AIR POLLUTION AND ECOSYSTEMS

Proceedings of an international symposium held in Grenoble (France) from 18 to 22 May 1987 and jointly organised by

- Commission of the European Communities, Brussels
- Ministère de l'Environnement, Paris

under the patronage of

Dr. Karl-Heinz NARJES, Vice-precident of the Commission of the European Communities

and

Monsieur Alain CARIGNON, Ministre Délégué, chargé de l'Environnement

This is Report No 7 in the Air Pollution Report Series of the Environmental Research Programme of the Commission of the European Communities, Directorate-General for Science, Research and Development. For more information concerning this series, please contact:

Mr H. Ott
CEC – DG XII/Gl
200, rue de la Loi
B – 1040 Brussels

Commission of the European Communities

AIR POLLUTION AND ECOSYSTEMS

Proceedings of an International Symposium
held in Grenoble, France, 18-22 May 1987

Edited by

P. MATHY

Commission of the European Communities,
Directorate-General for Science, Research and Development, Brussels, Belgium

D. REIDEL PUBLISHING COMPANY

A MEMBER OF THE KLUWER ACADEMIC PUBLISHERS GROUP

DORDRECHT / BOSTON / LANCASTER / TOKYO

Library of Congress Cataloging in Publication Data

CIP

Symposium on Effects of Air Pollution on Terrestrial and Aquatic Ecosystems
(1987: Grenoble, France)
Air pollution and ecosystems.

1. Air—Pollution—Environmental aspects—Congresses. I. Mathy,
Pierre, 1950– . II. Commission of the European Communities. III. Title.
QH545.A3S96 1987 574.5′222 87–24329
ISBN 90–277–2611–6

Publication arrangements by
Commission of the European Communities
Directorate-General Telecommunications, Information Industries and Innovation, Luxembourg.

EUR 11244

Published by D. Reidel Publishing Company,
P.O. Box 17, 3300 AA Dordrecht, Holland.

Sold and distributed in the U.S.A. and Canada
by Kluwer Academic Publishers,
101 Philip Drive, Assinippi Park, Norwell, MA 02061, U.S.A.

In all other countries, sold and distributed
by Kluwer Academic Publishers Group,
P.O. Box 322, 3300 AH Dordrecht, Holland.

Printed in The Netherlands

FOREWORD

In the concluding session of the symposium "Acid Deposition, a Challenge for Europe" held in Karlsruhe in September 1983, Dr. GINJAAR, the former Minister of Health and Environmental Protection of the Netherlands, emphasised the need for setting up a concerted research programme including the effects of air pollution on terrestrial and aquatic ecosystems. The Council of Ministers of the European Community in 1984 adopted a revision of the 3rd Community Programme on Environment comprising contract research and concerted action in the field of the effects of air pollution in ecosystems.

These research areas were also introduced in the 4th R-D Community Programme on Environment, adopted by the Council of Ministers in 1986 and are subject again to contract research and concerted action.

The Commission of the European Community is very concerned to increase the effectiveness of research projects carried out within the national programmes, and those undertaken at the Community level. The Commission tends to develop an integrated scientific approach, including not only the physico-chemical behaviour, the transport and the deposition of airborne pollutants but also the effects of these pollutants, in particular on living organisms and ecosystems. With regard to the specific issue of the effects, the Commission is trying to develop and strengthen a multi-disciplinary approach associating biologists, pathologists, ecophysiologists, and specialists of soil sciences, within the concerted action, as well as within coordinated research projects.

Progress in understanding the response of ecosystems to air pollution can only be achieved if one takes into account the network of interactions which govern the functioning of ecosystems.

The symposium confirmed the necessity to develop this kind of approach, emphasising the need to identify in each situation the limiting factor(s), biotic or abiotic, which could be responsible for permanent or temporary undesirable changes.

Air pollution is an increasing matter of concern which does not produce the same effects everywhere. Obviously, the effects depend on the type of pollution but there is strong evidence that they depend also on the state of the ecosystems and the type and the intensity of the other stresses to which the ecosystem is submitted. The contrasts between ecological situations and the diversity of pollution climates characterise Europe and these are further reasons for concertation and coordination.

Furthermore, it has been brought to light that the methodologies themselves have to be subject to concertation and coordination. A great deal of experimental facilities exist in Europe, some of them highly sophisticated and fully operational, others promising but needing further development, all of them with advantages and limitations.

The symposium confirmed that these methodologies should not be used in an isolated way but should be implemented within a single coordinated programme, associating the advantages of the various techniques in a complementary way.

Eight workshops were organised within the concerted action "Effects of Air Pollution on Terrestrial and Aquatic Ecosystems" since its launching in 1983. Major gaps in the knowledge were isolated at the occasion of these workshops and ways to develop coordinated research projects were indicated.

The symposium has summarized this concertation and coordination work and has provided recommendations concerning the establishment of the future research programmes.

The programme of the symposium covered research in the following fields :

. The atmospheric "pollution climates" in Europe : characterisation and classification ; mechanism of deposition ; fate of pollutants within the ecosystem and transfer to biota.
. The effects of atmospheric pollutants on agricultural crops.
. The effects of atmospheric pollutants on trees, forest ecosystems and semi-natural vegetation.
. The effects of atmospheric pollutants on aquatic ecosystems.
. Preventive and restoration measures.

The specific objectives of the symposium were :

1. to review the scientific knowledge, identify major gaps and recommend how these can be filled, in order to improve the understanding of eco-physiological processes which govern the responses of ecosystems to disturbances ;
2. to identify, as far as is feasible, critical "levels" and "loads" of air pollutants below which undesirable effects are unlikely to occur - information needed to guide environmental policy makers concerned with preventive measures. These "levels" and "loads" should be assessed taking note of interactive, and long-term cumulative effects ;
3. to recommend management rules and treatments which would aid the restoration of disturbed ecosystems.

The state of the knowledge in the most important research areas was presented at the occasion of critical reviews, in line with the objectives of the symposium.

A limited number of verbal communications and a permanent poster exhibition were presented to complement or illustrate the critical reviews.

For each session, rapporteurs were invited to prepare a general synthesis with recommendations. These syntheses are also included in the proceedings of the symposium.

We hope that the recommendations will be taken into consideration when the future research programmes will be designed and in particular the next R-D Community programme on Environmental Protection.

Finally, I would like to thank all those who actively participated in the organisation and success of the symposium. My particular gratitude goes to the reviewers who well achieved their task of summarising the available knowledge and to the rapporteurs who accepted the responsibility to draw conclusions and indicated some ways ahead.

P. MATHY

CONTENTS

Foreword v

OPENING SESSION

K.-H. NARJES, Vize-Präsident der Kommission der Europäischen Gemeinschaften, Belgium 2

K.-H. NARJES, Vice-President of the Commission of the European Communities, Belgium 5

K.-H. NARJES, Vice-Président de la Commission des Communautés Européennes, Belgique 8

A. CARIGNON, Ministre délégué, chargé de l'Environnement 11

C. BARTHOD, Direction de l'Espace Rural et de la Forêt du Ministère de l'Agriculture 15

KEY NOTES SPEECHES

Ecosystems and their response to airborne chemicals : the current situation in North America and Europe
E.B. COWLING, Associate Dean for Research, School of Forest Resources, North Carolina State University, Raleigh, North Carolina, U.S.A. 18

Overview of the U.S. National Acid Precipitation Assessment Program
P.M. IRVING, National Acid Precipitation Assessment Program, Washington, D.C., U.S.A. 39

SESSION I - POLLUTION CLIMATES IN EUROPE ; DEPOSITION IN ECOSYSTEMS

Report on the session
J. FUHRER, Swiss Federal Research Station for Agricultural Chemistry and Environmental Hygiene, Liebefeld-Bern, Switzerland ; and J. SLANINA, Netherlands Energy Research Foundation ECN, Petten, The Netherlands 50

Les climats de pollution en Europe : caractérisation et classification
G. TOUPANCE, Université Paris Val-de-Marne, Creteil, France 56

Deposition of pollutants on plants and soils ; principles and pathways
M. H. UNSWORTH and D. FOWLER, Institute of Terrestrial Ecology, Bush Estate, Penicuik, Scotland, United Kingdom 68

Critical loads for sulphur and nitrogen
J. NILSSON, The National Swedish Environmental Protection Board, Sweden 85

Regional ozone concentrations in Europe
P. GRENNFELT, Swedish Environmental Research Institute, Sweden ; J. SALTBONES, Norwegian Meteorological Institute ; and J. SCHJOLDAGER, Norwegian Institute for Air Research, Norway 92

SESSION II - EFFECTS OF AIR POLLUTION ON AGRICULTURAL CROPS

Report on the session
A.C., POSTHUMUS, University of Wageningen, The Netherlands; J.N.B. BELL, Imperial College, United Kingdom 104

Etat des connaissances sur l'impact des polluants de l'air sur les productions agricoles et approches méthodologiques
J. BONTE, Centre Départemental d'Etudes et de Recherches sur l'Environnement, Mourenx, France 109

Interactions between air pollutants and other limiting factors
T.A. MANSFIELD, P.W. LUCAS and E.A. WRIGHT, Department of Biological Sciences, University of Lancaster, Bailrigg, United Kingdom 123

Effects of ozone in ambient air on growth, yield and physiological parameters of spring wheat
J. FUHRER, A. GRANDJEAN, B. LEHNHERR, and W. TSCHANNEN, Swiss Federal Research Station for Agricultural Chemistry and Environmental Hygiene, Liebefeld-Bern ; and A. EGGER, Institute of Plant Physiology, University of Bern, Switzerland 142

Effects of atmospheric pollutants on grassland ecosystems
L. GRUNHAGE, U. HERTSTEIN, H.-J. JÄGER and U. DÄMMGEN, Institut für Produktions- und Okotoxikologie, Bundesforschungsanstalt für Landwirtschaft (FAL), Braunschweig, Federal Republic of Germany 148

Airborne pollutant injury to vegetation in Greece
C.D. HOLEVAS, Laboratory of Nonparasitic Diseases of the Benaki, Plant Pathology Institute, Kiphissia, Athens, Greece 154

Is plant resistance to air pollution related to plant resistance towards disease ?
G.P. DOHMEN, Gesellschaft für Strahlen und Umweltforschung, Institut für Toxikologie, Neuherberg, Federal Republic of Germany 158

SESSION III - EFFECTS OF AIR POLLUTION ON FOREST TREES AND FOREST ECOSYSTEMS

Part I

Report on the session
D.H.S. RICHARDSON, Trinity College, University of Dublin, Ireland ; O. QUEIROZ, Institut de Physiologie Végétale, C.N.R.S., Gif-sur-Yvette, France 164

Impact of air pollutants on above-ground plant parts of forest trees
G.H.M. KRAUSE, Landesanstalt für Immissionsschutz des Landes NW, Essen, Federal Republic of Germany 168

Direct effects of dry and wet deposition on forest ecosystems - in particular canopy interactions
P.H. FREER-SMITH, Biology Dept. University of Ulster, Newtownabbey, Northern-Ireland, United Kingdom 217

Tree responses to acid depositions into the soil. A summary of the COST workshop at Jülich 1985
E.-D. SCHULZE, Lehrstuhl Pflanzenökologie, Universität Bayreuth, Federal Republic of Germany 225

Experimentelle Untersuchungen zur Stoffauswaschung aus Baumkronen durch Saure Beregnung
K. KREUTZER, Lehrstuhl für Bodenkunde der Ludwig-Maximilians-Universität München, Federal Republic of Germany 242

Effect of artificial rain on soil chemical properties and forest growth
A.O. STUANES and B. TVEITE, Norwegian Forest Research Institute, Aas-NLH ; and G. ABRAHAMSEN, Institute of Soil Sciences, Agricultural University of Norway, Aas-NLH, Norway 248

Effects of sulphur dioxide on forest litter decomposition and nutrient release
P. INESON and P.A. WOOKEY, Merlewood Research Station, Institute of Terrestrial Ecology, Grange-over-Sands, Cumbria, United Kingdom 254

Les recherches sur le déperissement des forêts en France : structure et principaux résultats du programme DEFORPA
G. LANDMANN, I.G.R.E.F., I.N.R.A. Nancy, Champenoux, France 261

Part II

Report on the session
M. ASHMORE, Imperial College, London, United Kingdom 284

Note de synthèse sur les trois tournées en forêt effectuées le 20 mai 1987
M. BAZIRE, Inventaire Forestier National, France 288

Recent developments in the diagnosis and quantification of forest decline
J.N. CAPE, Institute of Terrestrial Ecology, Bush Estate, Penicuik, Midlothian, United Kingdom 292

Effects of H_2O_2-containing acidic fog on young trees
R.K.A.M. MALLANT and J. SLANINA, Energy Research Foundation ECN, Petten, The Netherlands ; and G. MASUCH and A. KETTRUP, Fachbereich Chemie, Universität-GH, Paderborn, Federal Republic of Germany 306

Effect of acidic mist on nutrient leaching, carbohydrate status and damage symptoms of Picea abies
K. MENGEL, M. TH. BREININGER and H.J. LUTZ, Institute of Plant Nutrition of the Justus Liebig-University, Giessen, Federal Republic of Germany 312

Part III

Report on the session
G. SELLDEN, University of Göteborg, Sweden 322

Methodological approaches : Part I : Experiments with open-top chambers : results, advantages and limitations
H.-J. JÄGER and H.J. WEIGEL, Institut für Produktions- und Ökotoxikologie, Bundesforschungsanstalt für Landwirtschaft, Braunschweig ; R. GUDERIAN, Institut für Angewandte Botanik, Universität Essen ; and U. ARNDT and G. SEUFERT, Institut für Landeskultur und Pflanzenökologie, Universität Hohenheim, Stuttgart, Federal Republic of Germany 327

Methodological aspects of the fumigation of forest trees with gaseous pollutants using closed chambers
T.M. ROBERTS, K.A. BROWN and L.W. BLANK, CEGB, Central Electricity Research Laboratories, Leatherhead, United Kingdom 338

The methodology of open-field fumigation
J.J. COLLS and C.K. BAKER, Nottingham University School of Agriculture, Loughborough, United Kingdom 361

SESSION IV - EFFECTS OF AIR POLLUTION ON AQUATIC ECOSYSTEMS

Report on the session
S.C. WARREN, Director, Water Research Centre, Medmenham, United Kingdom 374

Acid deposition and surface water acidification : the use of lake sediments
R. W. BATTARBEE, Palaeoecology Research Unit, University College, London, United Kingdom ; and I. RENBERG, Department of Ecological Botany, Umea University, Sweden 379

Acidification and ecophysiology of freshwater animals
J.H.D. VANGENECHTEN, Belgian Nuclear Research Center, S.C.K.-C.E.N., Biology Department, Mineral Metabolism Laboratory, Mol, Belgium 396

Reversibility of acidification
M. HAUHS, Universität Göttingen, Federal Republic of Germany 407

SESSION V - PREVENTIVE AND CURATIVE MEASURES

Report on the session

Part 1 : Preventive and curative measures for terrestrials ecosystems, in particular forests
G. LANDMANN, Centre de Recherches Forestières, I.N.R.A. Nancy, France 420

A.S. GEE, District Scientist, South Western District, Welsh Water Authority, Haverfordwest, Wales 422

Moyens préventifs et curatifs de lutte contre la pollution dans les ecosystèmes terrestres, en particulier les forêts
M. BONNEAU, I.G.G.R.E.F., Institut National de la Recherche Agronomique, Nancy, France 425

Fertilization as a tool to remove "New type" forest damages
R.F. HUTTL, Forestry Department, Kali und Salz AG, Kassel, Federal Republic of Germany 438

Judgement of the applicability of liming to restabilise forest stands - with special consideration of root ecological aspects
D. MURACH, Institute of silviculture, University of Göttingen, Federal Republic of Germany 445

The effects of land management on acidification of aquatic ecosystems and the implications for the development of ameliorative measures
M. HORNUNG, Institute of Terrestrial Ecology, Bangor Research Station, North Wales, United Kingdom 452

Practical preventive and curative measures for aquatic ecosystems
M. DICKSON, National Environmental Protection Board, Sweden 469

POSTER SESSION

THEME I - POLLUTION CLIMATES IN EUROPE ; DEPOSITION IN ECOSYSTEMS

Caractérisation de l'environnement atmosphérique en zone forestière. Station Laboratoire du Donon.
J.M. BIREN, C. ELICHEGARAY et J.P. VIDAL, Agence pour la Qualité de l'Air, Paris-la-Défense, France 480

Kleinräumige Verteilungsmuster der Stoffdeposition in naturnahen Waldökosystemen
W. BUCKING and R. STEINLE, Forstliche Versuchs- und Forschungsansalt, Baden-Württemberg, Abt. Botanik und Standortskunde, Federal Republic of Germany 486

Spatial distribution of wet and dry sulphur deposition in the United Kingdom
S. M. COTTRILL, Department of Trade and Industry, Warren Spring Laboratory, Stevenage, Herts, United Kingdom 493

The mont Lozere catchments : an observatory of the atmospheric pollution background level
J.F. DIDON-LESCOT, P. DURAND, F. LELONG, Laboratoire d'Hydrogéologie, Université d'Orléans ; and R. DEJEAN, Parc National des Cévennes, Florac, France 499

Palynologische und forstgeschichtliche Aspekte zur Versauerungsgeschichte von Schwarzwaldseen
K.H. FEGER und W. ZEITVOGEL, Institut für Bodenkunde und Waldernährungslehre der Albert-Ludwigs-Universität, Freiburg im Breisgau, Federal Republic of Germany 508

The pollution climate of the Fichtelgebirge, Northern Bavaria
J. FORSTER, Lehrstuhl für Hydrologie der Universität Bayreuth, Federal Republic of Germany 514

Relief effects on the deposition of air pollutants in forest stands - lead deposition as an example
V. GLAVAC, H. KOENIES, H. JOCHHEIM and R. HEIMERICH, University of Kassel, Federal Republic of Germany 520

Atmospheric deposition on semi-natural grassland vegetation
G.W. HEIL, B. HEIJNE and D. VAN DAM, "Acid rain" ecology, Department of Plant Ecology, University of Utrecht, The Netherlands 530

Atmospheric nitrogen deposition in a forest next to an intensively used agricultural area
W.P.M.F. IVENS, G.P.J. DRAAIJERS and W. BLEUTEN, Department of Physical Geography, State University of Utrecht, The Netherlands 536

Product analysis of the chemical/photochemical conversion of monoterpenes with airborne pollutants (O_3/NO_2)
K. JAY and L. STIEGLITZ, Kernforschungszentrum Karlsruhe, Federal Republic of Germany 542

Atmospheric inputs of cadmium to an arable agricultural system
K.C. JONES, C.J. SYMON, Department of Environmental Science, University of Lancaster ; and A.E. JOHNSTON, Soils and Plant Nutrition Department, Rothamsted Experimental Station, Harpenden, Hertfordshire, United Kingdom 548

Etude du déperissement forestier en Belgique. Premier Bilan d'une année de récolte des précipitations à la station pilote de Vielsalm
E. LAITAT et J. FAGOT, Faculté des Sciences agronomiques de l'Etat, Chaire de Biologie Végétale, Gembloux, Belgique 554

Atmospheric heavy metal deposition in Northern Europe measured by moss analyses
L. RASMUSSEN, Technical University of Denmark ; K. PILEGAARD, Riso National Laboratory, Denmark ; and A. RUHLING, University of Lund, Sweden 560

Cryptogamic epiphytes as indicators of air quality investigative methods and results from four selected areas in Portugal
C. SERGIO, Museu, Laboratorio e Jardim Botanico, Faculdade de Ciencias, Lisboa, Portugal ; and M.P. JONES, King's College, London, United Kingdom 566

Reaction and effect measurements in Bavaria with regard to forest damages
K. PFEIFFER and E. RUDOLPH, Bayer. Landesamt für Umweltschutz, München, Federal Republic of Germany 572

THEME II - DIAGNOSIS

Adenine and pyridine nucleotides, and carbohydrates in healthy and deceased needles : alterations in relation to vegetation period (May to October 1985) and needle age (1979 to 1985)
R. KEIL, T. BENZ, W. EINIG and R. HAMPP, Universität Tübingen, Institut für Biologie I, Tübingen, Federal Republic of Germany 584

Possible use of chlorophyll fluorescence in early detection of forest damages : influence of different criteria
F. BAILLON, X. DALSCHAERT and S. GRASSI, Chemistry Division, Joint Research Center, Ispra, Italy 590

Fluorescence as a means of diagnosing the effect of pollutant induced stress in plants
O. VAN KOOTEN, Lab. Plant Physiological Research, Agricultural University Wageningen ; and L.W.A. VAN HOVE, Department of Air Pollution, Agricultural University, Wageningen, The Netherlands 596

Rapid fingerprinting of damaged and undamaged conifer needles by soft ionization mass spectrometry and pattern recognition
H.-R. SCHULTEN, Fachhochschule Fresenius, Dep. Trace Analysis, Wiesbaden ; and N. SIMMLEIT, Institut Fresenius, Chemical and Biological Laboratories, Taunusstein-Neuhof, Federal Republic of Germany 602

Comparative investigation on the nutrient composition of healthy and injured spruces of different locations
W. FORSCHNER and A. WILD, Institut für Allgemeine Botanik der Johannes Gutenberg-Universität, Mainz, Federal Republic of Germany 604

Early diagnosis of forest decline - a pilot study
J.N. CAPE and I.S. PATERSON, Institute of Terrestrial Ecology, Edinburgh, United Kingdom; A.R. WELLBURN, J. WOLFENDEN and H. MEHLHORN, University of Lancaster, United Kingdom ; P. FREER-SMITH, University of Ulster, North-Ireland ; and S. FINK, Albert-Ludwigs-Universität, Freiburg, Federal Republic of Germany 609

THEME III - EFFECTS OF AIR POLLUTION ACTING ON ABOVE-GROUND PARTS OF PLANTS AND ECOSYSTEMS

Estimation of the plant-related resistances determining the ozone flux to broad bean plants (vicia faba var metissa), after long-term exposure to ozone in an open top chamber
J.M.M. ABEN, Environmental Research Department, N.V. KEMA, Arnhem, The Netherlands 616

Sulfur containing air pollutants and their effects on plant metabolism
L.J. DE KOK, F.M. MAAS, I. STULEN and P.J.C. KUIPER, Department of Plant Physiology, University of Groningen, Haren, The Netherlands 620

Analyse des mécanismes de transfert des particules à travers l'appareil stomatique
G. JULIENNE et P. LOUGUET, Laboratoire de Physiologie Végétale et d'Ecophysiologie Végétale Appliquée, UFR de Sciences et Technologie, Université Paris 12, Creteil, France 626

Open-top chamber studies to assess the effects of sulphur dioxide and acid rain, singly or in combination, on agricultural plants
H.J. WEIGEL, G. ADAROS and H.-J. JÄGER, Institut für Produktions- und Okotoxikologie, Bundesforschungsanstalt für Landwirtschaft, Braunschweig, Federal Republic of Germany 632

Ozone sensitivity of open-top chamber grown cultivars of spring wheat and spring rape
I. JOHNSEN, L. MORTENSEN, L. MOSEHOLM and H. RO-POULSEN, University of Copenhagen, Institute of Plant Ecology, and National Agency of Environmental Protection, Air Pollution Laboratory, Denmark 637

Effects of ambient air pollution on crop species in and around London
M.R. ASHMORE, A. MIMMACK and J.N.B. BELL, Department of Pure and Applied Biology, Imperial College of Science and Technology, London, United Kingdom 641

Effects of ozone and calcium nutrition on native plant species
M.R. ASHMORE, C. DALPRA and A.K. TICKLE, Department of Pure and Applied Biology, Imperial College of Science and Technology, London, United Kingdom 647

Secondary effects of photochemical oxidants on cereals : alterations in susceptibility to fungal leaf diseases
A. V. TIEDEMANN and H. FEHRMANN, Institut für Pflanzenpathologie und Pflanzenschutz der Georg-August-Universität, Göttingen, Federal Republic of Germany 653

Combined effects of ozone and acid mist on tree seedlings
M.R. ASHMORE, C. GARRETY, F.M. MCHUGH and R. MEPSTED, Department of Pure and Applied Biology, Imperial College of Science and Technology, London, United Kingdom 659

Does exhaust air from motorway tunnels affect the surrounding vegetation ?
S. BRAUN and W. FLUECKIGER, Institute for Applied Plant Biology, Schönenbuch, Switzerland 665

Rôle de la cuticule des plantes dans le transfert des xenobiotiques dans l'environnement
A. CHAMEL et B. GAMBONNET, DRF/LBIO/Biologie Végétale, Grenoble, France 671

Pollution atmosphérique et métabolisme respiratoire de l'épicea
A. CITERNE, J. BANVOY et P. DIZENGREMEL, Laboratoire de Physiologie Cellulaire Végétale, Université de Nancy I, Vandoeuvre-les-Nancy, France 678

Vergleichende Untersuchungen der Wachsmorphologie an Nadeln von Klonfichten und Altfichten
T. EUTENEUER und R. DEBUS, Fraunhofer-Institut für Umweltchemie und Okotoxikologie, Schmallenberg-Grafschaft, Federal Republic of Germany 684

Dépérissement forestier et précipitations acides dans les Pyrénées centrales
V. CHERET, J. DAGNAC et F. FROMARD, Laboratoire de Botanique et Biogéographie ; et D. GUILLEMYN, Institut de la Carte Internationale de la Végétation, France 690

Indirect effects of acid mist upon the rhizosphere and the leaves' buffering capacity of beech seedlings
S. LEONARDI and W. FLUCKIGER, Institute for Applied Plant Biology, Schönenbuch, Switzerland 697

Recherches sur les causes du dépérissement des forêts. Mise en place de tests d'exclusion au col du Donon (Vosges - France)
J.P. GARREC et M. BERTEIGNE, I.N.R.A, Centre de Recherches Forestières, Laboratoire d'Etude de la Pollution Atmosphérique, Champenoux, Seichamps, France 701

Effets des "pluies acides" sur les cuticules et les surfaces foliaires
J.P. GARREC et C. LAEBENS, I.N.R.A., Centre de Recherches Forestières, Laboratoire d'Etude de la Pollution Atmosphérique, Champenoux, Seichamps, France 704

Diffusions/Reaktionsmodell für SO_2 am Blattpfad
G.H. KOHLMAIER, F.W. BADECK, M. PLOCHL, K. SIEBKE, O. SIRE, and C.WIENTZEK, Insitut für Physikalische und Theoretische Chemie der Universität Frankfurt/Main, Federal Republic of Germany 710

Effets d'une pollution de l'atmosphère par l'Ozone et le Dioxyde de Soufre sur la croissance, la conductance stomatique et la teneur en composés phénoliques de 3 clones d'épicéas en chambre à ciel ouvert
P. MALKA, ADERA/CDERE. Contrat ELF ; D. CONTOUR-ANSEL et P. LOUGUET, Laboratoire de Physiologie Végétale et Ecophysiologie Végétale Appliquée, Université Paris XII ; et J. BONTE, Centre Départemental d'Etudes et de Recherches sur l'Environnement, Paris, France 717

Effets de la pollution atmosphérique sur le système sécréteur et la composition terpenique des aiguilles de picea abies
A. SAINT-GUILY, Laboratoire de Physiologie Cellulaire Végétale, Université de Bordeaux I, Talence, France 723

Direct effects of acid wet deposition on photosynthesis, stomatal conductance and growth of populus CV. Beaupre
P. VAN ELSACKER, C. MARTENS, and I. IMPENS, Department of Biology, University of Antwerp, Belgium 728

The uptake of atmospheric ammonia by leaves
L.W.A. VAN HOVE and E.H. ADEMA, Department of Air Pollution ; and W.J. VREDENBERG, Laboratory of Plant Physiological Research, Agricultural University Wageningen, The Netherlands 734

Investigating effects of air pollution by measuring the waterflow velocity in trees
J.H.A. VAN WAKEREN, H. VISSER and F.B.J. KOOPS, Environmental Research Department, N.V. KEMA, Arnhem, The Netherlands 739

Untersuchungen zum Kohlenhydratstoffwechsel an Fichte (picea Abies Karst.) Aus Klimakammerexperimenten und aus geschädigten Waldbeständen
K. VOGELS und T. LAMBRECHT, Universität GHS Essen, Institut für Angewandte Botanik, Federal Republic of Germany 743

Comparative investigations on the photosynthetic electron transport chain of spruce (picea abies) with different degrees of damage in the open air
A. WILD, B. DIETZ, U. FLAMMERSFELD, and I. MOORS, Institute for General Botany of the Johannes Gutenberg-University, Mainz, Federal Republic of Germany 754

Some effects of low levels of sulphur dioxide and nitrogen dioxide on the control of water-loss by Betula spp.
E.A. WRIGHT, Department of Biological Sciences, University of Lancaster, Lancashire, United Kingdom 760

Deposition of heavy metals and their distribution on needles and bark
V. CERCASOV, I. RENTSCHLER and H. SCHREIBER, Institut für Physik, Universität Hohenheim, Stuttgart, Federal Republic of Germany 766

Ammonia and pine tree dieback in Belgium
L. DE TEMMERMAN, A. RONSE, K. VAN DEN CRUYS and K. MEEUS-VERDINNE, Institute for Chemical Research, Ministry of Agriculture, Belgium 774

The effects of acid mist on conifer aphids and their implications for tree health
N.A.C. KIDD and M.B. THOMAS, Department of Zoology, University College, Cardiff, United Kingdom 780

Lichen and conifer recolonization in Munich's cleaner air
O. KANDLER, Botanical Institute, University of Munich, München, Federal Republic of Germany 784

THEME IV - FATE OF AIR POLLUTANTS IN SOIL. EFFECTS ON SOILS AND CONSEQUENCES FOR PLANTS

Long-term changes in the polynuclear aromatic hydrocarbon content of agricultural soils
K.C. JONES, J.A. STRATFORD and K.S. WATERHOUSE, Department of Environmental Science, University of Lancaster ; and A.E. JOHNSTON, Soils and Plant Nutrition Department, Rothamsted Experimental Station, Harpenden, Hertfordshire, United Kingdom 792

Mobilisation du fluor par acidolyse des sols
J.A. BARDY et C. PERE, A.P.P.A. Comité Bordeaux-Aquitaine, Laboratoire Municipal de Bordeaux, France 798

Apports atmosphériques et acidification des solutions de sol dans une forêt déperissante des Vosges (Est de la France)
F. GRAS, D. MERLET et J. ROUILLER, Centre de Pédologie Biolcgique, CNRS, Nancy, France 806

A six year nutrient budget in a coniferous watershed under atmospheric pollution
A. HAMBUCKERS and J. REMACLE, Ecologie Microbienne, Département de Botanique, Université de Liège, Belgium 814

A mathematical model for acidification and neutralization of soil profiles exposed to acid deposition
H.U. SVERDRUP and P.G. WARFVINGE, Departement of Chemical Engineering, Lund Institute of Technology, Chemical Centre; and U. VON BROMSSEN, Swedish Environmental Protection Board, Solna, Sweden 817

Le bassin versant du Strengbach à Aubure (Haut-Rhin, France) pour l'étude du dépérissement forestier dans les Vosges (programme DEFORPA) :
I - Equipement climatique, hydrologique, hydrochimique
D. VIVILLE et B. AMBROISE, Centre d'Etudes et de Recherches Eco-Géographiques, ULP, Strasbourg ; A. PROBST, Institut de Géologie, ULP, Strasbourg ; B. FRITZ, Centre de Sédimentologie et de Géochimie de Surface, Strasbourg ; et E. DAMBRINE, D. GELHAYE et C. DELOZE, Laboratoire "Sols et Nutrition des Arbres Forestiers", Institut National de la Recherche Agronomique, Centre National de Recherches Forestières (INRA-CNRF), Champenoux, Seichamps, France 823

Le bassin versant du Strengbach à Aubure (Haut-Rhin, France) pour l'étude du dépérissement forestier dans les Vosges (programme DEFORPA)
II - Influence des précipitations acides sur la chimie des eaux de surface
A. PROBST, Institut de Géologie, ULP, Strasbourg ; B. FRITZ, Centre de Sédimentologie et de Géochimie de la Surface, Strasbourg ; et B. AMBROISE et D. VIVILLE, Centre d'Etudes et de Recherches Eco-Géographiques, ULP, Strasbourg, France 829

Fate of mineral nitrogen in acid heathland and forest soils
J.W. VONK, D. BARUG, and T.N.P. BOSMA, TNO Technology for Society, Zeist, The Netherlands 835

The effect of atmospheric deposition, especially of nitrogen, on grassland and woodland at Rothamsted experimental station, England, measured over more than 100 years
K.W.T. GOULDING, A.E. JOHNSTON and P.R. POULTON, Soils and Plant Nutrition Department, Rothamsted Experimental Station, Herts, United Kingdom 841

Effects of ammonia deposition on animal-mediated nitrogen mineralization and acidity in coniferous forest soils in The Netherlands
H.A. VERHOEF and F.G. DOREL, Department of Ecology and Ecotoxicology, Free University, Amsterdam, The Netherlands 847

Investigations on the telluric microbiological features of forest dieback in Vosges
D. ESTIVALET and R. PERRIN, Station de recherches sur la flore pathogène dans le sol, Dijon ; and F. LE TACON, Laboratoire de microbiologie forestière CRF, Champenoux, Seichamps, France 852

The influence of acid rain on mycorrhizal fungi and mycorrhizas of Douglas fir (Pseudotsuga Menzies II) in The Netherlands
A.E. JANSEN, Agricultural University, Department of Phytopathology, Wageningen, The Netherlands 859

Lead contents and distribution in spruce roots
D.L. GODBOLD, E. FRITZ and A. HUTTERMANN, Forstbotanisches Institut, Universität Göttingen, Federal Republic of Germany 864

"New type" of forest decline, nutrient deficiencies and the "Virus"-Hypothesis
R.F. HUETTL, Institute of Soil Science and Forest Nutrition ; B.M. MEHNE, Institute of Forest Botany and Wood Biology, Albert-Ludwigs-University, Freiburg, Federal Republic of Germany 870

Effects of airborne ammonium on natural vegetation and forests
J.G.M. ROELOFS, A.W. BOXMAN and H.F.G. VAN DIJK, Laboratory of Aquatic Ecology, Catholic University, Nijmegen, The Netherlands 876

Acid rain studies in the Fichtelgebirge (NE-Bavaria)
R. HANTSCHEL, M. KAUPENJOHANN, R. HORN and W. ZECH, Institute of Soil Science and Soil Geography, University of Bayreuth, Federal Republic of Germany 881

Contribution 1 : Acidification research on Douglas fir forests in The Netherlands (Aciforn project)
P. EVERS, C.J.M. KONSTEN and A.W.M. VERMETTEN, ACIFORN group, Agricultural University, Department of Air Pollution, Wageningen, The Netherlands 887

Contribution 2 : Monitoring of root growth
A.F.M. OLSTHOORN, Dept of Silviculture and Forest Ecology, Agricultural University, Wageningen, The Netherlands 888

Contribution 3 : Monitoring of hydrological processes under Douglas fir
A. TIKTAK and W. BOUTEN, Department Physical Geography and Soil Science, University of Amsterdam, The Netherlands 891

Contribution 4 : Monitoring soil chemical parameters under Douglas fir
C.J.M. KONSTEN and N. VAN BREEMEN, Agricultural University, Dept. Soil Science & Geology, Wageningen, The Netherlands 896

Contribution 5 : Air pollution monitoring in a Douglas fir forest
A. VERMETTEN, P. HOFSCHREUDER and H. HARSSEMA, Agricultural University Wageningen, Department of Air Pollution, Wageningen, The Netherlands 903

Contribution 6 : Influence of air pollution on tree physiology
P. EVERS, P. CORTES, H. V.D. BEEK, W. JANS, J. DONKERS, J. BELDE, H. RELOU and W. SWART, Dorschkamp Research Institute for Forestry and Landscape Planning, Wageningen, The Netherlands 907

The influence of acid rain on fine root distribution and nutrition
B.U. SCHNEIDER and W. ZECH, Institute of Soil Science, University of Bayreuth, Federal Republic of Germany 910

Function of forest ecosystems under the stress of air pollution
W. GRODZINSKI, Department of Ecosystem Studies, Jagiellonian University and Institute of Fresh Water Biology, Polish Academy of Sciences, Krakow, Poland 918

Carbon assimilation and nutrition of trees in two Picea abies (L.) Karst. stands in Bavaria exhibiting apparent decline differences
R. OREN, E.-D. SCHULZE, U. SCHNEIDER, K.S. WERK and J. MEYER, Lehrstuhl für Pflanzenökologie, Universität Bayreuth, Federal Republic of Germany 919

Forest ecosystems research network FERN
A. TELLER , European Science Foundation 923

THEME V - EFFECTS OF AIR POLLUTION ON AQUATIC ECOSYSTEMS

Historical development and extent of acidification of shallow soft waters in The Netherlands
G.H.P. ARTS, Laboratory of Aquatic Ecology, Catholic University, Nijmegen, The Netherlands 928

Acidification of freshwaters in the Vosges mountains (Eastern France)
J.-C. MASSABUAU and B. BURTIN, Laboratoire d'Etude des Régulations Physiologiques, CNRS, Strasbourg ; et B. FRITZ, Centre de Sédimentologie et Géochimie de la Surface, CNRS, Strasbourg, France 934

Mercury pollution in Swedish woodland lakes
M. MEILI, University of Uppsala, Institute of Limnology, Sweden 940

Effects of hydrology on the reacidification of the limed lake Gardsjön
U. NYSTROM and H. HULTBERG, Swedish Environmental Research Institute, Gothenburg, Sweden 948

pH changes over two centuries in three Dutch moorland pools
H. VAN DAM, Research Institute for Nature Management, Leersum ; B. VAN GEEL, Hugo de Vries-Laboratory, University of Amsterdam, The Netherlands ; A. VAN DER WIJK, Centre for Isotope Research, Groningen, The Netherlands ; and M.D. DICKMAN, Dept. of Biological Sciences, Brock University, St. Catharines, Ontario, Canada 954

Documentation of areas potentially inclined to water acidification in the Federal Republic of Germany
A. HAMM, P. SCHMITT and R. LEHMANN, Bavarian Institute for Water Research Munich, Test Plant Wiedenbach ; and J. WIETING, Federal Environmental Agency Berlin, Federal Republic of Germany 960

Internal or external toxicity of aluminium in fish exposed to acid water
H.E. WITTERS and O.L.J. VANDERBORGHT, University of Antwerp, Biology Department, Wilrijk and Belgian Nuclear Energy Study Centre ; J.H.D. VANGENECHTEN and S. VAN PUYMBROECK, Belgian Nuclear Energy Study Centre (SCK/CEN), Biology Department, Mineral Metabolism Laboratory, Mol, Belgium 965

CLOSING SESSION - GENERAL CONCLUSIONS AND RECOMMENDATIONS

Conclusions générales
M. MULLER, Ministère de l'Environnement, Paris, France 972

Summing up of the symposium
P.J.W. SAUNDERS, Department of Environment London, United Kingdom 975

General conclusions and thanks
Ph. BOURDEAU, Commission of the European Communities, Brussels, Belgium 977

INDEX OF AUTHORS 979

OPENING SESSION

SEANCE D'OUVERTURE

EROFFNUNGSSITZUNG

ANSPRACHE

vom
Vizepräsidenten der Kommission der Europaïschen Gemeinschaften
K.-H. NARJES

Herr Minister, Herr Präfekt,
Meine Damen und Herren,

Im Namen der Kommission der Europaïschen Gemeinschaften, heiße ich Sie zu diesem Symposium in Bezug auf die "Auswirkungen der Luftverschmutzung auf terrestrische und aquatische Okosysteme", das gemeinsam von der Kommission und vom französischen Umweltministerium organisiert wurde, willkommen. Wir freuen uns sehr die Gastfreundschaft der Stadt Grenoble, von der Sie Bürgermeister sind, zu geniessen.

Mein Kollege Stanley Clinton Davis, der für die Umweltpolitik innerhalb der Kommission verantwortlich ist, bat mich Ihnen seine herzlichsten Glückwünsche zu übermitteln, sowie seine Abwesenheit zu entschuldigen, da er verpflichtet war, eine andere große Veranstaltung in München im Rahmen des Europaïschen Umweltjahres zu eröffnen.

Ich bin überzeugt, daß wir in den nächsten Tagen einen wichtigen Beitrag zu diesem Europaïschen Umweltjahr, dessen Symbol in der Uberschrift des Programmes dargestellt ist, leisten werden.

Sie sind, meine Damen und Herren, hier in Grenoble zusammengekommen, um die Ergebnisse von drei Jahren Gemeinschaftsforschung zu diskutieren und zu bewerten, und um daraus die Schlussfolgerungen zu ziehen, die wir Politiker benötigen, um unsere Entscheidungen zu treffen. Für die Kommission kann ich Ihnen zusichern, dass wir Ihre Schlussfolgerungen zur Kenntnis nehmen werden, um unsere Vorschläge für Rechtsvorschriften auf eine solide wissenschaftliche Grundlage zu stellen, denn dies ist das erklärte Ziel der gemeinschaftlichen Umweltforschung.

Innerhalb des Umweltforschungsprogramms nimmt Ihr spezielles Gebiet, die Erforschung der ökologischen Auswirkungen der Luftverschmutzung, nunmehr eine wichtige, wenn nicht die wichtigste Stellung ein. Dies war nicht immer so, und ich möchte daher hier kurz auf die Geschichte der Gemeinschaftsforschung in diesem Bereich eingehen.

Ende der siebziger Jahre kamen aus den skandinavischen Ländern, Canada und einigen Teilen der Vereinigten Staaten die ersten alarmierenden Berichte über Versauerung von Gewässern und Boden und ernste Folgen für das Leben in Flüssen und Seen. Wir haben dies in Mitteleuropa leider nicht ernst genug genommen.

Erst als zu Beginn der achtziger Jahre, sicher auch in Zusammenhang mit einer Reihe von klimatisch sehr ungünstigen Jahren, zuerst in der Bundesrepublik und dann in einigen Alpenländern zunehmend dramatische Schäden in den Wäldern beobachtet wurden, verstärkte sich der Druck der Offentlichkeit auf die Politiker in einem bis dahin nicht bekannten Ausmass. Der Ausdruck "Waldsterben" ging damals als Lehnwort in die französische Sprache ein. Auch die Gemeinschaft war gefordert, denn die grenzüberschreitende Natur der Luftverschmutzung liess rein nationale Massnahmen von vornherein als wenig aussichtsreich erscheinen, und einige Mitgliedsländer waren keineswegs von der Notwendigkeit eines raschen Handelns überzeugt.

Auch die Wissenschaft war nicht sehr gut auf diese Ereignisse vorbereitet ; viele, oft sich widersprechende Theorien für die Erklärung der Waldschäden wurden auf den Markt gebracht und "Schuldzuweisungen" an die einzelnen infrage kommenden Schadstoffe erteilt.

Ich war damals in der Kommission für die Umweltpolitik zuständig und in keiner sehr einfachen Lage. Ich sah als erstes eine Notwendigkeit, alle damals zur Verfügung stehenden wissenschaftlichen Erkenntnisse zu analysieren und zu bewerten und habe daher veranlasst, dass im September 1983 in Karlsruhe ein Symposium ausgerichtet wurde, dessen Titel "Saure Niederschläge, eine Herausforderung für Europa" bereits widerspiegelte, dass die Gemeinschaft vor einer grossen Aufgabe stand.

In der Folge konnten wir den Ministerrat davon überzeugen, uns zusätzliche Mittel für Umweltforschung zur Verfügung zu stellen, um damit ein europäisches Dringlichkeitsprogramm auf den Weg zu bringen. Nunmehr bin ich innerhalb der Kommission für die Forschung zuständig, und so ist es heute für mich eine Genugtuung, dass wir hier die Ergebnisse dieses Forschungsprogramms vorstellen können.

Lassen Sie mich, bevor ich näher auf das Forschungsprogramm eingehe, einige weniger erfreuliche Anmerkungen zum Fortschritt der Gesetzgebung der Gemeinschaft zur Minderung der Luftverschmutzung machen, ohne dabei auf Einzelheiten einzugehen. Die Kommission hat noch unter meiner Federführung dem Ministerrat eine Reihe von Direktiven vorgeschlagen mit dem Ziel einer erheblichen Reduzierung der Emissionen von Feuerungsanlagen und Kraftfahrzeugen. Obwohl einige von diesen, wie die Reduzierung des Schwefelgehalts von Heizölen, die Hürden des Rates genommen haben, sind die wichtigsten von ihnen immer noch nicht Gemeinschaftsrecht geworden. Dies ist betrüblich, und ich nütze die Gelegenheit für einen Appell an die Mitgliedsländer, sich bald zu einigen.

Nun zurück zur Forschung. Es war uns von Anfang an klar, dass die Kommission angesichts der spärlichen Mittel für Forschungsverträge in erster Linie eine Rolle der Koordinierung übernehmen sollte. Wir haben daher eine konzertierte Aktion im Rahmen von COST vorgeschlagen, der Norwegen, Schweden und die Schweiz beigetreten sind. Es freut mich besonders, auch die Wissenschaftler aus diesen Staaten hier begrüssen zu dürfen. Bei der Vergabe unserer eigenen Mittel haben wir in erster Linie versucht, gezielt Lücken in den verschiedenen nationalen Programmen zu schliessen, und die Verträge so zu streuen, dass wir ein breites Spektrum von Problembereichen überdecken. Es kam der Gemeinschaftsforschung sehr zustatten, dass auch einige der Mitgliedsländer ihre Forschungsanstrengungen erheblich steigerten ; zwischen diesen nationalen Programmen und der Kommission kam eine fruchtbare Zusammenarbeit zustande. Ich möchte mich darauf beschränken, das Programm DEFORPA (Dépérissement des forêts attribué à la pollution de l'air) des Gastlandes als einziges besonders zu erwähnen.

In der Zwischenzeit haben wir die Fortschreibung des Umweltforschungsprogramms auf den Weg gebracht, die vom Rat letztes Jahr beschlossen wurde, und es freut mich, dass viele von Ihnen auch in Zukunft mit der Kommission zusammenarbeiten werden.

Ich möchte den Ergebnissen des Symposiums hier nicht vorgreifen, meine Mitarbeiter, die an den zahlreichen spezifischen Arbeitssitzungen und Workshops teilgenommen haben, berichten jedoch von ermutigenden Fortschritten. Es scheint bereits jetzt möglich zu sein, die komplexen Mechanismen der Schadwirkungen auf einige wenige grundsätzliche Komplexe einzuengen und vielen der ursprünglichen Hypothesen eine Nebenrolle zuzuweisen. Der mühsame Weg einer quantitativen Beschreibung liegt allerdings noch vor uns.

Es hat sich aber auch neuer Forschungsbedarf ergeben. So scheint mir zum Beispiel sehr bedeutsam, dass im Grundlagenwissen über die Physiologie von Bäumen grosse Lücken bestehen. Sie bald zu schliessen ist eine wichtige Voraussetzung für eine schlüssige Erklärung von Schadsymptomen. Wir haben noch harte Arbeit vor uns.

Wir dürfen uns dabei nicht davon entmutigen lassen, dass das öffentliche Interesse an der Luftverschmutzung in letzter Zeit offensichtlich etwas zurückgegangen ist. Gerade die Debatte über die Kernenergie macht eine gründliche Bewertung der Folgen der Nutzung fossiler Brennstoffe um so notwendiger.

Es ist mir ein Bedürfnis, Ihnen, Herr Minister, und Ihren Mitarbeitern zu danken für die Mühe und harte Arbeit, die nötig waren, um dieses Symposium zusammen mit der Kommission auszurichten. Mein besonderer Dank gilt jedoch den Wissenschaftlern für ihre Beiträge zu dieser Veranstaltung, und wir sollten ihnen jetzt Gelegenheit geben, die eigentliche Arbeit zu beginnen. Ich wünsche Ihnen dazu viel Erfolg und einen angenehmen Aufenthalt in Grenoble.

Ich hoffe, möglichst viele von Ihnen heute abend auf dem Empfang der Kommission persönlich begrüssen zu dürfen.

OPENING ADDRESS

by

K.-H. NARJES

Vice-President of the Commission of the European Communities

Messieurs les Ministres, Monsieur le Préfet,
Ladies and Gentlemen,

On behalf of the Commission of the European Communities, I am happy to welcome you to this Symposium on the "Effects of Air Pollution on Terrestrial and Aquatic Ecosystems", organised jointly by the Commission and by the French Ministry of the Environment. We are extremely happy to enjoy the hospitality of this city of Grenoble of which you are mayor, Monsieur le Ministre.

My colleague, Stanley Clinton-Davis, responsible for environmental policy in the Commission, has requested me to convey to you his best regards. He asks you to excuse his absence on this occasion, as he is inaugurating another big event in Munich, within the framework of the European Year of the Environment.

I am convinced that during the next few days we shall make an important contribution to this European Year of the Environment, the symbol of which appears on the cover of the Programme of this Symposium.

You are meeting here in Grenoble, Ladies and Gentlemen, to discuss and to evaluate the results of three years of Community research and to draw conclusions based upon which we, the politicians, need to make decisions. I can assure you, that we in the Commission will take note of your conclusions which I am sure, provide a solid scientific basis for regulatory actions, this being a declared goal of Community research on the environment.

Within the EC Environment Research Programme, the particular area of the ecological effects of air pollution, which you are working on, has now an important role, perhaps the most important one. This was not the case in the past, and I would therefore like to use this opportunity to briefly sketch the history of Community research in this area.

At the end of the seventies the first alarming reports on the acidification of soil and surface waters and its serious consequence for living organisms in rivers and lakes reached us from Scandinavia, Canada and parts of the United States. Unfortunately, we in the Community did not take these reports seriously enough.

Only at the beginning of the eighties, a period of rather exceptional climatic conditions over a number of years, when increasingly dramatic damage to forests was observed, first in the Federal Republic of Germany and later in other alpine countries, did the pressure of the public opinion on politicians reach hitherto unknow dimensions. The term "Waldsterben" even became a technical term in French ! The problem was also a challenge for the Community. On one hand, pollution ignores national frontiers and, therefore, purely national counter measures were seen from the start as having little chance of success ; on the other hand, some Member Countries were not convinced of the need for rapid action.

The Community was in good company, because science was also ill prepared to meet this challenge. Many, often contradictory theories explaining forest damage were offered to the public and various pollutants were judged "guilty", often on rather flimsy evidence, of destroying the European forest.

I myself was at that time in charge of the environment policy in the Commission and my position was not an easy one. My first priority was to analyse and to evaluate all the relevant existing scientific knowledge and I, for that purpose, initiated the organisation of the Karlsruhe Symposium in September 1983 "Acid Rain, a Challenge for Europe", a title which indicates the importance of the task which the Community was facing.

After this we were able to convince the Council of Ministers to provide us with supplementary resources for research on the environment, permitting us to launch a major European effort on acid deposition. In the meantime I changed hats within the Commission and am now in charge of research, and it is, therefore, of particular satisfaction for me, that we can now present to you the results of this initiative.

Before dealing in more detail with the research programme, I have, unfortunately, to make a few general and less pleasant remarks on the progress made by the Community to produce legislation for reducing air pollution. Already under my responsibility, the Commission had proposed to the Council of Ministers a series of directives aiming at a considerable reduction in the emissions from fossil fuel burning installations and motor vehicles. Although a few of these directives have passed the hurdle of the Council, the most important ones have not yet become Community law. This is regrettable, and I would like to use this opportunity to appeal to the Member States, to urgently reach an agreement.

Now back to research. It was evident from the beginning, that given the scarce resources for financing research contracts, the Commission should primarily have a coordinating role. We had therefore proposed a concerted action within the COST framework, joined also by Norway, Sweden and Switzerland. And it is my great pleasure to welcome scientists from these countries to our meeting today. When committing our scarce financial resources, we attempted in the first place to close gaps in the various national programmes and to cover with our research contracts a broad range of problems. This Community research was also strengthened by the considerable increase in research expenditures in a number of Member Countries, resulting in mutually beneficial collaboration between these national programmes and that of the Community as a whole. Of the many such programmes, I wish to mention here only the programme DEFORPA (Dépérissement des forêts attribué à la pollution de l'air) of our host country.

Meanwhile, we were able to assure the continuation of the EC Environment Research Programme, which was approved by the Council last year, and which, I am happy to say, will permit many of you to collaborate with the Commission also in the future.

I do not want to anticipate the results of this Symposium, but my colleagues, who have had the opportunity to participate in numerous discussions and workshops, will report encouraging progress. It seems already possible to narrow down the complex mechanisms of air pollution to a few basic features, and to relegate many of the original hypotheses as being of secondary importance. The difficult task of quantitative description is, though, still ahead.

Need for new research has also become evident. It seems significant to me, for example, that there are major gaps in our basic knowledge of the physiology of trees. Closing these gaps rapidly, is a prerequisite for explaining tree damage. Much hard work is still before us.

In this work we should not be discouraged by the recent lessening of public interest in air pollution. The debate on nuclear energy makes a hard look at the exploitation of fossil fuels and its consequences even more necessary.

I want to thank you, Monsieur le Ministre, and your Colleagues for the care and hard work in organizing this Symposium together with the Commission. My particular thanks go to the scientists for their contributions to this event, and we should now give them the opportunity to get down to the essence of this meeting. I wish them success with their work and a pleasant stay in this beautiful town of Grenoble.

I hope to be able to welcome many of you personally this evening at the reception offered by the Commission.

ALLOCUTION D'OUVERTURE

par Monsieur K.-H. NARJES
Vice Président de la Commission des Communautés Européennes

Messieurs les Ministres, Monsieur le Préfet,
Mesdames, Messieurs,

Au nom de la Commission des Communautés Européennes, je vous souhaite la bienvenue à ce Symposium relatif aux "Effets de la pollution de l'air sur les écosystèmes terrestres et aquatiques", organisé conjointement par la Commission et par le Ministère français de l'environnement. Nous sommes très heureux de profiter de l'hospitalité de la ville de Grenoble dont vous êtes le Maîre, Monsieur le Ministre.

Mon collègue Stanley Clinton-Davis, responsable de la politique de l'environnement au sein de la Commission, m'a demandé de vous transmettre ses meilleurs voeux ; il vous prie d'excuser son absence due au fait qu'il s'est vu dans l'obligation d'ouvrir une autre grande manifestation à Munich, dans le cadre de l'Année Européenne de l'Environnement.

Je suis convaincu que nous apporterons pendant les jours qui suivent une contribution importante à cette Année Européenne de l'Environnement dont le symbole figure à l'entête du programme de ce symposium.

Vous êtes réunis, Mesdames et Messieurs, ici à Grenoble pour examiner et discuter les résultats de 3 ans de recherche communautaire et en tirer les conclusions dont nous, les hommes politiques avons besoin pour prendre des décisions. Je peux vous assurer que la Commission tiendra compte de vos conclusions pour préparer des propositions de directives basées sur de fermes connaissances scientifiques ce qui est le but évident des recherches communautaires en matière d'environnement.

Dans le programme de recherche Environnement il existe un domaine particulier concernant les effets écologiques de la pollution de l'air actuellement un important, sinon le plus important problème à étudier. Cela ne fut pas toujours le cas et je voudrais détailler ici brièvement l'histoire de la recherche communautaire.

A la fin des années 70, des pays Scandinaves du Canada et quelques régions des USA arrivent les premières indications alarmantes au sujet de l'acidification des eaux et des sols et des sérieuses conséquences pour les êtres vivant dans les cours d'eau et les mers. En Europe Centrale nous n'avons pas pris assez au sérieux ces avertissements.

Ensuite au début des années 80, sûrement en relation avec une série d'années au climat très défavorable, d'abord en République Fédérale d'Allemagne puis dans les pays alpins des dégâts dramatiques furent observés dans les forêts renforçant la pression du public sur les hommes politiques d'une façon jusqu'alors inconnue. L'expression "Waldsterben" est même passée dans le langage français.

Aussi l'action de la Communauté était favorisée du fait que la nature transfrontière de la pollution de l'air, dès le début, ne paraissait pas s'accomoder de mesures purement nationales et que certains Etats Membres étaient convaincus de la nécessité de s'entraider.

De plus les connaissances scientifiques n'étaient guère aptes à répondre à un tel évènement ; de nombreuses, et souvent contradictoires, théories sur les causes des dégâts aux forêts s'affrontaient et attribuaient les responsabilités à des polluants pris isolément.

J'étais à l'époque responsable de la politique de l'Environnement à la Commission et je me trouvais en situation délicate - En premier lieu je jugeais nécessaire d'analyser et de comparer toutes les connaissances alors disponibles et je décidais la réunion d'un symposium à Karlsruhe en septembre 1983 dont le titre "Pluies acides, un défi pour l'Europe" réflétait déjà que la Communauté se trouvait devant une mission importante.

Par la suite nous pûmes convaincre le Conseil des Ministres de nous donner les moyens nécessaires à la recherche en matière d'Environnement et de mettre en route d'urgence un programme européen. Actuellement je suis responsable de la recherche au sein de la Commission et c'est aujourd'hui pour moi une satisfaction de pouvoir prendre connaissance des résultats de ce programme de recherche.

Laissez moi, avant d'en venir au programme de recherche, faire quelques remarques sur l'avancement de la legislation communautaire sans entrer dans les détails.

La Commission, toujours sous ma responsabilité, a présenté au Conseil des Ministres une série de directives dans le but de réduire notablement les émissions des grosses installations de combustion et des véhicules à moteur. Bien que certaines d'entre elles, comme la réduction du contenu en souffre dans les liquides de combustion, aient franchi l'obstacle du Conseil, les plus importantes ne font pas encore partie du droit communautaire - ceci est désolant et je saisis l'occasion de faire un appel aux Etats à se mettre d'accord rapidement.

Revenons à la recherche. Il apparut clairement, dès le début, que la Commission, devait assumer principalement un rôle de coordination par suite de la faiblesse des moyens financiers disponibles pour des contrats de recherche. Nous avons donc lancé une action concertée dans le cadre du COST à laquelle la Norvège, la Suède et la Suisse ont adhéré. Il m'est agréable de pouvoir souhaiter la bienvenue aux scientifiques de ces pays. En utilisant nos moyens propres nous avons cherché d'abord à combler les lacunes dans nos connaissances et de répartir les contrats de façon à couvrir un large spectre dans les problèmes à résoudre.

Le programme communautaire est arrivé juste à point au moment où certains Etats Membres renforçaient énormément leurs moyens de recherche. Entre ces programmes nationaux et la Commission se sont établis des liens fructueux. Je voudrais me borner à souligner particulièrement l'importance du programme DEFORPA du pays qui nous accueille ici.

Entretemps nous avons préparé la continuation du programme de recherche en matière d'Environnement qui fut approuvée par le Conseil l'année dernière et je me réjouis que nombre d'entre vous travailleront avec la Commission dans l'avenir.

Je ne voudrais pas ici anticiper les résultats du Symposium que mes collaborateurs, qui ont participé à des réunions de travail et des sessions spécialisées, m'ont indiqué représenter des progrès importants. Il me semble possible de prévoir que les mécanismes complexes d'action des polluants puissent se ramener à quelques principes fondamentaux et que nombre des théories émises au début ne jouent qu'un rôle secondaire. L'établissement de relations quantitatives reste encore à faire.

Il existe d'autres besoins de recherche. Il me semble, par exemple, très significatif que de grosses lacunes subsistent dans les connaissances fondamentales de la physiologie des arbres. Ceci est une importante condition pour comprendre les symptômes des dégâts. Nous avons encore beaucoup de travail devant nous.

Nous ne devons pas nous décourager de ce que l'intérêt du public pour la pollution atmosphérique, autrefois évident, diminue quelque peu. En relation directe avec les débats sur l'énergie nucléaire, il est indispensable de faire une évaluation approfondie des conséquences de l'utilisation des combustibles fossiles.

Je dois vous remercier, vous Monsieur le Ministre et tous vos collaborateurs, pour la peine que vous avez prise et le travail nécessaires à la réalisation de ce Symposium avec la Commission. J'adresse mes remerciements particulièrement aux scientifiques pour leur contribution à cette manifestation et nous voudrions leur donner maintenant la possibilité de commencer à travailler. Je vous souhaite un bon succès et un bon séjour à Grenoble.

J'espère de pouvoir remercier personnellement nombre d'entre vous ce soir à la réception offerte par la Commission.

INTERVENTION DE MONSIEUR ALAIN CARIGNON
MINISTRE DELEGUE CHARGE DE L'ENVIRONNEMENT

Les problèmes posés par les pluies acides et autres formes de pollution atmosphérique à longue distance constituent l'une des préoccupations majeures des responsables chargés de la protection de l'environnement en Europe et en Amérique du Nord.

Pour ma part, j'en fais une des toutes premières priorités de mon action.

En effet, personne ne peut nier aujourd'hui l'existence de transports transfrontières et de transformations physico-chimiques des polluants rejetés dans l'atmosphère par les activités humaines, même si les mécanismes mis en jeu ne sont pas tous expliqués ni quantifiés. En outre, il est généralement admis que les dépots des produits acides ou oxydants ainsi formés peuvent, dans certaines conditions qui restent à préciser, contribuer à l'acidification très préoccupante de nombreux lacs en Scandinavie, au Canada et dans le Nord-Est des Etats-Unis, ainsi qu'aux dépérissements forestiers plus ou moins marqués qui sont observés dans la plupart des pays d'Europe et d'Amérique du Nord, surtout depuis le début des années 80. Enfin, n'oublions pas les effets possibles sur les productions agricoles et la santé humaine, et les dégradations, certaines celles-là, de nos constructions et de nos monuments historiques.

A la fois pour toutes ces raisons et en tenant compte de l'aspect international du problème, je suis persuadé que nous devons, au sein de l'Europe, poursuivre nos efforts dans la voie des réductions des émissions de polluants qui sont, au moins en partie, à l'origine des phénomènes observés, à savoir le dioxyde de soufre, les oxydes d'azote et les hydrocarbures.

L'APPORT DE LA RECHERCHE

Les remèdes préventifs (réduction des rejets polluants dans l'atmosphère) ou curatifs (restauration des écosystèmes perturbés) à mettre en oeuvre par les responsables de l'environnement dépendront évidemment du progrès d'un certain nombre de connaissances scientifiques.

Parallèlement à la mise en oeuvre de ces différents remèdes il convient de poursuivre et de développer les travaux de recherche pour mieux comprendre les causes et les mécanismes des altérations de nos écosystèmes terrestres et aquatiques.

Il faut préciser le rôle des différentes formes de pollution qui, si l'on excepte le cas de doses massives de polluants, ne sauraient être tenues pour seules responsables des dommages constatés, par exemple, parmi les causes possibles du dépérissement des forêts, il faut également envisager les agressions dues à des conditions climatiques extrêmes (gels, sécheresses...) ou à des parasites et micro-organismes divers (insectes, champignons, virus, ...). Il convient d'examiner l'influence néfaste de conditions écologiques précaires (surtout en montagne), voire de certaines pratiques sylvicoles mal adaptées aux conditions du milieu.

Il s'agit non seulement de délimiter la responsabilité de la pollution atmosphérique par rapport aux autres causes possibles, mais aussi d'identifier les composés chimiques réellement impliqués et de déterminer, dans la mesure du possible, leur origine, leurs concentrations et leurs doses critiques en dessous desquelles les effets indisérables ne se manifesteraient probablement pas.

C'est pourquoi, je suis particulièrement heureux d'accueillir à Grenoble ce symposium, organisé avec l'aide de mes services par la Commission des Communautés Européennes, et dont le but est justement de dresser le bilan de la première phase (1984-1986) des recherches effectuées sur ce sujet dans le cadre du programme de recherche et de développement de la Commission des Communautés Européennes dans le domaine de la protection de l'environnement.

L'EFFORT DE RECHERCHE DES COMMUNAUTES EUROPEENNES

Et je suis d'autant plus heureux de la présence à nos côtés du Dr. Karl-Heinz NARJES, Vice-Président de la Commission des Communautés Européennes, que cela me donne l'occasion de lui exprimer toute notre gratitude pour l'effort de recherche et de développement accompli par la CEE dans le domaine si important de la protection de l'environnement.

La France apprécie à leur juste valeur des résultats déjà obtenus et je suis certain que la conclusion des travaux de ce symposium le confirmera.

Au nom du gouvernement français, je vous remercie tout particulièrement, Monsieur le Vice-Président, pour le soutien que vous avez apporté, dès son lancement en octobre 1984, à notre programme national de recherche sur le "dépérissement des forêts attribué à la pollution atmosphérique" (programme DEFORPA), soutien que vous venez de lui renouveler pour la période 1987-1990. Ainsi, les recherches sur le dépérissement de nos forêts, dont les résultats les plus récents sont présentés ici même, bénéficient-elles des nombreux échanges d'informations et de la coordination des travaux que vous assurez au plan européen grâce aux actions concertées COST 612 et COST 611, cette dernière portant sur le comportement physico-chimique des polluants dans l'atmosphère.

DEFORPA : UN PROGRAMME DE RECHERCHE EXEMPLAIRE

Qu'il me soit permis, enfin de présenter un bref historique de la genèse du programme DEFORPA dont la conception scientifique et la conduite me paraissent exemplaires.

A la suite d'une enquête menée en 1982 auprès des services forestiers allemands, la RFA et la France (en ce qui concerne les Vosges où étaient apparus les premiers dommages) ont mis l'une et l'autre en place un réseau d'observation systématique de l'état sanitaire des forêts, selon des protocoles relativement proches.

Depuis 1984, le réseau français, géré par l'ONF, s'est régulièrement étendu. Monsieur le Directeur de l'Espace Rural et de la Forêt, présent à nos côtés, en dressera les grandes lignes et donnera les résultats les plus récents des observations effectuées.

Parallèlement à la mise en place de ce réseau, le Centre de Recherches Forestières de l'INRA à Nancy et le SRETIE, alertés par les observations des forestiers, ont impulsé une série de recherches sur les différents aspects du problème, avec la participation du CNRS et de nombreux laboratoires d'horizons divers, en particulier des laboratoires universitaires. L'intérêt des premiers résultats, l'exigence d'une évaluation scientifique rigoureuse et le besoin de coordonner les financements nécessaires, nationaux et européens, ont alors conduit le Ministère de l'Environnement (désigné comme coordonnateur), le Ministère de l'Agriculture et le Ministère de la Recherche à doter le programme DEFORPA d'un comité de direction inter-ministériel et inter-organismes ainsi que d'un comité scientifique placé sous la présidence du Professeur Pierre JOLIOT du Collège de France.

Le comité de direction, présidé par Monsieur Lucien CHABASON, Chef du

SRETIE, est responsable de l'organisation, de la coordination et du financement du programme. Le comité scientifique propose les orientations scientifiques, donne son avis sur les projets de recherche soumis au comité de direction et procède à l'évaluation des résultats.

Les recherches sont dirigées par Monsieur Maurice BONNEAU, Directeur de Recherche à l'INRA, assisté d'un groupe opérationnel de chercheurs spécialistes des disciplines impliquées dans le programme et travaillant tous vers le même but.

Je laisserai évidemment aux chercheurs le soin de présenter les résultats de leurs travaux mais je saisis l'occasion qui m'est aujourd'hui offerte pour remercier de leur engagement dans des recherches difficiles M. JOLIOT et M. BONNEAU ainsi que l'ensemble des chercheurs, au sein d'organismes scientifiques, d'universités, d'établissements publics et de laboratoires industriels qui apportent leur concours à ce programme.

LES REMEDES PREVENTIFS OU CURATIFS

Cela dit, au-delà de la recherche proprement dite, dans le domaine de la prévention et pour ce qui est des solutions techniques que nous pouvons actuellement impulser, plusieurs dossiers européens ont avancé réellement depuis quelques années.

En même temps, un certain nombre de questions attendent pour l'instant vainement des réponses concrètes. Ne pas les évoquer retarderait encore leur traitement : je vais aussi m'en expliquer.

La pollution automobile est le sujet sur lequel l'action européenne a été la plus visible. Dans ce domaine qui concerne à la fois la pollution acide et la pollution par les hydrocarbures, il était indispensable d'aboutir à une décision commune pour préserver l'intégrité du marché commun de l'automobile.

L'accord politique conclu à Luxembourg en 1985, qui a été solennellement confirmé à Bruxelles en mars dernier, représente pour l'Europe un grand pas en avant. Il permettra une réduction considérable de la pollution automobile dans des conditions économiques supportables pour les usagers.

Je regrette toutefois que cet accord n'ait pu être encore officialisé. Je souhaite que tout soit mis en oeuvre pour que cette affaire soit enfin réglée et pour qu'une décision soit rapidement prise sur deux dossiers encore en discussion : les rejets de particules de voitures diesel et la limitation de vitesse, dont l'aspect favorable pour l'environnement se double, bien sûr, d'un impact très positif sur la sécurité.

La France, pour sa part, cherche à faire progresser tous ces dossiers. J'aurai l'occasion de le dire de nouveau dans trois jours lors du prochain conseil des ministres européens de l'environnement.

J'accorde également une très grande importance à un autre dossier européen : le projet de directive sur les grandes installations de combustion.

Il s'agit en effet de la principale source de pollution acide en Europe : il serait utile que les états européens adoptent une réglementation stricte pour réduire cette pollution.

La France a soutenu l'ambitieux projet initial présenté par la Commission. Il n'a malheureusement pu être adopté par l'ensemble des états. Les discussions durent maintenant depuis trop longtemps alors qu'il est urgent d'aboutir. Je le dirai également à mes collègues dans trois jours : chaque état doit faire un geste de conciliation. Pourvu que chacun fasse un effort et que les mesures prises aient une réelle efficacité, la France, pour sa part, pourra accepter une décision allant moins loin que ce qu'elle espérait il y a quelques années.

J'ai beaucoup parlé de l'Europe, et c'est normal. Mais notre pays mène aussi une action nationale qu'il convient de souligner.

Dans notre programme de lutte contre les pluies acides, nous nous sommes fixés deux priorités : les oxydes de soufre et les hydrocarbures.

Pour ce qui est des oxydes de soufre, nous avons déjà obtenu de bons résultats puisque nos rejets ont été divisés par deux depuis 1980 grâce, en particulier, à la réalisation de notre programme électronucléaire. Mais nous voulons aller plus loin. C'est pourquoi nous encourageons le développement de la désulfuration, notamment sur les chaudières industrielles ou de chauffage urbain. Plusieurs réalisations sont en cours et nous devons continuer dans cette voie.

En ce qui concerne les hydrocarbures, la France a été, je crois, un des premiers pays à se fixer un objectif de réduction ; cet objectif est de moins 30 %. La réduction de la pollution automobile y contribuera, mais nous devons aussi agir vis-à-vis des sources fixes : d'ores et déjà j'ai pris des décisions visant à réduire l'évaporation dans les stockages d'hydrocarbures. Prochainement, des mesures seront prises dans les secteurs de la peinture automobile et du prélaquage. Des réglementations sont également à l'étude pour d'autres activités comme l'imprimerie.

Je souhaiterais, d'ailleurs, que la Communauté Européenne envisage de mener des actions dans ce domaine des sources fixes d'hydrocarbures.

CONCLUSION

Ces actions de réduction des pollutions, pour être indispensables, n'en seront pas moins longtemps insuffisantes.

Tant qu'il y aura des pollutions, c'est-à-dire tant qu'il y aura des activités humaines, nous devrons continuer à en surveiller les effets.

Nos efforts permettent cependant aujourd'hui d'affirmer qu'en Europe une démarche tout à fait positive est en cours. Il y manquait ces dernières années une véritable volonté de recherche scientifique : eh bien, ce symposium en est l'évidente manifestation, en 1987, année européenne de l'environnement, cet objectif est en passe d'être atteint.

Le programme DEFORPA va bénéficier d'un nouveau contrat européen de 8 MF et nous devons ici nous en féliciter tout particulièrement.

Dans le même temps, la Communauté Européenne s'apprête à arrêter ses décisions sur le volume et les priorités du prochain programme cadre de recherche : l'environnement y occupera une place importante. A l'échelle de l'Europe occidentale entière le programme EUROTRAC dans le cadre d'EUREKA, va être lancé. Il rassemble une centaine de laboratoires européens mobilisés pour mieux connaître le processus d'acidification à l'échelle du continent.

Ce symposium de Grenoble est un nouveau pas en avant vers une Europe scientifique de l'environnement en marche : notre pays y contribue activement.

EXPOSE

de Monsieur Christian BARTHOD
Direction de l'Espace Rural et
de la Forêt du Ministère de l'Agriculture

Messieurs les Ministres, Monsieur le Président, Monsieur le Préfet, Messieurs les Directeurs, Mesdames, Messieurs,

Je vous demande de bien vouloir excuser Monsieur François GUILLAUME, Ministre de l'Agriculture, chargé de la politique forestière, retenu par d'autres obligations.

Le réseau français d'observation du dépérissement des forêts a commencé à se mettre en place en 1983 dans les Vosges à l'initiative de l'Office National des Forêts. Depuis cette date, il s'est régulièrement étendu et permet de dresser l'état sanitaire d'une partie importante de la forêt de l'Est de la France.

En 1986, le réseau systématique de placettes d'observation de l'état sanitaire des arbres en forêt a concerné 13 régions administratives, avec au total 1.416 placettes du réseau à 16 km x 1 km, sans compter les placettes qui ont du être ajoutées dans certaines régions, en forêts gérées par l'ONF, pour améliorer l'échantillonnage avec alors une distribution de 8 x 1 km ou même 4 x 1 km.

Le Ministère de l'Agriculture assume la responsabilité du réseau de surveillance et de l'interprétation des quelques 260.000 données annuelles qui en découlent. De l'automne 1985 à l'automne 1986, on note une stabilisation de la santé du sapin dans l'ensemble des montagnes françaises, voire même, dans les Vosges au moins, une amélioration de sa vigueur, tandis que le jaunissement de l'épicéa persiste ou s'amplifie légèrement. La situation demeure certes préoccupante mais il n'y a pas lieu de céder au catastrophisme.

Sur un total de 1.416 placettes du réseau systématique à 16 x 1 km, couvrant une surface totale de 2.265.000 ha, 959 se trouvent dans les forêts gérées par l'ONF (soit 67,7 % du total) et 457 dans les forêts particulières (soit 32,3 % du total).

On constate dans l'ensemble, une aggravation de l'état des feuillus et une amélioration de l'état des conifères. Toutefois, les feuillus restent nettement moins endommagés que les conifères, autant par la perte de leur feuillage que par la décoloration de celui-ci.

L'aggravation de l'état des feuillus est particulièrement nette en Lorraine, Alsace, Bourgogne et Midi-Pyrénées, alors qu'elle est moins sensible en Franche-Comté, Rhône-Alpes, Nord-Pas de Calais, Champagne-Ardenne et pratiquement nulle en Haute-Normandie.

Sur l'ensemble de la France, c'est sur le chêne et les feuillus divers que les dommages paraissent augmentés, en passant respectivement de 5 % à 6,2 % et de 3,9 % à 5,5 % pour la proportion d'arbres ayant perdu plus de 25 % de leur feuillage.

Inversement, la situation du sapin pectiné et du pin sylvestre parait améliorée, le pourcentage des arbres ayant perdu plus de 25 % de leurs aiguilles passant respectivement de 18,6 % à 15,3 % et de 17,7 % à 14 %. L'épicéa est resté stable (10 % et 9,9 %). Le mélèze et les autres conifères sont trop peu nombreux pour que les résultats soient significatifs.

Du fait du comportement en sens contraire des feuillus et des conifères, la situation pour l'ensemble des essences est inchangée, la proportion des arbres ayant perdu plus de 25 % de leur feuillage étant de 8,4 en 1985 et de 8,3 % en 1986.

Le Ministère de l'Agriculture, très conscient des problèmes d'interprétation statistique des données issues du réseau bleu (notamment pour les comparaisons entre pays) a pris une part très active dans l'élaboration du protocole harmonisé de la CEE (réseau à grand maillage 16 km x 16 km) qui vient être adopté (6-7 mai 1987) dans le cadre du réglement "Protection des Forêts". Les observations 1988 seront assurées parallèlement sur tout le territoire national selon le protocole européen et sur les zones critiques de manière beaucoup plus détaillée selon le protocole actuellement en vigueur en France.

En collaboration avec le Ministère de l'Environnement et le Ministère chargé de la Recherche, le Ministère de l'Agriculture prend une part active à l'élaboration, au financement et au suivi du programme DEFORPA. Les forêts et leurs propriétaires sont en effet les principales victimes de ce phénomène encore mal connu.

Pour ce qui relève du domaine de compétence du Ministère de l'Agriculture, il apparaît prioritaire de mener un effort tout particulier dans trois directions :

- Caractériser de façon objective, quantitative et surtout spécifique les différents types de dépérissements, que les arbres extériorisent de façon voisine. Il s'agit de lancer une recherche sur la symptomatologie. La plupart des programmes de recherche ont trop négligé ce point capital.
- Amplifier l'approche dendroécologique qui consiste à mettre en évidence l'influence des fluctuations des conditions d'environnement sur l'accroissement des arbres, permettant, le cas échéant, de mettre en lumière des périodes de fragilisation des peuplements forestiers français dûes, par exemple, à des facteurs climatiques. Il s'agit d'un savoir-faire original des forestiers français qui représente une contribution appréciable aux recherches internationales.
- Rechercher des indicateurs biochimiques précoces du dysfonctionnement physiologique des arbres dépérissants.

Par ailleurs, les négociations entre la France et la RFA pour le montage du réseau européen de recherche sur la physiologie des ligneux (EUROSILVA) sont très avancées. Un projet détaillé, avec une liste de laboratoires concernés et un calendrier prévisionnel est actuellement à l'étude dans les ministères allemand et français chargés de la recherche. Le Ministère de l'Agriculture a participé aux négociations de ce projet dont certains aspects (physiologie des stress) pourront éclairer le problème du dépérissement. Mais ce réseau EUROSILVA n'est pas réductible au seul problème du dépérissement.

Pour mener une action efficace, l'effort de garder son sang froid et d'adopter une grande rigueur scientifique est indispensable. Les trois ministères concernés et le Comité Scientifique du Programme DEFORPA se rejoignent sur cette nécessité. Si la recherche forestière coordonne un très important effort dans et autour de son domaine de compétence, il est essentiel de souligner qu'on ne peut en rester là : la forêt n'est pas seule en jeu, d'où l'intérêt d'une approche intégrée et européenne, dont le présent symposium témoigne.

KEY NOTES SPEECHES

Chairman : T. SCHNEIDER

National Institute of Public Health
and Environmental Hygiene (The Netherlands)

ECOSYSTEMS AND THEIR RESPONSE TO AIRBORNE CHEMICALS: THE CURRENT SITUATION IN NORTH AMERICA AND EUROPE

ELLIS B. COWLING
Associate Dean for Research
School of Forest Resources
North Carolina State University
Raleigh, North Carolina 27695 USA

Summary

Human activities of many sorts are changing the chemical climate of the earth. Some of these changes are known to increase the productivity of terrestrial and aquatic ecosystems; but others induce stresses that change the current condition of specific plants, animals, and microorganisms or alter the health, productivity, genetic structure, and/or the geographic distribution of whole ecosystems. Recent research to determine the effects of airborne nutrient, acidic, toxic, growth-altering, and radiatively active airborne chemicals is leading to an improved understanding of how ecosystems develop and how they respond to stresses imposed by airborne chemicals and other environmental perturbations. This paper provides an overview of current and future changes in the chemical climate and the responses of ecosystems to such changes in North America and Europe.

1. INTRODUCTION

During the past several years, a series of major shifts have been taking place in both scientific and public debates about "acid rain", "air pollution", and "global climate change". All three issues are related because all three: 1) are manifestations of change in the chemical climate of the earth; 2) involve many of the same chemical emissions from combustion of fossil fuels; 3) involve many of the same photochemical transformations that occur during short-distance and long-distance transport of air pollutants; 4) are known or suspected to increase stress in terrestrial and aquatic ecosystems; and 5) involve increased risk of direct and indirect effects on human health, increased corrosion of engineering materials, accelerated damage to statuary and other cultural resources, increased haziness of the atmosphere, and other detrimental influences on society.

The objectives of this paper are to:

1) Briefly summarize evidence that the chemical climate of the earth is changing in ways that may affect the health, productivity, and stability of forests and other natural and managed ecosystems;

2) Summarize the current state of knowledge regarding known and projected responses of ecosystems to airborne chemicals. Since effects of airborne chemicals on forests are my special field of competence, many of the ideas outlined below use

forests as examples. These same ideas are also germane, however, to other terrestrial and aquatic ecosystems.

2. RECENT CHANGES IN THE CHEMICAL CLIMATE

Three general types of airborne chemicals are known or projected to affect terrestrial and aquatic ecosystems:

1) Essential nutrient substances that help living things grow;

2) Toxic or growth-altering substances that cause direct injury or inhibit normal development of living organisms; and

3) Radiatively active gases which can lead to general warming and other changes in the physical climate of the earth.

Beneficial Airborne Chemicals

The beneficial airborne chemicals that help ecosystems develop include the following: CO2, H2O, O2, NH3, NO3-, NH4+, PO4---, K+, S, SO4--, Mg++, Ca++, Mn++, Fe++, Cu++, Zn++, Mo++, BO4-, and Cl-. These substances include all 16 of the essential nutrient elements required by plants for normal growth and development. Nine other micronutrient elements are required for normal growth and development of many animals and some microorganisms; these include cadmium, chromium, fluorine, iodine, nickel, selenium, silicon, tin, and vanadium. All 25 of these essential nutrients are dispersed in the atmosphere and can be taken up readily through leaves and green shoots as well as through roots of plants. They are also ingested in food materials and assimilated by animals and microorganisms.

Injurious Airborne Chemicals

The injurious substances that inhibit growth or cause direct injury to forests and other ecosystems are of four general types:

1) toxic gases such as O3, SO2, NO, NO2, F, hydrogen peroxide (H2O2), peroxyacetyl nitrate (PAN), and peroxypropionyl nitrate (PPN) which can penetrate stomata and thus damage foliar organs or other plant parts after very brief periods of exposure (hours to days or weeks);

2) toxic metals such as Al, Pb, Hg, Cd, and Zn which are deposited from the atmosphere over many years or are gradually liberated from soils in toxic amounts and can inhibit growth or damage living tissues after relatively long periods of exposure (years to decades);

3) excess nutrient substances, especially all biologically available forms of nitrogen, which can be taken up by plants in greater-than-normal amounts and thus can alter normal patterns of growth and development over intermediate periods of time (weeks to months); and

4) growth-altering organic chemicals, such as ethylene, aniline, or dinitrophenol which can change the normal patterns of growth and development but for which the time period of action is unknown.

Climate-Altering Airborne Chemicals

The most important climate-altering airborne chemicals include H_2O vapor, CO_2, N_2O, NO_2, SO_2, CH_4, CCl_3F, CCl_2F_2, and O_3. All are energy absorbing gases which alter the radiative energy balance of the earth. Warming of the earth's surface is a result of a "blanketing" effect of water vapor, carbon dioxide, and the other trace gases listed above which absorb infrared radiation from the earth's surface and atmosphere, and reradiate a portion of that energy back to the earth causing the so-called "greenhouse effect" (McDonald, 1985).

Natural and Human Sources of Airborne Chemicals

All of these airborne substances -- beneficial, injurious, and climate altering -- are derived from a wide variety of natural and human sources. The natural sources include volatile emissions from decomposing plant and animal remains, volcanoes, and wild fires, sea spray from oceans and large lakes, wind-blown dust from arid regions, and biogentic particles such as spores and pollen.

The largest human sources include volatile waste products from combustion of fossil fuels in power plants, metal smelters, industrial boilers, transportation vehicles, and domestic and commercial space and water heating installations. They also include volatile emissions from decomposition or incineration of domestic, commercial, and industrial wastes, controlled burning of agricultural and forest residues, evaporation of solvents, liquid fuels, refrigerants, pesticides, and other volatile chemicals, and use of explosive devices in peace and war.

The relative contributions of natural and human sources of airborne chemicals vary greatly with the particular chemical and region of the world involved. For example, in North America and Europe, natural sources of ammonia and ammonium ion in the atmosphere are believed to be larger than human sources. In these same continental areas, however, human sources are about 20 times larger than natural sources in the case of sulfur oxides and about 10 times larger in the case of nitrogen oxides.

Ozone and some other photochemical oxidants such as hydrogen peroxide, formaldehyde, PAN, and PPN are natural constituents of the earth's atmosphere, but are present in comparatively low concentrations (up to about 30 ppb). In some urban-industrial areas, concentrations of ozone typically accumulate to 80-200 or more ppb. These high concentrations accumulate in the atmosphere as a result of photochemical transformations of volatile organic compounds and nitrogen oxides released from various human activities.

Some climate-altering chemicals such as CCl_3F and CCl_2F_2 are produced only by human activities but most radiatively active gases are both natural constituents and volatile waste products of human activities. For example, greatly increased amounts of CO_2 and CH_4 are accumulating in the global atmosphere because of increased use of fossil fuels, biomass burning, decay and burning of forests to make way for agriculture or human settlements, and decomposition of municipal and agricultural wastes (McDonald, 1985).

Time-Period of Change in the Chemical Climate

During the preindustrial period in which forests and other natural ecosystems evolved, the concentrations of these beneficial, injurious, and climate-altering chemicals in the atmosphere were relatively low. Forest trees, herbaceous plants, microorganisms, and animals all adapted their habits of growth and nutrition to a very dilute chemical environment. As the industrial revolution gathered momentum about the middle of the 19th century, however, human activities added more and more substances to those that circulate naturally among the air, water, soil, and living things.

During the past century, major increases in combustion of coal, oil, and other fossil fuels and production and use of volatile organic compounds have taken place in North America and Europe. As a result, equally major changes have taken place in emissions of sulfur and nitrogen oxides and volatile organic compounds (VOC). Since 1880, emissions of sulfur and nitrogen oxides have increased approximately 10 fold (Husar, 1986; Gschwandtner et al, 1986). Reliable data on changes in emissions of VOC are not available but are estimated by informed scientists to be of about the same order of magnitude (Altshuller, personal communication). Two of these three primary pollutants (SO2 and NOx) are toxic in their own right (as demonstrated by numerous observations and detailed investigations around major industrial and urban sources of sulfur and nitrogen oxides). Also, all three primary pollutants are the principal precursors of regionally dispersed secondary pollutants which have been rigorously proven in the case of ozone or are widely suspected in the case of acid deposition to be harmful to plants, animals, or microorganisms.

In the early days of industrialization, injury to forests and other vegetation (and even public health) were accepted as unavoidable side effects of early metal-smelting and combustion technologies. In time, however, these adverse effects became better understood and methods were devised to minimize such effects. Control technologies are now required by law in most of Europe, North America, and Japan. As a result, the known adverse effects of gaseous SO2, NOx, and F are now confined almost exclusively to the immediate vicinity of large sources of air pollution such as metal smelters, power plants, and industrial complexes. Thus, the center of gravity in research on the ecological effects of air pollution has shifted from intensive studies of acute effects of primary pollutants in the vicinity of large sources to extensive studies of chronic exposure to regionally dispersed secondary pollutants (such as ozone, hydrogen peroxide, and other photochemical oxidants) and to radiatively active gases which may alter the physical climate of the earth.

3. CURRENT KNOWLEDGE ABOUT EFFECTS OF AIRBORNE CHEMICALS

Present knowledge about effects of airborne chemicals on forests and other natural ecosystems has developed mainly in three distinct sets of scientific disciplines:

-- ecology and plant and animal nutrition with regard to the effects of beneficial nutrient substances that help organisms grow;

-- plant pathology, veterinary and medical science, and plant and animal physiology with regard to the effects of harmful substances that cause injury or inhibit normal growth and development of terrestrial and aquatic ecosystems; and

-- plant and animal geography, forest and wildlife ecology, and both physical and chemical climatology with regard to possible effects of global climate change on forests, agricultural, and aquatic ecosystems.

Many socalled "pollutant chemicals" such as sulfur and nitrogen oxides and certain heavy metals such as cadmium and zinc are also beneficial nutrients; that is, they are injurious at certain concentrations and beneficial at other concentrations. This is part of the reason why many scientists now prefer to use the term "airborne chemicals" rather then "air pollutants", "acid deposition", or "acid rain" to describe their area of special scientific interest.

General Responses of Ecosystems to Changes in the Chemical and Physical Climate

During all the millenia that followed the first development of life on our planet, microorganisms, plants, and animals adapted their habits of growth, nutrition, and metabolism to fit within the dominant physical and chemical features of their environment. The processes of natural selection determined which organisms survived and which organisms perished. If a given organism was well adapted to its environment, it thrived and reproduced. If it was not fit, it did not survive. Whenever changes occur in the physical climate or the chemical climate, new pressures of natural selection are applied to the population of surviving organisms (Johnson and Sharpe, 1982; Woodwell, 1970). These continuing processes of evolutionary change and adaptation have led to the development of the marvelously diverse communities of living organisms that we find everywhere over the land and in the surface waters of the earth.

These natural ecosystems, both terrestrial and aquatic, come as a gift from the evolutionary history of our planet. When human beings first appeared on the earth and the process of civilization began, our collective impacts on the processes of natural selection and evolution were hardly perceptible. But as we increased in numbers and particularly after we learned to:

-- harness the energy stored in fossil fuels, and

-- apply this energy to the processes of urbanization, industrialization, and intensive agriculture and forestry,

our collective impacts on the processes of natural selection and evolution became progressively more impressive. Most western and eastern peoples have considered that the diverse flora and fauna of the earth are ours to use as we see fit -- to manage carefully, that is within the sustainable productive capacity of the ecosystems in question, or to exploit carelessly, that is with only limited regard for the long term stability of the ecosystems themselves.

Today, the aquatic and terrestrial ecosystems in certain high elevation, industrial, and urban locations in North America, Europe, Japan, China, South Africa, and some developing countries are exposed to much heavier loadings of airborne nutrients, acidic, toxic, growth-altering, and climate-altering chemicals than were present during the preindustrial period in which these ecosystems evolved. In some locations, essentially all the nutrients needed to sustain some of these ecosystems are now provided from atmospheric sources. Never before in their evolutionary history have the natural and managed ecosystems of North America and Europe been "fed from above" to the extent that they are today!

Under these new conditions of continuing change in the chemical climate, additional pressures for further adaptation of plants, animals, and microorganisms are applied and have their influences within terrestrial and aquatic ecosystems. At present we have only limited experience and even less scientific evidence with which to identify regions where the rates of continuing change in the chemical climate are within the elastic limits of ecosystem resiliency and adaptability and where these rates of change will exceed those limits. Nowhere are these uncertainties more evident than in the forests of central Europe and in the lakes, streams, and forests of northern Europe and certain parts of eastern North America.

In theory, airborne chemicals in the form of gases, aerosols, and dissolved or suspended substances in air, cloud water, or precipitation could cause at least eight different types of effects on aquatic or terrestrial ecosystems, either alone or in combination with other stress factors:

1) visible symptoms of injury;
2) decreased growth, with or without visible symptoms of injury;
3) interference with normal reproductive processes;
4) decreased genetic diversity within a given species of organisms;
5) decreased productivity of whole ecosystems;
6) decreased regional productivity of ecosystems;
7) changes in species composition of an ecosystem; and
8) changes in the geographic distribution of an ecosystem.

Only the first three of these eight theoretical effects have been confirmed by observation and experiments in specific case studies of various aquatic or terrestrial ecosystems.

In both Europe and North America, most studies of the detrimental effects of air pollutants on forests and agricultural ecosystems have been made in the vicinity of strong point sources of SO_2, NO_x, and F (Woodwell, 1970; Carlson, 1974; Thompson, 1981; Legge et al, 1980; Treshow et al, 1967; Adams, 1962; Scurfield, 1960; Miller and McBride, 1975; Postel, 1984). Much less is known about effects of regionally dispersed airborne chemicals such as ozone or acid deposition (Berry and Hepting, 1964; Miller et al, 1963; Woodman and Cowling, 1987).

Concept of Multiple Stresses

There are seven general classes of stress factors that affect the health, productivity, species composition, genetic diversity, geographic distribution, and survival of ecosystems. These include:

1) Natural competitional stresses that occur whenever individual organisms compete with each other for limited supplies of growing space, food, solar radiation, water, and/or essential nutrients from the same ecological niche. Natural competitional stresses also include interactions among organisms of the same or different species which result from injurious chemicals (allelopathic chemicals) produced by certain organisms which adversely affect the survival and development of other organisms in the same environment. In any planted or naturally regenerated forest, for example, only a few trees survive to maturity; in fact more trees are inhibited in normal development or die of competitional stresses than all other stress factors combined.

2) Natural climate stresses that include freezing temperatures, drought, flooding, high winds, low humidity, extremes of heat and cold (especially rapid changes leading to sun scalding or frost injury and winter desiccation; mechanical damage by ice and snow; and finally, wild fire and/or controlled burning of vegetation.

3) Natural biological stresses that result in the impairment of normal physiological or ecological processes. These biotic stress factors include parasitic and pathogenic fungi, insects, nematodes, bacteria, viruses, viroids, plasmids, mycoplasmas, parasitic seed plants, and injurious animals such a porcupines and deer. In forests, the most important of these natural biotic stresses are induced by fungi and insects.

4) Natural chemical stresses that result from deficiencies (and occasional excesses) of essential nutrients or excesses of toxic soil chemicals such as aluminum. Nitrogen is the most common nutrient deficiency in forests and many other natural ecosystems.

5) Human disturbance stresses that are imposed by various advertent or inadvertent human activities include logging, burning, draining, flooding, and physical disturbance of soils leading to physical compaction, erosion, leaching of nutrients, or accumulation of toxic substances.

6) Air pollution stresses that occur whenever ecosystems are exposed to injurious concentrations of toxic gases such as SO_2, NO_x, ozone, or fluoride, toxic aerosol particles or coarse particulate matter, or dissolved or suspended chemicals in air, cloud water, or precipitation. Pollutant stresses can also occur as a result of the accumulation or mobilization of toxic substances in soils after wet or dry deposition from the atmosphere.

7) Global climate stresses that may occur as a result of increased accumulation of radiatively active gases in the

atmosphere. These stresses are projected to be manifest through a combination of general warming of the climate coupled with increased severity and longevity of droughts and wind, rain, or ice storms (Johnson and Sharpe, 1982).

Quality of Evidence Regarding Effects of Airborne Chemicals

At present a great disparity exists in the quality of scientific evidence about the detrimental effects of:

1) acute exposures to locally dispersed primary pollutants such as sulfur dioxide and fluoride which are emitted directly from point sources; and

2) chronic exposures to regionally dispersed secondary pollutants such as ozone and acid deposition which are formed from primary pollutants after dispersion in the atmosphere.

Steep gradients in pollutant concentrations in the case of most locally dispersed primary pollutants usually have made it fairly easy to establish a strong correlation between pollutant concentration and visible symptoms of damage and/or death of trees. When this strong correlative evidence was coupled with the results of a few controlled-exposure tests, little scientific uncertainty usually remained about what species of trees were affected, the source of the injurious substance(s), their chemical nature, or the concentrations that were injurious. Thus, sensible air-quality and/or forest-management recommendations could be formulated on the basis of straight forward relationships between the concentration of pollutant(s) in the air and the time period of exposure (dose) and the change in health or productivity of the forest (response). Only at the fringes of the area affected was there usually much uncertainty about cause/effect or dose/response relationships or about the role of competition, drought, frost, biotic pathogens, or other natural factors as predisposing or contributing stress factors.

Similarly neat cause-and-effect relationships have proven to be much more difficult to establish when chronic exposure to regionally dispersed secondary pollutants rather than locally dispersed airborne chemicals are involved. Often there are several reasons for this:

1) visible symptoms may be subtle or lacking entirely;

2) concentrations of airborne chemicals are often highly variable -- both in time and in space;

3) two or more airborne chemicals may occur simultaneously or sequentially and act additively, synergistically, or antagonistically; and, most important of all

4) rigorous scientific methods must be employed to distinguish the effects of airborne chemicals from those of natural stress factors which may act as predisposing, inducing, or contributing causal factors.

Most symptoms attributed to airborne pollutants are difficult to distinguish from those of natural stresses (Huttunen, 1984; Smith, 1981; Pinkerton, 1984). Gaseous pollutants often produce chlorosis or premature shedding of leaves -- symptoms similar to those induced by drought, frost, foliar diseases, and some nutrient deficiencies. Most postulated effects of acid deposition are indirect and thus similar to those induced by nutrient deficiencies and biotic pathogens that attack tree roots.

The two other reasons why it is difficult to detect airborne chemical stress are embodied in the principles of genetic diversity (Figure 1) and the relationship between forest productivity and growing stock in forests (Figure 2). In theory, it is possible that all individuals within a given species or population of trees might be uniformly susceptible or resistant to a given air pollutant. The preponderance of evidence from studies of both crop plants and forest trees indicates, however, that plant populations generally contain substantial genetic diversity in resistance and susceptibility to air pollutants (Scholz, 1984).

Figure 1 shows the hypothetical extent of this genetically controlled variation. Thus, only a portion are highly susceptible to air pollutants and these trees are distributed at random in most forests.

Figure 2 shows that forest productivity (usually measured as the volume of stemwood produced per unit area of land at a given age) is relatively constant over a wide range of tree densities or total forest volumes (Langsaeter, 1941). Thus, pollution-induced changes in growth or mortality rates of randomly distributed genetically susceptible trees are not likely to affect the productivity of a whole forest unless the total number of affected trees becomes so large that the forest becomes understocked.

It is possible that air pollution stress may simply increase mortality of pollution-sensitive trees enough that resistant trees will grow more rapidly because of their improved competitive advantage. For these several reasons it will be difficult to measure an impact on forest productivity even when many individual trees are injured or killed by airborne chemicals.

4. SCIENTIFIC APPROACHES IN THE DETERMINATION OF CAUSE AND EFFECT

The logical methods of determining cause-and-effect relationships dictate that a causal relationship can be inferred when there is a strong pattern of consistency, responsiveness, and/or a proven biological mechanism with the suspected causal factors (Mosteller and Tukey, 1977). In research on the effects of air pollution on ecosystems, consistency requires that injury or dysfunction symptoms in the ecosystem must be associated consistently with the presence of the suspected causal factor. Dose-response relationships are established through tests in which healthy plants or animals are exposed to various known concentrations of suspected airborne pollutants under controlled conditions that simulate conditions in the ecosystem. This relationship is called responsiveness. A mechanism is one or a series biological processes through which the suspected cause is related to the observed effect. In simple systems that involve only one or two causal factors, any two of these three patterns of linkage may be sufficient to infer cause. In more complicated systems, all three may be necessary.

Figure 1 Hypothetical relationship of inherent sensitivity of forest trees to pollution stress.

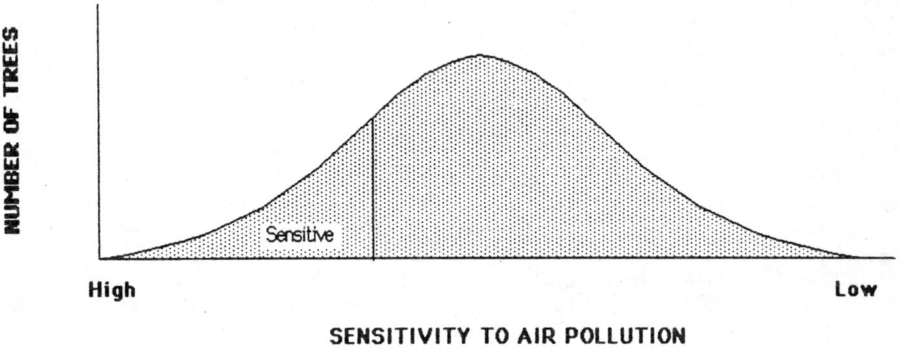

Figure 2 Hypothetical relationship of forest growth with forest biomass (Langsaeter 1941)

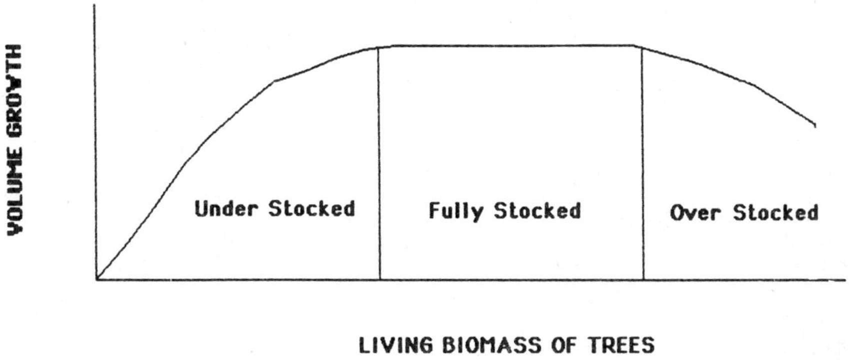

This notion of patterns of linkage is derived in part from a series of four rules for proof of causality which were first used by Robert Koch (1876) and later Louis Pasteur to establish the germ theory of disease. For more than a century, Koch's Postulates have defined both the procedural steps and standard for rigor in proof of causality in the medical, veterinary, and plant sciences.

Many scientists have recognized the need for adaptation of Koch's Postulates when applied to abiotic stress factors such as air pollution (Adams, 1962; Huttermann, 1984; Last et al., 1984; Cowling, 1985). These adaptations fulfill the consistency and responsiveness patterns of linkage with the following rules of proof:

Rule 1) The injury or dysfunction symptoms observed on individual organisms in the ecosystem must be associated consistently with the presence of the suspected causal factor(s);

Rule 2) The same injury or dysfunction symptoms must be duplicated when healthy organisms are exposed to the suspected causal factor(s) under controlled conditions;

Rule 3) Natural variation in resistance and susceptibility of organisms observed in the ecosystem must be duplicated when clones of these same organisms are exposed to the suspected causal factor(s) under controlled conditions.

The first two of these adaptations of Koch's Postulates must be fulfilled to draw a firm inference about cause-and-effect; the third increases the biological rigor of the second. These adaptations are useful mainly in studies of relatively simple systems involving one, two, or possibly three interacting causal factors.

Wallace (1978) has suggested a synoptic approach for diagnosis of complex problems involving three or more interacting causal factors -- initial surveys to identify key variables, multiple regression analyses to generate hypotheses, and field tests to verify the diagnosis achieved.

The four scientific approaches commonly used in research on the response of ecosystems to air pollutants are damage surveys, controlled-exposure tests, mechanisms tests, and risk assessments.

Damage or injury surveys quantify the number of organisms or geographical extent of economic damage or visible injuries to plants or animals (Anonymous, 1984; Weiss et al., 1985; Rezabec et al., 1986). One common method has been to examine changes in radial growth using increment cores from sample trees across a pollution gradient (Cook, 1985; Johnson et al., 1981; McLaughlin, 1985). In air pollution research, damage surveys are useful only when combined with reliable data on air quality. Rule 1 (above) requires that concentrations of airborne chemicals, sufficient to cause injury, are present consistently at the same time and place in which symptoms are observed. But such correlational evidence alone does not constitute proof of cause.

Controlled-exposure tests range from simple comparisons of effects on healthy plants or animals exposed to pollutant-free and pollutant-laden air, to complex experiments in which healthy organisms are exposed to various concentrations of one or more airborne chemicals and other stress factors. Most controlled-exposure tests are performed with small plants or animals in growth chambers, greenhouses, or open-top chambers; tests with mature trees are rare. Results are often expressed as dose-response relationships or as minimum exposure (threshold value) needed to induce symptoms of injury or growth responses.

Controlled-exposure tests fulfill the responsiveness requirement (rules 2 and 3) by demonstrating that the suspected pollutant(s) can induce injury. Before it can be concluded that air pollution does cause injury, however, the test results must be combined with supporting data that demonstrate consistency under field conditions (Rule 1).

Mechanism tests are used to evaluate the hypothetical pathways proposed during the past several years for action of airborne chemicals (Schutt and Cowling, 1985; USEPA, USFS, and NCASI, 1986). For example, ozone has been postulated to inhibit tree and crop growth by damaging cell membranes. This results in decreased photosynthesis and/or increased leaching of nutrients from foliage by acid rain. Similarly, atmospheric deposition of nitrogen has been postulated to predispose high-elevation red spruce foliage to injury by frost. Just as correlation alone is not a proof of cause, neither is demonstration of a plausible mechanism a proof of cause unless it is supported by tests for consistency and responsiveness under field conditions (Rules 1 and 2 or 3).

Risk assessments vary widely in objectives, methods, amount of effort required, and degree of certainty expected. They can be used to estimate the likelihood that one or more airborne chemicals will affect the health or productivity of ecosystems in a given region.

5. PRESENT KNOWLEDGE ABOUT EFFECTS ON FORESTS

Most European and North American studies of the detrimental effects of air pollutants on forests have been made near metal smelters and other major sources of sulfur and nitrogen oxides or fluoride (Scurfield, 1960; Adams, 1962; Treshow et al., 1967; Woodwell, 1970; Carlson, 1974; Legge et al., 1980; Thompson, 1981; Postel, 1984). Much less is known about effects of regionally dispersed airborne chemicals such as ozone or acid deposition (Pinkerton, 1984).

Although nearly 20 regionally important cases of change in condition of forest trees have been reported in Europe and North America (Cowling, 1986), in only a few cases are airborne chemicals considered probable or possible causal or contributing factor. Some of the most important cases are described below in order of decreasing quality of evidence that airborne chemicals may be involved.

Rigorous Proof of Cause

The studies that demonstrate the strongest proof of cause are those that involve white pine forests in the eastern United States and mixed conifer and hardwood forests in southern California. A number of other field and experimental studies have been conducted as well.

Eastern White Pine. Ozone and other photochemical oxidants have been shown to induce visible injury and decreased growth in randomly distributed white pine trees in various parts of the eastern United States and Canada (Berry and Ripperton, 1963; Berry and Hepting, 1964; Linzon, 1967; Dochinger and Seliskar, 1970; Hayes and Skelly, 1977; Dochinger, 1978; Benoit et al., 1982). Injured trees are randomly distributed in part because of genetic diversity in resistance and susceptibility to ozone. Ozone injuries in white pine trees include a wide variety of foliar symptoms such as chlorotic mottling and premature loss of needles, decreased growth of individual trees, and mortality. No studies have shown a loss in volume productivity of white pine forests.

Proof that these injuries were induced by ozone is based on a number of lines of evidence that include injury surveys with concomitant air quality measurements, dose-response studies with ozone, and tests with clones of individual trees observed to be resistant or susceptible in forests. Thus, all three adaptations of Koch's Rules have been satisfied in this case.

Southern California forests. Ozone and other photochemical oxidants have also been shown to cause visible injury, decrease growth of some trees, and induce changes in species composition of forests near Los Angeles, California (Parmeter et al., 1962; Cobb and Stark, 1970; Miller and Elderman, 1977; Ohmart and Williams, 1979; McBride and Miller, 1985; McBride et al., 1985). Premature loss of needles leading to decreased photosynthesis is the major visible symptom; mortality of ozone-weakened trees usually is caused by bark beetles and root-rotting fungi. The major tree species affected include ponderosa pine, Jeffrey pine, white fir, limber pine, incense cedar, and California black oak. Injured trees are distributed at random in part because of genetically controlled resistance and susceptibility to ozone. No studies have shown a loss in forest productivity.

The research approaches used to demonstrate that ozone and other oxidants are involved include injury surveys with concomitant air quality measurements and dose-response studies with ozone. Thus, both Rule 1 and Rule 2 (above) have been fulfilled.

Other field and experimental studies. A number of investigators have made field measurements or controlled-exposure studies which fulfill either Rule 1 or Rule 2 or 3 (above), but not both. In these cases, the results apply only to the specific field or experimental situations where the studies were made. We mention two such studies as examples. Field symptoms of oxidant damage to seedlings of tulip poplar, green ash, hickory, black locust, eastern hemlock, table mountain pine, Virginia pine, and pitch pine have been correlated with high concentrations of ozone in the Shenandoah National Park (Duchelle et al., 1982). Also, annual dry matter production was greater when ozone was removed from the air surrounding seedlings of four clones of trembling aspen in New York state (Wang et al., 1986).

Circumstantial Evidence of Causality

There are two regional cases with only circumstantial evidence of causality: one in central Europe and one in eastern U. S. high-elevation spruce forests.

Central European forests. Essentially all commercially important tree species in Germany and several other countries of central Europe have shown neuartige Waldschaden (new types of forest damages) beginning about 1979-82. The major tree species affected include Norway spruce, silver fir, European beech, and certain species of oak, maple, ash, birch, and alder (Blank, 1984; Postel, 1984). The most common symptoms include chlorosis and visible thinning of tree crowns, decrease in growth, decrease in root biomass, and changes in the shape and size of leaves (Schutt and Cowling, 1985). These symptoms are most commonly seen on trees that are at least 60 years old (Schutt et al., 1983) and are distributed at random in forest stands. Only a small portion of trees have died. No studies have shown a loss in forest productivity. A few controlled exposure tests with ozone plus acid mists (Prinz, 1985) and simulated acid deposition on Norway spruce (Huttermann, personal communication) reproduced only a few field symptoms. Thus, none of the adaptations of Koch's Postulates have been fulfilled. Airborne chemicals are suspected mainly because no single natural factor or combination of natural factors can explain the observed changes in forest condition.

High elevation spruce-fir forests. Visible injury, decreased radial growth, and widespread mortality of some high elevation red spruce trees have been observed in the Appalachian Mountains from Vermont to North Carolina (Johnson and Siccama, 1983; Bruck, 1987). Changes in condition include dieback of tree tops and branch tips on saplings and larger trees (Siccama et al., 1982); widespread synchronous decrease in radial growth beginning in the early 1960's (Johnson and Siccama, 1983; McLaughlin, 1985; Johnson and McLaughlin, 1986); and widespread mortality and decrease in live basal area in New York, New Hampshire and Vermont (Scott et al., 1984; Weiss et al., 1985).

Current field research includes injury surveys and tree ring studies with very few concomitant measurements of air quality. Mechanism tests and controlled exposure studies have just been started. The suspicion that airborne chemicals might be involved is based mainly on correlations between observed altitudinal gradients in abundance of visible symptoms and known altitudinal gradients of several airborne chemicals. None of Koch's Postulates has been satisfied.

Limited Evidence of Causality

There are several regional cases of forest injury for which there is little or no evidence of involvement of airborne chemicals.

Sugar maple forests. Visible injury and mortality of sugar maple and other hardwood trees has been observed in parts of Quebec and Ontario in Canada and in New England, New York, and Pennsylvania (Hibben, 1969; Robitaille, 1986).

Changes in forest condition include trees with branch and top dieback, decreased sap production, and increased mortality.

The only research to date involved injury surveys without concomitant air quality measurements. Water stress and other natural biotic and physical stress factors have been implicated in sugar maple decline. Air pollution is believed to be a primary or contributing causal factor mainly because no other more plausible explanation has been suggested.

<u>Low elevation coniferous forests</u>. Unexplained decreases in radial growth without visible symptoms have been detected in some trees in three areas:

-- low elevation red spruce in New York and several New England states (Hornbeck and Smith, 1985);

-- pitch pine and shortleaf pine in Pennsylvania and New Jersey (Johnson et al., 1981);

-- natural stands of loblolly, slash, and shortleaf pines in North and South Carolina, Georgia, Alabama, and Florida (Sheffield et al., 1985). Higher-than-expected mortality rates also have been detected in some southern pine forests.

All of these changes in growth were detected in surveys that were not designed to identify possible causal factors and with which there were no concomitant air quality measurements. Several natural stress factors are known which could explain these decreases in growth. Experimental tests for a possible causal or contributing role of airborne chemicals have only recently started.

6. CONCLUSIONS

The ideas outlined earlier in this paper may be summarized in the following ten points.

1) The chemical climate of the earth is changing as the result of many human activities. The most important of these activities are combustion of fossil fuels and decomposition, incineration, or volatilization of waste materials from many industrial and urban sources.

2) Some of these airborne chemicals have beneficial effects on the productivity of aquatic and terrestrial ecosystems but many others have detrimental effects on the health of individual organisms and may have detrimental effects on the productivity, genetic diversity, and geographic distribution of whole ecosystems.

3) Continuing release of radiatively active airborne pollutant chemicals may lead to a general warming of the earth's atmosphere and to increased severity of droughts and wind, rain, snow, and ice storms. These changes could have very far-reaching effects on the health, productivity, genetic structure, and geographic distribution of terrestrial and aquatic ecosystems throughout the world.

4) All living organisms are subjected to a wide variety of natural stresses during most stages of their growth and development. These stresses include several that are natural in origin -- competition with other organisms, diseases and insects, and the normal vagaries of both the physical and the chemical climate. In many industrial and developing regions

of the world, organisms are also subjected to additional stresses induced by airborne chemicals that include toxic gases and aerosols, excess nutrient substances, and acidic and growth-altering substances. They may also be subjected to chemically induced change in the physical climate of the earth. The responses of many living organisms to natural stresses are frequently very similar to their responses to airborne pollutant chemicals.

5) It is usually not too difficult to determine cause-and-effect relationships in the case of locally dispersed primary pollutant chemicals in the vicinity of strong point sources. It is usually much more difficult to determine cause-and-effect relationships in the case of regionally dispersed secondary pollutant chemicals such as acid deposition and ozone and other photochemical oxidants.

6) Rigorous scientific methods must be employed to distinguish the individual and combined effects of airborne pollutant chemicals from the individual and combined effects of many natural stress factors in both terrestrial and aquatic ecosystems.

7) In North America, ozone and other photochemical oxidants are the only regionally dispersed airborne pollutant chemicals that have been rigorously proven to cause detrimental effects on crops or forests. Hydrogen peroxide, excess nitrogen, and acid deposition leading to leaching of essential nutrients from foliage and from soils, are hypothetical mechanisms of action that are currently being tested in the research programs of government and industry in the United States, Canada, and many countries in Europe.

8) Many government and some industry groups in North America and Europe are expanding their programs of research on the responses of ecosystems to air pollutants to include the effects of chemically induced change in the physical climate.

9) Scientific uncertainty about cause-and-effect relationships between airborne pollutant chemicals and both forests and agricultural crops is leading to some hesitancy on the part of industry and government to increase pollution control efforts despite the far better known effects of many of the same pollutant chemicals on public health, atmospheric haze, materials damage, and acidification of aquatic ecosystems.

10) For all of the above reasons, it seems prudent that the United States and Canada, and various countries in Europe, earnestly continue their historical air-quality planning and evaluation procedures. In doing so, leaders of industry and government should recognize that:

a) Research is continuing and will continue to increase scientific and public understanding of air pollution and its direct and indirect effects on our society;

b) An integrated (multiple pollutant and regional) program for management of air-quality would have many advantages over the present single-pollutant and state-by-state and province-by-province methods of implementation;

c) Cooperation, consultation, and statesmanship is likely to produce more economically and scientifically sound management plans than legislatively mandated solutions; and

d) It is likely to take at least 10 years to implement a decision to make a significant change in the air quality in any large part of North America or Europe.

Ackowledgement

I am greatly indebted to James Woodman for many stimulating discussions and for his permission to repeat in this paper some of the ideas we have written about earlier.

REFERENCES

Adams, D.F. 1962. Recognition of the effects of fluorides on vegetation. J. Air Pollution Cont. Assn. 13:360-362.

Anon. 1984. Walderkrankung und Immissionseinflüsse. Report of Ministeriums fur Ernahrung, Land-wirtschaft, Umwelt und Forsten Baden-Württemberg. West Germany, 32 pp.

Benoit, L.F., J.M. Skelly, L.D. Moore and L.S. Dochinger. 1982. Radial growth reductions of Pinus strobus L. correlated with foliar ozone sensitivity as an indicator of ozone-induced losses in eastern forests. Can. J. For. Res. 12:673-678.

Berry, C.R. and G.H. Hepting. 1964. Injury to Eastern white pine by unidentified atmospheric constituents. For. Sci. 10:2-13.

Berry, C.R. and L.A. Ripperton. 1963. Ozone, a possible cause of white pine emergence tipburn. Phytopathology 53:552-557.

Blank, L.W. 1985. A new type of forest decline in Germany. Nature 314:311-314.

Bruck, R.I. 1987. The forest decline enigma -- is it air pollution related? Plant Disease 71: In Press

Carlson, C.E. 1974. Evaluation of sulfur dioxide injury to vegetation on federal lands near the Anaconda Copper Smelter at Anaconda, Montana. USDA Insect Disease Report 74-15, Missoula, MT, 7 pp.

Cobb, F.W. and R.W. Stark. 1970. Decline and mortality of smog-injured ponderosa pine. J. For. 68:147-148.

Cook, E.R. 1985. The use and limitations of dendrochronology in studying effects of air pollution on forests. In Proceedings, Effects of Acidic Deposition on Forests, Wetlands, and Agricultural Ecosystems. NATO Adv. Res. Workshop. Toronto, Canada.

Cowling, E.B. 1985. Critical review discussion papers - effects of air pollution on forests. J. Air Pollut. Control Assn. 35:916-919.

Cowling, E.B. 1986. Regional declines of forests in Europe and North America: The possible role of airborne chemicals. In: S.D. Lee, T. Schneider, L.D. Grant, and P.J. Verkerk, eds. Aerosols. Lewis Publishers, Chelsea, MI. pp. 1334-1348.

Dochinger, L.S. 1978. The impact of air pollution on eastern white pine: The chlorotic dwarf disease. J. Air. Poll. Cont. Assn. 18:814-816.

Dochinger, L.S. and C.E. Seliskar. 1970. Air pollution and the chlorotic dwarf disease of Eastern white pine. For. Sci. 16:46-55.

Duchelle, S.F., J.M. Skelly and B.I. Chevone. 1982. Oxidant effects on forest tree seedling growth in the Apalachian Mountians. Water, Air, Soil Pollut. 18:363-373.

Gschwandtner, G., K. Gschwandtner, K. Eldridge, C. Mann and D. Mobley. 1986. Historic emissions of sulfur and nitrogen oxides in the United States from 1900 to 1980. J. Air Poll. Cont. Assn. 35:139-149.

Hayes, E.M. and J.M. Skelly. 1977. Transport of ozone from the Northeast U.S. into Virginia and its effect on Eastern white pines. Plant. Dis. Rept. 61:778-782.

Hibben, C.R. 1969. Ozone toxicity to sugar maple. Phytopathology 59:1423-1428.

Hornbeck, J.W. and R.B. Smith. 1985. Documentation of red spruce growth decline. Can. J. For. Res. 15:1199-1201.

Husar, R. 1986. Manmade SOx and NOx emissions and trends for eastern North America. In: Acid Deposition: Long Term Trends. Nat. Academy Press, Washington, DC. pp. 48-92.

Huttermann, A. 1984. Die Anwendung der Koch'schen Postulate auf die Untersuchungen zur Ursachenforschung des Waldsterbens. Forstarchiv 55:45-48.

Huttunen,S. 1984. Interactions of disease and other stress factors with atmospheric pollution. In: Air Pollution and Plant Life. M. Treshow, ed. John Wiley & Sons, New York. pp. 321-356.

Johnson, A.H. and S.B. McLaughlin. 1986. The nature and timing of the deterioration of red spruce in the Northern Appalacian Mountains. In Acid Deposition Long-term Trends. Nat. Academy Press, Washington, DC. pp. 200-230.

Johnson, A.H. and T.A. Siccama. 1983. Acid deposition and forest decline. Env. Sci. Tech. 17:294-306.

Johnson, A.H., T.G. Siccama, D. Wang, R.S. Turner and T.H. Barringer. 1981. Recent changes in patterns of tree growth rate in the New Jersey pinelands: A possible effect of acid rain. J. Environ. Qual. 10:420-427.

Johnson, W.C. and D.M. Sharpe. 1982. Ecological consequences of a CO2-induced climatic change on forest ecosystems. Part 13 In Volume II. Environmental and social consequences of a possible CO2-induced climate change. U.S. Dept. Energy Rept. DOE/LY/10019-13. Washington DC. 31pp.

Koch, R. 1876. Die aetiologie der Milzbrand-Krankheit, begundet auf die entwicklungsgeschiachte des Bacillus Antracis. Beitr. Biol. Pflanz. 2:277.

Langsaeter, A. 1941. Om tynning i enaldret gan- og furuskog. Meddel. f. d. Norske Skogforsoksvesen 8:131-216.

Last, F.T., D. Fowler and P.H. Freer-Smith. 1984. Die Postulate von Koch und die Luftverschmutzung. Forst Central bladt, 103:28-48.

Legge, A.H., D.R. Jaques, G.W. Harvey, H.R. Krouse, H.M. Brown, E.C. Rhodes, M. Nasal, H.U. Schellhase, J. Mayo, A.P. Hartgerink, P.F. Lester, R.G. Amundson and R.B. Walker. 1980. Sulphur gas emissions in the Boreal Forest: The West Whitecourt case study. In: Water, Air, and Soil Pollution, 15:77-85.

Linzon, S.N. 1967. Ozone damage and semi-mature-tissue needle blight of eastern white pine. Can. J. Bot. 45:2047-2061.

McBride, J.R. and P.R. Miller. 1985. Responses of North American forests to photochemical oxidants. In: Proc. NATO Adv. Study Workshop. Toronto. In Press.

McBride, J.R., P.R. Miller and R.S. Laven. 1985. Effects of oxidant air pollutants on forest succession in the mixed conifer forest type of southern California. In: Air Pollutants: Effects on Forest Ecosystems. Acid Rain Foundation, St. Paul, MN. pp. 157-167.

McDonald, G.J. 1985. Climate change and acid rain. Mitre Corp., McLean, VA. 44pp.

McLaughlin, S.B. 1985. Effects of air pollution on forests. A critical review. J. Air Poll. Control Assoc. 35(5):512-534.

Miller, P.R. and M.J. Elderman. 1977. Photochemical oxidant air pollutant effects on a mixed conifer forest ecosystem: A progress report, 1976. U.S. Environ. Protection Agency Rep. EPA-600/3-77-104, Corvallis, OR.

Mosteller, F. and J.W. Tukey. 1977. Data Analysis and Regression. Addison-Wesley Pub. Co., Reading, MA, 588 pp.

Ohmart, C.P. and C.B. Williams, Jr. 1979. The effects of photochemical oxidants on radial growth increment for five species of conifers in the San Bernardino National Forest. Plant Dis. Rept. 63:1038-1042.

Parmeter, J.R, Jr., R.V. Gefa and T. Neff. 1962. A chlorotic decline of ponderosa pine in southern California. Plant Dis. Rept. 46:269-273

Pinkerton, J.E. 1984. Acidic deposition and its relationship to forest productivity. TAPPI 67:36-39.

Pinkerton, J.E. and A.S. Lefohn. 1986. Characterization of ambient ozone concentrations in commercial timberlands using available monitoring data. TAPPI 69:58-62.

Postel, S. 1984. Air pollution, acid rain, and the future of forests. Worldwatch Institute Paper 58, Washington, DC, 54 pp.

Prinz, B. Q.H.M. Krause, K.A. Jung. 1985. Untersuchunger der LIS essen zur Problematik Waldschden. In: Waldschden - Theorie und Praxis auf der suckenach Antworten. Olden Bouey Verlag, Mnchen, West Germany. pp. 143-193.

Rennie, P.J. 1985. Future scientific research programs and management approaches. In: Air Pollutants: Effects on Forest Ecosystems. Acid Rain Foundation, St. Paul, MN. pp. 347-365.

Rezabec, C.L., J.A. Morton, E.C. Mosher, A.J. Prey and J.E. Cummings. 1986. Regional effects of sulfur dioxide and ozone on Eastern white pine (Pinus strobus L.) in Eastern Wisconsin. Wisc. Dept. of Nat. Resources Report; Madison, WI. 20 pp.

Scholz, F. 1984. Wirken Luftverunreinigungen auf die genetische Struktur von Waldbaumpopulationen? Forstarchiv 55:43-45.

Schutt, P. and E.B. Cowling. 1985. Waldsterben, a general decline - symptoms, development, and possible causes. Plant Disease 69:548-558.

Schutt, P., W. Koch, H. Blachke, K.J. Lang, E. Reigber, H.J. Schuck and H. Summerer. 1983. So Stirbt der Wald. BLV Verlagsegesellschaft, Munich, West Germany, 128 pp.

Scott, J.T., T.G. Siccama, A.H. Johnson and A.R. Breisch. 1984. Decline of red spruce in the Adirondacks, New York. Bull. Torrey Botanical Club 111:438-444.

Scurfield, G. 1960. Air pollution and tree growth. Forestry Abstracts 21:339-347.

Sheffield, R.M., N.D. Cost, W.A. Bechtold and J.P. McClure. 1985. Pine growth reductions in the southeast. Resour. Bull. SE-83, USDA For. Serv. SE For. Exp. Sta., Asheville, NC, 112 pp.

Siccama, T.G., M. Bliss and H.W. Vogelmann. 1982. Decline of red spruce in the Green Mountains of Vermont. Bull. of Torrey Bot. Club 109:162-168.

Smith, W.H. 1981. Air Pollution and Forests. Springer-Verlag, New York. 379 pp.

Treshow, M., F.K. Anderson and F. Harner. 1967. Responses of Douglas-fir to elevated atmospheric fluorides. For. Sci. 13:114-120.

USEPA, USFS, and NCASI. 1986. Response of forests to atmospheric deposition: National research plan for the Forest Response Program. US Environment Protection Agency and US Forest Service. Washington, DC. 110 pp.

Wallace, H.R. 1978. The diagnosis of plant diseases of complex etiology. Ann. Rev. Phytopathology 16:379-402.

Wang, D., D.F. Karnosky and F.H. Bormann. 1986. Effects of ambient ozone on the productivity of Populus tremuloides Michx. grown under field conditions. Can. J. For. Res., 16:47-55.

Weiss, M.J., L.R. McCreery, I. Millers, J.T. O'Brien and M. Miller-Weeks. 1985. Cooperative survey of red spruce and balsam fir decline and mortality in New Hampshire, New York, and Vermont - 1984. USDA Forest Service Interim Report, NE, 130 pp.

Woodman, J.N. 1985. What the forest products industry is doing about pollution-induced forest health problems. TAPPI 68:34-37.

Woodman, J.N. and E.B. Cowling. Airborne chemicals and forest health. Env. Sci. Tech. 21:120-126.

Wolff, G.T. and P.J. Lioy. 1980. Development of an ozone river associated with synoptic scale episodes in the eastern United States. Env. Sci. Tech. 14:1258-1260.

Woodwell, G.M. 1970. Effects of pollution on the structure and physiology of ecosystems. Science 168:429-433.

OVERVIEW OF THE U.S. NATIONAL ACID PRECIPITATION ASSESSMENT PROGRAM

P.M. Irving
National Acid Precipitation Assessment Program
Washington, D.C., U.S.A.

Summary

The U.S. Acid Precipitation Act of 1980 (Title VII of the Energy Security Act of 1980, Public Law 96-294) established the Interagency Task Force on Acid Precipitation to implement a national program to increase the understanding of the causes and effects of acidic deposition. The National Acid Precipitation Assessment Program (NAPAP) was thus established to manage research efforts to reduce uncertainties for policymakers in the evaluation of impacts from acidic precipitation and associated pollutants. The program includes research within six federal agencies and is grouped in seven categories: 1) Emissions and Controls, 2) Atmospheric Chemistry, 3) Atmospheric Transport and Modeling, 4) Air Quality and Deposition, 5) Terrestrial Effects, 6) Aquatics Effects, and 7) Materials. The major research efforts of the Program include investigations to determine the relationship between emissions and deposition and the relationship between deposition and effects. The primary areas of research include: comprehensive inventories of emissions, studies in atmospheric chemistry, atmospheric modeling, deposition and air quality monitoring, and effects research on agricultural crops, plantation and natural forests, watersheds, acquatic systems and materials.

1. INTRODUCTION

The U.S. National Acid Precipitation Assessment Program (NAPAP) is a congressionally mandated, 10-year research program which coordinates research efforts of six agencies (NOA, EPA, USDA, DOE, DOI, and TVA) to reduce uncertainties which limit decisions of policymakers. The research program has increased its expenditures from $17 million in 1982 to over $85 million in fiscal 1987. The work is conducted in seven categories or Task Groups: I. Emissions and Controls, II. Atmospheric Chemistry, III. Atmospheric Transport and Modeling, IV. Air Quality and Deposition, V. Terrestrial Effects, VI. Aquatics Effects, VII. Materials. There are two major areas of scientific uncertainty that impact policy on acidic precipitation. These are: The relationship between changes in emissions to deposition and air quality at distant sensitive receptors (source/receptor uncertainties), current and long-term effects on sensitive receptors from ambient or hypothetical levels of acidic deposition and air quality (effects uncertainties). Within the scope of the National Program receptors include agricultural crops, plantation and natural forests, soils, aquatic systems, and materials.

2. SOURCE/RECEPTOR RELATIONSHIPS

Emissions

Emissions inventories of sulfur dioxide (SO_2), oxides of nitrogen (NO_x), volatile organic compounds (VOCs) and alkaline materials from various sources are required for an assessment of acidic rain precursors. NAPAP has inventoried historical data to estimate sulfur dioxide and nitrogen oxide emissions on a 5-year basis beginning in 1900 for each state (1). Emissions estimates of VOCs produced previous to or outside of NAPAP, extend back only to about 1940.

Historical patterns of emissions are important for assessing the role of acidic deposition in environmental change and for the development of policy. In general, SO_2 emissions for the nation as a whole were about the same in 1984 as they were in 1925 although SO_2 emissions peaked in 1972 and have declined by about 30% since that time. Anthropogenic emissions of NO_x and VOC on a national basis have increased over the last 50 to 60 years.

In the U.S. natural emissions of sulfur are small (only a few percent) relative to anthropogenic emissions on an annual national basis (2). There is considerable uncertainty about the relative quantities of nitrogen originating from humanmade and natural sources (2). Current estimates are that natural sources may contribute from 8 to 30 percent of the nitrogen emitted on a national basis over the course of the year. Natural emissions of volatile organic compounds and alkaline materials are poorly quantified.

In addition to emissions inventories, emissions projections are also important in assessing future environmental trends and for the development of policy. NAPAP is developing a set of models for use by policymakers which will project emissions and control costs for major emissions sectors given specifications about an array of engineering, economic, and regulatory assumptions.

Atmospheric Chemistry

Emissions of sulfur dioxide and oxides of nitrogen are chemically transformed in the atmosphere by oxidants to sulfuric and nitric acids. Oxidants are formed when volatile organic compounds interact with oxides of nitrogen in the presence of sunlight. One of the most important observations from atmospheric chemistry is that at least during some portions of the year the shortage of oxidants limit the production of sulfuric acid. As a result oxidation and deposition occur further from source areas during those oxidant limited periods.

For sulfur most of the pathways and rates are known and the research questions reflect the extent to which each of the reactions operate to produce sulfuric acid. But since the dominant oxidant (H_2O_2) is produced by VOC and NO_x from sources mostly different from those emitting SO_2, reduction of sulfate deposition may require control of sources not emitting SO_2.

For nitrogen, the pathways are not all known. Present understanding of atmospheric chemistry cannot account for the observations of nitrogen oxidation. One set of detailed observations near Philadelphia shows that the production of nitric acid is about double that which can be accounted for with known mechanisms. Thus for nitrogen, the research program focuses on the detailed reaction patterns and the measurement of ephemeral but reactivenitrogen species in the atmosphere.

NAPAP has a number of efforts underway to reduce and quantify these uncertainties in atmospheric chemistry. These include the development of instruments capable of more finely resolving the chemical species of nitrogen which exist in the atmosphere, and tracking the development outside of NAPAP of instruments to measure active but very short-lived oxidants, such as the hydroxy radical. NAPAP efforts also include additional field experiments to learn more about the importance and operation of major chemical pathways particularly in the aqueous phase. The integration of this information will take place in modeling efforts described below.

NAPAP has constructed an atmospheric model, the Regional Acid Deposition Model or RADM, which incorporates all processes thought to be important in acid deposition (3). Model components embody state of the art knowledge on meterological predictions, chemical transformations, dry deposition, and precipitation scavenging (the removal of pollutants leading to wet deposition). It provides predictions of acidic deposition over 80 x 80 km grids. A verified model will allow an evaluation of emissions control strategy options.

Deposition

Exposure of receptors to air pollution and acid deposition occurs via several routes: wet deposition (rain or snow), clouds, fog or mist, and dry deposition. NAPAP has developed a nationwide network of 150 wet deposition monitoring sites. As a result, the chemistry of rain and snow is fairly well characterized on a national basis although historical data are lacking. Analysis of trends in sulfate and nitrate deposition collected at the 19 stations with the longest records indicate decreases in deposition at most of these stations from 1978-1983 (4). This is probably due to climatological factors.

Exposure of receptors to pollutants in clouds is particularly important at high elevations where forests may be bathed in clouds for extended periods of time. These clouds contain not only higher acidity than rain, but also dissolved oxidants such as hydrogen peroxide which are biologically and chemically active. The chemistry of clouds is not directly reflected in wet deposition sampling. Technology to measure the chemical composition of clouds is being developed and has been deployed only in the past year at five locations in mountains of the eastern United States.

Dry deposition occurs when pollutants settle on or interact with receptors in the absence of rain or clouds. Deposition velocities are highly uncertain and vary with meteorological conditions such as wind velocity as well as with the type of surface (5). A thin film of water on a surface can greatly enhance the deposition of some pollutants, such as of sulfur dioxide.

Deposition velocities are best characterized for sulfur dioxide and ozone and are more uncertain for nitrogen compounds. Quantification of dry deposition, extrapolating individual measurements over broader areas, and developing uncertainty estimates is one of NAPAP's priorities. Preliminary data indicate that the relationship between dry and wet deposition is highly variable, both in space and time (Table I).

TABLE I: PRELIMINARY ESTIMATES OF ANNUAL AVERAGES OF SULFUR DEPOSITION AS SULFUR DIOXIDE, PARTICULATE SULFATE, AND WET SULFATE, AT THREE SITES OF THE DRY DEPOSITION RESEARCH NETWORK.

	SO_2-S Gas	SO_4^{2-}-S Particle	SO_4^{2-}-S Wet
	kg ha^{-1} y^{-1}		
Oak Ridge, TN			
Average	7.8	2.1	8.5
Standard Error	1.3	0.3	2.0
State College, PA			
Average	7.1	3.2	10.6
Standard Error	1.2	0.9	2.8
Whiteface Mt., NY			
Average	3.5	0.8	8.1
Standard Error	0.5	0.2	1.2

3. EFFECTS

The major areas of effects research within NAPAP are directed toward agricultural crops, forests, aquatic systems, and construction and cultural materials. Air pollution has other effects but these are largely outside of the scope of NAPAP research and are being studied through other organizations.

Effects research embodies investigations on the response of a resource to incremental changes in wet deposition and air quality (dose-response research) as well as evaluations regarding the extent of resource at risk (inventories).

Agricultural Crops

The effects of acidic rain on a number of important crops have been examined in greenhouse and field studies in which rain at different acidity levels is simulated. The available data, which includes replicated field studies of 13 cultivars for eight crops, indicates that there are no measurable and consistent yield responses from current levels of rain acidity (Table II).

Repeated investigations of the one crop cultivar ('Amsoy' soybean) which has been reported to be susceptible to acidic rain stress, have indicated that the response is not consistent from site-to-site or from year-to-year at a particular site. Preliminary experiments with two cultivars of field corn suggest that moisture stress may predispose one cultivar to damage from high levels of acidity in rain (pH = 3.0). These results and others suggest that environmental conditions and genetic characteristics which affect leaf surface characteristics may control the response of crops to acidic rain (6). Thus, results of acidic rain studies from specific, limited experimental locations should be interpreted cautiously until differences due to such factors as soil type, drought, and presence of other pollutants are understood. The sensitivity of untested crop species and cultivars must also be determined before final conclusions are made. Studies are being conducted within NAPAP to address these questions.

TABLE II: SUMMARY OF FIELD STUDIES TO EXAMINE CROP YIELD RESPONSE TO ACIDIC RAIN DOSE

Crop	Number of Cultivars studied	Total number of studies	Studies reporting reduced yield at at ambient pH[a]
Corn	2	8	0
Soybean	8	29	7[b]
Hay	2	2	0
Wheat	1	1	0
Tobacco	1	1	0
Potato	1	2	0
Oat	1	1	0
Beans (Snap)	1	2	0
Total	17	46	7

[a] Yield reduction from simulated rain of pH 3.8 to 4.5 compared to pH 5.6.

[b] One study at University of Illinois; 6 studies at Brookhaven National Lab. There were no cultivars which consistently exhibited a negative response to acidic rain.

Although there are some uncertainties, the available information indicates that decreases in current levels of sulfuric and nitric acids in rain will not result in any significant increases in crop yields. In fact, there is a possible economic benefit from sulfur and nitrogen input from acidic deposition because it may supplement fertilizer amendments on intensively managed land and is a supplement for nitrogen and sulfur in agricultural systems under low levels of management, such as grasslands. Even when the cost of lime needed to neutralize the input of H^+ to agricultural soils is considered, there is a net benefit from the sulfur and nitrogen input from acidic deposition.

The concentrations and exposure frequency of SO_2 and NO_2 gases, which have been reported to cause damage to crops in controlled fumigation studies, rarely occur in rural areas of the U.S. For a few local situations such as near an area of multiple sources, the combined exposures of sulfur dioxide and nitrogen dioxide may reach concentrations and frequencies that could reduce crop productivity.

In contrast, O_3 at ambient levels (35-70 ppb, daily 7-hr. average during the growing season) reduces the productivity of agricultural crops in many regions of the U.S. (7). The average gain in crop production of a 25% reduction in ozone concentration has been estimated to be approximately 2-3% and has a potential value on the order of one billion dollars. Crop, yield reduction from O_3 varies widely by species and cultivar (Figure 1), ranging from 1 percent or less for sorghum and corn, to 20 percent for potato to greater than 30 percent for alfalfa, in comparison to yields in charcoal-filtered air.

Environmental conditions such as temperature and moisture, and exposure characteristics such as dose rate and recovery interval between doses, are known to influence plant response to pollutants. These and other factors require further study to precisely quantify the productivity loss from acidic deposition and associated pollutants for a range of cultivars, environments, and exposure conditions.

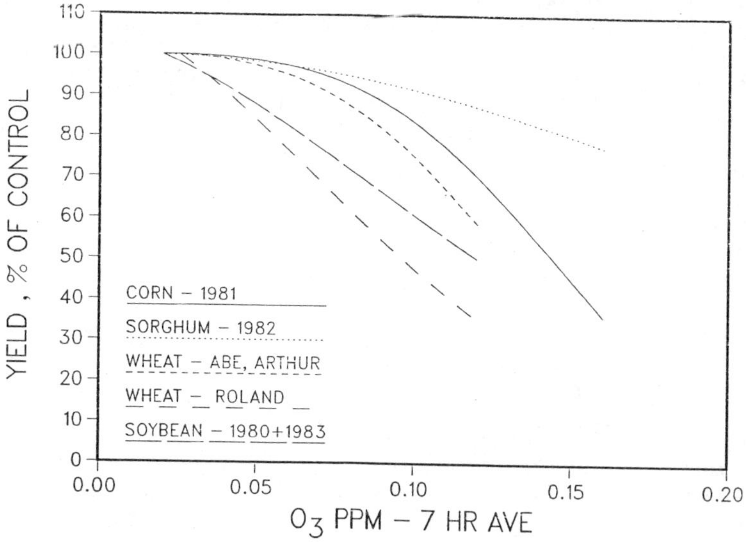

Figure 1. Ozone exposure-response functions for corn, sorghum, wheat, and soybean.

Forests

Like crops, the health of forests may be impacted by air pollution on a local scale due to intense point source emissions or on a regional scale from distant emissions from broad areas. In contrast to crops, not much is known about effects from current levels of pollutants on forest ecosystems. All forests experience some natural stresses such as plant competition, nutrient limitations, adverse weather, insects and disease. Some of these natural stresses increase with elevation. Factors altered by humans such as stand management, land use patterns, carbon dioxide, and possibly regional air pollution can also contribute to reduced forest health either alone or by interacting with natural factors.

There is currently no evidence that acidic deposition, SO_2, or NO_x ambient levels in the U.S. are responsible for observed changes in the nation's forest condition. Nor is it known whether atmospheric deposition potentially affects future forest productivity indirectly through impacts on soils. Ozone has been identified as the cause of regional scale forest damage in the mixed conifer forests of the San Bernardino Mountains near Los Angeles, California, and it may be responsible for damage in other regions of the country (8).

Increased concern about the health of U.S. forests has recently developed because of several unexplained changes. In the northeast, mortality and growth reductions are reported for red spruce particularly in the spruce-fir forests above cloudbase in the high elevations of the Appalachian Mountains. Other species such as eastern white pine and sugar maple are also experiencing decline symptoms for reasons that are as yet unclear. Likewise, forest inventory data in some areas of the southeast show growth reductions over the past several decades for commercial pine species, and causes are not certain. Concern also results from the declines of several tree species seen in the last decade or so in central European forests.

Consistent with these concerns, the overall objectives of the Forest Research Program are to:

1. Determine the effect on productivity and health of major tree species from exposure of both soils and foliage to controlled levels of acidic deposition and associated oxidants. Distinguish between soil-mediated effects).
2. Identify the major factors causing visible damage to forests above cloudbase in the high mountains of the eastern United States.
3. Initiate biochemistry, physiology and ecology studies to determine causal relationships between acidic deposition and related pollutants and forest conditions where productivity effects are demonstrable.
4. Establish a long-term monitoring program for the continued documentation and detection of new changes in forest conditions.

The research is organized and conducted regionally in major forest types -- eastern spruce-fir, southern pine, eastern hardwood, and western conifer. Field observations at Mt. Mitchell in the Southern Appalachians indicate that 15% of the spruce-fir stands have greater than 70% mortality; 4% have 30-70% mortality; and 76% have less than 30% mortality -- most of which is caused by balsam woolly adelgid on Fraser fir. Also, at Mt. Mitchell as well as other nearby mountains, a recent survey showed that no insects or pathogens other than the adelgid were present which could cause significant mortality in southern spruce-fir stands. Further, red spruce mortality did not correlate with elevation in this forest area.

A review of historical information available for the high-elevation spruce-fir forests in the South revealed that many past disturbances in these forests were caused by raiload logging, windstorms, slash fires, grazing and balsam woolly adelgid. These disturbances can change species composition, age distribution, site quality, and forest health depending upon time and location. Such background information is vital to understanding the role of air pollutants in forest health changes.

Initial experiments with simulated acid rain applied to tree seedlings show no foliar effects or above-ground dry weight loss at present ambient levels in eastern U.S., i.e. pH 4.2. Thus, direct foliar effects of acidic deposition to trees does not appear by itself to be a problem at the present time. However, cumulative effects and interactions of foliar, soil, and other stresses may lead to responses that would not otherwise be observed in short-term studies or under more favorable growing conditions.

In recent controlled laboratory studies with 3-year Fraser fir, sulfuric acid mist (pH 2.5) caused visible needle injury while nitric acid mist at the same pH did not. Ozone exposures (50 and 100 ppb) in combination with water stress treatments, over a 10-week period produced no significant interactive effects on the Fraser fir seedlings. Water stress alone reduced growth significantly, but there were no effects from ozone alone. Other research results have demonstrated consistent effects from O_3 on Liriodendron and Pinus seedling growth whereas no consistent effect from rain acidity has been observed (9, 10, 11).

Aquatic Effects

Knowledge regarding the relationship between acidic deposition and surface water chemistry has advanced significantly over the past few years. There is strong evidence that acidic deposition has affected some surface waters in the northeastern U.S. The NAPAP program is designed to increase knowledge of the extent, rate, and magnitude of effects of acidic deposition on aquatic systems on a regional scale.

Surveys conducted by NAPAP (Figure 2) describe the extent of acidic surface waters in regions considered potentially sensitive to acidic deposition (Table III) (12). The cause of acidity in any given lake may be due to natural factors, acidic deposition, or other human activities. The highest percentages of acidic lakes in the eastern United States were identified in the Adirondacks, the Upper Peninsula of Michigan, and Florida. No lakes sampled in the Western Lake Survey were acidic, however many have low acid neutralizing capacities (ANC).

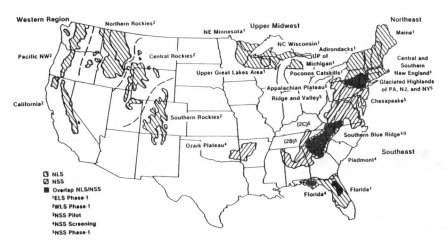

Figure 2. Regions of study for the U.S. National Surface Water Survey.

TABLE III: ESTIMATED TOTAL NUMBER OF LAKES (>4 $\leq$2000 ha) AND NUMBER AND PERCENTAGE OF LAKES CHARACTERIZED BY VALUES GIVEN FOR FAIR KEY VARIABLES.

		Northeast	Upper Midwest	Southern Blue Ridge	Florida
Total No. Lakes		7096	8502	258	2098
pH $\leq$5.0	No.	240	130	0	259
	%	3.4	1.5	0	12.4
ANC $\leq 0 \mu eq\ L^{-1}$	No.	326	148	0	463
	%	4.6	1.7	0	22
$SO_4^{2-} \geq 150 \mu eq\ L^{-1}$	No.	1846	608	22	846
	%	26	7.1	8.5	40.3
Extractable	No.	92	2	0	14
Al $\geq 150 \mu g\ L^{-1}$	%	2	0	0	1.5

Information is known about how acidic deposition interacts with individual biological, geological, and hydrological systems, but little is known about the interactions of these systems and their behavior within watersheds or in larger regions. For example, hydrologic processes are important because they determine the amount and nature of contact between incoming precipitation and watershed soils. If incoming precipitation runs overland or through large openings in the soil or rock, there may not be time for significant chemical reactions to occur between precipitation and the soil system, thus precipitation chemistry will be relatively unchanged as it enters a lake or stream.

Acidification, from whatever cause, changes surface water chemistry which affects fish and other biota over the long term (one or more years) or during episodes (lasting a few days). In concept, long-term acidification occurs gradually and affects the chemistry of the lake throughout the year. Episodic acidification occurs during a short period of time, perhaps contributing to long-term acidification. During this short period, the surface water biota may be damaged or lost for the rest of the year while the lake or stream chemistry returns to a "normal" state.

Episodic changes in surface water chemistry can be initiated during intense rainstorms in the warmer months which are dominated by sulfuric acid or by the spring melting of the snowpack in the Northeastern U.S. which is dominated by nitric acid. They may also be the result of surges in the activity of soil microbes which produce acidifying substances and which are more active when soils are moist. These episodic events may be more important in determining biological response than the longer, slower chronic acidification. In studies of four storm episodes in the northeastern and southeastern U.S., ANC and pH of streams decreased during the storm event; the depressions were more severe in the Northeast. Other studies demonstrated that stream discharge rate may be useful in predicting decreases in pH and ANC in response to episodes of H^+ input.

A study of 22 lakes in Maine examined the relationship between acidity and biological resources. The results indicated that for lakes with pH levels less than 6.0 the number of fish species and fish abundance are significantly correlated with lake pH (13).

NAPAP research results have indicated that the presence of alkaline substances in soils can minimize the influence of bedrock in determining the sensitivity of some ecosystems to acidic deposition and that periodic melting of snow throughout winter may lessen effects of snowmelt in spring. In other watersheds, bedrock alonee appears to be the controlling factor - the less reactive the bedrock, the more acidic a system will become under similar acidic deposition levels, regardless of soil cover.

For surface waters already acidified, mitigation strategies including liming and restocking are being investigated. Research demonstrated that the addition of lime to some acidic lakes can reduce mortality of fish populations.

Materials

A wide variety of materials, (metals, stone, paint, and others) are subject to damage from air pollution in addition to natural causes (light, heat, salt spray, freeze/thaw, etc.). The NAPAP goal is to determine whether current levels of acidic depositions, oxidants (O_3), or acidic gases (SO_2, NO_x) add a measurable or significant amount of damage and if so, the economic impact of that damage.

NAPAP materials research is conducted in highly controlled laboratory exposures to elucidate mechanisms, in open field exposures, to examine effects from ambient conditions, and in controlled field exposures where dosages of pollutants can be manipulated. The program encompasses 6 metals, two types of stone, and paint-substratum combinations. Research has progressed furthest for galvanized steel and for stone (limestone and marble). Significant research on paint has only been initiated recently. Preliminary results indicate that SO_2 gas has cumulative effects on limestone and marble and is much more damaging than SO_4^{2-} in rainwater (14).

REFERENCES

1. Historic Emissions of Sulfur and Nitrogen Oxides in the United States from 1900 to 1980, Volume I. Results: EPA-600/7-85-009a. April 1985. Volume II. Data: EPA-600/7-85-009b. April 1985.

2. Andreae, M.O. et al. 1985. Sulfur and nitrogen emissions. in; J.N. Galloway, ed., Biogeochemical Cycling of Sulfur and Nitrogen in Remote Areas. D. Reidel, Boston, MA.

3. National Center for Atmospheric Research. 1985. The NCAR Eulerian Regional Acid Deposition Model. Nat'l. Center Atmos. Res. Rep. NCAR/TN-256STR.

4. Schertz, T.L., and R.M. Hirsch. 1985. Trend Analysis of Weekly Acid Rain Data--1978-1983. U.S. Geol. Surv. WRI Rep. 85-4211.

5. Hicks, B.B. and J.A. Garland. 1983. Overview and suggestions for future research on dry deposition. Pages 1429-33 in: Pruppacher et al., eds., Precipitation Scavenging, Dry Deposition, and Resuspension. Volume 2. Elsevier Science.

6. Irving, P.M. 1986. Report on the Crop Response Workshop of the National Acid Precipitation Assessment Program. April 17-18, 1986. Chicago, IL. NAPAP, Washington, DC. 31 pp.

7. Heck. W.W., W.W. Cure, J.O. Rawlings, L.J. Zaragoza, A.S. Heagle, H.E. Heggestad, R.J.Kohut, L.W. Kress and P.J. Temple. 1984. Assessing impacts of ozone on agricultural crops II. Crops yield functions and alternative exposure statistics. J. Air Pollut Control Ass. 34:810-817.

8. McLaughlin, S.E. 1985. Effects of air pollutants on forests: a critical review. J. Air Pollut. Control Ass. 35:312-34.

9. Chappelka, A.H., B.I. Chevone, and T.E. Burke. 1985. Growth response of yellow-poplar (Liriodendron tulipifera, L.) seedlings to ozone, sulfur dioxide, and simulated acidic precipitation, alone and in combination. Environ. Exper. Bot. 25:233-244.

10. Jensen, K.F. 1985. Response of yellow poplar seedlings to intermittent fumigation. Environ. Pollut. 38:183-191.

11. Wood, T. and F.H. Borman. 1977. Short-term effects of a simulated acid rain upon the growth and nutrient relations of Pinus Strobus. L. Water, Air, Soil Pollut. 7:479-488.

12. Linthurst, R.A. et al. 1986. Characteristics of Lakes in the eastern United States. Volume I: Population Descriptions and Physico-chemical Relationships. EPA-600/4-86-007A. U.S. Environ. Prot. Agency, Washington, DC.

13. Haines, T.A., S.J. Pauwels, and C.H. Jagoe. 1985. Predicting and evaluating the effects of acidic precipitation on water chemistry and endemic fish populations in the northeastern United States. U.S. Fish Wildlife Serv. Eastern Energy Land Use Team Biol. Rep. 80(40.230.

14. Reddy, M.M., S. Sherwood, and B. Doe. 1985. Limestone and marble dissolution by acid rain. Proc. of 5th Internatl. Congress on Deterioration and Conservation of Stone. Sept. 25-27, 1985. Lausanne, Switzerland.

SESSION I

POLLUTION CLIMATES IN EUROPE : DEPOSITION IN ECOSYSTEMS

SESSION I

LES CLIMATS DE POLLUTION EN EUROPE : LES DEPOTS DANS LES ECOSYSTEMES

SITZUNG I

"POLLUTION CLIMATES" IN EUROPA : DEPOSITION IN OKOSYSTEMEN

Chairman : P.J.W. SAUNDERS, Department of Environment, United Kingdom

Rapporteurs : M.J. SLANINA, Energy Research Foundation, The Netherlands
J. FUHRER, Eidg., Forschungsanstalt für Agrikulturchemie und Umwelt hygiene, Switzerland

REPORT ON THE SESSION

J. Fuhrer
Swiss Federal Research Station for Agricultural Chemistry and Environmental Hygiene, Liebefeld-Bern, Switzerland

J. Slanina
Netherlands Energy Research Foundation ECN, 1755 ZG Petten, The Netherlands

The characterization of pollution climates in Europe, and the deposition of pollutants in terrestrial ecosystems were the topics discussed in Session I of this Symposium and during a Workshop held in Bern, Switzerland (27 - 30 April 1987). This report attempts to summarize some of the main results from these discussions and to indentify important needs for future research.

(1.) POLLUTION CLIMATES IN EUROPE - CHARACTERIZATION AND CLASSIFICATION

A complex relationship exists between an ecosystem and pollutants in the surrounding air. In order to better understand this relationship, the mixture of air pollutants must be characterized. The term "Pollution climate" is used to describe the mixture of pollutants that are found in a particular area. In different areas, different pollution climates can occur, and thus a different ecosystem response will result.

With respect to studies of effects of air pollution on terrestrial ecosystems, the pollution climate should be characterized on the basis of information on "key components". These key components, identified on the basis of their known effects on plants, soils etc., should be used as far as the present analytical methodologies permit. Such components could be: SO_2, NO_2, O_3 and inorganic ions in wet precipitation. On the other hand, improved measuring techniques need to be developed for substances such as HNO_3, peroxyacetyl nitrate, H_2O_2, NH_3 and others, before they can be used as "key components".

Routine measurements of the components used to characterize pollution climates should provide information relevant with respect to the evaluation of possible effects. Biologists involved in studies of air pollution effects must determine the necessary parameters. For instance, relevant parameters for ozone could be:

(a) average concentration for daylight hours,
(b) indication of frequency of episodes with concentrations above a certain threshold (e.g. above 60 ppbv),
(c) timing and duration of such episodes.

Information of minor relevance would be an average concentration on a daily or monthly basis. Any information on ozone should be related to the season of the year, i.e. it should cover the growth period of a specific crop. In the case of SO_2 and NO_2, annual 24-hr averages and one-hr peak values within 24 hours would be the relevant information.

At the present time, available information on pollutant concentrations from air pollution monitoring efforts in Europe is limited. As a consequence, only a few regions with distinctly different pollution climates can been identified. Photochemical oxidants, mainly ozone, determine the pollution climate at low altitudes in a region including S. Germany, Austria and Switzerland. This ist supported by recent data examined in the OECD project OXIDATE (1). The pollution climate of central Germany, Belgium and S. England is characterized by higher dry deposition of nitrogen and sulfur compounds. Wet deposition dominates a third region including N. Britain (2). This picture of pollution climates in Europe, however, ist incomplete. No classifications are possible for countries such as Italy, France, Spain, Greece and others. This is mainly because no measurements are taken, or because results from monitoring networks have not appeared in the international literature.

The definition of air pollution climates is further complicated by the effect of altitude. In a particular region, the pollution climate can be very different at different altitudes. Ozone, for example, shows different daily patterns at higher (e.g. >1000 m asl) and lower (e.g. >1000 m asl) altitudes. This is due to different ozone deposition and destruction during nighttime, which is high at lower and reduced at higher altitudes. In spite of different daily patterns of ozone concentrations, the 24-hr average can be very similar. This is not the case for SO_2 or NO_2. Under most circumstances, higher altitudes will be exposed to lower 24-hr average concentrations. Thus, in an area with high dry deposition of SO_2 and NO_2 at the low sites, the relative importance of ozone will increase with increasing altitude. Occult deposition (see below), by way of interception of cloud and fog droplets and of the formation of dew, also varies as function of region and altitude. The subdivision of regions according to altitude is particularly important in mountainous regions, e.g. Austria and Switzerland. Another difficulty with the characterization of pollution climates exists in areas exposed to generally low ozone concentrations, but where episodes with high concentrations can occur occasionally as a result of long-range transport (e.g. Scandinavia).

With respect to the characterization and classification of pollution climates, the following points should be considered in future research:

(a) Extension of national and international monitoring networks for gas concentration and atmospheric deposition of "key substances"; Relevant information should be stored in a central data bank.
(b) Identification of statistical parameters calculated from continous records of pollutant concentrations, that are relevant to the characterization of pollutant climates, must be emphasized.
(c) Studies on the effect of altitude on pollutant concentrations and deposition should be encouraged.

(2.) DEPOSITION IN ECOSYSTEMS

The response of individual plants or whole ecosystems to pollution climates largely depends on inputs via wet, dry and occult deposition.

Wet deposition (rain, snow) is relatively easy to collect and analyze. Thus, a good picture of spatial and temporal trends of ion concentrations and depositon rates is available. For instance, a change to higher NO_3^-/SO_4^{2-} ratios over the past years can be observed in many parts of Europe. More information is needed with regard to:

(a) variation of wet deposition with altitude,
(b) the origin of S and N background contributions, and
(c) improved models for regional deposition patterns.

Occult deposition, which refers to the capture of cloudwater by surface structures, has not been recognized as important process for the transfer of pollutants to the ground until recently. It is now known that various ionic and non-ionic compounds are enriched in cloudwater as compared to rainwater by a factor of up to 4 - 6, in the case of valley fogs by even more. At present, there are different systems of cloudwater collectors in use. Most of these need to be improved, e.g. in order to more efficiently exclude rainfall. It is not until optimized collection systems are available that cloudwater chemistry can be studied accurately, and cloudwater monitoring networks can be established. This would be necessary, however, because cloud and fog occurrence could be important in determining the pollution climate.

While some information is available on the chemical composition of cloudwater, very little is known about the rate of occult depositon. More direct measurements with sophisticated methods must be used to directly measure occult deposition to different types of surfaces. This would then allow to make models for a variety of situations.

Once cloudwater droplets are in contact with the surface, they can combine to form larger drops or they can be reduced in volume by evaporation of water. This latter process leads to a further enrichment of substances in the water remaining on the surface.

Dry deposition refers to inputs that occur during periods with no wet deposition. It includes the deposition of gases and particles of various sizes. Along the dry depositon pathway of a pollutant, a series of resistances determine its flux. The magnitude of these resistances depends on meteorology, type of surface cover, time of day etc. The resulting complexity makes it extremely difficult to measure rates of dry deposition directly in the field. Furthermore, occasional dew formation might affect deposition rates.

At the present time, best information is available in the case of dry SO_2 deposition to agricultural fields. Information on deposition of nitrogen oxides, ammonia and HNO_3, SO_2 deposition to forest canopies, and deposition of small particles is deficient.

Further work designed to improve our understanding of dry deposition will need to combine micrometeorological, physico-chemical and biological methods. Furthermore, in order to estimate regional differences in dry deposition, the controlling factors for dry deposition when surfaces are wet or covered with snow need to be determined. This should lead to improved accuracy of dry deposition estimates under a variety of conditions, and, ultimately, the development of models to predict dry deposition rates on a regional scale.

(3.) PATHWAYS AT THE SURFACE

After deposition onto surfaces and before reaching ultimate sinks in vegetative or soil compartments, pollutants follow different pathways. They can penetrate the plant cuticle or enter leaves and needles via the stomata, or they can reach the soil after being in contact with the surface of leaves, branches and stems (throughfall, stemflow) etc. Along these pathways, exchange processes occur. The phenomena controlling these exchange processes are not well understood. They can be the morphology, surface tension, chemical reactivity, cuticular permeability or other factors. Thus it is very difficult to predict the actual dose of a pollutant sensitive parts of terrestrial ecosystems are exposed to. For instance, rates of dry deposition measured by using surrogate surfaces, or derived from calculations on the basis of measurements of throughfall and stemflow may not accurately reflect the rate of dry deposition to the natural canopy, e.g. of a forest.

The understanding of surface exchange processes, together with a more complete knowledge of the overall ion cycling between soil, vegetation and air will be necessary in the future in order to be able to relate the pollution climate to the dose of pollutants vegetation and soils are exposed to.

(4.) CRITICAL LOADS OF POLLUTANTS

Loads of pollutants that are not likely to cause chemical changes leading to long-term harmful effects on most sensitive ecological systems, are below a "critical" level. Critical loads can be proposed on the basis of quality criteria for soils, groundwater and lakes, and the chemical properties of these receptors in a particular region of interest. For Scandinavia, the project "Critical Loads for Sulphur and Nitrogen", initiated by The Nordic Council of Ministers, proposed the following critical loads (3):

Protons: 10 - 20 keq H^+ km^{-2} yr^{-1}
Nitrogen: 10 - 20 kg N ha^{-1} yr^{-1}
Sulphur: 2 - 4 kg S ha^{-1} yr^{-1}

Although many uncertainties are associated with these estimated critical loads, they can lead to the evaluation of reductions in pollutant loads needed to protect sensitive parts of the environment. Based on the comparison of the above critical loads with estimated rates of atmospheric deposition, extensive reductions in many parts of Europe must be postulated. Consequences of overloading ecosystems with acidifying substances, can be observed as reduction in pH and cation content of forest soils in Scandinavia. In the same way, excessive nitrogen depositions contribute to alarming increases in the nitrate content of groundwater in the Netherlands.

The definition of critical loads is an important strategy to help decision makers to better understand the need for reductions in pollutant loads. This procedure should be applied to regions outside Scandinavia as well.

In conclusion, the work examined in this report demonstrates the need for a better understanding of pollution climates, deposition rates and surface exchange processes. An improved estimate of loads that are critical for sensitive ecological systems can result, as well as a better insight into the complex relationship between terrestrial ecosystems and air pollutants. However, the different problems should not be dealt with separately, rather, research efforts should include aspects of all different fields mentioned above. In consequence, researchers with an expertise in micrometeorology, chemistry or biology must interact with each other and participate in integrated studies.

REFERENCES

Greenfelt, P., Saltbones, J., Schjoldager, J. (1987), Oxidant data collection in OECD-Europe 1985-87 (OXIDATE). Draft report of the Norwegian Instiute for Air Research (NILU), Ref. 0-8535.

Last, F., Cape, J.N., Fowler, D. (1986), Acid rain - or "pollution climate"? Span 29/1, 3pp.

Nilsson, J., Ed. (1986), Critical loads for sulphur and nitrogen - Nordic Council of Ministers - Report 1986/11, 232 pp.

LES CLIMATS DE POLLUTION EN EUROPE: CARACTERISATION ET CLASSIFICATION

G. TOUPANCE
Université Paris Val de Marne,
94000, CRETEIL France

Summary: The concept of Pollution Climate is discussed by respect to the possible fields of application of the concept, to the chemical interactions of the pollutants, precursors and radical species in the atmosphere, to the spatial and temporal variability of the data, and to the various kinds of indicators of pollution levels needed by the plant physiologists. Results of the 1984 EMEP campaign are presented. Some guidelines for future measurements and for the edition of the results are suggested.

- 1 INTRODUCTION:

Le développement des dommages subis par la végétation conduit à s'interroger sur le rôle de la pollution de l'air dans ces phénomènes, dont l'expérience montre que la variabilité spatiale et temporelle est très grande. Comme pour toute relation dose-effet, seules les situations de pollution aigüe permettent une identification nette des causes: c'est le cas dans les régions de très fortes émissions de SO_2 (Silésie par exemple) où le dépérissement est observé depuis longtemps. La situation est beaucoup moins nette pour les faibles niveaux, heureusement les plus répandus, car l'expérimentation est beaucoup plus délicate. Il n'y a plus de facteur nettement dominant: les facteurs sont étroitement associés et il est difficile de discerner ceux qui sont fragilisants de ceux qui sont déclenchants, aggravants, ou même simplement des conséquences.

Ce n'est que par la réunion d'un grand nombre de données dans des situations variées et sur des périodes de temps importantes, que l'on peut espérer identifier des régularités permettant de séparer les effets éventuels de la pollution atmosphérique des autres effets, notamment géologiques, pédologiques et climatiques.

De même qu'il est possible de caractériser chaque région par un climât météorologique, il serait très utile de pouvoir caractériser chaque région par un "climât" relatif à chaque autre groupe de facteurs, en particulier la pollution.

- 2 CLIMAT de POLLUTION:

En météorologie, la notion de climât fait traditionnellement référence à la durée, par opposition à celle de temps: c'est l'ensemble des phénomènes météorologiques qui caractérisent pendant une longue période l'état moyen de l'atmosphère. Le climât diffère du temps, qui représente une combinaison éphémère et quelquefois exceptionnelle de ces phénomènes.

Malgré sa simplicité et son caractère apparent d'évidence, notamment du fait de l'expérience quotidienne de chacun, ce concept connaît de sérieuses variantes dans sa mise en oeuvre. Considérons par exemple la notion de climât météorologique: dans l'usage courant, cette notion est conçue en vue d'une typologie des habitats en fonction des contraintes que ceux-ci imposent à l'activité humaine, notamment à l'agriculture. Les facteurs principaux pris en compte sont la température, l'ensoleillement, la pluviométrie. On ne caractérise généralement pas la transparence atmosphérique car les effets biologiques directs sont négligeables; ce serait par contre un critère essentiel pour les astronomes: la climatologie générale et celle de l'astronome diffèrent considérablement par le poids qu'elles attribuent aux divers facteurs de l'environnement.

Il en est de même à propos de la pollution: il serait inopérant de définir un climât de pollution dans l'abstrait, sans explicitation précise de l'usage prévu, et donc sans une réflexion détaillée sur les facteurs à prendre en compte pour chaque type de problème, par exemple la production forestière, la production de céréales ou la modification des sols ou des lacs.

Cette remarque aux apparences triviales est centrale car les données actuelles sont fragmentaires et éparses: un effort important de métrologie devra être effectué: quels composés? où? quelle fréquence? quels types d'agrégation des résultats?

En particulier il importe de préciser dans quelle mesure la notion traditionnelle de climât, avec sa notion d'état moyen de l'atmosphère sur de longues périodes, est pertinente ou si elle doit aussi prendre en compte des échelles de temps plus brèves et représenter les situations de crise. Si oui à travers quels indicateurs?

Il est clair que la définition pertinente des composés à prendre en compte dans la définition des climats de pollution nécessite une coopération plus étroite entre physiologistes et physicochimistes.

- 3 QUELS COMPOSES RETENIR ?

Il est d'usage de distinguer les pollutions acides des pollutions oxydantes; de même la pollution acide est souvent ramenée à la seule pollution soufrée. Or, tous ces polluants sont en interaction étroite.

La figure 1 représente les pricipales voies de la chimie atmosphérique. L'accent y est surtout placé sur les oxydants mais les principales relations avec les composés soufrés sont représentées. Les polluants primaires sont: NO, SO2, Hydrocarbure (HC), CO. Les polluants secondaires, non directement émis mais qui se forment dans l'atmosphère à partir des précédents, sont NO2, O3, HNO3, H2SO4, et aussi H2O2, PAN, aldéhydes etc..

On note sur la figure 1: la spirale des hydrocarbures (en bas), les cycles de NOx et de l'ozone (au centre), la branche de H2O2 (au centre), les branches des nitrates (en haut) et des sulfates (à gauche), la branche des aldéhydes et du nitrate de péracétyle (PAN, en bas à droite), le rôle central des radicaux OH dans la chimie atmosphérique.

L'ozone est formé par photolyse de NO2; il peut en outre être apporté sur le site par advection depuis l'atmosphère libre. Consommé rapidement par NO, il ne peut s'accumuler que si NO est rapidement transformé en NO2 à travers la spirale des Hydrocarbures. Il oxyde SO2 dissous dans les nuages, gouverne l'une des 2 voies de formation de HNO_3 et est la principale source nette de radicaux OH dans le milieu.

SO_2 est oxydé en sulfate par OH en phase gazeuse et par H_2O_2 et O_3 dans les nuages. Du fait que les concentrations de ces trois espèces sont largement gouvernées par la chimie du couple NOx-Hydrocarbures, l'oxydation de SO_2 est étroitement couplée aux processus de pollution photooxydante: Si une émission de SO_2 est effectuée dans un milieu pauvre en NOx, en HC et en rayonnement, la vitesse d'oxydation est lente et SO_2 peut migrer loin de la source; dans le cas contraire SO_2 disparait rapidement et c'est sous forme de sulfate que le soufre est transporté. Des études de terrain ont montré que les concentration de SO_2 dans l'atmosphère et celles de sulfate dans les précipitations étaient anticorrelées (1).

NOx est oxydé en nitrate par réaction directe de NO_2 avec OH et par oxydation de NO_2 par O_3. La seconde voie est négligeable de jour du fait de la photolyse rapide de NO_3; Par contre c'est la voie principale d'oxydation de NO_2 la nuit. Il faut noter que la concentration de NO est toujours très faible en milieu naturel du fait de sa réaction rapide avec O_3.

On note en définitive une très grande interdépendance de tous les chainons réactionnels. On ne peut raisonner séparément en termes de pollution oxydante, acide, soufrée, azotée etc... Enfin, l'importance relative des diverses voies dépend de nombreux facteurs: intensité du rayonnement solaire (pour toutes les réactions photochimiques), état de l'atmosphère (plusieurs réactions clé ne se produisent qu'à des longueurs d'onde très courtes qui sont très affectées par l'absorpsion atmosphérique), présence d'eau liquide, température, débits d'émission des divers polluants primaires, apports par advection, dispersion, etc...

Caractériser le climat de pollution consiste à traduire cette complexité considérable par quelques indicateurs simples. Les polluants classiquement mesurés sont SO_2, NO, NO_2, hydrocarbures non méthaniques (NMHC), O_3, PAN depuis peu. On mesure aussi le dépôt humide (SO_4^{2-}, NO_3^-, H^+ et divers ions). En réalité on mesure ce qu'on sait mesurer à un coût acceptable. Il ne faut pas l'oublier et ne pas établir de relation causale excessive. Certains composés mesurés peuvent être des indicateurs utiles, mais pas nécessairement des agresseurs.

Plutot que SO_2 (lorsque sa concentration est faible) ou en tout cas NO, c'est souvent H_2SO_4 et HNO_3 qui sont agresseurs; Malheureusement on ne dispose pas de technique de mesure simple et satisfaisante en routine pour la phase gazeuse. De même on peut se demander si les radicaux présents dans une atmosphère polluée ne sont pas susceptibles d'attaquer directement la cuticule des végétaux: leur concentration est faible sans doute mais leur réactivité est très élevée. Certaines études sur les mécanismes d'oxydation de SO_2 dans les nuages (2) prennent aujourd'hui en compte la captation des radicaux OH et HO_2 par les gouttelettes et leur réactivité directe sur les espèces en solution. Malheureusement les radicaux sont très difficiles à mesurer directement.

Le PAN, pour lequel des techniques d'analyse commencent à être exploitées en routine dans plusieurs stations pourrait, à côté de son interet phytotoxique propre, donner des indications précieuses: c'est un composé très peu abondant dans le milieu naturel et qui est donc caractéristique d'un milieu pollué qui a subit une évolution photochimique importante. Les aldéhydes jouent aussi un rôle clé du fait de leur réactivité et de leur activité photochimque. Le peroxyde d'hydrogène, H_2O_2, est un composant majeur de la chimie atmosphérique et cependant encore peu connu. Ces composés secondaires seraient peut-être plus intéressants à déterminer dans une perspective d'effets écologiques que, par exemple, les hydrocarbures.

Il faut enfin effectuer quelques remarques sur la métrologie des polluants relativement à leur sensibilité et à leur spécificité. En effet, les niveaux des polluants en zone rurale ou naturelle sont généralement faibles alors que les appareils utilisés ont été conçus très largement pour la surveillance de la pollution urbaine. Les limites de sensibilité pratiques sont: 5 à 10 ppb pour SO_2 et 2 à 5 ppb pour NO et NO_2. C'est ainsi que dans une étude menée aux USA sur les climâts de SO_2 (3), les valeurs inférieures à 10 ppb n'ont pas été prises en compte car considérées par les auteurs comme non significatives. De même, le niveau de NO étant très généralement très inférieur à 3 ppb, les mesures de NO doivent être considérées commer non significatives. Par contre, les analyseurs NO_x donnent un signal avec HNO_3 et avec le PAN: ces composés constituent ainsi une part non négligeable et parfois prépondérante du terme NO_x - NO , lequel est toujours assimilé improprement à NO_2. Les relations dose-effet de NO_2 sur le terrain doivent en conséquence être établies avec circonspection.

- 4 <u>ECHELLE SPATIALE ET TEMPORELLE</u>:

Le terme climatologie de la pollution évoque l'élaboration de cartes des concentrations de polluants. Quels types de moyennes retenir? Dans quelle mesure faut-il prendre en compte des aspects plus qualitatifs ? on discutera ici à titre d'exemple le cas de l'ozone.

A partir des résultats publiés pour de nombreux sites (par exemple Fos-Berre (4), Bonn (5), Canada (6), Withe-Face mountain NY USA et Bodenmais RFA (7)), on peut établir une typologie des profils diurnes de O_3 en zone naturelle ou rurale (figure 2): 2 situations principales:

- sites placés au dessus de l'inversion nocturne ou réapprovisionnés en permanence par l'atmosphère libre sous l'effet des mouvements orographiques (fig. 2a): sommets, pentes dégagées etc... On n'observe pas de cycle diurne: le niveau de nuit comme de jour est le niveau de fond de la région, couramment 40 à 50 ppb en été en Europe continentale, parfois plus. Des pointes liées à des transports de masses d'air urbaines ou industrielles peuvent en outre se superposer au niveau de fond. Le niveau moyen sur 24 heures est égal ou supérieur au niveau de fond.

- sites de plaine (fig 2b): ils sont caractérisés par l'alternance d'une forte stabilité nocturne (convection très faible) et d'une forte convection le jour. Le cycle diurne est marqué: niveau nocturne nul ou très bas du fait de la consommation de O_3 par les émissions locales réductrices et par la déposition; niveau diurne pouvant être important du fait de production photochimique locale; pointes possibles liés à des transports. Le niveau moyen sur 24 heures est la plupart du temps inférieur au niveau de fond du fait des faibles valeurs nocturnes. Le niveau moyen durant le jour est par contre plus élevé que le niveau de fond.

Le climat de pollution dans ces deux situations est clairement différent et des indicateurs différents doivent peut-être être envisagés. Les mêmes chiffres n'ont pas la même signification selon le type de site.

La figure 3 (8) représente la valeur moyenne des concentrations de nitrate dans les précipitations en 4 points en Europe. Dans les régions périphériques (fig. 3a, 3b, 3c) les distributions sont très contrastées selon l'origine des masses d'air: Il est clair que les nitrates ont une origine continentale. Au contraire, au centre de l'Europe (fig. 3d) les niveaux moyens sont élevés pratiquement pour toutes les directions ce qui montre qu'au delà des émissions propres des pays concernés (parfois élevées, notamment pour SO2), il faudrait sans doute attribuer à cet effet omnidirectionnel, plus souvent qu'on ne le fait, une part des niveaux élévés de pollution systématiquement notés en Europe centrale. Il existe dans de nombreux pays d'Europe, RFA, PAYS BAS, SUISSE, etc, une forte densité d'agglomérations importantes dans toutes les directions: les zone rurales y sont pratiquement toujours exposées à des émissons industrielles ou urbaines plus ou moins lointaines. Ce n'est pas la cas à l'ouest, au nord ou au sud ouest de l'Europe; les climâts de pollution ont des caractéristiques différentes et les indicateurs doivent refléter cette diversité. Les effets observés peuvent être très directionnels, et parfois selon des relations complexes.

Citons enfin le cas de l'Espagne dans la région de Madrid (9): une zone de haute pression peut s'installer en été sur la péninsule pour plusieurs semaines, avec creusement durant le jour d'une faible dépression au centre: un régime de brise s'installe à l'échelle de l'ensemble de l'Espagne et globalement un transfert des polluants émis sur les côtes vers la zône de convergence, située dans la région de Madrid. Des épisodes sévères de pollution y sont régulièrement observés en fin d'été. Il s'agit d'un type de climât de pollution qui ne peut être décrit par de simples moyennes annuelles. On pourrait rattacher à cette catégorie de climât les zones soumises périodiquement à des transferts de panaches urbains ou industriels sur 50 à 200 km, notamment par effet de brises.

- 5 RESULTATS:

L'objet des remarques précédentes était de dresser les limites et les conditions de la définition des climâts de pollution. Il est clair qu'il ne saurait être question d'attendre des mesures idéales pour essayer d'évaluer l'état de l'environnement. Les mesures actuellement disponibles, notamment à travers le réseau EMEP permettent quelques éléments de discussion.

A partir des données 1984 du reseau EMEP (10), nous avons dressé des cartes de dépôt humide et d'exposition aux polluants gazeux. Le flux annuel de dépot humide de sulfate (figure 4) et de nitrate (figure 5) a été obtenu pour chaque station en multipliant la concentration moyenne annuelle des précipitations par le volume de celles-ci. Le sulfate pris en compte est le sulfate "non marin". Le maximum de la déposition humide est situé sur l'est et le sud est de l'Europe, avec un dépôt plus faible sur la vallée moyenne du danube. La déposition de nitrates concerne essentiellement le centre de l'Europe avec un maximum au centre sud.

La figure 6 représente la variation du rapport des concentrations (exprimées en équivalent ionique) de nitrate et de sulfate non marin de 1979 à 1985 à la station de Deuselbach (SW RFA) (11). L'augmentation régulière du rapport montre que les composés azotés jouent un rôle croissant dans la pollution acide sur ce site. Une observation analogue avait déjà été effectuée sur d'autres sites (8). Nous avons calculé la valeur de ce rapport pour 1984 sur l'ensemble de l'Europe (figure 7): on note que la contribution des nitrates est particulièrement marquée en Europe Occidentale et qu'elle peut par contre être considéré comme négligeable dans plusieurs pays d'Europe de l'est. Ces résultat recoupent ceux déjà publiés pour 1982 (8).

La figure 8 représente la concentration moyenne annuelle de SO2 gazeux: des niveaux élevés sont observés dans le sud de la RFA et en Yougoslavie. Par absence de données dans le sud de la RDA, de la Pologne, et en Roumanie, les valeurs EMEP ne permettent pas d'évaluer précisément l'étendue de cette zone vers l'est.

La figure 9 représente les données d'ozone disponibles pour l'été 1985 (avril à septembre), sous la forme de niveaux dépassés 2% du temps (98 percentile, d'après (12)). On note une croissance du niveau vers l'Europe du centre sud. On ne dispose que de trop peu de données sur NOx pour effectuer une relation entre O3 et NOx. On peut cependant noter une certaine coincidence entre les zônes de niveau élevé de O3 et celles de forte déposition humide de nitrates (fig. 5). Les quelques données disponibles sur NOx en RFA et en Scandinavie (10) vont aussi dans ce sens.

- 6 CONCLUSION:

Lors d'une réunion récente à BERNE (12), quelques points ont été mis en évidence:

La déposition humide semble actuellement convenablement évauée, sous réserve des fluctuations liées aux caractéristiques propres des points de prélèvement et des lacunes qui les séparent. Les données gagneraient cependant à être exprimées en terme de flux de déposition (nécessité de publier les concentrations et les volumes de pluie) et à bien distinguer déposition humide seule de déposition totale.

Les anions à évaluer sont $SO4^{2-}$, $NO3^{-}$, Cl^{-}, $HCO3^{-}$, $H2PO4^{-}$ et $HPO4^{2-}$; Les cation: Ca^{2+}, Mg^{2+}, K^{+}, Na^{+}, $NH4^{+}$, H^{+}, Mn^{2+}; L'ion Pb^{2+} est utile comme traceur de la pollution anthropogénique. Ces ions constituent, selon les physiologistes, le minimum nécessaire pour évaluer les effets de la déposition sur les écosystèmes. L'amélioration de la définition de protocoles standard de prélèvement et d'analyse a est souhaitée.

Par contre, l'évaluation du dépôt sec pose encore un problème sérieux car la déposition sur les végétaux est très complexe: aucune jauge physique ne semble à l'heure actuelle avoir été validée comme modèle acceptable du point de vue du physiologiste. Il semble actuellement préférable d'évaluer la déposition sèche sur les végétaux par calcul à partir des concentration en phase gazeuse, des données météorologiques et des facteurs de déposition espèce par espèce. Toutefois, la complexité du processus ne permet pas encore, selon nous, de dresser d'inventaire satisfaisant.

Pour le dépôt sec il est apparu en outre nécessaire d'affiner sensiblement les règles d'expression des résultats, notamment pour tenir compte des mécanismes d'assimilation des polluants gazeux, particulièrement efficaces lorsque les stomates sont ouverts (cycle diurne) et que le cycle végétatif est actif (cycle saisonnier). Il a été suggéré de fournir le données suivantes:

- SO_2 : Moyenne annuelle générale, moyennes journalières, moyennes annuelles pour chaque tranche horaire, valeur horaire maximale chaque 24h.
- NO_2 : idem
- O_3 : Moyenne annuelle des valeurs horaires diurnes (nuit exclue), nombre d'heures dans l'année ou le seuil horaire de 70 ppb a été dépassé (ou nombre de jours de l'année où le seuil horaire de 70 ppb a été dépassé au moins une fois). 200 mesures horaires supérieures à 70 ppb a été considéré comme représentant une situation sérieuse.

Un effort de développement de techniques analytiques spécifiques a enfin été souhaité, notamment pour HNO_3 et NH_3.

L'examen des figures montre que l'extension du réseau de surveillence aux pays actuellement peu générateurs de données est indispensable. La pollution atmosphérique est un phénomène par nature communautaire; certaines régions, notamment sur la façade occidentale bénéficient d'une forte alimentation en air peu pollué. Ce n'est évidemment pas le cas des pays du centre de l'Europe où une part des niveaux observés provient de l'effet cumulatif des rejets le long de la trajectoire des masses d'air. Un réseau suffisamment continu de stations permettrait de mieux décrire les climâts de pollution et de comprendre l'évolution des masses d'air au cours de leur déplacement.

- 7 BIBLIOGRAPHIE et REFERENCES:

(1) J.A. PENA, R.G. de PENA, V.C. BOWERSOX, J.F. TAKACS, 1982, Atmosph. Env., 16, No 7, p 1711-1715.

(2) C. SEIGNEUR, P. SAXENA, 1984, Atmosph. Env., 18, No 10, p 2109-2124.

(3) F.M. VUKOVICH, J. FISHMAN, 1986, Atmosph. Env., 20, No 12, p 2423-2433

(4) G. TOUPANCE, P. PERROS, 1985, Atmospheric Ozone, C. S. ZEREPHOS et A. GHAZI Edit., D. Reidel Pub. Co., p 820-824.

(5) B. SCHERER, R. STERN, 1981, Physico-Chemical Behaviour of Atmospheric Pollutants, B. VERSINO et H. HOTT edit., D. Reidel Pub. Co.,p561-571.

(6) R.P. ANGLE, H.S. SANDHU, 1986, Atmosph. Env., 20, No 6, p 1221-1228

(7) A.S. LEFOHN, V.A. MOHNEN, 1986, J. Air. Poll. Contr. Ass., 36, No 12, p 1329-1337.

(8) K. NODOP, 1987, Physico-Chemical Behaviour of Atmospheric Pollutants, G. ANGELETTI et G.RESTELLI edit., D. Reidel Pub. Co., p 520-528.

(9) M. MILLAN, M. NAVAZO, A.EZCURRA, 1987, Physico-Chemical Behaviour of Atmospheric Pollutants, G. ANGELETTI et G.RESTELLI edit., D. Reidel Pub. Co., p 614-626.

(10) J. SCHAUG, J. PACYNA, A. HARSTAD, T. KROGNES, J.E. SKJELMOEN, Février 1987, Rapport EMEP/CCC-Report 1/87, NILU, PO 64, N 2001, LILLESTROM, Norvège.

(11) D'après communication personnelle G. SCHMITT, 1987.

(12) P. GRENNFELT, J. SALTBONES, J. SCHJOLDAGER, Avril 1987, Rapport NILU 22/87, Référence O-8535, NILU, PO 64, N 2001, LILLESTROM, Norvège.

(13 COST 612 Workshop, "Definition of European Climates and their Perception by Terrestrial Ecosystems, Bern, Suisse, 27-30 avril 1987.

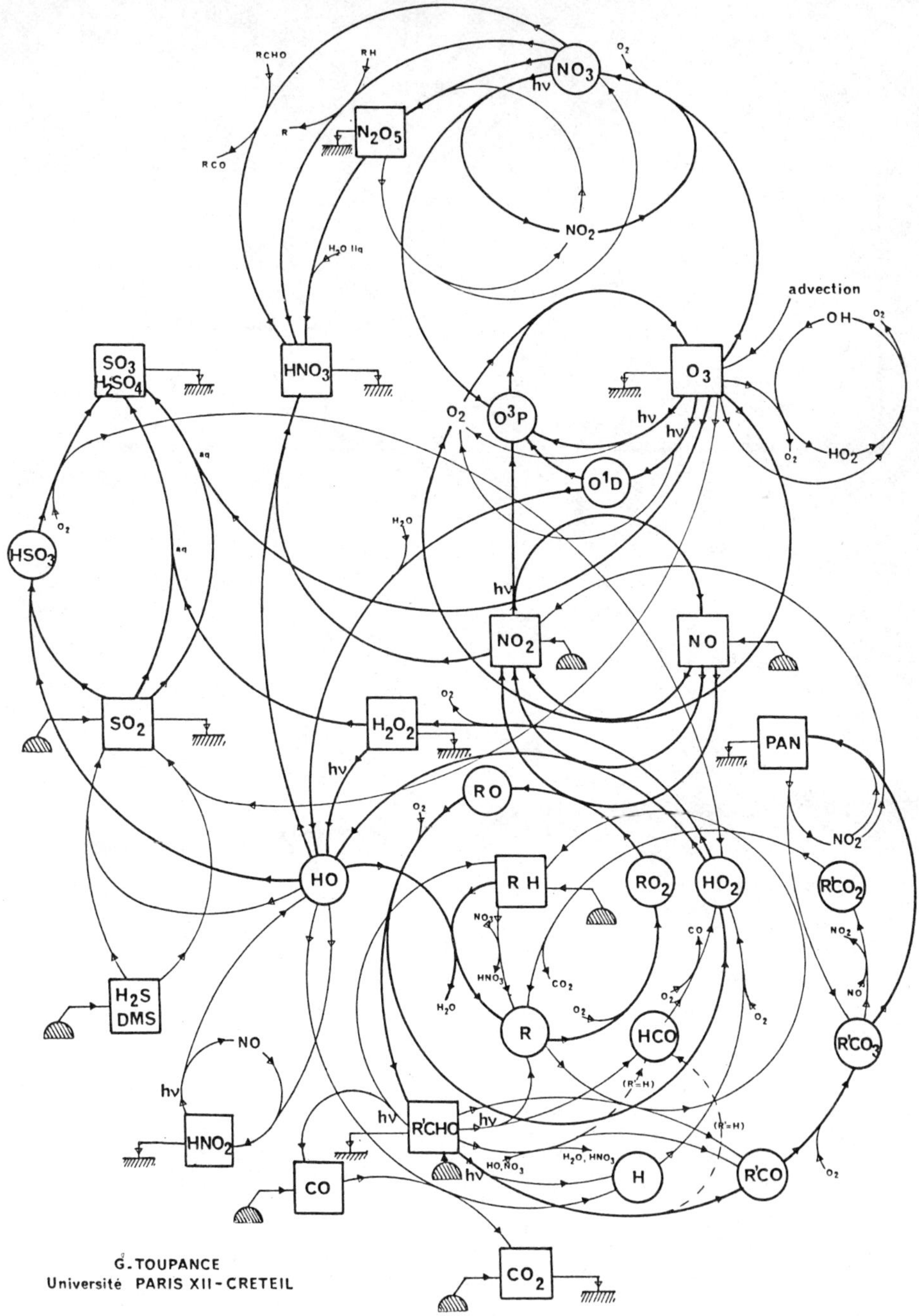

Figure 1 : Principales voies du métabolisme atmosphérique.

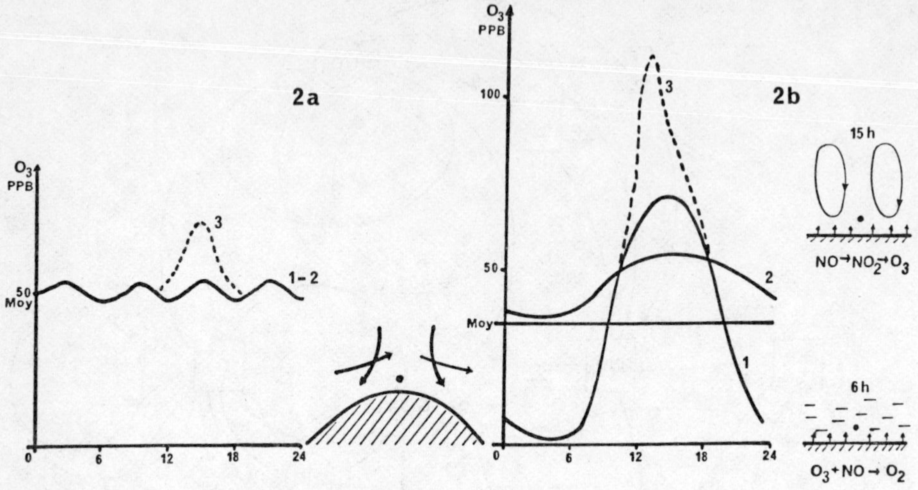

Figure 2 : Profils diurnes types d'ozone en zone rurale.
2a: station d'altitude. 2b: station de plaine.
(1- vent faible, 2- vent fort, 3- influence urbaine ou industrielle)

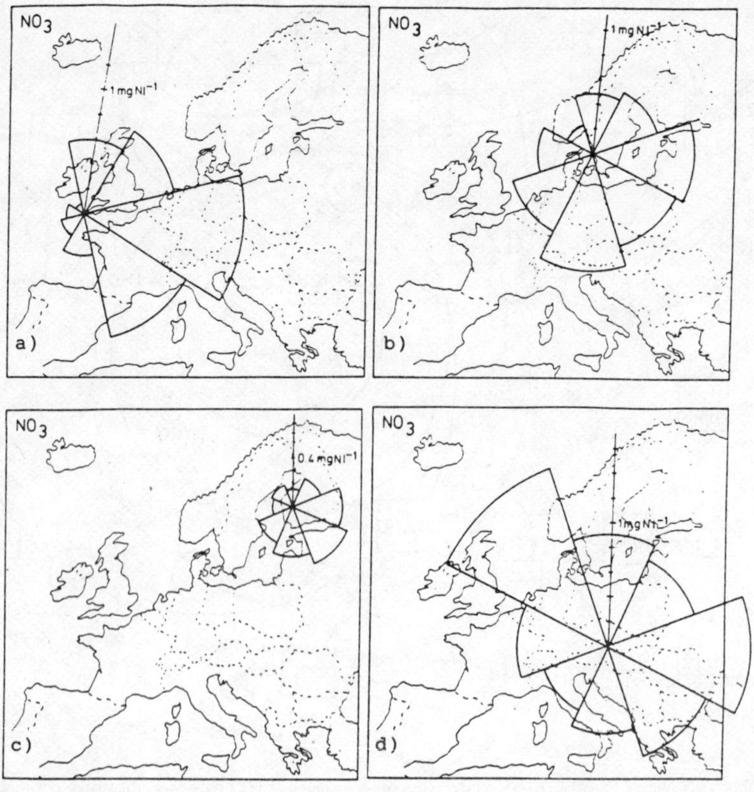

Figure 3 : Direction du vent et concentration de nitrate dans les pluies. D'après (8).

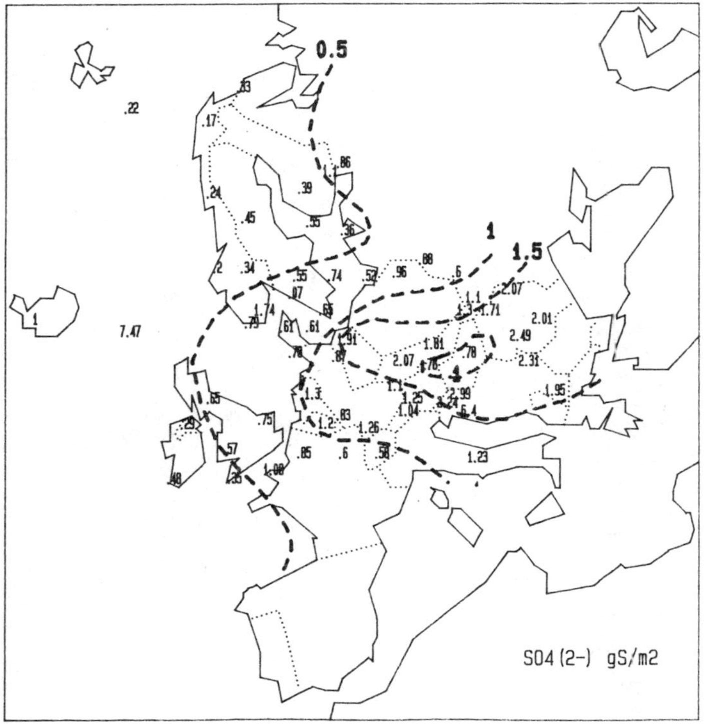

Figure 4 : Déposition de sulfate non marin par les précipitations en 1984. (gramme de soufre par an et par m^2).

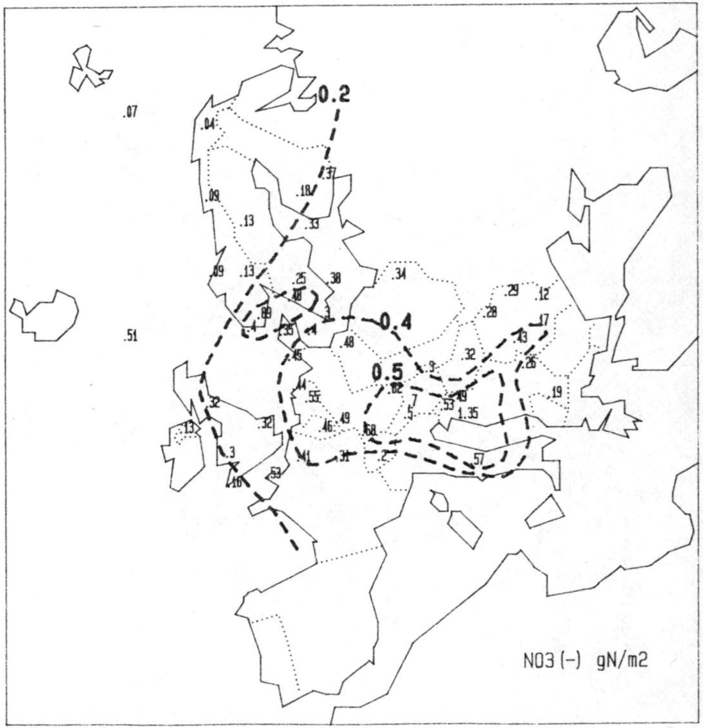

Figure 5 : Déposition de nitrate dans les précipitations en 1984. (gramme de N par an et par m^2).

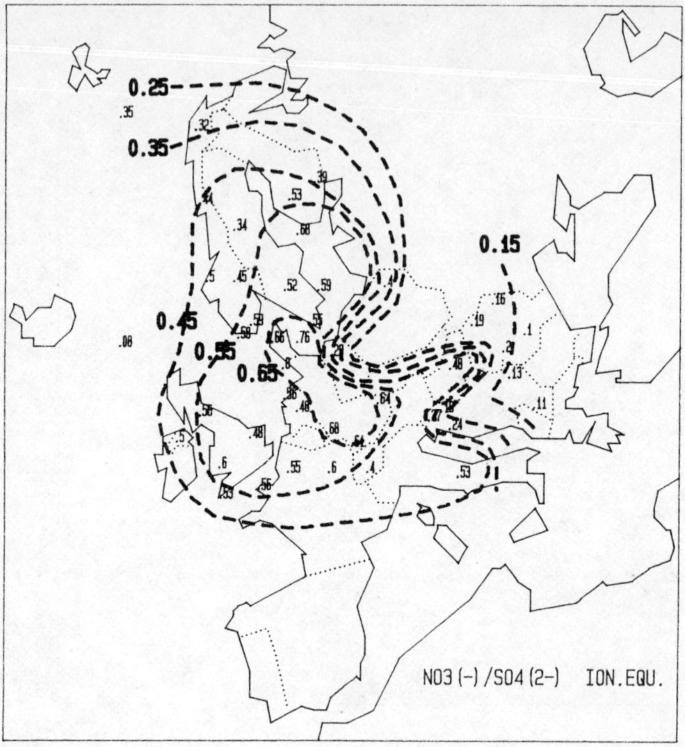

Figure 6 : Nitrate/Sulfate non marin (moyennes annuelles)
(Nitrate et Sulfate exprimés en ion équivalent)
(1984)

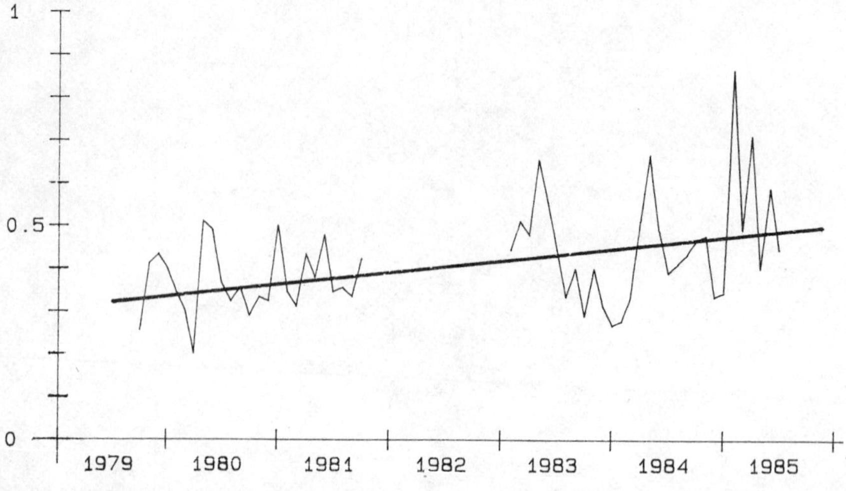

Figure 7 : Nitrate/Sulfate non marin (moyennes mensuelles, ion équivalent)
à DEUSELBACH (RFA), Septembre 1979-Juin 1985). (d'après (11))

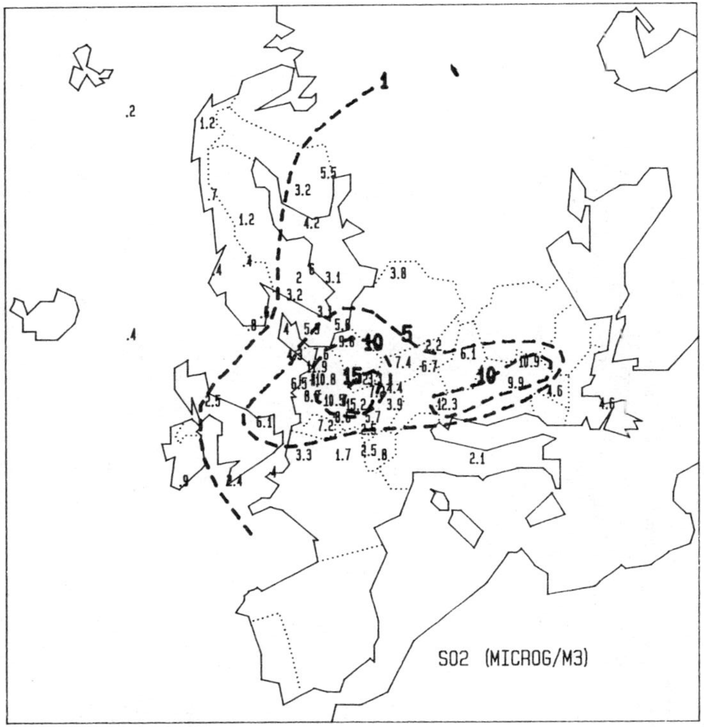

Figure 8 : Concentration moyenne annuelle de SO2 dans l'air. (1984).

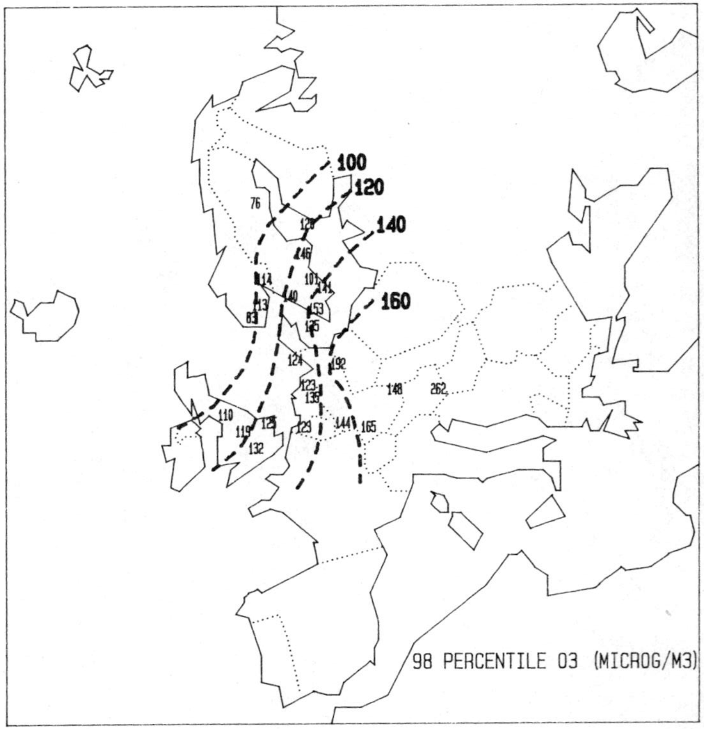

Figure 9 : Niveaux d'ozone dépassés 2% du temps. (98 percentile).
Période Avril-Septembre 1985.

DEPOSITION OF POLLUTANTS ON PLANTS AND SOILS; PRINCIPLES AND PATHWAYS

M.H. UNSWORTH & D. FOWLER
Institute of Terrestrial Ecology
Bush Estate
Penicuik
Midlothian EH 26 OQB
Scotland

Summary

There is considerable variation in the "pollution climate" in Europe, i.e. in the mixtures of pollutants found in various regions, and further monitoring of pollution climates is necessary to define relationships between deposition and effects of pollutants. Knowledge of rainfall chemistry has improved in recent years so that maps of rain acidity in Europe can be prepared, but the relative contributions from natural and anthropogenic sulphur are still unclear. Nitrate in rain has increased since the 1850's and is associated with a greater proportion of the acidity than in the past. Nitrogen deposition is also important because of its role in plant nutrition. Rainfall chemistry is strongly episodic. It is becoming clear that wet deposition in mountainous regions may be larger than previously estimated because of orographic enhancement and droplet scavenging. "Occult deposition", the deposition of soluble and insoluble materials in wind-driven cloud is also significant but has seldom been measured. Routine monitoring of cloud water chemistry is necessary as a first step, but is not sufficient for reliable occult deposition estimates. Rates of dry deposition are often strongly controlled by surface factors. Further dry deposition studies are needed for nitrogen-containing gases and for particles. Once deposited, pollutants may react with plant surfaces, they may be absorbed, and ion exchange may occur. These processes need to be studied in association with deposition measurement to improve our knowledge of the pathways by which pollutants move through ecosystems.

1. INTRODUCTION

It would be impossible in the space available for this paper to review the full range of our knowledge of the deposition of pollutants on plants and soils and of the subsequent pathways through ecosystems. Consequently, we have selected some topics where there have been new developments or where there is a need for further research, and they will be discussed under the main headings of wet deposition, occult (cloudwater) deposition, dry deposition, and surface phenomena. The emphasis of the review will be on the major pollutants associated with acidic deposition and photochemical oxidants; there is unfortunately no space to discuss in detail the deposition of other pollutants such as heavy metals and fluorine.

The relative importance of different forms of deposition in Europe depends greatly on the physical and pollution climates of the various regions. Physical climate alters deposition through, for example, the amount of rainfall, or the frequency of low cloud. The 'pollution climate' (Fowler et al., 1987) is a general term which describes the different mixtures of pollution that are found in various regions. Although the concentrations of pollution will vary from year to year and with season, pollution climate gives a general indication of the potential for different pollutants reaching terrestrial systems, much as physical climate gives a general indication of the likelihood of regions experiencing certain types of weather.

The combination of knowledge of physical and pollution climates is however still inadequate to define relationships between deposition and effects of pollutants in different parts of Europe until there is a clearer understanding of the links between deposition in its various forms, and the thresholds for plant response.

2. WET DEPOSITION

In the last decade there has been a substantial increase in the number of sites throughout Europe where rain chemistry is measured on a regular basis. Several different designs of wet-only rain collectors are now commonly used; although these devices minimise contamination of the collector when it is not raining, it becomes clear when identical collectors are run side by side that there are still problems over the reproducibility of such measurements, especially in the amount of rainfall collected. In cleaner parts of Europe where collector contamination is not such a problem, simpler collectors, without automatic lids, have been very usefully employed (Fowler & Cape, 1984). Rainfall chemistry is widely monitored throughout Europe, and the spatial and temporal patterns in wet deposition of the major ions are quite well defined on regional scales (EMEP 1985). However, there are still problems; for example, the wet deposition of sulphate in the high rainfall areas of western Britain and Norway dominates the sulphur inputs there, and greatly exceeds dry deposition. Yet the relative contributions of the sulphur inputs from natural and anthropogenic activity are still a matter for debate. Long range transport models incorporate a 'background' term to obtain satisfatory agreement between model predictions and measurements (Smith 1984). The 'background' is assumed to be a combination of natural sulphur emissions and components of anthropogenic emissions that have long residence times in the atmosphere. In these areas of large wet deposition, 'background' may represent 70% of the wet deposition (Barrett et al., 1984), and it is important to establish more clearly the origins of this material. There are also considerable difficulties in measuring amounts of snowfall and the associated chemistry, so that over much of northern Europe and at high altitudes in winter the amount of wet deposition in snowfall is poorly defined.

In discussing the variation in rain chemistry across Europe, acidity in rain is perhaps the least satisfactory ion to consider, as it is not conserved in the same way as the other major ions in rainfall. But acidity is, of course, of great public interest. Figure 1 shows that the areas where the annual average acidity in rain is largest range from the coast of Belgium eastwards through West and East Germany, into Poland and extending from the Baltic to central Germany (Fowler et al. 1987). In this region the annual volume-weighted acidity of precipitation is about pH 4.1 (80 μm H^+l^{-1}). To the west and north west, through France and the British Isles, there is a gradual decline in acidity, so that in north west Scotland, southwest France, and Portugal typical annual averages are pH 5.0.

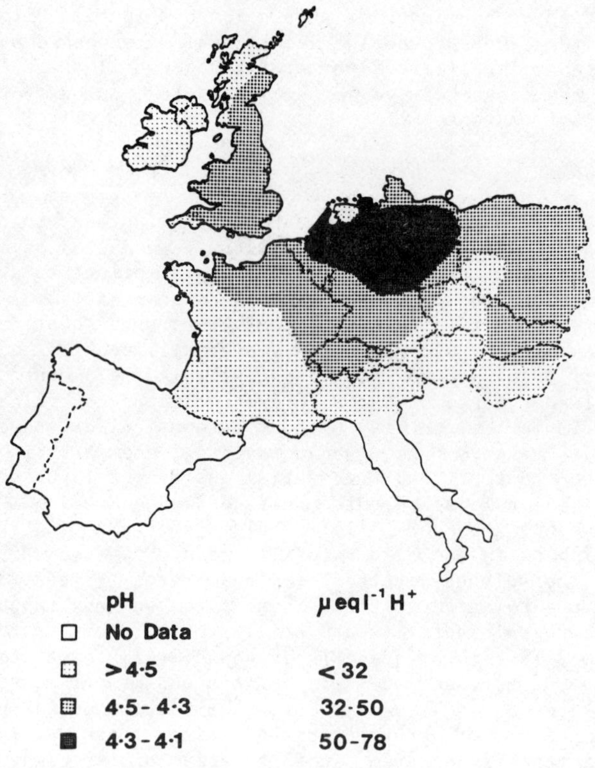

FIGURE 1. Annual mean acidity in precipitation (volume-weighed mean, 1978-83).

The major ions associated with acidity are sulphate and nitrate. The ratio of nitrate to non-marine sulphate has a maximum centred on West Germany and extending eastwards into East Germany, Czechoslovakia and Austria, and westwards into east and central France (Figure 2). Outside this region the relative importance of nitrate declines, indicating thatthe relative contributions of non-marine sulphate and nitrate to acidity throughout Europe differ (Fowler _et al_. 1987).

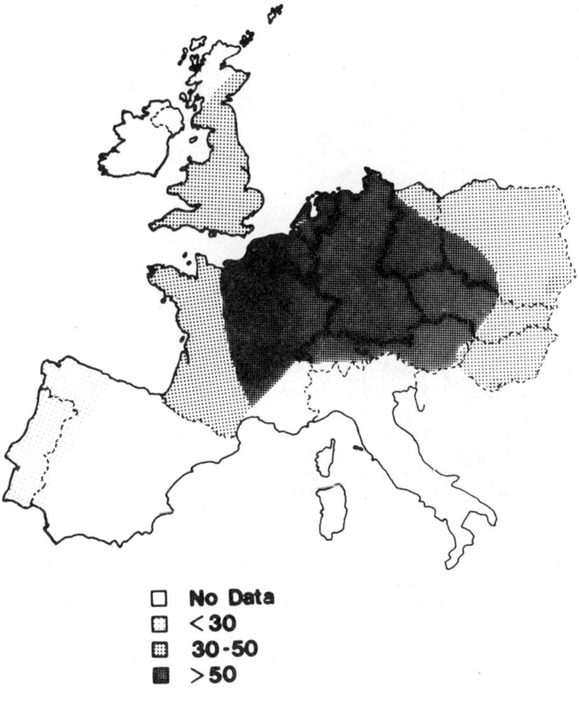

FIGURE 2. Ratio, expressed as %, of nitrate to sulphate equivalents in precipitation.

There is good evidence that the mean concentration of nitrate in rain has increased considerably throughout Europe in the last century. For example, at Rothamstead, UK, the mean concentration of nitrate in rain doubled between the 1850s and 1960, and an analysis of precipitation chemistry data from several sites in Europe indicates an approximately threefold increase in nitrate deposition between the 1890s and the late 1970s (Brimblecombe and Stedman 1982). Data from rural sites in Scotland show a marked increase in the importance of nitrate relative to non-sea sulphate during the four year period 1980 to 1983, and these data are consistent with analysis from Swedish stations which showed that nitrate and ammonium appeared to remain constant in rain between 1972 and 1984, whereas the sulphate concentration decreased by about 30% (Rodhe and Rood 1986). It is clear that a greater proportion of the acidity in rain over much of Europe is associated with nitrate than was the case in the past. The implications of the nitrogen deposition itself are also significant.

Another feature of the wet deposition climate is the variation in rainfall amount and composition between events. We can define episodicity as the tendency for a large proportion of the annual deposition to be associated with a small proportion of the annual rain events. For example,

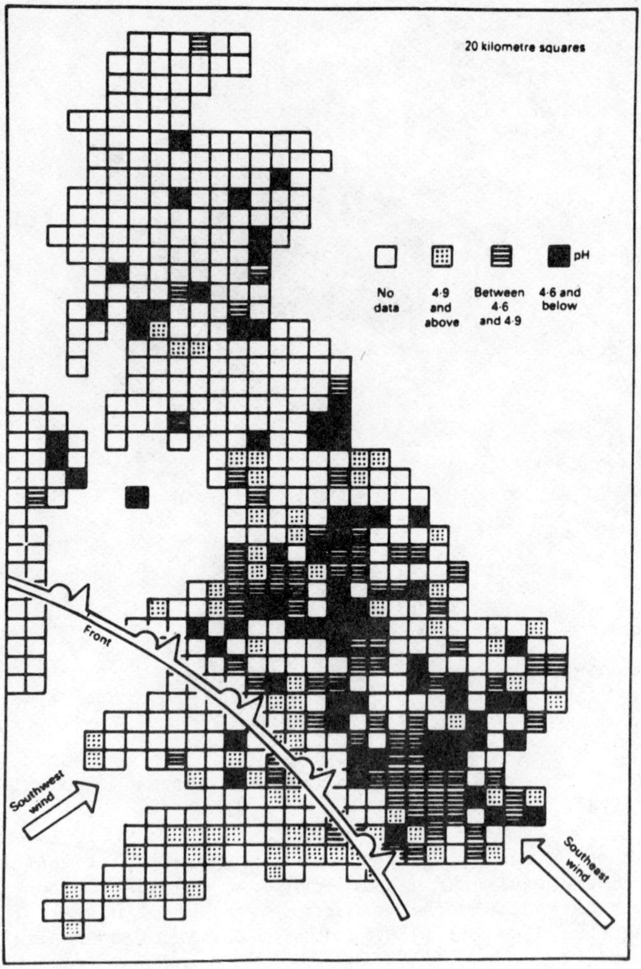

FIGURE 3. Mean pH of precipitation collected by school children on 21st January 1985.

Smith and Hunt (1978) identified parts of Scotland, Switzerland and southern parts of West Germany as those areas of Europe with the greatest episodicity for non-marine sulphate deposition. In these regions 30% of the wet deposited non-marine sulphate was deposited on less than 5% of the wet days. Different ions show different degrees of episodicity at the same site.

Although there are substantial numbers of rain chemistry collectors in Europe, there have been very few extensive analyses of rainfall chemistry as individual weather systems move across a region. An interesting example of the power of organizing surveys with large numbers of relatively simple observing devices was the UK 'Acid Drops' project (Baker et al. 1986). In this project, between January 14 and February 10 1985, thousands of children all over the United Kingdom monitored the acidity of rain and snow their own gardens using standardised collectors and indicator papers. The aim of the project was to collect detailed information about regional differences in rainfall acidity as weather fronts crossed the country. The results for January 21 are illustrated in Figure 3. A high pressure system was breaking down, and fronts carrying rain were moving into Britain from the southwest, whilst the east coast was still receiving air from Europe. The survey showed that rain in the southwest of the country was less acid than elsewhere, demonstrating the influence of long distance and local transport in the southeasterly air stream that covered most of the country. 1986). Figure 4 shows that calculations of nitrate in rain were within about a factor of 2 of observed concentrations, but only provided that the model calculations included (very uncertain) estimates of ammonia emissions and of the influence of ammonia on reaction rates (Hough 1986).

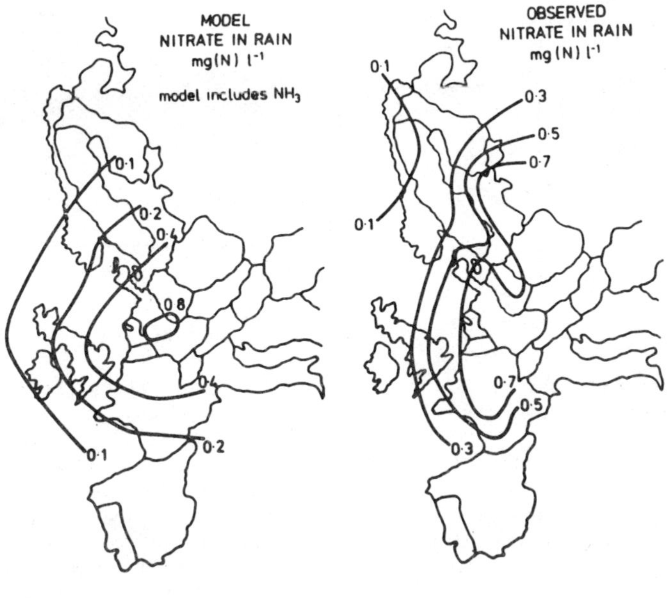

FIGURE 4. Comparison of observations and calculations of nitrate in rain.

Attempts to model the wet deposition of pollution in Europe are a good test of our knowledge of the appropriate physics and chemistry. Uncertainties in models of sulphate deposition were discussed earlier. More recently, models of nitrate deposition have been developed (Derwent

An important development in recent years has been the recognition that wet deposition in mountainous areas may be much larger than has been estimated previously by extrapolation from measurements of rainfall amount and chemistry made at convenient valley sites. It is well known that rainfall amounts increase with altitude and, although there is much less experimental evidence, concentrations of pollutants in rain may also increase with altitude. The amounts of such increases depend on rainfall type and on local orography. In the UK, particular attention has been paid to the orographic enhancement of rainfall amount and of wet deposition of pollutants. A large experimental research programme has been developed at Great Dun Fell, with collaboration from several research institutes and universities. The mechanism of orographic enhancement of rainfall (the "seeder-feeder" mechanism) is illustrated in Figure 5 (Fowler et al.

TYPICAL CONCENTRATIONS ($\mu eq l^{-1}$)

	Cap cloud	Rain 200m	Rain 800m
SO_4^{2-}	100-2000	42	60
NO_3^-	30-2000	45	65
H^+	10-1000	25	40

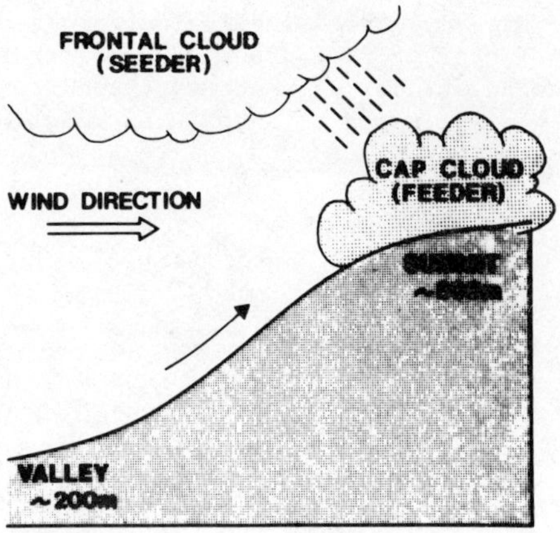

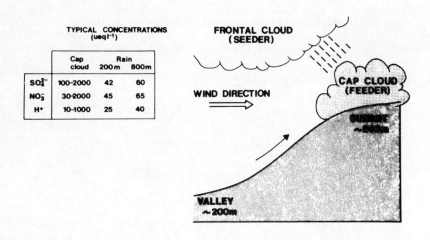

FIGURE 5. Seeder-feeder mechanism for enhanced rainfall concentrations of major ions in rain.

1987). Orographic cloud is formed when moist air is forced up a hillside and condensation occurs. Large concentrations of sulphate, nitrate and acidity are commonly observed in the droplets of the cap cloud. When rain then falls through the cap cloud from higher level frontal cloud, the falling raindrops scavenge the small cloud droplets in the cap cloud and this results in enhanced rainfall at the summit, and in an increase in chemical concentration of rainfall compared with that in the valley below.

TABLE 1.

RATIO OF CONCENTRATION AT SUMMIT/VALLEY FOR MAJOR IONS

	H^+	NH_4	Cl^-	NO_3^-	SO_4^{2-}	RAIN
x	2.9	3.1	2.9	2.3	2.2	2.0

Mean of 20 precipitation events measured at Great Dun Fell (1984-85) with measurements at 8 levels (244-847 m)

Table 1 illustrates the average ratio between the summit and the valley concentrations of major ions in rain at Great Dun Fell, showing that, with twice as much rain, and enhanced chemical concentrations, the average wet deposition of ions at the summit was typically 4 times larger than in the valley. Figure 6 shows the change in average concentration of sulphate in rain with altitude on Great Dun Fell during spring 1985. It is clear from this research that wet deposition estimates in many mountainous parts of Europe where the seeder-feeder mechanism operates are in error. In the higher mountains of central Europe the mechanisms of wet deposition differ from those operating in western Britain and Norway; rather smaller concentrations of major ions are found at these high altitude sites (Munzert 1987), but there is still considerable uncertainty over the amount of wet deposition.

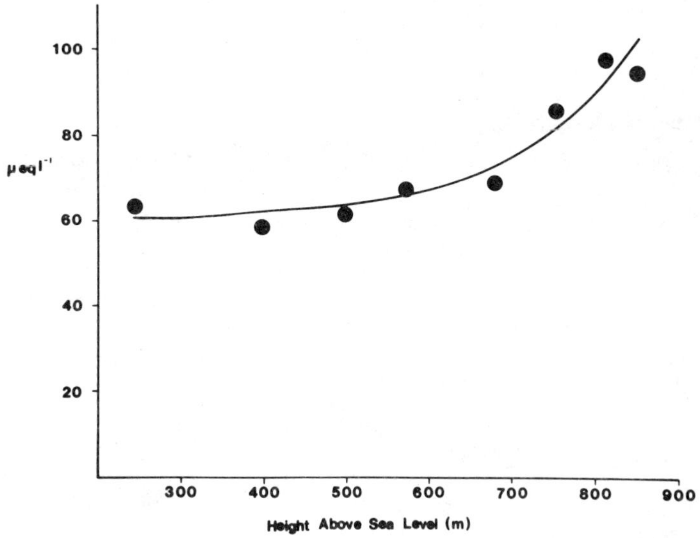

FIGURE 6. Variation of average concentration of sulphate in rain on Great Dun Fell.

The task of improving estimates of wet deposition in mountainous areas by direct measurements and through numerical models needs to be tackled before the relationships between deposition and effects on terrestrial and freshwater systems can be satisfactorily explored.

3. OCCULT DEPOSITION

It has long been recognized that vegetation in cloud captures small water droplets much more efficiently than nearby rain gauges. This input of water by cloudwater capture is therefore not accurately measured by rain gauges, and it has often been called "occult" (hidden) precipitation. The principles that we have discussed earlier concerning cloud formation on hills indicate that occult precipitation is likely to contain larger chemical concentrations than rainfall at the same site. It is not easy to collect cloudwater separately in the presence of rain, and so there are few reliable observations of the ratio of cloudwater to rainwater concentration. Table 2 shows this ratio for the major ions collected at 847 m on Great Dun Fell on occasions when we were reasonably confident of separating cloudwater from rainwater. These results suggest that the average concentration ratio is between 2 and 4; there are undoubtedly some occasions when the ratio is even larger, but the factor of 10 that is sometimes quoted is too large to be generally true.

TABLE 2.

AVERAGE RATIO OF CLOUD/RAIN WATER CONCENTRATIONS FOR MAJOR IONS*

H^+	NH_4^+	Cl^-	NO_3^-	SO_4^{2-}
3.9	2.4	2.6	2.8	2.0

* Mean of 11 precipitation and cloud events, measured at 847 m on Great Dun Fell 1985.

Before we can begin to estimate inputs of occult deposition to terrestrial ecosystems we need instruments for recording the frequency and density of cloud. As part of a CEC contract, Crossley et al. (1986) modified an automatic weather station at a hilltop site so that it recorded simultaneously the rates of rainfall and rates of cloudwater accumulation from adjacent rain gauges and cloudwater gauges respectively. Figure 7 shows that it was possible in this way to identify periods when there was cloud alone, rain alone, and mixed occurrences of the two.

At present we know far less about the chemical climatology of cloudwater than we do about the chemistry of rain in Europe. Because of the potential for damage that exists when vegetation is exposed to polluted cloudwater, there is a case of developing a network of European cloudwater monitoring sites in much the same way that there is a network of rainfall monitoring sites as part of the EMEP system.

Even when there are good records of the regional distribution of cloudwater chemistry, the frequency of cloud occurrence, episodicity etc. this will be only the starting point for estimating the amounts of occult deposition to different ecosystems. Unlike rainfall, the amount of cloudwater deposited depends strongly on physical factors such as windspeed and cloudwater content, and on surface factors such as canopy structure and

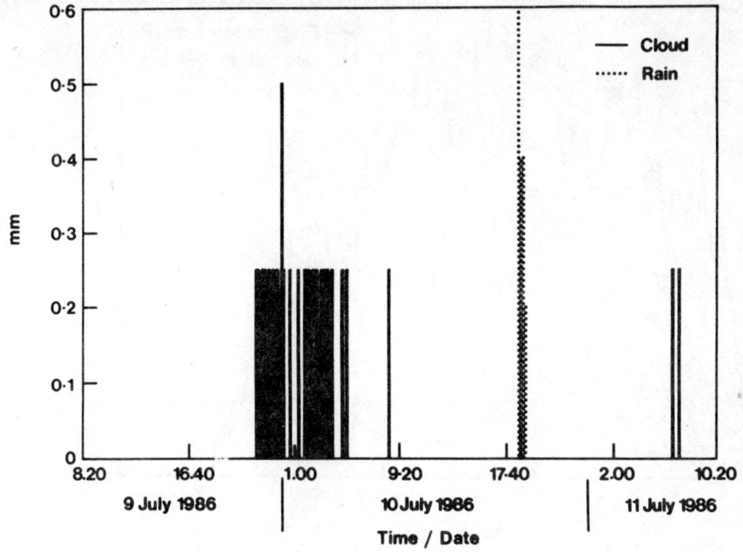

FIGURE 7. Example of the output versus time from tipping-bucket rainwater and cloudwater gauges at Castlelaw, Scotland. Over the 50 h period, the values at the top show that 219 mm of cloudwater were recorded and 50 mm rain.

species composition (Unsworth and Crossley 1987). There is an urgent need for direct measurements of occult deposition rates to different types of plant canopy. Such measurements need to be interpreted so that we can build generalised models that will enable us to estimate occult deposition on local or regional scales. Existing models such as those proposed by Shuttleworth (1978) and Lovett (1984) provide a useful framework for identifying the surface and atmospheric features that need to be measured, but these models make a number of assumptions that have not been fully tested by experiment, and it would be dangerous to believe that reliable estimates of occult deposition can be made from such models. Priority should be given to the development of experimental methods for measuring occult deposition, either by direct methods or from micrometeorological studies.

At times it is possible that cloud water may simultaneously be depositing on foliage and evaporating from it. In these circumstances Unsworth (1984) showed that chemical concentrations in the liquid film on foliar surfaces could be enhanced substantially over those occurring in the depositing cloud drops. The potential enhancement ratio is given by D/(D-E) where D is the deposition rate of cloudwater and E is the evaporation rate of pure water from the leaf. Foliar enhancement is likely to be a transient phenomenon, e.g. when the weather improves after a period of cloud, but it can also occur for prolonged periods if the cloud is suffiently thin to allow solar energy for evaporation to reach the

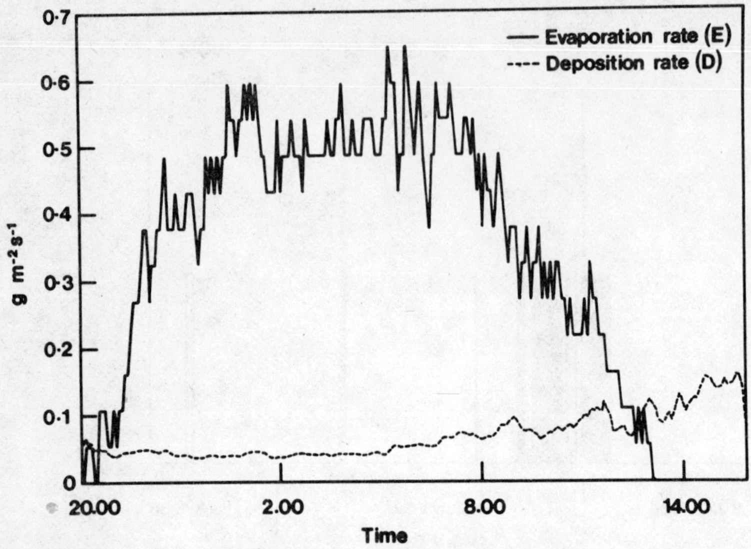

FIGURE 8 a) Calculations of rates of occult depositon D and evaporation E over a spruce forest at Castlelaw, Scotland.

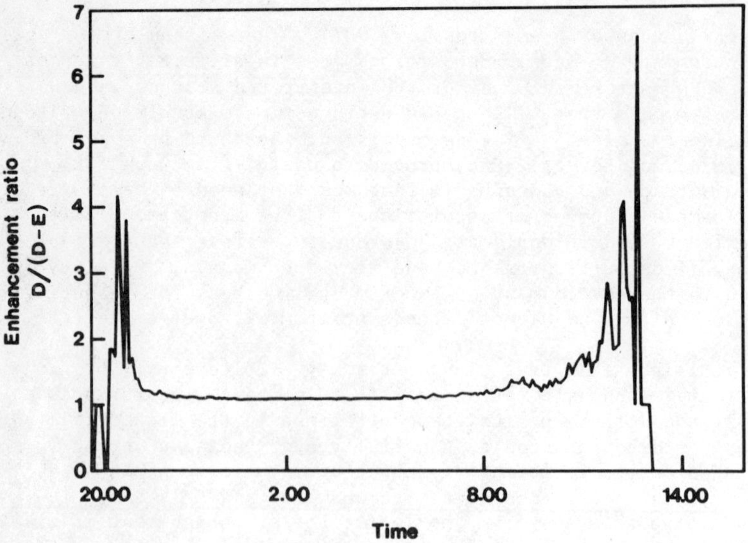

FIGURE 8 b) The potential enhancement ratio for foliar deposition, calculated for the conditons in Fig. 8 (a)

surface. Such situations are likely to occur at the top and bottom of cloud layers. It may be also significant that the chemical concentrations in cloudwater are more likely to be large in such situations than in the centre of a deep cloud layer. We know of no direct measurements of foliar enhancement during cloudwater exposure, but Figure 8 shows some calculations (Crossley et al. 1986) of the enhancement that would occur on the foliage of a spruce forest adjacent to our automatic weather station. For most of the time the enhancement ratio was about 1.1, but at the beginning and end of the cloud cover there was enough energy and the air was sufficiently unsaturated for the enhancement ratio to exceed 1.5, and to reach values of 4 to 6 for short periods. The significance for foliage of these brief exposures to highly concentrated foliar chemical solutions has not yet been studied.

The presence of orographic cloud on hills provides a stationary additional removal mechanism, whereby submicron particles become condensation nuclei, are converted into cloud water droplets and may be deposited either by scavenging by rain falling from above the cloud, or by turbulent deposition at the ground. This removal pathway for particles is much more efficient than dry deposition, discussed in the next section, and so ecosystems growing above cloudbase receive larger deposition of these particles than systems at lower elevation.

4. DRY DEPOSITION

The absorption of gaseous pollutants at the surface and the capture of particulate materials are forms of dry deposition. The importance of dry

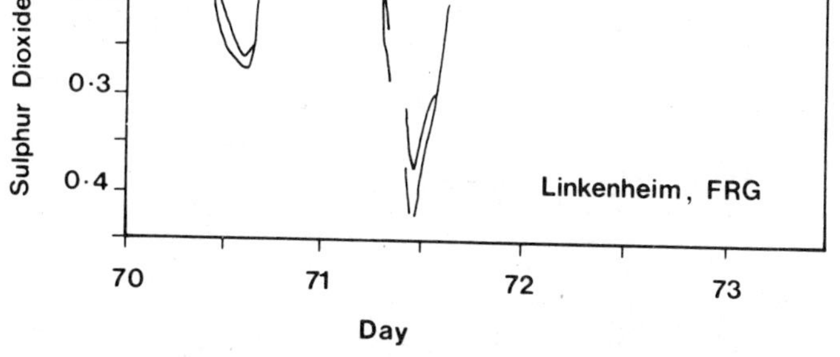

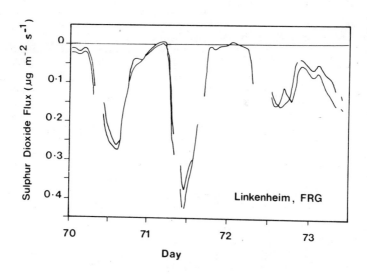

FIGURE 9. Eddy fluxes of sulphur dioxide measured with adjacent sensors over a grassy field near Linkenheim, FRG (data provided for Hicks (1986) by R.T. McMillen and D.R. Marr).

deposition should not be underestimated; in urban areas and in many industrial regions of Europe, dry deposition accounts for the major proportion of sulphur and nitrogen deposited on an annual basis.

Rates of dry deposition are often calculated as the product of atmospheric concentration and an appropriate deposition velocity v_d. Although this approach is attractive for large scale modelling exercises, it usually ignores the large amount of evidence collected over the last decade to show that v_d differs substantially between surfaces and often is controlled by physiological mechanisms that give v_d a substantial diurnal and seasonal variation. For example, Figure 9 shows measurements of the rate of dry deposition of sulphur dioxide to a grassy field near Linkenheim, FRG over a period of 3 days in spring 1985 (Hicks 1986). The main factor responsible for the large diurnal variation in deposition rate was the stomatal response of the grass; stomata were open widest in the middle of the day, leading to large deposition rates, and they closed at night, reducing deposition rates to almost zero. Pollutant gases may also be absorbed on the surfaces of leaves, in moisture such as dew on foliage, and on soils, so that it is fairly unusual for rates of dry deposition to be zero even when stomata are closed at night. The importance of stomatal responses in altering deposition velocity has been well demonstrated for SO_2 and O_3. A similar situation probably applies for NO_2, but difficulties in making sufficiently precise measurements have so far limited our knowledge of dry deposition of this gas. Highly reactive gases such as HNO_3 vapour are likely to have their deposition rates controlled mainly by atmospheric factors so that they deposit primarily on leaf surfaces and soils rather than through stomata. It is however dangerous to generalise in terms of reactivity or solubility; cuticular wax has evolved to be remarkably resistant to many physical and chemical stresses, whereas the structure of stomata and mesophyll tissue is designed to facilitate gas exchange. Consequently we should not be surprised that the deposition velocity of the soluble gas NH_3 to crop plants shows an order of magnitude variation between day and night (Aneja et al. 1986).

In spite of substantial advances in the precision of instrumentation over recent years, there have still been very few studies of dry deposition over surfaces where the factors controlling the deposition velocity can be fully analysed. There is a particular need for more work with nitrogen oxides and ammonia, and for SO_2 over forest canopies. We discuss in the next section some of the measurement problems that adsorption on foliage or storage within canopy air space can introduce in the interpretation of micrometeorological studies, and it seems most likely that a multi-disciplinary approach to field projects, combining micrometeorology, surface chemistry, and biology will prove most productive in providing clear explanations of observations. For estimating gaseous dry deposition over Europe there is a particular need to clarify the controlling factors when surfaces are wet with rain or dew, or are covered by snow.

There have been very few direct measurements of deposition rates of particles to natural surfaces, and our theoretical understanding of the processes involved is also rather limited. There are good physical reasons for believing that deposition velocity increases over rough surfaces and that v_d is also increased when the lower atmosphere is unstable (e.g. in hot dry weather). The situation is further complicated because deposition velocity increases rapidly with particle diameter in the important size range 0.1–1 μm. It will be difficult to make improved field measurements of the deposition rates of particles, but we are reaching a situation where lack of knowledge of this form of dry deposition is a limitation in our understanding of the amounts of sulphur and nitrogen compounds that are found in terrestrial and freshwater systems. This pathway is likely to be especially important in areas remote from major sources, where particulate sulphate and nitrate form a substantial proportion of observed sulphur and nitrogen concentrations in air.

5. PATHWAYS AT THE SURFACE

Once pollutants have been deposited on vegetation or soils they have in general a number of pathways towards ultimate sinks, i.e. sites where they are fixed, or at least remain for long periods. The main exception to this statement is ozone, which is not conserved, but which may alter surface affinities for other pollutants.

In considering the role of surface factors in controlling uptake, the simplest case to consider is wet deposition. In general, plants do not actively control their foliar uptake of pollutants deposited in rain or snow. For rain to interact with the leaf there must be a point of contact where the raindrop either becomes attached the leaf or is reflected. Exchange between the leaf and water droplets depends on the physical and chemical nature of the cuticle, and on the form of any epicuticular wax. Water drops on leaves may act as sources or sinks for inorganic and organic compounds. Many substances may be leached from within leaves, but material may also be taken up from solution by the leaf, probably by ion exchange processes. The kinetics of ion exchange and uptake may be relatively slow compared with the amount of time that water spends on plant surfaces during periods of rainfall. This could account for the observation (Hutchinson _et al._ 1986) that a forest canopy could still effectively neutralise rain after several days of almost continuous heavy rain. In resistance terms there would be a large resistance associated with the cuticle and controlling the rate of exchange between drops on the surface and a reservoir of ions within the leaf tissue. This interpretation implies that exchange phenomena will differ in experiments where different rainfall rates or drop sizes are used.

We need to know more about interactions between ions in their potential effects on plants. Skiba _et al._ (1986) showed that responses of plants to acid solutions with different anions were significantly different. At pH 3.5, significantly more Mg, K and Na was leached by sulphuric acid than by nitric acid. As rainfall chemistry is altering with time, it is perhaps surprising that interactions between ions on surfaces have not been explored more fully, in view of the interactions known to exist in the uptake of ions by roots from soil solutions, and in the responses of plants to mixtures of gaseous pollutants.

There have been many attempts at estimating pollutant deposition on forest canopies by comparing the chemistry of incoming rain with that of the throughfall and stemflow collected below the canopy. In a perfectly non-reactive, non leaching system, the throughfall and stemflow would contain dry- and occult-deposited material in addition to that which was contained in the incoming rain. However, the surface phenomena that we have just described make the system much more complex. The difficulty in interpretation arises because throughfall also contains substances leached from within the tree, and these may have been originally taken up from the soil where they originated in the soil minerals, or they may be re-cycled pollutants that moved from canopy to soil at an earlier time, or they may be materials that were absorbed from the atmosphere and stored temporarily within the tree. Substantial differences in the interpretation of similar results exist. Cape *et al*. (1987) studied annual budgets of sulphate in a Scots pine forest in central Scotland. They concluded that about one half of the excess sulphate found in throughfall originated as leaching from the forest canopy, and they suggested that this type of analysis could not be used to estimate dry deposition to the forest. On the other hand, Lindberg *et al*. (1986) studied deposition and throughfall in a mixed hardwood forest in the eastern United States, and concluded that there was little or no evidence of leaching of internal plant sulphate on an annual basis, and consequently they argued that it was possible to use throughfall measurements as an estimate of total deposition. Skeffington (1987) also argued that there was circumstantial evidence that most of the sulphate in throughfall below a Scots pine canopy in southern England was derived from dry deposition. It may be that these conclusions are not inconsistent when the different pollution climates are taken into account. Until there have been more detailed experiments using stable or radioactive tracers to precisely define sources it is unlikely that such differences of interpretation can be avoided.

6. CONCLUSIONS

We have reviewed the main forms of pollutant deposition and some of the factors that concern the fate of pollutant in ecosystems. There is a need for further research into variation of wet deposition with altitude (including rain and snow), and for the development of numerical models which will enable such research to be generalised on a European scale. Uncertainty over the origin of 'background' sulphate in European air is a major limitation in the utility of current long range transport models. Models for nitrogen deposition are at an even less satisfactory stage of matching known emissions and deposition with calculation.

The potential importance of cloudwater chemistry and of occult deposition for plant responses is clear. There need to be special efforts to monitor the chemistry of cloudwater, to understand the causes of episodicity, and measure rates of occult deposition. The potential for surface enhancement of acidity on vegetation in cloud has been demonstrated theoretically but needs to be assessed by direct measurement.

Dry deposition of sulphur dioxide and ozone are reasonably well understood, but there is scope to use this knowledge to produce generalised models appropriate to European conditions. More needs to be known about the dry deposition of nitrogen-containing gases and of particles.

Surface exchange phenomena for wet deposition and gases are not well understood. It seems unlikely that disagreements in the interpretation of throughfall chemistry will be resolved until more complete experiments have been attempted.

REFERENCES

ANEJA, V.P., ROGERS, H.H. and STAHEL, E.P. (1986). Dry deposition of ammonia at environmental concentrations on selected plant species. J. Air Pollut. Control Ass. 36, 1338-1341.

BAKER, J.M., THOMSON, C. and CAPE, J.N. 1986. Acid drops: educational initiatives in monitoring the acidity of rain (Occasional publication no. 12) Williton: Field Studies Council.

BARRETT, C.F., ATKINSON, D.H.F., CAPE, J.N., FOWLER, D., IRWIN, J.G., KALLEND, A.S., MARTIN, A., PITMAN, J.I., SCRIVEN, R.A. and TUCK, A.F. (1983). Acid Deposition in the United Kingdom. UK Review Group on Acid Rain. Warren Spring Laboratory, Stevenage.

BRIMBLECOMBE, P. and STEDMAN, D.H. (1982). Historical evidence for a dramatic increase in the nitrate component of acid rain. Nature, 298, 460-461.

CAPE, J.N. 1987. Chemical interactions between cloud droplets and trees. In: Proc. NATO Symp. Acid Deposition at High Elevation, Edinburgh. Unsworth, M.H. & Fowler, D. (eds.) in press, Elsevier.

CROSSLEY, A., MILNE, R. and UNSWORTH, M.H. (1986). Acid mist and tree injury. Final report, contract ENV 850, CEC Brussels.

DERWENT, R.G. (1986). Long range transport and deposition of acidic nitrogen species in northwest Europe. Nature, in the press.

EMEP (1984) Summary Report from the Chemical Coordinating Centre for the second phase of EMAP. Littlestrom: Norwegian Institute for Air Research, EMAP/CCC 2/84.

FOWLER, D. and CAPE, J.N. (1984). The contamination of rain samples by dry deposition on rain collectors. Atmos. Environ. 18, 183-189.

FOWLER, D., CAPE, J.N., JOST, D. and BEILKE, S. (1987). The air pollution climate of non-Nordic Europe. Water, Air and Soil Pollution (submitted).

GOLDSMITH, P., SMITH, F.B. and TUCK, A.F. (1984). Atmospheric transport and transformation. Phil. Trans. R. Soc. Lond B 305, 259-279.

HICKS, B.B., HOSKER, R.P. and WOMACK, J.D. (1986). Comparisons of wet and dry deposition derived from the first year of trial dry deposition monitoring. In: Symposium on Acid Rain, American Chemical Society.

HOUGH, A.M. (1987). Atmospheric chemistry at elevated sites. In: Proc. NATO Symp. Acid Deposition at high elevation. Unsworth, M.H. and Fowler, D. (eds.) in press, Elsevier.

HUTCHINSON, T.C., ADAMS, C.M. and GABER, B.A. (1986). Neutralization of acidic rain drops on leaves of agricultural crop and boreal forest species. Water, Air and Soil Pollution, 31, 475-484.

LINDBERG, S.E., LOVETT, G.M., RICHTER, D.D. and JOHNSON, D.W. (1986). Atmospheric deposition and canopy interactions of major ions in a forest. Science. 231, 141-145.

LOVETT, G.M. (1984). Rates and mechanisms of cloud water deposition to a subalpine balsam fir forest. Atmos. Environ. 18, 361-367.

MUNZERT, K. (1987). Chemistry of cloud water and precipitation at an Alpine mountain station. In: Proc. NATO Symp. Acid deposition at high elevation. Unsworth, M.H. and Fowler, D. (eds.) in press, Elsevier.

RODHE, H. and ROOD, M.J. (1986). Temporal evolution of nitrogen compounds in Swedish precipitation since 1955. Nature, 321, 762-764.

SHUTTLEWORTH, W.J. (1978) The exchange of wind-driven fog and mist between vegetation and the atmosphere. Boundary Layer Meteorol., 12, 463-489.

SKEFFINGTON, R.A. (1987). Do all forests act as sinks for air pollutants?: factors influencing the acidity of throughfall. In: Proc. UNESCO-IHP Symposium on "Acidification and water pathways", in the press.
SKIBA, U., PEIRSON-SMITH, G.J. and CRESSER, M.S. (1986). Effects of simulated precipitation acidified with sulphuric and/or nitric acid on the throughfall achemistry of Sitka spruce Picea sitchensis and heather Calluna vulgaris. Environ. Poll. B 11, 255-270.
SMITH, F.B. and HUNT, J. (1978). Meteorological aspects of the transport of pollution over long distances. Atmos. Environ. 12, 461-477.
UNSWORTH, M.H. and CROSSLEY, A. (1987). Capture of wind-driven cloud by vegetation. In Coughtrey, P.J., Martin, M.H., and Unsworth, M.H. (Eds.) Pollutant transport and fate in ecosystems. Blackwell, Oxford.
UNSWORTH, M.H. (1984). Evaporation from forests in cloud enhances the effects of acid deposition. Nature, 312, 262-264.

CRITICAL LOADS FOR SULPHUR AND NITROGEN

J.NILSSON
The National Swedish Environmental Protection Board

Summary
A group of Nordic experts has tried to draw conclusions on critical loads for sulphur and nitrogen. The critical load is defined as "The highest load that will not cause chemical changes leading to long-term harmful effects on most sensitive ecological systems".

Most soils, shallow groundwaters and surface waters would probably not be significantly changed by a load of 10-20 keq $H^{+}\cdot km^{2}\cdot yr^{-1}$ in areas with a low content of base cations in the deposition. The total deposition of hydrogen ions in southwestern Scandinavia is in the order of 100 keq $\cdot km^{-2}\cdot yr^{-1}$.

The long-term critical load for nitrogen is in the range of 10-20 kg $N\cdot ha\cdot^{1}-yr^{-1}$ in most forest ecosystems. In high productive sites it might be as high as 20-45 kg $N\cdot ha^{-}$ yr^{-1} in southern Sweden, and amounts to 30-40 $kg\cdot ha^{-1}\cdot yr^{-1}$ and even more over large areas in central Europe.

The current deposition of sulphur and nitrogen must be substantially reduced to keep the long-term changes in sensitive ecosystems within acceptable limits.

1. INTRODUCTION

Acid deposition constitutes a major environmental problem over large areas of Europe and North America. Its effects on various ecosystems, as well as on human health and material are manifold and complex.

Within the convention on long range transboundary pollutants most countries in these regions have agreed on a decrease of the emission by at least 30% in a 10 year period. A protocol on a corresponding agreement on the decrease of the emission of nitrogen is under discussion.

The purpose of this paper is to discuss scientifically defined critical loads for sulphur and nitrogen, especially in regard to their acidifying effects on forest soils, groundwater and surface water. In this context the critical load is defined:

"The highest load that will not cause chemical changes leading to long-term harmful effects on the most sensitive ecological systems".

The assessing of critical loads is a very complicated task. The sensitivity of different ecosystems is dependent on many factors and for this reason very few reports on the matter have been published.

This paper is a summary of the results published by a working group of Nordic experts (1).

2. CRITICAL LOADS FOR SULPHUR, (ACID DEPOSITION).

The deposition of sulphur compounds causes changes in the chemistry and biology of the soils. These changes include decreasing pH and alkalinity, elevated concentrations of soluble aluminium and increasing or decreasing contents of mineral nutrients in various parts of ecosystems.

Critical loads for acid deposition, forest soils.

To arrive at critical loads the following definition has been used:

"The soil should be protected from a long-term chemical change, due to anthropogenic impact, which cannot be compensated by natural soil processes. These chemical changes include decrease of pH as well as the mobilization of potentially toxic cations such as Al^{3+} and heavy metals. Such mobilization requires the presence of mobile strong acid anions as sulphate, nitrate and chloride".

The proton budget concept has been used. The total input of acidic substances should not exceed the alkalinity production produced by the weathering of base cations from primary minerals.

The main conclusions are:

- Assuming that there should be no reduction in the base saturation in a very long-term perspective (1000 years) the critical load for the deposition of strong acids on sensitive soils is not far from zero.

- Most soils in Scandinavia would probably not be damaged by a load of 10-20 keq H^+ $km^{-2} \cdot yr^-$ (Table 1). These values are also in accordance with the estimates for many soil types in central Europe. These figures on depositions of protons corresponds to critical loads of 200-400 kg sulphur $\cdot km^{-2} \cdot yr^{-1}$ in areas with a low content of base cations in the deposition (as in Scandinavia).

In areas with a high deposition of base cations the critical load may be higher without causing an excess leaching.

Table 1. Critical acid loads, at which in most soils there will be no decrease of pH or mobilization of inorganic cations (such as aluminium) in the soil solution of the rooting zone (from Nilsson 1986).

Reference	H^+ $keq \cdot km^{-2} \cdot yr^{-1}$	Area
Nilsson (1986)		
- median	20	Scandinavia
- range	0-75	Scandinavia
Åberg (1986)	20-30	Scandinavia
Fölster (1985) and Matzner (pers.comm)	0-60	FRG
Van Breemen (1986)	25	The Netherlands
Mulder (pers. comm)	0	The Netherlands
Paces (1985)	50-70	Czechoslovakia
de Vries et al (1986)	2-20	The Netherlands
Johnson et al (1985) (sulphate adsorption/ desorption included)	50-280	USA

- In soils with carbonates the alkali production can be very high (>500 $keq \cdot km^{-2} \cdot yr^{-1}$) and thus most acids, whether produced anthropogenically or naturally will be neutralized.

- The total deposition of hydrogen ions in southwestern Scandinavia is in the order of 100 $keq \cdot km^{-2} \cdot yr^{-1}$, which means that it will have to be reduced by about 80%, in order not to exceed the alkalinity production. In some areas of Europe the deposition exceeds 400 $keq \cdot km^{-2} \cdot yr^{-1}$.

- The conclusions in regard to critical loads for acid deposition are indirectly supported by the decrease in the pH of the soil that has been revealed in central Europe and Scandinavia. The decrease in pH during the last 20-50 years is often within the range 0.5-1 pH unit and even more, which may correspond to a decrease of the base saturation of about 50%. Both the humus layer and the mineral soil are affected.

Critical loads for acid deposition, groundwater

The long-term critical load of acidity is defined such that it should not exceed the produced alkalinity at the point of extraction. This may mean that acidity in the unsaturated zone should become neutralized during the time it takes for the water to reach the surface of the saturated zone. In some cases full neutralization of shallow groundwater may not be reached, particularly when the alkalinity production in the soil material is weak, because of coarse texture or high content of quartz.

Considering the varying alkalinity production rates in Scandinavian soils of precambrian rock origin, the critical load of acid for groundwater may be set to 10 $keq \cdot km^{-2} \cdot year^{-1}$ and even lower for the most sensitive soil types. In Table 2 critical loads are given for three sets of groundwater extraction systems, depending upon depth of wells. Five different aquifer materials are considered and two levels of groundwater recharge.

Table 2. Critical loads for groundwater in $keq \cdot km^{-2} \cdot year^{-1}$ (from Nilsson, 1986)

	Shallow wells 3 m		Medium wells 10 m		Deep wells 100 m	
Recharge $mm \cdot year^{-1}$	200	400	200	400	200	400
Silt	100	50	400	200	-	-
Fine sandy till	24	12	80	40	800	400
Medium sand	12	6	40	20	400	200
Coarse sand	3	-	10	5	100	50
Hard rock	3	-	10	5	100	50

It may be concluded that the wells in coarse sand - very frequently municipal water supplies even for fair sized communities - will not tolerate long-term acid deposition rates exceeding 10 $keq \cdot ^{-2} \cdot year^{-1}$ unless the aquifer is very deep. This acid deposition rate may therefore be taken as the critical load for acid deposition in Scandinavia with respect to municipal water supplies.

The conclusions drawn on the critical loads for groundwater are supported by studies on the quality and chemical changes in well waters in Sweden and other countries.

Critical loads for acid deposition, surface water

Criteria for critical loads of sulphur for surface water would be defined as:

> "The highest load that will not lead in the long-term (within 50 years) to harmful effects on biological systems, such as decline and disappearance of natural fish populations".

This means that the pH in clear water lakes should be higher than 5.3. It also implies that the alkalinity production of the total catchment area should be at least equal to the effective acid loading.

The main conclusions are:

- Analyses of data using empirical and processoriented models show that the anthropogenic contributions to the deposition of sulphur must be close to zero to maintain a positive alkalinity in sensitive aquatic ecosystems. At such a level there could still be lakes in some very sensitive areas with low or no alkalinity at a 100% reduction of the anthropogenic sulphur deposition. For example, a Norwegian study using an empirical model concludes that 3-7% of a sample of about 1000 lakes will still have zero alkalinity (pH 5.0-5.3) at 100% reduction.

- Observations from the very most sensitive lakes (Ca+Mg of 20 $ueql^{-1}$ in Norway would indicate a critical load of 20 $keq \cdot km^{-2} \cdot yr^{-1}$ wet deposition. Observations from similar lakes in Sweden would indicate a critical load of 15 $keq \cdot km^{-2} \cdot yr^{-1}$ wet deposition. With lower sensitivity (higher cation concentrations), higher critical loads are applicable. For example, in glacial till areas in Sweden the critical load of sulphur would be 40 $keq \cdot km^{-2} \cdot yr^{-1}$ total deposition.

- A working group (Canadian -USA) concluded based on wet deposition measured in 1980 that a loading of 40 keq $S \cdot km^{-2} \cdot yr^{-1}$ would serve to maintain surface water pH greater than 5.3 in clearwater lakes on an annual basis for basins having cation concentrations of 50 ueq/l or greater. More sensitive basins could, however, not tolerate this level of loading and maintain a pH greater than 5.3. A recent Canadian assessment concluds that a reduction of the deposition rates to about 20 keq $S \cdot km^{-2} \cdot yr^{-1}$ would be needed to assure no further losses of highly sensitive lakes to acidification and assure a recovery of presently acidified, less sensitive lakes (2).

3. CRITICAL LOADS FOR NITROGEN

Nitrogen has so far been a limiting factor in most forest ecosystems and also in some aquatic. There are however, indications that ecosystems can become nitrogen saturated. Such systems may leach considerable amounts of nitrate.

Any nitrogen input in excess of what organisms can utilize will in the long-term have an effect on the ecosystems. Two kinds of effects may be expected:

a) Acidification through nitrification and leaching of basic cations.

b) Eutrophication which means changes of population and an increased growth rate.

The aim in setting critical loads for nitrogen in the long-term (25-50 years) is to prevent forest ecosystems from becoming nitrogen saturated and so to minimize the leakage of nitrogen. Although nitrogen saturation is not a generally well-defined term, it may be defined as

> "Ecosystems where the primary production will not be further increased by an increase in the supply of nitrogen".

With a total long-term input of nitrogen less than the critical load value most of the potential harmful effects of an excess nitrogen input will be avoided or reduced. These impacts include

- terrestrial vegetation: nutrient inbalances, decreased frost hardiness, disturbances of mykorrhizza, excess nitrification, changed pest resistance, vegetational changes

- groundwater: acidification, increased concentration of nitrate

- surface water: short-term acidification and mobilization of aluminium

The critical load refers to a long-term input (25-50 years) of nitrogen to ecosystems. The input might thus be higher during some years, without bringing about a state of nitrogen saturation. As nitrogen so far has been a limiting factor for production in most forest ecosystems, the critical load has to be quantified to benefit from the short-term positive growth effect without venture the long-term productivity capacity of the system.

Using different concepts (see Table 3), each based on specific assumptions, the critical loads for many coniferous ecosystems is in the range 10-20 kg $N \cdot ha^{-1} yr^{-1}$, expressed as total deposition including both oxidized and reduced nitrogen compounds.

The total deposition of nitrogen compounds has now attained 30-40 $kg \cdot ha^{-1} \cdot yr^{-1}$ in many areas of central Europe. It exceeds 20 kg in the southern parts of Scandinavia. In extreme cases throughfall data indicate deposition more than 60 $kg \cdot ha^{-1} \cdot yr^{-1}$ in central Europe and above 40 $kg \cdot ha^{-1} \cdot yr^{-1}$ in south Scandinavia. Very high deposition has been recorded on slopes at the edge of forests and in the vicinity of areas with much animal farms and consequent manuring.

Table 3 Estimated critical loads for nitrogen based on different concepts (from Nilsson 1986)

Concepts	Critical load $kg\ N \cdot ha^{-} \cdot yr^{-1}$
Basic concept	
1. Net removal in forestry operation	
- low to medium productivity	5-20
- high productivity	20-45
Concepts for validation	
2. Input/output studies	10-15
3. Vegetational changes	10-20
4. The nitrogen productivity concept	15-35
5. Surface water	15
6. Nutrient balance in the soil	<20

In order to avoid long-term nitrogen saturation of sensitive ecosystems the total input of nitrogen has to be reduced by 30-50% in southern Sweden, and certainly even more in parts of central Europe. Based on an assessment of the potential risk of nitrogen saturation, the National Swedish Board of Forestry has recommended the foresters not to use nitrogen fertilizers in southern Sweden.

REFERENCES

More detailed information and references are to be found in

(1.) NILSSON, J., (1986) Critical loads for nitrogen and sulphur - Nordic Council of Ministers - Report 1986:11, 232 pp.

(2.) Assessment of the state of knowledge on the long-range transport of air pollutants and acid deposition - aquatic effects. - Research and monitoring coordinating committee, Canada (1986).

REGIONAL OZONE CONCENTRATIONS IN EUROPE

P. GRENNFELT[1)], J. SALTBONES[2)] and J. SCHJOLDAGER[3)]
[1)] Swedish Environmental Research Institute
[2)] Norwegian Meteorological Institute
[3)] Norwegian Institute for Air Research

Summary

Ozone data for April-September 1985 were collected from 24 monitoring stations in 9 countries in OECD-Europe. Frequency distributions, mean and extreme concentrations, trajectory sector distributions and occurrence of regional episodes are given together with a brief discussion of one oxidant episode.

1. INTRODUCTION

Episodes of high concentrations of ozone occur over north-western Europe every summer (1-4). During these episodes the concentrations of ozone may reach levels above ambient air quality guidelines over large regions. The ozone episodes are often associated with stagnant air masses over Europe, during which secondary pollutants such as photochemical oxidants and sulphate aerosol are formed.

The large scale oxidant phenomenon in Europe was first observed from measurements in southern England (5). Ozone concentrations above 200 μg m^{-3} were observed in connection in a high pressure situation with light easterly winds. The first evaluation of ozone data from several stations in Europe was made by Guicherit and van Dop (2). By examining data from a number of monitoring stations in six countries from Italy in the south to Sweden in the north they showed that high ozone concentrations occurred coincidentally at a number of stations. They showed regional build up and transport of oxidants in certain meteorlogical situations. Later, other studies are published presenting and evaluating ozone data from several countries in Europe (e.g. 3, 6).

Because of the regional scale of the photochemical oxidant problem, a successful control of oxidants will only be reached by joint efforts in several countries. At present, several research institutions are involved in modelling work on large scale oxidant formation and transport. Of particular interest is the Dutch-German PHOXA project (Photochemical Oxidant and Acid Deposition Model Application), and the work carried out in Norway within the framework of the European Monitoring and Evaluation Programme (EMEP) (7-8).

International bodies, e.g., ECE, EC and OECD, are concerned with control strategies. The OECD is currently carrying out a programme on "Control of Major Air Pollutants (MAP)" on an international scale. The programme is concerned with long range transport and large-scale formation of atmospheric pollutants, particularly photochemical oxidants and their precursors.

For the assessment of effects and for the model refinement and validation, a comprehensive data base on ambient air concentrations is necessary. Since no permanent or long term (i.e. several years) network yet exists in Europe for the monitoring of photochemical oxidants, Norway and Sweden offered during an OECD workshop in 1984, to collect hourly data on ozone, NO_2 and PAN from the European OECD countries, and redistribute the data to the participating countries. The project was presented for the OECD-Air Management Policy Group (AMPG) in March 1985, and received general support. It was agreed to include the project in the OECD MAP programme. The project is given the acronym OXIDATE and is planned to cover the years from 1985 through 1987.

For the first year ozone data were recieved from 24 monitoring stations in these nine countries: Austria, Belgium, Denmark, Federal Republic of Germany, Finland, The Netherlands, Norway, Sweden and United Kingdom. The monitoring stations are given in Figure 1 and Table 1. Institutions and contact persons for the ozone data collection are given in Appendix 1. Further details on the project organisation and results are given in the project report (9).

2. RESULTS AND DISCUSSION

Monthly mean values

The monthly mean values, given in Table 1, varied from 29 $\mu g\ m^{-3}$ (Ringamåla, Sweden) to 155 $\mu g\ m^{-3}$ (Illmitz, Austria). At most of the stations the monthly mean values varied between 50 and 90 $\mu g\ m^{-3}$. At Illmitz and Schauinsland (FRG) they were comparatively high for all months, 144-155 $\mu g\ m^{-3}$ at Illmitz and 97-110 at Schauinsland. At Gent, St. Kruiswinkel (Belgium) and Vindeln (Sweden), they were comparatively low (33-56 and 39-44, respectively). At 11 out of the 17 stations with monthly means from at least five months, the highest monthly mean occurred in May.

Extreme concentrations and exceeding of concentration limits

Ozone effects are often associated with peaks in ozone concentrations. Therefore, extremes in ozone concentrations are of importance. The hourly maxima and number of hours above certain concentrations are given in Table 2. At all stations but two, hourly ozone concentrations above 120 $\mu g\ m^{-3}$ were recorded. Four stations had hourly ozone concentrations above 240 $\mu g\ m^{-3}$. These were Illmitz (Austria), Gent St. Kruiswinkel (Belgium), Langenbrügge-Waldhof (FRG) and Jeløya (Norway).

From cumulative frequency distributions the 98 percentile was calculated. In Figure 2 the 98 percentile is given together with an estimation of concentration isopleths. The highest isopleths are only indicative because of the relatively small number of stations in the south and east. They show, however, a quite clear gradient with increasing concentrations from the northwest towards the southeast. Similar maps for the 90 and 95 percentiles are given in the project report.

Air trajectory sector distribution for elevated ozone concentrations

By air trajectory calculations it is possible to get indications of the geographical origin of the high ozone concentrations. Such calculations were made in the following way:

The days of the half-year period are distributed according to the back trajectory sectors. For a certain receptor point, one day is allocated to a 45° sector if the positions of the 96-h back trajectories at the 1000-hPa level arriving on that day are within the sector at least 50% of the time. Only trajectory positions between 150 km and 1500 km from the receptor point are considered. If this criterion is not satisfied for any of the eight 45° sectors, the day is called "undetermined". Trajectory roses are made both for all days in the half year period and for days on which the maximum 1-h ozone concentration exceeding certain levels.

In Figure 3 trajectory roses are given for Langenbrügge-Waldhof and Schauinsland. For these two stations the total half-year distribution of all days show a dominant transport from southwest, west and northwest. For days with high ozone concentrations the dominant trajectory direction for Langenbrügge-Waldhof was from the southwest and to a smaller degree from the southeast. For Schauinsland the high ozone concentrations occurred with trajectory directions from southeast and to a smaller degree from southwest and east.

It should be noted, that trajectory sector calculations are associated with many uncertainties, and great caution should be used when trying to indicate emission source areas. This is especially the case for high pressure situations often associated with high oxidant concentrations. Many of these days will have "undertermined" trajectory sectors.

Ozone episodes in the summer of 1985

The ozone data were analysed for episodes with high ozone concentrations occurring at several monitoring stations during the same period. If we define an "episode day" as a day when four or more stations have a maximum 1-h ozone concentration above 160 $\mu g/m^3$, six episodes were found in the summer of 1985 (Table 3).

One of the episodes, that of 12-14 July will be briefly discussed here. The discussion is based on the synoptic weather situation (10) back trajectories at 1000 hPa level and the air quality data (Figure 4 and Table 4). The weather was characterized by a high pressure area over the whole of central Europe except the British Isles. Southern Scandinavia was at the northern border of the high pressure area. The high pressure area moved to the east at the end of the period. Trajectory analyses for 13 July indicated transport from south-west to all stations except Illmitz. For 14 July the trajectories were clockwise with a main direction of between south and west.

The episode started on 12 July with high ozone concentrations at most of the stations on the European continent and in southern Scandinavia. The highest concentrations were measured at Illmitz, Austria (322 $\mu g/m^3$), Langenbrügge-Waldhof, Federal

Republic of Germany (242 µg/m^3), Rörvik, Sweden (214 µg/m^3) and Vavihill, Sweden (212 µg/m^3).

3. ADDITIONAL REMARKS

By collecting and evaluating ozone data for Europe it is possible to get a picture of the ozone situation. This study clearly shows that high concentrations often occur as episodes over large areas, but also that more short term and local ozone episodes may occur.

The examination of data reveals in many cases large variations of concentrations levels and patterns between the stations. Further work is needed to explain these variations. Local and mesoscale concentration variations of oxidant concentrations are well known and must be taken into account when data from different stations are compared.

There is a definite need to include more countries and measurement stations, in order to improve the understanding of the oxidant formation and transport. In particular, regional data from other European OECD countries are of interest, particularly France and Switzerland in which regional ozone monitoring is being carried out. Data from East Europe are also of great interest in future joint European measurement programmes. A dense network of ozone data will also be useful tool for estimating damage to crop, forests and natural flora in Europe.

REFERENCES

(1) Cox, R.A., Eggleton, A.E.J., Derwent, R.G., Lovelock, J.E. and Pack, D.H. (1975) Long range transport of photochemical ozone in north-western Europe. Nature, 235, 372-376.

(2) Guicherit, R. and van Dop, H. (1977) Photochemical production of ozone in Western Europe (1971-1975) and its relation to meteorology. Atmos. Environ. 11, 145-155.

(3) Schjoldager, J., Dovland, H., Grennfelt, P. and Saltbones, J. (1981) Photochemical oxidants in north-western Europe 1976-79. A pilot project. Lillestrøm, Norwegian Institute for Air Research (NILU OR 19/81).

(4) Grennfelt, P. and Schjoldager, P. (1984) Photochemical oxidants in the troposphere: A mounting menace. Ambio, 13, 61-67.

(5) Atkins, D.H.F., Cox, R.A. and Eggleton, A.J.E. (1972) Photochemical ozone and sulphuric acid aerosol formation in the atmosphere over southern England. Nature, 235, 372-376.

(6) Lübkert, B., Lieben, P., Grosch, W., Jost, D. and Weber, E. (1984) Oxidant monitoring networks. Oxidant monitoring data. Report from an International Workshop, 23-25 October 1984, Schauinsland, FRG. OECD Environment Directorate, Umweltbundesamt.

(7) Stern, R.M. and Builtjes, P.J.H. (1986) Long range modelling of the formation, transport and deposition of photochemical oxidants and acidifying pollutants. In: Fourth European Symp. on Physico-Chemical Behaviour of Atmospheric Pollutants, Stresa 23-25 September 1986.

(8) Hov, Ø., Stordal, F. and Eliassen, A. (1985) Photochemical oxidant control strategies in Europe: A 19 days case study using a Lagrangian model with chemistry. Norwegian Institute for Air Research (NILU TR 5/85).

(9) Grennfelt, P., Saltbones, J. and Schjoldager, J. (1987) Oxidant data collection in OECD-Europe 1985-87 (OXIDATE). April-September 1985. Lillestrøm. Norwegian Institute for Air Research (NILU OR 22/87).

(10) Weather Log (1985) Bracknell, Berkshire, British Meteorological Office.

<u>Table 1</u>. Monthly mean ozone concentrations (μg/m^3), April-September 1985. Station codes refer to Figure 1.

Station	name	April	May	June	July	Aug.	Sept.
A1	Illmitz	114	130	131	155	147	132
B1	Gent St. Kruiswinkel	33	37	33	56	37	38
DK1	Risø		87	80	64	58	50
D1	Brotajacklriegel	76	95	86	94		88
D2	Deuselbach	67	74	69	79	80	73
D3	Langenbrugge-Waldhof	74	79	59	83	78	54
D4	Schauinsland	97	109	98	110	98	104
D5	Westerland	68	85	83	76	75	75
SF1	Utö				80	86	73
NL2	Eibergen	54	61	49	51	37	32
NL3	Witteveen	23		68	57	46	38
N1	Birkenes				53	51	44
N2	Jeløya	71	78	63	60	58	41
N3	Langesund	63	68	69	42		
S1	Aspvreten	92	97	53	62	41	36
S2	Norra Kvill				57	60	53
S3	Ringamåla	80	95	70	60	41	29
S4	Rörvik	78	96	87	83	83	67
S5	Vavihill		106	75	68	65	74
S6	Vindeln				44	39	41
UK1	Bottesford	56	60	44	53	58	39
UK2	Harwell	86	78	64	66	59	67
UK3	Wray	70	72	63	57	59	51
UK4	Sibton	53	54	73		59	42

Table 2. Number of hours (h) and days (d) with hourly ozone concentrations exceeding 120, 160, 200, and 240 μg/m^3, and maximum hourly and daily ozone concentration (μg/m^3), April-September 1985.

Station	Number of hours and days								Maximum ozone concentrations (μg/m^3)	
	>120		>160		>200		>240			
	h	d	h	d	h	d	h	d	h	d
Illmitz	2226	168	994	123	405	65	152	32	446	197
Gent St. Kruiswinkel	77	15	38	7	17	4	1	1	253	120
Risø	107	22	19	4	3	1			210	146
Brotjacklriegel	314	41	8	3					174	109
Deuselbach	264	38	31	6					196	109
Langenbrugge-Waldhof	473	67	149	23	76	15	15	7	286	138
Schauinsland	1170	91	99	20	4	2			202	127
Westerland	101	19	4	2					166	105
Utö	26	7	1	1					198	120
Eibergen	81	17	18	7					181	99
Witteveen	88	16	17	6	3	1			217	112
Birkenes									115	70
Jeløya	53	12	22	5	14	5	1	1	266	118
Langesund	29	5							133	113
Aspvreten	272	30	62	6					198	173
Norra Kvill	13	2	3	1					194	100
Ringamåla	207	28	17	5	1	1			202	130
Rörvik	233	39	35	8	5	1			214	127
Vavihill	255	32	44	7	2	1			212	141
Vindeln									120	65
Bottesford	84	12	22	2	9	1			220	133
Harwell	121	23	16	4	1	1			206	136
Wray	51	11	4	1					176	113
Sibton	50	11	5	1					192	125

Table 3. Ozone episodes in north-western Europe, April-September 1985.

Time period	No. of days	No. of stations with ozone concentrations ≥160 μg/m^3	≥200	Maximum ozone conc. μg/m^3
26-28 May	3	9	3	242
3- 5 June	3	9	1	278
3- 6 July	4	8	3	242
12-14 July	3	11	4	322
24-26 July	3	7	3	330
29 Aug-1 Sept	4	10	3	320

Table 4. Maximum hourly ozone concentrations (μg/m^3) for the episode 12-14 July 1985.

Stations	July		
	12	13	14
Illmitz	250	322	262
Gent St.Kruiswinkel	85	187	
Risø	146	116	
Brotjacklriegel	150	158	144
Deuselbach	158	196	170
Langenbrugge	208	234	242
Schauinsland	176	194	166
Westerland	94	100	166
Utö	104	125	89
Eibergen	102	173	181
Witteveen	86	130	193
Birkenes			
Jeløya	83	73	60
Langesund	86	68	72
Aspvreten	80	72	112
Norra Kvill	108	101	133
Ringamåla			160
Rörvik	150	117	214
Vavihill	144	124	212
Vindeln	96	64	52
Bottesford	88	98	68
Harwell	98	132	107
Sibton	60	60	58

APPENDIX 1. Oxidant data collection in OECD-Europe 1985-87 (OXIDATE). List of contact persons and institutions, 1985.

Austria	Dr Ruth Baumann Umweltbundesamt Abteilung fur Lufthygiene Biberstrasse 11 A-1010 WIEN
Belgium	Dr. J. Beyloos Institute d'Hygiene et d'Epidemiologie 14, rue Juliette Wytsman B-1050 BRUXELLES
Denmark	Dr. Finn Palmgren Jensen Miljøstyrelsen, Luftforureningslaboratoriet Forsøgsanalg Risø DK-4000 ROSKILDE
Federal Republic of Germany	Dr. Rolf Sartorius Umweltbundesamt Bismarckplatz 1 D-1000 BERLIN
	Dr. Wolfgang Grosch Umweltbundesamt. Pilotstation Frankfurt Frankfurter Str. 135 D-6050 OFFENBACH
Finland	Mr. Heikki Lättilä Finnish Meteorological Institute (FMI) P.O. Box 50 SF-00810 HELSINKI
Netherlands	Dr. W.F. Blom Air Research Laboratory Rijksinstituut vor Volksgezondheit en Milienuhygiene Postbus 1 NL-3720 BA BILTHOVEN
Norway	Mr. Jørgen Schjoldager Norwegian Institute for Air Research (NILU) P.O. Box 64 N-20001 LILLESTRØM
Sweden	Mr. Peringe Grennfelt Swedish Environmental Research Institute (IVL) P.O. Box 47086 S-402 58 GÖTEBORG
United Kingdom	Dr. B. Sweeney Air Pollution Division Warren Spring Laboratory Gunnels Wood Road, Stevenage Herts SG1 2BX, ENGLAND

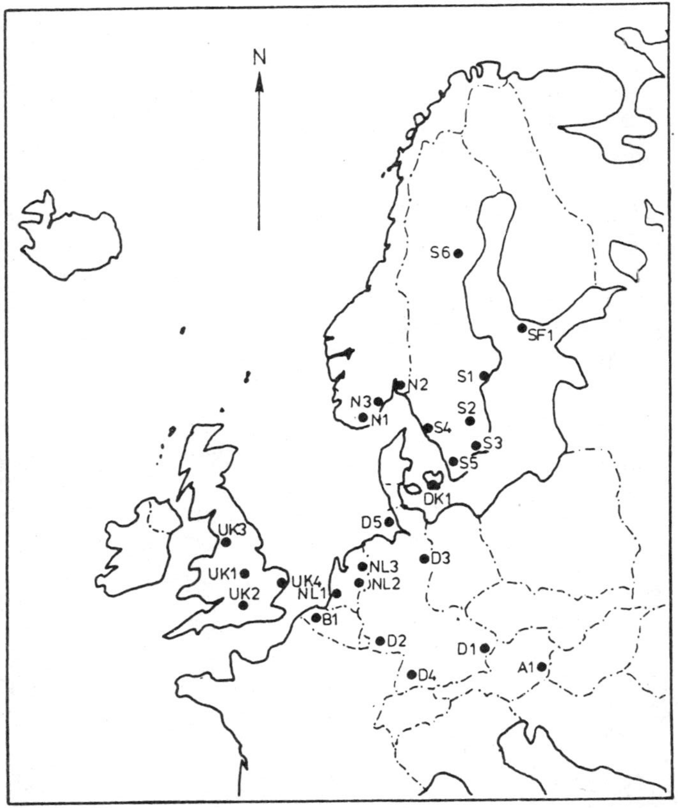

Figure 1 : Map of measurement stations in project OXIDATE 1985. (Station codes refer to Table 1 only).

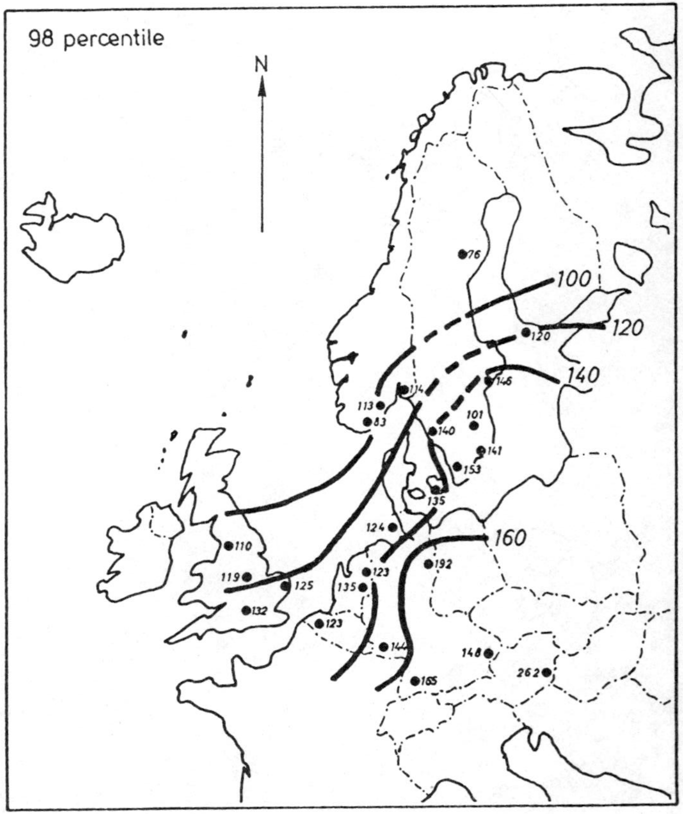

Figure 2 : 98 percentile of ozone concentrations ($\mu g/m^3$), April-September 1985.

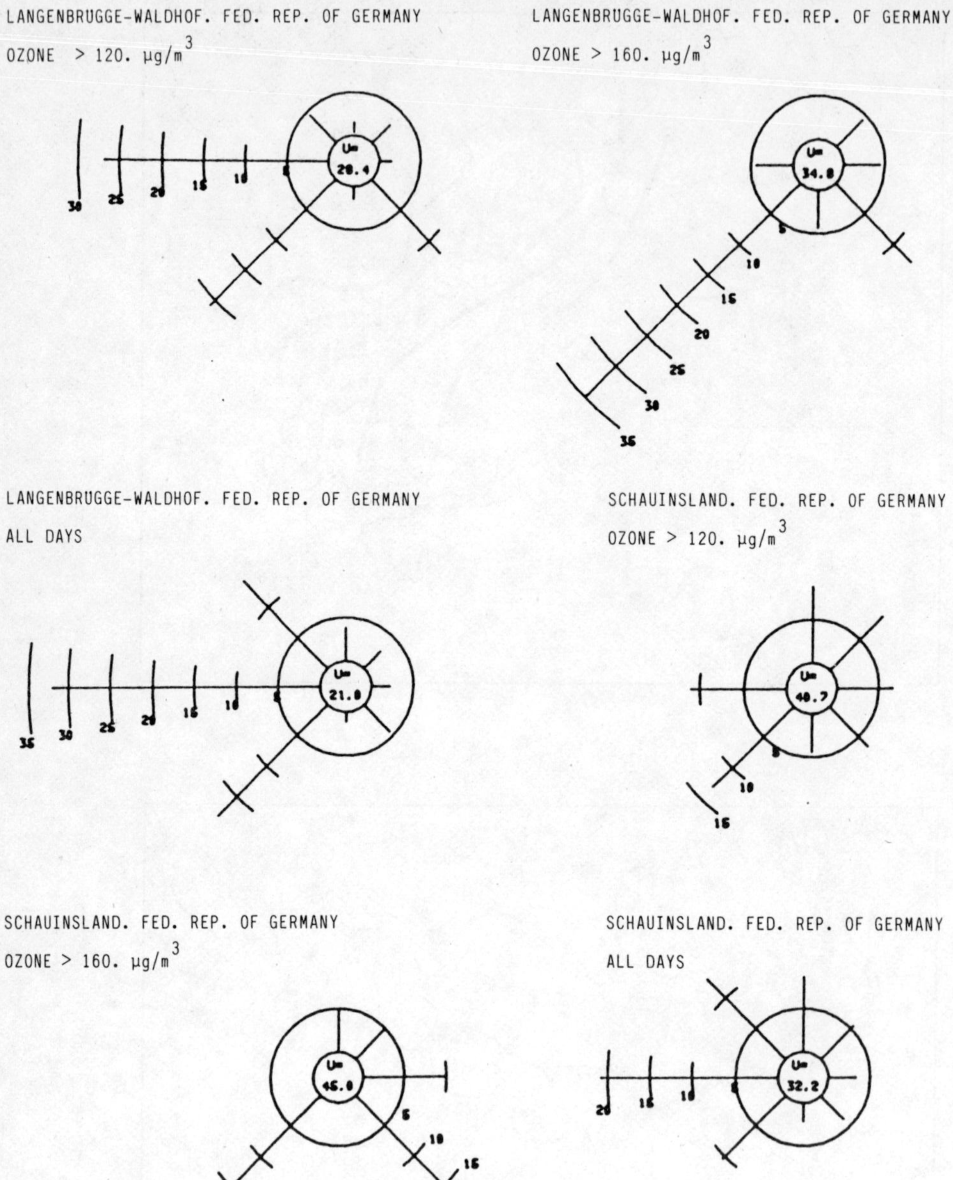

Figure 3 : Back trajectory sector distributions (%) for Langenbrügge-Waldhof ans Schauinsland (Federal Republic of Germany). U means percent of undetermined sectors.

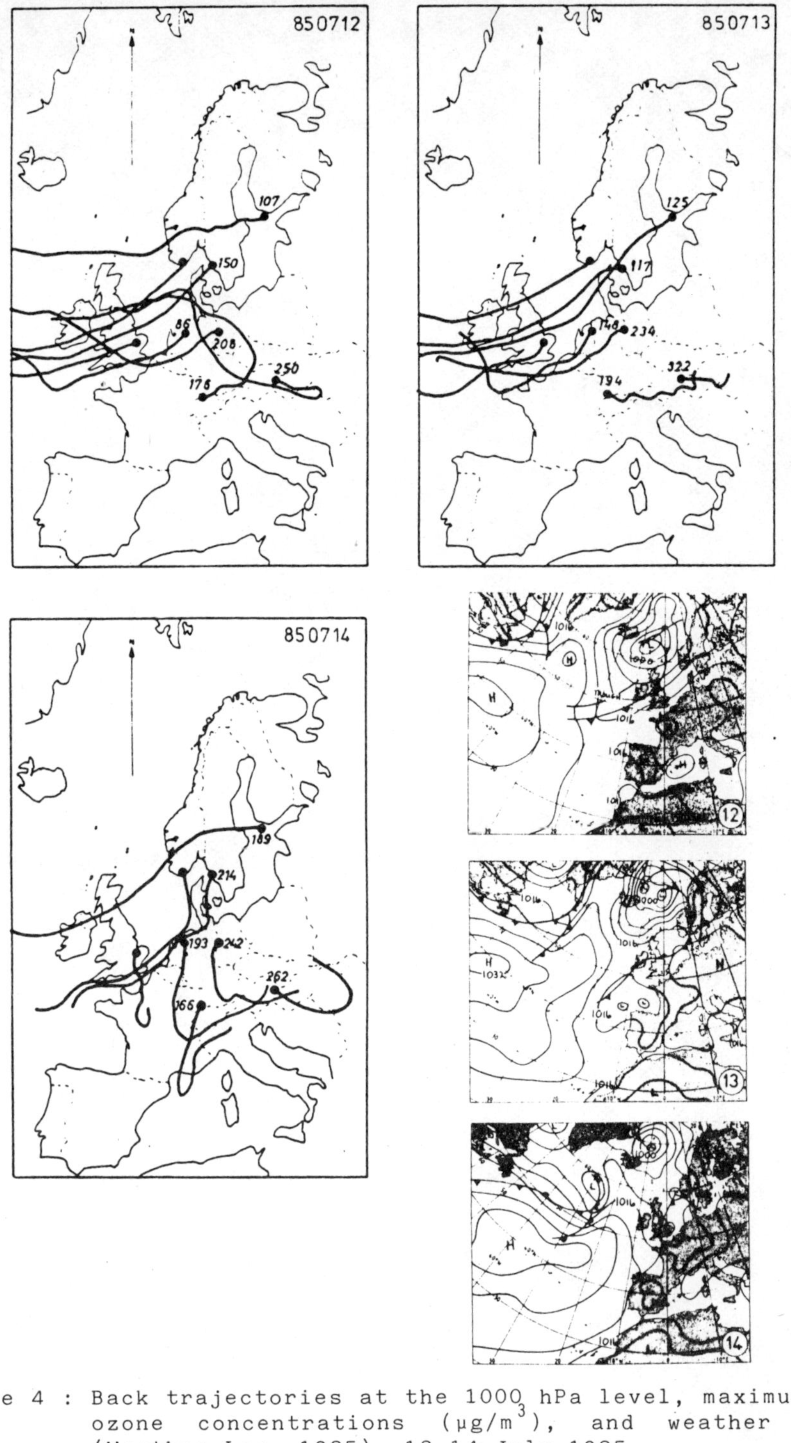

Figure 4 : Back trajectories at the 1000 hPa level, maximum 1-h ozone concentrations ($\mu g/m^3$), and weather maps (Weather Log, 1985), 12-14 July 1985.

SESSION II

EFFECTS OF AIR POLLUTION ON AGRICULTURAL CROPS

SESSION II

EFFETS DE LA POLLUTION DE L'AIR SUR LES CULTURES AGRICOLES

SITZUNG II

WIRKUNGEN VON LUFTSCHADSTOFFEN AUF LANDWIRTSCHAFTLICHE NUTZPFLANZEN

Chairman : Prof. F.T. LAST, United Kingdom

Rapporteurs : A.C. POSTHUMUS, University of Wageningen, The Netherlands
J.N.B. BELL, Imperial College, United Kingdom

REPORT ON THE SESSION

J.N.B. BELL & A.C. POSTHUMUS
Imperial College of Science & Technology, London
Research Institute for Plant Protection, Wageningen

The effects of air pollution on agricultural crops are currently attracting a relatively small amount of attention in Europe, compared with effects on forest trees, as evidenced by the programme of this conference. This is in stark contrast to research over the last 70 years in both Europe and North America, where emphasis has been placed on understanding the impact of air pollutants on the economic yield of major agricultural crops. The present situation is primarily the result of public concern in many European nations over the appearance and dramatic acceleration of the mysterious phenomenon of forest decline. However, it is difficult to escape the conclusion that a secondary cause of the switch in emphasis from crops to trees is the current surplus of many agricultural commodities within the EEC and the uncertainties concerning future agricultural policy.

Despite the present relative neglect of research into crops, most of the knowledge obtained over the years on the mechanisms of air pollution injury, at both the whole plant and physiological levels, has been based on experiments with such species. Most agricultural crops are annuals and, as such, are amenable for use as experimental material in fumigations designed to determine dose-response relationships. The major exception to this is the case of grassland species, where the longevity of the individuals concerned, as well as intense intra- and inter-specific competition in the sward present considerable difficulties in extrapolating data from fumigations to field conditions. A further advantage of working on agricultural crops, is that unlike trees, it is possible to work with mature individuals under realistic field conditions, thereby permitting a greater degree of confidence to be placed on observed effects being attributable to air pollution.

Over the years, there has been a progressive shift from the assessment of thresholds for adverse effects of air pollutants on agricultural crops on the basis of visible injury induced by short-duration exposure to high concentrations, towards fumigations with realistic doses throughout the life-cycle of the plant, accompanied by detailed measurements of responses in terms of growth, yield, and physiology. This increasingly realistic approach has resulted in the threshold long-term concentration of SO_2 accepted as potentially injurious to sensitive crop species being reduced by a factor of about ten over the last 20 years. At the same time realism has also been introduced into experiments by adopting fumigation and pollutant exclusion systems which have been progressively refined in order to minimise modifications of ambient microclimatic conditions. Early experiments used closed chambers, which often resulted in temperatures considerably elevated above the ambient. The introduction of open-top chambers about 15 years ago, followed by a range of improvements to the original designs, has substantially alleviated this problem and permitted experimentation under environmental conditions much more closely related to those of the field. Nevertheless, some differences from the ambient remain, particularly with respect to wind-speed, and thus questions are inevitably still raised about the validity of dose-response relations obtained using chambers, which are prompted by observations on occasions of a "chamber effect" on the performance of experimental plants. Such doubts have led to further experimental developments, notably fumigation and pollutant-exclusion systems in the absence of chambers where crops are subjected to either ambient or introduced pollutants, under otherwise normal field conditions. These systems, however, have their own disadvantages compared with chambers, principally problems of control of gas concentrations and possible interactions with ambient pollutants. A final experimental approach which also avoids problems arising from environmental modification is the exposure of standard cultures of plant material at sites with different pollutant concentrations, often along a transect representing a gradient of the latter ; this approach has been relatively little used, but can produce valuable insights into the effects of ambient pollution, although the results must be interpreted with caution due to other environmental differences between sites. Clearly, no single approach has all the advantages and a combination of techniques must be employed in a complementary manner if the effects of ambient air pollution on crops are to be elucidated, both with respect to their magnitude and causal mechanisms.

The most intensive investigation into air pollution effects on crops ever performed is now in its final stages. This is the National Crop Loss Assessment Network (NCLAN) in the U.S.A., which has adopted a rigorous common experimental protocol with standard open-top chambers in different parts of the country. Successive experiments have been performed over several years in which the most important American crops are grown to maturity in chambers ventilated with ambient or charcoal-filtered air or subjected to O_3 fumigation added onto ambient air for 7 hours per day at a range of concentrations ; in addition the plants are grown in outside chamberless plots in order to identify any influence of the chambers themselves. The NCLAN programme has yielded a substantial volume of data, which has resulted in the development of a series of functions, describing dose-response relationships. The information thus obtained has been incorporated into economic models to assess crop losses in monetary terms for the U.S.A. as a whole.

The success of the NCLAN programme has stimulated a similar European initiative in the form of the Open-Top Chamber Programme supported by the CEC Environmental Programme. This also employs a standard protocol, especially with respect to pollutant and microclimate monitoring, design of chamber to minimise differences from the ambient, crop species, and growth parameters measured. However, there are difficulties at the present time in producing a degree of standardisation comparable with NCLAN, arising essentially from the network being developed in different member states, at laboratories which have established different types of systems. Nevertheless, a successful pilot study has already been accomplished (1985-1986) at a number of locations in western and central Europe. The results of some of the experiments performed under this pilot study have been reported in this conference, and demonstrate that ambient air pollution in rural areas can have an adverse impact on crop growth.

Most of the experiments performed under the European Open-Top Chamber Programme compare crop performance in ambient and charcoal-filtered air. While this can demonstrate an adverse effect of air quality at a given location, it is difficult to determine the impact of the individual pollutants present, despite careful continuous monitoring of SO_2, NO_x and O_3 within and outside the chambers. Another multi-laboratory investigation, under the auspices of the CEC Environment Programme has recently been performed, also using a common protocol, and aimed at furthering the understanding of interactions between SO_2, NO_2 and O_3 with respect of crop growth. This involved growing Hordeum vulgare, Pisum sativum and Phaseolus vulgaris in different types of chambers, located either outdoors or indoors at 5 locations, for 28 days and fumigating with 100 μg m^{-3} O_3, 135 μg m^{-3} SO_2, and 96 μg m^{-3} NO_2, separately and in various combinations. While many common features emerged from this study, some differences were observed between locations, with additive, antagonistic and synergistic responses being recorded in some cases for the same pair of pollutants. These differences detected with the same concentrations of pollutants and a highly standardised crop culture protocol, point towards substantial interactions between pollutant response and climatic conditions, resulting from chamber and/or site differences ; in particular the siting of the chambers outside or under some degree of controlled environment conditions appeared to be particularly important in affecting plant response. This highlights the absolute necessity of the Open-Top Chamber Programme being accompanied by complementary controlled experiments designed to elucidate the nature of interactions between pollutants and other environmental factors taking place in the former. Such experiments have, in fact, been incorporated into the Fourth Environment Programme, in which the pilot Open-Top Chamber Programme has been developed substantially in scope and location. Such expansion of the programme has permitted the establishment of a denser network of experimental sites over much of Europe compared with the limited and somewhat heterogeneous distribution of the NCLAN locations in the U.S.A.

The data presented in this session suggest that the long-term average SO_2 concentration for reductions in growth of sensitive agricultural crops may lie between 40 and 50 μg m^{-3} on the basis of fumigation experiments, while in the presence of other pollutants in ambient air this may be reduced to 35 μg m^{-3} or lower. Physiological effects have now been demonstrated on a crop of 30 μg m^{-3} SO_2 + 20 μg m^{-3} NO_2 over 33 days, which have the potential to cause growth reductions under conditions of drought stress ; these are discussed in more detail later in this report. Filtration experiments in a rural area have shown reductions in wheat yield after exposure to ambient air containing mean daily O_3 concentrations of 70 $μg^{-3}$ between 09.00 and 17.00, for a 2 month period, during which time 13.4 % of all hourly means exceeded 100 μg m^{-3}. Interestingly, the NCLAN programme has indicated that wheat is one of the more resistant of the crops studied under its auspices, with a 7 hour seasonal mean O_3 concentration of 149 μg m^{-3} being needed to induce a 10 % yield reduction. This clearly indicates the dangers of extrapolating from NCLAN results to European field conditions and suggest that ambient O_3 pollution may be having a larger adverse impact on crops in many parts of Europe than predicted on the basis of North American research.

It has long been known that environmental factors can modify substantially the response of crops to air pollutants, although there remain large gaps in our knowledge of this subject, in particular the physiological mechanisms concerned. However, it is only recently that it has become apparent that the converse also applies, in that the response of plants to a wide range of environmental stresses, both biotic and abiotic, can be modified by air pollution. These environmental stresses are in themselves the cause of major economic losses to agriculture, varying in their importance across Europe, largely in accordance with prevailing climatic conditions. Thus it appears that air pollutants may be exerting hitherto unsuspected impacts on agricultural crops via such secondary pathways, but the nature and magnitude of these have so far scarcely been investigated.

A high degree of prominence was given in a number of papers in this session to the interactions between air pollutants and other environmental factors. Much of the developing interest in this subject is directed towards trees, in view of suggestions that forest decline results from complex interactions between air pollutants and climatic and nutritional stresses. Nevertheless, the findings of experiments on trees may well have relevance for agricultural crops and vice versa. It is noteworthy, that air pollutants may change the morphology of plants, with O_3, SO_2 and SO_2 + NO_2 often resulting in reductions in root/shoot ratios. Work reported in this session has demonstrated an increased rate of water loss from leaves of a grass species exposed to only 30 $\mu g\ m^{-3}$ SO_2 and 20 $\mu g\ m^{-3}$ NO_2 in combination. These observed responses to pollutants raise the possibility of ambient pollution increasing susceptibility to drought stress in agricultural crops. This problem is potentially greatest in southern European nations, where little research into air pollutant effects on vegetation has been carried out so far, with the exception of surveys around industrial point sources. In contrast, damage to crops by late frosts can occur in most parts of Europe. The potential significance of air pollution in this situation is being highlighted by current research demonstrating that O_3, SO_2 and SO_2 + NO_2 can predispose tree and agricultural crop species to frost injury, probably by influencing the "hardening" process occurring in the autumn.

The literature on the influence of mineral nutrition on crop responses to air pollution is complex and contains a considerable amount of contradictory evidence. In general, it would seem that such effects may be less important for agricultural crops, than for forests or wild herbaceous vegetation, in view of their subjection to regular controlled fertiliser applications aimed at maximising yield. However, the possible importance of nutritional effects on grassland species has been indentified in this session, with the demonstration that complex interactions can occur between calcium nutrition and O_3 sensitivity.

Biotic interactions with air pollutants represent a research area which has barely been investigated in Europe, despite a number of reports of changes in the incidence of fungal pathogens and insect pests at polluted locations. However, a number of research programmes have recently been initiated in order to elucidate the causal mechanisms of such changes. In the case of fungi, reports of increased virulence of "weak" pathogens of cereals in central Europe have been followed up by laboratory experiments which show that O_3 and O_3 + SO_2 fumigation can increase susceptibility of the host plant to the species concerned. Similarly, observations of increased aphid infestation in polluted areas are being confirmed by fumigation and filtration experiments which show that air pollution can stimulate the performance of a range of important pests of agricultural crops.

As research into the effects of air pollutants on agricultural crops progresses, it is becoming increasingly obvious how complex is the situation in the field, where plants are subjected to many other environmental stresses. This complexity is underlined by the fact that these stresses show interactions between themselves. Consequently there is a requirement for major research programmes which examine the response of agricultural crops to air pollutants in relation to interactions with other stresses, commencing with relatively simple experiments involving single pollutants and single stress factors, but progressing towards all combinations of these which are of potential importance in different parts of Europe. In many investigations, including those in both NCLAN and the Open-Top Chamber Programmes, other environmental stresses are often minimised by the provision of ample water supply and the control of pests and pathogens. It is now apparent that such programmes may result in serious underestimation of the impacts of air pollutants on crop performance by their removal or reduction of interactions with other forms of stress which would normally occur on occasions in the field.

It is the ultimate aim of nearly all research into the effects of air pollution on vegetation to provide information which can be used in legislation designed to prevent or minimise economic losses and ecological damage. The demonstration of complex interactions between pollutants and other environmental stresses which is emerging from recent research makes this task all the more difficult. In order to provide the required information the following recommendations are made for future research :

1) The identification of threshold levels of air pollutants for damage to crop plants, based on dose-response relationships covering pollutant concentrations which include those found in rural areas, over periods ranging from hours to years according to the crop under investigation.
2) The investigation of the impacts of mixtures of pollutants, characteristic of different agricultural regions, on a wide range of crop species, taking into account differences in response which may occur at different times of the growing season.
3) The interactions which may occur between pollutants and other environmental stresses, particularly with regard to the detection of threshold concentrations at which adverse effects of combined stresses take place, and the mechanisms responsible for them. Such studies should take into account the particular combinations of air pollutants and environmental stresses which occur in the principal agricultural regions of Europe.
4) The evaluation of controlled laboratory-based experiments under field conditions, using a range of techniques such as field fumigations, field pollutant exclusion experiments, and transect studies.

ETAT DES CONNAISSANCES SUR L'IMPACT DES POLLUANTS DE L'AIR SUR LES PRODUCTIONS AGRICOLES ET APPROCHES METHODOLOGIQUES

J. BONTE
Centre Départemental d'Etudes et de Recherches sur l'Environnement
LAGOR 64150 MOURENX

Résumé

Dès le début du siècle, des études sont entreprises pour évaluer les conséquences des activités industrielles sur la productivité des cultures et établir des relations dose-effet, effets aigus, visibles. Au cours des années soixante apparaît la notion de dégâts invisibles, notion qui traduit en fait le résultat de désordres métaboliques ressentis sur la croissance et la production.
Au début des années quatre vingt, le phénomène "pollution" atteint une échelle internationale. L'appréhension de ses effets sur la végétation prend une approche plus globale dans laquelle sont impliqués l'action combinée des polluants anthropogéniques et biogéniques mais aussi les stress climatiques et les atteintes parasitaires. L'évaluation des effets sur l'agriculture se coordonne à l'échelle nationale et l'on assiste à une fédération des équipes : aux Etats-Unis dans le National Crop Loss Assessment Network (NCLAN), en Europe à l'initiative de la Commission des Communautés Européennes un programme inspiré du NCLAN se met en place. Les techniques d'évaluation des effets se sont attachées à l'évolution des nuisances, elles s'orientent vers des dispositifs basés sur l'utilisation de chambres à ciel ouvert où les conditions de culture sont plus proches des conditions naturelles.

Summary

Since the beginning of the century studies were undertaken to evaluate consequences of industrial activity on crops and to establish dose-effect relationship. Acute and visible effects were only considered until sixties years when the invisible injury concept appears. This concept traduce results of metabolics perturbations having repercussions on growth and yield.
Since eighties years, the "pollution" phenomenon take an international scale. Understanding of its effets on plants take an inclusive approach including combinated actions of anthropogenic and biogenic pollutants but also climatic stresses and pathogens. Evaluation of effects on agriculture crops is coordinate at the national scale and we observe a teams federation : in United States the National Crop Loss Assessment Network (NCLAN), in Europe the Commission of the European Communities initiate a programm similar to the NCLAN.
These programms generalised the use of open top chambers where, at the present day, the better approach of natural conditions seems be reach.

1. INTRODUCTION

Les pluies acides et tout ce que ce terme inclut, ne se déposent pas sélectivement sur les écosystèmes forestiers et aquatiques. Si de nombreux indices révèlent que les forêts semblent sensibles à ce dépôt, il faut admettre, en regardant les courbes isodéposition (fig.1) de l'une des principales espèces polluantes que les cultures agricoles d'une grande partie de l'Europe Occidentale sont aussi soumises à des retombées atmosphériques importantes.

Quelle est l'incidence de tels dépôts sur les plantes agricoles ? Faut-il craindre dans les grandes régions agricoles un effet insidieux sur la qualité et la productivité des cultures ?

Pour les plantes pérennes et plus précisément les essences forestières le coût de la pollution résulte d'une série de déficits de production annuelle. L'une des questions qui se pose est de savoir si les dépérissements sont liés à cet effet cumulatif ou au franchissement des limites de tolérances de l'écosystème par l'un ou plusieurs des constituants des pluies acides.

Pour les plantes agricoles essentiellement annuelles la question se pose différemment. Le compte des déficits de production, éventuellement attribuables aux pluies acides, est effacé chaque année. Il serait, du fait de la variabilité de la production agricole, confondu avec celui des aléas climatiques.

En d'autres termes, si les conséquences d'un stress, quelle que soit son origine, peuvent être répercutées sur plusieurs années et s'accumuler voire s'aggraver chez les plantes pérennes, chez les plantes agricoles, chaque nouveau semis, dispose, aux conditions climatiques près, des mêmes chances de développement et de production que les cultures des années précédentes. L'objectif fondamental des pratiques culturales est justement d'optimaliser tous les paramètres de production pour, bien sûr, augmenter la productivité mais aussi en réduire les fluctuations. Ces pratiques ne peuvent toutefois exclure l'incidence de nombreux facteurs externes, aléatoires, parmi lesquels il faut rechercher un éventuel effet "pluies acides".

L'évaluation des effets de la pollution atmosphérique sur les cultures n'est pas une préoccupation nouvelle. En fixant précisément les conditions d'une expérience de fumigation artificielle sur de la luzerne, O'GARA dès 1922 a pu établir une relation étroite entre la concentration de dioxyde de soufre (SO_2) dans l'air, la durée de l'exposition et l'importance de la surface foliaire détruite à la suite du traitement. O'GARA avait constaté qu'on pouvait obtenir les mêmes effets en faisant varier le facteur temps ou le facteur concentration de telle sorte que leur produit soit constant.

Cette expérimentation lui a permis de classer une série de plantes en fonction de leur sensibilité et de constater que pour chacune d'elles, il existait, dans les conditions de l'expérience, une concentration de polluant supportable, selon lui, indéfiniment.

L'équation de O'GARA qui se rapportait à la manifestation de dégâts aigus, visibles à l'oeil nu, n'a plus qu'une valeur historique ; en revanche, la notion de dose, elle, est toujours d'actualité puisque la préoccupation des Physiologistes, des Agronomes et des Forestiers est toujours d'établir une relation dose (concentration x durée d'exposition) - effet (production qualité) capable d'établir une prévision des effets de la pollution de l'air et d'en mesurer les conséquences écologiques et économiques.

L'équation de O'GARA prévoyait qu'une culture de luzerne pouvait ug/m³. Il semble, dans l'état actuel de nos connaissances, que le seuil d'action de ce polluant sur la productivité des plantes sensibles se situe au-dessous de 50 ug/m³ en faisant abstraction des phénomènes de synergie qui peuvent apparaître avec des polluants ubiquistes comme l'ozone et les oxydes d'azote. On est frappé par l'apparente lenteur des progrès réalisés depuis O'GARA. La mise en équation de la dose et de l'effet suggère une relation univariable simple. En fait, il existe un nombre important de facteurs dans chacun des termes de cette relation.

La nature et l'amplitude des effets des polluants sur la végétation ne sont pas déterminées par la concentration seule. L'apparition et la sévérité des effets dépendent de la fréquence de l'exposition, de la durée des expositions, de la durée entre deux expositions, de l'amplitude des fluctuations de la concentration, de l'heure au cours de laquelle la pollution intervient dans la journée, de la séquence et du mode d'apparition de cette pollution, enfin du flux de polluant vers la plante qui est lui même modifié par les caractéristiques du couvert végétal et la structure des feuilles.

La réponse des végétaux intégrée dans le terme "effet" dépend d'un nombre encore plus important de facteurs. En plus des facteurs internes caractéristiques de l'espèce, de la variété, les facteurs externes, sol, climat, pratiques culturales, parasites, conditionnent de façon décisive les conséquences d'un épisode de pollution.

La difficulté pour résoudre ce problème est d'autant plus grande que son énoncé est en constante évolution :

- évolution de la pollution : au cours des années soixante dix, les émissions de produits soufrés ont considérablement augmenté. Focalisé par les effets aigus, visibles, des polluants, on s'est attaché, à l'aide de modèles mathématiques, à calculer les flux et des hauteurs de cheminées capables d'assurer une bonne dispersion des effluents et d'éviter aux cultures voisines les effets toxiques des fortes concentrations. En quelque sorte, on n'a pas supprimé le traitement mais simplement diminué la dose, augmentant ainsi l'aire de dispersion et favorisant les réactions secondaires ;
- évolution de la connaissance de la physicochimie de l'atmosphère et des techniques de dosage des polluants qui a permis de mieux définir la limite entre une teneur naturelle et les prémices d'un phénomène de pollution.

L'ozone polluant classé 1er par ordre d'importance aux Etats-Unis, pour ses effets sur l'agriculture, a été sous-estimé en Europe jusqu'au début des années 1980.

Les rares sites de mesure nous apprennent que les concentrations rencontrées sont du même ordre de grandeurs qu'aux Etats-Unis et quelquefois plus fortes ;

- évolution des techniques "d'investigation" et de la connaissance des effets de la pollution. Cette connaissance, très longtemps limitée aux dégâts visibles, ne cesse de progresser et fait reculer d'autant le seuil de concentration à partir duquel on fixe une action préjuciable
- enfin, bien que ce domaine soit lié au précédent, évolution de l'agriculture, de la productivité par la sélection génétique, de la technologie, des pratiques culturales et de la chimie des engrais. En quarante ans, la productivité du blé en France est passée de 20 à 100 Q/ha. Cette progression qui n'est pas sans influence sur le comportement et la sensibilité des plantes vis-à-vis des polluants, traduit aussi l'augmentation de la technicité de l'agriculteur et ses exigences vis-à-vis de la production végétale.

exigences vis-à-vis de la production végétale.

2. INFLUENCE DES DEPOTS HUMIDES SUR LES CULTURES

Le terme "dépôt humide" traduit le mélange complexe des composés chimiques qui est associé aux gouttes de pluie après dissolution, dans le nuage ou sous le nuage, lorsque la précipitation est amorcée. Les techniques à mettre en oeuvre pour appréhender l'influence de ce phénomène sur la productivité ainsi que l'exploitation des résultats font l'objet de belles controverses.

Ce sujet a été particulièrement abordé aux Etats-Unis où l'acidification des pluies concerne de grandes régions agricoles dans les Etats du Nord-Est. Des expérimentations, en plein champ, sur le maïs, le blé, la pomme de terre, le haricot, la luzerne, le tabac, l'avoine, indiquent que ces cultures ne sont pas affectées par l'acidité des pluies enregistrées pendant la période de végétation active. Dans ces régions, le pH moyen des pluies se situe entre 4,2 et 4,5. L'analyse des nombreuses études réalisées dans ce domaine (IRVING 1981) montre que la plupart des espèces agricoles ne subit pas de modification de productivité ou de croissance sous l'influence de précipitations artificielles dont l'acidité peut atteindre des valeurs très faibles : pH 3. Il semble que la réponse globale des cultures balance entre les effets positifs des apports de soufre et d'azote, indispensables au développement des plantes et les effets négatifs de l'acidité.

Ces premières études concernant les effets des dépôts humides sont plutôt rassurantes. Cependant, elles suggèrent la nécessité de pousser plus loin certaines investigations :

- le nombre de plantes testées est relativement limité, il est apparu que certains cultivars semblaient plus sensibles que d'autres ;
- des études exploratoires ont montré que des effets interactifs avec des facteurs naturels (stress hydrique, thermique, action des agents pathogènes) et des polluants gazeux pouvaient être attendus chez les plantes soumises à des précipitations acides ;
- enfin, et cela surtout a été démontré, dans les écosystèmes forestiers, les dépôts acides peuvent modifier les propriétés physico-chimiques des sols mal tamponnés et provoquer à long terme des altérations aux systèmes racinaires. Pour les sols agricoles, généralement bien tamponnés ou corrigés par des amendements, l'acidification des précipitations peut se traduire par des demandes supplémentaires de chaux. Cete conséquence ne semble pas avoir été évaluée.

3. INFLUENCE DES DEPOTS SECS SUR LES CULTURES

A l'interface entre le sol et l'atmosphère, le dépôt sec est constitué de différents polluants gazeux et/ou particulaires adsorbés ou absorbés par la couverture végétale ou le sol lui-même. Ce phénomène très difficile à évaluer semble le plus préoccupant à l'heure actuelle. Les polluants mis en cause sont toujours SO_2, NO_x, O_3, etc..., avec cependant un changement dans l'importance relative des espèces, une place prépondérante étant accordée aux polluants photochimiques, à l'ozone en particulier. Les informations acquises au cours des vingt dernières années ne sont donc pas inutiles. Elles constituent une base de connaissances et d'expérience qu'il est nécessaire d'adapter pour une perception plus globale du phénomène "pollution".

On conçoit devant l'importance des paramètres susceptibles d'influencer l'effet des polluants de l'air sur les cultures, que cette appréciation soit une entreprise difficile et soumise à de nombreuses critiques :

Un tel projet implique :

1) de bien connaître le mode d'occurrence et de co-occurrence des polluants : fréquence, durée, concentrations maximales et minimales, évolution chronologique, distribution géographique à l'échelle d'une région. En d'autres termes, le "climat pollution" qui caractérise cette région ;

2) de disposer d'un système d'expérimentation dans lequel les conditions de cultures s'approchent au plus près des conditions naturelles tout en permettant la simulation ou l'exclusion du phénomène "Pollution".

AVANCEMENT DES ETUDES AUX ETATS-UNIS

La meilleure approche semble avoir été réalisée aux Etats-Unis par le "National Crop Loss Assessment Network" (NCLAN).

Ce programme a été mis en place par l'Agence pour la protection de l'Environnement avec trois principaux objectifs :

1) définir une relation entre le rendement des principales cultures agricoles et différentes doses d'ozone, O_3, de dioxyde de soufre, SO_2, de dioxyde d'azote NO_2 et leurs mélanges[3] ;

2) évaluer les principales conséquences économiques résultant de l'exposition des cultures à O_3, SO_2, NO_2 ou leurs mélanges ;

3) progresser dans la compréhension des relations cause-effets qui déterminent la réponse des cultures aux polluants.

A l'heure actuelle les travaux sont essentiellement axés sur l'appréhension des phénomènes de pollution par l'ozone. Tenant compte du fait que les différences régionales des conditions de culture affectent la réponse des plantes aux polluants, les expériences du NCLAN sont conduites dans six terrains d'expérimentation, répartis dans quatre grandes régions agricoles. La comparaison des résultats est favorisée par une standardisation du matériel, des protocoles et une coordination de l'exploitation des résultats.

Choix des cultures

La sélection a porté sur les cultures économiques les plus importantes : maïs, soja, blé, coton, arachides, tomates et sorgho. Les cultivars les plus fréquemment utilisés dans la région ont été choisis et lorsque des informations étaient disponibles sur leur sensibilité relative, plusieurs cultivars d'une même espèce ont été retenus.

Dispositif expérimental :

C'est la technique des chambres à ciel ouvert "open top chamber" qui a été retenue dans ce programme. Ces chambres, cylindriques, d'environ 3 mètres de diamètre, en matériaux transparents, semblent présenter, dans l'état actuel de nos connaissances, la meilleure technique de simulation des effets de la pollution pour une approche agronomique.

Ces chambres sont balayées par de l'air débarrassé de ses polluants par passage sur un lit de charbon actif. L'air ainsi épuré peut être rechargé de doses croissantes de polluants. On obtient ainsi à partir de concentrations inférieures aux teneurs ambiantes une gamme de concentrations susceptibles de fournir des corrélations entre la dose et la productivité.

Afin de simuler les concentrations ambiantes des modulations peuvent être apportées au cours de la journée. L'addition d'ozone est généralement effectuée pendant 7 heures de 9h à 16h (séquence où la pollution ambiante est la plus élevée).

Concentrations utilisées et contrôle de la qualité de l'air : Le programme NCLAN dont l'objectif est l'étude des effets de O_3 anthropogénique repose sur la définition d'une concentration naturelle d'ozone au niveau des cultures. Concentration naturelle dont l'origine serait essentiellement liée à des transferts depuis la troposhère ou à des réactions photochimiques à partir de précurseurs biogéniques. Cette concentration "de référence" a été estimée à 25 $\pm$ 10 ppb. C'est par rapport à cette concentration que seront évaluées les pertes de production.

Un réseau de surveillance de la pollution oxydante et d'estimation des sources de précurseurs permet d'interpoler les concentrations d'ozone à l'échelle du comté, échelle où débute l'analyse comparée avec des paramètres de productivité agricole et à partir de laquelle on accède à une estimation des pertes économiques au plan national.

Résultats :

Ces résultats affinés et mis à jour au fur et à mesure de l'avancement du programme sont analysés en terme de perte de production (tableau 1 - figure 2).

Deux fonctions, linéaire et non linéaire ont été utilisées. On constate que selon les méthodes et les auteurs, la traduction de ces résultats en termes économiques peut varier considérablement, tableau 2.

Ce programme se poursuit avec pour objectif l'exploration de la sensibilité d'un plus grand nombre de cultivars et l'étude des interrelations avec le stress hydrique.

AVANCEMENT DES ETUDES EN EUROPE

De nombreuses équipes se consacrent à l'évaluation des effets des polluants gazeux sur les plantes d'intérêt économique. Dans l'ordre des priorités le dioxyde de soufre et les oxydes d'azote occupent toujours une place importante ; l'ozone apparaît dans les travaux les plus récents.

Ces études qui tendent toutes vers le même objectif : mesurer l'incidence de la pollution ambiante sur la productivité présentent la caractéristique d'être très difficiles à analyser globalement.

Chaque expérience utilise une méthode particulière :

- les techniques et les matériaux ont évolués, de façon dispersée, vers des dispositifs expérimentaux respectant mieux les conditions naturelles. Les chambres à ciel ouvert constituent une large part du matériel expérimental, mais les facteurs physiques externes inhérents aux types de matériaux se situent dans une grande plage de variation ;
- les doses appliquées et les temps d'exposition s'accordent bien avec les problèmes rencontrés mais, ils ne rendent pas compte du mode d'occurrence des pollutions. Pour des raisons pratiques, on applique généralement une concentration moyenne. On rencontre une multitude de doses correpondant à des concentrations et à des temps d'exposition très différents (tableau 3). Or, on sait définitivement que des pollutions épisodiques ont une plus grande influence que des pollutions séquentielles régulières ;
- les plantes utilisées, très dispersées dans les espèces et dans les variétés, constituent aussi une difficulté dans l'analyse de ces résultats ;
- enfin, les stades physiologiques au cours desquels la pollution est appliquée et les paramètres agronomiques retenus pour mesurer l'impact de la pollution ne sont standardisés.

Si cette variété de protocoles offre l'avantage, à travers les divergences constatées, de faire ressortir la variabilité des phénomènes à l'origine des dégâts, elle ne permet qu'une analyse approchée à l'échelle locale, à l'heure où les problèmes se posent à l'échelle internationale.

Une tentative de standardisation à l'initiative de la CCE montre que cette divergence persiste encore entre les équipes travaillant avec un degré supérieur de coordination.

Dans cette étude, la concentration, le temps d'exposition, les variétés, les protocoles expérimentaux et les paramètres de productivité ont été définis pour un petit nombre de Laboratoires.

L'étude comparative des résultats effectuée par ROBERTS et al (1985) montre que les effets de SO_2, O_3 et de leurs mélanges peuvent être fort variables d'une expérimentation à l'autre (figure 3). Ceci peut être attribué aux différentes techniques qui ont été utilisées par les laboratoires : soufflerie à Lancaster, phytotron à Wageningen, serre solaire à Leatherhead et chambre à ciel ouvert à Copenhague et Imperial College à Londres.

Ces différentes conditions de cultures ont une grande influence sur la croissance des témoins (tableau 4) et peuvent modifier la réponse aux polluants.

Ce constat des écarts qui peuvent naître entre des expérimentateurs pourtant animés de la volonté d'harmoniser leur protocole, a amené la CCE à poursuivre et étendre la recherche d'une coopération entre les équipes ayant les mêmes préoccupations.

Pour explorer l'opportunité d'établir en Europe un réseau d'expérimentation capable d'éprouver l'hypothèse selon laquelle les concentrations ambiantes de polluants atmosphériques sont susceptibles de réduire le rendement des cultures, la CCE a organisé en juillet 1984 à Bruxelles une réunion informelle suivie d'une réunion de travail à Edinbourgh en novembre. Cette réunion a permis de dégager les grandes lignes permettant l'accès à cet objectif :

- 1) établir un protocole expérimental commun en vue d'obtenir des résultats compatibles entre les états membres ;
- 2) utiliser des plantes d'intérêt économique local ;
- 3) inclure à ces expérimentations des stations de mesures de données physico-chimiques de l'atmosphère aussi complètes que possible ;
- 4) assurer une analyse statistique élaborée des résultats.

Les contacts suivants ont abouti à l'établissement d'un protocole commun adopté par 13 équipes scientifiques et testé dans une expérience probatoire.

Le programme de recherche et de développement dans le domaine de la protection de l'Environnement 1986-1990 devrait permettre de poursuivre ce projet et de renforcer la coordination entre les équipes impliquées.

BIBLIOGRAPHIE

- BAKER,C.K., COLLS,J.J., FULLWOOD,A.E., SEATON,G.G.R.(1986). Depresion of growth and yield in winter barley exposed to sulphur dioxide in the field. Newphytol 104,233-241.
- BELL,J.N.B.(1982).Sulphur dioxide and the growth of grasses. In Effets of gazeous air pollution in agriculture and horticulture. M.H. UNSWORTH and D.P. ORMROD Ed. BUTTERWORTHS.
- HECK,W.N., ADAMS,R.M., CURE,W.W., HEAGLE,A.S., HEGGESTAD,H.E.,

KOHUT,R.J., KRESS,L.W., RAWLINGS,J.O., TAYLOR,O.C.(1983). A reassessment of crop loss from ozone. Environ. SCi. Technol 17. 573.581

- HECK,W.W, CURE,W.W., RAWLINGS,J.O., ZARAGOZA,L.J., HEAGLE,A.S., HEGGESTAD,H.E., KOHUT,R.J., KRESS,L.W., TEMPLE,P.J.(1984). Assessing impacts of ozone on agricultural crops : II Crop yield fonctions and alternative exposure statistics. J.A. P.C.A. 34 810-816
- IRVING,P.M.,(1983). Crops in the acidic deposition phenomenom and its effects. Critical assessment review papers.vol.2, Effects sciences, 3-42-3-62 Report N°EPA-600/8-83-016 B EPA Washington D.C.
- O'GARA,(1922). Sulfur dioxide and fume problems and their solu tions ind. Eng. Chem. 14.744.
- ROBERTS,T.M., DARRALL,N.M., LEWIS,B.G., PROCTOR,M.V., BUNN,J., (1985). Effects of gas mixtures (SO_2 and O_3) on agricultural Crops Growth. Report to EEC by the CEGB EV3V.0888.UK(h).

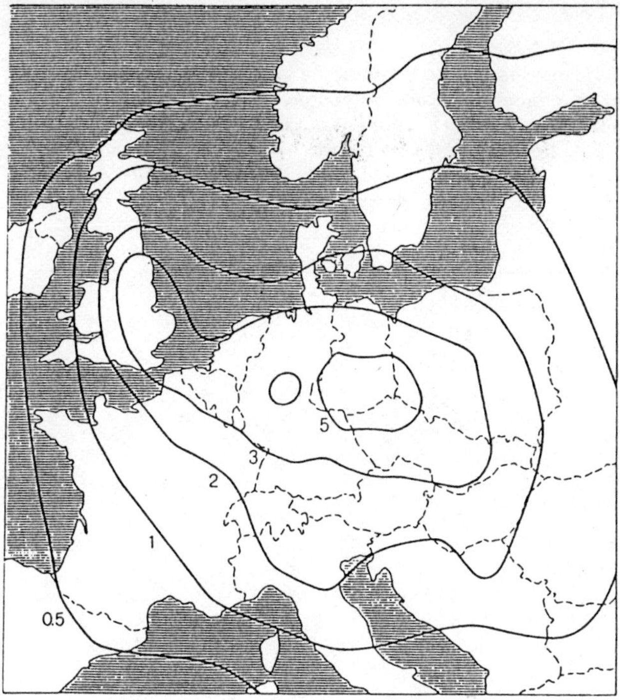

Figure 1 : Déposition sèche annuelle de soufre sur l'Europe en g (0.5 à 5) de SO_2 par m^2 et par an. Source anthropogénique (Source : UNECE EMEP Programme).

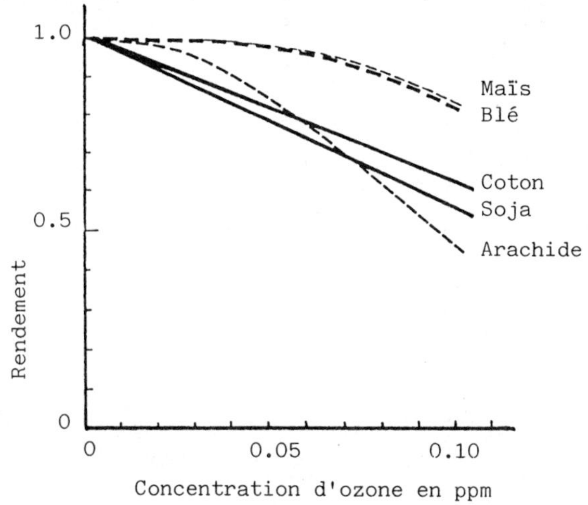

Figure 2 : Réponse relative de 5 cultures majeures à l'ozone évalué par le modèle WEIBULL (NCLAN 1983) HECK et al.

Tableau 1 **Prévisions des pourcentages de perte de rendement pour quatre concentrations moyennes saisonnières d'ozone (7h/jour) par rapport à la concentration naturelle (0,025 ppm) Modèle WEIBULL (NCLAN 1984)** HECK et al.

Espèces, Cultivars	Concentrations (ppm)			
	0.04	0.05	0.06	0.09
Orge				
'Poco'	0.1	0.2	0.5	2.9
Haricot				
'Calif. Light Red' (FP)	11.0	18.1	24.8	42.6
(PP)	1.7	4.4	9.1	38.4
Maïs				
'PAG 397'	0.2	0.7	1.5	8.1
'Pioneer 3780'	1.2	2.6	4.8	16.7
Coton				
'Acala SJ-2' 81 (I)	5.9	10.0	14.0	25.9
'Acala SJ-2' 81 (D)	1.4	2.7	4.2	10.1
'Acala SJ-2' 82 (I)	11.3	20.9	31.4	62.4
'Acala SJ-2' 82 (D)	4.5	10.4	19.1	57.4
'Stoneville 213'	4.8	9.9	16.4	42.2
Arachide				
'NC-6'	6.4	12.3	19.4	44.5
Sorgho				
'DeKalb 28'	0.8	1.5	2.5	6.5
Soja				
'Corsoy'e	5.6	10.4	15.9	34.8
'Davis' 81	11.5	18.1	24.1	39.0
'Davis" 82 (CA)	5.1	9.8	15.4	35.6
'Davis' 82 (PA)	1.8	4.8	10.2	43.7
'Essex'	4.7	8.2	12.0	
'Williams' 81	6.3	10.4	14.4	
'Forrest' (I)	1.3	2.8	5.0	15.3
'Forrest' (D)	7.1	11.5	15.7	27.2
'Williams' 82 (I)	5.6	9.9	14.5	29.1
'Williams' 82 (D)	6.7	10.9	14.9	25.9
'Hodgson' (FP)	10.3	16.6	22.4	
Tomate				
'Murrieta' 81	0.7	1.7	3.6	16.0
'Murrieta' 82	8.2	17.7		
Blé d'hiver				
'Abe'	3.1	6.2	10.2	26.7
'Arthur'	3.7	7.2	11.4	27.4
'Roland'	9.4	16.4	23.7	45.4
'Vona'	24.6	37.6	48.3	70.7

Tableau 2 Résultats des récentes estimations économiques des conséquences de la pollution aux U.S.A.

Etudes	Cultures	Bénéfices annuels du contrôle	Commentaires
Stanford Research Institut (1981)	Maïs, soja, luzerne et 13 autres cultures annuelles	1,8 Milliard $	Perte estimée pour 1980 dans 531 comtés
Shriner et al. (1982)	Maïs, soja, blé, arachides	3 "	Estimation en $ de 1978. Retour à un niveau de 50 μg/m^3 d'ozone. Utilisation des données du NCLAN
Adams et Crocker (1984)	Maïs, soja, coton	2,2 "	Dollar 1980. Retour à un niveau de 80 μg/m^3 d'ozone. Utilisation des données du NCLAN de 1980.
Adams, Crocker et Katz (1984)	Maïs, soja, blé, coton	2,4 "	Dollar 1980. Retour de 106 μg/m à 80 μg O_3/m^3. Utilisation des données du NCLAN de 1980, 1981 1982.
Kopp et al. (1983)	Maïs, soja, blé, coton arachides	1,2 "	Dollar 1978. Retour à un maxi horaire de 80 μg/m^3. Utilisation des données du NCLAN de 1980 et 1981.

Tableau 3 : MODIFICATION DU RENDEMENT DU RAY GRAS ANGLAIS EN POIDS SEC APRES EXPOSITION A DES POLLUTIONS CHRONIQUES DE DIOXYDE DE SOUFRE

Concentration de SO_2 ($\mu g\ m^{-3}$)	Durée en J	Modification du Poids Sec en %	Référence
43	173	- 68	Bell, Rutter et Relton (1979)
50	72	- 2 (N.S.)	Cowling et Koziol (1978)
50	77	- 5 (N.S.)	Lockyer, Cowling et Jones (1976)
50	87	- 3 (N.S.)	Cowling et Lockyer (1976)
55	85	- 3 (N.S.)	Cowling et Lockyer (1978)
62	42	+ 11 (N.S.)	Bell, Rutter et Relton (1979)
66	144	+ 8 (N.S.)	Bell, Rutter et Relton (1979)
100	77	- 3 (N.S.)	Lockyer, Cowling et Jones (1976)
106	194	- 24	Bell, Rutter et Relton (1979)
122	44	- 21	Bell, Rutter et Relton (1979)
122	44	- 25	Bell, Rutter et Relton (1979)
131	59	- 16 (N.S.)	Cowling, Jones et Lockyer (1973)
136	154	- 5 (N.S.)	Bell, Rutter et Relton (1979)
183	138	+ 1 (N.S.)	Bell, Rutter et Relton (1979)
191	180	- 50	Bell et Clough (1973)
200	77	+ 6 (N.S.)	Lockyer, Cowling et Jones (1976)
220	133	- 21	Bell, Rutter et Relton (1979)
312	28	- 34	Ashenden et Mansfield (1977)
312	28	- 18	Ashenden et Mansfield (1977)
343	62	- 40	Bell et Clough (1973)
347	61	- 21	Ayazloo et Bell (1981)
362	105	- 19	Ayazloo et Bell (1981)
367	131	- 8 (N.S.)	Ayazloo et Bell (1981)
380	64	- 24	Ayazloo, Bell et Garsed (1980)
400	72	- 4 (N.S.)	Cowling et Koziol (1978)
400	77	- 24	Lockyer, Cowling et Jones (1976)
423	108	- 43	Bell, Rutter et Relton (1979)
600	42	- 16	Bell, (in Horsman et al., 1979)
650	35	- 16	Horsman et al. (1979)
650	56	- 16	Horsman, Roberts et Bradshaw (1978)
650	56	- 18	Horsman, Roberts et Bradshaw (1979)
700	56	- 30	Horsman et al. (1979)

N.S. : non significatif à P = 0.05

D'après BELL (1982)

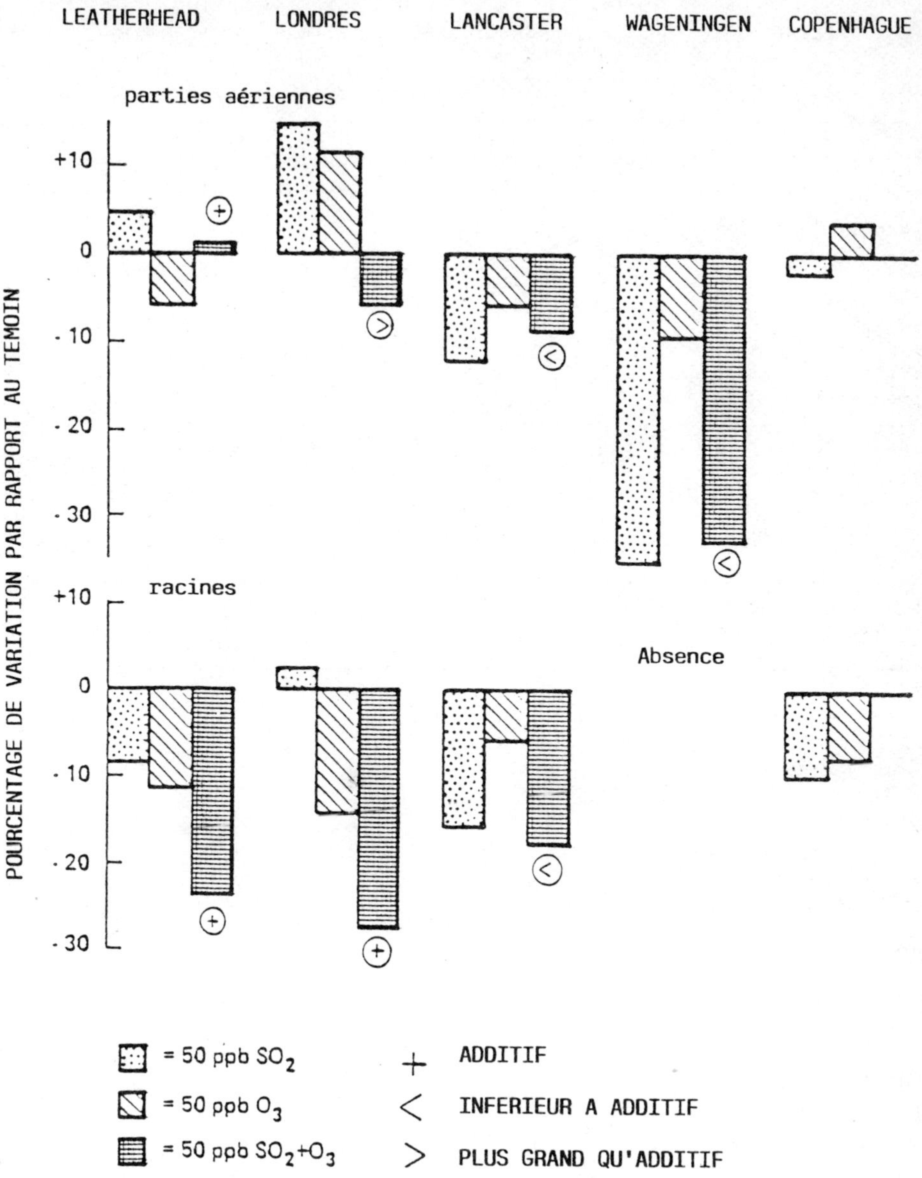

Figure 3 : **INFLUENCE DE 28 JOURS D'EXPOSITION à 50 ppb de SO_2, O_3 ou SO_2 + O_3 SUR L'ORGE D'HIVER cv. IGRI (ROBERTS et al 1986)**

TABLEAU 4 PARAMETRES DE CROISSANCE TESTES PAR 5 LABORATOIRES POUR L'ORGE D'HIVER cv. IGRI

ORGE	LANCASTER	WAGENINGEN	LEATHERHEAD	LONDRES	COPENHAGUE
Nombre de talles	7,92	4,56	6,32	4,23	-
Surface foliaire (cm^2)	220,80	185,30	131,13	123,70	215,28
Poids de feuilles sèches (g)	0,57	0,71	0,48	0,38	0,58
Poids sec par pied (g)	0,88	1,01	0,75	0,58	0,92
Poids sec par tige (g)	0,31	0,30	0,27	0,19	0,34
Poids des racines sèches (g)	0,11	-	0,43	0,20	0,48

INTERACTIONS BETWEEN AIR POLLUTANTS AND OTHER LIMITING FACTORS

T.A. MANSFIELD, P.W. LUCAS AND E.A. WRIGHT
Department of Biological Sciences, University of Lancaster,
Bailrigg, Lancaster LA1 4YQ, United Kingdom

SUMMARY

Several factors (climatic, biotic and edaphic) are considered which can modify the responses of plants to air pollutants. There is good evidence for the following:

(a) Exposure to SO_2, SO_2 + NO_2 or O_3 can increase the susceptibility of woody plants and herbs to frost. Mechanisms have not been fully explored but there is some evidence that the process of acclimation (hardening) can be affected.

(b) The form of the plant body, particularly shoot:root ratio, can be changed by exposure to pollution. This is likely to affect water relations, which may be further influenced by changes in stomatal behaviour. A reduced capacity for stomata to close under severe water stress has been detected after exposure to very low concentrations of SO_2 + NO_2 (e.g. 10 ppb of each).

(c) Attacks by both fungal and insect pathogens may be increased by exposure of plants to pollutants. New evidence suggests, for example, that the mean relative growth rate of aphids is directly related to SO_2 concentration between 0 and 110 ppb.

(d) The nutrient status of soil can exert a strong influence on the response of the whole plant to aerial pollution.

1. INTRODUCTION

Experimental fumigations of plants with air pollutants have usually been conducted under favourable conditions for growth. There are relatively few reports in the literature of experiments in which plants were exposed simultaneously to air pollutants and to one or more other adverse factors. We therefore have only a limited knowledge of the important topic of how the responses of plants to air pollutants are modified by the action of such factors. Nevertheless the evidence from a few well-designed experiments, supported by various observations in the field, suggests that interactions with other limiting factors may be very important. This is, consequently, a topic which must receive greater attention in future research.

This review is divided into sections dealing separately with different modifying agencies, namely (i) climatic, (ii) biotic and (iii) edaphic. It must, however, be emphasized that in the field many of these limiting factors may affect plants simultaneously, but evidence so far available usually applies to factors experienced singly. It is appropriate that such evidence should be collected and evaluated first before more complex situations can be considered.

2. CLIMATIC FACTORS

(a) Low temperatures. The responses of plants to temperature are very complex and it is important not to over-simplify the relationships between individual physiological processes and environmental temperature. It is, however, generally accepted that there are two main ways in which plants are damaged by low temperatures, namely "chilling injury" and "frost injury". The susceptibility to both kinds of injury differs enormously not only between species, but also in individual plants depending on their pretreatment. Plants that are potentially resistant to chilling or frost injury can be made susceptible if they receive particular kinds of pretreatment. The most fully investigated pretreatments involve environmental temperatures. Plants that are grown in high temperatures and are then placed suddenly in cold conditions are usually severely damaged. This damage can often be avoided if they are exposed to gradually decreasing temperatures, or for an appropriate period at an intermediate temperature. This process of acclimation or "hardening" is nearly always essential if chilling and/or frost injury are to be avoided, and the most frequent way in which other agents enhance cold injury is by interfering with acclimation.

Observations in the field suggesting that air pollutants reduce the frost resistance of woody plants extend back more than 60 years (10, 45). A full review of such evidence, much of it from the Nordic countries, has been provided by Huttunen (30). Many of the observations concern woody plants, the frost resistance of which follows a particular course during the year, and is closely related to the dormancy cycle. When extension growth is completed the plant enters a state known as "readiness for hardening" (37). From this point the process of hardening can proceed in phases which are governed by the temperatures and photoperiodic regime to which the plant is exposed. If the initial state ("readiness for hardening") is delayed for any reason the development of subsequent stages may be affected so that the plant is unprepared for the low temperatures in midwinter. In herbaceous plants the different phases of hardening are usually less clear but full hardening may also be dependent on the plant having reached the correct stage of growth.

Some specific examples of enhanced frost damage that have been attributed to pollution were given in the reviews by Huttunen (30) and Davison and Barnes (14). A few of these are shown in Table 1.

Table 1. Evidence of pollution/frost interactions from observations of trees in the field.

Authors	Species	Nature of evidence
Materna and Kohout (44) Materna (42) Materna (43)	*Picea abies*	SO_2 absorbed during the winter caused injury later, in the spring. Long-term mean SO_2 concentrations above about 7 ppb could increase frost sensitivity.
Havas (23) Havas and Huttunen (24) Huttunen (29) Havas and Huttunen (25) Huttunen (30)	*Pinus sylvestris*	Trees growing near urban/ industrial areas exposed to mixtures of pollutants (including SO_2, NO_x and HF) showed, visible lesions in spring after apparently undergoing mainly invisible injury during winter. This gradually produced injury is thought to involve membrane damage.
Huttunen, Karenlampi and Kolari (32)	*Picea abies*	In plants growing near a fertilizer factory the pollution appeared to affect physiological characteristics of cells and tissues required for the normal development of frost resistance, e.g. solute content.
Soikkeli and Tuovinen (52)	*Picea abies*	Ultrastructural damage in cells, particularly of chloroplasts, appearing in polluted plants over winter.
Huttunen, Havas and Laine (31) Cape and Fowler (9)	*Pinus sylvestris*	Changes in the water economy of trees in winter in polluted areas. Erosion of cuticular wax may be a contributory factor.

Despite these strong suggestions of a correlation between pollutants and frost injury, it is essential to acquire more precise information from experimental studies. In particular we need to know which pollutants are responsible, and some dose-response relationships. A conclusion from the review of

Davison and Barnes (14) was that very few critical experiments have been performed to determine the effects of individual pollutants or combinations of gases on frost and chilling resistance, and they pointed out that no experiments had used ozone, presumably because this pollutant does not normally occur in winter.

Some of the first controlled experiments were those of Keller (34). He found that when dormant seedlings of *Fagus sylvatica* were exposed to 50, 100 or 200 ppb SO_2 in winter, the concentrations of sulphur in the leaves that expanded in spring were significantly increased and the numbers of terminal buds found to be dead increased linearly according to the dose of SO_2 (Figure 1). In later studies Keller (35) exposed *Picea abies* to different SO_2 concentrations (25 to 225 ppb) from October to April, and the occurrence of late frost injury appeared to be correlated with dose of SO_2. The needle contents of ascorbic acid dropped due to fumigation, and the presence of this strong reductant seemed to be positively correlated with frost resistance.

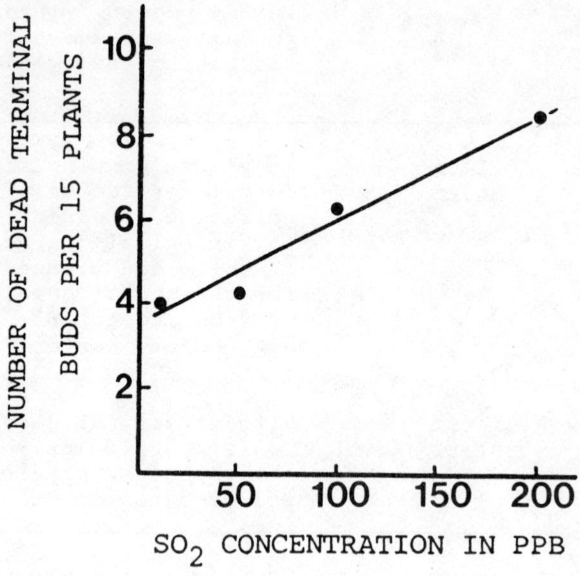

Figure 1. Numbers of dead terminal buds after fumigation of dormant seedlings of *Fagus sylvatica* with different concentrations of SO_2 from December to April. From Keller (34).

Davison and Bailey (13) fumigated the grass *Lolium perenne* for 5 weeks with 87 ppb SO_2, and then the plants were kept for 2 weeks in a temperature regime suitable for inducing frost hardening. When they were subsequently exposed to controlled sub-zero temperatures the pre-fumigated plants showed an appreciable reduction in frost resistance (Figure 2). Baker, Unsworth and Greenwood (1) found that winter wheat (Triticum aestivum, cv. Bounty) was more susceptible to natural frost in the field after fumigation in the open air with SO_2 (concentrations were variable over the treated area, but were in the region of 110-120 ppb in the centre). Severe frost injury occurred on the leaves of the fumigated plants after temperatures fell to -9°C in mid-January.

Freer-Smith and Mansfield (20) exposed dormant seedlings of Picea sitchensis to SO_2 and NO_2 separately and in combination, and then applied controlled frosts of different levels of severity in the range 0 to -15°C. There was poor survival of lateral buds in plants that had been exposed to 45 ppb SO_2 for 6 weeks then hardened at 6°C for 10 days in the absence of pollution, and afterwards cooled at a realistic rate to -5 or -10°C. the poor bud break in plants previously exposed to SO_2 and then cooled to -5°C meant that this treatment was less effective in meeting the chill requirements of the plants than in the clean air controls. In a second experiment, exposure to 30 ppb SO_2 and 30 ppb SO_2 + 30 ppb NO_2 caused small but consistent increases in frost injury on the needles of plants frozen to -5 and -10°C.

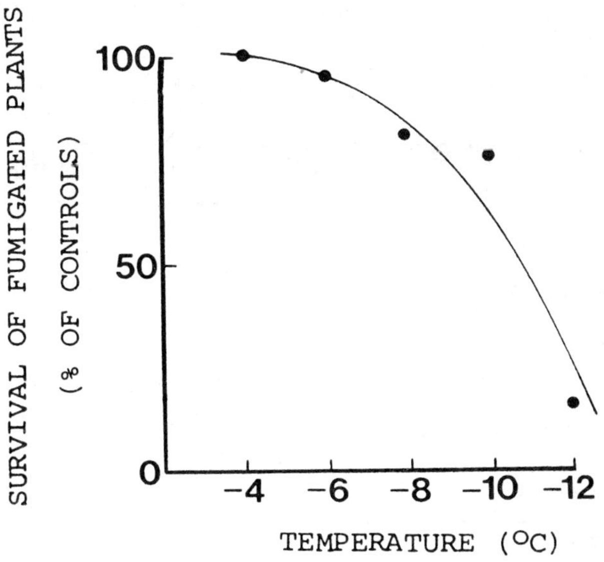

Figure 2. Survival of SO_2-treated Lolium perenne after exposure to sub-zero temperatures. Prefumigation with 87 ppb SO_2 lasted for 5 weeks. From Davison and Bailey (13).

As noted above, little attention has been given to the possibility that exposure to O_3 during the summer might enhance frost sensitivity during the following winter. Some recent experiments do, however, suggest that this is a phenomenon worthy of detailed investigation in the future. Brown, Roberts and Blank (8) fumigated ten different clones of 3-year-old Picea abies for 60 days from 19 July to 17 September with three different concentrations of O_3 of about 50, 100 and 150 ppb. These treatments produced no visible damage at the end of the fumigation period, but after an early natural frost down to -7°C three of the clones showed necrosis of the older needles which appeared to be related to the previous exposure to O_3 (Figure 3).

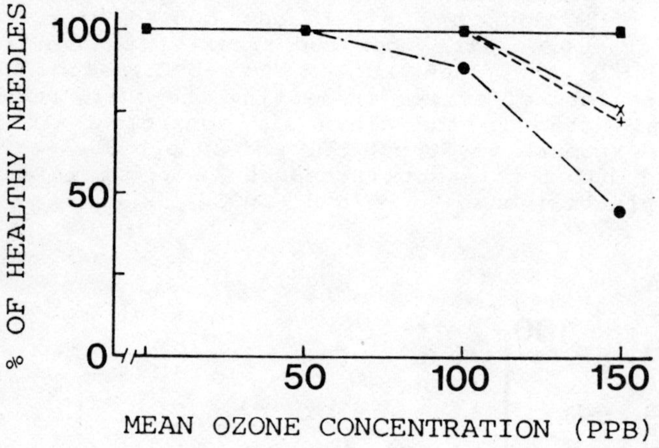

Figure 3. Effects of a summer fumigation with ozone on necrosis of the previous year's needles of Picea abies after a frost in November. ■ are the data for seven clones that appeared to be unaffected, and x, + and ● are for three clones that were sensitive. No damage was observed on the current year's needles. From Brown, Roberts and Blank (1987).

In a recent study conducted jointly between the University of Lancaster and the Institute of Terrestrial Ecology in Edinburgh, further evidence has emerged that summer exposure to O_3 may enhance frost sensitivity in the autumn (Cottam, Lucas and Sheppard, in preparation). Two-year-old seedlings of Picea sitchensis were exposed to 0, 70, 120 and 170 ppb O_3 for 7 hours per day for 86 days from 6 June to 30 August. They were then allowed to undergo a period of natural hardening in O_3-free conditions in the autumn, and on 10 November cuttings of the current year's growth were tested for frost hardiness at temperatures of -9, -12, -15 and -18°C. The cuttings from plants previously exposed to O_3 were more sensitive to frost, judging from the extent of damage to the needles. This experiment, like that of Brown et al. (8) strongly suggests that susceptibility to early frosts may be caused by exposure to O_3 during the summer. As we noted earlier, the development of frost resistance in woody plants is a physiological phenomenon associated with the dormancy cycle, and it is possible that the cellular injury caused by ozone interferes with this.

The evidence so far available strongly suggests that some combinations of pollution and cold stress may be very damaging to both herbaceous and woody plants. The two forms of stress clearly do not have to be coincident in time. Much more experimental research is needed to explore these types of interaction in detail, for they may be very important in forest decline and could also be critical for some field crops.

(b) Water deficits. The dry deposition of pollutants takes place directly into the above-ground portions of plants. It might, therefore, be expected that the main response would occur in the shoots and not in the roots, but this is not always the case because two pollutants (SO_2 and O_3) often bring about changes in the allocation of dry matter in the plant. These changes involve reduced translocation of newly manufactured photosynthates to the roots, and the eventual form of the plant may involve a major increase in shoot:root ratio. Such a shift in the balance of growth has clear implications for the resistance of plants to soil water stress. It has also been found that pollutants, especially SO_2, sometimes interfere with stomatal closure, so there could be quite a drastic interference with water relations if this effect occurred in combination with reduced root growth.

Long-term or intermittent exposures to doses of O_3 that are insufficient to cause visible injury have often been reported to alter dry matter allocation, or even to stimulate leaf production. Bennett and Oshima (2) found that Daucus carota (carrot) exposed intermittently to 190 or 250 ppb O_3 over a 108-day period showed marked increases in shoot:rot ratio against a background of declining dry weight of the plant as a whole. Later studies by Oshima, Bennett and Braegelmann (46) on Petroselinum crispum (parsley) showed that although the relative growth rate was first depressed by treatment with O_3, it recovered as the exposure continued. This and further studies (47) suggested that it is increased leaf growth that enables the relative growth rate to recover. A rise in the leaf area relative to the total dry mass of the plant provides a means of compensating for a drop in photosynthesis per unit leaf area such as that caused by O_3. Walmsley, Ashmore and Bell (54) showed for Raphanus sativus (radish) that the increase in leaf area in 170 ppb O_3 more than compensated for the reduced efficiency per unit leaf area caused by the O_3.

Okano et al., (48) labelled plants of Phaseolus vulgaris (dwarf bean) with $^{13}CO_2$ to study the translocation from individual source leaves, and provided conclusive evidence that a greater proportion of new assimilate was to the growing leaves at the expense of the root and partitioned stem. The effects of 200 ppb O_3 were large, for the labelled material moving into the stem and roots fell by only 28%.

The effects of SO_2, and particularly of mixtures of SO_2 and NO_2, have been found to be very similar to those of O_3 as far as the change in form of the plant is concerned. Jones and Mansfield (33) exposed seedlings of <u>Phleum pratense</u> (Timothy grass) to 120 ppb SO_2 for 40 days, and found that the first detectable response to the pollutant, appearing after 10 days, was a reduction in root growth. Changes in dry mass of the shoot appeared later, and eventually there were increases in leaf area ratio (LAR) in the polluted plants compared with the controls. LAR is the combined area of the leaves divided by the total dry mass of the plant, and it is an indication of what proportion of the body of the plant is committed to assimilation. The changes in response to SO_2, like those to O_3, can be regarded as compensatory, allowing the plant to overcome some of the effects of pollution, but at the expense of a change in form.

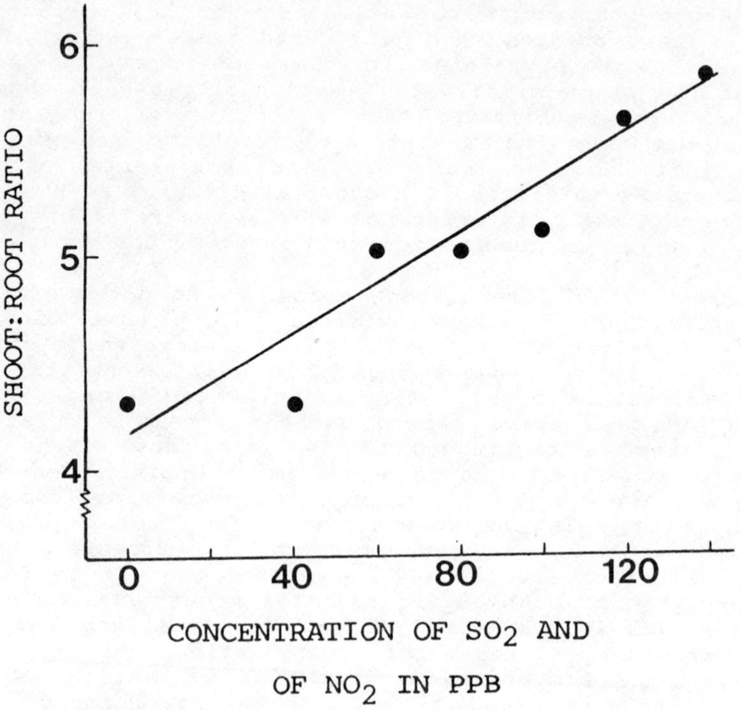

Figure 4. Changes in shoot:root ratio in seedlings of spring barley (cv. Patty) after fumigation for 2 weeks with mixtures of equal concentrations in ppb of SO_2 and NO_2. Fumigation began 3-4 days after germination. From Pande and Mansfield (49).

Our studies in Lancaster, mainly conducted on grasses, have shown that when SO_2 and NO_2 are applied simultaneously the effects on shoot:root ratio are more marked than those produced by SO_2 alone (e.g. 56). Figure 4 shows a dose-response relationship obtained using mixtures of SO_2 + NO_2 on spring barley. Freer-Smith (19) showed that mixtures of SO_2 + NO_2 caused comparable alterations in the allocation of dry matter in woody plants. If such effects occur in large trees in the field they could be very important in affecting water relations, particularly during periods of severe drought.

We have very sparse experimental evidence of the significance these changes in the form of plants have for the use and conservation of soil water. It is, however, difficult to believe that such changes are unimportant. Some preliminary data obtained in studies with grasses were presented by Mansfield, Davies and Whitmore (41). It was found that leaves of *Phleum pratense* from plants that had been exposed to 90 ppb SO_2 and 100 ppb NO_2 for 25 days showed an increase in specific leaf area (i.e. less dry mass per unit area), a higher water content, larger mesophyll cells and larger intercellular spaces. Studies of the water relations of excised leaves using a pressure chamber and a vapour pressure osmometer showed that the water potentials were similar in polluted and control plants but that the solute potentials were less negative in the polluted leaves. This meant that the turgors of the polluted leaves were lower than in the controls and consequently the maintenance of turgor during soil water stress might be more difficult for polluted plants.

Heggestad *et al.*, (26) obtained evidence of an interaction between low soil water potentials and realistic levels of O_3 in causing reductions in the yield of soybeans. Over two successive years in experiments in the NCLAN programme, a combination of soil moisture stress and ambient O_3 reduced yields by 25%. In irrigated plants the effect of ozone was much smaller (yield reduced by only 5%), and in clean air a yield reduction of only 4% was attributable to water stress. The cause of the O_3 × water stress interaction was not identified but the authors drew attention to the reduction in root growth caused by O_3 in soybeans. This study showed, however, that there is no simple relationship between O_3 and the responses of soybean to water stress, because at higher concentrations of O_3 (greater than 80 ppb for 7 hours a day) there was a *reduced* impact of water stress on yield. Such an effect may occur because higher doses of O_3 can cause stomatal closure and therefore can reduce water consumption.

The behaviour of stomata alongside other changes induced by air pollutants may be of great importance in determining pollutant/water stress interactions. There is much evidence in the literature that low concentrations of SO_2 can cause stomatal opening in some species. This topic has been fully reviewed by Mansfield and Freer-Smith (43) and Black (3). In the limited space available here some new evidence pointing to the possible significance of changes induced by mixtures of SO_2 and NO_2 will be considered.

Wright et al (57) described an experiment performed on the grass Phleum pratense which was designed to distinguish effects of SO_2 + NO_2 (30, 60 and 90 ppb of each gas) on root growth and on transpirational water loss. The consequences of differences in root growth were not displayed very clearly but the rate of use of soil water seemed to be substantially increased by the pollution treatment, and during a period in which water was withheld a clear pollution × water stress interaction emerged (Table 2).

TABLE 2

Effects of the mean relative growth rate (RGR) and net assimilation rate (NAR) for Phleum pratense exposed continuously for 40 days to SO_2 and NO_2 in combination, followed by a period of 23 days in clean air during which water was either supplied daily or was withheld.

Treatment	RGR $g\ g^{-1}\ d^{-1}$		NAR $g\ cm^{-2}\ d^{-1}$	
SO_2+NO_2 ppb	watered	unwatered	watered	unwatered
0 control	0.024	0.013	0.216	0.101
30	0.021	0.008	0.211	0.096
60	0.028	0.006*	0.183	0.076*
90	0.027	0.005*	0.162*	0.037*

* Indicates significant differences from control at the $P<0.05$ level.

Subsequent studies with two woody plants (Betula pendula and Betula pubescens) and further experiments with Phleum pratense have shown that changes in the ability of leaves to conserve water under conditions of severe stress may be mainly responsible for the interaction in Table 2. Further details of the experiments on Betula are included in the poster paper presented at this meeting by Dr. E.A. Wright. It seems likely that damage to cells in the epidermal layer, leading to malfunctioning of stomata, is mainly responsible for the increased rates of water loss. Black and Black (4, 5) found that the integrity and survival of epidermal cells was affected in leaves subjected to short exposures to SO_2. It seems likely that deposition of pollutants into epidermal cells from the substomatal cavity and the connecting intercellular spaces is involved, rather then uptake through the cuticle. The entry of pollutant molecules through cells involves a comparatively short route of low diffusive resistance, as indicated in the diagram in Figure 5.

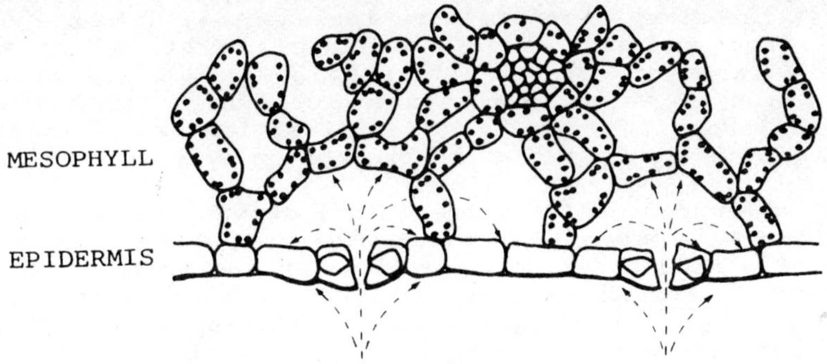

Figure 5. Routes for the uptake of pollutant molecules from the external atmosphere into the epidermal cells via their inner surfaces, together with some other locations for deposition.

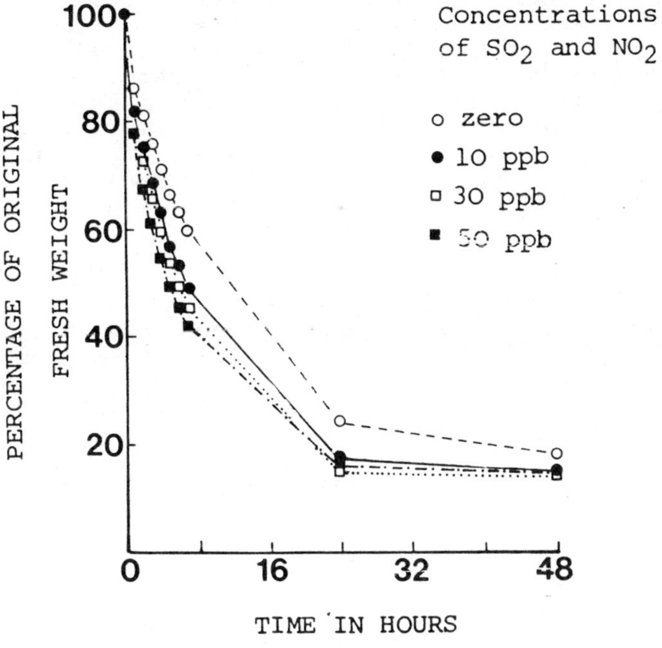

Figure 6. Decline in fresh weight over time of excised leaves of Phleum pratense grown for 33 days in clean air or the pollution treatments shown. Note that the concentrations are those of each gas (i.e. 10 ppb SO_2 + 10 ppb NO_2, etc). Each point on the graph represents the mean of nine replicate leaves all taken from different plants.

Our most recent experiments have included measurements of water loss from excised leaves of Phleum pratense after exposure to a range of concentrations of SO_2 + NO_2 beginning at 10 ppb. Even at this low concentration, which is of wide occurrence in Europe, the ability of excised leaves to retain water was signficantly less than in the controls (Figure 6).

It must be emphasized that these indications of a reduced ability of excised leaves to retain water cannot be regarded as evidence of a reduced efficiency of water use in the field. They do, however, suggest that more critical experiments are needed to evaluate pollution/water stress interactions. This is an area of research that needs to be given high priority in the future.

3. BIOTIC FACTORS

(a) Fungal pathogens. This complex topic was reviewed by Kvist (36) and space only permits a brief coverage here. There is not a simple answer to the question of whether pollutants predispose plants to fungal diseases. The changes in growth and physiology of plants caused by air pollution are clearly likely to have effects on host-pathogen relationships but few well-defined examples can be quoted. Complications arise because the pollutants may affect the pathogen as well as the host (51). Heagle (27) tentatively concluded that attack on plants by obligate fungi is most sensitive to SO_2, and that parasitism by nonobligate fungi shows a wide variation in response, i.e. it may be inhibited, stimulated or unaffected.

The lack of conclusive information in the literature does not indicate that this topic is unimportant. Fehrmann et al (17) drew attention to the fact that in central Europe, some cereal pathogens that were previously only of minor importance have recently caused epidemics of economic significance. They speculated that air pollutants have been responsible for these problems, by reducing the resistance of plants in some way. They described some experiments in which wheat and barley were first exposed to SO_2, O_3 or SO_2 + O_3, then inoculated with various pathogens. After inoculation the plants were kept in ambient air. The results suggested that prior exposure to the pollutants, expecially O_3 or SO_2 + O_3, could predispose the plants to injury. The effects were, however, different for different pathogens. More research is needed to explore these interactions in greater detail.

(b) Insect pathogens. Since the 19th century there have been several observations in the field suggesting that the presence of air pollutants may promote outbreaks of insect pathogens (18). More recently, experimental evidence has been forthcoming showing a good correlation between insect attack and particular pollutants. For example, rate of development and fecundity of the Mexican bean beetle (Epilachna varivestris) on soybeans was increased after the plants had been fumigated with SO_2 (28), and ambient air in London increased the growth rate of aphids (Aphis fabae) on broad

bean (16). Dohmen (15) later showed that the ambient air in Munich had the same effect on aphids (Macrosiphon rosae) on roses. There is no effect of low to moderate concentrations of SO_2 (e.g. 148 ppb) on the growth of aphids raised on artificial diets (16) and it therefore appears that the observed effects are mediated through changes in the host plants.

Warrington (55) has performed a careful dose-response study to define the effects of SO_2 on pea aphids (Figure 7). There was an apparently linear increase in mean relative growth rate up to a concentration of about 110 ppb, and thereafter a steep decline. The change in response at higher SO_2 concentrations may indicate that direct toxicity to the aphids occurred. It is interesting that the SO_2 concentration at which an effect first appeared is very low, and it seems likely that aphids in the field may be affected by SO_2 concentrations that are widespread in Europe.

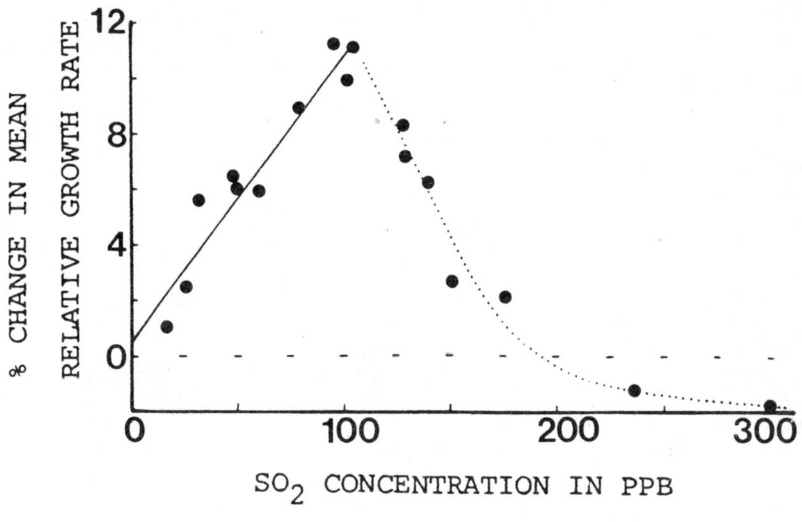

Figure 7. Effects of SO_2 on mean relative growth rate of pea aphids (Acyrthosiphon pisum). The aphids were allowed to feed uninterrupted for 4 days on garden peas grown for 20-22 days in the various SO_2 concentrations. Growth rate are % changes compared with controls in clean air (<5 ppb SO_2). From Warrington (55).

Plants exposed to SO_2 pollution show increases in soluble and total nitrogen and in free amino acids (e.g. 7, 11, 15, 21). It is well known that the performance of herbivorous insects is often directly correlated with amount of soluble nitrogen in the plants. McNeill *et al* (39) noted that apart from a general increase in amino acids, there were relative increases in important ones that would be expected to give nutritional benefit to aphids. McNeill and his colleagues are currently examining these matters in more detail by providing aphids with artificial diets containing different concentrations of specific amino acids.

Estimation of the economic significance of the increased performance of insect pathogens in polluted environments is complex because the pollutants also act directly on the plants. It is clear, however, that aphids are stimulated by low concentrations of pollutants (e.g. Figure 7), that are generally thought to have very little direct effect on crop yield. Thus economic losses brought about by the aphids themselves in polluted areas need to be explored in greater detail because they may constitute a major part of the overall effect in the field.

4. EDAPHIC FACTORS

There is quite a lot of evidence to suggest that the responses of plants to the uptake of pollutants by their aerial parts can be dependent on, or modified by, conditions in the soil. Particular emphasis has been given to the way in which the effects of SO_2 are altered by the sulphur status of the soil. There seems little doubt that when soils are deficient in sulphur, the uptake of SO_2 from the atmosphere can contribute usefully to the sulphur budget of plants (50). In experiments with controlled amounts of nutrients the uptake of sulphate from the medium has been shown to be reduced during fumigations with SO_2 (e.g. 53). The decrease in growth and yield caused by SO_2 may depend on the sulphur nutrition of the plants but the quantification of this dependency has proved difficult. A full review of the past literature was provided by Cowling and Koziol (12). McDonald (38) has stressed the importance of correctly controlled nutrition in future studies of the effects of air pollution. He suggests that some of the conflicting data published in the pollution literature can be explained in terms of the way in which plants are grown. These considerations are particularly relevant to the evaluation of effects of SO_2 and NO_x.

It is interesting to note that attention has been drawn to the importance of soil factors, and to the dearth of useful information on the subject, by senior authorities in air pollution in the past (e.g. 6, 22). Despite their pleas for more good experiments the state of the topic today is little improved. Other contributors to this meeting have covered some of the important questions concerning inputs of nitrogen and sulphur, and in this paper it is therefore appropriate merely to emphasize the need for continuing attention to this problem.

ACKNOWLEDGEMENTS. The research leading to the previously unpublished data reported here was supported by the UK Department of the Environment and the Commission of the European Communities.

REFERENCES

1 BAKER, C.K., UNSWORTH, M.H. and GREENWOOD, P. (1982). Leaf injury on wheat plants exposed in the field in winter to SO_2. Nature 299, 149-151.

2 BENNETT, J.P. and OSHIMA, R.J. (1976). Carrot injury and yield response to ozone. J. Am. Soc. Hort. Sci. 101, 638-639.

3 BLACK, V.J. (1985). SO_2 effects on stomatal behavior. pp. 96-117 of W.E. Winner, H.A. Mooney and R.A. Goldstein (eds) Sulfur Dioxide and Vegetation. Stanford University Press, Stanford, California.

4 BLACK, C.R. and BLACK, V.J. (1979). Light and scanning electron microscopy of SO_2-induced injury to leaf surfaces of field bean (Vicia faba L.). Plant, Cell and Environment 2, 329-333.

5 BLACK, C.R. and BLACK, V.J. (1979). The effects of low concentrations of sulphur dioxide on stomatal conductance and epidermal cell survival in field beans (Vicia faba L.) J. Exp. Bot. 30, 291-298.

6 BRANDT, C.S. and HECK, W.W. (1968). pp. 401-444 of A. C. Stern (ed) Air Pollution. Academic Press, New York.

7 BRAUN, S. and FLUECKIGER, W. (1985). Increased population of the aphid Aphis pomi at a motorway. Part 1, Field evaluation. Environ. Pollut. (Ser. A) 39, 183-192.

8 BROWN, K.A., ROBERTS, T.M. and BLANK, L.W. (1987). Interaction between ozone and cold sensitivity in Norway Spruce : a factor contributing to the forest decline in central Europe? New Phytologist 105, 149-155.

9 CAPE, J.N. and FOWLER, D. (1981). Changes to epicuticular wax of Pinus sylvestris exposed to polluted air. Silva Fennica 15, 457-458.

10 COHEN, J.B. and RUSTON, A.G. (1925). Smoke : a Study of Town Air. Edward Arnold, London.

11 COWLING, D.W. and BRISTOW, A.W. (1979). Effects of SO_2 on sulphur and nitrogen fractions and free amino acids in perennial ryegrass. J. Sci. Food Agric. 30, 354-360.

12 COWLING, D.W. and KOZIOL, M.J. (1982). Mineral nutrition and plant response to air pollutants. pp. 349-375 of M.H. Unsworth and D.P. Ormrod (eds). Effects of Gaseous Air Pollution in Agriculture and Horticulture. Butterworths, London.

13 DAVISON, A.W. and BAILEY, I.F. (1982). SO_2 pollution reduces the freezing resistance of ryegrass. Nature 297, 400-402.

14 DAVISON, A.W. and BARNES, J.D. (1986). Effects of winter stress on pollutant responses. pp. 16-32 of How are the Effects of Air Pollutants on Agricultural Crops Influenced by the Interaction with other Limiting Factors? Proceedings of CEC Workship within the framework of the Concerted Action 'Effects of Air Pollution on Terrestrial and Aquatic Ecosystems', Riso, Denmark.

15 DOHMEN, G.P. (1985). Secondary effects of air pollution: enhanced aphid growth. Environ. Pollut. (Ser. A) 39, 227-234.

16 DOHMEN, G.P., MCNEILL, S. and BELL, J.N.B. (1984). Air pollution increases Aphis fabae pest potential. Nature 307, 52-53.

17 FEHRMANN, H., von TIEDEMANN, A., BLANK, L.W., GLASHAGEN, B. and EISENMANN, T. (1986). Predisposing influence of air pollutants on fungal leaf attack in cereals and grape-vines. pp. 98-103 of How are the Effects of Air Pollutants on Agricultural Crops Influenced by the Interaction with other Limiting Factors? See reference (14) for details.

18 FLUECKIGER, W. and BRAUN, S. (1986). Effect of air pollutants on insects and host plant/insect relationships. pp. 79-91 of How are the Effects of Air Pollutants on Agricultural Crops Influenced by the Interaction with other Limiting Factors? See reference (14) for details.

19 FREER-SMITH, P.H. (1985). The influence of SO_2 and NO_2 on the growth, development and gas exchange of Betula pendula Roth. New Phytol. 99, 417-430.

20 FREER-SMITH, P.H. and MANSFIELD, T.A. (1987). The combined effects of low temperature and SO_2 + NO_2 pollution on the new season's growth and water relations of Picea sitchensis. New Phytol. in press.

21 GODZIK, S. and LINSKENS, H.F. (1974). Concentration changes of free amino acids in primary bean leaves after continuous and interrupted SO_2 fumigation and recovery. Environ. Pollut. (Ser. A) 7, 25-38.

22 GUDERIAN, R. (1977). Air Pollution. Springer-Verlag, Berlin.

23 HAVAS, P.J. (1971). Injury to pines growing in the vicinity of a chemical processing plant in northern Finland. Acta Forestalia Fennica 121, 1-21.

24 HAVAS, P.J. and HUTTUNEN, S. (1972). The effects of air pollution on the radial growth of scots pine (Pinus sylvestris L.). Biological Conservation 4, 361-368.

25 HAVAS, P.J. and HUTTUNEN, S. (1980). Some special features of the ecophysiological effects of air pollution on coniferous forests during the winter. In T.C. Hutchinson and M. Havas (eds). Effects of Acid Precipitation on Terrestrial Ecosystems, NATO Conference, Series 1 (Ecology) Plenum Publishing Corporation.

26 HEGGESTAD, H.E., GISH, T.J., LEE, E.H., BENNETT, J.H. and DOUGLAS, L.W. (1985). Interaction of soil moisture stress and ambient ozone on growth and yields of soybeans. Phytopathology 75, 472-477.

27 HEAGLE, A.S. (1982). Interactions between air pollutants and parasitic plant diseases. pp. 333-348 of M.H. Unsworth and D.P. Ormrod (eds) Effects of Gaseous Pollutants in Agriculture and Horticulture. Butterworths, London.

28 HUGHES, P.R., POTTER, J.E. and WEINSTEIN, L.H. (1982). Effects of air pollutants on plant/insect interactions: increased susceptibility of greenhouse grown soybeans to the Mexican bean beetle after plant exposure to SO_2. Environ. Entomol.-Ent. Soc. Amer. 11, 173-176.
29 HUTTUNEN, S. (1978). Effects of air pollution on provenances of Scots pine and Norway spruce in northern Finland. Silva Fennica 12, 1-16.
30 HUTTUNEN, S. (1984). Interactions of disease and other stress factors with atmospheric pollution. pp. 321-355 of M. Treshow (ed) Air Pollution and Plant Life. John Wiley and Sons, Chichester, U.K.
31 HUTTUNEN, S., HAVAS, P. and LAINE, K. (1981). Effects of air pollutants on wintertime water economy of the Scots pine (Pinus sylvestris L.). Holarctic Ecology 4, 94-101.
32 HUTTUNEN, S. KARENLAMPI, L. and KOLARI, K. (1981). Changes in osmotic values and some related physiological variables in polluted coniferous needles. Ann. Bot. Fenn. 18, 63-71.
33 JONES, T. and MANSFIELD, T.A. (1982). The effect of SO_2 on growth and development of seedlings of Phleum pratense under different light and temperature environments. Environ. Pollut. (Ser. A) 27, 57-71.
34 KELLER, T. (1978). Wintertime atmospheric pollutants - do they affect the performance of deciduous trees in the ensuing growing season? Environ. Pollut. 16, 243-247.
35 KELLER, T. (1981). Folgen einer winterlichen SO_2-Belastung die Fichte. Gartenbauwissenschaft 46, 170-178.
36 KVIST, K. (1986). Fungal pathogens interacting with air pollutants in agricultural crop production pp. 67-78 of How are the Effects of Air Pollutants on Agricultural Crops Influenced by the Interaction with other Limiting Factors? See reference (14) for details.
37 LARCHER, W. (1973). Physiological Plant Ecology, 2nd Edn. Springer-Verlag, Berlin.
38 McDONALD, A.J.S. (1986). Importance of mineral nutrition in studies of air pollutants and plant growth. pp. 38-49 of How are the Effects of Air Pollutants on Agricultural Crops Influenced by the Interaction with other Limiting Factors? See reference (14) for details.
39 McNEILL, S., BELL, J.N.B., AMINO-KANO, M. and MANSFIELD, P. (1986). SO_2, plant, insect and pathogen interactions. pp. 108-115 of How are the Effects of Air Pollutants on Agricultural Crops Influenced by the Interaction with other Limiting Factors? See reference (14) for details.
40 MANSFIELD, T.A. and FREER-SMITH, P.H. (1984). The role of stomata in resistance mechanisms. pp. 131-146 of M.J. Koziol and F.R. Whatley (eds) Gaseous Air Pollutants and Plant Metabolism. Butterworths, London.

41 MANSFIELD, T.A., DAVIES, W.J. and WHITMORE, M.E. (1986). Interactions between the responses of plants to pollution and other environmental factors such as drought, light and temperature. pp. 2-15 of How are the Effects of Air Pollutants on Agricultural Crops Influenced by the Interaction with other Limiting Factors? See reference (14) for details.
42 MATERNA, J. (1974). Einfluss der SO_2 - Immissionen auf Fichtenpflanzen in Wintermonaten. IXth Int. Tagung uber Luftverunreinigung und Forstwirtschaft, pp. 107-114. Marianske Lazne, Praha Czechoslovakia.
43 MATERNA, J. (1984). Impact of atmospheric pollution on natural ecosystems. pp. 397-416 of M. Treshow (ed) Air Pollution and Plant Life. John Wiley & Sons, Chichester, U.K.
44 MATERNA, J. and KOHOUT, R. (1963). Die Absorption des Schwefeldioxids durch die Fichte. Naturwissenschaften. Vol. 50. 407-408.
45 MUNCH, E. (1933). Winterschäden an immergrünen Gehölzen. Ber. Deutsch. Bot. Ges., 51, p.21.
46 OSHIMA, R.J., BENNETT, J.P. and BRAEGELMANN, P.K. (1978). Effect of ozone on growth and assimilate partitioning in parsley. J. Am. Soc. Hort. Sci. 103, 348-350.
47 OSHIMA, R.J., BRAEGELMANN, P.K., FLAGLER, R.B. and TESO, R.R. (1979). The effects of ozone on the growth, yield, and partitioning of dry matter in cotton. J. Environ. Quality 8, 474-479.
48 OKANO, K., ITO, O., TAKEBA, B., SHIMIZU, A., and TOTSUKA, T. (1984). Alteration of ^{13}C-assimilate partitioning in plants of *Phaseolus vulgaris* exposed to ozone. New Phytol. 97, 155-163.
49 PANDE, P.C. and MANSFIELD, T.A. (1985). Responses of spring barley to SO_2 and NO_2 pollution. Environmental Pollution (Ser. A) 38, 87-97.
50 PAUL, R. (1986). Sulphur dioxide uptake by plants according to the level of sulphur nutrition. pp. 104-107 of How are the Effects of Air Pollutants on Agricultural Crops Influenced by the Interaction with other Limiting Factors? See reference (14) for details.
51 SAUNDERS, P.J.W. (1966). The toxicity of SO_2 to Diplocarpon rosae Wolf. Causing blackspot of roses. Ann. Appl. Biol. 58, 103-114.
52 SOIKKELI, S. and TUOVINEN, T. (1979). Damage in mesophyll ultrastructure of needles of Norway Spruce in two industrial environments in central Finland. Ann. Bot. Fenn. 16, 50-64.
53 THOMAS, M., HENDRICKS, R., COLLIER, T. and HILL, G. (1943). The utilization of sulphate and sulphur dioxide for the sulphur nutrition of alfalfa. Plant Physiol. 18, 345-371.
54 WALMSLEY, L., ASHMORE, M.R. and BELL, J.N.B. (1980). Adaptation of radish *Raphanus sativus* L. in response to continuous exposure to ozone. Environ. Pollut. (Ser A) 23, 165-177.
55 WARRINGTON, S. (1987). Relationship between SO_2 dose and growth of the pea aphid, Acyrthosiphon pisum, on peas. Environ. Pollut. 43, 155-162.

56 WHITMORE, M.E. and MANSFIELD, T.A. (1983). Effects of long-term exposures to SO_2 and NO_2 on *Poa pratensis* and other grasses. Environ. Pollut. (Ser A) 31, 217-235.

57 WRIGHT, E.A., LUCAS, P.W., COTTAM, D.A. and MANSFIELD T.A. (1986). Physiological responses of plants to SO_2, NO_2 and O_3 : implications for drought resistance. Paper presented to CEC Workshop on Direct Effects of Dry and Wet Deposition on Forest Ecosystems, Lökeberg, Sweden. (in press).

EFFECTS OF OZONE IN AMBIENT AIR ON GROWTH, YIELD AND PHYSIOLOGICAL PARAMETERS OF SPRING WHEAT

J.Fuhrer, A.Grandjean, B.Lehnherr, A.Egger*) and W.Tschannen

Swiss Federal Research Station for Agricultural Chemistry and Environmental Hygiene, CH-3097 Liebefeld-Bern, and *)Institute of Plant Physiology, University of Bern, CH-3013 Bern, Switzerland.

SUMMARY

A spring wheat cultivar (Triticum aestivum, cv Albis) was grown near Bern (Switzerland) in open-top chambers and open-field plots and exposed daily to different ozone concentrations in ambient air. Seasonal 8-hr/day (0900-1700) ozone concentrations were 0.020 ppm in charcoal-filtered air, 0.035 ppm in nonfiltered air, 0.103 ppm in ozone-enriched air and 0.040 ppm in ambient air. Growth, yield and various physiological parameters were studied. Grain yield was the most ozone-sensitive parameter, much more sensitive than straw yield or growth. In filtered air, grain yield was 9% higher than in nonfiltered air, and 127% higher than in ozone-enriched air. Differences in grain yield appeared to be linked to direct effects of ozone on CO_2-fixation of flag leaves. Measurements of chlorophyll content and ethylene production revealed that stress from elevated ozone concentrations can affect processes associated with leaf senescence.

1. INTRODUCTION

Most previous studies on air pollution effects on agricultural crops in Switzerland were focused on the evaluation of effects in the vicinity of pollutant sources. Little attention was paid to possible chronic effects in more remote, rural areas. In these areas, ozone and other pollutants have been recorded occasionally. The limited data suggested that ozone may be the dominant pollution stress. This led to the first field-fumigation experiment. The aim of this experiment is to evaluate effects of ozone in ambient air on growth and yield of a spring wheat cultivar, and to describe changes in the physiology of flag leaves, which occur in association with ozone-induced alterations of growth and yield. The results from 1986 are included in this presentation. The experiment will be repeated in 1987 and 1988.

2. MATERIALS AND METHODS

2.1 Experimental site

The experiment was conducted at Oeschberg (N47°07'37''/E7° 37'01'') The site is located about 25 km northeast of Bern on the Swiss Central Plateau (see Figure 1). The elevation is 485 m a.s.l. The soil is low humus loam with a pH of about 6.5.

2.2 Cultural practices

A spring wheat cultivar (Triticum aestivum, cv. Albis) was sown with 16.7 cm row-spacing on 9 April 1986 (200 kg/ha). N-Fertilization was 46 kg/ha. Additional N was given on 6 May (29 kg/ha). Herbicide (4 May) and fungizide (4 July) was applied once. Open-top chambers were installed after the plants had emerged (6 May). Harvest date was 31 July inside the chambers and 8 August in the open-field plots.

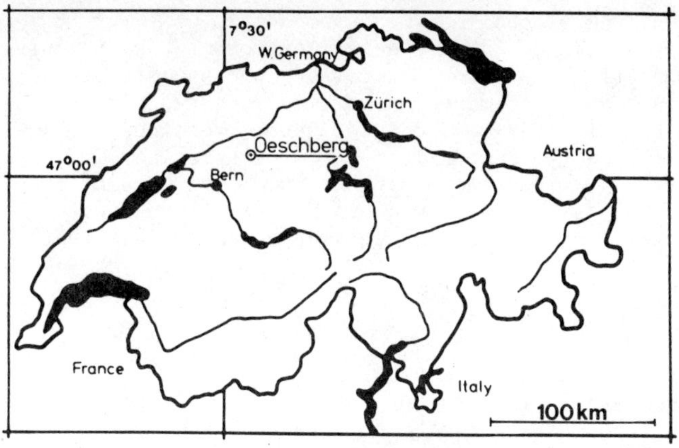

Figure 1: Location of the experimental site.

2.3 Pollutant exposure

Open-top chambers 1.5 m in diameter by 1.8 m in height were constructed according to the Institute of Terrestrial Ecology (Edinburgh) (2). Air was blown through a circular annulus at a rate of about 10 m^3/min. The annulus was fixed to the chamber wall 10 cm above the canopy. Ventilation was carried out day and night. One set of chambers was supplied with charcoal-filtered air (F) (Camcarb 2000, Camfil SA), four chambers received either nonfiltered (NF) or nonfiltered air with ozone added daily from 0900 to 1700 local time from 21 May to 31 July ($NF+O_3$). Ozone was generated by electrical discharge in dry air (Tecan ozone generator). A foam and fiberglass filter (Hi-Flo 3R-85, Camfil SA) was used to remove dust from air for the F-treatment. Open-field plots were used for comparison (A).

2.4 Environmental monitoring

Monitoring of environmental variables was performed at four points on a 15 min per plot time-share system. Measuring points were in ambient air and in one chamber of each different treatment. They were placed 30 cm above the canopy. Ozone was measured using a Dasibi 1003 AH analyzer. Ozone concentrations used in this paper are the 8-hr per day mean (0900 to 1700) for the growing period unless otherwise stated. Nitrogen oxides were measured using a Tecan CLD NO/NO_x analyzer. Gas samples were taken at each measuring point using an automated, computer-controlled valve and a vacuum pump. Climatic variables measured at all points included air temperature, soil temperature (at 5 cm depth), air humidity, and global radiation. Measurement data were recorded on tape with a Tecan Lx66 datalogger. At three different places in the field, soil water content was determined gravimetrically at 0-30 cm, 30-60 cm and 60-90 cm in triplicate twice a week. Rainfall was monitored in one chamber and one open-field plot. Wind speed and direction were recorded at one point.

2.5 Measurement of growth and yield

The length of the main shoot was measured biweekly during the whole experiment. For the determination of yield parameters, 50-cm row sections were harvested in each chamber or open-field plot for the analysis of yield components.

2.6 Physiological and biochemical measurements

Flag leaves were used for the determination of the following parameters. Chlorophyll content was determined in ethanolic extracts (5). Soluble protein was determined according to Bradford (1976) (1). Adaxial and abaxial leaf conductances were determined with a LiCor steady-state porometer (Lambda Inc.). Photosynthetic $^{14}CO_2$ uptake of attached leaves was measured in situ at 1600 to 2000 umol quanta m^{-2} sec^{-1} (400-700 nm) using the method of Shimishi (1969) (6,7). Ethylene production of leaf segments (9 cm) was measu-red as described earlier (3).

3. RESULTS

3.1 Microclimate and pollutant concentrations

Table 1 summarizes information on the microclimate and pollution concentrations in one chamber of each treatment and in one open-field plot. Daily means are given unless otherwise noted. Daily mean temperature of the air was higher by 2.3°C inside the chambers and the largest difference between chambers was 1.6°C. The temperature increase inside the chambers was largely due to the warming of the air inside the ventilation unit. Higher air temperature in the chambers was accompanied by higher soil temperature (+0.4°C) and lower relative humidity as compared to open-field plots. Chamber construction reduced global radiation by 15% and incident rainfall by 18%. The latter difference was compensated for by additional watering. Minimum moisture content of the soil was 25-30%. Ozone concentrations are given as either 24- or 8-hr/day means. In ambient air, 13.4% of all hourly means exceeded 0.050 ppm. 0.080 ppm was exceeded once. No major difference in NO and NO_2 was observed between the treatments.

The average diurnal pattern of hourly ozone concentrations in the different treatments is shown in Figure 2.

Table 1: Daily means for microclimatic parameters and pollutant concentrations from 1 June to 31 July 1986 in ambient air and in one open-top chamber of each treatment.

Parameter	(A)	(F)	(NF)	($NF+O_3$)
Temperature Air (°C)	19.6	21.8	23.4	22.0
Temperature Soil (°C)[1)]	19.4	20.1	19.7	19.6
Rel. humidity (%)	82.0	74.5	-	77.5
Global radiation (W/m^2)[2)]	340	261	301	299
Rainfall (mm)[3)]	308	252	-	-
Ozone (O_3) (24-hr/day, ppb)	27.3	12.3	23.4	46.5
Ozone (O_3) (8-hr/day, ppb)	38.3	20.1	35.3	103.4
Nitric oxide (NO, ppb)[4)]	1.8	1.4	1.7	1.5
Nitrogen dioxide (NO_2, ppb)	18.9	16.0	18.4	19.0

1) Measured at 5 cm depth; 2) Mean for daylight hours; 3) from 15 May to 31 July (measured in the center of the chamber); 4) NO and NO_2 from 17 June to 31 July. A: Ambient; F: Filtered; NF: Nonfiltered; $NF+O_3$: Ozone-enriched air.

3.2 Foliar injury

Foliar injury expressed as either chlorosis or necrosis was observed in the highest ozone treatment only ($NF+O_3$).

Figure 2:
Diurnal pattern of ozone concentrations in the different treatments.
A: Ambient air,
F: Charcoal-filtered air;
NF: Nonfiltered air;
$NF+O_3$: Ozone-enriched air.

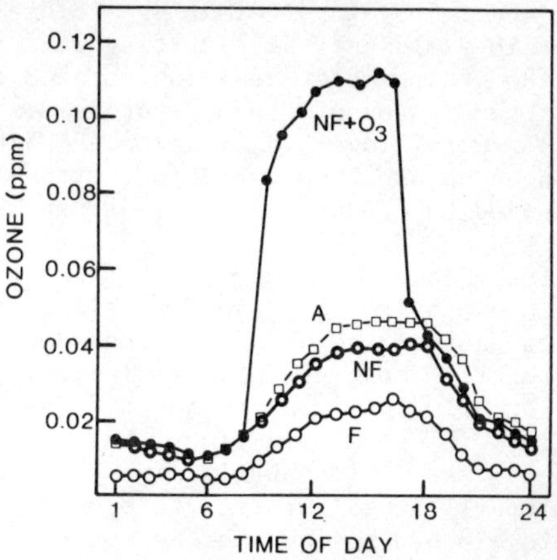

3.3 Growth and yield effects

In Table 2, results from growth and yield measurements are summarized. The response of plant height to different ozone exposures was very small. During their development, plants inside the chambers grew more rapidly than in the open-field plots (not shown). When ozone was reduced by filtration (F), grain yield and 1000-grain wt was increased by about 9% as compared to the unfiltered air treatment (NF), while straw yield only responded to an increase in ozone ($NF+O_3$). Differences in the various parameters between the four treatments were checked for statistical significance (α =0.05) using Duncan's multiple range test. The differences between open-field plots (A) and unfiltered air (NF) treatment was significant only for the height of the main shoot. Differences between unfiltered (NF) and filtered (F) air treatment were significant for grain yield and 1000-grain wt. Ozone-enriched air ($NF+O_3$) produced significant differences to nonfiltered air (NF) in the cases of grain yield, 1000-grain wt, number of seeds/head and plant height.

Table 2: Plant height, number of heads/m^2, straw yield, grain yield and 1000-grain wt of spring wheat exposed to selected ozone treatments in open-top chambers ($F,NF,NF+O_3$) or open-field plots (A)[1,2].

Treatment	O_3[3] (ppm)	Height (cm)	Heads (No/m^2)	Grains (No/head)	Straw (t/ha)	Grain (t/ha)	1000-grain wt (g)
A	0.038	103.5a	510a	37.4a	5.43a	5.54a	28.1a
F	0.020	108.9b	510a	40.6a	6.05a	6.68b	33.5b
NF	0.035	108.6b	520a	39.1a	6.17a	6.11a	30.7a
NF+03	0.103	97.9c	490a	29.1b	4.73a	2.94c	20.3c

1) Means followed by the same letter are not significantly different (Duncan's multiple range test, α =0.05); 2) 4 replicates per treatment; 3) Seasonal 8-hr/day mean concentration (0900–1700 local time).

3.4 Photosynthetic $^{14}CO_2$ uptake and leaf conductance

With increasing ozone concentration, uptake of $^{14}CO_2$ of attached flag leaves decreased (Figure 2A). In unfiltered air, $^{14}CO_2$ uptake was 10% lower than in filtered air. Similarly, grain dry weight per plant was reduced by 7%, indicating a close relationship between reductions in grain yield and rates of flag leaf photosynthesis.

The response of leaf conductance to increasing ozone was different from the response of photosynthesis. Both, adaxial and abaxial conductances were lower in filtered than in unfiltered air (Figure 2B). At enhanced ozone concentration, strongly reduced leaf conductance accompanied the reduction in $^{14}CO_2$ uptake.

3.5 Senescence

Chlorophyll loss and ethylene production were studied as processes associated with leaf senescence. Figure 4A shows the changes in chlorophyll content of flag leaves. In unfiltered and filtered air, chlorophyll loss started about 7 days after anthesis. This pattern was similar in open-field plots, but the plants outside the chambers contained less chlorophyll. In plants exposed to high ozone ($NF+O_3$), chlorophyll loss started before anthesis and was more rapid than in plants exposed to lower ozone concentrations.

Typically, ethylene production increases during the initial phase of chlorophyll loss. This pattern was confirmed by the analysis of flag leaves in open-field plots (Figure 4B). Ethylene production started to increase at the time of anthesis and reached a maximum rate

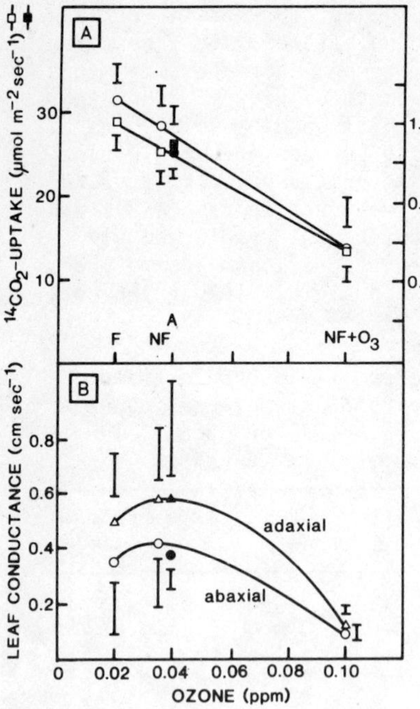

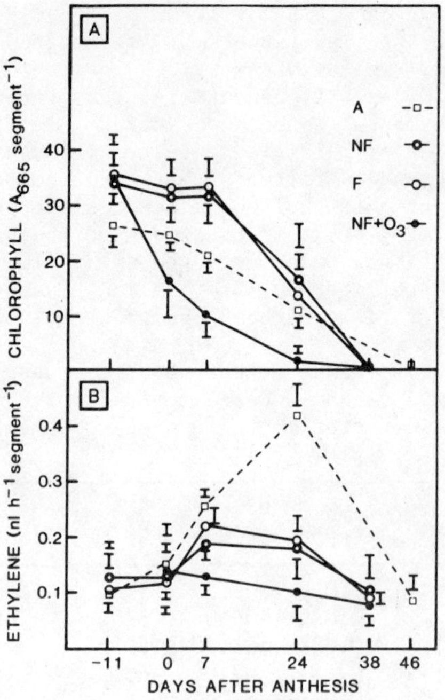

Fig. 2:
(A) $^{14}CO_2$ uptake of flag leaves and grain (seed) dry weight, and
(B) leaf conductance of flag leaves of wheat plants in response to different ozone treatments. Vertical bars: SEM.

Fig. 3:
(A) Time-course of chlorophyll content and (B) ethylene production of flag leaves of wheat plants treated with different ozone concentrations (symbols and concentrations as in Table 2). Vertical bars: SEM.

about 3 weeks later. Inside the chambers, maximum ethylene production was lower than in open-field plots, but the general time pattern in unfiltered and filtered air was similar to that in ambient air. In leaves exposed to elevated ozone, the post-anthesis increase in ethylene production was completely abolished.

4. DISCUSSION

Measurements in- and outside the open-top chambers revealed differences in the microclimate, which must be taken into account when using biological data obtained in chambers to predict pollution effects in the open field. Perhaps the most important difference exists for air temperature. The higher air temperature inside the chambers led to a more rapid growth and development. This could have been associated with a change in ozone resistance of the plants.

The results from yield determinations indicate a strong effect of ozone on grain yield, i.e. lower weight of individual kernels and a smaller number of kernels per head with increasing ozone pollution. Grain yield was significantly lower in unfiltered than in filtered air. In comparison, a study with different winter wheat cultivars in the midwestern USA with daytime ozone of 0.041 ppm revealed grain yield reductions of the same order as observed here, but in the case of the most sensitive cultivar only (4). This indicates the high ozone sensitivity of the spring wheat cultivar grown in the present study. However, as year-to-year variations in the yield response to ozone can be very high, the magnitude of the effect must be considered with care, and results from future experiments will have to be used for comparison.

The data included here suggest a direct relationship between photosynthetic CO_2 fixation and grain yield. Apparantly, ozone affects photosynthesis directly, and not via alteration of leaf conductance. Further investigations indicate that ozone reduces the amount of activatable RUBISCO present in the flag leaves (6). It is only at elevated ozone levels that yield can also be limited due to permature and accelerated leaf senscence.

5. ACKNOWLEDGMENTS

We wish to thank V. Lehmann, R. Perler and H. Shariatmadari for their help, and F.X. Stadelmann and C. Brunold for their support. The project is funded by the Federal Office for Education and Science (COST 612) and the Swiss National Science Foundation.

6. REFERENCES

(1) Bradford, M.M. (1976), Analytical Biochemistry 72, 248-254 (1976).
(2) Elphinstone, G.B., Fowler, D.Owen, G.H., Nicholson, I.A. (1983), Annual Report 1982, Institute of Terrestrial Ecology, 54-56.
(3) Fuhrer, J. (1982), Plant Physiol. 70, 162-167.
(4) Kress, L.W., Miller,J.E., Smith, H.J. (1985), Environ. Exp. Bot., 25, 211-228.
(5) Knudson, L.L., Tibbitts, T.W., Edwards, G.E. (1977), Plant Physiol. 60, 606-608.
(6) Lehnherr, B., Grandjean, A., Mächler, F., Fuhrer, J., J. Plant Physiol. (in press).
(7) Shimishi, D. (1969), J. Exp. Bot. 20, 381-401.

EFFECTS OF ATMOSPHERIC POLLUTANTS ON GRASSLAND ECOSYSTEMS

L. GRÜNHAGE, U. HERTSTEIN, H.-J. JÄGER and U. DÄMMGEN
Institut für Produktions- und Ökotoxikologie
Bundesforschungsanstalt für Landwirtschaft (FAL)
D-3300 Braunschweig, Bundesallee 50
Federal Republic of Germany

Summary

In order to investigate the effects of atmospheric pollutants on permanent grassland ecosystems the concentrations and depositions of SO_2 and of air-borne particulate matter have been measured. In addition to that a pilot study with open-top field chambers showed that white clover (which is sensitive to SO_2 and O_3) is affected seriously by unfiltered air. The amount of biomass production is a function of the N-supply (N-input by fertilizers, deposition of air-borne N-compounds, N-fixation by clover).

1. QUESTIONS, AIMS, AND METHODS

Atmospheric pollutants do not only affect yield and quality of crops and forages, they may also influence the structure, function and dynamics of the whole ecosystems. Therefore an investigation into short- and long-term effects of atmospheric pollutants on the compartments "vegetation" and "soil/soil water" of permanent grassland ecosystems, which represent about 40 per cent of the agriculturally used area in the Federal Republic of Germany, have been performed since 1983.

The plots investigated are situated in South East Lower Saxony and are influenced by three soft coal fired power stations close to the border to the German Democratic Republic. Our sampling stations at Rotenkamp, FAL (near Braunschweig), and Ettenbüttel are located 20, 35, and 55 km northwest of these sources, respectively, which emit approx. 150,000 tons of SO_2 per year. Up to 1983 all permanent grassland plots had been used as pastures, which were mown from time to time. The plots have different types of soils (Rotenkamp: humic gleysol on sandy-clayey fluviatile sediments over heavy clay; FAL: dystric-luvic cambisol on silty sandy cover-sediments overlying gravelly-sandy morainic deposits and glacifluviatile sands; Ettenbüttel:gleysol on sandy fluviatile sediments).

As it was difficult to distinguish between possible effects of the immission load and naturally occurring fluctuations and oscillations within the ecosystem compartments by investigating ambient plots only, open-top field chambers with charcoal-filtered and unfiltered air have been used since 1985 to detect changes caused by atmospheric pollutants.

The equipment used and the methods applied are described in detail elsewhere (1). First results of three years of investigation (1984 to 1986) are reported here.

2. CONCENTRATIONS AND DEPOSITIONS OF SO_2 AND OF AIR-BORNE PARTICULATE MATTER

Mean concentrations of SO_2 are between 35 and 50 µg/m³. They are much lower than all national or international threshold values (German TA Luft or EC guidelines), which were developed for the protection of human health and of vegetation. If we apply Stratmann's threshold function for very sensitive plants (2)

$$c_0 = 0.23\,t^{-0.17} \quad (c_0 \text{ in mg/m}^3;\ t \text{ in h}),$$

we find that during periods of vegetation the concentrations of SO_2 in air are considerably smaller than the threshold concentration for intervalls of one month (75 μg/m³). During winter months, however, the load is greater, so that very sensitive plants can be damaged in long term according to Stratmann's dose-response function (2):

$$c_0 = 0.23\,t^{-0.17} \quad (c_0 = \text{accumulated dose in mg·h/m}^3;\ t \text{ in h})$$

We know from experiments with O_3-sensitive tobacco (Bel W3) that our plots are exposed to phytotxic concentrations of air-borne oxidants from time to time. Thus we assume that sensitive grassland species e.g. clover (*Trifolium repens* L.) will be affected (3). In order to evaluate accute effects one needs more information about level, number, and temporal distribution of peak concentration of toxic gases.

Long term effects on ecosystems depend on the overall input from all forms of deposition, i.e. bulk deposition, deposition of aerosols and of gases. Table 1 gives the amounts of some chemical elements deposited into the grassland ecosystem at Rotenkamp. For S and N compounds there is a considerable input from the gas phase; for Pb and Cd the deposition of aerosols must not be neglected. Bulk-depositions of Pb and Cd are normal when compared to other rural regions in West Germany. Yet, the input of Cd (approx. 4 - 5 g/ha · a) exceeds the input from inorganic fertilizers (phosphates, 1 - 1.5 g/ha · a) significantly.

3. FIRST STEPS IN THE BALANCING OF CHEMICAL ELEMENTS

In order to establish a balance of chemical elements in grassland ecosystems it is necessary to consider the inputs of all types of depositions, the input from fertilizers and the outputs by leaching and harvesting. If leaching is not taken into account, one gets information on the load to soil and soil water (net input).

For the period dealt with in table 1 the net input of *sulphur* was approx. 40 to 60 kg/ha · a. Because the inputs from the atmosphere exceed the outputs by mowing considerably, a depletion of S in soils is very unlikely.

In contradiction to that the resources of the other major nutrients *potassium*, *magnesium*, *calcium*, and *phosphorus*, of which the inputs from air-borne particulates are comparatively less, will decrease by utilization in the course of time unless there is appropriate fertilizing.

The inputs of *cadmium* and *lead* are 2 - 4 g/ha · a and 200 - 300 g/ha · a, respectively.

Kloke (5) postulated tolerable upper limits for heavy metals in agricultural soils, i.e. 3 mg Cd/kg soil and 100 mg Pb/kg soil. If we assume that all inputs of heavy metals are accumulated in the top soil (0 - 5 cm), Kloke's limits for Cd will be exceeded in 366 years at the FAL, in 445 years at Rotenkamp, and in 107 years at Ettenbüttel; Pb concentrations will reach the upper limits in 119 years at the FAL, in 102 years at Rotenkamp, and in 80 years at Ettenbüttel. To us these calculations are an indication inviting us to investigate the *real load* to soils, as soils are the bases of all life. Yet, to calculate the real load it is necessary to find out *real deposition velocities* for all reactive gases considered as well as for aerosols and their compounds, not only for long periods but for *all important intervalls of time* in the course of the year.

4. RESULTS OF A PILOT STUDY USING OPEN-TOP CHAMBERS

Open-top chambers have been used to differentiate between "natural" influences on permanent grassland and those effects which are caused by gaseous air pollutants. For two years two pairs of open-top chambers (non-filtered/charcoal-filtered) have been operated at the FAL and at Rotenkamp. Thus we were able to compare unpolluted and polluted plots, which - apart from atmospheric influences - are subject to the same environmental conditions and their oscillations and fluctuations in the course of time. Amongst others we compared biomass production and the composition of species in either chamber. In particular we considered the developement of those species which are sensitive to certain gaseous pollutants and at the same time of special importance for the whole system.

At Rotenkamp we found a slight reduction in the production of above ground biomass in unfiltered air (cf. table 2): After a minor increase in 1985 (107.4 per cent compared to the filtered chamber) production dropped to 79.5 per cent in 1986 ($0.05 < P < 0.1$; Welsh test). During the same time both cover percentage and the contribution to yield of clover (*Trifolium repens* L.), which is sensitive to SO_2 and to O_3, became significantly less in the non-filtered chamber. At the end of the vegetation period in 1986 the clover had almost disappeared.

In contradiction to that we observed an increase in biomass production in unfiltered air at the FAL (cf. table 2. 1985: 139.2 per cent; 1986: 114.5 per cent; $0.05 < P < 0.1$; Welsh test). These plots had been fertilized with N, P, and K. It may be that this and/or water stress led to a sudden and general decrease of the cover percentage of clover and its contribution to yield in either chamber.

The differences between Rotenkamp and the FAL may be explained by the differences concerning the developement of clover and thus of the N supply. While at the FAL we had a better N supply in the non-filtered chamber (N from NO_x), we assume a better N supply in the filtered chamber at Rotenkamp caused by symbiontic N_2-fixation by clover. According to Klapp (6) a pasture containg clover can fix N in a range of 30 to more than 200 $kg/ha \cdot a$. The comparison of the N output with biomass supports this hypothesis (cf. table 2).

There is no significant relation between S concentrations in biomass and SO_2 concentrations in air. Also, biochemical/physiological investigations do not emphasize stress, which can be attributed to SO_2 only. We therefore assume that our observations at Rotenkamp have to be ascribed to combinations of gaseous pollutants, i.e. SO_2, NO_x, and O_3.

5. CONCLUSIONS

As a result of the investigations described we have started extendet field studies at the FAL using six pairs of open-top chambers (non-filtered /charcoal filtered). There we focus on N balancing and on the effects of atmospheric oxidants in combination with other air-borne pollutants on plants, microorganisms and soil/soil water. N balancing includes measurements of the input of gaseous NH_3 and particulte N compounds as well as the inputs from gaseous NO_x. At the same time we attempt to establish reasonable deposition velocities for the elements under investigation.

6. REFERENCES

(1) Grünhage, L., Hertstein, U., Dämmgen, U., Fleckenstein, J. & Jäger, H.-J. (1987). Auswirkungen von Depositionen saurer Luftverunreinigungen auf Grünlandökosysteme. Forschungsbericht des Bundesministeriums

für Forschung und Technologie. Forschungsprojekt 03-7320-4. Institut für Produktions- und Ökotoxikologie, Bundesforschungsanstalt für Landwirtschaft, Braunschweig.
(2) Stratmann, H. (1984). Wirkung von Luftverunreinigungen auf die Vegetation. LIS-Berichte *49*, 5 - 37.
(3) Cornelius, R., Faensen-Thiebes, A. & Meyer, G. (1985). Einsatz von Nicotinia tabacum L. Bel. W3. Staub-Reinhalt. Luft *45*, 59 - 61.
(4) VDI-Kommission Reinhaltung der Luft (1983). Säurehaltige Niederschläge - Entstehung und Wirkungen auf terrestrische Ökosysteme. Verein Deutscher Ingenieure. Düsseldorf.
(5) Kloke, A. (1980). Orientierungsdaten für tolerierbare Gesamtgehalte einiger Elemente in Kulturböden. Mitt. VDLUFA, H. 1-3, 9 - 11.
(6) Klapp, E. (1971). Wiesen und Weiden. Parey. Berlin, Hamburg.
(7) Dämmgen, U., Grünhage, L. & Jäger, H.-J. (1985). System zur flächendeckenden Erfassung von luftgetragenen Schadstoffen und ihren Wirkungen auf Pflanzen. Landschaftsökologisches Messen Auswerten *1*, 95 - 106.
(8) Umweltbundesamt (1986). Feststellung der Immissionsrate mit dem SAM-Verfahren. Monatsber. Meßnetz Umweltbundesamz 2/86, 36 - 48.

Table 1: Amounts of chemical elements deposited on permanent grassland at Rotenkamp (South East Lower Saxony, 4/85 - 4/86)

element	bulk deposition in g/ha · a	aerosol deposition in g/ha · a	dry deposition of gases in g/ha · a
Na	5.170		
K	1.250		
Mg	1.920		
Ca	10.100		
Al	1.750		
Pb	128	70 [1)]	
N			
- nitrate N	8.500	2.500 [1)]	17.000 [2)]
- ammonia N	8.000 [3)]		25.000 [4)]
S			
- sulfate S	9.700	2.300 [1)]	
- sulfite S	4.300		52.400 [5)]
Cl	16.400		
Zn	1.510		
Cd	3	1.5 [1)]	
Mn	140		
Fe	3.640		

1) Calculated from the element contents in aerosols using a deposition velocity of 0.2 cm/s.

2) Calculated from the NO_2 concentrations in ambient air using a deposition velocity of 0.5 cm/s.

3) Estimated from data given in (4)

4) Calculated from the NH_3 concentrations in ambient air using a deposition velocity of 0.7 cm/s.

5) Measured with surface active monitors, characteristic deposition velocity 1.0 cm/s (7,8).

Table 2: Above ground biomass production (in g/m^2), contribution to yield (in per cent) of grasses, herbs and legumes, S concentration in biomass (in mg/g dry wt.), and N output with biomass (in kg/ha) in charcoal filtered (CF) and non-filtered (NF) open-top chambers

FAL

	CF							NF						
	26.06.85	20.08.85	21.10.85	28.05.86	16.07.86	02.09.86	14.10.86	26.06.85	20.08.85	21.10.85	28.05.86	16.07.86	02.09.86	14.10.86
biomass production	437 ± 88	216 ± 35	85 ± 41	401 ± 101	262 ± 51	359 ± 54	141 ± 8	555 ± 130	305 ± 132	167 ± 148	453 ± 103	324 ± 105	376 ± 148	179 ± 68
contribution to yield														
- grasses	87.0	95.7	98.9	97.3	85.2	95.1	99.4	85.4	97.2	99.5	97.4	96.1	99.9	99.9
- herbs	10.2	4.3	1.1	2.7	14.8	4.9	0.6	5.0	2.8	0.5	2.6	3.9	0.1	0.1
- Trifolium repens	2.8	+ *)	+	-	-	+	-	9.6	+	-	-	+	-	-
S concentration in biomass	2.31	2.94	4.09	2.79	3.26	3.14	3.28	2.65	2.99	3.68	3.24	3.14	3.20	3.51
N output with biomass	97.9	60.1	29.6	109.4	78.5	111.2	54.3	154.3	85.1	59.5	124.5	97.3	142.1	65.2

Rotenkamp

	CF						NF					
	30.05.85	08.08.85	10.10.85	03.06.86	13.08.86	08.10.86	30.05.85	08.08.85	10.10.85	03.06.86	13.08.86	08.10.86
biomass production	744 ± 198	194 ± 76	80 ± 13	525 ± 138	214 ± 26	102 ± 10	755 ± 120	253 ± 61	85 ± 30	366 ± 85	236 ± 39	67 ± 2
contribution to yield												
- grasses	94.3	89.7	81.4	86.1	88.3	86.6	95.5	94.9	83.3	88.6	88.7	77.6
- herbs	4.8	5.2	6.4	9.3	6.0	6.6	3.6	3.2	10.6	11.4	11.3	22.4
- Trifolium repens	+	5.1	12.2	4.6	5.7	6.8	+	1.9	6.1	+	+	+
S concentration in biomass	2.03	2.70	4.21	2.23	3.70	4.69	2.03	2.90	4.03	2.54	2.86	5.29
N output with biomass	130.2	40.6	19.5	77.9	27.6	26.8	132.1	43.3	19.3	61.5	35.7	15.4

*) + ≙ < 1 %

AIRBORNE POLLUTANT INJURY TO VEGETATION IN GREECE

C.D. HOLEVAS
Laboratory of Nonparasitic Diseases of the Benaki
Plant Pathology Institute, Kiphissia, Athens, Greece

Summary

The photochemical complex formed in the atmosphere of some heavily urbanized districts of Greece constitutes a potential threat to crops grown in the surrounding agricultural land, but this topic has not been investigated so far.

Among the pollutants of industrial origin, fluorides have caused considerable injury to vegetation in a rural area, and some results concerning visual symptoms and fluoride content of plant tissues are given in this report.

1. INTRODUCTION

In Greece, a relatively heavy atmospheric pollution occurs mainly in the two metropolitan cities of Athens and Salonica, as well as in areas where various industries are operating. At present, however, urban pollution constitutes an increasingly important problem only in the Greater Athens Area, where special regulations hade to be imposed in an effort to reduce the burden of ambient pollution and prevent acute episodes in the city center and other parts of the region. On the other hand, in the Salonica area, although air quality is regarded poor as well, it has not reached untolerable limits.

Regarding single source air pollution, it can be said that there are several industrial operations all over the country which emit airborne phytotoxicants and cause pollution problems on a local scale. However, apart from fluorides, no other atmospheric pollutant has been evaluated, with regard to its impact on agricultural and other plant species.

2. URBAN AIR POLLUTION

The Greater Athens Area is less than 500 sq. km and has a population of more than three millions. This high concentration of population and the concurrent increase in industrial and other human activities together with a rapid build-up of motor vehicles in an area where meteorological conditions are conducive to a restricted air movement, brought about one of the most important air pollution problems existing in Greece today.

The first signs of the problem were already apparent in the early 1960's, mainly as a deterioration of visibility ; the famous blue sky and the clear horizon often lost their brightness by the development of a heavy cloud of smoke and dust, particularly after periods of atmospheric inversions. Quantitative observations commenced as early as 1962, and included measurements of sulphur dioxide and smoke ; since 1973, ambient levels of atmospheric pollutants are monitored in a net of permanent stations which are operated in the framework of a state project aiming at controling air pollution in the Greater Athens Area. The pollutants that are recorded in these stations are : particulate matter, sulphur dioxide, nitrogen dioxide ozone and hydrocarbons.

There is a variety of activities which are responsible for the atmospheric pollution occuring in the Greater Athens Area ; according to data collected so far they can be grouped as follows, in a decreasing order of contribution : automobile transportation (75 %), industrial processes (22 %) and space heating (3 %). Consequently, the atmosphere is polluted principally by exhaust gases from the combustion of organic fuels. This origin of air pollution and its association with sunshine and meteorological conditions producing air stagnation would suggest that the pollution complex which develops in the Athens Basin is photochemical in nature, with ozone as one of its major ingredients.

So far, studies of the air pollution arising in the Greater Athens Area have been designed to obtain basic information on pollutant levels and dispersion, to assess their effects on human health, archaeological monuments and other materials and to formulate remedial measures. The significance of this urban pollution for plants growing in the polluted area, as well as in the contiguous agricultural land has not been evaluated.

However, photochemical pollutants are phytotoxic and the ozone and perhaps other oxidants which are formed in the area of Athens are expected to affect adversely plants exposed to them. Though symptomatology of plants suffering from toxic effects of air pollutants is complicated by many other environmental interferences, it should be noted that there are several instances where plants in the polluted area were found to exhibit leaf chlorotic and necrotic patterns attributable to air pollution injury. A conspicuous and widespread pathological manifestation that has often been encountered is a sudden necrosis of pine needles in suburban parts of Athens ; affected needles, sooner or later were abscissed and the trees formed new growth. As this needle injury was observed in early summer and the situation of the trees improved later, it cannot be attributed solely to dry conditions ; the syndrome had many similarities to an ozone injury, and may have been induced by ozone generated in the urban atmosphere.

Furthermore, the impact of the atmospheric pollution arising in the Athens Basin may not be limited to vegetation in this area, taking into account that the nature of the pollutants favours their transfer and build-up in distant places. Therefore, adverse effects on agricultural crops grown close to this source of urban pollution shoud be considered as well. In this connection, plant damage should be seen also in the context of a reduced resistance to adverse physical factors such as water and low temperature stress. Under the climatic conditions of Greece a physiological stress imposed on agricultural plants by photochemical air pollutants and/or acidified rain, even in the absence of visible symptoms, may be proved of a greater significance for production and survival than an acute toxicity.

3. INDUSTRIAL AIR POLLUTION

The most serious problem of industrial air pollution damage to vegetation, in rural areas, has been observed in the district of Viotia, and it was caused by emissions of fluorides from a local aluminium production factory. It should be noted, however, that plant injuries by phytotoxic air pollutants, such as sulphur dioxide, fluorides, acidic fumes, particulates etc. have been recorded in some other industrialised regions as well, though not so extensive and destructive as in Viotia.

Fluorides rank among the most dangerous atmospheric pollutants, because of their high phytotoxicity and accumulation in plant tissues, even from relatively low concentrations in the air. Thus, they act as a cumulative poison on the plants directly, or on the grazing animals which consume plant food with a relatively high fluoride level. For this reason an investigation was undertaken to determine the extent and the severity of plant injury by airborne fluorides, and the data obtained are summarized in this report.

Occurence of fluoride plant injury

In addition to the Viotia region, visible injury to vegetation caused by airborne fluorides has been recorded in a district near Kavala in Northern Greece. In this district, however, very few observations have been made and so most of the information reported here refers to the Viotia area, where plants suffered severe injury. In this area, olive trees and grapevines are the main agricultural crops, and these two plant species served as guides for the detection of fluoride air pollution ; olive trees are evergreens and so their foliage constituted a continuous accumulation surface, whereas grapevines, due to their sensitivity to fluoride toxicity exhibited characteristic symptoms that helped in the diagnosis.

Visual symptoms in grape developed up to a distance of about 8 km from the factory, but increased fluoride levels were found much further, depending on distance, the topography and the prevailing winds. The influence of these factors on the accumulation of airborne fluoride by the leaves is reflected too, by the changes found in the rate of fluoride deposition on limed filter paper exposed at different distances and directions around the factory.

Visual symptoms

In general, toxicity symptoms were apparent in the leaves and were intensified as season progressed. Plant species differed greatly in their sensitivity to fluoride, but intensity of symptoms varied also among individuals of the same species.

Grapevine and apricot are the two species which suffered the most severe leaf necrosis ; it started at marginal and interveinal tissues which initially were water-soaked and dull green, but soon dried up and turned reddish brown in grapevine and gray brown in apricot. Apricot leaves showed a tendency to fold upwards and have wavy margins, several days before the appearance of necrotic tissues. Also, in this species nectotic areas of laminae separated easily and dropped out so that affected leaves became ragged at their edges. In some instances, grapevine leaves developed only necrotic spots and young berries had necrotic specks on their surface.

Olive trees exhibited leaf tip necrosis or chlorosis on a relatively small part of their foliage ; in very few instances, necrotic lesions were observed on inflorescences and young fruits on external portions of the trees.

Tip or marginal necrosis of leaves was observed on almond, plum, apple and quince ; it was not extensive and appeared in late season. Also, light leaf necrosis was recorded in pistachio, walnut, garlic and onion.

In pine, needles of the years growth became chlorotic and later necrotic at their tips.

Other species, viz. mulberry, plane tree and rose developed chlorotic mottles, without or very little necrosis, whereas others, viz. fig, broad bean and barley did not exhibit visible injury at all.

In the district of Kavala, gladiolus plants suffered severe leaf damage and their cultivation was abandoned. Also, in this area peach fruit of the Rubidoux variety developed symptoms like those of the suture red spot disease which is caused by fluoride absorption by the fruit.

Fluoride content of plant tissues

The presence of airborne fluorides resulted in an accumulation of relatively high fluoride levels in plant tissues. Leaves were the most vulnerable organ to fluoride build-up. Thus, petioles and shoot bark of grapevine contained 38 % and 10 %, respectively, of the amount of fluoride found in laminae. Fruits, also, contained relatively small amounts of fluoride ; mature berries from a vineyard with severe leaf burning caused by fluoride contained less than 3 p.p.m. F, on a fresh weight basis.

A great deal of variability was observed in the fluoride content of leaves of the same plant ; the age, the position and the portion of leaves analysed influenced their fluoride status. Mature leaves and those at exposed parts of the plant contained more fluoride. These factors affecting the concentration of fluoride in the leaves should always be taken into account when leaf analysis is used for detection of air contamination with fluorides, particularly in districts remote from a fluoride source, where fluoride content of plant tissues does not differ greatly from background levels.

Air fluoride content

A static technique was employed to detect fluorides present in the air. It was based on exposure of limed filter paper discs in different locations for various periods of time. The discs were sheltered in especially constructed boxes which were installed at different distances and directions around the fluoride source. The rates of fluoride accumulation on the discs differed considerably among localities. Also, a wide range of values corresponded to the same box for the different periods of exposure.

Effect of fluoride on olive trees

Although olive trees accumulated as high or higher leaf fluoride levels as grapevines and apricots, they did not exhibit severe injury as happened with these two sensitive plant species. It would appear that olive trees are tolerant to fluoride as far as visual injury is concerned. Nevertheless, the presence of such a strong toxicant in plant tissues at high concentrations could affect adversely plant productivity and hardiness. However, a direct assessment of crop performance of olive trees in the affected area was not possible, due to large fluctuations of yields caused by weather and other factors from year to year. Thus an attempt was made to obtain some indirect information by examining pollen germination in relation to fluoride. On the basis of these measurements, pollen of olive trees does not seem to be very sensitive to fluoride toxicity, and fluoride levels encountered in the polluted area are unlikely to cause a reduction of practical importance in the germination of pollen of this crop.

IS PLANT RESISTANCE TO AIR POLLUTION RELATED TO PLANT RESISTANCE TOWARDS DISEASE?

G.P. Dohmen
Gesellschaft für Strahlen- und Umweltforschung, Institut für Toxikologie, Ingolstädter Landstraße 1, D - 8042 Neuherberg, FRG

Summary

In order to study possible correlations between plant resistance towards disease and resistance to ozone, a number of fumigation experiments was carried out with different cultivars of wheat, barley, and maize, of which the average disease resistance is known. Only a poor correlation was observed between average plant susceptibility to disease and sensitivity to ozone. However, there is a very close correlation between ozone tolerance and wheat resistance to a certain pathotype of yellow rust. The knowledge of such relationships could offer one approach for research in pollutant effects on plants, as well as for breeding pollutant resistant plants.

When carrying out experiments on the effect of low doses of ozone on the disease potential of brown rust on wheat plants, I found that not only was disease development strongly inhibited on ozone fumigated wheat plants (1), but also that the wheat cultivar which is less sensitive to the fungus is also the cultivar with less visible injury after acute ozone treatment. Another case of correlation between resistance towards insect pests and air pollutants has been reported in the literature. The pine species, _Pinus nigra_, which is less attacked by insects is also less damaged by air pollutants (2).

In order to find out whether this relation between resistance to air pollution and resistance to parasites can be encountered more frequently, I carried out a number of fumigation experiments with different cultivars of wheat, barley, and maize (in the following only the results obtained with wheat will be dealt with). They were chosen according to their average resistance towards a number of fungal parasites as listed in the "Bundessortenliste". This publication by the "Bundessortenamt, Hannover" summarizes the investigations carried out to study the quality of certain crop cultivars. It gives, for example, relative values for yield, but also for susceptibility towards certain diseases. Eight different pathogenic fungi are listed for wheat, together with the relative resistance of each cultivar towards these diseases. An average resistance was calculated from these data. The cultivars with the highest and lowest relative average resistance respectively were used for the experiments.

The plants were grown in a greenhouse up to the three

leaf stage. Then they were transferred into closed fumigation chambers and kept for three days in clean air only. In order to produce visible symptoms the plants were fumigated - under artificial light - for eight hours per day at concentrations of 150-230 ppb. Each evening at the end of the fumigation period the plants were checked visually and samples were taken. A fumigation event lasted for five to eight days and was stopped when the plants showed visible symptoms.

Results: The original observations on differential sensitivity to ozone were made on the two wheat cultivars 'Kanzler' and 'Granada'. 'Kanzler' is very susceptible towards brown rust, whereas on 'Granada' the fungus hardly develops at all. Figure 1 shows the effect of an acute dose of ozone on these two wheat cultivars. 'Granada' was only slightly affected, 'Kanzler', however, showed severe necrosis and particularly the upper third of the older leaves had large damaged areas. In addition the growth of this cultivar was far more retarded by ozone than was that of 'Granada'. The symptoms developed in 'Kanzler' at an earlier stage of the fumigation period then they did in 'Granada'.

FIGURE 1

The two wheat cultivars 'Kanzler' and 'Granada' after acute ozone treatment. K: Kanzler; G: Granada; O_3: ozone treatment; cf: control.

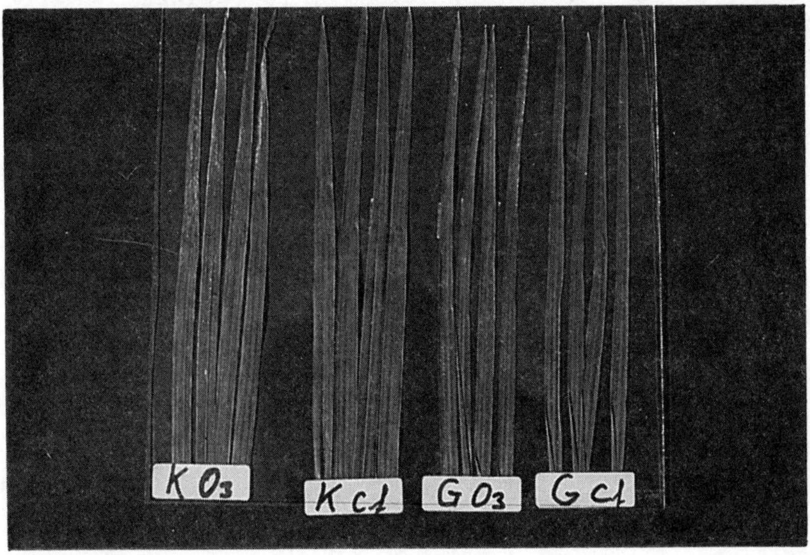

In the next experiment four other winter wheat cultivars were used, two with a high average susceptibility ('Basalt' and 'Götz') towards disease and two with a low sensitivity ('Kraka' and 'Kronjuwel'). The applied ozone concentrations did not cause very severe symptoms in any of these wheat cultivars. Both 'Basalt' and 'Götz' showed yellowing of the older leaves. In 'Götz' those had in addition some chlorotic mottling, whereas in 'Basalt' also in the younger leaves yellowing could be observed. The disease resistant cultivar 'Kraka' did not develop any symptoms, but the equally resistant 'Kronjuwel' showed a similar amount of damage to 'Götz', i.e. yellowing of older leaves with a little chlorotic mottling. Thus in 'Kronjuwel' there was no correlation between resistance towards air pollution and resistance to fungal diseases.

Two more experiments were carried out with four different cultivars of winter wheat each time. The results are summarized in table 1. Nine out of 13 cultivars showed a positive correlation between disease resistance and ozone tolerance, and four did not. Thus the average correlation between these two traits is only weak. However it appears that in certain cases a strong correlation can exist. That was in fact observed in the experiments with different wheat cultivars. Table 2 shows that there is a very close relationship between ozone tolerance and resistance towards pathotype "232" of yellow rust, <u>Puccinia</u> <u>striiformis</u>.

TABLE 1

Ozone sensitivity and average disease resistance in different wheat cultivars: The numbers are a measure of disease susceptibility (with "9" highest and "1" lowest susceptibility); a "(+)" indicates high and a "(-)" low ozone tolerance.

cultivar	more susceptible	more resistant
Albrecht		4.1 (+)
Basalt	5.9 (-)	
Bert		4.3 (+)
Götz	5.9 (-)	
Granada		4.4 (+)
Jubilar	5.6	(+)
Kanzler	6.3 (-)	
Knirps	(-)	4.0
Kraka		4.6 (+)
Kronjuwel	(-)	4.1
Merkur		4.1 (+)
Oberst	5.9 (-)	
Sorbas	(-)	3.8

Such experiments pose the question as to what could be responsible for the relationship. First one could think of anatomical and morphological factors. Leaves with a thick cuticle, many hairs, and small and well protected stomata are

TABLE 2

Ozone tolerance and resistance towards <u>Puccinia striiformis</u>, pathotype "232", in different wheat cultivars: The numbers are a measure of disease susceptibility (with "9" highest and "1" lowest susceptibility); a "(+)" indicates high and a "(-)" low ozone tolerance.

cultivar	more susceptible	more resistant
Basalt	9 (-)	
Bert		1 (+)
Götz	7 (-)	
Granada		3 (+)
Jubilar		2 (+)
Kanzler	5 (-)	
Kraka		1 (+)
Kronjuwel	7 (-)	
Merkur		2 (+)

usually quite well protected against many parasites; these properties also imply an advantage for plants under air pollutant stress, since they increase the 'boundary layer resistance'. Braun (3) observed significant morphological differences between spruce clones sensitive or tolerant to air pollutants. On a physiological level it is known that plants possess defence mechanisms against attack by parasites. Enzymes such as peroxidase or phenylalaninammoniumlyase can show enhanced activities which might lead to increased cell wall lignification. Hormonal changes may alter the physiological stage of cells thus making them more or less susceptible to parasite (or pollutant) attack. Plants with well functioning defence and repair mechanisms could also be more effective in repairing - or preventing - damage caused by air pollutants. On the other hand there are also more "incidental" events which determine tolerance towards ozone. Engel & Gabelman (4), for example, observed high resistance to ozone in some onion varieties. The main cause was found within the guard cells. In the resistant varieties the guard cells were sensitive to ozone and closed when damaged. Hence the other cells of the leaf did not come into contact with the pollutant and the variety was ozone tolerant. In this case ozone resistance is mainly controlled by a single dominant gene. But this seems to be the exception; in many other investigations it was found that several genes are responsible for the trait "tolerance" (5, 6). They are often additive and dominant over "sensitive".

As shown in this report, however, there is only a limited relationship between average resistance towards parasites and ozone. This is not necessarily surprising since plants often have very special properties which determine their susceptibility towards certain parasites (e.g. phytoalexines). It is not unusual for a plant cultivar resistant towards a certain pathotype of a parasite to be sensitive towards another pathotype of the same parasite species. Equally such distinct properties of the plant may also have a considerable influence on its sensitivity towards a pollutant. And this in fact appears to be the case for the strong correlation between ozone sensitivity and susceptibility towards the pathotype "232" of yellow rust, Puccinia striiformis.

The knowledge of such relationships could offer one approach to tackling the question of how exactly ozone affects plants, and how defence or repair mechanisms can work. From the genetic differences between the pathotype resistant and sensitive cultivars, a molecular biologist should be able to find the biochemical clues which also determine the plants' tolerance to ozone.

Such relationships could also be of some help in breeding pollutant resistant plant cultivars, but this should not be the main goal. Our primary aim should be to render resistance towards pollutants unnecessary.

REFERENCES

(1) DOHMEN; G.P. (1987). Secondary effects of air pollution: Ozone decreases brown rust disease potential in wheat. Environ. Pollut., 43:189-194.

(2) SIERPINSKI, Z. (1966). Schädliche Insekten an jungen Kiefernbeständen in Rauchschadengebieten in Oberschlesien. Archiv für Forstwesen, 15:1105-1114.

(3) BRAUN, G. (1977). Über die Ursachen und Kriterien der Immissionsresistenz bei Fichte (Picea abies [L.] Karst.), Teil 1: Morphologisch-anatomische Immissionsresistenz. Eur. J. For. Path., 7:129-152.

(4) ENGLE, R.L. and GABELMAN, W.H. (1966). Inheritance and mechanism for resistance to ozone damage in onion, Allium cepa L..Proc. Am. Soc. Hortic. Sci., 89:423-430.

(5) POVILAITIS, B. (1967). Gene effects for tolerance to weather fleck in tobacco. Can. J. Genet. Cytol., 9:327-334.

(6) HANSON, G.P., ADDIS, D.H. and THORNE, L. (1976). Inheritance of photochemical air pollution tolerance in petunias. Can. J. Genet. Cytol., 18:579-592.

SESSION III

EFFECTS OF AIR POLLUTION ON FOREST TREES AND FOREST ECOSYTEMS

SESSION III

EFFETS DE LA POLLUTION DE L'AIR SUR LES ARBRES ET LES ECOSYSTEMES FORESTIERS

SITZUNG III

WIRKUNGEN DER LUFTVERSCHMUTZUNG AUF WALDBÄUME UND WALDÖKOSYSTEME

PART I — Ière PARTIE — TEIL I

Chairman : Dr. H. OTT, Commission of the European Communities

Rapporteurs : Prof. D.H.S RICHARDSON, University Dublin, Ireland
Prof. O. QUEIROZ, Institut de Physiologie Végétale, Gif-sur-Yvette, France

REPORT ON THE SESSION

D.H.S. RICHARDSON AND O. QUEIROZ
Trinity College, University of Dublin, Ireland
and
Institut de physiologie vegetale, C.N.R.S., Gif-sur-Yvette, France

SUMMARY

Session III part 1 involved seven verbal presentations and was attended by more than 100 delegates. Six points were considered of key importance by the two nominated raporteurs of the session. Five of these identify areas where further research is required while the last is a warning to EC states that the current phase of forest recovery may be followed by a new cycle of forest decline. Some of the more interesting ideas which were presented at the session with regard to forest decline are also outlined.

1. MAIN POINTS

1. Research on air pollution in relation to forest trees is not as advanced as that on crop plants. However, with respect to trees, now is the time to start examining mechanisms in detail. To do this we must:

(a) know more about the basic physiology of forest trees. In particular, laboratory experiments and mechanistic studies need to be developed to examine the physiology of both cells and organs in forest trees. However such studies must be paralleled by experiments in the field.

(b) develop coordinated teams involving physiologists, soil scientists, chemists and foresters. More interaction with research groups examining crop plants in relation to pollution would be valuable. The the transfer of expertise and personnel from crop plant studies to forest problems should be strongly encouraged.

(c)concentrate on a few aspects possibly the interacting effects of pollutants with (i) drought (ii) frost hardiness (iii) nutrient translocation (iv) hormonal aspects. In each case investigate th mechanisms involved with the soil dimension receiving due emphasis.

2. Open top chambers should be used for more than confirming that ambient pollution levels at particular sites induce visible symptoms or growth responses. They should be employed to test mechanistic hypothesis evolved from rigorous experiments carried out in in phytotrons or controlled environment growth chambers. Complementary studies utilizing the two types of facilities should be encouraged as they have been in Japan. The use of clones is recommended to reduce individual variabliity of forest trees.

3. proton effects reqire to be examined more closely. Clear experiments seem not to have been designed to check whether known

buffering mechanisms are operative or effective in forest trees. Try and discover more about the compartmentalization of buffering in the cells of tree leaves and roots in relation to air pollution.

4. The differing views need to be reconciled with regard to the importance of enhanced nitrogen inputs in relation to pollution induced cation deficits resulting from (a) increased leaching (b) reduced uptake.

5. The long term dimension of pollutants in relation to trees should receive emphasis. For example enhanced nitrogen inputs apparently increase algal growth on trees. What are interactions with other epiphytes and their associated animals or indeed with the soil fauna and flora.

6. We must ensure that EC and governments are aware that current models of our understanding of forest decline include the possibility of a new cycle of forest damage. This may follow the current phase of forest recovery if a series of dry years occur. Only continued research and refining of ideas and damage mechanisms will eventually allow us to predict the likelihood of a new damage cycle or indeed to suggest measures to ameliorate its effects or delay its onset.

2. BACKGROUND

Three general themes were addressed in session III part 1. Firstly the effects of air pollutants on the above-ground and root systems of forest trees; secondly the leaching of substances from the canopy and soil system as a result of air pollutant exposure and finally effects on soil micro-organisms.

It is clear that the changing patterns of aerial emissions have complicated the assessment of forest damage. Small areas still exist where localised damage occurs as a result of high pollutant levels. However the trend towards release of emissions from high stacks has resulted in widespread pollution of low concentration. The effects of such emissions are more subtle and long term. Measuring these low levels accurately is technically difficult and it is now believed that both peak values as well as daily or annual means may be significant in terms of forest damage.

A detailed review of the levels and effects of the various air pollutants on forest trees was provided by Krause. It was clear that direct and indirect effects of individual pollants or pollutant mixtures was complex and could be ameliorated or exacerbated depending on both climatic and soil factors. For example one model accounting for forest decline that is currently being tested involves the following hypothesis. Exposure of the trees to elevated ozone levels results in increased permeability of cell membranes within leaves. In the presence of acid rain, leaching and loss cations including magnesium takes place. This results in magnesium deficienty which can affect protein synthesis, enzyme operation and metabolic balances. Leaves ultimately become yellow and may be shed. ozone can also affect carbohydrate redistribution from the leaves and this may account for observed changes on tree root systems where forest decline occurs.

The question of nutrient leaching was also addressed by Freier-Smith in his review of the effects of wet and dry deposition on tree canopies. He noted that while broad leaved trees and young conifers tend to buffer added acidity, older conifers acidify throughfall and stemflow. Leaching of calcium and other essential cations from the tree crown has serious implications for both the affected plant and the underlying soil environment. Two contributions dealt with the effects of simulated acid rain on spruce trees. The first by Kreutzer revealed that much of the hydrogen ion input was retained by canopy but that leaching of K, Ca, Mg, and Mn took place. The second study by Stuanes and coworkers indicated the possibility that tree growth rates could recover after irrigation ceased and that changes in soil chemistry were moderated in time. High soil aluminium levels during irrigation with the simulated acid rain may have been responsible for the reduced growth rates. However significant levels of Ca and Mg in the groundwater which was acidified for the irrigation experiments must be taken into account in interpretation of their results. These two ions have many important physiological roles including membrane stabilization that may ameliorate the effects of the high proton load.

In terms of soil chemistry, the paper by Schulze underlined the complex interactions that occur in soils underlying trees exposed to air pollution. Where there has been enhanced hydrogen ion input, there can be important consequences in the soils, especially a change in the Ca/Al ratio and upsets to mycorrhizal associations. He posed the question as to Why forests subjected to cation losses fail to conserve nutrient levels by growing more slowly as do trees on bogs etc?. Schulze proposed that enhanced inputs of nitrogen and ammonia throughout Europe might be a factor that caused the trees to retain unaturally high growth rates and hence nutrient imbalances which lead to the observed symptoms of yellowing and needle loss. However Zottl disputed this idea strongly. While he admitted that enhanced cation leaching resulted from pollution exposure, he believed that neither toxic metals, protons or aluminium toxicity could account for the recent forest decline. Furthermore while nitrogen deposition had undoubtedly increased, the input was small in relation to the overall nitrogen status of many soils and the fluxes involving micro-organisms in the forest soils. Unfortunately time did not permit a full discussion of the opposing viewpoints but they must be debated fully in the near future.

The reduced litter layer under many EC forests exhibiting forest decline could well reflect changes in the C/N ration of the litter following enhanced nitrogen inputs. In contrast under conditions of high sulphur dioxide input, an increase in litter layer can occur as has been found in some regions of Poland. The reported studies by Ineson and Wookey provided explanation for this. Decomposing pine litter experimentally exposed to sulphur dioxide in chamber fumigations and open air fumigations exhibited reduced respiratory activity. 90% of this activity was due to the activity of fungi which are generally sulphur dioxide sensitive especially at low pH. Enhanced leaching of cations from acidified litter was also noted and could have important implication for nutrient status soils under forests subject to acidic pollution.

3. CONCLUDING RECOMMENDATIONS IN RELATION TO SOME OF THE MAIN POINTS

1. Characterization of pollution/climate factor interactions.

Research is necessary to define how and why (a) pollution effects are modified by other factors, (b) pollution increases sensitivity to other stresses, (c) after-effects occur and the time-scale involved.

We need to know:

- whether both climatic stresses and pollution have a direct effect on the same metabolic targets
- or whether drought, cold, etc. affect targets not directly sensitive to pollutants but which interlink with pollution recovery mechanisms.

For example, mobilization of energy reserves and metabolite pools due to climatic stresses could weaken the capacity of a tree to respond to a pollution episode. Identifying these different modes of interaction will require metabolic and physiological experiments under controlled conditions in the laboratory as well as in open top chambers. Results on different clones or provenances from such studies could be important for tree breeders.

2. Effects ascribable to acidity.

Discussion will remain speculative and theoretical as long as no critical experiments are available to check pH changes within cell compartments of trees. Utilization of NMR may well yield valuable results as major advances in technology are being made in this area. The actual capacity of various cells in trees to resist pH change during pollution and/or stress need to be investigated. Such aspects as the malate/pyruvate system, the polyamines system, etc. in cell fractions need further research in this context.

3. Studies on gene expression.

A few results are now available showing that pollution by itself or when combined with frost can modify the protein patterns in spruce needles. Other results show that resistance varies in clones exhibiting different isozyme maps for several enzymes. Such approaches should be encouraged in order to obtain information on the steps in protein synthesis which are affected by particular pollutants. They may also help in an understanding of the selection pressures which operate in polluted forest populations.

4. Photosynthate translocation.

Studies should initiated both on photosynthate export and mobilization of reserve pools in old and young leaves during and after pollution and/or stress episides. These data with existing knowlege, would provide information that would allow a more integrated view of the response of trees to ambient air pollution.

5. The forest decline syndrome

One difficulty of assessing forest decline is to find a parameter that may be measured in mature trees that will reflect their health status. A review of the various ways to diagnose forest decline was presented in part 2 of this session. The paper by Krouse in the first part of session III provides a compilation of the tree species which are most sensitive to the major pollutants, fluorides, sulphur dioxide and ozone. A case might perhaps be made for selecting sensitive clones of one conifer and one deciduous species from each of the three lists. The six trees could be incorporated into a biological monitoring set. Each year a set of standard clones could be exposed at given sites throughout the EC as part of a monitoring network. This would be similar, in some ways, to the network that principally involves herbaceous plants and is currently used in Germany for pollution assessment. Symptoms on the six sensitive tree clones could be periodically recorded and related to measured parameters on mature trees. At the end of each season the plants could be returned to a central laboratory for physiological assessment and destructive analysis. Such a network would clearly not take account of the important soil dimension related to forest decline. However it might help to assign the forest decline syndrome to a number of distinct categories of causes. It might also lead to the development of a better parameters to diagnose pollution-induced damage both on the experimental clones and on mature forest trees.

IMPACT OF AIR POLLUTANTS ON ABOVE-GROUND PLANT PARTS OF FOREST TREES

GEORG H.M. KRAUSE

Landesanstalt fur Immissionsschutz des Landes NW,
Wallneyer Strasse 6, 4300 Essen 1, FRG

Summary

Impact of air pollutants on forest ecosystems is a well known phenomenon which has been observed for over 100 years in the case of sulfur dioxide and which includes other gaseous pollutants such as HF, NO_x, O_3, heavy metals and, more recently, inorganic acids as well as organic substances. With increasing efforts to reduce the air pollution burden, the pollution environment has changed considerably during the last decade. While sulfur dioxide emissions have stayed more or less constant and will be markedly reduced in the near future, secondary air pollutants such as photochemical oxidants have gained importance with respect to vegetation effects, due to increased emissions of nitrogen and organic compounds as precursor substances and the long range transport of these pollutants. Oxides of nitrogen and sulfur also undergo chemical reactions in the atmosphere to form acidic products which are considered increasingly deleterious to forest ecosystems. The impact of gaseous air pollutants on above-ground plant parts of trees is discussed with emphasis on low concentrations such as those present in forests remote from industrial areas. Special attention is given to air pollutant interactions.

1. INTRODUCTION

Reflecting on the possible causes of the new forest decline, it was once stated that nearly 2000 substances in the ambient air could be involved or should be considered for possible cause-effect relationships. Such a statement has little relevance to the evaluation of potential hazards from air pollution other than a general awareness of the risks associated with human activities. Because it is impossible to consider every single pollutant, it is therefore necessary to focus on our a priori knowledge of relevant pollutants and on their probability of occurrence with respect to space and time in those areas which are endangered.

The pollution situation has changed significantly in the last two decades, for example, in the Ruhr area over the last 15 years (1), in West Germany, and probably in other European countries. While in the 1950s and 1960s most pollution damage was observed in the vicinity of point sources and resulted frequently in acute injury, such cases are of diminishing relevance. The State Institute for Air Pollution Research in Northrhine

Westfalia (LIS) has dealt with over 500 requests to evaluate air pollution injury on vegetation between 1964 and 1986. The relative frequency of vegetation injury is shown in Figures 1 and 2. During the first decade, fluoride and sulfur dioxide induced injury accounted for nearly all cases. Some injury was caused by HCl and some by organics such as phenols or carriers used for pigmentation of cotton (2). Heavy metals were of minor relevance with respect to direct injury on above-ground plant parts, except for single cases during the early 1980s (3). Although incidences of fluoride and SO_2 damage decreased and accounted for less than 20 and 5%, respectively, of the observed injury cases during the last 15 years, injury believed to be caused by air pollutants increased by over 60% but was actually due to biotic or abiotic stress factors. This is indicative of the increased awareness of pollution problems which reflects a more intense awareness of the role of the environment in human life and is clearly associated with the ecological movement in the FRG. Must it be presumed, therefore, that air pollution problems become more and more a specious problem?

The purpose of this example is to point out that air pollution injury to vegetation has not only become more difficult to identify but more difficult to substantiate. In the case of local sources, the pollution factor is of greatest influence, but while it is still dominant on a regional scale, other stress factors reducing plant vitality gain importance (Tab. 1). In the case of new forest decline the pollution environment is still more complex, since former clean air areas are affected supra-regionally and these areas are influenced by various regional pollution sources in low concentrations with differing pollutant configurations. A good example can be given for the Egge-Mountains in Northrhine-Westfalia where, during periods of prevailing westerly winds, the pollution burden comes from the Ruhr area, while during periods of easterly winds it comes from Halle/Jena in the GDR and is associated with rather high SO_2 peaks, especially in winter (4). Still, in these regions, the air pollution factor is just one among many (i.e., climate, soil, pathogens, silviculture, etc.). Although one might anticipate a predisposition of vegetation to air pollution stress, within this noise of environmental stress factors, symptoms are no longer directly associable with air pollution injury and frequently are caused by nutrient deficiency or climate (Tab. 1).

What are now the most important gaseous pollutants affecting trees, especially with respect to potential impacts to forest ecosystems? I would like to focus this question on the Class 1 relationship defined by Smith (5), in which low doses of pollutants are prevalent and the impact is difficult to measure because it can also be stimulatory. Taking the emission parameters of potentially phytotoxic substances into consideration, such as the abundance of sources of a given type, stack height, or flue gas volume as well as the orographic or meterological conditions influencing pollution dispersion, emphasis must be placed on all compounds associated with combustion processes. Although to a certain extent these compounds are fluorides, primarily they are oxides of sulfur and nitrogen and some hydrocarbons, as well as the reaction products of these compounds. The compounds react with each other or with other media such as sunlight, rain or cloud-water in the formation of photochemical oxidants, organic reaction products and acids. These may be summarized as secondary air pollutants.

2. HYDROGEN FLUORIDE

Although reliable measurements of ambient fluoride concentrations are rare, fluoride levels in ambient air remote from sources are considered to be well below 1 $ug.m^{-3}$ (6). Even in the vicinity of urban areas, excluding the industrial agglomeration zone of the Ruhr, annual means rarely exceed 0.2 $ug.m^{-3}$ (7). Susceptible plants may be injured by concentrations as low as 0.8 $ug.m^{-3}$ (8,9), which are 10-1000 times lower than the atmospheric concentrations of other compounds discussed below. Fluoride is non-essential for plant life and is readily taken up by foliage from the atmosphere as well as from the soil by roots (10,37). It accumulates in leaves and needles, preferentially in needle tips or leaf margins (12) due to ready translocation (13,14,15), where the visible injury first occurs. Therefore, even low concentrations over longer periods can be harmful to long living organisms like conifers, inhibiting various metabolic processes such as glycolysis or glycogenesis (20), or undergoing reactions with trace elements to form certain chelates (18). In this context, results from Garrec et al. (21) are interesting because they found that F changes the nutrient balance in conifers. This was seen as a reduction in the Mg and Mn content of needles while Ca and F concentrations increased and N, P, and K remained constant.

Accumulation of F is dependent on factors such as dose, plant species, or plant to plant variation, apart from internal and external growth factors such as light and temperature, which increase the metabolic activity of the plant and thus, F uptake (10,16). Potential risks for vegetation are also increased, because F uptake during the night is high in comparison to SO_2 and HCl (17,18) and will occur, even during winter, presumably through the lenticells, thus resulting in increased F levels in spring (11,19).

In mixed forest stands deciduous plants generally accumulate more F than do conifers (22,23); conifers show greater variation in F accumulation (24) and F-tolerant species show higher F values than do sensitive ones (9). Sensitive tree species are <u>Pinus ponderosa</u> (25), <u>Abies nordmanniana</u> and deciduous species such as <u>Sorbus intermedia</u> and <u>Fagus sylvatica</u> which show severe leaf or needle injury after exposure to 1.3 ug $HF.m^{-3}$ for only several days (14) (Table 2).

Many studies have identified fluoride as the cause of severe tree mortality around industrial sources (26,27,28,29, and others), and it has frequently been observed that even if visible injury is absent, individual resistance against biotic or abiotic stress factors is reduced. Depending on the concentrations present, plant communities ultimately can be changed in composition and structure (30). Interactions with biotic factors were believed to be important in the greater area of the Mount Kitimat alumina reduction smelter, where infestations of saddleback-looper (<u>Ectropus crepuscularia</u>) and budworm (<u>Choristoneura orae Free</u>.) coincided well with fluoride deposition zones. Although no true cause-effect relationship was established, several hypotheses were put forward to explain the interaction: (1) F-induced changes in plant metabolism could have increased the attractiveness to insects, (2) resistance to the insects was reduced because of the latent F burden, or (3) there was selective toxicity of the F emissions to enemies of the attacking insects (24). Wenzel (31) and Thalenhorst (32) reported a similar indirect effect, an increased infestation of <u>Sacchiphantes abietis L.</u>, the gall aphid of spruce.

Fluoride accumulation in needles commonly represents the diagnostic feature for F-intoxication, although fluoride is leached from above-ground plant parts (13,34) and thus the F-content does not increase in proportion to dose (35). F-contents of needles showing characteristic symptoms of new forest decline are generally well below 10 ug F g^{-1} dry weight even for older needles (33), throughout FRG with little spatial variation, and have therefore to be considered insignificant. This applies too, despite the high phytotoxic potential of F and possible interactions with other pollutants such as NO_2 (36), NaF in soil, SO_2 etc. (37).

3. OXIDES OF SULFUR

Natural SO_2 concentrations in remote areas are well below 5 $ug.m^{-3}$ and annua means in Europe in clean air areas range between 5 and 25 $ug.m^{-3}$. However, since in Europe distances between rural and source areas are much shorter than in North America, concentrations frequently exceed 25 $ug.m^{-3}$ and are subject to great variability. The LIS is operating 65 monitoring stations within the TEMES network in Northrhine-Westfalia to automatically sample ambient air concentrations of SO_2, NO_2, NO, O_3, and suspended particles as well as meteorological data (202). Three of these stations are located in forested areas in Northrhine-Westfalia (Eifel, Egge-, and Rothaar-Mountains) and continuous monitoring has been carried out since 1983 (4). Figure 3 compares SO_2 concentrations between the Ruhr area and two of the sampling sites. Annual mean SO_2 concentrations at the sites are in the range of 30-40 $ug.m^{-3}$ and are 30 to 40% lower than those in the Ruhr area. Half hour means (max./month) are also lower in the remote areas. However, during the wintertime with easterly winds and high atmospheric stability, peaks of up to 1200 $ug.m^{-3}$, comparable to those of the Ruhr area, have been recorded (Fig. 3). In other areas with high elevations where forests are severely affected, such as the southern Black Forest, southern Bavaria, and parts of Switzerland, annual means are, on the other hand, well below 10 $ug.m^{-3}$ (38,39).

Sulfur dioxide is probably the best known phytotoxic pollutant and its deleterious effects on forest trees have been known for more than 100 years (40,41,42). Many excellent reviews on cause-effect have been published (18,43, 44,45). In most cases sulfur dioxide exposure was very much related to point sources and the identification of injury was possible due to the specific symptomatology, individual plant tolerance, S-accumulation on leaf tissue, and a clear foliar S concentration gradient with increasing distance from the source. Many publications have described SO_2 effects on woody plants, and the degree of susceptibility has been assessed by noting the magnitude and extent of foliar injury after exposure to SO_2 at defined concentrations. The sensitivity of woody species to SO_2 is provided in Table 3 (46,47).

Tree response to SO_2 is rather complex and involves steps such as pollutant uptake through gas phase interactions at leaf and stomatal surfaces (49,50,51). Factors influencing stomatal opening, such as light, leaf surface moisture, relative humidity, temperature, and soil moisture play a major role in plant sensitivity by limiting passive entry of SO_2 (52,53). Movement of stomata can be altered by SO_2 (54,55) which can lead to a possible increase in SO_2 uptake and transpiration with the potential for additional water stress (56). After contact of the SO_2 with wet cellular membranes, biochemical and/or physiological changes occur at the cellular level, which are of particular interest at low SO_2 levels, since they may serve as primary indicators of latent injury (57,58). These include interferences with enzyme activity (59), energy translocation (60), lipid biosynthesis (61), amino acid content and chlorophyll synthesis (62,63,64), thereby inhibiting photosynthetic processes (65,66,67). In addition, in remote areas during pollutant free periods (4), plant repair mechanisms (homeostasis) (68,69) are operative. The response of trees to SO_2 however, may not necessarily be deleterious because deposited sulfur can act as a fertilizer (70,71,72), but so far this has been shown only for agricultural crops. According to Keller (73,74,75), except for sites with extremely poor soil-sulfur content, beneficial effects on trees are unlikely because their S requirements are relatively low.

In reviewing the literature for SO_2 effects on forest trees, few studies are available in which SO_2 effects are described for the ambient concentrations which are encountered in remote areas. Frequently, SO_2 concentration measurements are even lacking and, therefore, no cause-effect relationships can be established. Furthermore, interference with other pollutants often present at the same time (48) is not given sufficient consideration.

In a major investigation in the vicinity of a SO_2 point source near Biersdorf, FRG, various tree species were exposed in the field at increasing distances from the source (46). Generally, growth was not affected at concentrations below 50 $ug.m^{-3}$ during the four year experiment. Other studies near point sources revealed similar results (76, compare also 22). In the vicinity of a sour gas processing plant in western Canada, changes in a conifer forest ecosystem were identified at ambient concentrations currently encountered in Europe (337,338). These included metabolic alterations such as changes in energy allocation (339). Bonte et al. (78) showed in an outdoor, non-chambered fumigation experiment that two species of larch had marked growth reductions when SO_2 concentrations were 60 $ug.m^{-3}$. Exposure of sensitive conifer species to a mean annual SO_2 concentration of 29 $ug.m^{-3}$ in a rural area in England caused growth reductions after two years, while more tolerant species performed without noticeable loss in vitality (79). In Finland and Czechoslovakia, field surveys were carried out in forests to assess possible effects of low levels of sulfur dioxide. According to these studies, summarized by Huttunen for a World Health Organization (WHO) criteria document (80), SO_2 concentrations ranged between 20 and 40 $ug.m^{-3}$ and resulted in elevated foliar S-levels and even chronic injury in the leaf tissues of conifers such as _Picea abies_, _Pinus sylvestris_, and in broadleaf species such as _Betula pubescens_. These effects were frequently observed in the springtime, being preceded by chronic injury during winter (86,87) and were especially pronounced after harsh winters, while plant parts covered by snow were unaffected (88). The decreased frost resistance appeared to be caused by increased electrolyte and sugar leakage from shoots (90,91). Extended field studies were also carried out by Materna in Czechoslovakia (82,83,84). As in the Finish studies, 20-40 $ug.m^{-3}$ of SO_2 was blamed for the degradation of large forested areas of _Picea abies_. However, according to Materna (85) the process of forest decline is slow and direct destruction of forests is not expected. Materna stresses on the other hand, that sensitivity of these ecosystems is very much influenced by other factors such as soil nutrient imbalances, climate, unusually harsh winters, or rapid changes in temperature or drought situations, which can trigger or intensify SO_2 effects. These studies raise the question of whether it is even possible to attribute observed effects to such low concentration air pollution stress, when pollution-free areas with comparable ecological features are not available as points of reference and, as stated in the beginning, the influence of numerous other stress factors cannot clearly be distinguished from the air pollution factor.

Another aspect frequently discussed is the role of peak episodes of sulfur dioxide and their possible influence on plant performance. At the experimental station of the LIS in Essen-Kettwig, Picea abies and Fagus sylvatica were fumigated continuously for one year either with 100 ug.m^{-3} SO_2, or with 60 ug.m^{-3} SO_2 accompanied by SO_2 peaks every 14 days at 500 ug.m^{-3}. While plants growing in the constant, high SO_2 regime showed slight growth reductions and diffuse needle discoloration at the end of the experiment, no visible injury occurred in the other group exposed to regularly occurring SO_2 peaks (Krause, unpublished data). Field exposure of Pinus sylvestris and Picea sitchensis to a fluctuating regime of sulfur dioxide with 100 ug.m^{-3} as the annual mean and peak concentrations of either 300-400 ug.m^{-3} or 750-900 ug.m^{-3} for mean durations of 20 and 5 h, respectively, in comparison to plants kept at a constant concentration of 100 ug.m^{-3} under controlled conditions, led to the conclusion that the main effects were attributable to mean, rather than peak, concentrations (256). However, this question has not definitively been answered because peaks tend to be seasonal, occurring in winter, in contrast to the low means during summer months.

Plant-pathogen interactions with SO_2 such as those described earlier for HF are possible but thus far have not been described for SO_2 and forest trees (126). Due to the antiseptic effect of sulfur dioxide, however, infestations of various fungi were reduced in the presence of fairly high SO_2 concentrations which, for the most part, exceeded 100 ug.m^{-3} as the annual mean (126).

To date, foliar S analyses of needles carried out in forest areas showing new decline symptoms in FRG have revealed that the foliar S contents, including those of older needles, rarely exceeded normal concentrations of 800-1200 ug.g^{-1} dry weight (33). These data were based on total foliar sulfur content and not on foliar sulfate-sulfur context which is a much better parameter for evaluating the influence of sulfur dioxide. Foliar sulfate-sulfur content should be taken into consideration in future chemical analyses of needles.

Besides physical and biological stress factors which predispose tree response (93), the pollution environment also consists of other pollutants occurring simultaneously or alternatively in various combinations. This factor will be discussed later in this paper.

4. OXIDES OF NITROGEN

Concentrations of NO_2 and NO are usually very much lower than SO_2 in remote areas, despite the fact that emissions of nitrogen oxides have increased considerably during the last 15 years in comparison to SO_2 emissions (94). For the monitoring sites in forested regions in Northrhine-Westfalia, NO_2 concentrations show annual means of 20-25 ug.m^{-3} while those of NO rarely exceed 5 ug.m^{-3} (4). These values are approximately 50% of those measured in the Ruhr area. Maximum daily averages are about 60-90 ug.m^{-3} (Figures 4 and 5). NO_2 concentrations at elevated locations like Brotjacklriegel in Bavaria or Schauinsland, Black Forest are generally below 20 ug.m^{-3} even during winter periods (95). NO_2 concentrations in a forested area in Sweden were below 2 ug.m^{-3} for the summer months and below 5 ug.m^{-3} during the winter, with maximum daily mean concentrations of 26 ug.m^{-3} (81). It seems, therefore, that the importance of nitrogen compounds is based more on their function as precursors for ozone formation than on their own phytotoxicity, which is comparatively low (96).

With respect to direct effects, nitrogen dioxide is of greater relevance than nitrogen monoxide, because the concentrations of NO_2 in the air are much higher than those found for NO (4,81), and NO_2 has a greater phytotoxic potential (203,204). Like the other gaseous compounds, NO_x can be taken up via stomata and the same interactions as discussed previously for sulfur dioxide apply in general for NO_x. Uptake rates are proportional to the concentration gradient between the atmosphere and the interior of the leaf and are regulated by stomatal resistance (96). Uptake is influenced by factors such as light (97) and is proportional to the transpiration rate. Generally, uptake is enhanced by factors influencing stomatal aperture such as high relative humidity or low CO_2 concentrations (99). Nitrogen contents of plants increase with increasing NO_x concentrations (98, 100,101) and plants can utilize atmospheric nitrogen as fertilizer (102); therefore, beneficial effects on growth performance should not be overlooked. Since soils in northern parts of North America and Europe are low in nitrogen, atmospheric deposition of N is likely to improve forest production (201), probably counteracting low concentration SO_2 stress, as suggested in Czechoslovakia (85), but this is a controversial point.

Not much is known about the impact of NO_x on cellular function. Interactions with ferredoxin have been discussed (104) but are questionable because NO would have to pass through several biomembranes before such reactions could take place (105). Due to the high reaction potential of nitrogen oxides, the primary site of action is believed to take place at the biomembranes. Another principal mechanism is the reaction of nitrogen oxides with the wet surfaces of membranes, leading to acid formation especially when the buffering capacity of the leaf is limited (108). Furthermore, accumulation of nitrite due to reduced nitrite reductase activity (109) has been associated with leaf injury (108,110) as a result of a decreased detoxification potential for nitrite (111). Similar mechanisms are thought to be involved in the relatively high sensitivity of roadside trees to NO_x in urban areas (112).

There are several reports in the literature describing acute or chronic injuries in the vicinity of nitrogen oxide sources such as fertilizer production plants (113,114) or ammunition plants (115), and various tree species such as *Pinus strobus*, *Liriodendron tulipifera* or *Pinus sylvestris* have been affected. The NO_x concentrations, however, were high in all these cases.

Undoubtedly, nitrogen input (dry and wet) in forested areas has increased considerably during the last decade. Kenk (116) reports N-deposition rates of 40 $kg.ha^{-1}.a^{-1}$ for certain areas of the Black Forest. Even higher loads were reported from agricultural areas in northern Germany with 80 $kg.ha^{-1}.a^{-1}$ (Werner, pers. comm., 117). Higher N-input leads not only to a greater growth increment in trees or stimulated growth such as that observed in *Pinus strobus*, which was accompanied by an increased N-concentration in the needles and roots (328, Kenk, cited in 118,119; Arovaara et al. cited in 120), but can also change the floristic composition in forests (121,122,123). Furthermore, frost hardiness of trees may be lowered (124) when nitrogen levels in needles exceed 1.8-2.0% of total needle N (Aronsson, cited in 120; compare also Davison et al. 125).

Generally, it can be stated that elevated deposition of nitrogen oxides on above-ground plant parts in concentrations discussed above can be beneficial, but certainly it is associated with definite risks such as reduced frost hardiness or increased potential for infection by biotic stresses due to softer leaf tissues, or unbalanced internal nutrient ratios (124). These risks are more relevant with respect to acidification processes of the soil.

5. PHOTOCHEMICAL OXIDANTS

Effects of photochemical oxidants on vegetation were first observed more than three decades ago in the greater Los Angeles area (127). The following discussion will focus on ozone, the most prevalent and most studied compound whose effects are best understood. Other photochemical oxidants such as Peroxyacetylnitrate (PAN) have a higher phytotoxicity than ozone but do not appear to be a problem for Europe since concentrations are very low (92,94). However, additional evaluation and research of PAN are warranted.

Ozone as a secondary air pollutant is formed under the influence of UV-light in photochemical reactions involving NO_x and reactive hydrocarbons. Chemical processes in the atmosphere are more or less well understood (128,129,130) and lead to a mixture of oxidizing air pollutants of which O_3 is the major component. Unlike the other pollutants in this discussion, O_3 is the only component which, because of its specific mechanisms of formation and decomposition (129), has been found in higher concentrations in areas of greater altitude (500 to 1500 m above sea level) remote from urban and industrial sources (131,132,133). In some regions of southern Germany and the German Democratic Republic (GDR), O_3 concentrations increased between 1967 and 1982 by 50 to 80% and by as much from 1981 to 1982 as in the two preceeding 7 year periods (134,135). In the past 25 years, O_3 concentrations have been found to increase even during the winter (136).

In 1980, the monthly mean O_3 concentration at the Schauinsland station in the southern Black Forest was 113 $ug.m^{-3}$, which is in the range of concentrations measured in the United States (132). In severely affected mountainous areas of Northrhine-Westfalia, O_3 concentrations are generally higher by a factor of 2.0 to 2.4 compared to the Ruhr area (Figure 6) and can reach monthly means of 100 $ug.m^{-3}$ during the summer, as in August 1984 (133). High O_3 concentrations, with peak values over 600 $ug.m^{-3}$ were registered under favorable weather conditions in 1976 (129).

The ozone concentrations in elevated areas of FRG discussed above are high enough to injure sensitive plant species (132,138). These levels have also been measured in other parts of Europe (139,140,141). Besides acute injury to sensitive plants, more resistant tree species such as fir and spruce may even be affected on the cellular level and in combination with other pollutants as well as site factors (132,137). The impact of ozone on vegetation has recently been well documented by Guderian et al. (142).

The effects of O_3 on plant metabolism can be described as follows: tree response to ozone starts with the diffusion of the gas from the atmosphere into the leaf through the stomata. Uptake is controlled by leaf resistances (143) as well as stomatal density and conductance (144,145), and is generally dependent on physical, chemical, or biological factors involving the transition between gas phase and liquid phase movement into the cells (69). In the liquid phase, ozone will readily undergo transformation, yielding a variety of free radicals which will then react with cellular components (69). Sensitivity of plants is modified, therefore, by any factors influencing stomatal aperture such as light (146,147), relative humidity (148), and soil moisture (150). Besides genetic (152,153) or developmental factors (154), soil fertility (151) or chemicals such as herbicides, fungicides (126) etc. can also influence stomatal aperature. When plants are unable to repair or compensate oxidant induced perturbations, injury on above-ground plant parts occurs and can be characterized for gymnosperms as destruction of the mesophyll cells leading to chlorotic bleaching of the needle tip or mottling of the youngest needles (155), frequently accompanied by an unspecific needle cast (156,157). Ozone injury in angiosperms starts with the destruction of the palisade parenchyma cells, resulting in bleaching and minute necrotic stiples in the intercostal areas (155).

However, ozone concentrations observed in Europe are not likely to be high enough to cause visible injury on dominant tree species and, therefore, effects at the cellular level are of special interest. The primary site of action for ozone is the cell membrane, leading to changes in membrane permeability of the plasmalemma as well as the chloroplasts, as shown by changes in flux of organic and inorganic plant metabolites (69,158,159,160,161). These phenomena have been associated with increased leaching of essential nutrients (161,178), accompanied or even enhanced by ozone induced weathering of the cuticle (162). Photosynthesis is very frequently reduced (164,165,166,167,168,192,327) either indirectly by closure of the stomatal aperture resulting in reduced CO_2 uptake (111,169), or directly by damage to the chloroplasts (170,171). Another effect is ozone induced reduction in assimilate translocation to roots with a resulting decrease in root size and fewer stored reserves, with the potential for increased sensitivity to frost, heat, and water stress (154,172,173,174). Other effects, such as reduction in chlorophyll content (168,169,175,192,327) or chlorophyll bleaching in the presence of light due to oxidation processes (111,168) have been observed. Biochemical perturbations are a further expression of subtle injury reactions, frequently indicating premature senescence which occurs under chronic O_3 pollution (142,258,284). These include the oxidation of sulfhydryl groups (176), and changes in the content of soluble sugars, starch, phenols, ascorbic acid, amino acids, and protein, as well as interferences with enzymes such as the nitrate- or nitrite-reductase involved in nitrogen metabolism (172,177,178).

Our knowledge is rather limited in terms of the many plant species indigenous to forest ecosystems (179). Ozone effects are much more difficult to evaluate than those of SO_2, because there are no point sources to allow observations along a concentration gradient. Furthermore, it is difficult to determine if tree responses are cumulative and the result of a number of influencing chronic stress factors, unless they can be traced back to specific biotic diseases or pollution exposure. In areas with chronic ozone exposure (and I think many areas of Europe should be ranked in this category), decline in vigor of trees and forest ecosystems is a commonly observed response (179,180,181,258). Symptoms of chronic decline differ markedly from acute visible injury and include, according to McLaughlin (180): (1) premature senescence with cast of older needles in autumn, (2) reduced assimilate storage in roots at the end of the growing season and reduced resupply capacity in spring, (3) increased reliance of new needles on self support during growth, (4) shorter new needles and thus reduced assimilate production, (5) reduced availability of photosynthates for homeostasis, and (6) premature cast of older needles.

The most complete study of ozone effects on forest ecosystems was done within the San Bernardino National Forest near Los Angeles, California where, due to specific climatic, orographic, and emission characteristics, elevated ozone levels of up to 580 $ug.m^{-3}$ were present (182). Tree species sensitive to ozone were listed as: Pinus ponderosa, Pinus jeffreyi, Abies concolor, Quercus velutina, Librocedrus decurrens and Pinus lambertiana. Foliar injury and premature leaf fall coincided with decreased rates of photosynthesis, reduced radial growth, tree height and seed production, as well as retarded nutrient retention (181,183). Injury to Pinus ponderosa occurred at concentrations of 100 to 120 $ug.m^{-3}$ for

24 h. However, sensitive tree species were not eliminated by the photo-oxidant burden, but by their O_3 induced predisposition to insect infestations such as bark beetles.

The other predominant tree species affected under ambient exposure of ozone is eastern white pine (Pinus strobus), as has been seen in many parts of the northeastern United States, such as the Appalachian region (184,185,186). White pine showed great variability in sensitivity and the trees were classified as sensitive, intermediate and tolerant based on visible injury, needle length and needle retention. Growth of sensitive and intermediate trees was 25 and 15% less, respectively, than tolerant trees but progressive growth reductions over the previous 24 year period was found in all sensitivity classes when O_3 concentrations had ranged between 100 and 140 $ug.m^{-3}$ on a recurring basis with peak episodes of 200 to 400 $ug.O_3\ m^{-3}$ (187).

Similar effects on white pine were found by McLaughlin et al. (180) in eastern Tennessee. Between 1962 and 1979, annual average growth was reduced by as much as 70% in sensitive species as compared to tolerant ones. The cause of these growth effects was attributed to chronic ozone exposure, frequently in the phytotoxic range (>160 $ug.m^{-3}$). In addition to growth reductions, premature senescence, and lower photosynthetic rate, perturbations in the processes of carbon allocation were also observed.

In other areas of the United States, such as the southern Sierra Nevada in Central California, and Indiana and Wisconsin, symptoms on western conifers have also been observed (183,188,189).

In another field study with filtered versus non-filtered air using open top chambers, Duchelle et al. (190) showed that summer mean concentrations of 86 $ug.m^{-3}$ produced visible injury at the end of the growing season on species such as Liriodendron tulipifera, Liquidambur styraciflua, and Fraxinus pennsylvanica, while those in filtered chambers showed no effect. Betula pendula, Fraxinus excelsior, and some species of Fraxinus americana were found to have ozone specific injury after one growing season in rural southeastern England using the same type of exposure system, while Fagus sylvatica and Quercus robur were more tolerant (191). According to Ashmore et al. (132), O_3 concentrations were markedly below summer means of 100 $ug.m^{-3}$ measured at the Schauinsland station in the Black Forest (94). On the other hand, fumigation experiments with Picea abies and Abies alba revealed that these major, European native species were relatively more tolerant because O_3 fumigation with 150 and 600 $ug.m^{-3}$ for 28 and 56 days, respectively, produced some visible injury. After 48 days of continuous fumigation with 150 $ug.m^{-3}$ O_3, unspecific chlorosis on the youngest needles was observed on Picea abies with light yellow discoloration of the upper parts of the crown. Ozone concentrations >150 $ug.m^{-3}$ produced marked mottling after 5 weeks and inhibited photosynthesis as well as transpiration (168). Other examples of the effects of experimental ozone fumigations are compiled in Table 4 (132) while the most complete listing of woody plant sensitivities to photochemical oxidants has been published by Davis and Wilhour (47) and a summary of this sensitivity ranking is given in Table 5.

Of all the pollutants discussed so far which are present in remote forested areas, ozone probably has the highest degree of phytotoxicity based on concentration and duration of exposure. It is, therefore, reasonable to assume that ozone is a major contributing factor in the decline phenomena observed in forests today.

6. EFFECTS OF ACIDIC DEPOSITION

Acidic deposition has increased substantially in most parts of the industrialized world, causing decreased pH values of precipitation. Much attention has been given to the soil effects caused by acid rain which influence the vitality of long living organisms such as trees by direct root damage or nutrient disturbances in the rhizosphere by changing the buffering capacity of soils (205). Apart from soil effects, wet acidic deposition in the form of rain, fog, dew, snow, or cloud water interception, may also have deleterious effects on above-ground leaf organs. The chemical constituents of wet deposition in remote areas are mainly ions of hydrogen, sulfate, and nitrate as well as ammonium, but other substances such as trace elements or organics scavenged from the atmosphere may be present (206).

The deposition of these substances to vegetation is dependent on the roughness of the response surface influencing deposition velocity and on the hydrophobic or hydrophilic properties of leaf organs, the latter being greatly influenced by the epicuticular waxes. Physical and chemical properties of waxy layers change with time and they become more hydrophilic during aging. This natural weathering process is influenced by the external environment such as wind, light, temperature, and nutrient status, among other factors (207,208,209, 210,211). Weathering processes can further be stimulated by air pollutants like sulfur dioxide, even at concentrations of 40 $ug.m^{-3}$ (annual mean) (212,213,335). Other pollutants such as ozone or acidic deposition act in a similar manner (33,214) and have also been associated with increased predisposition for fungal attack (215). Destruction can further be enhanced by deposition and evaporation of acidic substances on leaf surfaces, increasing the hydrogen ion concentration considerably by up to one pH-unit since acids such as H_2SO_4 and HNO_3, as well as their anhydrides, concentrate because of their limited volatilization (216). This can be especially important for conifers when water drops accumulate at needle tips. _Pinus strobus_, subjected to nightly acidic mist treatments (pH 2.6, 3.6, 4.6, 5.6) showed subtle but significant morphological changes in needles which were probably attributable to nutritional effects of nitrogen and sulfur compounds derived from the HNO_3 and H_2SO_4 in acidic mists (336). However, plants are able to buffer acidity on leaf surfaces by ion exchange (217,218,219,220,221), which explains the increased leaching of inorganic ions, proteins, amino acids, and carbohydrates by plants (222,223). Therefore, it is likely that acid deposition inhibits growth by excessive loss of nutrients and metabolites although healthy plants are able to compensate leaching losses by accelerated root uptake on suitable soils (224). The amount of nutrient and metabolic loss, however, will vary considerably from year to year with the precipitation pattern, so long-term observations are necessary to evaluate possible impacts on tree performance, especially since acidic deposition can also lead to an increased nutrient input (328). Experiments with simulated acid rain reveal a decrease in calcium and magnesium content of leaves (225,226,227) while no changes were observed for N, P, and K (226,228). Extensive leaching experiments were carried out under controlled conditions with _Picea abies_ and substantial leaching of magnesium, calcium, copper, chloride, sulfate, nitrate, and ammonia occurred at pH 4 using a simulated rain mixture close to the chemical composition of ambient rainfall. With decreasing pH, leaching was increased and elements like zinc and iron, not leachable at pH 4.0, were leached at pH 3.5 (162). No pH effect was found for sulfate and nitrate. The results indicated an ion exchange between cell wall-bound cations and the protons of the rain solution. Similar mechanisms have also been postulated by others (223,229,231).

Numerous publications have dealt with the question of leaf injury and yield loss caused by acid rain and visible injury has been observed in agricultural crops at pH <3.0 (230,231,232,233,234,235,236, and others). Most of these results, however, are questionable because pure diluted inorganic acid solutions were used to simulate acidic rain but were not comparable in chemical composition to ambient rain. Other results have shown beneficial effects of acidic rain on vegetation, leading to an increase in growth and yield (137,138,139, and others). Visible injury was found to occur on needles of Picea abies and Abies alba only when plants were fogged for 45 days consecutively for 3 h/d, at pH below 2.75 using a close to ambient fog solution (240). No effects were observed on growth, photosynthesis, or chlorophyll content when seedlings of Quercus rubra and Acer saccharum were exposed to acidic rain of pH 5.6 and 3.0 (12.5 mm of rain, twice a week, sulfate/nitrate ratio 2:1) for 2-3 months (327). In other experiments carried out by Onger and Teigen (241), visible injury occurred when acidic rain of pH 2.5 was applied to Picea abies at a constant temperature of 24°C. Temperatures below 18°C reduced injury symptoms markedly.

As yet, there is no clear answer to whether acidic deposition at ambient pH values normally present is deleterious to above-ground plant parts of forest trees, because long-term studies are lacking. Furthermore, no real answer can be given as to what extent internal and external growth factors influence foliar leaching. Since the amount of leachates increases with decreasing tree vitality, as observed in leaching studies (161,168), special attention should be given to soil-nutrient/plant interactions. The aspect of combined effects of acidic deposition with other pollutants will be discussed later.

7. ORGANIC COMPOUNDS

Not much information is available on the effects of organic air pollutants on vegetation although large quantities of organics (3 x 106 $t.a^{-1}$) are emitted (248), mostly due to the widespread use of solvents, fuels, and the emissions of other chemical processes. Nature itself is part of this pollution burden, since forests emit large quantities of hydrocarbons (HC's) commonly known as terpenes. These products can rapidly be consumed by fast reactions with OH radicals and are thus linked with the formation of photochemical oxidants under favorable weather conditions (242). Measurements of halogenated HC's were carried out by Frank (243) in the Swabien Alps, showing concentrations of 3.2 to 7.0 $ug.m^{-3}$ for freons and 0.4 to 1.3 $ug.m^{-3}$ for chloroethene, trichloroethene, and tetrachloroethene. Polycyclic aromatic hydrocarbon measurements (PAH's) were carried out from October 1983 to September 1984 by discontinuous measurements in the Egge Mountains and the Ruhr area (4). No significant differences were found between these regions, and PAH concentrations in the Egge Mountains were comparable with those in other remote areas.

PAH's and halogenated hydrocarbons (HHC's) are lipophilic substances which concentrate in the waxy layers of leaf surfaces and can also react with membrane lipids (243). This pathway has been applied for the determination of PAH's on vegetation by using the large waxy surface area of cale Brassica oleracea var. acephala as a bioindicator (244,245,246). HHC's can be activated by UV-light to reactive triplet-states or free radicals, and may decompose to highly toxic species such as phosgen, dichloroethyne, chloracetylchloride, and others, and interfere with plant metabolism

(243). Earlier experiments with PAN have also shown a remarkable relationship between light, PAN, and vegetation injury. It was shown that a wave length of 600 nm could produce PAN injury independent of light intensity, while in the presence of 660 nm and 700 nm no such injury was observed (247). Fumigation under controlled conditions in the laboratory with 14 $ug.m^{-3}$ tri- or tetrachloroethene for 5 hours in the presence of UV-light resulted in a marked change in the pigment content of Picea abies needles, most strongly attenuating chlorophyll a and B-carotene concentrations (243). Changes in plant pigments were also observed in field studies in which species of Picea omorica were exposed to the same organic compounds in the presence of sunlight (248). However, hardly any concentration measurements of the organics were made and the number of observations was limited. Pigments were also found to be changed in trees showing new decline syndrome versus healthy ones at a site in the Black Forest. This was especially true for substances like camphene or limonene which were markedly increased in injured Picea abies needles (249,250). Furthermore, a clear light dependency of injury in trees affected by the new forest decline has frequently been described (94 and others). It is interesting to note, however, that this light dependency of injury was also observed when Picea abies and Fagus sylvatica were subjected to ozone fumigation of >150 $ug.m^{-3}$ (168).

Extensive fumigation experiments with various organic substances were carried out by van Haut et al. (251) and Krause and Hoeckel (252). The concentrations of organic compounds which caused visible injury in horticultural and agricultural plants were mostly found to be in the mg-range and, hence, were incomparable to the much lower level concentrations encountered under ambient conditions (Tab. 7).

Recently, organic lead compounds found as constituents of wet deposition were associated with injury to above-ground plant parts of trees because they revealed a high toxic potential in cell culture systems (253, 254). However, as shown by Unsworth (255) and Krause (257), the results were questionable with respect to the chemical procedures as well as the temporal and spatial distribution of occurrence and, therefore, are not considered very significant in this context.

In essence, we know that HHC's and HC's have increased in concentration in rural areas of industrialized states, although quantification is yet rather crude, and further measurements, including concentrations in wet deposition, are necessary. Future research should also focus on determining phytotoxic potentials of organic compounds, giving special reference to long-term exposures using realistic concentrations and to possible interactions in the presence of ozone.

8. COMBINED EFFECTS OF AIR POLLUTANTS

The atmosphere usually contains a complex and dynamic mixture of pollutants occurring simultaneously or sequentially with great regional variation (259). Therefore, it is difficult to evaluate vegetation response in a given experimental design, since only limited pollution regimes out of many potential ones are reflected. Plant response to air pollutant mixtures can be additive, less than additive (antagonistic) or greater than additive (synergistic). The last response type has particularly led to great concern for forest ecosystems. The pollution situation in remote areas is mostly characterized by the presence of SO_2, NO_2, and O_3, as well as acidic deposition. Combinations of these pollutants are discussed shortly. No reference will be given to interactions with HF (260,261) since they are of minor relevance to the topic.

Uptake of pollutants, whether they are single or in combination, is higher in sensitive plant species than in resistant ones (292), due to a wider stomatal aperture in sensitive species (293). Antagonistic effects of mixtures such as NO_2 and SO_2, frequently observed as plant response, are probably related to increased stomatal resistance resulting in a decreased pollutant uptake (294,295). Similar effects are observed for combinations of SO_2 and O_3 where, in the presence of both gases, stomata tend to close tighter than with either of the compounds alone, although O_3 uptake is enhanced at first by SO_2-induced stomatal opening (262,263,264). Apart from other factors, uptake is further modified by increased relative humidity which leads to increased injury in the presence of both pollutants as compared to each pollutant alone (265,266). Other experiments show, on the other hand, that synergistic effects are not necessarily related to stomatal resistance (150). According to Elkiey and Ormrod (267,268) pollutant uptake from singly applied pollutants is considerably greater than from mixtures but greater phytotoxicity may result from a greater number of molecules being absorbed from mixtures than from single components (142). Combinations of NO_2 and O_3 reduced the transpiration rate similarly to O_3 alone, while no such effect was observed from combinations of NO_2 and SO_2 (296,297). Photosynthesis was inhibited by combinations of O_3/SO_2 (298,299), NO_2/O_3 (300), as well as NO_2/SO_2 (301,302). Other physiological changes in photosystems I and II were affected by combinations of NO_2/O_3 while NO_2/SO_2 influenced photosystem I only (303). Interferences in nitrogen metabolism occurred in the presence of SO_2 and NO_2 and it was concluded that the presence of SO_2 leads to an inhibition of nitrite reductase activity (304).

8.1 SO_2 AND O_3

Combined effects of SO_2 and O_3 show that injury symptoms are related more to ozone than to SO_2 (269,270,271); however, exceptions are also reported (272,273,274). Visible effects normally associated with PAN were found on Petunia leaves exposed to SO_2 and O_3 (275), while no clear response showing either single pollutant injury or a new type of injury was found for eastern white pine exposed to SO_2 and O_3 (276,277). As well, no clear answer with respect to a growth effect response could be given. While most herbaceous plants show additive or synergistic effects when exposed to SO_2 and O_3 (278,279,280, 281,282,283,284,285), woody species show slightly less than additive or even antagonistic responses in plant growth (142). Exposure of cottonwood, yellow poplar, and hybrid poplar to SO_2 and O_3 caused significant growth reductions but the effects were less than expected from responses to single pollutants (286,287). Fumigation of Ulmus americana with high concentrations of SO_2 (2 ppm/6 h) and O_3 (0.9 ppm/5 h) singly and in combination resulted in earlier visible injury and retarded leaf expansion in the combined treatment, but total effects at the end of the experiment could not be differentiated from single pollutant treatments (288). No growth effects were found when hybrid poplar was fumigated with a mixture of SO_2 and O_3 at a dose resembling ambient concentrations (<0.01 ppm O_3 and 0.4 ppm SO_2) while concentrations greater than ambient (0.06-0.08 ppm O_3, 0.06-0.11 ppm SO_2) resulted in reductions in growth, dry matter accumulation, and leaf longevity. However, no combined effect of both pollutants was observed and injury was mainly related to ozone (289).

Combined effects of both pollutants have been observed at the cellular level, however, which do not necessarily result in visible injury or growth reductions. Picea abies and Abies alba were exposed in open top chambers for two years at SO_2 concentrations of 30-50 $ug.m^{-3}$ and 15-30 $ug.m^{-3}$ during winter and summer, respectively. O_3 was applied during sunny days for only 12 h/d at concentrations of 80-100 $ug.m^{-3}$. It was shown that antioxidant formation (glutathione and vitamin E) increased significantly in the SO_2/O_3-treatment in comparison to single pollutant exposure (290). In controlled fumigation experiments with SO_2 and O_3, needles of Picea abies showed marked changes in the concentration of phenolic compounds and tannins in the combined treatment versus treatment with a single pollutant. No differences, however, were observed with respect to morphological changes in the single pollutant treatments while in the combined treatments morphological change was detectable (291).

In summary, chronic exposure to SO_2 + O_3 reduces growth of woody plants to varying degrees, but effects seem to be less than additive with yield reductions occurring at concentrations of about 0.05 ppm O_3 and 0.05 ppm SO_2 (142).

8.2 EFFECTS OF SO_2, NO_2 AND O_3 EXPOSURE, SINGLY AND IN COMBINATION

An excellent, recent review of the interactions of SO_2, NO_2, and O_3 by Guderian and Tingey (96) will be referred to in this section.

Plants show no specific visible injury pattern when exposed to SO_2 and NO_2. Combined exposure to both gases has sometimes produced O_3 related symptoms (305) or new symptoms (306,307). Freer-Smith (308) reported injury symptoms of deciduous trees exposed to a mixture of NO_2 and SO_2 to be similar to those produced by the compounds alone.

Combined effects of NO_2 and O_3 resulted in chlorotic mottling in Pinus taeda and were occasionally accompanied by tip-burn (309). Low levels of NO_2 and SO_2 increased early senescence in poplar trees, accompanied by premature leaf drop, irrespective of the addition of ozone (310). Guderian (311) reported that the combined effect of NO_2, SO_2, and O_3 was yellowing of needles of Picea abies, and when nutrient deficiency was present at the same time, the symptoms showed remarkable similarity with those of the new forest decline (94).

Visible injury symptoms have occurred in many ornamental and agricultural species fumigated with NO_2/SO_2 (269,307) but reactions are plant specific and therefore effects are not always observed (312). Injury symptoms appear to be related to low concentrations of both pollutants in the mixture while higher concentrations cause no combined effect (269,294,307,312). Similar results were obtained for combinations of NO_2 and O_3. Experiments carried out with poplar species fumigated with SO_2, NO_2, and O_3 singly and in combination revealed a less sensitive reaction to combinations of SO_2/NO_2 than to NO_2/O_3 or $NO_2/SO_2/O_3$. It seems that combinations of NO_2/O_3 are more important with respect to leaf injury than NO_2/SO_2 (311).

Continued exposure of agricultural plants (Phaseolus vulgaris, Lolium multiflorum, and Phleum pratense) to SO_2/NO_2 resulted in marked growth reductions in comparison to single applications of the same concentrations (297,313). Other agricultural plants such as Glycine max, Raphanus sativus, Poa pratense, Lolium prenne, Festuca rubra, Agrostis alba, Agrostis palustris, and Agrostis tenuis reacted similarly (306,318,319,321). However, low concentrations of both gases produced either slight (320) or no growth effects (315,316,317), and when applied singly, sometimes even a stimulatory response was seen (320). In general, this also applies to combinations of $SO_2/NO_2/O_3$ (305,322,323).

Few studies are available on the effects of pollutant combinations on trees. Summer fumigations of Populus nigra with SO_2 and NO_2 (0.11 ppm SO_2/NO_2) reduced plant growth more than single pollutant application, while no such effect was observed on dormant trees (308). Fumigations with chronic doses of SO_2 and NO_2 (weekly average 62 ppb NO_2/SO_2) over 2 years, singly and in combination, resulted at the end of the experiment in greater needle loss in Picea sitchensis and Pinus sylvestris, although NO_2 had initially enhanced needle retention in the latter species. Effects took nearly two years to develop and no seasonal response pattern was observed (324). Hight growth of Pinus taeda was reduced by two or three pollutant combination of $SO_2/NO_2/O_3$ in close to ambient concentrations (309). It was shown that effects of the three pollutant combination were similar to two pollutant combinations of O_3 + SO_2 or O_3+ NO_2, respectively. However fumigation of Platanus occidentalis with a combination of $SO_2/NO_2/O_3$ showed greater effects than with two-gas-mixtures without foliar injury (326). Other experiments revealed that growth reductions in Pinus strobus were more influenced by O_3 and/or SO_2 than NO_2 alone or combinations of NO_2 + O_3 or NO_2 + SO_2 (325). These findings contradict those of Mooi (310) where a greater decrease in growth occurred with the combination of either NO_2/O_3 or all three pollutants, than with SO_2/NO_2 (311).

8.3 INTERACTION OF GASEOUS POLLUTANTS AND ACIDIC RAIN

Interactions of gaseous pollutants and acidic deposition and their impact on forest ecosystems have been discussed only recently and, as yet, not many studies are available. Most research has focussed on interactions between ozone and acidic deposition in the form of either rain or fog.

Foliar leaching of essential nutrients such as magnesium, calcium, zinc, or copper, as well as nitrate and ammonium occurred in Picea abies when plants were fumigated consecutively with 200 or 600 ug m^{-3} O_3 and treated with acidic mist (pH 3.5) once a week (161,168,249). For most cations, leaching was dependent on O_3 dose and was further enhanced by low nutrient content of soils or a reduction in plant vitality prior to exposure (333). Although similar results were obtained in combined O_3 fumigation and fogging experiments, foliar leaching was further modified by additional frost treatments (329). Skeffington and Roberts (330), however, using a different methodological approach, found only increased nitrate leaching in Pinus sylvestris, while leaching of cations was not enhanced by O_3/acidic rain. Combined fumigation experiments with SO_2, O_3, and acidic rain over two years in open top chambers, using Picea abies, Abies alba, and Fagus sylvatica, revealed a marked increase in cation leaching when all three pollutants were supplied in concentrations approaching ambient levels (331). Combinations of ozone and acidic rain, however, had no particular pronounced effect on leaching. There were marked effects on the photosynthesis of red oak, sugar maple, and white pine at various low concentrations of ozone close to the ambient, but no such effects occurred for acidic deposition, nor for the combination of the two (332). Deleterious effects are assumed to be related to the reduced presence of mycorrhizae, which are markedly affected by acidic deposition and are probably modified in the presence of ozone due to changes in carbon allocation (331).

It is clear that pollutant mixtures should be given high priority in future research because the conflicting results from mixture studies are so difficult to interpret. Although combined effects of pollutants can be synergistic, especially at low concentrations, the results from experiments using artificial pollutant combinations should be addressed with care (96). It is absolutely essential, therefore, to produce more reliable information on mixtures, using exposure regimes which simulate the temporal and spatial variation of representative mean and peak concentrations which occur under ambient conditions.

9. RESUMEE

During the past 15 years dispersion of air pollutants over large areas due to increasing emissions of certain pollutants and control strategies, like the high-stack-policy, reveales a subtle but continuous burden for forest ecosystems in remote areas. Concentrations as such are generally low, apart from ozone and their effects seem likely to be considered negligible. However, it must be seen that we deal with combined chronic effects, since there is a clear temporal development, leading eventually to marked ecosystem disturbancies. This implies also for the soil as one of the ultimate sinks for air-born pollutants. Novel forest decline in forested areas of Europe has thus to be linked with the anthropogenic factor: air pollution. However, non-anthropogenic factors interact with this prevalent cause, modifying response in a positive or negative manner. The complexity of these interacting factors and their evaluation in respect to realistic cause-effect-relationships asks scientist for great discipline in future research to cope with this enormous task. The following statements recently compiled by Prinz (334) are probably some help in the right direction for future research in this field, although surely not new to you:

1. The definition of the type of tree decline taken as the basis for the planned research objective (tree species; site/soil factor; and symptomatology and its classification in the entirety of new forest decline, considering temporal and spatial development).
2. A description of the plant material used (age; origin; criteria for selection; time of removal from natural sites; site description; and pretreatment).
3. The definition of site factors relevant to the experiment (climate; soil; exposure design; technical details; etc.).
4. The definition of controls (description of control plants; location filter material).
5. The relevance of results gained (morphological; anatomical; biochemical; physiological).
6. Linkage of observed results to forest decline phenomena (as one constituent in a greater complex of actors; explanation for the whole phenomenon with reference to spatial and temporal occurrence of new forest decline, etc.).
7. The validity of results obtained with respect to foregoing knowledge and possible connection with known cause-effect relationships. (Experimental proof for and against current hypotheses; new explanations).
8. The proof of the relevance of newly gained results with respect to the "real world" ambient conditions (comparable dose; concentration; intensity with the given situation; evidence of time synchronous development of factor in question

with the real development of tree decline under consideration of pollutant and effects accumulation).

9. The consideration of questions still to be answered and recommendations for future approaches (evaluation of principal phenomena; improved statistical evaluation; comparability to other tree species; age classes; treatments or other experimental variables, etc.).
10. Personal evaluation of own results (new discovery; confirmation of the research of others; possibilities for final conclusions, etc.).

ACKNOWLEDGEMENTS

The author would like to acknowledge Dr. Allan H. Legge and Julie Lockhart of the Kananaskis Centre for Environmental Research, The University of Calgary, Calgary, Alberta, Canada for editing the English translation of this paper.

Table 1. Characterization of pollution environment on local, regional, and supra regional basis.

LOCAL — Pollution factor predominant

Injury by air pollution definable
- injury local, restricted to vicinity of emission source
- one specific pollutant causal agent
- in high concentration
- resulting in acute injury, giving specific symptoms
- easily detectable by methods of differential diagnosis
- subsequent measures for pollution reduction accomplishable by state law

REGIONAL — Pollution factor dominant

Injury by air pollution definable
- injury regional, limited to areas of industrial agglomeration
- mixture of pollutants interacting as causal agents
- high annual means of one or two pollutants (SO_2/NO_x)
- conspicuous symptoms
- interaction with other stress factors (biotic and abiotic)
- difficult to characterize by methods of differential diagnosis
- subsequent measures for pollution reduction accomplishable by federal law with knowledge of special compounds.

SUPRA REGIONAL — Pollution factor one among others

Injury by air pollution?
- injury in remote, rural areas
- mixture of pollutants from different regions interacting
- low annual means of phytotoxic compounds; peak concentrations?
- influence by photochemical oxidants
- unspecific symptoms, alien to air pollution
- strong influence of other stress factors (biotic and abiotic)
- hardly any method of differential diagnosis, except in complex, stepwise experimental procedures
- interaction between pathway, soil, and air
- measures for reduction of pollution burden only by federal law and interstate agreements (probability estimation)

Table 2. Relative sensitivity of woody species to fluorides

Very Sensitive Species	Sensitive Species	Less Sensitive Species
Abies alba	*Abies concolor*	*Chamaecyparis lawsoniana*
Larix occidentalis	*Abies grandis*	*Chamaecyparis nootkaensis*
Larix kaempferi	*Abies nordmanniana*	*Juniperus chinensis*
Picea abies	*Laris decidua*	*Taxus baccata*
Picea pungens	*Picea abies*	*Thuja occidentalis*
Picea omorica	*Picea glauca*	*Tsuga canadensis*
Pinus contorta	*Picea pungens*	*Ailanthus altissima*
Pinus mugo	*Pinus mugo*	*Alnus glutinosa*
Pinus ponderosa	*Pinus nigra*	*Amelanchier canadensis*
Pinus strobus	*Pinus sylvestris*	*Berberis thunbergii*
Pinus sylvestris	*Pseudotsuga menziesii*	*Betula papyrifer*
Pseudotsuga menziesii	*Taxus cuspidata*	*Betula pendula*
Acer negundo	*Thuja* spp.	*Cornus* spp.
Acer palmatum	*Acer campestre*	*Fraxinus velutina*
Carpinus betulus	*Acer platanoides*	*Ligustrum* spp.
Mahonia repens	*Acer saccharinum*	*Liquidambar styraciflua*
Prunus domestica	*Amelanchier canadensis*	*Parthenocissus quinquefolia*
Prunus armeniaca	*Betula* spp.	*Philadelphus coronarius*
Prunus persica	*Fagus sylvatica*	*Platanus* x *hybrida*
Sorbus scandica	*Fraxinus pennsylvanica*	
Syringa vulgaris	*Morus rubra*	*Platanus occidentalis*
Vaccinium vitis-idea	*Polygonum alpinum*	*Populus balsamifera*
Vitis vinifera	*Populus canadensis*-Hybr.	*Quercus* spp.
	Populus nigra	*Rocinia pseudoacacia*
	Populus tremuloides	*Salix* spp.
	Rhus glabra	*Sambucus nigra*
	Rhus typhina	*Sorbus aucuparia*
	Rubus idaeus	*Tilia americana*
	Salix spp.	*Ulmus americana*
	Sorbus aucuparia	*Ulmus parvifolia*
	Syringa vulgaris	*Ulmus pumila*
	Tilia cordata	

Table 3. Relative sensitivity of woody species to SO_2.

Very Sensitive Species	Sensitive Species	Less Sensitive Species
Salix nigra	*Pinus nigra*	*Populus balsamifera*
Ulmus parvifolius	*Abies balsamea*	*Populus canadensis*
Pseudotsuga menziesii	*Tilia americana*	*Abies grandis*
Pinus strobus	*Catalpa*	*Tilia cordata*
Pinus banksiana	*Prunus demissa*	*Pinus contorta*
Populus grandidentata	*Populus deltoides*	*Platanus acerifolia*
Acer negundo var. *interius*	*Picea engelmannii*	*Quercus rubra*
Pinus ponderosa	*Acer spicatum*	*Acer saccharinum*
Populus tremuloides	*Pinus resinosa*	*Acer saccharum*
Larix occidentalis	*Tsuga heterophylla*	*Thuja plicata*
Fraxinus americana	*Pinus monticola*	*Thuja occidentalis*
Betula papyrifera	*Ulmus americana*	*Picea glauca*
Juglans regia	*Fagus silvatica*	*Platanus* spp.
Ribes rubrum	*Carpinus betulus*	*Alnus* spp.
Ribes uva-crispa	*Malus domestica*	*Syringa vulgaris*
	Corylus avellana	*Salix* spp.
		Robinia pseudoacacia
		Betula spp.
		Prunus spp.
		Vitis vinifera
		Rhododendron spp.

Table 4: Examples of the effects of experimental ozone fumigations on trees.

Species	Exposure Conditions		Injury Type	Reference
Pinus strobus	30	48h	Foliar injury-enhanced in the presence of mists	193
	70	4h		
Pinus strobus	65	4h	Foliar injury	184
Pinus ponderosa	509h d^{-1}, 9 d		Foliar injury	194
Pinus taeda	50	6h d^{-1}, 28 d	Reduced height growth (18%)	195
	100	6h d^{-1}, 28 d	Reduced total dry weight (23%)	
Populus x *eruamericana* cv. Dorskamp	34	12h d^{-1}, 28 d	Foliar injury, premature leaf fall, reduced stem dry weight (21%)	196
Populus x *eruamericana* cv. Dorskamp cv. Zeeland	41	12h d^{-1}, 161 d	Increase in stem length bearing no leaves (40-fold in Dorskamp, 10-fold in Zeeland)	197
Pinus strobus (Eastern white pine) *Tsuga canadensis* (Hemlock)	100	6 h	Foliar injury	198
Larix decidua (European larch) *Larix leptolepis* (Japanese larch)	100	8h	Foliar injury	199
Acer saccharum (Bigtooth maple)	100	2h d^{-1}, 14 d	Slight foliar injury	200
Plantanus occidentalis	100	6h d^{-1}, 28 d	Reduced height growth (27%) Reduced total dry weight (61%)	195
Picea abies	250	8h	None	199
(Norway spruce)	150	28d	Foliar injury	168
Abies alba (Silver fir)	300	56d	None	168
Pinus sylvestris (Scots pine)	250	8h	Foliar injury	199
Fagus sylvatica (beech) 75 (shaded plants)	150	42d	Foliar injury	168

(from Ashmore, 132)

Table 5. Relative sensitivity of woody plants to ozone (from Davis and Wiehour shortened, 47)

Very Sensitive Species	Sensitive Species	Less Sensitive Species
Ailanthus altissima	*Acer negundo*	*Abies concolor*
Cotoneaster divaricatus	*Forsythia x intermedia* 'Lynwood Gold'	*Acer platonoides*
Cotoneaster horizontalis	*Laris kaempferi*	*Betula pendula*
Gleditsia triacanthos inermis	*Ligustrum vulgare*	*Buxus sempervirens*
Juglans regia	*Liquidambar styraciflua*	*Fagus sylvatica*
Larix decidua	*Philadelphus coronarius*	*Ilex aquifolium*
Ligustrum vulgare var. *pyramidae*	*Pinus strobus*	*Ilex crenata*
Liriodendron tulipifera	*Pinus sylvestris*	*Juglans nigra*
Pinus nigra	*Syringa vulgaris*	*Picea abies*
Platanus occidentalis	*Tsuga canadensis*	*Picea pungens*
Populus maximowiczii x trichocarpa		*Pieries japonica*
Sorbus aucuparia		*Pseudotsuga menziesii*
Spiraea x vanhouttei		*Quercus robur*
Symphoricarpos albus		*Quercus rubra*
		Robinia pseudoacacia
		Sophora japonica
		Thuja occidentalis
		Tilia americana
		Tilia cordata
		Viburnum x burkwoodii
		Viburnum carlesii

Table 6. No effect concentrations for various organic substances for *Lepidum sativum*, *Raphanus* sativ. *radicula*, *Trifolium pratense*, *Phaseolus vulgaris nana*, *Nicotiana tabacum* (BelW3), *Saintpaulia ionatha*, and *Petunia hybrida* after continuous fumigation for 14 days.

Name of Compound	Concentration ($mg.m^{-3}$)
Dichloro-methane	100 *
Toluene	60 *
Trichloroethane	100 *
Acetone	60 *
Xylene	160 *
Ethene	5 +
Methane	5 +
Butadiene 1,3	200 +
Butane	40 +
1,2-Dichloroethane	256 +
Ethanol	70 +
Ethyl-acetate	50 +
Ethane	31

* from (251)
+ from (252)

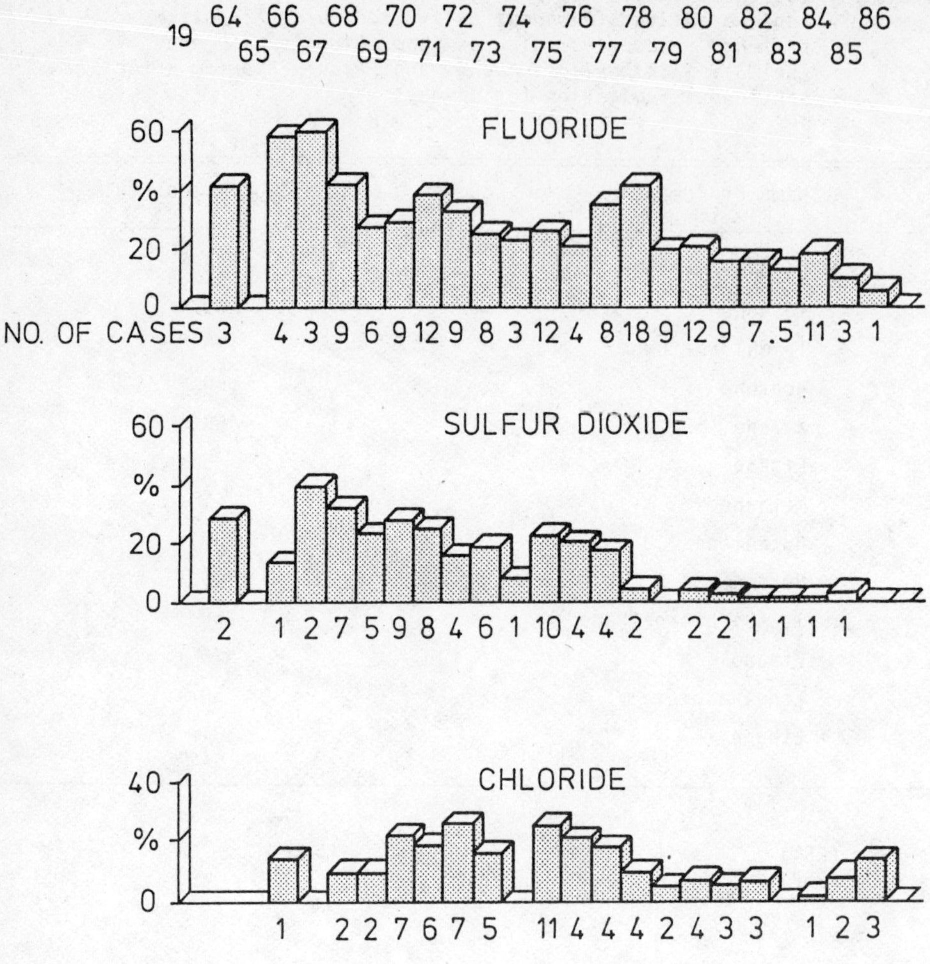

Figure 1. Relative frequency of air pollution injury on vegetation by fluoride, chloride, and sulfur dioxide in relation to total number of consultancy cases carried out by the LIS-ESSEN.

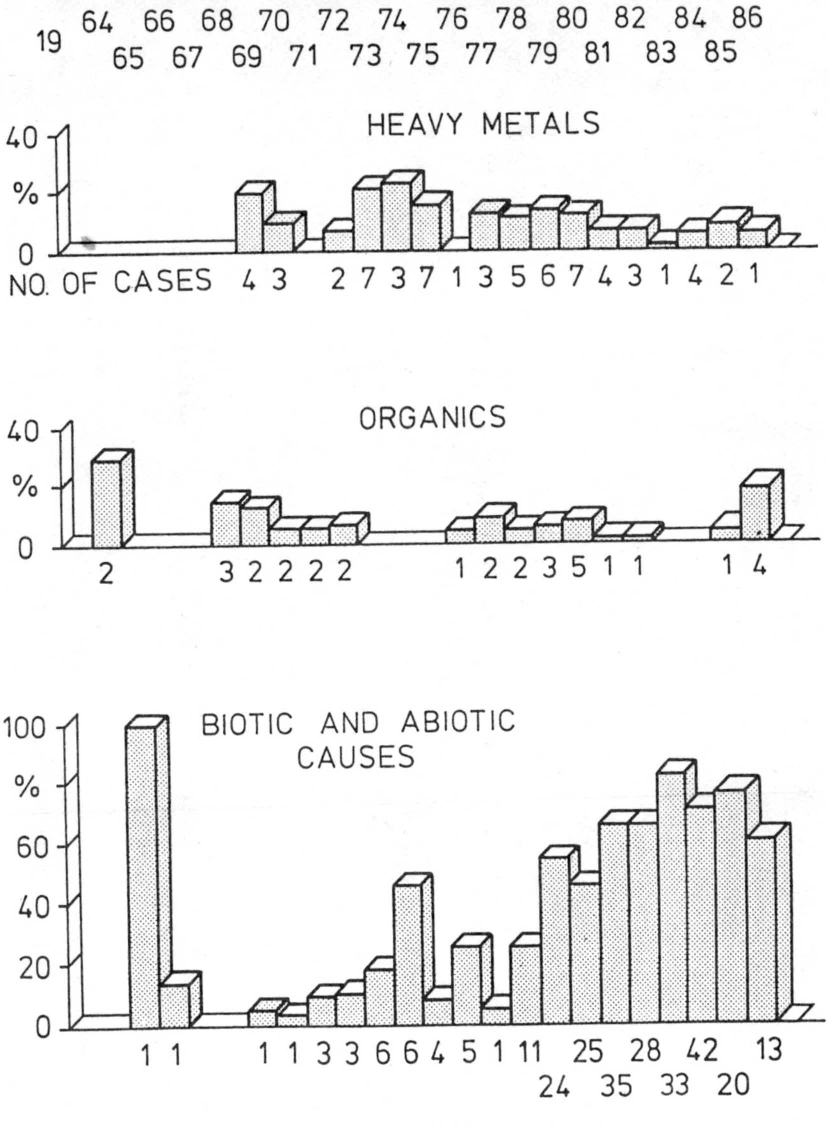

Figure 2. Relative frequency of air pollution injury on vegetation by heavy metals, organics, or non-anthropogenic cases in relation to total number of consultancy cases carried out by the LIS-ESSEN.

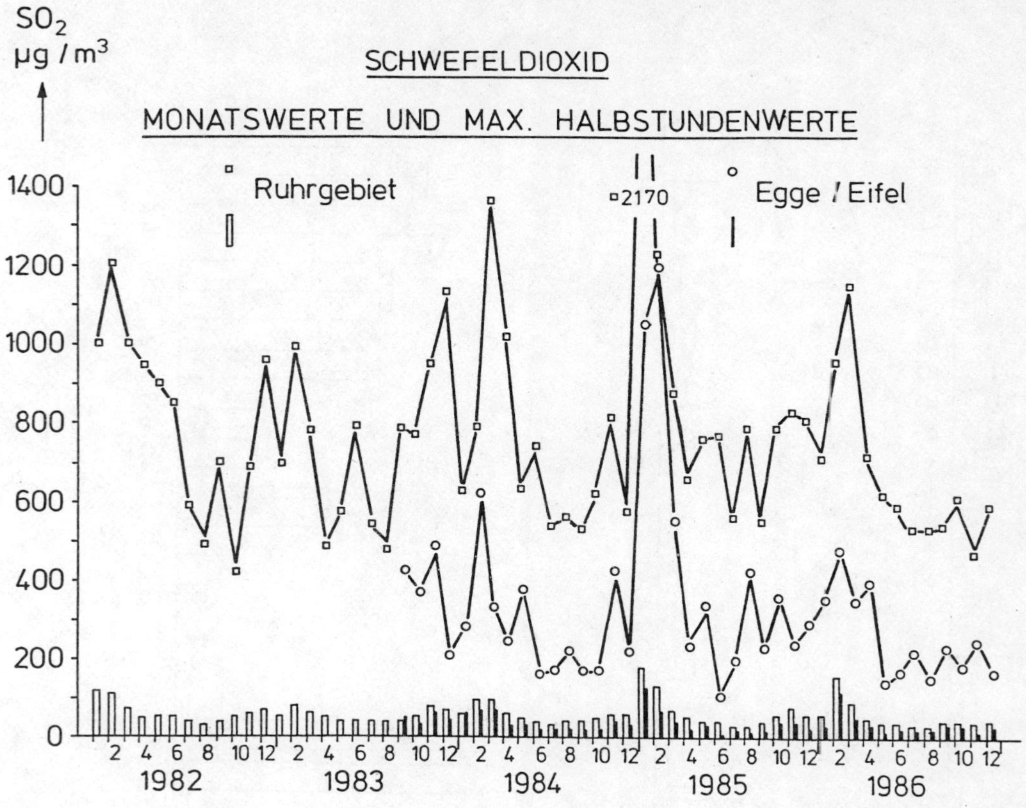

Figure 3. Monthly mean and maximum half hour mean SO_2 concentrations for the Ruhr area and remote forested areas in Northrhine-Westfalia (THEMES-network).

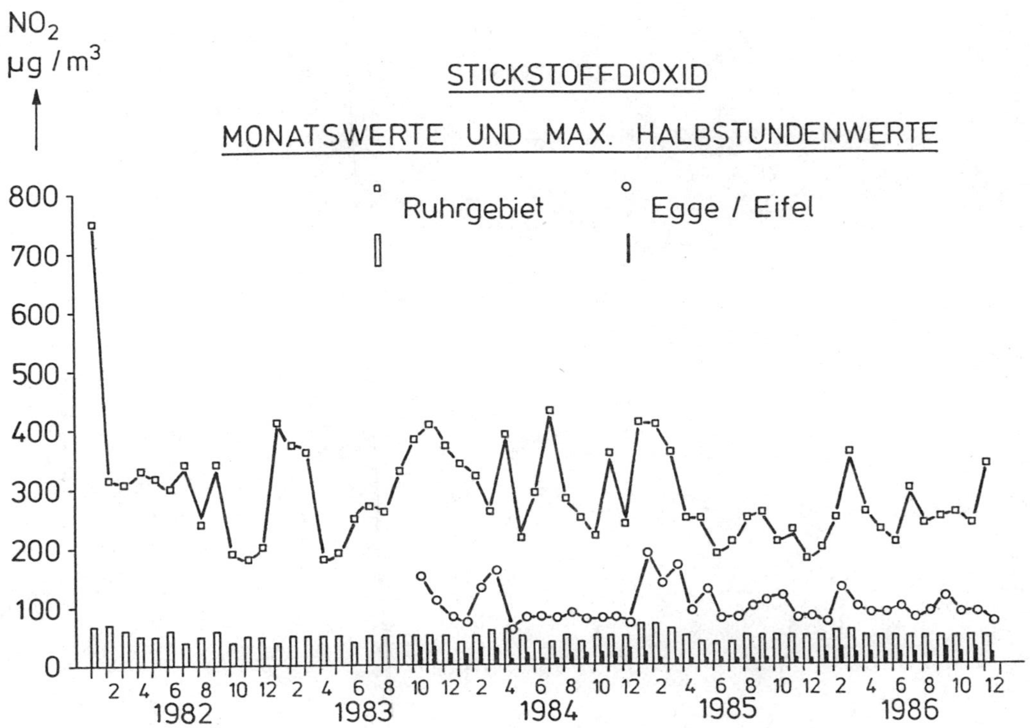

<u>Figure 4</u>. Monthly mean and maximum half hour mean NO_2 concentrations for the Ruhr area and remote forested areas in Northrhine-Westfalia (THEMES-network).

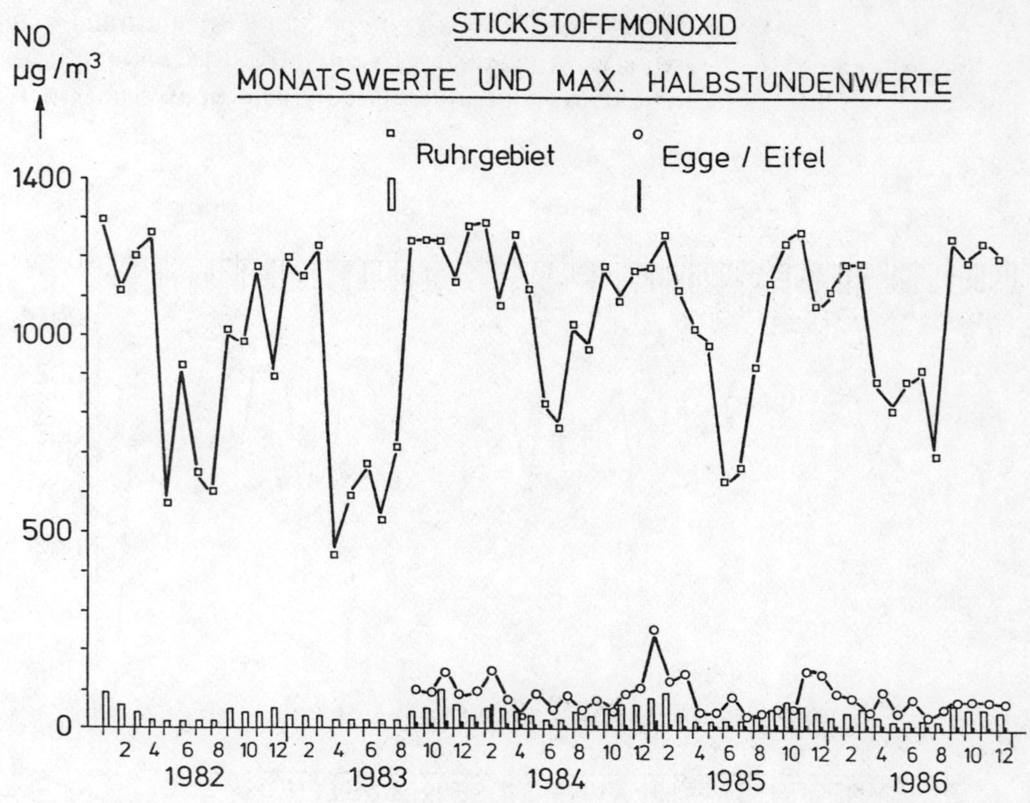

Figure 5. Monthly mean and maximum half hour mean NO concentration for the Ruhr area and remote forested areas in Northrhine-Westfalia (THEMES-network).

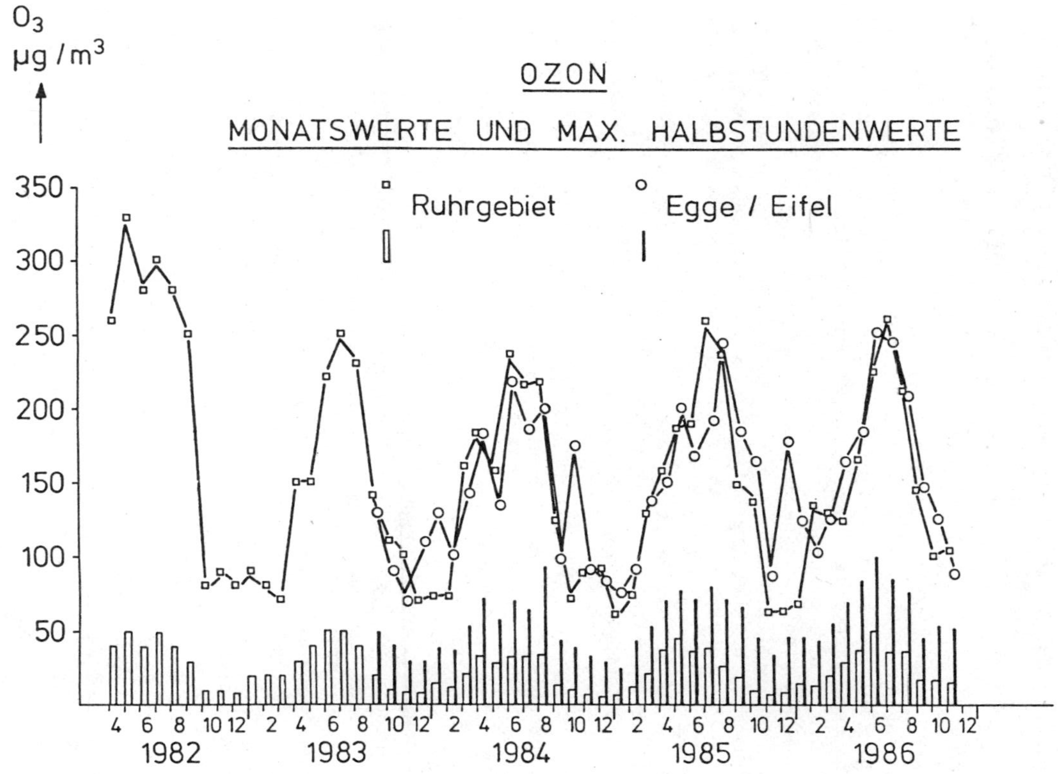

<u>Figure 6</u>. Monthly mean and maximum half hour mean O_3 concentration for the Ruhr area and remote forested areas in Northrhine-Westfalia (THEMES-network).

REFERENCES

(1) BUCK, M., IXFELD, H., and ELLERMANN, K. (1982). Die Veränderung der Immissionsbelastung in den letzten 15 Jahren im Rhein-Ruhr-Gebiet. Staub - Reinhalt. Luft 42, 51-58

(2) SCHÖNBECK, H. (1983). Untersuchungen über Vegetationsschäden in der Umgebung einer Stückfärberei. Ber. aus d. Landesanstalt für Bodennutzungsschutz des Landes NW, 4, 33-78.

(3) PRINZ, B., KRAUSE, G.H.M., and STRATMANN, H. (1979). Thalliumschäden in der Umgebung der Dyckhoff Zementwerke AG in Lengerich, Westfalen. Staub-Reinhalt. Luft 39, 457-462.

(4) PFEFFER, H.U., and BUCK, M. (1985). Meßtechnik und Ergebnisse von Immissionsmessungen in Waldgebieten. VDI-Berichte 560, 127-155.

(5) SMITH, W.H. (1974). Air pollution effects on the structure and function of the temperate forest ecosystem. Environ. Pollut 6, 111-129.

(6) YUNGHANS, R.S., and McMULLEN, T.B. (1970). Fluoride concentration found in NASN samples of suspended particles. Fluoride 3, 143-152.

(7) IXFELD, H., and ELLERMANN, K. (1981). Immissionsmessungen in Verdichtungsräumen. Bericht über Ergebnisse in Bielefeld, Bonn und Wuppertal im Jahre 1978. Schriftenr. Landesanstalt für Immissionsschutz des Landes NW, 53, 7-18.

(8) HILL, A.C. (1969). Air quality standards for fluoride vegetation effects. JAPCA 19, 331-336.

(9) WEINSTEIN, L.H. (1977). Fluoride and plant life. J. Occup. Med. 19, 49-77.

(10) GARBER, K., GUDERIAN, R., and STRATMANN, H. (1967). Untersuchungen über die Aufnahme von Fluor aus dem Boden durch Pflanzen. Qualitas Plant. Mater Vegetabilis 14, 223-236.

(11) KELLER, Th. (1978). Wintertime atmospheric pollutants - do they affect the performance of deciduous trees in the ensuing growing season. Environ. Pollut. 16, 243-247.

(12) LEDBETTER, M.C.R., MAVRODINEANU, R., and WEISS, A.J. (1960). Distribution of studies on radioactive fluorine-18 and stable fluorine-19 in tomato plants. Contrib. Boyce Thompson Inst. 20, 331-348.

(13) JACOBSON, J.S., WEINSTEIN, L.H., McCUNE, D.C., and HITCHCOCK, A.E. (1966). The accumulation of fluorine by plants. JAPCA 16, 412-417.

(14) GUDERIAN, R., van HAUT, H., and STRATMANN, H. (1969). Experimentelle Untersuchungen über pflanzenschädigende Fluorwasserstoff-Konzentrationen. Westdeutscher Verlag, Köln and Opladen; Forsch.Ber.d. Landes Nordrhein-Westfalen, Nr. 2017.

(15) KRONBERGER, W., HALBWACHS, G., and RICHTER, H. (1978). Fluortranslokation in Picea abies (L.) Karsten. Angew. Botanik 52, 149-154.

(16) BRANDT, C.J. (1981). Untersuchungen über Wirkungen von Fluorwasserstoff auf Lolium multiflorum und andere Nutzpflanzen. LIS-Berichte 14, 140 pp.

(17) BENEDICT, H.M., ROSS, J.M., and WADE, R.W. (1965). Some responses of vegetation to atmospheric fluorides. JAPCA 15, 253-255.

(18) GUDERIAN, R. (1977). Air pollution. Phytotoxicity of acid gases and its significance in air pollution control. Springer Verlag Berlin, Heidelberg, New York.

(19) McLEAN, D.C., and HANSEN, K.S. (1986). Responses of the deciduous plants to hydrogen fluoride during winter dormancy. JAPCA 36, 60-61.

(20) CHANG, C.W. (1975). Fluorides. In: J.B. MUDD and T.T. KOZLOWSKI, (eds.) Responses of plants to air pollution. Academic Press, New York, pp 57-97.

(21) GARREC, J.P., PLEBIN, R., and LHOSTE, A.M. (1978). The effect of fluoride on the mineral composition of polluted fir needles (Abies alba Mill.). Fluoride 11, 68-76.
(22) SIDHU, S.S. (1978). Patterns of fluoride accumulation in forest species as related to symptoms and defoliation. Paper 78-24.7, 16 p, 71st Annual Meeting, Air Pollution Control Assoc., Houston Texas.
(23) SIDHU, S.S.(1979). Fluoride levels in air, vegetation and soil in the vicinity of a phosphorous plant. JAPCA 29, 1069-1072.
(24) AMUNDSON, R.G., and WEINSTEIN, L.H. (1980). Effects of airborne F on forest ecosystems. In: Proceedings of the Symposium on Effects of air pollutants on mediterranean and temperate forest ecosystems Riverside, Calif. June 22-27, 1980, pp. 63-78.
(25) ADAMS, D.F., SHAW, C.G., GNAGY, R.M., KOPPE, R.K., MAYHEW, D.J., and YERKES, W.D. (1956). Relationship of atmospheric fluoride levels and injury indexes on gladiolus and ponderosa pine. J. Agr. Food Chem. 4, 64-66.
(26) TRESHOW, M., ANDERSON, K., and HARNER, F. (1967). Responses of douglas fir to elevated atmospheric fluorides. For.Sci. 13. 114-120.
(27) WHEELER, G.L. (1972). The effects of fluorine on the cycling of calcium, magnesium, and potassium in pine plantations of eastern North Carolina. Dissertation. North Carolina State University at Raleigh, 71 pp.
(28) CARLSON, C.E., GORDON, C.C., and GILLIGAN, C.J. (1979). The relationship of fluoride to visible growth/health characteristics of Pinus monticola, Pinus contorta, and Pseudotsuga menziesii. Fluoride 12, 9-38.
(29) FLÜHLER, H., KELLER, Th., and SCHWAGER H. (1981). Waldschäden im Walliser Rhonetal (Schweiz). Mitteilungen der Eidgenössischen Anstalt für das forstliche Versuchswesen 57, 361-499.
(30) GUDERIAN, R. and KÜPPERS, K. (1980). Responses of plant communities to air pollution. In: Proceedings of the Symposium on Effects of air pollutants on Mediterranian and temperate forest ecosystems. Riverside, Calif. June 22-27, pp 187-199.
(31) WENZEL, K.F. (1965), Insekten als Immissionsfolgeschädlinge, Naturwissenschaften 52, 113.
(32) THALENHORST, W. (1974). Untersuchungen über den Einfluß fluorhaltiger Abgase auf die Disposition der Fichte für den Befall durch die Gallenlaus Sacciphantes abietis (L.) Z. Pflanzenkrankh. Pflanzenschutz 81, 717-727.
(33) KRAUSE, G.H.M., and PRINZ, B. (1987). Neuere Untersuchungen zur Ursachenforschung der neuartigen Waldschäden. LIS-Berichte, in Vorbereitung.
(34) LEONE, I.A., BRENNAN, E., and DAINES, R.H. (1956). Atmospheric fluoride: its uptake and distribution in tomato and corn plants. Plant Physiol. 31, 329-333.
(35) KNABE, W. (1970). Natürliche Abnahme des aus den Immissionen aufgenommenen Fluors in Fichtennadeln. Staub-Reinhalt. Luft 30, 384-385.
(36) AMUNDSON, R.G., WEINSTEIN, L.H., van LEUKEN, P. and COLAVITO, L.J. (1982). Joint action of HF and NO_2 on growth, fluorine accumulation, and leaf resistance in Marcross sweet corn. Environ. Exp. Bot. 22, 49-55.
(37) KELLER, TH.(1980). The simultaneous effect of soilborne NaF and air pollutant SO_2 on CO_2-uptake and pollutant accumulation. Oecologia (Berl.) 44, 283-285.

(38) BUCHER, J.B. (1985). In: Nato Research Workshop: Effects of Acid Deposition and Air Pollution on Forests. Wetland and Agricultural Ecosystems, May 13-17, 1985 Toronto, Canada in press.
(39) OBLÄNDER, W. and HANSS, A. (1985). Landesanstalt für Umweltschutz Baden-Württemberg, Bericht 97/85.
(40) SCHROEDER, J. (1872-1873). Die Einwirkung der schwefligen Säure auf die Pflanzen. Tharander forstl. Jahrbuch 22, 185-239, 23, 217-267.
(41) WIELER, A. (1903). Über unsichtbare Rauchschäden. Z. Forst-Jagdwes. 35, 204-225.
(42) WISLICENUS, H.(1908). Sammlung von Abhandlungen über Abgase und Rauchschäden. In: Waldsterben im 19. Jahrhundert. VDI-Verlag, Düsseldorf, 1985.
(43) MUDD, J.B. (1975). Sulfur Dioxide. In: Responses of Plants to Air Pollution. J.B. MUDD and T.T. KOZLOWSKI, (eds.), Physiological Ecol. Mono. Series, Academic Press, Inc., New York, NY, pp. 9-22.
(44) HECK, W.W. and BRANDT, C.S. (1977). Effects on Vegetation: Native, Crops, Forest. In: Air Pollution. A.C. STERN, (ed.), Academic Press, New York, NY, pp. 157-229.
(45) ROBERTS, T.M., DARRALL, N.M., and LANE, P. (1983). Effects of gaseous air pollutants on agriculture and forestry in the UK. Adv.Appl. Bio. 9, 1-142.
(46) GUDERIAN, R. and STRATMANN, H. (1968). Freilandversuche zur Ermittlung von Schwefeldioxidwirkungen auf die Vegetation. III. Grenzwerte schädlicher SO_2-Immissionen für Obst- und Forstkulturen sowie für landwirtschaftliche und gärtnerische Pflanzenarten. Forschungsberichte des Landes NW, Nr. 1920, Westdeutscher Verlag, Köln 113 pp.
(47) DAVIS, D.D. and WILHOUR, R.G. (1976). Susceptibility of woody plants to sulfur dioxide and photochemical oxidants: a literature review. EPA 600/3-76-102, US Environmental Protection Agency, Corvallis, Oregon, USA.
(48) NICHOLSON, I.A., POWLER, D., PATERSON, I.S., CAPE, J.N., KINNAIRD, J.W. (1980). Continuous monitoring of airborne pollutants. In: D. DRABLOS and A. TOLLAN (eds.): Ecological impact of acid precipitation. Proc. Internat. Conf. Sandefjord, SASF-Project, Nisk, 1432 As-NLH, pp. 144-145.
(49) KODATA, M.and INOUE, T. (1972). Invading path of sulfur dioxide into pine leaves as revealed by microradioautoghraphy. J. For. Soc. 54, 207-208.
(50) JENSEN, K.R. and KOZLOWSKI, T.T. (1975). Absorption and translocation of sulfur dioxide by seedlings of four forest tree species. J. Environ. Qual. 4, 379-382.
(51) HALLGREN, J.E.(1978). Physiological and biochemical effects of sulfur dioxide zo plants. In: J.O. NRIAGU (ed.) Sulfur in the environment. Vol 2, John Wiley & Sons, New York, NY, pp. 163-210.
(52) MEIDENER, H. and MANSFIELD, T.A. (1968). Physiology of stomata. McGraw-Hill, England.
(53) McLAUGHLIN, S.B. and TAYLOR, T.A. (1981) Relative humidity: important modifier of pollutant uptake by plants. Science 211, 167-169.
(54) MAJERNIK, O., and MANSFIELD, T.A. (1970). Direct effect of SO_2 pollution on the degree of opening of stomata. Nature 227, 377-378.
(55) BISCOE, P.V., UNSWORTH, M.H., and RINCKNEY, H.R. (1973). The effects of low concentrations of sulfur dioxide on stomatal behaviour in Vicia faba. New Phytologist 72, 1299-1306.
(56) SUWANNAPINANT, W. and KOZLOWSKI, T. (1980). Effects of SO_2 on transpiration, chlorophyll content, growth and injury in young seedlings of woody angiosperms. Canad. J. Forest. Res. 10, 78-81.

(57) BUCHER, J.B. (1978). Zellwandabbauende Enzyme als Indikatoren von Schwefeldioxid-begasten Buchen- und Fichtensämlingen. Schweizerische Z. Forstwes. 129, 414-423.
(58) BUCHER-WALLIN, J.K., BERNHARD, L. und BUCHER, J.B. (1979). Einfluß niedriger SO2-Konzentrationen auf die Aktivität einiger Glykosidasen der Assimilationsorgane verklonter Waldbäume. Eur.J.For.Path. 9, 6-15.
(59) ZIEGLER, I. (1975). The effect of SO_2 pollution on plant metabolism. Residue Rev. 56, 79-105.
(60) BALLENTYNE, D.J.(1973). Sulfite inhibition of ATP formation in plant mitochondria. Phytochemistry 12, 1207-1209.
(61) MALHOTRA, S.S. and KAHN, A.A. (1978). Effects of sulfur dioxide fumigation on lipid biosynthesis in pine needles, Phytochemistry 17, 241-244.
(62) GODZIK, S., and LINSKENS, H.F. (1974). Concentration changes of free amino acids in primary bean leaves after continuous and interrupted SO_2 fumigation and recovery. Environ. Pollut. 7, 25-28.
(63) RAO, D.N. and LeBLANC, F. (1965). Effect of sulfur dioxide on the lichen algae with special reference to chlorophyll. Bryologist 69, 69-75.
(64) MALHOTRA, S.S. (1977). Effects of aqueous sulfur dioxide on chlorophyll destruction in Pinus contorta. New Phytologist 78, 101-109.
(65) BLACK, V.J. and UNSWORTH, M.H. (1979). Effects of low concentrations of sulfur dioxide on net photosynthesis and dark respiration of Vicia faba. J. Exper.Bot. 30, 473-483.
(66) KELLER, TH. (1981). Winter uptake of airborne SO_2 by shoots of deciduous species. Environ. Pollut. 26, 313-317.
(67) BRENNINGER, C. and TRANQUILINI, W. (1983). Photosynthese, Transpiration und Spaltöffnungsverhalten verschiedener Holzarten nach Begasung mit Schwefeldioxid. Europ.J. Forest. Path. 13, 228-238.
(68) GUDERIAN, R. and VAN HAUT, H. (1970). Nachweis von Schwefeldioxid-Wirkungen an Pflanzen. Staub-Reinhalt. Luft 30, 1-10.
(69) TINGEY, D.T., and TAYLOR, G.E.JR. (1982) Variation in plant response to ozone: a conceptual model of physiological events. In: UNSWORTH, M.H., and ORMROD, D.P. (eds.): Effects of gaseous air pollution in agriculture and horticulture. Butterworth Scientific, London, pp. 113-138.
(70) FALLER, N. (1970). Effects of atmospheric SO_2 on plants. Sulfur Inst.J. 6, 5-7.
(71) COWLING, D.W., JONES, H.P., and LOCKYER,D.R. (1973). Increased yield through correction of sulfur deficiency in ryegrass exposed to sulfur dioxide. Nature 243, 479-480.
(72) NOGGLE, J.C., and JONES, H.C. (1979). Accumulation of atmospheric sulfur in plants and sulfur supplying capacity of soils. EPA-600/7-79-109, US Environmental Protection Agency, Corvallis, Oregon, USA.
(73) KELLER, TH. (1982). The S-content in forest tree foliage fumigated during summer or winter. Europ.J.Forst. Path. 12, 399-406.
(74) KELLER, TH. (1983). Ökophysiologische Folgen niedriger, langdauernder SO_2-Konzentrationen für Waldbaumarten. Ges.f. Strahlen- und Umweltforschung mbH, München: SO_2 und die Folgen. GSF-Bericht A 3/83. pp. 31-47.
(75) KELLER, TH. (1984). Direct effects of sulfur dioxide on trees. In: Ecological effects of deposited sulfur and nitrogen compounds. Phil. Trans.R.Soc., London B 305, 317-326.
(76) LINZON, S.N. (1971). Economic effects of sulfur dioxide on forest growth. JAPCA 21, 81-86.

(77) ROBERTS, T.M. (1984). Effects of air pollutants on agriculture and forestry. Atmospheric Environm. 18,629-652.

(78) BONTE, J., deCORMIS, L., and TISNE, A. (1981). Etude des effects a long-terme dúne pollution chronique par SO_2. Ministry of Agriculture of France, INRA, Morlaas.

(79) GARSETT, S.G., and RUTTER, A.J.(1982). Relative performance of conifer populations in various tests for sensitivity to SO_2, and the implication for selecting trees for planting in polluted areas. New Phytologist, 92, 349-367.

(80) HUTTUNEN, S. (1985). The effects of SO_x on vegetation. WHO-Criteria document, 1987.

(81) SJÖDIN, A., and GRENNFELT, P. (1984). Regional background concentrations of NO_2 in Sweden. In: EC 3rd Symposium on Physico-chemical behaviour of atmospheric pollutants. Varese (Italy) 10.-12.4.1984, pp. 346-363.

(82) MATERNA, J. (1972). Einfluß niedriger Schwefeldioxidkonzentrationen auf die Fichte. Mitteil. forstl. Bundesversuchsanstalt, Wien 97, 219-231.

(83) MATERNA, J. (1978). Forstschäden in Fichtenbeständen in Abhängigkeit von Immissionseinwirkungen. X. Arbeitstagung forstl. Rauchschadens-sach-verständiger. IUFRO S 2.09, Ljubljana, 1978, pp. 241-352.

(84) MATERNA, J. (1981). Die Ernährung der Fichtenwälder im Riesengebirge. Lesnietvi (Forstwirtschaft) 27, 689-698.

(85) MATERNA, J. (1985). Luftverunreinigungen und Waldschäden. In: Symposium von Umweltschutz- eine internationale Aufgabe, Prag, 13.-15.3.1985, VDI-Verlag, Düsseldorf.

(86) HAVAS, P.J.(1971). Injury in pines growing in the vicinity of a chemical processing plant in northern Finland. Acta Forest.Fem. 121, 1-21.

(87) HAVAS, P.J., and HUTTUNEN, S. (1972). The effect of air pollution on the radial growth of Scots pine (Pinus sylvestris L.). Biol. Conserv. 4, 361-368.

(88) HAVAS, P.J., and HUTTUNEN, S. (1980). Some special features of the ecophysiological effects of air pollution on coniferous forests during winter. In: Hutchinson, T.C., and Havas, P.J. (eds.) Effects of acid precipitation on terrestrial ecosystems., M. Plenum Publ. Corp.

(89) HUTTUNEN, S. (1978). Effects of air pollution on provenances of Scots pine and Norway spruce in northern Finland. Silva Fenn. 12, 1-16.

(90) MICHAEL, G., FEILER, S., RANFT, H., and TESCHE, M. (1982). Der Einfluß von Schwefeldioxid und Frost auf Fichten (Piea abies L Karst.) Flora 172, 317-326.

(91) FEILER, S. (1985). Influence of SO_2 on membrane permeability and consequences on frost sensitivity of spruce (Picea abies L.). Flora 177, 217-226.

(92) MEYRAJM. H., HAHN, J., HELAS, G., and P. WARNECK (1984). Cryogenic sampling and analysis of peroxyacetyl nitrate in the atmosphere. In: EC 3th Symposium on Physico-chemical behaviour of atmospheric pollutants. Varese (Italy), 10.-12.4.1984, pp. 109-115.

(93) MANION, T.D. (1981). Tree disease concepts, Prentice Hall, 324.

(94) PRINZ, B., KRAUSE, G.H.M., and STRATMANN, H. (1982). Waldschäden in der Bundesrepublik Deutschland. LIS-Berichte 28.

(95) UBA (Umweltbundesamt) (1985, 1986) Monatsberichte aus dem Meßnetz.

(96) GUDERIAN, R., and TINGEY, D.T.(1987). Notwendigkeit und Ableitung von Grenzwerten für Stickstoffoxide. UBA-Berichte 1/87. Umweltbundesamt, Berlin.

(97) BENNET, J.H., and HILL, A.C. (1973) Absorption of gaseous air pollutants by a standardized plant canopy. JAPCA 23, 203-206.
(98) OMASA, K., ABO, F., NATORI, T., and TOTSUKA, T. (1980). Analysis of air pollutant sorption by plants. (3) Sorption under fumigation with NO_2, O_3 or NO_2+O_3. In: Studies on the effects of air pollutants on plants and mechanisms of phytotoxicity, Research Report form the National Institute for Environmental Studies Nr. 11, Yatabe-machi, Tsukuba, Ibaraki 305, Japan, pp. 213-224.
(99) SRIVASTAVA, H.S., JULLIFFE, P.A., and RUNECKLES, V.C.(1975). The effects of environmental conditions on the inhibition of leaf gas exchange by NO_2. Canad. J. Bot. 53, 475-482.
(100)ROGERS, H.H.Jr. (1975). Uptake of nitrogen dioxide by selected plant species. PhD Thesis. University of North Carolina, Chapel Hill. NC. USA.
(101)BENGTSON, C., BROSTROM, C.A., GRENNFELT, P., SKÄRBY, L., and TROENG, E. (1980). Deposition of nitrogen oxides to Scots pine (Pinus sylvestris L.). In: DRABLOS, D., and TOLLAN, A. (eds.) Ecological impact of acid precipitation, SNSF, Oslo, Norway, pp. 154-155.
(102)FALLER, N. (1972). Schwefeldioxid, Schwefelwasserstoff, Nitrosegase und Ammoniak als ausschließliche S- bzw. N-Quellen der höheren Pflanzen. Z. Pflanzenernährung, Düngung, Bodenkd. 131, 120-130.
(103)ZÖTTEL, H.W. (1986). Possible causes of forest damage in Germany. In: Atmospheric emissions and their effects on the environment in Europe with particular reference to the role of hydrocarbons. Coucawe, Report No. 86/61.
(104)HILL, A.C., and BENNET, J.H. (1970). Inhibition of apparent photosynthesis by nitrogen oxides. Atmos. Environm. 4, 341-348.(105)
(105)WELLBURN, A.R., WILSON, J., and ALDRIGE, P.H. (1980). Biochemical responses of plants to nitric oxides polluted atmospheres. Environ. Pollut (Series A) 3, 219-228.
(106)ROWLANDS, J.R., ALLEN-ROWLANDS, C.J., and GAUSE, E.M. (1977). Effects of environmental agents on membrane dynamics. In: LEE, S.D. (ed.): Biochemical effects of environmental pollutants. Ann. Arbor Scientific Press, Ann Arbor, Michigan, USA, pp. 203-246.
(107)MUDD, J.B., BANERJEE, S.K., DOOLEY, M.M. and KNIGHT, K.L. (1984). Pollutants and plant cells: Effects on membranes. In: KOZOIOL, M.J., and WHATLEY, F.R. (eds.) Gaseous air pollutants and plant metabolism. Butterworth Scientific Press. London, UK, pp. 105-116.
(108)ZEEVART, A.J. (1976), Some effects of fumigating plants for short periods with NO_2. Acta Bot. Neerl. 23, 345-346.
(109)HOGSETT, W.E., PETERMAN, J.J., HOLMAN, S.R., and TINGEY, D.T. (1981) Diurnal variation in foliar injury and nitrogen metabolism with NO_2 exposure. Plant Physiol. (Supplement) 67, 80-82.
(110)YONEYAMA, T., SASAKAWA, H., ISHIZUKA, S., and TOTSUKA, T. (1979). Absorption of atmospheric NO_2 absorption in leaves. Soil Sci. Plant Nutrit. 25, 267-1275.
(111)MOHR, H.D.(1983). Zur Faktorenanalyse des Baumsterbens. Biol. in unserer Zeit 13, 74-76.
(112)AMUNDSON, R.G., and McLEAN, D.C. (1982) Influence of oxides of nitrogen on crop growth and yield: an overview. In: SCHNEIDER, T., and GRANT, L. (eds.): Air pollution by nitrogen oxides, Elsevier Publishing Comp., Amsterdam, NL., pp. 501-510.
(113)KAWEKA, A.(1973). Changes of the leaves of Scots pine (Pinus sylvestris) due to the pollution of the air with nitrogen compounds. Ekologia Polska 21, 105-120.

(114)SCHOLL, G. (1975). Positive und negative Wirkungen von Stickstoffverbindungen im Einwirkungsbereich einer Düngemittelfabrik. Staub-Reinhalt. Luft 35, 201-205.
(115)SKELLY, J.M., MOORE, L.D., and STONE, L.L. (1972). Symptom expression of eastern white pine located near a source of oxides of nitrogen and sulfur dioxide. Plant Dis. Rep. 56, 3-6.
(116)KENK, G. (1986). Pers. communications.
(117)WERNER, H.(1987), Pers. communications.
(118)KRAUSE, G.H.M., ARNDT, U., BRANDT, C.J., BUCHER, J.B., KENK, G., and MATZNER, E. (1987). Forest decline in Europe: development and possible causes. Water, Air, and Soil Pollut. 31, 647-668.
(119)RÖHLE, H.(1985). Ertragskundliche Aspekte der Wald-Erkrankungen. Fostw. Cbl. 104, 225-245.
(120)ANDERSEN, B. (1986) Impact of nitrogen deposition. In: Nilsson, J. (ed.) Critical loads for sulfur and nitrogen. Nordisk Ministerrad, Milijo rapport 1986: 11, pp. 159-197.
(121)WITTIG, R., BALLACH, H.J., and BRANDT, C.J. (1985). Increase of number of acid indicators in the herb layer of millet grass-beech forest of the Westfalia Bight. Angewandte Botanik 59, 219-232.
(122)BALLACH, H.J., GREVEN, H., and WITTIG, R.(1985). Biomonitoring in Waldgebieten Nordrhein-Westfalens. Staub-Reinhalt. Luft 45, 567-573.
(123)TYLER, G.(1986). Probable effects of soil acidification and nitrogen deposition on the floristic composition of oak (Quercus robur). Department of Ecology, University of Lund, Sweden.
(124)NIHGARD, B. (1985). The ammonium hypothesis - an additional explanation to the forest dieback in Europe. Ambio 14, 4-16.
(125)DAVISON, A.W., BARNES, J.D.,and RENNER, C.J. (1987). Interactions between air pollutants and cold stress. Paper presented at 2nd International Symposium on Air Pollution and Plant Metabolism, April 6-9, 1987, GSF-München-Neuherberg, FRG.
(126)LAURENCE, J.A. (1981) Effects of air pollutants on plant pathogen interactions. J. Plant Diseases and Protection 87, 156-172.
(127)MIDDLETON, J.T., KENDRICK, J.B.Jr., and SCHWALM, H.W.(1950). Injury to herbaceous plants by smog or air pollution. Plant Dis.Reptr. 34, 245-252.
(128)FINLAYSON, B.J., and PITTS, J.N.Jr. (1976). Photochemistry of the polluted troposhere. Science 142, 111-119.
(129)BECKER, K.H., LÖBER, J., and SCHURATH, U. (1983). Bildung, Transport und Kontrolle von Photoxidantien. In: UBA Luftqualitätskriterien für photochemische Oxidantien. Bericht 5/83 pp. 3-132.
(130)BRUCKMANN, P. (1983). Bildung von Säuren und Oxidantien durch Gasphasenreaktionen, VDI-Berichte 500, 21-33.
(131)FRICKE, W. (1983). Großräumige Verteilung und Transport von Ozon und Vorläufern, VDI-Berichte 500, 55-62.
(132)ASHMORE, M.R., BELL, N., and RUTTER, J. (1985). The role of ozone in forest damage in W.Germany. Ambio 14, 81-87.
(133)PFEFFER, H.U. (1985). Immissionserhebungen in quellfernen Gebieten Nordrhein-Westfalens. Staub-Reinhalt. Luft 45, 187-193.
(134)WARMBT, W. (1979). Ergebnisse langjähriger Messungen des bodennahen Ozons in der DDR. Z. Meterol 29, 24-31.
(135)ATTMANNSPACHER, W., HARTMANNSGRUBER, R., and LANG, P. (1984). Langzeittendenzen des Ozons der Atmosphäre aufgrund der 1967 begonnenen Ozonmeßreihen am Meteorologischen Observatorium Hohenpeißenberg. Meteorol. Rdsch. 37, 193-199.
(136)FEISTNER, U., and WARMBT, W. (1984). Long-term surface ozone increase at Arkona. Internat. Ozone Symp. IAMAP, Halkidiki, Greece, Sep. 1984.

(137)ARNDT, U., SEUFERT, G., and NOBEL W. (1982). Die Beteiligung von Ozon an der Komplexkrankheit der Tanne (Abies alba Mill.) - eine prüfenswerte Hypothese. Staub-Reinhalt Luft 42, 243-247.

(138)ARNDT, U. (1985). Ozon und seine Bedeutung für das Waldsterben in Mitteleuropa. In: NIESLEIN, E., and VOSS, G. (eds.). Was wir über das Waldsterben wissen. Deutscher Instituts Verlag GmbH, Köln, ISBN 3-602-14158-6, p. 160-174.

(139)POSTHUMUS, A.C., and TONNEIJK, A.E.G. (1982). In: STENBING, L., and JÄGER, H.J. (eds.): Monitoring of air pollutants by plants. W. Junk Publ., The Hagne, NL.

(140)SKÄRBY, L., and SELLDON, G. (1984). Photochemical oxidants in the troposphere: a mounting menace. Ambio 13, 68-72.

(141)ARNDT, U., BARTTOLMEß, H., and SCHLÜTER, CHR. (1986). Allg. Forst.Z. 41, 3-9.

(142)GUDERIAN, R., TINGEY, D.T., and RABE, R. (1985). Effects of photochemical oxidants on plants. In: GUDERIAN, R. (ed.): Air pollution by photochemical oxidants. Springer Verlag, Berlin, Heidelberg, New York pp. 129-333.

(143)KRAUSE, C.R., and WEIDENSAUL, T.C. (1978). Ultra structural effects of ozone on the host-parasite relationship of Botrytis cinerea and Pelargonium hortorum. Phytopathology 68, 301-307.

(144)BUTLER, L.K., and TIBBITTS, T.W. (1979). Stomatal mechanisms determining genetic resistance to ozone in Phaseolus vulgaris L. J.Am.Soc.Hort.Sci. 104, 213-216.

(145)GASELMAN, C.M. and DAVIS, D.D. (1978). Ozone susceptibility for ten azalea cultivars as related to stomata frequency or conductance. J.Am.Soc.Hort.Sci 103, 489-491.

(146)DUNNING, J.A., and HECK, W.W. (1973). Response of pinto bean and tobacco to ozone as conditioned by light intensity and/or humidity. Environ.Sci.Technol. 7, 824-826.

(147)DUNNING, J.A., and HECK, W.W. (1977). Response of bean and tobacco to ozone: Effect of light intensity, temperature and relative humidity. JAPCA 27, 882-886.

(148)MILLER, C.A., and DAVIS, D.D. (1981). Effect of temperature on stomatal conductance and ozone injury of pinto bean leaves. Plant Dis. 65, 750-751.

(150)OLSZEYK, D.M., and TIBBITTS, T.W. (1981). Stomatal response and leaf injury of Pisum sativum L. with SO_2 and O_3 exposures. Plant Physiol.67, 539-544.

(151)HARKOV, R., and BRENNAN, E. (1980). The influence of soil fertility and water stress on the ozone response of hybrid poplar trees. Phytopathology 70, 991-994.

(152)HEAGLE, A.S. (1979). Ranking of soybean cultivars for resistance to ozone using different ozone doses and response measures. Environ.Pollut. 19, 1-10.

(153)HANSON, G.P., ADDIS, D.H., and THORNE, L. (1976). Inharitance of photochemical air pollution tolerance in petunia. Can. J. Genet. Cytol. 18, 579-592.

(154)BLUM, U.T., and HECK, W.W. (1980). Effects of acute ozone exposures on snap bean at various stages of its life cycle. Environ. Exp. Bot. 20, 73-85.

(155)SMITH, W.H. (1981). Air pollution and forest, interactions between air contaminants and forest ecosystems. Springer Verlag, Berlin.

(156)TRESHOW, M. (1970). Ozone damage to plants. Environ Pollut. 1, 155-161.

(157)EVANS, L.S., and MILLER, P.R. (1972). Ozone damage to Ponderosa pine: a histological and histochemical appraisal. Amer.J.Bot. 59, 297-304.
(158)HEATH, R.L., and FREDERICK, P.E. (1979). Ozone alteration of membrane permeability in Chlorella. Plant Physiol. 64, 455-459.
(159)HEATH, R.L. (1980). Initial events in injury to plants by air pollutants. Am.Rev.Plant Physiol, 31, 395-431.
(160)KEITEL, A., and ARNDT, U. (1983). Ozoninduzierte Turgeszenzverluste bei Tabak (Nicotiana tabacum Bel W 3) - ein Hinweis auf schnelle Permeabilitätsveränderung der Zellmembranen. Angew. Botanik 57, 193-204.
(161)KRAUSE, G.H.M., PRINZ, B., and JUNG, K. D. (1983). Forest effects in West Germany. In: DAVIS, D.D. (ED.) Air pollution and the productivity of the forest. Proc. of Symp. Izaak Walton League, Wash. 4.-5.10.1983, pp. 297-332.
(162)KRAUSE, G.H.M., JUNG, K.-D., and PRINZ, B. (1985). Experimentelle Untersuchungen zur Aufklärung der neuartigen Waldschäden in der BRD. VDI-Berichte 560, 627-656.
(163)SCHULTEN, H.R. (1987). Impact of ozone on high-molecular constituents of beech leaves. Naturwissenschaften in press.
(164)CARLSON, R.W. (1979). Reduction in the photosynthetic rate of Acer, Quercus and Fraxinus species caused by sulfur dioxide and ozone. Environ.Pollut. 18, 59-170.
(165)HEATH, R.L., FREDERICK, P.E., and CHIMIKLIS, P.E. (1982). Ozone inhibition of photosynthesis in Chlorella sorakiniana. Plant Physical. 69, 229-233.
(166)BLACK, V.J. ORMROD, D.P., and UNSWORTH, M.H. (1982). Effects of low concentration of ozone singly and in combination with sulfur dioxide on net photosynthesis rates of Vicia faba L. J. Experim.Bot. 33, 1302-1311.
(167)COYNE, P.J., and BINGHAM, G.E. (1981). Comparative ozone dose response of gas exchange in a Ponderosa pine stand exposed to longterm fumigation. JAPCA 31, 38-41.
(168)KRAUSE, G.H.M., and PRINZ, B. (1986). Zur Wirkung von Ozon und saurem Nebel auf phänomenologische und physiologische Parameter an Nadel- und Laubgehölzen im kombinierten Begasungsexperiment. Spezielle Berichte der Kernforschungsanlage Jülich, 369, 208-221.
(169)FAENSEN-THIEBES, A. (1983). Veränderung im Gaswechsel, Chlorophyllgehalt und Zuwachs von Nicotiana tabacum L. und Phaseolus vulgaris L. durch Ozon und deren Beziehung zur Ausbildung von Blattnekrosen. Angew.Botanik 57, 181-191.
(170)SAKAKI, T., KONDO, N., and SUGAHARA, K. (1983). Breakdown of photosynthetic pigments and lipids in spinach leaves with ozone fumigation: role of active oxygen. Physiol. Plant. 59, 28-34.
(171)LICHTENTHALER, H.K., and BUSCHMANN, C. (1984). Beziehungen zwischen Photosynthese und Baumsterben. Allg. Forst.Z. 39, 12-16.
(172)TINGEY, D.T., WILHOUR, R.G., and STANDLEY, C. (1976). The effect of chronic ozone exposures on the metabolic content of Ponderosa pine seedlings. Forest Sci. 22, 234-241.
(173)McLAUGHLIN, S.B., and McCONATHY, R.K. (1983). Effects of SO_2 and O_3 on allocation of 14-C-labelled photosynthate in Phaseolus vulgaris, Plant Physiol. 73, 630-635.
(174)HORSMAN, D.C., NICHOLLS, A.O., and CALDER, D.M. (1980). Growth resposes of Doctylis glomerata, Lolium perenne and Phalaris aquatica to chronic ozone exposures. Anst.J.Plant.Physial. 7, 511-517.
(175)REICH, P.B. (1983). Effects of low concentration of O_3 on net photosynthesis, dark respiration and chlorophyll contents in aging hybrid poplar leaves. Plant Physiol. 73, 291-296.

(176)MUDD, J.B., McMANUS, T.T., ONGUN, A., and McCULLOGH, T.E. (1971). Inhibition of glycolipid biosynthesis in chloroplasts by ozone and sulfhydryl reagents. Plant Physiol. 48, 335-339.
(177)BARNES, R.L. (1972). Effects of chronic exposure to ozone on soluble sugar and ascorbic acid contents of pine seedlings. Can.J.Bot. 50, 215-219.
(178)KRAUSE, G.H.M. (1987). Ozone induced nitrate formation in needles and leaves of Picea abies, Fagus sylvatica and Quercus robur. In preparation.
(179)SKELLY, J.M. (1980). Photochemical oxidant impact on Mediterranean and temperate forest ecosystems: Real and potential effects. In: MILLER, P.R. (ed.)Effects of air/pollutants on Mediterranean and temperate forest ecosystems. Riverside, Calif. 22. - 27.6.1980, pp. 38-50.
(180)McLAUGHLIN, S.B., McCONATHY, R.K., DUVICK, D., and MANN, L.K. (1982). Effects of chronic air pollution stress on photosynthesis, carbon allocation and growth of white pine trees. For.Sci. 28, 60-70.
(181)MILLER, P.R., TAYLOR, O.C., and WILHOUR, R.G. (1982). Oxidant air pollution effects on a western coniferous forest ecosystem. Corvallis, OR: U.S. EPA-600/D-82/276; 10 pp.
(182)MILLER, P.R., and ELDERMAN, M.J. (eds.) (1977). Photochemical oxidant air pollutant effects on a mixed conifer forest ecosystem: a progress report, 1976. Corvallis, OR.: U.S. EPA Report No. EPA-600/3-77-104.
(183)WILLIAMS, W.T. (1983). Tree growth and smog disease in the forests of California: Case history of Ponderosa pine in Southern Sierra Nevada. Environ Pollut. 30, 59-75.
(184)BERRY, C.R., and RIPPERTON, L.A. (1963). Ozone a possible cause of white pine emergence tipburn. Phytopath. 53, 552-557.
(185)BERRY, C.R., and HEPTING, G.H. (1964). Injury to eastern white pine by unidentified atmospheric constituents. For.Sci. 10, 2-13.
(186)DUCHELLE, S.F., SKELLY, J.M., SHARICK, J.L., CHEVONNE, B.I., YANG, Y.S., and NELLESON, J.E. (1983). Effects of ozone on the productivity of natural vegetation in a high meadow of the Shenandoah National Park of Virginia. J. Environm.Management 17, 299-308.
(187)GENOIT, L.F., SKELLY, J.M., MOORE, L.D., and DOCHINGER, L.S. (1982). Radial growth reductions of Pinus strobus L correlated with foliar ozone sensitivity as an indicator of ozone-induced losses in eastern forests. Can.J. For.Res. 12, 673-678.
(188)USHER, R.W., and WILLIAMS, W.T. (1982). Air pollution toxicity to eastern white pine in Indiana and Wisconsin. Plant Dis. 66, 199-204.
(189)REZABEK, C.L., MORTON, J.A., MOSHER, E.C., PREY, A.J., and CUMMINGS, J.E. (1986). Regional effects of sulfur dioxide and ozone on eastern white pine (Pinus strobus L.) in Eastern Wisconsin. Wisconsin Dept. of Natural Resources, Report 78255, 2/86, 20 p.
(190)DUCHELLE, S.F., SKELLY, J.M., and CHEVONNE, B.I.(1982). Oxidant effects on forest tree seedling growth in the Appalachian Mountains. Water, Air Soil Pollut. 18, 363-373.
(191)ASHMORE, M.R. (1984). Effects of ozone on vegetation in United Kingdom. In: GRENNFELD, P. (ed.): The evaluation and assessment of the effects of photochemical oxidants on human health, agricultural crops, forestry, materials and visibility. Swedish Environmental Research Inst., Göteborg, pp. 92-104.
(192)REICH, P.B., SCHOETTLE, A.W., RABA, R.M., and AMUNDSON, R.G. (1986). Response of soybean to low concentrations of ozone: I. Reductions in leaf and whole plant net photosynthesis and leaf chlorophyll content. J. Environ.Qual. 15, 31-36.

(193)COSTONIS, A.C., and SINCLAIR, W.A. (1969). Relationships of atmospheric ozone to needle blight of eastern white pine. Phytopathology, 59, 1566-1574.
(194)MILLER, P.R., PARAMETER, J.R., JR., TAYLOR, O.C., and CARDIFF, E.A. (1963). Ozone injury to the foliage of Pinus ponderosa. Phytopathology, 53, 1072-1076.
(195)KRESS, L.W., and SKELLY, J.R. (1982). Response of several eastern forest tree species to chronic doses of ozone and nitrogen dioxide. Plant Disease, 66, 1149-1152.
(196)MOOI, J. (1981). Influence of ozone and sulphur dioxide on defoliation and growth of poplars. Mittl. Forstl. Bundesversuch. Wien., 137, 47-51.
(197)MOOI, J. (1980). Influence of ozone on growth of two poplar cultivars. Plant Disease, 64: 772-773.
(198)HOUSTON, D.B. (1974). Response of selected Pinus strobus L. clones to fumigation with sulfur dioxide and ozone. Can. J. For. Res., 4, 65-68.
(199)DAVIS, D.D., and WOOD, F.A. (1972). The relative susceptibility of eighteen coniferous species to ozone. Phytopathology. 62, 14-19.
(200)HIBBEN, C.R.(1969). Ozone toxicity to sugar maple. Phytopathology 59, 1423-1428.
(201)ABRAHAMSON, G. (1984). Effects of acidic deposition on forest soil and vegetation. Phil.Trans.R.Soc. Lond. B. 305, 369-382.
(202)BUCK, M., AND PFEFFER, H.U. (1987). Air quality surveillance in the state Northrhine-Westfalia (FRG). LIS-Berichte Nr. 70, 21 p.
(203)ROWLAND, A., MURRAY, A.J.S., and WELLBURN, A.R. (1985). Oxides of nitrogen and their impact upon vegetation. Reviews on Environ. Health. Vol VII.
(204)TAYLOR, O.C., THOMPSON, C.R., TINGEY, D.T., and REINERT, R.A. (1975). Oxides of nitrogen. In: MUDD, J.B., and KOZLOWSKI, T.T. (eds.): Responses of plants to air pollution. Academic Press, Inc., New York, N.Y., pp. 121-139.
(205)ULRICH, B., MAYER, R., KHANNA, P.K. (1980). Chemical changes due to acid precipitation in a loess-derived soil in central Europe. Soil Science 130, 193-199.
(206)JACOBSON, J.S. (1984). Effects of acidic aerosol, fog, mist and rain on crops and trees. Phil.Trans.R.Soc.Lond. B 305, 327-338.
(207)SKOSS, J.D. (1955). Structure and composition of plant cuticle in relation to environment factors and permeability. Botanical Gazette 117, 55-72.
(208)SCHIEFERSTEIN, R.H., and LOOMIS, W.E. (1956). Wax deposits on leaf surfaces. Plant Physical. 31, 240-247.
(209)WHITZCROSS, M.I., and ARMSTRONG, D.J. (1972). Environmental effects on epicuticular waxes of Brassica napus L.. Aust.J.Bot. 20, 87-95.
(210)RENTSCHLER, I. (1973). Die Bedeutung der Wachsstruktur auf den Blättern für die Empfindlichkeit der Pflanzen gegenüber Luftverunreinigungen. In: Proc. 3rd. Internat. Clean Air Congress (IUAPPA), Düsseldorf 8.-12.10.1973, VDI-Verlag Düsseldorf, pp. A139-A142.
(211)KELLER, Th. (1986). Gehalte an einigen Elementen in den Ablagerungen auf Fichtennadeln als Nachweis der Luftverschmutzung. Allg. Forst und Jagdz. 157, 69-77.
(212)CAPE, J.N., and FOWLER, D. (1981). Changes in epicuticular wax of Pinus sylvestris exposed to polluted air. Silva Fen. 15, 457-458.
(213)CAPE, J.N. (1982). Contact angles of water droplets on needles of Scots pine (Pinus sylvestris) growing in polluted atmospheres. New Phytol. 93, 293-299.

(214)WENTZEL, K.F. (1985). Hypothesen und Theorien zum Waldsterben. Forstarch. 56, 5156.
(215)KAZDA, M., and GLATZEL, G. (1986). Schadstoffbelasteter Nebel fördert die Infection von Fichtennadeln durch pathogene Pilze. Allg. Forst Z. 41, 436-438.
(216)KLEMM, O., and FREVERT, T. (1985). Säure- und redoxchemisches Verhalten von Niederschlagswasser beim Abdampfen von Oberflächen. VDI-Berichte 560, 457-463.
(217)EATON, J.S., LIKENS, G.W., and BORMANN, F.H. (1973). Throughfall and stemflow in a northern hardwood forest. J. Ecol. 61, 495-508.
(218)McCOLL, J.B., and BUSCH, D.S. (1978). Precipitation and throughfall chemistry in the San Francisco Bay area. J. Environ. Qual. 7, 352-357.
(219)HOFFMANN, W.A., LINDBERG, S.W., and TURNER, R.R. (1980). Precipitation acidity: the role of forest canopy in acid exchange. J. Environ. Qual. 9, 95-100.
(220)CRAKER, L.W., and BERNSTEIN, D. (1984). Buffering of acid rain by leaf tissue of selected crop plants. Environ. Pollut. (Series A) 36, 375-381.
(221)SKIBA, U., PEIRSON-SMITH, T.J., and CRESSER, M.S. (1986). Effects of simulated precipitation acidified with sulfuric and/or nitric acid on the throughfall chemistry of sitka spruce Picea sitchensis and heather Caluna vulgaris. Environ. Pollut. (Series B) 11, 255-270.
(222)TUKEY, H.G. jr. (1970). The leaching of substances from plants. Ann. Rev.Plant Physiol. 21, 305-329.
(223)SCHERBATSKOY, T., and KLEIN, R.M. (1983). Response of spruce and birch foliage to leaching by acidic mists. J. Environ. Qual. 12, 189-195.
(224)MECKLENBURG, R.A., and TUKEY, H.B. Jr. (1964). Influence of foliar leaching on root uptake and translocation of calcium^{45} to the stems and foliage of Phaseolus vulgaris. Plant Physiol. (Lanc.) 39, 533-536.
(225)HINDAWI, I.J., REA, J.A., and GRIFFIS, W.L. (1980). Response of bush bean exposed to acid mist. Am.J.Bot. 67, 168-172.
(226)PLANPIED, G.D. (1979). Effect of artificial rain water pH and calcium concentrations on the calcium and potassium in apple leaves. Hort. Sci. 14, 706-708.
(227)KELLY, J.M., and STRICKLAND, R.C. (1986). Throughfall and plant nutrient concentration response to simulated acid rain tratment. Water, Air and Soil Pollut. 29, 219-231.
(228)PROCTOR, J.T.A. (1983). Effect of simulated sulfuric acid rain on apple tree foliage, nutrient content, yield and fruit quality. Environ. Exp. Bot. 23, 167-174.
(229)PARKER, G.G. (1983). Throughfall and stemflow in the forest nutrient cycle. Adv. Ecol. Res. 13, 56-120.
(230)WOOD, T., and BORMANN, F.H. (1974). Effects of an artificial acid mist upon the growth of Betula alleghaniensis. Environ. Pollut. (Series A) 7, 259-268.
(231)WOOD, T., and BORMANN, F.H. (1975). Increases in foliar leaching caused by acidification of an artificial mist. Ambio 4, 169-171.
(232)EVANS, L.S., CURRY, T.M., and LEWIN, K.F. (1981). Responses of leaves of Phaseolus vulgaris L. to simulated acid rain. New Phytol. 88, 403-420.
(233)EVANS, L.S., LEWIN, F.F., CONWAY, C.A., and PATTI, M.J. (1981). Seed yield (quantity and quality) of field-grown soybeans exposed to simulated acidic rain. New Phytol. 89, 459-470.

(234)EVANS, L.S., GMUR, N.F., and MANCINI, D.(1982). Effects of simulated acidic rain on yields of *Raphanus sativus*, *Lactuca sativa*, *Triticum aestivum* and *Medicago sativa*. Environ. Exp. Bot. *22*, 445-453.

(235)FERENBAUGH, R.W. (1976). Effects of simulated acid rain on *Phaseolus vulgaris* L. (Fabaceae). Am.J.Bot. *63*, 283-288.

(236)LEE, J.J., NEELY, G.E., PEERIGAN, S.C., and GROTHAUS, L.C. (1981). Effect of simulated sulfuric acid rain on yield, growth and foliar injury of several crops. Environ. Exp. Bot. *21*, 171-185.

(237)EVANS, L.S., and LEWIN, K.F. (1981). Growth, development and yield responses of pinto beans and soybeans to hydrogen ion concentration of simulated acidic rain. Environ. Exp. Bot. *21*, 103-113.

(238)IRVING, P.M. (1983). Acidic precipitation effects on crops: a review and analysis of research. J. Environ. Qual. *12*, 442-453.

(239)TROIANO, J., COLAVITO, L., HELLER, L., McCUNE, D.C., and JACOBSON, J.S. (1983). Effects of acidity of simulated rain and its joint action with ambient ozone on measures of biomass and yield in soybean. Environ. Exp. Bot. *23*, 113-119.

(240)KRAUSE, G.H.M., JUNG, K.D., and PRINZ, B. (1983). Neuere Untersuchungen zur Aufklärung immissionsbedingter Waldschäden. VDI-Berichte *500*, 257-266.

(241)ONGER, G., and TEIGEN, O. (1980). Effects of acid irrigation at different temperatures on seven clones of Norway spruce. Meddeleser fra Norsk Institutt for Skogforskning *36*, 3-38.

(242)CICCIOLI, P., BRANCALEONI, M., POSSANZINI, M., BRACHETTI, A., and DI-PAZO, C. (1984). Sampling, identification and quantitative determination of biogenic and anthropogenic hydrocarbons in forestal areas. In: 3rd E.C. Symposium on *Physico-chemical behaviour of atmospheric pollutants*, Varese (Italy) 10.-12.4.1984, pp. 134-145.

(243)FRANK, H. (1985). The new forest decline. Destruction of photosynthetic pigments by UV-activation of chloroethenes; a probable cause of forest decline. EPA-NEWESLETTER 04/06 1985, 7-13.

(244)HETTCHE, H.O.(1971). Pflanzenwachse als Sammler für polyzyklische Aromaten in der Luft von Wohngebieten. Staub.-Reinhalt. Luft *31*, 72-76.

(245)VAN HAUT, H. (1972). Nachweis mehrerer Luftverunreinigungskomponenten mit Hilfe von Blätterkohl (*Brassica oleracea acephala*) als Indikatorpflanze. Staub.-Reinhalt. Luft *32*, 109-111.

(246)BRORSTRÖM-LUNDEN, E., and SKÄRBY, L. (1984). Plants as monitoring samplers of air-borne PAH. In: 3th E.C. *Symposium on Physico-chemical behaviour of atmospheric pollutants*, Varese (Italy), 10.-12.4.1984, pp. 166-175.

(247)DUGGER, W.M. Jr., KAUKOL, J. READ, W.D., and PALMER, R.L. (1963). Effect of PAN on $C^{14}O_2$ fixation by spinach chloroplasts and pinto bean plants. Plant Physiol *38*, 468-472.

(248)FRANK, H., and FRANK, W. (1985). Chlorophyll bleaching by atmospheric pollutants and sunlight. Naturwissenschaften *72*, 139-141.

(249)JÜTTNER, F. (1986). Analyse stofflicher Veränderungen in Laub- und Nadelblättern immissionsgeschädigter Waldbäume. Spezielle Berichte der Kernforschungsanlage Jülich, FRG, *369*, 313-316.

(250)JÜTTNER, F.(1987). Jahresdynamik der Terpenkonzentration in Nadeln unterschiedlich stark immissionsgeschädigter Fichten (*Picea abies*). PEF-Bericht *10*, Kernforschungszentrum Karlsruhe, pp. 89-106 (ISSN No 0931-2749).

(251)VAN HAUT, H., PRINZ, B., and HÖCKEL, F.E. (1979). Ermittlung der relativen Toxioität von Luftverunreinigungen im LIS-Kurzzeittest - Verschiedene organische Komponenten und Ammoniak. Schriftenr. d. Landesanstalt für Immissionschutz des Landes NW 49, 29-65.
(252)KRAUSE, G.H.M., and HÖCKEL, F.E. (1987). Effects of organic air pollutants, NH_4 and H_2S on plants. In preparation.
(253)FAULSTICH, H., and STOURNARAS, C. (1985). Potentially toxic concentrations of triethyllead in 'Black Forest' rainwater samples. Nature 317, 714-715.
(254)FAULSTICH, H. (1986). Triäthylblei als mögliche Ursache der Waldschäden. Spezielle Berichte der Kernforschungsanlage Jülich, 369, 305-311.
(255)UNSWORTH, M., and HARRISON, R.M. (1985). Is lead killing German forests? Nature 317, 674.
(256)GARSED, S.G., and RUTTER, A.J. (1984). The effects of fluctuating concentrations of sulfur dioxide on the growth of Pinus sylvestris and Picea sitchensis (Bong.) Carr. New Phytol 97, 175-195.
(257)KRAUSE, G.H.M. (1986). Zusammenfassende Bewertung der Seminarergebnisse: Die Wirkung von Schwermetallen, einschließlich ihrer organischen Verbindungen. Spezielle Berichte der Kernforschungsanlage Jülich 369, 362-365.
(258)REICH, P.B., and LASSOIE, J.P. (1985). Influence of low concentrations of ozone on growth, biomass partitioning and leaf senescence in young hybrid poplar plants. Environ. Pollut. (Series A) 39, 39-51.
(259)LAST, F.T., and FOWLER, D. (1984). Koch's postulates and the aerial pollution environment. Forstw.Cbl. 103, 24-28.
(260)MANDL, R.H., L.H. WEINSTEIN, and KEVENY, M. (1975). Effects of hydrogen fluoride and sulphur dioxide alone and in combination on several species of plants. Environ. Pollut. 9, 133-143.
(261)MANDL, R.H., WEINSTEIN, L.H., DEAN, M. and WHEELER, M. (1980). The response of sweet corn to HF and SO_2 under field conditions. Environ. Exp. Bot. 20, 359-365.
(262)BECKERSON, D.W., and HOFSTRA, G. (1979). Response of leaf diffusive resistance of radish, cucumber and soybean to O_3 and SO_2 singly or in combination. Atmos. Environ. 13, 1263-1268.
(263)BECKERSON, D.W., and HOFSTRA, G. (1979). Stomatal responses of white bean to O_3 and SO_2 singly or in combination. Atmos. Environ. 13, 533-535.
(264)UNSWORTH, M.H., and BLACK, V.J. (1981). Stomatal responses to pollutants. In: JARVIS, P.G., and MANSFIELD, T.A. (eds.): Stomatal physiology. University Press, London, Cambridge, pp. 187-203.
(265)JENSEN, K.F., and ROBERTS, B.R. (1986). Changes in yellow poplar stomatal resistance with SO_2 and O_3 fumigation. Environ. Pollut. (Series A) 41, 235-245.
(266)KOBRINGER, J.M., and TIBBITTS, T.W. (1985). Effect of relative humidity prior to and during exposure on response of peas to ozone and sulfur dioxide. J. Amer.Soc.Hort.Sci 110, 21-24.
(267)ELKIEY, T., and ORMROD, D.P. (1981). Sulphite, total sulphur and total nitrogen accumulation by petunia leaves exposed to ozone, sulphur dioxide and nitrogen dioxide. Environ. Pollut. (Ser. A). 24, 233-241.
(268)ELKIEY, T., and ORMROD, D.P. (1981). Sorption of O_3, SO_2, NO_2 or their mixtures by nine Poa pratensis cultivars of differing pollutant sensitivity. Atmos. Environ. 15, 1739-1743.

(269) TINGEY, D.T., HECK, W.W., and REINERT, R.A. (1971). Effect of low concentrations of ozone and sulfur dioxide of foliage, growth and yield of radish. J. Amer.Soc.Hort.Sci. 96, 369-371.
(270) HEAGLE, A.S., BODY, D.E., and NEELY, G.E. (1974). Injury and yield responses of soybean to chronic doses of ozone and sulfur dioxide in the field. Phytopathology 64, 132-136.
(271) ELKIEY, T., and ORMROD, D.P. (1979). Ozone and/or sulphur dioxide effects on tissue permeability in petunia leaves. Atmos. Environ. 13, 1165-1168.
(272) GROSSO, J.J., MENSER, H.A., HODGES, G.H., and McKINNEY, H.H. (1971). Effects of air pollutants on Nicotiana cultivars and species used for virus studies. Phytopathology 61, 945-950.
(273) KENDER, W.J., and SPIERINGS, F.H.F.G.(1975). Effects of sulfur dioxide, ozone, and their interactions on 'Golden Delicious' apple trees. Neth. J. Plant Pathol. 81, 149-151.
(274) SHERTZ, R.D., KENDER, W.J., and MUSSELMAN, R.C. (1980). Foliar response and growth of apple trees following exposure to ozone and sulfur dioxide. J. Amer. Soc. Hort. Sci. 105, 594-598.
(275) LEWIS, E., and BRENNAN, E. (1978). Ozone and sulfur dioxide mixtures cause a PAN-type injury to petunia. Phytopath. 68, 1011-1014.
(276) COSTONIS, A.C. (1973). Injury to eastern white pine by sulfur dioxide and ozone alone and in mixtures. Eur. J. Forest Pathol. 3, 50-55.
(277) DOCHINGER, L.S., BENDER, F.W., BOX, F.O. and HECK, W.W.(1970). Chlorotic dwarf in eastern white pine caused by an ozone and sulphur dioxide interaction. Nature 255, 476.
(278) TINGEY, D.T., REINERT, R.A., WICKLIFF, C., and HECK, W.W. (1973). Chronic ozone or sulfur dioxide exposures, or both, affect the early vegetative growth of soybean. Can. J. Plant Sci. 53, 875-879.
(279) HEAGLE, A.S., HECK, W.W., RAWLINGS, J.O., and PHILBECK, R.B. (1983). Effects of chronic doses of ozone and sulfur dioxide on injury and yield of soybeans in open-top chambers. Crop Sci. 23, 1184-1191.
(280) TINGEY, D.T., and REINERT, R.A. (1975). The effects of ozone and sulphur dioxide singly and in combination on plant growth. Environ. Pollut. 9, 117-125.
(281) NEELY, G.E., TINGEY, D.T., WILHOUR, R.G. (1977). Effects of ozone and sulfur dioxide singly and in combination on yield, quality, and N-fixation of alfalfa. In: DIMITRIADES, B. (ed.): International conference on photochemical oxidant pollution and its control. EPA-600/3-77-001b, EPA, Research Triangle Park, North Carolina, pp. 663-673.
(282) HOFSTRA, G., and ORMROD, D.P. (1977). Ozone and sulphur dioxide interaction in white bean and soybean. Can. J. Plant Sci. 57, 1193-1198.
(283) REICH, P.B., AMUNDSON, R.G., and LASSOIE, J.P. (1982). Reduction in soybean yield after exposure to ozone and sulfur dioxide using a linear gradient exposure technique. Water, Air, Soil Pollut. 17, 29-36.
(284) REICH, P.B., and AMUNDSON, R.G.(1984). Low level O_3 and/or SO_2 exposure causes a linear decline in soybean yield. Environ. Pollut. 34, 345-355.
(285) ASHMORE, M.R. (1984). Modification by sulfur dioxide of the responses of Hordeum vulgare to ozone. Environ. Pollut. (Series A) 36, 31-43.
(286) JENSEN, K.F. (1981). Air pollutants affect the relative growth rate of hardwood seedlings. US Dep. Agric. For. Serv. Res. Paper NE 470.
(287) NOBEL, R.D., and JENSEN, K.F. (1980). Effects of sulfur dioxide and ozone on growth of hybrid poplar leaves. Am.J. Bot. 67, 1005-1009.

(288)CONSTANTINIDOU, H.A., and KOZLOWSKI, T.T.(1979). Effects of sulfur dioxide and ozone on Ulmus americana seedlings. I. Visible injury and growth. Can. J. Bot. 57, 170-175.

(289)REICH, P.B., LASSOIE, J.P., and AMUNDSON, R.G. (1984). Reduction in growth of hybrid poplar following field exposure to low levels of O_3 and (or) SO_2. Can. J. Bot. 62, 2835-2841.

(290)MEHLHORN, H., SEUFERT, G., SCHMIDT, A., and KUNERT, K.J. (1986). Effect of SO_2 and O_3 on production of antioxidants in conifers. Plant Physiol. 82, 336-338.

(291)KETTRUP, A., MASUCH, G., KICINSKI, H.G., and BOOS, K.S. (1987). Untersuchungen zur Schädigung von Jung- und Altfichten (Picea abies L.) und Jung- und Altbuchen (Fagus sylvatic L.) durch Ozon als Luftverunreinigung. Annual Report 1986 of BMFT-Project 03-7399-0, University of Paderborn.

(292)ELKIEY, T., and ORMROD D.P. (1981). Absorption of ozone, sulphur dioxide, and nitrogen dioxide by petunia plants. Environmental Exper. Bot. 21, 63-70.

(293)NATORI, T., and TOTSUKA T. (1984). Effects of mixed gas on transpiration rate of several woody plants. 1. Interspecific difference in the effect of mixed gas on transpiration rate, pp. 45-52. In: Studies on Effects of Air Pollutant Mixtures on Plants. Part 1. Research Report No. 65. The National Institute for Environmental Studies, Yatabe-machi, Tsukuba, Ibaraki 305, Japan.

(294)AMUNDSON, R.G., and WEINSTEIN L.H. (1981). Joint action of sulfur dioxide and nitrogen dioxide on foliar injury and stomatal behaviour in soybean. Environ. Qual. 10, 204-206.

(295)GUDERIAN, R., VOGELS, K., and MASUCH G. (1986). Comparative physiological and histological studies on Norway spruce (Picea abies Karst.) using climatic chamber experiments and field studies in damaged forest stands. 7th World Clean Air Congress & Exhibition. Sydney, Australia - 25-29 August 1986.

(296)NATORI, T., and TOTSUKA, T. (1984). Effects of mixed gas on transpiration rate of several woody plants. 2. Synergistic effects of mixed gas on transpiration rate of Euonymus japonica, pp. 55-61. In: Studies on Effects of Air Pollutant Mixtures on Plants. Part 1. Research Report No. 65. The National Institute for Environmental Studies, Yatabe-machi, Tsukuba, Ibaraki 305, Japan.

(297)ASHENDEN, T.W. (1979). Effects of SO_2 and NO_2 pollution on transpiration in Phaseolus vulgaris L. Environ. Pollut. 18, 45-50.

(298)FURUKAWA, A., YOKOYAMA M., USHIJIMA T., TOTSUKA T. (1984). The effects of NO_2 and/or O_2 on photosynthesis of sunflower leaves, pp. 89-97. In: Studies on Effects of Air Pollution Mixtures on Plants. Part 1. Research Report No. 65. The National Institute for Environmental Studies, Yataba-machi, Tsukuba, Ibaraki 305, Japan.

(299)YONEYAMA, T., and TOTSUKA, T. (1984). Enhancement of damage in sunflower plants by probable involvement of factors generated in the mixing of NO_2 and O_3. pp. 233-239. In: Studies on Effects of Air Pollutant Mixtures on Plants. Part 1. Research Report No. 65. The National Institute for Environmental Studies, Yatabe-machi, Tsukuba, Ibaraki 305, Japan.

(300)BELL, J.N., and MANSFIELD, T.A. (1974). Photosynthesis in leaves exposed to SO_2 and NO_2. Nature 250, 443-444.

(301)WHITE, K.L., HILL, A.C., and BENNETT, J.H. (1974). Synergistic inhibition of apparent photosynthesis rate of alfalfa by combinations of sulfur dioxide and nitrogen dioxide. Environ. Sci. Technol. 574, 476-478.

(302) CARLSON, R.W. (1983). Interaction between SO_2 and NO_2 and their effects on photosynthetic properties of soybean Glycine max. Environ. Pollut. (Series A) 32, 11-38.
(303) SUGAHARA, K., OGURA, K., TAKIMOTO, M., and KONDO, N. (1984). Effects of air pollutant mixtures on photosynthetic electron transport systems. In: Studies on Effects of Air Pollutant Mixtures on Plants. Part 1. Research Report No. 65. The National Institute for Environmental Studies, Yatabe-machi, Tsukuba, Ibaraki 305, Japan.
(304) WELLBURN, A.R., HIGGINSON, C., ROBINSON, D., and WALMSLEY, F. (1981). Biochemical explanations of more than additive inhibitory effects of low atmospheric levels of sulphur dioxide plus nitrogen dioxide upon plants. New Phytol. 88, 223-237.
(305) REINERT, R.A., and GRAY, T.N. (1981). The response of radish to nitrogen dioxide, sulfur dioxide, and ozone, alone and in combination. J. Environ. Qual. 10, 240-243.
(306) KLARER, C.I., REINERT, R.A., and HUANG, J.S. (1984). Effect of sulfur dioxide and nitrogen dioxide on vegetative growth of soybeans. Phytopathol. 74, 1104-1106.
(307) LUTTRINGER, M., and deCORMIS, L. (1975). Action synergique du dioxyde de soufre et des oxydes d'azote sur quelques vegetaus: etudes en conditions artificielles. C.R. Seances Acad. Agriculture France 61, 827-835.
(308) FREER-SMITH, P.H. (1984). The responses of six broadleaved trees during long-term exposure to SO_2 and NO_2. New Phytol. 97, 49-61.
(309) KRESS, L.W., SKELLY, J.M., and KINKELMANN, K.H. (1982). Growth impact of O_3, NO_2, and/or SO_2 on Pinus taeda. Environ. Monitoring Assessment 1, 229-239.
(310) MOOI, J. (1984). Wirkungen von SO_2, NO_2, O_3 und ihrer Mischungen auf Pappeln und einige andere Plfanzenarten. Forst- und Holzw. 39, 438-444.
(311) GUDERIAN, R., KÜPPERS, K., and SIX, R. (1985). Wirkungen von Ozon, Schwefeldioxid und Stickstoffdioxid auf Fichte und Pappel bei unterschiedlicher Versorgung auf Magnesium und Kalzium sowie auf die Blattflächte Hypogymnia physodes. VDI-Berichte 560, 657-701.
(312) BENNETT, J.H., HILL, A.C., SOLEIMANI, A., and EDWARDS, W.H. (1975). Acute effects of combinations of sulfur dioxide and nitrogen dioxide on plants. Environ. Pollut. 9, 127-132.
(313) ASHENDEN, T.W., and WILLIAMS, I.A.D. (1980). Growth reductions in Lolium multiflorum Lam. and Phleum pratense L. as a result of SO_2 and NO_2 pollution. Environ. Pollut. (Series A) 21, 131-139.
(314) WHITMORE, M.E., and MANSFIELD, T.A. (1983). Effects of long-term exposures to SO_2 and NO_2 on Poa pratensis and other grasses. Environ. Pollut. (Series A) 31, 217-235.
(315) LANE, P.I., and BELL, J.N.B. (1984). The effects of simulated urban air pollution on grass yield: Part I - Description and simulation of ambient pollution. Environ. Pollut. (Series B) 8, 245-263.
(316) LANE, P.I., and BELL, N.J.B. (1984). The effects of simulated urban air pollution on grass yield: Part 2 - performance of Lolium perenne, Phleum pratense, and Dactylis glomerata fumigated with SO_2, NO_2, and/or NO. Environ. Pollut (Series A) 35, 92-124.
(317) PETTITE, J., M., and ORMROD, D.P. (1984). Effects of sulphur dioxide and nitrogen dioxide on four Solanum tuberosum cultivars. Amer. Potato J. 61, 319-330.
(318) IRVING, P.M., and MILLER, J.E.(1984). Synergistic effect on field grown soybeans from combinations of sulfur dioxide and nitrogen dioxide. Can. J. Bot. 62, 840-846.

(319)GODZIK, S., ASHMORE, M.R., and BELL, J.N.B. (1985). Responses of radish cultivars to long-term and short-term exposures to sulphur dioxide, nitrogen dioxide, and their mixture. New Phytol. 100, 191-197.
(320)SHIMIZU, H., OKOAWA, T., and TOTSUKA, T. (1984). Effects of low concentration of NO_2 and O_3 alone and in mixture on the growth of sunflower plants, pp. 121-126. In: Studies on Effects of Air Pollutant Mixtures on Plants. Part 1. Research Report No. 65. The National Institute for Environmental Studies, Yatabe-machi, Tsukuba, Ibaraki 305, Japan.
(321)ELKIEY, T., and ORMROD, D.P. (1980). Response of turfgrass cultivars to ozone, sulfur dioxide, nitrogen dioxide, or their mixtures. J. Amer. Soc. Hort. Sci. 105, 664-668.
(322)SANDERS, J.A., and REINERT, R.A.(1982). Screening azalea cultivars for sensitivity to nitrogen dioxide, sulfur dioxide, and ozone alone and in mixtures. J. Amer. Soc. Hort. Sci. 107, 87-90.
(323)REINERT, R.A., and HECK, W.W. (1982). Effects of nitrogen dioxide in combination with sulfur dioxide and ozone on selected crops. In: SCHNEIDER, T., and GRANT, L. (eds.): Air Pollution by nitrogen Oxides. Elsevier Scientific Publishing Company, Amsterdam, The Netherlands, pp. 533-546.
(324)FREER-SMITH, P.H. (1985). The influence of gaseous SO_2 and NO_2, and their mixtures on the growth and physiology of conifers. In: KLIMO, E., and SALY, R. (eds.): Air pollution and the stability of coniferous forest ecosystems. Internat. Symp. 1.-5.10.1984, University of Agriculture Breno, CSSR, pp. 235-247.
(325)YANG, Y.S., SKELLY, J.,M., and CHEVONE, B.I. (1983). Effects of pollutant combinations at low doses on growth of forest trees. Aquilo Ser. Bot. 19, 406-418.
(326)KRESS, L.W., and Hinkelmann, K.H. (1982). Growth impact of O_3, NO_2 and/or SO_2 9on Platanus occidentalis. Agric. Environ. 7, 265-274.
(327)REICH, P.B. SCHOETTLE, A.W., and AMUNDSON, R.G. (1986). Effects of O_3 and acidic rain on photosynthesis and growth in sugar maple and northern red oak seedlings. Environ. Pollut. (Series A) 40, 1-15.
(328)REICH, P.B., SCHOETTLE, A.W., STROO, H.F., and AMUNDSON, R.G. (1986). Acid rain and ozone influence mycorrhizal infection in tree seedlings. JAPCA 36, 724-726.
(329)BOSCH, Chr., PFANNKUCH, E., REHFUESS, K.E., RUNKEL, K.H., SCHRAMEL, P., and SENSER, M. (1986). Einfluß einer Düngung mit Magnesium und Calcium, von Ozon und saurem Nebel auf Frosthärte, Ernährungszustand und Biomasseproduktion junger Fichten (Picea abies [L.] Karst.). Forstw. Cbl. 105, 218-242.
(339)SKEFFINGTON, R.A.,and ROBERTS, T.M. (1985). The effects of ozone and acid mist on Scots pine saplings. Oecologia 65, 201-206.
(331)SEUFERT, G.,and ARNDT, U. (1986). Beobachtungen in definiert belasteten Modellökosystemen mit jungen Bäumen. Allg. Forst Z. 41, 545-549.
(332)REICH, P.B., and AMUNDSON, R.G. (1985). Ambient levels of ozone reduce net photosynthesis in tree and crop species. Science 230, 560-570.
(333)KRAUSE G.H.M. (1987). Zur Wirkung von Ozon auf die Nitratbildung in Nadeln und Blättern von Picea abies [L.] Karsten, Fagus sylvatrice L. und Quercus robur L. In: Ministerium für Umwelt, Raumordnung und Landwirtschaft des Landes NW, Schwanstraße 3, 4000 Düsseldorf 30 (ed).: Symposium über Luftverunreinigungen und Waldschäden, Düsseldorf 29.-30.10.1986, pp. 27-31.

(334)PRINZ, B.(1987). Einführung und Überblick über den Themenbereich E: Oberirdischer Wirkungspfad, Streß und Stoffwechselphysiologie. Statusseminar: Ursachenforschung zu Waldschäden, Kernforschungsanlage Jülich, FRG 30.03. - 03.04 1987.

(335)CROSSLEY, A., and FOWLER, D. (1986). The weathering of Scots pine epicuticular wax in polluted and clean air. New Phytol. 103, 207-218.

(336)MAURICE, C.G., and CRANG, R.E. (1986). Increase in Pinus strobus needle transectional areas in response to acid misting. Arch.Environ.Contam.Toxicol. 15, 77-82.

(337)LEGGE, A. H., JAQUES, D. R., HARVEY, G. W., KROUSE, H.R., BROWN, H. M., RHODES, E. C., NOSAL, M., SCHELLHASE, H. U., MAYO, J., HARTGERING, A. P., LESTER, P. F., AMUNDSON, R. G., and WALKER, R. B. (1981). Sulphur gas emissions in the boreal forest: the West Whitecourt case study. I: Executive Summary. Water, Air, Soil Poll. 15, 77-85.

(338)LESTER, P. F., RHODES, E. C., and LEGGE, A. H. (1986). Sulphur gas emissions in the boreal forest: The West Whitecourt case study. IV: Air quality and the meterological environment. Water, Air, Soil Poll. 27, 85-108.

(339)HARVEY, G. W., and LEGGE, A. H. (1979). The effect of sulfur dioxide upon the metabolic level of adenosine triphosphate. Can J. Bot. 57 759-764.

DIRECT EFFECTS OF DRY AND WET DEPOSITION ON FOREST ECOSYSTEMS - IN PARTICULAR CANOPY INTERACTIONS.

P.H. Freer-Smith
Biology Dept. University of Ulster,
Shore Road, Newtownabbey, Co. Antrim
Northern Ireland. BT37 0QB U.K.

Summary

This paper is a synthesis and report on the workshop held in Lökeberg, Sweden in October 1986 which was organized in the framework of the Concerted Action "Effects of Air Pollution on Terrestrial and Aquatic Ecosystems" (COST 612) project, by the Commission of the European Communities, the Swedish University of Agricultural Sciences and the Swedish Environmental Research Institute. The meeting considered four themes each of which were introduced by an invited review and then contributed to by research presentations (1), the main points made and some general comments on each theme are given below. I will not review those papers concerned with methodology because this topic will be considered in Part III of the session on forest ecosystems.

1. EFFECTS OF DRY AND WET DEPOSITION ON TREE CANOPIES- IN PARTICULAR MINERAL AND NUTRITIONAL STATUS.

The review article reminded participants that nutrient and pollutant inputs to forest ecosystems are by **dry deposition** of gases and by **wet deposition** in rainfall and that these include the interception of aerosols and of mist in so called 'occult precipitation'. The forest canopy has an internal cycle of elements and, as water passes through the canopy it can gain or lose elements resulting in **foliar uptake** and **crown leaching** respectively. Recently it has been suggested that foliar uptake may be a major, or for some elements (Na, Cl, S) even the sole input to closed canopies. Data were presented showing that **throughfall** and **stemflow** may gain or lose protons giving increases or decreases of pH relative to **bulk precipitation**. Generally the neutralization of rainfall pH is less effective in winter, broadleaved species tend to neutralize the water which passes through their canopies as do young conifers, while after canopy closure older conifers may acidify throughfall and stemflow.

In contrast to the loss or gain of H^+ which cannot then be ascribed to damage, the crown leaching of nutrients is a component of many of the hypotheses which have been proposed as explanations of forest decline. It was suggested that it is now important to determine the significance of crown leaching and to consider whether it should be viewed with concern. Precipitation of low pH may give rise to greater crown leaching and we know that pine conserves it's foliar nutrients while spruce is more 'careless' in losing ions from the canopy. We have evidence for example, of Mg deficiency in a range of species, a deficiency which has previously only been a problem in nurseries. Finally, consideration was given to the model proposed by Bosch *et al* (2) which has linked O_3 damage to membranes which may be exacerbated by frost and leads to crown leaching, impairment of root growth and function and finally forest decline.

The presentations of experimental work which followed this theme reaffirmed many of the general statments made in the key paper and some went further in explaining the mechanisms responsible. Mengel *et al* showed that experimental applications of acid fog (pH 2.75) increased the quantities of K^+, Ca^{2+}, Mg^{2+}, Mn^{2+}, Zn^+ and carbohydrates leached from five-year old Norway spruce as compared to leaching from trees misted at pH 5.0. At the end of the seven-week experiment the elemental content of needles had not been decreased by the additional crown leaching in the acid misted trees. This finding is in agreement with other experimental evidence, but of additional interest were the repeated observations of needle blemish on the acid misted trees. The older needles discolouring to light greenish grey and the current year needles to reddish brown before being shed. Furthermore scanning electron micrographs showed that **acid fog** caused the **epicuticular wax** on needles to lose it's fine wax rod structure. Similar discolouration and changes of wax structure have been seen in some instances where *Picea abies* are in decline. Comments to this paper suggested that O_3 may have been present during misting and pointed out that reddening of spruce needles is often seen at low altitude while yellowing is more common at high altitude and has been associated with Mg^{2+} deficiency.

The paper presented by Seufert showed that results are now available from the open-top chamber system being run by Hohenheim University. This is a complex system with chambers in which fir, spruce and beech receive artificial rain as a mist at pH 4 and either control, O_3 alone (50-180 $\mu g\ m^{-3}$), SO_2 alone (30-40 $\mu g\ m^{-3}$) or both gases. SO_2 increased the proton and sulphate content of throughfall and it's Mg and Ca content and also the $SO_4{}^{2-}$, H^+, Mn^{2+}, Ca^{2+} and Mg^{2+} contents of the seepage water. The foliage contents of S and Mn are elevated and the Mg and Ca contents are decreased at times. Two important points emerged from this paper: Firstly, the effects described above were markedly greater for $SO_2 + O_3$ than they were for SO_2 alone; leading to the conclusion that S uptake as well as leaching is greater when O_3 is present giving accelerated turnover of S and of other elements. Secondly, as in the acid mist study described above, losses of the fine structure of epicuticular wax and **cracking** of the **wax plugs** in the stomatal antechambers were observed. The open-top chamber work suggested that for this symptom rainfall pH and SO_2 exposure were more important than O_3, and there remains some controversy as to whether the cracks are an artefact of specimen preparation to which polluted needles may be more susceptible. Work reported later in the workshop by Mcleod also showed that exposure to O_3 and acid mist can increase the leakiness of pine and spruce shoots to Mg, Ca, K, NO_3 & Mn but not P, although the foliar nutrient contents may not decrease. The experiments conducted by Johnson and his co-workers in open-top chambers did however identify decreases of S and Mg levels in *Picea abies* needles after exposure to $NO_2 + O_3$. Combinations of SO_2, NO_2 & O_3 increased the S content of foliage. Exposure to acid rain (2.4) also increased needle S concentrations and lowered foliar N and cation content. The effects detected in the Giessen, Hohenheim and Copenhagen studies are consistent and generally the finding that pollutants increase crown leaching seems conclusive.

Two papers were then presented on one of the few areas where the major cause of forest decline can be identified with a degree of certainty that is the southeast of The Netherlands (Roelofs & Van Dijk). In this area nitrogen deposition to forests as gaseous **ammonia** and **ammonium sulphate** in precipitation is 10 to 20 times the normal rates as a result of intensive stockbreeding and slurry spreading. The needles of Scots pine, Corsican pine and Douglas fir show very distinct yellowing and bleaching, and have very large nitrogen contents mainly as a result of the accumulation of the N-rich aminoacid arginine. These needles also undergo a **cation exchange** losing K, Mg, and Ca as they gain N, and fungal infection with *Sphaeropsis sapinea* is also associated with these changes. Large N deposition was also invoked as an explanation for the differences in the nutrient concentrations of spruce needles sampled regularly at three forest sites in Southern Sweden by Nihlgard. This paper provides data on seasonal and year-class differences in elemental content in relation to the elemental load of wet deposition but, as in many studies of this type, the conclusions are limited due to the unavailability of a true control site. In a second forest study in southern Sweden, Bergkvist, Folkeson & Olsson were able to describe the influence of tree species on throughfall, stemflow and soil solution as a result of analysis of these fractions from five adjoining stands of *Picea abies*, *Fagus sylvatica* and *Betula pendula*. Stemflow was more acidic than throughfall as in previous studies and was also more acidic in summer. The soil solutions had larger pH values beneath birch and beech than beneath spruce, and for spruce stemflow was always more acidic than bulk precipatition. Compared with spruce, the beech and birch showed less elemental flux but metal cations, particularly **aluminium**, and nitrate and sulphate all increased in concentration in precipitation with it's passage through the spruce canopy.

The total annual sulphate input to a *Pinus sylvestris* canopy from both wet and dry deposition, estimated from analysis of rainfall and by **eddy correlation** respectively, have been compared with the loss in stemflow and throughfall by Cape *et al*. Sulphate losses in stemflow plus throughfall exceeded inputs in rain. Of this 'excess' sulphate (3.16 g S m^{-2} $year^{-1}$) only half could be accounted for by dry deposition and by the interception of mist and fine rain. The remaining 'excess' lost from the canopy must be leached from the needles. This means that rates of dry deposition cannot be estimated from the difference between S inputs and losses from canopies, and further it seems important to understand why this amount of S is taken up only to be lost by crown leaching. We return to the question as to whether crown leaching should be viewed with concern?

2. INTERACTIVE EFFECTS OF DRY AND WET DEPOSITION ON STRUCTURE AND FUNCTION OF CANOPIES, IN PARTICULAR INTERACTIVE EFFECTS OF SO_2, NO_x AND O_3.

The key paper of this theme was presented by H. Ziegler and considered the entry of gaseous pollutants, aerosol particles and droplets into foliage. The unusual feature of the flux pathway into conifer needles is the plug of wax rodlets which accumulates with time in the antechambers of the stomatal pores. These plugs may be more permeable to CO_2 than to H_2O and their permeability to pollutants may be critical. Information on the development of cracks in these plugs in response to environmental stress was reviewed and experimental evidence was presented which showed that exposure of *Picea abies* to 150 $\mu g\ m^{-2}\ O_3$ causes aggregation of the wax rodlets in antechambers, while acid mist stimulates the cracking of the plug structure. The author emphasises the potential of such cracks as pathways for the uptake of aerosol particles and water droplets which carry large concentrations of pollutant molecules. The **permeability coefficients** of isolated cuticles to various gases were quoted, showing that the permeability of SO_2 is similar to that of O_2, but that cuticles are effectively impermeable to O_3 unless leaf veins or hairs are present. Of particular interest were the permeabilities obtained for NO_x. For *Abies alba* values of permeability were an order of magnitude greater than those obtained for the other gases and some NO_x appears to become permanently bound to the cuticle matrix giving an increase in mass and also increasing the cuticle's permeability to H_2O five-fold. The penetration and impacts of pollutants at the cellular level was also reviewed, with attention being drawn to the often synergistically interacting effects of pollutant combinations on photosynthesis and to the effects of sulphite on **phloem loading**. The extreme sensitivity of Strasburger cells which are considered to be responsible for sieve element loading may be responsible for the accumulation of starch in the needles of conifers which has been observed in the healthy upper parts of trees showing sub-top damage (3).

The results from the open-top system being run by the Univ. of Göteborg and The Swedish Environmental Research Institute confirm that O_3 exposure (84 $\mu g\ m^{-2}$) supresses the **net photosynthesis** of *Pinus sylvestris* by about 40 %, and show that this is accompanied by a decrease in transpiration of similar magnitude. Microscopic examination of pine and spruce needles from these chambers revealed that the chloroplasts of needles exposed to O_3 were smaller with denser stroma, suggesting that disrupted gas exchange is the result of damage to the photosynthetic apparatus. This ultrastructural damage progresses with disruption spreading to the cytoplasm and with the cells further from the epidermis gradually developing injury. Even earlier detection of injury to the photosynthetic apparatus of *Picea abies* was reported by Barnes and Davison, who were able to identify decreased chlorophyll fluorescence from needles after only 12 days of fumigation with 140 ppb O_3. This technique has considerable potential in diagnostic studies since it uses only small quantities of plant tissue (see 4), and the characteristics of fast response fluorescence also provide information on the site of injury. For the O_3 impacts on spruce this appears to have been electron transport on the donor side of PS II.

Wellburn and Wolfenden presented data which suggested that acidic precipitation may lead to lower vacuolar, rather than cytoplasmic, pH in young Sitka spruce. Recent techniques available for measuring intracellular pH were discussed. The effects of artificial acid precipitation (pH 5.6, 4.3 & 3.5) on the gas exchange of Norway spruce were reported by Van Elsacker and Impens. Leaf conductance of current year needles increases with decreasing pH, giving increases in **transpiration rates** which exceed those of net photosynthesis so that **water-use-efficiency** was 25% poorer in acid treated shoots. No distinction could be made between cuticular and stomatal conductance so these effects may be caused by cuticular changes or by alterations in the function of stomata. Less efficient water use would be critical for forest trees where periods of drought are experienced or when transient water deficit occurs in the upper parts of the canopy. However, a comprehensive study of the gas exchange of Norway spruce in the Fichtelgebirge Mountains has failed to detect differences in the gas exchange characteristics at two sites with differing proton loads (Zimmermann *et al*). At both sites impaired gas exchange could not be detected until needle chlorosis associated with nutritional changes occurred. Yellowing and subsequent depression of net photosynthesis were correlated with low Mg^{2+} and Ca^{2+} contents in needles.

Two papers described results obtained from O_3 and SO_2 fumigation chambers run by the Swiss Federal Institute of Forest Research at Birmensdorf. The first of these identified a decrease in starch translocation in Norway spruce resulting from exposure to 300 $\mu g\ m^{-2}$ O_3 and, although this observation is consistent with possible effects on strasburger cells, later experiments have not shown the same response. Decreases in **acid phosphatase** activity and increased **peroxidase activity** seem to be more repeatable effects and the potential of these as early indicators of damage was discussed. In the second paper an interesting new effect was described by Goerg-Günthardt & Keller: The stomata of spruce needles showed a covering of cuticule-like material which dissolves with difficulty in chloroform. This layer appears to be intermediate between polymerised cuticle and soluble wax and had only previously been observed on the older needles of forest grown material. The occurrence of this layer does not seem to be a pollutant response but is associated with water deficit.

3. INTERACTIVE EFFECTS OF AIR POLLUTION AND NATURAL STRESS FACTORS SUCH AS DROUGHT AND FROST.

The first paper of this theme reviewed the substantial body of evidence that plants respond to SO_2, NO_2 & O_3 by making changes of dry matter allocation which favour shoot growth at the expense of the roots; a response which could render plants less able to withstand water deficit. Although most of these data concern herbaceous plants they are consistent with observations of starch accumulation in the needles of declining conifers, with phloem loading being a possible initial site of action for both symptoms. Their influence on **partitioning of dry matter** is clearly only one possible mechanism by which pollutants may alter **susceptibility to drought**, and the data presented by Mansfield clearly identified effects of SO_2 + NO_2 on water loss from leaves of *Betula pendula*. The drying curves of excised leaves showed that the ability of polluted leaves to retain water was substantially decreased by exposure to as little as 20 ppb SO_2 + 20 ppb NO_2. Because of the rate of water loss in the drying curves the authors suggest that cuticular damage rather than disrupted stomatal action may be the cause of such effects, and point out that enhanced cuticular water loss after exposure to NO_2 is the effect which would be predicted from Lendzian's work on the H_2O permeabilities of excised fir cuticles (see Ziegler's above). The possibility that excessive water loss may be a major factor in forest decline is strongly supported by the observations of damage to epicuticular wax, to antechamber plugs and by the experimental evidence for alterations of gas exchange after exposure to acid mist and O_3 (see themes 1 & 2 above).

The second paper of this theme presented some evidence for interacting effects of SO_2 + NO_2 with those of **low temperature** on Sitka spruce. In spite of several reports of enhanced frost injury on trees growing in polluted forest areas this interaction is not yet clearly established for woody plants. However data published since the Lökeberg workshop indicate that interactions in the effects of O_3 and low temperatures could be of significance in forest decline (4). The study of two spruce stands close to the Belgium-German border was designed to establish criteria for the **early diagnosis** of forest decline, with much of the analysis again focusing on cuticular weathering and the cracking of stomatal wax plugs of spruce. However, Laitat also presented some interesting work in which artificial defolation (20, 40, 60 & 80 %) of four year old spruce clones was performed to reproduce a major symptom of forest decline. 60% of needles had to be removed before height growth and second order branch growth were influenced in the first season of **defoliation**. Interestingly, trees which had been 60 & 80 % defoliated also showed enhanced frost injury in the following winter, with both tree mortality and the percentage of shoots killed by frost increasing. This group are currently working on the effects of defoliation on drought resistance.

4. OCCURRENCE OF MICROPOLLUTANTS AND THEIR EFFECTS ON CANOPY STRUCTURE AND FUNCTION.

This was the shortest and most controversial of the themes considered in the meeting. In the first paper Garrec and Berteigne suggested that organic micropollutants present in the air could be responsible for some of the growth alterations and visible symptoms of forest decline. The compounds considered included **peroxyacetyl nitrate**, **hydrogen peroxide**, **ethylene**, **herbicides** and their residues, **aniline**, bitumen and tar vapours and **aromatic hydrocarbons and halocarbons**. It has been suggested that the burning of fossil fuel releases these compounds and that their atmospheric concentrations have risen in the last ten years (6). Many are reactive compounds and hydrocarbons certainly have a role in the formation of O_3 from NO_2. The oxidation products of hydrocarbons also have 'plant hormone' like structures.

Figge and his collegues have measured airborne organic compounds at two sites near Regensburg in West Germany using techniques based on gas chromatography. Their analysis has revealed that there are up to 150 organic compounds of anthropogenic origin at trace concentrations of down to a few ng m^{-3}, but our knowledge of the nature and occurrence of such compounds must still be regarded as limited. Forest conopies may be efficient filters of airborne organic molecules with needle cuticles being particularly permeable to lipophilic compounds. It is clear that interdisciplinary work will be required if progress is to be made in understanding the occurrence and impacts of such compounds. Their large number and small concentrations in air suggest that it may not be profitable to investigate these compounds individually, especially since, like those of other air pollutants, their effects may interact. The use of open-top chambers for selective filtration in combination with careful analysis as pioneered by Figge's group was a way forward suggested by the chairman of this session. In the final paper it was suggested that foliar samples could be collected from trees on a transect across Europe and analysed on return to the labortary for the presence of chemical toxins. It was unfortunate that this presentation was made at an early stage in the analysis of data from the project, and this approach was generally received with scepticism by participants of the workshop.

CONCLUSIONS

Exposure to acid mist, NO_x, SO_2 and O_3 generally increase crown leaching, although the foliar content of elements does not necessarily decrease. The S and N contents of needles can increase with increased deposition, but Mg, Ca and other cation concentrations may decrease in controlled experiments and in the forest. It is not clear whether the additional S, N and cation content in throughfall is supplied by interception deposition or by uptake via the roots followed by crown leaching. There is therefore still much to be learnt about the sources, mechanisms and cost (6) of the loss of ion from canopies.

There now seems good agreement that air pollutants and acid mist influence the rates at which the epicuticular wax structure of conifer needles is lost in the normal weathering processes. This effects wettability, may be important in crown leaching and could alter needle physiology. The importance of antechamber plugs and of the cracks which have been seen in them is not known. Controlled experiments can detect significant effects of realistic concentrations of O_3 and of acid mist on photosynthesis and on water-use-efficiency, but in the forest disruption of gas exchange has not been clearly identified prior to needle discolouration. Light and scanning microscope work has progressed to the stage where it can be valuable in early diagnosis of the causes of decline, and starch accumulation suggests that phloem loading may be an important site of action of pollutants.

The importance of interactions in the effects of pollutants and drought is now accepted, but work on trees still needs to be done. The observed effects of pollutants on epicuticular wax, water-use-efficiency and translocation of photoassimilates to roots are all effects where interactions with water deficit are clearly implicated. Interactions between low temperatures and pollutants are receiving attention because of the clear indications of their importance in the forest. The role of organic micropollutants in forest decline remains unclear, and in view of the large number and their atmospheric concentrations experiments using filtered versus unfiltered air are suggested.

Clearly the elemental composition and pH of throughfall and stemflow will influence the edaphic environment and probably the quality of water lost from the forest ecosystem. Changes in the soil and of tree root function will have effects on the shoots and vice versa and, although this workshop was primarily concerned with forest canopies, it is recognized that an understanding of forest productivity and health requires analysis of both the aerial and edaphic environments of forests.

REFERENCES

(1) Direct effects of dry and wet deposition on forest ecosystems - in particular canopy interactions. The proceedings of the CEC/COST Workshop held in Lökeborg Sweden, October 1986.

(2) BOSCH, C., PFANNKUCH, E., BAUM, U. & REHFUESS, K.E. (1983) Uber die erkrankung der fiche (*Picea abies* Karst) in der hochlagen des Bayerischen Waldes. Forstwiss Centralblatt. 102: 167-181.

(3) FINK Von S. (1983) Histologische und histochemische untersuchungen an nadeln erkrankter tannen und fichten im Südschwarzwald. Allgemeine Forst Zeitschrift 26/27: 660-663.

(4) DAVISON, A.W., BARNES, J.D. & RENNER, C.J. (1987) Interactions between air pollutants and cold stress. Proceedings of the 2 nd. international symposium on air pollution and plant metabolism held in Munich in April 1987.

(5) FRANK, H. & FRANK, W. (1985) Chlorophyll-bleaching by atmospheric pollutants and sunlight. Naturwisserschaften 72: 139-141.

(6) AMTHOR, J.S. (1986) An estimate of the 'cost' of nutrient leaching from forest canopies by rain. New Phytologist 102: 359-364.

TREE RESPONSES TO ACID DEPOSITIONS INTO THE SOIL A SUMMARY OF THE COST WORKSHOP AT JÜLICH 1985

E.-D. SCHULZE
Lehrstuhl Pflanzenökologie, Universität Bayreuth,
Postfach 10 12 51, D-8580 Bayreuth, FRG

Summary

The review is based on the results of the 1985 COST workshop at Jülich. The main pathway of tree responses to acid deposition are summarized in Fig. 1. It occurs, that leaching of nutrients causes a plant external nutrient cycle. Together with the overall proton and nitrate input, soil acidity and soil cation pools have changed. This has effects on root structure and performance which result in decreased nutrient uptake and needle ion deficits. But, the plant internal regulation of coping with nutrient deficits is additionally perturbed by the input of nitrogen which stimulates growth at low cation supply. This causes an inbalance of the needle nutrient status which in turn results in whole plant growth reduction and needle loss. The timing of the process contains a long-term component of sulphate inputs. More recently are nitrogen inputs. Both caused a predepostiton of the ecosystems. The short-term rapid increase of nutrient deficits, which have been observed as feature of forest decline, may be due to the soil physical process of soil-aggregation, which reduces the supply of cations in dry years.

1. INTRODUCTION

Since the initial hypothesis of Ulrich (1) on the destabilization of forest ecosystems due to nutrient losses and associated soil acidification, increasing evidence has been accumulated, that long-term trends exist of decreasing pH in the soil solution. For instance, in Swedish deciduous forests during the past 40 years the pH of the soil solution decreased by 0,24 pH units per 10 years, when the original pH was in the silicate buffer range (2). A comparison of these data with 60 year records (3) indicates that the main change in soil pH occured in the last decades (see also 4, 5, 5a).

Taking the changes of soil pH and associated changes in the buffer capacity of soils and changes in the nutrient pools as an historic fact, the present review will evaluate mechanisms by which plants respond to this stress situation in the soil compartment. These processes operate at the ecosystem level. They have a very slow time constant and they are not easily reversible. Therefore, the stress at the soil level has a different quality when compared with direct damages of leaves by atmospheric pollutants.

The following chapter is based on results of the COST workshop on "Indirect effects of air pollution on forest trees - root/rhizosphere interactions" held 1985 at Jülich, FRG (7) and associated literature which has been published since then. I regret that a complete treatment of the published material is not possible.

2. THE ENVIRONMENT

The first step in an ecological analysis is a detailed assessment of the environment. It is quite clear, that there is a large geographical variation of the atmospheric inputs of air pollutants (refer to session I of this conference). But, in order to illustrate interactions of whole plant performance with atmospheric depostitions and soil parameters, Fig. 1 refers to the situation in the Fichtelgebirge (NE Bavaria), taking for granted that ozone levels or nitrogen inputs may be higher and sulphur lower at other locations.

Fig. 1: The environmental conditions and the observed responses of Picea abies in the Fichtelgebirge (NE Bavaria, FRG)

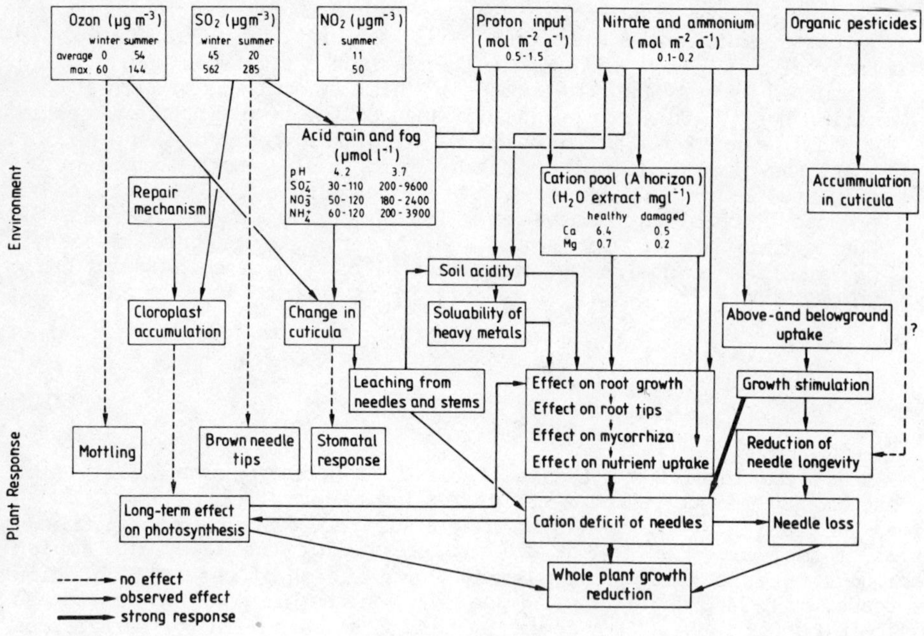

In a situation as exemplified by Fig. 1, there are no symptoms of direct damage of ozone or SO_2 (8,9). The SO_2 concentrations are high only in winter when the vegetaton is dormant. Based on theoretical considerations Heber et al. (12) postulated direct damages by SO_2 also for the Fichtelgebirge, but obviously repair mechanisms exist in the plants to maintain high rates of photosynthesis even in old needles (10). With respect to direct damage of pollutants the reader is referred to the review of Krause in this volume.

Leaching of nutrients from the forest canopy takes place at considerable rates (11, 11a), but canopy leaching by itself will not cause nutrient deficits, if the soil and root conditions allow a replacement of the leached nutrients. Leaching will establish a plant external nutrient cycle since the leached nutrients are again accessable to the roots for uptake. Nevertheless, canopy leaching will promote the soil acidification process, since the nutrients in the soil solution can also be washed out of the soil profile together with mobile anions such as nitrate. With respect to canopy leaching the reader is referred to the review of Kreutzer in this volume. The following text will therefore concentrate on the effects of low nutrient pools and of nitrogen inputs on plant performance.

3. IONICS EFFECTS ON ROOT TIPS AND MYCORRHIZA

Heavy metals have been accumulated in forest soils (38), which may become mobile with decreasing pH of the soil solution. Godbold and Hüttermann (13) and Schlegel et al. (14) give evidence that methylated heavy metals have toxic effects on roots and whole plant processes at concentrations which are likely to occur in the field (Fig. 2).

Fig. 2: Root elongation rates of sterile Picea abies seedlings grown for 7 days in nutrient solutions containing different concentrations of methyl-HgCl (after Godbold and Hüttermann (13)).

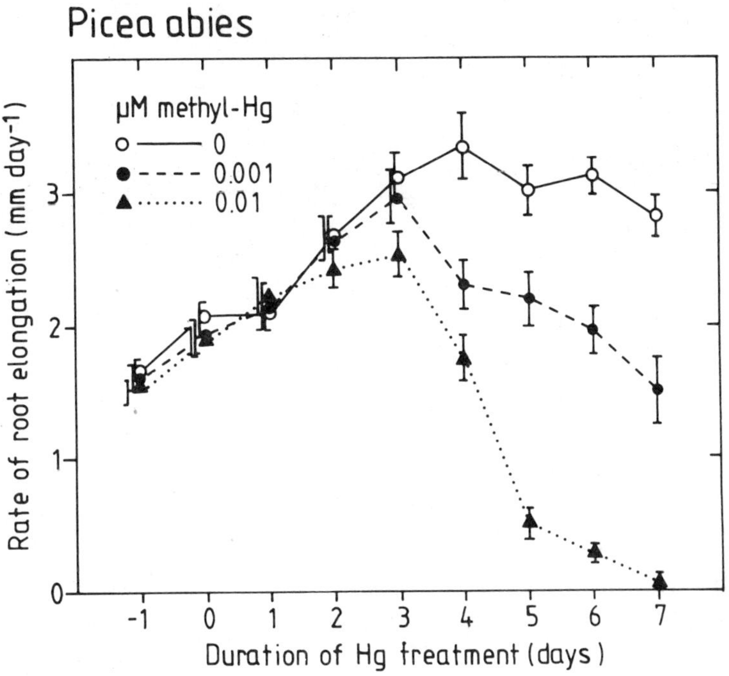

Concentrations of 1 nM of methylated mercury caused a depression of root growth after 3 days. These studies were made with sterile roots having no mycorrhiza, but free root tips are likely to occur also in natural soils during long-root formation and with nitrogen depositions. The significance of this observation is difficult to evaluate, since there are only few systematic surveys on concentrations of heavy metals and their organic binding forms in natural habitats (e.g. 15). Godbold and Hüttermann (13) summarize that, while it is unlikely that heavy metal toxidity is a dominating factor in forest decline, the heavy metal contents in forest soils are sufficiently high to be considered as a possible contributing factor. Even more important than direct heavy metal toxitidy may be ionic effects such as the inhibition of calcium uptake by lead or cadmium (16).

The change in soil pH is an expression of very pronounced alterations in soil chemistry (6). With decreasing pH units there is an increasing soluability of aluminum, which is concurrent with a reduction of the base saturation (Tab. I). Berholm (17) measured a linear increase of aluminum-ions from 0 to 0.5 mM when the pH decreased from 4.9 to 3.5 units. These aluminum concentrations in the soil solution correlate with reduced relative growth rates of spruce seedlings (18,19).

Fig. 3: Buffer systems and their pH ranges in soils (from Ulrich (6)).

Buffer substance	pH range	Main reaction product of lower ANC (chemical change in soil)
Carbonate buffer range		
$CaCO_3$	8.6 > pH > 6.2	$Ca(HCO_3)_2$ in solution (leaching of Ca and basicity)
Silicate buffer range		
Primary silicates	Whole pH scale (dominating buffer reaction in carbonate free soils pH > 5)	Clay minerals (increase of CEC)
Exchanger buffer range	5 > pH > 4.2	
Clay minerals		Nonexchangeable $n[Al(OH)_x^{(3-x)+}]$ (blockage of permanent charge, reduction of CEC
Mn-oxides		Exchangeable Mn^{2+} (reduction of base saturation)
Clay minerals		Exchangeable Al^{3+} (reduction of base saturation)
Interlayer Al $n[Al(OH)_x^{(3-x)+}]$		Al-hydroxosulfate (accumulation of acid in case of input of H_2SO_4)
Aluminum buffer range		
Interlayer Al Al-hydroxosulfate	4.2 > pH	Al^{3+} in solution (Al displacement, reduction of permanent charge)
Aluminum/iron buffer range		
As Al buffer range in addition: "Soil $Fe(OH)_3$"	3.8 > pH	Organic Fe complexes (Fe displacement, bleaching)
Iron buffer range		
Ferrihydrite	3.2 > pH	Fe^{3+} (Fe displacement, bleaching, clay destruction)

The experiments of Rost-Siebert (20) show that the concept of an aluminum toxidity must be modified such that not the aluminum itself, but the molar ratio of Ca/Al is of functional importance. Calcium ions can counterbalance the effect of aluminum, thus there is a linear decrease of root elongation in a broad range of Ca/Al conditions (Fig. 4). There is evidence that the mechanism is due to a competitive effect of 2- and 3-valence ions during cell wall formation (20). In many field situations it is the Ca rather than the Al which is changing the ratio (25).

Fig. 4: Root growth of Picea abies as related to the molar Ca/Al ratio of the cultural solution (after Rost-Siebert (20)).

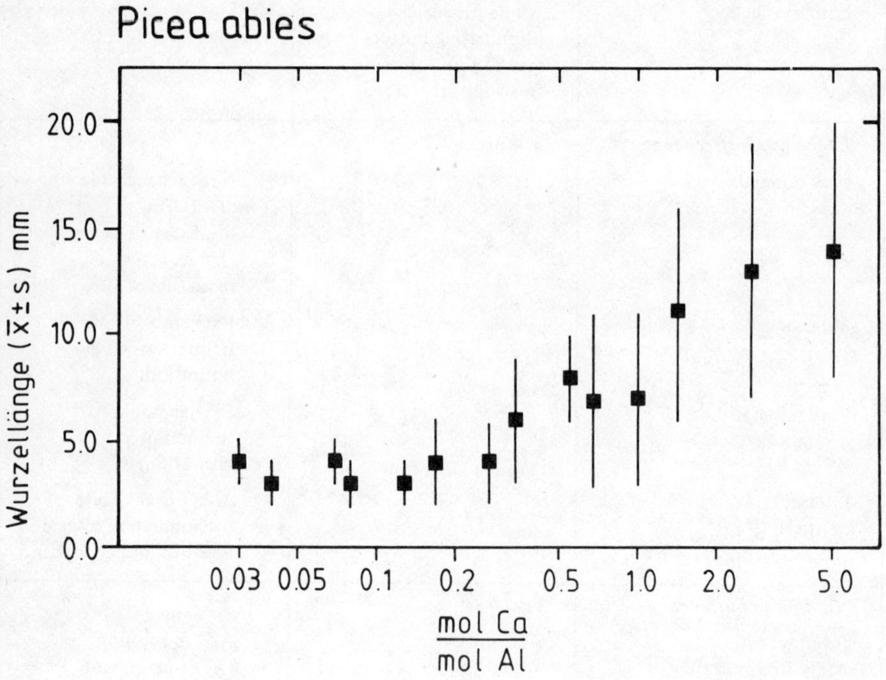

The importance of the Ca/Al-ratio has been varified for many laboratory and field conditions (21, 21a, 21b). A linear correlation was observed between the number of root tips in the field and the molar ratio of Ca/Al in the soil solution of plots showing various degrees of needle yellowing as symptom of forest decline (Fig. 5).

Fig 5: Number of root tips of Picea abies as related to the molar ratio of Ca/Al (after J. Meyer et al. (21)).

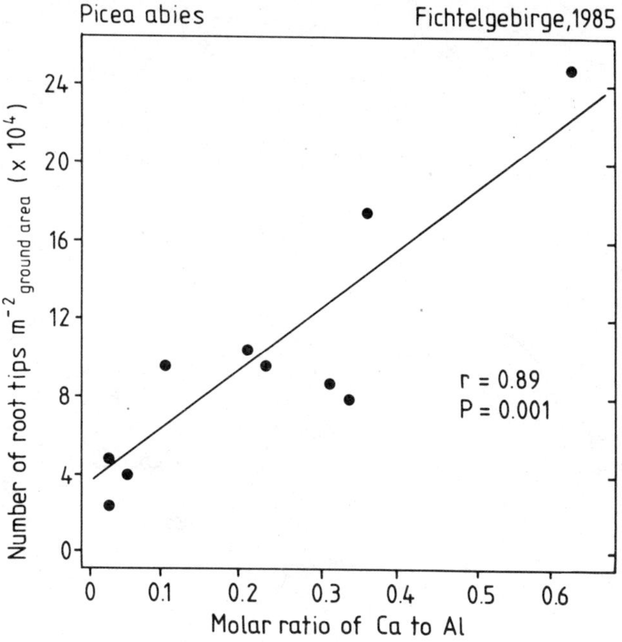

Referring to Fig. 1, there is experimental evidence that the historic changes in the buffer range of the soil solution had an effect on root development under natural conditions. This conclusions needs to be qualified by the observation, that root tip damage results in an increased root branching (20). There are earlier observations which show that the root tip number rather increases from 1.2 million to 5.3 million per ground area than decreases, when comparing spruce on eutrophic brown soil with trees on ion-humus-podsol (26).

The field situation is complicated by the interaction of root tips with mycorrhizal fungi. In a detailed review Kottke and Oberwinkler (22) conclude that the rate of mycorrhization is not affected by soil acidification, Ca/Al ratios or nitrogen inputs. Nevertheless, the number of mycorrhiza per dry weight of fine roots was higher in well growing than in poorly growing stands, but in all cases the degree of mycorrhization was 100 %. Therefore, the number of available root tips seems to be the rate limiting step. This is difficult to study since there is a feedback relation between root functioning and plant performance (Fig. 6) which results in allometric relations between numbers of root tips and fine root biomass at a given situation (23, 24), but which also can be perturbed by inputs of nitrogen (28).

Fig. 6: Interrelationships between tree canopy and root systems (from J. Meyer et al. (23)).

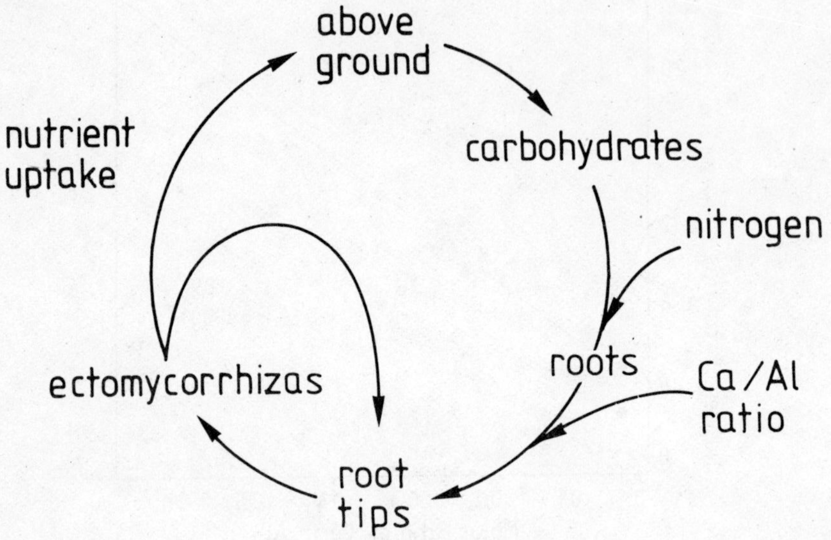

There is an additional factor which severely effects root development, namely the inpact of nitrogen (see review by Beese and Matzner (29)). F.H. Meyer (26) discussed the observation of increasing root tips with poor soils as an effect of soil nutrition rather than of acidification. Root elongation is stimulated by nitrogen (27, 28) and root tips may pierce through the mycorrhizal sheath under the impact of access nitrogen (22, 27). Under well fertilized conditons the number of root tips per root dry weight decreases (22). Also the lower root tip numbers of J. Meyer (23) when compared to the observations of F. H. Meyer (26) could be interpreted as an effect of nitrogen inputs. The effects of nitrogen on roots are manifold and depend on the nature of the NH_4/NO_3 ratio. The depositon of ammonium promotes soil acidification during uptake by roots (30) and nitrification (29). The effect may be more important than acidification by sulphate. In contrast to stable aluminum-sulphate formation, nitrate will leach through the soil horizon and promote a proportional leaching of cations.

It may be summarized (Fig. 1), that decreasing Ca/Al ratios and nitrogen deposition will decrease the amount of fine roots and the number of root tips per ground area. The effect on mycorrhiza is controversal. Kottke and Oberwinkler (22) conclude that mycorrhization is not affected, but may lag behind root tip elongation at nitrogen fertilization, whereas F.H. Meyer (28) observed changes in the degree of mycorrhization. It is difficult to extrapolate from laboratory experiments because of possible interactions with other ions. Ingrowth-cores of peat and sand show high mycorrhizal frequency even when fertilized with nitrogen in the field (29). J. Meyer et al. (27) observed a decrease of mycorrhizal frequency with increasing ammonium concentration, but this effect is associated with changes in pH and cation supply. Also, observations of decreasing mycorrhizal frequency in the presence of nitrate are generally confounded by pH and cation interactions (29).

4. EFFECTS OF ROOT AND MYCORRHIZAL DEVELOPMENT ON WHOLE-PLANT NUTRITON

Irrespective of the problem whether root tips or mycorrhiza are affected by soil chemical factors or nitrogen deposition, an overall correlation exists between the nutrient concentration in the soil solution and the ion content of the needles (31). Thus, soil acidification and reduced cation pools are one major pathway which lead to cation deficits of the needles (Fig. 1).

Fig. 7: Correlation between Mg-content of needles and the Mg-concentration in the soil solution (after Kaupenjohann et al. (31)).

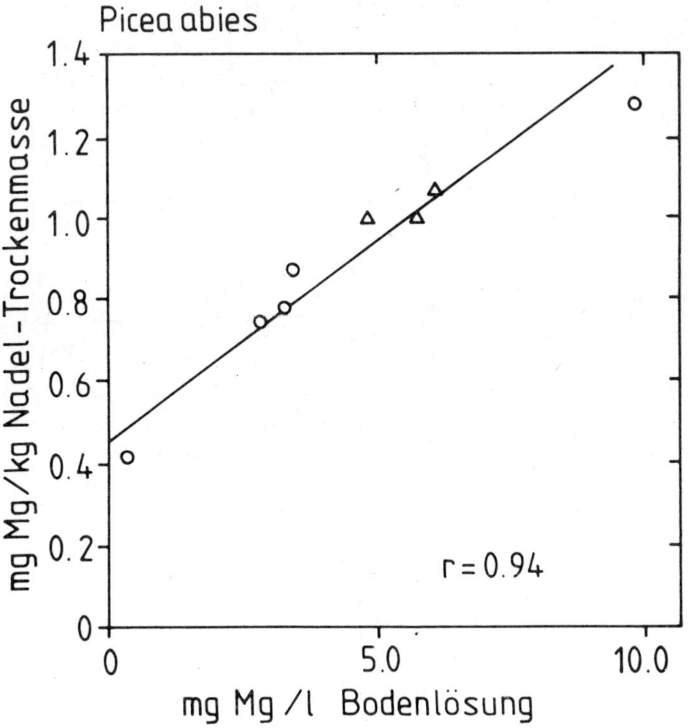

This overall correlation is supported by a correlation between leaf ion contents and the number of mycorrhizal root tips per leaf area (21). Furthermore a relation exists between the concentration of magnesium and calcium in the xylem and the needle contents (32).

Fig. 8: Correlations between leaf magnesium, calcium and aluminum contents and the number of ectomycorrhizas per leaf area (from J. Meyer et al. (21)).

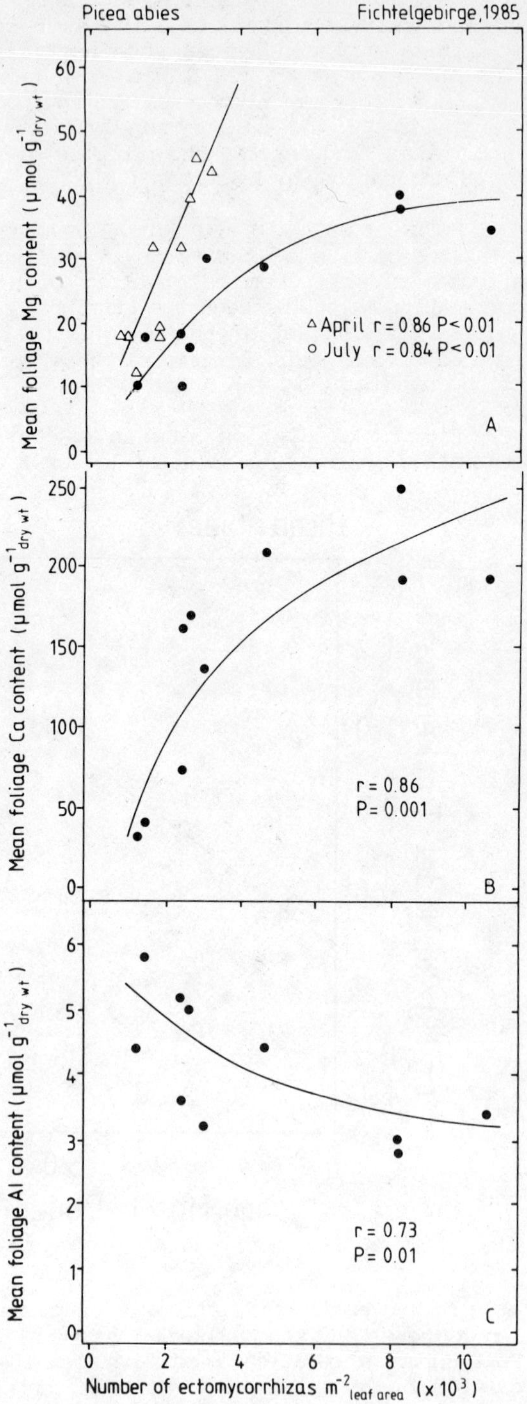

Although nutrient deficits in the soil and the effects of soil chemistry on root growth appear to be a reasonable explanation for observed nutrient deficits on silicate soils (Fig. 1), it remains uncertain why plants are not able to adjust their nutrient relations such that the rate of uptake balances the demand of growth. There are numerous natural situations, where low nutrient supply and soil acidity in the past did not result in nutrient deficits, e.g. growth of conifers on bogs, on serpentine, or dolomitic outcrops. In all these cases trees have a stunted appearance due to slow growth, but they show healthy green needles of consideralbe age. This indicates that an additional factor operates in the process of forest decline which perturbes the nutrient relations of the trees such, that a balanced growth is not prohibited. Although it is not yet possible to draw complete nutrient cycles, the overall balance sheet indicates that nitrogen has a key function in this process, which is different from direct damages by ammonium (33).

Tab. I: An estimation of the magnesium balance of Picea abies needles (after Schulze (34))

	magnesium use per needle $mg\ g^{-1}\ a^{-1}$	needle growth per ground area $g\ m^{-2}\ a^{-1}$	magnesium use per ground area $mg\ m^{-2}a^{-1}$
Mg-reallocation	0.13	450	58
Mg-uptake from soil	0.38	450	171
Mg-leaching	0.19	450	84
			---
		total uptake from soil	255
Assumption: reduce growth such that supply from soil is sufficient for healthy needles:			
Mg-incorporation from soil	0.90	234	211
Mg-leaching	0.19	234	44
			---
		total uptake from soil	255

The magnesium balance of Picea abies needles is based on a stand showing symptoms of magnesium deficiency by having 0.51 mg g^{-1} of magnesium in the current needles and 0.38 mg magnesium g^{-1} in the 1 year old needles. Even under conditions of magnesium deficiency there is a considerable uptake of magnesium from the soil which covers about 80 % of the total magnesium use including leaching. The amount would be sufficient to supply about half of the present growth rate of needles at a rate comparable to a healthy stand. There must be a factor that stimulates growth beyond the supply of magnesium.

Tab. II: An estimate of the nitrogen balance of healthy and damaged *Picea abies* stands. Deposition, demand and export were measured by independent methods and are expressed per ground area and year (g $m^{-2}a^{-1}$). (after Schulze (34)).

		healthy	damaged
N-deposition	NO_3-N	1.18	1.14
	NH_4-N	0.87	1.23
		----	----
	total	2.05	2.37
N-demand	needles	0.5	0.5 (consid. reallocation)
	wood	1.3	0.9
	root	0.1	0.1
		---	---
	total	1.9	1.5
N-export (conc x flow in 75 cm)		0.9	1.05
N-mineralization unkown			
Mg-supply		sufficient	insufficient 1/4 of N-deposition would balance Mg supply and avoid leaching from the profile

In the example shown in Tab. II about the same amounts of nitrate which are deposited on the soil are also leached out of the profile. This is not an exception. The deposition of nitrogen in Mid-Europe exceeds the demand for forest growth in most instances (29). As a consequence a decrease of the C/N ratio in the humus is observed over a broad range of soil types (35). Ammonium is quantitatively used during the path through the profile by processes of root uptake and nitrification. It is estimated in Tab. II that the deposition of nitrogen should decrease by 75 % in order to balance the supply of magnesium from the soil (Tab. I) and in order to avoid leaching of nitrate from the profile.

The deposition of nitrogen has multiple effects and pathways through the ecosystem. The trees gain nitrogen from uptake of NO_x through the stomata. There is also evidence that nitrogen enters in the liquid phase into the above-ground parts (leaf fertilization). Nitrate and NO_x will be reduced into amino acids in the needles. Part of this uptake will be leached again from the needles and added to the soil as reduced nitrogen. It is not possible so far to give exact numbers for these above-ground cycles of nitrogen, but it is quite clear that above-ground uptake and losses of nitrogen occur which are not under root control. In the presence of ammonium and nitrate in the soil solution, conifers appear to preferentially use ammonium as nitrogen source (36). In this case the nitrate of the soil sollution perculates through the profile (37) and promotes leaching. Since in the presence of nitrogen root tips and mycorrhizal development is reduced (26) also cation uptake will be decreased. Since conifers have little storage parenchyma all amino acids will be used for growth. The trees appear to grow at a rate that the nitrogen content of needles is, within certain limits, quite constant. Thus, the observed cation deficiency could be the result of nitrogen inputs. The use of nitrogen causes deficiency of those cations which are at a minimum in the soil solution. This could be manganese in the case of dolomite, potassium on limestone and magnesium on granite and schist.

5. ECOSYSTEM CONSEQUENCES

Fig. 9 shows a schematic presentation of how forest ecosystems become destabilized by acid depositions (6). Although this scheme was developed only with respect to acidification, it is quite obvious that the speed by which the system moves will be enhanced by nitrogen deposition. Since the resilience of the whole ecosystem is being altered, there is no rapid and simple way of curing the situation. Even if all immissions would stop immediately, the process of acidification will not recover at the same rates. It is also possible that it is the rate of change which has caused the main flush of forest die-back and the systems may be capable to stabilize at a low level of leaf area index as a consequence of recent rates of defoliation. It is very unlikely that large parts of Mid-Europe will become deforested.

Fig. 9: Resilience and stability of European forest ecosystems (from Ulrich (6))

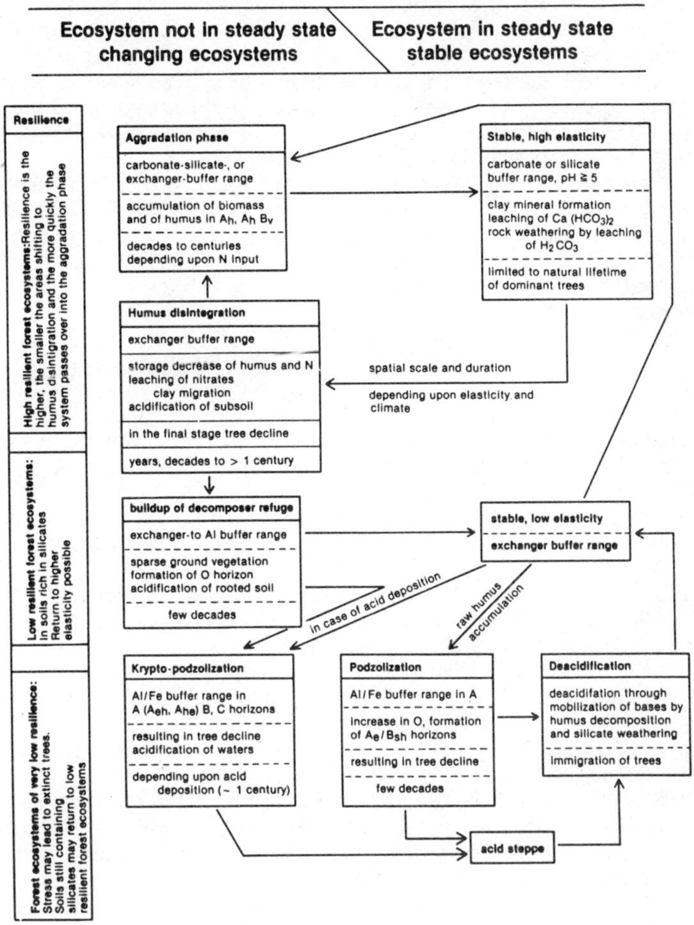

If acidification and nitrogen were the main components of causing forest decline it still needs to be answered why this phenomenon happened within such a short and defined period of time. Sulphate deposits occured since decades. They have started the acidification process but may have reached the same level of instability only some time in future, since sulphate is bound to other ions in the soil profile (6). Since about 1960 the immissions of nitrogen inreased due to industry, automobiles, city suage and agriculture (38). Between 1960 and 1970 the major constructions of suage plants took place which caused an increase of ammonium immissions. In addition between 1970 and 1980 the use of liquid fertilizer increased in agriculture as well as ammonium production increased due to mass animal production in Europe. The main step of forest decline followed these changes in nitrogen inputs. Nevertheless, short-term climatic factors accelerate this general scheme. Dry years promote the acidification and increase the nitrate concentration in the soil solution (38). In addition the soil physical processes of aggregation decrease the supply of cations to the root (39, 40).

It appears that long-term processes of acidification altered the stability of forest ecosystems. The addition of nitrogen, has accelerated this process and it has increased the demand for macro-nutrients in a situation of reduced supply. Short-term events of dry years have promoted and synchronized theses responses of trees to acid depositions.

It is to be expected in future that wet years will lead to a recovery due to soil physical processes (40). Also, following defoliation, the leaf area index has been reduced to a level that the demand for macro-nutrients for growth is again equilibrated with the supply. It should be pointed out however, that such responses of tree recovery do not represent a recovery of the ecosystem, since the same symptoms of damage will return as soon as the stand builds up the original leaf area, or as soon drought influences the availability of cations, or increases the concentration of nitrogen in the soil solution.

REFERENCES

(1) ULRICH, B. (1981). Destabilisierung von Waldökosystemen durch Akkumulation von Luftverunreinigungen. Forst- und Holzwirt 36, 525-532.

(2) FALKENGREN-GRERUP, U. (1986). Soil acidification and vegetation changes in deciduous forest in southern Sweden. Oecologia 70, 339-347.

(3) HELLBÄCKEN, L., TAMM, C.O. (1986). Changes in soil acidity from 1927 to 1982-1984 in a forest area of south-west Sweden. Scand. J. For. Res. 1, 219-232.

(4) BUTZKE, H. (1984). Untersuchungsergebnisse aus Waldböden Nordrhein-Westfalens zur Frage der Bodenversauerung durch Immissionen. Wissenschaft und Umwelt 2, 80-88.

(5) von ZEZSCHWITZ, E. (1982). Akute Bodenversauerung in den Kammlagen des Rothaar Gebirges. Forst- und Holzwirt 37, 275-276.

(5a) GEHRMANN, J., BÜTTNER, G., ULRICH, B. (1987). Untersuchungen zum Stand der Bodenversauerung wichtiger Waldstandorte im Land Nordrhein-Westfalen. Berichte des Forschungszentrums Waldökosysteme/Waldsterben, Reihe B, Vol. 4, 1-233.

(6) ULRICH, B. (1986). Stability, elasticity, and resilience of terrestrial ecosystems with respect to matter balance. In: E.-D. Schulze, H. Zwölfer (eds.). Potentials and Limitations of Ecosystem Analysis. Springer Verlag, Ecological Studies, Vol. 61, 11-49.

(7) COMMISSION OF THE EUROPEAN COMMUNITIES (1986). Indirect effects of air pollution on forest trees, root-rhizosphere interaction. Proceedings of the 1985 Jülich COST workshop. Directorate general for Science, Research and Development EAD 45.86 XII/ENV/24/86.

(8) LANGE, O.L., GEBEL, J., SCHULZE, E.-D., WALZ, H. (1985). Eine Methode zur raschen Charakterisierung der photosynthetischen Leistungsfähigkeit von Bäumen unter Freilandbedingungen. Forstwiss. Zentralbl. 104 g, 186-198.

(9) SCHULZE, E.-D., OREN, R., WERK, K., MEYER, J., ZIMMERMANN, R. (1986). Kohlenstoff-, Wasser- und Nährstoffhaushalt von Fichten stark belasteter Hochlagenstandorte auf Phyllit in NO-Bayern. In: Führ, F., Ganser, G., Kloster, G., Prinz, B., Stüttgen, E. (eds.) Wirkungen von Luftverunreinigungen auf Waldbäume und Waldböden. BMFT Statusseminar 1985, Jül-Spez-369, 106-117.

(10) ZIMMERMANN, R., SCHULZE, E.-D., OREN, R., WERK, K.S. (1987). Photosynthesis and Transpiration of green trees of Picea abies (L.) Karst.: A comparison of two sites with differens sulphate and proton input in Northern Bavaria, FRG. COST Workshop Göteborg 1986.

(11) HANTSCHEL, R., KAUPENJOHANN, M., HORN, R., ZECH, W. (1985). Wasser- und Elementtransport in unterschiedlich gedüngten, geschädigten Waldökosystemen. Z.dt.Geol.Ges. 136-473-480.

(11a) GODT, J. (1986). Untersuchungen von Prozessen im Kronenraum von Waldökosystemen und deren Berücksichtigung bei der Erfassung von Schadstoffeinträgen. Ber. d. Forschungszentrums Waldökosysteme/Waldsterben, Vol. 19, 1-265.

(12) HEBER, U., LAISK, A., PFANZ, H., LANGE, O.L. (1987). Wann ist SO_2 Nähr- und wann Schadstoff? Ein Beitrag zum Waldschadensproblem. Allg. Forstzeitschrift (in press).

(13) GODBOLD, D.L., HÜTTERMANN, A. (1986). The uptake and toxidity of mercury and lead to spruce (Picea abies Karst.) seedlings. Water, Air, and Soil Pollution 31, 509-515.

(14) SCHLEGEL, H., GODBOLD, D.L., HÜTTERMANN, A. (1987). Whole plant aspects of heavy metal induced changes in CO_2 uptake and water relations of spruce (Picea abies) seedlings. Physiol. Plantarum 69, 265-270.

(15) BUCHER, J.B. (1986). Chemical needle analysis as indicator of the polluted environment in Switzerland. Proceedings of the Jülich COST Workshop. Commission of the European Communities 13-15.

(16) FRITZ, E., GODBOLD, D.L., DICTUS, K., HÜTTERMANN, A., LARSEN, B. (1986). Mechanism of action of soil mediated stress factors: aluminum, heavy metals and excess nitrogen on spruce roots. Jülich COST Workshop, Commission of the European Communities 33-45.

(17) BERGHOLM, J. (1986). Soil and fine root chemistry in a SW Swedish spruce forest with different degrees of needle loss. Proceedings of the Jülich COST Workshop. Commission of the European Communities 119-125.

(18) GÖRANSSON, A. (1986). Effects of Al^{+3} on seedling growth and nutrient uptake of Norway spruce, Scots pine and European birch at steady state nutrition. Proceedings of the Jülich COST Workshop. Commission of the European Communities 46-55.

(19) ALXANDER, C., MILLER, H.G. (1986). Root health and soil microbial activity and the possible influence of pollutant inputs. Proceedings of the Jülich COST Workshop. Commission of the European Communities 170-180.

(20) ROST-SIEBERT, K. (1985). Untersuchungen zur H- und Al-Ionen-Toxidität an Keimpflanzen von Fichte (Picea abies, Karst.) und Buche (Fagus sylvatica, L.) in Lösungskultur. Ber.d. Forschungszentrums Waldökosysteme/Waldsterben, Vol. 12, 1-219.

(21) MEYER, J., SCHNEIDER, B.U., WERK, K., OREN, R., SCHULZE, E.-D. (1987). Performance of Picea abies (L.) Karst. at different stages of decline. V. Root tips and ectomycorrhiza development and their relation to above-ground and soil nutrients. Oecologia (in press).

(21a) BARTSCH, N. (1985). Ökologische Untersuchungen zur Wurzelentwicklung an Jungpflanzen von Fichte (Picea abies, (L.) Karst.) und Kiefer (Pinus sylvestris L.) Ber. d. Forschungszentrums Waldökosysteme/Waldsterben, Vol. 15, 1-231.

(21b) BERTILLER, M.B. (1986). Wachstums- und Schädigungsdynamik von Feinwurzeln als Funktion der bodenchemischen Bedingungen. Ber. d. Forschungszentrums Waldökosysteme/Waldsterben, Reihe A, Vol. 21, 1-167.

(22) KOTTKE, I., OBERWINKLER, F. (1986). Mycorrhiza of forest trees-structure and function. Trees 1, 1-24.

(23) MEYER, J., OREN, R., WERK, K.S., SCHULZE, E.-D. (1986). The effect of acid rain on forest tree roots. proceedings of the Jülich COST Workshop. Commission of the European Communities 16-30.

(24) MURACH, D. (1986). Soil chemistry and fine root growth of forest trees. Proceedings of the Jülich COST Workshop. Commission of the European Communities 95-107.

(25) BAUCH, J., SCHRÖDER, W. (1982). Zellulärer Nachweis einiger Elemente in den Feinwurzeln gesunder und erkrankter Tannen (Abies alba Mill.) und Fichten (Picea abies (L.) Karst.). Forstwiss. Cbl. 101, 285-294.

(26) MEYER, F.H. (1967). Feinwurzelverteilung bei Waldbäumen in Abhängigkeit vom Substrat. Forstarchiv 38, 286-290.

(27) MEYER, F.H. (1986). Root and mycorrhiza development in declining forests. Proceedings of the Jülich COST Workshop. Commission of the European Communities 139-151.

(28) MEYER, F.H. (1985). Einfluß des Stickstoff-Faktors auf den Mykorrhizabesatz von Fichtensämlingen in Humus einer Waldschadensfläche. Allg. Forstzeitschrift 40, 208-219.

(29) BEESE, F., MATZNER, E. (1986). Langzeitperspection vermehrten Stickstoffeintrags in Waldökosysteme: Droht Eutrophierung? Berichte des Forschungszentrums Waldökosysteme/Waldsterben, Reihe B, Vol. 3, 182-204.

(30) MARSCHNER, H., HÄNSSLING, M., Leisen, E.L. (1986). Rhizosphere pH of Norway spruce trees growing under both controlled and field conditions. Proceedings of the Jülich COST Workshop. Commission of the European Communities 113-118.

(31) KAUPENJOHANN, M., ZECH, W., HANTSCHEL, R., HORN, R. (1987). Ergebnisse von Düngungsversuchen mit Magnesium an vermutlich immissionsgeschädigten Fichten (Picea abies (L.) Karst.) im Fichtelgebirge. Forstw. Cbl. 106 (in press).

(32) OSONUBI, O., OREN, R., WERK, K.S., SCHULZE, E.-D., HEILMEIER, H. (1987). Performance of Picea abies (L.) Karst. at different stages of decline. V. Correlations with xylem sap concentrations of magnesium, calcium and potassium but not nitrogen. Oecologia (in press).

(33) NILGRÄD, B. (1985). The ammonium hypothesis - an additional explanation to the forest die back in Europe. Ambio. 14, 2-8.
(34) SCHULZE, E.-D. (1987). Wirkungen von Immissionen auf 30-jährige Fichten in mittleren Höhenlagen auf Phyllit. AFZ (in press).
(35) Von ZEZSCHWITZ, E. (1985). Qualitätsänderungen des Waldhumus. Forstw. Cbl. 104. 205-220.
(36) BEEVERS, L., HAGEMAN, R.H. (1983). Uptake and reduction of nitrate: Bacteria and higher plants. Encyclopedia Plant Physiol. N.S. 15 A, 351-369.
(37) BEESE, F. (1986). Nitrogen Cycling in temperate Forest Ecosystems. Göttinger Bodenkundliche Berichte 85, 195-211.
(38) ULRICH, B., MATZNER, E. (1983). Abiotische Folgewirkungen der weiträumigen Ausbreitung von Luftverunreinigungen. BMI Forschungsbericht 104 02 614.
(39) HANTSCHEL, R., KAUPENJOHANN, M., HORN, R., ZECH, W. (1986). Kationenkonzentrationen in der Gleichgewichts - und Perkolationsbodenlösung (GBL und PBL) - ein Methodenvergleich. Z. Pflanzenernähr. Bodenk. 149, 136-139.
(40) HORN, R., ZECH, W., HANTSCHEL, R., KAUPENJOHANN, M. (1987). The importance of soil aggregation to forest nutrition and decline phenomena. Oecologia (in press).

EXPERIMENTELLE UNTERSUCHUNGEN ZUR STOFFAUSWASCHUNG AUS BAUMKRONEN DURCH SAURE BEREGNUNG

K. Kreutzer
Lehrstuhl für Bodenkunde
der Ludwig-Maximilians-Universität München

Summary

Acid irrigations (H_2SO_4, pH 2.7) of a young spruce stand with an input of 366 mmol H^+/m^2 during 18 events from June to October 1985 caused a 6-7 times larger loss of Mn and Ca, a 4-5 times larger one of Mg, and a doubled one of K and NH_4, when compared with normal irrigations (pH 5.2). During the whole application period the acidified apoplast solution was a sink for NH_3, causing a higher input of NH_4 into the system. At the end of the observation period in November 1985 17 % of the proton input are retained in the canopy, mainly puffered by cation exchange. 15 % are consumed mainly by HCO_3-input of natural rain and 68 % did pass the canopy (54 % as H^+, 14 % as NH_4^+).

1. EINFÜHRUNG

Die Auswirkungen saurer Niederschläge auf die Kationenauswaschung aus den Baumkronen war in letzter Zeit häufig Gegenstand wissenschaftlicher Erörterungen (1, 2, 3, 4 u.a.). Nur wenige Forscher untersuchten das Problem jedoch experimentell im Freiland (5, 6, 7) und zwar mit freier Beregnung ohne wesentlich gestörtes Mikroklima (also z.B. nicht in Open-top- Kammern). Auch wurde bisher nicht genügend den dynamischen Prozessen der Protonenspeicherung und des Protonenverbrauchs in der Kronenschicht nachgegangen. Um diesen Mängeln abzuhelfen, richteten wir ein Freilandexperiment mit freier Beregnung über den Kronen ein. Die Untersuchungen sollten vor allem zu zwei Fragen einen Beitrag leisten:

- Wie stark beeinflußt ein definierter Mehreintrag an Protonen unter möglichst natürlichen Bedingungen die Kationenauswaschung?
- Welche Prozesse der Protonenpufferung spielen sich in der Krone ab und welches Ausmaß erreichen diese Prozesse im einzelnen?

2. MATERIAL UND METHODEN

Der Versuchsbestand ist eine sehr wüchsige, gesund aussehende, 13jährige Fichtenpflanzung auf saurer Braunerde. Über Einzelheiten des Bestandes, des Standortes und der Versuchstechnik informieren Kreutzer und Bittersohl (8). Die regionale luftchemische Situation ist als gering bis mäßig belastet einzustufen (9). Sie ist typisch für den ländlichen Raum im bayerischen Alpenvorland. Lediglich die Ammoniakgehalte der Atmosphäre sind aufgrund der intensiven Landwirtschaft angehoben. Die externe Staubimmission kann als minimal betrachtet werden, da der Versuchsbestand inmitten eines großen Waldareals sehr geschützt zwischen Altbeständen eingebettet ist.

Eingerichtet wurde die Überkronenberegnungsanlage im Jahre 1983 mit zwei Varianten: a) stark saure Beregnung (pH 2,7-2,8, H_2SO_4) und b) mäßig saure, "normale" Beregnung (pH 5-5,5). Die Beregnungen erfolgten während der frostfreien Zeit in den Jahren 1984 und 1985, zusätzlich zu den natürlich fallenden Regen (pH meist 4.7-5.2). Im folgenden werden die Ergebnisse der Beregnungssaison von 1985 dargestellt. Sie begann am 14. Juni und endete am 21. Oktober. In der Zwischenzeit wurden 18 Beregnungen zu je rund 10 mm mit insgesamt 174 mm Eintrag an den Meßplätzen durchgeführt. Davon gingen in den Kronen 67 mm (= 39 %) als Interceptionsverdunstung wieder verloren. Beprobt wurde das Niederschlagswasser oberhalb und unterhalb der Kronenschicht während der Beregnungssaison einschließlich einer Anschlußzeit vom 21. Oktober bis 29. November. Die Anschlußzeit dauerte so lange bis sich das pH im natürlichen Bestandsniederschlag nicht mehr wesentlich zwischen den Behandlungsvarianten unterschied. Die Beprobung erfolgte streng getrennt nach natürlichen Regen und Beregnungsereignissen, und zwar bei den Beregnungsereignissen immer jeweils, wenn der Abtropfvorgang vollständig abgeschlossen war. Die natürlichen Regen betrugen während der gesamten Beobachtungszeit 459 mm, die Interceptionsverluste 161 mm (= 35%).

Die Analyse der Wässer wurde mittels ICP, AAS, AA und IC durchgeführt. Der Ermittlung der Protonenkonzentrationen diente das pH und die Ionenstärke der Lösungen. Die Stoffflüsse wurden mit Hilfe der Wasserflüsse bestimmt.

Der Protoneneintrag betrug während der gesamten Beobachtungszeit durch saure Beregnung 366,0, durch "normale" Beregnung 1,1 und durch natürliche Regen 5,7 mmol/m². Die künstlichen Beregnungsereignisse dauerten in der Regel 1 h 20 min.

3. ERGEBNISSE

Die Tabelle 1 zeigt die Relation von Austrag zu Eintrag bei den Beregnungsereignissen. Es geht daraus hervor, daß die saure Beregnung im Vergleich zur "normalen" den Austrag von Mangan und Calcium auf das 6- bis 7fache, jenen von Magnesium auf das 4- bis 5fache und jenen von Kalium und Ammonium etwa auf das Doppelte erhöht. In absoluten Mengen gerechnet, ist der Mehraustrag von Calcium mit rund 25 mmol IE/m² am stärksten, es folgen die Mehrausträge von Kalium und Ammonium mit rund 8, dann jene des Mangans und Magnesiums mit rund 6; immer unter Berücksichtigung der Flußdifferenzen bei den normalen Beregnungen.

Tab. 1: Relation Austrag zu Eintrag bei den Beregnungsereignissen

	Mn	Ca	Mg	NH_4	K
Saure Beregnung pH 2,72	9,8	6,5	4,9	2,6	3,0
Normale Beregnung pH 5,2	1,5	1,0	1,1	1,1	1,7
sauer/normal	6,5	6,5	4,5	2,4	1,8

Die Abb. 1 veranschaulicht die Stoffflüsse oberhalb und unterhalb der Krone auf den Parzellen mit saurer Beregnung, und zwar getrennt nach a) sauren Beregnungsereignissen (die beiden linken Säulenpaare) und b) der gesamten Beobachtungszeit (die beiden rechten Säulenpaare).

Stoffeinträge u. -austräge in der Kronenschicht (mmol IE / m²)

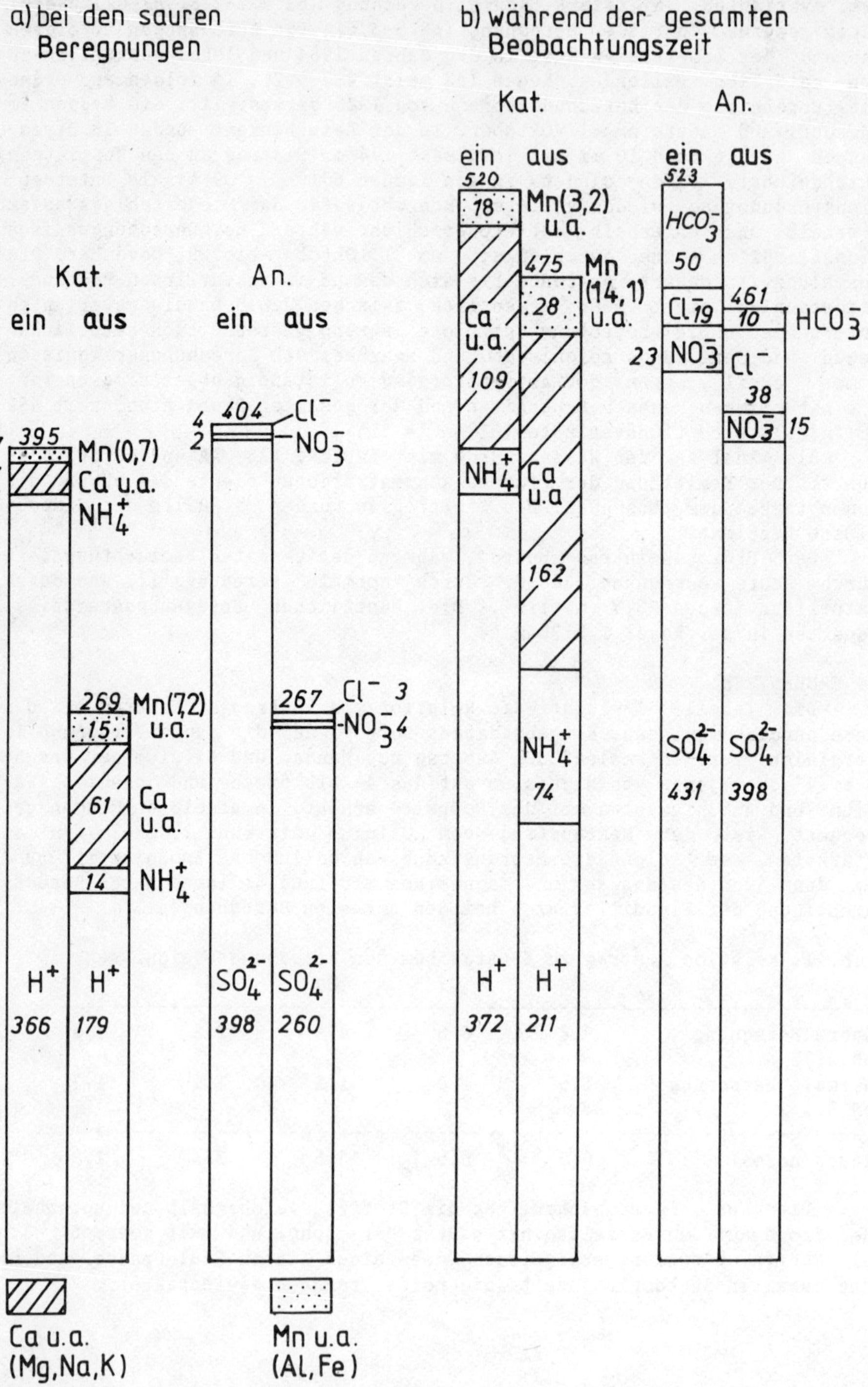

Offensichtlich wird die Hauptmasse der Kationen bereits bei den sauren Beregnungsereignissen mobilisiert und ausgewaschen. Ein großer Teil der eingetragenen Schwefelsäure bleibt jedoch in der Kronenschicht zurück und wird erst danach durch die natürlichen Regen eluiert.

Von besonderem Interesse ist bei dieser Retentions- und Austragsdynamik das Schicksal der Protonen. Darüber gibt Tab. 2 quantitativ Aufschluß:

Tabelle 2: Protonenbilanz

	bei sauren Beregnungen		während der gesamten Beobachtungszeit	
	mmol IE/m²	%	mmol IE/m²	%
Eintrag	366	100	372	100
Retention				
durch Kationenaustausch	44	11	53	14
als Säure	138	37	14	3
Konsumtion				
durch MnO_2-Lösung	6	2	11	3
durch HCO_3-Eintrag	-	-	40	10
durch NO_3-Aufnahme	-	-	8	2
Austrag				
als NH_4^+	9	2	53	14
als H^+	179	48	211	54
Rechnerische Unschärfe von Retention, Konsumtion und Austrag gegenüber dem Eintrag	+ 3 %		+ 5 %	

Zurückgehalten werden bei den sauren Beregnungsereignissen im Kronenraum rund 48% der eingetragenen Protonen: rund 11% durch Austausch gegen Ca, Mg und K und 37% als Säure gemeinsam mit Sulfatanionen, was vermutlich zu einer starken Versauerung des Apoplasten führt. In geringem Maße scheinen Protonen bei der Auflösung der im Apoplasten deponierten Manganoxide (11) verbraucht zu werden (2%).

Rund die Hälfte der eingetragenen Protonen passiert bei den sauren Beregnungen die Kronenschicht. Nur gering ist dabei der Anteil, der nicht direkt als H^+, sondern als NH_4^+ ausgetragen wird. Wir gehen dabei davon aus, daß die Austrags-Eintragsdifferenz von NH_4^+ dadurch zustande kommt, daß sich während der sauren Beregnung gasförmiges Ammoniak der Atmosphäre in den sauren Wasserfilmen löst und als NH_4^+ ausgewaschen wird. Die Zeit ist jedoch für eine verstärkte NH_3-Adsorption zu kurz, obwohl die tiefen pH-Werte begünstigend wirken.

Gegenüber den sauren Beregnungsereignissen erhöht sich während der gesamten Beobachtungszeit der direkte Protonenaustrag nur um 6 auf 54%, während der Austrag als NH_4^+ von 2 auf 14% ansteigt. Mit großer Wahrscheinlichkeit ist letzteres die Folge verstärkter Adsorption von atmogenem NH_3, denn die Entsäuerung des Apoplasten im Wege der Pufferung und des Säureaustrages erfolgt nicht abrupt (Gestützt wird diese Vermutung dadurch , daß das Versuchsglied mit normaler Beregnung insgesamt nur einen NH_4^+-Mehraustrag von 8.9 mmol IE/m² aufweist, während bei saurer Beregnung 53 mmol IE/m² Mehraustrag resultierte (7)).

Deutliche Veränderungen ergeben sich auch bei der Protonenkonsumtion. So verdoppelt sich annähernd die Auflösung von Manganoxiden in den Zeitabschnitten mit natürlichen Regen, erreicht aber insgesamt nur eine Pufferung von 3% der eingetragenen Protonen. Bedeutsamer ist dagegen die Pufferwirkung der mit natürlichen Regen eingetragenen Bikarbonationen (10%). Relativ gering ist mit 2% die Pufferung durch Vermetabolisierung von Nitrat.

Ebenfalls nur um 3% des Protoneneintrags steigt in den Zeitabschnitten mit natürlichem Regen die Pufferung durch Kationenaustausch an. Doch ergeben sich deutliche Verschiebungen in der Zusammensetzung der Kationenbilanz, wie Tab.3 zeigt. Auffällig ist vor allem, daß das Magnesium in der gesamten Beobachtungszeit eine Nettospeicherung in der Krone erfährt, während es bei den sauren Beregnungsereignissen eine deutliche negative Bilanz aufweist. Auch die Calciumbilanz ist während der gesamten Beobachtungszeit höher als bei den sauren Beregnungen, was bedeutet, daß in den Zeitabschnitten mit natürlichen Regen mehr Calcium ein- als ausgetragen wird. Umgekehrt verhält sich das Kalium, dessen erhöhter Austrag auch bei den natürlichen Regen andauert. Er liegt deutlich über den entsprechenden Werten bei den Parzellen mit normaler Beregnung (7).

Tabelle 3: Kationenbilanz in mmol IE/m²

	bei saurer Beregnung				während der gesamten Beobachtungszeit			
	Ca	Mg	K	Na	Ca	Mg	K	Na
Eintrag	4,8	1,8	5,4	5,0	43,1	23,6	21,4	20,3
Austrag	31,4	8,9	16,2	4,0	67,8	20,3	48,9	24,6
(E-A)	-26,6	-7,1	-10,8	+1,0	-24,7	+3,3	-27,5	-4,3

Durch die natürlichen Regen kommt es zu einer starken Auswaschung von anorganischen Anionen: Am Ende der Beobachtungszeit sind nur noch 3% des H^+-Eintrages in Begleitung von anorganischen Anionen im Kronenraum vorhanden. Möglicherweise liegt dieser Rest an den Membranen in Form von protonierten Aminogruppen vor, deren positive Ladung durch adsorbierte anorganische Anionen ausgeglichen ist. Umtauschvorgänge scheinen auch bei den Anionen abzulaufen: So kommt es in den Zwischenzeitabschnitten mit natürlichen Regen zu einem vermehrten Chloridaustrag, der durch eine entsprechende Sulfatspeicherung kompensiert wird (s.Abb.1).

Von dem gesamten Protoneneintrag verblieben am Ende der Beobachtungszeit 17% in den Kronen, während 68% dem Boden zugeführt wurden. 15% wurden konsumiert.

4. DISKUSSION

Die erhöhten Kationenausträge bei saurer Beregnung bestätigen bisherigen Befunde (1,5, 10). Daß es sich dabei zum größten Teil um eine Auswaschung aus dem Apoplasten handelt, ist sehr wahrscheinlich, da die partikuläre Immission wegen der Lage der Versuchsfläche eine nur untergeordnete Rolle spielt. Calcium und Magnesium liegen im Apoplasten z.T. als Brückenionen im Protopektin der Mittellamellen, z.T. als adsorbierte

Ionen vor und können hier von H^+-Ionen verdrängt werden. Mangan hingegen ist im Apoplasten wahrscheinlich als Braunstein deponiert (11), der bei Versauerung gelöst werden kann, so daß reduziertes 2-wertiges Mangan im Leachat erscheint.

Eine Minderung der Nährelementgehalte in den Nadeln ließ sich bei unseren Experimenten bislang noch nicht nachweisen. Grundsätzlich ist eine solche Nährstoffminderung jedoch nicht auszuschließen. Entscheidend ist dabei die Replazierungsrate aus dem Boden, die nach Mengel (mündl. Mitt.) bei jungen Fichten sehr rasch ablaufen kann. Besonders beim Kalium dürfte dieser Mechanismus eine besondere Rolle spielen, da, wie sich zeigte, die Kaliumauswaschung nach saurer Beregnung auch noch sehr stark bei natürlichen Regen weiterging.

Von Bedeutung ist, daß in der sauren Lösung des Apoplasten verstärkt atmogenes NH_3 adsorbiert wird. Ein Teil des dabei unter Abpufferung von H^+ gebildeten NH_4^+ gelangt im Wege der Auswaschung auf den Boden. Dort kann es bei der Aufnahme durch die Wurzel wie durch die Nitrifikation versauernd wirken. Daneben konkurriert es bei der Aufnahme sehr stark mit Magnesium und Calcium, so daß vor allem auf magnesiumschwachen Substraten Magnesiummangel induziert werden kann (7).

LITERATUR

(1) HOFFMAN, W.A., LINDBERG, S.E. and TURNER, R. R. (1980). Precipitation acidity: the role of the forest canopy in acid exchange. J. Environ. Qual. 9, 95-100.

(2) ZÖTTL, H.W. und MIES, E. (1983). Nährelementversorgung und Schadstoffbelastung von Fichtenökosystemen im Südschwarzwald unter Immissionseinfluß. Mitt. Dtsch. Bodenkdl. Gesellsch. 38, 429-434.

(3) BOSCH, C., PFANNKUCH, E., BAUM, U. und REHFUESS, K.-E. (1983). Über die Erkrankung der Fichte (Picea abies Karst.) in den Hochlagen des Bayerischen Waldes. Forstw. Cbl. 102, 167-181.

(4) MATZNER E., KHANNA, P. K., MEIWES, K. J., CASSENS-SASSE, E., BREDEMEIER, M. und ULRICH, B. (1984). Ergebnisse der Flüssemengen in Waldökosystemen. Berichte des Forschungszentrums Waldökosy./ Waldsterben. Bd. 2, 29-49.

(5) HORNTVEDT, R. (1979). Leaching of chemical substances from tree crowns by artificial rain. Mitt. IUFRO Tagung "Luftverunreinigung" Ljubljana, 115-125.

(6) ARNDT, U. und SEUFERT, G.(1985). Untersuchungen zur Beteiligung von Ozon und Schwermetallen an den Schäden unserer Waldbäume. BML-Forschungsbericht.

(7) KREUTZER, K. und BITTERSOHL, J. (1986). Stoffauswaschung aus Fichtenkronen (Picea abies (L.) Karst.) durch saure Beregnung. Forstw. Cbl. 105, H. 4, 357-363.

(8) KREUTZER, K. und BITTERSOHL, J.(1986). Untersuchungen über die Auswirkungen des sauren Regens und der kompensatorischen Kalkung im Wald. Forstw. Cbl. 105, H. 4. 273-282.

(9) KREUZIG, R. und KORTE, F. (1986). Luftchemische Charakterisierung des Standortes Höglwald. Forstw. Cbl. 105, H. 4, 292-295.

(10) SCHERBATSKOY, T. and KLEIN, R. M. (1983). Response of spruce and birch foliage to leaching by acidic mists. J. Environ. Qual. 12. 189-195.

(11) HENRICHFREISE, A. (1976). Aluminium- und Mangantoleranz von Pflanzen saurer und basischer Böden. Diss. Münster.

EFFECT OF ARTIFICIAL RAIN ON SOIL CHEMICAL PROPERTIES AND FOREST GROWTH

A.O. STUANES[1], G. ABRAHAMSEN[2] and B. TVEITE[1]
[1]Norwegian Forest Research Institute
P.O.Box 61, N-1432 Aas-NLH, Norway
[2]Institute of Soil Sciences
Agricultural University of Norway
P.O.Box 28, N-1432 Aas-NLH, Norway

Summary

Field plots, 150 m^2 in size, were watered 27 times, 50 mm each time in the period of July 1973 to September 1978. Groundwater adjusted with sulphuric acid to pH levels of 6, 4, 3, and 2.5, was applied to treatment plots. The experiment was in a homogenous stand of Norway spruce (Picea abies) planted in 1956 on a Cambic Arenosol soil. Tree growth measurements and needle and soil samples have been taken in the period 1973 to 1986. The relatively large changes in soil chemical properties created by applying strong acid to the soil were moderated within a relatively short period after the application of acid was terminated. No effect from the acid treatments were measured on tree height and basal area growth during the watering period. However, negative effects on growth from the acid treatments gradually developed after the watering was stopped. Maximum effects occurred two to five years later. Since then, recovery has been measured. The negative effects are difficult to explain by the nutrient status of the stand as measured by standard foliar analysis. High Al concentrations in the soil solution may have been important.

1. INTRODUCTION

During the period from 1972 to 1975, five field experiments were established in Norwegian forests to examine the effect of artificial acidification on soil properties and growth of trees. This paper describes the results from one such experiment. Preliminary results from this experiment have been published (1,2,3,4,5).

2. MATERIALS AND METHODS

At Nordmoen (11°6'E, 60°16'N) in southern Norway, 12 field plots, 150 m^2 in size, were established. The experiment (A-2 experiment) was in a homogenous stand of Norway spruce (Picea abies) planted in 1956 on a Cambic Arenosol soil (6,7). The average stand height in spring 1973 was 3.3 m, rising to 8.1 m in autumn 1986. The plots were watered 27 times, 50 mm each month, in the frost-free period from July 1973 to September 1978. The water was applied above the canopy by a sprinkler system. The plots received artificial precipitation in addition to an annual average of 742 mm of natural precipitation for the period in question (1). The pH and lime potential (LP) of bulk precipitation for the watering period are shown in Table I. The high LP of the pH 6 and pH 4 treatments compared to the not-watered treatment is due to higher amounts of Ca and Mg in the groundwater than in the precipitation. Equivalent amounts of precipitation with pH 4.2 and (Ca+Mg) = 7 µM and the total amount of SO_4-S supplied in the different treatments are also given (2). Groundwater, with the pH adjusted by sulphuric acid, was used as artificial rain, giving treatments of pH 6, pH 4, pH 3, and pH 2.5. In addition, areas between the watered plots were used for the not-watered treatment. Each treatment was replicated three times in a randomized design (6).

Table I. The pH and lime potential (LP) of bulk precipitation, the *equivalent amount of precipitation with pH 4.2 and (Ca+Mg) = 7 µM, and the total amount of SO_4 -S supplied in the different treatments. The numbers are from the watering period.

Treatment	n.w.	pH 6	pH 4	pH 3	pH 2.5
Bulk pH	4.36	4.46	4.26	3.61	3.16
Bulk LP	1.8	2.2	2.0	1.4	0.9
Precipitation (m)*	2.6	1.4	2.1	9.5	27.1
SO_4-S kg ha^{-1}	40	55	130	320	750

Tree height and girth increments were measured annually from autumn 1974. Foliar nutrient concentrations in both current and previous year needles were measured from autumn 1973, except for the years 1979 and 1980. Needle samples from eight trees per plot were taken from the third or fourth branch whorl from the top of the trees. Soil samples were taken from the O, E, Bs1, and Bs2 horizons in 1975, 1978, 1981, and 1984. Twenty subsamples were pooled by horizon for each plot. Exchangeable cations were determined after extraction with 1 M NH_4OAc at pH 7. Soil pH was measured in a 1:2.5 soil-water suspension. For further description of analytical methods, see Ogner et al. (8).

3. RESULTS AND DISCUSSION

Because the potential cation exchange capacity (CEC) was measured, the CEC did not show changes by treatments. CEC varies from 750 $mmol_c$ kg^{-1} in the O-layer to about 40 $mmol_c$ kg^{-1} in the Bs2-layer. Since CEC is not changed, the base saturation is a good measure for possible alterations in the pool of exchangeable cations in the soil. Variations in soil pH and base saturation with treatments and time are shown in Figure 1. In the years with watering, base saturation appeared to be higher for the pH 6 and pH 4 treatments compared with the not-watered treatment. This may be due to the higher lime potentials of these treatments (Table I).

The most pronounced effects on the soil were measured in

the O and E horizons (Figure 1). In these horizons the pH dropped for all treatments from 1975 to 1978. There were generally clear differences between the treatments in 1975 and 1978, with the two most acid treatments giving the lowest pH and base saturation. Since 1981, the differences between the treatments have become very small. An increase in soil pH was measured for the pH 3 and pH 2.5 treatments from 1978 to 1981. The soil acidification introduced by the watering therefore appears to be reversible. The pH in the most acid soil has increased and in the least acid soil has decreased. In general, base saturation has decreased during the entire period, but for the pH 3 and pH 2.5 treatments an increase has been measured in the O horizon after the watering stopped. However, there is a clear moderating effect from 1981 to 1984, and the differences between the treatments have significantly diminished. The soil pH and base saturation have also decreased in the not-watered treatment. This is most likely to be explained by extraction of nutrients from the soil to the tree biomass. This effect is very important since most of the N is most probably taken up as ammonium (2).

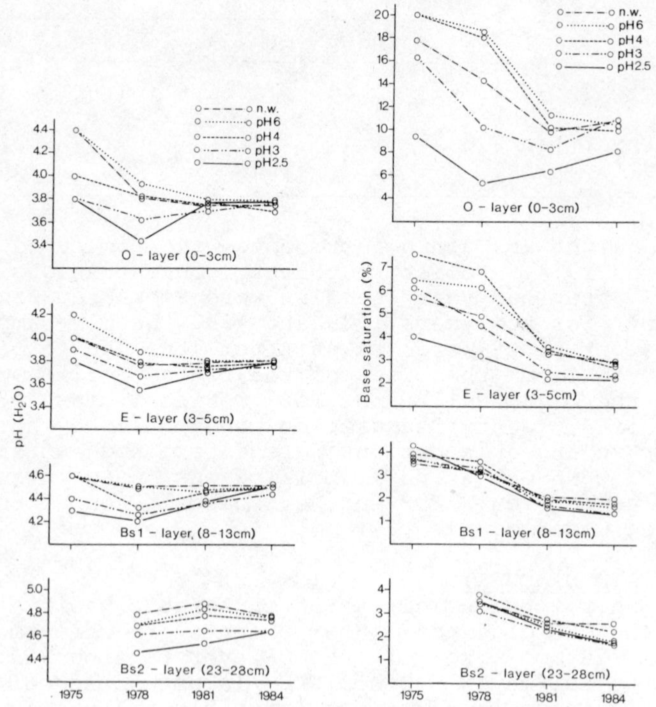

Figure 1. Effects of the different pH's in the artificial rain on soil pH and base saturation in various soil layers for the years from 1975 to 1984.

Negative effects from acid treatments on tree height and basal area growth developed gradually from 1978 (Figure 2). Before that, no significant effects of treatments were found

(5). Maximum effects occurred in the years 1980 to 1983, with an obvious recovery in the following years.

Certain time lags seem to exist between soil and tree growth effects. Even though the soil chemical properties in the most acidic treatments started to improve between 1978 and 1981, the tree growth did not start to improve before in 1983 to 1984.

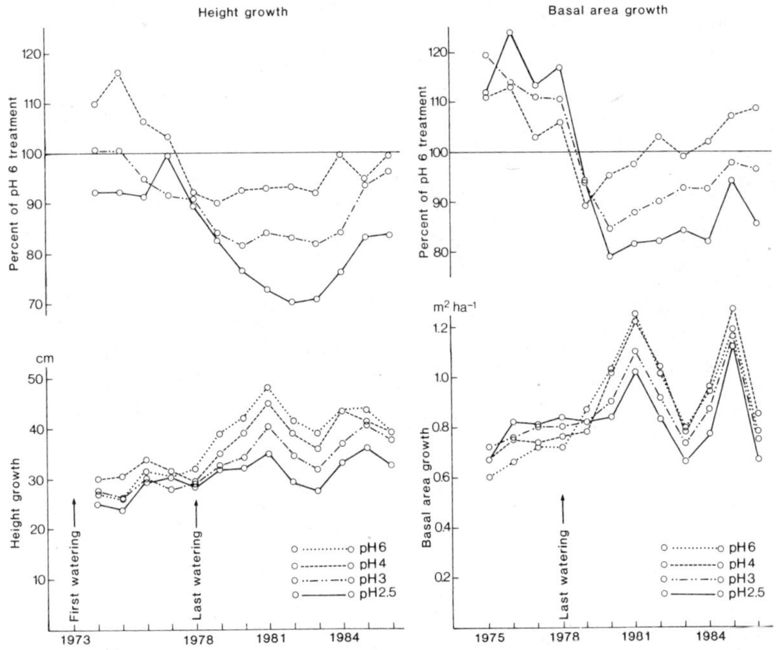

Figure 2. Height and basal area growth in absolute numbers and as percent of the pH 6 treatment.

In Figure 3 the foliar nutrient concentrations of N, P, K, Ca, and Mg are shown for the years 1973 to 1985. All nutrients showed a decreasing trend in the years with acid watering independent of treatment. The nitrogen content of the needles reached a maximum in 1983 and 1984, but the other nutrients showed the highest concentrations at the beginning of the experiment. Acid treatments lowered the Ca concentrations, while the Mg concentrations were apparently affected primarily by the pH 2.5 treatment.

The element concentrations in current year needles for 1985 are shown in Table II. The concentrations in 1978 are reported by Tveite (4). The nutrient status of the trees shows no obvious sign of deficiencies, though the N concentration is somewhat suboptimal. The pH 2.5 treatment still shows reduced Ca concentration. Raised needle concentrations for S, SO_4-S, and Al are found for the two most acidic treatments.

In conclusion, the relatively large changes in soil chemical properties created by the strong acid treatments have been moderated within a relatively short period after the

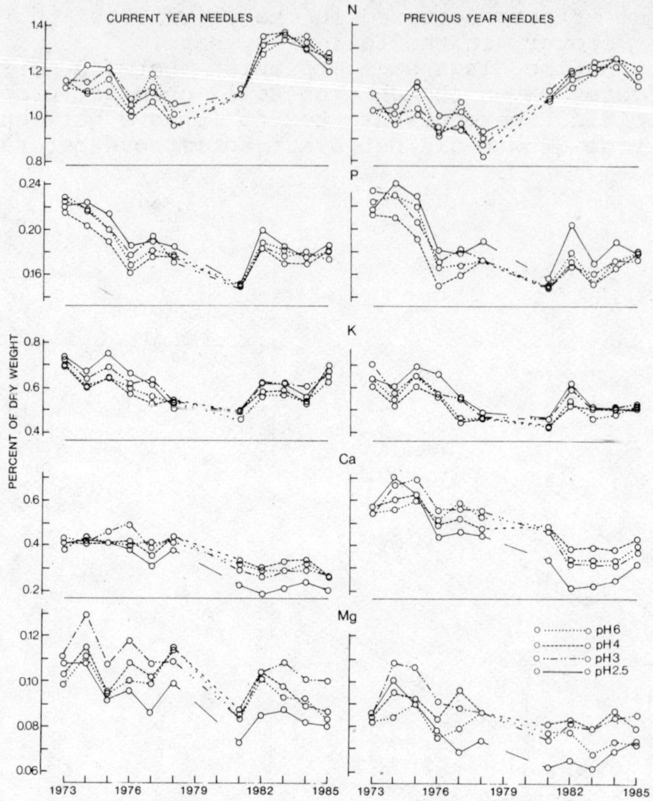

Figure 3. Foliar nutrient concentrations in current (left) and previous year needles (right) for the years 1973 to 1985.

Table II. Element concentrations (percent of dry weight) in current year needles for the different treatments sampled in November 1985.

Element	n.w.	pH 6	pH 4	pH 3	pH 2.5
N	1.30	1.26	1.28	1.24	1.20
P	0.19	0.18	0.18	0.17	0.19
K	0.64	0.62	0.69	0.65	0.67
Ca	0.26	0.26	0.26	0.28	0.20
Mg	0.086	0.083	0.086	0.100	0.080
S	0.096	0.096	0.094	0.099	0.100
SO_4-S	0.019	0.021	0.019	0.026	0.027
Mn	0.107	0.103	0.107	0.120	0.133
Al	0.0112	0.0111	0.0102	0.0125	0.0127
Zn	0.0030	0.0028	0.0029	0.0033	0.0028
Fe	0.0037	0.0037	0.0047	0.0042	0.0036
Cu	0.00026	0.00026	0.00025	0.00024	0.00025

applications terminated. No significant effect of the acid treatments was observed on tree height and basal area growth during the watering period. However, negative effects on growth from the pH 3 and pH 2.5 treatments gradually developed starting from the year the watering was stopped. Maximum effects occurred two to five years later. After that, significant recovery has been observed. The growth reduction cannot be explained by the nutrient status of the trees. We do not have any definite explanation for the growth reduction, but high Al concentrations may have been important. According to lysimeter experiments, Al concentrations above 0.4 mM may have prevailed in the soil solution over long periods (9).

REFERENCES

(1) ABRAHAMSEN, G. (1980). Impact of atmospheric sulphur deposition on forest ecosystems. In: Shriner, D.S., Richmond, C.R. and Lindberg, S.E. (eds.) Atmospheric sulfur depostion. Environmental impact and health effects. Ann Arbor Sciences Publishers, Inc. Ann Arbor, Michigan: 397-415.

(2) ABRAHAMSEN, G. (1987). Air pollution and soil acidification. Nato Advanced Research Workshop on The Effects of Acid Deposition on Forest, Wetlands and Agricultural Ecosystems. Toronto, May 12-17 1985. (in press).

(3) STUANES, A.O. (1980). Effects of acid precipitation on soil and forest. 5. Release and loss of nutrients from a Norwegian forest soil due to artificial rain of varying acidity. In: Drabløs, D. and Tollan, A. (eds.) Proc., Int. Conf. Ecol. Impact Acid Precip. SNSF-project, Norway 1980: 198-199.

(4) TVEITE, B. (1980). Effects of acid precipitation on soil and forest. 8. Foliar nutrient concentrations in field experiments. In: Drabløs, D. and Tollan, A. (eds.) Proc., Int. Conf. Ecol. Impact Acid Precip. SNSF-project, Norway 1980: 204-205.

(5) TVEITE, B. (1980). Effects of acid precipitation on soil and forest. 9. Tree growth in field experiments. In: Drabløs, D. and Tollan, A. (eds.) Proc., Int. Conf. Ecol. Impact Acid Precip. SNSF-project, Norway 1980: 206-207.

(6) ABRAHAMSEN, G., BJOR, K. and TEIGEN, O. (1976). Field experiments with simulated acid rain in forest ecosystems. SNSF-project, Norway, FR 4/76.

(7) STUANES, A. and SVEISTRUP, T.E. (1979). Field experiments with simulated acid rain in forest ecosystems. 2. Description and classification of the soils used in field, lysimeter and laboratory experiments. SNSF-project, Norway, FR 15/79.

(8) OGNER, G., HAUGEN, A., OPEM, M., SJØTVEIT, G. and SØRLIE, B. (1984). The chemical analysis program at The Norwegian Forest Research Institute, 1984. Norwegian Forest Research Institute, Aas, Norway. 27 pp.

(9) ABRAHAMSEN, G. (1983). Sulphur pollution: Ca, Mg and Al in soil and soil water and possible effects on forest trees. In: Ulrich, B. and Pankrath, J. (eds.) Effects of accumulation of air pollutants in forest ecosystems. D. Reidel Publishing Company: 207-218.

EFFECTS OF SULPHUR DIOXIDE ON FOREST LITTER DECOMPOSITION AND NUTRIENT RELEASE

P. INESON and P.A. WOOKEY
Merlewood Research Station, Institute of Terrestrial Ecology,
Grange-over-Sands, Cumbria LA11 6JU, United Kingdom

Summary

Sulphur dioxide gas is a potent inhibitor of respiration in decomposing pine litter at concentrations commonly encountered in rural air. Results from constant level and fluctuating level sulphur dioxide chamber fumigations, and from open air field fumigation systems, suggest that sulphur dioxide is readily dry deposited to forest litter, where it oxidises to form sulphuric acid. This in turn leads to a substantial drop in pH of the litter leading to enhanced leaching of cations, especially calcium and magnesium.

1. INTRODUCTION

Soil organisms are important in the maintenance of soil fertility, being responsible for many reactions which take place within the soil. In particular, they are central to the conversion of nutrients contained in soil organic matter into inorganic forms, thus completing an essential part of the nutrient cycling process. For example, the breakdown of organic nitrogen compounds to release ammonium and nitrate is the principal means by which non-leguminous trees can obtain this important element. The soil population also performs many other functions, such as the mineralisation of inorganically bound elements, maintenance of soil structure, degradation of pesticides and organic residues, and a host of reactions essential for the health of higher plants. The suggestion that air pollutants may impair forest soil biological activity is therefore a serious one, and one that has resulted in an increase in research effort into the assessment of the effects of wet deposition of air pollutants in recent years.

In marked contrast, there have been few studies examining the soil effects of dry deposited air pollutants, despite the fact that dry deposition is often the most significant deposition mechanism for pollutants in regions close to emissions sources. Indeed, several research groups have recently reported the major importance of dry deposition of sulphur dioxide in the acidification of areas in central Europe (1,2).

Microbial populations within soils act as effective uptake agents for many gaseous atmospheric pollutants (3,4), and gases such as sulphur dioxide have known anti-microbial properties. Field observations suggest that the distribution of certain fungi (noteably phylloplane species) is correlated with sulphur dioxide pollution (5,6,7). Air pollutants may affect soil/litter fungal and bacterial populations in a similar way and therefore suggested that gaseous atmospheric pollutants could adversely affect decomposer organisms, resulting in decreased microbial biomass and activity, thus slowing the rate of nutrient mineralisation. The impact of air pollutants on litter decomposition is considered to be particularly important because the litter layer lies at the interface between the atmosphere and the pedosphere and also marks the beginning of the decomposition process.

Little direct experimental evidence exists as regards the impact of air pollutants on soil micro-organisms, this lack of research reflecting some of the practical difficulties encountered when attempting to fumigate leaf litters and soils with low concentrations of pollutant gases. Until now studies examining the impacts of gaseous air pollutants have been mainly indirect in their approach, the solution products of the air pollutants being used in place of the gases themselves (8,9). These studies are based on the assumption that many gaseous air pollutants (such as sulphur dioxide and nitrogen dioxide) become anti-microbial only when in solution. Unfortunately, very little information exists at present on the concentrations of pollutant solution products encountered in soils in association with realistic concentrations of gaseous air pollutants. The results of solution product studies are therefore very difficult to relate to field conditions. The research presented here was performed using fumigation facilities constructed for higher plant work, and we describe some of the results of fumigations performed in chambers and in field fumigation. The principal pollutant we consider here is sulphur dioxide.

2. METHODS

Chamber fumigations

Needles were collected from the litter layer beneath a mature stand of *Pinus nigra* at Delamere Forest, U.K. (Grid Reference ST583689) the day before the experiment commenced. The litter was mixed well and placed on the soil surface in plant pots in each of five fumigation chambers at Imperial College Field Station. The fumigation chambers are fully described by Farrar *et al.* (10), and consisted of closed chambers maintained out of doors and designed for fumigation of trees with sulphur dioxide at constant and fluctuating levels. The pattern and degree of fluctuation were controlled by a microprocessor and monitored by a Meloy SA185 continuous sulphur analyser, enabling concentrations to be maintained in an interactive and pre-controlled fashion.

Litter, 15g, was placed on the soil surface within plant pots with ten pots in each chamber being treated in this way. Each pot was watered daily from above with tap water, incident precipitation being excluded. The sulphur dioxide concentration within the chambers was maintained according to the regime described in Table I. All chambers, with the exception of Chamber 1, were held at an overall mean SO_2 concentration between 30-40 parts per billion (ppb), despite the fact that the patterns of SO_2 concentrations in each chamber differed. Chamber 1 acted as a 0 ppb control. Chamber 2 was maintained at a relatively constant level of around 34 ppb, and experienced no days at 0 ppb. This chamber had occassional peaks over the 82 days of the fumigation, but the daily mean never exceeded 100 ppb. In marked contrast, Chamber 5 received quite high peaks of SO_2 on a regular basis, spending many days at 0 ppb in order to return the mean concentration to 37 ppb. Chambers 3 and 4 were effectively replicates, each receiving peaks in SO_2 concentration intermediate between

Chambers 2 and 5. After 82 days the litters were removed from the chambers and it is analysed for rate of carbon dioxide production and pH. CO_2 production is principally a measure of microbial respiration, and provides a useful index of microbial activity and rate of decomposition. CO_2 production was measured using infra-red gas analysis (11), and pH was determined in water suspension using a combination electrode.

Field fumigation

We exposed decomposing *Pinus sylvestris* needles to a range of environmentally realistic SO_2 concentrations in a field fumigation experiment The field fumigation system was designed and is operated by the Central Electricity Research Laboratories, Leatherhead, and a description is given in McLeod *et al.* (12). Briefly, the system consists of four replicate plots constructed in a 2.6 ha field at the Glasshouse Crops Research Institute, Littlehampton. The plots were designed to expose a large sample area to a spatially homogeneous SO_2 concentration, the four plots being used to provide a range of SO_2 concentrations from ambient to about 50ppb. *Pinus sylvestris* needles, collected in October 1985 from the L layer of a pure stand at Gisburn Forest, Lancashire (G.R. SD751589), were air-dried and used to construct 72 litter bags with 2mm nylon mesh. These bags each contained 5 grams of air-dry needles, the dimension of the bags being approximately 130x150mm. 18 such litter bags were placed out in each of the four field fumigation plots on 9/12/85. These litter bags were then left exposed until the sampling dates on 13/05/86 and 11/07/86. On each sampling date six litter bags were removed from each fumigation plot. Microbial respiration of the *Pinus sylvestris* needles was determined by infra-red gas analysis. Nutrient release from the needles was assessed by washing the litter bags with distilled water and analysing the resulting leachates. The following analyses were performed: pH and hydrogen ion concentration by combination electrode, calcium and magnesium ions by atomic absorption spectrophotometry, sodium and potassium by flame emission spectrophotometry, ammonium by auto-analysis (utilising the indophenol-blue colour reaction) and nitrate, sulphate and chloride by Dionex high-pressure liquid chromatography.

3. RESULTS

Chamber fumigations

Table 1 outlines both the experimental regime and also the effects of the treatments on litter respiration and pH. It is important to note that each chamber, with the exception of Chamber 1, was held at approximately the same overall mean SO_2 concentration and that treatment differences are a consequence of differences in the pattern of fumigation.

Field fumigation

Results for the two sampling dates show the same basic patterns of pollutant effect with the July data confirming those of the May sampling. Results from the July sampling are presented. Fig. 1 shows daily mean SO_2 concentrations for the four pollution plots during the fumigation period. Respirometry results (all determinations performed at 15^oC) for the four treatments show a marked reduction in microbial activity with increasing pollution (Fig. 2). An 18.5% reduction in microbial respiration was observed when sulphur dioxide exposure was increased from 7 to 14ppb. The chemical nature of needle leachates in relation to SO_2 exposure is summarised in Fig. 3. The results illustrate

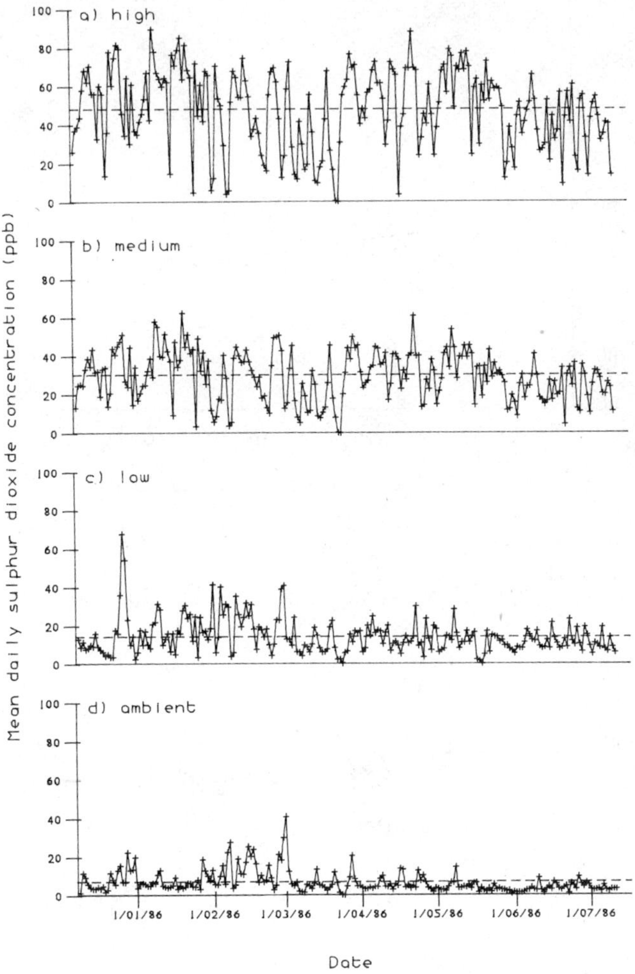

Figure 1. Mean daily SO_2 concentrations (ppb) for the four plots at the field fumigation site from 9/12/85 to 11/7/86. a), b), c) and d) refer to the pollution treatments in each plot. The broken lines show the mean SO_2 concentrations.

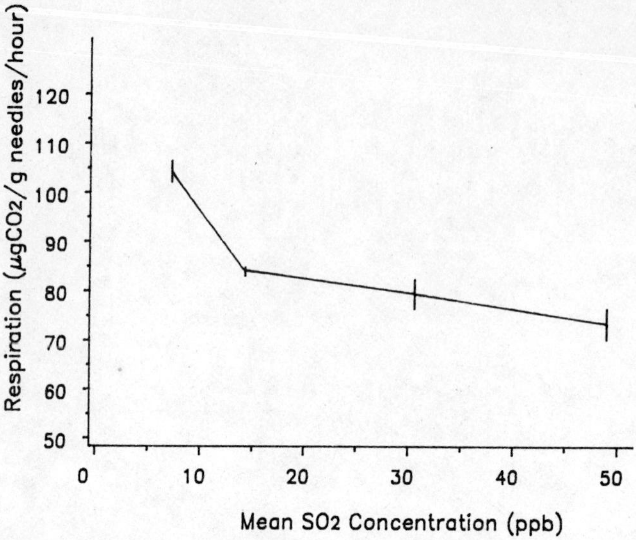

Figure 2. Microbial respiration for the field fumigation litter sampled on 11/7/86 (n=6). Mean ± 1 S.E. for each fumigation plot.

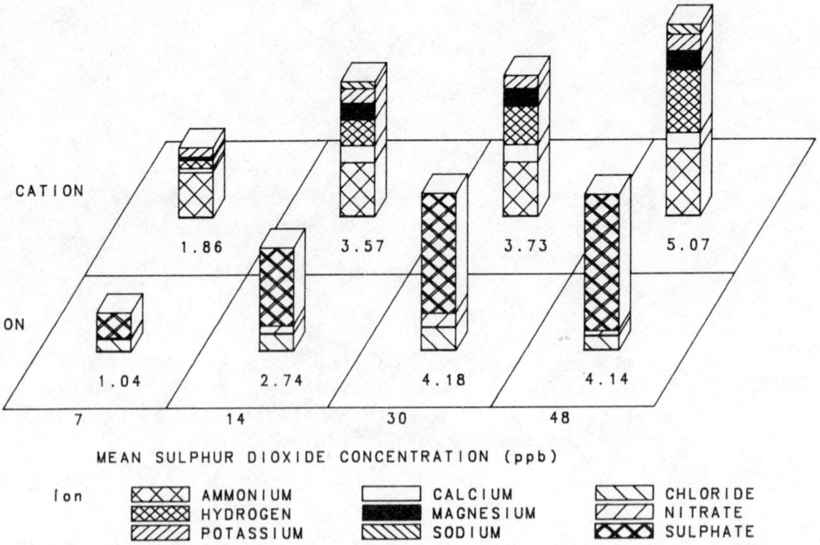

Figure 3. Leachate chemistry results for the field fumigation. Litter sampled on 11/7/86. All cations and anions converted to micro-equivalents per gram needles. The number at the base of each column gives the total micro-equivalents.

a marked increase in sulphate-sulphur and hydrogen ion concentration with increasing pollution exposure. The increase in leachate hydrogen with increasing pollution exposure reflects a drop in leachate pH from 5.1 in the control plot (7ppb SO_2) to 4.2 in the most highly polluted plot (48ppb SO_2). A very marked increase in leachate calcium and magnesium is apparent with increasing pollution exposure, six times more magnesium being found in leachates from the 14ppb plot as compared to the 7ppb control plot.

Table I. The effect of constant and fluctuating levels of SO_2 on the respiration and pH of P. nigra litter. Experiments were carried out over an 82 day period.

Chamber	Mean SO_2 concentration (ppb)	Maximum daily mean SO_2 (ppb)	Days at 0 ppb	CO_2 evolution (cm^{-3}/100g/h)	Final pH
1	0	0	82	7.3 ± 0.4	5.8
2	34	100	0	4.9 ± 0.5	5.3
3	35	127	82	6.0 ± 0.4	5.6
4	31	128	82	6.3 ± 0.3	5.3
5	37	178	82	7.6 ± 0.4	5.7

4. CONCLUSIONS

Sulphur dioxide at low concentration reduces the activity of litter micro-organisms, as reflected by measurements of respiration. Results from the chamber fumigations of Pinus nigra litter show the importance of pollution regime in determining pollutant impact. Fairly constant fumigation at around 38ppb (Chamber 2) could well be more damaging to decomposer organisms than fumigations with high peaks of SO_2 separated by periods of low or even zero SO_2 concentrations (Chambers 3,4 and 5). The after-fumigation litter pH values indicate that the greatest deposition of SO_2 occured to litter in Chamber 2 (Table I), and it thus indicates that constant fumigation with low concentrations allows greater deposition of the pollutant than fumigations with high magnitude fluctuations. Episodes with high SO_2 concentrations could result in rapid saturation of receiving surfaces with the pollutant, thus limiting further deposition.

Field fumigations provide a closer approximation to polluted environments than chamber fumigations. Again, with Pinus sylvestris litter, sulphur dioxide has markedly inhibited microbial respiration particularly over the range 7-14ppb SO_2 (see Fig. 2). Sulphur dioxide deposited onto the litter has subsequently solubilised under damp conditions and oxidised ultimately to form sulphate and hence sulphuric acid. It is believed that the intermediate oxidation states -sulphite, bisulphite and undissociated sulphurous acid- are the toxic agents. The increase in sulphate with increasing pollution has resulted in a marked increase in leachate hydrogen and hence a reduction in pH. Sulphate leaching and acidification have also led to enhanced divalent cation leaching from the needles, particularly of magnesium and calcium. All of these effects were noted after only 7 months of fumigation: the longer-term results of sulphur dioxide pollution remain to be evaluated.

In the European environment soil acidification and the leaching of important and often scarce nutrient cations as a result of dry-deposited acid gases remain a cause for concern (13). A reduction in microbial activity resulting from sulphur dioxide dry-deposition could also, in the longer term, result in reduced rates of nutrient mineralisation in forest soils with a progressive impoverishment of soil fertility and nutrient status.

ACKNOWLEDGEMENTS

The work described here was only made possible because of the use of fumigation facilities established and maintained by Imperial College and the Central Electricity Research Laboratories. We would particularly like to thank Drs T.W. Ashenden and A. McLeod for their support and collaboration in this respect.

5. REFERENCES.

(1) van BREEMEN, N., DRISCOLL, C.T. and MULDER, J. (1984). Acidic deposition and internal proton sources in acidification of soils and waters. Nature. **307**, 599-604.

(2) PACE, T. (1985). Sources of acidification in central Europe estimated from elemental budgets in small basins. Nature. **315**, 31-36.(3)

(3) CRAKER, L.E. & MANNING, W.J. (1974). Sulphur dioxide uptake by soil fungi. Environmental Pollution. **6**, 309-311.

(4) ABELES, F.B., CRAKER, L.E., FORRENCE, L.E. and LEATHER, G.R. (1971). Fate of air pollutants: Removal of ethylene, sulphur dioxide and nitrogen dioxide by soil. Science, **173**, 914-916.

(5) GRZYWACZ, A. and WAZNY, J. (1973). The impact of industrial air pollutants on the occurence of several important pathogenic fungi of forest trees in Poland. European Journal of Forest Pathology. **3**, 129-141.

(6) SAUNDERS, P.J.W. (1973). Effects of atmospheric pollution on leaf surface microflora. Pesticide Science. **4**, 589-595.

(7) BEVAN, R.J. & GREENHALGH, G.N. (1976). Rhytisma acerinum as a biological indicator of pollution. Enviromental Pollution. **10**, 271-285.

(8) GRANT, I.F., BANCROFT, K. & ALEXANDER, M. (1979). Sulphur dioxide and nitrogen dioxide effects on microbial activity in an acid forest soil. Microbial Ecology. **5**, 85-89.

(9) WODZINSKI, R.S., LABEDA, D.P. & ALEXANDER, M. (1978). Effects of low concentrations of bisulphite, sulphite and nitrite on micro-organisms. Applied and Environmental Microbiology. **35**, 718-723.

(10) FARRAR, J.F., RELTON, J. & RUTTER, A.J. (1977). Sulphur dioxide and the growth of Pinus sylvestris. Journal of Applied Ecology. **14**, 861-875.

(11) INESON, P. & GRAY, T.R.G. (1980). Monitoring the effects of acid rain and sulphur dioxide on soil micro-organisms. Society of Applied Bacteriology Tech. Ser. **15**, 21-26.

(12) McLEOD, A.R., FACKRELL, J.E. & ALEXANDER, K. (1985). Open-air fumigation of field crops: criteria and design for a new experimental system. Atmospheric Environment **19**, 1639-1649.

(13) ZOETTL, H.W. & HUETTL, R.F. (1986). Nutrient supply and forest decline in southwest Germany. Water, air and soil pollution **31**, 449-462.

LES RECHERCHES SUR LE DEPERISSEMENT DES FORETS EN FRANCE : STRUCTURE ET PRINCIPAUX RESULTATS DU PROGRAMME DEFORPA

G. LANDMANN
I.G.R.E.F.
I.N.R.A. NANCY
Champenoux, 54280 Seichamps, France

Résumé

Le programme français de recherche sur les causes du dépérissement des forêts (DEFORPA) a été élaboré en 1984, suite à l'observation, dans le massif des Vosges, de défoliations importantes, sur des résineux essentiellement. Le programme DEFORPA comprend 4 volets : 1) Evaluation des dommages, et relation avec les conditions écologiques. 2) Caractérisation du climat de pollution en zone forestière. 3) Action directe des polluants atmosphériques. 4) Effet de la pollution acide par l'intermédiaire du sol. L'évaluation provisoire des premiers résultats de recherche peut se résumer ainsi : une partie notable des dommages observés serait le fait d'écosystèmes de montagne fragilisés par l'Homme (vieillissement des peuplements, essences forestières acidifiantes, surdensité de gibier), et soumis à une série singulière de cycles d'années sèches depuis quelques décennies. Il n'en demeure pas moins que si l'action directe des polluants gazeux, aux niveaux où ils sont présents actuellement en zone forestière, ne paraît pas devoir être important, le rôle des polluants acides (H2SO4, HNO3) ou acidifiants (dépôts secs de SO2,NOx, NH4) pourrait être significatif, tant au niveau du feuillage (lessivage d'éléments nutritifs) qu'au niveau du sol (appauvrissement en éléments nutritifs, libération d'éléments toxiques), et rendre compte de certains symptômes typiques du dépérissement (jaunissements du feuillage).

Summary

The french research program on the causes of forest decline (DEFORPA) was established in 1984, following the observation of important defoliation in conifer stands in the Vosges mountains. The DEFORPA program is divided into 4 parts : 1) Damage assessment, and relationship with ecological conditions. 2) Characterization of the climate of pollution in the damaged forest areas. 3) Direct effects of air pollution on forest trees. 4) Indirect effects of acid pollution via the soil.
Initial evaluation of the first results may be summarized as follows : a significant part of the damage could be due to man's effects on the delicately balanced mountain ecosystems (poor forest management, introduction of soil acidifying forest species, high densities of game), coupled with the effects of an exceptionnal series of dry years during the last fifty years. Nevertheless, if direct effects of the gazeous pollutants, at the levels at which they occur in the forest areas, do not seem to be important, the influence of acid pollution (wet deposition, H2SO4, HNO3 and dry deposition, SO2, NOx, NH4) may be significant, both on the foliage (leaching of nutrients) as well as in the rhizosphere (leaching of nutrients, liberation of toxic elements), in the development of certain typical symptoms of forest decline (discolouration of the foliage).

1. INTRODUCTION

Le programme français de recherche sur le dépérissement des forêts a été élaboré en 1984 dans des conditions assez particulières : alertés par leurs collègues d'Outre-Rhin sur une dégradation préoccupante de l'état de santé du sapin pectiné, et à un moindre degré des autres essences forestières, les forestiers du Massif vosgien ont également observé, en 1983, des peuplements de diverses essences, sapin pectiné et épicéa commun principalement, présentant des déficits de feuillage importants. Sur la base des études engagées en République Fédérale d'Allemagne essentiellement, l'idée prévalait alors qu'aucun facteur ne pouvait, mieux que la pollution atmosphérique à longue distance, expliquer le phénomène observé. Le titre du programme "Dépérissement des forêts attribué à la pollution atmosphérique" (DEFORPA) ne signifie pas pour autant que tous les projets de recherche sont focalisés sur l'hypothèse "pollution". Au contraire, cela a été une attitude constante des responsables du programme français d'aborder le problème avec un esprit ouvert, car si l'existence de la pollution ne pouvait être niée, on ne pouvait pas non plus ignorer les cas de "dépérissements" de diverses essences, rapportés dans la littérature scientifique et les archives forestières, à des époques anciennes et plus récentes, et qui n'ont souvent été que mal élucidés (23,24,47,62,68, ...).

2. STRUCTURE ET CONTENU DU PROGRAMME DEFORPA

Le programme DEFORPA est clairement finalisé : il s'agit d'identifier les causes réelles du dépérissement actuel des forêts. Sont ainsi exclus de son champ d'action, à la différence de certains programmes nationaux, l'incidence de la pollution sur les écosystèmes aquatiques et les espaces cultivés, l'inventaire des émissions de polluants, la technologie et la stratégie de leur réduction.

En fonction de cet objectif, le programme DEFORPA a été divisé en quatre grandes parties :

2.1. Observation et symptomatologie des dommages ; relations avec les conditions écologiques (y compris climatiques)

La caractérisation fine du dysfonctionnement de l'écosystème étudié, qui, dans un contexte moins marqué par l'urgence apparente de la situation, aurait du précéder la formulation d'hypothèses et la définition d'une stratégie de recherches, ne doit pas, pour autant, être délaissée. Quelle valeur donner à des hypothèses qui ne rendent compte que d'aspects très marginaux du problème ?

2.2. Caractérisation de la pollution atmosphérique dans les zones forestières affectées par le dépérissement

L'effort de caractérisation du climat de pollution porte essentiellement sur le massif des Vosges, région où les dommages sont les plus importants. Le relief montagneux introduit une difficulté particulière dans les études de transport, de transformation et de dépôts dans les zones forestières considérées.

2.3. Etude de l'action directe des polluants atmosphériques

On entend par là l'action des polluants atmosphériques, à l'état gazeux mais aussi sous forme de pluies ou de brouillards, sur la partie aérienne des arbres. L'importance des études expérimentales (simulation, exclusion, ...) reflète l'idée, relativement admise en 1983, que les effets possibles de la pollution par l'intermédiaire du sol (cf 2.4) ne couvraient qu'une fraction des dommages observés *in situ*.

2.4. Etude de l'effet de la pollution acide par l'intermédiaire du sol

Parmi les modes d'action les plus vraisemblables de la pollution atmosphérique sur les arbres forestiers, on trouve l'acidification progressive des sols (pertes de cations, libération d'aluminium sous forme ionique dans les solutions du sol, ...) dont l'étude exige des techniques adaptées à la mise en évidence des processus et à la quantification des divers facteurs contributifs (apports atmosphériques, nature du peuplement, conduite sylvicole, ...). Quelle que soit l'origine exacte des déficiences observées, la pertinence et l'efficacité des mesures de fertilisation doivent être étudiées attentivement.

Des études spécifiques traitent du rôle possible des parasites racinaires ou d'une altération des mycorhizes dans le phénomène étudié.

Pour plus de détails sur les projets et les résultats détaillés, on se reportera aux principales publications mentionnées en bibliographie (74, 75).

3. ANALYSE DES PRINCIPAUX RESULTATS DU PROGRAMME DEFORPA

3.1. Evaluation et suivi des dommages. Symptomatologie. Relations avec les conditions écologiques

Comme la plupart des pays confrontés au problème de l'évaluation des dommages, la France a opté, en l'absence de connaissances sur la répartition spatiale des dommages, pour un réseau systématique de placettes d'observation qui couvre, en 1986, une partie significative du territoire français (figure 1). Ce réseau a pour objectif prioritaire de suivre l'évolution des dommages dans l'idée de confirmer (ou d'infirmer) l'idée d'un déclin de l'écosystème forestier. Les paramètres retenus sont la défoliation et l'altération de la couleur du feuillage, selon les classes en vigueur dans la plupart des pays européens. On admet généralement que cet objectif est correctement satisfait. Les résultats du suivi des dommages depuis 1983 peuvent se résumer ainsi (40) :

- dans les régions réputées "atteintes" en 1983 (Vosges, Jura), la défoliation ne s'est que faiblement agravée entre 1983 et 1984 pour se stabiliser ensuite, voire s'atténuer sensiblement (cas de l'épicéa dans le Jura).
- certains jaunissements du feuillage semblent constituer un fait nouveau ; c'est principalement le cas du jaunissement de l'épicéa dans les Vosges, en forte progression de 1983 à 1985, puis en légère baisse en 1986.
- une dégradation notable de l'état de santé des feuillus est observée en 1986 dans plusieurs régions. Le niveau général des dommages reste cependant faible.

Un second objectif est la détermination de la distribution spatiale des dommages. Le tableau 1 rassemble les régions forestières où une essence au moins présente un état jugé déficient sur la seule base de la défoliation. Un tel tableau, bien que riche d'enseignements, revêt inévitablement un caractère trompeur : comment, en effet, rendre compte de l'état sanitaire de diverses essences présentes dans des écosystèmes très différents les uns des autres (de plaine et de montagne , peuplements naturels et artificiels, gestion intensive ou extensive, ...)? Ajoutons que les données issues du réseau d'observation ne permettent pas de conforter l'idée, avancée ici ou là, d'une propagation des dommages au sein du territoire français. L'extension du réseau et l'attention croissante que portent forestiers et scientifiques à ce problème sont probablement à l'origine de cette "impression".

Figure 1

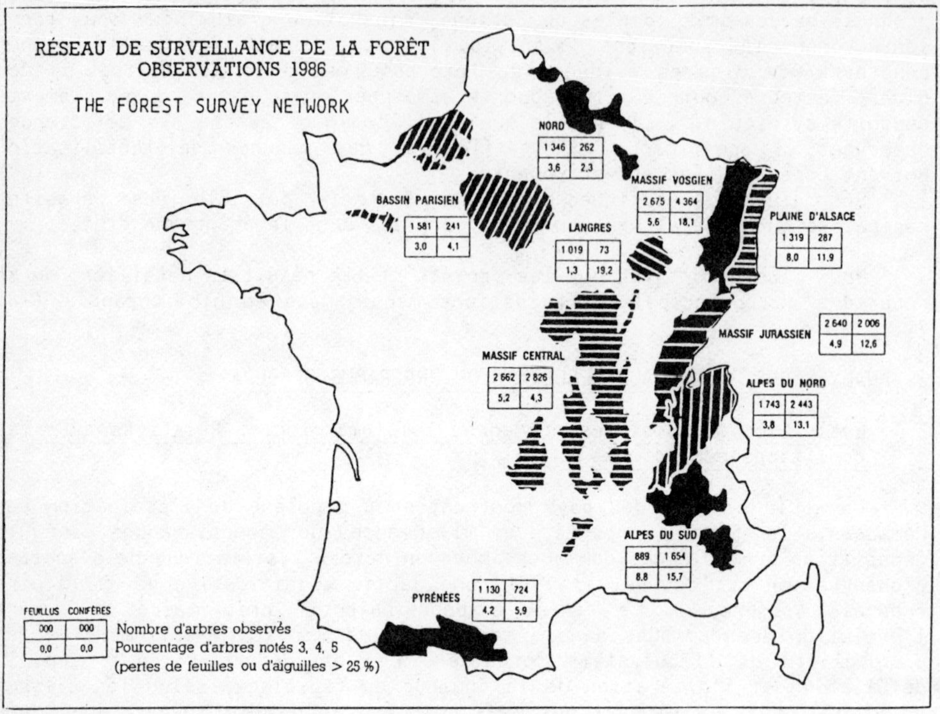

Un troisième objectif aurait pu être l'étude des relations entre l'intensité du dépérissement et les facteurs du milieu. Cette utilisation des données a été volontairement limitée en raison de la trop faible densité de placettes et de la faible fiabilité des paramètres stationnels disponibles. Les résultats parfois contradictoires des études spécifiquement conçues pour répondre à cette question justifient a posteriori cette attitude prudente. De façon de plus en plus claire, il s'est avéré indispensable de compléter "l'image" que donne le réseau d'observation du phénomène par :

a) une symptomatologie approfondie visant à cerner la pertinence des paramètres mesurés par rapport à d'autres estimateurs de la vitalité des arbres, au-delà du problème (réel !) de la précision des mesures. A ce titre, il s'est avéré en particulier que la relation entre la perte foliaire estimée et les paramètres de croissance était médiocre. Dans le cas du sapin dans les Vosges, seule une défoliation forte (40 %) entraîne systématiquement une perte d'accroissement (7,8), mais des "comportements" très variés, qui méritent d'être distingués, sont observés. Il faut dès à présent en conclure qu'un paramètre unique ne caractérise pas valablement la vitalité d'un arbre (notion qui se révèle très délicate à définir) et qu'une lecture trop littérale des données d'inventaire fournira une appréciation peu nuancée du phénomène étudié.

Tableau 1

Régions forestières pour lesquelles une essence au moins présente une proportion d'arbres ayant perdu plus de 25 % de feuilles ou d'aiguilles supérieure à respectivement 20 % (tableau 1 A) et 10 % (tableau 1 B)

(le tableau 1 B ne comprend que les régions complémentaires de celles figurant au tableau 1 A)

Forest regions for which at least one species presents a proportion of trees which have lost more than 25 % of their leaves or needles superior to respectively, 20 % (table 1 A) and 10 % (table 1 B)

(table 1 B only contains the regions complementary to those shown in table 1 A)

Tableau 1 A

RÉGION FORESTIÈRE / ESSENCE	SAPIN	EPICEA	PIN SYLVESTRE	CHÊNE	HÊTRE	AUTRES FEUILLUS (1)
Vosges						
Hautes Vosges gréseuses	[E] c	[C] b	[C] a	A a	[B] b	—
Vosges cristallines	[E] b	[D] b	[C] a	[C] b	[B] b	[B] a
Basses Vosges gréseuses	—	—	[E] c	[B] a	A b	—
Plaine d'Alsace						
PL de Haguenau	—	—	[E] c	[B] c		[E] c
Hardt	—	—	—	—	—	[E] a
Jura						
Haut Jura	[E] c	[C] b	—	—	A a	A a
Pente intermédiaire du Jura	[E] b	A a	—	—	A a	A a
1er plateau du Jura	A a	A a	—	[D] a	A a	[B] a
Alpes du Nord						
Chartreuse	[E] c	[E] a	—	—	A a	A a
Vercors	[E] c	[E] b	—	—	A a	A a
Alpes du Sud						
Préalpes de Haute-Provence ...	—	—	[E] a	—	—	[C] c
Préalpes de Digne	—	—	[E] a	—	—	[E] a

Tableau 1 B

Jura						
Petite montagne jurassienne ...	—	[C] a	—	[B] a	—	A a
2e plateau du Jura	B d	—	—	—	A a	—
Massif Central						
Cantal-Cézallier	—	[C] c	—	[B] a	[B] a	[B] d
Pyrénées						
Haute chaîne Pyrénéenne	[C] d	—	—	—	A b	—

☐ Etat de santé déficient (E pour les résineux ; D et E pour les feuillus).
Deficient state of health (E for conifers ; D and E for broadleaved species).

⁝⁝ Etat de santé intermédiaire (C et D pour les résineux ; B et C pour les feuillus).
Intermediate state of health (C and D for conifers ; B and C for broadleaved species).

— Nombre d'arbres insuffisant.
Number of trees insufficient.

(1) Autres feuillus que Chêne et Hêtre.
Broadleaved species other than the Oak, Beech.

Code de couleur : *Colour code :*
a = moins de 10 % d'arbres présentant une altération de la couleur du feuillage ; *less than 10 % of trees showing foliar discolouration.*
b = de 10 à 19 % d'arbres présentant une altération de la couleur du feuillage ; *from 10-19 % of trees showing foliar discolouration.*
c = de 20 à 39 % d'arbres présentant une altération de la couleur du feuillage ; *from 20-39 % of trees showing foliar discolouration.*
d = plus de 40 % d'arbres présentant une altération de la couleur du feuillage ; *more than 40 % of trees showing foliar discolouration.*

Code de défoliation ; *Defoliation code :*
A = moins de 5 % d'arbres ayant perdu plus de 25 % de feuilles ou aiguilles ; *less than 5 % of trees having lost more than 25 % of leaves or needles.*
B = de 5 à 9 % d'arbres ayant perdu plus de 25 % de feuilles ou aiguilles ; *from 5-9 % of trees having lost more than 25 % of leaves or needles.*
C = de 10 à 14 % d'arbres ayant perdu plus de 25 % de feuilles ou aiguilles ; *from 10-14 % of trees having lost more than 25 % of leaves or needles.*
D = de 15 à 19 % d'arbres ayant perdu plus de 25 % de feuilles ou aiguilles ; *from 15-19 % of trees having lost more than 25 % of leaves or needles.*
E = plus de 20 % d'arbres ayant perdu plus de 25 % de feuilles ou aiguilles ; *more than 20 % of trees having lost more than 25 % of leaves or needles.*

b) des estimations des surfaces "atteintes" et une compréhension plus fine de leur répartition spatiale sur des bases cohérentes avec le "sentiment" des forestiers, et en particulier avec la perturbation de la gestion forestière occasionnée par le phénomène. Jusqu'ici, cette question n'a pas été abordée sur un plan théorique mais par le biais de cartographies locales des dommages réalisées par les services forestiers (sur la base d'appréciations à l'échelle du peuplement) ou encore par l'Institut Géographique National (sur la base de photographies aériennes). Les cartes montrent, dans le cas des Vosges, une répartition spatiale des dommages plus marquée que ne le laissaient présager les relations "moyennes" entre les dommages et l'âge des peuplements ou l'altitude, établies sur la base des données du réseau d'observation. Il semble en particulier que ce dernier rende mal compte de la singularité des zones très touchées, relativement limitées en superficie.

c) des approches écologiques basées sur des échantillons stratifiés et des paramètres fiables. A l'image de la situation observée dans d'autres pays, des contradictions sensibles sont apparues entre les résultats des diverses études conduites sur la question de l'incidence des conditions écologiques sur l'intensité du dépérissement. Alors qu'une première étude portant sur 200 placettes de sapin (6) dans les Vosges concluait à une faible incidence du niveau de fertilité (apprécié par la flore), une étude ultérieure établissait des relations assez nettes pour 30 peuplements adultes d'épicéa et de sapin avec la garniture du complexe absorbant des sols (14). Une image cohérente se dégage, liant les déficits foliaires les plus marqués chez le sapin (essence résineuse dominante des Vosges), et les jaunissements les plus intenses chez l'épicéa, à des sols pauvres en magnésium et très alumineux (rapport Mg/Al très bas). Ces relations viennent d'être confirmées sur un échantillon de jeunes peuplements résineux des Vosges (41). A l'inverse, les défoliations observées dans le Jura (17) sur substrat calcaire, dans les Alpes (42) et les Pyrénées (20) sur moraines ou schistes, ne sont qu'en partie reliées à des déficiences nutritionnelles établies par des analyses foliaires. La déficience en potassium, répandue et parfois accusée, n'a pas pu être reliée à des caractéristiques édaphiques (17, 20). Sur un plan méthodologique, il semble que les échantillons systématiques ou aléatoires rendent mal compte des défoliations accusées à l'échelle d'un peuplement (cas peu fréquents), focalisant l'attention sur le cas des défoliations d'intensité moyenne des peuplements âgés (très représentés) pour lesquels un diagnostic de santé est très délicat. A l'inverse, les échantillons stratifiés évitent en partie cet écueil, mais sont souvent trop limités et peu représentatifs d'une situation régionale. L'échantillon de 2000 placettes observées dans les Vosges, dans le cadre de l'opération de télédétection (60) constitue probablement un compromis ; notons d'ailleurs que cette étude a confirmé l'incidence défavorable des roches-mères les plus pauvres sur l'état de santé des peuplements résineux. Les limites de reconnaissance fine des symptômes (confusion entre port des vieux arbres et défoliation, mauvaise perception des peuplements serrés, ...) réduisent toutefois la qualité du diagnostic. Cette remarque vaut plus encore dans le cas de la numérisation des photographies infra-rouge (5) et celui de l'utilisation des images satellitaires(61).

d) une étude de la croissance des peuplements forestiers des régions atteintes. L'étude dendroécologique du sapin dans les Vosges (200 placettes, 1200 arbres) (7, 8) n'a pas permis de conforter l'existence d'une dégradation importante de la productivité de cette essence. Certes on retrouve le résultat bien connu de la divergence ancienne, sur le plan de l'accroissement, entre arbres fortement et faiblement défoliés, mais, sur la base de la largeur d'aubier (critère déjà souvent cité comme pertinent pour caractériser la vitalité d'un arbre), ces divergences remontent à une époque encore plus ancienne, qui ne coïncide plus, comme la précédente, avec l'augmentation brutale de la pollution atmosphérique (figure 2). Par ailleurs, la comparaison de l'accroissement radial des arbres de même âge à diverses époques ne permet pas de reconnaître de tendance significative (figure 3). La reconstitution historique de la productivité des peuplements sur la base d'un échantillon figé est, il est vrai, très délicate. On présume toutefois qu'une évolution négative dramatique aurait été décelée. Il est cependant prudent de ne pas énoncer de conclusions définitives sur la seule base de telles études, ne serait-ce que parce que des phénomènes

Figure 2

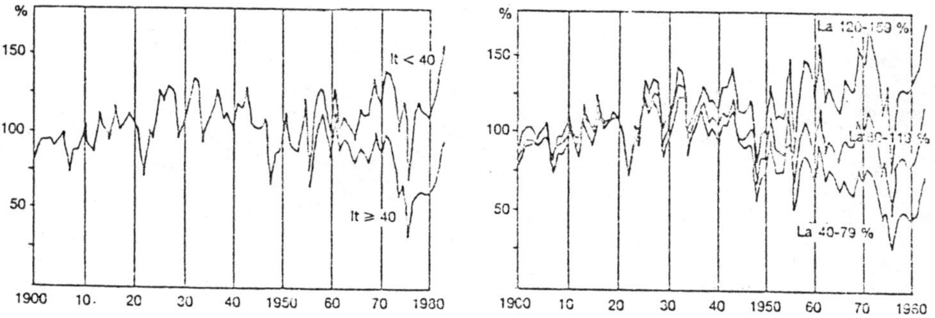

Evolution de la largeur corrigée des cernes, avec le temps, pour différentes sous-populations de Sapin dans les Vosges :
- deux sous-populations différant par le degré de défoliation (It : indice de transparence de houppiers)
- trois sous-populations différant par la largeur de l'aubier (La)

Evolution of the age corrected ring width, according to the date, for several fir sub-populations in the Vosges :
- two sub-populations differentiated by their degree of needles loss (It)
- three sub-populations differentiated by their sap-wood thickness (La)

Figure 3

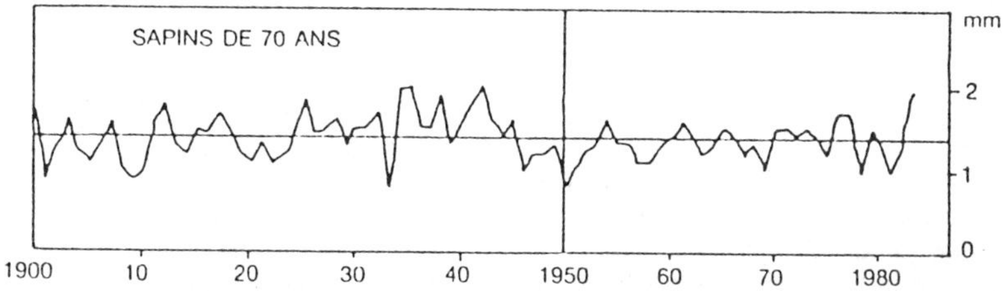

Largeur en (mm) de l'accroissement radial moyen des sapins de même âge (70ans) de 1900 à 1984

Mean annual ring width (in mm) of fir trees of the same age (70 years) from 1900 until 1984

d'émergence récente (le jaunissement du feuillage des épicéas par **exemple)** pourraient ne pas encore être décelées. Signalons encore un paradoxe : la période - supposée - de dégradation de l'état des houppiers de sapin dans les Vosges (1980-1984, par analogie aux observations réalisées en Bade-Wurtemberg)) puis la phase - observée - de jaunissement de l'épicéa (1983-1986) correspondent à partir de 1982 à un rétablissement remarquable de l'accroissement moyen du sapin vosgien (7, 8).

En résumé, si un réseau d'observation systématique peu dense permet de déceler des évolutions marquantes de l'état de santé des forêts, il n'en permet pas une bonne appréciation ; d'autres approches sont indispensables pour dépeindre de façon réaliste et consistante le phénomène étudié.

3.2. Etude du climat de pollution des Vosges

La connaissance très fragmentaire du climat de pollution des sites de montagne où étaient observés les dommages explique l'effort métrologique réalisé. Les mesures sont pour l'essentiel focalisées sur le site du Col du Donon (Vosges).

Une station permanente de mesure en continu (NOx, O3, SO2) fonctionne depuis début 1985 (29) (Tableau 2).

Le site du Donon présente un niveau moyen de pollution SO2 faible (environ 75 % des valeurs horaires en SO2 peuvent être considéres comme "nulles"), proche de celui de sites géographiquement voisins (Forêt Noire en particulier), avec des épisodes caractéristiques de pollution observés en hiver par vent de secteur Nord à Est (ex. : janvier 87, moyenne mensuelle, 69 µg/m3, plus forte moyenne journalière 299 µg/m3, plus forte moyenne horaire 375 µg/m3) et régime anticyclonique très stable. Pour l'ozone, on retrouve un profil journalier relativement peu marqué (29), déjà décrit dans d'autres sites de montagne à l'étranger, et des moyennes annuelles comparables à celles observées en Forêt Noire par exemple, de l'ordre de 60 µg/m3.

Tableau 2

	MOYENNE ANNUELLE	MOYENNE 24 H MAXIMUM	MOYENNE HORAIRE MAXIMUM	MOYENNES EN FONCTION VENT	
				NORD → EST	SUD → OUEST
SO_2 ($\mu g/m^3$)	15	200	340	38	5
NO_2 ($\mu g/m^3$)	12	56	105	26	8
NO ($\mu g/m^3$)	limite mesure	33	50	—	—
Ozone ($\mu g/m^3$)	58	152	200	—	—

Niveau de pollution gazeuse sur le site du Donon (01/04/85 au 31/03/86)
Levels of the gazeous pollutants measured at the Donon Pass (01/04/85 until 31/3/86

Deux campagnes (septembre 1985 et juin 1986) de caractérisation des aérosols (55) ont fourni, par vent de Sud-Ouest, des résultats comparables à ceux obtenus à la Station du Pic Wank (Garmisch-Partenkirchen) (59), tant du point de vue physique (concentration de l'ordre de 4500 particules/cm3) que de la composition chimique (2,5 µg/m3 d'ions sulfates). Par vent d'Est-Sud/Est, on observe une augmentation du nombre de particules (25 000/cm3), qui reste toutefois très en deçà des niveaux atteints sur des sites industriels (300 000/cm3 à Fos-sur-Mer) (54).

Le pluviomètre séquentiel à ouverture automatique (48) installé en 1986, a fourni des premiers résultats en période estivale, pour des précipitations orageuses et des précipitations associées aux circulations des principales familles de perturbations (Nord-Ouest, Ouest, Sud-Ouest) (28). On note l'opposition de la circulation d'Ouest avec la circulation Nord-Ouest : les concentrations moyennes en nitrates et en sulfates, sensiblement égales entre elles, sont de 2 mg/l pour les pluies de Nord-Ouest et 0,5 mg/l pour les pluies d'Ouest.

Une étude focalisée principalement sur les précipitations neigeuses (22) en mars et avril 1985 a confirmé le caractère acidifiant de la neige, avec un pH variant de 3,05 à 5,26. Les valeurs faibles sont enregistrées par vents de secteur Nord à Sud-Est. L'espèce acide majoritaire est $H2SO4$, et non $HNO3$ comme cela a été observé lors de collectes en Champagne.

Les autres équipements (analyseur de PAN (56), collecteurs de rosée (49) et de brouillard (71) sont à des stades variables de mise au point. La "Station laboratoire du Donon" s'articulera autour d'une tour de mesure qui sera construite en 1987 au sein d'un peuplement adulte de sapin et d'épicéa dans une zone forestière présentant les symptômes typiques du dépérissement. La conception de la Station est proche de celle de réalisations similaires d'autres pays (République Fédérale d'Allemagne, Suisse, Etats-Unis notamment), et repose sur la mesure en continu des paramètres physico-chimiques de l'air à proximité immédiate des arbres et aux différents niveaux du peuplement (12).

L'étude des trajectoires des masses d'air arrivant au Donon (67) a fourni, grâce à un modèle simple, des résultats très comparables à ceux obtenus avec le modèle d'Eliassen, utilisé dans le cadre du programme EMEP (figure 4 et tableau 3) ; bien que les cadrans Sud-Est et Sud-Ouest (vents dominants) soient plus pauvres en émissions que les cadrans Nord-Ouest et Nord-Est, le territoire national est à l'origine de 40 % environ de la pollution $SO2$ arrivant au Donon.

Un modèle tri-dimensionnel de mésoéchelle (19) est développé dans le but de cerner de façon la plus réaliste possible les dépôts acides (composés soufrés et azotés). Ces dépôts résultant d'interactions non-linéaires complexes entre facteurs météorologiques et paramètres microphysiques et physico-chimiques, il s'avère nécessaire d'intégrer à "parité" ces divers processus. Des modules chimiques ont fait l'objet d'analyse de sensibilité dans des situations précipitantes typiques (nuages d'origine maritime ou continentale).

La responsabilité des oxydants d'origine photochimique dans le dépérissement des forêts ($O3$ en particulier) étant parfois avancée , une investigation, associant campagne de terrain et modélisation, a été engagée sur la base d'un modèle proche, dans sa conception, du modèle EKMA (26, 10). Le but final est de cerner la contribution régionale aux concentrations d'ozone mesurées au Donon.

En résumé, l'état de l'environnement atmosphérique en zone forestière montagneuse, largement méconnu en 1984, est à présent mieux cerné. La description doit être poursuivie et complétée par la poursuite des travaux engagés. S'il est clair que la pollution générale au Donon est assez faible, l'existence de pointes élevées pouvant survenir pendant des périodes appréciables doit être intégrée aux schémas expérimentaux de l'étude de l'action des polluants (3.3.1). D'autre part, une meilleure connaissance des dépôts humides et des dépôts secs en site montagneux devient un objectif prioritaire si l'on souhaite confronter leur distribution spatiale (échelle régionale et interrégionale) à celle du dépérissement, ou du moins à certains de ses aspects (problèmes nutritionnels), dans lesquels ces dépôts pourraient avoir une part significative.

Figure 4

Contribution de chaque pays aux concentrations d'un produit passif arrivant au Donon (sans tenir compte des données d'émission)

Each countries contributions to concentrations of a passive product arriving at the Donon Pass (without taking into account emission data)

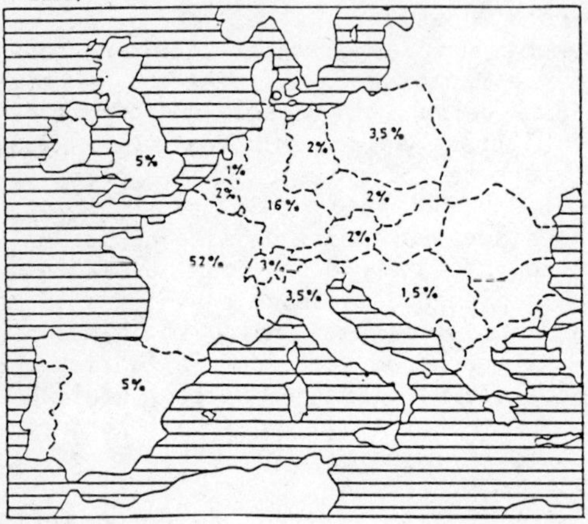

Tableau 3

Comparaison des contributions de chaque pays aux concentrations en SO_2 au Donon, d'après le modèle EMEP (European Monitoring and Evaluation programme) et d'après les contributions calculées sur la base d'un maillage 1° x 1°

Comparison of each countries contribution to the SO_2 concentrations at Donon Pass, according to EMEP (European Monitoring and Evaluation Programme) as well a contribution calculated on a grid 1° x 1°

	EMEP	$1/r^2$
France	37.5	38.4
RFA	31.7	22.0
Royaume Uni	7.2	10.5
Italie	3.2	3.8
RDA	5.0	5.9
Suisse	2.3	1.1
Belgique	3.2	4.0
Tchécoslovaquie	2.9	5.4
Espagne	2.3	1.0
Pays-Bas	1.4	1.1
Pologne	1.3	4.1
Luxembourg	0.9	1.3
Yougoslavie	0.7	0.6
Autriche	0.4	0.9

3.3. Etat des recherches sur les différentes causes possibles du dépérissement

3.3.1. Action directe des polluants sur l'appareil aérien

Plusieurs expérimentations ont été mises en place en 1985 et 1986 pour tester l'hypothèse d'une action directe des polluants sur l'appareil aérien. Si la littérature est riche de résultats expérimentaux démontrant l'effet nocif de SO_2, NO_x, O_3 sur divers végétaux, il est vrai aussi que relativement peu de travaux ont été conduits sur les arbres forestiers avec des niveaux de polluants comparables avec ceux observés en zone rurale.

Toutes les expérimentations sont conduites sur de jeunes épicéas de trois mêmes provenances (Gérardmer-Vosges, Lac de Constance-RFA, Istebna-Pologne); Il s'agit d'expériences en chambres à ciel ouvert, d'exclusion au Donon (Vosges) et de simulation à Montardon (Pyrénées Atlantiques) où sont reproduits les niveaux de SO_2 et O_3 (seuls ou en combinaison) enregistrés au Donon.

Dans les chambres à ciel ouvert du Donon, les mesures de croissance et le suivi physiologique et biochimique n'ont pas permis d'identifier une influence significative de la pollution ambiante. Ont été mesurés : l'activité photosynthétique (38), la résistance stomatique (46), la composition en phénols (46), l'activité enzymatique (21). Un effet "chambre" et des différences entre clones sont par contre relevés pour plusieurs de ces paramètres. L'expérimentation de Montardon est trop récente pour avoir fourni des résultats (16).

Les expérimentations en conditions contrôlées ont montré une interaction significative entre pollution par SO_2 et sécheresse édaphique, notamment dans l'activité photosynthétique et dans la teneur en eau des jeunes plants ou semis (57). SO_2 seul (80 ppb) provoque des réponses à cinétique plus lente et de plus faible amplitude (altération de la protéosynthèse par exemple) sur l'épicéa que sur des espèces herbacées (57).

L'action combinée d'une sécheresse édaphique et de brouillards acides (pH 3,4) a mis en évidence des différences sensibles entre clones principalement en ce qui concerne l'économie en eau, alors que l'activité photosynthétique et la régulation stomatique ne semblent que peu ou pas affectées par l'application de brouillard acide (3).

Il faut toutefois signaler que ces expérimentations (combinaison sécheresse et S02, sécheresse et brouillard acide) n'ont pour l'instant été conduites qu'en période estivale, alors que les pointes de S02 et les brouillards sont essentiellement hivernaux dans les Vosges.

Les premiers résultats de l'étude de la perméabilité cuticulaire des feuilles sous l'action des pluies acides et de l'ozone (35), conduite sur des cuticules isolées de lierre, de laurier et de philodendron, ne semblent pas indiquer qu'il y ait un effet important de l'ozone ou de l'acide sur les perméabilités cuticulaires à l'eau et aux ions. Cela ne préjuge pas des modifications éventuelles de la perméabilité membranaire des cellules.

Les nombreuses mesures faites in situ sur des arbres de différents âges et de différentes classes de dommages ont permis de dresser un portrait physiologique et biochimique intéressant. On observe ainsi, chez les arbres dépérissants, des variations d'activité des enzymes (57), une augmentation des sucres libres, du tryptophane, de la putrescine (70), une diminution des protéines solubles (57) et de l'arginine (70). La composition terpénique est d'une grande variabilité (63) et ne semble pas être un critère de caractérisation à retenir.

On peut toutefois s'interroger sur l'aptitude de telles observations à nous renseigner sur une cause déterminée, sachant qu'il semble extrêmement délicat de trouver des indicateurs de stress spécifiques. Tout au plus peut-on espérer tirer un "parallèle" entre certaines des modifications observées in situ et celles qui apparaîtraient dans les chambres à ciel ouvert ou dans des expérimentations contrôlées conduites en conditions réellement proches de celles observées in situ. Sans doute faudrait-il aussi limiter les mesures in situ à des paramètres très précis, en référence à des modalités d'actions (établies) de tel ou tel polluant très "en amont" dans la machinerie métabolique. Il est intéressant de noter, dans le cas des arbres étudiés in situ, que les différences établies entre individus sains et dépérissants sont le plus souvent plus nettes chez les sujets jeunes, ce qui peut s'interpréter de diverses manières. Cela pourrait correspondre à l'observation d'une maladie unique chez les jeunes arbres (jaunissement du feuillage) alors que les aspects sont plus variés et les distinctions moins nettes sur les vieux sujets.

3.3.2. Effets des polluants par l'intermédiaire du sol

Une meilleure connaissance de la dynamique des éléments minéraux dans les écosystèmes forestiers soumis à la pollution atmosphérique, tout particulièrement dans un contexte de sols très acides et désaturés, est indispensable pour vérifier l'hypothèse selon laquelle les sols s'acidifieraient et développeraient des carences ou des toxicités qui pourraient jouer un rôle important dans le dépérissement observé.

L'équipement, en 1985 et 1986, d'un petit bassin versant (80 ha entre 860 m et 1 100 m) à Aubure (Vosges) sur le plan hydrogéochimique et biogéochimique, répond à ce besoin.

Seuls des résultats préliminaires sont disponibles (2,25) :

- le pH moyen des pluies n'est pas très bas (4,3) mais on note une acidité plus marquée des précipitations de printemps (pouvant descendre à pH 3,7 avec des concentrations de 350 μ eq/l en sulfates et 170 μ eq/l en nitrates)

- les eaux à l'exutoire sont sensiblement neutres (ph 5,5 à 6,5) avec une alcalinité très faible en présence de sulfates abondants représentant 60 % de la charge anionique

- les eaux à l'exutoire sont très différentes de celles d'un bassin versant voisin, couvert de prés, qui présentent une forte alcalinité, atteignant 50 % de la charge anionique.

La mesure des flux d'éléments minéraux à travers les différents compartiments de l'écosystème forestier n'en sont qu'à leur début. Des premiers essais de récolte de pluviolessivats, prélevés sous épicéas à l'automne 1986, ont donné des concentrations en sulfate 5 à 10 fois plus élevées que dans les pluies hors couvert. Les dépôts secs, dont l'estimation est extrêmement délicate, sont probablement pour beaucoup dans cette différence.

L'étude de l'aluminium dans les solutions de deux sols vosgiens (39) a confirmé l'existence de pics d'aluminium dans les eaux gravitaires, lors des premiers épisodes pluvieux après une sécheresse, alors que la concentration moyenne est basse (0,5 à 1 mg/l pour l'horizon de surface). On a également observé une corrélation très forte entre les concentrations en Al et NO3, soulignant le risque de dommages aux racines par excès de nitrification entraînant des pointes de concentration en aluminium.

Ces projets s'inscrivent, dans le cas des Vosges, dans un contexte forestier où les manifestations de troubles nutritionnels, principalement sur le magnésium et le calcium (14, 15), se sont très nettement amplifiées depuis 1983. Rappelons aussi que les sols des stations où les peuplements sont les plus dépérissants présentent souvent une garniture de complexe absorbant très défavorable à la nutrition des arbres (14, 15). Par ailleurs, les résultats de travaux récents conduits à l'étranger (33, 66) confortent l'idée d'une contribution significative des dépôts humides et secs au lessivage des mêmes ions (Mg, Ca, ...) à partir du feuillage.

3.3.3. Action possible d'agents pathogènes

Les recherches concernent uniquement le rôle que pourraient jouer certains champignons et la flore du sol. Ne sont pas pris en compte : les virus, les viroïdes, les mycoplasmes et les rickettsies dont la présence a été signalée ou supputée par des chercheurs à l'étranger.

Il semble clair que les agents pathogènes classiques (Pythium, Fusarium et Rhizoctonia) ne sont pas impliqués dans le dépérissement observé (44). Il existe par contre, dans les sols vosgiens étudiés, une microflore ayant un effet dépressif sur la nutrition et la croissance des plants, sans provoquer de nécroses racinaires (44). L'état des mycorhizes d'arbres sains et dépérissants a également fait l'objet d'observations : il semble que l'état de dépérissement puisse être associé à une dérive qualitative du statut mycorhizien plutôt qu'à une dégradation complète des mycorhizes (44). La caractérisation des mycorhizes, outre qu'elle se heurte à des difficultés méthodologiques importantes, ne fournit, pas plus que l'observation des houppiers, d'indications immédiates sur les causes de la dégradation.

Parmi les champignons des grosses racines observés, seule Armillaria obscura semble pouvoir jouer un rôle significatif. On observe toutefois que le grand développement de ce champignon chez les arbres ne s'observe en régle générale qu'en phase ultime de dépérissement (27), observation en accord avec les caractéristiques d'un parasite d'équilibre et saprophyte. Sa participation possible à une érosion progressive du système racinaire est très incertaine.

Globalement, peu d'indices suggèrent une forte participation d'agents biotiques au dépérissement.

3.3.4. Incidence des conditions climatiques sur le dépérissement des forêts

S'il est généralement admis que les conditions climatiques peuvent contribuer au dépérissement des forêts, on conclut le plus souvent qu'elles ne peuvent être considérées comme une cause dominante (51, 72). Si, en dernière analyse, chacun s'accorde à reconnaître que les données disponibles sont de qualité médiocre (des données fiables de suivi de régime hydrique des sols in situ manquent presque totalement) et ne permettent pas de conclure sans ambiguités, les arguments invoqués pour énoncer la conclusion sus-mentionnée sont les suivants (18, 43) :

. l'étendue du phénomène actuel dépasse celle des dépérissements passés
. les dégâts sont apparus de façon synchrone sur une très vaste étendue
. la distribution des dommages ne coïncide pas avec celle que provoqueraient les extrêmes climatiques
. il n'y a pas coïncidence de la défoliation avec un épisode climatique singulier.

Si ces arguments paraissent raisonnables, leurs fondements scientifiques sont néanmoins fragiles. Il faut souligner, dans le contexte du dépérissement des forêts, le double danger d'observations approximatives des dommages et une utilisation imprudente des connaissances actuelles (très lacunaires) sur le fonctionnement de l'écosystème forestier. Les quelques résultats provisoires issus de divers projets du programme DEFORPA qui ont trait directement (69), ou indirectement (7, 8) au climat permettent de souligner cette fragilité :

. sur la base du tableau 1, on pourrait logiquement conclure à un phénomène grave et inédit affectant plusieurs essences dans plusieurs régions françaises. Il faut cependant remarquer que les conséquences de ces "déficiences" restent jusqu'ici limitées : dans plusieurs régions, les services gestionnaires ne considèrent pas cette situation comme nouvelle ou grave ; dans les Vosges, où les symptômes de dépérissement, rapportés en 1983-1984, ressemblent beaucoup à ceux décrits il y a 35 ans par FOURCHY (34), les cas de récolte prématurée de peuplements forestiers restent peu nombreux. Il faut insister sur les limites des observations au sol (40), forcément récentes et donc impropres à des comparaisons historiques. Un examen des courbes de croissance (figure 5 par exemple), autre approche possible, montre que la crise (de croissance) "actuelle" a débuté en fait en 1973 et trouve son paroxysme en 1976. A partir de 1981, on observe un rétablissement, fortement confirmé en 1984-85 (données partielles ne figurant pas sur les courbes) et nettement visible sur le terrain à partir de 1985 (reprise de la croissance apicale) (40). Deux autres crises, presque comparables en durée et en intensité, peuvent être repérées : 1916-1925, centrée sur 1922, et 1943-1951, centrée sur 1948. On peut d'ailleurs penser qu'elles ont été plus profondes (que ne le laisse deviner la courbe) car nombre d'arbres (les plus touchés) ont certainement disparu depuis (7, 8). On peut aussi signaler un dépérissement du chêne qui a sévi dans diverses régions françaises au cours des trois mêmes périodes (68, 62, 9).

. l'idée de la propagation des dommages ne peut être vérifiée en France ; rappelons en particulier que les cas de dépérissement du sapin en Normandie (4) et dans les Pyrénées (20) sont antérieurs à celui signalé dans les Vosges). Est-il raisonnable de prétendre reconstituer une dissémination à travers l'Europe ?

Figure 5

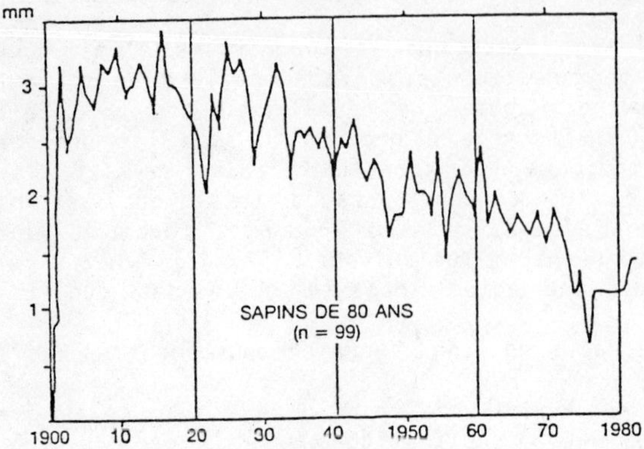

Largeur (en mm) de l'accroissement radial moyen des sapins ayant 80 ans en 1984

Mean annual ring width (in mm) of fir trees which were 80 years old in 1984

le tableau 1 rappelle que si plusieurs essences peuvent être touchées dans une région, cela est loin d'être le cas général, et des "cas de figures" très variées sont signalés (40). Dans les Vosges, l'étude de télédétection a confirmé que les dommages sur les diverses essences ne coïncident pas dans l'espace : les résineux sont plus atteints en altitude, l'inverse est observé pour les feuillus (60). Les seuls écosystèmes globalement affectés (arbres adultes de diverses essences, semis) sont ceux où domine le symptôme jaunissement (41).

. l'idée, très répandue, selon laquelle une sécheresse entrainerait surtout des dégâts à basse altitude, en épargnant sensiblement les arbres dominants (18, 43) ne peut être acceptée sans discussion. A ce sujet, mentionnons encore FOURCHY (1951) décrivant les conséquences de la période de sécheresse 1942-1949 : "dans les Basses Vosges (Vosges gréseuses), habituellement si humides, de nombreux sapins ont séché depuis 3 ans, sur les plateaux ou vers les crêtes, et ce sont souvent les sujets les plus gros et les plus beaux qui ont péri les premiers, ..." (34).

. de même que les symptômes visibles ne sont apparus, semble t-il, qu'à l'issue de la série climatique sèche 1943-1949 (34), il n'est pas forcément étonnant que les symptômes sur le sapin en Forêt-Noire ou dans les Vosges ne se soient manifestés qu'au début des années 1980, à l'issue d'une période de déficit pluviométrique. Les "arrière effets" décrits par les écophysiologistes semblent (quoique les connaissances soient encore limitées sur ce point) pouvoir expliquer certains décalages, comme par exemple l'apparition du dépérissement du chêne à Tronçais en 1978 (9). Les données météorologiques de Strasbourg montrent des déficits pluviométriques plus ou moins longs et marqués : 1917-1921, 1943-1949, 1961-1964, 1969-1976. Les trois périodes les plus longues coïncident avec les crises visibles sur les courbes de croissance ; les plus importantes (1943-1951, 1973-1981) se sont soldées par des dégâts visibles rapportés dans la littérature, celle de 1961-1964, sans répercussion apparente sur les courbes, semble pourtant avoir eu un rôle décisif dans les situations actuelles des peuplements (45). L'analyse des données météorologiques sur 36 postes répartis dans les Vosges, sur la période 1961 à 1984, a confirmé l'existence des deux dernières périodes de sécheresse en ajoutant l'année 1983 (non repérable sur les courbes) de sécheresse moins intense que 1961, 1964, 1971, 1976, mais correspondant à des stress hydriques pour toutes les stations vosgiennes étudiées (69). On peut supposer que ce stress, de courte durée, a pu se solder par une défoliation "supplémentaire" sans conséquences sur la croissance. De façon intéressante, la période 1977 à 1984 est marquée par une très forte fréquence des systèmes perturbés dans toutes les Vosges (69).

Il est évident que la portée de certains de ces résultats dépasse nettement le cadre géographique des Vosges (les travaux récents parus en R.F.A. ne les démentent pas(1, 36)). L'importance des séries d'années contrastées, bien plus que les caractéristiques des années individuelles, déjà cernée avec pertinence par FOURCHY (34) est une notion que retrouvent les spécialistes de la dynamique des populations animales ; en zone de montagne, on peut mentionner le cas des tétraonidés (31, 32).
Si la série de périodes sèches depuis 40 ans est réellement unique depuis que l'on dispose de données météorologiques (1880 à Strasbourg), il ne faut pas s'étonner, a priori, de dérèglements d'une ampleur " exceptionnelle". Sur le long terme, le parallèlisme entre la courbe des "indices de croissance" (accroissement dégagé de l'influence de l'âge) et les variations lentes de la température au niveau mondial et de la pluviométrie est pour le moins troublant (figure 6) (7, 8).

Figure 6

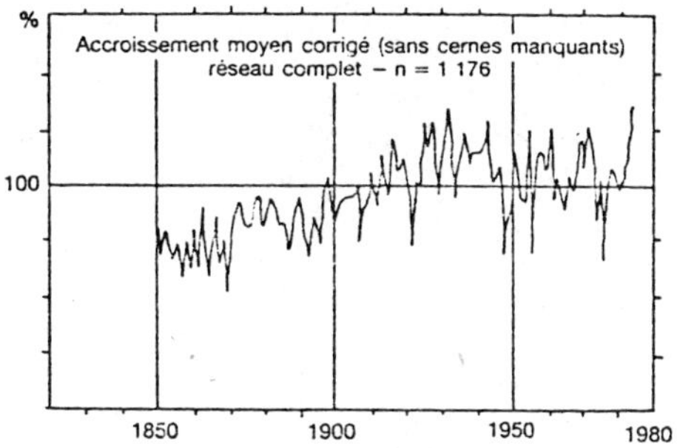

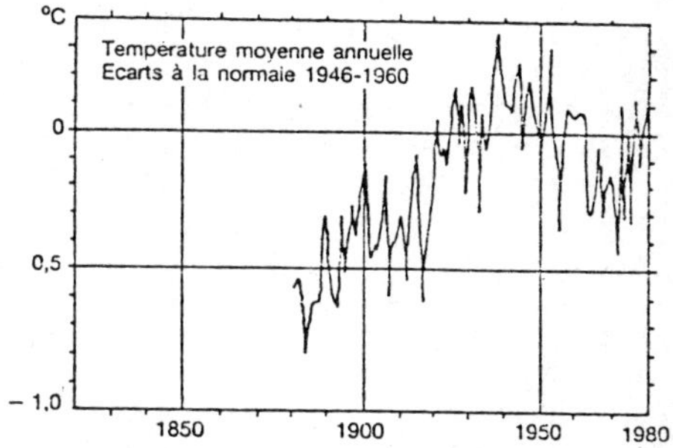

Largeur corrigée des cernes du Sapin (indépendante de l'âge) de 1850 à 1980 et évolution, pour la même période, de la température moyenne annuelle de l'hémisphère Nord

Age corrected annual ring width, according to the date, from 1850 until 1980, for the fir in the Vosges, and evolution, during the same period, of the mean annual temperature of the North hemisphere

3.3.5. Incidence de l'état actuel des forêts sur les dommages

Dans une Europe forestière profondément marquée par l'Homme, on est, a priori, frappé du peu d'importance donnée, dans la littérature sur le dépérissement des forêts, à l'influence possible de la gestion passée et présente des peuplements forestiers. Les évaluations de la situation à cet égard, et les conséquences qu'on en tire pour la sylviculture future, paraissent sur certains points assez contradictoires (13). Alors qu'en Suisse, une influence notable du vieillissement des forêts de montagne est rejetée (53), une mise en rapport des données sur les dommages avec les caractéristiques de peuplement a conduit, en Bade-Wurtemberg, à l'idée selon laquelle "les symptômes de la maladie sont d'autant plus exprimés que les houppiers sont exposés à l'air et par là même aux polluants atmosphériques" (65). L'idée d'une responsabilité majeure de la structure actuelle des peuplements dans les dommages observés est généralement réfutée sur la base de la répartition des dommages qui toucheraient aussi, voire surtout, les peuplements gérés de façon "naturelle" (peuplements mélangés, étagés), et concernent les peuplements feuillus parfaitement en station (51).

Les analyses conduites sur les données du réseau d'observation français sont restées prudentes pour les raisons déjà évoquées (cf. 3.1). Dans le cas des Vosges, il s'est révélé en outre que la répartition même des peuplements au sein du massif, et les inévitables corrélations "moyennes" entre roche-mère, altitude, âge des peuplements, ..., étaient de nature à suggérer une hiérarchie artificielle des facteurs, que des études plus précises risquaient de mettre en défaut. C'est du reste ce qui semble se dégager d'un certain nombre de projets du programme DEFORPA :

- les peuplements entrouverts sont, quelle que soit l'essence, les plus touchés (60) mais ce sont aussi les plus âgés. Dans le contexte vosgien, on trouve en particulier nombre de sapinières entrouvertes depuis longtemps, par suite des échecs de régénération, dues probablement à des phénomènes d'allélopathie dans certains cas (30), et souvent à de fortes densités de gibier. Faut-il s'étonner de la situation précaire de tels peuplements ?

- des observations spécifiques menées sur des "couples" de peuplements (sain/dépérissant) à faible distance les uns des autres montrent que le peuplement le plus dense est généralement le plus touché (45). L'influence négative de houppiers peu développés a été mentionnée par d'autres auteurs (64).

- pour le sapin, l'observation réalisée par télédétection sur placettes comportant cette essence, suggère que les peuplements purs sont plus atteints que les peuplements mélangés (60). Cette même observation est faite sur les 200 placettes étudiées par ailleurs (8), alors même que le pourcentage de feuillus (hêtre) n'y varie qu'entre 0 et 30 %.

A ces observations, ajoutons le fait que des essences productives et acidifiantes comme l'épicéa, conduisent, dans le cadre d'une sylviculture "normale" à des exportations probablement du même ordre que celles disponibles. Pour les Ardennes primaires, NYS et al (52) chiffrent à "386, 760, 104 et 250 kg l'exportation de K, Ca, Mg et Mn pour deux révolutions d'épicéa alors que les réserves du sol ne sont que de 270, 370, 80, 110 kg/ha respectivement pour ces mêmes éléments à l'état échangeable". La fragilité de tels écosystèmes est souvent soulignée (58). Il est vrai aussi que la conséquence négative envisagée était une diminution globale et progressive de la fertilité plutôt qu'une crise brutale au sein de l'écosystème.

En résumé, et bien que certains de ces éléments doivent encore être étayés, il ne semble pas que l'incidence de la sylviculture passée et actuelle soit sans incidence notable. A tout le moins, des dommages qui pourraient rester relativement modestes semblent exacerbés par des conditions sylvicoles défavorables.

4. QUE CONCLURE SUR LES CAUSES DU DEPERISSEMENT DES FORETS ?

A mesure que les recherches progressent, la synthèse et l'évaluation des résultats deviennent de plus en plus difficiles. Conclure devient dès lors une gageure. Mais il faut essayer ...

Un consensus est acquis sur l'idée que le dépérissement est un phénomène complexe et multiforme, différent de région à région. Rappelons que le terme "dépérissement" reste usité par commodité, bien qu'aujourd'hui, plus encore qu'hier, manquent les éléments scientifiques permettant d'augurer une disparition pure et simple de l'écosystème forestier sur des régions entières, et sa transformation en steppe. Remarquons ensuite que si certains "cas régionaux" sont bien documentés (jaunissement de l'épicéa), d'autres cas attendent, non pas des explications, mais encore des hypothèses.

Pour rendre compte des phénomènes observés à l'échelle de la France (ceux rapportés ici et d'autres qui ont échappé à la surveillance), il semble déraisonnable de voir en toute défoliation ou en toute altération de couleur du feuillage un phénomène anormal. Ainsi, peut-on envisager que les problèmes observés dans certaines régions calcaires soient essentiellement la conséquence, sur des écosystèmes fragiles, des conditions climatiques très particulières des décennies récentes. Cela est probablement le cas également pour les défoliations d'intensité moyenne, de caractère diffus, des peuplements vosgiens. Quand toutefois ces crises sont amplifiées, ou plus exactement suivies, dans le cas des Vosges, par des problèmes de nutrition sur des sites ayant des caractéristiques édaphiques communes et très typées, le risque de "dérapage" augmente. La nouveauté et la gravité des divers troubles nutritionnels recensés sont difficiles à apprécier : carences en Mg, Ca sur substrat cristallin(14, 15), en K sur calcaires, schistes, moraines (20, 41), Mn sur calcaire (16, 41) ; la seule classification selon les corrélations avec les critères édaphiques, appréciée selon les méthodes conventionnelles, parait insuffisante. L'apparition brutale de toute une palette de carences ne peut raisonnablement être liée qu'à un facteur climatique (excédents pluviométriques hivernaux et printaniers, gels ?). Comment expliquer autrement, la même année (1985), les jaunissements sur épicéa dans diverses montagnes cristallines du territoire français (15), du hêtre sur calcaire en plaine (41), du Pin cembro dans le Mercantour (49), et même des arbres fruitiers de diverses régions ?

Au facteur climat, dont l'étude circonstanciée devrait, si elle est conduite à une échelle suffisante (l'Europe !) apporter des compléments aux éléments proposés pour le cas français, et permettre de "prévoir", c'est-à-dire d'éviter aux forestiers et scientifiques d'autres "surprises", il faudra ajouter une vision plus réaliste de l'incidence de la sylviculture actuelle ; si certains éléments avancés ici devaient se confirmer, les conséquences logiques devront - aussi - en être tirées.

La pollution par SO_2 ne semble pas devoir être dans les conditions des massifs montagneux français une cause déterminante, sinon (et cela peut être significatif) pour aggraver des sécheresses précoces, exceptionnelles par ailleurs (il faut cependant évaluer avec attention les conséquences physiologiques de pointes hivernales).

Les dépôts secs et humides (éventuellement accentués par la présence d'ozone) paraissent par contre, au niveau de pollution actuel, avoir une action significative sur le lessivage au niveau du feuillage des arbres et du sol. Ces points doivent être approfondis. Une attention particulière doit, dans ce contexte, être portée au cas des apports azotés, en augmentation tendancielle. Les éléments scientifiques sur cette question sont encore insuffisants (51). On peut cependant craindre la sensibilité d'écosystèmes très fragiles à une offre supplémentaire d'azote.

Le dépérissement des forêts a mis en relief de grandes lacunes dans la science forestière. Si de nouvelles capacités de recherche devaient être créées pour pallier, à long terme, à ces déficiences, elles devraient intégrer à "parité" les domaines sus-mentionnés, en particulier les réactions des peuplements forestiers au climat et aux interventions sylvicoles, et les cycles biogéochimiques.

(1) ABETZ, P. (1985). Ein Vorschlag zur Durchführung von Wachstumsanalysen im Rahmen der Ursachenerforschung von Waldschäden in Südwestdeutschland. Allg. Forst.-u. J.-Ztg, 156, 9/10, 177-187.

(2) AMBROISE, B., FRITZ, B., PROBST, A., VIVILLE, D. (1987). Bilan hydrogéochimique d'un petit bassin versant des Vosges en relation avec le dépérissement forestier. In "Dépérissement des Forêts...*" Vol. 3, 215-232.

(3) AUSSENAC, G., CLEMENT, A., GUEHL, J.M., (1987). Influence d'une application combinée de brouillards acides et d'une sécheresse édaphique sur la capacité photosynthétique, la régulation stomatique et le pool anionique des tissus foliaires de l'épicéa commun. In "Dépérissement des Forêts...*" Vol. 3, 41-50.

(4) BARTHOD, C., LE BOZEE, G. (1979). Le dépérissement du sapin en Normandie. ENGREF, D.D.A. de l'Orne, rapport.

(5) BECKER, F., PAUL, P. (1987). Détection des zones atteintes dans les forêts vosgiennes et étude des corrélations avec les zones de brouillard. In "Dépérissement des Forêts...*" Vol. 1, 43-64.

(6) BECKER, M. (1985). Le dépérissement du sapin dans les Vosges. Quelques facteurs liés à la détérioration des cimes. Revue Forestière Française, XXXVII, 4, 281-287.

(7) BECKER, M. (1987). Etude écologique et dendroécologique du dépérissement du sapin dans les Vosges. In "Dépérissement des Forêts...*", Vol 1, 83-96.

(8) BECKER, M. (1987). Bilan de santé actuel et rétrospectif du sapin (Abies alba Mill.) dans les Vosges. Etude écologique et dendrochronologique. Annales des Sciences Forestières (à paraître).

(9) BECKER, M., LEVY, G. (1983). Le dépérissement du Chêne. Les causes écologiques (exemple de la forêt de Tronçais) et premières conclusions. Revue Forestière Française, XXXV, 5, 341-356.

(10) BELIN, C. (1987). I. Validation de la méthode CDM/CIF pour les composés légers en trace dans l'air. II. Campagne de mesures des composés organiques en trace dans l'air "Région Donon France". In : "Dépérissement des Forêts...*" Vol 2, 187-206

(11) BERTEIGNE, M., ROSE, C. (1987). Test d'exclusion en chambre à ciel ouvert. In "Dépérissement des Forêts...*" Vol 3, 5-12.

(12) BIREN, J.M., ELICHEGARAY, C., VIDAL, J.P. (1987). Caractérisation de l'environnement atmosphérique en zone forestière. Station laboratoire du Donon. In "Dépérissement des Forêts..." vol 2, 111-118.

(13) BONNEAU, M. (1987). Moyens préventifs et curatifs de lutte contre la pollution dans les écosystèmes terrestres, en particulier les forêts. In "Effects of air pollution on terrestrial and aquatic ecosystems" Grenoble, 18-22 Mai 1987 (à paraître).

(14) BONNEAU, M., LANDMANN, G., ADRIAN, M., 1987. Le dépérissement du sapin pectiné et de l'épicéa commun dans le massif vosgien est-il en relation avec l'état nutritionnel des peuplements ? Revue Forestière Française XXXIX, 1, 5-11.

(15) BONNEAU, M., LANDMANN, G. (1987). Analyses foliaires. In "Dépérissement des Forêts...*", Vol 1, 187-204

(16) BONTE, J., CANTUEL, J., MALKA, P. (1987). Simulation de l'action du dioxyde de soufre et de l'ozone, seul ou en mélange, sur l'épicéa en chambre à ciel ouvert. In "Dépérissement des Forêts...*"Vol 3, 13-40.

(17) BRUCKERT, S., BOUN SUY TAN (1987). Etude du dépérissement des forêts de l'étage montagnard du Jura en relation avec les sols calciques ou acides du Karst, l'importance de leurs éléments nutritifs et l'état nutritionnel des peuplements. In "Dépérissement des Forêts...*" Vol 1, 133-148.

(18) BURSCHEL, P. (1985). Waldschäden-Forstwirtschaft-Witterung. Betrachtung zu zwei publikationnen aus der chemischen Industrie. All. Forstztg, 3, 43-46, 49.
(19) CHAUMERLIAC, M., ROSSET, R. (1987). Modélisation mésoéchelle des dépôts acides. In "Dépérissement des Forêts...*", Vol 2, 133-148.
(20) CHERET, V., DAGNAC, J., FROMARD, F. (1987). Le dépérissement du sapin dans les Pyrénées luchonnaises. Revue Forestière Française. XXXIX 1, 12-24.
(21) CITERNE, A., DIZENGREMEL, P. (1987). Effets du dépérissement sur la respiration des aiguilles d'épicea. In "Dépérissement des Forêts...*, Vol 3, 125-140.
(22) COLIN, J.L. (1987). Echantillonnage de pluie et de neige par un collecteur séquentiel. Mise en évidence et comparaison de certains facteurs acidifiants. In "Dépérissement des Forêts...* Vol 2, 55-66.
(23) COWLING, E.B. (1985). Comparison of regional decline of forest in Europe and North America : a possible role of airborne chemical. In "Air pollutants : Effets on Forest Ecosystems" Acid rain Found, St Paul, Minn. pp. 217-234.
(24) CRAMER, H.H., CRAMER-MIDDENDORF, M. (1984). Untersuchungen über Zusammenhänge zwischen Schadensperioden und Klimafaktoren in mitteleuropaïschen Forsten. In : "Waldschadensproblematik. Pflazenschutznahrichten, Bd 37, 55, H.2
(25) DAMBRINE, E., NYS, C., RANGER, J. (1987). Etude du fonctionnement biogéochimique d'écosystèmes forestiers soumis à différentes perturbations exogènes - Bassin versant d'Aubure, Vosges. In "Dépérissement des Forêts...*" Vol 3, 233-238.
(26) DECHAUX, J.C. (1987). Elaboration d'une stratégie de lutte contre les oxydants photochimiques et les pluies acides. In "Dépérissement des Forêts...*" Vol 2, 163-186.
(27) DELATOUR, C. (1987). Intervention d'Armillaria obscura dans le dépérissement du sapin (Abies alba). In "Dépérissement des Forêts...*" Vol. 1, 211-216.
(28) DEREXEL, Ph., MASNIERE, P. (1987). Etude des précipitations collectées au Col du Donon. Premiers résultats. In "Dépérissement des Forêts..* Vol 2, 47-54.
(29) DRACH, A., TARGET, A. (1977). Exploitation des données de la station du Donon. In "Dépérissement des Forêts...* Vol 2, 5-14.
(30) DRAPIER, J. (1985). Les difficultés de régénération naturelle du Sapin (Abies alba Mill) dans les Vosges. Etude écologique. Revue Forestière Française XXXVII 1, 45-55.
(31) EIBERLE, K., MATTER, J.F. (1984). Witterungsverlauf und Auerhuhnbestand. Schweizerjäger, 69, 16, 776-782.
(32) EIBERLE, K., MATTER, J.F. (1985). Zur Bedeutung einiger Witterungselemente für das Birkhuhn (Tetrao tetrix) im Alpenraum. Allg. Forst.-u. J.-Ztg 156, 6-7, 101-105.
(33) EVERS, F.H. (1987). Wasser-, nadel- u.bodenanalytische Begleitungtersuchungen zu den Versuchen mit oben offenen Kammern u. dem Luftmessprogramm beim Edelmannshof-Welzheimer Wald. PEF Kolloquium 10-12 Mära 1987 (à paraître).
(34) FOURCHY, P. (1951). Sécheresse, variations climatiques et végétation. Revue Forestière Française, 32, 1, 47-55.
(35) GARREC, J.P., LAEBENS, C. (1987). Etude de la perméabilité cuticulaire des feuilles sous l'action des pluies acides et de l'ozone, et son rôle dans le dépérissement des forêts. in "Dépérissement des Forêts ...* Vol 3, 175-194.
(36) GERECKE, K.L. (1986). Zuwachsuntersuchungen an vorherrschenden Tannen aus Baden-Württemberg. Allg. Forst.-u.J.- Ztg 157, 3/4, 59-68.

(37) GIROMPAIRE, L., RAMEAU, J.C. (1987). Recherche de relations entre le dépérissement et les conditions écologiques dans les Vosges. In "Dépérissement des Forêts...*" Vol 1, 167-186.
(38) GOUNOT, M. (1987). Echanges gazeux in situ d'épicéas dans le massif du Donon. In "Dépérissement des Forêts...*" Vol 3, 87-104.
(39) GRAS, F. (1987). L'aluminium dans les sols acides forestiers et leurs solutions : exemple de deux sols sur grès du massif vosgien. In "Dépérissement des Forêts...*" Vol 3, 195-214.
(40) LANDMANN, G. (1987). Observation au sol de l'état sanitaire des forêts : quels enseignements peut-on tirer des résultats disponibles. In : "Recherches en France sur le dépérissement des Forêts Programme DEFORPA, 1er rapport." Ed. ENGREF, 17-26.
(41) LANDMANN, G. (1987). à paraître
(42) LANDMANN, G., DAMBRINE, E. (1987). A paraître
(43) LEHRINGER, S. (1985). Waldschäden - Forstwirtschaft - Witterung ; Ergebnisse eines AFZ. Kolloquiums, All. Forstzeitsch. 997-1000.
(44) LE TACON, F. et al. (1987). Aspects microbiologiques du dépérissement des forêts : pathogènes des fines racines et mycorhizes. In "Dépérissement des Forêts...*" Vol. 1, 205-210.
(45) LEVY, G., BECKER, M., Le dépérissement du Sapin dans les Vosges - Rôle primordial de déficits d'alimentation en eau (à paraitre)
(46) LOUGUET, P., MALKA, P., CONTOUR-ANSEL, D. (1987). Variations de la résistance stomatique et de la teneur en composés phénoliques de feuilles d'épiceas soumis à une pollution contrôlée par le dioxyde de soufre et l'ozone en chambres à ciel ouvert. In "Dépérissement des Forêts...* Vol 3, 105-124.
(47) McLAUGHLIN, S.B. (1985). Effects of air pollution on forests : a critical review. J. Air. Poll. Cont. Assn. 35, 511-533.
(48) MASNIERE, P. (1987). Mise au point d'appareils d'échantillonnage séquentiel de pluies et de prélèvements de brouillard. In "Dépérissement des Forêts...*" Vol 2, 105-110.
(49) MATHIEU, J. (1987). Mise au point d'un collecteur de rosée. In "Dépérissement des forêts...*" Vol 2, 81-84.
(50) MAURIN, V. (1986). Le dépérissement du Pin cembro dans le Mercantour. Rapport de stage B.T.S.A. Productions Forestières, ENITEF, 37p, + annexes.
(51) MOOSMAYER, H.U. (1986). Stand der Forschung über das Waldsterben und Fortschritte seit 1984. All. Forstzeitsch, 1150-1153
(52) NYS, C., RANGER, D., RANGER, J. (1983). Etude comparative de deux écosystèmes forestiers, feuillu et résineux des Ardennes primaires françaises. III. Minéralomasse et cycle biologique. Ann. Sci. Forestières 40, 1, 41-66.
(53) OTT, E. (1985). Wie ist die Frage der Überalterung für unsere Schweizer Gebisgswälder zu beurteilen ? Schweiz. Zeit. für Forstwesen, 136, 11, 931-944.
(54) PERRIN, M.L., BOURBIGOT, Y., MADELAINE, G. (1982). Fine particles measurements around and industrial area. Aerosol Science and Technology (2), 165.
(55) PERRIN, M.L., FRAMBOURT, C. (1987). Mesure de la concentration et de la granulométrie et nature chimique des aérosols à la station du Donon. In "Dépérissement des Forêts...*" Vol 2, 15-28.
(56) PERROS, P., PIGEON, A., TOUPANCE, G., TSALKANI, N. (1987). Mesure du PAN. Mise au point d'un appareil de mesure in situ. In "Dépérissement des Forêts...*" Vol 2, 67-80.
(57) PIERRE, M., SIEFFERT, A., SAVOURE, A., QUEIROZ, O., CORNIC, G., HUBAC, C., MACREZ, V. (1987). Effets de la pollution et synergie sécheresse et pollution. In "Dépérissement des Forêts...*" Vol 3, 59-86

(58) RANGER, J., BONNEAU, M. (1986). Effets prévisibles de l'intensification de la production et des récoltes sur la fertilité des sols en forêts. II. Les effets de la sylviculture. Revue Forestière Française, XXXVIII, 2, 105-123.
(59) REITER, R., POTZL, K., SLADKOVIC, R. (1984). Détermination of the concentration of chemical main and trace elements in the aerosol of at a North-Alpine pure air mountain station. Arch. Met. Geoph. Biocl., Ser. B 35, 1-30
(60) RETEAU, F. (1986). Etude par télédétection du dépérissement des forêts. Utilisation de la photographie aérienne sur le massif des Vosges. Mémoire ENITEF, 54p + annexes
(61) RIOM, J. (1987). Télédétection : estimation des dommages. In : "Dépérissement des Forêts...*", Vol 1, 33-42.
(62) ROL, R. (1951). Le dépérissement des chênes. Revue Forestière Française 3, 11, 707-709.
(63) SAINT GUILY, A. (1987). Effet du dépérissement sur le métabolisme terpénique de Picea abies. In "Dépérissement des Forêts...*" Vol 3, 161-174.
(64) SCHLÄPFER, R. et al (1985). Der Gesundheitszustand des Waldes im Revier Schaffhausen. Schweiz. Zeit. Für Forstwesen., 136, 1, 1-18
(65) SCHÖPFER, W., HRADETZKY, J. (1984). Der Indizienbeweis : Luftverschmutzung massgebliche Ursache der Walderkrankung. Forstw. Cbl. 103, 231-248
(66) SEUFERT, G., ARNDT, U. (1986). Beobachtungen in definiert belasteten Modellökosystemen mit jungen Waldbaümen. Allg. Forstztg 545-549
(67) STRAUSS, B., CUILLER, G. (1987). Trajectographie à l'échelle synoptique In "Dépérissement des Forêts...*" Vol 2, 119-132.
(68) TURC, L. (1927). Note sur le dépérissement du chêne pédonculé dans les forêts du plateau nivernais. Rev. des Eaux et Forêts, 65, 11, 561-565.
(69) VIGNAL, C. (1987). Conditions climatiques et dépérissement forestier dans les Vosges. In "Dépérissement des Forêts...*" Vol 1, 65-82
(70) VILLANUEVA, V.R. (1987). Etude comparée du métabolisme cellulaire entre épicéas sains et épicéas dépérissants. In "Dépérissement des Forêts...*" Vol 3, 141-161.
(71) ZEPHORIS, M. (1987). Mise au point d'un capteur de gouttelettes de brouillard. In "Dépérissement des Forêts...*" Vol 2, 85-104.
(72) Bundesminister für Forschung un Technologie (1985). Umweltforschung zu Waldschäden 3. Bericht.
(73) Forschungsbeirat Waldschäden/Luftverunreinigung der Bundesregierung und der Länden (1986). 2. Bericht. Literaturabteilung des Kerforschungszentrums, Karlsruhe.
(74)* Dépérissement des forêts attribué à la pollution atmosphérique, 1987. Programme DEFORPA - Etat des recherches à la fin de l'année 1986. Ministère de l'Environnement - INRA
(75) Les recherches en France sur le dépérissement des forêts, 1987 - Programme DEFORPA, 1er rapport. 88p - Ed. ENGREF

SESSION III

EFFECTS OF AIR POLLUTION ON FOREST TREES AND FOREST ECOSYSTEMS

SESSION III

EFFETS DE LA POLLUTION DE L'AIR SUR LES ARBRES ET LES ECOSYSTEMES FORESTIERS

SITZUNG III

WIRKUNGEN DER LUFTVERSCHMUTZUNG AUF WALDBÄUME UND WALDÖKOSYSTEME

PART II | IIème PARTIE | TEIL II

Chairman : Prof. R. IMPENS, Faculté des Sciences Agronomiques de l'Etat, Gembloux, Belgium

Rapporteur : Dr. M. ASHMORE, Imperial College, London, United Kingdom

REPORT ON THE SESSION

M.R. ASHMORE
Dept. of Pure & Applied Biology
Imperial College of Science & Technology
Silwood Park
Ascot
Berks SL5 7PY
U.K.

Until recently, pollution damage to forests was largely due to individual sources, which affected trees in their immediate vicinity. An example of this type of problem was seen in our visit to the Grand-Chartreuse forest during this symposium. Dieback of conifers, and replacement by beech and other deciduous species had occurred in the vicinity of a small cement works situated at the bottom of a steep valley. Visible leaf injury could be observed close to the works, while analysis of leaf sulphur concentrations had shown significantly increased concentrations in the area around the works. Thus inspection of macroscopic damage in the area, together with a simple chemical test, had allowed the problem to be diagnosed as being due to SO_2 emissions from the works, with the spatial variation in symptoms in relation to the source being a key element in the diagnosis.

One should not underestimate the problems of making a successful diagnosis in such cases. Other chemical, physical and biological stresses may cause similar symptoms, while the sensitivity of vegetation to individual pollutants may be influenced by the presence of other pollutants and other environmental stresses. Nevertheless there is no doubt that these problems are trivial in comparison with those which we confront in trying to diagnose the cause of the forest damage now affecting many parts of Europe.

This new damage is different in character; it is widespread, and often shows no clear geographical relationship with possible sources of pollution. The macroscopic symptoms observed are rather non-specific, often being responses typical of a range of stresses, or being manifestations of some form of nutrient deficiency. The time-scale over which pollutant exposure affects the target organism may be considerable. Furthermore, to date we have no clear morphological, biochemical or physiological responses which are consistently and specifically associated with the visible injury. In traditional terminology, we are dealing with chronic pollution injury rather than acute injury. It is important to note, however, that this is not a firm distinction, since in both cases pollution is one of many environmental factors influencing tree health; the key difference lies in the scale of the spatial and temporal variation in pollution levels relative to that of other environmental variables.

Assessments of forest health based on visible appearance have concentrated primarily on crown structure and density, and foliar chlorosis. Since such symptoms can be caused by different types of stress, or indeed develop as part of the natural ageing process, they have little diagnostic value in isolation. The alternative which has been proposed is to replace them by a set of symptoms based on morphological, physiological and biochemical parameters which can be specifically related to individual pollutants, pollutant combinations or combinations of pollutant exposure and other stresses.

However, before any symptom can usefully be employed in the diagnosis of forest decline phenomena, it is essential to demonstrate a consistent and unique association with a stress factor.

The presentations at this symposium illustrated the range of different methods which are currently being used to establish that specific plant responses are linked to pollutant exposure. These methods can conveniently be grouped into the following categories:-

(1) Comparison of individual trees at the same site showing different levels of visible damage. This approach eliminates many of the confounding environmental variables. However it can only demonstrate associations between visible injury and other symptoms at different levels of biological organisation, and cannot prove a link between these and possible causes.

(2) Comparison of symptom expression at different sites. This is essentially an epidemiological approach, in which large numbers of sites are assessed, and correlations sought between the intensity of observed symptoms and possible causal agents; such studies require intensive input and an assurance that all relevant factors are included, but can only provide associations which need further testing in laboratory and/or field. A more useful approach in terms of establishing cause-effect relationships may be to examine trees along a known gradient of pollution levels. Such an approach can be most valuable when specific sources of a pollutant exist, and variations in pollutant concentration are large in relation to those of other factors, but it is less useful for secondary pollutants, such as O_3, which spread in high concentrations over large areas.

(3) Comparison of temporal variation in pollutant exposure and symptom expression in the field. This approach is another type of epidemiological approach, which necessitates a large programme of continuous monitoring of many variables, in order to establish statistical associations. However, the known links between pollutant concentrations and climatic conditions make it unlikely that unequivocal associations can be obtained. Nevertheless this type of study, if extended over several years, may provide valuable insights into how different plant responses interrelate through time, and how episodes of high pollution concentrations, or of other abnormal environmental stresses influence subsequent symptomology.

(4) Laboratory studies of plant response to known pollutant exposures. These experiments can demonstrate a link between a specific pollutant and a symptom unequivocally, but cannot demonstrate that the symptom is uniquely linked to the pollutant. Furthermore, since pollutant exposures are frequently unrealistic, exposure conditions are often unlike those in the field, and physical constraints restrict their application to seedlings, they cannot guarantee that the same response will actually be observed in the field.

(5) Experimental manipulation in the field. This approach has received relatively little attention, although its potential may be considerable, since it can allow hypotheses to be tested in the field. Applications of specific fertilisers, for example, have been used to identify macroscopic symptoms associated with a particular nutrient deficiency; the same approach might also allow physiological or morphological symptoms associated directly with the development of nutrient deficiencies to be distinguished from those associated with pollutant

exposure or other stresses. However field techniques suitable for pollutants are poorly developed; although chamber enclosures may be used to test the effects of ambient air, we cannot evaluate the impact of individual pollutant gases, while manipulation of precipitation acidity is also problematical.

Each of these approaches has clear methodological or philosophical disadvantages, and there is to date no established technique for evaluating new diagnostic procedures. Thus, although many promising results have been obtained, the intractable problem of fully testing their significance on relevant spatial and temporal scales makes it difficult to provide firm scientific conclusions from this session.

What are the prospects for future development in this field? Among the factors to be considered in developing strategies for improved diagnosis are the following:-

(1) There are dangers in relying too much on diagnostic tests which provide, for example, a quantitative measurement of an individual metabolic component, since these may lend a false air of objectivity to the diagnostic exercise. The health of a tree at any point in time is the integrated result of numerous interacting factors acting over many years, and it may be naive to expect to quantitatively summarise these processes with a few laboratory tests. Field pathologists like good medical doctors, rely on experience, and even instinct, as much as laboratory tests in developing their diagnosis. Professional foresters are often rather cynical about scientific activities and perhaps with some reason; their intimate knowledge of a local area, often gained over many years, provides essential background information against which to judge whether symptoms are new or abnormal. Information and assessment based on subjective judgements is too easily dismissed because it cannot readily be assimilated using current scientific methodology; more attention should be paid to methods of integrating it with the more objective measurements.

    Secondly, it is essential that any new method of assessment should provide a quantitative measure of the health of the whole organism, and not of components which may be irrelevant to the ultimate course of a tree's decline. It may thus be argued that assessments based on macroscopic symptoms which result from long-term declines in vitality should continue to occupy a central role in health evaluation, once a tree has passed into a phase in its decline in which they become manifest. However, these are currently based only on an assessment of the vitality of the above-ground part of the tree. The central role of root health in many hypotheses of forest decline emphasises the importance of developing techniques for the assessment of vitality of the below-ground organs, in order to provide a more complete measure of the health of the whole tree.

(2) It is important not to become too restrictive in the range of pollutants being examined. Few studies, whether in the laboratory or the field, consider pollutants other than SO_2, NO_x, O_3, NH_3 and the acidity of wet deposition, but other substances may yet prove to be of major importance. For example, recent studies have demonstrated the phytotoxicity of hydrogen peroxide in mists, while the importance of a range of organic molecules for plant health is still uncertain. Thus there is a danger that because symptoms cannot be linked to the major pollutants their importance will not be properly recognised.

(3) It is essential to recognise that diagnosis cannot be based purely on observation of specimen trees at one point in time. My dictionary

defines diagnosis as "identification of a disease by investigation of its symptoms and history". Thus knowledge of the time course of symptom expression is essential to understand relationships between pollutant exposure and symptom development, interrelationships between different types of symptom, and to identify symptoms which are characteristic of different stages in the decline of an individual tree. Both laboratory and field studies should incorporate this element of time if they are to be of maximal value.

Analysis of tree ring growth may provide valuable insights into the time-course of disease development; for example, German studies have demonstrated that trees which have only recently developed visible symptoms actually started to decline in growth relative to their healthy neighbours over thirty years ago. This historical record provided by tree rings may contain other valuable information, such as in changes in wood morphology or chemical composition through time, which may be more specifically related to pollutant expsoure.

(4) Our knowledge of the way in which the normal range of biotic and abiotic stress factors influence tree physiology is limited. The lack of background information on the natural variability in an individual symptom makes it difficult to devise field evaluation programmes and assess its potential value. There is a need for more basic research on forest eco-physiology, but this should be closely coordinated with the processes of symptom identification and evaluation.

(5) The complexity of the problem and the range of methodological approaches being used means that coordination of research, collaboration between groups and integration of the information acquired become of critical importance. Integration, however, means more than data collation; it is essential also to develop conceptual models to provide the framework through which the significance of individual observations to our understanding of the overall decline process can be assessed and the role of individual stress factors in producing varying symptomology in different locations can be interpreted.

In summary, the value of both reductionist and holistic approaches to this problem need to be recognised. Each provides information of central importance to developing a more complete diagnosis of forest decline. The scientific challenge is to find methods of integrating information obtained at very different levels of biological organisation in a form which will lead to clearer understanding, rather than greater confusion.

NOTE DE SYNTHESE SUR LES TROIS TOURNEES EN FORET EFFECTUEES LE 20 MAI 1987

M. BAZIRE
Inventaire Forestier National

Ces trois tournées effectuées aujourd'hui devaient nous montrer divers aspects des massifs forestiers proches de Grenoble :

d'une part dans les Préalpes calcaires (étages jurassique et surtout crétacé) des massifs du Vercors et de la Chartreuse, à l'Ouest et au Nord de Grenoble ;

d'autre part, dans le massif cristallin de Belledonne, précédé de collines jurassiques, à l'Est.

Malgré leur proximité géographique, et par suite, une relative similitude climatique, les conditions écologiques y sont assez différentes.

I - VERCORS - CHARTREUSE

Vercors et Chartreuse ont plusieurs traits communs certes : formations géologiques en "synclinaux perchés" où le facies urgonien forme d'impressionnantes falaises. Mais les calcaires plus ou moins gréseux dominent dans le Vercors qui offre en outre un relief nettement plus tabulaire. En Chartreuse, les barres calcaires du Tithonique (Jurassique supérieur) et du Valangenien (Crétacé inférieur) coopèrent avec les calcaires urgoniens pour dessiner l'ossature du paysage, beaucoup plus fortement disséqué par de profondes vallées. Sur les pentes dominent les marnes et calcaires marneux du crétacé inférieur, donnant des sols moins superficiels et plus riches.

La pluviosité est nettement moins élevée surtout en été dans le Vercors qu'en Chartreuse, à altitude égale, alors que la situation de ce bastion isolé en bordure du sillon rhodanien l'offre à la violence des vents du secteur Ouest.

Aussi les forêts du Vercors sont-elles plus fragiles qu'en Chartreuse, avec un aspect général souffreteux. Dans les deux massifs, cependant, les stations forestières appartiennent à la vaste série de la hêtraie-sapinière mêlée d'épicéa, où le hêtre peut devenir envahissant, notamment sur les sols rocailleux et les sols bruns calciques ou calcaires de pente, dans la zone des brouillards : Hêtraie à Tilleul et Acer opalus
Hêtraie à Asperula odorata.

Le sapin est à son optimum sur les ubacs à sols profonds, entre 800 et 1 300 m, tandis qu'ailleurs et surtout à plus haute altitude, l'épicéa domine.

1.1 - Tournée dans le Vercors

Dans le Vercors, la première station, en forêt communale de Méaudre, aménagée depuis 1726, près du Col de la Croix Pernin vers 1200 m, a montré une belle futaie jardinée de sapin et d'épicéa, pourvue d'une abondante régénération naturelle, surtout en sapin. Des indices de dépérissement sont cependant visibles sur les sapins (jaunissement) ; les épicéas y offrent leur facies "Vercors" légèrement "déguenillé", à cime claire qui les a toujours caractérisés dans ce massif. C'est notamment le cas des pessières sur Lapiaz. Le sol, sur dalle calcaire fissurée, reste très superficiel et de faible "capacité en eau".

La seconde forêt visitée, au-dessus d'Autrans, dans le massif des Clapiers, aménagée depuis 1867 en futaie jardinée résineuse, se trouve à plus haute altitude. Elle offre des aspects variés, mais souvent moins "bien venants". La tornade de novembre 1982 a créé une vaste trouée de chablis dans le haut du versant et ainsi favorisé ensuite la prolifération de l'Ips Typographus contre lequel la lutte par piégeage avec phéromone est activement menée. Ici encore, l'aspect très médiocre ou nettement dépérissant de certains arbres, sapins et surtout épicéas, paraît devoir être imputé à leur isolement brutal, aux dégâts provoqués par la tempête dans le système racinaire des arbres restés debout, mais tout de même fortement ébranlés, et enfin aux récentes sécheresses (1983, 1985, 1986).

1.2 - Tournée en Chartreuse

En Chartreuse, nous visitons une partie de la forêt domaniale de la Grande Chartreuse, aménagée en futaie jardinée de sapin, épicéa avec une forte proportion de hêtre, qui s'y montre souvent envahissant. C'est une forêt nettement vieillie dans son ensemble, mais en voie de rajeunissement.

Les stations pratiquées au cours de la visite ont offert divers aspects caractéristiques de cette forêt de plus de 8 000 ha. Ici encore quelques trouées de chablis et attaques d'Ips Typographus, notamment près du Col de Porte.

Surtout une zone de dépérissement très accentuée puisque les épicéas et les sapins ont pratiquement disparu tandis que le hêtre parvient encore à subsister, mais avec un aspect nettement dépérissant en fin de saison. C'est le fond de la Vallée du Guiers mort proche de la cimenterie de la Perelle où les observations du C.D.E.R.E. effectuées depuis 6 ans ont révélé une importante pollution soufrée. Les analyses foliaires pratiquées sur des échantillons prélevés systématiquement tous les 50 m en altitude suivant différents transects, ont permis de tracer des courbes d'isoteneurs en soufre dans les aiguilles et de révéler des teneurs maximales de l'ordre de 7 ‰ , alors que la teneur normale est inférieure à 2 ‰. Il s'agit donc ici de pollution de proximité impliquant essentiellement le SO_2. Les analyses récentes sur sapin montrent une teneur des aiguilles en soufre, de 2,2 ‰ dans la zone polluée et de 1 % dans le reste du massif, ce qui est proche du niveau naturel.

Notons encore, au voisinage, la fréquence du gui (Viscum album) sur les sapins, qui souffrent en outre d'être à trop basse altitude (500 m) environ).

La visite d'une placette d'observation de l'état sanitaire des arbres ("réseau bleu") ne montre pas d'évolution sensible de la perte des aiguilles des sapins et épicéas entre 1985 (1ère année d'observation) et 1986 ; il a été constaté une réduction du jaunissement.

Les "placettes rouges" installées sciemment dans un peuplement déjà dépérissant montrent une nette aggravation de la perte des aiguilles, au Col de Porte, où la proportion des arbres ayant perdu plus de 25 % de leur feuillage passe de 46 % en 85 à 62 % en 86.

II - MASSIF DE BELLEDONNE

Nous abordons ici la partie externe de la zone cristalline de la chaîne alpine : Mécaschistes et grès donnent des sols acides, tandis que vers l'axe du Massif de Belledonne, les roches cristallophylliennes basiques, comme les amphibolites, donnent des sols plus riches en calcium et potassium.

Dans leur ensemble, les sols sont, ici, plus profonds que dans les massifs calcaires, avec une meilleure capacité en eau ; on passe des sols brunifiés mésotrophes vers 1 000 m d'altitude à des sols podzoliques vers 1 500 m.

Le hêtre est nettement défavorisé par l'acidité générale des sols et les forêts sont à base de sapin et d'épicéa, mêlés d'érable sycomore, de bouleau et de tremble, sauf à basse altitude et aux adrets, domaine de la hêtraie subthermophile et acidocline.

- Tournée dans la forêt communale de Sechilienne et dans la forêt de Premol.

La tournée débute au lac de Luitel, tourbière d'altitude (1 100 m environ), la plus ancienne réserve naturelle de France, à l'entrée de la forêt communale de Sechilienne.

Le circuit traverse la série de futaie jardinée de sapin-épicéa puis une série feuillue (ancien taillis) en voie d'enrésinement. On peut observer des arbres atteints de jaunissement (sapins surtout) et quelque peu dégarnis en cime. La futaie jardinée a été malmenée par une tempête en 1974, qui a obligé à exploiter environ 30 % du matériel sur pied ; si la partie basse se régénère abondamment, la partie haute est en cours de dépérissement et attaquée par des bostryches. Il a fallu recourir aux plantations d'épicéa.

La deuxième partie de la tournée, en forêt de Premol, étagée sur près de 1 200 m d'altitude (490 - 1 680) se limite à la 2ème série (le Clos du Verney) traitée en futaie jardinée résineuse à épicéa dominant et sapin. L'altitude moyenne ici de 1 500 m, explique la prépondérance de l'épicéa. La saison de végétation est courte et le diamètre d'exploitabilité n'est atteint qu'à un âge avancé. Les arbres souffrent des conditions climatiques, et aussi, ici, du sol superficiel et relativement sec de cet adret escarpé. La situation, à l'Est de Grenoble, favorise l'influence de la pollution venant de la zone industrielle.

RESUME DES OBSERVATIONS EFFECTUEES SUR
LES PLACETTES DU "RESEAU BLEU"

- Vercors : 30 placettes - 720 arbres

  en fait 693 arbres.

  Ont perdu plus de 25 % de leur feuillage en 1986 :

  Sapin : 24,9 % Epicéa : 26,1 %

  Tous résineux : 24,8 % - Toutes essences : 18,6 %

  En 1985, notations analogues. Toutes essences : 18 %

  Jaunissement : 27 % en 1985, 18 % en 1986.

- Chartreuse : 20 placettes - 480 arbres

  en fait 477 arbres.

  ont perdu plus de 25 % de leur feuillage en 1986 :

  Sapin : 41,5 % Epicéa : 26,3 %

  Tous résineux : 33,6 % - Toutes essences : 20,1 %

  En 1985, notations analogues - Toutes essences : 21 %

  Jaunissement : 26 % en 1985, 12 % en 1986.

- Belledonne-Sud : 24 placettes - 576 arbres

  en fait 599

  ont perdu plus de 25 % de leur feuillage en 1986 :

  Sapin : 5,8 % Epicéa : 4,7 %

  Tous résineux : 5 % - Toutes essences : 5,2 %

  En 1985, notations analogues : toutes essences : 4 %

  Jaunissement : 11 % en 1985, 6 % en 1986.

On constate donc une légère aggravation ou la stabilité de la perte des aiguilles, d'une façon générale, et une réduction nette du jaunissement. L'effet de la pollution de proximité en Chartreuse, et des conditions difficiles de sols en Vercors est très marqué.

RECENT DEVELOPMENTS IN THE DIAGNOSIS AND QUANTIFICATION OF FOREST DECLINE

J.N. CAPE
Institute of Terrestrial Ecology
Bush Estate, Penicuik, Midlothian EH26 OQB, UK.

Summary

The objectives for the development of tests for the early diagnosis and quantification of forest decline are defined. The ideal characteristics of such tests are then discussed in terms of factors affecting the variability of responses and their specificity to given causes, eg. air pollution. The practical constraints for applying such tests in the field are detailed, in particular the sampling strategy. A brief description of possible diagnostic tests is then given, classified as microscopical, whole-leaf, leaf surface, physiological, non-destructive biochemical and destructive biochemical. Finally, the stages of development of suitable tests are summarized, and some suggestions given as to analytical methods appropriate for field use. The importance of collaborative work between research groups is emphasized.

1. INTRODUCTION

Within the last decade there has grown an increasing awareness of widespread disturbances to forest ecosystems which have been described as a new type of forest decline. Symptoms have been observed in most European countries, and in many cases the damage has been attributed either directly or indirectly to the influence of air pollutants. Many and various have been the hypotheses suggested to account for the recent appearance of similar symptoms in different tree species over wide geographical areas, and some of these may be tested in laboratory and field experiments. However, this review is concerned with field measurements which may provide additional information on the extent, and possibly also the causes, of this new type of forest decline.

The extent of the problem has been assessed by visual inspection of forests from the ground, using trained observers. Factors included in the assessment are crown form, leaf/needle retention and leaf/needle colour. National surveys based on such observations have been operating in many countries now for several years, and the recent CEC directive on national inventories of forest health (1) will extend this type of survey over the whole European Economic Community. Beyond the EEC, the United Nations Economic Commission for Europe is in the process of formulating protocols for surveying forest health (2). However, this type of survey relies subjective assessments (albeit by trained personnel) which have shown bias in the past (3,4), and which may well show bias in future surveys.

Furthermore, a survey based on visible symptoms alone gives little or no information on the causes of the symptoms, in view of the non-specific nature of these responses.

What is required, therefore, is a series of tests to provide

i) an objective quantitative assessment of 'damage' in areas with trees showing visible symptoms and

ii) an objective quantitative assessment of 'latent damage' in areas with no visible symptoms.

In the second case, the 'latent damage' may be irreversible (ie. indicating the start of a one-way progression through visible damage to tree death) or reversible (ie. indicating a state of stress from which the tree may recover, but which may lead to irreversible decline). The definition of 'reversibility' in this case is rather vague, but the distinction between reversibility and irreversibility probably occurs for processes which operate on a time scale of years, rather than months or days. It may also be applied to visible symptoms insofar as some of these (eg. crown form) are effectively irreversible whereas others (eg. foliar yellowing) may disappear over a period of months or if treatment is applied.

A third objective for such tests would be the identification of the causal agent(s) and mechanisms involved, but it must be made clear that in order to show cause and effect field measurements can at best show correlations which may be relevant. The only way to establish cause/effect relationships for certain is through the formulation and testing of hypotheses by direct experiment, difficult as this may be for forest trees. It is an interesting philosophical point to discuss how one treats hypotheses which by their very nature preclude experimental testing on time and financial scales accessible to science!

2. DIAGNOSTIC TESTS

If we admit the possibility that such diagnostic tests exist (and it is unlikely that a single test will provide the solution), what would be their characteristics? To be useful in the context of a survey, there should ideally be a degree of independence of the tests with different tree species. At first sight, this appears an unreasonable constraint, yet the visible symptoms are similar for several conifer species.

Within a species, any diagnostic test should ideally be independent of:

- genotype
- tree age
- stage of development (season of year)
- time of day (weather, eg. sunshine)
- position on tree of sample taken.

If the test is to give an indication of cause, it should ideally depend on specific factors such as:

- air pollution (gases or wet acidity)
- soil acidification (nutrient availability)
- drought stress (which includes soil factors)
- cold stress/frost hardiness
- pathogen stress, etc.

It would be unrealistic to expect any test to meet these ideal requirements, but useful tests may still be possible if these requirements are relaxed. For example, if the response to a given test is well established as a function of tree age, or time of year, it may be possible to 'normalise' a response relative to known patterns. Alternatively, sampling and testing may be restricted to a fixed time of year, or stratified by tree age, as happens already in existing surveys.

The ideal independence from genetic variation refers more to the qualitative nature of the response than the quantitative nature. We must accept that individuals and populations of a species may differ in their response to a given stress or stresses. On the negative side, such variability may prove to be so great as to swamp any 'signal' related to visible symptoms. On the positive side, it may allow us to identify sensitive or tolerant gentoypes, thereby permitting an assessment of the potential for damage within a given population, and the possibility of active selection for future forest management.

The most critical requirement for a test is its independence of short-term variability in the environment, on a time-scale from minutes to days. 'Real-time' responses to the physical environment, such as the stomatal response to mid-day water stress or the rapid biochemical responses to a short-term fumigation, should not mask the longer term response. In many cases the capacity for short-term response (eg. enzyme acitivity) may have developed in response to chronic stress, so that the magnitude of the 'real-time' response is greatly altered, and may be useful as a diagnostic indicator. Episodes of 'acute' pollutant or environmental stress may be recorded permanently (eg. as cell damage). Chronic stress may be recorded by the accumulation of certain metabolites or by a progressive change in the relative amounts of certain components of the plant. It is these cumulative or permanent responses that are likely to prove most useful for diagnosis.

3. PRACTICAL CONSTRAINTS

If a diagnostic test is to be useful under field conditions as part of a survey, then the method must be inexpensive and rapid, and involve the minimum of preparation before storage and transport to a laboratory. It may be possible to adapt laboratory techniques for field use, and instances of this type of approach will be described later. More usually the requirement in the field is to collect a sample and preserve it for later analysis. This may entail the transportation to and from a field site of 'dry ice' (solid CO_2) or liquid nitrogen, and in all cases speed is essential.

However, before material can be prepared for analysis, a sample must be taken. The sampling strategy may be divided into two levels: which tree, and where on the tree? In the first case, the sampling strategy should be little different in principle from any ecological sampling, and should ensure an unbiased selection of a number of trees from a population. Inevitably, problems arise with trees which may require the inclusion of a stratification regime for sampling which would not be appropriate for other plants. For example, it may be practically very difficult to obtain comparable samples of leaf material from 30 year-old trees and 120 year-old trees because of inaccessibility. An age stratum (which would be desirable in any case) may therefore include an element of sample availability. In view of what is observed in terms of visible symptoms it may also be necessary to stratify the sampling explicitly in terms of edge or non-edge trees, dominance, site aspect, elevation etc.

The practical problem (largely avoided by visual inspection from the ground) of physically obtaining sample material from, say, the mid or upper crown of a tree is not to be underrated. Many methods have been used (shooting, turntable ladders, semi-permanent towers, telescopic pruners) but all have drawbacks. This problem is likely to be the severest constraint on the applicability of any diagnostic test to trees growing in a real forest.

The sampling strategy within a tree is also important; height within the crown, needle age class (for conifers), leaf type (sun or shade), aspect, order of lateral branches etc. Again, there may be practical constraints on accessibility.

The timing of the sample may be all important for many biochemical tests, both within the day and within the year, as noted earlier. This may pose additional difficulties if results are to be compared over large geographical areas where stages of development occur at different times. What is the 'correct' time basis for comparison - a calendar date or a physiological 'date' such as budbreak?

4. IDENTIFICATION AND DEVELOPMENT OF DIAGNOSTIC TESTS.

All the foregoing presupposes the existence of tests which may be used for diagnosis. There are many possibilities, ranging from the non-specific but relatively straightforward to the specific but complex. A compromise has to be sought dictated by the resources available. In most cases, a potential test will not have been applied under field conditions. If measurements obtained from the application of a given test appear to correlate well with present descriptions based on visible symptoms, then that test may be useful without further development for the quantification of decline. More usually, we would seek information on the cause of the observations based on the specific nature of the test. In this case a test must have been intensively studied under controlled conditions to investigate its dependence on external factors and the specific nature of its response, as discussed earlier. A technique which shows a pollutant response in the laboratory may be influenced by so many other factors under field conditions as to be valueless for diagnosis. Conversely, a test which appears to show a pollutant-related response in the field must be investigated experimentally to demonstrate that the response is in fact caused by pollution.

5. POTENTIALLY USEFUL DIAGNOSTIC TESTS.

There is a wide range of tests which have been, or may be, used for diagnostic purposes. These may be classified:

- Changes visible only under the microscope.
- Changes in 'whole leaf' properties.
- Changes to external surfaces of leaves.
- Changes in physiological responses.
- Changes in biochemistry, measurable externally (non-destructive).
- Changes in biochemistry, measurable internally (destructive).
- Changes in root systems.

Some of the techniques in each of the classifications will be described briefly, with comments on suitability or otherwise as potential diagnostic tests.

5.1 Changes visible under the microscope

Microscopy may be used to investigate both the external surface (usually by scanning electron microscopy, SEM) or the ultrastructure of cells and internal tissue (by light microscopy or transmission electron microscopy, TEM). The use of SEM to study leaf surfaces has a long history, and even pollutant-related studies go back over nearly 20 years (5). The technique has been applied to field-grown mature trees (6,7), and differences attributed to air pollution have been found. However, the structure of epicuticular waxes is also dependent upon non-pollutant factors such as physical climate, water stress and nutrient status (8), making such a test rather non-specific. More importantly, the technique requires a large number of samples to be examined and 'scored', and the scoring criteria (7) are not strictly objective. Although the technique is attractive inasmuch as visible changes are apparent, it may not be practically useful on a scale appropriate for diagnosis unless image analysis techniques can be developed to score large numbers of micrographs automatically.

Microscopic investigation of cell ultrastructure has been more concerned with qualitative diagnosis than quantitative diagnosis, and so may be more useful as a specific indicator of causes than for estimating the overall extent of damage. Responses to individual pollutants and stresses are known (eg. 9,10) and may indicate other tests appropriate for quantitative study. For example, Fink (11) has recently shown the accumulation of starch in Norway spruce needles from areas showing visible symptoms, in contrast to needles from areas with no visible symptoms. This may form the basis of an objective chemical determination.

5.2 Changes in 'whole leaf' properties.

One property of foliage which may be useful as an indicator is the water content of foliage, expressed as a dry weight: fresh weight ratio. For conifers, variability may be reduced by comparing such ratios for different year classes (12). For broad-leaved species, the specific leaf area may similarly be useful, as this is known to respond to air pollutants (13), but great care must then be used in sampling to distinguish, for example, shade leaves.

Elemental composition may also be regarded as a 'whole leaf' property, and has been widely studied. In this case there is a large amount of data against which field data may be compared, and many of the visible symptoms of forest decline (discolouration) have been associated with deficiencies of Mg^{2+}, K^{+}, Mn^{2+} or other metal ions (14). However, for diagnostic purposes the sampling regime may prove difficult in view of the known variability with leaf age and position in crown (15). For conifers, comparison of the elemental composition of different needle ages from the same branch may be used to remove some of the inherent variability.

5.3 Changes to external surfaces of leaves.

The visible (by SEM) changes in epicuticular wax structure discussed in 5.1 may be studied quantitatively by the behaviour of water droplets on leaf surfaces. The contact angle of water droplets has been shown to well with visible changes (16), and also to be a potential diagnostic test (17), in view of its ease of use. However, like SEM studies, it is not necessarily a specific test, and only addresses the outermost surface of the leaf. The total amount of epicuticular wax may also prove useful as an indicator of stress, especially when comparing conifer needles of different ages from the same tree. Care must be taken in interpreting such measurements, however, as the amount of wax per unit area may vary less with age than the amount expressed as a proportion of the needle dry weight. As the next stage, the chemical composition of the epicuticular wax may be studied, and there are indications (18) that changes in certain components may be related to air pollutant stress.

One of the earliest diagnostic tests for the effects of sulphur dioxide on trees was devised by Härtel (19) in 1953, and is based on the chemical properties of the wax and needle composition. The turbidity of a boiling water extract of conifer needles was shown to increase with proximity to a point source of sulphur dioxide. Later studies (20) showed that this response was complex, and related to the availability of calcium ions in the extract. Nevertheless, the test is simple to apply (if not to interpret) and may yet be useful.

The cuticle proper may also be of use, although its isolation may prove too time-consuming for general use. Cuticle thickness may be determined microscopically, but recent measurements (21) suggest that the mass of the isolated cuticle and its permeability may be altered in response to air pollutants. Much development would be required for these properties to be of use in diagnosis.

There is, however, a technique which combines the properties of cuticle and wax (and may involve physiological responses also) which has been shown to be strongly dependent on pollutant exposure. The loss of water through the leaf surface (loosely termed 'cuticular transpiration') has been shown to be enhanced for excised leaves exposed to air pollutants under controlled conditions (22) and in the field (23). This test is worthy of further evaluation and development as it involves a physiologically meaningful measurement and is straightforward to use.

5.4 Changes in physiological response

Changes in photosynthesis in response to air pollutants are well known in the laboratory. However, in the field such responses are only likely at unusually large pollutant gas concentrations, and may persist for only a few minutes. In such circumstances what may be useful diagnostically is not the actual response but the capacity for response. This approach, measuring photosynthetic capacity at light and CO_2 saturation, has been applied in the field (24) and shows promise as a diagnostic test given the relatively small variability observed with sampling height and the correlation with visible symptoms.

5.5 Changes in biochemistry - non-destructive measurements.

There are several techniques which may be applied externally to foliage to determine the biochemical status of the leaf. The following is not an exhaustive list, but indicates the types of techniques which are available, and which in general have not yet been developed for use under field conditions.

5.5.1 Reflected light.

Remote sensing of light reflectance from vegetation may be helpful in assessing visible damage symptoms, and multi-spectral scanners and thematic mappers on aircraft and satellites have been used to correlate images with known visible symptoms. For forest stands of a single tree species there may be good correlation with damage estimates made at ground level (25,26) using ratios of spectral bands, but the discrimination between different damage classes is likely to be poor relative to the 'noise' introduced in mixed forest, and the validation of the technique under such conditions is very labour-intensive. It is nevertheless an attractive proposition in terms of the areas which may be covered. The sampling problems become part of the stratification required for assessing the spectral response in terms of tree species, age, altitude etc. This technique should be seen as a simple extension of the present type of survey.

5.5.2 Chlorophyll fluorescence

The stimulated fluorescence of leaves has been used in the laboratory for the investigation of photosynthetic efficiency, and the technique is now being developed for field use. This may be combined with remote sensing passively (by correcting fluorescence for reflectance) or actively using a pulsed laser source from an aircraft (27), but development is still in the early stages. Induced fluorescence has been used under field conditions, either using a modulated light source or by making simultaneous measurements at 2 wavelengths. The latter technique (28) removes much of the internal heterogeneity by taking ratios, but both techniques are more easily applied to broad-leaved species than to conifers. As equipment is developed for field use, this type of measurement may prove useful empirically, but may not be very specific.

5.5.3 Electron paramagnetic resonance.

This technique is included insofar as it involves relatively little preparative work and may show some specificity towards gaseous pollutants. The equipment required is not suitable for use in the field, but samples may be brought to the laboratory for analysis. Although preliminary results (29) refer to a relatively small sample, the discrimination of different damage classifications suggest that this technique is worthy of further evaluation as a potential diagnostic test, relating to free radical activity within the photosynthetic process.

5.5.4 Nuclear magnetic resonance.

The use of high field nmr for phosphorus (31p) is also a relatively simple technique for the determination of intercellular pH (30), but again requires non-mobile equipment for analysis of samples. The technique has not been applied (to my knowledge) to field studies, and before being considered as a potentially useful test in the field, extensive studies must be made of its dependence on the parameters listed in section 2 above, and on the time-scale of the pH response. The same comments apply to epr (section 5.5.3).

5.5.5 Emissions of small organic molecules.

The emission of ethene (ethylene) in response to acute stress is a well-known response of plants. Other organic compounds (ethane, propane, propene, ethanol, methanal (formaldehyde), ethanal (acetaldehyde) etc.) may also be emitted under certain conditions (31). Recently, Rennenberg (32) has drawn attention to the experimental difficulties associated with making such measurements, in that in vivo responses may be very different from in vitro responses. For diagnostic purposes, the response time-scale for emissions may be too short to be generally useful in the field, unless methods can be developed to integrate gas concentration measurements over days or weeks. However, little is known of the long-term responses of plants, and further development may identify individual compounds or groups of compounds which can be used for diagnosis. The methods for gas sampling will also need to be developed for use in the field, but the recent availability of a portable gas chromatograph using a photo-ionization detector makes such developments more likely.

5.5.6 Emissions of monoterpenes.

Measurement of monoterpenes emitted from plants have been included here in view of their alleged role as amplifiers of ozone damage and relation to visible symptoms (33). Similar properties may also be observed with ethylene (34), but have not been demonstrated for trees under field conditions. Experimental methods for routine field measurements have not yet been developed.

5.6 Changes in biochemistry - destructive measurements.

Most of the following possible analyses cannot be carried out in the field, but require transportation of the sample to the laboratory. However, some sample preparation and preservation will be required in the field, and the development of any test for field use must include a study of the stability of the analyte during transport and storage. This may require the use of liquid nitrogen, with the inherent problems of supply and transportation between field site and laboratory.

5.6.1 Enzymes and substrates/precursors.

Although many measurements have been made of the activity of enzymes such as superoxide dismutase, peroxidase, glutathione reductase, nitrate and nitrite reductase, etc (35) as part of laboratory experiments, there has been relatively little work on field measurements. Part of the problem perhaps arises from the interpretation of such measurements from field material. To what extent does the activity of an enzyme system depend upon recent acute exposure to pollutant gases rather than to chronic low-level exposure? Where there is a clear correlation of enzyme acitivity with visible damage, we may assume that the measured response tends to the longer-term and may be a useful indicator (35), but is unlikely to be quantitatively related to the degree of damage once visible injury is present (37). Such enzyme assays may also not be specific for individual pollutants, except perhaps for nitrate reductase response to NO_x.

In addition to enzyme activity, the concentrations of substrates, precursors and related compounds may also change in response to pollutant stresses. Amounts of ascorbate, glutathione, and α-tocopherol may be elevated when compared with control plants in fumigation experiments, and similar results have been observed in the field. Not surprisingly, the variability of these concentrations may be very great, and it may be difficult to separate acute and chronic responses under field conditions. Development of these biochemical markers as diagnostic tests will require a better understanding of the dynamics of their response to pollutant and other stresses.

5.6.2 Pigments

Bleaching and loss of pigments is responsible for visible symptoms, and it would follow that early changes in pigment amounts might be a useful indication of changes occurring before visible symptoms appear. Pigments are difficult to extract and analyse quantitatively, but recent developments in high performance liquid chromatography (hplc) now permit good quantification. Although absolute determinations are more difficult, pigment ratios are readily determined, and these may in any case be better as indicators as the use of ratios may eliminate unwanted variation. Both chlorophylls and carotnoids have been studied (35). Recent results for Norway Spruce (17) suggest that violaxanthin:antheraxanthin ratios may be related to the occurrence of visible damage in the field.

5.6.3 Other metabolites.

It was noted earlier that starch grains may accumulate in conifer needles under certain conditions. More generally, carbohydrate metabolism may be altered by pollutant stress in ways which may be useful diagnostically. Field studies on spruce have shown lower levels of monosaccharides and acid-soluble polysaccharides in healthy trees compared with those showing visible damage (38). These results are particularly encouraging in that there appears to be only a small seasonal variation in the response.

Amino-acids may also be useful markers (35). Tryptophan has been shown to accumulate in damaged trees (39), and where exposure to pollutant ammonia occurs, amino acids may accumulate in tissue to a very great extent (40). Although responses of amino-acid concentrations are well documented for laboratory experiments, detailed field studies are required.

Polyamines are another group of compounds whose concentrations have been shown to increase in line with visible damage in the field (41) and in response to gaseous pollutants in the laboratory (42). Small RNAs have also recently been reported to be found in larger concentrations in visibly damaged spruce trees, independent of season, needle age and branch position (43).

5.6.4 Buffer capacity

The effects of gaseous air pollutants on plant cell buffering capacities was suggested as a fruitful research area 3 years ago (44) and non-destructive methods were described earlier. Recent results, where the buffer capacity of homogenised spruce needles was determined by titration (17), show differences which are related to the incidence of visible damage and air pollutant exposure. These differences are most apparent when comparing tissue of different ages, and the technique is still to be evaluated under controlled conditions to determine whether or not there is a response to specific pollutants.

5.7 Changes in root properties.

Roots, and the effects of soil acidification, have been studied in some detail, but not usually with a view to quantitative diagnosis. There is certainly a role for root analysis, but at present mainly for qualitative diagnosis. It is difficult to see how the quantitative aspect can be addressed, given the heterogeneous nature of the growing medium (soil) when compared with the homogeneous nature of the atmosphere, and the complex sampling difficulties associated with foliar analysis. However, just because roots are difficult to sample should not mean that this area be omitted from quantitative diagnostic research. It is perhaps in this area above all that research initiatives need to be taken, not just of roots themselves, but also of the associated mycorrhizal fungi. There are already indications of long-term changes in mycorrhizal populations in the Netherlands (45) which may be related to air pollution, and there is a clear need for an experimental approach on a wider scale.

6. CONCLUSIONS

The development of a potential diagnostic test goes through a number of distinct stages, starting with a requirement that the response to a given test be quantitatively related to the degree of damage. As a guiding principle, responses which are cumulative or long-term are to be preferred to short-term responses. This may mean that the capacity for a particular response under ideal conditions is a more appropriate measure than the actual response under field conditions. An example of this for photosynthesis was described earlier (section 5.4).

Once a potential test has been identified it must be evaluated for the tree species of interest both in the laboratory and in the field. Laboratory studies should examine the specificity of the test to applied pollutant stresses, and interactions with other variables such as climate. Field studies should examine the sources and magnitude of variability introduced by sampling from populations of different age and genotype, and interactions with other factors normally controlled in the laboratory such as climate, soil type and pathogens.

Laboratory studies would normally be performed with clonal stock, and it must be shown that field responses from non-clonal populations are qualitatively similar to those obtained in the laboratory. We should be able to address questions such as; 'are the trees in this forest damaged because of exposure to large concentrations of air pollutants, or because the trees are of a genotype sensitive even to small concentrations of pollutants?'.

In moving tests from the laboratory to the field we must decide whether to take the apparatus to the forest, or bring samples to the laboratory. In either case there are logistical problems of transport and storage to be considered.

The next stage would be to compare results from more than one field site, as a means of evaluating the geographical variability. Genotype, microclimate, soil type and air pollution will all vary, but if a test is to be useful for diagnostic purposes, the responses should be comparable over a wide geographical range. At this stage the collaboration of two or more research groups may become necessary in order to gain access to suitable sites and facilities.

A single test is very unlikely to provide useful information related to all the visible symptoms currently included in field surveys, and several tests should be evaluated together, where possible. In the past, measurements of several properties of trees from the same field site have often been made in isolation. The results are then not comparable because, for example, samples have been taken at different times of year, or from different trees. Valuable information on the covariation of different physiological or biochemical tests may be gained by applying them to the same samples. This is another area where collaboration between research groups can be of great value. Comparison of results which are' internally' normalised for variation resulting from sampling strategy may show relationships which would otherwise be missed.

A second general principle which may prove useful in moving from the laboratory to the field is to make use of measurement ratios. This is already done for many routine tests when results are expressed as a proportion of dry weight, but other bases for expressing results may be more appropriate. The difference between amounts of epicuticular wax expressed relative to dry weight or surface area were mentioned earlier. Ratios of measurements may be internal (eg. the ratios of pigment concentrations, or of chlorophyll fluorescence wavelengths) or, for conifers especially, may compare different age classes of needles. If long-term changes in properties are to be used for diagnosis, then a single sample may contain several years' information. Changes from one year's foliage to the next may also give information on relative rates of change, which may be less sensitive to genetic variability, for example, than absolute values (16).

If ratio measurements are to be used, however, we must take care in statistical analysis of the results that our data are (approximately) normally distributed, so that statistical tests of significance apply. In some cases it may be necessary to develop suitable statistical analyses in order to extract the maximum amount of useful information from the hard-won data.

The steps in the development of a potential diagnostic test need not follow those given above. There may be good grounds for 'screening' possible tests by a pilot study at a large number of sites and under widely varying conditions, as has been attempted recently (17). We are still in the early stages of development of any generally applicable objective quantitative diagnostic tests, but it is to be hoped that our search for a better means of defining and understanding the new type of forest decline will be successful so that appropriate remedial action can be taken. This will only happen with the active collaboration of research groups within Europe and world-wide.

REFERENCES

(1) COMMISSION OF THE EUROPEAN COMMUNITIES (1986). On protection of the Community's forests against atmospheric pollution (Council regulation 3528/86). Official Journal of European Communities L326, 2-4.

(2) UNECE EXECUTIVE BODY FOR CONVENTION ON LONG-RANGE TRANS-BOUNDARY AIR POLLUTION (1986). Elements of a draft manual: Methodologies and criteria for harmonised sampling, assessment, monitoring and analysis of the effects of air pollution on forests in the ECE region. UNECE Draft publication.

(3) KRAUSE, G.H.M., ARNDT, U., BRANDT, C.J., BUCHER, J., KENK, G. & MATZNER, E. (1986). Forest decline in Europe: development and possible causes. Water,Air & Soil Pollution 31, 647-668.

(4) INNES, J.L., BOSWELL, R., BINNS, W.O. & REDFERN, D.B. (1986). Forest Health and Air Pollution: 1986 Survey. Forestry Commission Res. and Devt. Paper 150. Forestry Commission, Edinburgh.

(5) BYSTROM, B.G., GLATER, R.B., SCOTT, F.M. & BOWLER, E.S.C. (1968). Leaf surface of *Beta vulgaris* - electron microscope study. Botanical Gazette 129, 133-138.

(6) HUTTUNEN, S. & LAINE, K. 1983. Effects of air-borne pollutants on the surface wax structure of Pinus sylvestris needles. Annales Botanica Fennici 20, 79-86.
(7) CROSSLEY, A. & FOWLER, D. (1986). The weathering of Scots pine epicuticular wax in polluted and clean air. New Phytologist 103, 207-218.
(8) JEFFREE, C.E., BAKER, E.A. & HOLLOWAY, P.J. (1976). Origins of the fine structure of plant epicullticular waxes. In: Microbiology of Aerial Plant Surfaces, C.F. Dickinson & T.F. Preece (eds) pp 119-158, Academic Press, London.
(9) SOIKELLI, S. (1981). A review of the structural effects of air pollution on mesophyll tissue of plants at light and transmission electron microscope level. Savonia 4, 11-34.
(10) FINK, S. (1983). Histologische und histochemische Untersuchungen an Nadeln erkrankter Tannen und Fichten in Südschwarzwald. Allgemeine Forst Zeitschrift 26/27, 660-663.
(11) FINK, S. (1987). Histological and cytological changes caused by air pollutants. In: Proc. 2nd Int. Symp. Air Pollution and Plant Metabolism, Neuherberg, L.W. Blank, N.M. Darrall, S-Schulte-Hostede & A.R. Wellburn (eds).
(12) MEHLHORN, H., FRANCIS, B., PATERSON, I.S., CAPE, J.N., WOLFENDEN, J. & WELLBURN, A.R. (1987). Dry weight: fresh weight ratio as an easy to measure biochemical indicator of tree health. Poster presented at 2nd Int. Symp. Air Pollution and Plant Metabolism, Neuherberg, April 1987.
(13) WHITMORE, M.E. & MANSFIELD, T.A. (1983). Effects of long-term exposures to SO_2 and NO_2 on Poa pratensis and other grasses. Environmental Pollution A 31, 217-235.
(14) ZÖTTL, H.W. & HÜTTL, R.F. (1986). Nutrient supply and forest decline in south-west Germany. Water,Air & Soil Pollution 31, 449-462.
(15) PETERS, J. & MAURER, W. (1985). Monitoring environmental impacts of forest ecosystems using spruce needles - investigations on representative sample collection programs. In: Air Pollution and Plants, C. Troyanowsky (ed) p 217. VCH Verlag, Weinheim (FRG).
(16) CAPE, J.N. (1983). Contact angles of water droplets on needles of Scots pine (Pinus sylvestris) growing in polluted atmospheres. New Phytologist 93, 293-299.
(17) CAPE, J.N., PATERSON, I.S., WELLBURN, A.R., WOLFENDEN, J., MEHLHORN, H., FREER-SMITH, P. & FINK, S. 1987. Early diagnosis of forest decline - results of a pilot study. (in press). Institute of Terrestrial Ecology, Huntingdon.
(18) CAPE, J.N. (1986). Effects of air pollution on the chemistry of surface waxes of Scots pine. Water, Air & Soil Pollution 31, 393-399.
(19) HÄRTEL, O. (1953). Eine neue Methode zur Erkennung von Raucheinwirkungen an Fichten. Zentralblatt für die gesante Forst-und Holzwirtschaft 72, 12-31.
(20) FUCHSHOFER, H. & HÄRTEL, O. (1985). Zur Physiologie des Trübungstests, einer Methode zur Bioindikation von Abgaswirkungen auf Koniferen. Phyton 25, 277-291.
(21) ZIEGLER, H. (1987). Interactive effects of different air pollutants on structure and function of tree leaves. In: Proc. CEC Workshop Direct Effects of Dry and Wet Deposition on Forest Ecosystems, Lökeberg, Sweden. CEC, Brussels (in press).

(22) MANSFIELD, T.A., WRIGHT, E.A., LUCAS, P.W. & COTTAM, D.A. 1987. Interactions between air pollutants and water stress. In: Proc. 2nd Int. Symp. Air Pollution and Plant Metabolism, Neuherberg, L.W. Blank, N.M. Darrall, S. Schulte-Hostede & A.R. Wellburn (eds).
(23) CAPE, J.N. & FOWLER, D. (1981). Changes in epicuticular wax of Pinus sylvestris exposed to polluted air. Silva Fennica 15, 457-458.
(24) FÜHRER, G. (1987). Rapid gas exchange measurements in the field: photosynthetic capacity of cut spruce twigs. In: Proc. 2nd Int. Symp. Air Pollution and Plant Metabolism, Neuherberg, L.W. Blank, N.M. Darrall,S. Schulte-Hostede & A.R. Wellburn (eds).
(25) ROCK, B.N. & VOGELMANN, J.E. (1986). Use of TMS/TM data for mapping of forest decline damage in the northeastern United States. In: Proc. IGARSS 86 Symp. pp 1405-1410. ESA Publication SP-254.
(26) KOCH, B. & KRITIKOS, G. (1984). Integrative investigation of forest damage detection based on airborne multispectral scanner data. In: Proc. Integrated Approaches in Remote Sensing pp 109-113. ESA Publication SP-214.
(27) KLIFFEN, C. (1987). Early diagnosis of air pollution injury by estimating the chlorophyll fluorescence. In: Proc. 2nd Int. Symp. Air Pollution and Plant Metabolism, Neuherberg. L.W. Blank, N.M. Darrall, S. Schulte-Hostede & A.R. Wellburn (eds).
(28) STRASSER, R.J. (1987). Reported in early diagnosis workshop, ibid.
(29) STEGMANN, H.B., RUFF, H.J. & SCHEFFLER, K. (1985). Photosignal II as an indicator for damage to forests - EPR spectroscopic investigation of spruce needles. Angew. Chem. Int. Ed. Engl. 24, 425-426.
(30) WOLFENDEN, J. & WELLBURN, A.R. (1986). Cellular readjustment of barley seedlings to simulated acid rain. New Phytologist 104, 97-109.
(31) BUCHER, J.B. (1984). Emissions of volatiles from plants under air pollution stress. In: Gaseous Air Pollutants and Plant Metabolism, M.J. Koziol & F.R. Whatley (eds). pp 399-412. Butterworths, London.
(32) STAHL, K. & RENNENBERG, H. (1987). Early diagnosis by emission of volatile hydrocarbons. In: Proc. 2nd Int. Symp. Air Pollution & Plant Metabolism, Neuherberg. L.W. Blank, N.M. Darrall, S. Schulte-Hostede & A.R. Wellburn (eds).
(33) WAGNER, E. (1987). Reported in early diagnosis workshop, ibid.
(34) MEHLHORN, H. & WELLBURN, A.R. (1987). Stress ethylene formation determines plant sensitivity to ozone. Nature 326, (in press).
(35) DARRALL, N.M. & JÄGER, H.J. (1984). Biochemical diagnostic tests for the effect of air pollution on plants. In: Gaseous Air Pollutants and Plant Metabolism, M.J. Koziol & F.R. Whatley (eds). pp 333-349. Butterworths, London.
(36) KRINGS, B. & RENNENBERG, H. (1987). Superoxide dismutase activity in spruce trees. In: Proc. 2nd Int. Symp. Air Pollution & Plant Metabolism, Neuherberg. L.W. Blank, N.M. Darrall, S. Schulte-Hostede & A.R. Wellburn (eds).
(37) HEISKA, E. & HUTTUNEN, S. (1987). Seasonal variation of superoxide dismutase (SOD) activity in Scots pine and Norway spruce needles in northern Finland. ibid.
(38) VILLANEUVA, V.R., SANTERRE, A., MONCELON, F. & MARDON, M. (1987). Biochemical markers in polluted Picea trees III. Pollution and carbohydrate metabolism. ibid.

(39) VILLANEUVA, V.R., MONCELON, F., SANTERRE, A. & MARDON, M. (1987). Biochemical markers in polluted Picea trees II. Pollution, tryptophan metabolism and growth. ibid.

(40) ROELOFS, J.G.M. & VAN DIJK, H.F.G. (1987). Effects of airborne ammonium on the nutritional status of pine needles. In: Proc. CEC Workshop Direct Effects of Wet and Dry Deposition on Forest Ecosystems, Lökeberg, Sweden. CEC, Brussels (in press).

(41) VILLANEUVA, V.R., MARDON, M., MONCELON, F. & SANTERRE, A. (1987). Biochemical markers in polluted Picea trees I. Is putrescine a useful pollution biological marker? In: Proc. 2nd Int. Symp. Air Pollution and Plant Metabolism, Neuherberg. L.W. Blank, N.M. Darrall, S. Schulte-Hostede & A.R. Wellburn (eds).

(42) BORLAND, A.M. & ROWLAND, A.J. (1987). Pollution and polyamines, ibid.

(43) KÖSTER, S.,BEUTHER, E. & RIESNER, D. (1987). Small RNAs originating from healthy and diseased spruces (Picea abies L.) ibid.

(44) NIEBOER, E., MACFARLANE, J.D. & RICHARDSON, D.H.S. (1984). Modification of plant cell buffering capacities by gaseous air pollutants. In: Gaseous Air Pollutants & Plant Metabolism, M.J. Koziol & F.R. Whatley (eds). pp 313-330. Butterworths, London.

(45) ARNOLDS, A. (1985) (ed). Veranderingen in de Paddestoelenflora (Mycoflora). Wetenschappelijke Mededeling K.N.N.V. 167, K.N.N.V., Hoogwoud, NL.

EFFECTS OF H_2O_2-CONTAINING ACIDIC FOG ON YOUNG TREES.

Ronald K.A.M. MALLANT, Jacob SLANINA,
Energy Research Foundation ECN, P.O. box 1, 1755 ZG Petten,
The Netherlands.

Georg MASUCH, Antonius KETTRUP,
Fachbereich Chemie, Universitaet-GH, D-4790 Paderborn,
Federal Republic of Germany.

Summary

Intercepted clouds and fog are considered to play an important role in the forest die-back. As a result, there is a growth of the number of plant effect studies in which fog is involved. Most of these experiments suffer from some shortcomings in that sense that:
a. acidity is considered to be the only damaging factor,
b. most exposure-chambers are not suited to operate at 100% relative humidity, leading to substantial changes in the chemical composition of the fog, due to condensation or evaporation.

The experiments presented in this paper were set up with the intention to give a realistic simulation of the physics and chemistry. Along with the description of the exposure facility, some results of experiments in which young trees were exposed to hydrogen peroxide-containing fog are presented. Experiments have shown microscopical changes in the tissue of plants that are exposed to sub-ppm levels of H_2O_2 in fog, during six weeks.

1. INTRODUCTION

In several papers the effect of acidic fog or mist on plants is described. WOOD and BORMANN (1) exposed yellow birch seedlings to artificial acid mists, and observed that pH values had to drop to 3 to cause foliar tissue damage, while significant growth decreases were first encountered at pH 2.3. Exposure took place once a week, for six hours, over 15 weeks.

MAURICE and CRANG (2) observed an increase in the transectional areas of eastern white pine needles with decreasing pH. Though pH values went as low as 2.6, the nutritional effects of the nitrogen and sulfur compounds in the acid mists apparently exceeded the adverse effects of acidity, although it was doubted that such benificial effects would persist if duration of exposure is increased.

KRAUSE et al. (3) exposed spruces to O_3 in concentrations of 0; 200 and 600 microgram/m^3. Twice a week, the plants were exposed to acidic fog of pH 3.5. Increased leaching of several cations, nitrate and sulfate was observed at O_3 concentrations of 200 microgram/m^3. For most ions, leaching increased with decreasing pH. Similar experiments were performed by SKEFFINGTON and ROBERTS (4), who used O_3-concentrations of 0; 100; 200 and 300 microgram/m^3. The fog applied was generated by spraying a diluted mixture of sulfuric and nitric acids of pH 3. After 56 days, in which fog was applied during a few minutes (twice daily), and at high O_3 concentrations, 'chlorotic mottle' was observed in case of Scots pine while Norway spruce mainly showed a red-brown discolouration. The level of nutrients in the needles increased as function of O_3 concentration, which is contradictionary to the hypothesis of PRINZ et al. (5). In the experiments reported by CRANG and McQUATTIE (6) one-year-old yellow poplar seedlings were exposed to fog 4h/day each afternoon, for a period of 8 weeks. The pH values ranged from 5.6 to 2.6. In addition, some plants were fumigated with either 0.1 ppm of O_3 or NO_2. The symptoms observed were a decrease of mesophyll air space, and an increase in the frequency of occurrence of 'dense inclusion bodies, possibly tannins'. A less clear dependency of pH occurred when the plants were fumigated with O_3 or NO_2.

More literature exists on experiments in which the effects of acidic fog are studied. All these experiments have in common that acidity is considered to be the only factor causing the adverse effects of fog. In most of these experiments, the pH range at which effects become significant is about 3-3.5. GROSCH and SCHMITT (7) report that for a station in the Taunus mountains (FRG), the probability that the pH of fog water is less than 4 is 50%, while pH values below 3.5 occur in 10% of the cases. So, highly acidic fog occasionally does occur in forested regions, and a direct effect of acidity alone, or in combination with O_3 can be expected.

Field observations have shown that fog and intercepted clouds often contain appreciable amounts of unstable and possibly phytotoxic compounds. These compounds are peroxides, sulfite and hydroxymethanesulfonic acid (HMSA). Most measurements were made in the United States (8-10). Hydrogen peroxide is a highly soluble oxidant that is photochemically formed in the gas phase, primarely if the radicals formed by the photolysis of O_3 are not completely scavenged by reactions with NO and NO_2 (11). Sulfite is formed by the dissolution of SO_2 in water. The concentration depends strongly on the pH of the water. If neutralizing compounds (e.g. NH_3) are available, the concentration of sulfite in fog droplets or (dew-) water layers can amount to 1-10 ppm (12),(13). HMSA forms by the reaction of formaldehyde and sulfite. Measurements at the ECN site in Petten in the Netherlands have shown that HMSA concentrations of 1-20 ppm occur in dew and fog water.

The adverse effects of H_2O_2 have already been investigated (13-16,18). Young spruces and beeches suffered from severe cell deformation after six weeks in which they were exposed for 3 hours/day to acidic fog (pH 4) that contained maximal 5 ppm H_2O_2. In this paper, results of further experiments are reported.

2. EXPERIMENTAL

The fog chambers used in the experiment were five 0.6 m^3 glass boxes, illuminated by 15.000 Lux of fluorescent and incandescent light. They were ventilated at a rate of 12 m^3/hr with filtered air (figure 1). The incoming air is humidified to avoid more or less complete drying of the droplets, that can occur even if liquid water concentrations, as used in our experiments, are high relative to ambient values. We used approximately 2 g/m^3, natural fogs and clouds contain 4-10 times less liquid water.

It should be obligatory for researchers reporting their experiments with fog to make mention of the complete set of experimental conditions that may influence the chemical composition of the fog such as it will be present on the plants. These conditionss are: relative humidity and temperature of the air, the air exchange rate and the amount of water sprayed per unit time or per m^3 of air.

Heat fluxes between the fog chambers and the surrounding air must be minimized, since minor temperature changes have drastic effects on relative humidity. These fluxes are avoided by maintaining a temperature equilibrium between the exposure facility and the laboratory. A column filled with 0.8 cm Raschig rings through which the air was fed, was used to humidify and thermostate the air entering the boxes. Details of the design of similar packed-bed humidifiers are given in (17).

In the experiments that are presented here, H_2O_2 concentrations were different in all boxes. One box served as a reference. The H_2O_2 was added in a concentrated form to a main flow of a stock solution at a point a few centimeters from a pressurized nebulizer. Concentrations and flow rates were chosen such as to result in the desired H_2O_2 concentrations in the various exposure boxes. We have chosen this configuration because a concentrated H_2O_2 solution has good stability. The composition of fog water is given in table I. Further details on the methods used to generate the fog are given in (16). To check if evaporation still took place in the experiment, despite the precautions taken to avoid this, fog was sampled from the boxes using small glass cyclones (16). The sample was subsequently analysed and compared with the original spray solution. Analytical methods as well as techniques used to prepare the needles and leaves for examination with light microscopy and scanning and transmission electron microscopy are described earlier (16).

Three-year old test plants; spruces (Picea abies Karst.) and beeches (Fagus sylvatica L.), were exposed to fog containing 0, 200, 700, 2000 and 5000 ppb of H_2O_2. Exposure lasted 6 weeks, in which the plants were exposed for 3 hours per day. After this period the plants were prepared for examination.

3. RESULTS

The composition of the sampled fog was compared with that of the spray solution, both for the chemically stable components, as well as for H_2O_2. Thus, effects of evaporation or decay of H_2O_2 are revealed. As an example, the concentrating factor for Na^+ is given in figure 2. During the exposure, factors of 1.6 have been observed. As a rule, these factors are considerably closer to unity. Figure 2 also shows the variation in H_2O_2-concentration in the spray solution and in the sampled fog water.

Many histological changes were observed, both in beeches as well in spruces. Most of these changes appear even at the lower concentrations (200-700 ppb), although they may not yet be statistically significant (i.e.: $P < 0.05$). A rather qualitative presentation of these histological changes is given in this presentation, more detailed information on the observed effects on trees are reported elsewhere (14-16, 18).

One very illustrative symptom observed in the spruce needles, is the continuous increase of tannins as function of the H_2O_2-concentration (figure 3). The vacuoles in the reference series are transluctent, containing few tannin bodies. Exposure to 200 ppb H_2O_2 results in a higher concentration of tannin bodies. The diameter of these bodies has decreased with respect to the reference series. They form a web-like structure that fills the vacuole. Further increase of the H_2O_2-concentration leads to higher density of bodies, that have decreasing primary diameters. Chemical analyses of the tannin has shown that, apart from a change in the absolute amount, there is also a change in the ratio of the concentrations of the different compounds that form the tannin, i.e., shikimic acid, benzoic acid, picein, esculin, catechin, etc. (18).

4. CONCLUSION

An experimental facility is presented in which the effect of fog on plants can be studied. Fog water composition can be controlled and monitored to enable experiments with unstable components.

Experiments have shown that fog containing 700 ppb H_2O_2 has significant effects on the foliar tissue of beeches. Although this concentration is not excessively high, the integrated exposure time of the experiment probably exceeds general ambient values. Observation of effects has been restricted to changes in plant anatomy. More sensitive techniques, e.g. measuring physiological parameters, would probably reveal effects at lower exposure doses.

It can be concluded that H_2O_2 may play an important role in forest die-back. The experiments reported here stress the need for research in which atmospheric chemistry and physics are integrated with effects studies.

5. REFERENCES

(1) WOOD, T., and F.H. BORMANN (1974) Environ. Pollut. 7:259
(2) WOOD, T. and F.H. BORMANN (1975) Ambio 4(4):169
(3) KRAUSE, G.H.M., K.D. JUNG and B. PRINZ (1985) VDI Berichte 560:627
(4) SKEFFINGTON, R.A. and T.M. ROBERTS (1985) VDI Berichte 560:747
(5) PRINZ, B., G.H.M. KRAUSE and H. STRATMANN (1982) Landesanstalt fuer Immisionsschutz des Landes Nordrhein-Westfalen, Essen, FRG, Rep. No 28
(6) CRANG, R.E. and C.J. McQUATTIE (1986) Can. J. Bot. 64:1237
(7) GROSCH, S. and G. SCHMITT (1985) VDI Berichte 560:313
(8) DAUM, P.H., S.E. SCHWARTZ and L. NEWMAN (1983) Precipitation Scavenging, Dry Deposition, and Resuspension, 1, Elsevier Science Publishing Co., Inc. p.31
(9) KADLECEK, J., S. McLAREN, N. CAMAROTA, V. MOHNEN and J. WILSON (1983) ibid. (10), p.103
(10) MUNGER, J.W., C. TILLER, and M.R. HOFFMANN (1986) Science 231:247
(11) KLEINMAN, L.I. (1986) J. Geophys. Res. 91(D10):10889
(12) SLANINA, J., R.K.A.M. MALLANT, G. MASUCH and A. KETTRUP (1985) proc. of the Cost Action 61Abis meeting, Petten, The Netherlands, Dec. 9-10
(13) SLANINA, J. and R.K.A.M. MALLANT (1986) presented at the "Kolloqium Aktuelle Aufgaben der Messtechnik in der Luftreinhaltung", Heidelberg
(14) MALLANT, R.K.A.M., J. SLANINA, G. MASUCH and A. KETTRUP (1986) AEROSOLS, Research, Risk Assessment and Control Strategies, Ed. Lee, Schneider, Grant & Verkerk, Lewis Publishers, Inc., p.901-908
(15) MASUCH,G., A. KETTRUP, R.K.A.M. MALLANT and J. SLANINA (1985) VDI Berichte 560:761
(16) MASUCH,G., A. KETTRUP, R.K.A.M. MALLANT and J. SLANINA (1986) Int. Journ. of Environmental Analytical Chemistry, 27:183
(17) MALLANT, R.K.A.M. (1986) paper presented at the NATO advanced research workshop 'Acid deposition processes at high elevation sites' Edinburgh, Sept. 1986.
(18) KETTRUP, A., H.G. KICINSKI, G. MASUCH, R.K.A.M. MALLANT, J. SLANINA, proc. of the 7th World Clean Air Congress, Sydney Australia, August 1986, Vol.IV, pp. 228-253

fig. 1 schematic drawing of exposure facility (2 boxes shown). Aspirated air is humidified and thermostated by a packed-bed humidifier in which the water is passed via a microcomputer controlled heater. Fog is sprayed seperately in each box. Water is supplied to the plants from a reservoir by glass-fibre ribbons protruding from each container.

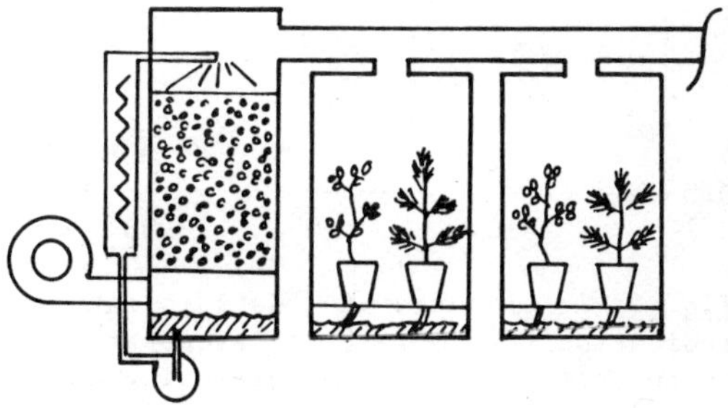

Table I Composition of fog water before spraying (micromol/l)

Na^{+}	743	K^{+}	45	Mg^{2+}	90
Ca^{2+}	77	NH_4^{+}	359	H^{+}	99
SO_4^{2-}	375	NO_3^{-}	106	Cl^{-}	785
H_2O_2 0; 6; 21; 59 or 147					

Table II Changes in tissues of beech trees exposed to fog containing 0.7 or 5 ppm H_2O_2. Reference experiment (R) without H_2O_2. Values between parenthesis are the number of leaves in each test.
Significant ($P < 0.05$) increase or decrease denoted by ++ respectively --. Insignificant ($P \geq 0.05$) changes denoted by + or -.

parameter	R -> 0.7	R -> 5	
weight:			(7)
fresh	-	-	
fresh (specific)	--	-	
dry	-	-	
dry (specific)	--	--	
water content	-	-	
water content (%)	+	+	
thickness (absolute):			(20)
total	--	--	
upper epidermis	--	--	
palisade mesoph.	--	--	
spongy mesoph.	-	-	
lower epidermis	-	--	
thickness (% of total):			(20)
upper epidermis	--	--	
palisade mesoph.	-	-	
spongy mesoph.	++	++	
lower epidermis	+	-	
stomata:			(40)
number	++	++	
degree of opening	+	++	

parameter	R -> 0.7	R ->5	
tissue area per unit length of transverse section (abs):			(10)
total	--	--	
upper epidermis	--	--	
palisade mesoph.	--	--	
spongy mesoph.	-	--	
lower epidermis	-	--	
interc. space	-	++	
tissue area per unit length of transverse section (%):			(10)
upper epidermis	-	--	
palisade mesoph.	-	--	
spongy mesoph.		--	
lower epidermis	+	-	
interc. space	+	++	
palisade mesophyll cells:			
size	--	-	(10)
number	--	--	(25)
(perc. of) vascular bundle:			(15)
phloem	+	--	
xylem	+	++	

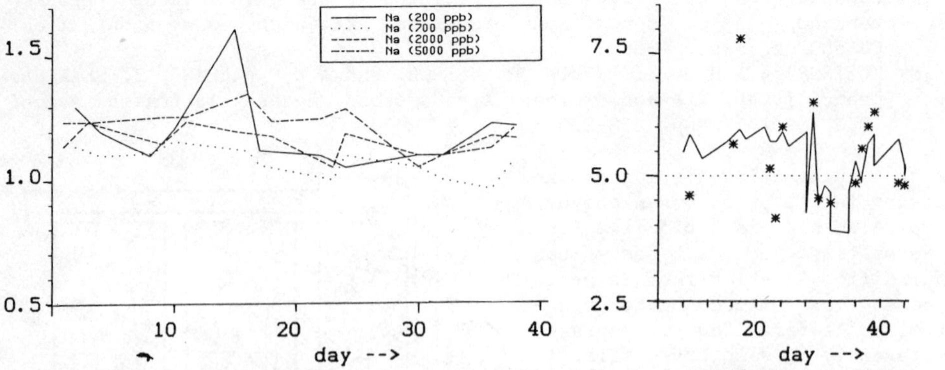

fig. 2 left: ratio of $[Na^{+}]$ in sampled fog water and the stock solution, indicating the effect of evaporation of fog in the exposure facility. Concentrating factor is given for the 0.2; 0.7; 2.0 and 5.0 ppm series. right: graph showing the variation in the H_2O_2 concentration in the solution before spraying (drawn line), and in the sampled fog water (indicated by asteriks). Dotted line: 5 ppm level (5 ppm series only).

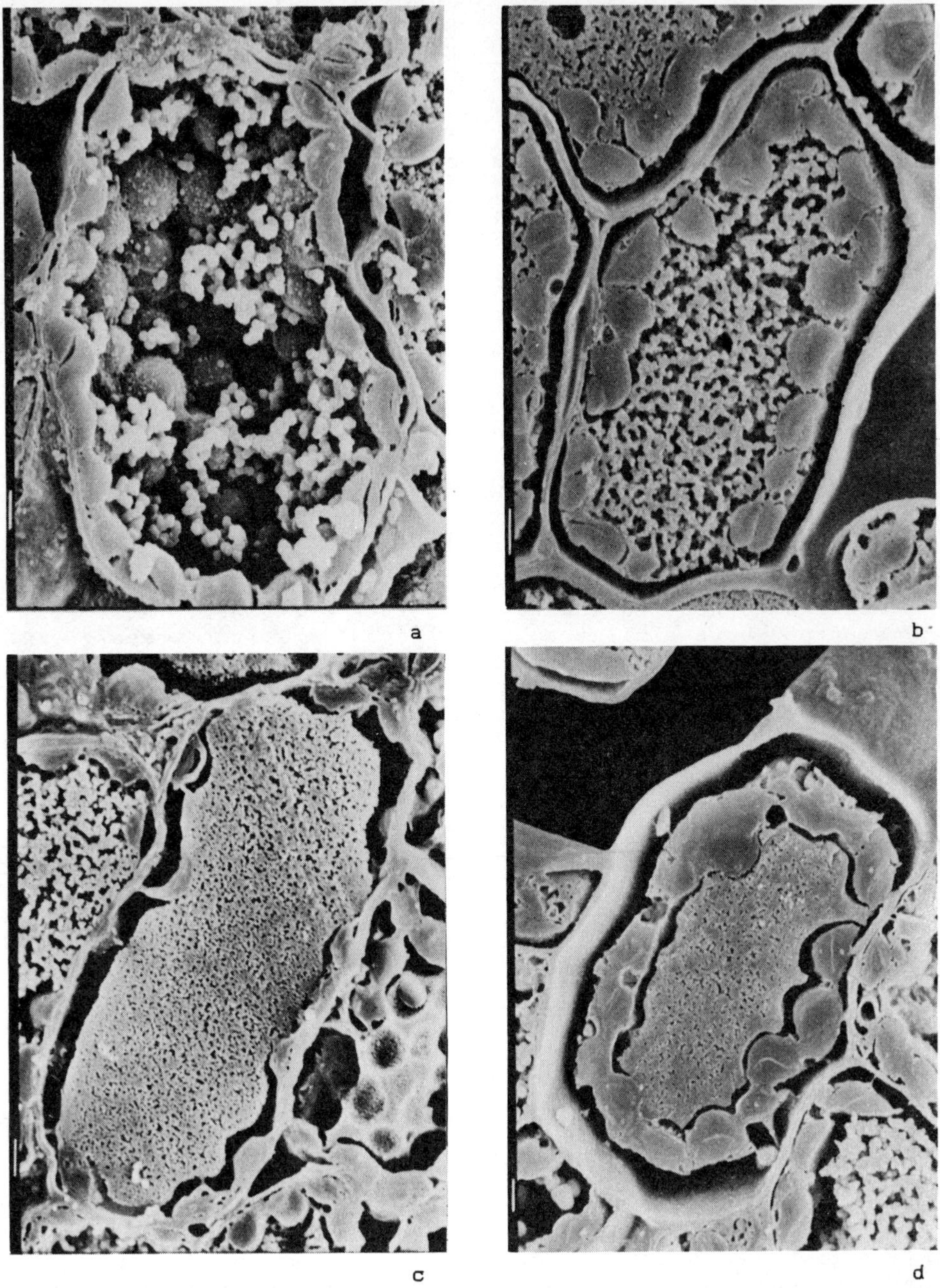

fig. 3 Primary needle mesophyll cells of Norway spruce with accumulations of tannin corpuscles in the central vacuoles.
a: reference series; b,c, and d: after exposure to fog containing respectively 0; 0.2; 0.7 and 5 ppm H_2O_2 (bar = 5 micrometer).

EFFECT OF ACIDIC MIST ON NUTRIENT LEACHING, CARBOHYDRATE STATUS AND DAMAGE SYMPTOMS OF PICEA ABIES.

K. MENGEL, M.Th. BREININGER and H.J. LUTZ
Institute of Plant Nutrition of the Justus Liebig-University
63oo Giessen, Fed. Rep. Germany

Summary

Experiments with young spruce trees (Picea abies) showed that acidic mist (pH 2.75) leached much higher rates of cations than a control mist (pH 5.oo). Nevertheless the cation concentration in the needles was hardly affected by this higher leaching rates. Exposure to acidic mist led to higher leaching of sugars and reduced the starch content of the youngest needles to about one half compared with needles exposed to the control mist.
Acidic mist damaged the wax layer and the stomatal wax props of the needles of the youngest shoot. Some of these needles turned redish brown and then were shed. These symptoms were similar to those found in spruce trees grown under natural conditions and showing decline. Needle loss in the treatment with acidic mist was much higher than in the control treatment.

1. INTRODUCTION

From the various air pollutants studied during the last decade acidic mist has gained in importance. According to Waldmann (1) modern forest decline in Central Europe is not clearly related to the concentrations of SO_2, NO_x and ozone in the atmosphere. A significant correlation (r = o.651), however, was found between forest damages and the frequency of mist (2). Mist droplets having a diameter from 1 to loo nm appear to be a strong sink for SO_2 and NO_x, because of their large relative surface. In these droplets the S and N oxides may be oxidized and by dissolving in water they form strong acids such as HNO_3 and H_2SO_4. For this reason pH in mist droplets may be 1 to 2 pH units lower than in rain drops, mist droplets often attaining pH levels lower than 3. In USA Munger et al. (3) found pH levels as low as 2.2 in mist.

Our experimental work is focussed on the effect of acidic mist on Norway spruce. Decline of this species is common in the area of Giessen particularly on soils derived from basalt. The symptoms mainly appear in older trees (3o to 5o years old). They are characterized by a redish brown colour of the needles which is not only seen in the older needles but appears also in the needles of the youngest shoots. Green needles lack brilliancy. Premature needle shedding is frequent. Eventually needle losses are so heavy that the trees die.

2. EXPERIMENTATION

Experiments were carried out with five years old trees of Picea abies in 1985 and 1986. In one treatment the trees were periodically exposed to an acidic mist, pH 2.75 in the 2. treatment trees were treated with the control mist, pH 5.oo. The drops dropping from the trees were collected and analyzed for inorganic ions and carbohydrates. Further details of the experimental design were reported by Mengel et al. (4).

3. LEACHING EFFECTS

The leaching rates obtained for K^+, Na^+, Mg^{2+}, Ca^{2+}, Zn^{2+}, and Mn^{2+} by acidic mist were several times higher than those of the control mist. In cases in which the trees had received a higher K fertilizer rate also higher K^+ leaching rates occurred. The total leaching of K^+, Ca^{2+}, Mg^{2+}, and Mn^{2+} collected in the acidic treatment during the experimental period of six weeks (1985) and eight weeks (1986) amounted only to 3 to 6% of the quantity of the corresponding element present in the needles. Only in the case of Zn^{2+} the leached quantity was two to ten times higher than the amount on Zn in the needles.

At the end of the experimental period needles were analyzed for the various nutrients. These data are shown in Tab. I.

Tab. I Element concentration in the needles at the end of the experimental period as related to the pH of the mist. K_o = without K fertilizer, K_1 = with K fertilizer. Asterisks between the columns denote significant differences; ** at the 1%, *** at the o.1% level.

	Youngest shoot pH 2.75		Youngest shoot pH 5.oo	Older shoots pH 2.75	Older shoots pH 5.oo
Potassium, K_o	2.96	***	2.2o	2.46	2.18
mg g^{-1} DM, K_1	6.49	***	5.79	4.88	4.66
Calcium, K_o	3.45		3.16	1o.8	1o.4
mg g^{-1} DM, K_1	4.15	***	3.24	1o.7	1o.8
Magnesium, K_o	1.6o		1.36	o.81	o.87
mg g^{-1} DM, K_1	1.75		1.7o	o.88	o.9o
Manganese, K_o	379		387	612	568
/ug g^{-1} DM, K_1	416		424	612	599
Zinc, K_o	14.7	**	9.1	34.3	4o.6
/ug g^{-1} DM, K_1	14.3		12.8	39.5	39.o

Leaching by acidic mist had not resulted in a decrease of the element in question. Even not for Zn^{2+}, although this element had been leached in amounts which were several times higher than the amounts present in the needles. This finding shows that the elements leached by the mist were easily replenished. It is particularly surprising that the exposure to the acidic mist led to a highly significant increase in the K^+ concentration in the needles of the youngest shoot. This was the case for both K^+ treatments and was found in the 1986 experiments (shown here) and in the 1985 experiments (not shown here). It is noteworthy that the K^+ concentrations in the needles of the K_o treatment were very low so that K^+ deficiency symptoms should have appeared.

4. CARBOHYDRATE STATUS

Fig. 1 shows the amounts of carbohydrates leached by the mist. The first sample taken, one week after the beginning of the exposure to the mist does not show any difference between treatments, but later a very clear and significant differentiation was observed. The curve for the acidic mist is characterized by a steep linear slope providing evidence that almost equal amounts of carbohydrates (sucrose + glucose + fructose) were leached from one sample date to the other. In the control treatment (pH 5.oo) these increments were very low. Only at the 1. leaching date (left point in Fig. 1) a high leaching rate was obtained. It is speculated whether the carbohydrates leached at the 1. date mainly originated from surface carbohydrates and not so much from carbohydrates synthetized in the needles during the experimental period.

Starch concentration found in the needles of the youngest shoot was twice as high in the control treatment as compared with the acidic mist treatment (see Tab. II).

Tab. II Starch concentration in the needles of the youngest shoot at the end of the experimental period as related to the mist treatment. K_o = without K fertilizer, K_1 = with K fertilizer; *** significant difference at the o.1% level. mg starch g^{-1} DM.

	pH 2.75	pH 5.oo
K_o	8.52	19.7***
K_1	8.18	17.o***

$^{14}CO_2$ assimilation experiments carried out at the end of the experimental period with three trees per treatment yielded no major difference in the CO_2 assimilation capacity.

5. SYMPTOMS

Needle shedding was much higher in the treatment with acidic mist as compared with the control treatment. This trend was the same in the K_0 and in the K_1 treatment as can be seen from Fig. 2. In 1985 five weeks and in 1986 two weeks after beginning with the mist treatment first visible symptoms occurred at the needles of the youngest shoot. The needles turned reddish brown and were partially shed. This symptom was found with all trees (14 per treatment) exposed to acidic mist. The symptom is very similar if not identical with that found at the needles of damaged spruce trees growing in the forest.

Brown needles from our 1986 experiment were analyzed for fungi attack. No fungi, also no Lophodermium was found.

Scanning electromicroscopic photos carried out by Dr. S. Fink, Freiburg revealed that the acidic fog had degradated the cuticle of the still green needles of the youngest shoot to a considerable extent. This is shown in photo 1 (1985 experiment). In 1986 an analogous finding was obtained as can be seen from photo 2 in which the stomatal props are shown. The upper part shows the treatment with acidic mist the lower the control treatment. It is evident that the stomatal props exposed to the acidic mist show symptoms of dissolution.

6. DISCUSSION

Our experimental results are in agreement with those of other authors (5, 6) who also found high leaching rates of various cation species when acidic mist was applied. Whether such leachings may lead to nutrient deficiency, however, is doubted, since the lost nutrients are quickly replenished. Krivan et al. (7) analyzed numerous needle samples from various forests of Southern Germany with damaged and healthy spruce trees. They could not find a relationship between the needles of damaged trees and the cation concentration in the needles.

From Fig. 1 it is clear that in the control treatment there was hardly any leaching of carbohydrates except for the 1. leaching date at which presumably mainly superficially adhering carbohydrates were leached. The high carbohydrate leaching rates obtained with acidic mist at the later dates are a clear indication that the mist increased the permeability of the needle tissue. It remains to be shown whether this increase in permeability also involves the plasmalemma or whether it is mainly restricted to the apoplast and the cuticula.

The remarkable impact of the acidic mist on the starch content indicates that the carbohydrate metabolism was affected, probably not by a decrease in CO_2 assimilation but rather by carbohydrate leaching and presumably by higher respiration rates.

The damage in the wax layer and in the stomatal props as shown in photo 1 and 2 equals the damage recently found in spruce needles which were exposed to the natural stress in the forest (8). Magel and Ziegler (9) reported, that young spruce trees exposed for four months to the natural stress of an area known for forest decline (Dreisessel, Bayer. Wald) showed melted wax props.

We assume that the entry of H^+ into the needle tissue

affects the synthesis of the wax layer and of the stomatal wax props. Since the wax is a protection against fungi attack and water loss needles with a degraded wax layer may be more prone to fungi infection and water stress. The increased leaching of carbohydrates from needles exposed to acidic mist (see Fig. 1) may be due to a more permeable cuticle. Such degraded cuticles should impair the water retention and make the trees more susceptible to water stress. This assumption is in good agreement with the observation that forest decline in particular increases during dry years. The reddish brown colour of needles observed by us in the treatment with acidic mist is considered as a consequence of drying out of the needles which is followed by needle shedding. This pattern of decline is in accordance with the observations made at spruce trees grown under natural conditions.

REFERENCES

(1) WALDMANN, G. (1985). Zur Anreicherung von Säuren im Baumkronenbereich. Allg. Forst- u. J.-Ztg. 156, 2o4-21o.

(2) MAIER, G. (1984). Aufnahme und Auswertung von Fichtenbeobachtungsflächen in Bayern. Schriftenreihe der Forstw. Fak. München u. der Bayr. Forstl. Forschungsberichte 62.

(3) MUNGER, J.W., JAKOB, D.J., WALDMANN, I.M and HOFFMANN, M.P. (1983). Fog water chemistry in an urban atmosphere. J. Geophy. 88, 51o9.

(4) MENGEL, K., LUTZ, H.J. and BREININGER, M.Th. (1987). Auswaschung von Nährstoffen durch sauren Nebel aus jungen intakten Fichten (Picea abies). Z. Pflanzenernähr. Bodenk. (in press)

(5) SCHERBATSKOY, T. and KLEIN, R.M. (1983). Response of spruce and birch foliage to leaching by acidic mists. J. Environ. Qual. 12, 189-195.

(6) SKEFFINGTON, R.A. and ROBERTS, T.M. (1985). The effects of ozone and acid mist on Scots pine saplings. Oecologia 65, 2o1-3o6.

(7) KRIVAN, V., LÜTTGE, U. and SCHALDACH, G. (1986). Profile von Makro- und Mikromineralnährstoffen in gesunden und kranken Fichten (Picea abies (L.) Karst.) auf verschiedenen Standorten in Süddeutschland. Angew. Bot. 6o, 373-389.

(8) LAITAT, E. (1986). Defoliation in Norway spruce and resistance to frost and drought. Workshop on direct effects of dry and wet deposition on forest ecosystems. Kungälv, Sweden, October.

(9) MAGEL, E. and ZIEGLER, H. (1986). Einfluß von Ozon und saurem Nebel auf die Struktur der stomatären Wachspfropfen in den Nadeln von Picea abies (L.) Karst. Forstw. Cbl. 1o5, 234-238.

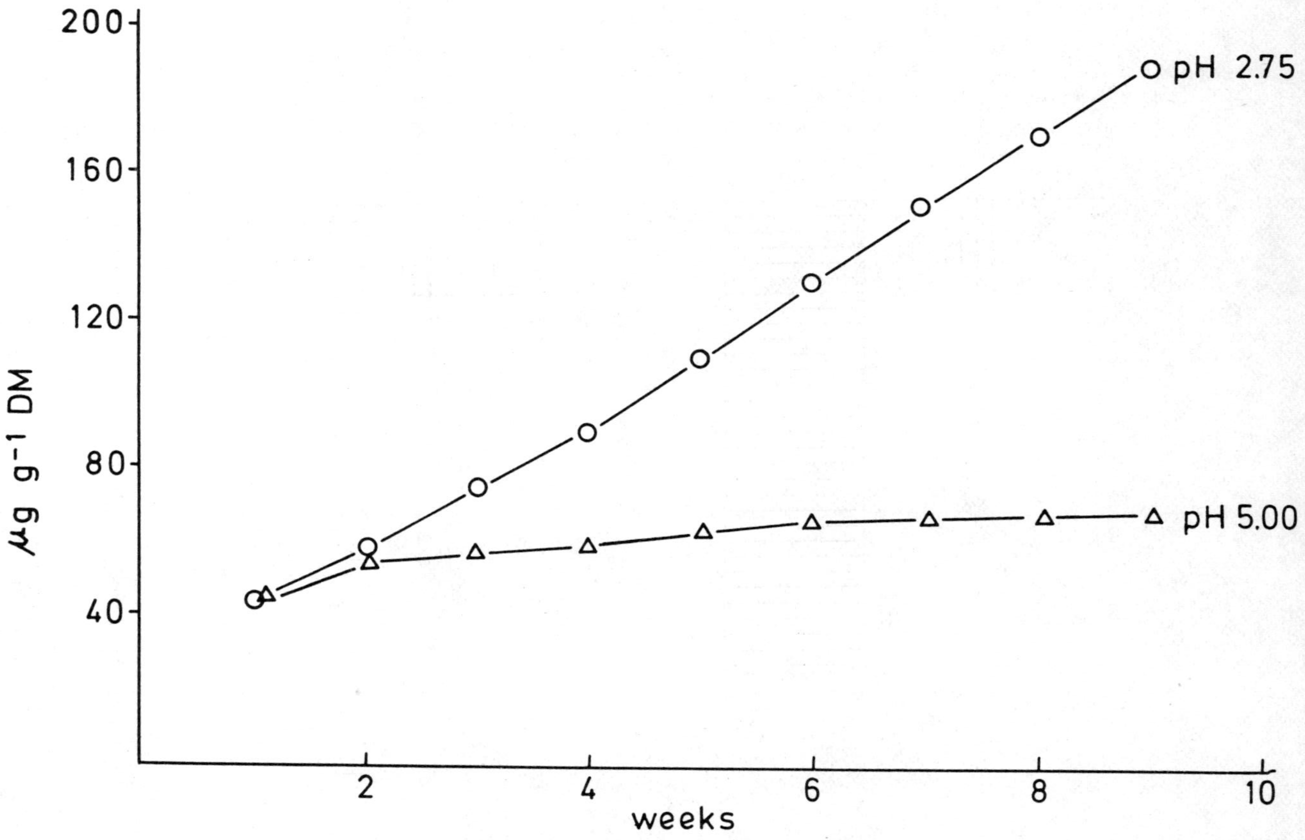

Fig. 1 : Cumulative leaching of carbohydrates (glucose + fructose + sucrose) out of Picea abies as related to mist pH (pH 2.75 and pH 5.oo).

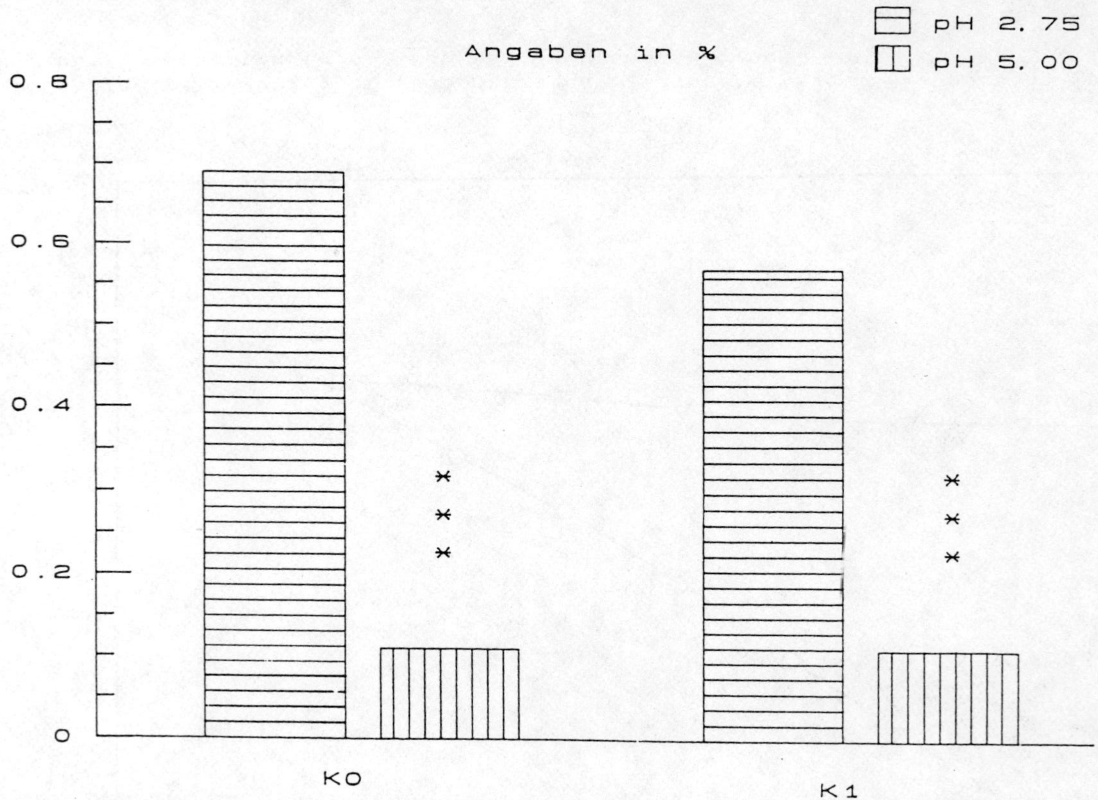

Fig. 2 : Needle shedding as related to the pH of the applied mist (pH 5.00 and pH 2.75) *** on columns denotes a significant difference at the 0.1 % level between the pH treatment. K_o = without K^+ fertilizer application, K_1 = with K^+ fertilizer application. Needle shedding in % of the total needles of the tree (= 100 %).

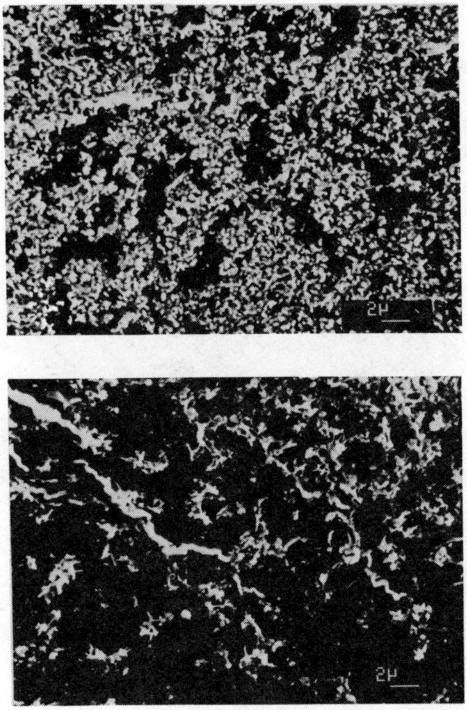

Photo 1 Scanning electromicrograph of the cuticle of a green needle from the youngest shoot. Upper part with control mist (pH 5.oo), lower part with acidic mist (pH 2.75) (Micrograph: S. Fink, Freiburg).

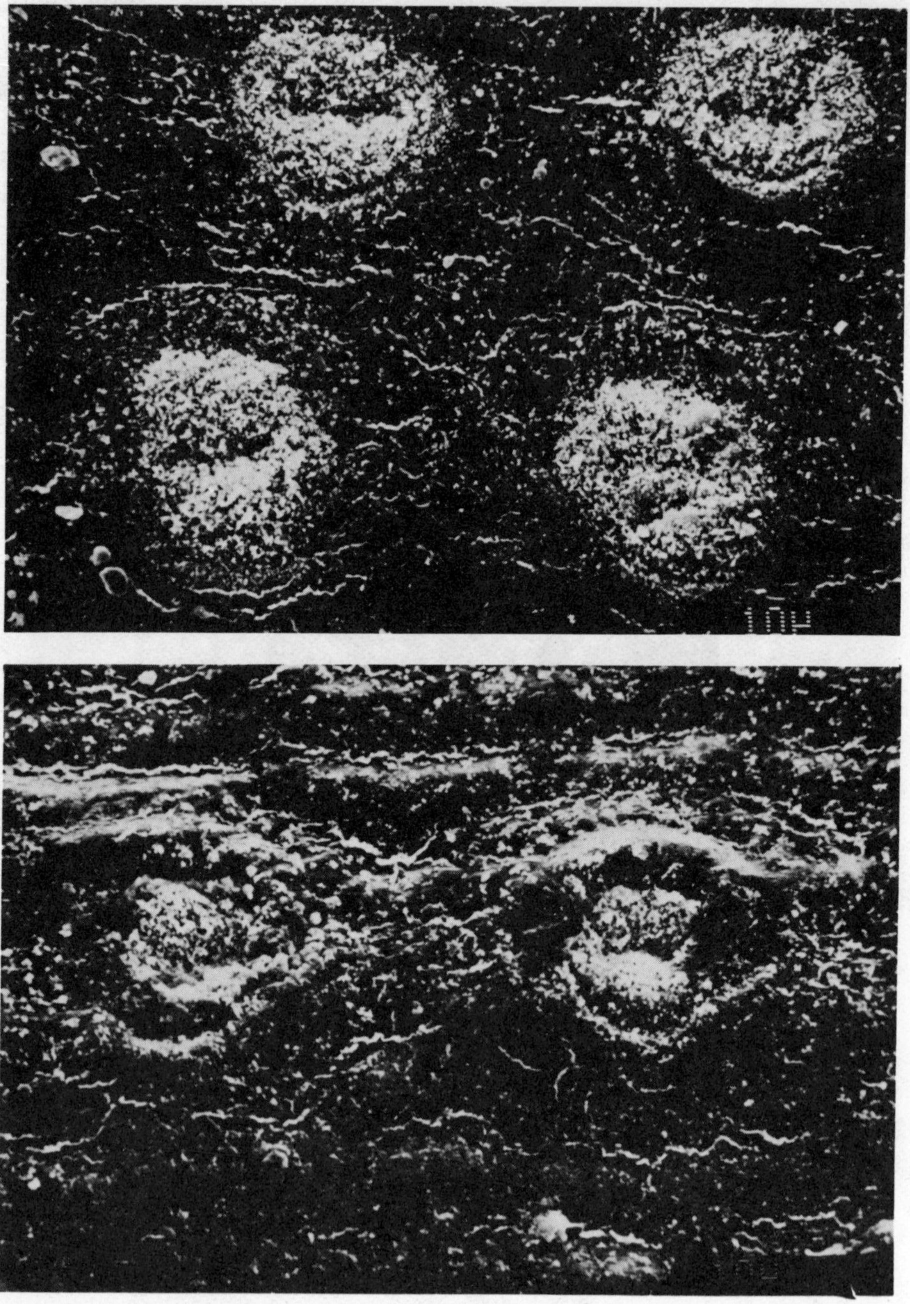

Photo 2 Scanning electromicrograph of the stomatal props. Upper part control mist (pH 5.oo), lower part acidic mist (pH 2.75).
(Micrograph: S. Fink, Freiburg).

SESSION III

EFFECTS OF AIR POLLUTION ON FOREST TREES AND FOREST ECOSYSTEMS

SESSION III

EFFETS DE LA POLLUTION DE L'AIR SUR LES ARBRES ET LES ECOSYSTEMES FORESTIERS

SITZUNG III

WIRKUNGEN DER LUFTVERSCHMUTZUNG AUF WALDBÄUME UND WALDÖKOSYSTEME

PART III — IIIème PARTIE — TEIL III

Chairman : Prof. R. IMPENS, Faculté des Sciences Agronomiques de l'Etat, Gembloux, Belgium
Rapporteur : Dr. G. SELLDEN, University of Göteborg, Sweden

REPORT ON THE SESSION

Sellden, G and Wallin, G

Institute of Botany, Department of Plant Physiology, University of Göteborg, Carl Skottsbergs gata 22, S-413 19 Göteborg, Sweden

In the following, points made in the three papers presented in session III, part III are discussed in relation to the objectives of the symposium. The papers are referred to by roman numbers.

One objective of the symposium was to identify, as far as feasible, critical "levels" and "loads" of air pollutants below which undesirable effects are unlikely to occur. To do this test systems have to be employed since information from the field is difficult to interpret. Test systems are defined by a set of properties, such as species, number of trophic levels, mode of exposure, dose and influence of other stress factors. As testsystems are used to predict effects of pollutants in ecosystems, the representativity of the test system will be of importance for the predictive capacity of the test.

A central problem in toxicology is the extrapolation from laboratory tests to field situations. Environmental standards today are mostly based on single species toxicity tests carried out in the laboratory. Those few studies made to validate results obtained with single species tests show that it is difficult, if not impossible, to extrapolate from one level of biological organisation to another (Kimball & Levin 1985, Cairns 1984, Cairns 1986). As Thornley (1980) points out, information on one level can be used to explain functions on a higher level. Information from the lower level can however not be used to predict functions on the higher level. While a working system requires intact subsystems, the opposite is not necessarily true.

Awareness of the limitations of single species test to predict effects in ecosystems has led to the employment of more complex testsystems, ie the ecological realism of the test systems has been increased. By using multispecies test systems, systems with several species at different trophic levels, it has been possible to study effects on interactions between species (Blank 1983, van Voris et al 1985, Cairns 1986). By increasing the ecological realism it is therefore possible to obtain further information about effects of pollutants. However, results from multispecies test systems have seldom been validated why it is not possible to judge if interactions occuring in test situations are representative of interactions occuring in ecosystems (Smies 1983, Tebo 1985). Thus, in principle, most multispecies testsystems suffer from the same limitations as single species tests.

As mentioned earlier test systems are defined by a set of properties and that the predictive capacity of the test system depends on these properties. Different issues will pose different demands on the test systems. This will influence the design of the system. One property that will bias the results is the pollution climate.

If the aim of the testing is to determine effects of single pollutants on plants qualitatively a certain amount of information can be obtained without knowledge about interactions with other pollutants and the internal dose. If, however, predictions of effects on ecosystems are to be attempted knowledge about effects of realistic internal doses are essential.

To obtain exposure conditions which are representative of ambient conditions several aspects have to be taken into account. Different technical solutions to this problem exists and the following four different designs of fumigation systems are rewied in papers I,II and III:

- o closed indoor chambers
- o closed outdoor chambers
- o open-top chambers (OTC)
- o open-field fumigation systems

Compared with field situations, the facilities needed to manipulate the fumigation lead to changes of the environmental conditions. All chambers affect the climate regardless of design (I,II,III). The temperature is increased and the movement of air is changed, the latter becoming a function of ventilation geometry and spacing of plant material. If fans are run continously dew formation is inhibited. Rain interception is changed by the presence of frustums in open-top chambers and totally abolished in closed chambers. Depending on the quality of wall material, solar radiation intensity and spectrum are also changed.

To simulate realistic pollution climates the following aspects have to be considered:

- o concentration frequency distribution
- o vertical distribution
- o composition
- o deposition

Plants in the field are never exposed to constant amounts of pollutants since concentrations constantly fluctuate. This variability can be expressed as a concentration frequency distribution (III). The fluctuations have to be assesed as they affect the internal dose and thus the response of the plant. To a large extent, pollutants enter plants through stomata. Thus a larger total upt ke can be expected when high stomatal conductances coincide with peaks of pollutant concentrations.

As the distribution frequency will affect the internal dose it is desirable that the distribution frequency of the added pollutant has similar characteristics as that of the ambient pollutant (III). From paper III, fig 2, it is clear that the only way to increase the concentration of a pollutant and at the same time simulate the distribution frequency of the ambient situation is to increase the ambient concentration by a constant factor.

Plants in the field are furthermore exposed to multitude of pollutants, a large number of which are unknown. One major drawback with laboratory test systems is therefore the difficulty in simulating realistic pollution climates. This particular problem is avoided by the use of outdoor test systems where ambient air can be used.

In the field the deposition of pollutants is resticted by a number of resistances (II, fig 1). The within-canopy resistance is determined by crop density, morphology and wind speed. Transfer of pollutants to leaf surfaces is determined by the boundry layer resistance which is determined by morphology and turbulence. The deposition of pollutants and thus the dose will to a large extent depend on these resistances. In chambers the high air flow, needed to keep temperatures down to acceptable levels, lowers these resistances substantially causing corresponding changes of the pollution uptake. The change of resistances may also lead to an altered assimailate production. Open-field fumigation systems are designed to leave environmental factors, such as within-canopy resistances, unchanged. However, even in this fumigation system the release geometry of the dispersion system will alter the exposure conditions somewhat compared with the field situation (III).

All changes of environmental conditions will lead to changes in the growth of plants. However a major gap in present knowledge is that it is not known how the response of a plant to a pollutant is influenced by different environmental conditions. This leads to difficulties in comparing results obtained by research groups using different fumigation systems. It should however be possible to assess the effect of climatic conditions on the sensitivity of plants to pollutants if the climate of the different systems presently in use could be accuratley determined. It should however be remebered that as the plants grow the turbulence within the chamber, and thus the microclimate, will change why climatic conditions have to be assessed repeatedly.

Conclusions

The first objective of the symposium was "to rewiev the scientific knowledge, identify major gaps and recommend how these can be filled , in order to improve the understanding of ecophysiological processes which govern the response of ecosystems to disturbances".

According to Peters (1980) ecology is defined as "that branch of science which predicts the distribution of biomass, production and kinds of organisms". The science of ecotoxicology is defined by adding in polluted environments. The problem with ecotoxicological research today is however that there exists no unfying theory on which to base predictions. The root of this problem lies in the inability of ecologists to define ecosystem function, ie to name critical variables and how these should be measured. One reason for this inability may be that ecology today is based too much on untestable concepts than on testable theories (Peters 1980). The increasing employment of test systems with a high ecological realism can be seen as a function of the awareness of the inability of single species test to predict effects on ecosystems but also as a function of the lack of theory on which to base predictions. Blanck (1983) states that "aquatic ecotoxicology is characterized by a lack of theory to predict ecosystem events, and by a multitude of test systems, concepts and protocols to generate estiamtes and statements on environmental hazard". In terrestial ecotoxicology the situation is similar.

The second objective of the symposium was to identify critical levels and loads. It will never be possible to fullfill neither the first objective nor the second unless testsystems which allow predictions are developed. However, this is certainly easier said than done.

A short-time aim would be to facilitate comparisons of results. To be able to do this it is necessary that a consensus on common research protocols is reached, especially regarding on how to describe doses.

A more long-term aim would be to construct test systems that allow predictions. One way to proceed would be to start validating results obtained with present test system and to determine the predictive capacities of those systems.

Papers presented in session III, part III

I. Jäger,H.J. Experiments with open-top chambers; results, advantages and limitations.

II. Roberts,T.M., Brown,K.A. & Blank,L.W. Methodological aspects of the fumigation of forest tees with gaseous pollutants using closed chambers.

III. Colls,J.J. & Baker,C.K. The methology of open-field fumigation.

References

Blanck, H. (1983) On the impact of long-chained aliphatic amines on photosynthesis and algal growth in ecotoxicological test systems. Thesis, Univ of Göteborg, Sweden.

Blanck, H (1984) Species dependent variation among aquatic organisms in their sensitivity to chemicals. Ecol. Bull. 36:107 119.

Cairns, Jr. J. (1984) Are single species toxicity tests alone adequate for estimating environmental hazard. Environmental Monitoring and Assessment 4:259-273.

Cairns, Jr. J. (1986) Multispecies toxicity testing: a new information base for hazard evaluation. Curr. Practices in Environ. Sc. & Engng. 2:3649.

Kimball, K.D. & Levin S.A. (1985) Limitations of laboratory bioassays: the need for ecosystem-level testing. BioScience 35:165-171.

Peters, R.H. (1980) Useful concepts for predictive ecology. Synthese 43:257-269.

Smies, M. (1982) On the relevance of microecosystems for risk assessment: some considerations for environmental toxicology. Ecotoxicology and Environmental Safety 7:355-365.

Tebo, Jr L.B. (1985) Technical considerations related to tGhe regulatory use of multispecies toxicity tests. In Multispecies toxicity testing. Ed. Cairns Jr, J. Pergamon Press.

Thornley, J.H.M. (1980) Research strategy in plant sciences. Plant, Cell and Environment 3:233-236

Voris van, P., Tolle, D.A., Arthur, M.F. & Chesson J. (1985) Terrestial Microcosms: applications, validation and cost-benefit analysis. In Multispecies toxicity testing. Ed. Cairns Jr, J. Pergamon Press.

METHODOLOGICAL APPROACHES : PART I : EXPERIMENTS WITH OPEN-TOP CHAMBERS: RESULTS, ADVANTAGES AND LIMITATIONS

H.-J. Jäger*, H.J. Weigel*, R. Guderian**, U. Arndt*** and G. Seufert***

*Institut für Produktions- und Ökotoxikologie, Bundesforschungsanstalt für Landwirtschaft, Bundesallee 50, D-3300 Braunschweig, **Institut für Angewandte Botanik, Universität Essen, Universitätsstraße 1, D-4300 Essen 1
***Institut für Landeskultur und Pflanzenökologie, Universität Hohenheim, Postfach 700562, D-7000 Stuttgart 70

Summary

To answer the question how air pollutants affect plants and what are the causes of the effects and the quantitative relationships between exposure and response, different approaches, i.e. different experimental methods are required. In experimental studies the influence of the experimental apparatus on the plants reaction to air pollutants has to be minimized. With this respect open-top chambers which were developed in the early 1970's in the U.S.A. are an experimental method which has significant advantages, e.g. over closed chambers and open-atmosphere approaches. This presentation summarizes the fundamental designs of various types of open-top chambers and the modifications introduced to improve the chambers performance. There are two principle ways to use open-top chambers for air pollution effect studies: i) by comparing plants in chambers supplied with filtered and non-filtered ambient air it is possible to judge, if ambient loads of pollutants affect plant growth and ii) by injecting pollutants into the chambers it is possible to examine cause-effect and dose-response relationships. The chambers can be placed over plants grown in the soil or filled with potted plants. Depending on the chamber design used small differences in microclimatic parameters (light quality and quantity, temperature, relative humidity) and the growth performance of plants in chambers and plots without chambers have been observed. These differences are described and critically evaluated.

1. EINLEITUNG

Die Auswahl von Methoden für eine bestimmte Untersuchung von Immissionswirkungen auf Pflanzen hängt primär von der Versuchsfrage ab, sekundär von wirkungsbestimmenden und objektabhängigen Parametern. Im vorliegenden Zusammenhang erscheint eine Aufteilung der gesamten Immissionswirkungen auf Pflanzen und terrestrische Ökosysteme auf die folgenden Fragebereiche vorteilhaft:

- Ermittlung der Wirkungsweise von Einzelkomponenten und Immissionsgemischen
- Ermittlung der Schädigung nach Art, Intensität und räumlicher Ausdehnung
- Ermittlung der Schädigungsursache
- Ermittlung von quantitativen Zusammenhängen zwischen Immissionen und Wirkung / Dosis-Wirkungsbeziehungen
- Ermittlung von Grundlagen für Abhilfemaßnahmen am Pflanzenstandort.

Jeder dieser Fragenkomplexe besteht aus einer Vielzahl von Detailfragen. Zu ihrer Beantwortung wird ein differenziertes Instrumentarium in Form unterschiedlicher Versuchsanordnungen benötigt, auch deshalb, weil Boden und Klima als äußere sowie Alter und Entwicklungsstadium der Pflanze als innere Faktoren die Wirksamkeit einer gegebenen Immissionsbelastung stark variieren.

Besondere Beachtung beanspruchen bei der Auswahl der Expositionssysteme die Immissionsverhältnisse. Die gegenwärtige Immissionssituation in Mittel- und Westeuropa ist gekennzeichnet durch das Vorherrschen komplexer Immissionstypen, relativ niedriger Immissionskonzentrationen und großräumiger Belastungen. Es besteht somit die Notwendigkeit, Schadstoffangebot und Schadstoffwirkung mit verschiedenen Kriterien zu bewerten und der Zwang, zur Ermittlung von Immissionswirkungen verschiedenartige Expositionssysteme zu verwenden.

Aus der Sicht des praktischen Immissionsschutzes geht es letzten Endes darum, immissionsbedingte Abweichungen auf den einzelnen Organisationsstufen im Ökosystem von der "Norm", d.h. von der unbelasteten Kontrolle zu ermitteln. Als "Norm" kann nur die Pflanze bzw. die Pflanzengemeinschaft gelten, wie wir sie in den agrarischen, forstlichen und natürlichen Ökosystemen vorfinden. Hiernach wären also Versuchsbedingungen zu wählen, mit deren Hilfe sich die Immissionswirkungen an Pflanzen unmittelbar in ihren jeweiligen Biotopen ermitteln lassen.

Mit den Anfang der siebziger Jahre synchron im Boyce Thompson Institute for Plant Research in Yorkers, N.Y. (1) sowie im U.S. Department of Agriculture / U.S. Environmental Protection Agency (2) entwickelten open-top Kammern steht eine Methode zur Verfügung, die wesentliche Vorzüge von Expositionen in geschlossenen Kammern einerseits und in der freien Atmosphäre andererseits in sich vereinigt. Seitdem sind verschiedene Fortentwicklungen und Veränderungen des ursprünglich für landwirtschaftliche Kulturpflanzen entwickelten Kammertyps erfolgt, so daß heute Kammern zur Verfügung stehen, die auch für Untersuchungen an älteren Bäumen geeignet sind (Abb.1). In der Regel wurde das Kammerdesign von den einzelnen Forschergruppen den jeweiligen Fragestellungen angepaßt. Die Ausführungen der Kammern reichen daher von einfachen, tragbaren Systemen bis hin zu festinstallierten Versuchsanlagen (Abb.2). Zu den essentiellen Bausteinen einer open-top Kammer gehören jedoch nach wie vor folgende Komponenten: das Kammergestell oder -gerüst; eine möglichst lichtdurchlässige Kammerbespannung; ein Gebläse zur Ventilation der Kammer; Staub-, Partikel- und Aktivkohlefilter zur Reduzierung der Schadstoffkonzentration der Umgebungsluft und Vorrichtungen zur Verteilung und Vermischung der in die Kammer eingeblasenen Zuluft. Die Möglichkeiten und Grenzen dieses Expositionssystems sollen nachfolgend aufgezeigt und bewertet werden.

2. BAU UND AUSFÜHRUNGEN

Während es sich bei der Mehrzahl der bisher für Versuchszwecke eingesetzten open-top Kammern um zylindrische Konstruktionen wie beim ursprünglichen Kammertyp handelt (1,2), sind auch hexagonale (3), oktagonale (4,5,6) und quadratische bzw. rechteckige Ausführungen in Gebrauch (7,8). Das gemeinsame Merkmal all dieser Kammern ist das oben offene Ende, durch das die in den unteren Kammerteil eingeblasene Luft entweichen kann. Im Hinblick auf den Einsatz von open-top Kammern für Untersuchungen an Bäumen sind die Kammerausmaße z.T. beträchtlich vergrößert worden. Während die Dimensionen kleiner Ausführungen bei ca. 1,30 m^2 Nutzfläche bzw. 2,85 m^3 Volumen liegen (9,10), sind für Untersuchungen an Wald- oder Obstbäumen größere Kammern eingesetzt worden, deren Ausmaße bis ca. 19,5 m^2 bzw. 75 m^3 reichen (8,11,12).

Abb. 1
Schematische Darstellung einer open-top Kammer der Universität Hohenheim aus dem Schwäbisch-Fränkischen Wald in Baden-Württemberg / BRD

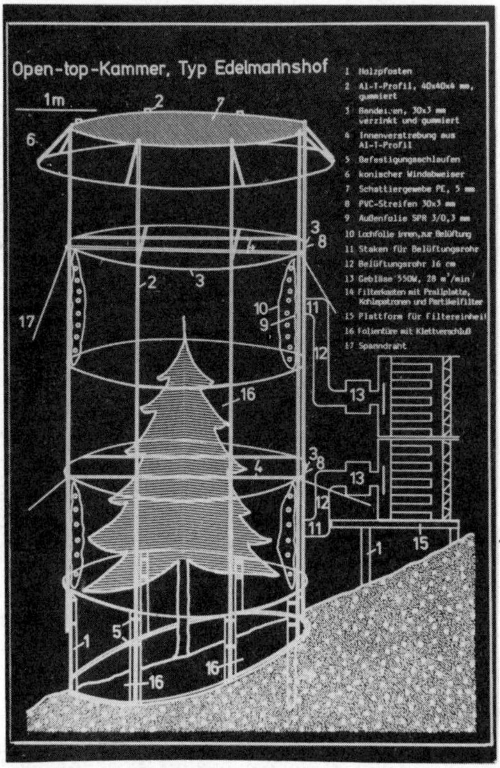

Die Dimensionen des am weitesten verbreiteten Kammertyps nach Heagle et al. (2) und Mandl et al. (1) liegen zwischen 7-8 m^2 bzw. 17-18 m^3. Da es besonders bei hohen Windgeschwindigkeiten in der Außenluft zu einem Fremdlufteinfall in die Kammern kommen kann, wurden als Neuerungen kegelförmige Verengungen des oberen, offenen Kammerrandes (frustum, cone) oder ins Kammerinnere ragende Ränder (baffle, lip) eingeführt (3,8,13,14,15,17). Diese Vorrichtungen sorgen außerdem durch verstärkte Wirbelbildung für eine bessere Durchmischung der Kammerluft, verhindern jedoch z.T. das Eindringen des natürlichen Niederschlages (s.u.). Um bei Untersuchungen zur Wirkung nasser Depositionen (Regen, Nebel) den Eintritt des Umgebungsniederschlages zu verhindern, können die Kammern ca. 30-50 cm oberhalb des oberen, offenen Randes mit einer Dachkonstruktion versehen werden (semi-open-top chamber), die aus dem gleichen Material wie die Kammerbespannung besteht (13,16,17,19). Als Material für das Kammergerüst findet wegen seines geringen Gewichtes und seiner hohen Korrosionsbeständigkeit meist Aluminium (1,2,3,8,13,18), seltener galvanisiertes Stahlrohr Verwendung (11,15). Das Wandmaterial für die Kammerbespannung kann aus Polyvinylchlorid-, Teflon- oder Polyäthylenfolien aus Hart-PVC (Novolux[R], Corulux[R]), Polycarbonat (Lexan[R]), Fiberglas oder Glas bestehen.

Abb. 2
Gesamtansicht der open-top Begasungsanlage des Instituts für Produktions- und Ökotoxikologie der Bundesforschungsanstalt für Landwirtschaft in Braunschweig / BRD

Sämtliche der aufgeführten Materialien verändern zu einem gewissen Teil den Lichteinfall in die Kammern (s.u.)

3. LUFTFÜHRUNG/FILTERUNG

Die Luftzufuhr in die open-top Kammern erfolgt je nach Kammertyp in der Weise, daß die Pflanzenbestände entweder von der Seite und/oder vom Boden her mit Zuluft versorgt werden. In einigen Untersuchungen sind jedoch Konstruktionen gewählt worden, die einen von oben nach unten gerichteten Luftstrom erzeugen (9,10). Die für Freilandverhältnisse eher atypische Luftführung von der Seite wird beim zylindrischen Kammertyp nach Heagle et al. (2) und Mandl et al. (1) durch den unteren, doppellagigen Teil der Kammerbespannung erreicht, der einen die Kammer ringförmig umlaufenden, an der Innenseite gelochten Luftkanal bildet. Während junge Bestände daher mehr von seitwärts-oben beströmt werden, ist bei älteren Pflanzenbeständen die Luftzufuhr eher von seitwärts-unten gegeben. Alternativen, die der sich verändernden Höhe eines Pflanzenbestandes Rechnung tragen, sind gelochte Rohrleitungen, die die Kammer ringförmig umlaufen und in der Höhe verstellbar sind (10,21). Daneben sind Ausführungen bekannt, bei denen die Luft durch ringförmig am Boden (5,15) oder parallel zwischen den Pflanzenreihen verlaufenden (3), gelochten Röhren in die Kammer eingeblasen wird, so daß die Luft die Bestände von unten nach oben durchströmt. Die auf diese Weise in die Kammer eingeblasene Zuluft entweicht durch den oberen, offenen Teil. Diese Art der Luftführung durch die Kammer ist eine Grundvoraussetzung dafür, daß ungereinigte Fremdluft nur in geringem Ausmaß eindringen kann. Die Gebläseleistung bzw. der Luftwechsel pro Zeiteinheit durch die Kammer bestimmen neben der

Windgeschwindigkeit der Außenluft und den oben erwähnten baulichen Vorkehrungen (frustum, lip) das Ausmaß des Fremdlufteintrittes. Als Gebläse werden elektrisch angetriebene Radial- oder Axialgebläse unterschiedlicher, an die Kammerdimensionen angepaßter Leistung eingesetzt. Der Luftwechsel pro Minute, der über die Windgeschwindigkeit in der Kammer den aerodynamischen Widerstand oder Grenzflächenwiderstand an der Nadel- oder Blattoberfläche mitbestimmt, schwankt in weiten Bereichen und liegt zwischen Extremwerten von 1,3-fach bis 10-fach (1,2,3,10,13). Beim zylindrischen Kammertyp nach Heagle et al. (2) und Mandl et al. (1) liegen die Luftwechsel pro Minute in der Regel zwischen 2,0-fach (1,17,18) bis 4,0-fach (2,5,8,21), was einer Gebläseleistung von ca. 2000 - 4500 m^3/h entspricht. Die Windgeschwindigkeit innerhalb der Kammern ist im allgemeinen in horizontaler Richtung relativ gleichmäßig, während sie zum oberen, offenen Ende hin leicht abnimmt. Die Häufigkeit des Luftwechsels durch die Kammer beeinflußt zudem die Lufttemperatur in der Kammer, da hohe Luftwechselraten eine übermäßige Erwärmung verhindern.

Die Filterung der Umgebungsluft zur Reduktion atmosphärischer Schadstoffkonzentrationen geschieht bei allen Systemen mit Hilfe von Aktivkohle und der Kohle vorgeschalteter Staub- und Partikelfilter. Während unbehandelte Aktivkohle für Ozon ein effektives Filtermaterial darstellt, muß zur Erzielung hoher Filtereffizienzen für saure Schadgase wie Schwefeldioxid, Stickstoffdioxid oder Fluorwasserstoff imprägnierte, d.h. mit chemischen Methoden behandelte Aktivkohle eingesetzt werden. Als Imprägnierungsmittel finden alkalisch reagierende Verbindungen, z.B. Kaliumhydroxid oder Kaliumcarbonat Verwendung (17,18). Die Kohle wird dabei in Form von Patronen (15,18), Platten (2,3,17) oder auch als loses Schüttgut eingesetzt.

Während die Filtereffizienz der Kohle selbst durch Faktoren wie Kontaktzeit mit der ungereinigten Umgebungsluft und Verschmutzungsgrad (= Standzeit) bestimmt wird, ist die Filtereffizienz oder Ausschlußeffizienz (exclusion efficiency) einer kompletten open-top Kammer z.B. bei Betrieb mit gefilterter oder ungefilterter Umgebungsluft zusätzlich davon abhängig, wie hoch der Anteil des Fremdlufteinfalls in die Kammer ist. Infolgedessen sind die in der Literatur gemachten Angaben zur Filter- bzw. Ausschlußeffizienz von open-top Kammern relativ uneinheitlich und können nur vor dem Hintergrund vergleichbarer Windgeschwindigkeiten interpretiert werden. So berichten einige Autoren über durchschnittliche Ausschlußeffizienzen für Schwefeldioxid, Stickstoffdioxid und Ozon von 50%-60% (3,9,18), während andere vor allem für Ozon Werte zwischen 70%-90% fanden (1,2,15,21, 22). Allgemein nimmt die Ausschlußeffizienz mit steigender Windgeschwindigkeit ab. Da die Filtereffizienz selbst von imprägnierter Aktivkohle für Stickstoffmonixid sehr gering ist (< 10 %), kann dieses Gas mit den oben beschriebenen Methoden nicht aus der Atmosphäre entfernt werden. Gute Filtereigenschaften für Stickstoffmonoxid besitzt hingegen ein mit Kaliumpermanganat imprägniertes Aluminiumoxid (Purafil^R), das in einer chemischen Reaktion Stickstoffmonoxid zu Stickstoffdioxid aufoxidiert, welches danach mit Hilfe alkalisch imprägnierter Aktivkohle entfernt werden kann.

4. KLIMA

Open-top Kammern wurden als ein Expositionssystem entwickelt, das die für das pflanzliche Wachstum entscheidenden Umwelt- bzw. Klimabedingungen nur unwesentlich beeinflussen sollte. Die bisher beobachteten Abweichungen vom Freilandklima betreffen hauptsächlich die Faktoren Licht, Temperatur, relative Feuchte, Wind und Niederschlag. So ist durch Verwendung der oben beschriebenen Materialien zur Kammerbespannung der Anteil der photosynthetisch aktiven Strahlung, der in die Kammer gelangt, geringer als im

Freiland, wobei die Lichtverluste meist zwischen 15%-20% liegen (1,2,3,9, 11,13,23,24) und nur bei Verwendung des sehr teuren Materials Teflon geringer sein dürften. Der Lichtverlust ist dabei tages- und jahreszeitabhängig und z.B. als Folge des flacheren Sonnenstandes im Frühjahr und im Herbst größer als im Sommer (9,23,24,25). Da einige der Wandmaterialien die UV-Strahlung zurückhalten, muß auch mit Veränderungen der Lichtqualität in den Kammern gerechnet werden.

Die Auswertung der Temperaturverhältnisse in open-top Kammern der verschiedensten Ausführungen zeigt, daß die durchschnittliche Lufttemperatur in den Kammern ca. 0,5-2,0°C höher liegen kann als im Freiland (1,2,3,13,18,23,24,26). Messungen der Blatttemperaturen in den Kammern zeigen ebenfalls, daß die Werte dort ca. 0,5-2,0°C höher sein können als unter Freilandbedingungen (24). Vergleichsmessungen der relativen Luftfeuchte zwischen Kammer und Freiland ergaben, daß z.B. bei hohen Außentemperaturen die Feuchte ca. 5%-10% niedriger sein kann als außerhalb der Kammern (24). Auch über geringfügig höhere Feuchtewerte (ca. 5%) während der Nachtstunden wird berichtet (7). Häufig wurden jedoch keine Unterschiede zwischen Freiland und open-top Kammern festgestellt (3, 9,11). Die relative Feuchte in den Kammern ist zudem vom Transpirationsverhalten des jeweiligen Pflanzenbestandes abhängig. Bei Langzeituntersuchungen z.B. mit Bäumen besteht derzeit noch eine gewisse Unsicherheit, inwieweit Temperaturerhöhungen von durchschnittlich 1°-2°C in den Kammern langfristig den Metabolismus der Pflanzen und damit auch deren Reaktion auf Immissionseinflüsse verändern. Die geringen Temperaturabweichungen könnten vor allem in den kühleren Monaten des Jahres, z.B. während der Frosthärtung oder während der Knospenruhe von Bedeutung sein.

Auf die Windverhältnisse in open-top Kammern wurde z.T. schon bei der Besprechung der Luftführung eingegangen (s.o.). Ein grundsätzliches Problem bei allen Kammertypen ist die durch das Gebläse bedingte kontinuierliche und gleichmäßige Luftbewegung. In dieser Hinsicht weichen die Kammerbedingungen von den Verhältnissen im Freiland ab, die durch wechselnde Windrichtungen und -geschwindigkeiten gekennzeichnet sind. Bei kontinuierlichem Gebläsebetrieb führt die konstante Luftbewegung u.a. dazu, daß die nächtliche Taubildung in den Kammern unterbleibt (9). Welche Folgen die dadurch bedingte fehlende Benetzung der Blätter für die Aufnahme leicht wasserlöslicher Gase wie Schwefeldioxid hat, ist noch ungeklärt. Bei einigen Kammerausführungen muß zudem berücksichtigt werden, daß Pflanzen in unmittelbarer Nähe des Lufteintrittes stärker beströmt werden als der Rest eines Bestandes, d.h., daß es zu Randeffekten kommen kann.

Ein weiteres Problem stellt der in die Kammern gelangende Niederschlag in Form von Regen oder bei ganzjährigen Untersuchungen in Form von Schnee dar. Vor allem bei Verwendung der oben beschriebenen Vorrichtungen zur Reduktion des Fremdlufteinfalls (frustum, lip) kommt es zu ausgeprägten "Regenschatten" (rain shadows), d.h. zu ungleichmäßigen Niederschlagsverteilungen (1). Der natürliche Niederschlag kann daher sowohl bei Verwendung von Gefäßkulturen als auch bei Untersuchungen an im anstehenden Boden gewachsenen Pflanzen zur Wasserversorgung nicht ausreichen. Bei ganzjährigen Untersuchungen an Bäumen müssen die Folgen einer fehlenden oder ungenügenden Schneedecke, z.B. die Gefahr einer Frosttrocknis bei tiefen Bodentemperaturen und hoher Sonneneinstrahlung bedacht werden.

5. EINSATZMÖGLICHKEITEN/ANSÄTZE

Zur Untersuchung der Auswirkungen von Luftverunreinigungen auf Pflanzen gibt es zwei grundsätzlich verschiedene Wege, open-top Kammern einzusetzen:
a) Wachstumsvergleiche in gefilterter und ungefilterter Umgebungsluft;

b) Kontrollierte Zudosierung atmosphärischer Schadstoffe (trockene und nasse Deposition).

Der Vergleich der Auswirkungen gefilterter (CF=charcoal filtered) und ungefilterter (NF=non-filtered) Umgebungsluft ermöglicht Aussagen über die potentielle Belastung der pflanzlichen Leistungsfähigkeit durch die vorherrschende Immissionssituation als Folge der aktuellen Luftqualität am Ort der Untersuchung (9,18). Da nur eine Kontrolle und eine Behandlungsvariante untersucht werden, sind generelle Aussagen über Belastungs- und Wirkungszusammenhänge jedoch nur schwer möglich, d.h. Schlußfolgerungen über die Konsequenzen einer Verbesserung oder Verschlechterung der Luftqualität sind nur bedingt zu ziehen. Während dies vor allem für annuelle Pflanzen gilt, können bei Dauerkulturen (z.B. Koniferen) mehrjährige Wachstumsvergleiche in gefilterter und ungefilterter Luft dazu beitragen, eine Immissionswirkung zu belegen. Weiter verfolgt werden sollte der Versuchsansatz (34), mehrere Abstufungen der Filterung einzuführen, d.h. teilweise gefilterte Luft in die Kammern einzublasen. Dadurch sind möglicherweise schlüssigere Aussagen über Immissions- und Wirkungszusammen hänge zu erarbeiten. Im Rahmen des amerikanischen NCLAN-Programmes (National Crop Loss Assessment Network), bei dem open-top Kammern zur Ermittlung der Auswirkungen von Ozon auf landwirtschaftliche Kulturpflanzen eingesetzt wurden (27,28), wurde der ursprüngliche Ansatz,mit gefilterter und ungefilterter Umgebungsluft zu arbeiten, ebenfalls verändert. Durch proportionales Zudosieren eines Vielfachen (z.B. des 1,3-1,9-fachen) der jeweils herrschenden Ozonkonzentration der Umgebungsluft zu den NF-Varianten und Vergleich mit den CF-Varianten wurde es möglich, Dosis-Wirkungsbeziehungen zwischen Ertrag und Schadgas zu erarbeiten (27). Eine solche Vorgehensweise wurde durch das Vorherrschen von Ozon als dominierende Luftschadstoffkomponente erleichtert, und dürfte auf europäische Verhältnisse nicht ohne weiteres übertragbar sein.

Die Zufuhr von gereinigter Umgebungsluft in open-top Kammern gestattet es, prinzipiell jeden relevanten gasförmigen Luftschadstoff unter kontrollierten Bedingungen zuzudosieren. In dieser Hinsicht unterscheiden sich open-top Kammern nicht von geschlossenen Expositionssystemen. Dabei ist es möglich, sowohl die Einzelwirkung von Schadgasen (13,15,29,30,31) als auch deren Kombinationswirkungen zu untersuchen (15,32,33). Die bisher durchgeführten Untersuchungen zeigen, daß sich reproduzierbare Gaskonzentrationen in den Kammern einstellen lassen. Während jedoch die Schadgase in der Mehrzahl der Fälle in konstanten Konzentrationen und als Einzelgase appliziert wurden, mangelt es noch an Untersuchungen in open-top Kammern zur Kombinationswirkung von Schadgasen sowie zur Wirkung fluktuierender, d.h. den natürlichen Immissionsgegebenheiten folgenden Schadgaskonzentrationen (15). Open-top Kammern sind durch Einbau entsprechender Vorrichtungen wie Regen- oder Nebeldüsen auch zu Untersuchungen der Wirkungen nasser Depositionen wie z.B. saurer Niederschläge einsetzbar (15,17,19). Zur Erstellung von Stoffbilanzen und zur Untersuchung der Auswirkungen trockener und nasser Depositionen auch auf Böden werden open-top Kammern mit Lysimetern eingesetzt (15,17). Damit ist es prinzipiell möglich, z.B. forstliche Modellökosysteme über mehrere Jahre in open-top Kammern zu untersuchen (15).

Die kontrollierte Zudosierung von Schadstoffen in open-top Kammern gestattet es daher, sowohl Dosis-Wirkungsbeziehungen zu erarbeiten als auch, durch Einbeziehung physiologischer und biochemischer Untersuchungen an belasteten Pflanzen, Wirkungsmechanismen aufzuklären und damit Wirkungskriterien zu einer möglichst frühen Diagnose von Schäden zu finden.

6. ZUSAMMENFASSENDE BEWERTUNG

Aufgrund der bisher vorliegenden Erfahrungen, vorwiegend mit landwirtschaftlichen Kulturpflanzen, lassen sich die Möglichkeiten und Grenzen von open-top Kammern folgendermaßen zusammenfassen:

a) Durch Filterung der Umgebungsluft, d.h. durch Ausschluß der natürlichen Luftbelastung ist es möglich
   - vor Ort, d.h. am belasteten Standort, Wachstumsvergleiche in gefilterter und ungefilterter Umgebungsluft auch über längere Zeiträume durchzuführen und
   - durch Zudosierung verschiedener Schadgase und Schadgasgemische sowie durch Ausbringung nasser Depositionen ein breites Spektrum reproduzierbarer Schadstoffbehandlungen zu simulieren.

b) Open-top Kammern können sowohl über natürlichen Beständen aufgestellt als auch mit Gefäßkulturen beschickt werden.

c) Im Vergleich mit allen anderen Kammersystemen sind die Abweichungen der Umwelt- bzw. Klimabedingungen als geringfügig anzusehen und gestatten damit auch langfristige Untersuchungen an Dauerkulturen. Diese Abweichungen sind zu dokumentieren, und bei der Bewertung der Wirkungsergebnisse zu berücksichtigen.

d) Zur Ermittlung immissionsbedingter Pflanzenschäden ist - verglichen mit anderen Verfahren - der Kostenaufwand als angemessen anzusehen.

e) Wie jedes andere Kammersystem auch bedingen open-top Kammern ein gewisses Maß an Künstlichkeit.

f) Vor allem bei Langzeituntersuchungen mit Dauerkulturen führen die geringen Abweichungen des Mikroklimas in den Kammern von den Freilandbedingungen möglicherweise zu kumulativen Effekten auf Wachstum und Metabolismus. Die Langzeitfolgen dieser geringen Klimaabweichungen für die Reaktion von Pflanzen auf immissionsbedingte Streßsituationen sind noch zu wenig bekannt.

g) Durch unspezifische Filterwirkung derzeit gebräuchlicher Filtersysteme können andere als die zur Untersuchung anstehenden Schadstoffe zurückgehalten werden und die Interpretation von Ursache und Wirkung erschweren.

h) Trotz erfolgversprechender Ansätze sind der Größe von open-top Kammern bei Untersuchungen an älteren Bäumen Grenzen gesetzt, da z.B. Luftführung und Klimaverhältnisse bei sehr großen Kammern immer weiter von den Freilandbedingungen abweichen.

i) Einige Ausführungen von open-top Kammern simulieren in ungenügender Weise die für Freilandverhältnisse typischen vertikalen Schadgasgradienten.

Während geschlossene Kammern oder Küvetten ein Höchstmaß an Reproduzierbarkeit im Hinblick auf die Schadstoffbehandlung gewährleisten, dabei aber gleichzeitig die Wachstumsbedingungen für die Pflanzen stark modifizieren, erlauben kammerlose Freilandbegasungssysteme zwar ein natürliches Pflanzenwachstum, begrenzen jedoch den Einsatz und die Reproduzierbarkeit von Schadstoffbehandlungen. Open-top Kammern stellen einen Mittelweg zwischen diesen beiden "extremen" Expositionssystemen dar. Bei guter Reproduzierbarkeit der Schadstoffbelastungen und geringer Modifikation der Umweltfaktoren ist der Einsatz von open-top Kammern zur Erforschung der Wirkung von Luftschadstoffen auf Pflanzen weiterhin empfehlenswert.

LITERATUR

(1) Mandl,R.H.,Weinstein,L.H.,McCune,D.C. and Keveny,M. (1973). A cylindrical open-top field chamber for exposure of plants to air pollutants in the field. Journal of Environmental Quality 2, 371-376.

(2) Heagle,A.S.,Philbeck,R.B. and Heck,W.W.(1973). An open-top chamber to assess the impact of air pollution on plants. Journal of Environmental Quality 2, 365-368.

(3) Buckenham,A.H.,Parry,M.A.,Whittingham,C.P. and Young,A.T. (1981). An improved open-topped chamber for pollution studies on crop growth. Environmental Pollution (Series B) 2, 275-282.

(4) Seelinger,T.,Wichmann,A.,Schweckendiek,J. and Bornkamm,R. (1986). Simulation natürlicher O_3-Belastung in open-top chambers mittels einer rechnergesteuerten Begasungsanlage. Technische Universität Berlin, unpublished paper.

(5) Skärby,L.,Berglind,B.,Grennfeld,P.,Sellden,G. and Wallin,G.(1986). Design of open-top field chambers for studies of effects of air pollutants on conifers. Commission of the European Communities,DOC OTC 12/86. Second European Open-Top Chamber Workshop "Environmental management in open-top chambers", September 1986, Freiburg/FRG.

(6) Durrant,D. and Willson,A. (1986). The forestry commission pattern of open-top chambers for air pollution research. Commission of the European Communities, DOC OTC 12/86. Second European Open-Top Chamber Workshop "Environmental management in open-top chambers". September 1986, Freiburg/FRG.

(7) Brewer,R.F.(1978). The effects of present and potential air pollution on important San Joaquin Valley crops:sugar beets. Final Report to the California Air Resources Board for project A6-161-30.

(8) Heck,W.W.,Clore,W.J.,Leone,A.I.,Ormrod,D.P.,Pool,R.M. and Taylor,O.C. (1985). Multi-year research plan for the Lake Erie Generating Station (LEGS) grape study. Vol.II. - Technical report. NY State Board on Electrical Generation Sitting and the Environment. Department of Public Service, Albany, New York.

(9) Roberts,T.M.,Bell,R.M.,Horsman,D.C. and Colvill,K.E. (1983). The use of open-top chambers to study the effects of air pollutants, in particular sulphur dioxide, on the growth of ryegrass Lolium multiflorum L. Part I. Characteristics of modified open-top chambers used for both air-filtration and SO_2-fumigation experiments. Environmental Pollution (Series A) 31, 9-33.

(10) Ashmore,M.R.,Bell,J.N.B. and Mimmack,A. (1986). Specification for open-top chambers. Commission of the European Communities, DOC OTC 12/86. Second European Open-Top Chamber Workshop "Environmental management in open-top chambers", September 1986, Freiburg/FRG.

(11) Kats,G.,Olszyk,D.M. and Thompson,C.R. (1985). Open-top experimental chambers for trees. Journal of the Air Pollution Control Association 35, 1298-1301.

(12) Mandl,R.H.,Laurence,J.A. and Kohut,R.J.(1985). Development and testing of new open-top chambers for exposing large perennial plants to air pollutants in the field. Phytopathology, in press.

(13) Hogsett,W.E.,Tingey,D.T. and Holman,S.R.(1985) A programmable exposure control system for determination of the effects of pollutant exposure regimes on plant growth. Atmospheric Environment 19, 1135-1145

(14) Davis,J.M. and Rogers,H.H.(1980). Wind tunnel testing of open-top field chambers for plant effects assessment. Journal of the Air Pollution Control Association 30, 905-908.

(15) Seufert,G. und Arndt,U. (1985). Open-top Kammern als Teil eines Konzeptes zur ökosystemaren Untersuchung der neuartigen Waldschäden. Allgemeine Forstzeitung 1/2, 13-18.

(16) Jäger,H.J.(1986). Specification for open-top chambers. Commission of the European Communities, DOC OTC 12/86. Second European Open-Top Chamber Workshop "Environmental management in open-top chambers", September 1986, Freiburg/FRG.

(17) Krause,G.H.M.(1986). Persönliche Mitteilung.

(18) Weigel,H.J.,Adaros,G. and Jäger,H.J.(1987). An open-top chamber study with filtered and non-filtered air to evaluate the effects of air pollutants on crops. Environmental Pollution, in press.

(19) Johnston,J.W.,Shriner,D.S. and Abner,C.H.(1986). Design and performance of an exposure system for measuring the response of crops to acid rain and gaseous pollutants in the field. Journal of the Air Pollution Control Association 36, 894-899.

(20) Ashmore,M.R.,Bell,J.N.B. and Dalpra,C.(1980). Visible injury to crop species by ozone in the United Kingdom. Environmental Pollution (Series A) 21, 209-215.

(21) De Temmerman,L. and Istas,J.R.(1986). Specification for open-top chambers. Commission of the European Communities, DOC OTC 12/86. Second European Open-Top Chamber Workshop "Environmental management in open-top chambers", September 1986, Freiburg/FRG.

(22) Bonte,J.(1986). Specification for open-top chambers. Commission of the European Communities, DOC OTC 12/86. Second European Open-Top Chamber Workshop "Environmental management in open-top chambers" September 1986, Freiburg/FRG.

(23) Olszyk,D.M.,Tibbitts,T.W. and Hertzberg,W.M.(1980). Environment in open-top field chambers utilized for air pollution studies. Journal of Environmental Quality 9, 610-615.

(24) Weinstock,L.,Kender,W.J. and Musselman,R.C.(1982). Microclimate within open-top air pollution chambers and its relation to grapewine physiology. Journal of the American Society of Horticultural Sciences 107, 923-929.

(25) Unsworth,M.H.(1986). Principles of microclimate and plant growth in open-top chambers. Commission of the European Communities, DOC OTC 12/86. Second European Open-Top Chamber Workshop "Environmental management in open-top chambers". September 1986, Freiburg/FRG.

(26) Heagle,A.A.,Philbeck,R.B.,Rogers,H.H. and Letchworth,M.B.(1979). Dispensing and monitoring ozone in open-top field chambers for plant effects studies. Phytopathology 69, 15-20.

(27) Heck,W.W..,Taylor,O.C.,Adams,R.,Bingham,G.,Miller,J.,Preston,E. and Weinstein,L.H.(1982). Assessment of crop loss from ozone. Journal of the Air Pollution Control Association 32, 353-361.

(28) Heck,W.W.,Cure,W.W.,Rawlings,J.O.,Zaragosa,L.J.,Heagle,A.S., Heggestad,H.H.,Kohut,R.J.,Kress,L.W. and Temple,P.J.(1984). Assessing impacts of ozone on agricultural crops: I.Overview. Journal of the Air Pollution Control Association 34, 729-735.

(29) Collvill,K.E.,Bell,R.M.,Roberts,T.M. and Bradshaw,A.D.(1983). The use of open-top chambers to study the effects of air pollutants, in particular sulphur dioxide, on the growth of ryegrass Lolium multiflorum L. Part II. The long-term effect of filtering polluted urban air or adding SO_2 to rural air. Environmental Pollution (Series A) 31, 35-55.

(30) Murray,F.(1984). Responses of subterranean clover and ryegrass to sulphur dioxide under field conditions. Environmental Pollution (Series A) 36, 239-249.

(31) Lee,E.H.,Heggestad,H.H. and Bennett,J.H.(1982). Effects of sulphur dioxide fumigation in open-top field chambers on soil acidification and exchangeable aluminium. Journal of Environmental Quality 11, 99-102.
(32) Heagle,A.S.,Heck,W.W.,Rawlings,J.O. and Philbeck,R.B.(1983). Effects of chronic doses of ozone and sulphur dioxide on injury and yield of soybeans in open-top field chambers. Crop Science 23, 1184-1191.
(33) Johnston,J.W. and Shriner,D.S.(1986). Yield response of Davis soybeans to simulated acid rain and gaseous pollutants in the field. New Phytologist, 103, 695-707.
(34) Oshima,R.J.(1978). The impact of sulphur dioxide on vegetation: a sulphur dioxide-ozone response model. California Air Resources Board Final Report, Agreement No. A6-162-30. Sacramento.

METHODOLOGICAL ASPECTS OF THE FUMIGATION OF FOREST TREES WITH GASEOUS POLLUTANTS USING CLOSED CHAMBERS

T.M. Roberts, K.A. Brown and L.W. Blank
CEGB, Central Electricity Research Laboratories,
Kelvin Avenue, Leatherhead, UK.

SUMMARY

This paper reviews the development of outdoor and indoor closed chambers which have been used in studies of the effects of air pollutants on trees. Large well-ventilated outdoor closed chambers have been designed which optimize pollutant flux and minimise changes in environmental conditions. Differences in intensity and spectral composition of solar radiation, air, soil and leaf temperatures and relative humidity are reviewed with reference to the large geodesic (Solardome) chambers currently in use at CERL (Leatherhead, UK). Changes in these parameters are within the range achieved in open-top chambers.

Indoor chambers have been developed for pollutant effect studies under repeatable environmental conditions or when pollutant-climatic interactions are under investigation. Problems to be overcome include selection of light sources to give realistic intensity and spectral composition and adequate temperature control in view of the high heat load generated by the lights. Adequate air-flow over the experimental material should be ensured particularly when subchambers are used. Removal of NO can be achieved by additional filters and is especially important when O_3 fumigations are being carried out in both indoor and outdoor chambers. Recent technical developments are reviewed with reference to the controlled-environment chambers at the GSF (Munich). Closed chambers are particularly useful in the study of pollutant mixtures or fluctuating concentrations when a low constant control value is required. Studies on the effects of O_3 and acid mist on Norway spruce have largely been carried out in closed chambers as have studies on interactions with pests and pathogens where an enclosed system is essential. Nevertheless, it is important to maintain a range of technical approaches to the study of air pollution and forest health.

1. INTRODUCTION

There are at least five basic methodological approaches available for studying the effects of gaseous pollutants on trees. These include field transects, pollutant exclusion chambers, controlled pollutant exposures using closed or open-top chambers, open-air fumigation systems and laboratory chambers (which sometimes have facilities for modifying climatic factors as well as pollutant levels). Each of these methods has its advantages and drawbacks, and the choice of system will reflect the specific question under investigation and may be influenced by the resources available.

The closed chambers used in air pollution studies on trees can be broadly divided into outdoor and indoor systems. The ideal outdoor chambers should allow the experimental conditions to be directly related to field conditions with only minimal differences in pollutant flux, environmental and edaphic factors. The ideal indoor chambers should allow control of environmental conditions as well as pollutant levels and thereby facilitate repeatable studies of these factors and their possible interactions. Improvements in chamber design were largely made with studies on agricultural crops in mind. For tree studies, particular problems arise from the relatively large size of each plant and the need to study many trees due to the genetic diversity between individuals. Long-term experiments are essential in understanding the response of trees to air pollutants but problems may arise through restricted root growth and poor nutrition.

This paper reviews the development of closed chamber systems, both indoor and outdoors, for studies of air pollution effects on trees and summarises some recent results relating to forest decline in Europe. Particular reference will be made to the outdoor Solardome fumigation chambers at CERL (Leatherhead, U.K.) and the indoor controlled-environment chambers at the GSF (Munich, West Germany).

2. OUTDOOR CLOSED CHAMBERS

2.1 Technical Considerations

The design of outdoor chambers should minimise differences in climatic or edaphic factors and ensure that the pollutant flux can be related to field conditions. However, when plants are enclosed in chambers their atmospheric and soil environments are always different from the field. In early studies, these differences clearly altered both the pollutant flux and the sensitivity of plants to the pollutant treatments so it is often not possible to use the results of these chamber studies to predict field effects. Open-top and closed outdoor chambers have been designed to minimise chamber-field differences, largely by increasing the rate of air exchange and turbulence. With the recent emphasis on interactions between pollutant-climate-soil factors it is important that these chamber effects be adequately described.

The major environmental features that influence plants and are likely to be altered by chambers are light, temperature, water, and wind speed. Chambers decrease light by shading from the framework, reflection or absorption by the covering material particularly in the ultra-violet range of the light spectrum. Light energy absorbed by plants and the soil surface is mostly used to heat the absorbing material or evaporate water. The heat is then transferred by convection, conduction and radiation to heat the air or soil. Temperatures and humidity in chambers are thus strongly influenced by the radiant energy input, wind speed over the absorbing surfaces and rate of air changes through the chamber.

In modern chambers, the flux of pollutant is maximised and the heat load minimised by increasing the rate of air flow to 2-4 air changes per minute and/or adding internal mixing fans giving >0.5m s^{-1} over the plant surface. In the field the flux of pollutants can be limited by the above-canopy resistance which depends upon air turbulence and surface roughness of the vegetation (Figure 1). The within-canopy resistance to flux is determined by plant density, morphology and air turbulence. Transfer to the leaf or needle is then limited by the boundary layer resistance which is determined by morphology and air turbulence. In modern chambers these resistances are all minimised by the high air flow or turbulence, the use of spaced plants in many studies and the general practice of blowing air through the canopy. Deposition will also be

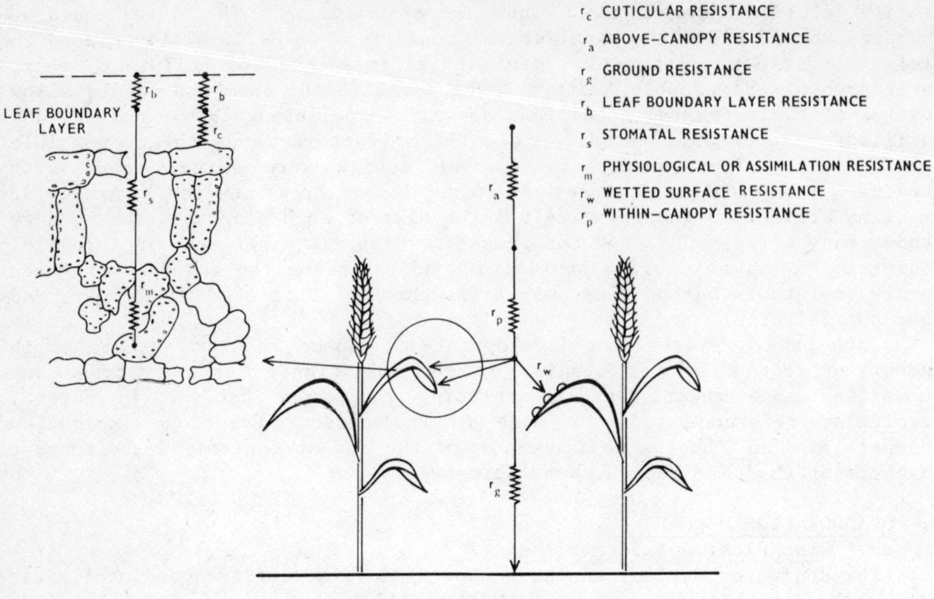

FIG. 1 PATHWAYS OF POLLUTANT TRANSFER TO VEGETATION

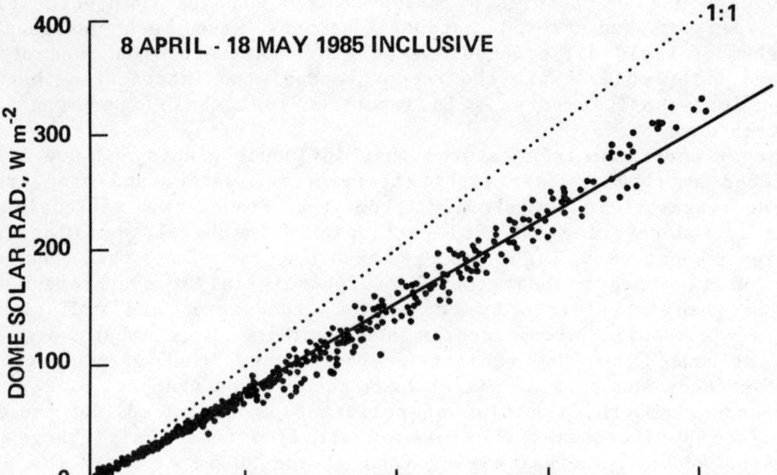

Figure 2. Mean Hourly Solar Radiation (PAR. 400-700nm) for April 8th - May 18th 1985 incl. in Domes and in the Open.

affected by the extent and duration of surface wetness which is generally less in chambers than in the open.

Uptake of a pollutant may also be partly determined by the stomatal resistance. Stomatal opening is under physiological control and may vary with light intensity and quality, vapour pressure deficit and leaf water potential. Exposure to pollutants may also cause a change in stomatal aperture. Within the leaf a further mesophyll resistance controls the rate of absorption into the cells particularly for gases of low solubility (eg. COS or NO). Any change in the factors which determine stomatal opening (such as vapour pressure deficit or leaf water potential) within chambers could change the rate of gas uptake (i.e. the effective dose rate).

2.2 Radiation Conditions

Solar radiation is always reduced in outdoor chambers due to absorption and reflectance by the glass or plastic covering. Within the photosynthetically-active range (PAR, 400-700nm) there is little difference in the transmissivity of glass or plastics which is generally around 90% of incident radiation. Shading by the chamber supports and reflectance increases the light loss in this spectral range to 20-25%. The PAR in glass-covered geodesic Solardomes at Leatherhead (UK) was reduced by about 25% independently of intensity or sun angle (Figure 2). Light absorption by PVC sheets, plexiglass or polyesters differs from glass outside the visible range. Polyesters and PVC sheet do not transmit radiation below 400nm, compared to 320nm for normal glass or 280nm for plexiglass. However, special glasses (Sanalux) are now available which allow transmission of UV light (down to 280nm) and this has recently been installed in the Solardomes at Leatherhead. The plastics absorb more than glass in the short infra-red (1600-2500nm), but less in the far infra-red. Conseqently, poorly-ventilated plastic chambers show higher daily amplitudes of temperature than do glass chambers. Plastics may also become translucent with age. Teflon has good transmissivity and low reactivity with pollutants but is too fragile for chamber walls. This problem does not occur with some of the new fluoroplastics which may find a use in future outdoor chamber designs.

2.3 Improvements in Pollutant Flux in Closed Chambers

The earliest chambers designed by O'Gara and subsequently modified, e.g. by Thomas (1), consisted of a celluloid chamber (about 1.5 metres square and 1.5 metres high) with air forced in at the top and vented at the bottom, at 0.5 to 1.0 air changes per minute (ac/min).

These chambers were used extensively by Katz and Lathe (2) to study the effects of SO_2 episodes on conifers around the Trail smelter in British Columbia. Fumigations were performed at 0.25 to 5ppm for 9 hours to 70 days on natural stands (3-30 years old) and on transplanted seedlings (4-10 years old) in portable celluloid chambers (about $5m^3$ volume). Larch was most sensitive, pines intermediate and firs were resistant. Visible injury was most pronounced following fumigation during the early summer, whereas little injury occurred during the winter or in summer periods of low relative humidity or low soil moisture status. The authors concluded that conifers could tolerate 0.25 ppm of SO_2 for an indefinite period without visible injury. The dose-response relationships for acute injury produced in these chambers are now known to be incorrect as the low air-flow in the chambers ($<$ 1 air change per minute) produced a low SO_2 flux to the needle surface, increased air temperatures, decreased relative humidity and decreased CO_2 concentrations.

A chamber design which has been used extensively in outdoor as well as laboratory studies was developed by Heck et al (3). These chambers were about $0.5m^3$ in volume. Air injected at the top passed through a perforated false ceiling, to equalise flow, down through the canopy and out through the floor of the chamber at a rate of one change every two minutes. Similar designs had been used earlier in Germany but with even lower air flows (eg. van Haut and Stratmann (4)). A number of important studies on pollutant interactions were published in the early 1970's using such systems. However, Cowling and Jones (5) in a similar set of chambers found that two air changes per minute were necessary to achieve a deposition velocity for SO_2 to ryegrass swards which approached values reported under field conditions (0.5-0.1 cm s^{-1}). Other chambers used for tree studies, in which the air-flow was less than satisfactory, included those of Houston and Stairs (6) (0.2 ac/min) and Zahn (7) (1.3 ac/min).

The importance of maintaining sufficient wind speed through the chamber to ensure realistic rates of pollutant flux was understood by Bennett and Hill (8), who developed a closed recirculating environmental chamber which closely reproduced field conditions. The point was re-emphasised by Unsworth and Mansfield (9). As a result, a number of studies have been carried out more recently in closed chambers with improved air-flow which indicate that long-term exposures can produce yield reductions in the absence of visible injury. Keller (10) investigated the concept of "latent" injury in a series of experiments in which *Pinus sylvestris* (Scots pine) and *Picea abies* (Norway spruce) clones were exposed to 50, 100 and 200 ppb SO_2 for 10 weeks to 9 months in outdoor at Birmensdorf (Switzerland). Yield reductions and premature needle loss were observed at the two higher SO_2 levels. Indications of metabolic changes, such as reduced CO_2 uptake and enhanced peroxidase levels, were also observed at the lowest level. Ten-week exposures of Norway spruce clones in the summer or winter reduced both wood production and root length in the following growing season. Similar "carry-over" effects occurred after a winter fumigation of *Fagus sylvatica* (beech) in that many terminal buds failed to open in the following spring. Latent effects were observed at 50ppb continuous exposure in this study. However, there are indications that effects could occur at lower concentrations as the air-flow through the chambers (<0.2m s^{-1}) was not adequate to minimise changes in environmental parameters. For example, air temperatures up to 37°C were recorded in the chambers so a screen had to be drawn over the chambers during bright sunlight. The temperature increase would lower the relative humidity so stimulating stomatal closure and thereby reduce the rate of pollutant uptake.

Farrar et al (11) and Garsed et al (12) carried out a series of continuous exposures of young trees to a constant SO_2 concentration (about 60ppb) in similar outdoor closed chambers (1.5 air changes per minute). Exposures lasting 24 to 62 weeks produced small reductions in growth of Scots pine associated with variable changes in needle elongation and root development. Exposures of four deciduous species for 15 to 71 weeks produced reductions in leaf area and early leaf senescence but no consistent effects on growth.

Various types of chambers have been developed in the last decade to maximise pollutant flux and minimise changes in environmental conditions. Open-top chambers force air through the crop or trees at 2-3 air changes per minute and out through the top of the chamber. Closed chambers have been developed with improved air-flow (2ac/min) and additional internal mixing fans as in the constant-stirred tank reactor system (CSTR) of Heck et al (13) or the Solardome system of Ashenden et al (14) and Lucas et al (15).

2.4 Environmental Conditions in Well-Ventilated Chambers

Pollutant flux in chambers has been maximised by the high air-flow which results in a low leaf boundary layer resistance. This also has implications for the energy balance of the leaves. At low boundary layer resistances, transpiration varies over a wide range with changes in vapour pressure deficit with relatively small effects on leaf temperature. Consequently, leaf temperature in well-ventilated chambers should be similar to the air temperature.

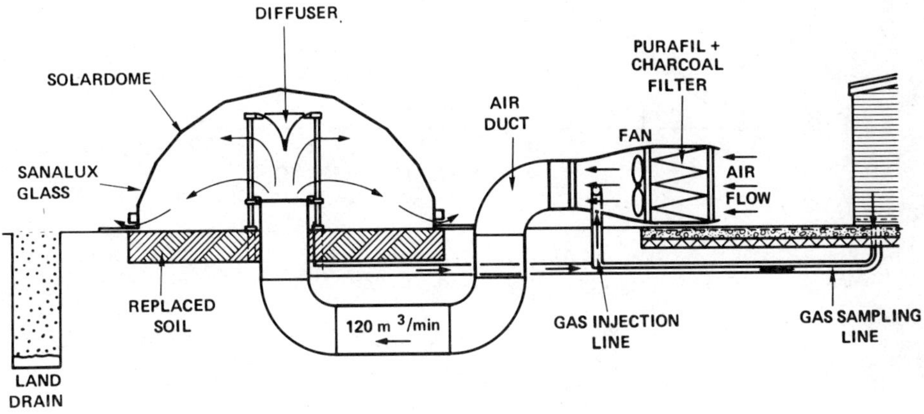

Figure 3. Cross-Section of One of the Eight CERL Solardome Outdoor $SO_2/NO_x/O_3$/Acid Mist Fumigation Chambers

The effect of large outdoor chambers on the microclimate of the test plants can be seen from the data collected from the Solardome chambers at CERL in which the air flow was increased to 4 ac/min.

Each of the eight chambers was constructed of glass and aluminium and had a volume of 25 m^3 (experimental floor area of 9 m^2). Each chamber was ventilated with 100 m^3 min^{-1} by a 5 hp internal fan which pushed the air into each chamber through an underground duct (0.7 m^2 cross-section area). There were four air changes per minute and the wind velocity at crop height was approximately 0.5 m s^{-1} (Figure 3). Each filter contained Sutcliffe Speakman Grade 207C copper-activated charcoal. Each filter had a surface area of 15.8 m^2 and was 2 cm deep. Residence time of the air stream in the charcoal was 0.15 seconds. The photosynthetically-active radiation (PAR - 400-700nm) was reduced by about 25% irrespective of sun angle or intensity (Figure 2). The air temperature in the chamber was about 1°C above ambient over a range from -2°C to 24°C (Figure 4). This implies that 4 ac/min was adequate to remove the radiant energy. The temperature of barley leaves in the chamber was the same as the air temperature. Outside the chamber, the leaf temperature was less than the air temperature below 5°C due to radiant cooling and up to 5°C above at >15°C due to radiant heating following stomatal closure. Consequently the leaf temperature in the chamber was 1-2°C above the outside leaf temperature at <5°C but up to 5°C lower at

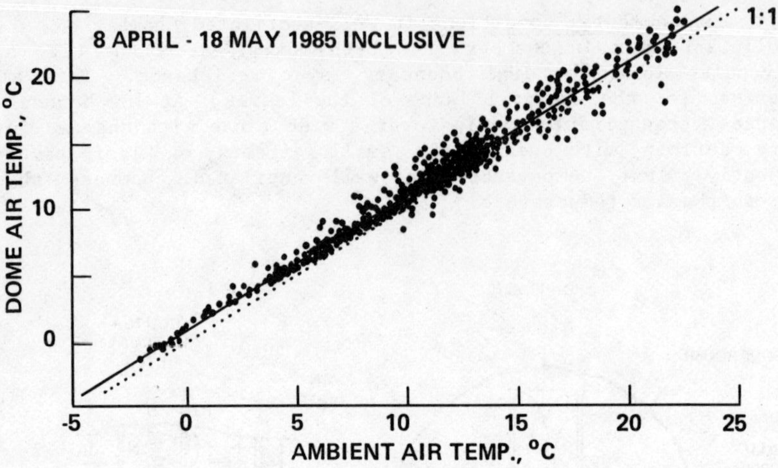

Figure 4. Mean Hourly Air Temperature (°C) for April 8th - May 18th 1985 incl. in Domes and in the Open

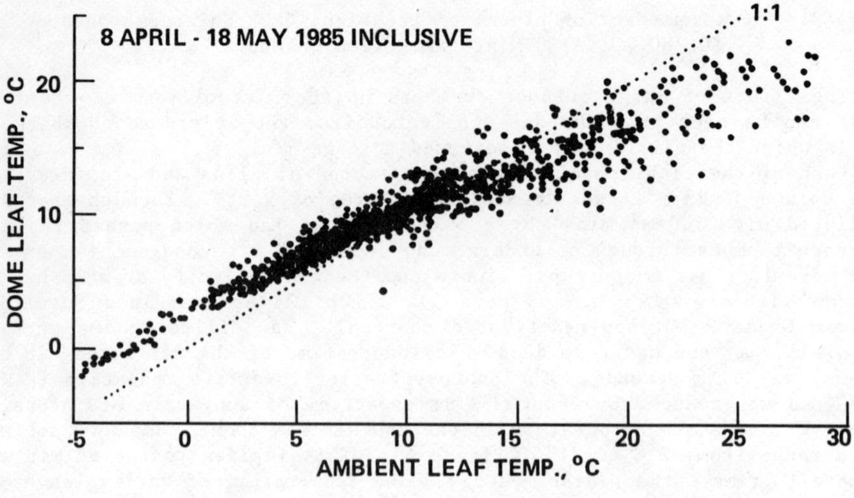

Figure 5. Mean Hourly Barley Leaf Temperatures (°C) for April 8th - May 18th 1985 for the Domes and in the Open.

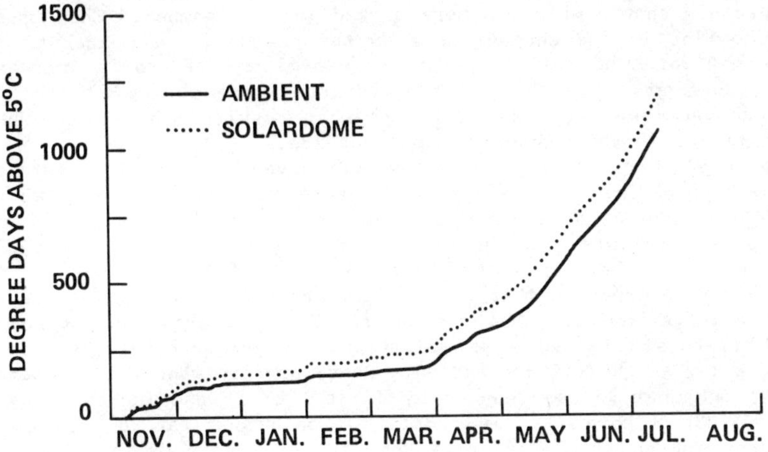

Figure 6. Degree days (above 5°C) Cumulated from Nov. 6th 1984 to July 16th 1985 for the Domes and in the Open

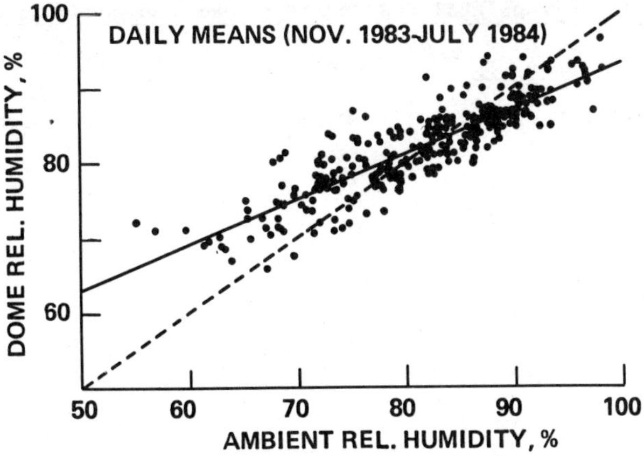

Figure 7. Mean Daily Relative Humidity (Nov 1983 - July 1984) for the Domes and in the Open

>15°C (Figure 5). The increase at lower temperatures might infer that pollutant-cold stress interactions would be underestimated in chambers. However, this may not be so as cold injury is also determined by the rate of temperature change which is more marked in the chamber. The absence of radiant cooling in the chamber is also the reason for the lack of dew or frost formation. The soil temperature (0-5cm) was generally increased by 0.5-1°C. However, when the air temperature dropped below 0°C the higher rate of convective heat loss in the chamber resulted in periods when the soil was frozen in the chamber but not outside.

The development of crops or trees over a year is largely determined by the degree day. The 1°C average temperature increase in the chambers can make a difference of 7-10 days in the cumulated degree days >5°C over a growing season (Figure 6). This is the main reason for early ripening of cereals in the chambers and early budbreak of conifers.

It might be expected that the 1°C temperature increase would result in a lower relative humidity in the chamber. However, this was not so as the charcoal filters retained some moisture at night and this was released during the day as the filters warmed. Consequently, when the chambers were empty, the chamber RH was reduced by 5% at >85% RH and increased by about 5% at <60% RH (Figure 7). In addition, the transpiration of a maturing cereal crop increased the relative humidity even more during the daytime. Consequently the daytime vapour pressure deficit in the chamber during an experiment on winter cereals was up to 25% less than outside.

2.5 Comparison of Closed and Open-Top Chambers

A comparison of open-top chambers and well-ventilated solardome chambers reveals that the changes in environmental parameters, when compared to the field, are rather similar (Table 1). Solar radiation is reduced slightly more in closed chambers but changes in air temperature, leaf temperature and relative humidity do not show major differences between the two types of fumigation system.

Table I Summary of Environmental Changes in Closed and Open-Top Chambers

Chamber Type	Open-Top	Open-Top	Closed	Closed
Location	New York State	Glasgow	Lancaster	Leatherhead
Air Changes/min.	3.7	3	2	4
Radiation (PAR as %of Open)	76-90%	80%	80%	75%
Air Temperature	+0.4 to 3.7°C	+1 to 2°C	+1 to 2°C	+1°C
Leaf Temperature	+1 to 5°C	±0	--	±0
Relative Humidity	-5 to 10%	--	--	+5%(day)
Vapour Pressure Deficit	+ about 5%	--	--	-5%(day)
Author	Weinstock et al (16)	Fowler and Cape (unpubl)	Lucas et al (15)	Roberts (unpubl)

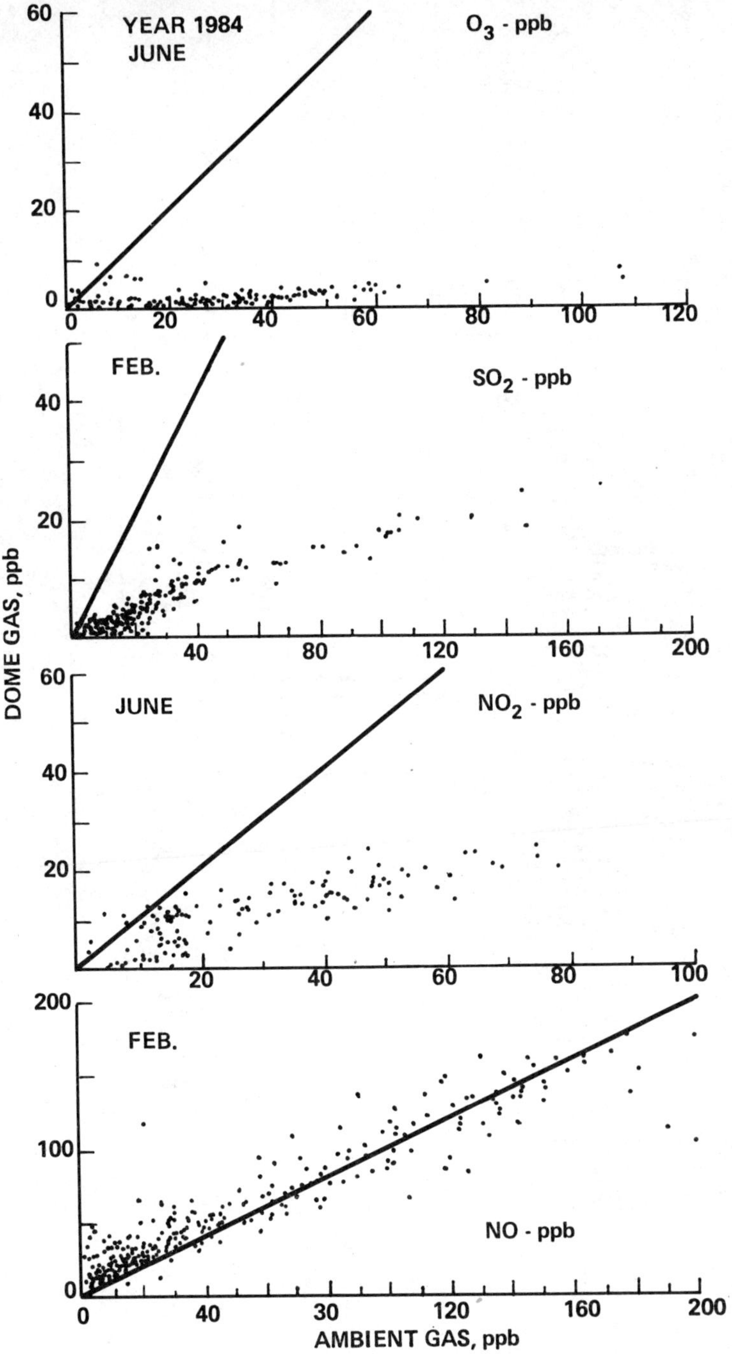

Figure 8. Effect of Activated Charcoal Filters on Pollutant Concentrations in the Solardomes (for selected 10 day periods in June or February 1984 - hourly mean values)

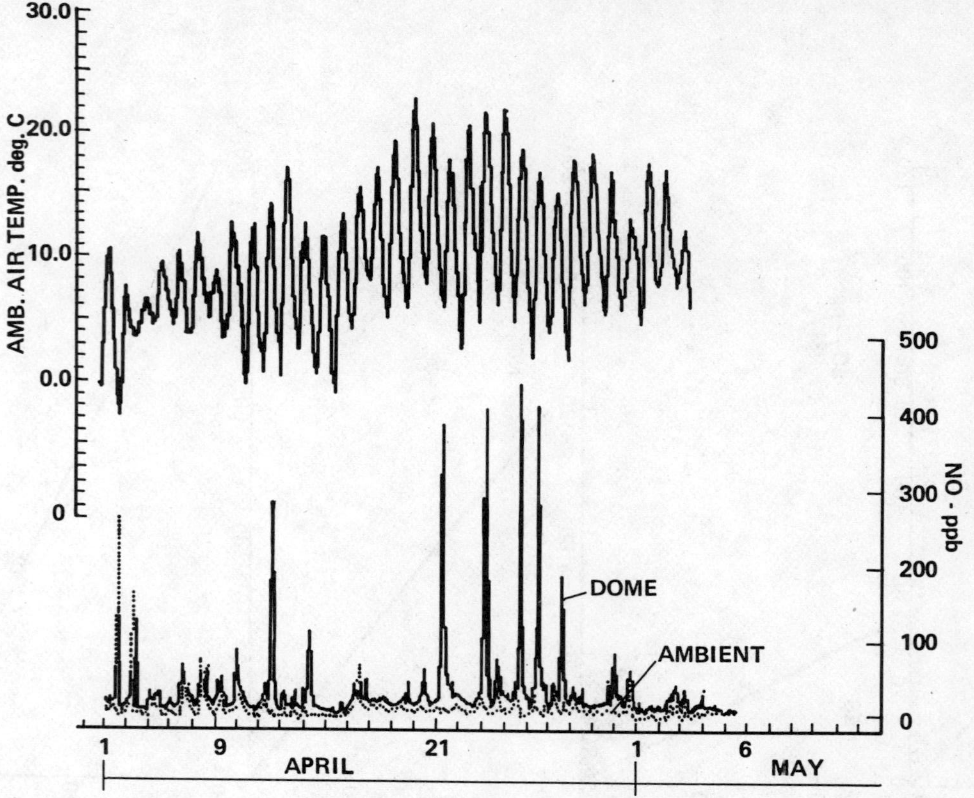

Figure 9. Effect of Air Temperature on Release of NO from Charcoal Filters in April/May 1984 (after filters had been used overwinter)

Open-top chambers have been extensively used to compare plant growth in filtered and unfiltered air. This has proved satisfactory for situations in which ozone is the main concern as O_3 episodes generally coincide with low wind speeds when air intrusion through the open-top will be small. However, overwinter studies in Europe when SO_2 and NO_2 are the main pollutants, have found intrusion a serious problem which can only be solved by the addition of a perforated roof (Lane, 17).

Open-top chambers were originally designed to allow normal wetting of crop and soil by rain. However, with the addition of sloping frustrum to reduce air intrusion, the rain shadow often results in the need for additional watering. In studies of the combined effects of gaseous pollutants and acid deposition, OTC's have to be converted into closed chambers by the addition of a roof (Krause, pers. comm., Seufert and Arndt, (18)) or by a sliding cover activated during rainfall (Johnston et al, (19)).

2.6 Studies of Pollutant Mixtures Using Closed Chambers

Closed chambers have been widely used for studies of the effects of pollutant mixtures and/or fluctuating levels when low control values are required. Representative data on the filtering efficiency of large activated charcoal filters is shown for the CERL solardomes in Figure 8. Ozone, SO_2 and NO_2 levels were reduced by 95%, 85% and 75% respectively. However, NO levels were, slightly higher in the chamber following large NO peaks or during periods of high NO_2 levels. In addition, filters operated overwinter released NO in the spring when the air temperature exceeded the previous highest value (starting at about 15°C) (Figure 9).

Whitmore and Freer-Smith (20) reported the effects of exposure to 62ppb SO_2 and/or NO_2 from March until August in Solardome chambers on six broadleaf tree species. The SO_2 treatment reduced growth in three of the six species whereas NO_2 produced small growth increments. The SO_2 + NO_2 mixture produced more-than-additive effects on four of the six species. Kress and Skelly (21) used CSTR's in a study of the effects of O_3 and/or NO_2 on 10 eastern forest species over 28 days which showed yield reductions in 8 species between 50 and 100ppb O_3 whereas the interactions with 100ppb NO_2 were generally additive or less-than-additive. Garsed, Mueller and Rutter (22) used closed chambers in a comparison of constant and fluctuating SO_2 concentrations for 650 days on pine seedlings which showed that growth was reduced by similar amounts irrespective of the frequency or duration of peaks.

In 1983 experiments were started in the CERL solardomes to investigate the effect of ozone, acid mists and environmental factors on conifers. The aim of the work was to evaluate the substantive hypothesis of Prinz, Krause and Stratmann (23) in which O_3 and acid mist combine to cause some of the symptoms of the Waldschaden currently affecting many species of tree in Central Europe. In brief, this hypothesis states that exposure to O_3 made conifer needles more permeable and thus susceptible to leaching by acid mists. Leaching reduces the needle nutrient concentrations, as the tree is unable to replace the leached nutrients quickly enough due to reduced root growth resulting from inhibition of photosynthesis by O_3. Needle nutrient deficiencies are a characteristic but not universal symptom of the Waldschaden, especially in Norway spruce. Krause et al (24) have provided some evidence from laboratory experiments that O_3 increases the leaching of nutrients from various tree species, but not that needle nutrient contents are reduced.

Exposure of Norway spruce saplings to between 50 and 150ppb O_3 for 60 days in 1984 in the CERL Solardomes produced some chlorotic mottle and

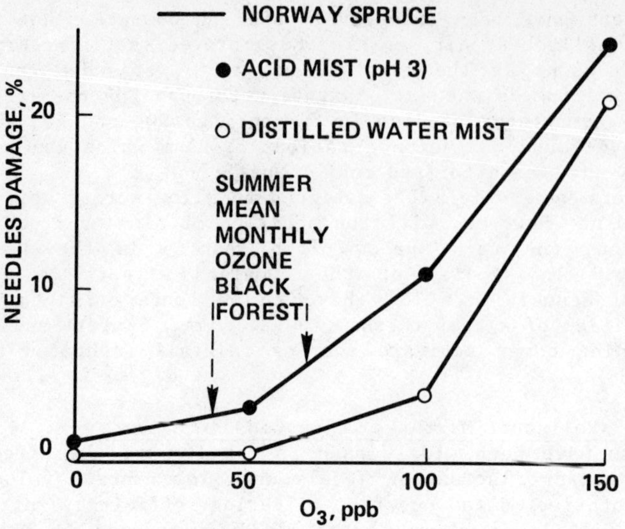

Figure 10. Effect of O_3 and Acid Mist (pH 3) on Needle Damage in Norway spruce after 60 day Exposures in 1984

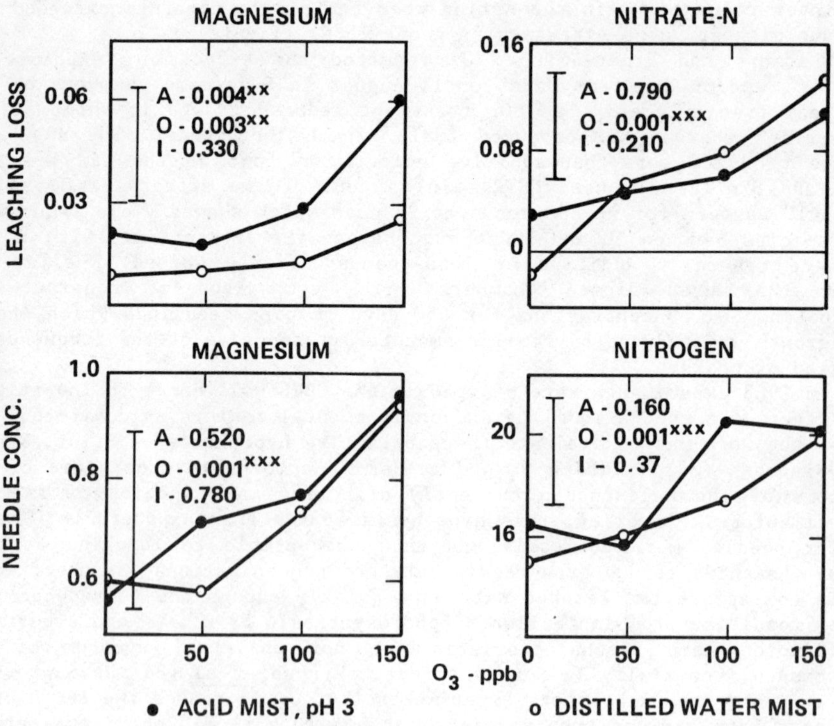

Figure 11. Effect of O_3 and Acid Mist (pH 3) on Nitrogen and Magnesium in Needles and Leachate (with pH 3 H_2SO_4 for 24 hours) from Norway spruce.

tip necrosis on the youngest shoots of Norway spruce and the necrosis was increased by spraying with pH3 acid mist (Figure 10). The hypothesis of Prinz et al (23) predicts that treatment with O_3 and leaching by acid mist should reduce needle nutrient concentrations. On the contrary, O_3 significantly increased the Mg and N content of Norway spruce needles, with other elements showing the same trend (Figure 11). Acid mist also had no effect. However, evidence was obtained that O_3 increased the leakiness of needles. A shoot was taken from each tree at the end of the experiment and leached with pH3 sulphuric acid. O_3 appeared to increase the leakiness of shoots to Mg, Ca, K, NO_3 and Mn but not P. Previous treatment with acid mist also increased the leakiness of most of these elements (25).

The increased N content and leachable nitrate-N highlights a problem which arises when O_3 is added to charcoal-filtered air. Ozone will react with NO to produce NO_2 and perhaps HNO_3 vapour which have a higher deposition velocity. The increased needle N content and nitrate in leachate may, therefore, result from NO_2/HNO_3 deposition rather than the ozone treatment (Brown and Roberts, in prep.). Similarly, the enhanced deposition of N species may also contribute to the visible injury attributed to O_3. NO should therefore by removed by special filters or the ozone fumigation should only be carried out during low NO periods. The use of charcoal-filtered air as a control in studies of interactive effects of pollutants on trees has also been questioned by the results of Seufert and Arndt (18) who established that the deposition of about 20ppb SO_2 to Norway spruce was more than doubled when 40-60ppb O_3 was added.

Closed chambers will also be used extensively in the study of interactions between pollutants and environmental factors. Brown et al (26) found that 100-150ppb O_3 for 60 days increased the cold sensitivity of 3 of 10 Norway spruce clones. A similar interaction with cold stress has also been reported by Davison et al (27). Lucas et al (15) have described the development of an air cooling system which can lower the air temperature by 4-8°C in Solardome chambers. Closed chambers are also ideal for studying the mechanisms for interactions between pollutants, pests and pathogens (28, 29).

3. INDOOR CHAMBERS

In an ideal outdoor fumigation system plants are subjected to various pollutants/pollutant levels without alteration of climatic conditions, thus effectively allowing a direct comparison with plant material grown under ambient conditions. With this methodological approach, however, it is not possible to repeat fumigation experiments under identical climatic conditions, or to study pollutant: climate interactions. When addressing these two objectives, it becomes necessary to use closed laboratory systems with adequate climate as well as pollutant control.

Closed indoor fumigation systems with various levels of climatic control, from passive control by housing subchambers in climate rooms (chamber-in-chamber) to actively controlled systems, have been extensively used for laboratory studies of the effect of pollutants. Various designs have been described in the literature (eg. van Haut and Stratmann (4); Heck et al (13); Bennett and Hill, (8); Aiga et al (30)).

Most of these systems have initially been used for crop studies, and experimental work with trees has only recently become a major focus of scientific interest. Although crop fumigation systems may easily be modified to accommodate small trees, the climatic requirements for a realistic experiment with forest trees may require substantial modifications. The temperature regime, appropriate to high altitude forests, for example, may differ substantially from lowland agricultural areas. This may cause problems for long-term studies on trees when winter conditions have to be reproduced. An additional drawback of existing indoor as well as outdoor closed systems is the lack of dew or frost formation. Taking these technical problems and the recent emphasis on pollutant-climate-soil interactions into account it is important that chamber characteristics, technical performance, and the chamber effects be adequately described.

The indoor closed fumigation system at the GSF, Munich/FRG is, at present, technically the most advanced facility (Payer et al (31)) and two years experience with short as well as long-term exposure experiments indicates what can be achieved and highlights the areas where particular care is required.

The GSF system consists of 4 identical environmental chambers, which are effectively large climatically-controlled rooms. Each of the four walk-in chambers (floor area approx. 10 m^{-2}) can either be used directly for fumigation experiments, or can be used to house 4 subchambers. Trees in each subchamber are then exposed to the same climatic conditions but may be fumigated differently. The system is described in detail by Payer et al. (31), and the main features of the large chambers include:

Climatic Range

Air temperature	-20°C to +40°C
Rel. humidity	20 to 95%± 10% (gradient depending on temperature)
Light regime	500 to 120,000 ± 3,000 lux (with a horizontal gradient of ± 10%; spectral characteristics similar to standard daylight)
Air movement	0.05 to 0.7 m s^{-1} horizontal air flow; recirculating system (max. purge/make-up 750 m^{-3} h^{-1}) with a circulation rate of 3,000 to 23,000± 500 m^{-3} h^{-1} (equivalent to 1 to 12 air changes min^{-1}). When subchambers are being used, their internal wind speed is considerably lower (0.03 to 0.1 m s^{-1}).

Exposed plants can be subjected to various pollutants, eg. SO_2, NO_x, O_3, singly or in combination. An additional feature, important for long-term studies, is the availability of controlled root compartments to prevent unrealistic temperature regimes.

A major problem in indoor chambers is the generation of an adequate light climate with high light intensities as they frequently occur at field sites (Figure 12). Light intensities at elevated field sites may often be well in excess of 80 klux (Table II), which is clearly higher than what is used in most indoor chambers (eg. Guderian et al (33): max 40klux). The use of realistic light intensities is particularly important in relation to forest decline as the frequent symptom (Mg-deficient chlorosis) only occurs on branches exposed to direct sunlight. The large amount of excess heat produced by powerful light systems, however, poses a serious technical problem as it will interfere with the temperature control especially when cold winter days with a high rate of illumination are to be simulated. This problem of heat generation is of particular importance for chamber-in-chamber facilities with low air flow.

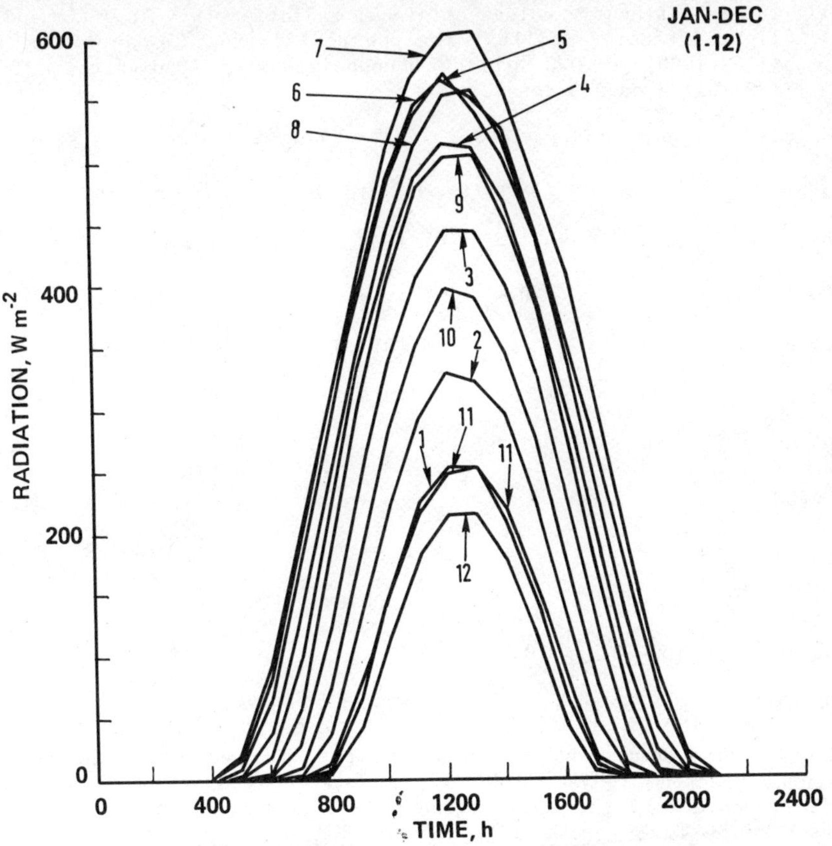

Figure 12. Monthly Average Total Radiation (1966-75) at Hohenpeissenberg/FRG (Baumgartner et al, 1985)

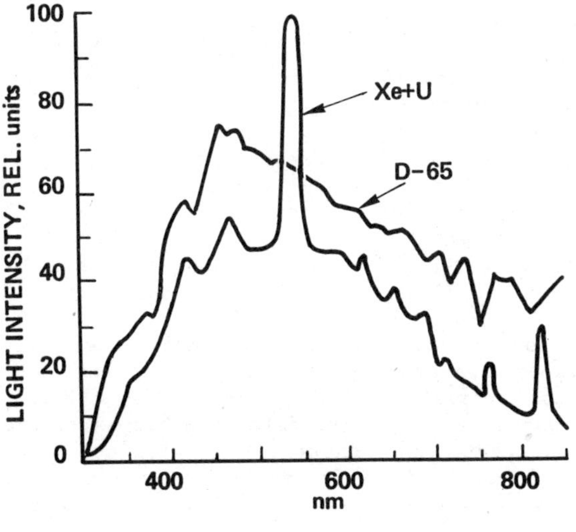

Figure 13. Transmission Spectrum of the Mixed Lamp System (Xenon; Ultra lamps) in the GSF Chambers Compared with the CIE D-65 Standard Sunlight Spectrum

Table II 10 highest hourly values of global radiation (S & D; Wm^{-2}) and light intensity (E; Lux) for the period from 15 May to 15 July (1960-1980) at DWD station Hohenpeissenberg (Baumgartner *et al* (32) based on DWD data).

Time (hours)	11.00		12.00		13.00	
Date	S+D	E	S+D	E	S+D	E
11.06.75	-	-	1045.3	109441	-	-
17.06.76	-	-	-	-	1031.4	107986
24.06.62	-	-	1017.5	106530	-	-
29.05.72	-	-	1014.7	106239	-	-
31.05.69	1000.8	104784	-	-	-	-
22.06.68	-	-	-	-	1000.8	104784
01.07.74	-	-	-	-	1000.8	104784
01.06.69	-	-	1000.8	104784	-	-
13.06.76	1000.8	104784	-	-	-	-
17.06.63	-	-	998.0	104493	-	-

It is also important to select light sources which produce a near-natural spectrum (Gaastra, (34)). A detailed comparison between the CIE D-65 standard sunlight spectrum and the spectrum at the GSF chambers was presented by Payer *et al* (35) (Figure 13). The selective filtering properties of material frequently used for chambers (perspex, glass etc) have already been discussed in the context of closed outdoor systems. Similar problems occur in indoor chambers when chamber-in-chamber systems are being used (Guderian *et al* (33); Payer *et al* (35)).

The use of chamber-in-chamber systems, with generally comparatively small subchambers of a volume of less than 1 m^{-3} (Guderian *et al* (33); Payer *et al* (35)) can also create problems with the rate of air flow and with gas dispersion. Most large climate chambers use recirculating air systems with a limited rate of purge and fresh air intake (GSF system: max 750 m^{-3} h^{-1} per chamber). Subchambers with different pollutant levels, using once-through systems sucking air from the climate-chamber, are therefore severely restricted in the volume of air available. Care must be taken to ensure that a decrease in the rate of air exchange does not result in an air movement insufficient to overcome the aerodynamic resistance of leaf surfaces, a problem extensively addressed by Unsworth and Mansfield (9).

The results of the first longer-term tree exposure experiment with the GSF system, involving various research groups, have recently been published (Forstwissenschafliches Centralblatt 105, 1986). The climatic and pollutant exposure program, based on monitoring data from high altitude forest sites in the Bavarian Forest, was carried out without major problems, so target and actual data were generally well in line (Figure 14). Although the minimum air temperatures (-17°C) were the lowest

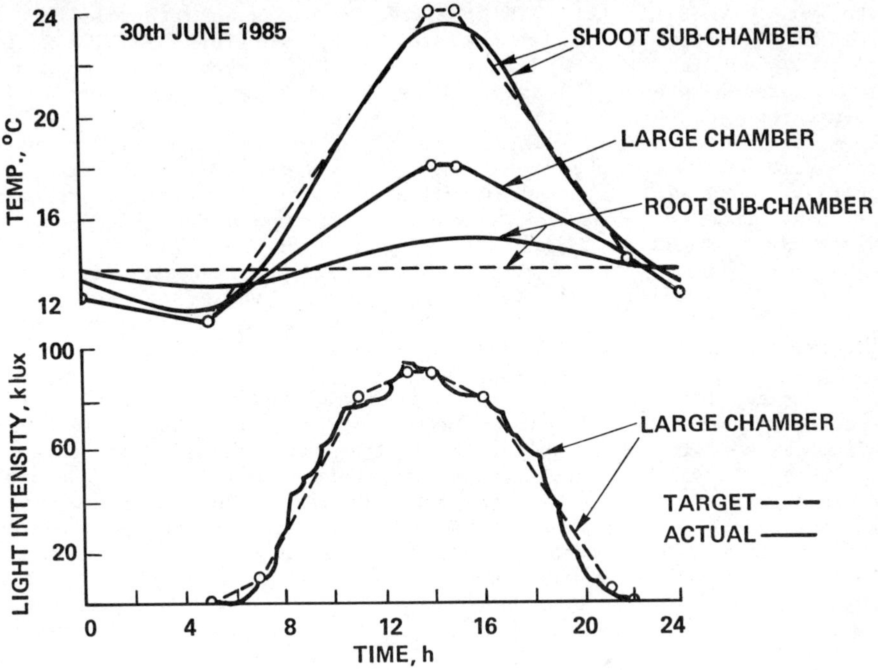

Figure 14. Daily Variation in Illumination (Klux) and Air Temperature in the GSF Chambers (30.6.1985; chamber-in-chamber approach with temperature controlled root compartment)

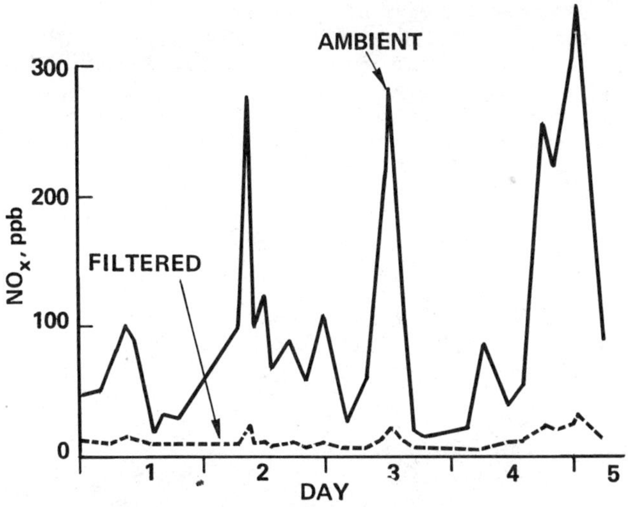

Figure 15. Air Filter Efficiency in the GSF chambers (charcoal plus permanganate filters) During Episodes of High NO_x Concentrations

achieved in indoor fumigation work it should be noted that winter air temperatures at elevated forest sites may be well below that achievable even with the GSF system. The analysis of 20 years meteorological data from the Bavarian Forest, which forms the basis for this GSF climatic program, does indeed show long-term (1960-1980) mean daily minimum air temperatures between -12°C and -20°C with daily extremes as low as -29°C ((Baumgartner et al (32)).

For indoor fumigation facilities which use air recirculation systems, great care has to be taken regarding the fresh-air filter efficiency. In contrast to once-through systems with short retention times, intrusion of pollutant gases can pose a serious risk to experimental plants. The selection of effective filters is of particular importance in urban and industrial areas, where elevated NO_x levels necessitate effective NO filters. The problem of high urban NO_x levels, where peak concentration may be in excess of 500ppb can largely be overcome by fitting permanganate filters, in addition to charcoal, as shown in Figure 15 for the GSF chambers.

Closed fumigation systems are now frequently being used for experimental studies on forest trees (eg. spruce, pine, beech) in order to establish the role of air pollutants in the recent forest decline reported throughout Central Europe. Indoor fumigation systems have become an important tool in studies on the interaction of pollutants with soil and climate. Bosch et al (36) suggested an interaction between elevated ozone concentrations, acid mist, nutrient deficiency and severe climatic conditions (sharp frost) may cause damage to spruce on acid soils at high altitude forests in S. Germany. This explanatory hypothesis for the most common type of forest decline was the basis of a 5 month "pilot-experiment" in the GSF chambers (Rehfuess and Bosch, (37)).

During this first experiment, which used the chamber-in-chamber approach, 4-year-old clonal Norway spruce were exposed to 16 stress combinations (normal winter temperature or an episode of severe forest; low or elevated O_3 concentrations; misting with water of pH5.6 or pH3.0; Mg and Ca fertilised soil or nutrient poor soil), with the exposed plant material being subsequently examined by various research groups. Bosch et al (38) could demonstrate, that foliar leaching of the key elements Mg and Ca was increased by application of acid mist (pH3.0) and particularly by a combination of acid mist and elevated ozone concentration (varied between 25 and 75ppb). Foliar deficiences of Mg and Ca were, however, restricted to the nutrient poor soil. Magel and Ziegler (39) found changes in needle wax structure, i.e. a disturbance of wax plug formation in current year needles. It appeared that ozone treatment caused a fusion of the normally fine wax tubes, and that acid mist application leads to the development of cracks in the wax plugs. Photosynthesis and stomatal conductance were also found to be affected (Selinger et al (40)). In all non-fertilised plants exposed to ozone, maximum carboxylation capacity, maximum net photosynthetic rate, apparent quantum yield, and stomatal regulation were found to be reduced considerably. Fertilizer application, (Mg and CA) led to an increase in photosynthetic capacity and reduced effects of pollutant treatments.

Foliar injury, however, was not induced by any of the pollutant combinations. This agrees with observations reported by Guderian et al (33), who fumigated spruce with O_3 (100ppb for 8 hours d^{-1}) in an indoor chamber for four months. In a subsequent indoor chamber experiment over 5½ months, Guderian (33) found discolouration of current years needles could be induced by fumigating Mg- and Ca- deficient spruce with a combination of SO_2 (40ppb continuously) and O_3 (90ppb-100ppb from

8.00-16.00 hours plus 45-50ppb from 16.00-8.00 hours). The discolouration, however, was restricted to trees exposed to high light intensities of max. 40,000 lux compared with a second group receiving only a maximum of 15,000 lux. Jurat et al (41), using modified growth chambers for tree fumigation, found an increase in needle yellowing of three year old spruce grown in acid soils (pH 3.64 and 3.90) with poor Ca and Mg supply after 7 weeks of continuous SO_2 fumigation (55 ppb). They also report a reduced rate of photosynthesis after fumigation both for trees grown in these acid soils and for a third group grown in standard soil (pH 6.29) with good Ca and Mg supply; the combination of SO_2 fumigation and acid soil, however, caused a larger reduction in net photosynthesis than each factor on its own.

4. CONCLUSIONS

It is important to maintain a variety of technical approaches to the study of air pollution effects on trees. Open-air fumigation systems have been developed to the point where pollutant concentrations can be maintained at target values over 10-25m diameter area. Close monitoring of crop responses, pests, pathogens and environmental factors has begun to reveal the complex interactions which can occur under field conditions. However, outdoor closed or open-top chambers must be used to compare growth in filtered and unfiltered air. It is also difficult to use open-air fumigation systems to study the effects of single gases or pollutant mixtures due to the difficulty of finding a site free of air pollutants.

Closed chambers must, therefore, be used for studies of pollutant mixtures and comparison of constant or fluctuating levels where a low constant control value is required. Closed chambers will also be necessary for studies where an enclosed environment is essential - as in the case of studies on interactions between pollutants, wet deposition, pests and pathogens. Closed chambers have also been developed to assess the interaction between pollutants and cold stress. It is essential to maintain adequate air-flow through the chamber to maximise pollutant flux and minimise changes in environmental conditions. Large well-ventilated closed chambers do not appear to modify environmental conditions more than open-top chambers.

Indoor controlled-environment chambers must be used when pollutant treatments need to be repeated under identical experimental conditions or in studies of pollutant-climate interactions. When designing the light system, realistic light intensities and a near-natural spectrum are clearly major objectives. The resulting heat load can generate problems in controlling temperature in recirculating systems requiring an increased cooling capacity to achieve temperature regimes comparable to field sites, particularly when natural winter conditions are to be simulated. Again adequate air-flow should be maintained, especially when subchambers are used. Effective filters for NO removal are important when O_3 fumigations are being carried out, particularly in recirculating systems.

5. ACKNOWLEDGEMENTS

The authors thank Drs S.H.D. Payer, J. Grace and N.M. Darrall for asistance in preparing this manuscript.

6. REFERENCES

(1) THOMAS, M.D., (1951). Gas damage to plants. Annual review of Plant Physiology, 2, 293-322.

(2) KATZ, M. and LATHE, F.E. (1939). Effect of Sulphur Dioxide on Vegetation. National Research Council of Canada, Ottowa.

(3) HECK, W.W., DUNNING, J.A. and JOHNSON, H., (1968). Design of a Simple Plant Exposure Chamber. National Centre for Air Pollution Control. U.S. Dept. Health, Education and Welfare, Washington (USA).

(4) VAN HAUT, H. and STRATMANN, H., (1960). Experimentelle Untersuchungen uber die Wirkung von Schwefoldioxyd auf die Vegetation. Forschungsberichte des Landes Nordrhein-Westfalen. NR. 884.

(5) COWLING, D.W. and JONES, L.H.P., (1978). Sulphur and amino acid fractions in perennial ryegrass as influenced by soil and atmospheric supplies of sulphur. IN. Sulphur in Forages. (Brogan, J.C., Ed.) pp. 15-31. Anas Foras Taluntais, Dublin.

(6) HOUSTON, D.B. and STAIRS, G.R., (1973). Genetic control of SO_2 and O_3 tolerance in eastern white pine. Forest Science, 19, 267-271.

(7) ZAHN, R., (1961). Wirkung von Schwefeldioxyd auf die Vegetation, Ergebnisse aus Begasungsversuchen. Staub, 21, 56-60.

(8) BENNETT, J.H. and HILL, A.C., (1973). Absorption of gaseous air pollutants by a standardised plant canopy. JAPCA., 23, 203-206.

(9) UNSWORTH, M.H. and MANSFIELD, T.A., (1980). Critical aspects of chamber design for fumigation experiments on grasses. Environmental Pollution (A), 23, 115-120.

(10) KELLER, T. (1985). SO_2 effects on tree growth. IN. Sulphur Dioxide and Vegetation. (Winner, W.E., Mooney, H.A. and Goldstein, R.A., Eds.). pp250-263. Stanford University Press, Stanford.

(11) FARRAR, J.F., RELTON, J. and RUTTER, A.J., (1977). SO_2 and growth of Pinus sylvestris. Journal of Applied Ecology, 14, 861-875.

(12) GARSED, S.G., FARRAR, J.F. and RUTTER, A.J., (1979). The effects of low concentrations of SO_2 on the growth of four broad-leaved tree species. Journal of Applied Ecology, 16, 217-226.

(13) HECK, W.W., PHILBECK, R.B. and DUNNING, J.A. (1978). A Continuous Stirred Tank Reactor (CSTR) System for Exposing Plants to Gaseous Air Pollutants. USDA. Agricultural Research Service, Publication No. ARS-S-181, Washington.

(14) ASHENDEN, T.W., TABNER, P.W., WILLIAMS, P., WHITMORE, M.E. and MANSFIELD, T.A., (1982). A large-scale system for fumigating plants with SO_2 and NO_2. Environmental Pollution (B), 3, 21-26.

(15) LUCAS, P.W., COTTAM, D.A. and MANSFIELD, T.A. (1987). A large-scale fumigating system for investigating interactions between air pollution and cold stress on plants. Environmental Pollution, 43, 15-28.

(16) WEINSTOCK, L., KENDER, W.J. and MUSSELMAN, R.C., (1982). Microclimate within open-top air pollution chambers and its relation to grapevine physiology. J. Amer. Soc. Hort. Sci., 107 (5), 923-929.

(17) LANE, P.I., (1983). Ambient Levels of Sulphur and Nitrogen Oxides in the UK and their Effects on Crop Growth. Ph.D. Thesis, Imperial College, London.

(18) SEUFERT, G. and ARNDT, U. (1986). Beobachtungen in definiert belasteten Modellökosystemen mit jungen Waldbäumen. Allgemeine Forst Zeitschrift, 22, 545-548.

(19) JOHNSTON, J.W., SHRINER, D.S. and ABNER, C.H., (1986). Design and performance of an exposure system for measuring the response of crops

to acid rain and gaseous pollutants in the field. JAPCA., 36, 894-899.

(20) WHITMORE, M.E. and FREER-SMITH, P.H., (1982). Growth effects of SO_2 and/or NO_2 on woody plants and grasses observed during the spring and summer. Nature, 300, 55-57.

(21) KRESS, L.W. and SKELLY, J.M., (1982). Response of several eastern forest tree species to chronic doses of O_3 and NO_2. Plant Disease, 66 (2), 1149-1152.

(22) GARSED, S.G., MUELLER, P.W. and RUTTER, A.J., (1982). An experimental design for studying the effects of fluctuating concentrations of SO_2 on plants. IN. Effects of Gaseous Air Pollution in Agriculture and Horticulture. (Unsworth, M.H. and Ormrod, D.P., Eds.) pp. 474-475.

(23) PRINZ, B., KRAUSE, G.H.M. and STRATMANN, H., (1982). Waldschäden in der Bundesrepublic Deutschland. LIS Berichte., 28, Essen.

(24) KRAUSE, G.H.M., JUNG, K.D. and PRINZ, B. (1985). Experimentelle Untersuchungen zur Aufklärung der neuartigen Waldschäden in der Bundesrepublik Deutschland. VDI Berichte, 560, 627-656.

(25) SKEFFINGTON, R.A. and ROBERTS, T.M., (1985). Effects of O_3 and acid mist on Scots pine and Norway spruce - an experimental study. VDI Berichte, 560, 747-760.

(26) BROWN, K.A., ROBERTS, T.M. and BLANK, L.W., (1987). Interaction between O_3 and cold sensitivity in Norway spruce - a factor contributing to forest decline in Central Europe? New Phytologist, 105, 149-155.

(27) DAVISON, A.W., BARNES, J.D. and RENNER, C.J., (1987). Interactions between air pollutants and cold stress IN. Proc. 2nd International Symposium on Air Pollution and Plant Metabolism, Munich (in press).

(28) BRAUN, S. and FLUCKIGER, W., (1985). Increased populations of the aphid, Aphis pomi, at a motorway. Environmental Pollution, 39(A), 183-192.

(29) DOHMEN, G.P., (1987). Secondary effects of air pollution - ozone decreases brown rust disease potential in wheat. Environmental Pollution, 43, 189-194.

(30) AIGA, I., OMASA, K., and MATSUMOTO, S., (1984). Phytotrons in the National Institute for Environmental Studies. Res. Rep. Natl. Inst. Environ. Stud., Japan, 66, 133-154.

(31) PAYER, H.D., BLANK, L.W., BOSCH, C., GNATZ, G., SCHMOLKE, W., and SCHRAMEL, P. (1986). Simultaneous exposure of forest trees to pollutants and climatic stress, Water, Air and Soil Pollution, 31, 485-491.

(32) BAUMGARTNER, A., MAYER, H. and KONIG, C., (1985). Report on meteorological data from the Bavarian Forest: Evaluation for use in Environmental Chambers, GSF Internal Report.

(33) GUDERIAN, R., KUPPERS, K and SIX, R., (1985). Wirkungen von Ozon, Schwefeldioxid und Stickstoffdioxid auf Fichte und Pappel bei unterschiedlicher Versorgung mit Magnesium und Kalzium sowie auf die Blattflechte Hypogymnia physodes, VDI - Berichte 560, Düsseldorf, 657-701.

(34) GAASTRA, P., (1970). Climate rooms as a tool for measuring physiological parameters for models of photosynthesis systems, in Prediction and Measurements of Photosynthetic Productivity, Proc. of the JBP/PP Technical Meeting, Trebon 14-21, Sept. 1969.

(35) PAYER, H.D., BOSCH, C, BLANK, L.W., EISENMANN, T. and RUNKEL, K.H., (1986). Beschreibung der Expositionskammern und der Versuchsbedingungen bei der Belastung von Pflanzen mit Luftschadstoffen und Klimastress. Forstw. Cbl. 105, 207-218.

(36) BOSCH, C., PFANNKUCH, E., BAUM, U. and REHFUESS, K.E., (1983). Uber die Erkrankung der Fichte (Picea abies (L.) Karst) in den Hochlagen des Bayerischen Waldes,. Forstw. Cbl. 102, 167-181.

(37) REHFUESS, K.E. and BOSCH, C., (1986). Experimentelle Untersuchungen zur Erkrankung der Fichte (Picea Abies (L.) Karst.) auf sauren Boden der Hochlagen: Arbeitshypothese und Versuchsplan. Forstw. Cbl 105, 201-206.

(38) BOSCH, C., PFANNKUCH, E., REHFUESS, K.E., RUNKEL, K.H., SCHRAMEL, P. and SENSER, M., (1986). Einfluss einer Düngung mit Magnesium und Calcium, von Ozon und saurem Nebel auf Frosthärte, Ernahrungszustand und Biomasseproduktion junger Fichten (Picea abies (L.) Karst). Forstw. Cbl. 105, 218-229.

(39) MAGEL, E. and ZIEGLER, H., (1986). Einfluss von Ozon und saurem Nebel auf die Struktur der stomatären Wachspfropfen in den Nadeln von Picea Abies (L.) Karst. Forstw. Cbl. 105, 234-238.

(40) SELINGER, H., KNOPPIK, D. and ZIEGLER - JONS, A. (1986). Einfluss von Mineralstoffernährung, Ozon und saurem Nebel auf Photosynthese-Parameter und stomatäre Leitfahigkeit von Picea Abies (L.) Karst. Forstw. Cbl. 105, 239-242.

(41) JURAT, R., SCHAUB, H., STIENEN, H. and BAUCH, J., (1986). Einfluss von Schwefeldioxid auf Fichten (Picea abies (L.) Karst.) in verschiedenen Bodensubstraten. Forstw. Cbl 105, 105-115.

THE METHODOLOGY OF OPEN-FIELD FUMIGATION

J.J. COLLS & C.K. BAKER
Nottingham University School of Agriculture,
Sutton Bonington, Loughborough LE12 5RD, England.

Summary

Field fumigation systems are designed to determine the effects of increased pollutant concentrations on the development, growth and economic yield of agricultural crops. They offer improved simulation of field growing conditions and ambient pollutant exposure than either controlled environment or open-top chamber experiments. Nevertheless there are many ways in which such field fumigation can fail to meet its design objectives. The main ways - pollutant time variations, spatial homogeneity, continuity and statistical validity - are examined in more detail and the implications for the design and interpretation of such experiments are discussed.

1. INTRODUCTION

It cannot be often that a review has so few relevant papers on which to base a coverage of the field. Therefore, instead of attempting to relate the results obtained by different field fumigation systems working with different species, we are going to concentrate on the design and operational constraints which are implicit in the concept of open-field fumigation. These constraints affect the validity of all such experiments and hence the interpretation of their results.

The most straight-forward type of experimental design is one in which all parameters are fixed, one variable is controlled at a particular value, and a response is observed. In pollutant effects work such a design is only appropriate if the variable (which could be the pollutant concentration) is large enough in value for the effects due to it alone to be much greater than any other effects due, for example, to unnatural lighting or soil conditions.

The emphasis in pollutant effects work has moved away from short-term investigations of the acute response to high concentrations. One of the questions to be answered now concerns the biological and economic impact of long-term exposures of both agricultural crops and natural ecosystems to regional ambient air pollutants (1). The concentrations of these pollutants are normally less than the levels at which acute, visible plant damage occurs of the type identified in short-term exposures (2, 3). The low levels and rates of damage have two important consequences

a) Quantitative investigations of plant effects should be made over a period of the same order as the productive cycle; several years for trees and a growing season for annual crops.

b) An experimental design must be used which can distinguish the pollutant effects from either normal biological variability or from spurious effects due to the experimental system itself. The simple comparison between a 'treatment' and a 'control' may not be adequate if the control is not itself at field conditions. If, for example, a particular yield change is found on an experimental cereal crop that is planted at low density, there is no a priori justification for assuming that the same yield change will be experienced on a crop grown at field density.

Recognition of these ideas has led to increased interest in the design of experiments which have minimum influence on the agronomic or environmental conditions of the plants. The two main avenues of development have been open-top chambers and field fumigation systems.

The principle of open-top chambers is to control only the air supply to the plants, so that pollutants can be removed or added but other environmental factors are left unchanged (4). This is hard to put into practice. Firstly, the chamber itself changes the solar radiation intensity and spectrum, the air temperature and the humidity of the plants' environment throughout the depth of the canopy (5, 6). Secondly, the very fact that the air supply is controlled changes the air flow regime above and within the canopy. Maximum windspeeds and turbulence intensity are both reduced, and the vertical profile of windspeed (and hence of pollutant supply) becomes an arbitrary product of the ventilation geometry. All these factors affect the growth of plants to some degree; it has not yet been shown whether such unavoidable environmental modifications make a quantitative difference to pollutant sensitivity.

Field fumigation systems are designed to add a specific pollutant, or combination of pollutants, to the natural air flow. They should leave all environmental factors, including windspeed profile within the canopy, unchanged. Both controls and treatments should be grown using normal crop husbandry, and the treatment should experience no adjustment other than in the concentration of pollutant. These conditions cannot be satisfied by chambers (whether open or closed), by short-term experiments or by the use of potted plants. This degree of rigour is required because we are not seeking merely qualitative indications, but quantitative responses which may eventually serve as inputs to economic cost/benefit analyses. Thus we need to be reasonably certain that a 5% change in a relevant parameter during a trial would be reflected by a corresponding change in the same parameter in a field crop.

2. SYSTEM DESIGNS

Each field fumigation system that has been reported has been based on a different design and control approach. The main variations are in the release geometry (whether from a line source upwind of the crop, or point sources within the crop) and in sophistication of control (whether gas is released at constant rate or at a variable rate depending on the dispersion conditions).

The French Ministry of Agriculture system has been described by DeCormis et al. (7), Bonte (8), and Bonte et al. (9). They used a network of vertical pipes to release sulphur dioxide among various tree species. The SO_2 flow rate was constant, so that the actual concentrations depended on windspeed and turbulence. Concentration increased in the downwind direction within the experimental area, since more upwind sources were included. A single treatment was compared with an equivalent, unfumigated control.

Lee, Preston & Lewis (10) developed the ZAPS (Zonal Air Pollution System). This was used to fumigate prairie grassland with SO_2 distributed from perforated horizontal pipes. Again the release rate was maintained constant, giving little control over concentration. The research is notable for its comprehensive coverage of the impact of a toxic pollutant on different aspects of an ecosystem (11).

Miller et al. (12) described a system at Argonne based on parallel release pipes suspended in an east-west direction above the crop - soybeans in this case. Fumigation was carried out on occasions when the wind direction was north or south. Four (and subsequently five) of these piped plots were used for SO_2, and later for mixtures of NO_2 and SO_2 (13). The results are compared with those from control plots away from the fumigation plume trajectory.

These three systems all attempt to apply a spatially uniform concentration of pollutant to a crop, with little control over the concentration. Greenwood et al. (14) report a more sophisticated system at Nottingham University in which the gas concentration (SO_2) can be controlled. Gas is released from one or two of the four 20m perforated horizontal pipes defining a square. The SO_2 concentration at the centre of the plot is measured continuously and output fed back via motorized needle valves to adjust the SO_2 release rate. The system is controlled by a small computer, which also uses the wind direction to decide from which pipe or pair of pipes the gas should be released. A variable wind direction can be exploited to improve uniformity of fumigation - in fact the standard deviation of SO_2 concentrations across the central 10 m x 10 m experimental area was ± 7 ppb on an average concentration over the fumigation period of 100 ppb (15).

This feedback approach to concentration control has also been used at the Central Electricity Research Laboratory (16). In this case a circle of high-level SO_2 sources was supplemented by distributed low-level sources for fumigating winter cereals with SO_2. A larger system in which gas is released at two heights on one circle is now being deployed for tree fumigation (17).

Fumigation systems have also been devised in which a linear gradient of the fumigant gas is established (18). Plants at different locations will then experience different concentrations, and in principle a simple method would exist for examining the variation of response to concentration. In practice the uncontrolled nature of the fumigation may make interpretation difficult.

There are now fumigation systems that can achieve reasonable spatial uniformity, and temporal control, of the fumigant concentration. The operation and management of such systems is nevertheless beset by uncertainties. We shall go on in this paper to look at some of the difficulties and limitations which impinge on the interpretation of results.

3. FUMIGATION PHILOSOPHY

The most fundamental decision concerning the experimental design is that of the imposed fumigation regime. This decision is not straight-forward, and has implications for both the management and interpretation of the experiment. In the ambient atmosphere, pollutant concentrations are never constant. They vary continuously, responding to changes in source strength, wind direction, dispersion conditions, deposition velocity and other parameters. The variability is often expressed in terms of the concentration frequency distribution; this may be log-normal or a close approximation to it (19). The exact characteristics of the distribution may have implications for the ultimate effects of the pollutant on the plant (20, 21). To demonstrate this more clearly, we have constructed an exactly log-normal distribution (Table 1) from 288 values (these could be, for example, 5-minute averages over a period of 24 hours) at 16 concentrations. The three diurnal variations shown in Figure 1 are constructed from the same 288 points and hence have the same frequency distribution and average value. The plant response to these time series will depend partly on the coincidence between peaks of pollutant concentration and variations in stomatal conductance. Thus the daytime peak (a) has an average concentration during daylight (0600 - 1800) of 24.4 ppb, compared with 4.1 ppb for the night-time peak (b) The third graph (c) shows a quasi-random variation which is more probable in practice than either of these extremes. In experiments, results from fumigated plots are often compared with those from nearby control (ambient) plots. This comparison will be made more realistic if the frequency distribution of the imposed fumigation has similar characteristics to that of the ambient pollutant. However, this is not easy to arrange. In Figure 2, we have compared the effects of three possible fumigation regimes: constant addition (CA) in which the ambient concentration is elevated by a constant amount (say 40 ppb); constant resultant (CR) in which a continuously varying fumigation is applied so as to maintain the total concentration steady at 53.8 ppb - fumigation ceases when the ambient concentration rises above 54 ppb; and constant factor (CF) in which the fumigation is varied so as to raise the concentration to 3.86 times the prevailing ambient concentration. The ambient air pollutant (AA) is represented by the frequency distribution of Table 1 . All of these fumigations are 'equivalent' in the sense that they give the same long-term average concentration of 54 ppb. However, they clearly have very different effects on the frequency distribution. Although the CR method is conceptually convenient, it destroys any resemblance between the fumigation regime and its ambient counterpart. There is a significant difference between CA and CF methods in practical control terms. With a CA system, the ambient concentration is measured, then the pollutant release rate fixed to give the required addition. That addition is then independent of ambient concentration changes, varying only with dispersion conditions. A CF system, on the other hand, requires a continuously varying emission rate to match a varying ambient concentration, even when dispersion conditions are steady. This difference may be important if, as is generally the case, one gas analyser is time-shared between several treatments. Then it may be 20 minutes before the treatment concentration can be checked and the flow reset; the CF system will inevitably suffer from larger drift away from the target concentration during this period. The benefit of the CF system, which offsets the more difficult control, can be seen in Fig. 2. It is the only way that we can simulate both the ambient variation and frequency distribution without corrupting the latter at low concentrations.

4. INTERMITTENCY

We have seen the possible effects of different pollutant time-series. A more dramatic disturbance to the fumigation regime is brought about by intermittency - the inability of the system to supply the required concentration of pollutant to the plant at any given time. Consider three different causes of intermittency - system availability, low wind-speed failure, and plume meander.

System Availability

Fumigation systems for annual crops may have to be installed on the field after crop emergence. This can lead to an initial period of non-fumigation while the system is set up and made operational. There will be a corresponding period at the end of the growing season if the system must be dismantled and removed before harvest, although this lost time will normally be much shorter. The significance of these lost time periods may be greater for a spring-sown crop with a 5 month growing season than for a winter cereal with a 9 month season, and may be negligible if trees are being fumigated for several years.

Low windspeed failure

Fumigation in the required concentration range of 10 - 100 ppb is normally achieved in two stages - a predilution in ambient air from the pure gas down to the release concentration range of 0.1 - 1% followed by further dilution down to the target concentration by natural aerodynamic turbulence above the canopy. There is thus an acutely phytotoxic gas concentration available at canopy height; as windspeed falls and the gas flow-rate control comes below its minimum turn down ratio, the only way of avoiding over-fumigation is to shut down at a pre-set windspeed - about 1 m s^{-1} in practice. On gently convective days the windspeed may hover for hours around the shut-off speed, producing erratic system operation and never achieving stable fumigation. The problem is compounded if a 'soft start' is incorporated to minimize the possibility of over-fumigation whenever the system starts up. The intermittency due to this factor could be reduced by running at the lowest possible release concentration. This could be achieved by having a large air flow rate or by extending the dynamic range through the use of a two-stage flow controller.

Plume meander

Gas release occurs from multiple point sources around, or within and around, the experimental area. The spatial uniformity of fumigation across that area depends strongly on the dispersal of the released gas. This is controlled by two processes of different time scales - the steady long-term broadening of the plume and the erratic short-term variations caused by changing wind directions. Applications of dispersion theory are dominated by the far-field approximation, in which the short-term direction variability tends to a stable average, and plume width is controlled by long-term broadening (22, 23). This works well on a scale of 100 m - 10 km. In the near-field, however, (0.1 - 100 m) the plume itself remains quite narrow, and is switched in different directions rapidly and erratically (24). When the downwind concentration is measured with a typical gas analyser having a 95% response time of around 100 s, these rapidly alternating sequences of high-concentration plume and low-concentration ambient air are effectively integrated and averaged to give the mean concentration over that period. The plants in the experimental area also receive this pulsating exposure; we know nothing about the effects of such relatively high-frequency variations as compared to a steady level. The pulsations can be minimized by increasing the number (i.e. reducing the spacing) of the point sources, and by increasing the downwind distance between the gas release points and the plants. In this respect a system having release points outside the experimental area should give a more continuous fumigation than one in which the release points are distributed over the area.

These three examples show how the applied fumigation may be at variance either with the spatial uniformity of ambient pollution or with the intended concentration. We are not suggesting that these imperfections necessarily invalidate the experimental results - we simply do not know sufficient at present about the comparative effects of time-varying and steady pollutant concentrations. It is clear, however, that any design of system should strive to minimize these imperfections until we do know more about the plant response. We are currently using the Sutton Bonington fumigation system to apply three different concentrations of SO_2 to 3 different plots on different time cycles (continuous, 3 days on / 3 days off, one day on / 5 days off). Overall, the plots will all receive the same dose (concentration x time) over each contiguous 6-day period; this will enable us to discriminate the effects of one of the types of intermittency discussed above.

5. VERTICAL PROFILES

The metabolic reactions that lead to assimilate production in each leaf of a plant are partly controlled by the leaf's physical environment - radiation flux density, saturation vapour pressure deficit, air temperature, windspeed and CO_2 concentration (25). These parameters vary in both absolute and relative magnitude through the height of the canopy. At each height, the effects of the physical environment will depend on the plant material - for example, leaves of different ages may have very different inputs to the plant. One of the strengths of field fumigation systems is that they make no adjustment at all to the vertical profiles of these parameters. However, this is not necessarily true of the fumigating gas. This achieves a vertical profile which depends on the release geometry, dispersion conditions and flux to the canopy. It will in general be different in the ambient canopy and the different treatment canopies; it will also vary over the area of each treatment. Again, we do not know what the implications of such variations are for plant effects. In the absence of such an understanding, the general rule for field fumigation system designers must be 'if in doubt - emulate'. The more closely that field conditions are reproduced, the higher the chance of getting relevant results. This principle applies in areas other than pollution research. For example, it has been shown that the effects of an increased ultra violet component in solar radiation are qualitatively different in field experiments to the effects found in controlled environments (26).

6. MULTIPLE POLLUTANTS

In the discussion so far we have examined the case of fumigation with one pollutant. In some instances, a single pollutant dominates the phytotoxic ambient gases. The fumigation system can then be tailored to the measurement, release and control of that gas alone. In general, however, more than one gas is prevalent. For example, although sulphur dioxide can occur almost alone, it is often accompanied by nitric oxide, nitrogen dioxide and ozone in variable proportions (27). If sulphur dioxide is the only fumigant but other gases are present, then we must expect additional effects. Evidence from both open and closed-chamber experiments suggests that such effects will vary in both sign and magnitude according to the relative and absolute concentrations of phytotoxic gases in the mixture (28). Hence although the non-fumigated gases occur in equal concentrations on all the experimental plots, their effects on each plot may be synergistic, neutral or antagonistic depending on the concentration of fumigant gas. It is normal to monitor the concentrations of these significant non-fumigant gases; however, such measurements are of limited practical value and may in the end only be used to justify unforeseen differences between treatments. A further complication is caused by different diurnal variations. As an exaggerated example, the two sequences in Figure 3 have identical average values of non-fumigant gas. In one case, the peaks of non-fumigant gas [3] occur before and after, but not during, the midday period when the ambient concentration (and hence fumigant concentration) is a maximum. In the second case [4] the peaks are coincident. Consequently the effects can only be understood in the light of phase data - statistical summaries are not sufficient.

7. STATISTICAL LIMITATIONS

The final area of fumigation system design that we will discuss here is that of statistical limitations. We shall touch briefly on two aspects -interactions and replication - which both impose restrictions owing to the large scale and high cost of field fumigation.

The design of fumigation experiments may involve control plots (ambient air) and a series of about 3 identical treated plots, each of which is fumigated with a different gas concentration. It is desirable to locate these plots so that the control or lowest-concentration plots receive negligible fumigation even when the highest-concentration plot is directly upwind. In practice this can result in excessive lengths of gas dispensing and sample lines; in addition it becomes difficult to keep the plots within an area of homogeneous soil type. Consequently the plot spacing is reduced to a manageable compromise that will give an acceptably small interaction; for example, at Sutton Bonington 2 control plots experienced mean fumigation concentrations of 4 and 10 ppb of SO_2, compared with concentrations on the experimental plots of 50, 100 and 200 ppb.

The high degree of variability associated with biological experiments has led to a general acceptance that treatments need to be replicated to achieve statistical confidence in the results, and that the degree of replication must be increased if smaller differences between treatment and control are to be demonstrated at a given level of confidence. The high cost of field fumigation work makes it very unlikely that true replication can be funded. Thus the investigator with funding for 3 fumigated plots is more likely to opt for 3 treatments, no replicates than for 1 treatment, 3 replicates. If these treatments are carefully planned, then much insight can still be gained from rank and trend analyses. It is no substitute for replicates to take subsamples from the same treatment, since these are statistically dependent. The only way in which confidence can be increased is to repeat the experiment, preferably on a different soil; natural weather variation will provide a further indication of consistency.

CONCLUSION

Field fumigation offers the greatest likelihood of any experimental system of determining correctly the quantitative response of crops to pollutant stress. However, many factors must be optimized in order to achieve realistic fumigation. In some respects, essential understanding is still lacking of aspects that are vital to this optimization process. Sub-experiments on the effects of long-term (days) gaps in fumigation, short-term (seconds) pulsations and the importance of vertical profiles are all required. Every effort should be made at the design stage to reduce such shortcomings, and detailed measurements of, for example, spatial variations should be made so that departures from the design ideal can be adequately described.

ACKNOWLEDGEMENT

The authors would like to acknowledge financial support from the Natural Environment Research Council, Ministry of Agriculture Fisheries and Foods and Department of the Environment.

REFERENCES

(1) ROBERTS, T.M. (1984) Effects of air pollutants on agriculture and forestry. Atmos. Env. 18 629-652.

(2) JACOBSON, J.S and HILL, A.C. (eds) (1970) Recognition of Air Pollution Injury to Vegetation: A Pictorial Atlas. Air Pollution Control Association, Pittsburgh, Penn.

(3) TAYLOR, H.J., ASHMORE, M.R. and BELL, J.N.B. (1986) Air Pollution Injury to Vegetation. Institute of Environmental Health Officers, London.

(4) HEAGLE, A.H., BODY, D.E. and HECK, W.W. (1973) An open top field chamber to assess the impact of air pollution on plants. J. Env. Qual. 2 365-368.

(5) OLSZYK, D.M., TIBBITTS, T.W. and HERTZBERG, M.W. (1980) Environment in open-top field chambers utilized for air pollution studies. J. Env. Qual. 9 610-615.

(6) OLSZYK, D.M., KATS, G. DAWSON, P.J., BYNEROWICZM, A., WOLF, J. and THOMPSON, C.R. (1986) Characteristics of air exclusion systems vs. chambers for field air pollution studies. J. Env. Qual. 15 326-334.

(7) deCORMIS, L., BONTE, J. and TISNE, A. (1975) Experimental method for the study of the effects on vegetation of sulphur dioxide applied continuously at sub-necrotic dose. Poll. Atmos. N66 103-107.

(8) BONTE, J. (1977) Effects of long-term fumigations with low concentrations of sulphur dioxide on field crops. Report on Contract No. 72-35 (PA 101). I.N.R.A. Morlaas.

(9) BONTE, J., deCORMIS, L. and TISNE, A. (1981) Etude des effets a long terme d'une pollution chronique par SO_2 a concentration sub-necrotique (50μg SO_2 m^{-3}). Final Report on Contract No. 79-03 (PA77 - 156). I.N.R.A. Morlaas.

(10) LEE, J.J., PRESTON, E.M. and LEWIS, R.A. (1978) A system for the experimental evaluation of the ecological effects of sulphur dioxide. In: Proceedings of the 4th Joint Conference on Sensing of Environmental Pollutants, American Chemical Society, Washington D.C., pages 49-53.

(11) LAUENROTH, W.K. and PRESTON, E.M. (eds) (1984) The effects of SO_2 on a grassland : a case study in the northern Great Plains of the United States. Springer, New York.

(12) MILLER, J.E., SPRUGEL, D.G., MULLER, R.N., SMITH, H.J. and XERIKOS, P.B. (1980). Open-air fumigation system for investigating sulphur dioxide effects on crops. Phytopathology 70 1124-1128.

(13) IRVING, P.M., XERIKOS, P.B. and MILLER, J.E. (1981) The combined effect of sulphur dioxide and nitrogen dioxide gases on the growth and productivity of soybeans. Argonne National Laboratory Report No. ANL 81-85, Part III, 17-22.

(14) GREENWOOD, P., GREENHALGH, A., BAKER, C. and UNSWORTH, M. (1982) A computer-controlled system for exposing field crops to gaseous air pollutants. Atmos. Env. 16 2261-2266.

(15) COLLS, J.J., BAKER, C.K. and SEATON, G. (1986) Computer control of SO_2 fumigation in a study of crop response to pollution. In: Computer Applications in Agricultural Environments, eds Clark, Gregson, Saffel. Butterworths, London.

(16) McLEOD, A.R., FACKRELL, J.E. and ALEXANDER, K. (1985) Open-air fumigation of field crops : criteria and design for a new experimental system. Atmos. Env. 19 1639-1649.

(17) SHAW, P.J.A. (1986) The Liphook forest fumigation experiment : description and project plan. Central Electricity Generating Board Report No. TPRD/L/2985/R86.

(18) SHINN, J.H., CLEGG, B.R. and STUART, M.L. (1977) A linear gradient chamber for exposing field plants to controlled levels of air pollutants. UCRL Reprint No. 80411. Lawrence Livermore Laboratory, U.S.A.

(19) FOWLER, D. and CAPE, J.N. (1982) Air pollutants in agriculture and horticulture. In: Effects of gaseous air pollution in agriculture and horticulture. Unsworth and Ormrod (eds) Butterworths, London.

(20) MALE, L.M. (1982) An experimental method for predicting plant yield response to pollution time series. Atmos. Env. 16 2247-2252

(21) JACOBSON, J.S. and McMANUS, J.M. (1985) Pattern of atmospheric sulphur dioxide occurrence : an important criterion in vegetation effects assessment. Atmos. Env. 19 501-506.

(22) U.S. ENVIRONMENTAL PROTECTION AGENCY - Guideline on air quality models (1978) EPA-450/2-78-027.

(23) HINRICHSEN, K. (1986) Comparison of four analytical dispersion models for near-surface releases above a grass surface. Atmos. Env. 20 29-40.

(24) GRIFFITHS, R.F. (1984) Aspects of dispersion of pollutants into the atmosphere. Sci. Prog. Oxf. 69 157-176.

(25) MONTEITH, J.L. (1973) Principles of Environmental Physics. Arnold, London.

(26) TERAMURA, A.H. (1986) Cited in Chemical and Engineering News, 64 (47) 28-29.

(27) MARTIN, A. and BARBER, F.R. (1981) Sulphur dioxide, oxides of nitrogen and ozone measured continuously for 2 years at a rural site. Atmos. Env. 15 567-578.

(28) ORMROD, D.P. (1982) Air pollutant interactions in mixtures. In: Effects of gaseous air pollution in agriculture and horticulture. Unsworth and Ormrod (eds). Butterworths, London.

Table I

Hypothetical log-normal cumulative frequency distribution of pollution concentration.

Concentration/ppb	% < Conc.	Concentration /ppb	% < Conc.
0.5	1.0	9.0	59.0
0.75	2.4	13.0	70.5
1.05	4.8	18.5	81.0
1.50	9.0	26.0	88.0
2.2	15.0	37.0	93.0
3.1	23.5	53.5	96.5
4.4	34.0	74.0	98.1
6.3	46.0	100.0	99.0

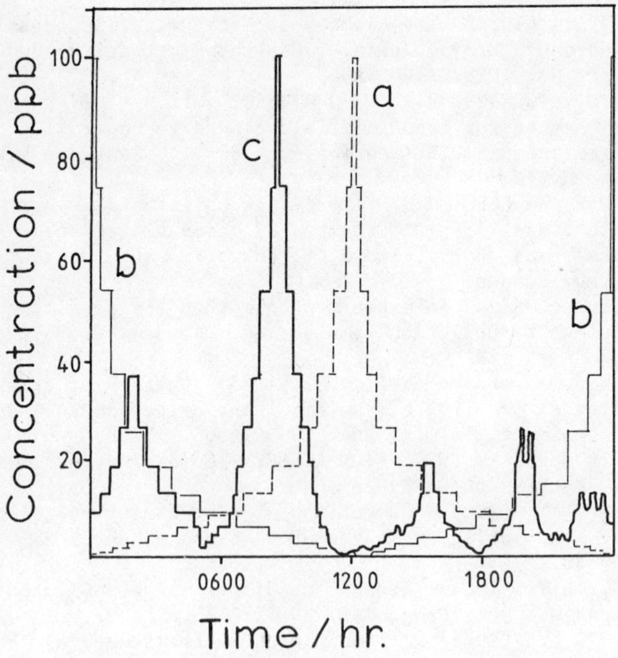

FIGURE 1

Hypothetical diurnal variations of a pollutant gas having the frequency distribution of Table 1.

a) Single day-time peak

b) Single night-time peak

c) Quasi-random variation

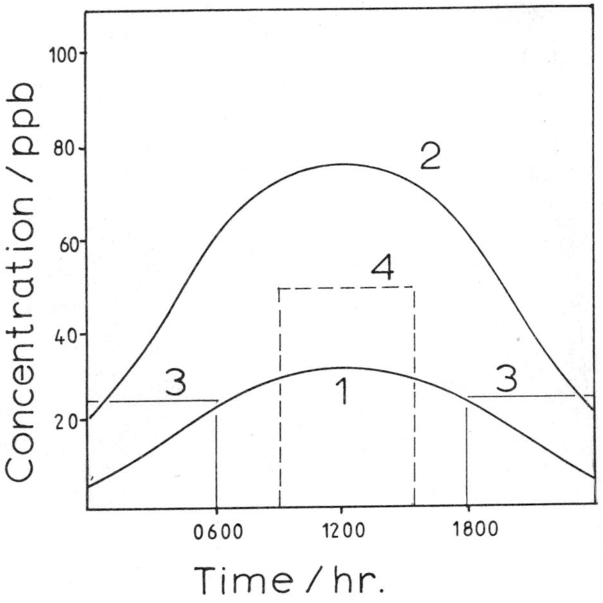

FIGURE 3.

Showing the significance of the phase relationship between the ambient gas concentration (1), its fumigated counterpart (2) and the occurrence of a second phytotoxic, but not fumigated, gas (3 or 4)

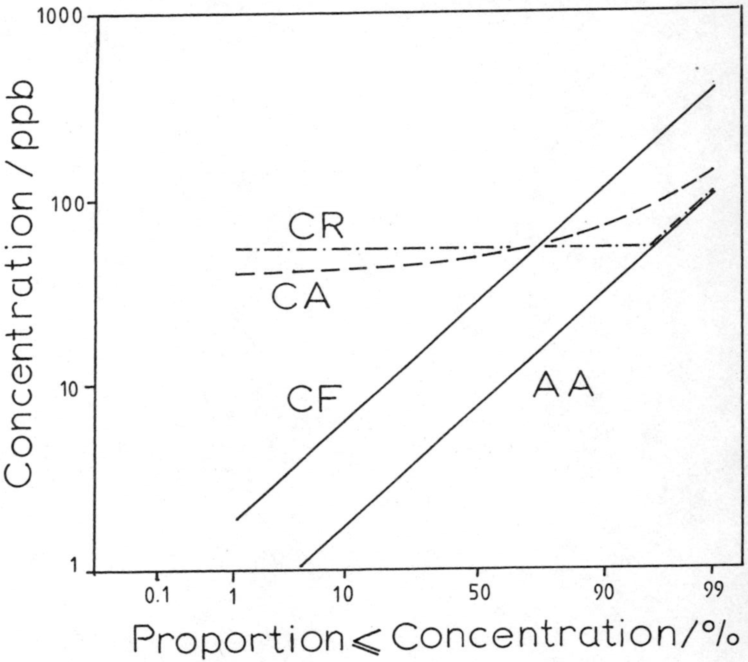

FIGURE 2.

The effect on the natural frequency distribution (AA) of different fumigation regimes.

CR - constant resultant

CA - constant addition

CF - constant factor

SESSION IV

EFFECTS OF AIR POLLUTION ON AQUATIC ECOSYSTEMS

SESSION IV

EFFETS DE LA POLLUTION DE L'AIR SUR LES ECOSYSTEMES AQUATIQUES

SITZUNG IV

WIRKUNGEN DER LUFTVERSCHMUTZUNG AUF AQUATISCHE OKOSYSTEME

Chairman : Dr. A. TOLLAN, Norwegian Water Resources and Energy Administration, Norway

Rapporteur : Dr. S.C. WARREN, Water Research Center, United Kingdom

REPORT ON THE SESSION

S C WARREN
Director, Water Research Centre, Medmenham

1. THE HISTORY AND CAUSES OF SURFACE WATER ACIDIFICATION IN EUROPE

The study of fossil diatoms in sediment cores from lakes shows that acidification of surface waters began in parts of the UK as much as 150 years ago. In Europe acidification began later following the spread of industrialisation.

The rate and extent of acidification depends on total acid deposition (loading) and catchment sensitivity (the content of weatherable silicate and carbonate minerals in the soils and rocks) but current evidence suggests that :

Acidification is confined to areas where the pH of rain averages less than 4.7. There is no evidence of recent (post 1800) acidification of clear water lakes where the pH of rain averages 4.9 or more.

Evidence from lake sediment cores does not support the view that changes in land management, other than afforestation, have brought about the acidification of lakes in Great Britain. The effects of afforestation are, however, probably due in part to their greater capacity to capture atmospheric acids and increase acid loading. Studies in Scotland and Wales have shown that forestry does increase the acidity of run-off water and evidence from Llyn Brianne suggests that streams in forested catchments experience more intense acid episodes without a great reduction in mean pH. There are reports that afforestation has caused soil acidification in Germany and The Netherlands but it is difficult to distinguish effects due to the capacity of forests to remove acidic materials from the atmosphere from those due to the trees alone. The absence of acidified lakes in afforested catchments in regions of low deposition and the experience of acidified lakes in unafforested catchments support the conclusion that :

Forests, although they may exacerbate and enhance the effects of acid deposition, are neither necessary nor sufficient to cause the acidification of lakes.

Recommendations :

In order to provide guidance for land use policy and practice in regions where rainfall averages pH 4.7 or below :

1a) A better understanding is needed of the mechanisms by which deciduous and coniferous forests and other vegetation
- acidify soils and surface waters, irrespective of acid deposition
- increase the capture of acids from the atmosphere
- quantitative estimates are desirable of the relative importance of these two factors.

1b) Methods need to be developed and tested for preventing or ameliorating the adverse effects of forest planting and other land use changes.

Atmospheric inputs of inorganic nitrogen compounds are generally retained and accumulated in forest ecosystems. There is, however, evidence that some forest systems are becoming saturated with nitrogen and that, as a result, nitrate is leaking from them. Removal or destruction of the forest can also result in the release of some of the accumulated nitrogen as nitrate, which then contributes to soil and water acidification. Waldsterben may therefore increase soil and surface water acidity in the affected catchment.

Recommendation :

1c) In order to improve forecasts of the secondary effects of Waldsterben on surface waters and, if necessary, to guide rehabilitation policy and practice, quantitative estimates are needed of the potential nitrate pool which might be released by land use changes, whether deliberate or accidental, in sensitive catchments. These should be used to estimate the potential acidifying effects on catchments of different types of vegetation.

Palaeolimnological evidence is available only for lakes, most of which do not experience short term fluctuations in pH. Fast flowing upland streams, on the other hand, can experience rapid temporary drops in pH and rises in aluminium concentration, particularly during snowmelt in the spring and during the early autumn rains.

Streams may thus experience temporary changes in chemistry including lower pH which do not occur in an associated lake. The diatom record in the lake sediment will therefore give no indication of the extremes of pH reached in, say, a stream feeding the lake.

In general a stream may well experience acid episodes, and perhaps suffer a decline in fish population, even when the pH of an associated lake still supports a diatom population characteristic of a rather higher pH. The pH profile inferred from the diatom record may thus lag behind the extremes experienced by the stream. Conversely, if deposition falls, the stream may continue to experience occasional damaging episodes after the lake pH and its diatom population records show a recovery. The magnitude of this hypothetical lag is uncertain.

2. BIOLOGICAL EFFECTS OF ACIDIFICATION

Falling pH affects biota at each trophic level and results in altered species composition, reduced species number and altered relative abundance. Fish are nearly always absent in waters of pH less than 5 but a decline in the population often occurs when the mean pH is greater than this. At least two species of fish (the Mudminnow (Umbra pygmaea) and Tilapia (Oreochromis mozambicus) seem able to tolerate lower pH values.

Acidic episodes can be severe enough to cause fish kills but knowledge of the tolerance of ecosystems to episodes of varying severity and duration is poor. The status of a stream cannot necessarily be inferred from the average values of its pH and alkalinity.

Recommendations :

2a) In order to be able to predict the biological consequences of changes in acid deposition we need a better understanding of the response of ecosystems to short term fluctuations, as well as to long term changes, in pH and alkalinity.

2b) In order to be able to detect the effects of changes in water acidification at an early stage we need more sensitive, perhaps sub-lethal, biological indicators.

3. POSSIBLE MECHANISMS FOR THE BIOLOGICAL EFFECTS OF ACIDITY ON FISH

Rapid acidification causes loss of sodium through the branchial epithelium, which can itself be lethal. If the water is acidified more slowly the electrolyte loss is much lower, but in most experiments measurements have not been made for long enough periods to establish whether low level losses continue for long periods and, if so, their significance. The presence of calcium ions greatly reduces electrolyte loss and affords some protection for the fish but during acid episodes stream flow increases, temperature falls and other transient changes in chemistry occur, which may cause additional stress.

The effects of exposure of fish to low pH are broadly understood, but little is known about the effects of additional stress factors which usually occur at the same time as low pH, nor about their effects on other aquatic organisms.

Recommendation :

3a) In order to improve our understanding of the response of aquatic ecosystems to acidification laboratory studies of the response of fish to acid stress should be carried out using acidification rates and, ideally, changes in temperature typical of those which occur naturally during actual acid events.

Aluminium is implicated in the effects of low pH on trout. At pH 5.2-5.5 aluminium hydroxide precipitates on the gills and causes suffocation. At lower pH values the aluminium in ionic form seems to be toxic.

The toxicity of aluminium is modified in the presence of organic matter (humic acids) and by the presence of calcium. Life stages vary in their sensitivity to pH and aluminium concentration. The environmental conditions to which fish have previously been subjected greatly affect their response to low pH and perhaps also to aluminium. Changes in species composition and abundance caused by acidification may reduce the supply of food and further weaken the fish.

Laboratory studies may provide a better understanding of the importance of some of these factors but so many variables are involved interactively in the response of aquatic organisms to acidification that it is difficult to predict reliably and accurately the response of a particular ecosystem to an event.

Recommendations :

3b) In order to improve our ability to forecast the effects of reductions in deposition, laboratory and field studies of the response of ecosystems to acidification and to acid episodes should be carried out.

3c) The results should also help to identify measures which might be taken to mitigate the impact of acidification and acidic episodes.

4. REVERSIBILITY OF ACIDIFICATION

In order to forecast the effects of reducing emissions a model is needed to describe the way in which a catchment responds to acid deposition. The most commonly used model is MAGIC (Model of Acidification of Groundwater in Catchments).

Retrospective predictions of surface water acidification made by MAGIC using historical data have been compared with the pH record inferred from diatom remains in dated core samples. The agreement for pH is poor. In general MAGIC seems to predict a more regular fall in pH with time whereas the diatom record more often shows a relatively constant pH before a fairly sharp fall to a new level. The agreement for alkalinity is a little better but until we have a convincing explanation for the discrepancies between the methods, confidence in them will be restricted. It must be borne in mind that sediment cores record the history of a lake but MAGIC is explicitly designed to perform long term simulations of changes in soilwater and streamwater chemistry. The possible sources of error in sediment core analysis are described by Davis and Smol in 'Diatoms and Lake Acidity' edited by Smol, Battarbee, Davis and Merilainen but :

in general diatom records are likely to give a reasonably accurate indication of trends and the timing of change. Reconstruction of actual pH values is more difficult although probably reliable within reasonable limits.

MAGIC is a valuable tool but it has a number of shortcomings :

* **it uses estimated historic regional emissions as a surrogate for acid deposition, data for which are, of course, lacking**
* **it contains no hydrological element and so cannot take account of changes in intensity and frequency of episodes and cannot, therefore predict the onset of biological perturbation**
* **it lumps soil chemistry on a catchment basis and thus does not address 'fine scale' heterogeneity.**

Recommendations :

4a) The errors inherent in inferring pH trends from sediment cores should be quantified as far as possible and limits placed on the range of pH/time curves which might be consistent with the fossil record.

4b) Efforts should be made to refine the method so as to minimise these errors as far as possible.

4c) MAGIC should be further tested against palaeolimnological, historical and experimental (eg the RAIN project) data sets to assess its usefulness as a predictive tool.

d) Attempts should be made to infer actual trends in local deposition from sediment cores, perhaps by using soot particles or other means. If successful these will provide more accurate and realistic input data for MAGIC to see if better agreement could be obtained.

The response of a catchment to reductions in sulphur deposition depends on the rate of release of base cations from weathering, the sulphate absorption properties of the soil and the size of the pool of soil sulphate. At one extreme sensitive catchments with very thin soil covering will be in equilibrium with deposited acid, will have a very small reserve of sulphate ion and the water base cation levels will respond relatively rapidly to any change in deposition. At the other extreme insensitive catchments have thick soil coverings carrying a high load of previously deposited sulphate and are not in equilibrium with deposited acid. These catchments will continue to leach sulphate even if no further acid is deposited and some of them may, for practical purposes, be irreversibly acidified.

Because of its treatment of soil chemistry MAGIC is more likely to be successful in predicting the response to changes in deposition of catchments with a thin or negligible soil covering than of those with deeper soils with more than one horizon. In predicting the response of a catchment to changes in emissions much will depend on how accurately local deposition can be inferred from emissions. Doubts will remain about the detailed accuracy of predictions of MAGIC and similar models until they have been better validated and, if necessary, improved.

The episodes to which streams will still be subject may delay the recovery of the stream ecosystem compared with that of a lake, but this will depend on the turnover time of the lake.

Even if we could forecast the effects of emission reductions on water chemistry, the response of ecosystems to stream chemistry, particularly to acid episodes, is not yet well enough understood to support reliable forecasts of the effects of such reductions, or of other measures on ecosystems.

Recommendation :

4e) An examination of the recent fossil record from a sensitive catchment in a region in which acid deposition has been falling for some years, might yield evidence for a change in lake chemistry which could be compared with retrospective MAGIC predictions.

ACID DEPOSITION AND SURFACE WATER ACIDIFICATION: THE USE OF LAKE SEDIMENTS

RICHARD W. BATTARBEE, Palaeoecology Research Unit,
University College London,
26 Bedford Way, LONDON WC1H OAP, UK.

and

INGEMAR RENBERG, Department of Ecological Botany,
Umeå University, S-901 87
Umeå, Sweden.

Summary

A summary of the palaeolimnological techniques used in lake acidification studies are presented. Sediments reconstruct trends in lake chemistry and biology, in catchment history and in atmospheric contamination. Comparisons of trends in space and time both within and between lakes allow alternative hypotheses for the acidification of lakes to be tested. Although there are situations where catchment changes have caused lake acidification sediment studies show that clear water lakes with pH values <5.0 have only become common during the last few decades in areas of high acid deposition. Repeat sediment sampling can be used as an effective means of monitoring lake response to sulphur-reduction strategies.

1. Introduction

In studies of surface water acidification we need to differentiate between acid and acidified lakes, to assess the extent and timing of acidification, and to evaluate alternative causes for acidification. None of these can be achieved without reliable historical information. In the absence of accurate continuous documentary chemical and biological records for acid lakes the historical record that can be derived from the analysis of lake sediments is of fundamental importance. Lake sediments not only contain a diatom record that can be used for pH reconstruction (1) but also other physical, chemical and biological records that can be used to reconstruct aspects of lake and catchment ecology as well as the history of atmospheric contamination (Fig. 1). Such an approach is indirect and requires calibration and interpretative skill but the combined analysis of a full range of parameters can be used to document environmental trends and to test alternative hypotheses (2). This approach is being pursued in the United Kingdom under the auspices of the Department of Environment (3), it is the basis of the palaeolimnology project within the Surface Water Acidification Programme (SWAP) being carried out in Norway, Sweden and the UK (4), and has been used effectively in North America in the PIRLA (Palaeoecological Investigation of Recent Lake Acidification) project (5).

Lake sediments have two unique advantages over other sources of historical information: they usually provide a continuous record that can be extended back through the whole post-glacial period, and they constitute a record that is potentially available at all lacustrine sites. But there are also problems which vary in seriousness from site to site. These include sedimentological complexities, hiatuses, missing or disturbed sequences, problems of dating accuracy and temporal resolution, questions of microfossil preservation, and difficulties in distinguishing between alternative sources of similar material.

In this paper we review the use of paleolimnological techniques in studies of lake acidification and draw upon data both published and unpublished from a range of European countries, and where appropriate from North America. Almost all the information acquired in this way is dependent on the establishment of accurate sediment chronology for the relevant time period and the most important technique used for this purpose is 210Pb dating (6,7).

2. Trends in lake chemistry, especially pH

The most useful technique for reconstructing pH trends in acid lakes is diatom analysis (8,9,10) although both cladoceran assemblages (11) and chrysophytes (12) have also been used. Many acidified lakes in western and northern Europe follow the sequence exemplified by Gårdsjön in S.W. Sweden. Planktonic diatoms such as Cyclotella kützingiana decline first when the pH falls below about 5.8. This is followed by a decline in circumneutral periphytic diatoms such as Achnanthes minutissima and Anomoeoneis vitrea which are often replaced by acidophilous taxa such as Eunotia veneris and Tabellaria flocculosa. In many cases acidobiontic diatoms, especially T. quadriseptata, and T. binalis are found in abundance in the uppermost sediment indicating a decline in pH to less than 5. In extreme circumstances with pH values as low as 4.0 the diatom assemblage may consist only of Eunotia exigua, a species that appears to be tolerant of very high acidity and aluminium levels (13,14).

Since pH and factors co-varying with pH usually explain most of the variation in diatom distribution in dilute freshwaters, past pH values can be inferred with confidence from diatom assemblages, and a number of simple linear regression models have been developed for this purpose (15,16,17,8,9,10). Most of these models are based on a classification of diatoms into pH groups (18). Fig. 2 shows an example for Gårdsjön using the Index B approach of Renberg & Hellberg (17). The advantages and limitations of these methods have been discussed by Battarbee (1), Davis et al.(19), Charles & Norton (20), and Battarbee & Charles (21). Most recently Birks (22) has proposed a new statistical method which is more suited to the non-linearity of pH-diatom distributions, an approach which is being validated at present. In general all models show very similar pH trends although pH values can vary by about +/- 0.3 pH unit.

Although many of the same diatom species occur in almost all acid and acidified lakes within a region, there can be substantial floristic differences between lakes of similar pH in response to variations in other environmental factors such as water colour, lake morphometry, and available microhabitats. The importance of colour has been especially stressed by Davis et al. (23) who show that diatom assemblages can be used to reconstruct quantitative trends in TOC as well as in pH. Considerable research effort is now being given to the reconstruction of water quality parameters other than pH.

3. Trends in lake biology

Lake sediments contain a wide range of fossil material derived from different communities and habitats within the lake including diatoms, chrysophytes, chlorophyta, aquatic macrophyte pollen and seeds, cladocera, rotifera, and chironimidae.

3.1. Diatoms

Diatoms are the algae most completely represented in the sediments and their remains are usually used as environmental indicators (see above). However, they are also an important group of primary producers. The upper sediments of very acid lakes in areas of high acid deposition characteristically show decreasing proportions of planktonic diatoms (Fig. 2, 17,24,25), consistent with the observation that this group of diatoms is rare in waters with pH <5.8 (9,10). Smith has recently shown that Asterionella formosa held in culture at a pH of 5.5 grows well in water of this pH with aluminium concentrations less than 100 ug l-1, but higher concentrations, especially in waters with low (< 100 ueq l-1) calcium levels, are lethal (26). Despite the decline in plankton populations, non-planktonic diatoms (especially epilithic and epiphytic forms) are abundant in acid lakes, although the species present are substantially different from those in less acid situations. Calculations of diatom fluxes to the sediments of acidified lakes show no consistently downward trends during acidification. These data, although uncorrected for biomass differences between species, indicate that there is a compensatory shift in diatom production from planktonic habitats to littoral and benthic ones, but no major decline in overall diatom productivity.

3.2. Chrysophyceae

The siliceous cysts and scales of chrysophytes are also often present in abundance in lake sediments. Unfortunately many of the cysts can not yet be identified so the potential of this part of the fossil record has not been fully realised. On the other hand scales from species in the Mallomonadaceae family can be identified and considerable use of them has already been made in acidification studies (12,27). These scaled chrysophytes are mainly planktonic forms. Consequently in lakes with pH <6.0 they can be better indicators of open-water conditions than diatoms. Stratigraphic analyses of a core from Deep Lake, an acidified Adirondack lake (12) showed the replacement of a circumneutral flora dominated by Mallomonas crassisquama by the acidobiontic forms M. hindonii and M. hamata. Similar results were obtained from an analysis of Big Moose Lake (27).

3.3. Other primary producers

In addition to diatoms and chrysophyceae, microfossil remains of other primary producers also occur in lake sediments including some species of chlorophyceae, especially Pediastrum, algal and bacterial pigments, and the pollen, spores, and macrofossils of some aquatic plants (28,14). More studies of these remains in the sediments of acid lakes are required.

3.4. Cladocera

Cladocera form populations in the zooplankton (eg Bosmina, Daphnia) and in the littoral zone (the Chydoridae). Their chitinous parts are preserved well in sediments. Few cores from acidified lakes have been analysed but some tentative generalisations can be made. One feature appears to be a decline in Daphnia ephippia concentrations at the time of acidification (29). This finding is consistent with the observation that Daphnia populations are rare in lakes with low pH (30,29). Evidence for changes in Bosmina populations is conflicting. Arzet et al. (31) report an abrupt decline in the concentration of Bosmina longispina in the uppermost sediments of Grosser Arbersee (Bavaria, FRG) whereas Dayton (pers. comm.) has shown that Bosmina increases during the acidification of the Round Loch of Glenhead in south-west Scotland.

Chydorid populations are more diverse and changes in the composition of the sediment assemblage seem to parallel other indicators of acidification (31,32,11). At Nedre Mălmesvatn in southern Norway increases in Alonella nana and Alonopsis elongata and decreases in Alona affinis and Chydorus piger occurred at the same time as the increase in acidophilous diatoms (32). At other sites, such as Grosser Arbersee and Pinnsee (31), these species seem to be less sensitive to pH changes suggesting that other factors such as habitat change, predator-prey relationships, as well as water chemistry may be important. More studies of these important groups are required both in terms of modern ecology and biostratigraphy. Several new sites are being studied as part of the SWAP programme (4).

3.5. Chironomidae

Chironomidae occupy benthic habitats and leave well preserved larval head capsules in the sediment (33). There have been a few studies of these in the sediments of acid lakes. Henriksen et al. (34) and Henriksen & Oscarson (35) have described the changes in Gårdsjön sediments during the period in which pH declined from about 6 to 4.5. In particular the acidification coincided with an increased abundance of Sergentia, Psectrocladius and Protanypus. In a comprehensive study of the sediments of Achterste Goorven, a strongly acidified pool in the Netherlands, Dickman et al.(14) described a change from Psectrocladius psilopterus assemblage to a P.platypus assemblage and a substantial increase in the number of Chironomus head capsules in the upper 10 cm.

3.6. Other animal remains

There are few direct records of fish history in lake sediments, although in one of our SWAP projects we are trying to evaluate whether there is a relationship between changes in cladoceran morphology (eg body size, antenna length) and recorded fish presence and absence (4,36). Rotifer eggs are also preserved in lake sediments but so far attempts to use them in acidification studies have failed (May pers. comm.).

4. Trends in catchment characteristics

Water quality and the flora and fauna of lakes are influenced by their catchments. In studies of lake acidification it is necessary to reconstruct the history of catchments in order to separate possible catchment influences from atmospheric ones. For recent times the study of documentary records and local interviews are the most useful techniques (37,38), but changes in the character of the catchment can also be discerned from the geochemical, mineral magnetic, pollen and charcoal records of the lake sediment and from changes in the texture and accumulation rate of the sediment itself.

Battarbee et al. (39) have shown how sediment accumulation rates rapidly increased as a result of soil erosion caused by deep catchment ploughing prior to afforestation in Galloway. Accumulation rates of lakes without such disturbance showed no change. Soil inwash can also be indicated by increases in the concentration of cations especially Na and K in sediments (40,41), by changes in magnetic mineralogy, and by changes in loss on ignition values. In many British cases the impact of such events on biota in the lake can be clearly observed most often by depressions in the proportions of the spores of the submerged macrophyte *Isoetes lacustris* and sometimes by a sudden response of the diatom phytoplankton as the inwash releases nutrients to the lake for a short period (42).

The most common technique for indicating vegetation and land-use change in sediments is pollen analysis, although in some situations the method is not very precise. Pollen grains are derived from areas beyond the immediate lake catchment and many important pollen groups can only be identified to family or genus levels. Nevertheless much useful information can be gained from the trends indicated chiefly in order to test hypotheses about land-use change and acidification (see below). For example, pollen analysis at a range of sites in the Galloway hills, south-west Scotland shows no increase in the *Calluna*/Gramineae ratio during the period of lake acidification. In association with other evidence (2) these data discount the hypothesis that acidification may have been due to land abandonment and the regeneration of *Calluna* heathland. Over longer periods of time within the post-glacial period pollen analysis is essential in evaluating the causes of limnological change (43,44,45,46). Before modern problems of pollution, soil and vegetation disturbances, whether caused by natural (climatic or geological) or anthropogenic factors, were the main influences on freshwater ecology.

5. Trends in atmospheric contamination

Lake sediments record changes in atmospheric contamination in a variety of ways. In some cases the relative contributions of catchment and atmospheric sources are not clear and in others post-depositional processes modify the stratigraphic accuracy of the record. However, the range of independent techniques available is large enough for many of these doubts and ambiguities to be resolved. The most important techniques for acidification studies are trace metal and sulphur geochemistry, polyaromatic hydrocarbon (PAH) analysis, carbonaceous particle analysis and magnetic mineral analysis.

5.1. Trace metals

The trace metal record in lake sediments derives partly from the catchment and, since the 18th century, partly from the atmosphere. The increased atmospheric flux is especially reflected by large increases in recent sediments in the concentration of Pb, Zn, Cd, and other elements. Pb is widely regarded to be the most stable element. Unlike Zn and Cd which can be mobilised from sediments in acidified lakes (47), the pattern of Pb can be used as a general surrogate for historical air pollution related in a direct way to fossil fuel combustion.

Fig. 3 shows a Pb and Zn profile for L. Tinker in the southern Highlands of Scotland. Background concentrations prior to about 1800 AD are very low. From 1800 onwards there is a steep increase in values reflecting the historic rise in atmospheric contamination. Maximum values occur in post-1940 sediments, but declining values characterise the most recent sediments. Depressions in the general pattern (eg at 17 cm) are due to dilution by occasional soil inwash events. The sub-surface decrease in both elements occurs at about the same time. Because of this and because Loch Tinker has a relatively high pH of 5.9 the declines in both elements are probably related to a real decrease in atmospheric contamination at this site over the last two decades or so. A contrasting situation is shown in Fig. 4 where the trace metal curves for Gårdsjön in South-west Sweden show that the beginning of significant contamination occurred over a century later than in Scotland and where declines in Zn and Cd concentrations are related more to the remobilisation of these elements as the lake acidified to its present pH of 4.5 than to the decline in atmospheric contamination indicated by the small Pb decline.

5.2. Total sulphur

Although the aquatic and terrestrial components of lake-catchment systems are capable of supplying a considerable sulphur flux to sediments, it is intuitively probable that elevated SO_2 emissions and the related increased SO_4 concentrations in lake waters in areas of high acid deposition might cause an increase in the concentration of total S in lake sediments in parallel with the increases in trace metal values. There is now good evidence that sulphate reduction at the sediment surface is an important process in removing sulphur from the water column and in generating alkalinity. It can also account for the permanent accumulation of S in sediments (48,49,50). However, the extent to which a total S profile from a sediment core accurately reflects the historical trend of S deposition is a matter of some debate since S can be remobilised by chemical diagenesis and transported through the sediment pore water in response to vertical diffusion gradients (51). Consequently the S profile can vary from lake to lake in response to both input history and post-depositional changes.

Fig. 3 shows a S profile from L. Tinker. The trends in the upper part are similar to those of Pb from the same core and probably reflect increased S deposition. However, the lower part of the profile has relatively high values indicating that catchment sources of sulphur are also important at this site. Some acidified sites show no S elevation in the upper sediment levels suggesting that sulphate reduction is not an important process at such sites and that sulphate is exported from the lake basin at a rate equal to its input.

5.3. Carbonaceous particles

Carbonaceous particles derived from fossil fuel combustion are found in considerable quantities in recently deposited sediments (52,53). In Swedish lakes the annual accumulation of coarse spheres (> 5 um diameter) in varved sediments follow the same main pattern as statistical data for coal and oil combustion over the last two centuries (Fig. 5, 54). It is likely that carbonaceous particle records in non-varved sediments also accurately reflects the deposition history of these pollutants and high concentrations are found in the recent sediments of all acidified lakes so far studied both in Sweden and in the UK (54). Detailed SEM and EDS analysis shows that most of the particles have a morphology and chemical structure indicating that they are mainly derived from combustion of fuel oil (55). Alumino-silicate fly ash particles also occur in lake sediments (56) that are derived from high temperature coal combustion but techniques have not yet been developed to quantify them effectively.

6. Causes of acidification - hypothesis testing

Whilst the above techniques can be used separately and together to describe changes in lakes through time they can also be used to test alternative hypotheses for the causes of lake acidification. This can be achieved in two main ways, first, by comparison between sites in which site characteristics are systematically varied and second, by reference to within site trends in which lake responses to catchment or atmospheric perturbations are monitored. In ideal circumstances these two approaches can be combined. For example, in Galloway the hypothesis that lakes with pH <5.5 are acidified due to post-war afforestation can be simply disproved by showing that adjacent afforested and non-afforested catchments are acidified and that acidification of afforested catchments occurred well before planting (57,25). This does not imply, of course, that afforestation is not an acidifying influence. Indeed studies of sediment cores from Loch Fleet, also in Galloway, show that acidification took place at this site in the 1970's a decade or so after afforestation (42). This is consistent with the hypothesis of Harriman & Morrison (58) that the forest effect becomes significant 10-15 years after planting as the forest canopy closes.

A unique advantage of the palaeoecological approach is the possibility of studying lake/catchment relationships in the past when acid deposition from industrial sources was minimal. In this way the so-called "land-use" hypothesis can be effectively tested for catchments known to be sensitive to acidification. For example, Jones et al. (59) have shown that the wholescale development of Calluna heathland in the catchment of the Round Loch of Glenhead 4-6000 years ago did not cause the pH of the lake to fall below about 5.5, whereas a rapid decline in pH since about 1850 has occurred coincidentally with the rise in SO_2 emissions in the UK. Renberg has also used this approach to assess the effect of recent spruce (Picea) forest expansion in south Sweden. Diatom analysis of sediments before and after the natural immigration of spruce into north Sweden 3000 years ago showed that spruce caused no acidification (60).

The acid deposition hypothesis can also be tested (2). We can try to disprove it by showing that there is no evidence for atmospheric contamination at a particular site in question. In all cases so far studied this hypothesis has not been disproved since elevated trace metal concentrations and high carbonaceous particle concentrations are found at all acidified sites.

We can also attempt to disprove this hypothesis by demonstrating that the timing and spatial distribution of acidified sites is inconsistent with the hypothesis (61). Fig. 6 shows a map of sites in Europe for which dated diatom analyses are available. So far it can be seen that acidified sites only occur within the zone of high acid deposition, similarly sensitive sites outside this zone are not acidified. Dates for the onset of acidification between regions largely coincides with those expected on the basis of industrial history, with the UK showing evidence for the earliest acidification and with Finland the most recent. Norway, the Netherlands, and West Germany have intermediate dates.

7. Conclusion

Sediment core studies show beyond doubt that lakes with pH <5.0 were exceptionally rare during the post-glacial period. They have become common only in the last few decades and only in areas of high acid deposition.

Attempts to disprove the acid deposition hypothesis have so far failed. Catchment afforestation can have a substantial additional influence on a lake but so far it is not known whether this is independent of acid deposition. Present evidence suggests that the forest effect is related to the enhanced trapping of acid pollutants by forest canopies and that forests in "clean" areas impose little acidification influence. In the UK we are hoping to find afforested sites on sensitive geology in an area of low acid deposition where this effect can be assessed. The palaeolimnological method is well suited to many of the problems posed by the acid rain issue, but much further work is required to provide new insights and to refine present knowledge. Statistical models relating diatoms and other microfossils to water chemistry need continuing development and additional sites need to be included in the ongoing survey work. In particular we need to know more about the recent and long-term history of brown-water lakes and we need to study the limnology and palaeolimnology of non-acidified soft-water lakes in areas of low acid deposition both as a control for acidified lakes and as a reference for the future.

Palaeolimnology also has a role to play in the future since as sediments accumulate a continuing integrated record of lake response to environmental improvement is formed. Repeat sediment sampling of lakes with previously analysed sediment cores is probably one of the most efficient means of monitoring the effectiveness of sulphur reduction strategies in the future.

References

(1) Battarbee, R.W.: 1984, 'Diatom analysis and the acidification of lakes', Phil. Trans. R. Soc. 305, 451-477.

(2) Battarbee, R.W., Flower, R.J., Stevenson, A.C., and Rippey, B.: 1985, 'Lake acidification in Galloway: a palaeoecological test of competing hypotheses', Nature, 314, 350-352.

(3) Fritz, S.C., Stevenson, A.C., Patrick, S.T., Appleby, P.G., Oldfield, F., Rippey, B., Darley, J., & Battarbee, R.W. 1986: Palaeoecological evaluation of the recent acidification of Welsh lakes. 1: Llyn Hir, Dyfed. Research Paper No. 16. Palaeoecology Research Unit, University College London.

(4) Battarbee, R.W., and Renberg, I.: 1985, 'Royal Society Surface Water Acidification Project (SWAP) Palaeolimnology Programme' Working Paper No. 12, Palaeoecology Research Unit, University College London, 18 pp.

(5) Charles, D.F. et al.: 1986, 'The PIRLA project (Palaeoecological Investigation of Recent Lake Acidification): Preliminary results for the Adirondacks, New England, N. Great Lake States, and N. Florida', Water, Air and Soil Pollution, 30, 355-366.

(6) Appleby, P.G., and Oldfield, F.: 1978, 'The calculation of lead-210 dates assuming a constant rate of supply of unsupported 210Pb to the sediment', Catena, 5, 1-8.

(7) Appleby, P.G., Nolan, P.J., Gifford, D.W., Godfrey, M.J., Oldfield, F., Anderson, N.J., & Battarbee, R.W.: 1986, '210Pb dating by low background gamma counting' Hydrobiologia 143, 21-27.

(8) Davis, R.B. and Anderson, D.S.: 1985 'Methods for pH calibration of sedimentary diatom remains for reconstructing history of pH in lakes', Hydrobiologia, 120, 69-87.

(9) Charles, D.F.: 1985, 'Relationships between surface sediment diatom assemblages and lakewater characteristics in Adirondack lakes', Ecology, 66, 994-1011.

(10) Flower, R.J.: 1987 'The relationship between surface sediment diatom assemblages and pH in 33 Galloway lakes: some regression models for reconstructing pH and their application to sediment cores' Hydrobiologia 143, 93-104.

(11) Krause-Dellin, D. & Steinberg, C.: 1986, 'Cladoceran remains as indicators of lake acidification' Hydrobiologia, 143, 129-134.

(12) Smol, J.P., Charles, D.F., and Whitehead, D.R.: 1984, 'Mallomonadacean microfossils provide evidence of recent lake acidification' Nature, 307, 628-630.

(13) Renberg, I.: 1985, 'A sedimentary diatom record of severe acidification in Lake Blåmissusjön, N. Sweden, through natural soil processes' In Smol, J.P., Battarbee, R.W., Davis, R.B. & Merilainen, J. (eds.) Diatoms and lake acidity W. Junk, Dordrecht, pp 213-220.

(14) Dickman, M.D., van Dam, H., van Geel, B., Klink, A.G., & van der Wijk, A.: 1987, 'Acidification of a Dutch moorland pool, a palaeolimnological study' Archiv fur Hydrobiologie (in press).

(15) Nygaard, G.: 1956, 'Ancient and recent flora of diatoms and chrysophyceae in Lake Gribso. Studies on the Humic, Acid lake Gribso', Fol. Limnol. Scand., 8, 32-94.

(16) Merilainen, J.: 1967, 'The diatom flora and the hydrogen-ion concentration of the water', Ann. Bot. Fenn., 4, 51-58.

(17) Renberg, I., and Hellberg, I.: 1982, 'The pH history of lakes in southwestern Sweden, as calculated from the subfossil diatom flora of the sediments', Ambio, 11, 30-33.

(18) Hustedt, F.: 1937-1939, 'Systematische und ökologische Untersuchungen über den Diatomeen-Flora von Java, Bali, Sumatra', Arch. Hydrobiol.(Suppl.) 15 and 16.

(19) Davis, R.B. & Smol, J.P.: 1986, 'The use of sedimentary remains of siliceous algae for inferring past chemistry of lake water - problems, potential and research needs' In Smol, J.P., Battarbee, R.W., Davis, R.B., & Merilainen, J. (eds.) Diatoms and lake acidity, W. Junk, Dordrecht, pp 291-300.

(20) Charles, D.F., and Norton, S.A.: 1985, 'Paleolimnological evidence for trends in atmospheric deposition of acids and heavy metals', in Atmospheric Deposition: Historic Trends and Spatial Patterns, National Academy Press, Washington, D.C., USA, Chapter 9.

(21) Battarbee, R.W. & Charles, D.F.: 1987, 'The use of diatom assemblages in lake sediments as a means of assessing the timing, trends, and causes of lake acidification' Progress in Physical Geography (in press).

(22) Birks, H.J.B.: 1987, 'Methods for pH - calibration and reconstruction from palaeolimnological data: procedures, problems, potential techniques' Unpublished manuscript, 11pp.

(23) Davis, R.B., Anderson, D.S., and Berge, F.: 1985, 'Loss of organic matter, a fundamental process in lake acidification: palaeolimnological evidence', Nature, 316, 436-438.

(24) Renberg, I., Hellberg, T. & Nilsson, M.: 1985, 'Effects of acidification on diatom communities as revealed by analyses of lake sediments' Ecological Bulletins 37, 219-223.

(25) Flower, R.J., Battarbee, R.W. & Appleby, P.G. 1987: The recent palaeolimnology of six acid lakes in Galloway, south-west Scotland. Diatom analysis, pH trends, and the role of afforestation. Journal of Ecology in press.

(26) Smith, M.A.: 1987, 'An hypothesis for the physiological response of planktonic diatoms to lake water acidification' Unpublished manuscript, 10pp.

(27) Smol, J.P.: 1986, 'Chrysophycean microfossils as indicators of lakewater pH' in Smol, J.P., Battarbee, R.W., Davis, R.B., and Merilainen, J. (Eds.) Diatoms and lake acidity, W. Junk and Sons, The Hague, The Netherlands, pp. 275-287.

(28) Renberg, I., and Wahlin, J-E.: 1985, 'The history of the acidification of Lake Gårdsjön as deduced from diatoms and Sphagnum leaves in the sediment', Ecological Bulletins 37, 219-223.

(29) Brakke, D.F., Davis, R.B., & Kenlan, K.H.: 1984, 'Acidification and changes over time in the chydorid Cladocera assemblage of New England lakes' In Hendrey, G.R. (ed.) Early biotic responses to advancing lake acidification Ann Arbor, Boston, pp 85-104.

(30) Nilssen, J.P.: 1980, 'Acidification of a small watershed in southern Norway and some characteristics of acidic aquatic environments' Int. Revue ges. Hydrobiol. 65, 177-207.

(31) Arzet, K., Krause-Dellin, D., and Steinberg, C.: 1986, 'Acidification of selected lakes in the Federal Republic of Germany as reflected by subfossil diatoms, cladoceran remains and sediment chemistry', in Smol, J.P., Battarbee, R.W., Davis, R.B., and Meriläinen, J. (Eds.) Diatoms and Lake Acidity, W. Junk, The Netherlands, pp 227-250.

(32) Davis, R.B., Norton, S.A., Hess, C.T., and Brakke, D.F.: 1983 'Palaeolimnological reconstruction of the effects of atmospheric deposition of acids and heavy metals on the chemistry and biology of lakes in New England and Norway' Hydrobiologia, 103, 113-123.ton, pp 85-104.

(33) Frey, D.G.: 1964, 'Remains of animals in Quaternary lake and bog sediments and their interpretation' Arch. Hydrobiol. Beih. Ergebn. Limnol. 2, 1-114.

(34) Henriksen, L., Olofsson, J.B. & Oscarson, H.G.: 1982, 'The impact of acidification on Chironomidae (Diptera) as indicated by subfossil stratification' Hydrobiologia 86, 223-229.

(35) Henriksen, L. & Oscarson, H.G.: 1985, 'History of the acidified lake Gårdsjön: The development of chironomids' Ecological Bulletins 37, 58-63.

(36) Sandoy, S. & Berge, F. In preparation.

(37) Patrick, S.T. & Stevenson, A.C. 1986: Palaeoecological evaluation of the recent acidification of Welsh lakes: III. Llyn Conwy & Llyn Gamallt, Gwynedd. Research Paper No 19. Palaeoecology Research Unit, University College London.

(38) Patrick, S.T. 1987: Palaeoecological evaluation of the recent acidification of Welsh lakes. Research Paper No. 21, Palaeoecology Research Unit, University College London, 47 pp.

(39) Battarbee, R.W., Appleby, P.G., Odell, K. & Flower, R.J.:1985, '210Pb dating of Scottish lake sediments, afforestation and accelerated soil erosion' Earth Surface Processes and Landforms 10, 137-142.

(40) Mackereth, F.J.H. 1966: Some chemical observations on post-glacial lake sediments. Philosophical Transactions of the Royal Society, London, B 250, 165-213.

(41) Engstrom, D.R. & Wright, H.E. 1984: Chemical stratigraphy of lake sediments as a record of environmental change. In Haworth, E.Y. & Lund, J.W.G. (eds.) Lake sediments and environmental history, Leicester University Press, Leicester, pp 11-67.

(42) Anderson N.J., Battarbee R.W., Appleby, P.G., Stevenson, A.C., Oldfield, F., Darley, J., & Glover, G. 1986: Palaeolimnological evidence for the recent acidification of Loch Fleet, Galloway. Research Paper No 17, Palaeoecology Research Unit, University College London.

(43) Pennington, W., Haworth, E.Y., Bonny, A.P. & Lishman, J.P. 1972: Lake sediments in Northern Scotland. Philosophical transactions of the Royal Society, London, B 264, 191-294.

(44) Digerfeldt, G., 1972: The post-glacial development of Lake Trummen. Folia Limnol. Scand., Vol 16, 1-104.

(45) Whitehead, D.R., Charles, D.F., Jackson, S.J., Reed, S.E. & Sheehan, M.C.: 1986, 'Late-glacial and Holocene acidity changes Adirondack (N.Y.) lakes. In Smol, J.P., Battarbee, R.W., Davis, R.B. & Merilainen J. (eds.) Diatoms and Lake Acidity W. Junk, Dordrecht, pp 251-274.

(46) Jones, V.J. & Stevenson, A.C. & Battarbee, R.W.: in preparation.

(47) Norton, S.A.: 1986, 'A review of the chemical record in lake sediment of energy related air pollution and its effects on lakes' Water, Air and Soil Pollution 30, 331-346.

(48) Schindler, D.W.: 1986, 'The significane of in-lake production of alkalinity' Water, Air and Soil Pollution 30, 931-946.

(49) Cook, R.B. & Schindler, D.W.: 1983, 'The biogeochemistry of sulfur in an experimentally acidified lake' Ecological Bulletins 35, 115-127.

(50) Rudd, J.W.M., Kelly, C.A. & Furutani, A.: 1986, 'The role of sulfate reduction in long term accumulation of organic and inorganic sulfur in lake sediments' Limnol. Oceanogr. 31, 1281-1291.

(51) Holdren, G.R., Brunelle, T.M., Matisoff, G. & Whalen, M. 1984: 'Timing the increase in atmospheric sulphur deposition in the Adirondack Mountains' Nature 311, 245-248.

(52) Griffin, J.J. & Goldberg, E.D.: 1981, 'Sphericity as a characteristic of solids from fossil fuel burning in a Lake Michigan sediment' Geochim. Cosmochim. Acta 45, 763-769.

(53) Renberg, I., and Wik, M.: 1984, 'Dating recent lake sediments by soot particle counting', Verh. Internat. Verein. Limnol., 22, 712-718.

(54) Wik, M., Renberg, I. & Darley, J.: 1987, 'Sedimentary records of carbonaceous particles from fossil fuel combustion' Hydrobiologia 143, 387-394.

(55) Parker, J. 1987: Unpublished B.Sc dissertation, University College London.

(56) Vizzard, R.N.: 1986, 'Investigation of selected particles within Loch Fleet sediments' Loch Fleet Project - Interim report No. 6, NCB Research

(57) Flower, R.J. & Battarbee, R.W.: 1983, 'Diatom evidence for recent acidification of two Scottish lochs', Nature, 305, 130-133.

(58) Harriman, R. & Morrison, B.R.S.: 1982, 'The ecology of streams draining forested and non-forested catchments in an area of central Scotland subject to acid precipitation' Hydrobiologia 88, 251.

(59) Jones, V.J., Stevenson, A.C. & Battarbee, R.W. 1986: Lake acidification and the "land-use" hypothesis: a mid-post-glacial analogue. Nature, 322, 157-158.

(60) Renberg, I.: 1987, Unpublished manuscript.

(61) Battarbee, R.W. & Charles, D.F.: 1986, 'Diatom-based pH reconstruction studies of acid lakes in Europe and North America: a synthesis.' Water, Air and Soil Pollution 30, 347-354.

(62) EMEP.: 1984, 'Co-operative programme for monitoring and evaluation of the long-range transmission of air pollutants in Europe', Summary report from the chemical co-ordinating centre for the second phase of EMEP, EMEP/CCC-Report 2/84, Norwegian Institute for Air Research, 120 pp.

(63) Arzet, K., Steinberg, C., Psenner, R., and Schulz, N.: 1987, 'Diatom distribution and diatom inferred pH in the sediment of four alpine lakes', Hydrobiologia 143, 247-255.

(64) Battarbee, R.W., and Flower, R.J.: 1985, 'Palaeoecological evidence for the timing and causes of lake acidification in Galloway, Southwest Scotland', Working Paper No. 8, Palaeoecological Research Unit, University College London, 79 pp.

(65) Berge, F.: 1975, 'pH-forandringer og sedimentasjon av diatomeer i Langtjern', Norges Teknisk Naturvitenskapelige Forskningsrad, Intern. rapport 11, 1-18.

(66) Berge, F.: 1979, 'Kiselalger og pH i noen insjoer i Agder og Hordaland', SNSF Prosjektet IR 42/79, 64 pp.

(67) Berge, F.: 1985, 'Relationships of diatom taxa to pH and other environmental factors in Norwegian soft-water lakes', Ph.D. Dissertation, University of Maine, Orone, Maine, USA.

(68) Davis, R.B. and Berge, F.: 1980 'Atmospheric deposition in Norway during the last 300 years as recorded in SNSF lake sediments II. Diatom stratigraphy and inferred pH', in Drablos, D. and Tollan, A. (Eds.) Ecological impact of acid precipitation, Proceedings of an International Conference, Sandefjord, Norway. SNSF Project, Oslo, Norway, pp 270-271.

(69) Haworth, E.Y.: 1985, 'The highly nervous system of the English Lakes: aquatic ecosystem sensitivity to external changes, as demonstrated by diatoms', FBA Annual Report, 53, 60-79.

(70) Miller, U.: 1973, 'Diatoméundersokning av bottenproppar fran Stora Skarsjön, Ljungskile', Statens Naturvardsverk Publikationer, 7, 43-60.

(71) Simola, H., Kenttamies, K., and Sandman, O.: 1986, 'The recent pH-history of some Finnish headwater and seepage lakes, studied by means of diatom analysis of 210-Pb dated sediment cores', Aqua Fennica, in press.

(72) Smol, J.P., Battarbee, R.W., Davis, R.B., and Merilainen, J. (Eds.):1986 Diatoms and lake acidity, W.Junk and Sons, The Hague, The Netherlands, 307 pp.

(73) Steinberg, C., Arzet, K., and Krause-Dellin, D.: 1984, 'Gewasserversauerung in der Bundesrepublik Deutschland im Lichte paleolimnoligischer Studien', Naturwissenschaften, 71, 631-633.

(74) Tolonen, K., and Jaakkola, T.: 1983, 'History of lake acidification and air pollution studied in sediments in South Finland', Ann. Bot. Fenn., 20, 57-78.

(75) Tolonen, K., Liukkonen, M., Harjula, R., and Patila, A.: 1986, 'Acidification of small lakes in Finland documented by sedimentary diatom and chrysophycean remains' in Smol, J.P., Battarbee, R.W., Davis, R.B., and Merilainen, J. (Eds.) Diatoms and lake acidity, W. Junk and Sons, The Hague, The Netherlands, pp 169-200.

(76) van Dam, H., Survmond, G., and ter Braak, C.J.F.: 1981, 'Impact of acidification on diatoms and chemistry of Dutch moorland pools', Hydrobiologia, 83, 425-429.

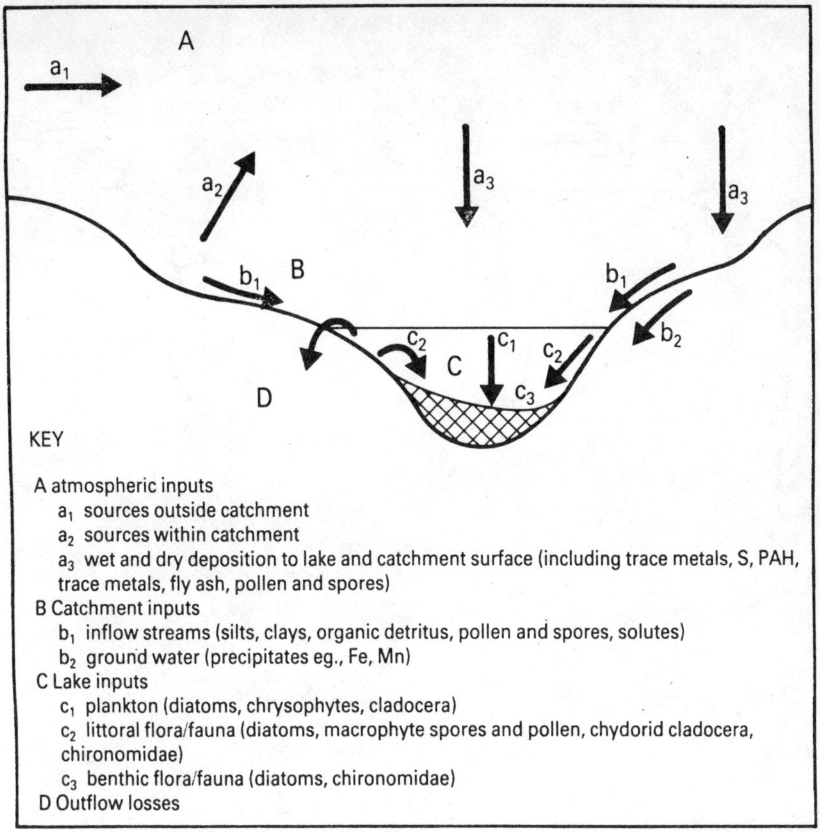

Figure 1 : Diagram to show the main sources of material found in lake sediments.

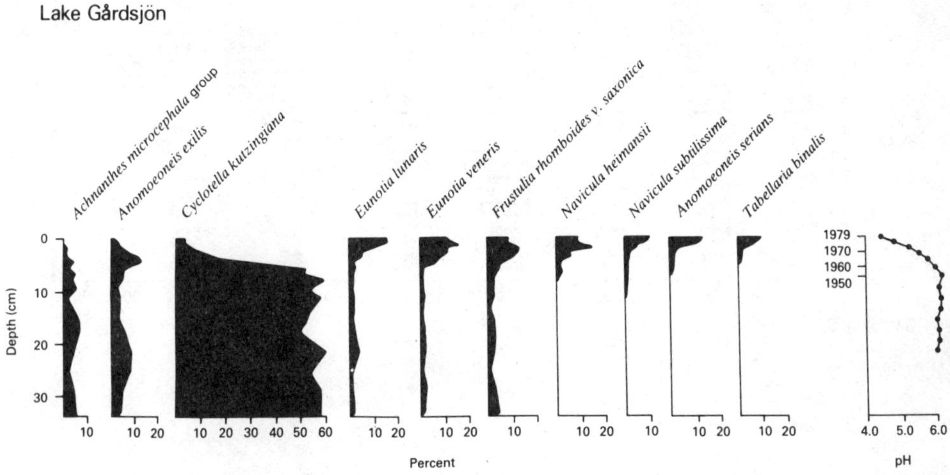

Figure 2 : Diatom diagram for Gardsjon, South-west Sweden.

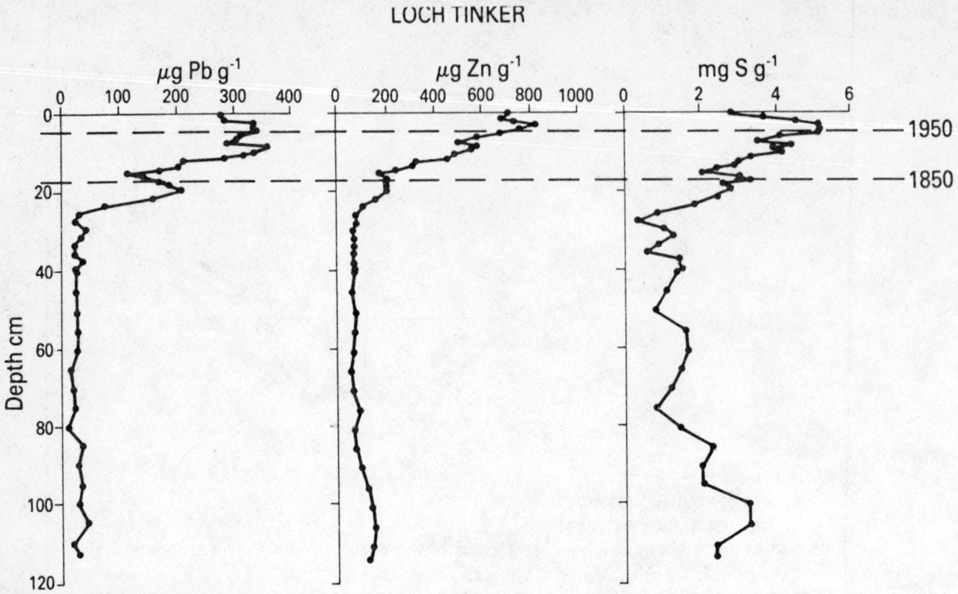

Figure 3 : Pb, Zn, and S trends for Loch Tinker, Scotland.

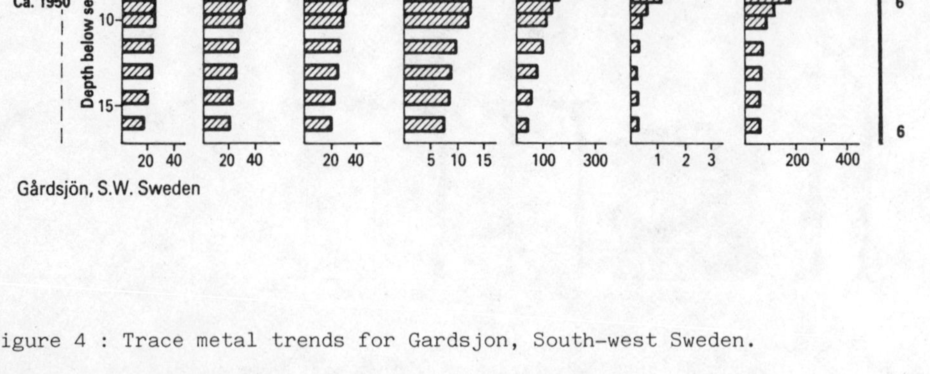

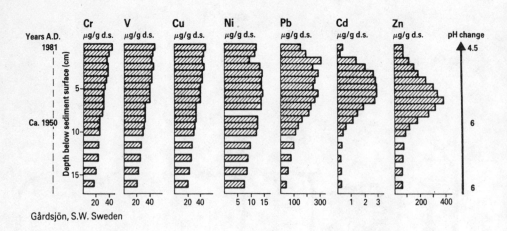

Figure 4 : Trace metal trends for Gardsjon, South-west Sweden.

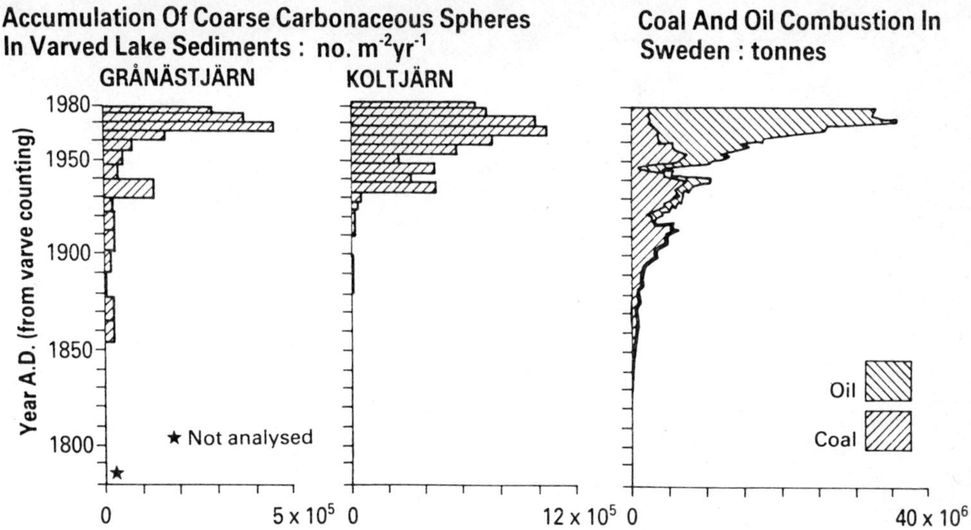

Figure 5 : Trends in carbonaceous particle concentration for two Swedish lakes compared with coal and oil combustion history (from ref. 54)

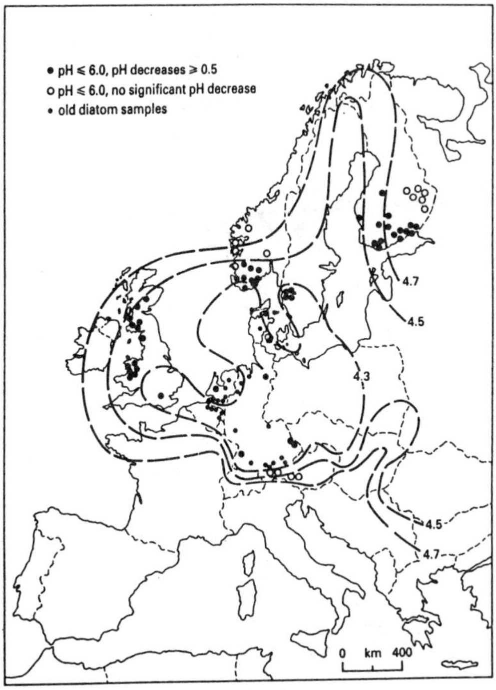

Figure 6 : Distribution of sites where recent (post 1800) acidification has occurred based on diatom data. Information from references 42, 63, 31, 1, 64, 61, 21, 65, 66, 67, 68, 32, 14, 62, 57, 25, 3, 69, 70, 17, 71, 72, 73, 74, 75, 76.

ACIDIFICATION AND ECOPHYSIOLOGY OF FRESHWATER ANIMALS

Jef H.D. VANGENECHTEN

Belgian Nuclear Research Center, S.C.K.-C.E.N.
Biology Department, Mineral Metabolism Laboratory
B-2400 Mol, Belgium

Summary

Ionic regulatory failure is supposed to be the principal mechanism determining death of animals in acid soft waters. Respiratory failure plays an important role if Al is present. Under specific physico-chemical conditions Al leads to mucus clogging at the gills and this hampers normal gas diffusion. This paper focusses attention on the many uncertainties in the mode of toxic action of acidified water. Especially the combined effects of high acidity, high metal concentration (esp. Al) and the presence of other cations (esp. Ca) and of natural organic acids are poorly understood.

Our knowledge on the ecophysiology of acid stress momentarily turns out to be inadequate to understand the responses on an ecosystem level. A more detailed understanding of the physiological basis of the acid toxicity stress would improve our ability to predict the impact of acidification as well as the impact of reversal of acidification.

To conclude, an overview of the gaps of knowledge and possible research priorities are given.

1. INTRODUCTION

Acidification of lakes and streams and the concomitant changes in water chemistry, can lead to adverse effects on aquatic biota. Our knowledge on acid-induced losses of fish populations and of other aquatic organisms, has increased rapidly during the past decade.

This paper is mainly based on data presented at an international symposium on "Ecophysiology of Acid Stress in Aquatic Organisms" held in Antwerp in 1987 (1). This symposium pointed to the need of a more advanced knowledge of the toxicity mechanisms to understand and predict the impact of water acidification. We are indeed often not able to predict the effects on the biota although we posses carefully registered chemical data.

This text reviews existing knowledge of acid stress physiology and assesses the merits and limitations of the recent data. Gaps or weaknesses of knowledge in this field of research, and a selection of recommandations for future research are eventually listed.

2. BIOTIC RESPONSES TO ACID STRESS

Not only the much emphasized decimation of fish populations, but all levels of biota are affected by acidification (2,3,4,5). The most obvious changements are : altered species composition and reduced species number. Species losses occur in many taxonomic groups : the diversity of most groups decreases as lakes acidify (4, and references therein). Concomitantly an increase in the number of tolerant individuals has been seen, perhaps partly due to a decreased predative pressure.

3. ABIOTIC CONDITIONS

To understand the basic physiological problems in aquatic animals under acid stress, we need to know more about the water quality conditions we are dealing with. Of course there is the acid pH (< pH 5.5), but acidified waters may be generally characterized by having low Ca content, low ionic strength and increased aluminium concentrations. For a detailed description of these abiotic conditions I refer to the paper of Wright in this volume (6). But not only these abiotic conditions are important, there is also the acidification rate. Acid waters are the result of a slow acidification process which has taken decades to reach the values which we are measuring now. Next to this slow acidification rate there is the periodical acidification which occurs during springtime snowmelt and/or during heavy rain in autumn. Under these conditions, pH may drop by 0.4-0.5 pH units in days or weeks and may remain low for short-time periods ; the Al concentrations increase concomitantly. These episodic events have regularly been described to lead to high mortality in fishes. The mortality in these conditions is related to the rate of pH change (7) and to the time of exposure to combined stress of low pH and high dissolved aluminium concentrations (8). However, more knowledge is needed on the speciation of aluminium during these short-term acid episodes in realistic conditions of decreased pH, higher water flow, possible resolubilisation of sediment-adsorbed aluminium etc.

4. THE PHYSIOLOGY OF H^+ (PROTON) STRESS

Exposure of aquatic animals to acid water leads to net losses of ions from the animal to the water. These reactions are studied in most detail in fishes (9), but some data are available also for invertebrates (10).

Net losses of Na and Cl in fish are typically largest during the first hours (0-12 h) immediately following acute pH stress. Wood and McDonald (11) in their work with trout, point to the fact that these initial ion losses are a good predictor of eventual mortality. This initial "shock" phase of large ion losses may kill by setting up intolerable fluid volume disturbances, hemoconcentration and circulatory collapse (9). In the later recovery phase, after the initial shock phase, zero-balance is slowly regained within 24-48 hours. However, this return to zero-balance may be apparent more than realistic. Net loss measurement procedures may not be sensitive enough to detect very small net losses. These losses may occur over very long periods of time leading to a further slow decrease in blood ion concentrations (12). As most of the experimental work confronts animals with acid pH for time periods for maximally 7 days, a further decrease of the blood ion content may be overlooked.

The detected net losses of Na and Cl in the initial shock phase invariably result from a combination of influx inhibition and efflux stimulation. Influx inhibition may be explained by a competitive action of H^+ on the Na^+/H^+ uptake mechanism. The high efflux rate on the other hand, appears to be the main cause of the extreme net ion losses. This is believed to be due to the opening of paracellular channels in the branchial epithelium (13,14). It is further believed that Ca ions work protectively at this stage by binding to gill surface ligands, thereby stabilizing the apical membranes and the integrity of the tight junctions on the paracellular channels (15). This could explain why ion-losses are less when Ca concentrations in the ambient water are raised (13,16,17).

The recovery phase which follows the initial shock phase reflects the return of efflux and influx to control levels. This adaptive response to acid exposure is seen in most fish species studied up to now, but the physiological basis is poorly known.
It may result from a hormonal response, but other mechanisms may be involved which reduce paracellular permeability. It is indeed the restoration of the ion efflux towards normal values which is primarily responsible for the return of net ion losses to pre-acidification values. However, stimulation of active ion uptake has been reported to occur in trout (17) and in the mudminnow (18).

Very few physiological investigations carried out momentarily go beyond this point and most research only describes physiological reactions observed during the first days immediately following acid exposure. Long term experiments carried out with Tilapia who were exposed for over 2 months to pH 4, reveal an activation of the prolactin cells and show prolactin synthesis (19). During this chronic exposure, prolactin secretion remained high and this was believed to maintain the normal branchial permeability and plasma electrolyte levels.

The matter of acid-induced hormone production is clearly a complex one. Balm and coworkers (20) exposed Tilapia for 5 days to a low pH regime (pH 3.5). This resulted in an increased sensitivity of the cortisol producing interrenal cells to ACTH and to a product with alpha-MSH immunoreactivity. However, no increase in plasma cortisol levels could be detected. Plasma cortisol was neither significantly altered in brown trout exposed to high Ca acidic water. But when these fish where exposed to low calcium acidic water, cortisol appeared to be elevated, but the response was variable and was only significant after 7 days (21).

Much more research is needed on the endocrinological response to acid stress especially if we are interested in understanding processes of recovery or differences in sensitivity and also in reproductive success.

5. GAPS OF KNOWLEDGE IN H^+ (PROTON) STRESS

(1) The acidification rate determines the quantity of ion loss in laboratory experiments (19,20). Rapid acidification (minutes), a method used by most investigators, was shown to overestimate the effect of the final pH. Such rapid acidification may even cause morphological damage to the branchial epithelium and is of less ecological significance. Future physiological experiments have to take into account these recent results. Indeed, the most rapid acidification in nature during snowmelt for example takes days or weeks.

(2) We are just beginning to learn what the role of hormones may be in counteracting the pH-induced ion-flux changes in the gill epithelium. If we want to understand the differences of H^+ sensitivity among different species, we need to know more about the hormonal action.

(3) If passive diffusion of ions over the gill epithelium is the major cause of net ion loss and eventual death, we need to learn much more about the opening and closing of the tight junctions. Especially the influence of Ca and of hormones on tight junctions needs further investigation.

6. THE PHYSIOLOGY OF COMBINED H^+ AND ALUMINIUM STRESS

Losses of fishes and fish kills have often been described to occur at higher water pH's than anticipated from the known effects of acidity alone. Increased concentrations of Al have been recognized as being responsible for this phenomenon (22,23). Al is abundant in rocks and soil, soluble at low pH and extremely toxic to fish at around pH = 5.0. The aqueous chemistry of Al is complex but is critical to estimate the Al toxicity to water animals.

In this respect, the following Al fractions seem important :
1) the ion-exchangeable, in Driscoll's (24) original terminology : labile (inorganic) monomeric Al = Al^{+++}, $Al(OH)^{++}$, $Al(OH)_2^+$, AlF^{++}, $Al(SO_4)^+$
2) the organically bound fraction e.g. the non-labile (organic) monomeric aluminium which passes an ion-exchange resin
3) a fraction remaining on filters : precipitation products formed by polymerisation, adsorption on existing particles and co-precipitation with humic matter.
The presence of the different fractions is pH-dependent. Although the chemistry of Al is known in detail as far as equilibrium conditions of aqueous Al of synthetic gibbsite, kaolinite etc. are concerned (24,25), this may not exactly be the chemical knowledge needed by the toxicologists and physiologists. Indeed, these equilibrium conditions refer to aged solutions of Al and are determined in laboratory conditions. Such situations may not always be realistic especially in the light of the short-term events of Al-increase during acid episodes and the presence of natural organic acids.

Leivestad et al. (26) for example described a higher Al solubility and a different pH-dependency under naturally realistic conditions than discussed in literature.

For toxicological and physiological work, a constant and accurately defined speciation of Al is necessary. This implies among others an accurate control of the ambient pH. Hutchinson and coworkers (27) in their work on the survival of yolk-sac fry of lake trout elegantly illustrated the importance of pH in Al-toxicity. Survival after 14 days exposure decreased from 100% in pH 4.6 to 52% if 50 µg Al/l was added, whereas at pH 4.8 more than 300 µg Al/l was needed to yield a decrease in survival to 70%. This difference of 0.2 pH units for the survival of the animals thus illustrates the complexity with which researchers are faced when studying the problem of Al-stress. In this respect it seems to be vital to work with open, flow-through systems, so that Al speciation has changed as little as possible during contact with the animals.

From toxicological (27) and physiological (11,26,28,29,30,31) work, the responses of fish towards combined acid and Al stress can be summarized as follows :

1) If pH is such that H^+ acts as a sublethal stress and Al is present in a labile monomeric form, the Al is toxic and causes an increased mortality : this is noticed for example at pH 4.4-4.8 and pH 4.2 for fry of lake trout (Salvelinus namaycush) and brook trout (S. fontinalis) respectively. The mode of toxic action seems to be predominated by a failure in the ionoregulatory ability. In fact labile Al exacerbates the ionoregulatory problems already observed with acid pH alone (11,28,29).

2) At higher pH values (between pH 5.1-6.0 but the limits vary) Al toxicity has been described to occur in solutions in which Al exists mainly in a non soluble, polymeric form (27,30,31). The mode of toxic action in this case is explained as being due to physical obstruction of

the gill function by the polymeric and colloidal Al forms leading to mucus formation, hindering of normal O_2 and CO_2 exchange and resulting in hypoxemia and acidosis (e.g. 11). Ionoregulatory failure seems to be of minor importance under these conditions.

Hutchinson and coworkers (27) and Leivestad and coworkers (26) failed to demonstrate the above mentioned toxicity of "polymeric" Al solutions. In Hutchinson's experiments this may be due to the fact that the Al test-waters were not "aged" before they were used and thus the formation of polymeric and colloidal Al species was not completed. In Leivestad's experiments (26) however, polymeric Al was actually measured by filtration (0.05 μm filter), but this Al form showed lower toxicity to eggs and larvae of Atlantic salmon than the labile Al form.

These examples clearly illustrate that our knowledge on the chemistry of these "polymeric" solutions and on the influence of organic and inorganic constituents in the water, are totally insufficient to explain the observed toxicity patterns.

The relative importance of the two toxic mechanisms of Al e.g. ionoregulatory failure (point 1) and the respiratory problems (point 2), depend on water pH, Ca and the animal species. Wood and McDonald (11) speculate on the mechanisms of Al and H^+ toxicity in the context of the branchial micro-environment, more specifically on paracellular ion loss, mucus formation (see also 32) and Al complexation. They point to the fact that a great deal more work is needed on the nature of the gill surface to confirm or disprove these ideas. Aluminium speciation at the gill surface is likely to be complex partly because of the mucus layers but also because of intricate pH conditions (NH_3 loss for example, 33).

3) At lethal H^+ stress (thus very low pH), addition of Al may increase survival time of eggs and fry of salmonids (27). This beneficial effect of Al at very low pH (pH 4.4 in the above mentioned example), has also been observed in trout and in the waterfly Daphnia magna and was confirmed by ion flux measurements in both animal species (11,34).

The physiological basis for this (short-term) beneficial effect of Al on survival at low pH is poorly studied and hence poorly understood. Although this effect appears to be of a short-term nature, it may have important ecological applications. During episodic short-term pH-decreases in the water after snow melt or after heavy rainfall, the water quality conditions may become lethal in as far as H^+ concentrations are concerned. However the Al present in the water may temporarily prolong the actual survival of the animals long enough to overcome this period of extreme H^+ stress. A more thorough understanding of the physiological basis may help to know which water quality conditions are required for this effect to be of ecological relevance.

7. LACKS OF KNOWLEDGE IN PHYSIOLOGY OF COMBINED H^+ AND AL STRESS

Far less is known about the physiology of combined acid and Al stress then about pure acid stress.

1) Physiologists and toxicologists need more "realistic" data on the speciation of Al in natural waters especially in regard with presence of organic matter and the interaction between sediment and water in periodical or episodic acidification events.

2) The interaction of Al on the gill micro-environment is poorly known. Especially the induction of mucus formation and the processes determining metal complexation in the mucupolysaccharides need more attention. Nothing is known about the mechanisms of Al-induced ion losses and current ideas on the opening of tight junctions have not been examined in detail.

Moreover the role of Ca in this process is completely unclear.
3) In current ideas Al toxicity is situated on the external gill epithelium. Evidence on a possible assimilation of Al and of an internal toxic action on enzymatic reactions for example in the epithelial cells is completely lacking (35).
4) We do not know anything at this moment on the possible role of hormones in sensitivity or tolerance of aquatic animals towards Al stress in acid water. However many investigators experienced for example seasonal differences in sensitivity towards these stressors.

7. IMPAIRMENT OF OTHER PHYSIOLOGICAL MECHANISMS

Most data on the physiological responses of aquatic organisms, deal with those mechanisms which were believed to be the cause of death of the animals, e.g. the ionoregulatory failure, the respiratory problems and the secondary effects (hemoconcentration, circulatory collaps). But what about the physiological basis of differences in sensitivity in various life stages for example. What are the physiological mechanisms responsible for often observed decreases in reproductive success, of decreased growth rate of skeletal deformities (36), of abnormal gill development (37) ? What about effects on immunology, on sensitivity to infections, on the endocrinology ? Some data are available on these responses but they mostly do not go beyond the descriptive stage and they do not provide much information on the physiological basis of the observed response.

Chronic sublethal effects however may have adverse effects on the whole community in the long run and these effects may be overlooked in the short-term laboratory experiments that have been performed up to now.

8. EARLY WARNINGS OF ACID STRESS

A thorough knowledge of sublethal effects may help to detect unfavourable water quality conditions with much more accuracy and with higher sensitivity than we can do with the "lethal toxicity" data available now. Physiological parameters may be used to set safer (acceptable) limits on the process of acidification of surface waters. They will be of particular interest when defining a "critical load" on pollution i.e. the amount of pollution which nature can withstand. And such data are highly needed by decision makers.

Examples of possible early warning have been ample in the contributions to the symposium. A possible early sign of acidification may be the attempt of organisms to escape the unfavourable conditions as was observed by Raddum and Fjelheim (38) in the mayfly *Baetis rhodani*. Several studies have shown that plasma glucose in fish increased rapidly during acid stress and thus may be an excellent indicator of sublethal stress (20,21). However, plasma glucose may be elevated in response to a wide range of stressors (39) and natural variation in it may be large (40). Therefore the hyperglycaemic response may not be specific enough to serve as an indicator of sublethal acid stress alone. Other examples of early signs may be tests on osmoregulatory abilities (41), measurements of blood osmolality, gill histology and oocyte atresia (42).

9. ECOSYSTEM APPROACH

The conference pointed to the need for a holistic ecosystem approach. Laboratory experiments only give indications on the (physiological) reactions of organisms in particular conditions. But on an ecosystem level even small changes in survivorship, growth, reproduction, etc. can lead to the extinction of the population over several tens of generations.

10. WHY ARE SOME SPECIES MORE TOLERANT THAN OTHERS ?

The problem of tolerance and sensitivity of aquatic animals has been addressed in several papers (12,43,44,45,46,47,48,49). Acidification is a well-known stressor of freshwater animals, yet species can differ in their tolerance of acid stress. There exists variance in sensitivity not only between species, but even between populations within species (10,45) and between life-cycle stages of individuals. Based on literature data and original work on invertebrates subject to pollution by acid minewaters, Maltby and coworkers (43) concluded that low pH can act as a selection pressure, favouring more tolerant genotypes and, by having a differential effect on different size/age classes, influencing life-cycle evolution.

A rather unexpected way by which differences in tolerance may be explained is the capacity of the nymphs of the mayfly *Leptophlebia cupida* to increase its number of chloride cells by increasing its frequency of moulting. Moulting is the only way a nymph is able to increase the number of chloride cells (44). An increase in the numbers of epithelial chloride cells, generally recognized as the sites of ion transport, is also demonstrated in fish (50) under acid stress.

In experiments using *Hyalella azteca* (amphipods) collected from acidic and from circumneutral lakes, France (45) tested their ability to gain tolerance or to lose it. Resistance and tolerance to low pH was neither lost by the tolerant amphipods exposed to neutral water for 10 days, nor readily gained by the non-tolerant amphipods exposed to sublethal low pH for a similar duration.

Jernelov and co-workers (51) found that the most acid-tolerant individuals of chironomids had the highest hemoglobin content. They hypothesized that the acid-neutralizing capacity rather than the oxygen binding capacity of this red blood pigment, gave a selective advantage to the red individuals. Their hypothesis is intriguing, the more because it is known that fish kept at low pH in laboratory conditions also increase their hemoglobin, hematocrit and red blood cell count (52).

In a very elegant experiment M. Havas (53) tested Jernelov's hypothesis using the water flea *Daphnia*. *D. pulex* was considerably more acid tolerant than *D. magna*, but in neither of these species did hemoglobin enhance survival at low pH. Hemoglobin thus provides no ecological advantage to *Daphnia* in acidic environments.

An extreme tolerance against acidity has been described for the East American mudminnow *Umbra pygmea* which inhabits acid waters ($pH < 4.0$) in the Netherlands (46). This tolerance is believed to be the result of various morphological and physiological adaptations. The relatively small surface area of the gill lamellae may be an efficient adaptation to reduce branchial ion leakage in acid water. Furthermore, the swim-bladder can be used for airbreathing so that water flow over the gills for respiratory purposes can be minimized (48). Whole body sodium exchange studies (49) revealed that Na^+ influx was stimulated at pH 4.5 as compared to pH 7 and that Na^+ efflux was not influenced by ambient pH. Very well adapted chloride cells were observed in the gills of the fish adapted to pH 4.5 and these cells may be the morphological correlate of the enhanced Na^+ uptake rates observed.

The physiological adaptation of the increased Na influx at acid pH in the mudminnow seemed cortisol dependent (49). In this respect, the mudminnow differs from another acid-resistant fish e.g. tilapia. The latter species establishes a positive Na balance at low pH by a reduction in Na^+ efflux (18). This regulatory ability of the integumental permeability to ions may be accomplished by prolactin (19).

Another important issue adressed the acclimation of fishes to low pH and elevated Al concentration. Chronic exposure of tilapia to pH 4 for example, increased its sensitivity to water with pH 3.5, as was illustrated by measurements of plasma electrolyte losses (19). In rainbow trout, a chronic exposure to pH 5.0 showed a gradual recovery of the initial high ion-losses, but the fishes remained extremely sensitive to a subsequent Al stress (200 µg/l) (54). Very high ion losses reappeared after Al addition and plasma NaCl declined rapidly while the haematocrit increased, leading to death within the next days.

11. CONCLUSIONS

The conclusions of the international symposium on "Ecophysiology of Acid Stress in Aquatic Organisms" are summarized in table 1 under the form of recommendations for future research.

Table 1. Research priorities on ecophysiology of acid stress in aquatic organisms (55).

1. Detection of sublethal physiological responses is needed to determine the critical load on acid pollution.
2. More data are needed on the influence of acid stress on the mechanistics of the ionoregulatory system. But there is a need for research on other possible sensitive mechanisms such as respiration, endocrinology and reproduction, as well as on behaviour.
3. There is a critical need for better understanding of the Al speciation under realistic conditions (organic substances) and its toxicity.
4. More ecophysiological research on aquatic plants, such as macrophytes, mosses, algae as well as on bacteria in acidified water is needed.
5. Experiments have to be performed in ecologically relevant conditions ; parallel experiments have to be undertaken in the laboratory and in the field.
6. A multidisciplinary approach is strongly recommended in order to come to a whole community and ecosystem understanding of acid stress effects.

REFERENCES

1) WITTERS, H. and VANDERBORGHT, O. (1987). Ecophysiology of acid stress in aquatic organisms. H. Witters and O. Vanderborght, Eds. Annls. Soc. r. Zool. Belg. 117, supplement 1, 472 pp.
2) ALMER, B., DICKSON, W., EKSTROM, C., HORSTROM, E., and MILLER, U. (1974). Effects of acidification on Swedish lakes. Ambio 3, 30.
3) EILERS, J.M., LIEN, G.L. and BERG, R.G. (1984). Aquatic organisms in acidic environments : a literature review. Wisc. Dep. Nat. Resour. Tech. Bull. 150, 18 pp.
4) MILLS, K.H. and SCHINDLER, D.W. (1986). Biological indicators of lake acidification. Water, Air and Soil Pollution 30, 749-789.
5) OKLAND, J. and OKLAND, K.A. (1986). The effects of acid deposition on benthic animals in lakes and streams. Experientia 42, 471-486.
6) WRIGHT, R.F. (1987). Effects of acidification on aquatics : chemical aspects. In : "Effects of air pollution on terrestrial and aquatic ecosystems", Proc. of an international symposium, Grenoble, 18-22 May 1987, CEC Luxemburg.
7) LACROIX, G. and TOWNSEND, D. (1987). Responses of juvenile atlantic salmon to episodic increases in the acidity of some rivers of Nova Scotia, Canada. In (1), 297-307.

8) GAGEN, C. and W. SHARPE (1987). Influence of acid runoff episodes on survival and net sodium balance of brook trout (Salvelinus fontinalis) confined in a mountain stream. In (1), 219-230.
9) WOOD, C.M. 1987. The physiological problems of fish in acid waters. In : Acid toxicity and aquatic animals. Ed. R. Morris, D.J.A. Brown, E.W. Taylor and J.A. Brown. Society for Experimental Biology, Seminar Series, Cambridge University Press, Cambridge. In press.
(10) VANGENECHTEN, J.H.D., WITTERS, H. and VANDERBORGHT, O.L.J. (1987). Laboratory studies on invertebrate survival and physiology in acid waters. In : Acid toxicity and aquatic animals. Ed. R. Morris, D.J.A. Brown, E.W. Taylor and J.A. Brown. Society for Experimental Biology, Seminar Series, Cambridge University Press, Cambridge. In press.
(11) WOOD, C.M. and D.G. McDONALD (1987). The physiology of acid/aluminium stress in trout. In (1), 399-410.
(12) VANGENECHTEN, J.H.D., WITTERS, H., VAN PUYMBROECK, S., VANDERBORGHT, O.L.J. and J.N. CAMERON (1987). The acid-base and electrolyte balance during acid stress in two species of acid tolerant catfish : Ictalurus nebulosus and I. punctatus. In (1), 265-275.
(13) McDONALD, D.G. (1983). The interaction of environmental calcium and low pH on the physiology of the rainbow trout, Salmo gairdneri. I. Branchial and renal net ion and H^+ fluxes. J. exp. Biol., 102, 123-140.
(14) MARSHALL, W.S. (1985). Paracellular ion transport in trout opercular epithelium models osmoregulatory effects of acid precipitation. Can. J. Zool. 63, 1816-1822.
(15) McWILLIAMS, P.G. (1983). An investigation of the loss of bound calcium from the gills of the brown trout, Salmo trutta, in acid media. Comp. Biochem. Physiol., 74A, 107-116.
(16) McDONALD, D.G., HOBE, H. and WOOD, C.M. (1980). The influence of calcium on the physiological responses of the rainbow trout, Salmo gairdneri, to low environmental pH. J. exp. Biol. 88, 109-131.
(17) McDONALD, D.G., WALKER, R.L. and WILKES, P.R.H. (1983). The interaction of environmental calcium and low pH on the physiology of the rainbow trout, Salmo gairdneri. II. Branchial ionoregulatory mechanisms. J. exp. Biol. 102, 141-155.
(18) FLIK, G., VAN DER VELDEN, J.A., SEEGERS, H.C.M., VAN DER MEIJ, J.C.A., KOLAR, Z. and WENDELAAR BONGA, S.E. (1987). Water acidification and bidirectional sodium flow in the tilapia Oreochromis mossambicus and the East American mudminnow Umbra pygmaea. Submitted for publication. From (19).
(19) WENDELAAR BONGA, S.E., FLIK, G. and BALM, P.H.M. (1987) Physiological adaptation to acid stress in fish. In (1), 243-254.
(20) BALM, P.H.M., LAMERS, A. and WENDELAAR BONGA, S.E. (1987) Regulation of pituitary-interrenal activity in Tilapia acclimating to low pH conditions. In (1), 343-352.
(21) EDWARDS, D., BROWN, J.A. and WHITEHEAD, C. (1987). Endocrine and other physiological indicators of acid stress in brown trout. In (1), 331-342.
(22) GRAHN, O. (1980). Fish kills in two moderatly acid lakes due to high aluminium concentration. In : Ecological impact of acid precipitation. D. Drablos and A. Tollan, Eds., pp. 310-311. SNSF Project.
(23) HENRIKSEN, A., SKOGHEIM, O.K. and ROSSELAND, B.O. (1984). Episodic changes in pH and aluminium speciation kill fish in a Norwegian salmon river. Vatten 40, 255-263.

(24) DRISCOLL, C.T. (1980). Chemical characterization of some dilute acidified streams in the Adirondack Region of New York State. Ph.D. Thesis, Cornell University, Ithaca, N.Y.
(25) SEIP, H.M., MULLER, L. and NAAS, A. (1984). Aluminium speciation : comparison of two spectrophotometric analytical methods and observed concentrations in some acidic aquatic systems in Southern Norway. Water, Air and Soil Pollution 23, 81-85.
(26) LEIVESTAD, H., JENSEN, E., KJARTANSSON, H. and XINGFU, L. (1987). Aqueous speciation of aluminium and toxic effects on Atlantic salmon. In (1), 387-398.
(27) HUTCHINSON, N., HOLTZE, K., MUNRO, J. and PAWSON, T. (1987). Lethal responses of salmonid early life stages to H^+ and Al in dilute waters. In (1), 201-217.
(28) DALZIEL, T., MORRIS, R. and BROWN, D. (1987). Sodium uptake inhibition in brown trout, Salmo trutta exposed to elevated aluminium concentrations at low pH. In (1), 421-434.
(29) ORMEROD, S., WEATHERLEY, N., FRENCH, P., BLAKE, S. and JONES, W. (1987). The physiological response of brown trout, Salmo trutta to induced episodes of low pH and elevated aluminium in a Welsh hill stream. In (1), 435-447.
(30) BAKER, J.P. and SCHOFIELD, C.L. (1982). Aluminium toxicity to fish in acidic waters. Water, Air and Soil Pollut. 18, 289-309.
(31) NEVILLE, C.M. (1985). Physiological response of juvenile rainbow trout, Salmo gairdneri, to acid and aluminium-prediction of field responses from laboratory data. Can. J. Fish. Aquat. Sci. 42, 2004-2019.
(32) LINNENBACH, M., MARTHALER, R. and GEBHARDT, H. (1987). Effects of acid water on gills and epidermis in brown trout (Salmo trutta) and in tadpoles of the common frog (Rana temporaria L.). In (1), 365-374.
(33) WRIGHT, P. and RANDALL, D. (1987). The interaction between ammonia and carbon dioxide stores and excretion rates in fish. In (1), 321-329.
(34) HAVAS, M. and LIKENS, G.E. (1985). Changes in ^{22}Na influx and outflux in Daphnia magna (Straus) as a function of elevated Al concentrations in soft water at low pH. Proc. Natl. Acad. Sci. USA 82, 7345-7349.
(35) WITTERS, H., VANGENECHTEN, J.H.D., VAN PUYMBROECK, S. and VANDERBORGHT, O.L.J. (1987). Internal or external toxicity of aluminium in fish exposed to acid water. In : Effects of air pollution on terrestrial and aquatic ecosystems. Proc. of an international symposium, Grenoble, 18-22 May 1987, CEC Luxemburg.
(36) CUMMINS, C. (1987) Factors influencing the occurence of limb deformities in common frog tadpoles raised at low pH. In (1), 353-364.
(37) JAGOE, C., HAINES, T. and BUCKLER, D. (1987) Abnormal gill development in atlantic salmon (Salmo salar) fry exposed to aluminium at low pH. In (1), 375-386.
(38) RADDUM, G. and FJELLHEIM, A. (1987). Effects of pH and aluminium on mortality, drift and moulting of the mayfly Baetis rhodani. In (1), 77-87.
(39) DONALDSON, E.M. (1981). The pituitary interrenal axis as an indicator of stress in fish. In : Stress and fish. Pickering A.D. (Ed.), pp. 11-47. Academic Press, London.
(40) HILLE, S. (1982). A literature review of the blood chemistry of rainbow trout, Salmo gairdneri Rich. J. Fish. Biol. 20, 535-569.
(41) HAUX, C., HOGSTRAND, C., MAGE, A. and OLSSON, P. (1987). Physiological parameters in fish toxicology : a method to interprete sublethal effects observed in the labortory in terms of ultimate survival of the fish in the field. In (1), 11-20.

(42) McCORMICK, J., JENSEN, K., LEINO, R. and STOKES, G. (1987). Fish blood osmolality, gill histology and oocyte atresia as early warning acid stress indicators. In (1), 309-319.
(43) MALTBY, L., CALOW, P., COSGROVE, M. and PINDAR, L. (1987). Adaptation to acidification in aquatic invertebrates ; speculation and preliminary observations. In (1), 105-115.
(44) BERRILL, M., ROWE, L., HOLLETT, L. and HUDSON, J. (1987). Response of some aquatic benthic arthropods to low pH. In (1), 117-128.
(45) FRANCE, R. (1987) Differences in H-ion sensitivity among Hyalella azteca populations : an illative hypothesis invoking natural selection. In (1), 129-137.
(46) LEUVEN, R., WENDELAAR BONGA, S. and HAGEMIJER, W. (1987). Effects of acid stress on the distribution and reproductive success of freshwater fish in Dutch soft waters. In (1), 231-242.
(47) ROSSELAND, B. and SKOGHEIM, O. (1987). Differences in sensitivity to acidic soft water among strains of brown trout (Salmo trutta L.). In (1), 255-264.
(48) DEDEREN, L., WENDELAAR BONGA, S. and LEUVEN, R. (1987). Ecological and physiological adaptations of the acid-tolerant mudminnow Umbra pygmaea (De Kay). In (1), 277-284.
(49) FLIK, G., KOLAR, Z., VAN DER VELDEN, J., SEEGERS, H., ZEEGERS, C. and WENDELAAR BONGA, S. (1987). Sodium balance in the acid resistant East American mudminnow, Umbra pygmaea (De Kay). In (1), 285-294.
(50) LEINO, R.L. and McCORMICK, J.H. (1984). Morphological and morphometrical changes in chloride cells of the gills of Pemephales promelas after chronic exposure to acid water. Cell and Tissue Res. 236, 121-128.
(51) JERNELOV, A., NAGELL, B. and SVENSON, A. (1981). Adaptation to an acid environment in Chironomous raparius (Diptera, Chironomidae) from smoking Hills, N.W.T., Canada. Holarctic Ecology, 4, 116-119.
(52) NEVILLE, C.M. (1979). Sublethal effects of environmental acidification on rainbow trout (Salmo gairdneri). J. Fish. Res. Bd. Can. 36, 84-87.
(53) HAVAS, M. (1987). Does hemoglobin enhance the acid-tolerance of Daphnia. In (1), 151-165.
(54) WITTERS, H., VANGENECHTEN, J., VAN PUYMBROECK, S. and VANDERBORGHT, O.L.J. (1987). Ionoregulatory and haematological responses of rainbow trout, Salmo gairdneri to chronic acid and aluminium stress. In (1), 411-420.
(55) BARTH, H., HIGLER, B., JOHANNESSEN, M., OTT, H., VANDERBORGHT, O.L.J., WENDELAAR BONGA, S. and WOOD, C. (1987). Some future research priorities on ecophysiology of acid stress in aquatic organisms. In (1), 459.

Reversibility of Acidification

Hauhs,M. *

1. Introduction

Acidification of surface waters has occurred in a number of regions in Europe and the regional distribution of acid surface waters provides strong empirical evidence for a causal link to acid deposition. Regions with acidified waters such as southernmost Norway are characterized by acid soils and weathering resistant bedrock mineralogy (*Wright, 1983a*). Since most of the acid soils in Europe date back from times before acid deposition it is necessary to evaluate the qualitative and quantitative changes that were introduced to these ecosystems by acid deposition.

The acid-base status of soils is controlled by a number of key factors such as base saturation, silicate weathering, Al mobilization, CO_2 solubility, and S and N dynamics. Simple soil chemical models demonstrate that the introduction of mobile, strong acid anions to sensitive, naturally acid soils is sufficient to initiate the transfer of acidity from the soil to surface waters (*Reuss and Johnson, 1985*). Thus the links between acid deposition and water acidification have been established both at the empirical and process-oriented levels (*Reuss et al., 1987*).

The SO_2 emissions in Europe, the main precursor to acid deposition have been decreasing since the early 1980s. In addition most countries in Europe have agreed on a 30% reduction in SO_2 emissions relative to the 1982 level. The other main acidifying agent NO_3 is derived from NO_x and NH_3 emissions. The absolute and relative importance of nitrogen in the acidity of the deposition is expected to increase.

Any decision on mitigation strategies for affected aquatic ecosystems requires knowledge about the expected response to a reduced load from the atmosphere. The presumable gradual and delayed response to the increase in deposition over the last 150 year, however, was given only little scientific attention. Records of lake acidification have to be reconstructed from the sediments by methods such as diatom analysis (*Jones et al, 1986*). Therefore a discussion of reversibility of water acidification is hampered by the lack of experience about the time aspects especially at the level of soil chemical processes.

This article is a progress report from a literature review on reversibility that is financed by the COST 612 programme and is to be completed by end of 1987. It is restricted to the reversibility

Institute for Soil Science and Forest Nutrition
Büsgenweg 2, D-3400 Göttingen, Federal Republic of Germany

of surface water chemistry and provides a short summary about the empirical, experimental, and modelling evidence on the subject. I will not consider in-lake processes.

2. Soil chemical key processes for water acidification

In this section the key processes that control the flux of acidity from soils into surface waters will be reviewed with respect to their rates of response to a change in deposition. Water acidification can be defined as a loss in alkalinity where the alkalinity of soil solution and surface waters can be defined in the two ways that are represented by the two sides of the following equation ([ ] denote molar concentrations):

$$2[Ca^{2+}] + 2[Mg^{2+}] + [Na^{+}] + [K^{+}] - [SO_4^{2-}] - [NO_3^{-}] - [Cl^{-}] =$$

$$[HCO_3^{-}] + [OH^{-}] - [H^{+}] - \sum Al^{+}$$

The Al-terms stand for the sum of all inorganic, monomeric, positively-charged Al compounds. A rearrangement of this equation yields the statement of ionic balance. Organic anions, Mn, NH_4, and CO_3^{2-} can be neglected for this purpose.

All terms on the left side of the above equation are independent of the CO_2 partial pressure (pCO_2) whereas the right side summarizes all compounds that change when the soil solution is degased from its CO_2. Thus alkalinity is in contrast to pH independent of pCO_2 and can be used to access the transport of acidity from the soil into the surface water (*Reuss and Johnson, 1985*).

The ecological effects of water acidification especially to fish depend on pH and Al levels in surface waters. As a close relation exists between pH and Al solubility the pH is the appropriate variable to investigate the links between water chemistry and biota. The two key variables alkalinity and pH are coupled by the fact that in natural waters Al and HCO_3 are mutually exclusive. They coexist only in a narrow range around zero alkalinity at which the pH is highly sensitive to alkalinity changes. Therefore H^+- and Al^{3+}-levels that can be harmful to fish occur only at a negative alkalinity.

These definitions have been introduced to distinguish between water acidification as a loss in alkalinity and its ecological effects in systems switching from positive to negative alkalinity. Acid deposition is unlikely to have caused ecological effects to acid waters that had negative alkalinities already prior to acid deposition nor to systems where the initial alkalinity was much higher than the loss caused by acid deposition.

The reversibility of acidification of aquatic ecosystems depends on the time aspect of the processes involved. This will be demonstrated by the effect of acid deposition to an undisturbed, sensitive watershed as exemplified by the results from Sogndal (Norway), a site than has artifically been acidified (*Wright et al,1986*).

We will assume that base cations † in the soil are in equilibrium with biomass cycling (uptake into biomass is equal to mineralisation) and the levels of NO_3 and SO_4 in the leachate are small as typical of large areas in Northern Scandinavia. In this case the release of base cations from silicate weathering (an input flux to the soil) is equal to their export in the streamwater. It will also control the availability of these ions at the exchange sites. The accompaning anion typically is HCO_3 and its source the dissolution of CO_2 produced by respiration in the soil.

The seasonal variation of CO_2 production in soils (with a maximum during the vegetation period) and the flux of water through the soil (with maximum flow and shallowest flow path ways in snowmelt) causes minimum alkalinity to occur during highflow especially snowmelt events. Such a seasonal pattern is exemplified by the data from White Oak Run (Virginia,USA) (*Cosby et al,1985*).

Silicate weathering is the ultimate source for base cations in soils and this release rate bounds the long term net proton supply that the system is able to buffer. In ecosystems that are in equilibrium with respect to biomass accumulation alkalinity in streamwater is a measure of sensitivity towards acid deposition.

Soils with low silicate weathering rates are widespread on the granitic bedrock of Scandinavia. These soils are acid and show low base saturation. The streamwater typically is at low but positive alkalinity levels. The above concept of acidity transfer implies that Al compounds present in the soil solution will precipitate when this solution emerges as streamwater and degases. That is why in the absence of mobile strong acid anions alkaline waters can drain acid soils (i.e: Sogndal, Norway or Petersburg, Alaska) (*Wright et al,1986; Johnson,1981*).

The changes that can be introduced to such systems by acid deposition group under two aspects:

I. intensity controlled: Due to the exchange characteristics of acid soils any increases in ionic strength will cause a small but instantaneous decrease in soil solution alkalinity. This effect is controlled by the amount of mobile strong acid anions and does not require any change in base saturation nor is it dependent on a proton input. The so-called mobile anion concept thus describes the widespread occurrence of water acidification in sensitive areas of Scandinavia (*Seip, 1980*).

† The cations Ca,Mg,K, and Na will be termed base cations as they are usually stored in the soil together with weak acid anions such as silicates and exchange surfaces of clay minerals. An exception to this are soils that contain acid salts as described by *Prenzel (1983)*.

II. capacity controlled: The long-term effect of proton deposition in connection with the input of SO_4 and NO_3 ions is a buffering reaction at the exchange sites that leads to a net decrease in base saturation. This soil acidification will last until a new base saturation is reached that again is in equilibrium with a lower soil solution alkalinity. Such a net depletion phase may last years to decades and it is not possible to judge from streamwater output alone whether a system has achieved a new steady state. The capacity effects can only be accessed by complete ion budgets (*Ulrich,1984*).

Any ecosystem receiving acid deposition will experience both phases. The regional difference in acidification is that highly sensitive watersheds (i.e.: in Scandinavia) facing a low to moderate acid deposition load (S-input: 10-25 $Kg.ha^{-1}.yr^{-1}$) may reach a negative alkalinity and loose their fish populations already during phase I. This initial response will be followed by a chronic capacity-controlled acidification phase as exemplified by an artificial acidification experiment at Sogndal (Norway) (*Wright et al,1986*).

A quantification of the capacity effect requires the separation of changes in base cations and silicate weathering. In the light of the time scale of such soil chemical changes this implies the combination of ion budget studies and repeated soil inventories (*Fölster,1985 Matzner,1987*). The silicate weathering rate also is essential to any estimate of reversibility of water acidification. At low weathering rates a reduction in base saturation may be 100% reversible after a decrease in acid load but may require several centuries for such a recovery.

The input of mobile strong acid anions such as sulfate and nitrate is involved in both the intensity and capacity effect. Therefore processes involved in the turnover and mobility of these anions will affect the response to a change in deposition. In most of the acidified sensitive sites of Scandinavia and North America an equilibrium exists between total input (dry and wet) and output of sulfate in streamwater (*Christophersen and Wright,1980; Wright,1983b; Hultberg,1986; Rochelle et al,1987)*. The soil storages of sulfur can vary between an amount equivalent to one or several decades of input. In soils where most of this total storage is organically bound a fast response of sulfate output to a reduction in deposition can be expected.

In the deeply weathered soils typical of middle Europe S-accumulation is often dominated by mineral forms such as adsorption/desorption or precipitation/dissolution procceses. The nature of the accumulated sulfate is still under discussion (*Prenzel,1983 Khanna et al,1987, Dise and Hauhs,1987*), however, a pH-dependence of this process has been demonstrated (e.g.: *Nodvin,1986*). The long-term soil solution monitoring at Solling (West Germany) showed a accumulation phase that was followed by a net sulfate release from the soil concurrent with a drop in soil solution alkalinity (*Khanna et al,1987*). Here the further sulfate output with the leachate will rather depend on the chemical status of the soil than on future deposition rates. The acid deposition to such sites has to be reduced before a breakthrough of the critical pH that controls sulfate accumulation into the surface waters occurs. Otherwise water acidification for such streams may be irreversible.

In contrast to sulfate nitrate deposition is rarely in excess of the biological demands of forest ecosystems. Moreover nitrate often was and in some cases still is the growth-limiting factor of European forests. Nitrate is strongly implicated in the internal nutrient cycling and undisturbed forest ecosystems are usually tight with respect to nitrate leaching *Vitousek and Melillo (1979).* That is why the input in deposition and the output in streamwater from forest ecosystems are poorly correlated (*Nilsson, 1986*). A disturbance of forest ecosystems such as the forest decline in middle Europe, however, will affect the nitrate utilization and thus export rates. This is indicated by the pattern of streamwater nitrate levels in space and time in middle Europe (*Hauhs and Wright, 1986*). A spruce forest at Hils (West Germany), a site with severe forest dieback, shows nitrate saturation (deposition input equals streamwater output) at a concentration level of 500 μeq.L^{-1}. That is much higher than sulfate levels in any acidified Scandinavian stream. To reverse such a nitrate-related stream acidification the status of the forest ecosystem clearly plays a role which is at least equally important as the current deposition rates. The nature of forest decline in middle Europe is still too poorly understood to evaluate implications on runoff chemistry at regional scales.

3. Reversibility

The few examples on reversibility of water acidification group into empirical, experimental, and modelling evidence. Most of these examples have been discussed at a workshop of the COST 612 working group in Grimstad, Norway in 1986 and will be shortly summarized below.

3.1. Empirical evidence

Such examples require a reduction of SO_4-deposition (as a surrogate of total acid input) from levels that affected surface water alkalinity.Because of the possibly delayed response the reduction should have lasted for several years. *Dillon et al (1986)* report about a rapid recovery in lake chemistry after the SO_2 emissions in the Sudbury area (Canada) decreased from 1.41 x 10^6 t.yr^{-1} (1973-78) to 0.68 x 10^6 t.yr^{-1} (1979-85). The recovery in SO_4,pH,and Al has occurred in a number of intensively studied lakes close to the emission source. Comparative pH inventories (1974 and 84) also demonstrated a recovery for a larger group of lakes in the Sudbury area.

SO_2 fumigation in Sudbury has widely damaged the vegetation and caused extensive soil erosion. Therefore the catchments have large proportions of bare rock surfaces and became hydrologically more responsive. The water retention time of the lakes, however, is 3-4 years. The improvement in water chemistry can to a large degree be explained by a dilution model (*Schindler,1986*).The results from Sudbury will thus have only limited relevance for systems where direct pathways into the surface waters are less important.

Another example of a reduced acid load stems from the west coast of Sweden (*Forsberg et al,1986*).Here SO_4 concentrations in wet deposition have been declining since the late seventies. *Forsberg et al* quantify this reduction for two stations with 15% and 22%. In this area, however, dry deposition can be as high as the wet deposition (*Hultberg,1985*). Lakes in this area experienced a reduction in sulfate levels of up to 35% (*Forsberg et al*).As a consequence of that pH increased. But the data give no conclusive evidence about the related input and output fluxes.The highest values of sulfate concentrations in both precipitation and lakewater to which the reduction relates coincide with several hydrological extreme years.Low precipitation and a low level of groundwater table in this spell of dry years might have contributed to the high sulfate levels in the mid seventies (*Bringmann,1986*). An oxidation of reduced sulfur concurrent with the drop in groundwater table was described by *Hultberg and Wendblad (1980)* and might have contributed to the maximum SO_4 levels.

3.2 Experimental evidence

Since 1984 acid deposition has been reduced by 70% for a whole headwater catchment at Risdalsheia,an acidified area at southernmost Norway. This was achieved by means of two roofs that extend over the boundaries of the two enclosed catchments. One of these catchments serves as a control and receives recycled ambient acid rain. The soils at Risdalsheia are thin, patchy,and poorly developed.

The results from the first two years of treatment are given by *Wright,1986 and Wright et al.1986*. The exclosure of acid rain resulted in a decrease in streamwater nitrate and sulfate. The increase in alkalinity, however, is much smaller. That is because of the following reasons:

The seasalt inputs to the control catchment were relatively small during the treatment period. The amount of artifically added seasalts for the catchment receiving clean rain has been adjusted to the long-term average at this site. Therefore the reduction in the runoff concentration of sulfate and nitrate is partly offset by a higher chloride concentration. An intensity-controlled response due to a reduced ionic strength has thus not been detectable up to now.

The treated catchment shows in contrast to the control catchment a net output of SO_4 and H (output: 176% and 167% of input levels respectively) and a net retention of Na and Mg (output: 83% and 67% of input levels respectively). The result may be a build up of the pool of exchangeable base cations in the soil.

The organic anions are buffering against a pH increase.

These data may indicate that a change in base saturation is necessary before a significant improvement in runoff pH takes place. In the preceeding recovery phase the concentrations of base cations in the streamwater will be below the presumeably steady state levels. Thus systems that respond during the acidification phase primarily to the intensity factors may in a recovery phase be delayed by capacity controlled soil changes.

Another whole ecosystem experiment was conducted in a Norway spruce stand that received additional artificial acidity at three deposition levels (*Stuanes,1987*). A period of recovery over several years illustrates the role of capacity factors: A gradual increase in soil pH and base saturation is probably due to the input of base cations from weathering in excess of uptake and leaching rates.

3.3 Modeling evidence

At present the heuristic role of models in acidification research is to integrate between process-level research and the understanding of long-term catchment response. A set of models that describe the various time scales of freshwater acidification was reviewed by *Reuss et al (1986), Cosby (1986), and Kämäri (1987)*. Currently the so-called MAGIC model (Model of Acidification of Groundwater In Catchments) seems to be most widely used. It is a process-oriented conceptual model that explicitly considers long-term capacity effects such as changes in base saturation. This feature enables the model to simulate chemical reversibility. No complete data set of a history of catchment acidification, however, is available for a rigorous test of the model.

Small, hydrologically responsive catchments that are sensitive to acidification will show a relative fast response to changes in deposition and may be able to provide the necessary data sets. The MAGIC model uses a simple description of catchment hydrology and is not suited to describe the interaction between acidification and hydrology at short time scales. Since hydrology is especially important in small catchments (*Hauhs,1987*) the application to such data sets is hampered by the inherent long time scale of the model (*Wright and Cosby,1987*). A solution to this problem can be either the development of a refined hydrology module for the MAGIC model or the incorporation of capacity factors into short-term models of the Birkenes type (*Christophersen et al,1983*).

In areas with deeper soils and a less responsive catchment hydrology application of the MAGIC model is less problematic. *Hornberger et al (1986)* avoid the problem of missing long-term data by replacing time by space within a regional model application. Under high input of acid deposition as at Solling (West Germany) sulfate retention can no longer be described by an adsorption/desorption model (*Khanna et al,1987*). Here the MAGIC model can serve as a useful tool to inspect alternative concepts of the sulfate turnover. This is also the case for the results on N-retention in stressed forest ecosystems that will probably emerge from the investigation of the forest decline in Central Europe.

4. Conclusions

All available information about reversibilty of water acidification shows that the response to changes in deposition is hysteretic. This behavior can be explained at the process level: Due to the current understanding acidification is dominately controlled by cation exchange and sulfate retention characteristics whereas deacidification depends largely on silicate weathering (*Cosby,1986)*. Since the emissions of acidifying substances has been changing in most countries of Western Europe (reduced SO_2, increased NO_x and NH_3) more evidence on this subject can be expected for the near future. Two scenarios seem to be especially important.

1. Sensitive, acidified areas receiving low to moderate acid deposition (typical southern Scandinavia): Here the acidified catchments will in most cases be in an equilibrium with the input S-fluxes (input H^+-fluxes). Thus sulfate levels in streamwater will respond to a decreasing input. Improvements in pH, however, may be delayed by years to decades due to low silicate weathering rates and the necessary changes in base saturation.

2. Unsensitive, deeply weathered areas receiving high deposition loads (typical within the belt of maximum SO_2 emissions over Central Europe). These sites have not achieved an equilibrium with current deposition levels and base saturation may be still declining (*Hauhs and Dise,1987)*. The dynamics of the sulfur and nitrogen pools that have been accumulated from the historical deposition within catchment soils may interfer with the role of current and future deposition rates. More knowledge about the accumulation mechanisms is needed before predictions are possible, but the few examples of long-term monitoring sites show the risk that acidification of such sites may take so long as to be practically irreversible.

References

BRINGMAN,I. (ED.): 1986, Monitor 1986. pp. 180, Swedish Environmental Protection Board, Box 1302 S-17125 Solna, Sweden.

CHRISTOPHERSEN,N. AND WRIGHT.R.F,,1980: Sulfate budget and a model for sulfate concentrations in streamwater at Birkenes, a small forested catchment in southernmost Norway. Water Resour. Res., 17: 377-389.

CHRISTOPHERSEN,N., SEIP,H.M., AND WRIGHT,R.F.,1982: A model for streamwater chemistry at Birkenes, Norway. Water Resour. Res.,18: 977-996.

CHRISTOPHERSEN, N., JOHANNESSEN, M., AND SKAANE, R.: 1986, Testing a soil-oriented charge balance equilibrium model for freshwater acidification. pp. 169-176, in : Effects of air pollution on terrestrial and aquatic ecosystems (COST 612), Workshop on reversibility of acidification, Grimstad, Norway (preprint).

COSBY, B.J., Hornberger,G.M., Galloway,J.N., and Wright,R.F.: 1985, Modeling the effects of acid deposition: Assessment of a lumped parameter model of soil water and streamwater chemistry. Water Resour. Res.,21: 51-63.

COSBY, B.J.: 1986, Modelling reversibility of acidification with mathematical models. pp. 137-148, in : Effects of air pollution on terrestrial and aquatic ecosystems (COST 612), Workshop on reversibility of acidification, Grimstad, Norway (preprint).

DILLON,P.J., REID,R.A., AND GIRAD,R.: 1986, Changes in the chemistry of lakes near Sudbury, Ontario following reduction in SO_2 emissions. Water, Air, and Soil Pollution, 31: 59-65.

FORSBERG, C., MORLING, G., AND WETZEL,R.G.: 1986, Indications of the capacity for rapid reversibility of lake acidification. Ambio, 14: 164-166.

FÖLSTER, H.: 1985, Proton consumption rates in Holocene and present-day weathering of acid soils. pp. 197-209, In: Drever,J.I. (ed.) The chemistry of weathering, D.Reidel Publ. Comp., Dordrecht The Netherlands.

HAUHS,1987: Water and ion movement through a minicatchment at Risdalsheia, Norway (final Report). Acid Rain Research Report, Norwegian Institute for Water Research (NIVA), Box 333 Blindern, Oslo (in press).

HAUHS,M. AND WRIGHT,R.F.: 1986b, Regional pattern of acid deposition and forest decline along a cross section through Europe. Water Air and Soil Pollution, 31: 463-474.

HAUHS,M. AND DISE,N.: 1987, Depletion of exchangeable base cations in an acid forest soil at Lange Bramke, West Germany. Geoderma, (submitted)

HORNBERGER, G.M., COSBY, B.J., AND GALLOWAY, J.N.: 1986, Modeling the effects of acid deposition: Uncertainty and spatial variability in estimation of long-term sulfate dynamics in a region. Water Resour. Res., 22: 1293-1302.

HULTBERG,H. AND WENBLAD,A.: 1980, Acid groundwater in southwestern Sweden. pp. 270-271, in: D.Drabløs and A.Tollan (eds.) Ecological impact of acid precipitation. Proceedings of an Internat. conf., Sandefjord, Norway. Norwegian Inst. for Water Research, Oslo, Box 333.

HULTBERG, H.: 1985, Budgets of base cations, chloride, nitrogen and sulphur in the acid lake Gårdsjön catchment, Sweden. Ecological Bull. (Stockholm), 37: 133-154.

JONES,V.J., STEVENSON,A.C., AND BATTÁRBEE,R.W.: 1986, Lake acidification and the land-use hypothesis: a mid-post-glacial analogue. Nature, 322: 157-158.

KÄMÄRI.J: 1987, Prediction models for acidification. Proceedings of the Symposium on Acidification and Water Pathways. Bolkesjø Norway. 405-416, Norwegian National Committee for Hydrology, Box Majorstua Oslo, Norway.

KHANNA,P.K. , Prenzel, J.,Meiwes, K.J., Ulrich, B., and Matzner, E.: 1987, Dynamics of sulfate retention by acid forest soil in an acid deposition environment. Soil Sci. Soc. Am. J., **51**: 446-452.

KHANNA, P.K., WEAVER, G.T. AND BEESE, F.: 1986, Effect of sulfate om ionic transport and balance in a slightly acidic forest soil. Soil Sci. Soc. Am J., 50: 770-776.

NILSSON, J. (ED.): 1986, Critical loads for Sulphur and nitrogen. Nordic Council of Ministers Rapport 11, p. 188. Stockholm

NODVIN, S.C., DRISCOLL, C.T., AND LIKENS, G.E.: 1986, The effect of pH on sulfate adsorption by a forest acid soil. Soil Science, 142: 69-75.

MATZNER, E.: 1987, Rates of Na, K, Ca, and Mg release by weathering derived from long term flux monitoring in two forest ecosystems of Germany. pp. 10-12, in: Moldan, B. and Paces, T. (eds.) Geomon - Internat. workshop on geochemistry and monitoring in representative basins. Geological Survey Prague, 118 21 Praha 1, Czechoslovakia.

PRENZEL, J.: 1983, A mechansim for storage and retrieval of acid in acid soils. pp 157-170, In: B.Ulrich and J.Pankrath (eds.) Effects of accumulation of air pollutants in forest ecosystems, D.Reidel Publ. Comp., Dordrecht The Netherlands

ROCHELLE, B.P., ROBBINS CHURCH, M., AND DAVID, M.B.: 1987, sulfur retention at intensively studied sites in the U.S. and Cananda. Water, Air, and Soil Pollution 33: 73-83.

REUSS, J.O., CHRISTOPHERSEN, N., AND SEIP, H.M.: 1986, A critique of models for freshwater and soil acidification. Water, Air, and Soil Pollution 30: 909-930.

REUSS, J.O. AND JOHNSON, D.W.: 1985, Effects of soil properties on the acidification of water by acid deposition. J. Environ. Qual., 14: 26-31.

SCHINDLER, D.W.: 1986, Recovery of Canadian lakes from acidification. pp, 11-22. in : Effects of air pollution on terrestrial and aquatic ecosystems (COST 612), Workshop on reversibility of acidification, Grimstad, Norway (preprint).

SEIP, H.M. 1980, Acidification of freshwater sources and mechanisms. pp 358-365, in: D.Drabløs and A.Tollan (eds.) Ecological impact of acid precipitation. Proceedings of an Internat. conf., Sandefjord, Norway. Norwegian Inst. for Water Research, Oslo, Box 333.

STUANES, A.: 1987, Effects of artificial rain on soil chemical properties and forest growth. Proceedings : Effects of air pollution on terrestrial and aquatic ecosystems, May 1987 Grenoble (in press).

ULRICH, B.: 1984, Stability and destabilization of central European forest ecosystems - A theoretical, data based approach. In: Cooley, J.H. and Golley, F.B. (eds.) Trends in ecological research for the 1980s. Plenum Publ. Corp.

VITOUSEK, P.M. AND MELILLO, J.M.: 1979, Nitrate losses from disturbed forests: Patterns and mechanisms. Forest Sci., 25: 605-619.

WRIGHT, R.F. AND COSBY, J.B.: 1987, Use of a process oriented model to predict acidification at manipulated catchments in Norway. Atmospheric Environment 21: 727-730.

WRIGHT, R.F.: 1983a, Acidification of freshwaters in Europe. Water Quality Bull., 8: 137-143.

WRIGHT, R.F.: 1983b, Input-output budgets at Langtjern, a small acidified lake in southern Norway. Hydrolobiologia, 101: 1-12.

WRIGHT,R.F. , Gjessing,E., Christophersen,N., Lotse,E., Seip,H.M., Semb,A., Sletaune,B., Storhaug,R. and Wedum,K.,1986: Project RAIN: changing acid deposition to whole catchments. the first year of treatment. Water, Air and Soil Pollution 30: 47-64

WRIGHT,R.F.: 1986, RAIN-project results after two years of treatment. pp, 23-40. in : Effects of air pollution on terrestrial and aquatic ecosystems (COST 612), Workshop on reversibility of acidification, Grimstad, Norway (preprint).

SESSION V

PREVENTIVE AND CURATIVE MEASURES

SESSION V

MESURES PREVENTIVES ET CURATIVES

SITZUNG V

VORSORGE-UND WIEDERHERSTELLUNGSMASSNAHMEN

Chairman : Prof. Dr. H.W. ZOTTL, Universität Freiburg, Federal Republic of Germany

Rapporteurs : Dr. G. LANDMANN, Institut National de la Recherche Agronomique, Nancy, France

Dr. A. GEE, Welsh Water Authority, United Kingdom

REPORT ON THE SESSION

Part 1 : Preventive and curative measures for terrestrials ecosystems, in particular forests

Rapporteur : G. LANDMANN, Centre de Recherches Forestières, I.N.R.A. Nancy (F)

This topic was covered by three papers : a general review by BONNEAU dealt with the three main types of measures which may be considered (genetic measures, sylvicultural measures and fertilization), and two communications ; the first one, by HUETTL, dealt with the possibility to correct acute nutritional disturbances ; the second one, by MURACH, focused on the effect of fertilization on root status.

BONNEAU emphasized the great difficulty to define correctives measures for a "disease" whose causes are not yet well known. He based his ideas on a scheme in which, beside extreme climatic events (droughts, frost, ...), the acid deposition (wet and dry) plays a significant role, especially by depletion of essential nutrients in the soil and mobilization of ions e.g. $Al3+$, which may be toxic or at least reduce nutrient uptake. In the regions where direct effect of $SO2$ or excessive atmospheric nitrogen input are supposed to be the main causes of the observed disturbances, correctives measures must be specifically defined.

- Genetic measures do not seem to be a convenient way of dealing with the problem, because of the long delay needed to obtain resistant varieties and the uncertainties about the adaptation of these to the ecological conditions of the areas to be reafforested. This method should only be persued in heavy damaged forest areas where the pollution load will remain high even with considerable reduction of the emissions.

- Sylvicultural measures will only have the expected effects in the long term, by changing the structure of the forest stands. It is, in fact, very difficult to propose a general "rule", because conclusions of the existing studies which have been focussed on the susceptibility of the forest stands towards the present decline are somewhat contradictory. The need of thinning young stands is emphasized by nearly all the sylviculturists, but there are still questions, for example about the way to conduct the regeneration.

- Fertilisation could, if considering the supposed causes, be the most rapid way to mitigate damage and to avoid further deterioration of the soil :

. in response to progressive acidification ot the soil, a "compensative" fertilisation may improve the chemical status of the soils. Improvement of the soil status and also, in some cases, in the mycorrhizal status of the trees, has been noticed when old fertilization trial sites have been re-examined. A positive effect on the growth of the trees and on the presente state of the foliage is not always recorded.

. in response to severe nutritional disturbances (mostly Mg and K) fast soluble fertilizers have lead to rapid greening up (see HUETTL), essentially in rather young stands. The results are less obvious in areas heavily loaded by $SO2$.

In conclusion, fertilisation is thought to be the only way to improve the state of health of the forests, as long as the effect of a significant reduction ot the emissions is not observable. To date, three decades (or more) of acidification have caused a deterioration of the soils of some areas, which may be corrected only by a moderate liming. However a "compensative" and "ameliorative" fertilization should not be practised in a uniform way, but should be based on a precise diagnosis of the nutrient supply.

HUETTL described a positive and rapid response to fast soluble sulphate based fertilizers (greening up within one vegetation period) on young conifer stands showing typical Mg or K deficiencies, respectively on cristaline and on moranic derived soils.

MURACH's paper dealt more specifically with the effects of fertilisation on root growth : beside the often described effects on the soil chemistry or on the foliage content, this approach is supposed to give a complementary view on short and long term effects on the stability of forest stands. A spruce stand (105 years) and a beech stand (135 years) were studied. In the spruce stand, fertilization (mostly Mg and Ca) promoted the development of slower growing flat root systems, whilst in the beech stand, root growth was unaltered though development was greater in the mineral horizons.

The three papers and the resulting discussions highlighted areas where understanding of the subject is lacking :

- the role of forest composition and structure to their relative vulnerabilities to the current decline is not clear. To explain the discrepancies between existing studies, one could fall back on the characteristics of the different regions, but it is without doubt that the methods used in the studies play an important role. At the same time as collating data from large scale studies, it would be very desirable to have data from localized studies (including information in the forest structure, yield, nutrition, changes in use, ...) specifically aimed to answer particular questions.

- the effects of fertilizers on root systems is not well known and should be made the object of detailled investigations. It will be necessary to evaluate, not only the short term effects, but also the changes in the medium and long term, towards stabilising the stands. These studies should be made because, of the very slow effect that fertilizers applied to the upper horizon have on enriching the depleted lower horizons.

- the impact of applied fertilizers, especially that of moderate liming, on water quality, even if thought insignificant by some scientists who have studied this subject, should be the object of further studies, especially in the light of the large scale application of fertilizers which are now on route or being proposed. Likewise, an evaluation of existing results on the effects of fertilizer application on the fauna and flora of the soil should be made.

A S GEE
District Scientist, South Western District,
Welsh Water Authority, Haverfordwest, Wales

1. TERRESTRIAL ECOSYSTEMS, PARTICULARLY FORESTS

There is now a wealth of information demonstrating that the acidification of terrestrial ecosystems, including forests, is due to a number of interacting forces; Climate, local conditions as well as atmospheric pollution are important factors. As a result of the greater degree of management of agricultural land compared with forests, it was stated that measures to counter acidification are easier and more effective in the former. However, there was little discussion of this ascertion, the symposium presentations concentrating on forests.

Measures to prevent or cure the acidification of forest ecosystems can be divided into three main categories:

(i) Genetic - whereas there may be some local value in selecting stocks for increased resistance to acidification, it is unlikely that these methods can have any significant regional benefit.

(ii) Forest management practices - in many areas, traditional forest practices have not provided resistance to acidification. However, imaginative combinations of species of tree, together with changes in thinning and planting policies could be useful.

(iii) Fertilisation - in general, there is no evidence that the addition of calcium alone can be universally beneficial. However, where there is a deficiency, for example in magnesium or potassium, the addition of fertiliser containing these elements has been shown to be effective. The addition of soluble fertiliser in Bavaria has had a beneficial effect when targetted on particular nutrient deficiencies. Indeed, many of the symptons of acidification are those of nutrient deficiency. Acid fogs and ozone, for example, increase the turnover of nutrients in leaves

It can be concluded that where soils are increasing in acidity and where soil water aluminium concentrations are high, the addition of calcium and magnesium is useful. Where nutrients are deficient, depending on local conditions and the degree of forest dieback, the addition of nutrients has been used effectively. It should be noted however, that, in some cases, the addition of fertiliser can lead to increased leaching of nitrate and further acidification.

Recommendation: Further work is necessary to determine the scope for effective genetic manipulation of stocks to suit local conditions, including acidification. More importantly, there is likely to be greater benefit from experimenting with combinations of forest management practices, including deficiency-targetted fertiliser and nutrient additions. These studies should include measurement of the responses of trees during the growing season and monitoring of the effects on runoff waters.

2. AQUATIC ECOSYSTEMS

The amelioration of acidification of aquatic ecosystems can be achieved in addition to reducing deposition, by two principal means:

(i) Land Management

Land use is particularly important in acid sensitive regions where the underlying geology and soils afford little buffering capacity for the incoming acidity.

Vegetation has an important effect on the acidity of surface waters. There is an increasing amount of information on the effects of vegetation on the fate of atmospheric pollutants. Different plant species, e.g. grasses or mixtures of hardwood and softwood trees, affect the chemistry of throughfall and stemflow in different ways. Where soils are deficient in base cations, nitrogen fixing species can acidify runoff waters, particularly in Winter when biological uptake is low. Vegetation also affects water yield through evapotranspiration and can result in significant concentrations of deposited acidity.

Other land mangement activities can also acidify soils and runoff, e.g. addition of nitrate fertiliser to acid soils; harvesting of crops can result in removal of bases, whilst clear-felling of trees results in a rapid loss of nitrate; land drainage results in more rapid runoff, oxidation of sulphur and nitrogen compounds and reduction in sulphate retention.

Clearly, changes in land use will affect current levels of equilibrium. If carried out sensitively then they can limit or even prevent acidification. In contrast, if carried out inappropriately, such activities can seriously exacerbate acidification.

Recommendation: Although on an European scale, land management is not an important factor in determining the acidity of surface waters, on a local scale it can be crucial. Where local conditions dictate, full consideration ought to be given to the implications of land use management, as well as liming, in order to avoid, reduce or overcome the acidification of streams and lakes.

In Wales and Scotland, streams in conifer afforested catchments in acid sensitive areas are much more acidic and contain elevated levels of toxic aluminium than those in adjacent moorland catchments. As part of a major research study, the Llyn Brianne Project is providing valuable information on the ways in which land use affects water quality and ecology. Leading from this study, collaboration between interest parties has led to the adoption of planting guidelines for forest developments to reduce or prevent acidification.

Recommendation: Alternative land preparation and planting techniques (including novel mixtures of species) should be evaluated with a view to ameliorating acidification in base poor soils.

(ii) Liming

Whereas lake liming has been practiced successfully in Scandinavia and elsewhere for several years, the liming of running waters has received less attention. Techniques and equipment have been developed to lime rivers and streams but there are considerable practical difficulties and costs in ensuring that headwaters are adequately dosed during periods of peak flows when acidity levels are greatest.

In order to prevent aluminium from leaching into streams lime needs to be added to land, with best results achieved if lime is applied to wetland areas.

Despite the practical success of liming in certain circumstances, it was not generally considered that it is an universally applicable nor practicable solution to the problem of regional acidification.

Recommendation: Further work is necessary on failsafe systems for lime dosing flowing waters in remote areas. The general ecological consequences of liming catchments need further attention.

MOYENS PREVENTIFS ET CURATIFS DE LUTTE CONTRE LA POLLUTION DANS LES ECOSYSTEMES TERRESTRES, EN PARTICULIER LES FORETS

M. BONNEAU
I.G.G.R.E.F.
Institut National de la Recherche Agronomique, Nancy

Résumé

Plusieurs causes interviennent dans le dépérissement des forêts : accidents climatiques, pollution atmosphérique, conditions stationnelles.

La sélection de variétés résistantes à la pollution n'est pas un moyen convenable car il faut réduire la pollution et non chercher à adapter la nature à un environnement pollué. On n'est pas encore sûr des méthodes de traitement sylvicole des peuplements qui pourraient leur permettre de résister à la pollution.

La fertilisation paraît pouvoir être un moyen efficace, mais coûteux, à condition d'être adaptée aux conditions de nutrition réelles des peuplements. Le chaulage systématique des sols acides ne peut être efficace que si l'on corrige en même temps d'éventuelles carences nutritives établies par analyse foliaire et analyse du sol. Il faut aussi choisir les peuplements à fertiliser en fonction de leur âge et de leur degré de dépérissement.

Summary

Several causes are responsible for forest decline : climatic accidents, air pollution, site conditions, particularly soil conditions.

The selection of air pollution-resistant trees is not a convenient way : pollutant emissions should be reduced instead of trying to adapt the nature to a polluted environment. The silviculturists are not yet sure of the silvicultural treatments wich could make forests resistant to air pollution.

Though expensive, fertilization seems to be a good way of limiting the decline, but has to be adapted to the present nutrition conditions of the forest stands. Systematic liming of acid soils can be efficient only if eventual nutrition deficiencies are corrected at the same time, after soil and foliar diagnosis. The stands to be fertilized should be chosen according to their age and degree in decline.

1. INTRODUCTION

Malgré l'importance et la qualité des recherches mises en oeuvre depuis quelques années, les causes du dépérissement des forêts restent mal connues,

On peut cependant, pour simplifier, retenir trois grands groupes de facteurs dont les effets sont additifs.

- Les accidents climatiques et notamment la sécheresse. Il semble bien que, depuis 1947, des épisodes de sécheresse longs et intenses interviennent à des intervalles de temps plus rapprochés que dans la période antérieure. Les recherches qui ont été conduites dans les Vosges (5) et qui

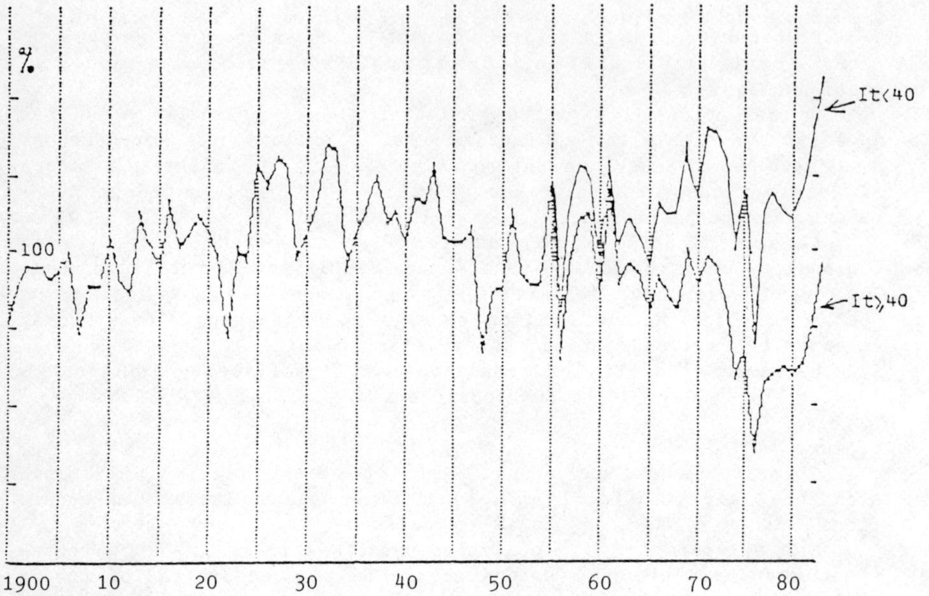

Figure 1 : Indice de croissance (dégagée de l'influence de l'âge) selon l'indice de transparence des houppiers It ; jusqu'à 40%, pas de différences significatives.

ne contredisent pas d'autres résultats acquis en Allemagne ou en Suisse, montrent que les arbres les plus atteints aujourd'hui ont vu leur vitalité (traduite par la largeur des accroissements) commencer à baisser en 1949-50, plus nettement en 1969, puis à partir de 1972 jusqu'en 1976 (figure 1). Il semble que la sécheresse de 1959 ait donné le choc fatal à bon nombre d'entre eux. Dans certaines régions, des froids intenses ont pu être le facteur déclenchant du dépérissement (23).

La pollution atmosphérique. Une forte chute de productivité en 1960 ne coïncide pas seulement avec la forte sécheresse de 1959, mais aussi avec le début d'une période de fort accroissement de la pollution atmosphérique, cause qu'il faut donc retenir comme agent vraisemblable de dommages aggravant ceux de la sécheresse.

L'action de l'ozone, pour les forêts d'Europe au moins, ne semble pas la plus importante. Des pointes de pollution par SO_2, même si la teneur moyenne annuelle de l'air reste faible, semblent par contre avoir une action physiologique importante, et notamment renforcer l'effet de la sécheresse (28).

Mais c'est surtout les pollutions acides (H_2SO_4, HNO_3) ou acidifiantes (dépôts secs de SO_2, NOx, NH_4) qui semblent jouer un rôle déterminant, d'abord par leur action de lessivage d'éléments nutritifs à partir des feuillages, ensuite en contribuant à l'appauvrissement du sol en éléments nutritifs et à l'augmentation du niveau d'éléments potentiellement toxiques (aluminium, manganèse...).

La pollution n'agit pas seulement en appauvrissant les écosystèmes forestiers mais parfois en déséquilibrant la nutrition, comme dans certaines régions où l'apport exogène d'azote atmosphérique est très élevé.

- Les conditions stationnelles. Il est tout à fait visible que le dépérissement n'est jamais uniformément réparti, sauf peut-être dans les régions où la pollution est très aiguë. Les conditions stationnelles interviennent donc pour sensibiliser les peuplements aux facteurs du dépérissement. Il est bien évident que les arbres sont plus exposés aux accidents de sécheresse sur les sols superficiels et plus sujets à des déséquilibres nutritionnels ou à des carences sur les sols pauvres.

Sur les sols les plus pauvres, le déroulement normal de l'activité sylvicole, en exportant avec le bois des éléments nutritifs en quantité souvent plus grande que celle qui serait compatible avec un équilibre des apports au sol et des pertes par les écosystèmes, est aussi un facteur d'acidification. On peut dire, dans ce sens, que même une sylviculture normale intervient dans le dépérissement en contribuant à appauvrir l'écosystème en éléments nutritifs.

Nous admettrons que ce schéma rapide des causes du dépérissement est convenable ; sinon il serait tout à fait illogique de discourir sur les remèdes d'un mal dont les causes sont inconnues.

Nous admettrons aussi, sans nous attarder sur ce qui est pour la plupart d'entre nous une évidence, que la réduction des émissions de polluants de toutes catégories constitue le remède le plus logique et aussi le plus efficace, le plus durable à long terme, à la fois pour rétablir la santé des forêts et pallier les inconvénients de la pollution dans de nombreux autres domaines.

Nous n'envisagerons donc ici que les mesures de protection et de réhabilitation des écosystèmes forestiers que le sylviculteur est lui-même capable d'appliquer en forêt, avec évidemment le concours technique d'organismes compétents et éventuellement l'aide financière des états. Normalement, si l'on est décidé à réduire la pollution à la source, ces mesures n'ont de sens qu'en tant que mesures transitoires et donc si elles ont un effet rapide.

Nous parlerons successivement :

- des mesures génétiques
- des mesures sylvicoles proprement dites, c'est à dire de la conduite des peuplements
- de la fertilisation en insistant plus particulièrement sur ce dernier thème.

2. MESURES GENETIQUES

La voie génétique, à priori, ne constitue pas une solution du problème. Sur le plan éthique, comme sur le plan logique, il faut éviter les agressions extérieures que les écosystèmes forestiers subissent du fait de l'action de l'homme, plutôt que d'accepter cette agression comme une fatalité et d'y adapter un nouveau matériel végétal.

Cette remarque faite, on peut cependant évoquer et commenter diverses voies génétiques possibles :

a) sélectionner des espèces ou des variétés plus résistantes à la pollution. C'est une voie en principe peu intéressante car elle est longue, elle implique souvent une sélection en ambiance de pollution assez courte et à concentration élevée plutôt que vis-à-vis d'une pollution faible et de longue durée qui est le plus généralement celle que subissent aujourd'hui les forêts des zones rurales. On ne peut sélectionner que vis-à-vis d'un petit nombre de polluants qui sont ceux d'aujourd'hui, mais qui pourront changer dans quelques décennies. Cette sélection ne garantit pas que les espèces ou variétés sélectionnées, une fois installées en forêt, ne se montreront pas excessivement sensibles à des aléas climatiques ou parasitaires (26). Cette voie n'a de sens, peut être, que pour des forêts très proches de gros centres industriels ou urbains qui risqueraient de rester des sources de pollution notable malgré les efforts de réduction des émissions.

b) sélectionner en forêt et multiplier par voie végétative, à l'intérieur des espèces qui peuplent déjà les massifs forestiers, des provenances ou des individus qui se seraient montrés, pendant un temps suffisamment long après le début du dépérissement, à la fois résistants à ce dépérissement et bien adpatés aux conditions écologiques. Mais on se dirige alors vers une sylviculture clonale qui a l'inconvénient d'appauvrir le potentiel génétique des forêts et de réduire l'adaptabilité de l'ensemble des peuplements à des accidents ultérieures de nature inconnue. Cette voie est cependant déjà entamée en R.F.A. (12) et peut être dans d'autres pays.

c) un peu en marge de la mesure précédente les généticiens ont également la préoccupation de réaliser des plantations conservatoires pour sauver des provenances intéressantes qui seraient menacées de disparition par dépérissement aux endroits où elles sont naturellement implantées (12).

La voie génétique, inopportune en général vis-à-vis des dégâts dus à la pollution, ainsi qu'il a été indiqué plus haut, risque aussi, par sa longueur de mise en oeuvre, de n'apporter une solution qu'après que la réduction des émissions aura déjà produit ses effets bénéfiques.

3. MESURES SYLVICOLES

Les observations sur les relations entre dépérissement et sylviculture sont souvent contradictoires. On a observé que les peuplements pleins et homogènes souffraient moins du dépérissement, ce qui a d'ailleurs été un argument pour étayer la thèse de l'intervention des polluants atmosphériques dans le dépérissement (25, 21). De même, KELLER et IMHOF (19) observent que des hêtraies éclaircies par le bas sont moins sensibles au dépérissement que celles qui ont subi des éclaircies par le haut. Par contre, SCHÜTZ (26) indique que "des sapins provenant de peuplements étagés possèdent à âge égal une meilleure vigueur que ceux issus d'un environnement régulier. Ceci démontre l'effet favorable d'éclaircies pratiquées à temps, c'est à dire en jeunesse". Par ailleurs, tout le monde s'accorde à dire que, dans les peuplements déjà soumis au dépérissement, les extractions d'arbres doivent être aussi limitées que possible pour éviter une ouverture de la forêt qui facilite la circulation des polluants.

Cependant, maintenir des peuplements très fermés risque de nuire à leur faculté de surmonter sans dommage les périodes de sécheresse. Des comparaisons, faites dans les Vosges, de peuplements dépérissants et sains situés dans des conditions écologiques très semblables, montrent que ceux qui ont surmonté sans dommage la crise de sécheresse de 1959-60 sont ceux qui avaient bénéficié quelque temps auparavant d'éclaircies assez fortes (LEVY, communication personnelle).

L'idéal serait (26) d'intervenir très précocément et très énergiquement dans les jeunes peuplements, par des interventions faibles mais fréquentes, de manière à les amener très vite à une densité moyenne, avec un développement important des cimes qui doivent cependant assurer un couvert complet, et ensuite de ne plus intervenir. Dans le hêtre les éclaircies devraient cesser vers 70 ans (19).

Les régénérations doivent être conduites de manière à éviter la multiplication des lisières, exposées à l'action des polluants ; c'est-à-dire qu'il ne faut pas céder à la tentation de purger les peuplements des arbres dépérissants au fur et mesure de leur apparition, ce qui conduit à réaliser des trouées anarchiques. Il faut au contraire s'efforcer de réaliser des opérations ordonnées dans l'espace (26). A la limite, ce souci semble pouvoir être contradictoire avec la sylviculture des peuplements de haute- montagne qu'on a plutôt tendance à traiter par des coupes de faible surface.

La création de peuplements mélangés peut aussi être appréciée de manière différente suivant le point de vue qu'on adopte. D'une part les peuplements mélangés sont réputés mieux résister aux agressions diverses, notamment parasitaires, mieux prospecter le sol, être plus stables. Il est logique de penser que dans un peuplement mixte feuillus-résineux, les dépôts secs acides ou acidifiants sont plus faibles puisque la masse foliaire est réduite en hiver. Mais, à l'inverse, le mélange risque de créer des hétérogénéités de structure qui favorisent la circulation et le dépôt des polluants atmosphériques dans les cimes.

Certains auteurs (21) tiennent également pour souhaitable une réduction de la densité de cervidés, car les blessures dues au frottis compromettent la résistance des arbres à d'autres attaques.

Au total, il semble que les sylviculteurs soient encore très partagés sur la manière de conduire les peuplements et des observations minutieuses sont certainement encore nécessaires pour arriver à une certaine sûreté de jugement et parvenir à surmonter les dilemmes qui ne manquent pas de se présenter au sylviculteur. Les mesures à prendre seront certainement différentes selon les régions, le relief, l'altitude, l'intensité de la pollution. De toute façon, elles ne peuvent produire un effet qu'à très long terme car il faut beaucoup de temps pour modifier la structure des peuplements.

4. FERTILISATION

Si l'on se réfère aux causes invoquées pour expliquer le dépérissement, la fertilisation pourrait être un moyen efficace et relativement rapide de préserver les peuplements forestiers d'une dégradation plus forte, et même de restaurer la vigueur de certains d'entre eux.

On peut considérer la fertilisation de deux manières :

- comme une lutte contre l'acidification des sols et le retour à une acidité plus faible ; il s'agit dans ce cas d'une fertilisation calcique ("chaulage de compensation" ou "d'assainissement" selon GUSSONE (13))
- comme une amélioration générale de la nutrition, destinée à conférer aux arbres un fonctionnement physiologique optimal qui leur permettrait de mieux résister à des agressions externes, en particulier celles des polluants atmosphériques, mais aussi la sécheresse et les froids intenses.

Bien évidemment ces deux types de fertilisation ne s'excluent pas.

4.1. Fertilisation conçue comme une lutte contre l'acidification du sol

Une forte acidification du sol, en réduisant l'activité biologique, en diminuant la disponibilité en calcium, en entraînant éventuellement des épisodes de toxicité aluminique, est nuisible à la vie des peuplements et peut être une des causes du dépérissement. On peut donc penser qu'un apport d'engrais calcique est susceptible de stopper le dépérissement et même d'améliorer l'état de santé des peuplements fertilisés.

Les essais mis en place dans le cadre spécifique du dépérissement sont généralement trop récents pour avoir pu fournir des résultats. Aussi les recherches ont-elles plutôt eu comme objectif d'étudier des essais anciens et de voir si les placeaux ayant reçu un amendement calcique avaient moins souffert du dépérissement que les autres.

Les résultats ne sont pas extrêmement concluants. Par exemple, KENK et al (20) ne notent qu'une faible amélioration de l'état de santé d'un peuplement d'épicéa de l'Odenwald après un apport calcique et phosphaté (figure 2). Nous-même, sur un essai de fertilisation de sapins de 80 à 100 ans dans les Vosges, installé en 1970, avons noté que la détérioration de l'état de santé depuis cette date a été aussi importante sur les placeaux ayant reçu du calcaire que sur les témoins. DERÔME et al (9), en analysant l'effet d'apports de calcaire sur 96 essais de Finlande, trouvent qu'après 25 ans le sol a été amélioré (taux de saturation en bases plus élevé, aluminium échangeable moins abondant, acidité échangeable moins forte), mais l'accroissement en volume a plutôt diminué.

Si l'on applique, au lieu d'amendement calcique, un amendement calco-magnésien, les résultats sont généralement meilleurs (10). Ainsi KENK et al (20) notent dans l'essai de l'Ödenwald une moindre perte d'aiguilles dans les traitements ayant reçu du kalimagnésia en plus de la fertilisation calcique et phosphatée. Cependant, ALDINGER (1), sur d'autres essais, ne note qu'une diminution du jaunissement mais non de la défoliation.

Il semble que, d'une manière générale, l'apport de calcaire seul n'ait pas été très bénéfique, même sur des sols très acides, si les peuplements ne souffraient pas d'une carence alimentaire en calcium.

Il semble cependant difficile d'admettre l'idée que le maintien d'un état calcique convenable, et mieux une resaturation du sol, soient sans effet à long terme. RITTER et al (24) notent, dans les essais de Pfalzgrafenweiler et de Ziefele en Forêt Noire, que l'état des mycorhizes et la vitalité des racines sont meilleurs dans les placeaux chaulés que dans les témoins. Il est difficilement pensable que les arbres ne tirent pas à la longue profit de ce meilleur état racinaire.

Si une carence magnésienne s'ajoute à l'acidité du sol, l'apport de calcaire magnésien est probablement bénéfique, mais son effet est lent. Dans des essais des Vosges où un apport calco-magnésien ($CO_3Ca + CaO + MgO$) a été effectué, la nutrition magnésienne n'a pas encore été sensiblement améliorée après 2 ans.

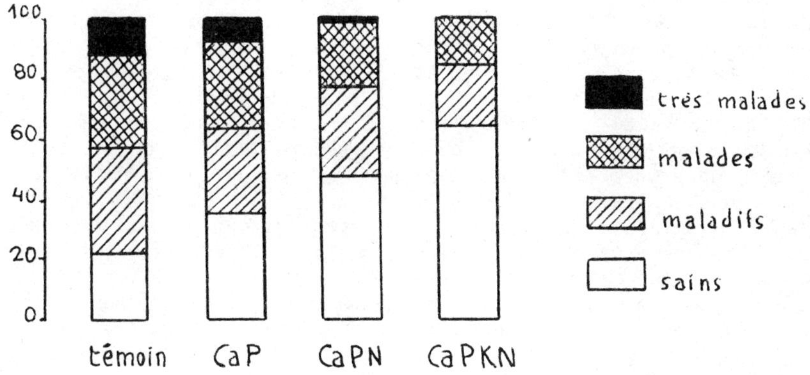

Figure 2 : Pourcentage d'arbres sains et dépérissants dans différents traitements d'un essai sur épicea dans l'Odenwald (d'après KENK et al., 1984) dans le traitement Ca, P, K, N, K est apporté sous forme de kalimagnésia : il y a donc aussi apport de Mg).

4.2. Fertilisation conçue comme une amélioration de la nutrition

Il a été montré dans de nombreuses régions que le dépérissement allait souvent de pair avec de fortes carences minérales en cations (Mg, K, Ca, Zn, Mn) et avec une dégradation, depuis une vingtaine d'années, de la teneur des aiguilles en ces éléments (7, 32, 15, 30, 31, 6, 8, 17). BAULE (2, 3, 4) montre que, sur un essai déjà ancien, une fertilisation apportant du magnésium et du calcium sous forme soluble (N, P, K, Mg) conduit au maintien d'un état de santé convenable alors qu'avec une fertilisation uniquement azotée et phosphatée la défoliation des épicéas est très importante.

De nombreux essais ont donc été mis en place dans un passé récent pour étudier les effets d'une correction de la nutrition. Ils utilisent généralement des engrais très solubles afin que l'effet sur l'alimentation minérale de l'arbre soit rapide. Ainsi, dans les Fichtelgebirge, en Bavière, un apport systématique de 5 kg de kalimagnésia par arbre (sulfate double de Mg et K) a permis d'obtenir une spectaculaire augmentation de la masse foliaire de sapins très dépérissants (29). HÜTTL et ZÖTTL (16), montrent, dans une série d'essais diagnostiques en Forêt Noire que, après des analyses foliaires, des analyses de sol et une fertilisation avec des engrais solubles de composition ajustée aux carences minérales constatées, de jeunes plantations d'épicéa jaunissantes reverdissent rapidement, en une saison de végétation (tableau I).

Dans les régions à très forte pollution, et malgré des essais déjà anciens qui semblaient montrer l'effet bénéfique de fertilisations complètes (y compris apport d'azote) (27), la situation ne semble pas encore très claire ; dans une publication récente (FIEDLER (11) estime que des recherches sont encore nécessaires pour "tester l'effet encore peu clair de la fertilité chimique du sol sur l'intensité des dommages". NEBE (22) indique qu'en R.D.A., une fertilisation appliquée avant dépérissement a légèrement réduit les pertes de croissance qui, à partir de 1972, ont précédé l'apparition de la défoliation. Dans les régions fortement polluées par SO_2, la carence magnésienne est d'ailleurs beaucoup plus légère que dans les régions à pollution plus diffuse (Europe occidentale) où sont apparus les "nouveaux dommages". Dans le premier cas, sur des porphyres acides, ce sont surtout les associations N + Ca ou N + P, qui ont eu le plus d'effet. Dans le cas de peuplements fertilisés alors que la pollution avait déjà provoqué des chutes de croissance, l'effet de la fertilisation a été presque nul (figures 3 et 4).

Il est par ailleurs certain que, dans les quelques régions où un apport extérieur d'azote très élevé sous l'influence de pratiques agricoles et d'élevage très intensives conduit à un déséquilibre très important de la nutrition ($N/P > 20$), la fertilisation des peuplements devra revêtir un aspect particulier, le but étant de valoriser cet apport d'azote très élevé et de rééquilibrer la nutrition par une alimentation très abondante en autres éléments (y compris oligo-éléments).

5. DISCUSSION - CONCLUSION

La fertilisation apparaît comme le seul moyen efficace d'améliorer l'état de santé de la forêt pendant la période transitoire qui va de l'époque actuelle à une époque future où les émission de polluants seraient très réduites. Les autres mesures possibles, génétiques ou sylvicoles, sont inadéquates, incertaines ou à effet trop lent.

Il faut distinguer :

1. Une fertilisation calcique généralisée à toutes les forêts sur sol acide ; son but est de réduire ou de stabiliser l'acidité de ces sols

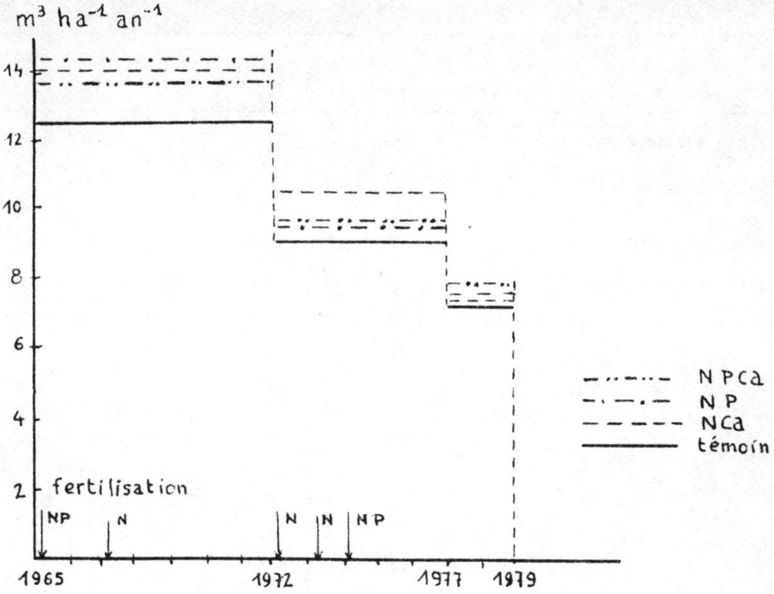

Figure 3 : Essai sur porphyre quartzeux avec dégâts de pollution 7 ans après l'application de la fertilisation (d'après NEBE, 1986).

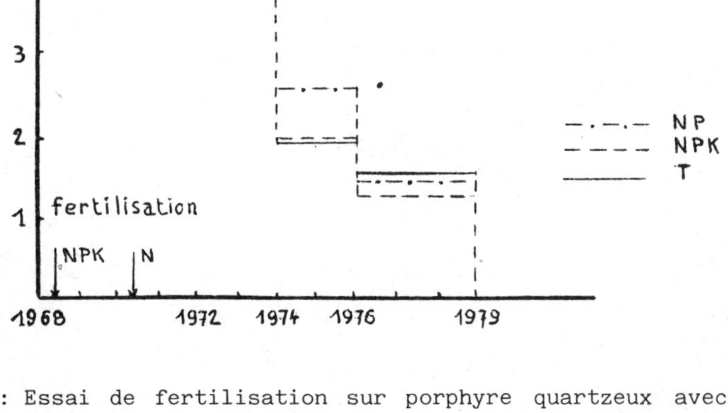

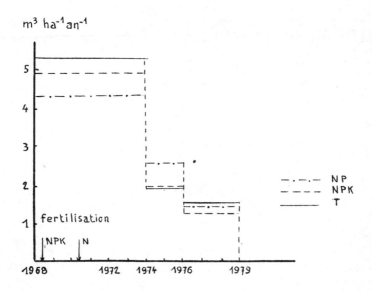

Figure 4 : Essai de fertilisation sur porphyre quartzeux avec dégâts de pollution déjà existants lorsque la fertilisation a été appliquée.

	N (1)	P (1)	K (1)	Ca (1)	Mg (1)	Mn (2)
Sapin adulte Vosges non fertilisé (6)	1.40	0.16	0.70	0.35	0.070	1.300
Epicea adulte Vosges non fertilisé (6)	1.45	0.16	0.75	0.19	0.045	1.200
Epicea plantations Vosges non fertilisées (G. LANDMANN)	1.89	0.23	0.74	0.33	0.040	-
Epicéa Jura non fertilisé	1.24	0.12	0.50	0.66	0.090	340
Sapin adulte Pyrénées non fertilisé (8)	1.29	0.21	0.30	0.41	0.151	828
Epicea plantation Staufen (16)						
- non fertilisé	1.37	0.21	0.99	0.32	0.054	600
- fertilisé	1.68	0.21	0.96	0.46	0.083	635
Epicea plantation Dischingen (16)						
- non fertilisé	1.45	0.14	0.45	0.59	0.130	900
- fertilisé	1.48	0.14	0.65	0.69	0.130	1.270
Epicea plantation Immendingen (16)						
- non fertilisé	1.01	0.12	0.83	0.68	0.067	3
- fertilisé	1.03	0.12	0.89	0.63	0.071	20
Epicea plantation Baden-Baden (32)						
- non fertilisé	1.60	0.20	1.00	0.09	0.041	450
- fertilisé	1.85	0.22	0.86	0.19	0.061	460

Tableau I : Composition foliaire de quelques peuplements dépérissants de Sapin et d'Epicea montrant la diversité des insuffisances d'alimentation minérale et la possibilité de les corriger par fertilisation.

(1) en % de la matière sèche
(2) en ppm

qui a fortement augmenté au cours des trente dernières années ; son effet sur l'état de santé des peuplements ne sera pas spectaculaire mais elle est probablement nécessaire à long terme car on arrive à des acidifications extrêmes.

D'après DERÔME et al (9) des doses de 2000 à 3000 kg de calcaire par ha seraient suffisantes pour 30 ans, ce qui correspond à la neutralisation d'un apport de protons de 1,3 à 2 $kg.eq.ha^{-1}.an^{-1}$. Des doses supérieures ne sont pas souhaitables, car elles pourraient avoir des conséquences secondaires fâcheuses pour les peuplements (blocage d'oligoéléments, difficultés d'alimentation en potassium), les écosystèmes (perte d'azote par minéralisation de la matière organique) ou l'environnement (augmentation de la teneur des eaux en nitrates). De plus, des apports de calcaire seul ne devraient pas être décidés avant qu'une analyse de l'état de la nutrition ait permis de déterminer si l'apport d'autres éléments est nécessaire.

2. Une fertilisation a effet plus immédiat, visant à corriger des déséquilibres nutritifs et utilisant surtout des engrais solubles (une fertilisation calcique préalable est cependant souhaitable en sol acide) et qui doit être précédée d'une investigation très minutieuse de l'alimentation minérale des peuplements. Les types de fertilisation à pratiquer sont très divers en fonction des conditions locales, par exemple : fertilisation calcique et magnésienne sur les grès et certains granites très pauvres des Vosges ou de Forêt Noire ; potassique dans la région préalpine du sud-ouest de l'Allemagne sur sols limoneux, sur les schistes des Pyrénées Centrales de la région de Luchon ; probablement potassique, phosphatée et azotée dans certaines régions du Jura et des Alpes ; phosphatée dans les zones d'apport azoté très élevé ; peut-être fertilisation N+Ca+K dans les régions à très forte teneur de l'air en SO_2. Si de l'azote doit être apporté, une partie devrait l'être sous la forme nitrique car l'absorption de NO_3^- facilite celle des cations et neutralise un H^+ (18). Sur les sols acides, la fertilisation idéale consisterait probablement en une fertilisation à base de calcaire finement broyé complétée par des engrais solubles apportant K ou Mg suivant les cas. Il est probable que cette fertilisation devrait être renouvelée, car, si le complexe absorbant est très désaturé, l'effet de l'apport de calcaire et de magnésium est peu durable (14, 20, 1).

Cette fertilisation idéale est cependant coûteuse et il faut s'interroger sur l'opportunité d'une application systématique. Des peuplements trop âgés ou trop endommagés ne sont plus susceptibles de valoriser économiquement les frais de fertilisation. Pour ces peuplements il vaut mieux réserver l'apport d'engrais à une nouvelle génération, en l'appliquant au début par pied d'arbre pour éviter une trop forte concurrence de la végétation spontanée, puis en la renouvelant sur toute la surface quelques années plus tard (10). Les jeunes plantations et les peuplements adultes d'âge moyen et dont le degré de dépérissement est encore limité semblent les plus capables de rentabiliser les frais de fertilisation. NEBE (22) estime que, dans les zones très polluées par SO_2, l'effet de la fertilisation sur les peuplements adultes est si faible qu'il vaut mieux réserver le financement disponible à une fertilisation d'amélioration des plantations nouvelles avec enfouissement des engrais par labour.

Il faut donc conclure que la fertilisation ne peut se révéler utile que si elle est pratiquée très judicieusement en fonction des conditions de peuplement et des conditions écologiques générales. Des opérations systématiques de type uniforme sont contrindiquées. Il faut raisonner l'intérêt de la fertilisation et sa nature à l'échelle de chaque parcelle de forêt.

REFERENCES

(1) ALDINGER, E. (1986). Wirkungen älterer Kalkungen in Fichten-Tannen Beständen des Buntsandstein-Schwarzwaldes. I.M.A. Querschmittsemi--nar "Restabilisierungsmassnahmen-Düngung" Kerforschungszentrum Karlsruhe 15/16 avril 1986, 64-72.

(2) BAULE, H. (1983). Wie weit kann Düngung gegen Walderkrankungen helfen ? Presse-Information Kali und Salz A.G., 30 august 1983.

(3) BAULE, H. (1984). Forstpflanzenernährung und Walderkrankungen. Arbeits-technisches Merkheft der Waldarbeit n°48.

(4) BAULE, H. (1984). Zusammenhänge zwischen Nährstoffversorgung und Walderkrankungen. Allg. Forst. Zeitschr. 30/31.

(5) BECKER, M. (1987). Etude écologique et dendroécologique du dépérissement du sapin dans les Vosges. in "Dépérissement des forêts attribué à la pollution atmosphérique. Etat des recherches à la fin de l'année 1986". Ministère de l'Environnement et l'I.N.R.A.- Nancy, Vol. 1, 83-96.

(6) BONNEAU, M., LANDMANN, G., ADRIAN, M., 1987. Le dépérissement du sapin pectiné et de l'épicea commun dans le massif vosgien est-il en relation avec l'état nutritionnel des peuplements ? Revue Forestière Française, XXXIX (1) 5-11.

(7) BOSCH, C., PFANNKUCH, E., BAUM, U., REHFUESS, K.E., 1983. Über die Erkrankung der Fichte (Picea abies Karst.) in den Hochlagen des Bayerischen Waldes. Forstw. Cbl. 102 (2), 167-181.

(8) CHERET, V., DAGNAC, J., FROMARD, F. (1987). Le dépérissement du sapin dans les Pyrénées luchonnaises. Revue Forestière Française XXXIX (1) 12-24.

(9) DERÔME, J., KUKKOLA, M., MALKÖNEN, E. (1986). Forest liming on mineral soils. Results of finnish experiments. National Swedish Environment Protection Board. Report 3084, 107 p.

(10) EVERS, F.H. (1984). Lässt sich das Baumsterben durch Walddüngung oder Kalkung aufhalten ? Der Forst-und Holzwirt 39 (4) 75-80.

(11) FIEDLER, H.J. (1986). Forstdüngung gegen negative Immissionswirkungen im Mittelgebirgsraum. Arch. Naturschutz u. landsch. Forsch. Berlin, 26 (2), 117-131.

(12) GELDBACH, J., SCHRÖDER, S., WIDMAIER, Th. (1987). Nachkommenschaftprüfung von Fichtenbeständen des Schwarzwaldes (Herkunftgebiete 840-08 und 840-09) mit den Zielen : 1) Verbesserung der Immissionstoleranz, 2) Erhaltung der Genresources geschädigter autochtoner Hochlagenbestände (Genbank). Statuskolloquium des P.E.F. 10-12 märz 1987. Kernforschungszentrum Karlsruhe.

(13) GUSSONE, A. (1987). Kompensationskalkungen und die Anwendung von Düngemitteln im Walde. Der Forst. und Holzwirt 42, 6, 158-163.

(14) HILDEBRAND, E.E. (1986). Zustand und Entwicklung der Austauscheigenschaften von Mineralböden auf Standorten mit erkrankten Waldbeständen. Forstw. Cbl, 105, 1, 60-76.

(15) HÜTTL, R., ZÖTTL, H.W. (1985). Ernährungzustand von Tannenbeständen in Süddeutschland. Ein historischer Vergleich. Allg. Forst. Zeitschr. 40, 1011-1013.

(16) HÜTTL, R.F., ZÖTTL, H.W. (1986). "Neuartige" Waldschäden und diagnostische Düngungsversuche. Premier Colloque Scientifique des Universités du Rhin Supérieur "Recherches sur l'environnement dans la région". Strasbourg, 27-28/6/1986.

(17) HÜTTL, R.F. (1987). "Neuartige" Waldschäden, Ernährungsstörungen und Düngung. Allgem. Forst. u. Jagdzeitschr. n°12, 289-299.

(18) ISERMANN, K. (1985). Diagnose und Therapie des "neuartigen Waldschäden" aus der Sicht der Waldernährung. Vortrag anlässlich des V.D.F. Kolloquiums "Waldschäden" am 18-20 juni 1985 in Goslar (Harz).

(19) KELLER, W., IMHOF, D. (1987). Zum Einfluss der Durchforstung auf die Waldschädenuntersuchungen in Buchen-Durchforstungsflächen der E.A.F.V.. Schweiz. Z. Forstwes. 138, 21-38.

(20) KENK, G., UNFRIED, P., EVERS, F.H., HILDEBRAND, E.E. (1984). Düngung zur Minderung der neuartigen Waldschäden. Auswertungen eines alten Düngungsversuchs zu Fichte im Buntsandstein-Odenwald. Forstw. Cbl. 103, 307-320.

(21) MAYER, H. (1984). Kann ein naturnaher Waldbau die Auswirkungen des Waldsterbens im Gebirge mindern ? Schweiz. Z. Forstwes. 135, 7, 613-618.

(22) NEBE, W. (1986). Zum quantitativen Nachweis von Ertragsminderung und zur Düngung in immissionsgeschädigten Fichtenbeständen. Cbl. ges. Forstwesen 135, 7, 613-618.

(23) REHFUESS, K.E., BOSCH, Chr. (1986). Experimentelle Untersuchungen zur Erkrankung der Fichte (Picea abies (L) Karst.) auf sauren Böden der Hochlagen : Arbeitshypothese und Versuchsplan. Forstwiss. Cbl 105, 4, 201-206.

(24) RITTER, Th., KOTTKE, I., OBERWINKLER, F., 1987. Vergleichende Untersuchungen zur Vitalität von Mykorrhizen nach Düngungsmassnahmen. 3 Statuskolloqium des P.E.F. 10-12 märz 1987 Kernforschungszentrum Karlsruhe.

(25) SCHÖPFER, W., HRADETSKY, J. (1984). Der Indizienbeweiss : Luftverschmutzung als massgebliche Ursache der Walderkrankung. Forstwiss. Cbl. 103, (4-5), 231-248.

(26) SCHÜTZ, J.P. (1984). Mesures sylvicoles immédiates et à long terme face au dépérissement des forêts. Journal Forestier Suisse, 135, 4, 307-319.

(27) TRILLMICH, H.D., UEBEL, E. (1983). Ergebnisse von langjährigen Forstdüngungsversuchen in einem Rauchschadengebiet. Beitrage f. d. Forstwirtschaft 17, 3, 123-130.

(28) VAN PRAAG, H.J., WEISSEN, F. (1986). Foliar mineral composition, fertilization and dieback of Norway spruce in the Belgian Ardennes. Tree physiology, 1, 169-176.

(29) ZECH, W. (1983). Kann Magnesium immissionsgefährdete Tannen retten ? Allg. Forstzeitschr. 9/10 p. 237.

(30) ZECH, W., POPP, E. (1983). Magnesiummangel, einer der Gründe für das Fichten-und Tannensterben in N.O. Bayern. Forstw. Cbl. 102, 50-55.

(31) ZÖTTL, H.W., MIES, E. (1983). Die Fichtenerkrankung in Hochlagen des Südschwarzwaldes. Allg. Forst. u. Jagdz. 154, 110-114.

(32) ZÖTTL, H.W. (1986). Nährelementversorgung mitteleuropaischer Wirtschaftwälder. Symposium "Möglichkeiten und Grenzen der Sanierung immissionsgeschädigter Waldökosysteme" Wien, nov. 1986.

FERTILIZATION AS A TOOL TO REMOVE "NEW TYPE" FOREST DAMAGES

R.F. HUETTL
Forestry Department, Kali und Salz AG,
P.O. Box 10 20 29, D-3500 Kassel, FRG

SUMMARY

A considerable number of investigations indicates that the so called new type of forest damages are often associated with nutritional disturbances. Clear correlations between the site specific substrate chemistry and the actual nutritional status of the trees were found. To explain the sudden and wide spread appearance of the forest declines beside natural stress factors the effect of various air pollutants and their derivatives are discussed. Diagnostic fertilization trials proved that a fast and sustained revitalization of declining forest stands characterized by disturbed nutrient supply is possible. Inspite of the improved phenotyp revitalization of fertilized trees is indicated by foliar analyses. Furthermore soil analyses reveal a considerable improvement of the chemical soil status due to the application of fast soluble sulphate based fertilizers (neutral salts). Base saturation was elevated, nutrient availability increased and pH-values did not change. Based on these and similar results fertilization of stands damaged by atmospheric deposition is subsidized by 80% in the Federal Republic of Germany. At present approximately 100,000 - 150,000 ha of declining forests are fertilized annually.

1. INTRODUCTION

In Central Europe "new type" of forest damages are observed since one dacade. Particularly in coniferous trees but also in deciduous stands beside overall unspecific foliage losses various decline types can be found. Hereby spatial and temporal distribution and above all natural and atmospheric deposition related site conditions must be differentiated. From manyfold observations of foliar discoloration symptoms it has become evident that nutritional disturbances are frequently associated with these declines. It should be noted that not the decline symptoms but the rather sudden and wide spread appearance of the damages must be addressed as truely "new". Disturbed nutrient supply in declining forest stands was found by Zech and Popp (1), Bosch et al. (2), Zoettl and Mies (3), Huettl (4), Hauhs (5), Reemtsma (6), Isermann (7) and v. Bredow et al. (8) in the Federal Republic of Germany, by Flueckiger et al. (9) in Switzerland, by Bonneau (10) in France, by Roelofs et al. (11) and van den Burg (12) in the Netherlands, by ÖDB (13) in Austria, by Delecour and Weissen (14) in Belgium.

2. HYPOTHESES

For the explanation of the observed nutritional disturbances various hypotheses have been suggested (cf. 15):

2.1 Indirect impacts:

Accelerated soil acidification due to increased acid deposition leads to a depletion of essential nutrients in the soil. In very acid soils toxic ions like Al^{3+} are mobilized damaging the roots/mycorrhizae and/or inducing nutrient uptake antagonisms (Mg/Al, Ca/Al; e.g. 16,17).

Increased N deposition leads to nutrient imbalances in the soil inducing deficiencies (e.g. 18, 19, 20).

Decreased input of alkaline substances due to the drastic reduction of particle emission and the historic change from cole to oil/gas burning causes nutrient deficiencies (21).

Climatic and other "natural" stress factors play a major role in the cause of this phenomenon (22, 23).

2.2 Direct impacts

Air pollutants e.g. SO_2 and O_3 damage foliar tissue. Via precipitation nutrient deficiencies can be induced due to increased leaching of mobile or exchangeable nutrient elements (K, Zn, Mg, Mn, Ca) from the foliage (24, 25, 26).

Acid precipitation particularly acid fog, dew and cloud water causes erosion of foliar waxes and elevated nutrient leaching leading to deficiencies (27).

Precipitation events with high NH_3/NH_4 concentrations lead to nutrient imbalances due to direct N uptake from the atmosphere (11).

Various authers believe, however, that all or some of these factors act additively or synergistically resulting in site and species specific decline types.

3. DIAGNOSTIC FERTILIZATION

Following two types of recent nutritional disturbances in Norway spruce stands (*Picea abies*) are presented. Via diagnostic fertilization the revitalization potential of declining trees is demonstrated.

3.1 Mg deficiency

At higher elevation sites with base poor parent materials Mg deficiency (tip yellowing of older needle year classes) is a characteristic symptom of the "new type" of forest decline. In those stands there is generally a rather wide variety of vitality ranging from relatively healthy looking trees to heavily damaged trees.
A typical example of this decline type is trial Staufen 6. The experimental area is a westernly exposed slope at 650 m a.s.l. in the Southern Black Forest. The deeply weathered acid brown earth is derived from base poor cristalline rock (solifluction debris of granite/gneiss). The low Mg/Al and Ca/Al ratios indicate the insufficient availability of particularly Mg and also Ca at this site (Table I). Due to the high Al^{3+} and H^+ soil contents base saturation is extremely low. Needle analyses data of the 12-yr-old Norway spruce stand coincide well with the substrate chemistry as well as with the different degrees of decline symptoms (Table II). The Mg contents of current needles are in all cases deficient, Ca values are low, Zn too is not sufficiently supplied in trees with very pronounced symptoms. Despite Mg and Ca also Zn values show a marked gradient i.e. decreasing element contents with increasing visible symptoms. P and K contents are optimal, N supply is good to sufficient. In order to improve the nutrient supply at this site various fast soluble Mg fertilizers were applied in spring 1985 (cf. Table III). When comparing the needle analyses data of the various control plots of fall 1985 (Table III) with the data of winter 1984 (Table II) a pronounced annual variation can be observed for most elements probably due to variing climatic conditions (less precepitation in 1984 than in 1985; in 1985 a long dry period in late summer and early fall was followed by heavy precipitation events and mild temperatures in late fall). The relatively stable P, K, and Mg contents underline the optimal supply of P and K and the definite Mg deficiency. The needle analyses results also indicate the remarkable influence of the Mg fertilizers on the Mg supply. Within only one vegetative period the fertilized trees had greened up considerably and needle losses were clearly reduced. The Mg fertilizer application has obviously increased overall nutrient uptake except for the anyhow well supplied elements P and K. In the following winter up to 30% of the control trees died due to frost impact whereas none of the fertilized trees was damaged.

Table I: Staufen 6. Chemical soil status in spring 1985 (NH_4Cl extraction; exchangeable cations at soil pH); 20-30 cm soil depth, before fertilization.

Mg^{2+}	Ca^{2+}	K^+	Al^{3+}	H^+	Mg/Al	Ca/Al	K/Ca	base sat.	pH
μval g^{-1}					mol			%	($CaCl_2$)
0.4	1.4	1.5	68	11.5	0.009	0.03	2.1	4.0	3.7
(preliminary threshold values Huettl in prep.)					(0.05-) 0.1	0.1 (-1)	0.5 (-1)	(15-20 ?)	(4-5 ?)

Table II: Staufen 6. Element contents of current needles (top whorl) of Norway spruce; samples winter 1984 (n=10).

damage class (foliar discoloration)	N	P	K	Ca	Mg	Mn	Zn	Fe	Al
	mg g^{-1} d.m.					μg g^{-1} d.m.			
(deficiency threshold values)	12-13	1.1-1.2	4.0-4.5	1.0-2.0	0.7-0.8	<20 (-80)	<13	(17)	0
moderately damaged	14.0	2.8	11.6	1.8	0.5	420	17	32	85
heavily damaged	14.0	2.1	9.0	1.5	0.3	510	11	34	105
very heavily damaged	12.6	2.5	9.1	1.3	0.2	370	9	34	100

Table III: Diagnostic fertilization trial Staufen 6. Element contents in current needles (top whorl); samples fall 1985 (n=10); fertilization (150 g/tree) in spring 1985.

plot	N	P	K	Ca	Mg	Mn	Zn	Fe	Al
	mg g^{-1} d.m.					μg g^{-1} d.m.			
control	17.7	2.6	9.2	2.6	0.3	690	20	48	135
Kieserite (27% MgO)	18.0	2.1	7.4	3.8	0.7	970	29	52	140
control	16.2	2.0	7.7	3.3	0.3	940	21	38	110
trial fertilizer I (18% MgO, 12% K_2O)	19.0	2.3	9.0	4.9	0.8	1250	38	49	155
control	17.9	2.4	8.8	4.0	0.4	780	29	40	150
trial fertilizer II (18% MgO, 12% K_2O, 7% CaO)	19.3	2.3	8.9	5.1	0.8	1450	30	67	190

Trial Freiburg 9, too, represents the common decline type associated with Mg deficiency. The well growing 24-yr-old Norway spruce stand is again located in the Southern Black Forest at 970 m a.s.l. on an acid brown earth derived from gneiss. When first investigated in fall 1983 the stand showed typical Mg deficiency symptoms coinciding with the insufficient Mg needle contents (0.6 mg g^{-1} d.m.). In spring 1984 various fertilizers were applied resulting in a quick revitalization response of the fertilized trees (Table IV, F_2). Again discoloration symptoms had disappeared almost completely within only one growth period. Due to the application of N, Ca, Mg, and the micronutrients Mn and Zn all element contents were considerably elevated. Only K contents did not change probably because of the optimal geogenic K supply of the parent material. Furthermore a sustained revitalization effect was found when fertilized trees were investigated three vegetative periods after fertilization (Table IV, F_1) . From this observation a pronounced change of the chemical soil status had to be expected. Indeed element contents of exchangeable cations in the fertilized soil show a remarkable improvement indicating a clear correlation between soil chemistry and nutritional status of the trees (Table V). Especially Mg and Ca soil contents were elevated, Al^{3+} adsorption was reduced and thus base saturation was improved. pH values were not influenced.

Table IV: Diagnostic fertilization trial Freiburg 9. Element contents in current needles (top whorl) of Norway spruce; fall sampling (n=10); fertilization in spring 1984.

plot	sample year	N	P	K	Ca	Mg	Mn	Zn
		mg g^{-1} d.m.					$\mu g\ g^{-1}$ d.m.	
control	1984	13.3	1.8	7.7	1.5	0.6	890	20
F_2*	1984	15.0	1.7	7.8	3.3	0.9	1270	42
F_1**	1986***	16.8	2.4	7.6	2.2	1.0	-	-

* F_2: trial fertilizer III (10% K_2O, 15% MgO, 10% CaO), 1500 kg ha^{-1}; calcium nitrate, 650 kg ha^{-1}; $MnSO_4$ (7%) and $ZnSO_4$ (4%) solution, 50 l ha^{-1}.

** F_1: trial fertilizer III, 1500 kg ha^{-1} ; calcium nitrate, 650 kg ha^{-1}.

*** data from Zoettl et al. in prep.

Table V: Diagnostic fertilization trial Freiburg 9. Chemical soil status (NH_4Cl extraction) in spring 1985.

plot	soil depth (cm)	exchangeable cations at soil pH Mg^{2+}	Ca^{2+}	K^+	Al^{3+}	H^+
		$\mu val\ g^{-1}$				
control	0-10	1.2	2.0	2.1	53.9	11.8
	20-30	1.0	1.0	1.5	53.1	11.1
F_1	0-10	+235*	+100	+23	-18	-3
	20-30	+ 12	+ 59	+ 4	-27	-14

plot	soil depth (cm)	Mg/Al	Ca/Al	K/Ca	base sat.	pH
		mol			%	($CaCl_2$)
control	0-10	0.033	0.055	2.1	7.7	3.7
	20-30	o.027	0.027	3.0	5.2	3.9
F_1	0-10	0.140	0.136	1.3	16.1	3.8
	20-30	0.045	0.061	2.1	8.2	4,0

* differences are given in percent to control (=100%)

3.2 K deficiency

At a variety of sites K deficiencies have recently been observed. E.g. Zoettl and Huettl (28) found a dramatic reduction of K contents via historical comparison of needle analyses in declining Norway spruce stands on moranic material in Southwestern Germany. Insufficient K supply has also been detected in declining stands on sandy (29) and on calcerous soils(4, 8, 9).

In fall 1983 the 64-yr-old Norway spruce stand at Ebnat 1 in NE Baden-Wuerttemberg was marked by K deficiency symptoms (lemon yellow discoloration of older needle year classes). Symptoms are well related to the nutritional status (Table VI). The stand grows on jurassic limestone coverd by a shallow layer of acidified loam as can be seen from the chemical soil data (Table VII). The extremely poor K availability becomes evident from the very low K/Ca ratio. Also in this case the application of a fast soluble K fertilizer provoked a pronounced revitalization. Two years after nutrient supplementation K supply was remarkably improved (Table VI).

Table VI: Diagnostic fertilization trial Ebnat 1. Element contents in current needles (top whorl) of Norway spruce; fall sampling (n=10); Kalimagnesia (30% K_2O, 10% MgO), 900 kg ha^{-1} in spring 1984.

plot	sampling year	N	P	K	Ca	Mg
		mg g^{-1} d.m.				
control	1983	13.4	1.6	3.5	4.5	1.2
fertilized	1985	14.7	1.4	6.5	6.3	1.2

Table VII: Ebnat 1. Chemical soil status (NH_4Cl extraction) in spring 1984 before fertilization.

soil depth (cm)	exchangeable cations at soil pH				
	Mg^{2+}	Ca^{2+}	K^+	Al^{3+}	H^+
	µval g^{-1}				
0-10	4.2	43	1.6	37	11
20-30	3.5	107	1.0	0	5
	Mg/Al	Ca/Al	K/Ca	base sat.	pH
	mol			%	($CaCl_2$)
0-10	0.171	1.748	0.074	50	4.0
20-30	-	-	0.019	96	4.9

4. FROM DIAGNOSIS TO THERAPY

Based on these and similar experiences from recent and older fertilization and liming trials focused on mitigating or removing forest damages due to atmospheric deposition impact fertilization in declining forest stands is now subsidized by 80% of total costs (fertilizers, application) in the Federal Republic of Germany.
In forestry practice generally between "compensative" and "ameliorative" fertilization is differentiated. Compensative fertilization is determined as the application of Mg limes (generally 20-30 dt ha^{-1}) to neutralize future acidic deposition on forest soils. Ameliorative fertilization is focused on the specific application of nutrients particularly in stands revealing nutritional disturbances. This is done with various fast soluble Mg and K fertilizers appliing 5 to 10 dt ha^{-1}. In the rare cases of insufficient P and N supply these elements are fertilized. Recently it was suggested strongly to combine the two above described treatment types to a more

effective site and species specific fertilization based on symptomatology, foliar and soil analysis (30, 31, 16).
Since 1983 approximately 200,000 ha of declining forests have been limed and/or fertilized in West-Germany. Due to the establishment of various state programs probably 100,000 - 150,000 ha will be fertilized annually within the next 5-10 years. Similar practices are being carried out in Poland, CSSR, GDR, USSR and are investigated (or planned) in the Netherlands, Sweden, Danmark, Belgium, France, Switzerland, Austria, Canada, and the USA.

5. CONCLUSIONS

Investigations into the "new type" of forest decline indicated that nutritional disturbances are frequently involved in the causes of this phenomenon. Via diagnostic fertilization trials the revitalization potential of declining forest stands was documented. Thus, site and stand specific fertilization can be used as a tool to mitigate and remove "new type" forest declines associated with disturbed nutrient supply.

6. ACKNOLEDGEMENT

The presented research work was carried out at the Institute of Soil Science and Forest Nutrition, Albert-Ludwigs-University, Freiburg, FRG under Prof.Dr. H.W. Zoettl. I want to thank Prof.Dr. Zoettl for his excellent guidance and the Federal Ministery of Science and Technology, Bonn for financial support.

7. REFERENCES

(1) Zech, W. and Popp, E.: 1983, Forstw. Cbl. 102, 50.
(2) Bosch, C., Pfannkuch, E., Baum, U., Rehfuess, K.E.: 1983, Forstw. Cbl. 102, 167.
(3) Zoettl, H.W. and Mies, E.: 1983, Allg. Forst- u. Jagdz. 154, 110.
(4) Huettl, R.F.: 1985, Freiburger Bodenkundl. Abh. 16, 195 p.
(5) Hauhs, M.: 1985, Ber. Forschungsz.Waldoekosysteme/Waldsterben Univ. Goettingen 17, 206 p.
(6) Reemtsma, J.B.: 1986, Naehrelementversorgung von Fichtenbeständen in Niedersachsen (manuscript).
(7) Isermann, K.,: 1987, in Ende,P.: 1987, Allg. Forstz. 42, 285.
(8) Bredow,v., B., Buggert, A., Eckhoff , A., Hollstein, B., Neumann, M., Schindel, R., Weber, A., Zech, S., Glavac,V.: 1986, Allg. Forstz. 41, 551.
(9) Flueckiger, W., Braun,S., Flueckiger-Keller, H., Leonardi, S., Asche, N., Buehler, U., Lier, M.: 1986, Schweiz. Zeitschr. f. Forstw. 137, 917.
(10) Bonneau, M.: 1986, Analyses Foliaires, in: Recherches sur le dépérissement de forêts attribué à la pollution atmosphérique (rapport).
(11) Roelofs, J.G.M., Kempers, A.J., Hondijk,A.L.F.M., Jansen, J.: 1985, Plant and Soil 84,45.
(12) Van den Burg, J.: 1987 in: Ende, P. (cf. (7)).
(13) ÖDB: 1986, Duengung - ein Weg zu gesunden Waeldern, 23 p.
(14) Delecour, F., Weissen, F.: 1986, Peuplements d'epiceas - Croix Scaille Forêt communale de Gedinne (rapport).
(15) Huettl, R.F.: 1986, Forest decline and nutritional disturbances. IUFRO-World-Congress in Ljubljana, Yugoslavia, Sept. 7-21, 1986 (in press).
(16) Ulrich, B.: 1986, Forstw. Cbl. 105, 421.
(17) Murach, D.: 1984, Goettinger Bodenkundl. Ber. 57, 126 p.
(18) Nihlgard, B.: 1985, Ambio 14 , 2.
(19) Mohr, H.: 1986, Biologie in unserer Zeit 16, 83.
(20) Friedland, A.J., Gregory, R.A., Karenlampi, L., Johnson, A.H.: 1984, Can.J.For.Res.14,963.
(21) Schenck, G.O.: 1986, Zur Beteiligung photochemischer Prozesse an den photodynamischen Lichtkrankheiten der Pflanzen (Waldsterben) (in press).
(22) Cramer, H.: 1986, Zusammenhaenge zwischen Schadensperioden und Klimafaktoren seit Mitte des letzten Jahrhunderts (in press).
(23) Rehfuess, K.E., Bosch, C.: 1986, Experimentelle Ueberpruefung der Auswirkungen eines Witterungsstresses in Expositionskammern (in press).

(24) Arndt,U., Seufert,G., Nobel,W.: 1982, Staub- Reinhaltung Luft 42, 243.
(25) Prinz,B., Krause, G.H.M., Strahtmann, H.: 1982, LIS-Ber. 28, 154 p.
(26) Guderian,R.,Kueppers, K., Six, R.: 1985, VDI-Ber. 560, 657.
(27) Mengel, K., Lutz,H.J., Breininger,M.Th.: 1987, Auswaschung von Naehrstoffen durch sauren Nebel aus jungen intakten Fichten (Picea abies) (in press).
(28) Zoettl, H.W., Huettl, R.F.: 1985, Allg. Forstz. 40, 197.
(29) Evers, F.H.: 1987, in: Ende,P. (cf. (7)).
(30) Gussone, H.A.: 1987, Der Forst- u. Holzwirt 42, 158.
(31) Horn,R., Zech,W., Hantschel,R., Kaupenjohann, M., Schneider,B.U.: 1987, Allg. Forstz. 42, 300.

JUDGEMENT OF THE APPLICABILITY OF LIMING TO RESTABILISE FOREST STANDS - WITH SPECIAL CONSIDERATION OF ROOT ECOLOGICAL ASPECTS

D. MURACH
Institute of silviculture, University of Goettingen, West-Germany

Summary

The results of fine root inventories in 1983 to 1985 in limed and unlimed plots of a spruce and beech stand of the Solling project in West-Germany are presented. The data show that for both tree species the liming treatment is visible in the chemical composition of the fine roots. The response of the fine root growth to the fertilisation is different among the species. In opposite to the beech the fine root turnover in the spruce stand is decreased. Spruce is forming a more shallow root system on the limed plots caused by better rooting in the humus layers to the charge of the mineral soil horizons. In the beech stand liming caused a better rooting in the strongest acidified upper mineral soil layers. In both stands the total fine root biomass is not significantly changed by fertilisation. The consequences for the damage symptoms of the shoots are discussed.

1. Introduction

An increase of the acid deposition within the last decades has caused a decrease in base saturation or an accumulation of acids in the soil. This is the main reason for the long term increasing soil acidification.
The consequences are nutrient losses and the appearance of potential toxic cations in the soil solution, which can lead to a restricted growth of the roots (6). Thus, the disordered nutrient and water supply of the trees in connection with further stress factors can lead to visible damage symptoms at the shoots. This knowledge has early lead to the claim for a compensation liming(7).
With the analysis of the effects of liming experiments the stress is put upon questions concerning soil and needle chemistry.There are less investigations on the effects on the root growth of trees.
The following presentation of single results of a 6 years fine root investigation in the frame of research in forest decline in the Federal Republic of Germany should deliver some further data to this question.

2. Stands and method

In a 105 years old spruce stand (Picea abies Karst.) and a nearby 135 years old beech stand (Fagus silvatica L.) in the Solling there were taken soil samples by the auger method with 10 replications at each of the 10 dates in the period from 1983 to 1985. After dissection of the soil core the soil slices were overfloated with demineralized water and washed out by the low pressure method (5). Fine roots (less than 2 mm diameter) longer than 1 cm were separated under the microscope into living and dead rootlets by morphological characteristics and then their dry matter was destinated (5).

These investigations were executed on a limed and an unlimed plot in the spruce and beech stand. Both stands are growing on a podsolic brown earth derived from loes over red sandstone. The time, kind and quantity of the fertilizer is to be seen in table I.

Tab.I: Fertilisation treatments in the Solling stands

date	fertilizer	nutrients in kg/ha							
		NH4-N	NO3-N	Ca	K	Na	Mg	Mn	Cl
June 1973	calcium ammonium nitrate + KCl	166	150	115	170	24	4		186
Okt. 1975	calcium silicate (smelter ash)			1188	30	12	360	16	
Dez. 1980	dolomit lime			772	5	2	456	1	

The effects of the fertilization are evident with the exchangeable cations mainly in the upper mineral soil horizon (s. tab. II). In these horizons the higher amounts of Ca and Mg at the cation exchange capacity to the charge of H, Al, Fe and Mn are especially obvious. In the lower soil the chemical soil conditions on the limed and unlimed spruce- and beech-stands are assimilating.
The increase of the pH-value by the liming is not very distincted in the upper soil. On the spruce stand the pH-value in $CaCl_2$ is increasing from 2.9 to 3.2 and on the beech stand only from 3.2 to 3.4. In the humus horizon the differences are greater, but they also show a great variation.

3. Results of the fine root investigations

The vertical distribution of the fine roots on the 4 plots is to be seen in figure 1. In the average of the investigation period the spruce stands show an obviously better rooting in the humus horizon of the limed plot, especially in the Of horizon. This is accompanied by less fine root biomass in the upper mineral soil. In the lower soil there are no significant differences. The differences in the root growth in the humus horizon of the spruce plots are especially obvious during periods with higher precipitation. Compared to

Tab. II: Exchangable cations in the different plots

stand	depth (cm)	CEC (µmolIE/g)	X^S_H	X^S_K	X^S_{Ca}	X^S_{Mg} (%)	X^S_{Fe}	X^S_{Mn}	X^S_{Al}
spruce unlimed	0-10	133	16	1	2	1	13	3	62
	10-20	79	1	1	1	1	1	6	88
	30-40	55	0	3	1	0	0	3	93
spruce limed	0-10	138	12	1	12	8	7	4	56
	10-20	76	2	2	4	3	1	6	82
	30-40	51	0	3	2	2	0	3	88
beech unlimed	0-10	108	15	1	3	1	5	5	69
	10-20	74	3	2	2	1	1	7	84
	40-50	42	0	2	0	0	0	1	94
beech limed	0-10	112	12	1	11	7	4	3	61
	10-20	80	3	2	3	2	1	4	85
	40-50	55	0	3	1	1	0	1	92

the drier spring and summer period 1983 (precipitation May to Oct. 1983: 360 mm) with 100 mg fine root biomass per 100 ml soil on the unlimed plot and 180 mg per 100 ml soil on the limed plot, the differences in the year 1984 (precipitation May to Oct 1984: 570 ml) are twice as great (130 mg/100 ml and 280 mg/100 ml respectively).
There is the assumption, that in drier periods the lower K/Ca- ratios in the soil solution of the limed plot restricts the growth promoting effect of higher pH-values. This is supported by the chemical analysis of the fine roots. The Ca/K-ratios of the living fine roots in the Of-horizon in 1983 is 3.4 in the limed plot compared to 1.8 in the unlimed plot as a mean of six sampling dates. In the Oh-horizon the mean values are 2.3 and 1.6 respectively. Furthermore, much better Ca/K ratios in the fine roots are evident during the periods with better root growth in the humus horizon of the limed spruce stand. By the end of June 84 they are about 2.0 in the Of horizon.
The better fine root growth in the humus layer is compensated by the decreased fine root concentrations in the upper mineral soil. The liming treatment in the Solling area at least has caused a more shallow root system but not an increased total fine root biomass. In both spruce plots they are in the range of 2500 kgs/ha.
So the conclusion may be drawn that the spruce is reacting stronger on a pH-value increase in the humus layer than on the increase of the Ca/Al-ratio in the mineral soil, which is in the range of 1 to 2 according to analysis of watery extractions. Thus, this ratio is 5 times higher than in lysimeter solutions of the unlimed plot.
One of the reasons for the missing positiv reaction on the improved Ca/Al-ratios may be caused by the accumulation of Al in the fine roots. With increasing age the Ca/Al-ratio in the

fine roots is decreasing due to the accumulation of aluminum in the fine root cortex. This could be shown in field investigations with the comparison of young and old rootlets (s. tab. III). Young growing rootlets from inventories (stand Hils) respectively from ingrowth core samplings with partially limed soil (stand Solling) show similar Ca-concentrations as the elder rootlets. But the K-contents are significantly higher and the Mn-, Fe- and especially Al-concentrations lower.

Tab.III: Comparison of the element concentration in young and old fine roots in two spruce stands

stand	age	treat-ment	P	K	Ca	Mg	Mn	Fe	Al
			mg/g						
Hils	young	-	2.8	7.1	.5	.2	.04	.4	.5
	old	-	1.5	1.3	1.4	.2	.06	.4	1.2
Solling	young	limed	2.7	7.5	4.2	1.5	.46	.9	.7
	young	unlimed	2.5	5.5	1.5	.5	.32	.8	.4
	old[1]	unlimed	1.1	2.5	1.3	.5	1.1	1.8	12.3
	old[2]	unlimed	.8	1.7	.2	.2	.10	2.3	25.5

[1]: living fine roots [2]: dead fine roots

On reason of the accumulation of Al in the fine roots there are nearly no differences between the Ca/Al-ratios in the unlimed and limed plots with increasing soil depth (s. fig. 3).

This might be the reason that the negative influence of unfavourable Ca/Al-ratios on the fine root growth is not compensated by superficial liming. Unfavourable Ca/Al-ratios result in a increased rooting of the organic influenced upper soil horizons, where Al is complexed and therefore detoxicated (6).

Transferring this principle also on other stands, this would mean that on all noncalcareous soils, where the exchanger is covered with Aluminum and anions from immission enable the mobility of cation acids in the soil, the root system of Al-sensible trees like the spruce is tending to produce a flat root system. Of significant importance should be the presence of anions in the soil, which is, except from nitrate, mainly caused by air pollution within areas far from the coast. Nitrate itself is naturally appearing in the soil in an important quantity only within phases of decoupled ion cycling. Thus, its appearance is restricted by time and space. Therefore, the ion concentrations on stands with reduced anthropogene influences should be relatively low. The accumulation of damaging elements in the fine roots is rather delayed under such conditions. On the other hand the delivery of dissolved nutrients is also slowed down. There is the possibility, that in such stands the involvement of mycorrhizas in short nutrient cyclings, which is discussed in the literature (8),is getting a special importance. These aspects should be more considered in the present discussion about the damaging influence of high nitrate concentrations on the mycorrhiza and on the fine root systems of the trees. When supposing that the acidification of fine roots is showing a limiting factor for the root growth of the spruce, the positive effect of increased Ca-concentrations in the soil solution of limed plots should foremost be a prolongation of the durability of fine roots, as the obtainment of a damaging Ca/Al-ratio in the fine roots takes a longer period.

In order to check such effects in field experiments it is of an advantage that an increased durability of the fine roots is correlated to a reduced fine root turnover, which is easier to pick up.

The calculations for 1983 showed,in fact,essentially higher fine root turnover in the limed plot, with values of 2.500 $kgs \cdot ha^{-1} \cdot a^{-1}$ for the unlimed and 1.100 $kgs \cdot ha^{-1} \cdot a^{-1}$ for the limed stand. This is corresponding to the increament investigations of BAUCH et al. (1) on these plots. On the limed plot BAUCH et al. could determine an about 10 % higher basal area increment.

The calculation of the fine root turnover for this year was facilitated by the heavily restricted decomposition of the dead roots in the course of the dry year until the rewetting period in autumn (see figure 2). By the accumulation of the dead fine roots the turnover could be calculated by a single difference formation and, thus, it is charged with less variation than calculations by summing up of repeated difference formations.

A further positive effect of the liming is a better nutrient supply of the stands (see figure 3). While there are no differences with N and P, the living fine roots of the limed spruce plot show slightly reduced concentrations of Mn and Fe. In contrary to this the Mg- and Ca-concentrations are obviously, the K-concentrations only slightly increased. Essentially, these results are corresponding to microanalytic analysis in fine roots, which were executed by Bauch et al. on the same plot (1). Needle analysis show, that the better Ca- and Mg-supply of the fine roots in the limed plots correlates with higher contents of these elements in the needles (3).

The effect of the liming treatments on the chemical composition of the fine roots within the beech plots are less obvious than with the spruce. A significant increase of the Ca- and Mg-contents of the living roots is only valid for the upper soil. With strong saisonal and spacial variation there are sometimes great differences in the N-contents with higher concentrations in the limed plot. With the other elements there are no significant differences.

In contrary to spruce a reduction of the fine root turnover is not evident, as this is essentially lower than with the spruce even in unlimed stands. At least, in 1983 a accumulation of dead fine roots comparable with the spruce stand couldn't be observed. With this tree species, the liming treatment is promoting the rooting of the most

acidified layers Oh and 0-5 cm. In contrary to earlier investigations in the same stand, which were executed by GOETTSCHE (2) in 1971/72, the rooting in these horizons has been getting worse in comparison with the other horizons. Thus the beech is obviously differing from the spruce as far as the reaction of their fine root growth on increasing soil acidification is concerned. The spruce earlier tends to a shallow root system than the beech and is therefore reacting more sensible on increasing Al-concentrations in the soil. But the beech preferably reacts on decreasing pH-values. These results of field investigations fit in with the results of hydroculture experiments (6).
As with the spruce the total fine root mass per hectar is not increased.
If these results are interpretated in view to the damage symptoms at the shoots, which are appearing in connection with the present forest damages, it is possible that after a surficial liming the yellowing of the spruce needles which is caused by Mg deficiency is decreasing. But new symptoms due to induced K-deficiency are possible. As far as the losses of the needles are concerned, there shouldn't appear improvements. On the contrary on stands with a worse water budget it is to calculate that these symptoms could be intensified. In contrary, a positive effect on the growth might be observed when the water supply is sufficient.

REFERENCES

(1) Bauch, J., Stienen H., Ulrich B. and Matzner E. (1985): Einfluß einer Kalkung bzw. Düngung auf den Elementgehalt in Feinwurzeln und das Dickenwachstum von Fichten aus Waldschadensgebieten. AFZ 43, 1148-1150

(2) Goettsche, D. (1972): Verteilung der Feinwurzeln und Mykorrhizen im Bodenprofil eines Buchen- und Fichtenbestandes im Solling. Diss. Reinbeck

(3) Matzner, E. (1985): Auswirkungen von Düngung und Kalkung auf den Elementumsatz und die Elementverteilung in zwei Waldökosystemen im Solling. AFZ 43, 1143-1147

(4) Meyer, F.H. (1985): Einfluß des Stickstoff-Faktors auf den Mykorrhizabesatz von Fichtensämlingen im Humus einer Waldschadensfläche. AFZ 40, 208-219

(5) Murach,D. (1984): Die Reaktion der Feinwurzeln von Fichte (Picea abies Karst.) auf zunehmende Bodenversauerung. Goettinger Bodenkundliche Berichte, Bd. 7

(6) Rost-Siebert,K. (1985): Untersuchungen zur H- und Al-Ionen-Toxizität an Keimpflanzen von Fichte (Picea abies,Karst.) und Buche (Fagus silvatica, L.) in Lösungskultur. Ber. d. Forschungszentrums Waldökosysteme/Waldsterben; Bd.12

(7) Ulrich,B. (1971): Grundsätzliches zur Forstdüngung. Forst- und Holzwirt 26, 147-148

(8) Walter, H. (1975): Besonderheiten des Stoffkreislaufs einiger terrestrischer Ökosysteme. Flora 164, 169-184

FIG. 1: VERTICAL DISTRIBUTION OF THE FINE ROOTS

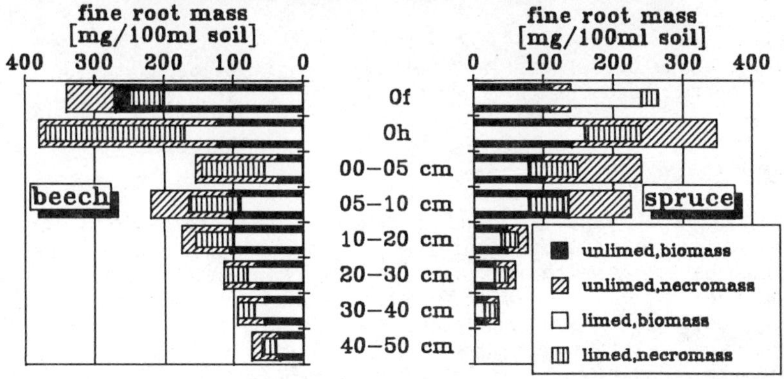

FIG. 2: SAISONAL DEVELOPEMENT OF THE FINE ROOT NECROMASS IN THE SPRUCE STANDS

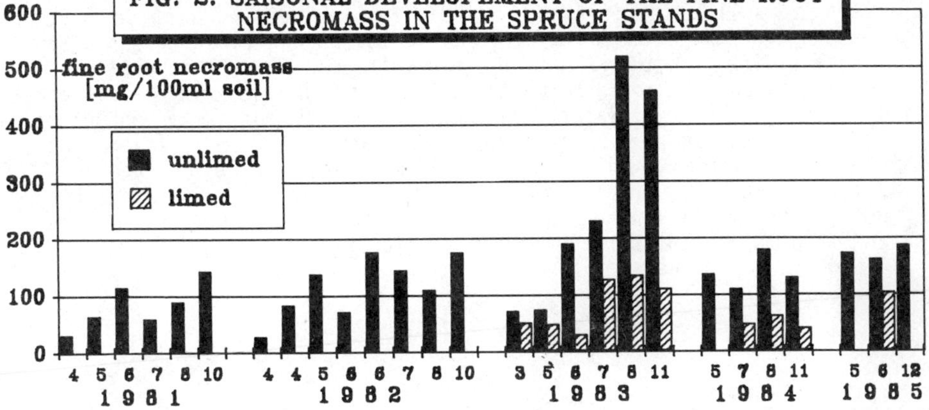

FIG. 3: ELEMENT CONCENTRATION IN THE LIVING FINE ROOTS

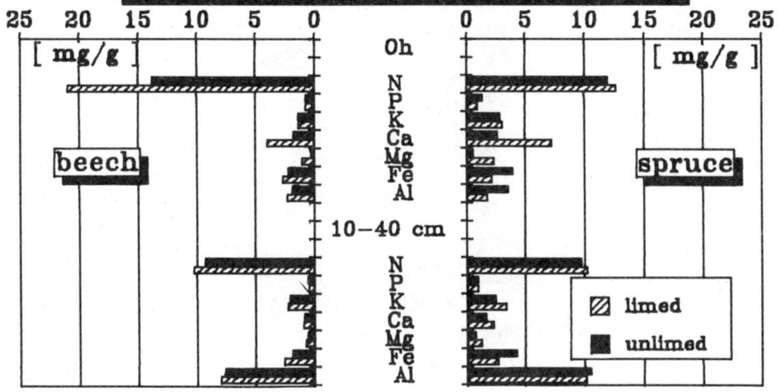

THE EFFECTS OF LAND MANAGEMENT ON ACIDIFICATION OF AQUATIC ECOSYSTEMS AND THE IMPLICATIONS FOR THE DEVELOPMENT OF AMELIORATIVE MEASURES

M. HORNUNG
Institute of Terrestrial Ecology, Bangor Research Station
Penrhos Road, Bangor, Gwynedd, LL57 2LQ North Wales, UK

Summary

The impact of land use, and management on factors which influence the acidity of aquatic ecosystems is discussed. Those aspects of land management which increase anion leaching, whether by increasing atmospheric inputs or within-system generation, or modify catchment hydrology will have the most rapid impact on drainage water chemistry. Sites with acid soils and massive, base poor bedrock will be most sensitive to such changes. Vegetation change influences atmospheric inputs, the formation of nitrate within the soil, organic matter type and rate of turnover and the amount of precipitation which reaches streams. Nitrogen fertilizers increase rates of soil acidification and leaching. Cropping removes base cations and acidifies soils; nitrate formation and leaching can also increase after cropping. Drainage of soils can lead to oxidation of sulphides and organic compounds of S and N; it also modifies catchment hydrology, water pathways and the mix of water entering lakes and streams. Following any given land use change, several of these mechanisms will operate simultaneously, and interact. Afforestation in the british uplands is used to illustrate the complexity of the changes and interactions, and to examine land use and land management strategies for ameliorating their impact.

1. INTRODUCTION

Land use, and management, influence several of the interacting factors which control the chemistry of aquatic ecosystems. Thus, they produce changes in vegetation, and therefore atmosphere-vegetation interactions and inputs; they modify chemical, physical and biological properties and processes which obtain in soils, and therefore soil-water interactions and drainage water chemistry; they alter hydrological pathways in catchments and the mix of waters draining into streams and lakes. These changes in soil-water interactions and catchment hydrology can be important factors in the acidification of aquatic ecosystems. The impact of land management on any given aquatic system will depend on the management practices themselves and the soils, drift and bedrock geology, and hydrology of the catchment; to this extent, the impacts will be site specific.

In some situations the influence of land management on the acidity of aquatic ecosystems may act in addition to the effects of acidic atmospheric deposition. In other cases land use, and atmospheric pollution may interact to enhance the overall impact on aquatic systems. Land use may be the dominant factor producing acidification of surface waters or it may play a minor role compared with that of acidic deposition. Whatever the situation at a given site, it is important that the role of land use, and specific management practices, is identified and evaluated. The development of measures to ameliorate, or reduce the acidification of aquatic systems should consider land use and management options in addition to direct treatment of the aquatic system. The identification of possible management options necessitates an understanding of the impact of land use on the processes and mechanisms controlling soil-water interactions and catchment hydrology.

In the following sections, mechanisms are discussed whereby land management practices can produce increased leaching and acidification of soils, and increased acidity, and/or aluminium concentrations, in drainage waters. The acidification of soils will have a longer term influence on drainage water chemistry and may influence the ability of given soils to buffer atmospheric inputs of acidity. Soils with low to moderate base saturation, and low cation exchange capacity are most sensitive to further acidification (1, 2, 3). The most rapid impacts on drainage water chemistry will result from processes, or management practices which increase the mobile anion loading (4) or modify the catchment hydrology. Areas with acid, low base saturation soils and massive, base poor bedrock, and low conductivity waters, will be most sensitive to the effects of increased anion loading and hydrological changes. On this type of site the increased anion load, input from the atmosphere or generated within the soil, can lead to acidification of drainage waters and mobilisation of aluminium. The impact of any increased anion loading will, however, be influenced by the sulphate adsorption properties of the soils (5) and by any retention of nitrate inputs within the catchment (6).

2. VEGETATION CHANGE

2.1 The impact on atmospheric inputs

The transfer of elements, as gases, small particles and occult deposition, from the atmosphere to terrestrial ecosystems is strongly influenced by the nature of the vegetation canopy; in particular, the aerodynamic roughness and surface resistance of the canopy (7). These two characteristics vary widely between vegetation types. Land use, and mangement, can therefore influence the rate of transfer of elements from the atmosphere, and total deposition, by changing or modifying the vegetation of a given site. It should be noted that the gas, particulate and occult deposition is in addition to precipitation inputs, which are not canopy dependent.

In this context, the most important vegetation changes brought about by land management are the replacement of grassland, or moorland, by conifer forests and the change of tree species from hardwoods to conifers. Coniferous forest canopies have a much greater aerodynamic roughness than shorter, smoother grass or moorland vegetation. As a result, the afforestation of moorland areas can result in a large increase in inputs to a site of both pollutant and non-pollutant elements and ions from the atmosphere; the "filtering effect" or "enhanced capture" of forest canopies. The magnitude of the increase will vary with specific site conditions such as ambient atmospheric chemistry, altitude, aspect, occurrence of cloud and mist cover.

A number of methods have been used in an attempt to quantify the additional input following afforestation, for example comparative catchment studies (8) and calculations based on data from shielded and filter-gauge type precipitation collectors (9). Catchment studies in western Britain (8) show sulphate outputs from afforested catchments to be between two and three times greater, and chloride outputs roughly double, those from adjacent moorland catchments. Studies at Lake Gardsjon, Sweden, also showed a relationship between sulphate output from catchments, and the proportion of forest cover, Table I (10). A number of clearfelling studies have also reported reductions in sulphate and chloride outputs following felling, eg. Table II (11). Together, these data provide a measure of the likely change in atmospheric inputs as a result of afforestation.. It is interesting that at most sites in western Britain the increase in sulphate inputs seems to be greater than the increase in chloride.

	F1	F2	F3
% Norway spruce, Scots pine and 20 year clearcuts	93	82	42
Mean SO_4-S in runoff 1979-1981 kg. ha^{-1} yr^{-1}	28.8	26.2	19.8

Table I. Mean annual SO_4-S outputs from three catchments with different proportions of forest cover at Lake Gardsjon, Sweden

[Data from Hultberg (10)]

	SO_4-S	Cl
Felled	38.8	60.9
Unfelled, control	52.9	118.0

Table II. Sulphate-S and chloride outputs in drainage from felled and unfelled forest plots in Cumbria, England; kg. ha^{-1} yr^{-1}

[Data from Adamson <u>et al</u>., (11)]

The deposition onto forest canopies at a given location will be influenced by tree species, height and spacing. In general, the deposition is less onto broadleaved canopies. For example, in the Solling Forest, West Germany total S deposition to beech forest was 45-51 kg S ha^{-1} y^{-1}, to spruce forest 80-86 kg S ha^{-1} y^{-1}, and to bare soil 23 kg S ha^{-1} y^{-1} (12). Similarly, Nys <u>et al</u>. (13) calculate that S inputs to oak-dominated broadleaved woodland in north eastern France were between 38 and 42 kg S ha^{-1} y^{-1} and to spruce plantations between 56 and 64 kg S ha^{-1} y^{-1}. There is also a considerable variation between deposition onto different coniferous species. Recent work in west Wales and eastern Scotland suggests a considerably greater deposition onto larch than spruce canopies; at the relatively unpolluted welsh site total inputs of sulphur to the larch are roughly double those to spruce while at the more polluted scottish site they are upto five times higher (Reynolds & Cape, pers comm.).

Thus, afforestation of moorland with conifers in upland Britain, the invasion of abandonned farmland by forest in Scandinavia or the change from hardwood forest to coniferous plantations in France, Belgium and Germany could increase inputs of atmospherically derived elements to soils. On sensitive sites, particularly where acid or shallow soils,overlie massive, base poor bedrock, this could increase the acidity and aluminium concentrations of drainage waters. In some areas it may be advisable to limit the amount of afforestation, or to maintain hardwood forests rather than replace them with conifer plantations, to prevent acidification of waters. Alternatively the affected soils could be limed to increase their base status and pH; on some sites, however, this can result in a reduction in tree growth (cf. 14). The choice of species used in plantations should also include a consideration of the impact on atmospheric inputs.

2.2 **Nitrogen fixing species**

Nitrogen fixing species increase the input of nitrogen to terrestrial ecosystems and the production of nitrate in soils. This can be an important acidification mechanism and in systems dominated by these species nitrate may dominate leaching mechanisms. Haynes (15) has reviewed soil acidification by agricultural crops such as clover species (Trifolium sp.), lucerne (Medicago sativa L.) and soybeans (Glycine max L.). He reports a number of published examples of reductions in soil pH beneath clover pastures of between 0.5 and 1.0 unit over 20-30 years (16, 17). There are no data on the impacts of this acidification on drainage waters. The most important nitrogen fixing tree crops in temperate areas are the alders (Alnus sp.). Van Miegroet and Cole (18, 19) compared nitrification, nitrate fluxes and leaching mechanisms below red alder and Douglas fir stands in north western USA. In the alder stands nitrification released upto 4500 mol H+ ha^{-1} yr^{-1} while the input via precipitation was 320 mol H+ ha^{-1} yr^{-1}. The upper 30 cm of the soil beneath the alder had been acidified, compared with that below the Douglas fir. Nitrate comprised 75% of the total anion flux at 40 cm in the alder system compared with c. 0.6% below the Douglas fir. In the system studied by van Miegroet and Cole increased acidity generated beneath the alder had leached base cations from the upper soil but had been effectively buffered within the system. Preliminary results from a study in the Pennine uplands, northern England parallel those of van Miegroet and Cole (Iles, pers. comm.).

In agricultural systems, lime is usually added to offset the enhanced acidification due to the legumes. The increase in proton production and excess nitrate below alder on acid soils could produce acidification of drainage waters, and/or raised aluminium concentrations, and aquatic ecosystems. While more research is needed, it would seem unwise to plant extensive stands of alder on sensitive, or acidic soils, especially near water courses unless ameliorative liming was also carried out.

2.3 **Changes in organic matter and decomposition**

The nature of the soil organic matter at a given site reflects the interaction of vegetation, climate and soil chemical and physical properties. Alteration of vegetation as a result of land use or management may change the form of organic matter. Thus, the replacement of broadleaved forests, or coppice woodland by coniferous plantations leads to replacement of mull or moder humus by mor and an acid forest floor (eg. 23). Similarly, the invasion of abandonned farmland by coniferous forest or Calluna heath in Norway has resulted in the development of an acid surface humus (24). Rosenquist (24) has suggested that this development of acid humus layers beneath coniferous forest, or heath, vegetation has played an important role in the acidification of surface waters in Norway.

The development of acid organic matter can give rise to an increased production of organic acids. Cronan and Atkins (25) found that organic acids accounted for some 25% of the acidity in the drainage water from a coniferous forest floor. In podzols or acid brown soils, with vertical movement of water, the organic acids are usually precipitated or degraded in the lower horizons and do not, therefore, influence drainage water acidity. However, where the soils are very shallow over bedrock and/or there is significant lateral water movement through the organic horizons to streams, the organic acids may contribute to drainage water acidity. Even in deeper soils there may be considerable water movement through the surface layers during storms.

The acid humus may also acidify input precipitation as a result of ion exchange between neutral salts and the soil (22). In peats and peaty surfaced soil this mechanism gives rise to the so called 'sea salt events' (26). This mechanism can give rise to extremely acid episodes in streams but is unlikely to produce a permanent increase in drainage water acidity.

In some areas vegetation management by fire may provide a means of preventing the invasion by coniferous forest or Calluna. The ash would also have a neutralising effect on drainage waters. Where planting of conifers, as a deliberate policy, will produce a change to a more acid humus then liming could be used to counteract any acidifying effect, if necessary. The possible adverse effects of liming on tree growth (14) must, however, be borne in mind.

The change in humus type, following vegetation change, partly reflects a change in the rates of organic matter decomposition and turnover. There is also a considerable variation in these rates between conifer crops, and between deciduous trees. Thus, larch (Larix sp.) litter breaks down quicker and produces more nitrate than spruce (Picea sp.) litter (27). In a current study at Llyn Brianne, in mid-Wales, soil drainage waters from below hybrid larch contain significant higher levels of nitrate than those from below Sitka spruce; the increased levels of nitrate seem to be driving enhanced mobilization of aluminium. This effect will add to the greater atmospheric inputs of both sulphate and nitrate below larch, as compared with spruce. Clearly, the choice of tree species, for use in afforestation, must be made with care.

2.4 **Water usage**

The vegetation of a given site can have an important influence on the amount of the incoming precipitation which reaches the ground and drainage waters. In the uplands of Britain, afforestation of grassland or moorland leads to a reduction in water yield from catchments of between 15 and 25% (28). The majority of the increased water loss in this instance results from enhanced evaporation of incoming precipitation from the canopy, usually referred to as interception. Transpiration is also slightly higher from the forest than the moorland. In less windy, and less humid sites the difference in drainage water quantity between grassland and forest may be mainly due to the greater transpiration loss from the forest. The replacement of broadleaved woodland by coniferous plantations can also lead to a reduction in the quantity of drainage water (eg. 23).

The increased evapo-transpiration following afforestation, or replacement of broadleaved trees by conifers, can have important impacts on soil and drainage water chemistry. The evaporation from the canopy increases solute concentrations of input solutions to the soil. Increased transpiration raises solute concentrations in the soil solution. Some ion exchange, and mineral dissolution reactions in soils are concentration dependant (29). Thus, in acid soils with very low base saturation, c. 10%, increasing the concentration of mobile anions, particularly sulphate, in solution can significantly increase aluminium concentrations in soil waters. Thus, Johnson and Reuss (29) calculate that, in a soil with 5% calcium saturation, increasing solution sulphate concentrations from 25 to 250 ueq l^{-1}, increases the total charge associated with aluminium species, in solution, from 5.4 to 39.3 ueq. l^{-1}. This mechanism could influence the acidity and aluminium concentrations of drainage waters. The mechanism may, however, have little influence during heavy rainfall and storm events. On these occasions, evaporation from the canopy is reduced and water may move rapidly through the soil via macro-pores, or over the surface. Once again, in sensitive sites, the impact on a given catchment, and its aquatic ecosystems, could be controlled by limiting the area afforested; or by retaining broadleaved forest in preference to coniferous plantations. Alternatively, ameliorative additions of lime would increase soil pH and base saturation.

3. FERTILIZERS

Nitrogen fertilizers are commonly applied as an ammonium salt, eg. ammonium nitrate or phosphate, or urea. The oxidation of the ammonium to nitrate within the soil generates protons and can be an important acidification mechanism. Excess nitrate, whether added as nitrate or produced by oxidation of added ammonium, over and above plant requirements also produces enhanced leaching. Gasser (20) estimates, in agriculture soils, that oxidation of NH_4 plus nitrate leaching, as a result of the addition of N fertilizers, will generate 4-6 kg $H+$ ha^{-1} yr^{-1}. This is considerably greater than reported inputs from atmospheric sources in western Europe.

In most agricultural systems the impact of the fertilizer-driven acidification, and leaching, on drainage water acidity will be insignificant. Agricultural crops generally require relatively high pH and base saturation and, where necessary, this is maintained by additions of lime. The net result, in these situations, of the increased proton generation and nitrate leaching is enhanced outputs of base cations to drainage waters. On acid soils, however, the addition of fertilizers can lead to a marked acidification of drainage waters. The use of N fertilizers on acid soils will mainly take place in intensive forest systems. Wiklander (21) for example, reports acidification of drainage waters following addition of fertilizers to acid, sphagnum peats in Sweden in connection with energy cropping forestry. Even the addition of the fertilizer as the neutral salt $Ca(NO_3)_2$ produced a significant acidification; this was probably due to mobilisation of protons from exchange sites on the organic matter (cf. 22). At the Swedish site,

Wiklander (21) found that the acidifying effect of the fertilizers could be neutralised by a simultaneous addition of ground limestone; dolomite was not suitable as it did not enter solution quickly enough. On acid soils, such as peats or podzols, it may be advisable to limit the use of N fertilizers within sensitive catchments unless there is a simultaneous application of ground limestone.

4. CROPPING

The removal of agricultural and forest crops involves a net removal of base cations from sites and is therefore an acidifying process (30, 31, 32). Breeusma and de Vries (30) suggest that this mechanism produces the equivalent of c. 0.5-2.0 k mol H+ ha^{-1} yr^{-1}, a similar amount to H+ inputs in precipitation in western Europe. The mechanism may produce an acidification of soils but this will have long term implications for drainage water chemistry rather than short term or, more particularly, event related effects (31). In agricultural systems, the acidification will be counteracted by additions of lime on naturally base poor soils. In forest systems, however, where liming is not part of routine land management, there may be permanent acidification of sensitive soils.

Removal of the vegetation also produces a change in the microclimate at the soil surface, stops root uptake and may stop the production of allelopathic chemicals. As a result, there can be a large increase in nitrification, following crop removal (eg. 33), and H+ production. Van Breeman <u>et al</u>. (34) calculate that this mechanism resulted in an acidification rate of 6.1 k mol H+ ha^{-1} yr^{-1} in the Hubbard Brook deforestation study. Breeusma and de Vries (30) suggest that in agricultural systems acid production from this process is between 3 and 14 k mol H+ ha^{-1} yr^{-1}. The most important effect on drainage water chemistry results from the large flux of nitrate which can occur after harvesting. At Hubbard Brook this was balanced by increased calcium mobilization by leaching and weathering. On acid soils, however, it can produce an acidification of soil drainage waters and mobilization of aluminium. At a study in north Wales there was a flux of c. 70 kg NO_3-N ha^{-1} yr^{-1} from the C horizon in each of the two years after felling (35). In the second year the pH of soil drainage waters was reduced by c. 0.5 of a pH unit and aluminium concentrations increased by c. 0.2 mg l^{-1}. The impact of this on the aquatic ecosystems in any given catchment can be controlled by limiting the area clearfelled in any one year. The output of nitrate is generally reduced again once a groundflora develops and root uptake begins again.

5. DRAINAGE AND FORESTRY PLOUGHING

Drainage of soils for agricultural or forestry purposes influences soil processes and catchment hydrology. The main influence on soil processes results from the change from anaerobic to aerobic conditions, following the drainage. The most dramatic impact on drainage water acidity is found following drainage of acid sulphate soils (36, 37). Drainage of these soils results in oxidation of sulphides, contained in the soil materials, to produce sulphuric acid (38, 39); this can lead to strong acidification of drainage waters. In an experimental study, Trafford <u>et al</u> (39) report drainage waters of pH 2.9. Potentially acid sulphate soils can occur in marine derived deposits, as in the Netherlands, eastern England and central Finland or in lake deposits (40). Gosling and Baker (41) provide an example of the potential impact on aquatic ecosystems of drainage of sulphide containing soils. They report the complete loss of a fish community, freshwater mussels and most aquatic macrophytes at a broadland site in eastern England when the pH of the waters dropped to between 3.0 and 3.4.

Potentially acid sulphate soils are usually drained for agricultural purposes. In these cases, lime is added to raise the soil pH but this may only influence a relatively shallow, near surface zone. If the drainage lowers the water table significantly below the zone incorporating lime, acidic drainage waters may still result. The impact on drainage water acidity can be limited by maintaining a relatively high water table, but still below the rooting zone, combined with liming. This can be difficult to achieve as there are often large seasonal variations in the water table. The most acid drainage waters occur as the water table rises, in autumn or after dry periods, mobilizing the oxidation products.

Drainage, and deep forestry ploughing, of peats and peaty surfaced soils can result in increased mineralization of organic N and S compounds with the production of ammonia, nitrate and sulphate (42, 43, 44, 45). There may also be an increase in the production of organic acids. The generation of sulphate, nitrate and organic acids can result in the acidification of drainage waters. The impact on any given site will, however, depend on the peat chemistry, depth of drainage ditches and the nature of any mineral material exposed in the ditch floors (21). If base rich mineral material is exposed then any increased acidity in the peat drainage may be neutralised. There are few published studies with which to assess the magnitude of the imapct of peat drainage on water acidity. Wiklander (21) found no significant effect following draining of acid sphagnum peat in Sweden. A recent study in west Wales, however, showed a marked increase in sulphate levels and a drop in pH, compared with a control stream, after drainage of peats and peaty surface podzols. When the drainage waters from the peat entered ditches and forestry plough furrows, in the podzols, aluminium was also mobilised (Gee [Welsh Water Authority], pers. comm.). In sensitive catchments, the drainage of peats may be undesirable. In such catchments, the impact of peat drainage could be reduced by controlling the proportion of the area drained. Alternatively, lime could be added at the time of ploughing.

Drainage, and forestry ploughing, can modify site hydrology by increasing the proportion of streamwater derived from near surface soil layers (46) and by increasing the rate of runoff in storms (47). In acid soils, the soil-derived waters are considerably more acid and contain several times as much aluminium, as waters from deeper soil layers, or groundwater (48). Whitehead et al. (49) have used a modelling approach to demonstrate the potential importance of a change in the balance in a stream between soil-derived and groundwaters.

Alternative methods of ground preparation for forestry are being explored in the United Kingdom, to reduce the impact on streamwater acidity, eg. screef ploughing, contour ploughing and mole drainage. The screef ploughing and mole drainage will reduce surface disturbance, and the exposure and drying of organic horizons. Contour ploughing is aimed to keep the drainage waters on site longer to allow some buffering. Early results from a site in west Wales, where contour ploughing has been used, are promising; to-date there has been no significant change in drainage water acidity.

Drainage of peats also seems to influence the retention of SO_4 input from the atmosphere. A number of studies have shown that peatlands can be important in controlling sulphate concentrations, and fluxes in drainage waters (eg. 50, 51, 52, 53). The input sulphate is reduced to hydrogen sulphide in anaerobic, waterlogged peat and is then incorporated into organic molecules (54, 55, 56). These organic sulphur compounds can be oxidised, to produce sulphate, if the peat dries out as a result of drainage or in dry periods. Prior to drainge, however, most peatlands seem to act as net sinks for sulphate input from the atmosphere. Following drainage little, or non, of the sulphate is retained in the now aerobic upper peat layers. The net effect is to increase the flux of sulphate through the system and this may contribute to the acidification of drainage waters.

6. A CASE STUDY - AFFORESTATION IN THE BRITISH UPLANDS

In the previous sections of the paper, I have considered a number of mechanisms whereby land management can influence the acidity of aquatic ecosystems. The mechanisms were each considered separately. In reality, however, several mechanisms operate simultaneously, following a given change in land use, and interact to determine the overall impact on drainage water acidity. The role of each mechanism, and the nature of any interactions should be identified when developing land use and management strategies to prevent or ameliorate acidification of aquatic systems. Afforestation in the british uplands will be used to illustrate the complexity of the changes in ecosystem processes, brought about by a change in land use. I will also examine the development of land use and management strategies to limit the impact of these ecosystem changes on drainage water acidity. The comments are largely based on preliminary, unpublished results from studies in progress in Wales, and on the hypothesis being examined in these studies, particularly the Llyn Brianne project.

6.1 Background

There is now considerable evidence that, in certain parts of the British uplands, afforestation of grassland or moorland catchments with conifers results in increased acidity, and aluminium concentrations, in streams and lakes (57, 58, 59, 60). The changes in streamwater chemistry consequent upon afforestation have also been linked to reductions in fish stocks and changes in freshwater invertebrate faunas (60, 61, 62, 63). The acidification of streams is most marked where the afforestation takes place on acid soils overlying massive and/or base poor bedrocks. In this context, the most widespread acid soils are brown podzolic soils, podzols, stagnopodzols, stagnohumic gley soils and peats (64), and their equivalents. These soils often have a pH of 5 throughout the profile and a low base saturation, 15%; the stagnopodzols, stagnohumic gleys, peats and some podzols also have a peaty surface horizon. The massive base-poor rocks include acid igneous rocks, gneisses, some schists, slates, mudstones and grits. The magnitude of the increase in acidity, and aluminium concentrations, varies within this range of soil-rock combinations and with the chemistry of atmospheric inputs. A large increase in aluminium concentrations may also occur where there has been little impact on water acidity (65). A recent regional study of stream and river water chemistry in Wales, by the Welsh Water Authority, has shown a highly significant correlation between dissolved aluminium concentrations and proportion of catchment afforested (Stoner, pers comm.). The difference in acidity, and aluminium chemistry, between afforested and moorland catchments, on similar soils and bedrock, is greatest during storm events (57, 63).

6.2 **Changes resulting from afforestation**

Afforestation in these upland areas results in a number of interacting changes in the ecosystem processes, which have been discussed individually in the previous sections of the paper. Some of these changes result primarily from the influence of ground preparation, prior to planting, and others from the growth of the tree crop. The ground preparation usually involves ploughing and drainage (66, 67). This produces drying and oxidation of peaty horizons and peats with a potential production of ammonia, nitrate and sulphate, and acidification of drainage waters. Retention of sulphate, from atmospheric inputs, by the peats will decline. The plough furrows and ditches increase the rate of removal of surface waters; Robinson (47) reports a reduction in the time to peak flows and an increase in the height of flood peaks. The proportion of acid near surface, soil derived water in streams is thought to increase (cf. section 5).

As the tree crop grows and the canopy developes, evapo-transpiration losses increase. Solute concentrations in inputs to the ground are increased as a result of evaporation from the canopy; increased transpiration produces a further increase in solute concentrations in soil waters (cf. sections 2.1 & 2.4). Atmospheric inputs increase as a result of the increased transfer onto the forest canopy, compared with the moorland or grassland canopy. In higher altitude areas, the increase in occult deposition, with its higher solute concentrations than rainfall, may be particularly important (68). The interaction of the increased atmospheric inputs, some with high solute concentrations, with the concentration effects due to increased evapo-transpiration produce increased acidity and solute concentrations in soil waters; aluminium concentrations may be increased by 300%, compared to those in waters from moorland soils (65).

Growth of the tree crop also produces further drying of peats and peaty horizons, which adds to that due to ploughing and drainage. In peats, the increased decomposition of organic materials, plus the accumulation of base cations in the tree crop and forest floor, produce an acidification and a reduction in base saturation (69). Acidification of surface horizons of acid brown soils and brown podzolic soils have also been reported (70).

6.3 **Preventative and ameliorative measures**

6.3.1 Land use strategies

If the link between afforestation and acidification of surface waters is accepted as proven, it could be considered desirable to limit further afforestation of the catchments of streams thought sensitive to acidification. Such a policy would necessitate the definition, and identification, of sensitive waters, or those areas of the uplands where sensitive waters occur (60). The Welsh Water Authority is exploring a classification of the sensitivity of streams to acidification based on hardness. The amount of afforestion the Authority would see as acceptable in given catchments is then suggested, based on the sensitivity class (Table III). The proportion of planting acceptable assumes present forest management practices, present levels of emission of acidic pollutants, and the absence of ameliorative additions of lime, for example.

MEAN HARDNESS ($mg\ l^{-1}$ $Ca\ CO_3$)	ACCEPTABLE CONIFER AFFORESTATION
12	No planting other than small catchments drained by first order streams having a negligible effect on receiving water course.
12-15	Upto 30% of catchment, subject to planting guidelines in Mills (71) and any more recent Forestry Commission guidelines for planting in acid-sensitive areas.
15-25	Upto 70% of catchment, subject to planting guidlines in Mills (71) and any more recent Forestry Commission guidelines for planting in acid-sensitive areas.
25	Observation of planting guidelines contained in Mills (71).

Table III. Suggested guidelines for acceptable conifer afforestation in upland Wales (Stoner, pers. comm.)

A second approach, which is being developed, is combining data on soils, drift and solid geology in an attempt to identify those areas of the uplands, where surface waters sensitive to acidification are likely to occur (cf. 60). The aim is to rank soil-rock combinations in terms of the sensitivity of their drainage waters to acidification, and increase in aluminium levels, as a result of afforestion. For example, on slates and mudstones in the welsh uplands, the increase in acidity and aluminium in drainage waters is greater on stagnopodzols than on brown podzolic soils. Planting could be avoided on the most sensitive soil-rock combinations; this may provide a basis for planning planting within catchments. A preliminary map of Wales will soon be available showing the distribution of the various soil-rock sensitivity classes.

6.3.3 <u>Management strategies</u>

The further extension of forestry in areas with sensitive waters may necessitate the development of methods of forest management which will reduce the impact on the acidity of the aquatic systems.

The development of alternative management practices requires an understanding of the relative roles of the changes in ecosystem processes, brought about by afforestation, in the acidification of waters. If the changes in hydrology, as a result of ploughing and drainage, are the dominant mechanism, then new methods of ground preparation may be needed. A series of experiments are, therefore, examining the potential of contour ploughing, screef ploughing and turf planting, and stopping drainage ditches short of streams. The aim is to reduce surface disturbance and to keep the drainage waters on site longer to allow further time for buffering. As noted earlier (section 5), preliminary results are encouraging and suggest a reduction in the impact of the site preparation activities on drainage water acidity. It may also be possible eventually to provide guidelines for planting within catchments. For example, it may be advisable to leave some peat areas unplanted and undrained so that they continue to act as a partial sink for atmospheric inputs, and for SO_4, No_3 and aluminium input in drainage waters from adjacent drained and planted areas (cf. 53).

If the dominant mechanisms causing surface water acidification result from tree growth, eg. increased atmospheric deposition and evapo-transpiration, then ameliorative treatment of soil and/or waters may be necessary. Liming of soils could be carried out to increase pH and base saturation, and to provide a reservoir of soluble carbonates. The whole of the catchment area could be limed or selected areas. The liming must, however, be able to buffer the drainage water during storm events when residence times are short. Even without drainage and ploughing, the upland streams respond very rapidly to rainfall events. Rapidly soluble carbonates, eg. ground limestone, should be used (21). The lime should also be concentrated in contributing areas, if only selected parts of the catchment are limed. A good knowledge of the catchment hydrology is necessary to plan this liming.

Amelioration of lake water acidity with lime is now widely practiced (Dickson, this volume). Liming of flowing waters is a more difficult problem; any additions must be flow related to maintain a target pH. On some sites it has been suggested that it may be possible to buffer the acid, soil derived waters by bringing them into contact with more base rich materials before they enter the stream. These base-rich materials could be lower soil horizons, drift or bedrock, or imported carbonates or other base-rich rocks. Deeper drainage ditches, or specially dug sumps could be used to expose relatively base-rich material which occurs in situ on the site. This method would only be applicable on certain sites, particularly those with deeper drift deposits or weatherable material at depth. The use of imported material in sumps would be more widely applicable. Whether in situ or imported material were used, as the buffering material, the method must be shown to adequately buffer drainage waters during high flow, storm events.

Treatment of soils and/or waters could also be used if the dominant mechanisms, causing the acidification of the aquatic systems, were hydrological, ie. as a result of ploughing and drainage. In practice, a combination of possible ameliorative measures is needed to cover the range of site types which is encountered. On some sites, modification of ground preparation techniques may be sufficient to reduce the impact, of afforestation on surface water acidity, to acceptable levels. Elsewhere, exposure of relatively base-rich drift materials in ditches and sumps may provide adequate buffering. Other sites may necessitate the use of lime to offset the impact of the change in land use; for example, there are sites in Wales where acidification of streams has occurred following afforestation without ploughing or drainage, and where there is no base-rich material at depth. Part of a catchment could be afforested, the least sensitive soil-rock combinations, and lime applied to all or part of the unplanted area. If this approach is used, it is essential that the contributing areas are limed.

6.3.3 Amelioration of acidity of aquatic systems due to existing plantations

Addition of buffering materials, seems the only feasible approach on these sites. Major modification of drainage systems is only really feasible following clear felling. If the use of sumps, whether or not filled with imported materials, proves successful, it may be possible to construct these in existing forests. The comments made above on the addition of carbonates will apply here. It is essential to lime the contributing areas, especially all those operative during storms, if only parts of the catchment are limed. The additions of lime to the contributing areas should be large enough to cope with rapid flows during storms. Rapidly soluble carbonates should be used.

ACKNOWLEDGEMENTS

The author is grateful to John Stoner and Alun Gee of the Welsh Water Authority for helpful discussions and permission to quote unpublished data: colleagues in the Institute of Terrestrial Ecology, particularly Brian Reynolds and Paul Stevens, for valuable comments and discussions. The work in Wales is part funded by the UK Department of the Environment and the Welsh Office.

REFERENCES

(1) MCFEE, W.W. (1980). Sensitivity of soil regions to long-term acid precipitation. In: Atmospheric sulphur deposition: environmental impact and health effects, edited by D.S. Shriner, C.R. Richmond, and S.E. Lindberg, 405-505. An Arbor, U.S.A., Ann Arbor Science.

(2) WANG, C. and COOTE, D.R. (1981). Sensitivity classification of agricultural land to long term acid precipitation in Eastern Canada. (L.R.R.I. contribution 98) Agriculture Canada Research Board.

(3) WIKLANDER, L. (1974). The acidification of soil by acid precipitation. Grundforbattring 26, 155-164.

(4) CRISTOPHERSEN, N. and SEIP, H.M. (1982). Model for streamwater chemistry at Birkenes, Norway. Water Resources Research 18, 977-996.

(5) JOHNSON, D.W. and REUSS, J.O. (1984). Soil-mediated effects of atmospherically deposited sulphur and nitrogen. Phil. Trans. R. Soc. (Lond.) B. 305, 383-392.

(6) HEMOND, H.F. and ESHLEMAN, K.N. (1984). Neutralization of acid deposition by nitrate retention at Bickford Watershed, Massachusetts. Water Resources Research 20, 1718-1724.

(7) FOWLER, D. 1984. Transfer to terrestrial surfaces. Phil. Trans. Roy. Soc. (Lond.) B. 305, 281-297.

(8) REYNOLDS, B., HORNUNG, M., STEVENS, P.A. and NEAL, C. (In press). Input-output budgets for selected catchments in mid-Wales: some considerations for selecting representative basins. Proceedings of workshop on, "Geochemistry and Monitoring in Representative Basins". Prague, Czechoslovakia, May 1987.

(9) LAKHANI, K.H. and MILLER, H.G. 1980. Assessing the contribution of crown leaching to the element content of rainwater beneath trees. In: Effects of Acid Precipitation on Terrestrial ecosystems. Editors, T.C. Hutchinson and M. Havas. (Volume 4 NATO Conference Series). New York, U.S.A.: Plenum Press.

(10) HULTBERG, J. (1985). Budgets of base cations, chloride, nitrogen and sulphur in the acid Lake Gardsjon catchment in SW Sweden. Ecol. Bull. (Stockholm) 37, 133-157.

(11) ADAMSON, J.K., HORNUNG, M., PYATT, D.G. and ANDERSON, A.R. (In press). Changes in solute chemistry of drainage waters following the clearfelling of a Sitka spruce plantation. Forestry.

(12) MAYER, R. and ULRICH, B. (1978). Input of atmospheric sulphur by dry and wet deposition to two central European forest ecosystems.

Atmospheric Environment 12, 375-377.
(13) NYS, C., STEVENS, P.A. and RANGER, J. (In press). Sulphur nutrition of forests examined using a sulphur budget approach. In: Field Methods in Terrestrial Nutrient Cycling Studies. Editors, P. Ineson and A.F. Harrison. Elsevier Applied Science Publishers.
(14) DEROME, J., KUKKOLA, M. and MALKONEN, E. (1986). Forest liming on mineral soils. Results of Finnish experiments. National Swedish Environmental Protection Board, Report 3084.
(15) HAYNES, R.J. 1983. Soil acidification induced by leguminous crops. Grass and Forage Science 38, 1-11.
(16) WILLIAMS, C.H. and DONALD, C.M. (1957). Changes in organic matter and pH in a podzolic soil as influenced by sub-terranean clover and super-phosphate. Australian Journal of Agriculture Research 8, 179-189.
(17) WILLIAMS, C.H. (1980). Soil acidification under clover pasture. Australian Journal of Experimental Agriculture and Animal Husbandry 20, 561-567.
(18) MIEGROET, H. VAN. and COLE, D.W. (1984). The impact of nitrification on soil acidification and cation leaching in a red alder ecosystem. Journal of Environmental Quality 13, 586-590.
(19) MIEGROET, H. VAN. and COLE, D.W. (1985). Acidification sources in Red Alder and Douglas Fir Soils - importance of nitrification. Soil Sci. Soc. Am. J. 49, 1274-1279.
(20) GASSER, J.K.R. (1985). Processes causing losses of calcium from agricultural soils. Soil Use and Management 1, 14-17.
(21) WIKLANDER, G. (1984). Effects of energy forest silviculture on water pH in peatland environments. In: Ecology and Management on Forest Biomass Production Systems. Ed. K. Perttu. Dept. of Ecol. and Environ. Res., Swedish University of Agricultural Sciences Report 15, 101-112.
(22) WILKANDER, L. (1975). The role of neutral salts in the ion exchange between acid precipitation and soil. Geoderma 14, 93-105.
(23) NYS, C. and RANGER, J. (1985). Influence de L'espece sur le fonctionnement de l'ecosysteme forestièr. Le cas de la substitution d'une essence resineuse a une essence feuillue. Science du Sol, 203-216.
(24) ROSENQVIST, I.T. (1981). Importance of acid precipitation and acid soil in freshwater lake chemistry. Vann 4, 402-409.
(25) CRONAN, C.S. and AIKEN, G.R. (1985). Chemistry and transport of soluble humic substances in forested watersheds of the Adirondack Park, New York. Geochemica et Cosmochimica Acta 49, 1697-1705.
(26) HORNUNG, M. (1984). Precipitation-canopy and soil-water interactions. In: Acid Rain. Conference papers, Institute of Water Pollution Control and Institute of Water Engineers and Scientists. Birmingham, UK.
(27) CARLYLE, J.C. and MALCOLM, D.C. (1986). Nitrogen availability beneath pure spruce and mixed larch and spruce stands growing on a deep peat. 1. Net N mineralization by field and laboratory incubations. Plant and Soil 93, 95-114.
(28) CALDER, I.R. and NEWSON, M.D. (1979). Land use and upland waer resources in Britain - a strategic look. Water Resources Bulletin 15, 1628-1639.
(29) REUSS, J.O. and JOHNSON, D.W. (1986). Acid deposition and the acidification of soil and waters. Ecological Studies V 59. New York, U.S.A.; Springer Verlag New York Inc.

(30) BREEUSMA, A. and DE VRIES, W. (1984). The relative importance of natural production of H+ in soil acidification. Netherlands Journal of Agricultural Science 32, 161-163.
(31) NILSSON, S.I., MILLER, H.G. and MILLER, J.D. (1982). Forest growth as a possible cause of soil and water acidification: an examination of the concepts. Oikos 39, 40-49.
(32) ROWELL, D.L. and WILD, A. (1985). Causes of soil acidification: a summary. Soil Use and Management 1, 32-33.
(33) VITOUSEK (1981). Clear-cutting and the nitrogen cycle. In: Terrestrial Nitrogen Cycles. Ecosystem Strategies and Management Impacts. Edited by R.E. Clark and T. Rosswall. pp. 631-642. Ecological Bulletins (Stockholm) 33.
(34) BREEMAN, N. VAN, DRISCOLL, C.T. and MULDER, J. (1984). The role of acidic deposition and internal proton sources in acidification of soils and waters. Nature 307, 599-604.
(35) STEVENS, P.A. and HORNUNG, M. (In prep.). Nitrate leaching from a felled Sitka spruce plantation in Beddgelert Forest, north Wales. Submitted to Soil Use and Management.
(36) BREEMAN, N. VAN. (1982). Genesis, morphology and classification of acid sulphate soils in coastal plains. In: Acid Sulphate Weathering. Edited by J.A. Kittrick, D.S. Fanning and L.R. Hossner. S.S.S.A. Special Publication 10.
(37) BRINKMANN, R. and PONS, L.J. (1973). Recognition and prediction of acid sulphate soil conditions. In: Acid Sulphate Soils. Edited by H. Dost. Proceedings of Int. Symp. on Acid Sulphate Soils. Wageningen; International Institute for Land Reclamation and Improvement. Publication 18.
(38) BLOOMFIELD, C. (1972). The oxidation of iron sulphides in soils in relation to the formation of acid sulphate soils, and of ochre deposits in field drains. J. Soil Sci. 23, 1-16.
(39) TRAFFORD, B.D., BLOOMFIELD, C., KELSO, W.I. and PRUDEN, G. (1973). Ochre formation in field drains in pyritic soils. J. Soil Sci. 24, 453-460.
(40) ERVIO, R. (1975). Cultivated sulphate soils in the drainage basin of the river Kyronjoki. Journal of the Scientific Agricultural Society of Finland. 47, 550-561.
(41) GOSLING, L.M. and BAKER, S.J. 1980. Acidity fluctuations at a broadland site in Norfolk. Journal of Applied Ecology 17, 479-490.
(42) PIISPANEN, R. and LAHDESMAKI, P. (1983). Biogeochemical and geobotanical implications of nitrogen mobilization caused by peatland drainage. Soil Biol. Biochem. 15, 381-383.
(43) URBAN, N.R. and BAYLEY, S.E. (1986). The acid-base balance of peatlands: a short term perspective. Water, Air and Soil Pollution 30, 791-800.
(44) FIEDLER, H.J. and THAKUR, S. DEV. (1984). Gehalt und Transformation des Schwefels in Boden von Waldokosystem. Archiv fur Naturschutz und Landschaftsforschung 24, 17-35.
(45) KENTTAMIES, K. (1980). The effects on water quality of forest drainage and fertilization in peatlands. In: The influence of man on the hydrological regime with special reference to representative and experimental basins. IAHS - AISH Publication No. 130.
(46) MILLER, H.G. (1985). The possible role of forests in streamwater acidification. Soil Use and Management 1, 28-30.
(47) ROBINSON, M. (1986). Changes in catchment runoff following drainage and afforestation. Journal of Hydrology, Netherlands 86,

71-84.
(48) HORNUNG, M., ADAMSON, J.K., REYNOLDS, B. and STEVENS, P.A. 1986. Influence of mineral weathering and catchment hydrology on drainage water chemistry in three upland sites in England and Wales. Journal of the Geochemical Society, London, 143, 627-634.
(49) WHITEHEAD, P.G., NEAL, C. and NEALE, R. (1986). Modelling the effects of hydrological changes in streamwater acidity. J. Hydrology 84, 353-364.
(50) BRAEKKE, F.H. (1981). Hydrochemistry of high altitude catchments in South Norway. 1. Effects of summer droughts and soil-vegetation characteristics. Reports of the Norwegian Forest Research Institute 36(8), 1-26.
(51) WIEDER, R.K. and LANG, G.E. (1986). Fe, Al, Mn and S chemistry of Sphagnum peat in four peatlands with different metal and sulphur input. Water, Air and Soil Pollution 29, 309-320.
(52) SKEFFINGTON, R.A. (1981). Tillingbourne Catchment - interim report. Report No. RD/L/2083N81. Leatherhead, UK; Central Electricity Research Laboratories.
(53) URBAN, N.R. and BAYLEY, S.E. 1986. The acid-base balance of peatlands: a short-term perspective. Water, Air and Soil Pollution 30, 791-800.
(54) BROWN, K.A. (1985). Sulphur distribution and metabolism in waterlogged peat. Soil Biol. Biochem. 17, 39-45.
(55) BROWN, K.A. and MACQUEEN, J.F. (1985). Sulphate uptake from surface water by peat. Soil Biol. Biochem. 17, 411-420.
(56) BROWN, K.A. (1986). Formation of organic sulphur in anaerobic peat. Soil Biol. Biochem. 18, 131-140.
(57) HARRIMAN, R. and MORRISON, B.R.S. (1982). Ecology of streams draining forested and non-forested catchments in an area of central Scotland subject to acid precipitation. Hydrobiologia 88, 251-263.
(58) STONER, J.H. and GEE, A.S. (1985). Effects of forestry on water quality and fish in Welsh rivers and lakes. Journal of the Institute of Water Engineers and Scientists, 39, 27-45.
(59) REYNOLDS, B., NEAL, C., HORNUNG, M. and STEVENS, P.A. (1986). Baseflow buffering of streamwater acidity in five mid-Wales catchments. J. of Hydrology 87, 167-185.
(60) WELLS, D.E., GEE, A.S. and BATTERBEE, R.W. (1986). Sensitive surface waters - a UK perspective. Water, Air and Soil Pollution 31, 631-668.
(61) STONER, J.H., GEE, A.S. and WADE, K.R. (1984). The effects of acidification on the ecology of streams in the upper Tywi catchments in west Wales. Environmental Pollution (Series A) 35, 127-157.
(62) ORMEROD, S.J., MAWL, G.W. and EDWARDS, R.W. (1987). The influence of forests on aquatic fauna. In: Environmental aspects of plantation forestry in Wales. Edited by J.E.G. Good, pp. 37-49. (ITE Symposium No. 22). Grange-over-Sands, UK; Institute of Terrestrial Ecology.
(63) HARRIMAN, R., MORRISON, B.R.S., CAINES, L.A., COLLEN, P. and WATT, A.W. (1987). Long-term changes in fish populations of acid streams and lochs in Galloway south west Scotland. Water, Air and Soil Pollution 32, 89-112.
(64) AVERY, B.W. (1980). Soil Classification of England and Wales. Soil Survey Technical Monograph No. 14. Harpenden, UK; Soil Survey of England and Wales.

(65) HORNUNG, M., REYNOLDS, B., STEVENS, P.A. and NEAL, C. (In press). Increased acidity and aluminium concentrations in streams following afforestation: causative mechanisms and processes. Proceedings of symposium on "Acidification and Water Pathways". Bolkesjo, Norway, May 1987.
(66) THOMPSON, D.A. (1979). Forest Drainage Schemes. Forestry Commission Leaflet 72. London; HMSO.
(67) THOMPSON, D.A. (1984). Ploughing of forest soils. Forestry Commission Leaflet 71. London; HMSO.
(68) UNSWORTH, M.H. (1984). Evaporation from forests in cloud enhances the effect of acid deposition. Nature (London) 312, 262-264.
(69) BOGGIE, R. and MILLER, H.G. (1976). Growth of Pinus contorta at different water-table levels in deep blanket peat. Forestry 49, 123-131.
(70) HORNUNG, M., STEVENS, P.A., and REYNOLDS, B. (1987). The effects of forestry on soils, soil waters and surface water chemistry. In: Environmental aspects of plantation forestry in Wales. Edited by J.E.G. Good, pp. 25-36. (ITE Symposium No. 22). Grange-over-Sands, UK; Institue of Terrestrial Ecology.
(71) MILLS, D.H. (1980). The management of forest streams. Forestry Commission Leaflet No. 78. London, UK; HMSO.

PRACTICAL PREVENTIVE AND CURATIVE MEASURES FOR AQUATIC ECOSYSTEMS

W. DICKSON
National Environmental Protection Board, Sweden

Summary
Next to emission reduction preventive measures like base additions seem acceptable to save the biology of sensitive aquatic ecosystems. During the last decade lake liming has been used as a chemical treatment on a large scale in Sweden. However, as acidic deposition in the major parts of Europe far exceeds the weathering rate of basic substances, soil acidification proceeds and increases the leaching of toxic elements to the surface run off. Therefore more liming activities are needed on land to stop the leaching of these substances and to stop the acid surges which are now destroying some of the biology of limed lakes.

The effects of atmospheric pollution on the corrosion of materials have been widely experienced during the last fifty years. The phenomenon is worldwide.

During the last two decades there has been much concern regarding the effects of atmospheric deposition of strong acids on surface water quality. Surface water acidification by acidic deposition has been reported in Scandinavia, Central Europe, U.K. southeastern Canada and the northeastern Uniters States.

During the last decade groundwater acidification with respect to health effects and corrosion costs has attracted some interest, whereas the effects of soil acidification because of atmospheric pollution has gained the major public awareness because of the huge economic and ecological values at stake.

Several cost-benefit analyses have been performed through the years to evaluate the costs of reducing emissions versus the benefit to ecology and materials. These analyses have normally been carried out by economists and modellers.

When the ecologists have not been able to offer the modellers the accurate values for the environment, e.g. the costs of the damage of spruce forest at risk or the total length of cables dug down in low-buffered mineral soils, or the economic value of a unique strain of a lake spawning sea trout, the economists have left these figures out and put their values as zero.

With the present knowledge of the environment and the general awareness among scientists it has become possible to give far better estimates on our environment and also to show that the benefits of emission reduction often far exceed the costs. (This is not a goal per se, however, the "ecological optimum" level is probably not zero emission).

CHEMICAL TREATMENT

Next to emission reductions the second most practical preventive measure against aquatic acidification is the chemical addition of bases to neutralize the acidified ecosystems. When the surface water acidification problems became obvious and generally accepted in Scandinavia 15 years ago, several offers were made by oil companies of free chemical bases as remedies to the lake water acidification.

The approach that has been used to mitigate aquatic effects of acidic deposition is treatment of acidic waters by base application. A variety of materials has been proposed or used as neutralizing agents including limestone, calcite ($CaCO_3$), lime ($Ca(OH)_2$), quicklime (CaO), olivine ($MgFeSiO_4$), dolomite ($CaMg(CO_3)_2$), sodium bicarbonate ($NaHCO_3$), soda ash (Na_2CO_3), fly ash and industrial slags. Of these materials, limestone and dolomite have clearly been the chemicals of choice for neutralization efforts.

Several countries with aquatic acidification problems now have liming programs with the aim of subsiduing private persons, sport fishing clubs or local authorities for liming costs. In Sweden a national program started in 1977 with grants from the State of SEK 10 million. The grants for 1987 are SEK 100 million (Fig. 1) amounting to 85 % or 100 % of the total liming costs. Also in Norway, liming subsidies are available (NOK 11 million in 1987).

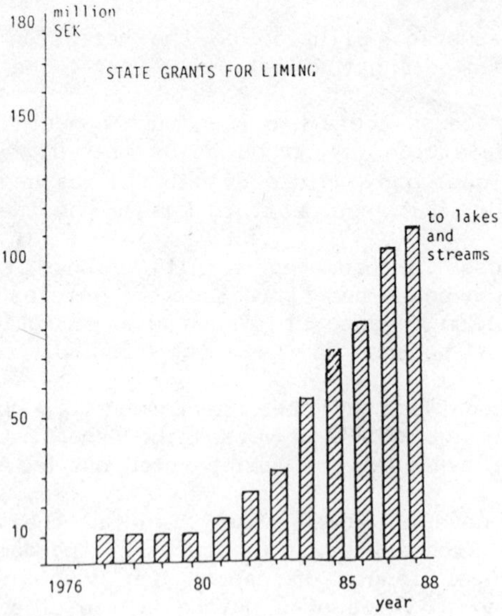

Figure 1. The state grants for liming surface waters in Sweden increased from SEK 10 million 1977 to SEK 100 million 1987, with addition of specific amounts to follow up and national planning.

Because liming implies manipulation of already stressed systems several national programs operate to evaluate the effects and to optimize the liming activities (1,2,3,4,5,6,7,8,9,10)

Practical experiences

In Sweden some 4000 lakes and several hundred streams have been limed during the last decade, Fig. 2. The chemical goal is to reduce the toxic levels of protons, aluminium and other metals in the water-levels which increase through acidification - and to restore the alkalinity of the water to values above 50 ueq/l.

A highly acidic water of pH 4.5 contains 30 ueq/l of strong acids and about as much of weak acids from aluminium etc. Theoretically only 3 grams of $CaCO_3$ per m^3 is needed to reduce the toxic levels. To increase the alkalinity to 100 ueq/l another 5 grams per m^3 is needed.

Application of much higher doses, in reality, shows that it is not difficult either chemically, economically or ecologically to detoxify acid water. Several courses of action are available: Different measures, fractions and doses, different ways, places and times of application.

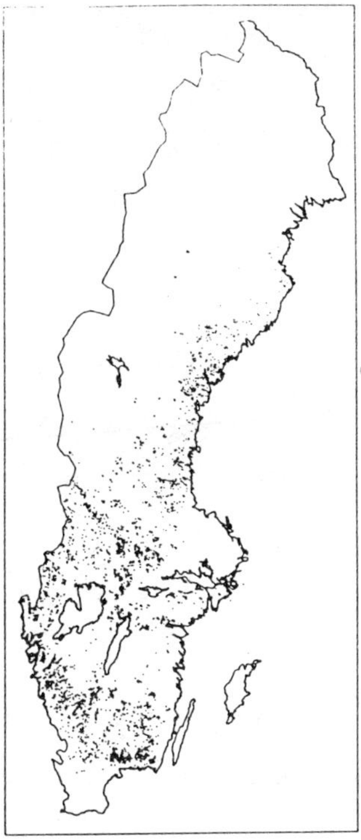

Fig. 2 Around 4000 lakes and several hundred streams have been limed the last ten years in Sweden. The liming activities are concentrated to the south and central parts because of high acid deposition and of slowly weathering bedrock.

Measures

The most economical means of lake liming seems to be limestone powder of 0-0.5 mm, with about 90 % of the amount in fractions less than 250 u or possibly somewhat coarser. Very fine powder gives a high momentaneous dissolution rate but very little long-term dissolution effects and is more expensive (4,11). For stream liming with dosing equipment where more instantaneous dissolution is necessary fractions of limestone, $CaCO_3$ of 0-0.1 mm with 90 % < 100 ɥ are recommended. If dolomite is to be used powders of 0-0.1 and 0-0.05 mm respectively are recommended (12). Several different stream liming equipments with advanced dosing and alarming systems in order to increase the reliability are now available on the market.

Doses

Lakes with extremely acid water and short retention time need more limestone to keep the water healthy than do less acid waters with a long turnover time.

Table 1. Recommended doses of $CaCO_3$ (grams per m^3 lake volume) (12)

pH before liming	Turn over times years <0.5	0.5-3	>3
4,5	75	50	25
5,0	60	40	20
5,5	45	30	15

The practical spreading in lakes is recommended to littoral bottoms along the shores (12). Even doses of 20 tonnes per hectare results in a 90 % disolution efficiency (11).

Ecological effects

The effects of liming on the ecosystem illustrate the reversibility of acidification.

chemically: pH, alkalinity and calcium levels increase. Aluminium and other metal levels in the water volume decrease, but increase on sediment surfaces. Nitrate levels normally decrease, as phytoplankton and denitrification activities increase (13).

Phosphorus levels increase (1,13,14) as well as selenium (15) and the decomposition of organic material (16).

biologically: Phytoplankton, zooplankton and bottomfauna diversity increases (13).

Whitemosses decrease and are replaced by the former flora: Lobelia, Littorella, Isoetes etc (17).

Fish reproduction and status improves. Crayfish and molluscs populations increase (18, 14).

Positive: The effects illustrated above are regarded as positive, when they reflect the natural unpolluted lake systems.

Negative effects

The metal enrichment in limed lakes becomes similar to that of natural alkaline lakes in acidified areas. But as limed lakes will reacidify after some time the metals precipitated on lake bottoms will dissolute and potentially reach toxic levels if acidification goes fast (19)(20). Limestone application directly into the lake or stream water precipitates toxic metals as Al or Cd within the lake water body. During wintertime, however, aluminium stays toxic to aquatic organisms even though pH is above 6 due to slow precipitation and flocculation rates (21, 22). Therefore "Ecological" liming has to include more applications on land to stop the leaching of toxic substances from the acidified soil. Today acid surges from the catchment area may destroy much of the valuable bottom fauna in the littoral zone of limed lakes because the surrounding land had not been limed. (23). Fig. 3.

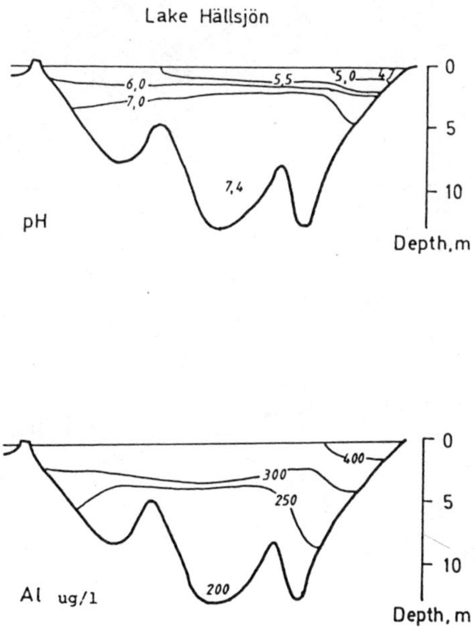

Fig. 3 Additions of limestone merely to the lake volume does not neutralize the acid input via inflowing water. In Lake Hällsjön the surface water pH fell from above 7 to less than pH 5 because of acid surges although the lake volume was limed recently before. High toxic aluminium levels were observed in the biotic zone. The phenomenon is regularly appearing in limed lakes all over Scandinavia (23) during high flow periods.

CONCLUSION

Base addition to lakes and streams is a possible method of keeping poorly buffered waters alive. Next to emission reductions it seems to be an ecologically acceptable way to improve the water quality. However, further liming operations are recommended on land in order to decrease acid and toxic surges polluting the surfacae waters from the acidified catchment soils. A widened soil and wetland liming will increase soil and ground water pH and also stabilize the leaching of micronutrients to the surface waters, at a level more similar to natural conditions.

REFERENCES

(1) FISKERISTYRELSEN, STATENS NATURVÅRDSVERK (1981) Kalkning av sjöar och vattendrag 1977-1981. Information från Sötvattenslaboratoriet Nr 4. 201 pp. (Sweden)

(2) MILJÖVERNDEPARTEMENTET (1985). Kalking av surt vann. Kalkningsprosjektet slutrapport 1985. 145 pp. (Norway).

(3) DILLON P.J. et al (1979). Acidic lakes in Ontario, Canada: Characterization, extent and responses to base and nutrient additions. Arch. Hydrobiol. Beih. Ergebn. Limnol. 13. pp. 317-336.

(4) ACID RAIN INFORMATION CLEARINGHOUSE (1985). Liming acidic waters. Environmental and Policy Concerns. Proceedings of a conference, 82 pp. (USA)

(5) U.S. DEPARTMENT OF THE INTERIOR. FISH AND WILDLIFE SERVICE (1986). Extensive Evaluation of lake liming, restocking strategies, and fish population response in acidic lakes following neutralization by liming. 117 pp. (USA)

(6) BOOTH, G.M. et al. (1986). Liming in Ontario: Short term biological and chemical changes. WASP. 31. pp. 709-720. (Canada).

(7) HOWELLS G.D. and BROWN, D.J.A. (1986). Techniques for acidity mitigation. WASP, 31, pp. 817-825.

(8) HULTBERG, H and ANDERSSON, I.B. (1982). Liming of acidified lakes: Induced long-term changes. WASP. 18. pp. 311-331.

(9) INSTITUTE OF FRESHWATER RESEARCH, Drottningholm (1984). 19 acidification and liming articles by Scandinavian researchers 202 pp.

(10) LESSMARK, O. and THÖRNELÖF E. (1986). Liming in Sweden. WASP. 31. pp. 809-815.

(11) ALENÄS, I. (1987). Kalkningsprojekt Härskogen 1976-1986, IVL L86/201 (In Swedish) (in press).

(12) STATENS NATURVÅRDSVERK (1987). Allmänna råd för kalkning av sjöar och vattendrag (in press).

(13) HÖRNSTRÖM, E. and EKSTRÖM, C. (1986). Acidification and liming effects on phyto- and zooplankton in some Swedish west coast lakes. Naturvårdsverket Rapport 1864. 108 pp.

(14) RADDUM, G.G. et al (1986). Liming the acid lake Hovvatn, Norway: A whole-ecosystem study. WASP. 31. pp. 721-763.

(15) LINDSTRÖM, K (1983) Selenium as a growth factor for plankton algae in laboratory experiments and in some Swedish lakes. Hydrobiologia 101 pp. 35-48.

(16) GAHNSTRÖM, G (1987). Metabolic processes in sediments of acidified and limed lakes with special reference to lake Gårdsjön SW Sweden. Dr. Thesis. 101 pp.

(17) SANGFORS, O. and GRAHN, O. (1987). A comparative study of macrophytes in lake Gårdsjön during acid and limed conditions. SNV publication. Gårdsjöprojekt II (in press).

(18) NYBERG, P. et al (1986). Effects of liming on crayfish and fish in Sweden. WASP. 31. pp. 669-687.

(19) MINISTRY OF AGRICULTURE (1982). Acidification today and tomorrow 231 pp. (Sweden).

(20) ANDERSSON, P and BORG, H (1986). Effekter av sjökalkning på kadmium i vatten, sediment och organismer. Naturavårdsverket Rapport 3119, 14 pp.

(21) DICKSON, W (1983). Liming toxicity of aluminium to fish, Vatten, 39. pp. 400-404.

(22) KARLSSON-NORRGREN, L. et al. (1986) Acid water and aluminium exposure: gill lesions and aluminium accumulation in farmed brook trout, Salmo trutta L. Journ. Fish. Dis. 9, pp. 1-9

(23) HASSELROT, B. et al. 1987. Response of limed lakes to episodic acid events in south-western Sweden. WASP. 32. pp. 341-362.

POSTER SESSION

THEME I : POLLUTION CLIMATES IN EUROPE ; DEPOSITION IN ECOSYSTEMS

THEME I : ENVIRONNEMENT CHIMIQUE DE L'ECOSYSTEME ET DEPOTS

THEMA I : "POLLUTION CLIMATES" IN EUROPA ; DEPOSITION IN ÖKOSYSTEMEN

CARACTERISATION DE L'ENVIRONNEMENT ATMOSPHERIQUE EN ZONE FORESTIERE
Station laboratoire du Donon

J.M BIREN, C.ELICHEGARAY, J.P.VIDAL
Agence pour la Qualité de l'Air
PARIS-LA-DEFENSE
FRANCE

Summary

As a part of the French Research program DEFORPA*, a monitoring and experimental air pollution station is implemented actually in the Vosges forest, where trees are observed in various stage of decline.

The objectives of this station are to determine the level of acidic deposition with regard to the heigth of vegetation, and to set up an area for experimental studies in the field.

This station is constitued of a 44 meters metallic tower for continuus measurement of acidic atmospheric compounds and meteorological parameters at 4 levels of altitude near the trees, monitoring equipments beeing located at ground level in a technical building.

The French Air Quality Agency is conducting the implementation of the station, funds have been provided by Commission of the European Community, French Environmental Ministery, French Air Quality Agency.

1. DESCRIPTION GENERALE DE L'OPERATION

L'implantation de la Station laboratoire du Donon s'inscrit dans le cadre des études destinées à caractériser l'environnement atmosphérique et les depôts acides en zone forestière, engagées en exécution du programme de recherche DEFORPA* mis en oeuvre par le Ministre de l'Environnement.

Le principe de base retenu pour la conception de cette station est la surveillance continue de l'environnement atmosphérique à **différentes hauteurs par rapport à la végétation** et au voisinage d'arbres dépérissants. Cette station est en outre destinée à constituer une structure d'accueil pour l'étude d'un écosystème forestier perturbé.

Le Ministère de l'Environnement a confié la réalisation de la station du DONON à l'Agence pour la Qualité de l'Air. La mise en place de cette station bénéficie d'un soutien financier de la **Commission des Communautés Européennes (Direction Générale de l'Agriculture)** à hauteur de 300 000 ECUS (environ 2 MF), en complément des financements du **Ministère de l'Environnement** (1 MF) et de l'**Agence pour la Qualité de l'Air** (1,6 MF).

* Dépérissement forestier attribué à la pollution atmosphérique.

Les coûts de fonctionnement de la Station seront couverts par le Ministère de l'Environnement et l'Agence pour la Qualité de l'Air.

La Station est localisée sur le territoire de la Commune de GRANDFONTAINE à environ 2 km au sud du col du DONON dans une parcelle forestière de l'Office National de Forêts, touchée par des dépérissements (n° 62 - Etoile 1). Ce site, à environ 750 m d'altitude, est situé sur la pente Nord-Ouest d'un vallon peuplé de sapins et d'épicéas d'une hauteur moyenne de 29 à 34 m, il comprend une placette d'observation de l'état sanitaire du massif.

2. DESCRIPTION DE LA STATION

La station a été conçue pour constituer une structure évolutive adaptée aux objectifs scientifiques du programme DEFORPA et provoquer le minimum d'atteinte au site (aucun abattage d'arbre prévu). L'élément central de la station est constitué d'un pylône métallique, du type autostable, d'une hauteur totale de 46,50 m au pied duquel est implanté le local technique.

Le pylône est constitué d'une ossature tubulaire contreventée, réalisée en 3 tronçons à base carrée, solution qui minimise les effets aérauliques, et dispose de 4 plates-formes de travail échelonnées entre le sol et 10 m environ au-dessus de la cîme des arbres. Les plates-formes de travail sont desservies par un ascenseur accessible depuis le local technique (charge utile de l'ascenseur : 250 kg).

3. EQUIPEMENT SCIENTIFIQUE

Il s'agit d'une part de l'équipement permanent destiné au programme scientifique de base (Tableau I), et d'autre part des équipements complémentaires qui seront implantés par des équipes de recherche pour la réalisation de programmes particuliers.

Dans l'attente de la définition des équipements complémentaires, n'est ici décrit que l'équipement scientifique de base de la station.

DISPOSITIF D'ECHANTILLONNAGE DE L'AIR

Différentes options techniques ont été initialement envisagées, cependant les critères de fiabilité, de servitudes, d'entretien, d'une station située en zone forestière isolée ont conduit à opter pour les prélèvements d'air au moyen de tubes d'échantillonnage avec retour au sol et analyse chimique par batterie d'analyseurs implantés dans un local technique climatisé. Cette solution nécessite qu'un soin tout particulier soit apporté à la conception du système d'échantillonnage, en particulier pour les polluants très réactifs tels que l'ozone. Des essais ont été réalisés à l'Institut de Recherches Chimiques Appliquées (IRCHA) à la demande de l'Agence pour la Qualité de l'Air pour définir ce système d'échantillonnage.

MESURES PHYSICO-CHIMIQUES

Les mesures chimiques sont effectuées par des analyseurs automatiques implantés dans le local technique et reliés de manière séquentielle aux lignes d'échantillonnages par des électrovannes.
Ce choix instrumental a résulté d'un nécessaire compromis entre les qualités métrologiques souhaitables du fait des niveaux de pollution à mesurer, et le besoin d'équipements automatiques (station en site isolé).

La station est équipée en outre d'un pluviomètre à ouverture pluvio-commandée pour l'analyse chimique ultérieure en laboratoire des précipitations. La station du Donon constitue l'une des 18 stations du REseau NAtional de MEsure des Retombées Atmosphériques (RENAMERA) actuellement implanté en France à l'initiative du Ministère chargé de l'Environnement, avec la participation technique et financière de l'Agence pour la Qualité de l'Air.

Les paramétres météorologiques sont mesurés au moyen d'instruments météo, disposés sur les plates-formes de travail du pylône, ces mesures comprennent la vitesse et la direction du vent, et la mesure du rayonnement solaire au sommet du pylône, une mesure de pression au sol, des mesures d'humidité et de température aux quatre niveaux (Tableau I).

ACQUISITION - TRAITEMENT DES DONNEES

Un microordinateur implanté dans le local technique acquiert et traite les données et pilote le cycle de mesures aux différentes plates-formes. Les données acquises sont transférées par lignes téléphoniques puis archivées au poste central de l'Association de Surveillance de la Pollution Atmosphérique en Alsace, à Strasbourg (ASPA), chargée de la gestion de la Station.

Les données feront notamment l'objet de traitements et de présentations graphiques pour l'édition de bulletins mensuels récapitulatifs des évènements survenus.

4. MODALITES, ETAT D'AVANCEMENT DES TRAVAUX

L'Agence pour la Qualité de l'Air assure la maitrise d'ouvrage de la construction de la station.

Les travaux de réalisation de la Station du DONON ont été confiés au Cabinet d'Architecture J.J. RIZZOTTI et à l'entreprise P.L. MAITRE, sélectionnés à la suite d'un appel d'offre.

La Direction Départementale de l'Equipement est chargée de la conduite de l'opération.

Le marché de construction de la station a été signé le 25 novembre 1986.

Après construction des éléments en atelier, le montage à l'aide d'un hélicoptère et l'assemblage du pylône sont prévus au cours du printemps 1987, le complet achèvement des infrastructures de la Station étant actuellement programmé pour septembre 1987.

L'Association de Surveillance de la Pollution Atmosphérique en Alsace contribue à la mise en place de la Station et en assurera la gestion.

Mesures	+ 7m	+ 16m	+ 30 m	+ 44 m
<u>Météorologiques</u>				
Vent (1D) (vitesse-direction)				X
température	X	X	X	X
humidité	X	X	X	X
pression	X			
rayonnement				X
<u>Polluants</u>				
SO	X	X	X	X
NO	X	X	X	X
O	X	X	X	X
CO	X	X	X	X
Aérosols				X

<u>Tableau 1</u>: principales mesures physico-chimiques

(le CO_2 est mesuré à titre d'indicateur de l'activité biologique des arbres)

STATION DU DONON (ETOILE 1)

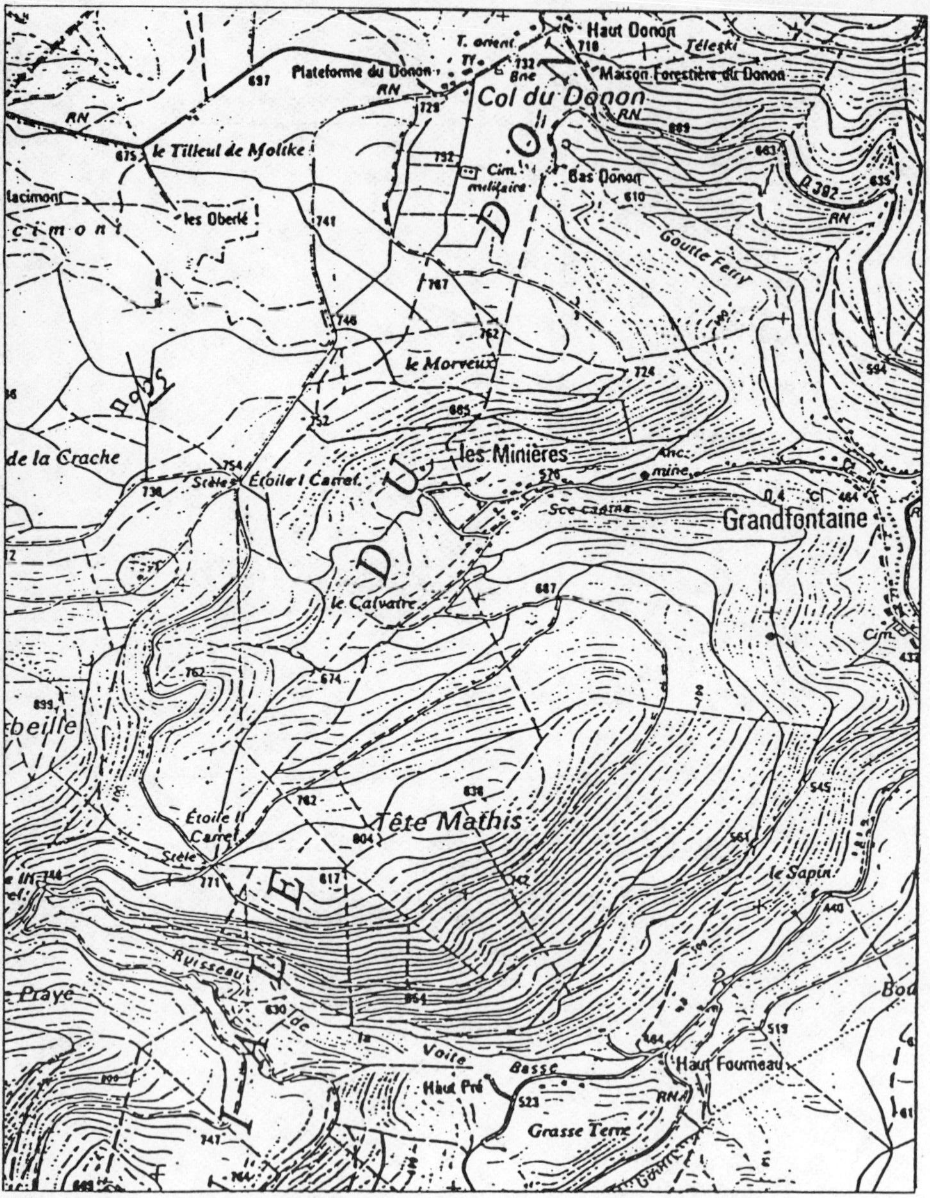

Figure 1 : Localisation géographique de la Station du DONON

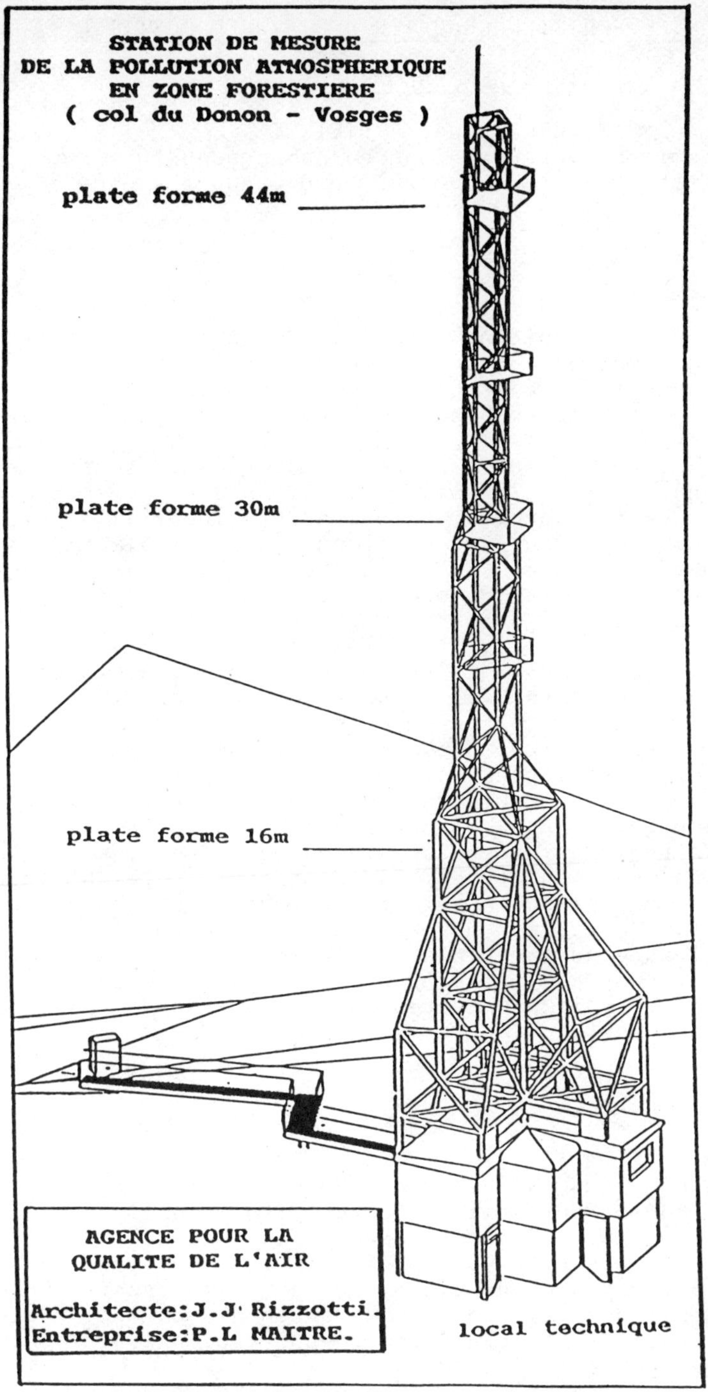

Figure 2 : Station du DONON

KLEINRÄUMIGE VERTEILUNGSMUSTER DER STOFFDEPOSITION IN NATURNAHEN WALDÖKOSYSTEMEN

W.BÜCKING und R.STEINLE
Forstliche Versuchs- und Forschungsansalt
Baden-Württemberg, Abt. Botanik und Standortskunde

Zusammenfassung

In den Versuchsgebieten Schönbuch (südlich Stuttgart) und Feldberg (südlicher Schwarzwald, bei Freiburg) wurde die Stoffdeposition in Nadel- und Laubholzbeständen mit ständig offenen Niederschlagssammlern (bulk-sampling-Verfahren) untersucht.

Beide Gebiete weisen bei sehr unterschiedlichen Niederschlagshöhen (750-1730 mm) vergleichsweise geringe bis mittlere Deposition von Säurebildnern auf (Tab. I,II).

Die Deposition nimmt von Freiland über Laubholzbestände zu Fichtenbeständen stark zu (Beispiel Schönbuch). Bergahorn-Bestände (Beispiel Feldberg) verhalten sich in Bezug auf die Deposition mit der Kronentraufe ähnlich wie Buchenbestände.

Während bereits in einschichtigen, geschlossenen Beständen die Niederschlagsdeposition variiert (Tab. I), wirkt sich auch die Öffnung des Kronendaches auf die Deposition deutlich aus (Tab. II). Schon kleine Lücken lassen die Niederschlagsdeposition auf dem Freiland vergleichbare Werte absinken. Aus dem räumlich differenzierten Depositionsmuster können langfristig im Naturwald kleinräumige Unterschiede in der Belastung der Puffersysteme und im Nährstoffangebot entstehen. In Hanglagen gleicht der laterale Wassertransport solche Unterschiede jedoch aus Sickerwasser, mittels Saugkerzen gewonnen, zeigt im Beispiel Feldberg nur geringe Variation auf kleinem Raum. Im Schönbuch werden große Unterschiede zwischen dem Chemismus des Sickerwassers unter dem Buchen- und dem Fichtenbestand deutlich.

1. Einleitung

In der Ursachen- und Wirkungsforschung zu den neuartigen Walderkrankungen kommt der Ermittlung von Stoffdepositionen im Wald große Bedeutung zu. Bisherige Ergebnisse zeigen, daß die Freilanddeposition (abgesehen vom Umfeld lokaler Quellen) zwar relativ gering variiert und vor allem von der Niederschlagshöhe abhängt, die Depositionsraten in Waldbeständen jedoch von zahlreichen Faktoren beeinflußt werden und sowohl regional als auch lokal große Schwankungen aufweisen.

Bei bisherigen Untersuchungen dienten als Forschungsobjekte überwiegend gleichaltrige und gleichförmige Waldbestände die für entsprechende Bestandestypen vergleichbare Werte liefern.

Es wurde im Projekt "Schönbuch" geprüft, wie stark die

Depositionswerte innerhalb solcher Bestände variieren. In Naturwäldern ist zusätzlich mit Baumartenmischungen, vielschichtigen Bestandesstrukturen und mit ungleichmäßiger Auflösung geschlossener Bestände zu rechnen. Wie sich die Deposition in so strukturierten Wäldern verhält, ist noch wenig bekannt und wurde daher im Feldberggebiet (Naturwaldreservat "Napf") untersucht.
Um zu prüfen, wie sich Stoffdepositionen auf das Bodenwasser in Abhängigkeit von Bestand und Standort auswirken, wurde das Sickerwasser analysiert.

2. Methode und Untersuchungsgebiete

2.1. Sammeleinrichtungen

Freilandniederschlag und Kronendurchlaß ("Kronentraufe", als abtropfender und durchfallender Niederschlag) wurden mit Niederschlagssammlern "Münden" (100 cm^2 Auffangfläche, Aufstellhöhe (Oberkante Sammeltrichter) 100 cm; im Bestand je 10-16) gewonnen (1). Dieser ständig offene Sammmler ermöglicht die Bestimmung der Niederschlagsmengen und Niederschlagsdepositionen ("bulk-sampling-Verfahren"; vgl. 3).

Saugkerzen: Zur Gewinnung von Bodenwasser aus dem ungesättigten Boden (Saugspannung 600 m Wassesrsäule, pF = 2,8 = Porenäquivalentdurchmesser bis ca. 5 µm) wurden Keramik-Saugkerzen "Poröse Masse" 80), auf Plexiglasrohre montiert, verwendet (1, 7). Je fünf Einzelkerzen ("Batterie") einer Bodentiefe waren über ein Vakuumreservoir untereinander verbunden. Weitere Hinweise zur Probenahme (1).

2.2. Analytische Methoden

Folgende Parameter, Elemente oder Verbindungen wurden untersucht: pH-Wert: elektrometrisch mit kombinierter Glaselektrode; Kalium (K^+), Natrium (Na^+), Calcium (Ca^{++}): flammenphotometrisch; Magnesium (Mg^{++}): atomabsorptionsspektrometrisch; Nitrat (NO_3^-: photometrisch mit Natriumsalicylatmethode, seit 1983 ionenchromatographisch; Ammonium (NH_4^+): photometrisch mit Indophenolmethode; Sulfat (SO_4^{2-}): nach Winkler, Kontrollbestimmungen potentiographisch, seit 1983 ionenchromatographisch.

2.3. Auswertung

Die Analysedaten der Niederschlagsproben wurden mit den jeweiligen Volumina zu gewogenen, die Saugkerzen-Proben zu arithmetischen Mittelwerten verrechnet. Die Datenverarbeitungsprogramme prüfen die Plausibilität jeder einzelnen Analyse über die Ionenbilanz.

2.4. Untersuchungsgebiete und Probeflächen

Das Untersuchungsgebiet "Schönbuch" (1,2), Teil des württembergischen Keuperberglandes, ist südlich von Stuttgart gelegen. Die Jahresmitteltemperatur beträgt etwa 8,6° C. Die Versuchsfläche "Fichte", in 450 m Meereshöhe, ist mit 60 jährigem Bestand bestockt (N= 1041/ha, G= 60,4 m^2/ha, Höhenrahmen 25 - 30m). Der Bestand steht auf einer Lehmdecke über Knollenmergel, der Bodentyp ist eine Pseudogley-Parabaunerde. Die Buchen-Versuchsfläche (495 m NN) ist 80jährig (N= 275/ha, G= 267 m^2/ha, Höhenrahmen 20-35 m). Der Bestand stockt auf Lias-Verwitterungsdecke, der Bodentyp ist eine Pseudogley-Parabraunerde.

Das Untersuchungsgebiet "Feldberg" (5,6) liegt in West-Südwest-Orientierung in 1350 m Meereshöhe (Jahresmitteltemperatur etwa 3,2° C) im Naturwaldgebiet (Bannwald) "Napf", an der Grenze des Waldes zur heute offenen Hochweide des Feldberges (1493 m), überragt jedoch noch deutlich um rund 100 m westlich vorgelagerte Gipfel. Der untersuchte Fichten-Bestand ist vor 120 Jahren bis 150 Jahren auf ehemals beweideten Flächen entstanden, daher noch relativ homogen. Stellenweise beginnt jedoch die Verlückung. In der Bodenvegetation dominieren Beersträucher (Vaccinium myrtillus). In Tälchen und in beginnnenden Talmulden stocken Gruppen und Kleinbestände des Bergahorns, eine im Urwald dieser Höhenlage stark beteiligte Baumart. Der Bestand steht auf tiefgründigen Hochlagen-Lehmböden.

3. Ergebnisse und Diskussion

Die Meßergebnisse sind in Tab. I und II dargestellt. Beide Gebiete weisen nur vergleichsweise geringe bis mittlere Stoffdeposition auf; Höchstwerte liegen in Südwestdeutschland in Nadelholzbeständen bei 3,3 kg/ha H^+ bzw. 200 kg/ha • a SO_4^{2-} und 120 kg/ha• a NO_3^- (4). Während der Fichten-Bestandsniederschlag vom Schönberg zum Napf auf das Vierfache ansteigt, nimmt die Protonendeposition nur um das Zweieinhalbfache zu. Die Deposition steigt vom Freiland über Laubholzbestände zu Fichtenbeständen an. Bergahorn-Bestände verhalten sich in Bezug auf die Deposistion mit der Kronentraufe ähnlich wie Buchenbestände, in dem in ihnen die Deposition gegenüber dem Freiland in geringem Maße erhöht wird. Während auch in geschlossenen Beständen die Niederschlagsdeposition variiert (Beispiel Schönbuch), wirkt sich die Öffnung des Kronendaches ebenfalls auf die Deposition deutlich aus (Beispiel Feldberg). Schon kleine Lücken lassen die Niederschlagsdeposition auf dem Freiland vergleichbare Werte absinken. Hieraus folgt, daß aufgelichtete und aus verschiedenen Baumarten gemische Bestände ein sehr differenziertes Depositionsmuster aufweisen; offene Stellen erhalten mehr Licht und weniger Nähr- und Schadstoffe. Im Buchenwald tritt der Stammabfluß mit hohen Niederschlagsmengen und Stoffbelastungen als weiterer kleinräumig differenzierender Faktor hinzu (1). Daraus können langfristig im Naturwald mosaikartig verteilte Unterschiede in der Belastung der Puffersysteme, aber auch im Nährstoffangebot entstehen.

In hängiger Lage gleicht der laterale Wassertransport solche Unterschiede jedoch aus; Sickerwasser zeigt beim Beispiel Feldberg nur geringe Variationsbreite auf kleinem Raum.

Bemerkenswert ist der Unterschied der Sickerwasserbefrachtung mit Inhaltsstoffen unter Fichte und Buche bei vergleichbaren Standortsverhältnissen, der im Projekt Schönbuch offenkundig wird. Besonders hervorzuheben sind die sehr unterschiedlichen Nitratkonzentrationen. Auch unter naturnaher Laubholz-Bestockung können jedoch die Nitratpegelwerte im Sickerwasser ansteigen, wie weitere Untersuchungen auf lehmüberdeckten Kalkstandorten der Schwäbischen Alb zeigten: unter Buchen-Eichen-Bestockung finden wir ein Konzentrationsniveau von 81 - 94 mg NO_3^- /l, unter Fichte von 110 bis 179 mg/l. Die Nitratkonzentrationen im Bodenwasser sind nicht nur bestimmt von der Deposition, sondern auch von ökosysteminternen Faktoren.

Die chemische Zusammensetzung des Sickerwassers unter Fichte im Versuchsgebiet Feldberg ist eher mit dem Buchensickerwasser des Schönbuchs zu vergleichen: die pH-Werte (über alle Tiefen 4,65-4,84) entsprechen dem Austauscher-Pufferbereich des

Bodens (1). Die Sulfatkonzentrationen sind bei größerem Eintrag in vergleichbarer Größenordnung, das Chlorid ist geringer konzentriert, das Nitrat (bei einem doppelt so großen Eintrag) aber 10 bis 100 fach erhöht. Unter den Kationen sind Kalium und Natrium sowie Ammonium in ihrer Größenordnung vergleichbar, Calcium und Magnesium dagegen - der geringeren Ausstattung des Substrats entsprechend - in deutlich verminderter Konzentration anwesend.

Summary

Small-scale distribution patterns of material deposition in semi-natural forest ecosystems

In experimental areas Schönbuch (south of Stuttgart) and Feldberg (Southern Black Forest, near Freiburg), material deposition was measured by bulk sampling techniques. Both areas show relatively small to medium deposition rates (Table I, II),but very different precipitation amounts (750 and 1730 mm). Deposition increases from the open area over broad-leafed tree stands to spruce stands (example Schönbuch). Mountain maple stands (example Feldberg) can be compared with beech stands. While in one-layered closed stands (Schönbuch) precipitation deposition varies in general only to a small extent, opening of the crown cover (Feldberg) has a great effect on deposition. Even small gaps diminish deposition to rates comparable to the open field. In addition to stem flow, deposition of beech, small-scale deposition pattern related to natural forest structure and species distribution may cause differences in impact of buffer systems and in nutrient disposition. In steep areas, lateral water transport equalizes such differences, as can be seen from seepage water analyses in the Feldberg area. In the Schönbuch experimental area, spruce stand seepage water is much more impacted by sulfuric acid, nitrate, calcium, magnesium, and other element concentrations than beech stand seepage water.

Literatur

(1) BÜCKING, W., EVERS, F.H., KREBS, A. (1986): Stoffdeposition in Fichten- und Buchenbeständen des Schönbuchs und ihre Auswirkungen auf Boden- und Sickerwasser verschiedener Standorte. In Einsele, G. (Hrsg): Das landschaftsökologische Forschungsprojekt Naturpark Schönbuch. Wasser- und Stoffhaushalt, Bio-, Geo- und Forstwirtschaftliche Studien in Südwestdeutschland. DFG-Forschungsbericht, 271-324, Weinheim (Verlag Chemie).

(2) BÜCKING, W., KREBS, A. (1986): Interzeption und Bestandesniederschlag von Buche und Fichte im Schönbuch. In: Einsele, G. (Hrsg.): Das landschaftsökologische Forschungsprojekt Naturpark Schönbuch. Wasser- und Stoffhaushalt, Bio-, Geo- und Forstwirtschaftliche Studien in Südwestdeutschland. DFG Forschungsbericht, 113-131, Weinheim (Verlag Chemie).

(3) DVWK 1984: Ermittlung der Stoffdeposition in Waldökosysteme Regeln zur Wasserwirtschaft 122, 6 S. Hamburg und Berlin (Parey).

(4) EVERS, F.H. (1985): Ergebnisse niederschlagsanalytischer Untersuchungen in südwestdeutschen Nadelwaldbeständen. Mitt. Verein Forstl. Standortskunde Forstpflanzenzüchtung 31, 31-36.

(5) JAKOB, R., REINHARDT, W. (1984): Struktur und Entwicklungsdynamik von Bannwäldern in Baden-Württemberg. DFG-Abschlußbericht. Unveröff. Mskr., 54 S + 29 S Anhang. Forstl. Versuchs- u. Forschungsanstalt Bad.-Württ., Freiburg i. Brg.

(6) LOPEZ, J.A. (1985): Analyse der Struktur und Dynamik eines naturnahen Fichten-Tannen-Buchenwaldes im Hochschwarzwald (Bannwald Napf Naturschutzgebiet Feldberg). Schriftenr. Waldbau-Inst. Univ. Freiburg i. Br., 5, 282 S.

(7) MEIWES, K.J., HAUHS, M., GERKE, H., ASCHE, N., MATZNER, E., LAMMERSDORF, N. (1984): Die Erfassung des Stoffkreislaufs in Waldökosystemen. Konzept und Methodik. Ber. Forschungszentrum Waldökosysteme/Waldsterben 7, 68-142.

Tabelle I: Deposition, Niederschlags- und Sickerwasserchemismus, Schönbuch (1979 - 1983)
rate of deposition, composition of precipitation and seepage water, Schönbuch area

	Volumen mm	pH	H^+	SO_4^{2-}	Cl^-	NO_3^-	NH_4^+	K	Na	Ca	Mg
			kg /ha · a								
			(Gewogener Mittelwert mg / l)								
Freiland Fichte	796	-	0,37	22,8	7,7	18,1	4,9	2,4	3,9	9,4	0,9
	-	4,33	(0,05)	(2,9)	(1,0)	(2,3)	(0,62)	(0,3)	(0,5)	(1,2)	(0,1)
Fichte 1	449	-	0,46	72,5	10,9	20,0	3,8	18,5	3,8	15,8	2,3
	-	3,99	(0,10)	(16,1)	(2,4)	(4,46)	(0,84)	(4,12)	(0,9)	(3,53)	(0,52)
Fichte 2	481	-	0,69	110,04	8,7	21,9	3,4	16,4	3,7	14,1	2,1
	-	3,84	(0,14)	(23,0)	(1,8)	(4,55)	(0,71)	(3,41)	(0,8)	(2,93)	(0,44)
Fichte 3	433	-	0,46	54,0	9,3	20,4	3,4	16,3	3,1	12,2	2,3
	-	3,98	(0,11)	(12,5)	(2,2)	(4,71)	(0,78)	(3,76)	(0,7)	(2,82)	(0,52)
Sickerwasser Fichte											
(Saugkerzen) 30 cm	-	3,97	(0,10)	(89,2)	(6,9)	(25,6)	(0,06)	(1,0)	(4,73)	(18,5)	(6,1)
(Saugkerzen) 60 cm	-	4,62	(0,02)	(19,2)	(13,4)	(7,0)	(0,04)	(0,8)	(10,0)	(43,8)	(17,8)
Freiland Buche	779	-	0,21	21,5	5,8	11,1	5,4	4,4	3,6	4,4	0,8
	-	4,56	(0,03)	(2,9)	(0,8)	(1,5)	(0,72)	(0,6)	(0,5)	(0,6)	(0,1)
Buche 1	539	-	0,23	31,8	7,8	22,3	6,3	13,8	3,6	11,6	1,6
(ohne Stammabfluß)	-	4,38	(0,04)	(5,9)	(1,5)	(1,4)	(1,16)	(2,6)	(0,7)	(2,2)	(0,3)
Buche 2	528	-	0,24	21,8	6,6	20,6	4,0	11,0	2,6	7,8	0,9
(ohne Stammabfluß)	-	4,35	(0,05)	(4,1)	(1,3)	(3,9)	(0,75)	(2,1)	(0,5)	(1,5)	(0,2)
Buche 3	555	-	0,26	25,4	6,6	19,9	4,1	13,5	2,7	7,0	1,11
(ohne Stammabfluß)	-	4,35	(0,05)	(4,6)	(1,2)	(3,6)	(0,73)	(2,4	(0,5)	(1,3)	(0,2)
Sickerwasser Buche											
(Saugkerzen) 30 cm	-	5,96	(0,001)	(12,3)	(3,6)	(0,1)	(0,02)	(0,4)	(1,8)	(6,1)	(1,5)
(Saugkerzen) 60 cm	-	6,35	(<0,001)	(4,7)	(6,5)	(0,1)	(0,03)	(0,3)	(2,4)	(3,8)	(1,1)

Tabelle II: Deposition, Niederschlags- und Sickerwasserchemismus, Feldberg (1986)
rate of deposition, composition of precipitation and seepage water, Feldberg area

		Volumen	pH	H^+	SO_4^{2-}	Cl^-	NO_3^-	NH_4^+	K	Na	Ca	Mg
		mm		kg /ha · a								
				(Gewogener Mittelwert mg / l)								
Freiland		1909	-	0,43	39,8	14,0	29,0	9,1	4,7	6,3	4,3	1,1
		-	4,65	(0,02)	(2,1)	(0,7)	(1,5)	(0,48)	(0,2)	(0,3)	(0,2)	(<0,1)
Fichte		1982	-	1,35	83,2	30,3	46,5	5,3	18,5	9,0	13,8	2,5
		-	4,17	(0,07)	(4,2)	(1,5)	(2,4)	(0,26)	(0,9)	(0,5)	(0,7)	(0,1)
Bestandes-lücken		1493	-	0,47	41,3	13,0	27,7	6,0	8,7	7,2	5,2	1,2
		-	4,50	(0,03)	(2,8)	(0,9)	(1,9)	(0,40)	(0,6)	(0,5)	(0,4)	(<0,1)
Bergahorn		1779	-	0,58	63,4	15,0	28,3	5,9	20,1	6,5	9,5	2,1
		-	4,49	(0,03)	(3,6)	(0,8)	(1,6)	(0,33)	(1,1)	(0,4)	(0,5)	(0,1)
Saugkerzen 30 cm	(1)	-	4,65	(0,02)	(3,5)	(2,4)	(5,3	(0,05)	(1,2)	(0,7)	(0,8)	(0,3)
	(2)	-	4,73	(0,02)	(4,2)	(0,8)	(14,2)	(0,04)	(1,4)	(0,6)	(1,0)	(0,2)
	(3)	-	4,60	(0,03)	(4,2)	(1,7)	(10,3)	(0,28)	(1,2)	(0,5)	(0,8)	(0,3)
	(4)	-	4,76	(0,02)	(4,8)	(1,2)	(4,0)	(0,10)	(2,2)	(0,8)	(0,4)	(0,3)
60 cm	(1)	-	4,72	(0,02)	(3,1)	(0,8)	(9,1)	(0,03)	(0,4)	(1,2)	(0,7)	(0,2)
	(2)	-	-	-	-	-	-	-	-	-	-	-
	(3)	-	4,84	(0,02)	(6,4)	(1,1)	(4,3)	(0,04)	(1,5)	(0,5)	(0,7)	(0,2)
	(4)	-	4,75	(0,02)	(6,9)	(2,0)	(6,4)	(0,03)	(2,2)	(1,0)	(1,0)	(0,4)

SPATIAL DISTRIBUTION OF WET AND DRY SULPHUR DEPOSITION IN THE UNITED KINGDOM

STEPHANIE M. COTTRILL
Department of Trade and Industry
Warren Spring Laboratory, Gunnels Wood Road,
Stevenage, Herts, United Kingdom

Summary

Assessment of the impact of acid deposition on sensitive areas of the United Kingdom is being assisted by information from a new monitoring network of over fifty sites. These measurements are providing, for the first time, a detailed picture of the deposition of major ions in precipitation. Kriging has been used to interpolate non-marine sulphate concentrations over the country from which wet deposition has been estimated. The relative importance of wet and dry sulphur deposition pathways has been investigated using statistical models. Trends in emissions and deposition provide a historical context for the present situation.

1. INTRODUCTION

Since January 1986 a network of over fifty precipitation composition monitoring sites has been operational throughout the United Kingdom, providing comprehensive coverage of the country (1). The data obtained from this network will provide a valuable input to assessments of the impact of acidic deposition on natural ecosystems. The rainwater samples are analysed for all major ions but the discussion of wet deposition in this paper will be confined to the spatial distribution of non-marine sulphate.

2. DATA SET

In order to ensure their representativity, sites in the network were established in accordance with the criteria drawn up by the United Kingdom Review Group on Acid Rain (2). The resulting data were subjected to quality assurance procedures as described by Heyes et al (3). The data presented in this paper, however, are still provisional.

After the application of these criteria, annual precipitation weighted mean non-marine sulphate data were available for 46 sites for 1986. The spatial distribution of this ion was investigated using the geostatistical technique of kriging, which has previously been applied to air pollution data sets (eg 4,5). One advantage of this method is that it provides an estimate of confidence limits on interpolated values. Semi-variograms were calculated for four main axes but in this paper, for simplicity, a non-directional semi-variogram has been employed to calculate kriged maps and the corresponding standard error. A straight line model for the semi-variogram was found to be adequate.

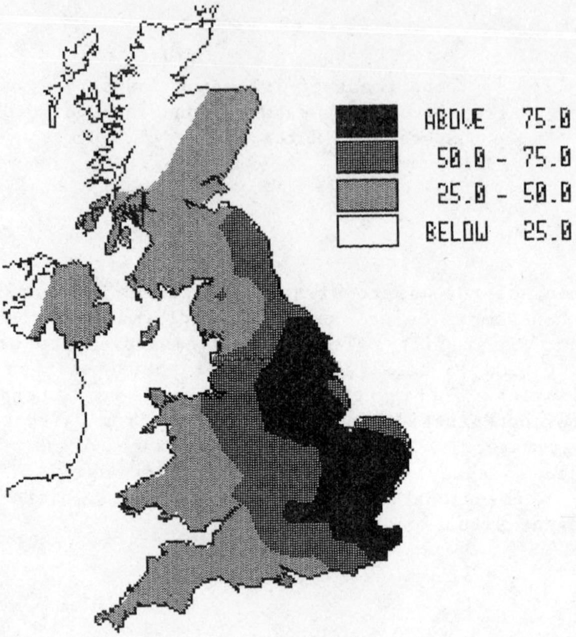

Figure 1. Precipitation Weighted Annual Mean Non-Marine Sulphate Concentration (μeq l^{-1}), 1986

3. SPATIAL DISTRIBUTION OF NON-MARINE SULPHATE CONCENTRATION

Figure 1 shows the spatial distribution of non-marine sulphate concentration. The smallest values are found in the north-west of Scotland and the west of Northern Ireland. The largest concentrations occur in the south-east of England, the East Midlands and East Anglia. The north-west to south-east gradient in concentration is clearly shown, the difference being approximately a factor of five. The spatial pattern of non-marine sulphate concentration over the United Kingdom has remained fairly constant when compared with the much more limited data sets available for 1978-1980 (6) and 1981-1985 (7). However the concentrations of non-marine sulphate in 1986 are somewhat smaller than those averaged over the earlier periods.

Figure 2 shows the associated standard error map. Over most of the country the standard error is < 10 μeq l^{-1}. Areas where this value is exceeded mostly reflect the fact that not all the monitoring sites were operational for more than 90% of the first year; a situation which will be rectified in future years.

4. SPATIAL DISTRIBUTION OF WET NON-MARINE SULPHATE DEPOSITION

An estimate of wet non-marine sulphate deposition for 1986 has been calculated by combining the kriged annual average concentration field with 30-year average rainfall on a 20 x 20 km grid. The resulting deposition field is shown in Figure 3. In using this method it is necessary to assume

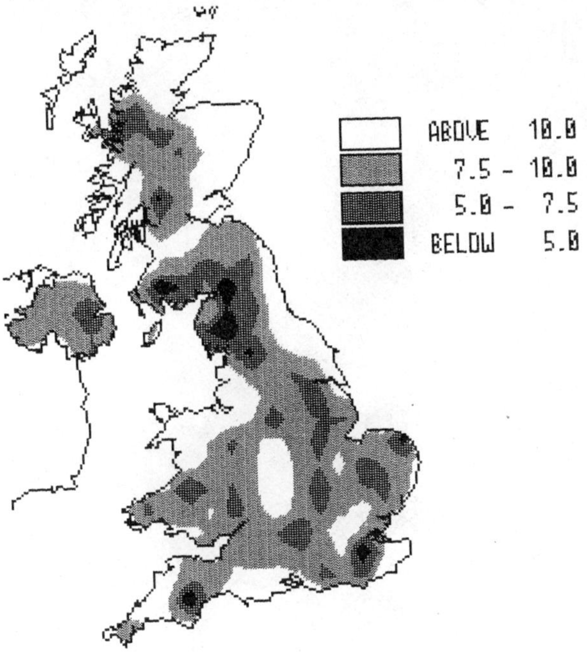

Figure 2. Estimated Standard Error of Annual Mean Non-Marine Sulphate Concentration ($\mu eq\ l^{-1}$), 1986

that rainfall composition does not vary with altitude and this will not be the case in some areas, eg where mountain tops are frequently shrouded by orographic cloud. In such areas deposition may be significantly under-estimated (7) with implications for acidic inputs to upland catchments but only a small effect on the national sulphur budget. The map indicates maximum values in the high rainfall regions of the country including mountainous areas of Scotland, northern England and Wales, where calculated deposition exceeds 1.0 g S m^{-2} $year^{-1}$. Comparison of these values with the limited data, calculated in the same way, for earlier periods (6,7) provides some evidence of a contraction in the areas of high deposition.

5. COMPARISON OF WET AND DRY DEPOSITION OF SULPHUR

The relative contribution of wet and dry pathways to sulphur deposition in 1983 over the country is shown in Figure 4. The dry and wet deposition fields are derived from an Eulerian, statistical model designed for calculating long-period average concentrations in air and rain. The model is based on that described by Perrin (8), but has been enhanced by including estimates of the contributions to sulphur in air and rain from European sources. To take account of possible oxidant limitation near to major SO_2 sources, a skip distance (9) within which SO_2 is not removed in rain, has been incorporated in the model. Figure 4 shows that sulphur

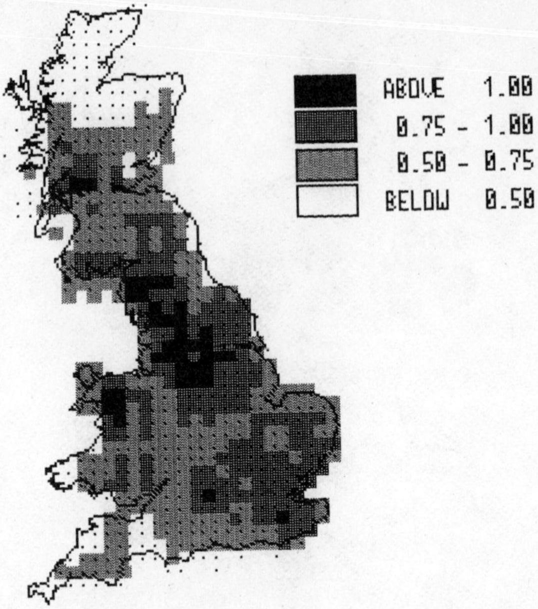

Figure 3. Wet Deposited Non-Marine Sulphate (g S m^{-2}), 1986

deposition in the Midlands and the south of the country is dominated by dry deposition which in some areas accounts for more than 75% of total sulphur deposition. In the higher rainfall areas of the north and west the reverse is true with wet deposition dominating.

6. TRENDS IN SO_2 EMISSIONS

Anthropogenic sulphur is mainly emitted in the form of SO_2, and emissions can be calculated using fuel consumption data and emission factors appropriate to the process and fuel concerned. In the United Kingdom maximum annual SO_2 emissions ($\sim$3 million tonnes as S) occurred in the sixties and early seventies and emissions have generally declined since then. Present SO_2 emissions are about 40% lower than the maximum; this is mainly due to the marked reduction in emissions from low and medium level sources (10). This decrease in United Kingdom SO_2 emissions is reflected by the decrease in the dry deposition of SO_2 over the country (6,7) most of which arises from indigenous sources.

The nature of wet deposition is more difficult to interpret as non-marine sulphate in United Kingdom precipitation may be regarded as arising from three sources - anthropogenic sulphur emissions in the United Kingdom, similar emissions in the European Continent, and an unattributable "background". A further complication is the fact that wet deposition from sources close to a receptor may vary less than proportionally with changes in emission. Clearly the data available to date are inadequate to

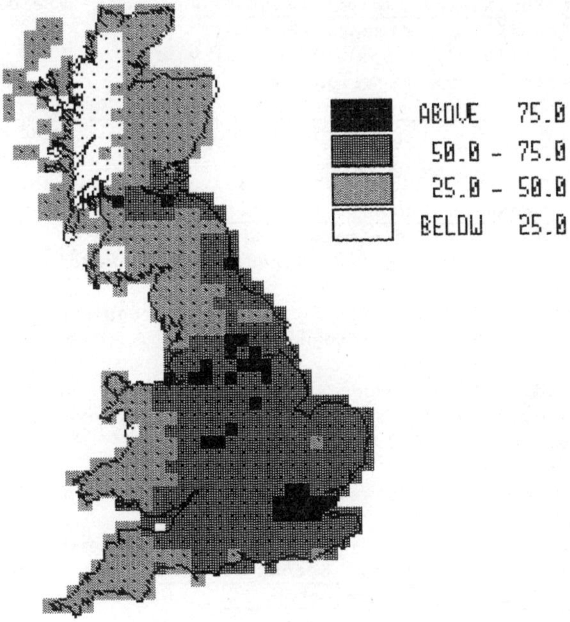

Figure 4. Dry Deposition as a Percentage of Total (Wet + Dry) Sulphur Deposition, 1983

distinguish reductions in wet non-marine sulphate deposition attributable to reductions in SO_2 emissions particularly in view of the large variability due to changes in weather. Nevertheless the observations are consistent with indications from modelling studies (7) that over the country only a varying fraction, typically 10-50%, of non-marine sulphate deposition in rain is directly attributable to indigenous SO_2 emissions and that this contribution has decreased over the past decade.

7. ACKNOWLEDGEMENTS

The funding of this research by the United Kingdom Department of the Environment is gratefully acknowledged as is the advice of the United Kingdom Review Group on Acid Rain and the willing assistance of site operators throughout the country.

8. REFERENCES

(1) DEVENISH, M. (1986). The United Kingdom precipitation composition monitoring networks. Stevenage: Warren Spring Laboratory, LR 584 (AP).
(2) RGAR (1986). United Kingdom Review Group on Acid Rain. Future Acid Deposition Monitoring in the United Kingdom. Stevenage: Warren Spring Laboratory.

(3) HEYES, C.J., IRWIN, J.G. and BARRETT, C.F. (1985). Acid deposition monitoring networks in the United Kingdom. In Advancements in Air Pollution Monitoring Equipment and Procedures. Freiburg im Breisgau, Federal Republic of Germany, 155-168.

(4) ZOLITZ, R. (1985). Spatial validity of the EMEP monitoring net: Geostatistical Investigations of the 1981 sulphate deposition in Europe. In Advancements in Air Pollution Monitoring Equipment and Procedures. Freiburg im Breisgau, Federal Republic of Germany, 200-210.

(5) EYNON, B.P. and SWITZER, P. (1983). The variability of rainfall acidity. Canadian Journal of Statistics, 11, 11-24.

(6) RGAR (1983). United Kingdom Review Group on Acid Rain. Acid Deposition in the United Kingdom. Stevenage: Warren Spring Laboratory.

(7) RGAR (1987). United Kingdom Review Group on Acid Rain. Acid Deposition in the United Kingdom 1981-1985. Stevenage: Warren Spring Laboratory.

(8) PERRIN, D.A. (1986). Modelling the transport and removal of sulphur dioxide emissions in the United Kingdom. Stevenage: Warren Spring Laboratory, LR 560 (AP).

(9) FISHER, B.E.A. (1985). The modelling process. National Society for Clean Air Workshop, Warwick. Brighton: National Society for Clean Air.

(10) EGGLESTON, H.S. and McINNES, G. (1985). United Kingdom emission inventories. In Proc. Second Annual Acid Deposition Emission Inventory Symposium, Charleston SC. Washington DC: EPA, Report 600/9-86-10, 245-260.

LES BASSINS VERSANTS DU MONT LOZERE : UN OBSERVATOIRE DU BRUIT DE FOND DE LA POLLUTION ATMOSPHERIQUE

THE MONT LOZERE CATCHMENTS : AN OBSERVATORY OF THE ATMOSPHERIC POLLUTION BACKGROUND LEVEL

J.F. DIDON-LESCOT[+], R. DEJEAN[++], P. DURAND[+] et F. LELONG[+]
+ Laboratoire d'Hydrogéologie, CNRS UA 724, Université d'Orléans, France
++ Parc National des Cévennes, 48400 FLORAC

Résumé

Les bassins versants de recherche du Mont Lozère (département de Lozère, France) se trouvent dans un secteur assez peu influencé par l'homme (Parc National des Cévennes, classé réserve mondiale de la Biosphère) et sont de ce fait propices à suivre les évolutions quasi naturelles d'écosystèmes terrestres éloignés de sources de pollution industrielle.

Les précipitations abondantes (1900 mm/an) ont une composante méditerranéenne importante. Elles sont en général acides (pH moyen de 4,5), épisodiquement basiques (poussières sahariennes), et les teneurs en sulfates peuvent exceptionnellement être élevées. On ne décèle pas actuellement d'effet de dépérissement sur la végétation.

De ce fait, le site du Mont Lozère, bien équipé et suivi de façon permanente depuis 1981, peut jouer le rôle d'observatoire du bruit de fond de la qualité de l'air et servir de référence pour des secteurs nettement plus perturbés.

Abstract

The Mont Lozère hydrological research basins (France) are situated in a country (Parc National des Cévennes, one of the worldly MAB sanctuary) with little anthropic perturbation and are therefore propitious to quantify the geochemical trends of terrestrial ecosystems in nearly natural conditions.

Wet precipitations (1900 mm per year) are high and originate partly from mediterranean area ; they are generally acid (mean annual pH value near 4.5, with rare basic, Sahara originated occurrences), SO_4 is the most prevalent anion, nitrogen concentrations are low and till now no noticeable forest damage has been noted.

Consequently the well instrumented Mont Lozere basins which have been continuously monitored since July 1981 are good observatories of the atmospheric pollution background level and they give reference values for comparison with perturbated basins.

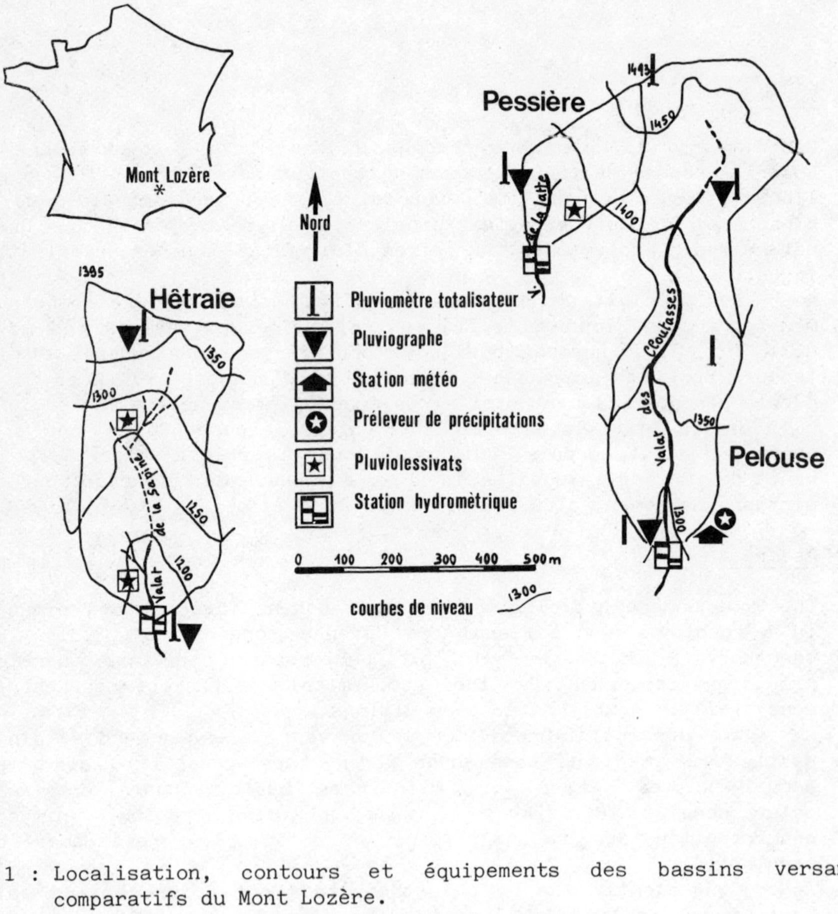

Fig. 1 : Localisation, contours et équipements des bassins versants comparatifs du Mont Lozère.

INTRODUCTION

Les bassins versants expérimentaux du Mont Lozère sont le fruit d'une collaboration menée entre l'Université d'Orléans (UA 724 du CNRS) et le Parc National des Cévennes (PNC) depuis 1981 dans le but général de comprendre les échanges entre l'eau, le sol et la végétation dans trois sites proches et très semblables, différents surtout par l'occupation des sols (2 forêts peu ou pas exploitées, de résineux, de feuillus et une pelouse pâturée), et peu marqués par l'influence de l'homme (1,2).

Les bilans hydrologiques et hydrochimiques réalisés sur 5 ans de 1981 à 1986 fournissent de précieux renseignements sur la dynamique des écosystèmes (2) et sont une base de référence indispensable dans la perspective d'aménagements ultérieurs ou de comparaisons.

A ce titre, le programme de recherche bénéficie d'un soutien pour l'étude sur bassin versant des effets des "pluies acides" (contrat CEE-DEFORPA en 1985-86, et Ministère de l'Environnement) en comparaison avec le bassin forestier d'Aubure, secteur dépérissant des Vosges (3).

LE CADRE EXPERIMENTAL ET LE DISPOSITIF D'ETUDE

Le Mont Lozère qui culmine à 1702 m, en bordure sud du Massif Central français (département de Lozère) est un massif granitique formé de plateaux d'altitude supérieure à 1000 m, fortement entaillé par le réseau hydrographique (Lot au Nord, Tarn au Sud).

Situé en zone centrale du PNC, ce territoire est soumis à une réglementation particulière qui permet d'y protéger plus particulièrement la nature, tout en encourageant un certain type d'agriculture de montagne (élevage ovin et bovin, exploitations forestières) et le tourisme rural. La densité de population est très faible. Une campagne de prospection d'uranium est actuellement en cours à 4 km des bassins versants. Le secteur est intégralement pris dans la Réserve mondiale de la Biosphère créée en 1985.

A 100 km de la Méditerranée, le Mont Lozère se présente comme un promontoire largement ouvert au Sud. Le climat est de type montagnard à caractère méditerranéen bien marqué (4). Les vents dominants et les précipitations associées à l'Aigoual (à 34 km au SW) se retrouvent légèrement atténués sur le Mt Lozère. Les précipitations restent abondantes, de l'ordre de 1900 mm en moyenne et fortement variables d'une année à l'autre et l'enneigement est irrégulier.

La végétation distingue chacun des 3 bassins. Pour le premier, c'est une forêt de hêtres (Fagus silva tica) d'environ 90 ans ; pour le second, une forêt d'épiceas (Picea excelsa) de 60 ans et pour le troisième une prairie à Nard (Nardus stricta), pâturée par un troupeau ovin transhumant. La croissance des arbres est lente, et il faut signaler une attaque de parasites (Dendrochtonus sp.) sur l'épicéa, nécessitant une éclaircie sévère qui sera pratiquée dès l'été 1987. Des parcelles témoins dans une zone non touchée sont en cours d'aménagement. Les placettes d'observations mises en place par l'ONF ne montrent pas de signe de dépérissement dans le secteur du Mont Lozère. Les dispositifs de mesure, largement décrits par ailleurs (1,5) sont représentés sur la figure 1. Un réseau de 5 pluviographes et 7 tubes totalisateurs permet d'apprécier la variabilité spatiale des quantités des précipitations. Pour le contrôle de la qualité, on dispose depuis février 1987 d'un doublet de collecteurs à la station de la Vialasse (à l'aval du bassin de pelouse), dont un pluviomètre à ouverture limitée aux périodes de pluies, exempt de dépôts secs, l'autre restant ouvert en permanence (dépôts secs et humides). Les échantillons sont récupérés après chaque épisode ; le pH, la conductivité et l'alcalinité sont rapidement mesurées au laboratoire de Genolhac. Les

Tableau 1 : Hauteur des précipitations, directions des vents et concentrations chimiques moyennes.
(1) moyenne pondérée par les hauteurs ; (2) moyenne des pluies ; (3) moyenne des neiges.

	Précipitations (mm)	Direction des vents lors de pluies %				pH moyen	Cl^- mg/l	SO_4^{2-} mg/l	NH_4^+ mg/l N
		N	S	W-NW	autres				
							1.34 (2)	3.09	0.28
1981-82	1580	1.4	62.4	35.3	0.9	4.2	1.30 (1)	2.70	0.31
							0.93 (3)	4.02	0.44
1982-83	2390	4.6	79.5	15.8	0.1		0.88	2.40	0.14
							1.29	7.45	0.27
1983-84	1476	4.9	64.5	25.3	5.3		1.16	5.29	0.21
							1.46	6.88	0.33
							0.88	3.28	0.30
1984-85	2173	2.9	72.9	24.0	0.2		0.79	2.99	0.23
							0.94	3.84	0.36
							1.53	5.32	0.17
1985-86	1418	1.0	66.8	28.1	4.7	4.6	1.37	3.60	0.16
							1.05	3.47	0.18

Tableau 2 : Mont-Lozère. Apports atmosphériques annuels moyens ($\overline{M}$) de 1981 à 1986 sur la pelouse (P), la hêtraie (H) et la pessière (E) en kg/ha/an, et dispersion autour des moyennes (σ, maxi, mini).

$kg.ha^{-1}.an^{-1}$		Ca	Mg	K	Na	Cl	SO_4	Si	N
$\overline{M}$	P	13	2.2	2.6	11.6	19.4	59.5	0.7	4.5
	H	12	2.1	2.5	10.9	18.3	57.4	0.7	4.2
	E	13	2.2	2.4	11.5	19.3	59.5	0.7	4.5
σ interannuel		6.4	0.4	0.8	3.4	2.6	16.9	0.8	2.5
Mini		6.0	1.8	1.8	8.4	16.8	41.0	0	2.3
Maxi		21	2.7	3.7	16.3	22.9	80.3	2	8.1

autres analyses sont effectuées soit sur place, NH_4^+ en particulier, soit à Orléans pour les cations (absorption atomique) ou SO_4^{2-} (néphélométrie).

Le même protocole est adopté pour les pluviolessivats prélevés sous les frondaisons. Des bougies poreuses pour collecter les eaux du sol doivent être mises en place dès la fonte des neiges pour comparer la composition des eaux ainsi recueillies à celle obtenue à l'aide de plaques lysimétriques ou extraite sur échantillons humides par centrifugation.

Des études conduites entre 1982 et 1985 n'ont pas montré de variation spatiale nette de la composition chimique des précipitations (5,6), le site de la Vialasse est considéré comme représentatif des apports aux 3 bassins.

APPORTS AUX BASSINS PAR VOIE ATMOSPHERIQUE

1) Niveaux de concentration des précipitations totales (sèches et humides)

Le tableau 1 donne les teneurs moyennes annuelles des différents éléments imputables à la pollution atmosphérique, correspondant à 180 épisodes de précipitations totales, neigeuses et pluvieuses, allant de juillet 1981 à juin 1986 ; il s'agit de valeurs moyennes pondérées par les hauteurs (précipitations totales) ou des moyennes arithmétiques simples (pluies, neige). On constate que les valeurs restent en général faibles. 80 % des SO_4^{2-} se situent entre 0 et 5 mg/l (SO_4) et 5 % dépassent 10 mg/l (avec de très rares pointes à 30 mg/l). Pour NH_4^+, 83 % des valeurs sont comprises entre 0 et 0,5 mg/l-N (4,4 % > 1 mg/l) ; l'anion NO_3^{2-} a été analysé au cours du premier cycle de mesures (1981-82), et les teneurs obtenues ont toujours été faibles (de 0 à 0,26 mg/l-N). Les pH, quant à eux, sont nettement acides, 70 % d'entre eux étant compris entre 3,5 et 5,0. Notons que les bonnes corrélations obtenues sur le cycle 1982-83 (5) entre pH et SO_4^{2-} et SO_4^{2-} et NH_4^+ disparaissent sur la période 1981-86.

Les concentrations fluctuent d'un épisode à l'autre en fonction de l'origine des masses d'air. Le site est soumis à une alternance de vents du NW, S et N dont les fréquences sont respectivement de 36,8, 35,6 et 25 % sur 5 ans ; mais les précipitations sont largement liées aux apports méditerranéens (flux de Sud) et plus modestement par les épisodes océaniques. Quelques essais pour tenter de caractériser le niveau de concentration par le type de temps sur le cycle 84-85 (6) ont donné des résultats incertains. Les apports d'origine saharienne y sont particulièrement nets au printemps et ont comme effet de diminuer l'acidité des pluies (pH compris entre 6 et 6,35, augmentation de Ca^{2+} et H_4SiO_4), tout comme en Corse (7) où le phénomène semble nettement plus dominant.

Il n'y a pas de pollution apparente en chlorures ; en effet les valeurs du rapport Na/Cl fluctuent entre 0,8 et 1,2 et sont proches de la valeur (0,88) caractéristique de l'eau de mer (9). Par contre, les pH et les concentrations mesurées en sulfates donnent, compte tenu des hauteurs de précipitations, des flux acides entrant dans les bassins relativement importants (40 à 80 $kg.ha^{-1}.an^{-1}$ de SO_4^{--}, selon les années et selon les bassins). D'après Katz et al. (8), l'acidification des eaux de surface et les dégâts sur l'environnement peuvent apparaître quand les flux apportés par les précipitations dépassent 20 $kg.ha^{-1}.an^{-1}$ de sulfate. Cependant les apports azotés sont très faibles, ils se présentent essentiellement sous la forme de NH_4^+, et l'on ne note pas de symptôme important de dépérissement des forêts sur le Mont Lozère. Les sources potentielles de pollution sont :

- le secteur minier et industriel d'Alès, où les émissions de SO_2 sont jugées faibles (10),

Tableau 3 : Apport en solution par voie atmosphérique. Comparaison des apports mesurés en France avec quelques valeurs mondiales. Valeurs exprimées en kg.ha^{-1}.an^{-1}. Les teneurs de sulfates sont exprimées en soufre (S) et celles des espèces azotées en azote (N).

Ca	Mg	K	Na	Cl	SO_4^-	NH_4^+	NO_3	H mm	Site	Durée	Référence
5.0	1.0	2.2	5.0	7.5	35	2.2	3.3	1500-2000	COWETTA (USA)	69-76 2 ans	(11)
2.2	0.6	0.9	1.6	6.2	12.7		6.7	1300	HUBBARD BROOK (USA)		(12)
5.4	1.2	1.6	3.7	5.4	10	2.9	3.0		VELEN (Suède)	68-74 5 ans	(13)
					19		11		ARDENNES (Belgique)		(14)
3.2	0.36	0.51	0.75	2.2	12	5.3	4	736	RURAL	5-7 ans	(15)
7.4	1.1	1.7	2.2	25.1.	19.6	7.5.	5.5	812	Industriel (Tchécoslovaquie)		
4.4	0.70	2.9	2.4	5.8	22.3	4.6	13.5	1000	TRENGBACH (Vosges)	1 an 85-86	(3)
18.1	2.6	3.7	11.9	19.5	17.3	2.3		1413-2390	Mt LOZERE	85-86	présente étude

Tableau 4 : Mont Lozére. Bilan entrée-sortie simplifié sur 5 ans (1981-86) en Kg/ha

Kg/ha	Pelouse	Hêtraie	Pessière
Cl	+ 2,8	+ 2,1	- 0,7
SO_4	+20,4	+20,7	+ 8,1

- les complexes urbains et industriels de Marseille-Fos-Berre-Sète,
- la zone de Barcelone.

2) Apports en phase soluble ("bulk precipitations")

Environ 95 % des pluies ont été analysées pour la plupart des éléments, avec des incertitudes sur les apports annuels estimés à 5-10 % pour les cations, et de 20 % pour Cl^-, $SO_4 2-$ et $NH_4 +$ (1). Le tableau II regroupe les résultats des 5 cycles traités. On constate que si la variabilité spatiale est faible (les bassins étant proches), la variabilité interannuelle est, elle, plutôt forte surtout pour Ca^{2+}, Na^+, K^+ et S. Elle intègre à la fois les variations de concentration liées aux processus de dissolution et de lame d'eau précipitée qui peut passer du simple au double sur la période étudiée.

Par rapport à d'autres sites français et étrangers (3,11 à 15), les apports de sulfates du Mont Lozère sont relativement importants et se rapprochent des valeurs des Vosges ; par contre, les apports d'azote sont très largement inférieurs.

3) Les apports sous forme solide

En l'absence de mesure directe (mesures mises en oeuvre depuis Février 1987) on a tenté d'évaluer les précipitations sèches par 2 méthodes, les bilans entrées-sorties (E-S) de Cl^- et $SO_4 2-$ et l'étude des pluviolessivats.

Bilans de Cl^- et $SO_4 2-$

Les ions Cl^- étant absents de la roche-mère et peu impliqués dans les cycles biologiques, un bilan $E - S < 0$ peut être le signe d'une sous-estimation des entrées et par conséquent de la présence de dépôts secs (1). Les moyennes des bilans sur 5 ans (tableau 4) sont faiblement positives pour la pelouse et la hêtraie, faiblement négatifs dans la pessière. Eu égard aux incertitudes sur ces bilans, l'hypothèse de bilans équilibrés est plausible. Notons toutefois que les années à bilan négatif dans la pessière correspondent aux cycles bien arrosés, ce qui a été déjà signalé (14). Si l'on admet que la période d'observation porte sur une série d'années de pluies déficitaires - d'environ 100 mm en moyenne interannuelle - on peut s'attendre en année normale à une exportation globale notable ($E - S < 0$) dans le bassin enrésiné, fait indiquant des dépôts secs de Cl^- non négligeables.

Le même raisonnement peut s'appliquer au bilan de $SO_4 2-$ en supposant que les quantités de soufre libérées par l'altération du substrat granitique soient négligeables et que la fixation directe de SO_2 atmosphérique par la végétation soit sensiblement la même pour les trois bassins. On observe pour cet élément une accumulation moyenne interannuelle nette pour les trois bassins étudiés, mais l'accumulation est de 10 $kg.ha^{-1}$ inférieure sous forêt résineuse. Cet écart pourrait être dû aux entrées atmosphériques solides occultes (non mesurées) car on sait que les végétations résineuses à organes foliaires finement divisés captent plus intensément les aérosols (5).

Les pluviolessivats

Dix neuf épisodes pluvieux, étudiés en 1982-83 (5) et un suivi continu depuis 1986 en cours de traitement permettent de dégager les faits suivants :

- la charge chimique de la pluie augmente notablement après la traversée de la frondaison, surtout sous pessière et dans les eaux d'égouttage des troncs. L'évaporation au niveau du feuillage ne peut à elle seule expliquer cette augmentation ; celle-ci concerne tous les éléments mais elle est moins nette pour l'azote. En ce qui concerne l'acidité, la variation du pH au contact des végétaux dépend de la nature de ceux-ci : dans 70 % des épisodes, le pH des pluviolessivats est inférieur (pessière) ou supérieur (hêtraie) à celui de la pluie. La concentration en $SO_4$2- augmente en moyenne de 50 % mais avec une grande variabilité d'un épisode à l'autre. Ainsi, à l'occasion de pluies riches en $SO_4$2-, la différence peut s'annuler, ce qui tendrait à indiquer un apport essentiellement sous forme dissoute.

Les deux types d'approche convergent pour montrer que les précipitations sèches sont plus importantes dans le bassin enrésiné, mais qu'elles restent d'intensité réduite de sorte que tout essai de quantification indirecte est hasardeuse.

CONCLUSION ET PERSPECTIVES

Les apports par voie atmosphèrique de polluants sur les bassins versants du Mont Lozère sont variables selon les éléments ;

- modestes pour la pollution azotée (et négligeable pour NO_3^-) :
- relativement importants, et assez prochcs des valeurs des Vosges pour SO_4^{2-}.
- le pH est également assez bas.

Ces caractères traduisent probablement un léger enrichissement des masses d'air, d'origine surtout méditerranéenne lors du passage au-dessus de centres industriels où la combustion de soufre (SO_2) semble prépondérante par rapport à la pollution de type NO, NO_x. Mais le niveau global de pollution est modéré et peut s'assimiler à un bruit de fond avec exceptionnellement quelques pointes nettement supérieures. A ce titre, le site du Mont Lozère peut servir de secteur de référence peu marqué par la pollution atmosphérique.

La poursuite du programme sur le Mont Lozère permettra de mieux comprendre le devenir de SO_4^{2-} et H^+ dans ces écosystèmes, de vérifier s'il y a piégeage de sulfate dans le sol, de vérifier les conditions de piégeage de ce composé dans les écosystèmes, et de suivre leurs incidences éventuelles sur les cycles biogéochimiques des éléments minéraux et sur l'état sanitaire de la végétation. Actuellement celle-ci paraît peu affectée par la pollution atmosphèrique mais des sites assez proches come l'Aigoual ainsi que certaines placettes du réseau d'observation "Hautes Cévennes" montrent certains symptômes de dépérissement forestier.

BIBLIOGRAPHIE

(1) DUPRAZ, C. (1984). Bilan des transferts d'eau et d'éléments minéraux dans trois bassins versants comparatifs à végétations contrastées (Mont Lozère, France). Thèse Doc. Ing. Univ. d'Orléans, 363 p.

(2) LELONG, F. et al. (1987). Bilans hydrologiques (1981-86) et hydrochimiques (1981-85) annuels des trois bassins comparatifs du Mont Lozère (France). Rapport final DEFORPA, 2 p.

(3) PROBST, A., VIVILLE, D., AMBROISE, B. et FRITZ, B. (1987). Bilan hydrogéochimique d'un petit bassin versant des Vosges en relation avec le dépérissement forestier. Rapport final DEFORPA.

(4) ASCENSIO, M. (1981). Département de Lozère, in An. Clim. du Lang. Roussilon, p. 5-74.

(5) WEDRAOGO-DUMAZET, B. (1983). Modification de la charge au cours du transit à travers trois écosystèmes distincts du Mont Lozère (hêtraie, pessière, pelouse). Thèse 2ème cycle Univ. d'Orléans, 147 p.

(6) DIDON, J.F. (1985). Contribution à l'étude de la variabilité spatio-temporelle des pluies sur le Mont-Lozère. Campagne 84-85. DEA Sciences de l'eau, Univ. de Montpellier, 79 p.
(7) LOYE-PILOT, M.D. et al. (1986). Impact of saharian dust on the rain acidity in the mediterranean atmosphere. 4th Eur. Symposium on physico chemical behaviour of atmospheric polluants. Stresa, Italy, 11 p.
(8) KATZ, B.G., BRICKER, O.P., KENNEDY, M.M. (1985). Geochemical mass balance relation ships for selected ions in precipitation and stream water, Catoctin Mountains, Maryland. Am. Journ. of Sci., 285, p. 931-962.
(9) MEYBECK, M. (1983). Atmospheric inputs and river transports of disolved substances. Proceed. of the Hamburg. Symposium I.A.H.S. Publ. n° 141.
(10) A.M.P.A.D.I. (1984). La pollution atmosphérique en Languedoc-Roussillon. Résultats des mesures de 1983. Rapp., 25 p.
(11) SWANK, W.T. et DONGLASS, J.E. (1985). Nutrient flux in undisturbed and manipulated forest ecosystems in the southern appalachian mountains. AISH, Symposium Tokyo, pp. 445-456.
(12) BORMANN, F.H. et LIKENS, G.E. (1979). Pattern and process in a forested ecosystem. HUBBARD BROOK, New Hampshire (U.S.A.).
(13) ANDERSON-CALES, U.M. et ERIKSSON, E. (1979). Mass balance of dissolved inorganic substances in three representative basins in Sweden. Nordic hydrology, pp. 99-114.
(14) BULDGEN, P. et al. (1984). Biochimie de deux bassins versant de l'Est de la Belgique. Physico-Géo., n° 9,pp. 47-59.
(15) PACES, T. (1985). Sources of acifification in Central Europe estimated from elemental budgets in small basins. Nature, vol. 315, pp. 31-36.
(16) FOSTER, I.D.L. (1980). Chemical yields in rnnoff, and denordation in a small arable catchment, East Devon, England, J. of hydrol., vol. 47, pp. 368-394.

Palynologische und forstgeschichtliche Aspekte zur Versauerungsgeschichte von Schwarzwaldseen

K.H. Feger und W. Zeitvogel
Institut für Bodenkunde und Waldernährungslehre
der Albert-Ludwigs-Universität
D-7800 Freiburg im Breisgau

Zusammenfassung

Rekonstruktionen der pH-Wertgeschichte mittels fossiler Diatomeen und anderer Bioindikatoren in Seesedimenten dokumentieren eine Absenkung des pH-Werts im Herrenwieser See, während der pH-Wert im Feldsee hingegen unverändert geblieben ist. Die radiometrische Datierung der Sedimentkerne macht deutlich, daß die Versauerung des Herrenwieser Sees bereits 1-2 Jahrhunderte vor Beginn der Industrialisierung begonnen hatte. Pollenanalysen im Sediment des Herrenwieser Sees sowie forstgeschichtliche Studien verdeutlichen den Wandel in der Vegetationszusammensetzung innerhalb der letzten 300 Jahre. Intensive Waldnutzungen, besonders die großflächigen Kahlhiebe des 18. Jahrhunderts, resultierten in der Umwandlung des natürliche Tannen-Buchenwalds in überwiegend reine Fichtenforste. Diese nutzungsbedingte Veränderung der Vegetationszusammensetzung, die im Nordschwarzwald besonders stark war, stand in enger Wechselwirkung mit der rasch fortschreitenden Vernässung und Versauerung der Böden. Diese Entwicklungen haben offensichtlich auch zu einer allmählichen Versauerung der Oberflächengewässer in diesem Gebiet geführt. Diese nutzungsbedingte Versauerung wird heute von den möglichen Auswirkungen eines verstärkten atmogenen Säureeintrags überlagert, die bislang nicht voneinander getrennt werden können. Der pH-Wert des im Südschwarzwald gelegenen Feldsees hat sich aufgrund anderer natürlicher Bedingungen und einer anderen Nutzungsgeschichte seit mehreren Jahrhunderten nicht verändert.

Summary

The reconstruction of pH-history by means of fossil diatoms and other bioindicators has documented a decrease in the pH-value of Lake Herrenwies, while there was no change in the pH-value of Lake Feldsee. The radiometric dating of sediment cores makes evident that the acidification of L. Herrenwies had started already 1-2 centuries before the beginning of industrialization. Pollen analysis in the sediment of L. Herrenwies as well as the evaluation of forest history demonstrate the change in the composition of vegetation cover during the last 300 years. Intensive silvicultural practices, especially the clearcuts of the 18th century, resulted in the alteration of the natural forest association of beech and fir in nearly pure spruce plantations. This alteration of vegetation cover due to land-use, which was strongest in the northern part of the Black Forest, was mutually related to a progressive waterlogging and acidification of the soils. These tendences have apparently also led to a gradual acidification of surface waters in that area. Acidification due to land-use changes are recently superimposed by the possible effects of an increased atmospheric deposition. So far it is not possible to separate the effects from each other. The pH-value of L. Feldsee located in the southern part of the Black Forest has not been changed for many centuries due to different land-use history and different natural conditions.

1. Einleitung und Problemstellung

Mehrere regionale Untersuchungen der aktuellen chemischen Zusammensetzung von Oberflächengewässern in bewaldeten Einzugsgebieten des Schwarzwalds fanden in bestimmten Gebieten pH-Werte zwischen 4 und 5 sowie Al-Konzentrationen bis zu 1000 $\mu g.L^{-1}$ (1, 2, 3, 4). Solche Gewässer sind besonders in den perhumiden Hochlagen des Nordschwarzwaldes anzutreffen (Jahresniederschlag >2000 mm). Der geologische Untergrund (Mittlerer Buntsandstein) ist extrem basenarm. Die zwergstrauchreichen Fichtenforsten stocken auf stark sauren Böden mit oberfächennahen Stauhorizonten (Stagnogleye, Bändchen-Staupodsole). Gewässer in Gebieten mit basenreicheren Böden (Braunerden) mit meist Gneis aus Ausgangsgestein weisen weit höhere pH-Werte zwischen 6 und 7 und nur sehr geringe Al-Konzentrationen auf.

Es stellt sich deshalb die Frage, wie die niedrigen pH-Werte im Buntsandstein-Gebiet des Nordschwarzwalds erklärt werden können. Sind diese Gewässer von Natur aus so sauer? Oder haben Veränderungen in den Einzugsgebieten (Landnutzung, Vegetationszusammensetzung) oder die gestiegenen atmogenen Einträge der letzten Jahrzehnte eine Versauerung der Gewässer in diesem Gebiet bewirkt? Da aus der Analyse des aktuellen Gewässerchemismus nicht entschieden werden kann, ob und wann ein Gewässer versauert ist und länger zurückliegende hydrochemische Meßreihen fehlen, kann nur ein paläolimnologischer Ansatz zur Lösung dieses Problems herangezogen werden. Durch die Analyse von Sedimenten stehender Gewässer lassen sich nämlich Veränderungen im See bzw. seinem Einzugsgebiet erkennen (5, 6).

2. Material und Methoden

Ungestörte Sedimentkerne wurden aus zwei Seen im Schwarzwald mittels eines BORG-Gravitationslots (7) entnommen. Um ungestörte Einzelproben auch aus den obersten, wenig verfestigten Sedimentbereichen zu gewinnen, erfolgte die Unterteilung der Kerne in 1-cm-Abständen sofort nach Kernentnahme am Ufer. Teilproben wurden der Pollenanalyse, der radiometrischen Datierung sowie der chemischen Analyse zugeführt. Eine detailierte Darstellung der Methodik erfolgt in (8, 9). Die Kerne wurden mit ^{137}Cs, ^{210}Pb und im Falle des Herrenwieser Sees auch mit ^{14}C datiert.

Tabelle I. Vergleich der aktuellen chemischen Zusammensetzung des Feldsees (Südschwarzwald, Gneis-Anatexit, Braunerden) und des Herrenwieser Sees (Nordschawrzwald, Mittlerer Buntsandstein, Podsole, Stagnogleye). Mittelwerte aus 28 Probenahmen an den Ausflüssen der Seen im Zeitraum Juni 1984 bis Juni 1985.

Parameter	Einheit	Feldsee	Herrenwieser See
pH-Wert		6.3	4.2
Leitfähigkeit (20°C)	$\mu S.cm^{-1}$	19.6	47.6
UV-Extinktion λ=254 nm	$E.m^{-1}$	1.0	12.0
Färbung λ=436 nm	$E.m^{-1}$	0.2	2.0
DOC	$mg.L^{-1}$	1.2	8.4
Na^+	$\mu eq.L^{-1}$	40	25
K^+	"	5	22
Ca^{2+}	"	106	69
Mg^{2+}	"	26	28
HCO_3^-	"	81	0
SO_4^{2-}	"	58	154
Cl^-	"	19	36
NO_3^-	"	24	32
Al	$\mu g.L^{-1}$	16	610

Tab. I gibt einen Überblick über die aktuelle chemische Zusammensetzung beider Gewässer. Beim Feldsee handelt es sich demnach um einen fast neutralen, oligotrophen Klarwassersee. Für die von uns gemessenen hydrochemischen Kennwerte ergeben sich, soweit analytisch überhaupt vergleichbar, keine Veränderungen gegenüber den Untersuchungen von ELSTER zu Beginn der fünfziger Jahre (10). Der Herrenwieser See dagegen ist ein stark saurer, braungefärbter dystropher See mit einem breiten *Sphagnum*-Verlandungsgürtel. Für diesen See gibt es keine älteren hydrochemischen Untersuchungen.

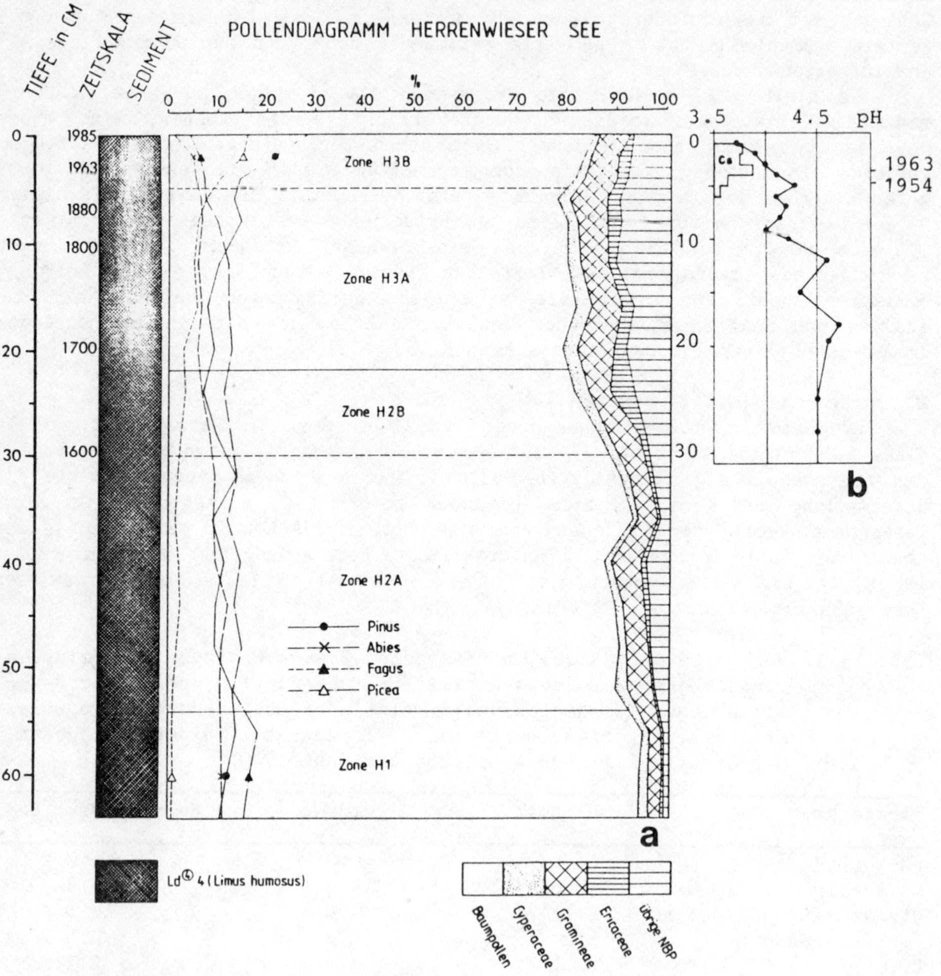

Abb. 1 a. Stratigraphische Gliederung, Datierung und Pollenverteilung im Sediment des Herrenwiesern Sees (Summenkurven der relativen Pollenhäufigkeiten von Baumpollen, Cyperaceen, Gramineen, Ericaceen und der übrigen Nichtbaumpollen; innerhalb der Baumpollen sind die Hauptbaumarten Kiefer, Tanne, Buche und Fichte als Einzelkurven dargestellt).

b. Rekonstruierter Verlauf des pH-Werts im Herrenwieser See aus der Analyse der fossilen Diatomeen im Sediment nach (11).

Für beide Schwarzwaldseen liegen jedoch Rekonstruktionen der pH-Wertgeschichte aus der Analyse fossiler Diatomeen und anderer Bioindikatoren in den Seesedimenten vor (11, 12). Nach diesen Ergebnissen fiel der pH-Wert im Herrenwieser See deutlich, während der pH-Wert im Feldsee über lange Zeit unverändert blieb (Abb. 1b). Für eine Interpretation dieser Befunde erscheint es deshalb von Interesse, diese mit den Ergebnissen unserer palynologischen Befunde in Beziehung zu setzen. Von zentraler Bedeutung ist dabei die zeitliche Einordnung der Versauerung des Herrenwieser Sees.

3. Ergebnisse und Diskussion

3.1. Herrenwieser See

In Abb. 1a ist die stratigraphische Gliederung, Datierung und Pollenverteilung im Sediment des Herrenwieser Sees dargestellt. Das Sediment besteht aus einem sehr homogenen Material von gallertartiger Konsistenz, das sich nach (13) als "Limus humosus" oder als "Lebermudde" (nach 14) klassifizieren läßt. Die Ergebnisse der ^{210}Pb-Datierung zeigen eine sehr geringe jährliche Sedimentakkumulation und einen gleichmäßigen Sedimentationsverlauf. Die Sedimentationsrate beträgt in der Schicht 1-2 cm 1.2 mm.a^{-1}, nimmt mit der Tiefe verdichtungsbedingt jedoch stark ab. In der Schicht 6-7 cm beträgt die Sedimentationsrate nur noch 0.4 mm.a^{-1}. Die ^{137}Cs-Messung ergibt ein Maximum bei 2-3 cm und dürfte damit die Jahreszahl 1963 als Maximum der Freisetzung dieses Radionuklids aus den oberirdischen Atombombenversuchen markieren, was sich auch mit der ^{210}-Datierung deckt. Die ^{137}Cs-Verteilung stimmt recht gut mit der von STEINBERG et al. (11, 12, vgl. Abb. 1b) gefundenen überein. Allerdings dürfen die Sedimentationsraten der obersten Schichten nicht auf tiefere Sedimentbereiche extrapoliert werden, was zu einer starken Unterschätzung des Sedimentalters führen würde. Die Tiefenstufe 6-7 cm entspricht nach dem ^{210}Pb-Alter bereits dem Jahr 1880. Somit dürften die obersten ca. 8 cm den Zeitraum der Industrialisierung und der gesamte Kern die letzten 1000 bis 1500 Jahre entsprechen. Ersteres wird durch eine Zeitmarke in 5-6 cm Tiefe (ca. 1910) bestätigt. Leicht erhöhte Werte für Trockendichte, Al, K und Mg sprechen für eine kurzzeitig erhöhte Zufuhr von mineralischem Detritus (8). Für diese Zeit ist der Bau des Forstwegs am Westufer des Sees belegt. Ortientierende ^{14}C-Messungen stellen die Schichten um 30 cm etwa in das Jahr 1600, während die Schichten um 65 cm wahrscheinlich bereits um die Zeitwende abgelagert wurden.

Das Pollendiagramm verdeutlicht die starken Veränderungen in der Vegetationszusammensetzung innerhalb der letzten 300 Jahre, was die Waldnutzungsgeschichte des Gebiets um Herrenwies widerspiegelt (vgl. 15). Intensive Nutzungen der Wälder, besonders die großflächigen Kahlhiebe des 18. Jahrhunderts, resultierten in der Umwandlung des natürlichen Tannen-Buchenwalds (Pollenzonen H1 und H2A) in die heutigen Fichtenforste (Zone H3B), was in den Tiefenverläufen der Hauptbaumarten Tanne, Buche, Kiefer und Fichte deutlich zum Ausdruck kommt. Bereits vor diesen Kahlhieben wurden die Wälder durch die zahlreichen "Waldgewerbe" (Köhlerei, Pottaschesiederei, Harzerei, Glashütten) in zunehmendem Maße devastiert, was ebenfalls zu einem Zurückdrängen von Buche und Tanne und zu einer Begünstigung von Tanne und Kiefer führte. Auch der Einfluß der Weidewirtschaft kommt im Pollendiagramm durch den Rückgang der Baumpollen und den Anstieg der Gräser gut zum Ausdruck. Beweidet wurden besonders die für diesen Zweck kahlgeschlagenen Hochflächen. Aber auch die Waldweide war weit verbreitet. Die starke Zunahme der Ericaceen im 17. und 18. Jahrhundert spiegelt die zunehmende Auflichtung und Verheidung der Landschaft wider. Die nutzungsbedingte Veränderung der Vegetationszusammensetzung, die im Nordschwarzwald besonders stark war, hatte aufgrund der dortigen, von Natur aus basenarmen Böden und Gesteine weitreichende ökologische Konsequenzen. Sie stand in enger Wechselbeziehung mit einer Degradierung der Böden. So führte der Weidebetrieb der perhumiden Hochlagen zu Bodenverdichtungen und Vernässung weiter Flächen. Die Umwandlung des einstigen Tannen-Buchen-Mischwalds in zwergstrauchreiche Fichtenforste sowie der nutzungsbedingte Nährstoffentzug, besonders durch die weit verbreitete Streunutzung, resultierten in einer starken Bodenversauerung (Podsolierung) und Ortsteinbildung (17, 18). So schreibt MÜNST (17) bereits zu Beginn dieses Jahrhunderts: "... *(es) ist der Schluß berechtigt, daß der größte Teil der Ortstein- und Missebildungen des oberen Murgtales auf die großen Waldverwüstungen zurückzuführen sind, welche im frühen Mittelalter begonnen und im Jahre 1800 mit dem Waldbrand im Freudenstädter Forst ihr Ende gefunden haben, und daß nur ein Teil als die natürliche Folge klimatischer Einwirkungen sich darstellt. Bei der allgemeinen Kalkarmut der Schwarzwaldböden hat sich die Raubwirtschaft bitter gerächt. Waren ehedem die Laubhölzer durchweg den Nadelhölzern reichlich beigemischt, so schlugen die Wiederanbauversuche nach dem Brande völlig fehl infolge der Rohhumusbildung und der Versaurung des Bodens. Die ungünstigen Folgen reiner Nadelholzbestände dauern bis heute an. Fichte, Forche und Tanne vermögen der Versaurung des Bodens keinen Einhalt zu tun und die Ortsteinbildung nimmt auch jetzt noch ihren Fortgang.*"

Diese Entwicklungen haben offensichtlich auch zu einer allmählichen Versauerung der Oberflächengewässer geführt. Der Chemismus vieler Gewässer im Nordschwarzwald wird heute deutlich durch Ionenaustausch- und Komplexierungsprozesse in den stark sauren Oberböden in Verbindung mit lateralem oberflächennahen Abfluß bestimmt (2, 3, 8). Der pH-Wert im Herrenwieser See muß nach der Diatomeen-Rekonstruktion (11, 12) mit Werten unter pH 5 natürlicherweise bereits sehr tief gewesen sein. Stellt man die pH-Wert-Geschichte (Abb. 1b) dem Pollendiagramm (Abb. 1a) gegenüber, so zeigt sich, daß der ab ca. 12 cm Tiefe deutlich erkennbare pH-Rückgang bereits über 100 Jahre vor der Industrialisierung begann. Die Versauerung des Herrenwieser Sees ist somit auf die nutzungsbedingte Versauerung des Einzugsgebiets und nicht auf den Eintrag anthropogener Luftschadstoffe zurückzuführen.

3.2. Feldsee

Der Anteil an mineralischem Detritus im Feldseesediment ist wesentlich höher als im Sediment des Herrenwieser Sees, was dem Seetypus als oligotrophem Klarwassersee entspricht. Das Sediment läßt sich als "Feindetritusmudde" klassifizieren und ist durch mehrere Fazieswechsel gekennzeichnet. Für das Feldseesediment ergibt sich eine deutlich höhere Sedimentakkumulation, weshalb der Kern nur die letzten ca. 400 Jahre umfaßt. Als durchschnittliche Sedimentationrate der obersten 20 cm lassen sich 1.9 $mm.a^{-1}$ angeben.

Das Pollendiagramm, das aus Platzgründen hier nicht widergegeben ist, spiegelt eine im Vergleich zum Herrenwieser See wesentlich geringere Veränderung in der Vegetationszusammensetzung wider. Die Fichte war aufgrund ihres östlich des Feldbergs natürlichen Vorkommens bereits schon früher in höheren Anteilen vertreten. Im Gegensatz zum Nordschwarzwald ist im Südschwarzwald der Anteil an Laubwald auch heute wesentlich höher. So finden wir im Einzugsgebiet des Feldsees einen naturnahen Bergahorn-Buchen-Mischwald. Im Südschwarzwald war die Waldnutzung in den vergangenen Jahrhunderten weit weniger intensiv als im Nordschwarzwald (19). Aufgrund der basenreicheren, lehmig verwitternden Böden aus Gneis haben sich auch kaum Podsole ausgebildet. Diese naturräumlichen und nutzungsgeschichtlichen Unterschiede dürften den Grund darstellen, warum sich der pH-Wert im Feldsee im Gegensatz zum Herrenwieser See in den letzten Jahrhunderten nicht geändert hat.

4. Zusammenfassung und Schlußfolgerungen

Die aktuellen tiefen pH-Werte vieler Gewässer im Nordschwarzwald können nicht das Ergebnis der atmogenen Säureeinträge der letzten wenigen Jahrzehnte sein, wie von STEINBERG et al. (11, 12) angenommen wurde. Wie die Rekonstruktion der Versauerungsgeschichte des Herrenwieser Sees deutlich zeigt, war der See mit pH-Werten unter 5 bereits vor mehreren Jahrhunderten natürlicherweise sauer. Die dokumentierte Versauerung begann bereits 1-2 Jahrhunderte vor Beginn der Industrialisierung infolge der damaligen, tiefgreifenden anthropogenen Eingriffe in die Waldökosysteme, die zu großflächigen Waldverwüstungen, Fichtenauffostungen und starken Bodendegradierungen (Vernässung, Versauerung) führten. Diese Auswirkungen werden durch die in den letzten Jahrzehnten gestiegenen atmogenen Stoffeinträge überlagert. Um die Auswirkungen der natürlichen, langfristig ablaufenden sowie der nutzungsbedingten Versauerung von den der möglichen Versauerung durch den Eintrag von Luftschadstoffen zu trennen, ist ein öko- und Hydrosphäre verbindender Ansatz erforderlich.

Daß der Feldsee im Vergleich zum Herrenwieser See nicht versauert ist, kann nicht mit unterschiedlichen aktuellen atmogenen Einträgen erklärt werden. Der Stoffeintrag mit dem Niederschlag ist in beiden Gebieten annähernd gleich und im Vergleich zu anderen Gebieten in Mitteleuropa eher gering (8). Vielmehr müssen hierzu Unterschiede in den Gewässereinzugsgebieten (Böden und ihre Ausgangsgesteine, Vegetationszusammensetzung, hydrologische Fließwege) sowie Unterschiede in der Nutzungsgeschichte herangezogen werden.

5. Danksagung

Das Forschungsprojekt "Versauerung und Schwermetalleintrag in Seen des Schwarzwalds" wurde gemeinsam von der Kommission der Europäischen Gemeinschaften und dem Land Baden-Württemberg durch das "Europäische Forschungszentrum für Maßnahmen der Luftreinhaltung (PEF)" am Kernforschungszentrum Karlsruhe gefördert. Besonders danken wir Herrn Prof. Dr. G. Lang (Systematisch-Geobotanisches Institut der Universität Bern) für die Betreuung der Pollenanalyse, Herrn G. Brahmer für die Mithilfe bei den oft schwierigen Geländearbeiten, sowie Herrn Dr. A. Mangini (Institut für Umweltphysik der Universität Heidelberg für die Durchführung der radiometrischen Datierungen.

Literatur

(1) ZöTTL, H.W.; FEGER, K.H. und BRAHMER, G.: 1985, KfK-PEF Berichte 2, 285-302.
(2) ZöTTL, H.W.; FEGER, K.H. und BRAHMER, G.: 1985, Naturwissenschaften 72, 268-270.
(3) FEGER, K.H. und BRAHMER, G.: 1986, Water, Air Soil Poll. 31, 257-265.
(4) SCHOEN, R. und KOHLER, A.: 1984, UBA Materialien 1/84, 58-59.
(5) ZüLLIG, H.: 1956, Schweiz. Z. Hydrol. 18, 5-143.
(6) BATTARBEE, R.W.; FLOWER, R.J.; STEVENSON, A.C. und RIPPEY, B.: 1985, Nature (London) 314, 350-352.
(7) AXELSON, V. und HAKANSON, L.: 1978, J. Sediment. Petrol. 48, 630-633.
(8) FEGER, K.H.: 1986, Freiburger Bodenkundl. Abh. 17, 1-253.
(9) ZEITVOGEL, W.: 1985, Diplomarbeit (unveröffentlicht) Institut für Bodenkunde und Waldernährungslehre, Universität Freiburg i.Br.
(10) ELSTER, H.J.: 1955-1962, Arch. Hydrobiol. Suppl. 22, 24, 25.
(11) STEINBERG, C.; ARZET, K. und KRAUSE-DELLIN, D.: 1984, Naturwissenschaften 71, 631-634.
(12) STEINBERG, C. und ARZET, K.: 1984, UBA Materialien 1/84, 136-147.
(13) TROELS-SMITH, J.: 1955, Danmarks Geologiske Undersøgelse IV/3, 10, 41-73.
(14) MERKT, J.; LüTTIG, G. und SCHNEEKLOTH, H.: 1971, Geol. Jb. 89, 607-623.
(15) HASEL, K.: 1944, Forschungen zur dt. Landeskunde 45, Leipzig.
(16) HAUSBURG, H.: 1967, Mitt. Ver. Forstl. Standortskunde und Forstpflanzenzüchtung 17, 3-22.
(17) MüNST, M.: 1910, Mitt. d. Geol. Abt. d. Königl. Statist. Landesamtes 8, Stuttgart.
(18) RIEK, G.: 1953, Neues Jb. Geol. Paläontol. Abh. 97, 402-462.
(19) BRüCKNER, J.: 1970, Dissertation Universität Freiburg i.Br.

THE POLLUTION CLIMATE OF THE FICHTELGEBIRGE, NORTHERN BAVARIA

J. FÖRSTER
Lehrstuhl für Hydrologie der Universität Bayreuth
Postfach 101251, D-8580 Bayreuth

Summary

The Fichtelgebirge, a forested upland region in Northern Bavaria (FRG), has been experiencing an intense forest decline in recent years. Therefore, characteristics of pollutant input (type, origin, pathways, variability) are of special interest for research in this area. The concentrations of the most abundant inorganic ions and some organic micropollutants (Polycyclic Aromatic Hydrocarbons (PAH) and Hexachlorocyclohexanes (HCH)) in precipitation and atmospheric particulate matter were measured over short sampling intervals. Additionally, data on the concentrations of gaseous pollutants (SO_2, NO, NO_2, O_3) could be obtained from a station on the forest site. Almost independent of duration and type of a storm event, the graphs of most of the pollutants' concentrations in rainwater form a U-shaped curve, starting with pH values as low as 3.4. Immission of PAH (tracers for anthropogenic incineration processes) and acidic anions (SO_4^{2-}, NO_3^-) shows a winter maximum. Values are especially high when long-lasting, stable inversion-layers reduce the atmospheric mixing height. Concentrations of most of the analyzed gaseous and particulate phase pollutants are higher when air masses are coming in from E (Czechoslovakia) or N (GDR), whereas the wet deposition pollutant load is mainly associated with advection from westerly directions.

1. INTRODUCTION

The Fichtelgebirge is a mainly forested upland region in Northeast Bavaria with altitudes between 500 and 1050 m above sea level (Fig. I).

The area has been experiencing an intense forest decline in recent years with the percentage of damaged trees (mainly fir and spruce) having reached more than 50 % (1).

As the impact of atmospheric pollutants is considered to be one of the main reasons for the dieback of forests in central Europe (2), the characteristics of pollutant input are of special interest for research in this area: type and variability of immission and deposition have to be known when the reactions of the ecosystem are to be investigated. Knowledge about the atmospheric pollutants' origin and pathways is of interest for effective pollution control strategies.

Fig. I: Location of Research Area

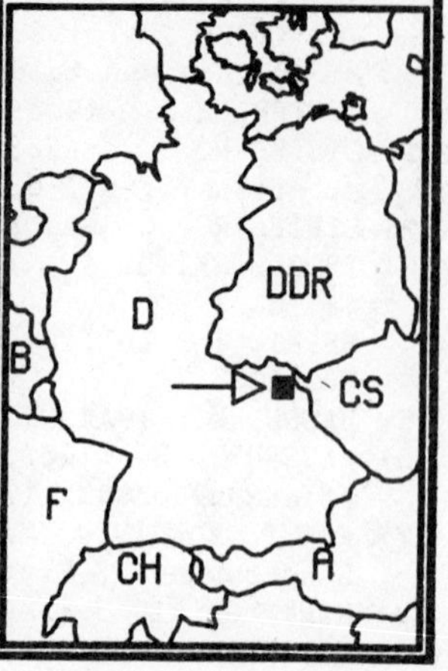

2. SET-UP OF THE INVESTIGATION

Precipitation water as well as particulate aerosols were analyzed for the most important inorganic ions (SO_4^{2-}, NO_3^-, Cl^-, NH_4^+, H_3O^+, Mg^{2+}, Ca^{2+} Na^+, K^+) and some organic micropollutants, polycyclic aromatic hydrocarbons (PAH) and hexachlorocyclohexanes (HCH). The PAH serve as tracers for pollutants from anthropogenic incineration; their main sources in central Europe are traffic (fossil-fuel fired engines), residential heating and steel production (smelters) (3). Abbreviations used in the Figures are as follows: Flu = Fluoranthene; BaP = Benzo(a)-pyrene;, IcdP = Indeno(cd)pyrene; BghiP = Benzo(ghi)perylene. From the group of the HCH, γ-HCH (Lindane) has been widely used as a pesticide.

The organic micropollutants cover a wide range of physico-chemical properties, hence allowing an estimation of the behaviour of other organic compounds of interest.

Mean pollutant concentrations over periods of weeks or months do not allow clear conclusions to be drawn on their relevance for phytotoxicity because peak values are leveled out. Therefore, it was the aim of this study to measure atmospheric aerosols and precipitation water over short sampling intervals.

3. EXPERIMENTAL

The two sampling sites, Wülfersreuth and Oberwarmensteinach (elevation 680 m and 750 m, respectively), are both situated on clearings inside forests.

Precipitation water was collected by parallel sets of totalisators. Glass funnels and bottles were used for sampling water for the organic analysis, polyethylene for the inorganic analyses. Athmospheric particulate matter was sampled on quartz fibre filters by means of a High-Volume Sampler. Sampling was performed discontinuously between July 1985 and April 1986.

Anions were analyzed by ion chromatography, cations by AAS and flame photometry; H_3O^+ was calculated from pH. HCH were quantified using gas chromatography, while PAH were analyzed by HPLC and HPTLC.

Pollutant gas concentration data were obtained from the Bavarian State Agency for Environmental Protection who run a measuring station directly on the Oberwarmensteinach forest site.

4. CHARACTERISTICS OF PRECIPITATION CHEMISTRY

Considering the relatively high annual precipitation of about 1200 mm in the Fichtelgebirge, it becomes evident that wet deposition is an important potential pathway for the input of pollutants into the terrestrial ecosystem. The sum parameters for the acid and ion content of precipitation, pH and conduc-

Fig.II: Substance Concentrations in Precipitation Water Wuelfersreuth Nov. 5.-6., 1985

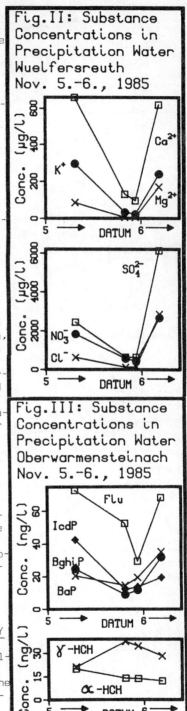

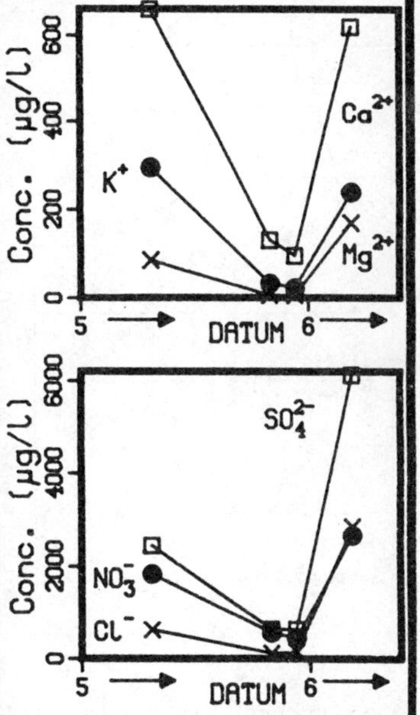

Fig.III: Substance Concentrations in Precipitation Water Oberwarmensteinach Nov. 5.-6., 1985

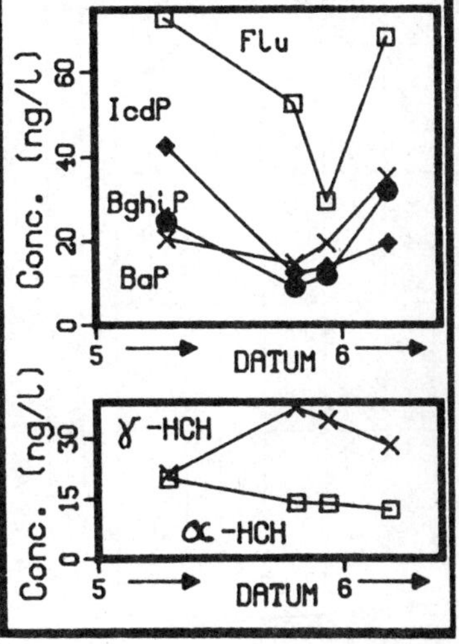

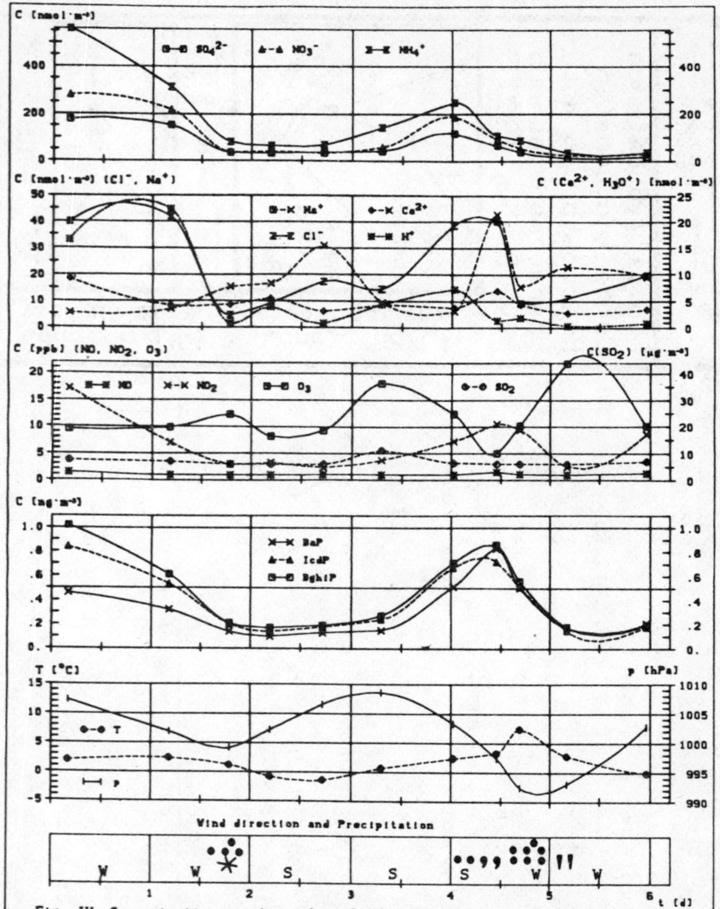

Fig. IV: Concentrations in Aerosol in Relation to Meteorological Conditions
Sampling Site and Time: Oberwarmensteinach, 31. Oct. - 7. Nov. 1985

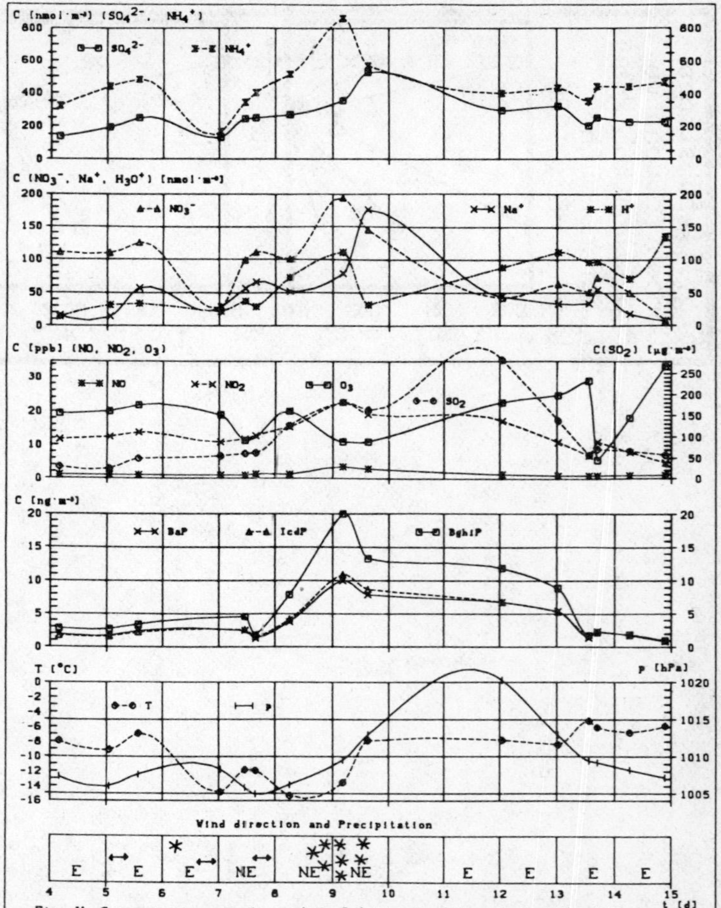

Fig. V: Concentrations in Aerosol in Relation to Meteorological Conditions
Sampling Site and Time: Oberwarmensteinach, 4. - 16. February 1986

tivity, vary between pH 3.4 and 6.0 with typical values of pH 4.0 - 4.5, whereas electrical conductivity readings had values between 6 and 121 µS cm^{-1}.

The precipitation water pollutographs of all substances investigated - except HCH - most typically form a U-shaped curve, for single showers as well as for cyclonic fronts passing through (Fig. II and III): starting with high pollutant concentrations, the values usually decrease until they reach a low constant level. Towards the end of the rainfall event, the concentrations increase again. The final values in most cases are lower than the initial values, but occasionally are higher. This final increase is observed less frequently than the decrease at the start; the latter occasionally is preceded by an additional increase.

The general type of pollutograph described is almost independent of the season and the duration of a single precipitation event. While high peak concentrations were recorded in all seasons, the trough in the middle of the event tends to be lower in summer than in winter.

HCH concentrations show a pattern different to all other substances investigated (Fig. III): They remain fairly constant during a single precipitation event but exhibit a clear seasonal differentiation with a maximum in early summer and a minimum in the cold season.

5. AEROSOL CHEMISTRY

Both aerodynamic resistance and specific surface area of forests are high, hence forest ecosystems are very susceptible to dry deposition which is difficult to be assessed exactly (4). However, concentrations in the air furnish a good representation of the aerosol pollution stress on a forest.

Figs. IV and V show examples for results from aerosol monitoring periods. The values (marker points) are centered to the middle of the sampling intervals which were of varying length according to the weather situation. The points have been connected to pollutographs using third-order polynomials. Fig. IV: The passage of a cold front from NNW (2.11.) and a Biskaya-Low's warm (5.11., morning) and cold fronts (5./6.11.) led to a clear decrease of acidic ion and PAH concentrations. The combined effects of air mass change and washout cannot be separated. Fig. V documents a mid-winter pollution climate with air mass advection from E and NE. On 9./10. Feb. 1986, a stable, low-lying inversion layer led to a smog episode; absolute maxima were recorded for most of the pollutants despite gas and particle scavenging by the falling snow.

Fig.VI: Mean Charge Proportions of single Ions

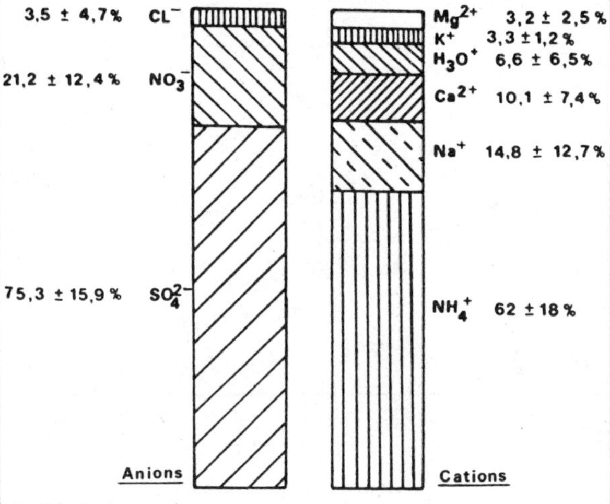

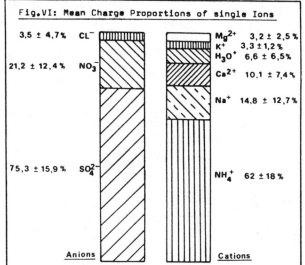

The inorganic ions analyzed in the project account for an average of 41 % of the total particle mass. The mean charge proportions (Fig. VI) indicate that SO_4^{2-} and NO_3^- are almost neutralized by soil-born cations and NH_4^+.

To calculate the directionally differentiated

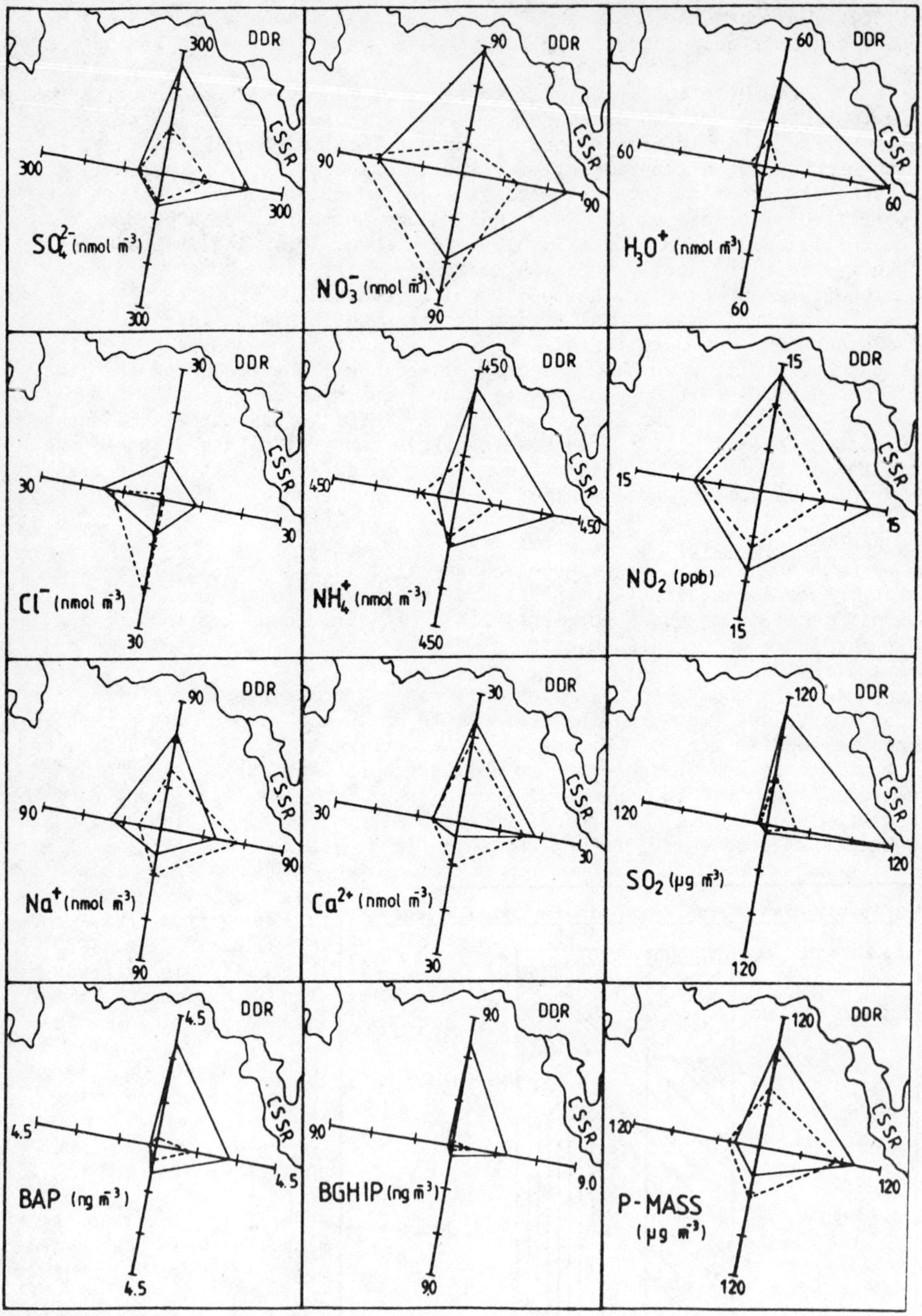

Fig. VII: Directionally Differentiated Concentration Means
---: Summer (T > 0°C); ——: Winter (T < 0°C)

concentration means (Fig. VII), each sampling interval had to be put into a 90°-directional class. The four sectors' approximate association with source regions for medium-range transport (up to 500 km) is as follows for ideally straight trajectories: 55° - 144°: Czechoslovakia; 145° - 234°: southern FRG; 235° - 324°: Central and W FRG including the Ruhr and Rhein-Main industrial regions; 325° - 54°: GDR. Three independent sources of information for the classification of each sampling interval were used: weather charts, radio-sonde readings and wind direction on the sampling site. For the substances that mainly originate from anthropogenic incineration (SO_4^{2-}, SO_2, PAH), the diagrams clearly show wintertime immissions to be much higher than summer values. Advection from N and E causes higher input than from S and W. Free acidity in atmospheric particles (H_3O^+) follows this trend although the second important acidic anion, NO_3^-, distinctly differs with summertime peak concentrations from W and S.

This differentiation was confirmed by principal component analyses of the aerosol data together with meteorological variables: SO_4^{2-}, NH_4^+, H_3O^+, the soil-born cations and the PAH all highly load the first principal component (PC) together with low temperature (-T), high pressure and wind direction parameters (N and E, but no W winds), whereas NO_3^- is grouped with Cl^- on a separate PC.

6. RESUME AND CONCLUSIONS

A: As precipitation events are frequent with westerly winds, the main proportion of pollutants' wet deposition comes from this direction, but there is no distinct directional difference for pollutant concentrations in precipitation water. Ion and PAH pollutographs are typically U-shaped.

B: The lower atmosphere's gas and particle phase pollutant load (the potential for dry deposition) is highest in winter and for advection from N and E.

C: There are no major emittants in FRG territory N and E of the Fichtelgebirge and the population density is low. Therefore, the main air pollution sources are presumably located in GDR and Czechoslovakia (steel and chemical industry, coal-fired power plants).

D: Air masses from W and S have a relatively higher proportion of traffic-born pollutants (e.g. NO_3^-) than those from E and N.

E: Stable wintertime inversion layers may enhance pollution and lead to severe smog episodes. This high potential for phytotoxicity should be considered when planning future botanical investigations on forest damage.

REFERENCES

(1) VDI-KOMMISSION REINHALTUNG DER LUFT (ed.) (1985). Die Waldschadenssituation in der Bundesrepublik Deutschland. VDI-Verlag, Düsseldorf.

(2) SCHOLZ, F. and LORENZ, M. (1984). Schadensursachen und Wirkungsmechanismen bei den Waldschäden. Allgemeine Forstzeitschrift 39/1984, 1275-1278.

(3) GRIMMER, G. (ed.) (1983). Environmental Carcinogens: Polycyclic Aromatic Hydrocarbons. Chemistry, Occurrence, Biochemistry, Carcinogenicity. CRC Press, Boca Raton/Florida.

(4) PRUPPACHER, H. R., SEMONIN, R. G. and SLINN, W. G. N. (eds.) (1983). Precipitation Scavenging, Dry Deposition, and Resuspension (2 Vols.) Elsevier, New York.

RELIEF EFFECTS ON THE DEPOSITION OF AIR POLLUTANTS IN FOREST STANDS - LEAD DEPOSITION AS AN EXAMPLE

V. GLAVAC, H. KOENIES, H. JOCHHEIM and R. HEIMERICH
University of Kassel

Summary

The stemflow influenced microhabitats beneath old beech trees can be used as an indicator for air pollution degree. The analysis of heavy metal accumulation rates in OH/Ah soil horizons of four forest districts in North Hessen (Baunsberg, Isthaberg, Langenberg and Hoher Meißner) allows the following conclusions:

- The immission of heavy metals rises with increasing altitude. This can be mainly attributed to dry deposition. Due to a greater air mass flux at higher sites a greater amount of air pollutants are deposed.
- In the investigated region the western slopes are more contaminated than the eastern slopes.
- The scattering of the measured values is mainly caused by the different soil humus contents. Nevertheless it is shown by partial correlation analysis that altitude and not humus content is the decisive factor for lead accumulation.

1. INTRODUCTION

The symptoms of forest damage can be observed very often at mountain ridges, hilltops and upper slope sites. In many regions of Central Europe this is a wide spread and well-known geoecological phenomenon. It could be assumed that mountain forests are exposed to higher pollutant input rates. Direct effects of ozone and atmospheric acids on the leaf surface would be more intensive and indirect influence of soil nutrient impoverishment would be stronger marked. Therefore an increase of air pollutant contamination following an altitudinal gradient could be expected.

The verification of this hypothesis is very difficult. The number of air pollution measurement stations is too small and the surveying periods are too short. In addition, there are no methods for reliable measurements of dry deposition available at present.

The manifold sink features and deposition processes in forest ecosystems cannot be reproduced artificially with technical equipment. Moreover field investigations have to overcome several difficulties because there are many factor groups that make a realistic measurement of deposition rates hardly feasable, such as:

- the permanent variation in the meteorological conditions
- the diversity of pollutant sources and their spatial distribution
- the varying atmospheric dispersion of air pollutants
- the long range air transport routes
- the different chemical and physical composition of air pollutants
- the distinct reversible and irreversible transformation of air pollutant compounds, and
- the versatile small space concentrations of air pollutants within a forest stand.

Due to these facts a direct approach of deposition measurement in a given area would require a dense and expensive network of measurement stations and long observation periods, otherwise results would be too uncertain.

The direct instrumental measurement of air pollutants can be avoided by means of snow cover analysis. During a longer frosty and dry weather period the surface of a permanent snow layer can be a suitable receptor. This indirect approach was chosen by some authors (14, 10, 7, 3). They found that the heavy metal content of the snow surface did rise with increasing altitude.

The third possibility to make this geoecological phenomenon evident offers the humus layer. The forest soil is a long-term sink for air pollutants, a kind of a 'ecosystem-memory'. It can be regarded as a gathering instrument, especially for trace elements. They are adsorbed, precipitated as insoluble compounds or chelated in the organic soil matrix. Their amount indicates the long-term atmospheric input. To show the correlation between the increase of heavy metal accumulation, as a consequence of enhanced input rates, and the altitude hitherto was attempted in Europe as well as in the USA (11, 12, 5, 6).

The aim of this report is to prove the effects of relief factors (altitude and aspect of sloping) on the rates of air pollutant deposition in forest ecosystems by the example of atmospheric lead input and its accumulation in forest soils.

The smooth-barked beech tree (Fagus sylvatica) leads 14% of its total crown precipitation to the ground as stemflow water, which is highly enriched with dry deposited air pollutants and leachates. Depending of the tree's crown projection area and the precipitation level 5 to 10 thousand l/yr run down the trunk to the foot of the tree. This way particular 'microhabitats' (4, 1, 2) are created beneath old trees, as a kind of 'atmospheric dustbin' caused by an enlarged mass flux (9, 8).

Soil samples (n= 480) were taken directly at the trunk footing area from the transit OH/Ah horizon (Fig. 1).

The 2N HCl-extract concentrations of lead, cadmium, copper, zinc, chrome and nickel were measured by atomic absorbtion. In addition pH(H_2O), pH(KCl) and humus content(%) were determined.

The sample plots (n= 52), in each case 10 microhabitats, were selected at different altitudes and aspects of sloping in north-hessian forest districts near Kassel (Baunsberg,

Isthaberg, Langenberg and Hoher Meißner) in the 'Mittelgebirgslandschaft' of North Hessen with the highest hilltops from ca 400 to 750 m.

2. RESULTS

In this paper only data on lead are represented. Mainly by automobile and industrial emission its aerosols are widely distributed and therefore lead suits for comparative research. The low mobility of lead causes its very long residence time in the organic soil horizons, namely ca 5,000 yr (13).

The effect of altitude on lead accumulation in old beech stands at the western slope of the Baunsberg is shown in Fig. 2.

The variation of lead accumulation along an altitudinal gradient at the Isthaberg is given in the Fig. 3 and 4. The greater pollutant load at the western side (luff), open to the industrial agglomerations of the Ruhrgebiet, is obvious. On the eastern side (lee) an irregularity is caused by air turbulences.

The difference between the western and the eastern slope of the Langenberg is shown in Fig. 5. In this case the samples were not collected at an elevation profile but randomized in the district. In this case the diverging aspects of sloping were classified into two main (west and east) directions.

Finally the correlation between altitude and Pb-accumulation for the western slopes of the Hohe Meißner is shown in Fig. 6. Analogous to Langenberg the sample plots are dispersed in the landscape. Therefore the correlation of the altitude/Pb-contamination is not so tight as by an elevation profile.

The means of measured lead concentrations in four mentioned districts are represented in Fig. 7.

Beneath orographic factors the soil humus content influences the extent of lead accumulation (Fig. 8).

Although the elevation differences of the investigated forest stands are relatively small (100 to 300 meters), the humus content rises with increasing altitude (except at the Baunsberg). Nevertheless the partial correlation analysis attest the decisive influence of a l t i t u d e on lead accumulation (without organic matter, west: r= 0.6690, east: r= 0.6510; without altitude, west: r= 0.0393, east:r=0.1608).

The effects of both environmental factors are made evident in Fig. 9 for western and in Fig. 10 for eastern slopes. In both cases staggered lead concentrations occur.

Considering the small altitudinal differences between lower and upper slopes more than 100% higher lead deposition on upper sites cannot be explained only by wet but to the greater part by higher dry deposition. It is probably the effect of greater air mass flux at wind exposed mountain sites (Fig. 11).

REFERENCES

(1) GERSPER, P.L. and HOLOWAYCHUK, N. (1970). Effects of stemflow water on a Miami soil under beech trees: 1. Morphological and physical properties. 2. Chemical properties. Soil Sci. Amer. Proc. 779-794.

(2) GERSPER, P.L. and HOLOWAYCHUK, N. (1971). Some effects of stemflow from forest canopy trees on chemical properties of soil. Ecology, 52, 691-702.

(3) GLATZEL, G., KAZDA, M. and LINDEBNER, L. (1986). Die Belastung von Buchenwaldökosystemen durch Schadstoffdeposition im Nahbereich städtischer Ballungsgebiete: Untersuchungen im Wienerwald. Düsseldorfer Geobot. Kolloq. 3, 15-32.

(4) GLAVAC, V., KRAUSE, A. and WOLF-STRAUB, R. (1970). Über die Verteilung der Hainsimse (Luzula luzuloides) im Stammabflußbereich der Buche im Siebengebirge bei Bonn. Schriftenr. f. Vegetationskunde, 5, 187-192.

(5) GLAVAC, V., KOENIES, H. and PRPIC, B. (1985). Zur Immissionsbelastung der industriefernen Buchen- und Buchen-Tannenwälder in den Dinarischen Gebirgen Nordwestjugoslawiens. Verhandlungen d. Ges. f. Ökologie, 15. Jahrestagung in Graz 1985 (in print).

(6) GLAVAC, V. (1986).Die Abhängigkeit der Schwermetalldeposition in Waldbeständen von der Höhenlage. Natur und Landschaft, 61, 43-47.

(7) GODT, J. and LUNKENBEIN, H. (1983). Höhenzonale Abhängigkeit der Schwermetallbelastung im Teutoburger Wald. Mitteilg. Dtsch. Bodenkdl. Gesellsch., 38, 203-208.

(8) JOCHHEIM, H. (1985). Der Einfluß des Stammablaufwassers auf den chemischen Bodenzustand und die Vegetationsdecke in Altbuchenbeständen verschiedener Waldgesellschaften. Berichte des Forschungszentrums Waldökosysteme/Waldsterben, 13, 226 p.

(9) KOENIES, H. (1982). Über die Eigenart der Mikrostandorte im Fußbereich der Altbuchen unter besonderer Berücksichtigung der Schwermetallgehalte in der organischen Auflage und im Oberboden. Diss. GhK, reprint: Berichte des Forschungszentrums Waldökosysteme/Waldsterben, 9, 289 p., 1985.

(10) KORFF, H.C., SCHRIMPFF, E., BRANDNER, I. and JANOCHA, F. (1980). Einfluß der Orographie auf die räumliche Verteilung von Schadstoffen im Steigerwald und im Fichtelgebirge. Bayreuther Geowissenschaftliche Beiträge über Oberfranken, 1, 39-56.

(11) REINERS, W.A., MARKS, R.H. and VITOUSEK, P.M. (1975). Heavy metals in subalpine and alpine soils of New Hampshire. Oikos, 26, 264-275.

(12) SICCAMA, Th.G. and SMITH, W.H. (1980). Changes in Lead, Zinc, Copper, Dry Weight, and Organic Matter Content of the Forest Floor of White Pine Stands in Central Massachusetts over 16 Years. Environmental Science and Technology, 14, 54-56.

(13) SMITH, W.H. (1984). Pollutant Uptake by Plants. In: TRESHOW, M. (ed.): Air Pollution and Plant Life. New York

(Wiley), 417-450.
(14) SCHRIMPF, E. (1980). Zur zeitlichen und räumlichen Belastung des Fichtelgebirges mit Spurenmetallen. Natur und Landschaft, 55, 460-462.

ACKNOWLEDGEMENTS

We would like to thank the Bundesministerium für Forschung und Technologie (BMFT) and the Projektleitung Biologie, Ökologie und Energie of the Kernforschungsanlage Jülich for financial support.

Fig. 1 Soil sample places

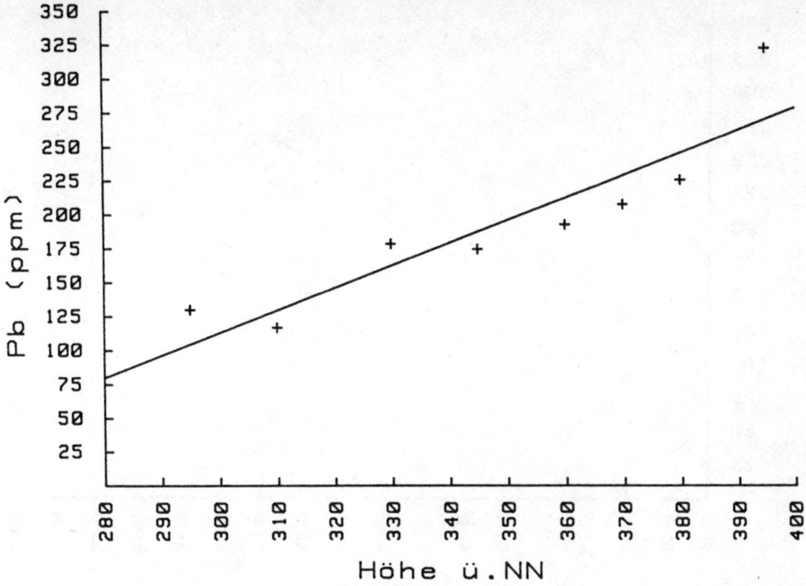

Fig. 2 Average rates of lead concentration (n= 5) in 2N HCl extracts from stemflow influenced OH/Ah horizons of 8 old beech stands at the western slope of the Baunsberg (r= 0.90456**) following an altitudinal gradient

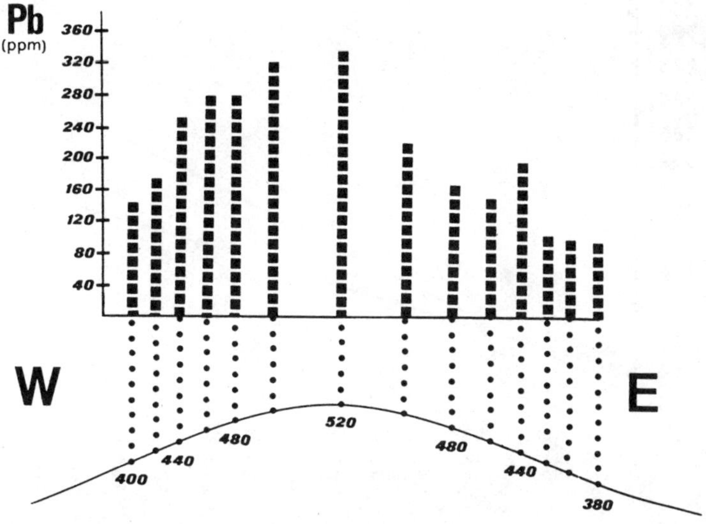

Fig. 3 Average rates of lead concentration (n= 10) in 2N HCl extracts from stemflow influenced OH/Ah horizons of 14 old beech stands at the Isthaberg along an altitudinal gradient

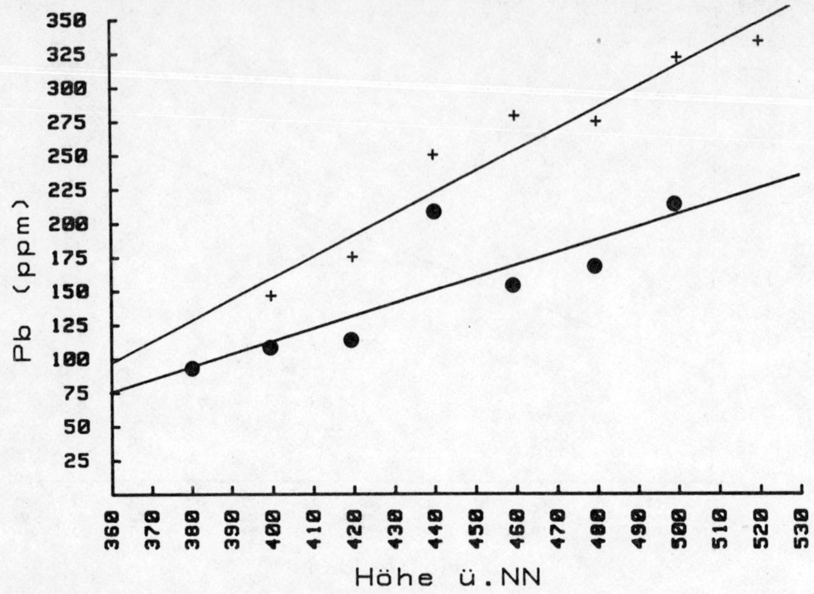

<u>Fig. 4</u> Average rates of lead concentration (n= 10) in 2N HCl extracts from stemflow influenced OH/Ah horizons of 7 old beech stands at the western (cross, r= 0.96295***) and eastern (point, r= 0.82486*) slope of the Isthaberg following an altitudinal gradient

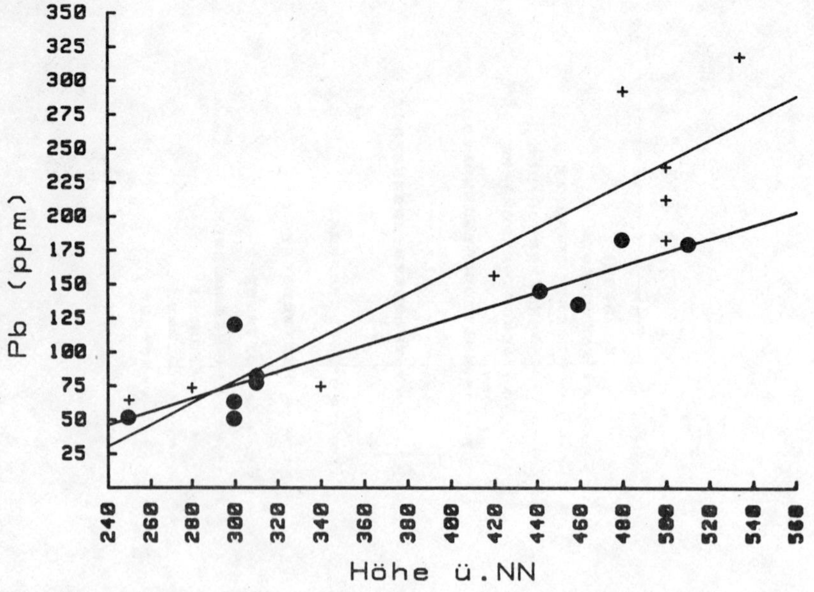

<u>Fig. 5</u> Average rates of lead concentration (n= 10) in 2N HCl extracts from stemflow influenced OH/Ah horizons of 10 old beech stands, at the western (cross, r= 0.91004***) and eastern (point, r= 0.92326***) slope of the Langenberg following an altitudinal gradient

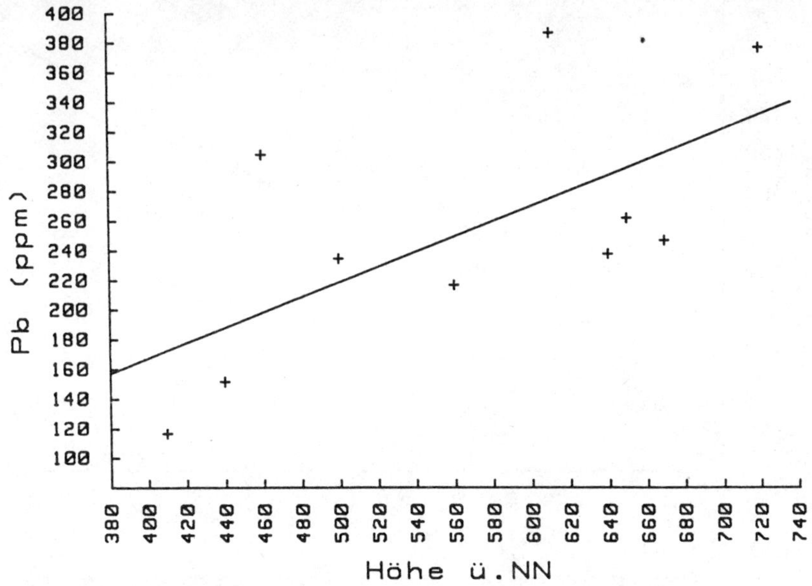

Fig. 6 Average rates of lead concentration (n= 10) in 2N HCl extracts from stemflow influenced OH/Ah horizons of 10 old beech stands at the western slope of the Hohe Meißner (r= 0.64398*)

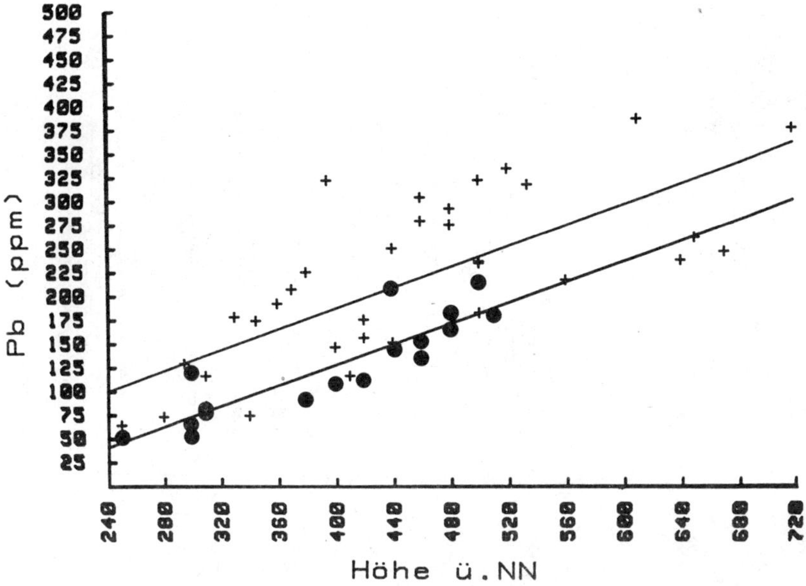

Fig. 7 Representation of the summerized results of all investigated sample plots. The increase of lead accumulation along the altitudinal gradient (west: cross, r= 0.72981***; east: point, r= 0.88250***)

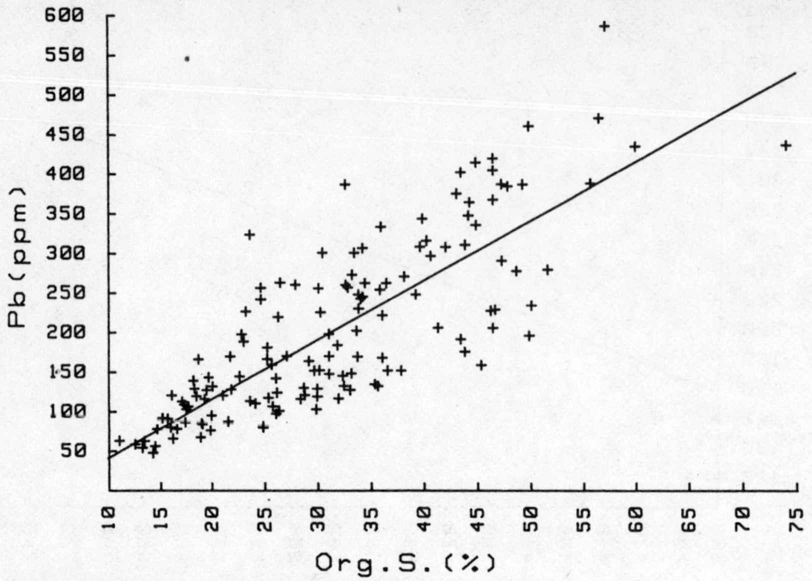

<u>Fig. 8</u> Dependence of lead concentration of soil humus content, western slope of the Isthaberg as an example (r= 0.80420***)

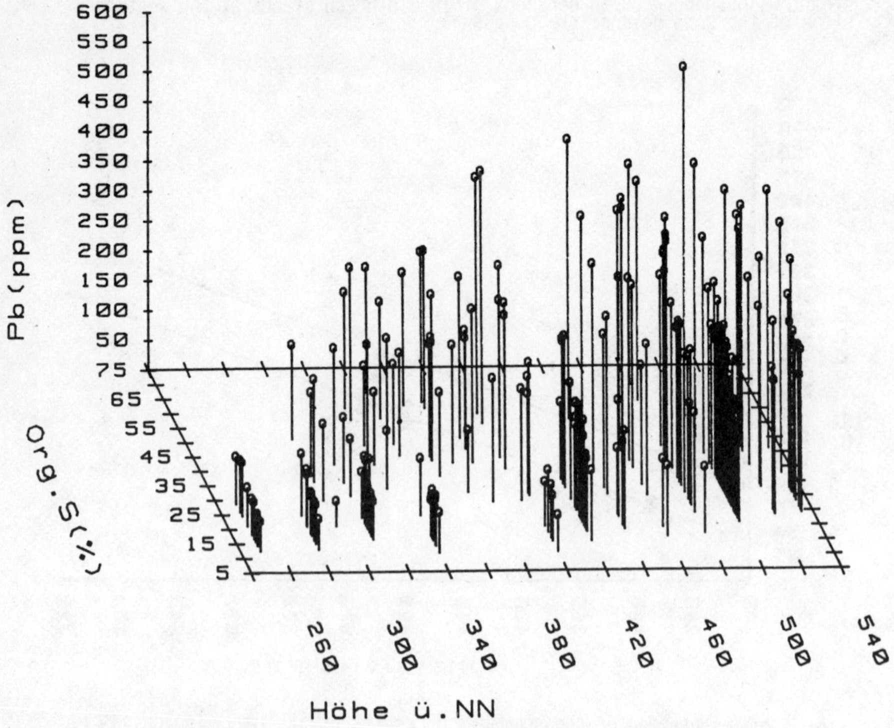

<u>Fig. 9</u> Lead concentration of (n= 210) stemflow influenced soils of 25 old beech stands at the western slopes of the Baunsberg, Isthaberg and Langenberg in dependence of altitude and humus content

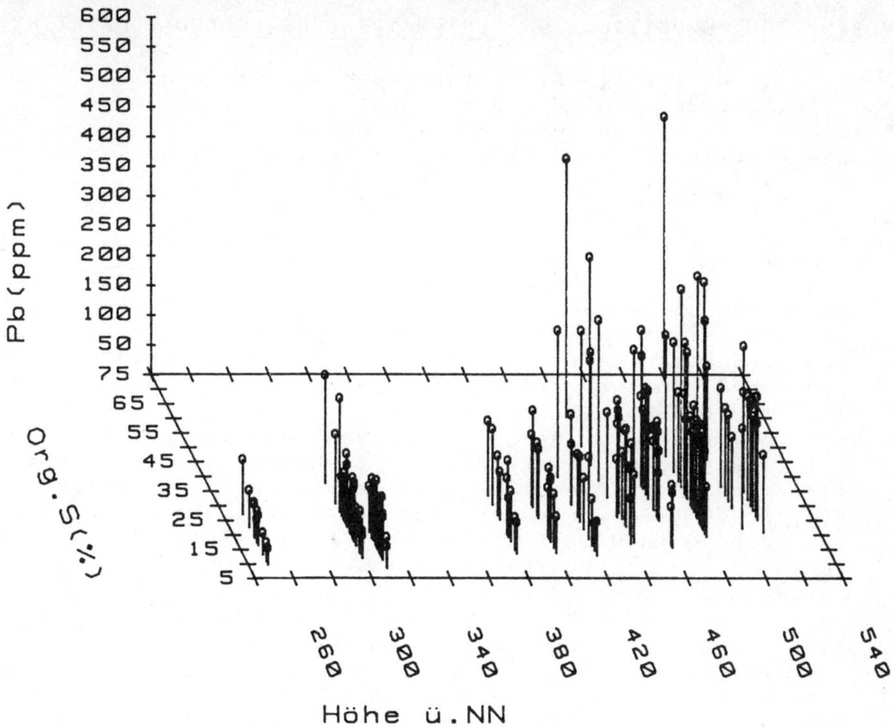

Fig. 10 Lead concentration of (n= 170) stemflow influenced soils of 17 old beech stands at the eastern slopes of the Isthaberg and Langenberg in dependence of altitude and humus content

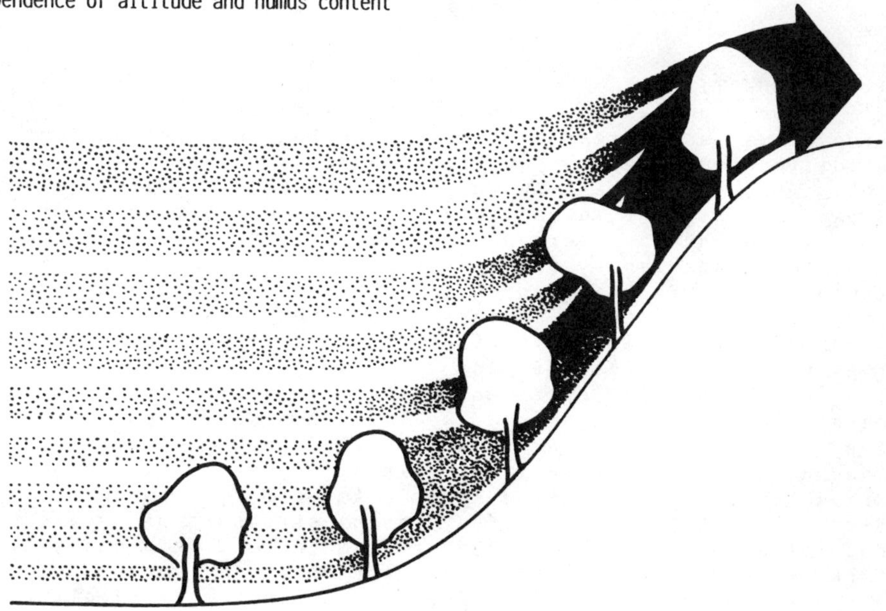

Fig. 11 Relief effects on air movement and dry deposition in forest stands, a possible explanation

ATMOSPHERIC DEPOSITION ON SEMI-NATURAL GRASSLAND VEGETATION

G.W. Heil, B. Heijne and D. van Dam
'Acid rain' ecology, Department of Plant Ecology
University of Utrecht
Lange Nieuwstraat 106, 3512 PN Utrecht, The Netherlands.

Summary

Vegetation is a considerable sink for atmospheric deposition. Analogous to throughfall and stemflow measurements in forests, deposition measurements were performed in two grasslands during a growing season. The two grasslands were situated adjacent to each other, and the canopy of one of the grasslands was cut once in July. In the uncut grassland, dry deposition of sulphate increased significantly as a result of the filtering effect of the canopy structure (LAI) of this vegetation (ca 0.8 m in height). Sulphate throughfall deposition amounted to approximately 5 times the bulk deposition during maximum standing crop, whilst ammonium throughfall deposition was significantly less than the bulk, although, a comparably higher amount of ammonium throughfall deposition could be expected. Probably, the low amount of ammonium throughfall deposition resulted from assimilation by the vegetation. The effect of atmospheric ammonium deposition on the species composition of the vegetation is discussed.

1. INTRODUCTION

Air pollution affects the species composition of ecosystems as a result of acidification and eutrophication. Acidification of e.g. aquatic ecosystems, as a result of air pollution, is a well-known phenomenon (1). Earlier studies showed a significant correlation between the amount of SO_2 deposition and the decrease of some terrestrial plant species in The Netherlands (2). Eutrophication as a result of atmospheric NH_3/NH_4^+ deposition. changes the plant species composition of (semi-)natural ecosystems, because it gives advantage to relatively fast growing plant species (3).

Characteristics of the deposition-surface (roughness length) have little effect on wet-deposition and on dry-deposition resulting from sedimentation. The sum of wet-deposition and sedimentation is called bulk deposition. The roughness length strongly determines, however, the amount of dry-deposition resulting from impaction, phoresis and diffussion (4). Thus, the canopy structure of vegetation determines to a large extent the amount of dry deposition. Experiments showed the co-deposition of SO_2 and NH_3. The opposite pH-dependent behaviour of these two air-pollutants mutually stimulates their deposition on (wet) surfaces (5). In this paper the relation between grassland canopy structure and amount of atmospheric deposition will be shown.

2. STUDY AREA

The study has been carried out in the nature reserve "Komgronden-reservaat" near Waardenburg, The Netherlands. The nature reserve is surrounded by agricultural areas with temporarily high emissions of NH_3 due to manuring. The grasslands of the nature reserve are not manured. Phytosociologically the grassland vegetation can be classified as Arrhenatherion elatioris (6). In the grassland two plots of 5 x 20 m^2 were used for deposition measurements. One plot was cut once during the growing season in the beginning of July.

3. METHODS

Analogous to throughfall and stemflow measurements in forests, a method for throughfall measurements has been employed, which produces deposition fluxes (7). For the measurements a system of half-open pipes with a capturing surface of 165 cm^2 were used, which could be placed within the canopy of the grassland without disturbance of its structure. The pipes were covered with polyethylene gauze, with mesh size of 0.02 mm, to prevent contamination with plantmaterial & dirt. The pipes were slanted to bottles to collect the samples. To keep the samples photochemically stable, the collecting bottles were placed in holes in the ground and covered with black plastic sheets. Bulk deposition was measured above the vegetation. All series were replicated five times. Throughfall and bulk deposition-samples were collected every fortnight from June to September 1986. The samples were colorimetrically analysed on SO_4^{2-} and NH_4^+ (Skalar auto-analyser). The canopy structure of the two grassland types was determined once a month using five replicates each sample time. The plants were clipped at soil surface. The vegetation samples were used for measurements of total Leaf Area Index (LAI), being the total surface area of standing leaves above the sample plot per unit ground area. Dead leaves on the soil surface that developed later in the growing season were excluded from LAI measurements. LAI was taken as an approximation of roughness length. The results were statistically analysed using ANOVA (8). The calculations spanned periods of 28 days.

4. RESULTS

Figure 1 shows the LAI-values of both grasslands. It is obvious that after cutting one plot LAI-values significantly differed. Both grasslands possess approximately the same LAI-value at the end of the growing season, as a result of regrowth in the cut plot and subsequently the simultaneous die off of the above-ground vegetation in both plots.

In Figure 2 the relationship between amount of SO_4^{2-} and NH_4^+ in the bulk precipitation is shown. The deposition of the two ions is significantly correlated with each other ($P < 0.05$). There is one relatively high datapoint of ammonium, however. This is caused during a manuring period in the surrounding agricultural areas. This correlation between the deposition of sulphate and ammonium seems to confirm other results regarding the co-deposition of the precursors of ammonium and sulphate, NH_3 and SO_2 respectively.

Fig. 1 LAI-values (m^2/m^2), being the total surface area of leaves standing above the sample plot.

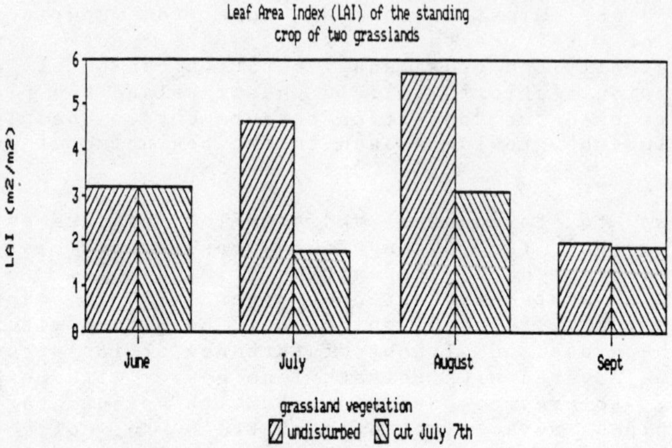

Fig. 2 Relationship between amount of sulphate and ammonium in the bulk precipitation.

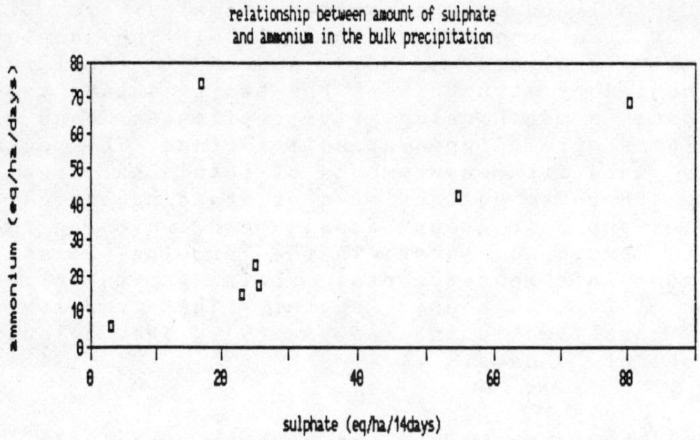

Figure 3 shows the amount of sulphate in the throughfall of the two grasslands and of the bulk. It is evident that the amount of sulphate in the throughfall of the grassland vegetation with the highest LAI-values is much higher than that in the bulk deposition ($P < 0.01$). At each measurement date the amount of deposition is significantly correlated with the the LAI-value of this vegetation ($P < 0.05$)(see also Fig 1). After the vegetation has partially died off, the amount of sulphate is the same in throughfall and in the bulk. The throughfall in the cut grassland shows no higher amount of sulphate (Fig. 3).

Fig. 3 Amount of sulphate (eq/ha) in throughfall and the bulk.

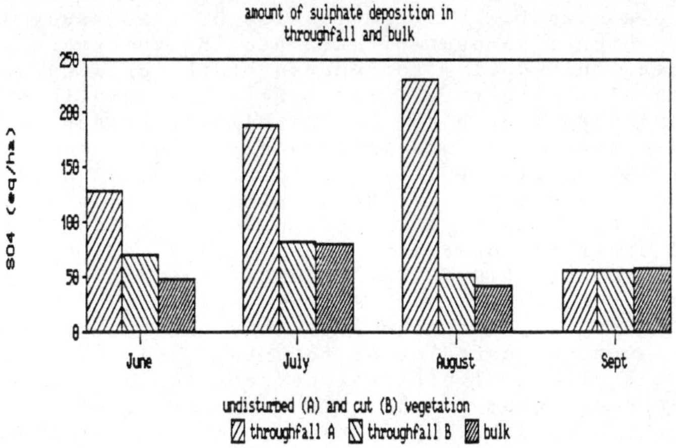

Contrary to the amount of sulphate throughfall, however, the amount of ammonium in the throughfall of both grasslands is less than in the bulk (Fig. 4). This can be ascribed to canopy-exchange i.e. assimilation of ammonium by green leaves of the vegetation canopy in exchange for other cations e.g. potassium.

Fig. 4 Amount of ammonium (eq/ha) in throughfall and bulk.

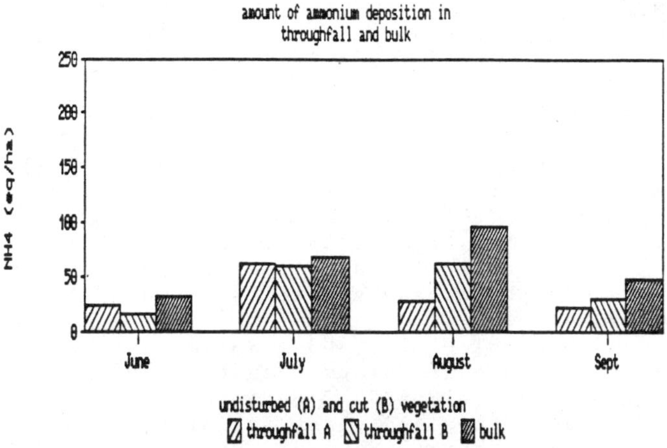

5. DISCUSSION

It is often assumed that the interceptive properties of grasslands are constant, but the results of this study show a significantly higher amount of sulphate in the throughfall relative to the bulk during the course of the growing season. Thus, one should be carefull for general assumptions with respect to roughness length and boundary layer transport.

Canopy-exchange processes influence the ionic composition of throughfall deposition (9). Laboratory experiments indicate that sulphate is not involved in assimilatory canopy-exchange processes (10). Contrary to sulphate, however, ammonium can be exchanged for other cations, such as H^+ and K^+, in the living above-ground biomass (10). Our results show no correlation between LAI and ammonium throughfall deposition as a result of the assimilation of ammonium by the plant parts. We demonstrated such a correlation for LAI and sulphate in throughfall and also a significant correlation between amount of sulphate and ammonium in the bulk deposition. This shows that the rates of deposition for sulphate and ammonium are equal. Consequently, estimates of the amount of canopy-exchange of ammonium can be deduced from measurements of sulphate and ammonium in the throughfall.

In grasslands, as in other ecosystems, changes in ammonium availability regulate species composition. Increased ammonium availability as a result of atmospheric deposition will result in changed competitive relations between the species (11), and this will lead ultimately to changes in species composition (3). The observed correlation between the amount of deposition of the precursor of sulphate and the decrease of some plant species in The Netherlands (2) may actually be the result of the co-deposition of the precursor of ammonium.

ACKNOWLEDGEMENTS

We are grateful to Wim de Mol for his help in field and laboratory.

REFERENCES

(1) OVERREIN L.N.,SEIP H.M. and TOLLAN A. (1981). Acid precipitation effects on forest and fish. Final report of the SNSF-project 1972-1980. Oslo.

(2) DAM D. VAN, DOBBEN H.F. VAN, BRAAK C.F.J. TER and WIT T. DE (1986). Air pollution as a possible cause for the decline of some phanerogamic species in The Netherlands. Vegetatio 65, 47-52.

(3) HEIL G.W. and DIEMONT W.H. (1983). Raised nutrient levels change heathland into grassland. Vegetatio 53, 113-120.

(4) HOSKER R.P. and LINDBERG S.E. (1982) Review: atmospheric deposition and plant assimilation of gasses and particles. Atmospheric Environment Vol. 16(5)889-910.

(5) ADEMA E.H., HEERES P. and HULSKOTTE J. (1986). On the dry deposition of NH_3, SO_2 and NO_2 on wet surfaces in a small scale windtunnel. Proceedings 7th World Clean Air Congress, Australia Sydney Vol.2, 12-18.

(6) WESTHOF V. and HELD A.J. DEN (1969).

Plantengemeenschappen in Nederland. Thieme-Zutphen.
(7) HEIL G.W. and DAM D.VAN (1986). Vegetation structures and their roughness lengths with respect to atmospheric deposition. Proceedings of the 7th World Clean Air Congress 1986. Sydney, Australia Vol. 5, 16-21.
(8) SOKAL R.R. and ROHLF F.J. (1979). Biometry:principles and practice of statistics in biological research.
(9) ULRICH B. (1983) Interaction of forest canopies with atmospheric constituents. In: B. Ulrich and J. Pankrath (eds.) Proceedings of a workshop held at Götingen, West Germany (1982). D. Reidel publishing company, Dordrecht.
(10) KLAASSEN R.K.W.M. (1987). Atmosferische depositie en canopy-exchange. Doctoral report, Department of Plant Ecology, University of Utrecht.
(11) HEIL G.W. and BRUGGINK M. (1987). Competition for nutrients between Calluna vulgaris (L.) Hull and Molinia caerulea (L.) Moench. Oecologia (Berlin) in press.

ATMOSPHERIC NITROGEN DEPOSITION IN A FOREST NEXT TO AN INTENSIVELY USED AGRICULTURAL AREA

W.P.M.F. IVENS, G.P.J. DRAAIJERS and W. BLEUTEN
Department of Physical Geography, State University of Utrecht, The Netherlands

Summary

Atmospheric deposition in a forest next to an intensively used agricultural area is measured by sampling throughfall, stemflow and open-field precipitation. Within 1 year 20 rain storms were sampled on an event basis. Concentrations in initial throughfall and dry and wet deposition fluxes to the soil of Ca, K, Mg, Na, NH4, CL, NO3, SO4 are given. Spatial variability op pH is shown. Concentrations in throughfall showed a large temporal variability. Maximum concentrations of NH4 and SO4 exceeded respectively 7 and 5 meq/l. Total nitrogen deposition on Douglas firs was for about 87% caused by dry deposition. Annual nitrogen fluxes to the forest floor were mainly in the form of NH4. Throughfall pH and stemflow pH showed an increase at sample sites next to ammonia sources, probably caused by neutralisation by ammonia.
Possible effects of the observed phenomena on the forest ecosystem are given.

1. Introduction

Besides natural sources, atmospheric nitrogen compounds originate from combustion engines (traffic), industry, livestock farms (pigs, dairy cattle, poultry) and the spreading of animal manure on arable land and pastures. The emission of nitrogen by agricultural activities mainly takes place in the form of ammonia. In the Netherlands high amounts of ammonia are emitted by agricultural sources. For some agricultural areas the ammonia emission has been estimated to be more than 200 kg/ha (cultivated area) per year (1). In consequence, a high atmospheric deposition of ammonia and ammonium occurs. The average contribution of ammonia to the total acid deposition in the Netherlands has been estimated on 32% (2). In areas next to farms or intensively manured land, this percentage probably will be higher. The deposition of NHx in a forest can cause the following effects:

- disturbance of the tree-physiology by excessive stomatal uptake of ammonia and/or ammonium (3);
- acidification of the soil by nitrification of ammonium (4);
- eutrophication of the soil and/or groundwater (5).

Besides these effects, ammonia can promote the deposition of other pollutants as SO2 due to its basic properties (6,7).

This paper deals with the atmospheric deposition in a forest next to an agricultural area with high ammonia emissions and its possible effects on the forest ecosystems.

2. Method

Trees act as effective sinks for atmospheric pollutants. Especially in perods without precipitation, trees will trap large amounts of gasses and aerosols in comparison to other surfaces due to their large foliage area. The amounts to be trapped in these sinks depend on atmospheric concentrations, tree morphology and physiology, and (micro) meterological conditions during the rainless period. The material accumulated on the tree surface during the rainless period will be washed off during rainfall. The water dripping from the canopy is called throughfall and the water running along the trunk is called stemflow. The composition of throughfall and stemflow strongly depends on the amount of dry deposition in the period preceding the rainstorm and to a lesser extent on the composition of the rain.

Using these phenomena, the atmospheric deposition in a forest was studied by sampling throughfall, stemflow and "open field" precipitation during rain storms. The sample sites were located in a forest east from a 30 km. wide agricultural area, with very high breeding-cattle densities ("Gelderse Vallei")(Fig. 1). On a line (5 km. long) from the forest edge into the forest 9 sample sites were installed in Douglas fir (Pseudotsuga menziesii) stands. At two locations the stemflow of Beech (Fagus Sylvatica) was sampled. "Open field" precipitation was collected simultaneously at locations close to the line of the forest sample sites. The age of the sampled trees ranges from 37 to 62 years. They all grow on poor, sandy soils.

Figure 1. Situation of the sample sites.

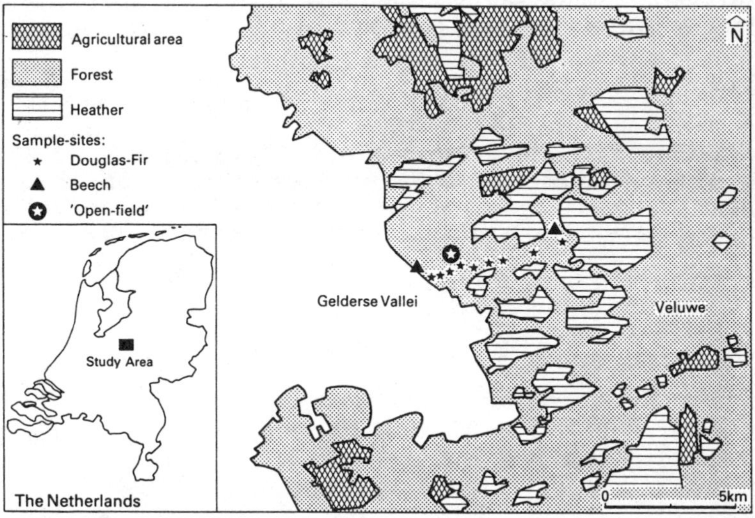

Throughfall was caught by narrow gutters (opening width = 0.01 m). The gutters were placed at 2 m above the forest floor. Gutters were exclusively placed under the canopy. No gutters were placed in open spots in the forest. To get a sample representative for the whole canopy-throughfall, the catch-area of the gutters was doubled towards the outside of the canopy. At each sample site two gutters with a joint catch-area of 0.094 m2 were placed perpendicular to each other under one tree. The endings of the two gutters were attached to the same reservoir. Stemflow was caught by a spiral cord around the trunk at breast height. The ends of the cord were attached to a reservoir. The precipitation in the "open-field" was caught by means of a funnel (0.040 m2) at 1.5 m above the ground.

To avoid contamination by e.g. insects, the water was filtered (mesh-width = 250 μm) before entering the reservoir. Only opaque reservoirs were used to avoid light penetration. Concerning the ions considered in this research, the used materials (principally polyethene) showed to be chemically inert. The funnels and reservoirs were cleaned with deionized water before each rainstorm. The gutters were cleaned once a month.

Within 36 hours after the ending of the rainstorm the samples were stored dark at 5 degrees C. The samples were analysed on acidity (pH) within 24 hours of sampling. Ion concentrations were measured within 7 days of sampling. Samples were analysed by colorometry for ammonium, sulphate, nitrate, chloride, magnesium and calcium; by flame-photometry for potassium and sodium. All chemical analyses were done at the Laboratory of Physical Geography of the State University of Utrecht.

To get a better understanding of the temporal variability of the dry deposition and the resulting concentrations in throughfall and stemflow, each rainstorm was sampled separately. Between October 1st 1985 and October 1st 1986 20 rainstorms were sampled. These storms accounted for 13% of the annual precipitation.

The water dripping from the canopy/trunk in the beginning of a rainstorm is called the initial throughfall/stemflow. The highest concentrations will occur in the initial water. Therefore the initial throughfall and stemflow was collected separately from the remainder. The initial volume was 275 ml, corresponding to 2.9 mm throughfall.

3. Throughfall concentrations

Table I shows the ion concentrations in initial throughfall and in "open field" precipitation. Average initial throughfall concentrations of ammonium, nitrate, sulphate, chloride and sodium are 8 to 11 times larger, potassium and magnesium concentrations more than 15 times larger than mean concentrations measured in rain water. Initial throughfall appears to be less acidic than rain water.

Table I Concentrations in initial throughfall and "open field" precipitation (μmol/l).

	initial throughfall			open field precipitation		
	mean	s.d.	max.	mean	s.d.	max.
CA^{2+}	166.3	123.0	1190	28.3	27.8	140
K^+	186.7	96.2	550	11.8	11.3	49
Mg^{2+}	106.1	84.9	390	6.1	10.6	43
Na^+	488.7	371.3	3330	40.9	24.8	104
NH_4^+	2338.1	1040.0	7110	198.9	138.7	756
Cl^-	606.8	431.8	4050	62.5	38.6	155
NO_3^-	659.4	454.5	2900	70.5	44.0	179
SO_4^{2-}	840.2	518.4	2590	69.6	48.4	190
H^+	6.4	7.8	50	36.5	52.1	251
pH	5.2	---	6.6	4.4	---	6.4

The high concentrations in throughfall will be principally due to dry deposition on the canopy in the dry period preceding the rainfall. The relatively high concentrations of potassium and magnesium in the initial throughfall suggest that leaching of these ions from the needles contributes substantially to the concentration in the throughfall. Leaching of these ions is also found at Picea abies (8).

The initial throughfall is strongly dominated by ammonium and to a lesser extent by sulphate. Concentrations of more than 3 meq/l ammonium and 2 meq/l sulphate are found several times, meaning that the needle surfaces have been exposed to at least the same concentration levels. Maximum concentration levels of these two ions are even substantially higher.

4. Fluxes to the forest floor

In table II the annual deposition fluxes to the forest soil directly under the tree is given. These fluxes are calculated from throughfall, stemflow and "open field" precipitation fluxes. In these calculations the effect of canopy exchange and of dry deposition directly on the forest floor are not taken into account. One should be aware of the fact that to some extent these figures will be larger than the average deposition fluxes for the whole forest, because the method applied did not include open spots in the forest.

Table II Contribution of the dry and wet deposition to the annual ionflux directly beneath the tree canopy. (kg/ha/year)

	dry deposition		wet deposition		total deposition	
	mean	s.d.	mean	s.d.	mean	s.d.
CA^{2+}	11.8	7.2	6.8	1.4	18.1	7.1
K^+	24.1	8.0	2.9	0.3	27.0	8.0
Mg^{2+}	5.5	2.6	1.1	0.2	6.6	2.6
Na^+	32.9	19.7	6.8	0.6	39.7	19.7
NH_4^+	143.4	46.6	18.3	0.2	161.7	46.6
Cl^-	51.9	30.1	22.6	0.3	74.5	30.1
NO_3^-	149.3	46.0	23.2	1.5	172.5	46.0
SO_4^{2-}	240.3	102.2	36.4	1.7	276.7	102.2
N-NH4	111.6	36.2	14.2	0.2	125.8	36.2
N-NO3	33.8	10.4	5.2	0.3	39.0	10.4
S-SO4	80.1	34.1	12.1	0.6	92.2	34.1

The total nitrogen input to the soil is 164.8 (± 37.7) kg/ha/year. It can be concluded that dry deposition is the main deposition process for the major compounds. About 72% of the total nitrogen input to the soil is due to dry deposition of nitrogen compounds on the tree. Ammonium counts for 76% of the total nitrogen input.

The maximum deposition fluxes of all ions (except for nitrate and potassium) occur at the sample site at the forest edge. Probably this is caused by an increased turbulency of the air at the forest edge and by a horizontal flux into the forest as additional to the vertical flux.

5. Acidity of throughfall, stemflow and rain

Acidity of stemflow of Douglas fir shows a remarkable spatial variation (Fig. 2). The pH of stemflow decreases significantly with increasing distance (x) from the ammonia source area (pH = 6.48 - 0.34 log(x); r = .93; p <.005). Stemflow of Beech shows a corresponding pattern. Previous research (9) in another area showed also higher pH values of Beech-stemflow at sites next to agricultural sources of ammonia. Higher stemflow-pH in the direct surroundings of intensive cattle breeding can be caused by higher deposition of ammonia at these sites. Ammonia will (partly) neutralize the acid-compounds in the water film on the trees.

Figure 2. pH of throughfall, stemflow and open field precipitation.

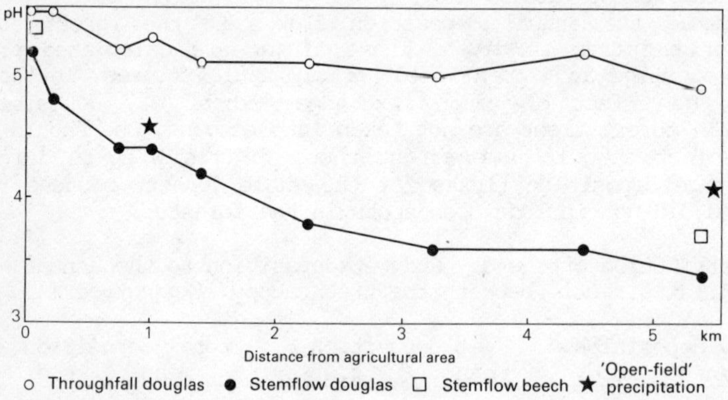

The pH of open field precipitation at the west side of the transect is slightly but significant (p<.01, one-tailed), higher than at the east side of the transect. This might be caused by some dry deposition of ammonia in the funnel.

Although less pronounced, throughfall of Douglas fir also shows a significant decrease of pH with increasing distance to the forest edge (pH = 5.92 - 0.11 log (x); r = .77; p <.005). Throughfall-pH is much higher than stemflow-pH. This might be caused by a larger proton-buffering in the canopy exchange processes.

6. Discussion

There appears to be a large temporal variability of accumulation of dry deposition on the tree canopy resulting in a large variability of throughfall concentrations. Therefore sampling on event basis or even sequential sampling within an event is recommended in assessing the direct impact of peak concentrations on trees.

Measured peak concentrations of ammonium and sulphate in initial throughfall exceed concentrations that effect the cuticula of needles (10). The measured initial concentrations are averages of 2.9 mm throughfall. It may be expected that in the water film, existing on the needles at the beginning of a rain storm, even higher concentrations will occur. This implicates that in the studied forest effect threshold values are exceeded frequently.

Measured ion fluxes to the soil may lead to high acidification rates and/or eutrophication of the nutrient poor forest soils. Deposition of ammonia compounds will play a very important role in forests ecosystems next to agricultural ammonia sources.

High ammonia deposition next to agricultural areas appears to increase the pH of water dripping from the trees during rainfall. It is feasible that the epiphytic lichen flora on these trees will be affected by this change in pH.

7. Acknowledgements

This research is partly financed by the Dutch Priority Programme on Acidification (Coordination: National Institute of Public Health and Environmental hygiene).

References

(1) BUIJSMAN, E; MAAS, H and ASMAN, W.A.H. (1984) Een gedetailleerde ammoniak emmissiekaart van Nederland. IMOU, V 84-20.

(2) ANONYMOUS (1986). Acidification in the Netherlands. Effects and policies. Ministry of Housing, Physical Planning and Environment.

(3) EERDEN, L.J.M. VAN DER (1982). Toxicity of ammonia to plant. Agriculture and Environment, 7, p. 223-235.

(4) BREEMEN, N. VAN; BURROUGH, P.A.; VELTHORST, E.J.; DOBBEN, H.F. VAN;WIT, T. DE; RIDDER, T.B. and REYNDERS, H.F.R. (1982). Soil acidification from atmospheric ammonium sulphate in forest canopy throughfall. Nature, vol. 299.

(5) ROELOFS, J.G.H.; CLASQUIN, L.G.M.; DRIESSEN, J.M.C. and KEMPERS, A.J. (1983). De gevolgen van zwavel- en stikstofhoudende neerslag op de vegetatie in heide en heidevenmilieus. In: Zure regen: oorzaken, effecten en beleid, E.H. ADEMA and J. VAN HAM (Red.), PUDOC, Wageningen.

(6) ADEMA, E.H. (1987). De betekenis van waterfilms op bladeren voor de droge depositie van NH_3 en SO_2. In: Acute en chronische effecten van NH_3 (en NH_4^+) op levende organismen: proceedings van de studiedag "Effecten van NH_3 op levende organismen" gehouden op 12 december 1986, Faculteit der Wiskunde en Natuurwetenschappen, Katholieke Universiteit, Nijmegen/red: A.W. BOXMAN en J.F.M. GELEEN.

(7) BUIJSMAN, E. (1982). Zure regen: atmosferische processen. IMOU, V 82-29.

(8) ULRICH, B (1983). Interaction of forest canopies with atmospheric constituents: S02, alkali and earth alkali cations and chloride. In: Effects of accumulation of air pollutants in forest ecosystems, 33-45, B. ULRICH and J. PANKRATH (Eds.); D. Reidel Publ. Co.

(9) IVENS, W.P.M.F.; DRAAIJERS, G.P.J. and BLEUTEN, W. (1986). Ruimtelijke en temporele variabiliteit van de atmosferische depositie in bossen. AD 1986-1, Vakgroep Fysisch Geografie, Rijksuniv. Utrecht.

(10) EERDEN, L.J.M. VAN DER and WIT, A.K.H. (1987). Effecten van NH_3 en NH_4 op planten en vegetaties; relevantie van effectwaarden. In: Acute en chronische effecten van NH_3 (en NH_4^+) op levende organismen: proceedings van de studiedag "Effecten van NH_3 op levende organismen" gehouden op 12 december 1986, Katholieke Universiteit, Nijmegen / red: A.W. BOXMAN en J.F.M. GELEEN.

PRODUCT ANALYSIS OF THE CHEMICAL/PHOTOCHEMICAL CONVERSION OF MONOTERPENES WITH AIRBORNE POLLUTANTS (O_3/NO_2)

K. JAY and L. STIEGLITZ
Kernforschungszentrum Karlsruhe

Summary

The monoterpenenes α-pinene, ß-pinene, camphene and limonene were ozonized in the gas phase and irradiated in air in the presence of NO_2. By ozonolysis following substances were identified:

with α-pinene: 1-acetyl-2,2,3-trimethylcyclobutane, norpinonaldehyde, verbenone, 1-hydroxypinocamphone, pinonaldehyde, 2,2-dimethyl-cyclobutyl-1-acetate-3-acetaldehyde, norpinonic acid, pinonic acid.

with ß-pinene: 6,6-dimethyl-[3.1.1]-bicyclohep-3-en-2-one, nopinone, 3-hydroxynopinone, 3-ketonopinone.

with camphene: camphenilone, cis and trans camphenylan aldehyde, 6,6-dimethyl-ε-caprolactone-2,5-methlyene.

with limonene: 1-acetyl-4-methylcyclohex-3-ene, 3-methylhepta-2,6-dione, ß-isopropenyl- δ -acetyl-n-valeric aldehyde, ß-isopropenyl-δ-acetyl-n-valeric acid, ß, δ -diacetyl-n-valeric acid.

At the photochemical reaction with NO_2 were detected:

with α -pinene: 1-isopropenyl-4-methylcyclohexadiene-2,4, pinol, α -pinene oxide, campholene aldehyde, cis and trans pinocamphone, 2-norbornanone-5,5,6-trimethyl, pinonaldehyde, norpinonaldehyde, 4,4,5-trimethylcyclopentan-1-one-3-acetaldehyde.

with ß-pinene: 1-isopropenylcyclohex-3-ene, nopinone, cis and trans myrtanal, bicyclo-[2.2.1]-heptane-7,7-dimethyl-1-carboxaldehyde.

with camphene: cis and trans camphenylan aldehyde, camphenilone.

with limonene: dihydrocarvone, 2-methyl-5-acetyl-cyclohexan-1-one, 2-(cyclohexan-5-yl-1-one-2-methyl)-proponal.

The bicyclic lactone resulting from gas phase ozonolysis of camphene suggests that dioxiranes may participate as reactive intermediates in agreement with the ozone/olefine reaction mechanism postulated by Dodge and Arnts in 1979. Dioxiranes are powerful oxgen donors. The may possess phytotoxic properties.

With α -and ß-pinene the main product of the NO_2/terpene reaction without irradiation is a nitroterpene of empirical formula $C_{10}H_{15}NO_2$.At NO_2 concentration down to 6 mg/m^3 products from the addition of atomic oxygen to the terpenes are predominant, while at the low NO_2 concentration of 600 µg/m^3 mainly the ozonolysis products were detected.

1. Einleitung

Die Vegetation gibt beträchtliche Mengen organischer Substanzen in die Atmosphäre ab. Die wichtigsten emittierten Stoffe sind Isopren und Monoterpene (1). Während für die Brutto-Umsetzungen von Terpenen mit Ozon und die photochemischen Reaktionen mit NO_x aus Laborversuchen Modelle vorliegen, die den Umsatz von Kohlenwasserstoff, O_3 und NO_x beschreiben lassen, war über die Struktur der entstehenden organischen Produkte bisher wenig bekannt (2,3). Ziel dieser Arbeit ist die Identifizierung der Umsetzungsprodukte von Terpenen und Schadgasen (O_3/NO_2) als Beitrag zu der Frage, inwieweit diese Produkte eine Rolle bei der Schädigung von Waldökosystemen spielen können (4). Die primären Umsetzungsprodukte von α -Pinen, ß-Pinen, Camphen und Limonen mit Ozon und die Photolyseprodukte dieser Terpene in Luft/NO_2 wurden größtenteils identifiziert und die Struktur dieser Substanzen durch unabhängige Synthese bestätigt (synthetisierte Substanzen sind auf den Abbildungen ohne Klammern wiedergegeben).

2. Experimentelle Anordnung

Die Photolyse und die Ozonolyse wurde in einem 9,8 l fassenden, thermostatisier- und evakuierbaren Glaskolben durchgeführt. Als Lichtquelle wurde eine durch Duranglas gefilterte 2500 Watt Xenonhochdrucklampe verwendet. Die Reaktionsprodukte wurden durch Umpumpen des Reaktorvolumens mit einer Metallbalgenpumpe auf Mikroaktivkohlefilter gesammelt und mit Hilfe der Kapillargaschromatographie-Massenspektrometrie analysiert.

3. Ergebnisse

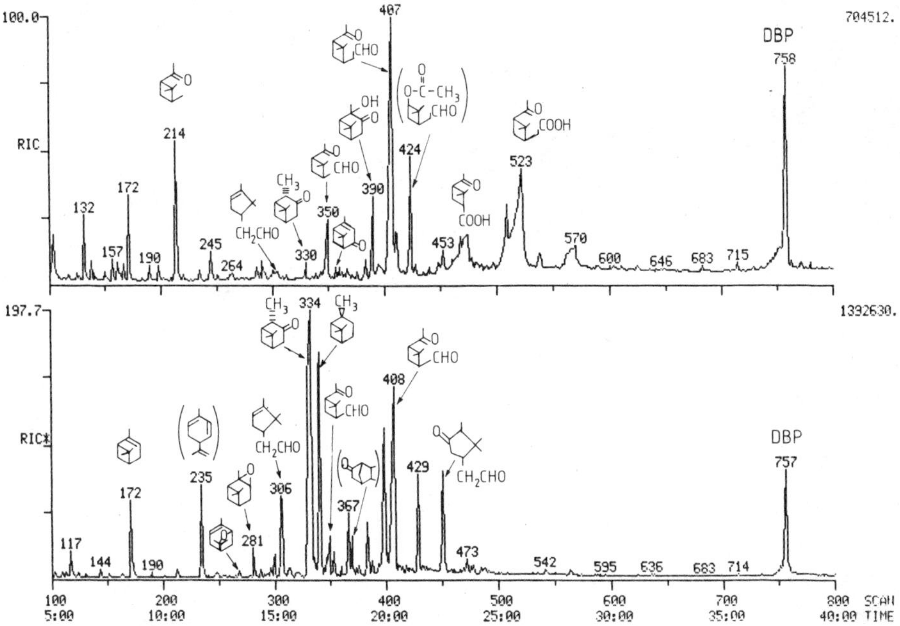

Abb. 1: Umsetzung von α-Pinen (20 vpm) in der Gasphase (Luft) oben: Dunkelreaktion mit Ozon, unten: Photolyse mit NO_2 (50 vpm)

Bei der Ozonolyse von α-Pinen (Abb. 1) wird hauptsächlich Pinonaldehyd erhalten, demgegenüber entstehen bei der photochemischen Umsetzung mit NO_2 (50 vpm) in erster Linie Addukte von $O(^3P)$. Bei NO_2 Konzentrationen kleiner 0.3 vpm (= 600 mg/m^3) überwiegen die Ozonolyseprodukte (Abb. 2)

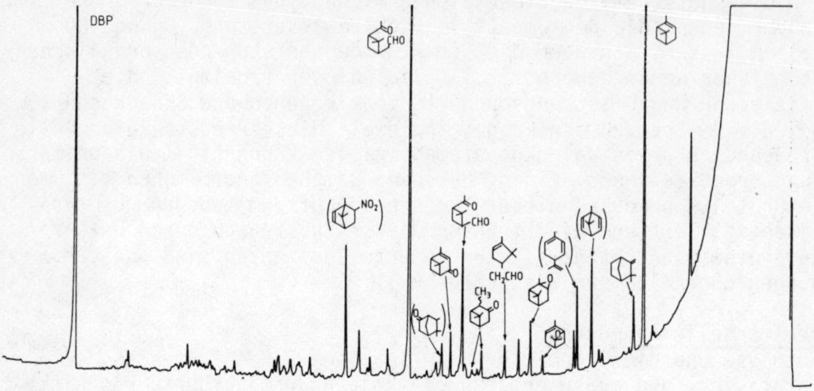

Abb. 2: Gaschromatogramm der Umsetzungsprodukte von α-Pinen (0.02 vpm) mit NO_2 (0.03 vpm)hν:30'.

Die Gaschromatogramme der Ozonolyse und der photochemischen Umsetzung von β-Pinen, Limonen und Camphen sind in Abb. 3-5 wiedergegeben. Diese Terpene unterscheiden sich in ihrem Verhalten prinzipiell nicht von α-Pinen.

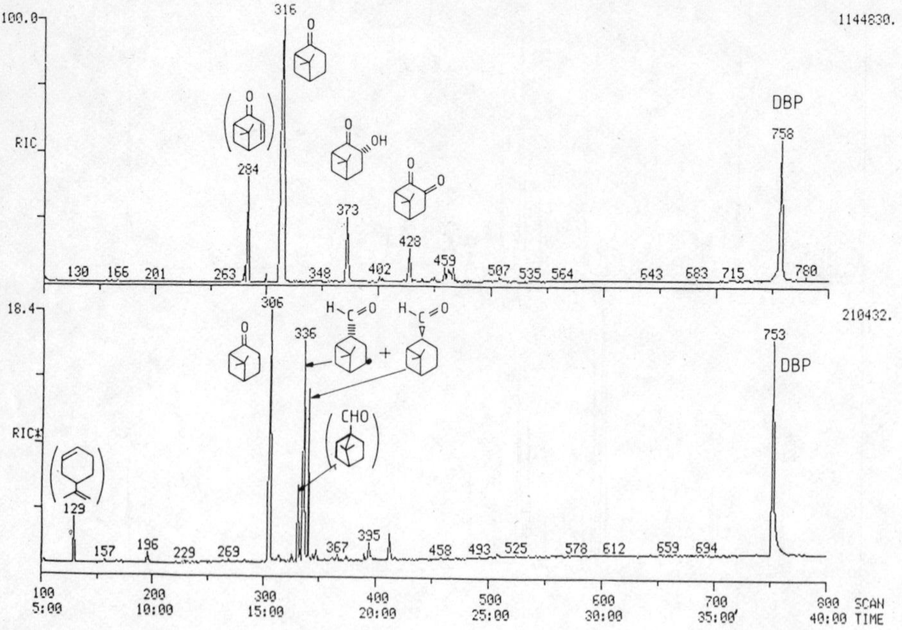

Abb. 3: Umsetzungsprodukte von β-Pinen (20 vpm) in der Gasphase (Luft) oben: Dunkelreaktion mit Ozon, unten: Photolyse mit NO_2 (50 vpm)

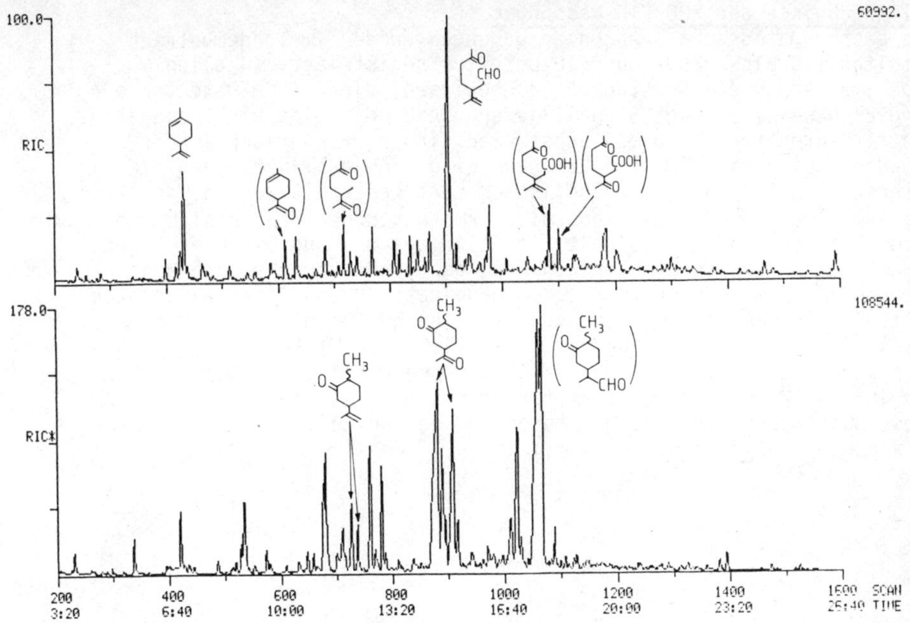

Abb. 4: Umsetzungsprodukte von Limonen (20 vpm) in der Gasphase (Luft) oben: Dunkelreaktion mit Ozon, unten: Photolyse mit NO_2 (50 vpm)

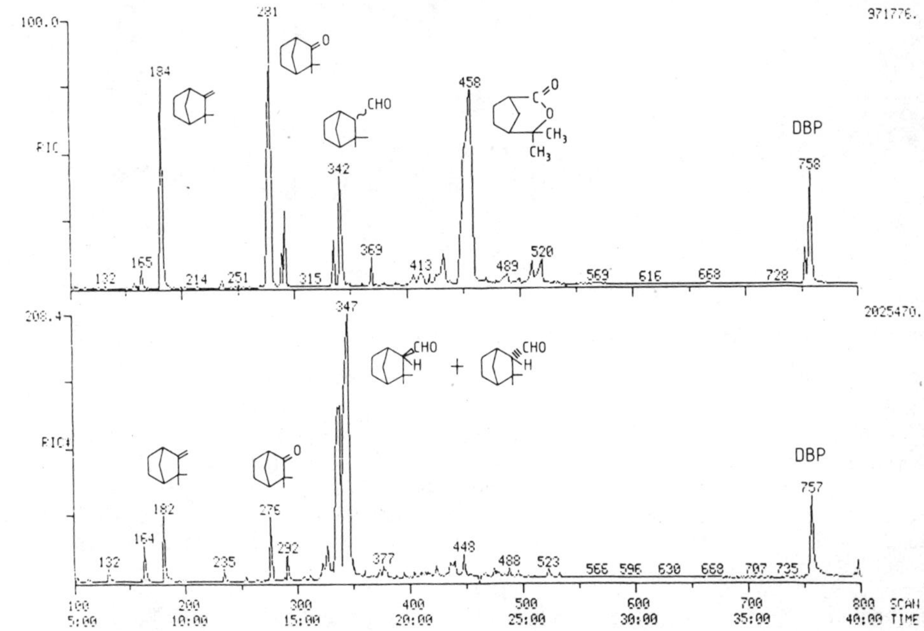

Abb. 5: Umsetzungsprodukte von Camphen (20 vpm) in der Gasphase (Luft) oben: Dunkelreaktion mit Ozon, unten: Photolyse mit NO_2 (50 vpm)

3.1 Ergebnisse der ECD-FID Messungen

Die gaschromatographischen Untersuchungen wurden im Tandembetrieb simultan mit einem ECD- und FID-Detektor registriert. In allen Experimenten wurden Substanzen nachgewiesen, die im ECD-Detektor ein vergleichsweise starkes Signal liefern (Abb. 6), also elektronegative Substituenten tragen. Diese Substanzen sind photochemisch leicht abbaubar. Ein Teil dieser Substanzen wird auch gebildet, wenn die untersuchten Terpene in synthetischer Luft bestrahlt werden, der kein NO_2 zugemischt wurde. Es liegt die Vermutung nahe, daß es sich dabei um Autooxidationsprodukte handelt. Im Falle des α - und ß-Pinens zeigen massenspektroskopische Untersuchungen des größten Peaks, daß Mononitroterpene der Summenformel $C_{10}H_{15}NO_2$ gebildet wurden. Diese Substanzen entstehen entweder durch eine Addition von NO_2 an die Doppelbindung, gefolgt von einer Umlagerung mit anschließender Wasserstoffabspaltung oder aber duch H-Abstraktion aus dem Terpen gefolgt von NO_2 Addition. Der in Abb. 2 und 6 gemachte Strukturvorschlag ist noch hypothetisch. Diese Substanzen sind die Hauptprodukte der Dunkelreaktion von α-bzw. ß- Pinen mit NO_2/synth. Luft.

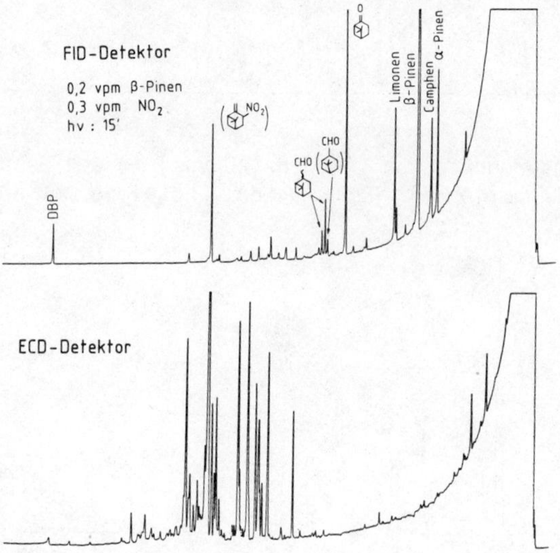

Abb. 6: Gaschromatogramme der Umsetzungsprodukte von ß-Pinen mit NO_2/hv

3.2 Ozonolyse von Camphen

Die Ozonolyse von Camphen in Lösung ist eingehend untersucht (5). Unter gewöhnlichen Ozonolysebedingungen entsteht hauptsächlich das Lacton wenn in Chloroform ozonisiert wird und Camphenilon bei der Ozonolyse in Methanol. Beide Substanzen werden auch bei der Gasphasenozonolyse als Hauptprodukte erhalten. Ihre Struktur wurde durch unabhängige Synthese bestätigt.

Während in Lösung die Umlagerung über das Sekundärozonid erfolgt, entsteht das Lacton bei der Gasphasenozonolyse von Camphen wahrscheinlich über ein Dioxiran, in Übereinstimmung mit dem von Dodge und Arnts 1979 postulierten Mechanismus der Ozon/Olefin Reaktion (6).

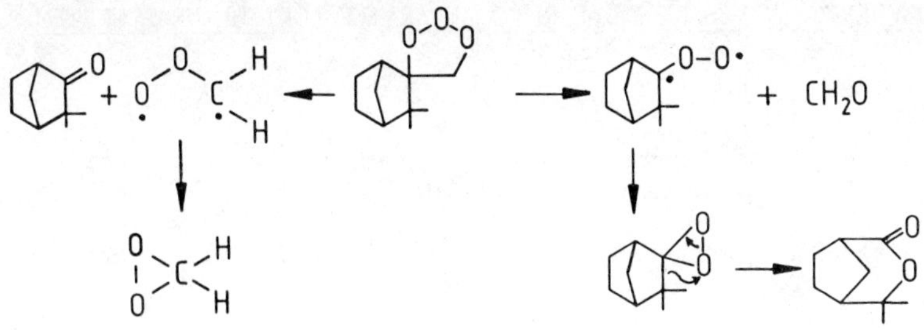

Der Grundkörper wurde von Lovas und Suenram 1977 als Produkt der Tieftemperaturreaktion von Ozon mit Ethylen identifiziert (7). Dimethyldioxiran ist in Lösung beständig und Gegenstand eingehender Untersuchungen (8).

Die Bildung des Lactons bei der Gasphasenozonolyse von Camphen zeigt, daß Dioxiranen auch bei Raumtemperatur in der Gasphase eine reale Existenz zukommen solte, auch wenn es sehr reaktive und daher kurzlebige Substanzen sind.

Während Dioxirane relativ unbekannte Substanzen sind, zeigt die Chemie dieser Verbindungsklasse, daß sie sehr starke Oxidationsmittel sind, im besonderen sind sie starke Sauerstoffdonatoren (8).

In diesem Zusammenhang erwähnenswert sind die Arbeiten von Arnold (9) und Darley et al. (10) aus dem Jahre 1959, die bei der Gasphasen-Ozonolyse von Olefinen kurzlebige phytotoxische Substanzen nachwiesen und sie als Zwitterion bzw. Olefin-Ozon Komplex beschreiben. Bei diesen Substanzen könnte es sich um die entsprechenden Dioxirane gehandelt haben, deren Existenz zu jener Zeit noch unbekannt war.

Es bleibt zu untersuchen, ob Dioxirane bei der Chemie der Luftverunreinigungen eine Rolle spielen und einen Beitrag zur Schädigung von Waldökosystemen leisten.

Literatur

(1.) T. E. Graedel, Rev. Geophys. Space Phys. 17, 937(1979)
(2.) L. A. Hull, Atmos. Biog. Hydrocarbons, Vol. 2, p. 161-186 (1981)
(3.) W. E. Schwartz, P. W. Jones, C.I.Riggle, D. F. Miller Report EPA - 650/3-74-011, August 1974
(4.) G. H. Kohlmaier et al., Allgem. Forst- und Jagdzeitschrift, Aug. 1983, Sonderband Waldsterben III
(5.) Ph. S. Bailey, Ozonation in Org. Chem. Academic Press New York 1978, p. 170
(6.) M. C. Dodge, R. R. Arnts, Int. J. of Chem. Kin. 11, 399 (1979)
(7.) F. J. Lovas, R. D. Suenram, Chem. Phys. Lett. 51, 453 (1977) J. Am. Chem. Soc. 100, 5117 (1978)
(8.) R. W. Murray, R. Jeyaraman J. Org. Chem. 50, 2847 (1985) J. Am. Chem. Soc, 108, 2470 (1986)
(9.) W. N. Arnold, Int. J. Air Poll. 2,167 (1959)
(10.) E. F. Darley, E. R. Stephens, J. T. Middelton, Ph. L. Hanst, Int. J. Air Poll. 1,155 (1959)

ATMOSPHERIC INPUTS OF CADMIUM TO AN ARABLE AGRICULTURAL SYSTEM

K.C. JONES[1], C.J. SYMON[1] and A.E. JOHNSTON[2]

[1]Department of Environmental Science, University of Lancaster, Lancaster LA1 4YQ, UK.

[2]Soils and Plant Nutrition Department, Rothamsted Experimental Station, Harpenden, Hertfordshire AL5 2JQ, UK

Summary

Soils and wheat grain samples collected and stored since the mid/late-1800s to the present have been analysed recently for Cd. The samples from the Broadbalk continuous wheat experiment at Rothamsted Experimental Station (U.K.) were selected to quantify changes in the soil Cd content due to atmospheric deposition. Soil plough layer Cd increased by ∿50%. The average increase was 1.9 μg Cd kg^{-1} $year^{-1}$, which is equivalent to 5.4 g Cd ha^{-1} $year^{-1}$. However, the data indicate an accelerated rate of increase in soil Cd in more recent decades. The Cd concentration in the wheat grain does not give such a clear temporal picture. It is concluded that any concomitant rise with time will be masked by (i) seasonal fluctuations in Cd uptake, (ii) additional variability introduced by growing several different cultivars over the years, and (iii) changes in soil properties, which may modify Cd uptake.

1. INTRODUCTION

Agricultural soils receive inputs of Cd from soil amendments, such as phosphate fertilisers and sewage sludge, and from atmospheric deposition. Currently there is concern in many countries that increases in the soil Cd burden may increase the risk of adverse effects on human health. This is because there is a positive relationship between soil and plant Cd concentrations (although this clearly differs with soil type and crop species), and because for the general population plant-based foodstuffs provide the major source of dietary Cd (1,2). Consequently, the average daily intake of Cd (which is currently about 20-40 μg day^{-1} in Europe)(3) may, in the long term, approach the World Health Organisation's suggested tolerable intake of ∿70 μg day^{-1} for an adult (4). The European Commission has recently acted to restrict the soil-crop plant-man movement of Cd by setting a limit of 3 mg Cd kg^{-1} soil for arable land receiving applications of sewage sludge (5). Additional action is being considered to limit the amount of Cd added to soils in phosphate fertilisers as part of a wider ranging strategy aimed at reducing the dispersal of Cd into the environment.

To date attention has focussed on quantifying the increases in soil Cd through inputs of sewage sludge and phosphate fertilisers (2,6,7). However, on the basis of a mass-balance approach adopted to quantify inputs and outputs of Cd to Danish soils, Tjell et al. (8) predicted that atmospheric inputs may also make a substantial contribution to a projected national increase in the soil plough layer Cd burden, of

~0.6% per annum. Indeed, a recent soil inventory for the UK indicated atmospheric deposition and phosphate fertilisers to be the major sources of Cd to agricultural land nationally (9).

This paper, together with the more detailed report elsewhere (7), gives what is probably the most comprehensive investigation to date of temporal trends in soil Cd. This has been made possible by retrospective analysis of a set of archived soils collected from long-term experimental plots at Rothamsted Experimental Station in the U.K. Samples have been collected and stored periodically since the beginning of the Classical Experiments in the 1840/50s. The control plots have only received Cd inputs from the atmosphere, whilst others have received known amounts of fertilizers, which has enabled the investigation of time-trends resulting from phosphate additions (6,7). This paper deals specifically with atmospheric inputs of Cd to a continuous arable cropping system with winter wheat, the Broadbalk Experiment at Rothamsted. In addition to reporting temporal trends in the Cd content of the soil plough layer (0-23 cm), samples of wheat grain from the experiment have also been analysed for Cd and these results are reported and briefly discussed.

2. STUDY LOCATION AND EXPERIMENTAL METHODS

Rothamsted (grid reference TL 120137) is a 'semi-rural' location in western Europe 42 km north of central London (7). The surface soil of Broadbalk is a neutral or slightly calcareous silty clay loam and contains ca. 25% clay and 1.0 - 2.6% C. The collection, storage and analysis of the soils has been described previously (7); soils were sampled on eleven separate occasions between 1846 and 1980; wheat (*Triticum aestivum* L.) grain has been harvested and stored annually. For this study grain from the years 1877-1881, 1922-1926, 1936-1940, 1968-1970 and 1979-1984 were subsampled and pooled (ca. 1 kg) and analysed by neutron activation analysis. Several wheat varieties have been grown: Red Rostock, 1877-81; Red Standard, 1922-26 and 1936-40; Cappelle Desprez, 1968-70; Flanders 1979-84. Following irradiation (0.5 g powdered sample for 7 h at 100 kW) the samples were dissolved in H_2SO_4/HNO_3 and CdS precipitated. The CdS was then washed, dissolved and Cd ammonium phosphate precipitated and weighed. Samples and reference materials (NBS 1567 - wheat flour and 1568 - rice flour) were counted for 10000 - 12000 s and the 527 KeV γ peak of ^{115}Cd and the 336 KeV γ peak of ^{115m}In used to measure Cd.

3. RESULTS AND DISCUSSION

Soils

Soils from the unmanured 'control' plots show a marked increase in Cd concentration between 1846 and 1980, from 0.51 to 0.77 mg kg^{-1} (Fig. I). The average annual increase is approximately 1.9 µg Cd kg^{-1}, which is equivalent to an average increase in the soil plough layer Cd burden of 5.4 g ha^{-1} $year^{-1}$ over the last hundred years or so (2870 t soil ha^{-1}). Data from the other Experiments at Rothamsted suggest the average increase in soil plough Cd is lower at 1.2 µg kg^{-1} $year^{-1}$ (equivalent to 3.2 g ha^{-1} $year^{-1}$)(7).

A recent study has assumed a present-day atmospheric input of 3g Cd ha^{-1} $year^{-1}$ to U.K. agricultural land (9). This is similar to the average increase observed on the Rothamsted plots. However, this measured increase results from the balance between inputs and outputs of Cd. Therefore, it is likely that the average atmospheric input over the last century has been somewhat higher than 3.2 g ha^{-1} $year^{-1}$. Also, because anthropogenic Cd emissions are thought to have increased over

Figure I : Changes in soil plough layer Cd at Broadbalk

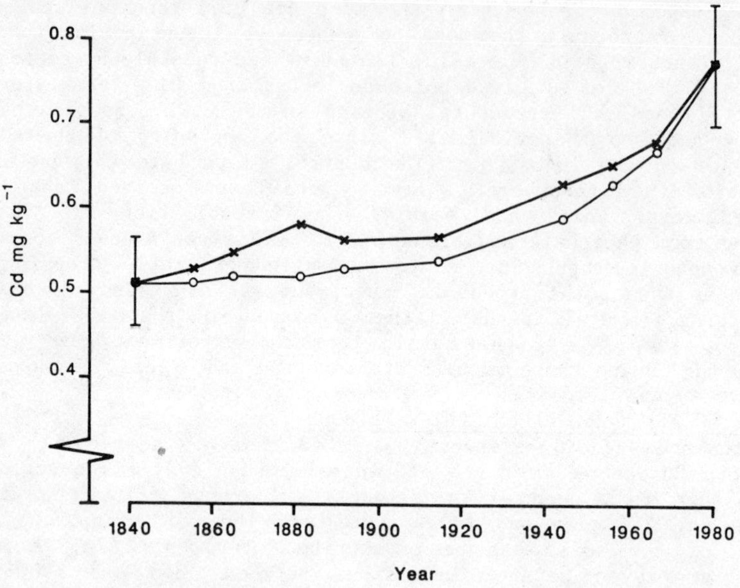

the last hundred years or so (10), it is probable that atmospheric Cd inputs at Rothamsted have varied through time. This is reflected by the data for Broadbalk, where a progressive increase in the soil Cd concentration over the last 130 years is observed, with the greatest change occurring over the last twenty years or so.

In order to reliably predict future changes in soil Cd levels, it is necessary to establish the rate of previous changes in soil Cd in relation to past inputs. However, without information on Cd emissions and their subsequent effect on soil Cd concentrations this is difficult. The major contemporary sources of Cd to the atmosphere in the UK are refuse incineration, non-ferrous metal production, iron and steel production and fossil fuel combustion (9). Because the trend in Cd levels over the last 130 years is reasonably well known for the Broadbalk plot, it seemed worthwhile to attempt to predict the change in soil Cd levels using the only suitable data available, namely Nriagu's (10) estimated global anthropogenic Cd emissions. This was achieved by apportioning the net input of 750 g Cd ha^{-1} at Broadbalk between 1846 and 1980 on the basis of the change in global anthropogenic Cd emissions (10) for each of the ten years for which a soil sample is available. The predicted soil Cd level was then determined by adding the estimated net Cd input to the plot since 1846 to the existing background soil burden. Because natural inputs of Cd to the atmosphere are likely to have remained constant over this time they have been ignored for the purposes of this calculation (7). According to Fig. I, which shows both the predicted and observed soil Cd levels on the untreated Broadbalk plot, the predicted concentrations follow the same pattern of increase as the observed levels, although they are consistently lower. Obviously this is only a crude method of estimating retrospective changes in soil composition, but the agreement is encouraging. Throughout all of these calculations it has not been necessary to quantify losses of Cd from the soil plough layer because the concentrations observed are those following any losses. A negligible amount (∿1%) of the soil Cd burden would be taken up by crops and removed at harvest. Loss of Cd due to soil mixing across plot boundaries is likely to be unimportant at this site. Of the two remaining possibilities, leaching and runoff, the former will be more important, as the experimental plots are on level ground.

Wheat grain

Samples of the wheat grain from plots 2 and 8 at Broadbalk were analysed for Cd. In addition to the atmospheric input discussed above, the soils of plot 2 have received 35 t (wet weight) ha^{-1} $year^{-1}$ of farmyard manure (FYM), whilst plot 8 received additions of various fertilisers, including 33 kg ha^{-1} y^{-1} granular superphosphate (19% P_2O_5). It has been estimated that Cd inputs from this latter source have been an additional 2 g ha^{-1} $year^{-1}$ (7). There is also some evidence to suggest that FYM is a significant source of Cd to the Rothamsted plots (7). These additional long-term Cd inputs to the soils need to be considered together with several other factors when interpreting the wheat grain data (Fig. II). Firstly, there are considerable seasonal differences in Cd uptake. For example, grain samples for the years

Figure II : Changes in wheat grain Cd at Broadbalk

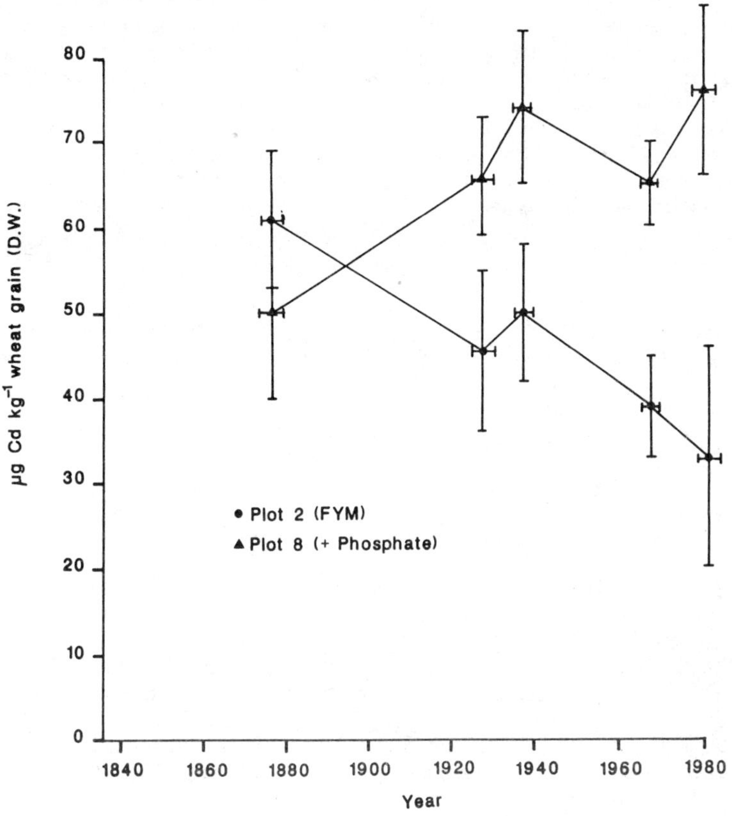

1979-84 were analysed separately and ranged from 16 - 50 ng Cd g^{-1} (dry weight) for site 2 and from 63 - 90 ng g^{-1} for plot 8 (it should be noted, however, that these ranges do not overlap). Secondly, it is likely that Cd uptake will vary between wheat cultivars (11). Clearly, therefore, a large scatter may be expected for these analyses, which may mask any underlying trend. However, there does appear to be a generally upward trend in Cd content of grain grown on plots receiving NPK fertilizers. The declining concentration in grain grown on FYM-treated soils is at variance with the much larger increase in soil Cd burden on this plot compared to that on the NPK-treated plot. However, whilst there has been no change in the organic matter content of the NPK-treated soils during the last 100 years, that on the FYM-treated soil has gradually increased until there is now 2.5 times as much organic matter in this soil as in fertilizer-treated soil. It has been shown in a previous study (6) that organic matter derived from permanent grass retained Cd in the surface horizons of the soil. Further work would be needed to show whether extra organic matter binds Cd sufficiently strongly to lessen its uptake by cereals and other food crops.

4. CONCLUSIONS

Atmospheric deposition is a significant source of Cd to UK agricultural soils and may result in long-term increases in the soil plough layer Cd burden. The rate of increase in soil Cd from this source will be dependent on the balance between inputs (i.e. atmospheric deposition rate) and outputs (via leaching, runoff and crop offtake). Although crop offtake is of negligible importance in a soil mass balance for Cd, it is of significance in terms of human dietary exposure to Cd. On a national basis several recent studies (e.g. 2,3) have recognised the importance of increasing soil Cd levels on human dietary intake of Cd, particularly since cereal-based foodstuffs constitute a major source of dietary Cd. However, the limited data presented here suggest that any relationship between increasing soil Cd and crop Cd may be masked by other factors which may vary in response to agronomic practices.

REFERENCES

(1) MINISTRY OF AGRICULTURE, FISHERIES AND FOOD (1983). Survey of cadmium in food: first supplementary report. Food Surveillance Paper No. 12. HMSO, London.

(2) DAVIS, R.D. (1984). Cadmium - a complex environmental problem. Part II. Cadmium in sludges used as fertilizer. Experientia 40: 117-126.

(3) HUTTON, M. (1982). Cadmium in the European Community: a prospective assessment of sources, human exposure and environmental impact. Monitoring and Assessment Research Centre Technical Paper No. 26, University of London.

(4) World Health Organisation (1972). Technical Report Series, No. 505, Geneva.

(5) Council Directive 86/278/EEC.

(6) ROTHBAUM, M.P., GOGUEL, R.L., JOHNSTON, A.E. and MATTINGLY, G.E.G. (1986). Cadmium accumulation in soils from long-continued applications of superphosphate. J. Soil Sci. 30: 147-153.

(7) JONES, K.C., SYMON, C.J. and JOHNSTON, A.E. (in press). Retrospective analysis of an archived soil collection II. Cadmium. Sci. Total Environ.

(8) TJELL, J.C., HANSEN, J.A., CHRISTENSEN, T.H. and HOVMAND, M.F. (1981). Prediction of cadmium concentrations in Danish soils. In: Characterization, Treatment and Use of Sewage Sludge, P. L'Hermite and H. Ott (Eds). Reidel Pub. Co., Dordrecht, Holland, pp. 652-664.

(9) HUTTON, M. and SYMON, C.J. (1986). The quantities of cadmium, lead, mercury and arsenic entering the U.K. environment from human activities. Sci. Total Environ. 57: 129-150.

(10) NRIAGU, J.O. (1979). Global inventory of natural and anthropogenic emissions of trace elements to the atmosphere, Nature 279: 409-411.

(11) KJELLSTROM, T., LIND, B., LINNMAN, L. and ELINDER, C.-G. (1975). Variation in the cadmium concentration in Swedish wheat and barley. Arch. Environ. Health 30: 321-328.

ETUDE DU DEPERISSEMENT FORESTIER EN BELGIQUE PREMIER BILAN D'UNE ANNEE DE RECOLTE DES PRECIPITATIONS A LA STATION PILOTE DE VIELSALM (x)

E. LAITAT & J. FAGOT
Faculté des Sciences agronomiques de l'Etat
Chaire de Biologie Végétale
Avenue du Maréchal Juin 2A
B 5800 GEMBLOUX

Summary

The Vielsalm station is part of a four stations network established to study the acid rains in Wallonia. Various research projects are currently being carried out, one of which deals with measuring concentrations and depositions in open field, stemflow and throughfall. The input-output research is done on five areas on a plateau : one open field area and four areas situated at 200 m intervals in a moderately declined spruce stand. Half of these forest area was fertilised with a suspension of calcium-magnesium, the other half being the control area. Results of the first year of these plots are presented concerning volume - pH - NO_3-N - SO_4-S - Mn and Pb. Rules of computing are also described.

1. INTRODUCTION

Le Département de Biologie végétale de la Faculté des Sciences Agronomiques de l'Etat a mis progressivement en place, depuis avril 1984, un réseau d'observation du dépérissement forestier orienté sur l'étude de l'écosystème pessière (1). Il est actuellement constitué de 4 stations de niveau d'atteinte gradué.

En raison d'une différence dans la symptomatologie de l'épicéa (Picea abies Karst.), marquée selon l'âge du peuplement (2), chaque station du réseau d'étude des précipitations se compose de trois parcelles : une en clairière, une sous couvert forestier jeune et une sous couvert forestier plus âgé.

Dans un premier temps, nous avons choisi avec grand soin nos parcelles et éprouvé diverses techniques d'échantillonnage. Après validation, nous avons commencé à accumuler les données.

Dans la station de Vielsalm, les paramètres écologiques et de pollution atmosphérique sont représentatifs des conditions générales de croissance des forêts résineuses de l'Est de la Belgique : le niveau moyen de pollution est très faible. La station de Vielsalm, objet de nombreuses études complémentaires, de parasitologie, de télédétection, de pédologie, d'analyse de la qualité de l'air, de chambres de simulation (open top chambers) et de fertilisation, est notre station pilote. Elle bénéficie d'une infrastructure renforcée. Elle est composée d'une parcelle en clairière et de quatre parcelles sous couvert forestier, dont 2 en peuplement âgé (± 90 ans) et 2 en peuplement jeune (± 30 ans). Les quatre parcelles

---

(x) Recherche subventionnée par la Communauté Economique Européenne et le Ministère de la Région wallonne pour l'agriculture et l'environnement.

Figure 1 VIELSALM
COMPARAISON DES PRECIPITATIONS MOYENNES
PEUPLEMENTS JEUNES - PEUPLEMENTS AGES
THROUGHFALL

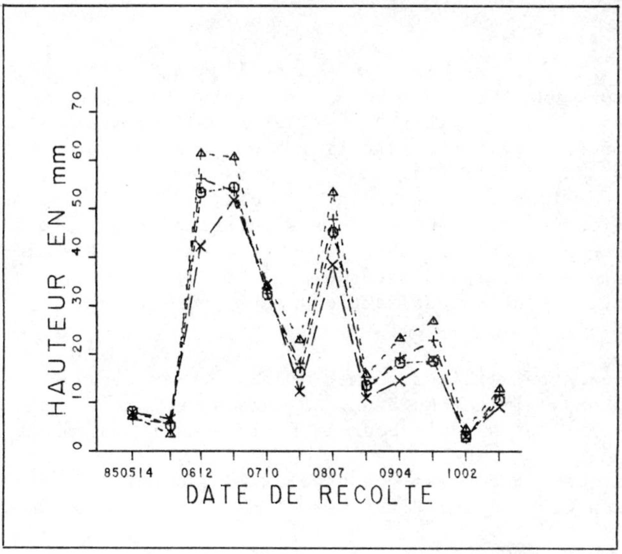

Figure 2 VIELSALM
COMPARAISON DES pH MOYENS
PEUPLEMENTS JEUNES - PEUPLEMENTS AGES
THROUGHFALL

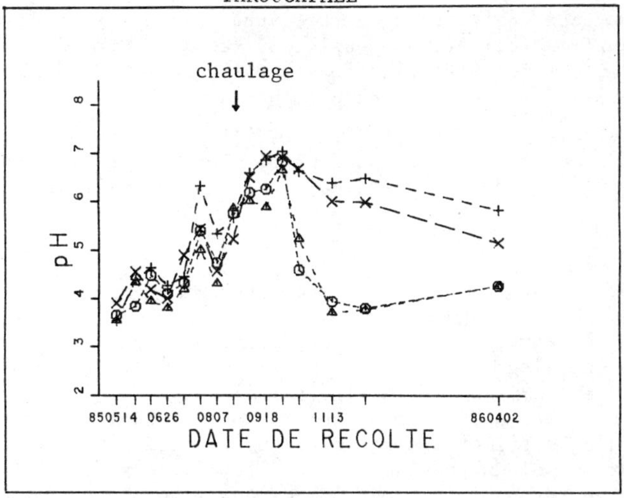

PPLT JEUNE TEMOIN
PPLT AGE TEMOIN
PPLT JEUNE CHAULE
PPLT AGE CHAULE

forestières, distantes au plus de 200 m, sont à l'origine des écosystèmes jugés totalement comparables deux à deux. Ces parcelles sont intégrées dans un programme de fertilisation forestière. La moitié du dispositif a été pulvérisée par voie aérienne d'une suspension de particules de 70 µm de carbonate double calcium-magnésium (55-45).

2. MATERIEL ET METHODE

Afin d'uniformiser les techniques d'échantillonnage, nous nous sommes alignés sur les recommandations du DWVK (3). 15 pluviomètres par parcelle récoltent les dépôts totaux en clairière et les pluviolessivats (throughfall) sous couvert forestier. Par parcelle, 5 collecteurs récoltent l'eau de ruissellement le long des troncs.

Chaque quinzaine, 15 paramètres, dont le volume, le pH, la conductivité et la concentration en anions et cations significatifs de l'équilibre ionique, sont mesurés pour chaque échantillon récolté. Les techniques d'échantillonnage, les conditions, matériel et méthodes d'analyse mis en oeuvre ont déjà fait l'objet de publications antérieures(2-4).

3. RESULTATS

L'analyse des eaux recueillies permet d'établir les flux au sein de l'écosystème. Nous nous limiterons à une analyse du volume, du pH, du NO_3-N, du SO_4-S, du Mn^{++} et du Pb^{++} pour la période du 1 mai 85 au 2 avril 1986.

De façon classique, la conversion du volume d'eau récoltée (exprimé en L) en hauteur de précipitation (exprimée en mm) est fonction de la surface de récolte.

Dans le cas des eaux de ruissellement le long des troncs, cette conversion s'effectue par la formule suivante :

$$fQ = \frac{G}{g} Q = h \qquad (a)$$

où Q est le volume des précipitations recueillies le long des troncs (exprimé en L)
g est la surface terrière de l'arbre échantillon (exprimée en cm2)
G est la surface terrière du peuplement (exprimée en m2)
h est la hauteur des précipitations attribuées à l'écoulement des eaux le long des troncs du peuplement (exprimée en mm)

Selon cette relation, la totalité des précipitations imputées à l'écoulement le long des troncs est, pour un peuplement donné, proportionnelle au rapport de la surface terrière du peuplement à celle de l'arbre échantillon.

3.1. Le volume

La figure 1 permet de visualiser l'importance des eaux de lessivage du feuillage (throughfall) dans l'écosystème pessière . Des 681 mm de pluie tombés sur le couvert, 86 %, soit 586 mm arrivent au sol, dont 98 % après lessivage du feuillage et 2 % par ruissellement le long des troncs (respectivement 573 et 13,5 mm). La figure 1 illustre de plus la très grande similitude des deux peuplements jeunes choisis ainsi que la précision des méthodes d'échantillonnage et de calcul utilisées : 583 mm de précipitation arrivent au sol du peuplement jeune témoin, dont 22,5 mm en stemflow et 561 mm en throughfall, pour 589 mm en ce qui concerne le peuplement jeune chaulé, dont 24 mm pour le stemflow et 565 mm pour le throughfall. Nous pouvons ainsi confirmer notre hypothèse de comparabilité des peuplements.

Préc. 681 pH 5,4 NO_3-N 0,84 a / 5,95 b SO_4-S 2,47 a / 16,43 b Mn^{++} 15,80 a / 129,05 b Pb^{++} 10,99 a / 113,94 b

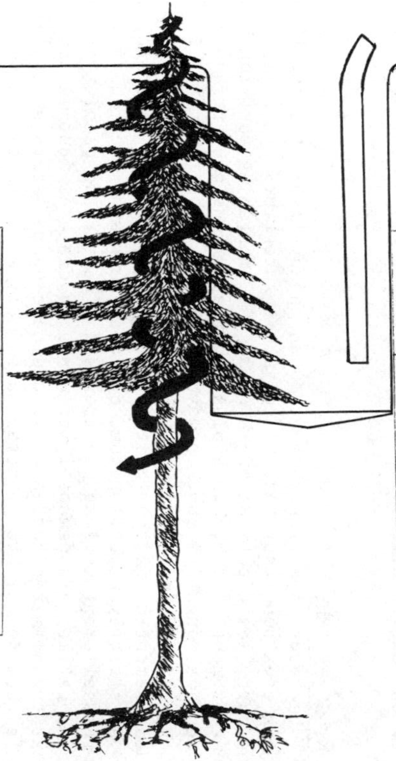

	Stemflow			
	PJ		PM	
	T	C	T	C
Préc.	22,5	24	3,15	4,5
pH	3,8	3,9	4,8	4,90
NO_3-N (a)	1,94	1,61	1,79	2,39
(b)	0,20	0,20	0,03	0,20
SO_4-S (a)	25,49	26,86	18,18	25,16
(b)	2,84	3,82	0,61	0,30
Mn^{++} (a)	1231,16	1082,67	620,26	705,39
(b)	123,10	127,02	22,43	12,78
Pb^{++} (a)	4,32	2,49	3,73	4,10
(b)	0,37	0,44	0,07	0,03

	Throughfall			
	PJ		PM	
	T	C	T	C
Préc.	561	565,5	639	525
pH	4,8	5,6	4,7	5,4
NO_3-N (a)	2,55	1,45	3,61	3,20
(b)	10,78	6,19	18,83	12,74
SO_4-S (a)	7,42	9,17	8,78	9,84
(b)	29,85	37,65	45,83	39,51
Mn^{++} (a)	362,02	161,39	342,64	260,38
(b)	1623,68	644,36	1789,68	1312,86
Pb^{++} (a)	4,78	14,59	39,58	19,55
(b)	179,28	63,27	170,82	107,08

Figure 3. : Précipitations moyennes en zone forestière entre le 1 mai 1985 et le 2 avril 1986 dans la station de VIELSALM

3.2. Le pH

La figure 2 rend compte de l'évolution du pH des précipitations. Le pH moyen des pluies en clairière, au cours de la période d'observation, est de 5,4. Lors de leur entrée dans l'écosystème, les eaux s'acidifient au contact du feuillage et de l'écorce.

Notons la différence dans l'évolution de ce paramètre entre les peuplements témoins, âgés et jeunes. Alors que, dans ce dernier, les eaux de stemflow sont en moyenne 1,6 unités et celles de throughfall 0,6 unités plus acides que les précipitations en clairière; en peuplement âgé, les eaux de stemflow et de throughfall ont la même valeur de pH et sont respectivement 0,6 et 0,7 unités plus acides que les pluies.

La figure 2 est également révélatrice de certains effets du chaulage des forêts. La pulvérisation d'une suspension de chaux magnésienne a été effectuée le 21 août 1985. Il faudra attendre le 2 octobre pour que les effets de l'épandage se manifestent au sol. Des valeurs voisines de pH 7 seront enregistrées à cette époque et voisines de pH 6 au cours de tout l'hiver. Notons que seules les eaux de throughfall sont fortement influencées par ce mode de fertilisation des forêts par voie aérienne.

Alors que les eaux de throughfall, tant en peuplement jeune qu'en peuplement âgé, sont de l'ordre de 0,7 à 0,8 unités plus basiques, les eaux de stemflow ont une acidité voisine en peuplement âgé ou jeune, chaulé et témoin.

3.3. NO_3-N, SO_4-S, Mn, Pb

Nous observons un enrichissement ou un appauvrissement sélectif selon les ions des eaux pénétrant dans l'écosystème forestier.

D'une part, les dépôts de NO_3-N et SO_4-S, imputables au throughfall, sont respectivement 2 et 2,3 fois plus importants sous couvert qu'en clairière et ceux de manganèse sont en moyenne 10 fois plus importants.

La contribution du stemflow dans les dépôts totaux est faible, en raison essentiellement du facteur multiplicatif f (a). Il se pourrait, d'un point de vue strictement écologique, que cette extrapolation du volume d'eau de ruissellement le long des troncs à l'hectare de peuplement nous amène à sous-estimer grandement l'influence de ces eaux dont l'aire d'influence est très limitée certes, mais capitale pour l'arbre puisqu'elle se limite à la zone de colonisation racinaire. Il est intéressant de remarquer, respectivement pour NO3-N et SO4-S, que les eaux de throughfall sont en moyenne 3,2 et 3,5 fois plus concentrées que les eaux de clairière et que le stemflow l'est de l'ordre de 2,3 et 10 fois plus. Enfin, en ce qui concerne le manganèse, les facteurs d'enrichissement des eaux de throughfall et stemflow par rapport aux eaux de précipitations en clairière sont respectivement, en moyenne, de l'ordre de 17 et 57.

D'autre part, les concentrations en plomb sont révélatrices de l'absorption sélective d'un ion par l'écorce. En moyenne, le throughfall est 3 fois plus concentré en plomb que les eaux de pluie et contribue à des dépôts du même ordre de grandeur que ceux de clairière. Par contre, la concentration des eaux de stemflow est réduite au tiers de celle des eaux de pluie. Environ 1 g de plomb par hectare et par an serait retenu sur l'écorce.

La figure 3 reprend la hauteur de précipitation (en mm) au cours de la période d'observation, le pH moyen et la concentration moyenne par quinzaine en mg/l pour NO_3-N et SO4-S et en μg/l pour Mn^{++} et Pb^{++}, ainsi que les dépôts en kg/ha/an pour NO3-N et SO_4-S et en g/ha/an pour Mn^{++} et Pb^{++}. Ces données sont relatives aux précipitations en clairière et aux stemflow (SF) et throughfall (TF), en peuplement jeune (PJ) ou mature (PM), chaulé (C) ou témoin (T).

BIBLIOGRAPHIE

(1) IMPENS, R., LAITAT, E. (1984). Belgian research on Acid Deposition and the sulphur cycle. Ed. Vanderborgh, pp. 141-148.

(2) LAITAT, E., IMPENS, R. (1985). Surveillance du dépérissement des forêts en Belgique. Pollution atmosphérique, Jan.-Mar., pp. 16-23.

(3) DVWK (1984). Regeln für Wasserwirtschaft, 122, P. Parey.

(4) LAITAT, E., DELCARTE, E., LENELLE, Y. (1986). Le dépérissement des forêts : une stratégie d'étude. Cm Chimie 4, pp. 13-17.

ATMOSPHERIC HEAVY METAL DEPOSITION IN NORTHERN EUROPE MEASURED BY MOSS ANALYSES

L. RASMUSSEN, Technical University of Denmark
K. Pilegaard, Risø National Laboratory, Denmark
Å. Rühling, University of Lund, Sweden

Summary

In a joint Nordic project the atmospheric heavy metal deposition as reflected in the concentrations in moss, was monitored in Denmark, Finland, Norway and Sweden in 1985. The aims of the project were to assess the regional and temporal variability of the atmospheric deposition of arsenic, cadmium, chromium, copper, iron, lead, nickel, vanadium and zinc. In this presentation only the data for nickel and lead are given. For all metals the concentration levels are higher in southern Fennoscandia than in the northern parts of the area. Some local emission sources are clearly revealed as hot spot areas upon the general background levels. Comparisons with similar investigations in Sweden every 5th year since the late 60's showed a distinct reduction in concentrations during the period.

1. INTRODUCTION

Several studies have shown that surveys of the metal concentrations in moss may be a valuable means of identifying sources of airborn pollution and of mapping the metal deposition. The atmospheric heavy metal deposition was in that way measured in 1985 in a joint Nordic project in Denmark, Finland, Norway and Sweden. The project was a continuation and follow-up of a comparable Danish-Swedish study from 1980 (1). It was initiated by the Steering Body of Environmental Monitoring in the Nordic Countries under the Nordic Council for Ministers.

The objectives of the project were to find the regional background levels of heavy metal concentrations in moss and to locate emission sources important on a regional scale. The findings are compared with similar investigations in Sweden in 1968-69, 1975 and 1980.

2. MATERIAL AND METHODS

Samples of two different moss species, *Hylocomium splendens* and *Pleurozium schreberi*, have been collected at 185 localities in Denmark, 534 in Finland, 522 in Norway and 839 in Sweden (Fig. 1). The samples were taken in openings in coniferous forests or young plantations, not exposed directly to throughfall precipitation. The sampling sites were located at least 300 m from main roads and populated areas, and at least 100 m from any road or single house. On each sample site 5-10 subsamples were taken within a 50 x 50 m area and mixed in the same bag. The concentrations of arsenic, cadmium, chromium, copper, iron, lead, nickel, vanadium and zinc were measured in the samples by atomic absorption, neutron activation or ICP techniques after intercalibration of the analytical procedures. In this presentation only the results for nickel and lead are given.

3. RESULTS AND DISCUSSION

The data are presented in the form of contour maps over northern Euro-

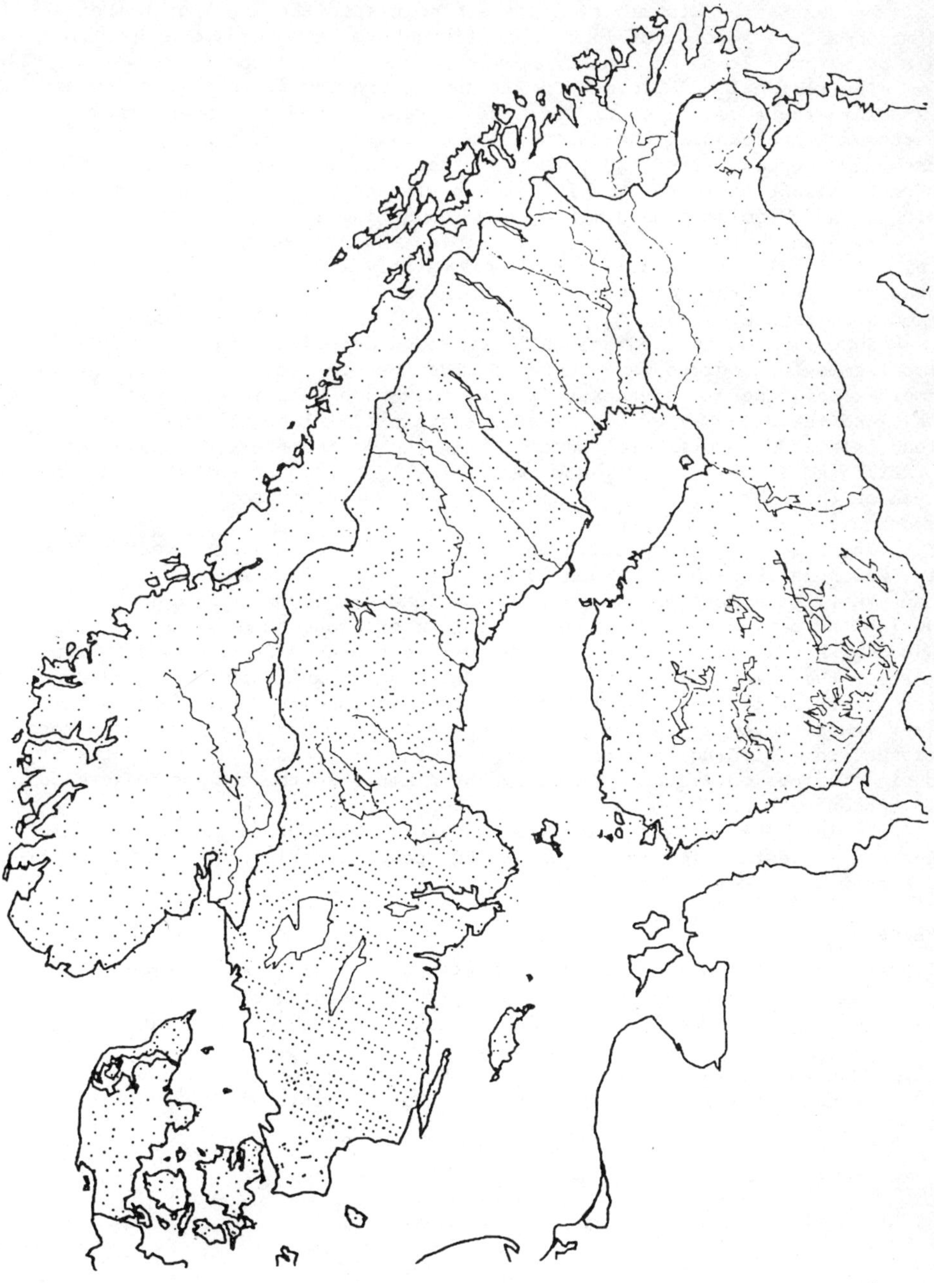

Fig. 1. Location of sampling sites.

pe (Fig. 2 and 3). For lead the highest concentration levels were found in samples from the southern parts of Fennoscandia. The levels decreased towards the north to a minimum in the northernmost areas of Fennoscandia. This concentration distribution pattern points to the conclusion that the atmospheric deposition of lead is mostly due to long-range transport of air pollutants from the densely populated areas of Europe.

The regional background deposition pattern for arsenic, cadmium and vanadium also showed a steep decreasing gradient towards the northern parts of Fennoscandia, whereas the concentrations of nickel (Fig. 3), chromium, copper, iron, and to some extent zinc, showed a weaker gradient. On a local but still relatively large scale enhanced values were found for nickel and chromium around mining and industrial activity centres.

In areas with no forests, like alpine and agricultural regions, metals originating from soil dust showed relatively high concentration values. That was true for arsenic, chromium and copper, and, especially for iron and vanadium.

Superimposed on the regional background deposition pattern a few big local emission sources were revealed. The most important were the metal smelter at Rönnskär, northeast Sweden, the Sauda-Odda area in western Norway, and the Pori-Harjavalta area in southwest Finland. In the case of lead, elevated levels were found also in the most densely populated and trafficated areas (Fig. 2). However, not all of the concentration maxima seen on the maps could be explained from the present knowledge of emission sources.

In Fig. 4 a retrospective comparison with the similar investigation in Sweden in 1968-69, 1975 and 1980 is made (2). It is evident from the maps that the atmospheric deposition of lead has decreased significantly during the period. The decrease has been most pronounced in southwestern Sweden, whereas little or no changes were detectable in the north. Since most of the atmospheric deposited lead in background areas is descharged from traffic, it is no doubt the reduction of lead in petrol during the last years, which is reflected in decreasing concentration levels in moss tissue. This finding is clear, in spite of an increased traffic intensity during the period, which presumably has diminished the observed decrease in concentrations.

It is the intention that this study will be a precursor of a common European investigation including all European countries. The next investigation will be carried out in 1990.

REFERENCES

(1) MILJØSTYRELSEN and STATENS NATURVÅRDSVERK (1983). Moss analyses as a means of surveying the atmospheric heavy-metal deposition in Sweden, Denmark and Greenland in 1980. National Swedish Environment Protection Board Bulletin SNV PM 1670.

(2) RÜHLING, Å. and TYLER, G. (1984). Recent changes in the deposition of heavy metals in northern Europe. Water, Air, and Soil Pollution 22: 173-180.

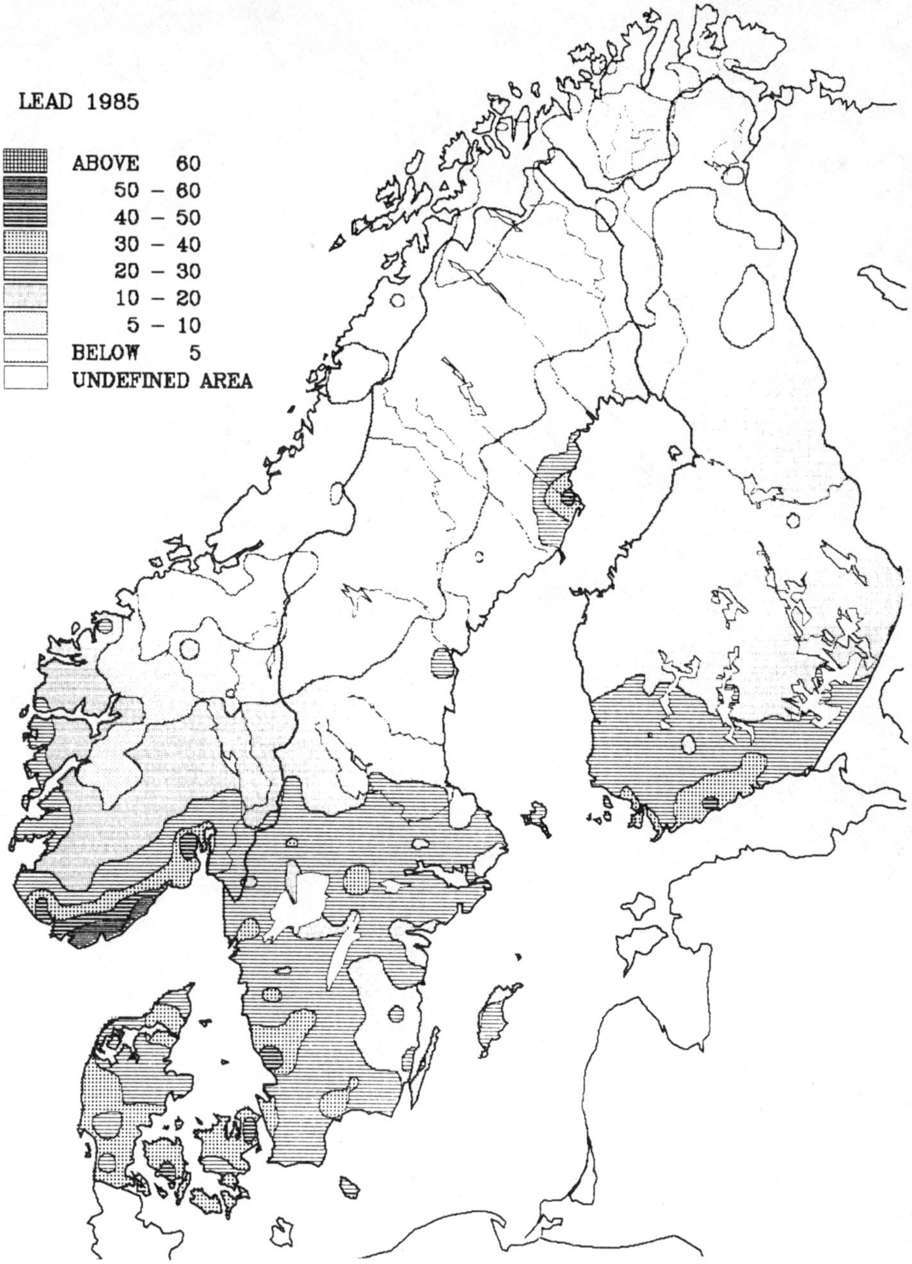

Fig. 2. Concentration of lead in moss.

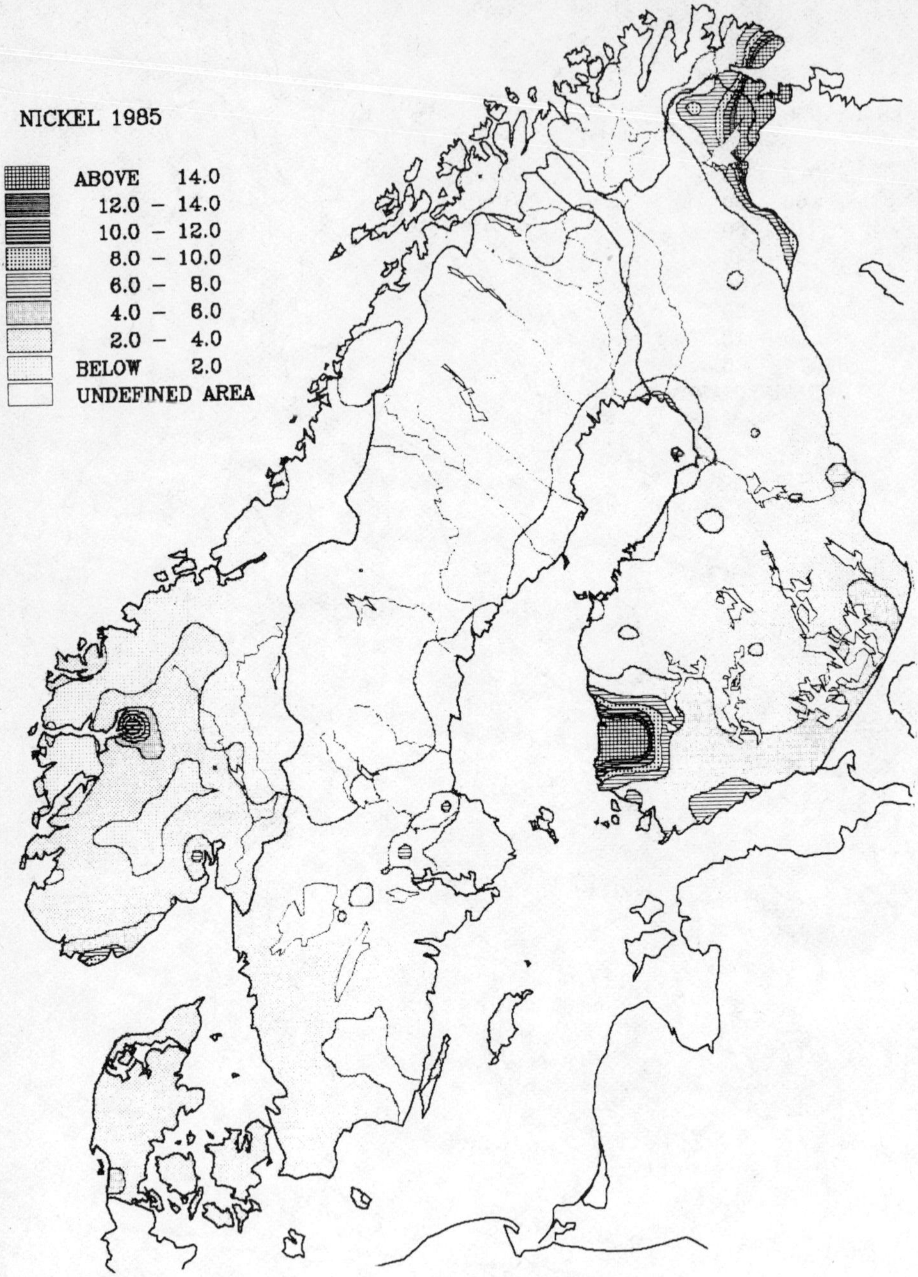

Fig. 3. Concentration of nickel in moss.

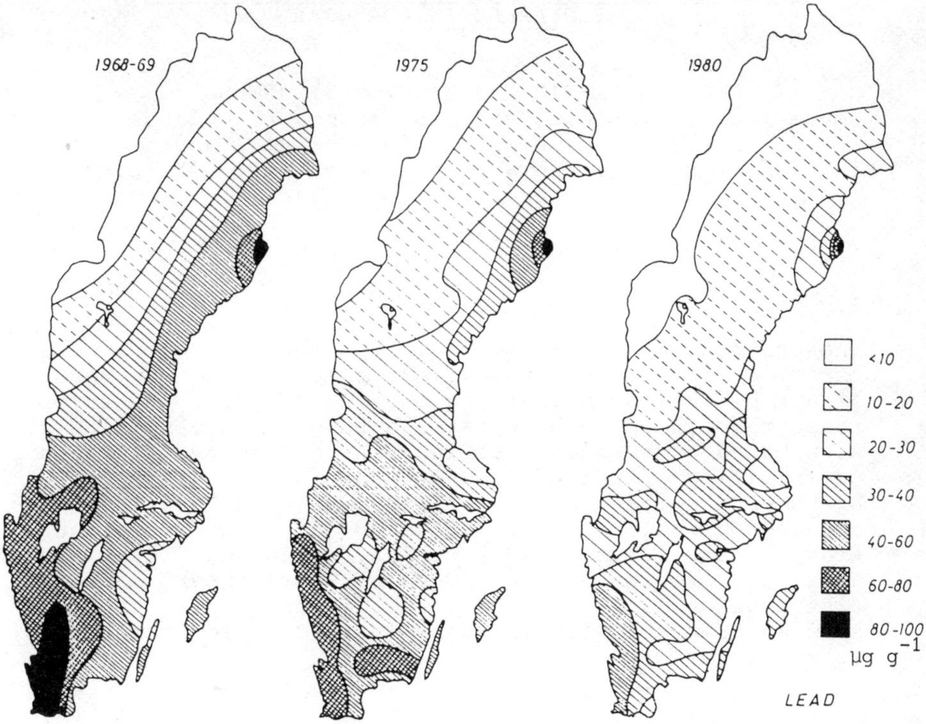

Fig. 4. Concentrations of lead in moss in Sweden 1968-69, 1975 and 1980.

CRYPTOGAMIC EPIPHYTES AS INDICATORS OF AIR QUALITY INVESTIGATIVE METHODS AND RESULTS FROM FOUR SELECTED AREAS IN PORTUGAL

CECÍLIA SÉRGIO and MAURICE P. JONES
Museu, Laboratório e Jardim Botânico, Faculdade de Ciências 1294 Lisboa Codex Portugal and
King's College, Hortensia Road, London SW10 OQX, U.K.

Summary

The present work follows up earlier investigations of alterations in the epiphytic flora composition as a result of air pollution in four regions in Portugal.With several research methods such as permanents quadrats, mapping studies based on species distribution, diversity,alterations of fertility of epiphytes and transplant experiments it has been possible to monitor affected areas and where increased pollution levels may be anticipated. A completed study of the Tagus estuary region, including the Lisbon area, was presented with levels of SO_2 (winter means)and distribution maps of some of the differentially sensitive species. Some baseline results in threatened areas are also provided.

1. INTRODUCTION

Pollution mapping studies based on the distribution,diversity, abundance and luxuriance of epiphytic species have been carried out almost all over the world, mainly during the past twenty five years.

Much of the extensive literature on this topic is summarised and discussed by bryologists (ANDO 1984, RICHARDSON 1981) and also by lichenologists (HENDERSON 1987) and, generaly,epiphytic vegetation is considered to be valuable in the estimation of air pollution levels (HAWKSWORTH and ROSE 1970, 1976, HOFFMANN 1974, CRESPO 1977, DERUELLE 1978).

In addition to mapping studies, phytosociological data have also been used in quantitative surveys of areas affected by pollution.LE BLANC and DE SLOOVER (1970) using number of species, frequency-cover, and resistant factor species, provided a clear picture of polluted sites, and lead to the establishement of the Index of Atmospheric Purity (IPA). Many other ecologists have shown that these methods are useful in monitoring levels of pollution, as SÉRGIO and SIM-SIM (1985), with the investigation of the epiphytic flora around Lisbon and the Tagus estuary.

The rich epiphytic flora of the mediterranean forest in Portugal, in areas previously undermaged by air pollution can provide a good opportunity to observe the impact of the establishement of new industrial areas, as the important industrial complex of the Sines area (JONES et al. 1981).

Another objective of these studies is to produce a record of the present day epiphytic flora of important areas in Portugal, and to detect regions with climax vegetation. These sites are important to provide a constant supply of new propagules of epiphytic species, that have became more and more rare in Europe and Mediterranean area.

2. OUTLINE OF RESEARCH METHODS AND SELECTED SITES

Sines Industrial area-Alentejo (Fig.1-area A)

During the early construction phase of the Sines complex, which began in 1977, in order to produce a record of the existing epiphytic vegetation, marked quadrats (20x20 cm traced,on acetate) have been established on suitable tree species. An attempt was made to sample regulary within the general area, using a superimposed grid (more than 70 sites by 1986). At each site 3 quadrats were selected and marked. The vegetation was recorded by tracing the outlines of the thalli on plastic sheeting. Two further examinations, have been made at each site, since 1977, with new plastic overlays to detect any changes in the dynamics of the epiphytic vegetation.

In order to produce a more complete record of the lichen flora, collections have been made and a list of species made and used for mapping studies (JONES 1983).

Monitoring of air pollution in urban and industrial areas in the Tagus estuary (Fig.1-area B)

The investigation of the epiphytic flora in the Lisbon area and Tagus estuary has been conducted since 1979 by the present senior authors (SÉRGIO et al. 1981, 1985).

During the years 1979-1984, 200 sites in the Lisbon area, were sampled for the total epiphytic flora (lichens and bryophytes), with quantitative analysis for determination of the IAP values for the area. At each site more than ten trees were investigated. The main species involved are *Olea europaea* and *Ulmus* spp. wich are either crop trees or are planted as ornamentals.

By comparison of the composition of the epiphytic flora with some SO_2 physico-chemical measurements (winter mean µgr/m^3), an adaptation of some accepted monitoring scales (HAWKSWORTH and ROSE 1970, CRESPO et al. 1981) was possible, and the mapping of differents polluted zones carried out.

This scale of toxitolerance for epiphytes in the Lisbon area (SÉRGIO and BENTO-PEREIRA 1981), however, has made it possible to extend it to other areas with similar bioclimatic conditions.

In addition, notes on elevation and phorophyte conditions (pH value of tree bark and its water content) were recorded at each site.

A change in the bryophytic flora as a result of increasing pollution emissions has been subsquently observed (SÉRGIO 1981, 1987).

Tortula laevipila, a bryophyte species with wide distribution and which produces spores and/or gemmae at different levels of SO_2, was transplanted along with their substrates, from unpolluted sites to sites that were more or less ecologically similar, but with different SO_2 values (SÉRGIO 1987). This study provided evidence of injury caused by increased pollution levels.

The Atlantic coast area of Leiria to Figueira da Foz (Fig. 1- area C)

For the determination of the IAP values, mapping studies based on species distribution, diversity, abundance and fertility of epiphytes have been carried out since 1984 in an threatened area by the establishement of a Coal-Fired Power Plant. This is an important region in the Central part of Portugal, where only some regional urbanized areas are located and one of the most important features are the *Pinus pinaster* forests, cultivated in Portugal from the 14th century.

The Tagus valley-Ribatejo to Spain (Fig. 1- area D)

For the same objectives, to produce a record of the present-day epiphytic flora of the area, and to make the AEI for of a new Coal Fired Power Plant in the Abrantes region in the Tagus valley, a study is being conducted in an area of cultivated plains with also a large exploited woodland area of *Pinus pinaster* and *Eucalyptus globulus*. There are some interesting areas with "montado" with *Quercus suber* (cork trees) and *Olea europaea* (olive trees) still present.

3. RESULTS

The investigations outline above have produced a number of significant results, some of which have been refered to in passing.

1. Distribution patterns attributable to point pollution sources and conturbation situations have been found for several species. (Fig. 2, Fig. 3, Fig. 4)
2. An increase in the species diversity of epiphytic communities corresponding to decreasing SO_2 levels. (Fig. 2, Fig. 3, Fig. 4)
3. A comparison of the toxitolerance of some widely distributed species by comparing with their distribution, in Portugal, in response to SO_2 levels, with known distributions patterns elsewhere on a global scale.
4. Changes in the morphology and reproductive strategy in some species in response to pollution levels. (Fig. 2-3)
5. The base-line studies have contributed significantly to the knowledge of the diversity, distribution and dynamics of the epiphyte population. This will serve as a base for future surveillance and forms a response to an international desire for information on the distribution of these threatned and often sensitive species.

REFERENCES

(1) ANDO, H. and MATSUO, A. (1984). Applied Bryology in SCHULTZE-MOTEL, Advances in Bryology 2. Vaduz.

(2) CRESPO, A., MANRIQUE, E., BARRENO, E. and SERIÑA, E. (1977). Valoracion de la contaminacion atmosferica del area urba-

na de Madrid mediante bioindicadores (líquenes epifítos). Anal. Inst. Bot. Cavanilles 34, 1: 71-94.
(3) CRESPO, A., BARRENO, E., SANCHO, L.G. and BUENO,A.G.(1981) Establecimiento de una rede de valoración de pureza atmosférica en la provincia de La Coruña (España) mediante bioindicadores liquénicos. Lazaroa, 3: 289-311.
(4) DERUELLE, S. (1978). Les lichens et la pollution atmosphérique. Bull. Ecol. 9, 2: 87-128.
(5) HAWKSWORTH, D. L. and ROSE, F. (1970). Quantitative scale for estimating sulphur dioxide air pollution in England and Wales using epiphytic lichens. Nature 277: 145-148.
(6) HAWKSWORTH, D. L. and ROSE, F. (1976). Lichens air pollution monitors. London.
(7) HENDERSON, A. (1987). Literature on air pollution and lichens xxv. Lichenologist 19, 2: 205-210.
(8) HOFFMANN, G. R. (1974). The influence of a paper pulp mill on the ecological distribution of epiphytic cryptogams in the vicinity of Lewiston, Idaho and Clarkston, Washington. Environ. Pollut. 7: 283-301.
(9) JONES, M. P., CATARINO, F.M., SÉRGIO, C. and BENTO-PEREIRA, F. (1981). The Sines industrial complex monitoring programme: a preliminary report. Environmental Monitoring and Assessment 1: 163-173.
(10) JONES, M. P. (1983). Epiphytic macrolichens of the Sines area, Alentejo, Portugal. Revista de Biologia, 12:313-325.
(11) LE BLANC, F. and DE SLOOVER, J. (1970). Relation between industrialization and the distribution and growth of epiphytic lichens and mosses in Montreal. Can. J. Bot. 48: 1485-1496.
(12) RICHARDSON, D. H. S. (1981). The Biology of Mosses.Oxford.
(13) SÉRGIO, C. and BENTO-PEREIRA, F. (1981). Líquenes e briófitos como bioindicadores da poluição atmosférica.I.Utilização de uma escala qualitativa para Lisboa. Bol. Soc.Broteriana, 54, 2ªSér.: 291-303.
(14) SÉRGIO, C. and SIM-SIM, M. (1985). Estudo da poluição atmosférica no estuário do Tejo. A vegetação epifítica como bioindicadora. Portug. Acta Biol. (B) 14: 213-244.
(15) SÉRGIO, C. (1987). Epiphytic bryophytes and air quality in Tejo estuary. Acta Botanica Hungarica. (in press).

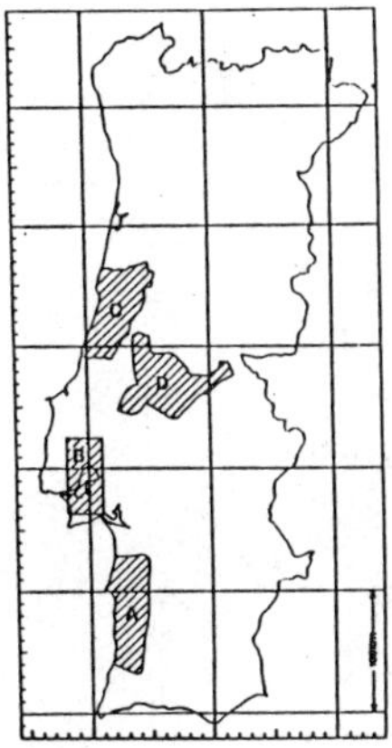

Fig. 1- Location of the selected areas of the epiphytic studies in Portugal (A,B,C,D). UTM-10x10 km.

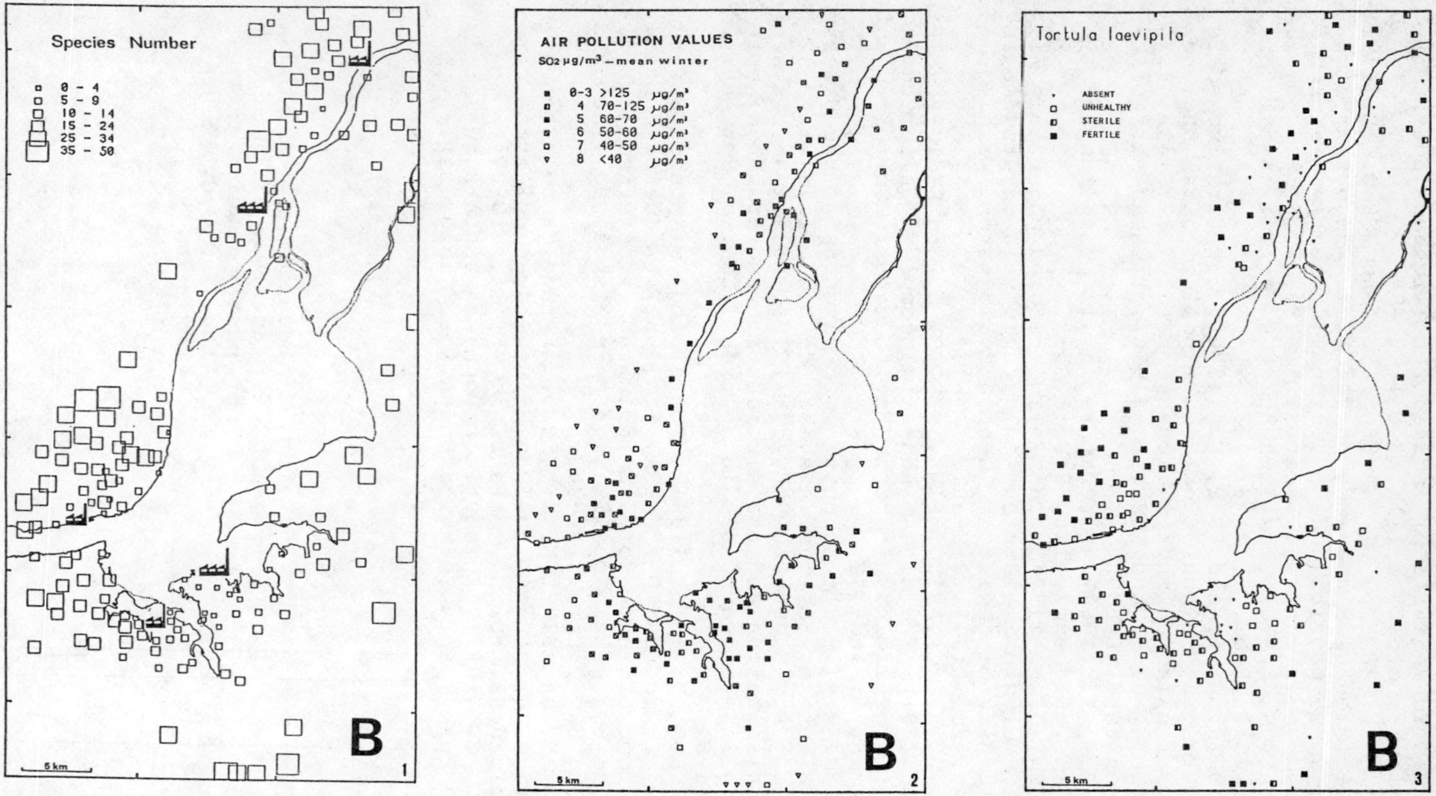

Fig. 2- Area B. 1- Highest number of epiphyte species; 2- SO_2 levels (µg/m3, mean winter) correlated with the epiphytic flora; 3- Occurence of a differentialy sensitive species to air pollution.

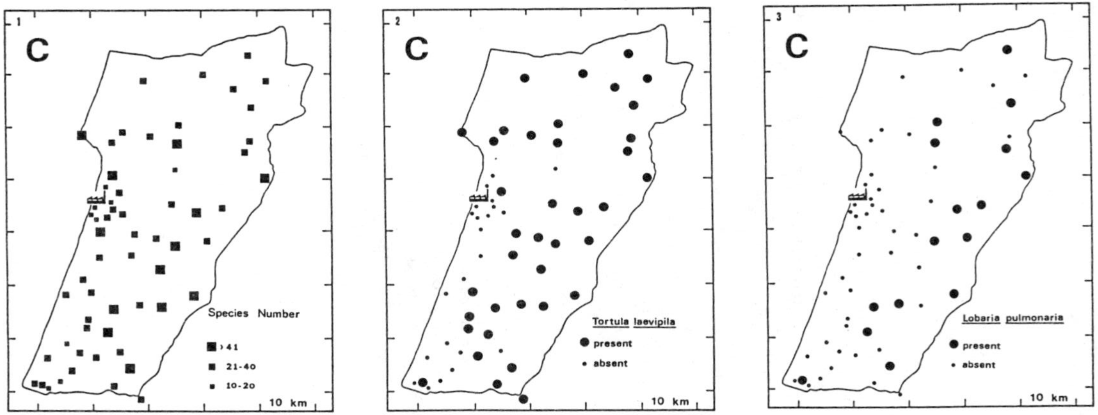

Fig. 3- Area C. 1- Highest number of the epiphyte species; 2- Occurence of <u>Tortula laevipila</u>; 3- Occurence of <u>Lobaria pulmonaria</u> a climax element.

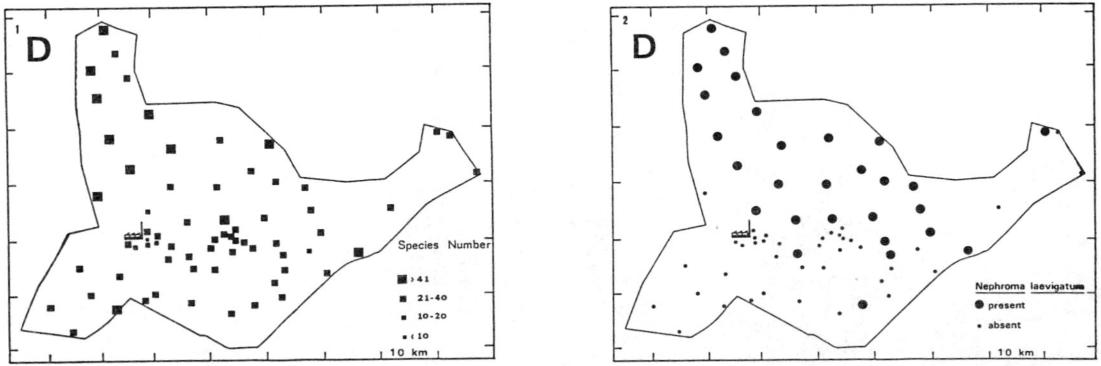

Fig. 4- Area D. 1- Highest number of epiphyte species; 2- Occurence of <u>Nephroma laevigatum</u> a preclimax element.

REACTION AND EFFECT MEASUREMENTS IN BAVARIA WITH REGARD TO FOREST DAMAGES

Karl Pfeiffer
Ernst Rudolph
Bayer. Landesamt für Umweltschutz, München

Summary

Reaction and effect measurements since 1974 have included the following investigations:

- Bioindicator network of spruces on natural locations for determination of S- and F-concentrations in needles
- Bioindicator network of epiphytic mosses for determination of heavy metals
- Rainwater network for determination of conductivity, of H^+-, Ca^{++}, NH_4^+, SO_4^{--}, NO_3^--concentrations and input during the vegetation period
- Exposition of metal plates for determination of acid corrosion rates
- Exposition of elastomeres for determination of photooxidative changes in tensile force
- Evaluation of rainfall data between 1976 and 1981

Allin vestigations are based on a network of uniform structure, but different density.

The main results are:

- In spruce needles S-content has diminished since 1977, except in high pollution areas (north-east Bavaria).
- In mosses, the development of heavy metal content has varied since 1979, but on the whole there seems to be a slight decrease
  Results of rainwater investigations show that the acid components correlate with S-concentration in needles of bioindicator spruces
- The same is true for corrosion rates of metal plates
- Tensile force of elastomeres is best correlated with height above sea level
- Amount and distribution of rainfall differ from long-time average. There have been deficits in some regions, and, particularey during April and May, all over the country
- Distribution of forest damages frequently does not correspond to possible influence factors.

1. Vorbemerkung

Das Bayerische Landesamt für Umweltschutz führt seit 1974 wirkungsbezogene Erhebungen durch, die neben den Belastungsgebieten gemäß Bundes-Immissionsschutzgesetz ab dem Jahr 1977 auf ganz Bayern ausgedehnt wurden.

Hierfür sind Bioindikatornetze nach einem festen Raster (Grundabstand 16 km x 16 km) eingerichtet worden, die nach Bedarf - insbesondere in den Belastungsgebieten verdichtet sind.

Zusätzlich bestehen ein Regenwassermeßnetz und Expositionsstellen für die Untersuchung von Wirkungen an Materialien, die in einem Raster von 64 km x 64 km angeordnet und identisch mit den Probenahmeorten der Bioindikatiornetze sind.

Die Niederschlagsdaten der Jahre 1976-1981 im Vergleich zu den langjährigen Mittelwerten wurden für die Probenahmeorte der Bioindikatornetze erhoben und ausgewertet. Außerdem wurden Dauerbeobachtungsflächen der Bodenvegetation eingerichtet.

Alle Ergebnisse sind EDV-gespeichert, durch Überlagerung der Daten sind kombinierte statistische Wertungen möglich.

2. Bioindikatornetz Standortfichten

Flächendeckend eingerichtet seit 1977, das Netz setzt sich grenzüberschreitend seit 1983 nach Österreich fort. Probenahme ist jeweils im Herbst von einjährigen Fichtennadeln, Ermittlung der Schwefelgehalte mittels Roentgen-Fluoreszenzmethode.

Das Verteilungsbild der Belastung ist in Bayern seit Einrichtung des Netzes ziemlich konstant.

Es werden nur auf relativ kleiner Fläche S-Akkumulationen erreicht, die als phytotoxisch gelten (ab 1750 µg/g TS). Im weitaus überwiegenden Teil Bayerns werden die natürlichen S-Grundgehalte (-1300 µg/g TS) seit mehreren Jahren nicht mehr überschritten.

Der Trend der Akkumulationswerte ist rückläufig, im Landesdurchschnitt um ca. 21 %, er ist in den einzelnen Landschaftsräumen unterschiedlich verlaufen. Ein starker Einbruch der S-Gehalte ist im Jahr 1980 zu verzeichnen, der nicht mit der Luftbelastungssituation erklärbar ist.

Ab 1983 stellten sich die Werte in den meisten Landschaftsräumen auf konstant niedrigem Niveau ein. Eine Ausnahme bilden die Landschaftsräume der stark von Fremdimmissionen betroffenen Mainniederung, Spessart, Nordost-Bayern. (vgl. Abbildung 1)

3. Bioindikatornetz Moose

Netz flächendeckend eingerichtet seit 1981 im Anschluß an ein Forschungsprojekt des Lehrstuhles für Hydrologie der Universität Bayreuth.

Probenahme jeweils im Herbst von epiphytischem Moos (Hypnum cupressiforme) in Anlehnung an das Raster des Bioindikatornetzes Standortfichten. Ermittlung der Metallgehalte mittels dreier Analysentechniken (AAS, NAA, ICP-AES).

Die Landesmittelwerte entwickelten sich von 1978-1983 wie folgt (Angaben in µg/g lufttrockener Probe):

Element	1978	1981	1982	1983
Blei	58,40	38,60	44,80	30,20
Cadmium	0,57	0,58	0,45	0,40
Chrom	3,85	6,40	3,75	4,10
Eisen	1300,00	1547,00	1111,00	1112,00
Kupfer	12,13	14,20	10,03	10,94
Mangan	220,00	257,80	218,20	232,50
Nickel	5,80	5,90	5,20	4,80
Zink	43,20	72,20	65,50	62,30
Quecksilber	-	0,39	0,19	0,14
Titan	-	40,40	80,00	70,40
Vanadium	-	9,20	7,22	7,28
Aluminium	-	-	1096,00	1016,00
Antimon	-	-	0,37	0,34
Arsen	-	-	0,98	0,92
Magnesium	-	-	1417,00	1296,00
Selen	-	-	0,43	0,36

Das regionale Verteilungsbild für die einzelnen Elemente ist uneinheitlich, für die Mehrzahl der Elemente sind Schwerpunktbereiche in Mittel- und Oberfranken und im nördlichen Unterfranken erkennbar. Bei der Mehrzahl der untersuchten Elemente ist die Tendenz rückläufig.
(vgl. Abbildung 2 als Darstellungsbeispiel)

4. Regenwasseruntersuchungen
Netz eingerichtet seit 1981 mit einem Rasterabstand 64 km x 64 km in Anlehnung an das Bioindikatornetz Standortfichten. Probenahme in 14tägigen Abständen während der Vegetationszeit.

Untersuchungsparamter: Niederschlagsmenge, pH-Wert, Protoneneintrag, Leitfähigkeit, Sulfat-Nitrat-Ammoniumgehalt und -eintrag, Calciumgehalt und -eintrag.

Menge des Niederschlags bestimmt im wesentlichen die Höhe des Protoneneintrages.

Beim pH-Wert ist insgesamt im Vergleich zum biologischen Neutralpunkt (5,6) eine Absenkung von 1 -. 1,5 Stufen gegeben.

Das Verteilungsmuster der Sulfatgehalte zeigt Übereinstimmungen mit dem der S-Gehalte in Fichtennadeln. Die Verteilung der Gehalte von NH_4 und NO_3 weist einen statistisch signifikanten Zusammenhang auf.

Die Werte der Einträge lassen, abgesehen vom Ammoniumeintrag, eine deutliche Abhängigkeit von der Niederschlagsmenge erkennen.
(vgl. Abbildung 3 als Darstellungsbeispiel)

5. Wirkungsuntersuchungen an Metallplättchen

Netz seit 1982 eingerichtet, Rasterabstand 64 km x 64 km Exposition je 3 Monate, jeweils fortlaufend mit anschließender Feststellung der korrosionsbedingten Gewichtsabnahme.

Das Verteilungsmuster der Metallkorrosion ist ähnlich mit dem der Sulfatgehalte der Niederschläge (Signifikanz 99 %) und damit auch mit dem der S-Gehalte in Fichtennadeln.

6. Wirkungsuntersuchungen an Elastomeren

Netz seit 1983 eingerichtet, Rasterabstand 64 km x 64 km Exposition von Naturkautschuk unter 20 % Zugdehnung in 4facher Wiederholung.

Expositionszeit je 3 Monate ganzjährig. Es wird die Restreißfestigkeit gemessen.

Das Verteilungsmuster des Reißfestigkeitsverhaltens bestätigt sich seit mehreren Jahren. Stärkste Wirkungen im wesentlichen in den ostbayerischen Mittelgebirgen.

Inwieweit diese Wirkungen auf Witterungseinflüsse zurückzuführen sind, ist noch nicht quantifizierbar. Die Reißfestigkeit nimmt mit zunehmender Höhenlage deutlich ab. Abgesehen von einem möglichen Einfluß von Oxidantien können Temperatureffekte hier wirksam sein. Die relativ geringe Reißfestigkeitveränderung im Alpenvorland entspricht nicht der Verteilung des dortigen Waldschadensbildes.

7. Abweichung der Jahres-, Monatsniederschläge von den langjährigen Mittelwerten in den Jahren 1976-81 an den Probenahmepunkten des Bioindikatornetzes.

In bestimmten Landschaftsräumen ist über den gesamten Erhebungszeitraum ein Defizit erkennbar. Übereinstimmungen mit dem ausgeprägteren Auftreten stärkerer Waldschäden (Schadensstufe 2 und 3) zeichnen sich ab.

Es ist ein deutliches Defizit in allen Landschaftsräumen zu Beginn der Vegetationszeit (Mai) von 1976 bis 1981 zu verzeichnen.

(vgl. Abbildungen 4, 5)

8. Folgerungen

Es lassen sich als Folgerungen ziehen:

- Schadensbild der Waldbestände und Bild der landesweiten S-Kontamination sind häufig nicht identisch. Statistisch auswertbare Zusammenhänge sind nur in Gebieten mit hoher S-Belastung gegeben.
- Das Verteilungsbild der Schwermetallgehalte in Moosen läßt keine Ähnlichkeit mit der Schadensverteilung bei den Waldbeständen erkennen.
- Reißfestigkeitsraten der Elastomeren stimmen ebenfalls weniger mit der Verteilung des Schadensbildes der Waldbestände überein.

- Die Niederschlagsbilanz der Jahre 1976 - 1981 ist in einigen Landschaftsräumen defizitär, eine Übereinstimmung mit dem Verteilungsbild stärkerer Schädigungen an Koniferen deutet sich an. Das Niederschlagsdefizit zu Beginn der Vegetationszeit in allen Landschaftsräumen läßt klimatische Ursachen für die Schädigung der Wälder nicht ausgeschlossen erscheinen.
- Aus diesen statistisch ausgerichteten Erhebungen läßt sich ableiten, daß es eine monokausale Begründung für das Schadensbild wohl nicht gibt. Einer weiteren Vertiefung der Untersuchungen soll ein Netz von Expositionsstellen mit Fichten genetisch gleicher Eigenschaften (Klonfichten) in identischem Bodensubstrat (Einheitserde) dienen.

Literatur

Rudolph,E: Immissionswirkungen an Pflanzen, Schriftenreihe Landesamt für Umweltschutz, Naturuschutz und Landschaftspflege, Heft 9, 1978

Lufthygienische Jahresberichte, 1978-83, Schriftenreihe des Landesamtes für Umweltschutz

Knabe, W.: Immissionsökologische Waldzustandserfassung. Ergebnisse und ihre Bedeutung für die Forstwirtschaft in Nordrhein-Westfalen. Mitt. der LÖLF, Sonderheft, 1982

Rudolph, E.: Funktionale Zusammenhänge zwischen Kenngrößen des Regenwassers und Wirkungskriterien, VDI-Bericht 500, S. 231, 1983

Luckat, S.: Beziehungen zwischen der Korrosionsrate von Stahl und den Immissionsraten verschiedener Schadstoffe, Staub 34, S. 205, 1974

Optimierung gebräuchlicher Bioindikationsmethoden mit besonderer Berücksichtigung der Repräsentanz und Reproduzierbarkeit der Ergebnisse. Unveröffentlichte Ergebnisse des UBA-Forschungsberichtes Nr. 106 07 016/03 des LfU, 1982, Teil 1

Kennel, E.: Waldschadensinventur Bayern 1983 - Verfahren und Ergebnisse - Schriftenreihe der Forstwiss.Falkultät der Uni München und der Bayer. Forstl.Versuchs- und Forschungsanstalt Heft 57, 1983

Herrmann, R. und Thomas, W.: Über die räumliche und zeitliche Verteilung von einigen atmosphärenbürtigen Umweltgiften in Bayern. Abschlußbericht zum Forschungsvorhaben vom 30.09.1980

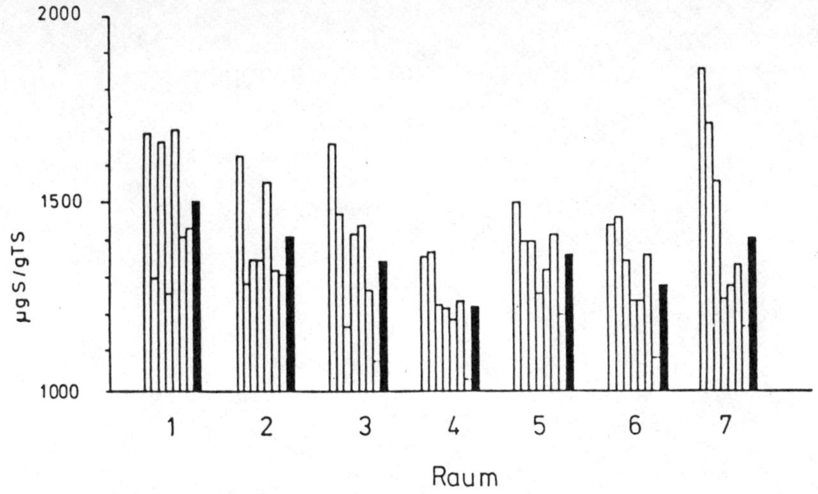

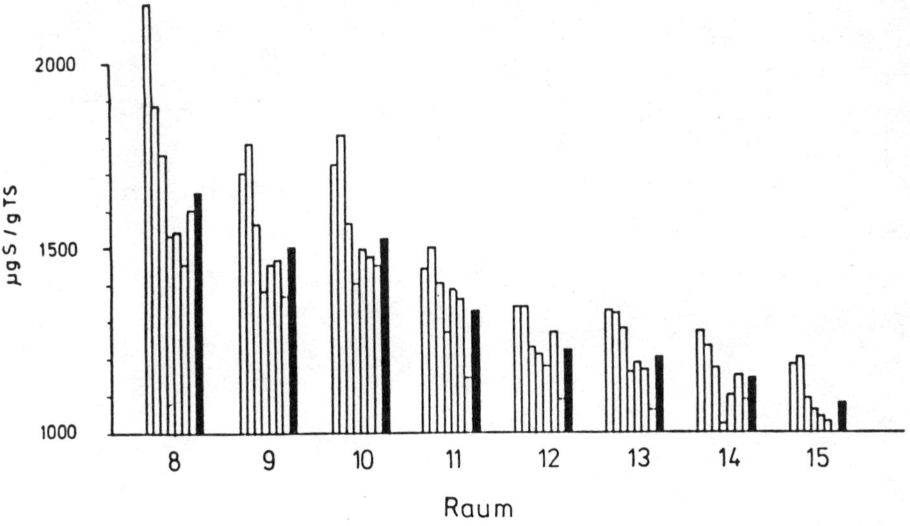

1	Rhein-Main-Niederung	10	Oberpfälzer Wald
2	Spessart	11	Bayer. Wald
3	Rhön	12	Tertiärhügelland, Iller-Lechplatten und Donautal
4	Fränkische Platten	13	Schwäbisch-Bayer. Schotterplatten und Altmoränen
5	Fränkisches Keuper-Lias-Land	14	Schwäbisch-Bayer. Jungmoränen und Molassevorberge
6	Fränkische Alb	15	Bayer. Alpen
7	Obermain Schollenland		
8	Frankenwald und Fichtelgebirge		
9	Oberpfälzer Becken und Hügelland		

Abb. 1 Änderung der S-Gehalte in Fichtennadeln 1977-1983 in den Hauptlandschaftsräumen

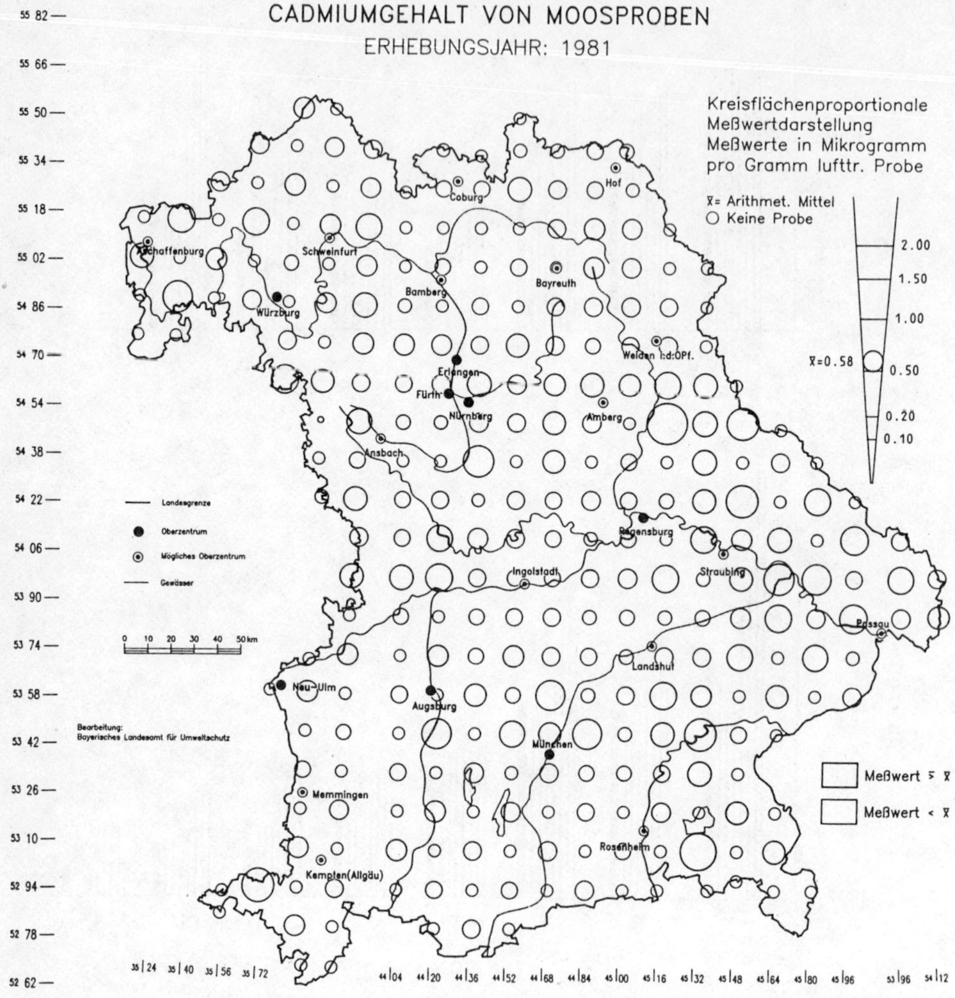

Herausgeber: Bayerisches Staatsministerium für Landesentwicklung und Umweltfragen

Abb. 2 Bioindikatornetz Moose

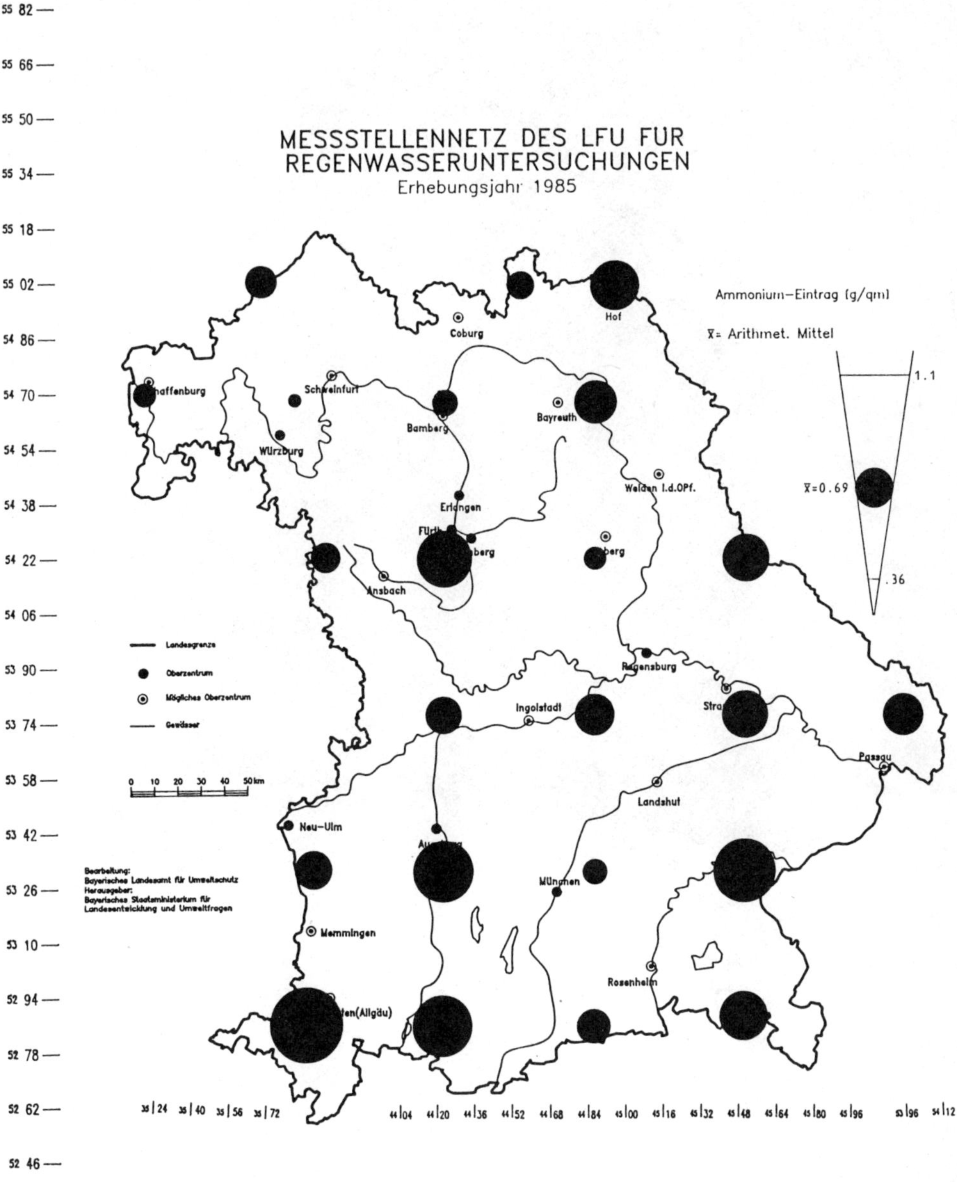
MESSSTELLENNETZ DES LFU FÜR
REGENWASSERUNTERSUCHUNGEN
Erhebungsjahr 1985
Ammonium-Eintrag (g/qm)
X̄= Arithmet. Mittel
1.1
X̄=0.69
.36
Coburg
Hof
Bamberg
Bayreuth
Würzburg
Erlangen
Ansbach
Ingolstadt
Regensburg
Passau
Landshut
Neu-Ulm
München
Memmingen
Rosenheim
Landesgrenze
Oberzentrum
Mögliches Oberzentrum
Gewässer
0 10 20 30 40 50km
Bearbeitung:
Bayerisches Landesamt für Umweltschutz
Herausgeber:
Bayerisches Staatsministerium für
Landesentwicklung und Umweltfragen

Abb. 3

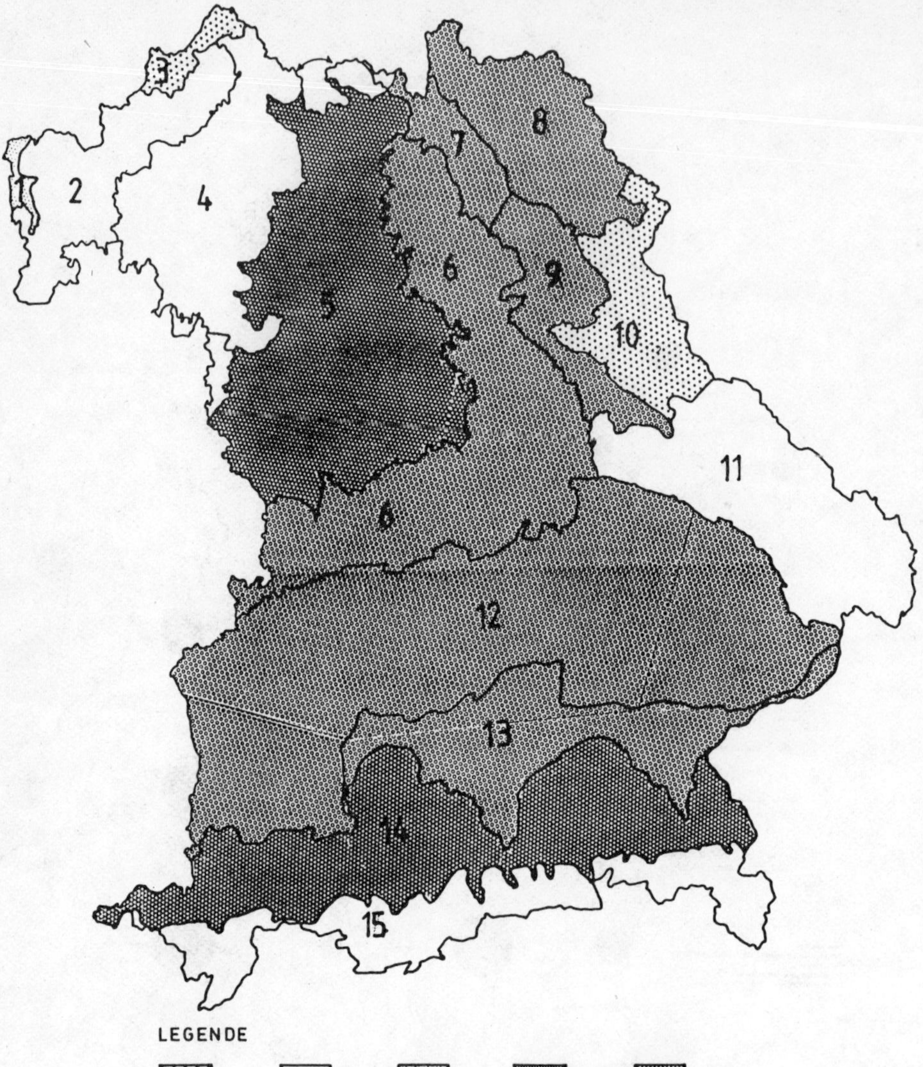

Abb. 4 Abweichung der Jahresniederschläge 1976 - 1981 vom langfristigen Mittel in den Hauptlandschaftsräumen

ANTEIL DEUTLICHER SCHÄDEN IN DEN WUCHSGEBIETEN BAYERNS

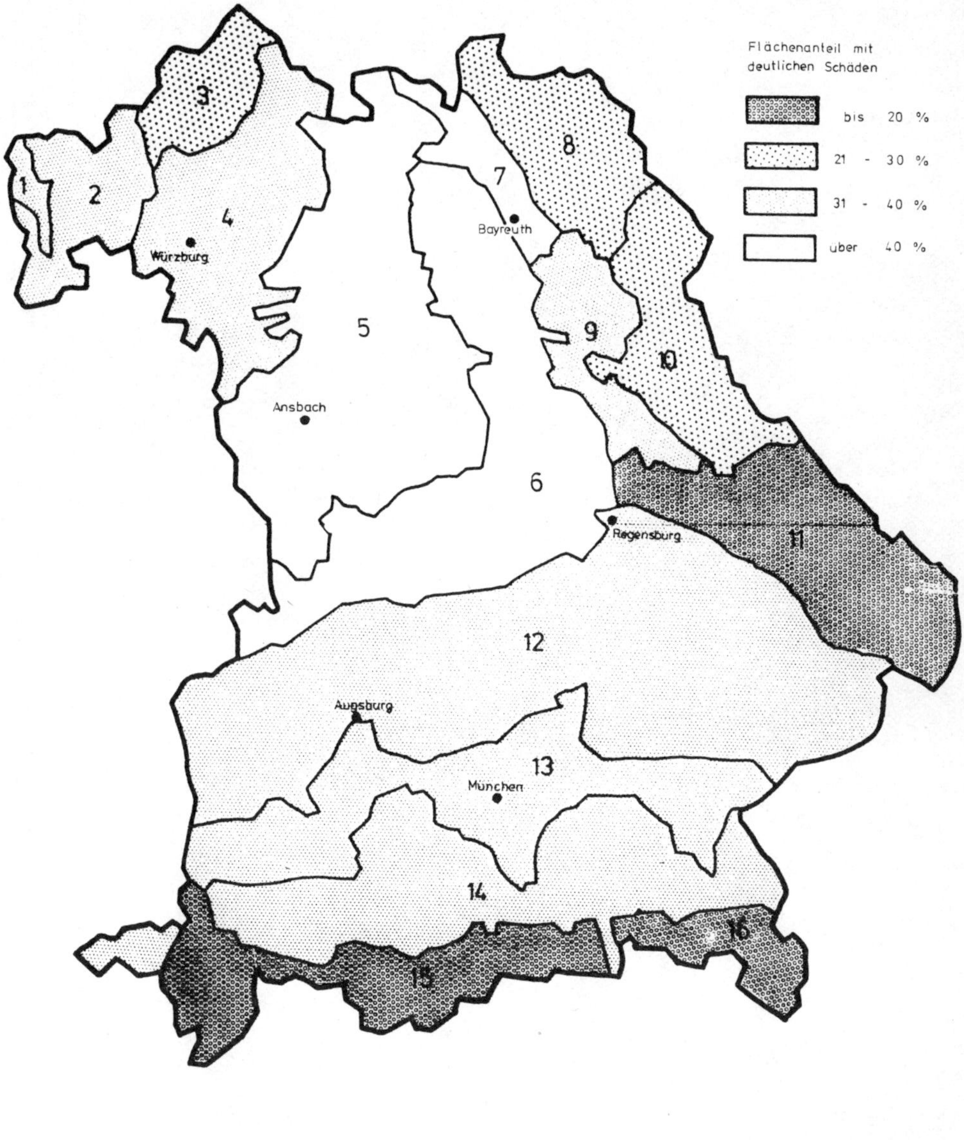

Abb. 5 Verteilung der Gebiete mit deutlichen Schäden im Jahr 1985
Quelle : Forstl. Forschungsberichte München 1986

THEME II : DIAGNOSIS

THEME II : DIAGNOSE

THEMA II : DIAGNOSE

ADENINE AND PYRIDINE NUCLEOTIDES, AND CARBOHYDRATES IN HEALTHY AND DECEASED NEEDLES: ALTERATIONS IN RELATION TO VEGETATION PERIOD (MAY TO OCTOBER 1985) AND NEEDLE AGE (1979 TO 1985)

R. Keil, T. Benz, W. Einig and R. Hampp
Universität Tübingen, Institut für Biologie I,
Auf der Morgenstelle 1, D-7400 Tübingen

Summary

Lyophilized spruce needles (Picea abies; location: Kälbelescheuer, southern Black Forest, 900 m a. sea level) were analyzed for their contents of adenine nucleotides (ATP, ADP: AdN), reduced (NADH, NADPH) and oxidized (NAD, NADP) pyridine nucleotides, and of carbohydrates (starch, sucrose, fructose, glucose). The metabolite levels were related to needle age, vegetation period, and the degree of damage (classified as 0 (control) and 2). Samples were taken of all distinct parts of a twig and analyzed individually. This was necessary as the levels of metabolites largely depended on the respective position of the needles.

The results were as follows:

1) With increasing needle age there was a significant decrease in the total AdN-pool, combined with an increase of the redox levels of both pyridine nucleotide systems (NADH/NAD; NADPH/NADP). This was true for both healthy and damaged needles.

2) The ratios of ATP/ADP, NADH/NAD and NADPH/NADP of deceased needles were higher compared to those of healthy ones.

3) Accumulation of starch from May to June was evident in control needles but largely absent in those from damaged trees. In contrast, levels of sucrose (about 150 nmol/mg dry weight), glucose and fructose (between 5 and 40 nmol/mg dry weight) indicated some dependence on the vegetation period but no significant differences with respect to the level of decease.

1. INTRODUCTION

Deceased Norway spruce in elevated sites of the southern Black Forest (Kälbelescheuer; West Germany) shows symptoms of bleaching combined with the loss of older needles. As part of a research program directed by the "Projekt Europäisches Forschungszentrum für Maßnahmen zur Luftreinhaltung" (PEF) investigations were performed in order to add to our understanding of the causal sequence. These were of physiological, biochemical, virological/bacteriological/mycological, and histological nature. Our part in this program is the analysis of metabolite pools which are involved in the cellular energy and carbohydrate metabolism, and which possibly can be used as biochemical indicators for the decease before optical changes are evident.

In this contribution we present data for adenine nucleotides (ATP, ADP: AdN), carbohydrates (starch, sucrose, fructose, glucose) and reduced (NADH, NADPH) as well as oxidized (NAD, NADP) pyridine nucleotides. The metabolite levels or ratios are related to needle age, vegetation period (May to October 1985) and the degree of damage.

2. MATERIAL AND METHODS

The trees were 30 to 50 years old Norway spruce (Picea abies L. Karst), classified as 0 (control) and 2 (levels of decease: 0 to 3). Location: Kälbelescheuer and Haldenhof, southern Black Forest, 900 m above sea level. Dominant air pollutant: Ozone.

Needles were taken from the 7th whorl, freeze stopped in liquid nitrogen, and stored at -80°C until use. For analysis they were homogenized (Dismembrator, Braun-Melsungen, F.R.G.: -20°C), and lyophilized. Adenine nucleotides were determined by luminometry (1), carbohydrates by end point determination (2,3). For the assay of pyridine nucleotides "enzymatic cycling" techniques were employed (4).

3. RESULTS AND DISCUSSION

ADENINE NUCLEOTIDES

Figs. 1 and 2 show the levels of ATP, ADP and the ratio of ATP/ADP in relation to the needle age and the date of sample collection. The values are the average contents of needles from S-, SN-, and N- positions of the respective twig (5).

ATP levels were always highest in the youngest needles and largely independent of the level of tree damage. In contrast, ADP was lower in class 2 needles. As a consequence, ATP/ADP ratios were significantly higher in class 2 needles. Together with the general increase of this ratio with needle age (both class 0 and 2), this data could be taken as evidence for accelerated senescence (6).

PYRIDINE NUCLEOTIDES

In parallel to the increase of the ratio, ATP/ADP, there was also a needle age-dependent increase of the redox level of the pyridine nucleotide system (Figs. 3 to 6). The ratios NADH/NAD (Fig. 5) and, less consistent, NADPH/NADP (Fig. 6) were higher in needles from class 2 trees. This observation could be related to enhanced activities of catabolic pathways. It thus could constitute additional evidence for accelerated processes of senescence in these needles (7).

CARBOHYDRATES

The amount per dry weight of starch, sucrose, glucose and fructose is illustrated by Figs. 7 to 10. While we observed accumulation of starch in control needles (May to June), this increase was virtually absent in class 2 samples (Fig. 7). The increase in starch seems typical for healthy needles during that time of the year (8). Its absence in class 2 needles should thus be due to a less efficient system of light-dependent carbon dioxide fixation. From June onwards all samples showed a steady decrease in their starch content. Time course and differences healthy/class 2 tree were less conclusive for sucrose and the hexoses. Sucrose levels were rather constant, independent of needle class or sampling date. The amounts of glucose (Fig. 9) and fructose (Fig. 10) exhibited considerable fluctuations on a much lower level compared to sucrose.

4. CONCLUSIONS

The decreased accumulation of starch in class 2 needles is surely the result of a less efficient system of photosynthetic CO_2-fixation. Such a decreased efficiency can be due to both a reduced number of light harvesting protein complexes and an impaired light activation of CO_2 converting enzymes. As these enzymes are Mg-dependent, the low levels of soil and tissue Mg, which are typical for the investigated location, should be involved in the observed effects.

The increased ratios of ATP/ADP, NADH/NAD and NADPH/NADP are possibly indicative for processes of accelerated senescence. As we observed a good correlation between these ratios and the degree of damage, such ratios could possibly be used as early indicators of an altered metabolic situation which, on the longer run, can lead to bleaching and, finally, needle loss.

REFERENCES

(1) HAMPP, R. (1985). ADP, ATP: luminometric method. In Methods of Enzymatic Analysis. Ed. Bergmeyer, H.U., Vol. VII, 370-379, 3. ed. Verlag Chemie Weinheim.

(2) BEUTLER, H.O. (1984). Starch. In Methods of Enzymatic Analysis. Ed. Bergmeyer, H.U., Vol. VI, 2-10, 3. ed. Verlag Chemie Weinheim.

(3) OUTLAW, W.H. and TARCZYNSKI, M.C. (1984). Sucrose. In Methods of Enzymatic Analysis. Ed. Bergmeyer, H.U., Vol. VI, 96-103, 3. ed. Verlag Chemie Weinheim.

(4) HAMPP, R. et al. (1984). Determination of compartmented metabolite pools by a combination of rapid fractionation of oat mesophyll protoplasts and enzymic cycling. Plant Physiol. 75, 1017-1021.

(5) LICHTENTHALER, H. et al. (1985). Untersuchungen über die Funktionsfähigkeit des Photosyntheseapparates be Nadeln gesunder und geschädigter Koniferen. Kernforsch. Zentrum Karlruhe, KfK-PEF-2, 81-105.

(6) MALIK, N.S.A. and THIMANN, K.V. (1980). Metabolism of oat leaves during senescence. Plant Physiol. 65, 855-858.

(7) MEYER, R. and WAGNER, K.G. (1986). Nucleotide pools in leaf and root tissue of tobacco plants: Influence of leaf senescence. Physiol. Plant. 67, 666-672.

(8). ERICSSON, A. (1979). Effects of fertilization and irrigation on the seasonal changes of carbohydrate reserves in different age-classes of needles in 20-year-old scots pines. Physiol. Plant. 45, 270-280.

ACKNOWLEDGEMENT. The work presented in this paper was financed by a grant from the "Projekt Europäisches Forschungszentrum für Maßnahmen zur Luftreinhaltung" (PEF; 84/043/1A).

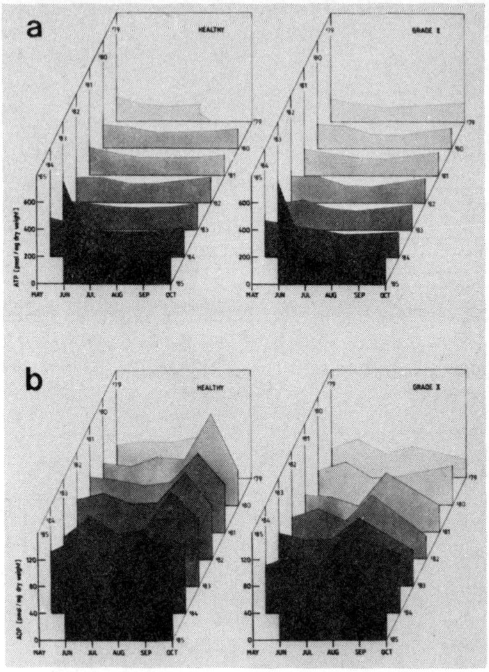

FIG. 1. ATP (A) AND ADP CONTENT (B) OF HEALTHY (LEFT) AND DAMAGED (RIGHT) NEEDLES OF NORWAY SPRUCE IN RELATION TO NEEDLE AGE AND VEGETATION PERIOD.

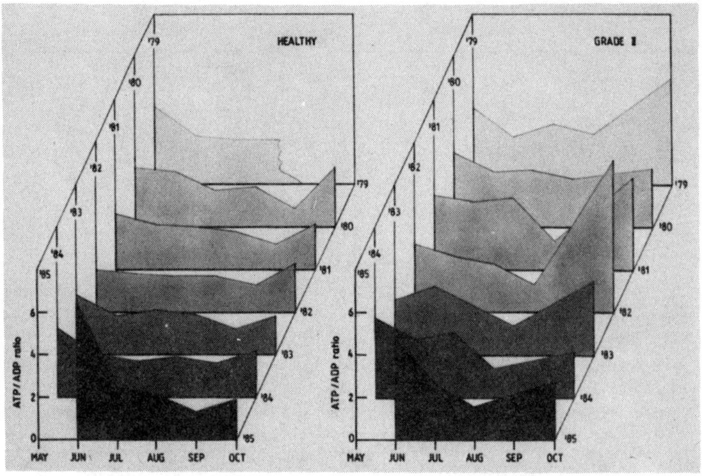

FIG. 2. ATP/ADP RATIOS OF HEALTHY (LEFT) AND DAMAGED (RIGHT) NEEDLES OF NORWAY SPRUCE IN RELATION TO NEEDLE AGE AND VEGETATION PERIOD.

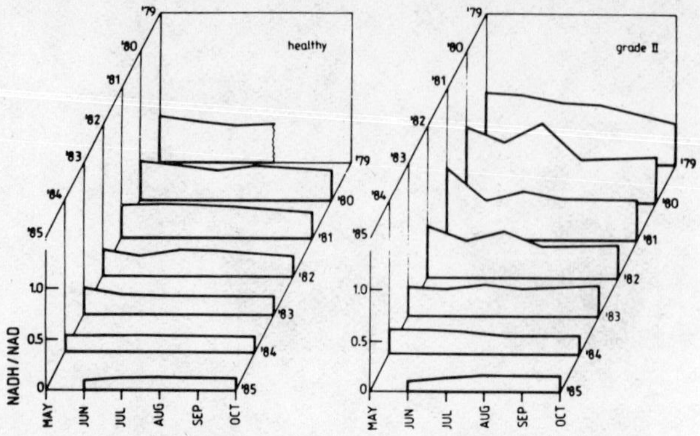

FIG. 3. NADH/NAD RATIOS OF HEALTHY (LEFT) AND DAMAGED (RIGHT) NEEDLES OF NORWAY SPRUCE IN RELATION TO NEEDLE AGE AND VEGETATION PERIOD.

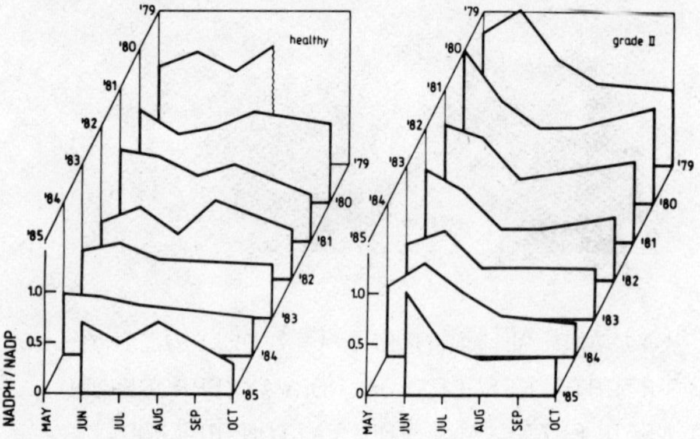

FIG. 4. NADPH/NADP RATIOS OF HEALTHY (LEFT) AND DAMAGED (RIGHT) NEEDLES OF NORWAY SPRUCE IN RELATION TO NEEDLE AGE AND VEGETATION PERIOD.

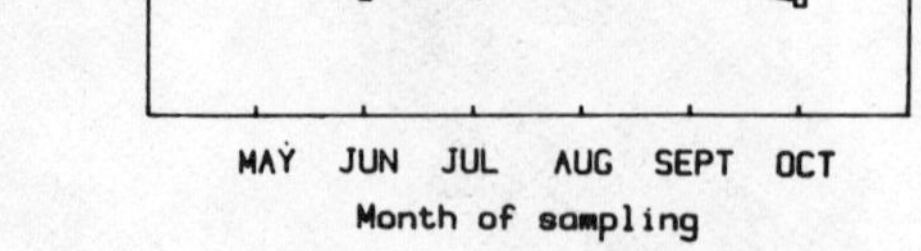

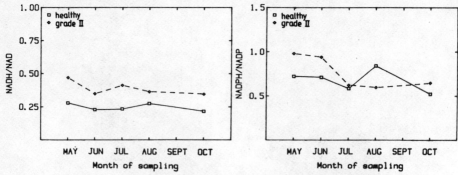

FIG. 5. TIME COURSE OF NADH/NAD RATIOS OF NORWAY SPRUCE NEEDLES. MEAN VALUES (N = 25) OF EACH SAMPLING DATE. VALUES OF CURRENT YEAR NEEDLES ARE NOT INCLUDED

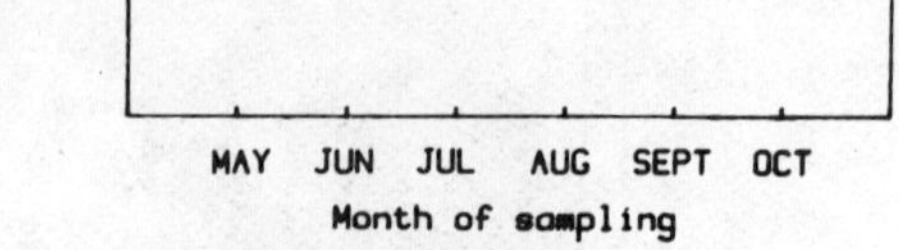

FIG. 6. TIME COURSE OF NADPH/NADP RATIOS OF NORWAY SPRUCE NEEDLES. MEAN VALUES (N = 25) OF EACH SAMPLING DATE. VALUES OF CURRENT YEAR NEEDLES ARE NOT INCLUDED.

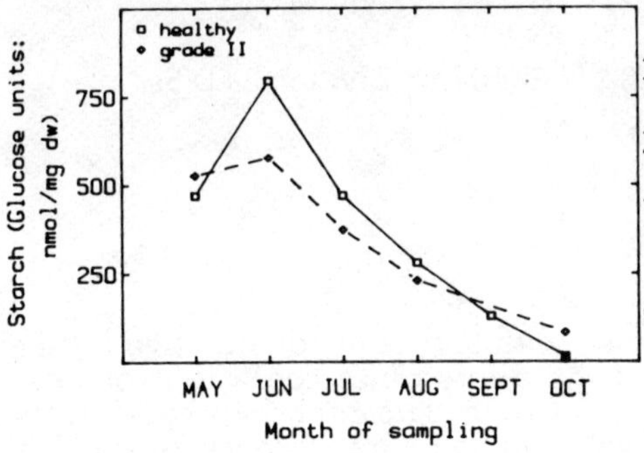

FIG. 7. TIME COURSE OF STARCH CONTENT OF NORWAY SPRUCE NEEDLES. MEAN VALUES (N = 25) OF EACH SAMPLING DATE. VALUES OF CURRENT YEAR NEEDLES ARE NOT INCLUDED.

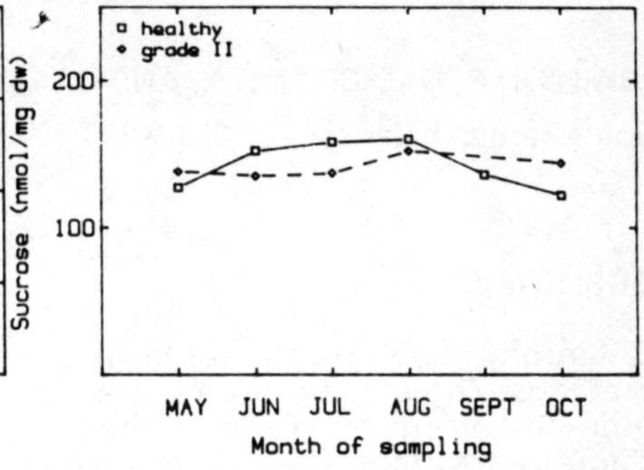

FIG. 8. TIME COURSE OF SUCROSE CONTENT OF NORWAY SPRUCE NEEDLES. MEAN VALUES (N = 25) OF EACH SAMPLING DATE. VALUES OF CURRENT YEAR NEEDLES ARE NOT INCLUDED.

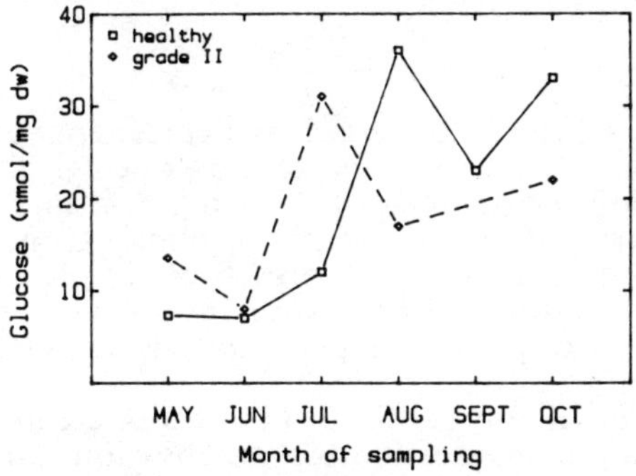

FIG. 9. TIME COURSE OF GLUCOSE CONTENT OF NORWAY SPRUCE NEEDLES. MEAN VALUES (N = 25) OF EACH SAMPLING DATE. VALUES OF CURRENT YEAR NEEDLES ARE NOT INCLUDED.

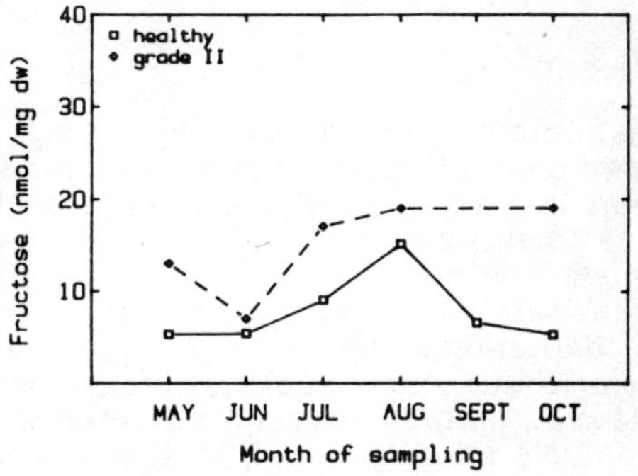

FIG. 10. TIME COURSE OF FRUCTOSE CONTENT OF NORWAY SPRUCE NEEDLES. MEAN VALUES (N = 25) OF EACH SAMPLING DATE. VALUES OF CURRENT YEAR NEEDLES ARE NOT INCLUDED.

Possible use of chlorophyll fluorescence in early detection of forest damages: Influence of different criteria.

BAILLON F., DALSCHAERT X. AND GRASSI S. - Chemistry Division - ED 5A
Joint Research Center - 21021 ISPRA (VA) - ITALY

SUMMARY

Chlorophyll fluorescence has been analysed with Spruce needles by two different methods. Those allow the calculation of the fluorescence coefficient or vitality index Rfd on one hand and, on the other hand, of the two quenching coefficients of fluorescence (photochemical quenching, qQ, and non photochemical one, qE). The influence of different parameters on fluorescence has been studied. They are linked to the conditions of measurement, ligth intensity and time of dark adaptation or to the green material as age of needles and genetic features. Moreover, the first results obtained in a study on magnesium deficiency, give support to the possible use of the second method in early detection of forest damages.

RESUME

La fluorescence de la chlorophylle a été mesurée sur des aiguilles d'Epicéas par deux méthodes différentes. Ces deux méthodes d'analyse permettent de calculer d'une part le coefficient de fluorescence ou index de vitalité Rfd, et d'autre part, les coefficients de quenching de la fluorescence (quenching photochimique, qQ, et quenching non photochimique, qE). L'influence sur la fluorescence de plusieurs paramètres liés soit aux conditions de mesure, intensité lumineuse et temps d'adaptation à l'obscurité, soit au materiel végétal lui-même, âge des aiguilles et caractèristiques génétiques, a été étudié.
De plus, les premiers résultats obtenus lors d'une étude sur l'effet d'une carence minérale en magnésium, laissent envisager des possibilités d'utilisation de la deuxiéme méthode d'analyse pour la détection précoce des dommages forestiers.

INTRODUCTION

Ligth energy absorbed by the different foliar pigments is transferred to the reaction centers, and then transformed in energetic substrates (ATP and NADPH) during the photochemical reactions of photosynthesis.During this process, a part of this energy is dissipated into heat, but also reemitted as ligth from the photosystem 2. This process is called fluorescence. The re-emitted ligth has a lower energetic level than that of the absorbed ligth. Therefore, the chlorophyll fluorescence is emitted at a higher wavelength (peaks at 685 and 730 nm).
Fluorescence is an indicator of photosynthesis, and its intensity depends on the efficiency with which ligth energy is transformed within the photosynthetic apparatus. Moreover, photosynthesis is a sensible parameter of physiological activity in case of plant stress.
The aim of this work is to evaluate the possibilities of fluorescence use in natural conditions to detect, before the apparition of any visible damages, anomalies in the physiological working of plants, and specially for a use in early detection of forest damages. To begin with, a study is carried out with healthy Spruces to determine the influence of phenological stages of the tree and of environmental conditions on chlorophyll fluorescence measurements. Secondly, the effect of magnesium mineral

deficiency is studied with Spruces. During this work, two different methods of fluorescence measurements are investigated.

PLANT CULTURE

Four years' old Spruces (Picea abies L.) were grown in 1 L. pot.
Non cloned Spruces provided from the forestry station of ORINO (ITALY).
Cloned Spruces belong to the clone 0665 producted by AFOCEL-EST in LESSEUX (FRANCE).

In another experiment, Spruces (non cloned) were grown in a complete hydroponic solution (HOAGLAND half strength) during two months, and then in a magnesium deficient solution.

Two months before the beginning of the experiments, the plants were placed in a controlled chamber:

light period: 16H, $100W/m^2$, 20°C, 45%RH.
dark period: 8H, 15°C, 70%RH.

METHODS

Chlorophyll fluorescence displays an induction kinetics, called the Kautsky effect, upon illumination of dark pre-adapted leaf (Fig.1). Fluorescence rises quickly (in ms range) to a maximum peak, Fm, and then decreases slowly (in min. range) to a steady-state value, Fs.
Two kinds of instruments are used to detect fluorescense signals emitted by plant material.

Fig.1

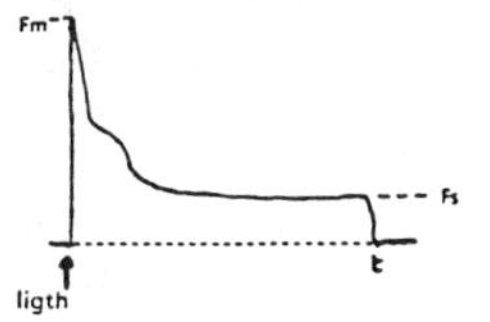

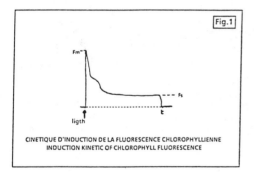

CINETIQUE D'INDUCTION DE LA FLUORESCENCE CHLOROPHYLLIENNE
INDUCTION KINETIC OF CHLOROPHYLL FLUORESCENCE

Fig.2

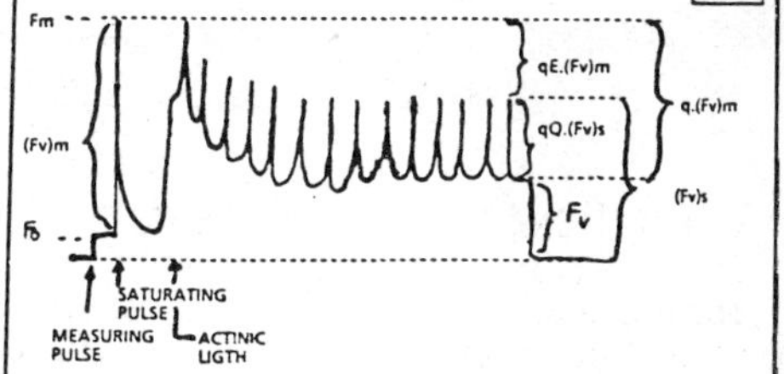

DEFINITION DES COEFFICIENTS DE QUENCHING DE LA FLUORESCENCE
DEFINITION OF THE COEFFICIENTS OF FLUORESCENCE QUENCHING

The first unit is built by HANSATECH to be used in photosynthetic applications with LD2 oxygen electrode unit. The fluorescence signal is detected by a photodiode (>700nm) and converted into an electrical signal proportional to the light received (<500nm).This signal (Kautsky effect) is then amplified to a level suitable to be recorded (signal 1). It allows the calculation of the florescence coefficient or vitality index: $R_{fd} = \frac{(Fm-Fs)}{Fs}$, (LICHTENTHALER, 1986).

The second unit was developped by Schreiber U. (University of Würzburg). The PAM chlorophyll fluorometer is based on a new pulse modulation measuring principle. This system uses a special measuring ligth for fluorescence excitation (<670nm) and only the resulting fluorescence signal (>700nm) is amplified. Using the saturating pulse method (Fig.2), new informations can be obtained on the photosynthetic capacity of plants by the fluorescence quenchings.Indeed, the photochemical quenching q_Q reflects the rate of electrons transport in the thylakoid membrane, and the non photochemical quenching q_E informs on the energization of the thylakoid membrane, i.e. on the protons gradient, and indirectly on the ATP synthesis . The coefficients of quenching are calculated from the induction kinetic relationship:

$$q_Q = \frac{[(Fv)s - Fv]}{Fv} \qquad q_E = \frac{[(Fv)m - (Fv)s]}{(Fv)s}$$

(SCHREIBER AND BILGER, 1985).

RESULTS

1 - INFLUENCE OF DARK ADAPTATION TIME

The coefficient of fluorescence, Rfd, increases rapidly for dark adaptation time in the range 0 and 15 minutes. After 30 minutes, it reaches a constant value, depending on the kind of needles (Fig. 3). The quenching coefficients of fluorescence reach a constant value more rapidly, after 15 minutes.

Fig. 3

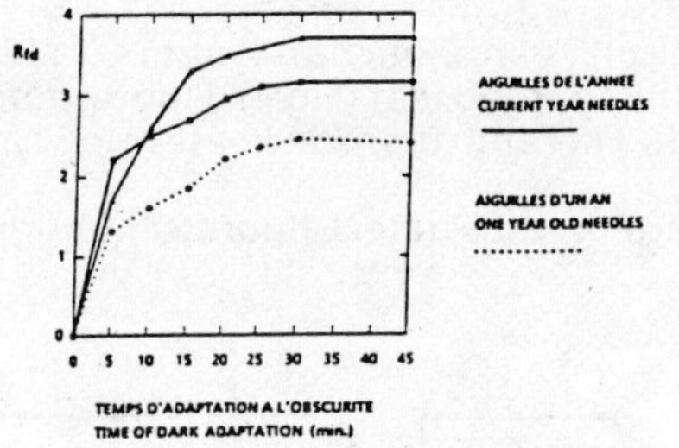

INFLUENCE DU TEMPS D'ADAPTATION A L'OBSCURITE SUR LA FLUORESCENCE
INFLUENCE OF THE TIME OF DARK ADAPTATION ON FLUORESCENCE

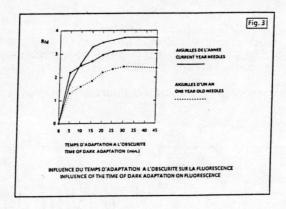

2 - INFLUENCE OF LIGTH INTENSITY

The effect of ligth intensity on the coefficient of fluorescence; Rfd; (Fig. 4) can be compared to its effect on photosynthesis activity (Fig. 5); i. e. apparition of a saturation plateau at a value around 300 μE/m2.s in the case of current year needles and at 500 μE/m2.s with one year old needles.

Fig. 4

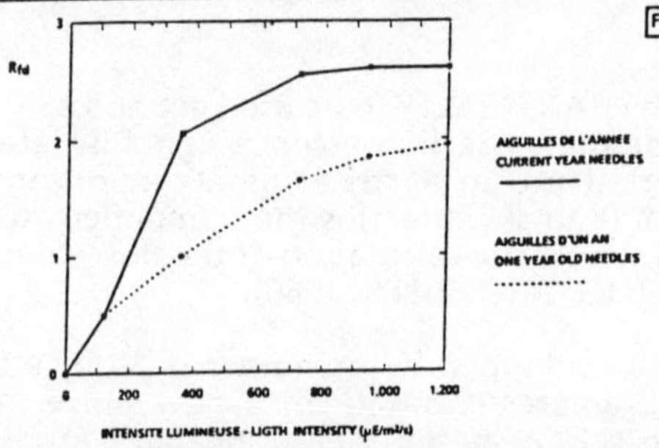

COEFFICIENT DE FLUORESCENCE (Rfd) MESURE SUR EPICEA EN FONCTION DE L'INTENSITE LUMINEUSE.
FLUORESCENCE COEFFICIENT (Rfd) MEASURED ON SPRUCE AT VARIOUS LIGTH INTENSITIES.

Fig. 5

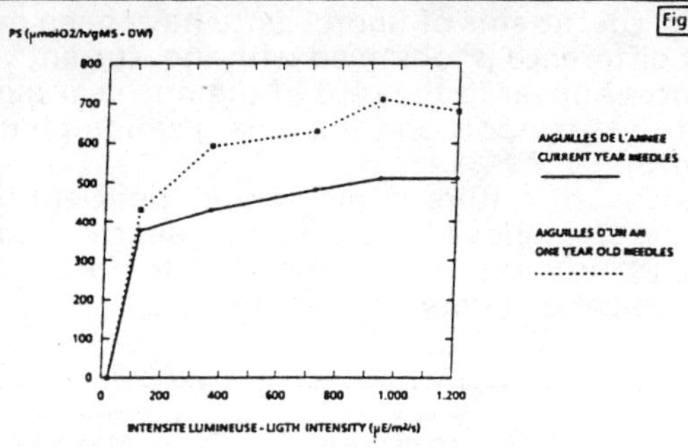

INTENSITE PHOTOSYNTHETIQUE (PS) DE L'EPICEA EN FONCTION DE L'INTENSITE LUMINEUSE.
PHOTOSYNTHETIC ACTIVITY (PS) OF SPRUCE AT VARIOUS LIGTH INTENSITIES.

3 - INFLUENCE OF THE NEEDLES AGE

The coefficient Rfd of current year needles is higher than the one of one year old needles (Tabl. 1), at the opposite of photosynthetic activity (Fig. 3, 4 and 5), whereas no significant difference is observed on the two quenching coefficients (Fig. 6). This confirms that they are qualitative indicators of the working capacity of the photosynthetic apparatus (SCHREIBER and BILGER, 1985). The photochemical quenching q_Q decreases rapidly (q_Q = 0,5 to 0,6), and then increases more slowly (q_Q = 0,9). This corresponds to an initial rapid reduction of the electrons transport chain followed by a slow reoxidation. At the opposite, the non photochemical quenching q_E increases rapidly (in 1min., 0,7) and then decreases slowly (0,5), corresponding to the membrane energization relaxation as the Calvin cycle and correlated use of ATP and protons gradient set in.

An important variability of values is observed with the non cloned trees compared to the cloned ones. Presently, the experiments are made with cloned Spruces.

Tabl. 1: VARIATION DU COEFFICIENT DE FLUORESCENCE EN FONCTION DE L'AGE DES AIGUILLES.
VARIATIONS OF THE COEFFICIENT FLUORESCENCE WITH NEEDLES AGE.

A - AIGUILLES DE L'ANNEE
CURRENT YEAR NEEDLES

B - AIGUILLES D'UN AN
ONE YEAR OLD NEEDLES

Rfd	NON CLONES	CLONES
A	2,11 + 0.59	2,41 + 0,45
B	1,53 + 0.48	1,82 + 0,37

Fig. 6

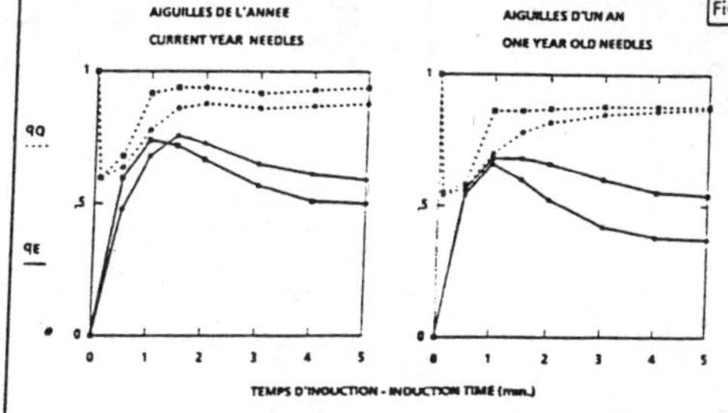

COEFFICIENTS DE QUENCHING DE LA FLUORESCENCE MESURES SUR DES AIGUILLES D'EPICEA
QUENCHING COEFFICIENT OF FLUORESCENCE MEASURED ON SPRUCE NEEDLES

4 - EFFECT OF MAGNESIUM MINERAL DEFICIENCY

After one month of culture in magnesium deficient hydroponic solution, no visible symptoms have been observed on the needles of different age. Moreover, the photosynthesis measurements (oxygen evolution) and the foliar pigments analysis do not show a significant difference compared to the controls. It can be noticed that no differences are found between the trees grown in complete hydroponic solution and those grown in soil (Fig. 6 and 7). Therefore, there is no influence of the nutrition conditions.

The quenching coefficients of fluorescence have been measured on these needles. No significant difference is observed with the current year needles. On the other hand, differences appear in the case of the one year old needles. The reoxidation phase of electrons transport and the energization phase of thylakoïd membrane are slower (Fig. 8).

After two months of culture in magnesium deficient solution, there are visible symptoms on the two kinds of needles. The quenching coefficients are modified in all cases (Fig. 9), but the most important feature is a non relaxation of the membrane energization (slow phase of q_E).

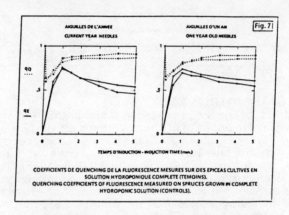

Fig. 7

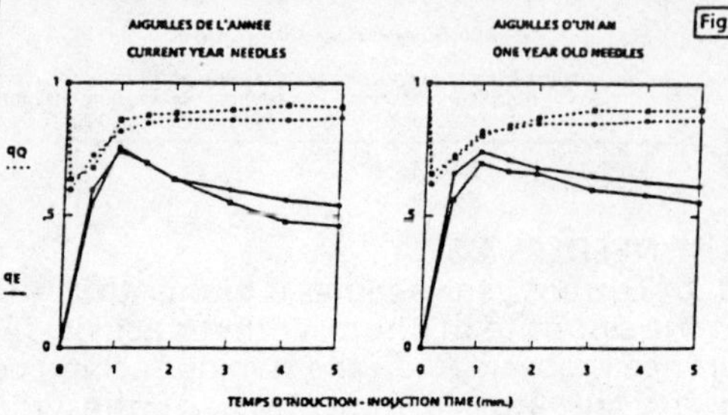

COEFFICIENTS DE QUENCHING DE LA FLUORESCENCE MESURES SUR DES EPICEAS CULTIVES EN SOLUTION HYDROPONIQUE COMPLETE (TEMOINS).
QUENCHING COEFFICIENTS OF FLUORESCENCE MEASURED ON SPRUCES GROWN IN COMPLETE HYDROPONIC SOLUTION (CONTROLS).

Fig. 8

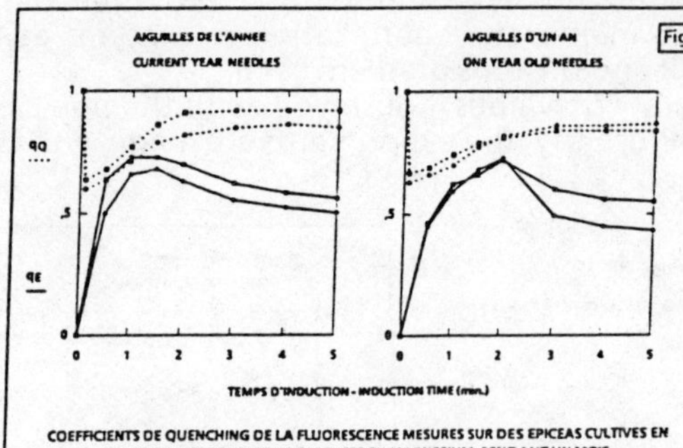

COEFFICIENTS DE QUENCHING DE LA FLUORESCENCE MESURES SUR DES EPICEAS CULTIVES EN SOLUTION HYDROPONIQUE CARENCEE EN MAGNESIUM, PENDANT UN MOIS.
QUENCHING COEFFICIENTS OF FLUORESCENCE MEASURED ON SPRUCES GROWN IN MAGNESIUM DEFICIENT HYDROPONIC SOLUTION , FOR ONE MONTH.

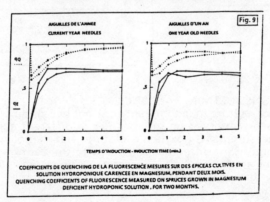

Fig. 9

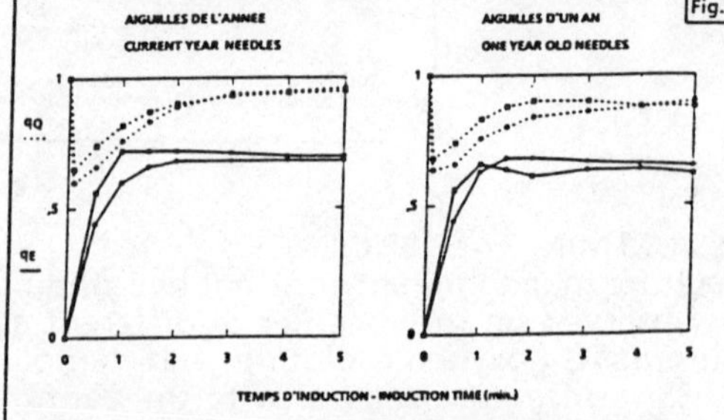

COEFFICIENTS DE QUENCHING DE LA FLUORESCENCE MESURES SUR DES EPICEAS CULTIVES EN SOLUTION HYDROPONIQUE CARENCEE EN MAGNESIUM, PENDANT DEUX MOIS.
QUENCHING COEFFICIENTS OF FLUORESCENCE MEASURED ON SPRUCES GROWN IN MAGNESIUM DEFICIENT HYDROPONIC SOLUTION , FOR TWO MONTHS.

CONCLUSION

The first results on the influence of ligth intensity, dark adaptation time and the age of needles confirm those already obtained by others authors (LICHTENTHALER, 1986; SCHMUCK, 1986). The one year old needles of Spruce are firstly affected by a magnesium mineral deficiency, as this has also been showed by FINK (1987). The fluorescence and particularly the quenching coefficients allow the detection of a physiological disfunction before seeing any visible damages on the needles.The protons gradient through the thylakoid membrane, indirectly the ATP synthesis and the Calvin cycle are the first sites reached. This confirms that magnesium is involved in the activation of RuBP carboxylase enzyme (ANDERSON, 1987).
The results obtained with the method allowing the calculation of the two coefficients of fluorescence quenching are more homogeneous. Presently, the experiments continue with this method, and others stress conditions are studied.

REFERENCES

ANDERSON L., 1987; Stromal reaction to air pollutants; 2nd International Symposium on Air Pollution and Plant Metabolism, Munich; April 6-9.

FINK S., 1987; Histological and cytological changes caused by air pollutants; 2nd International Symposium on Air Pollution and Plant Metabolism, Munich; April 6-9.

LICHTENTHALER H. K., 1986; Laser induce chlorophyll fluorescence of living plants; Proceedings of IGARSS'Symposium, Zürich; Sept. 8 - 11.

SCHMUCK G., 1986; Vergleichende Untersuchungen zur Erfassung der Vitalität von Baümen: Ein Beitrag zur Waldschadensforschung.; Karlsr. Beitrag. Pflanzenphysiol., 15, 1-173.

SCHREIBER U. and BILGER W., 1985; Rapid assessment of stress effects on plant leaves by chlorophyll fluorescence measurements.; NATO Advanced Research Workshop, Sesimbra, Portugal; Oct.

ACKNOWLEDGEMENTS

Grateful thanks to Drs. OMENETTO and ROSSI for their help and advices during the experiments described, and to the european community for the financial support.

FLUORESCENCE AS A MEANS OF DIAGNOSING THE EFFECT OF POLLUTANT INDUCED STRESS IN PLANTS

O. van Kooten[1] and L. W. A. van Hove[2]
[1]Lab. Plant Physiological Res., Agricultural Univ. Wageningen, Gen. Foulkesweg 72, 6703 BW Wageningen, the Netherlands. [2]Department of Air Pollution, Agricultural Univ., P. O. Box 8129, 6700 EV Wageningen, the Netherlands.

Summary

Recent developments in chlorophyll fluorescence studies have led to the application of the so called fluorescence-light-doubling technique (1). The technique is based on frequently changing the light intensity during the measurement of the variable fluorescence or the so called "Kautsky" curve (2). This enables us to deconvolute the measurement in a photochemical (q_Q) and a non-photochemical (q_E) part (3). Further manipulation of the actinic light intensity and the ambient CO_2 and O_2 concentrations enhances our possibilities of probing the plant cell at different metabolic and molecular levels (4). Long term exposures to relatively low levels of NH_3 (50 and 100µg/m^3) caused small visual effects in poplars and no visual effects in bean plants. Maximum photosynthesis rates, as measured by the CO_2 consumption in a leaf chamber (see Van Hove these proceedings), exhibited only small differences between the exposed plants and the controls. However, the fluorescence measurements did reveal a correlation between the maximum photosynthesis rate and the photochemical dependent fluorescence quenching (q_Q) in the steady-state illumination. Leaf development differences caused by the NH_3 exposure in *Phaseolus vulgaris* L. and *Populus euramericana* L. could be detected with the fluorescence curves. Short time exposure to a high concentration of a secondary pollutant (> 5ppm NO_x) revealed a stronger response in the NH_3 treated plants than in the control plants (5).

Introduction

Intensive livestock breeding in the Netherlands is concentrated in areas with predominantly sandy soils. The emission of ammonia (NH_3) in these areas, caused by NH_3 volatilization of animal manure, amounts up to $76 \cdot 10^3$ tons NH_3 a year or 53 percent of the total NH_3 emission in the Netherlands (6). There is increasing evidence that this high NH_3 emission contributes significantly to the serious dieback of forests and to the strong decline in plant species in these areas (7).

To investigate the effects of long-term exposure of plants to gaseous ammonia we have employed the light-doubling-fluorescence technique (1) and compared the results with CO_2 fixation rate measurements in a special designed leave cuvette (see Van Hove, these proceedings). Young poplar cuttings were exposed to 0, 50 or 100 µg $NH_3 \cdot m^{-3}$ and their fluorescence response during the transition from a dark to a light adapted state was followed in time. After eight weeks of exposure the light dependent photosynthesis rate was determined in a gas exchange leaf cuvette (see Van Hove, these proceedings). During the exposition it was evident that the exposed plants grew faster than the controls. In the fluorescence response of the young leaves one can observe the effect of an age difference between leaves of the same internodium but exposed to different concentrations of ammonia. When the leaves are fully developed this age difference disappears but a significant difference in electron transport rate and in the energy charge of the chloroplasts remains. These values are determined through the two quenching components of the fluorescence in the light adapted state (1). These results indicate that the level of photosynthesis rate in the exposed plants is slightly higher than in the controls. Preliminary results of maximum CO_2 fixation rate measurements seem to corroborate these observations.

Previous experiments with poplar plants and with beans showed a heightened sensitivity to short term exposures of high concentrations of NO_x (> 5ppm) for the ammonia exposed plants (5). This could be in line with the results mentioned above. It has been measured that the plants assimilate large quantities of the gaseous ammonia (Van Hove, these proceedings). Thus one can consider the control plants to be nitrate deficient, while this deficiency is relieved (at least in part) in the exposed plants. Such plants grow faster and are in general more susceptible to external stress factors.

We conclude that in poplars, a nitro-phylic plant, the exposition of ammonia as a <u>sole</u> pollutant in relatively low doses can be used by the plant as a source of extra reduced nitrogen. It enhances the electron transport rate in fully developed chloroplasts and lowers the (Gibbs) free-energy of ATP in the stroma during illumination. This is a strong indication for a higher turn over rate of the Calvin cycle in the light in fully developed chloroplasts, which is substantiated by preliminary results of direct photosynthesis measurements.

<u>Materials and Methods</u>

Poplar plants, i. e. *Populus euramericana* L., and beans, i. e. *Phaseolus vulgaris* L., were fumigated and treated as described by Van Hove in these proceedings. Fluorescence curves were measured starting the second week of ammonia exposition. Plants were taken out of the fumigation chambers and dark adapted for 30 minutes at 25 C and 70% relative humidity. Leaves were clamped in a laboratory build cuvette, positioning the single end of a quadruforcated light guide at 3 mm from the leaf and enabling a flow of moistened gas (2% O_2, 320 ppm CO_2 plus N_2) to pass at a rate of 80 ml/min. past the cuticula of the bottom side of the leaf. This cuvette is carefully designed not to damage the leaf. The low oxygen concentration in the gas is meant to eliminate Mehler reactions in the chloroplasts. In the absence of Mehler reactions it must be possible to relate electron transport rate to photosynthesis rate even at low CO_2 concentrations in the chloroplasts (8). With poplar the leaf of the 10th internodium was used and with bean one of the first triadic leaves was used. The actual fluorescence measurement proceeded in the following manner: first a low intensity (< 10 $mW{\bullet}m^{-2}$, 650 nm) pulsating measuring light beam illuminates the leaf through the light guide. Through the same light guide a photodiode can monitor returning light, i. e. emitted by the leaf, with a wavelength > 700nm and pulsating with the same frequency and phase as the measuring light beam. Due to this design the detected light can only be caused by chlorophyll fluorescence. The low intensity of the measuring beam combined with the dark adapted state of the leaf ensures a maximum quantum yield of photochemistry for the light absorbed by the plant's pigments. Thus the intensity of the light re-emitted as fluorescence is minimal and we denote this minimal level of fluorescence detected at the start of the measurement Fo (3). After detection of Fo an actinic light source (≈ 25 $W{\bullet}m^{-2}$, 650 nm) induces a rise in fluorescence due to a lowering of the quantum yield for photochemistry. This new changing level of fluorescence intensity is denoted by **Fv** and its development in time is known as the "Kautsky" curve (9). After 1 s of actinic illumination a third light source (1000 $W \cdot m^{-2}$, white light < 700 nm) illuminates the leaf for 700 ms. As a consequence of this bright illumination the electron transport chains in the chloroplast's thylakoids are completely saturated and all plastoquinones between photosystem 2 (PSII) and photosystem 1 (PSI) are reduced including the primary and secondary quinone acceptors of PSII, i. e. Q_A and Q_B respectively (10). Because of the virtual stop of photochemistry during this bright illumination the quantum yield becomes nil. The fluorescence level is now maximal and we denote it as **Fmax**. After this saturating illumination the fluorescence will fall off to the level Fv induced by the continuous actinic lamp. Three seconds later the high intensity illumination of 700 ms is repeated, but this time the level of saturated fluorescence remains below Fmax due to a mechanism known as energy dependent fluorescence quenching (q_E). We call this level of variable but saturated fluorescence **Fvs** (1, 3). During the first minute this bright illumination is repeated every 3 seconds. After the first minute the repetition rate is 0.1 Hz. This is continued until both Fv and Fvs seem to have reached a steady-state level, i. e. after a total illumination time of 6 minutes. Subsequently the actinic light is turned off and the development of Fv and Fvs in the dark is monitored for one minute. The result of such measurements is illustrated in figure 1.

In order to relate the measured fluorescence curves to photosynthetic performance of the chloroplasts it is necessary to calculate the photochemical dependent quenching (q_Q) and the energy dependent quenching (q_E) from Fv and Fvs (1).

$$q_E(t) = (Fmax - Fvs(t))/(Fmax - Fo)$$

$$q_Q(t) = (Fvs(t) - Fv(t))/(Fvs(t) - Fo)$$

q_E is believed to be linearly related to the pH gradient across the thylakoid membrane (11); q_Q reflects the redox state of the quinone pool, which is directly correlated to the electron transport rate in the thylakoid membranes (8). Due to the low O_2 pressure in the gas mixture the electron transport rate can be correlated with the actual CO_2 reduction rate.

Results and conclusions

In figure 1 we see averaged fluorescence measurements and their calculated q_Q and q_E curves for different periods of fumigation. Fig. 1A reveals fluorescence from poplar leaves, which had been in the fumigation chambers for 2 weeks. Each curve is an average of 7 measurements on different plants. The standard deviation of these points is below 10% in all cases. The main difference in the fluorescence curves can be detected in the first minute of illumination. When we look at q_Q and q_E at the left it is evident that also in the last minutes of illumination there is a significant difference in q_Q. The q_Q of the control plants is highest, while the plants exposed to 100 $\mu g/m^3$ have the lowest level of q_Q. For q_E the reverse can be said to be true. This leads to the conclusion that in young leaves of control plants electron transport is higher and the energy charge is lower than in similar leaves of fumigated plants. The effect seems to be concentration dependent. A high electron transport rate and a low energy charge in the light is typical for young leaves. Upon aging the electron transport rate diminishes and the

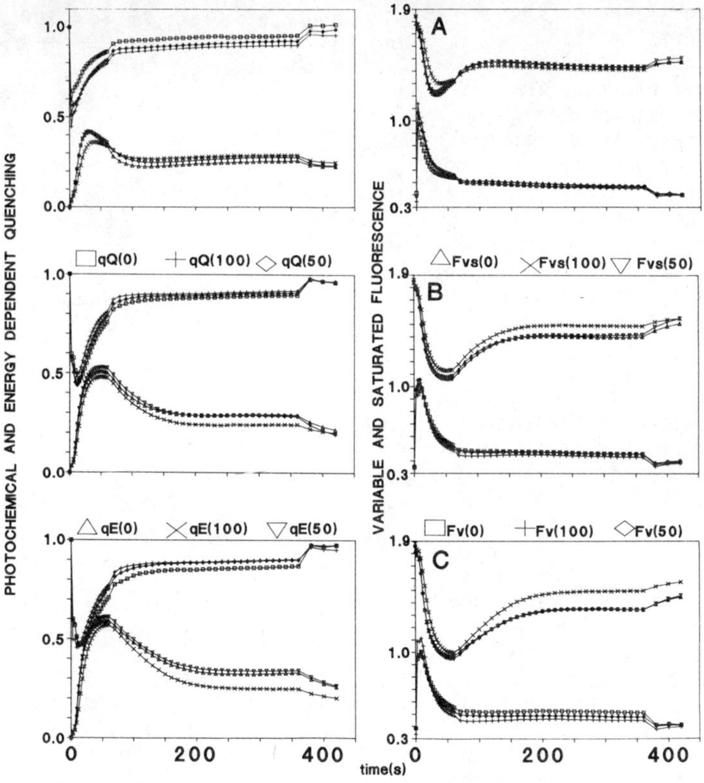

Figure 1. The variable (Fv) and saturated (Fvs) fluorescence on the right, the photochemical (q_Q) and the energy dependent (q_E) quenching on the left of fumigated poplar leaves (*Populus euramericana* L., 10th internodium). At t=0 s the actinic light is turned on and it is switched off 360 s. The ammonia concentration of the fumigation chambers the plants were taken from are in $\mu g \cdot m^{-3}$ and are given in brackets.
Every point is an average of 7 measurements. A: are plants that have been in the chambers for two weeks. The leaves are still very young and small. B: plants that have been exposed for 5 weeks and the leaves are fully developed. C: plants that have been exposed for 7 weeks. Here each point is an average of 5 measurements.

energy charge in the light rises due to a lower turnover rate of the Calvin cycle. Thus we conclude that in the first stage of leaf development the fumigated plants grow faster in order to explain the age difference between leaves of the same internodium. In fig. 1B the same measurements are depicted, but now after 5 weeks of fumigation. As can be seen there are hardly any differences in the first minute of illumination nor in q_Q. The leaves are fully developed and this is reflected in the equivalence of the electron transport rate. Only q_E in the steady-state is seen to be substantially lower in the case of the 100 $\mu g \cdot m^{-3}$ exposed plants. This indicates that the metabolic rate of the Calvin cycle is higher in these plants. In fig. 1C, which are curves measured after 7 weeks of treatment, this tendency seems to be enhanced. The q_E is clearly much lower for the 100 $\mu g \cdot m^{-3}$ plants, but now the q_Q is also lower in the control plants. The latter indicates diminished photosynthesis rates, in the control plants and consequently a higher Calvin cycle activity in the case of a higher ammonia exposition. These effects suggest a faster development in the first two weeks and a retarded ageing in the last weeks of NH_3 fumigated plants within the time period under study.

In figure 2 we take a closer look at these tendencies. Fig. 2A reveals the development of q_Q as measured in the steady-state, i.e. when the actinic illumination is turned off at t=360 s, for the three different ammonia concentrations. Here also the values are averages of 7 measurements. There is a clear tendency for q_Q of the control plants to decline throughout the period of treatment, while q_Q of the fumigated plants stays more or less constant. The development of q_E in the steady-state is less clear (fig. 2B). The spread in the measurements is much larger, but one can detect a slight rise in the control plants and a slight diminution in the case of the 100 $\mu g \cdot m^{-3}$ exposed plants. Wether this is statistically significant has not been analyzed up till now.

We have determined values of q_Q and q_E in the steady-state and subsequently the maximum photosynthetic rate (v_max) for the same leaf. In figure 3 a correlation can be seen between q_Q and the maximum photosynthesis rate. These are not averaged measurements, implying that q_Q in the steady-state is a rather good quantitative value for the photosynthetic capacity of the plant. The spread in the q_E

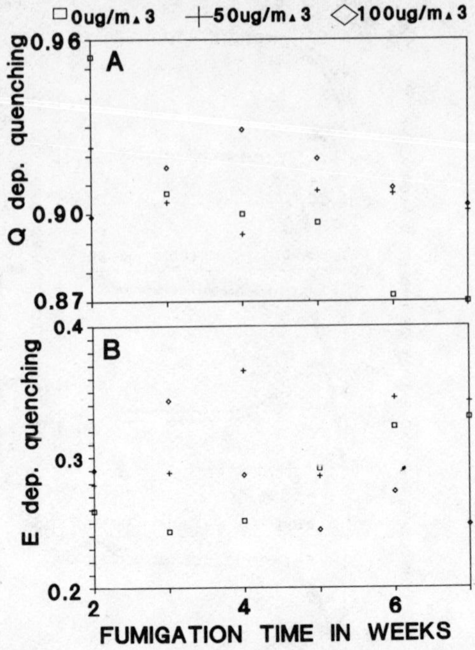

Figure 2. Development of the quenching components, i.e. q_Q and q_E, during exposition of poplar to different concentrations of ammonia. The q values were determined at the end of the 6 min. illumination period. The values of the first 5 weeks are averages of 7 measurements. 6 Weeks is an average of 6. And 7 weeks is an average of 5 measurements.

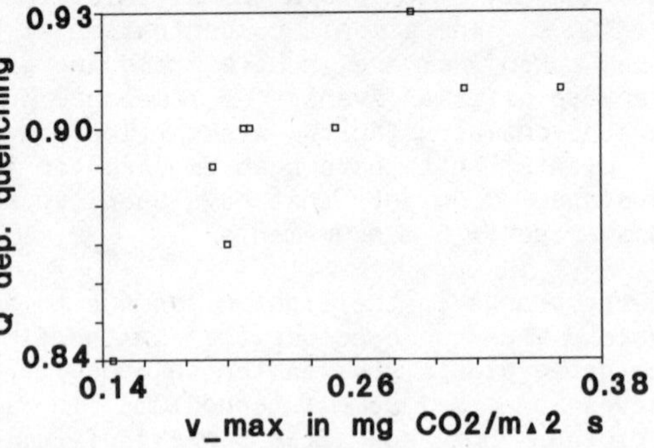

Figure 3. Q dependent quenching in the steady-state of poplar leaves after 8 weeks in the fumigation chambers, compared to the maximum photosynthesis rate measured in the same leaf the next day. A correlation between the two values is evident. These are single measurements.

measurements is much larger (data not shown).

Other measurements with poplars and beans have revealed that the fumigated plants respond to an exposition of > 5 ppm NO_x for 5 hours by raising the q_E and lowering q_Q in the steady-state (data not shown). The control plants do so to a much lesser extent. The effect seemed to be reversible and disappeared within two weeks after the extra exposure. However these measurements should be repeated, since the number of measurements were too few in order to make these results statistically reliable (5).

Discussion

The light-doubling-fluorescence technique is clearly fit to give us a qualitative diagnosis of the stress situation a plant is in. Before a photosynthesis curve from a leaf was measured, the fluorescence response of that same leaf was determined. Inspection of this fluorescence curve gave us an indication of how fast the leaf would respond to changing light intensities in the gas exchange cuvette. A quantitative correlation between q_Q in the steady-state and the maximum photosynthesis rate can be seen in fig. 3. Our results indicate that gaseous ammonia functions as a fertilizer for poplar plants. How plants react to ammonia in combination with other pollutants such as SO_2 is now under study and will be published elsewhere.

Acknowledgements

We would like to mention the invaluable assistance of mr. R.M. Bouma of our electronics department in upgrading the measuring equipment to a level necessary for the experiments. This research was made possible by a grant from the Ministry of Housing, Physical Planning and Environment of the Netherlands and is part of the Dutch Priority Programme on Acidification.

References

(1) Schreiber, U., Schliwa, U. and Bilger, W. (1986) Photosynth. Res. 10, 51-62.
(2) Kautsky, H. and Hirsch, A. (1934) Biochem. Z. 274, 422-434.
(3) Krause, G.H. and Weis, E. (1984) Photosynth. Res. 5, 137-157.
(4) Stitt, M. (1987) in "Progress in Photosynthesis Research" Vol. III (Biggens, J. ed.), pp. 685-692, M. Nijhoff Pub., Dordrecht, the Netherlands.
(5) Van Kooten, O., Lahey, G. and van Hove, L.W.A. (1987) Acta Bot. Neerl., in the press.
(6) Buijsman, E., Maas, H. and Asman, W. (1985) in: Een gedetailleerde ammoniak emissiekaart van Nederland. Rapportnummer 41. Ministry of Housing, Physical Planning and Environment, The Hague.
(7) Boer, W.J.M. den and Bastiaans, H. (1984) in: Verzuring door atmosferische depositie (vegetatie). Ministry of Agriculture and Fisheries, Ministry of Housing, Physical Planning and Environment, The Hague.
(8) Schreiber, U. and Bilger, W. (1987) NATO ASI series, of a symposium in Sesimbra, Portugal. in the press.
(9) Schreiber, U. (1984) Photosynth. Res. 4, 361-373
(10) Schreiber, U. (1986) Photosynth. Res. 9, 261-272
(11) Briantais, J.M., Vernotte, C., Picaud, M. and Krause, G.H. (1980) Biochim. Biophys. Acta 591, 198-202

RAPID FINGERPRINTING OF DAMAGED AND UNDAMAGED CONIFER NEEDLES BY SOFT IONIZATION MASS SPECTROMETRY AND PATTERN RECOGNITION

H.-R. SCHULTEN and N. SIMMLEIT*
Fachhochschule Fresenius, Dep. Trace Analysis,
Dambachtal 20, D-6200 Wiesbaden
* Institut Fresenius, Chemical and Biological Laboratories,
Im Maisel 14, D-6204 Taunusstein-Neuhof

Summary

Field desorption and pyrolysis field ionization mass spectrometry have been used to characterize chemical differences between damaged and undamaged plant materials. Molecular-specific information about the composition of spruce needles and their epicuticular wax is obtained from the applied methods. Rapid screening of numerous samples can be carried out. The mass spectra are evaluated statistically by pattern recognition techniques and compared with the spectra of defined treated needles (sulfuric acid, ozone). Using this new approach, it appears feasible to determine the causes for the environmental damage of plant leaves.

1. INTRODUCTION

Direct and indirect attacks of atmospheric pollutants on plants can occur principally at the two interphases atmosphere/leaf and soil water/roots. Considering the visible leaf damages, leaching etc., it seems important to investigate the cuticular layer and the whole plant leaves in order to monitor the attack of atmospheric pollutants at the surface of the receptor plant and to detect early alterations in the structure of the whole leaf (1). The outer envelope of a conifer needle consists of epicuticular wax which should be the primary target of air pollutants. Leaf wax as well as the whole leaf material are complex biomaterials which should be first analysed by rapid screening methods. Therefore, spectroscopic methods such as FTIR, carbon-13-NMR, field desorption (FD) and field ionization (FI) mass spectrometry (MS) are used for an universal characterization of untreated biomaterials.

2. MATERIALS AND SITE DESCRIPTION

Spruce needles of defined age were taken every two months from 32 Picea abies trees at the Kleiner Feldberg, Taunus, F.R.G. The Taunus mountains are exposed to acidic precipitation and gaseous air pollutants which have been transported from other areas. In particular, the older trees show visible damages such as discolouring and substantial needle loss.

3. METHODS

The epicuticular wax of spruce needles is extracted with chloroform in 15 s at room temperature. The extract is filtered to remove the permanently on the leave surface adsorbed aerosols particels (2). The concentrated wax is analysed by FD/FIMS, FTIR and carbon-13-NMR (3-5). Dried spruce needles are pulverised and analysed by temperature-programmed pyrolysis (Py)-FIMS (6) and Curie-point-GC-EI/FIMS (7). Determinations of elemental compositions of FI ions are carried out with high resolution photoplate records. The Py-FI mass spectra are statistically evaluated by the ARTHUR software package.

4. RESULTS AND DISCUSSION

Epicuticular Wax Epicuticular raw wax has been chemically characterized by FTIR, carbon-13-NMR (5) and FDMS (4). It consists mainly of aliphatic compounds such as estolides, alkylesters, n-fatty acids and secondary alcohols (4,8). The wax of green, yellowish and brownish needles could not be distinguished by FTIR and carbon-13-NMR. But FI mass spectra showed clear differences indicating a premature ageing of the wax due to environmental pollution (8). However, these alterations could not be simulated by application of sulfuric acid and heavy metal cations to the wax (5).

Spruce needles The Py-FI mass spectra consists of 600 mass signals which are derived from molecular ions of thermal degradation products. It could be shown that Py-FI mass spectra represent the major constituents of spruce needles such as polysaccharides, mono- and dilignol building blocks, cutin, polyphenols, lipids and antioxidants (7). The Py-FI mass spectra between green, yellowish and brownish spruce needles are very different, indicating a reduction of membrane lipids, tocopherols and sitosterols and an increase of stilbenes from green to yellowish needles.

The mass spectral patterns of spruce needles fumigated with ozone or irrigated with sulfuric acid show mainly changes in the lignin structure and a reduction of antioxidant signals. Their pattern is not similar to the mass spectra of the damaged needles from the Taunus mountains and hence, it is concluded that neither sulfuric acid nor ozone alone are responsible for the observed needle damages.

CONCLUSION

Soft ionization mass spectrometry was successfully applied for rapid screening of plant leaves. By comparison of the mass spectral patterns slight differences e.g. in the lignin structures of various spruce needle samples can be detected. This alterations could be related to the impact of defined air pollutant mixtures, if enough treated material would be available.

REFERENCES

(1) H.-R. Schulten, H.H. Rump, N. Simmleit und R. Müller (1986) Untersuchungen über immissionsbedingte Veränderungen ausgewählter Pflanzenstoffe in und an Koniferennadeln. UBA-Texte **18**, 13-23.

(2) N. Simmleit, H.H. Rump und H.-R. Schulten (1986) Nichtabwaschbare Aerosolteilchen auf den Oberflächen von Koniferennadeln. Staub Reinhalt. Luft **46**, 256-258.

(3) H.-R. Schulten, K.E. Murray and N. Simmleit (1987) Natural waxes investigated by soft ionization mass spectrometry. Z. Naturforsch. **42c**, 178-190.

(4) H.-R. Schulten, N. Simmleit and H.H. Rump (1986) Soft ionization mass spectrometry of epicuticular waxes isolated from coniferous needles. Chem. Phys. Lipids **41**, 209-224.

(5) N. Simmleit and H.-R. Schulten (1987) On the impact of acidic precipitation on the epicuticular wax of spruce needles. Proc. Int. Conf. Acid Rain, Lisbon, Sept. 1987, in press.

(6) H.-R. Schulten and N. Simmleit (1986) Impact of ozone on high-molecular constituents of beech leaves. Naturwissenschaften **73**, 618-620.

(7) H.-R. Schulten, N. Simmleit and H.H. Rump (1986) Forest damage: characterization of spruce needles by pyrolysis field ionization mass spectrometry. Intern. J. Environ. Anal. Chem. **27**, 241-264.

(8) H.-R. Schulten, N. Simmleit and K.E. Murray (1987) Differentiation of wool and spruce waxes affected by environmental influences using fingerprinting by soft ionization mass spectrometry. Fresenius Z. Anal. Chem **327**, 235-238.

COMPARATIVE INVESTIGATION ON THE NUTRIENT COMPOSITION OF HEALTHY AND INJURED SPRUCES OF DIFFERENT LOCATIONS

W. FORSCHNER and A. WILD
Institut für Allgemeine Botanik der Johannes Gutenberg-Universität, D-6500 Mainz, Federal Republic of Germany

Summary

Investigations on the content of cations (Ca^{2+}, K^{+}, Mg^{2+} and Mn^{2+}) were carried out in spruce trees at two exemplary locations in the Taunus montains (Königstein) and in the Hunsrück mountains (Hattgenstein) in the course of the vegetation period in 1985 and 1986.

The magnesium content decreased at both locations with increasing damage and with the needle age. The potassium content was likewise reduced in Königstein; unequivocal differences in the manganese and calcium content could not be detected here. In contrast to this, clearly damaged trees in Hattgenstein showed lower manganese and calcium contents than slightly damaged trees, but no reduction in their potassium contents.

Elution owing to membrane damage may be assumed to be the cause of the lower magnesium content in the needles of damaged spruce. Since an alteration of thylakoids and other chloroplast structures occurs at an early stage in damage and the decrease of the magnesium content is associated with a reduction of the chlorophyll content, the photosynthetic membranes appear to be directly affected.

1. INTRODUCTION

The investigation of the chemical elements was performed in the context of a project on the physiological, biochemical and cytomorphological characterization of spruce trees with different degrees of damage. Besides photosynthesis, the water balance and nitrogen assimilation were investigated. In addition, light microscopic and electron microscopic investigations were undertaken.

The investigations were carried out on spruce at two exemplary locations in the Taunus mountains (Königstein) and in the Hunsrück mountains (Hattgenstein) in West Germany. Detailed descriptions of both locations were made by (1).

2. MATERIAL AND METHODS

In Königstein needles were taken from 80 years old spruce trees (Picea abies) of different damage levels. During the measurement period a subdivision of the trees into three damage groups was made on the basis of the characterization of the harvested branches and needle samples with regard to loss of needles, necrotic areas of the needles and pigmentation. In order to arrive at comparable investigation material the two year old needles of a branch of the first order from the 7th whorl of the spruce crown were combined per harvesting date.

At the Hattgenstein location the investigations were carried out as a pair comparison according to (2), since slightly damaged trees and trees of the same age (about 20 years) with clearly visible damage were directly contiguous. Five tree pairs were selected for the measurements on the basis of visible characteristics in spring 1985 and spring 1986 (3). In 1985 two year old needles were investigated whereas in 1986 two and three year old needles were tested.

The disintegration of the washed and dried needles was performed with sulphuric acid under high pressure. The measurements of the cations (K^+, Ca^{2+}, Mg^{2+}, Mn^{2+}) were carried out with an AAS.

3. RESULTS

At both locations clear differences in the content of cations could be found in needles of spruces with different degrees of damage.

Königstein location (Fig. 1): The magnesium content decreased with increasing damage and with the needle age. The potassium content was likewise reduced, unequivocal differences in the manganese and calcium content could not be detected here.

Hattgenstein location (Fig. 2): Spruces with clearly visible damage showed a striking decrease of the magnesium content, in comparison to slightly damaged trees (pair comparison). A reduction of the values with the needle age could also be registrated at this location. Furthermore, clearly damaged trees had lower contents of manganese and calcium than slightly damaged ones, a reduction with the needle age could not be observed here. In the postassium content unequivocal differences could be found.

Similar results, as documented here for the two year old needles, could be obtained from the investigations of the three year old needles. The direct comparison of both needle generations in the year 1986 showed ecpecially a higher content of magnesium in the younger needles. In the manganese, calcium and potassium content significant differences could not be proved between both needle generations.

4. DISCUSSION

Both the comparison of the different damage groups at Königstein in the Taunus mountains and the pair comparison at the Hattgenstein location in the Hunsrück mountains showed a decrease in the magnesium content with increasing damage and with the needle age. Investigations on the chlorophyll content also showed an unequivocal decrease in more severe damage to the trees at the different harvest dates

(4). Elution owing to membrane damage may be assumed to be the cause of the lower magnesium content. The photosynthetic membranes appear to be directly affected, since an alteration of the thylakoids and other chloroplast structures occurs at an early stage in damage (5). Benner and Wild (6) also found a reduced photosynthesis in clearly damaged spruces.

The results of our physiological and cytomorphological investigations on spruce from different locations consistently show the presence of early damage to cell membranes (7, 8, 9). This membrane damage already occurs to a substantial extent in the green needles of damaged spruce. In the genesis of this membrane damage anthropogenic air pollutants appear to play a major role.

REFERENCES

(1) WILD, A., BODE, J. (1986). Physiologische, biochemische und anatomische Untersuchungen von immissionsbelasteten Fichten verschiedener Standorte. In: Wirkungen von Luftverunreinigungen auf Waldbäume und Waldböden, S. 166-181. Spezielle Berichte der Kernforschungsanlage Jülich - Nr. 369, ISSN 0343-7639.

(2) SCHÜTT, H.P., BLASCHKE, H., HOQUE, E., KOCH, W., LANG, K.J., SCHUCK, H.J. (1983). Erste Ergebnisse einer botanischen Inventur des "Fichtensterbens". Forstwiss. Cbl. 102, 158-166.

(3) BODE, J., KÜHN, H.P., WILD, A. (1985). Die Akkumulation von Prolin in Nadeln geschädigter Fichten (Picea abies [L.] Karst.). Forstwiss. Cbl. 104, 353-360.

(4) DÜBALL, S., WILD, A.: (1987). Investigations on the activity of glutamine synthetase, the content of free ammonium, soluble protein and chlorophyll in spruce trees with different degrees of damage in the open air. Planta, in press.

(5) JUNG, G., WILD, A. (1987). Electron microscopic studies of spruce needles in connection with the occurence of new kinds of tree damage. I. Investigations of the mesophyll. J. Phytopath., in press.

(6) BENNER, P., WILD, A. (1987). Measurements of photosynthesis and transpiration in spruce trees with various degrees of damage. J. Plant Physiol., in press.

(7) WILD, A. (1987). Licht als Streßfaktor bei Holzgewächsen. In: Klima und Witterung im Zusammenhang mit den neuartigen Waldschäden. Symposium Gesellschaft für Strahlen- und Umweltforschung München 1986, in press.

(8) WILD, A. (1987). Physiological and cytomorphological investigation of spruce-needles exposed to air pollution. In Sonderheft: Pflanzenphysiologische Untersuchungsergebnisse als Beitrag zur Waldschadensforschung. Allg. Forstz. (AFZ), in press.

(9) WILD, A. (1987). Licht als Streßfaktor bei Waldbäumen. Naturwiss. Rundschau, in press.

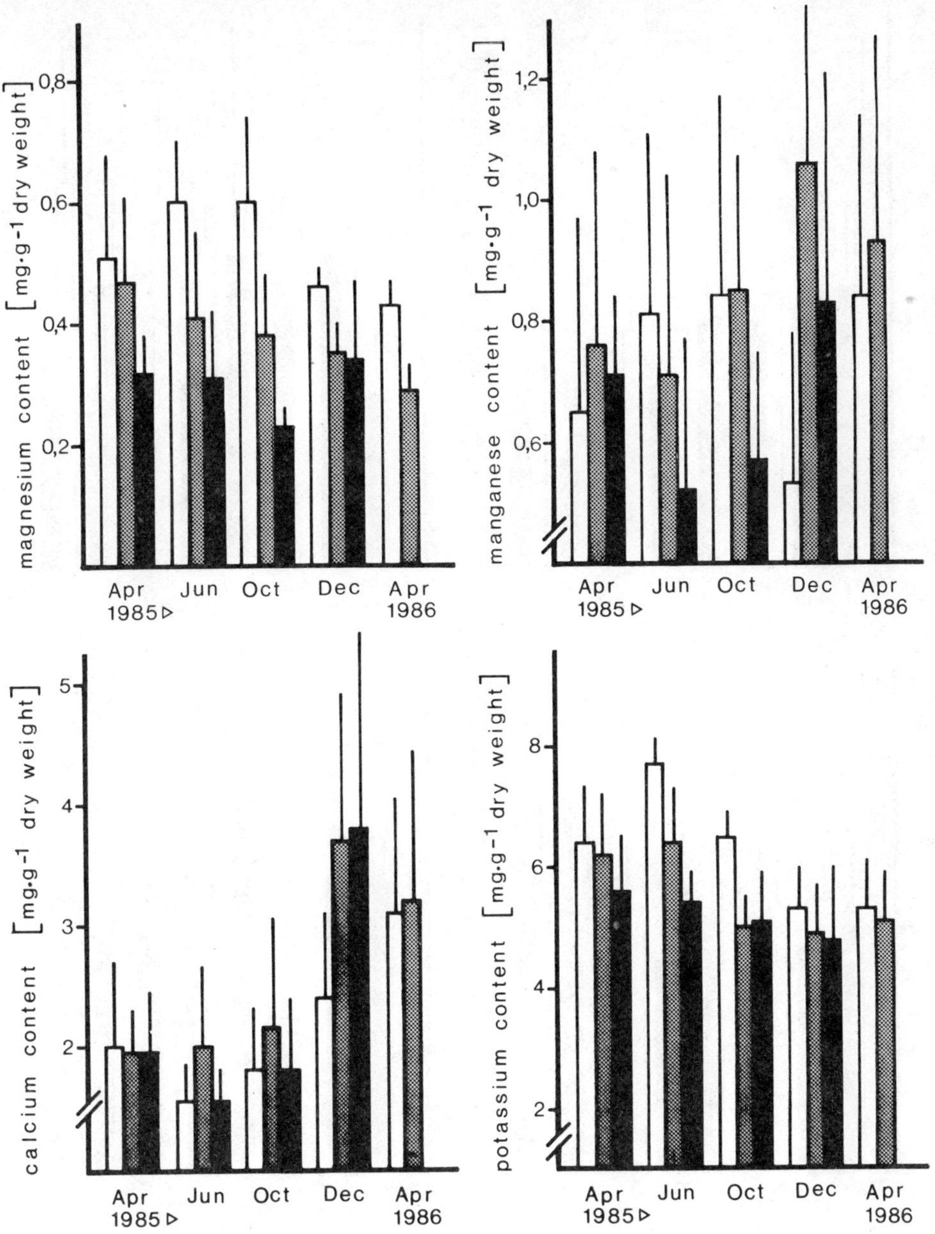

Fig.1: Content of cations of spruce needles at the Königstein location (Taunus mountains).

Damage groups: 1 2 3

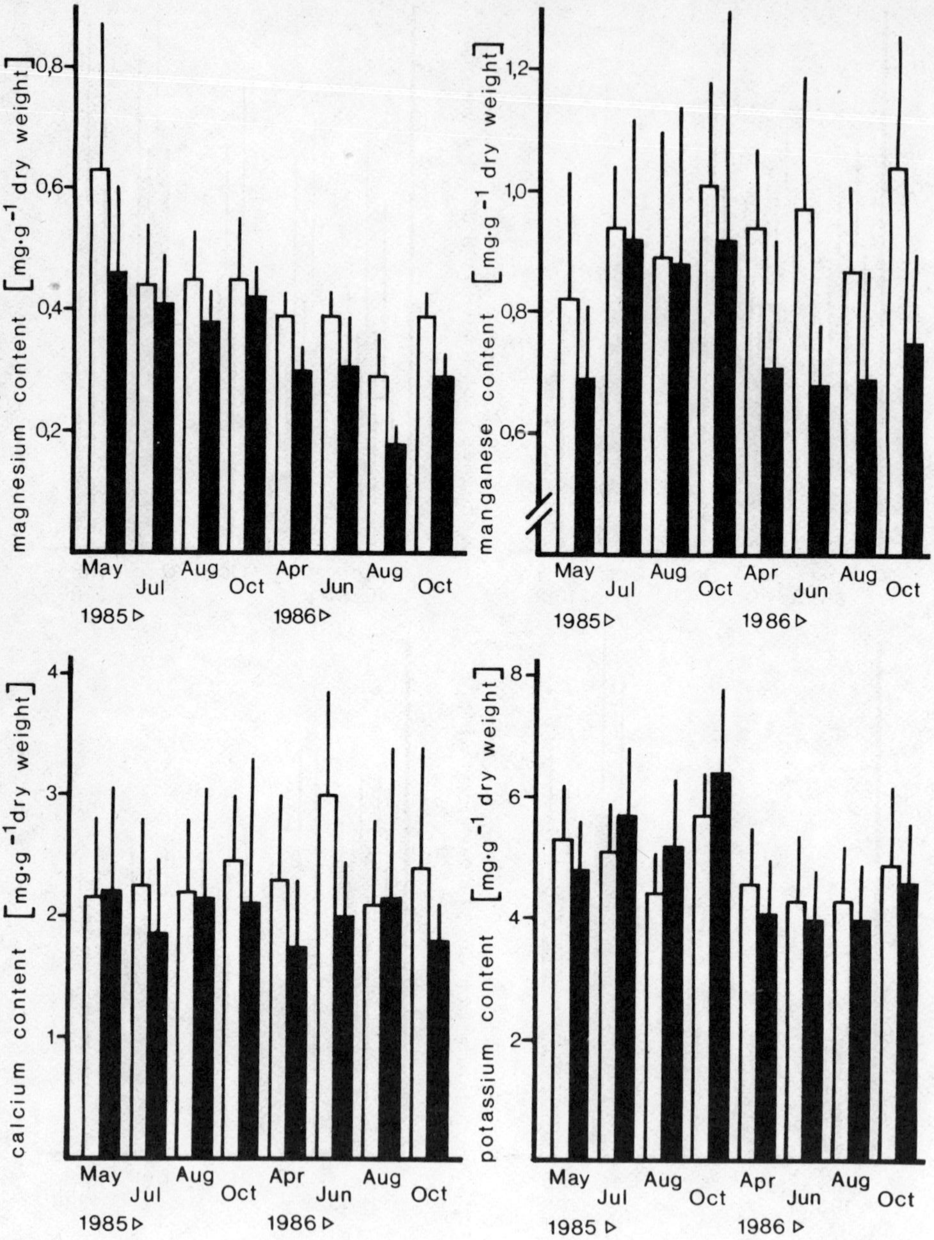

Fig. 2: Content of cations of spruce needles at Hattgenstein location (Hunsrück mountains).

Pair comparison: ☐ slightly damaged trees ■ clearly damaged trees

EARLY DIAGNOSIS OF FOREST DECLINE - A PILOT STUDY

J.N. CAPE & I.S. PATERSON
Institute of Terrestrial Ecology, Edinburgh, U.K.
A.R. WELLBURN, J. WOLFENDEN & H. MEHLHORN
University of Lancaster, U.K.
P. FREER-SMITH
University of Ulster, N. Ireland, U.K.
S. FINK
Albert-Ludwigs-Universität, Freiburg, Federal Republic of Germany.

Summary

A pilot survey to investigate possible diagnostic tests for the study of forest decline was undertaken during 1986. Samples were taken from sites from S. Germany to N. Scotland, and a wide variety of tests performed. Preliminary results from the most promising diagnostic tests are presented here.

1. INTRODUCTION

The extent of forest decline is usually measured by the presence of visible symptoms such as yellowing or crown density, which may be difficult to estimate. The aim of this project was to identify objective tests which could be used to estimate the degree of damage before symptoms were visible. A number of potentially useful tests (biochemical, physico-chemical and histological) were developed in the laboratory and then applied to samples from 3 tree species (*Picea abies*, *Pinus sylvestris*, *Fagus sylvatica*).

2. METHODS

During 1986, samples were taken from 12 sites along a transect from S.W. Germany to N.E. Scotland. This gave a very wide range of pollution climates, and the criterion for success in this initial study was that there should be a clear distinction between sites with and without visible damage symptoms. There are obvious large differences in soil type, genotype and other factors which will confound detailed interpretation, but any potentially useful test should at least give a clear indication of sites where damage is known to occur.

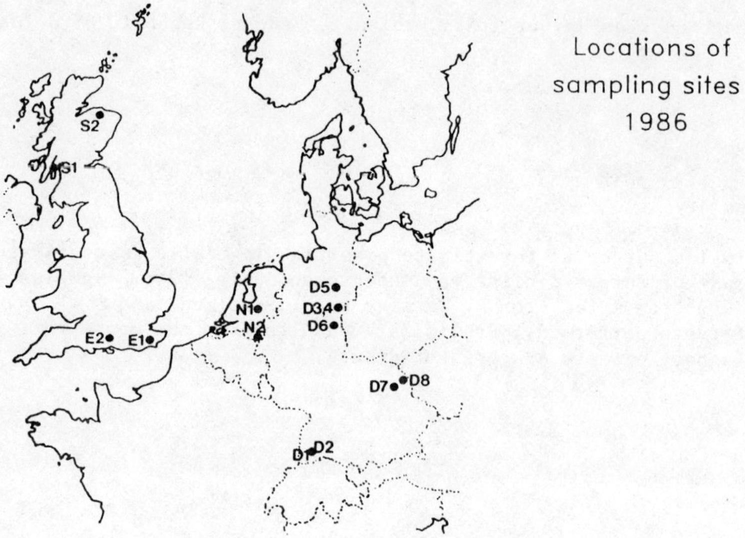

The tests which were applied to the plant material are as follows:

I.T.E.: dry weight/fresh weight ratios, nutrient analysis (K, Ca, Mg, S, N), amounts of surface wax and chloroform-insoluble material, Hartel turbidity test and contact angle of water droplets.

Lancaster: buffer capacity of leaf tissue, pigment analysis, emissions of hydrocarbons, modulated fluorescence and concentrations of α- tocopherol.

Ulster: water relations and photosynthesis.

Freiburg: histology and histochemistry.

3. RESULTS

The results presented here all refer to Norway spruce (*Picea abies*).

3.1 Contact angle of water droplets.

Shoots were sent by express post to the laboratory for analysis. A droplet of deionised water (0.2   ) was placed on the abaxial needle surface and the contact angle measured using a protractor graticule in a binocular microscope. Up to 80 measurements were made for each year class at each site. The results show small but significant differences between sites for current-year (1986) needles. Site differences increase greatly for two-year-old (1984) needles, as surface properties are affected by the different environments. The observed pattern showed large changes from year 0 to year 2 at British sites (no visible damage) compared with German sites (visible damage) is in contrast to the observations on Scots pine in Britain (Cape 1983) where needles from the polluted sites showed the larger changes, but the morphology of surface waxes is different in these two species.

Contact angle of water droplets

Norway spruce current year needles

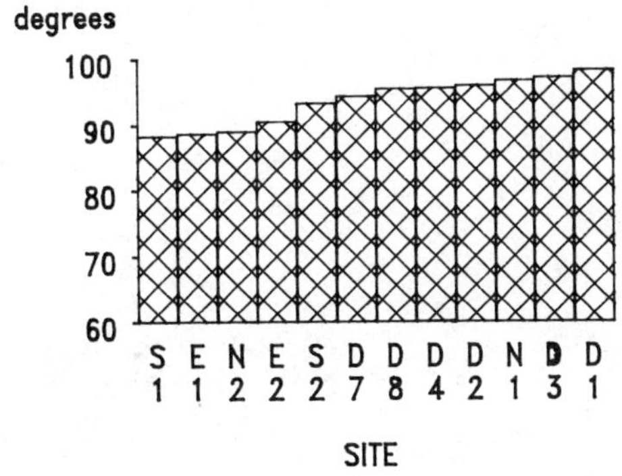

Norway spruce 2-year-old needles

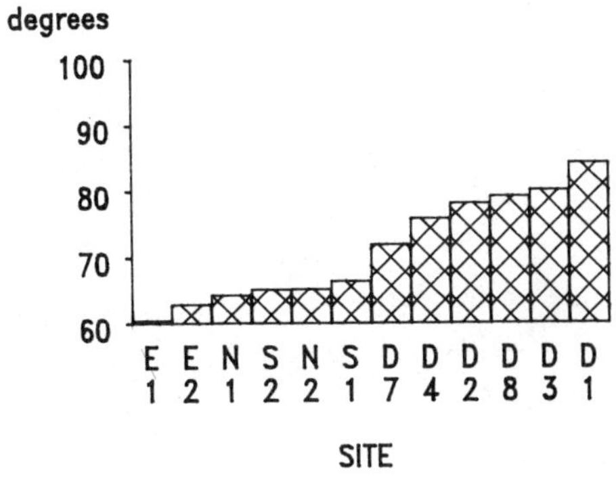

3.2 Buffering capacity.

Material was transported to the laboratory in liquid nitrogen, where 2 grams of needles were homogenised in 80 ml distilled water and titrated against 0.05 M HCl and 0 05 M NaOH. Buffer capacity (μmol H or OH/pH unit/g) was calculated from the slope of the titration curve. In general, alkaline buffering increases with needle age whilst acidic buffering decreases. These age-related differences are most pronounced at sites showing visible damage. Current-year needles show a greater variation in buffer capacity between sites than do 2-year-old needles. Thus at many German and Dutch sites showing damage, young needles appear to have a greater ability to buffer against acid, and a reduced alkaline buffering, when compared with the relatively undamaged British sites. This shift in buffer capacities is not maintained with age.

Buffer capacity increase at pH 8.5
from year 0 to year 2 in Norway spruce.

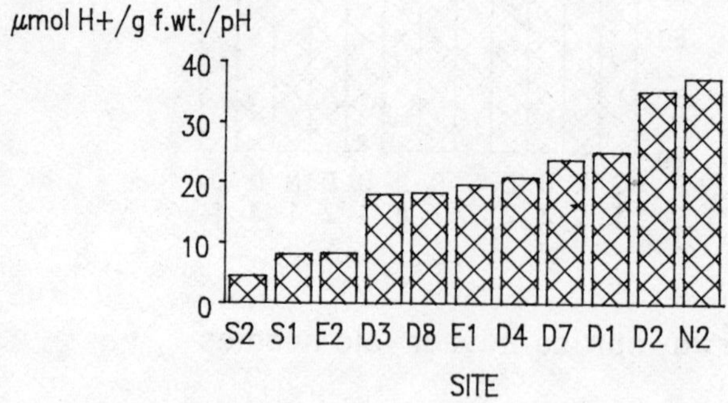

3.3 Pigment Ratios : Violaxanthin/Antheraxanthin.

Needles for pigment analysis were stored in liquid nitrogen. 0.5 grams material was homogenised in 5 ml methanol and filtered. Pigments were extracted in chloroform and separated by HPLC. The ratio of violaxanthin to antheraxanthin in 2-year-old needles shows significant variation between sites, being lowest at British sites and highest at German and Dutch sites. Increases in this ratio may be indicative of exposure to oxidants, which bring about the oxidation of antheraxanthin to violaxanthin (Yamomoto *et al*, 1962).

Violaxanthin : Antheraxanthin Ratio
Norway spruce 2-year-old needles

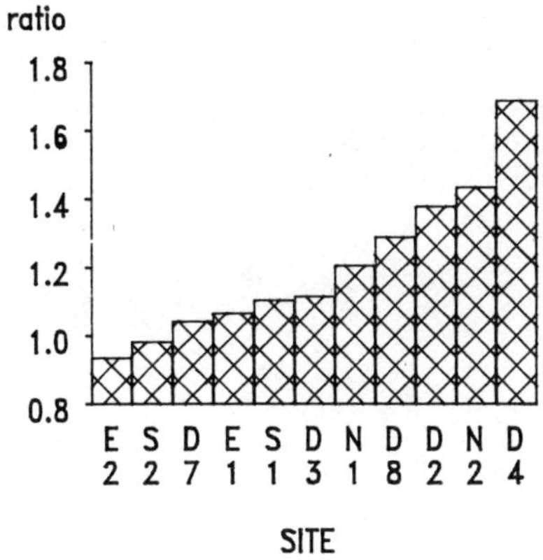

3.4 Histology

Single twigs from each of the conifer samples were first placed for 24 hours in the dark in order to allow dissolution and translocation of assimilation starch in the chloroplasts. Then segments of each year's needles were fixed, embedded, sectioned and stained for visualization of cell walls and starch. The sections were evaluated with regard to cell necrosis (mesophyll cells or sieve cells) and pathological starch accumulation. Three types of appearance could be distinguished:

(i) healthy needles with intact cells and without starch accumulation;

(ii) needles with slight damage to some mesophyll cells and local starch accumulation but intact vascular bundle;

(iii) needles without necrotic mesophyll cells but with phloem collapse and uniform starch accumulation in the mesophyll.

Whereas type (ii) seems to be indicative of direct impact of gaseous pollutants, type (iii) represents mainly the reaction to mineral deficiencies (Mg, K, Ca).

4. ACKNOWLEDGEMENTS

We are indebted to all those who helped in the selection of suitable sampling sites. The Ontario Ministry of the Environment (Canada) assisted with the field work. This project was partly funded by the Commission of the European Communities. This poster was also presented at the 2nd International Symposium on Air Pollutants and Plant Metabolism, Neuherberg, April 1987.

REFERENCES

CAPE, J.N. (1983). New Phytologist, 93 pp 293-299.

YAMOMOTO, H.Y. NAKAYAMA, T.O.M. & CHICHESTER, C.O. (1962). Arch. Biochem. Biophys., 97 p. 168.

THEME III : EFFECTS OF AIR POLLUTION ACTING ON ABOVE-GROUND PARTS OF PLANTS AND ECOSYSTEMS

THEME III : EFFETS DE LA POLLUTION DE L'AIR AGISSANT AU NIVEAU DES PARTIES AERIENNES DES ARBRES ET DES ECOSYSTEMES

THEMA III : WIRKUNGEN DER LUFTSCHADSTOFFE AUF OBERIRDISCHE TEILE VON BAUMEN UND OKOSYSTEMEN

ESTIMATION OF THE PLANT-RELATED RESISTANCES DETERMINING THE OZONE FLUX TO BROAD BEAN PLANTS (VICIA FABA VAR METISSA), AFTER LONG-TERM EXPOSURE TO OZONE IN AN OPEN TOP CHAMBER

J.M.M. ABEN
Environmental Research Department, N.V. KEMA,
P.O. Box 9035, 6800 ET Arnhem, The Netherlands

Summary

The ozone uptake and the transpiration rate of broad brean plants, previously exposed to a realistic ozone concentration in an Open Top Chamber during fourteen days, are measured in a whole-plant gas-exchango chamber. From the ozone uptake rate the total conductance of the plant is derived, and from the transpiration rate the stomatal conductance is derived. Relating the total conductance to the stomatal conductance shows that the adsorption of ozone at the epidermal layer and the diffusion of ozone through the cuticula are negligible. It is also shown that the uptake and turnover by the mesophyll cells does not limit the ozone uptake. This means that the measured ozone uptake can be adequately explained by the ratio of the ozone exposure concentration and the sum of the boundary layer and stomatal resistance for ozone.

1. INTRODUCTION

Knowledge of the plant-related factors determining the flux of a gaseous pollutant to the plant, and of their relative importance, can be used in studies concerning:

- the relation between the real pollutant dose, defined as the total amount of the pollutant taken up during exposure, and the response of the plant
- the development of explanatory crop growth models which take the influence of air pollution into account
- the modelling of atmospheric deposition velocities of gaseous pollutants above canopies.

Several authors (1, 2, 3) reported that the ozone flux to a plant is totally determined by the exposure concentration and the sum of the boundary layer and stomatal resistance. Taylor et al. (4) reported the existence of a residual or internal resistance which became the predominant resistance at exposure concentrations above 600 $\mu g.m^{-3}$, a very high concentration, at least for The Netherlands. All these studies were performed on plants grown and exposed under controlled conditions, and only short exposure times were used. The aim of this study is to estimate the plant-related resistances determining the ozone flux to plants grown under approximated field conditions and simultaneously exposed to a realistic ozone concentration.

2. THEORY

The factors determining the flux of air pollution to a plant can be modelled like "electrical resistances". Such a resistance model is given in figure 1.

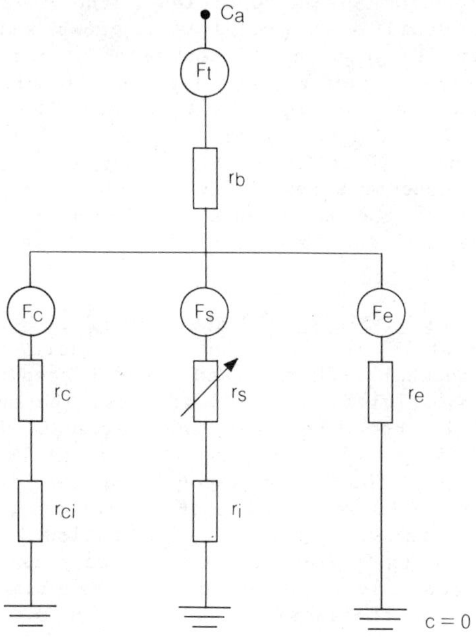

Figure 1 Resistance model for the flux of a gaseous air pollutant to a plant

F_t total flux to the plant
F_c flux through the cuticula
F_s flux through the stomates
F_e flux to the epidermis
C_a ambient pollutant concentration
r_b boundary layer resistance
r_c resistance against diffusion through the cuticula
r_s resistance against diffusion through the stomates
r_e resistance against adsorption to the epidermis
r_i, r_{ci} resistance against uptake and turnover by the mesophyll cells

According to (5) it follows from the model that:

$$(C_a/F_t-r_b)^{-1} = (r_s+r_i)^{-1} + (r_c+r_{ci})^{-1} + r_e^{-1}.$$

Thus, if $(C_a/F_t-r_b)^{-1}$, the total conductance of the plant, is plotted against $(r_s+r_i)^{-1}$, the conductance of the pathway through the stomates, a straight line with a slope of unity should result. The intercept of this line equals $(r_c+r_{ci})^{-1}+r_e^{-1}$, the combined conductance of the non-stomatal pathways (c_{ns}).

If relating values for F_t, C_a, r_b and r_s are known, the value for r_i can be found by iteration; r_i is varied until the slope becomes 1. The value of the intercept at this r_i-value is the combined conductance of the non-stomatal pathways.

3. EXPERIMENTAL SETUP

To estimate the values of the ozone transfer resistances for plants grown and exposed under field conditions, the following experimental setup was chosen. Broad bean plants were grown and exposed to ozone during fourteen days in an open top chamber. The exposure concentration amounted to 120 $\mu g.m^{-3}$ during the day and 30 $\mu g.m^{-3}$ at night. At the end of the exposure the ozone flux to the plant and the water release by the plant were measured with a whole-plant gas-exchange chamber at several irradiances (0-300 $W.m^{-2}$) to vary the stomatal resistance. The carbon-dioxide concentration in the chamber ranged between 300 and 340 ppmv, depending on the irradiance. The chamber temperature and humidity ranged between 20 and 25°C, and 50 and 70% respectively, also depending on the irradiance.

4. CALCULATION OF FLUX DENSITIES AND RESISTANCES

In the steady-state situation the flux density equals the product of the air flow through the chamber and the difference between inlet and outlet concentration, divided by the leaf area. The outlet concentration for ozone has to be corrected for the degradation of ozone at the chamber wall (6). The total resistance for water vapour is calculated from the flux density for water vapour and the difference in water vapour concentration between the stomatal cavity and the chamber air, applying Ohm's law. The water vapour concentration in the stomatal cavity is assumed to be the saturated water vapour concentration at leaf temperature. The average leaf temperature of the plant is estimated from the air temperature and the sensible heat flux density, the latter being derived from the energy balance. The net absorbed total radiation, one of the terms in the energy balance, is estimated from the incident radiation, according to (7). The boundary layer resistance for water vapour is calculated in the same way as the total resistance for water vapour, using a wetted model plant, prepared from filter paper. The stomatal resistance for water vapour equals the difference between total and boundary layer resistance. The stomatal resistance for ozone is derived from that for water vapour by multiplication with the ratio of the diffusion coefficients of ozone and water vapour in air.

5. RESULTS

As can be seen from table I a negative r_i-value and zero conductance for the non-stomatal pathways were derived for practically all experiments from the relation between total conductance of the plant and the conductance of the pathway through the stomates. The correlation coefficient for this relation was mostly higher than 0.99 (data not shown).

The negative r_i-values are likely to have no physical meaning. They should rather be interpreted as corrections of too high values for the stomatal resistance, probably due to overestimations of the leaf temperature. Small errors in the estimated leaf temperature give rise to large errors in the stomatal resistance, due to the exponential relation between temperature and the saturated water vapour concentration.

Table I Internal resistance (r_i) and the conductance of the non-stomal pathways (c_{ns})

exp. numb.	$r_i (s.m^{-1})$	$c_{ns} (mm.s^{-1})$
1	- 39	0.0
2	- 9	0.0
3	- 52	0.0
4	- 29	0.0
5	- 42	0.0
6	- 30	0.0
7	- 83	0.0
8	- 35	0.0
9	+ 1	0.0
10	-101	0.0
11	- 91	0.0
12	- 83	0.0
13	- 97	0.0
14	- 95	0.0
15	- 52	0.0

6. CONCLUSION

Taking into account that the mean value of the internal resistance is only 20% of the mimimum stomatal resistance (at maximum light intensity) and that negative r_i-values have no physical meaning, one may conclude that the stomatal resistance is the only plant-related factor determining the ozone uptake by broad bean plants grown under approximated field conditions and simultaneously exposed to a realistic ozone concentration.

REFERENCES

(1) RICH, S., WAGGONER, P. and TOMLINSON, H. (1970). Ozone uptake by bean leafs. Science. Vol. 169, 79-80.

(2) OMASA, K., ABO, F., NATORI, T. and TOTSUKA, T. (1979). Studies of air pollutant sorption by plants. II Sorption under fumigation with NO_2, O_3 or $NO_2 + O_3$. J. Agric. Meteorol. Vol. 35, 77-83.

(3) UNSWORTH, M.H. (1981). In: Plants and their atmospheric environment (Grace, J., Ford, E.D. and Jarvis, P.G., eds.), Blackwell Scientific Publications, Oxford.

(4) TAYLOR, G.E., TINGEY, D.T. and RATSCH, H.C. (1982). Ozone flux in Glycine max (L) Merr.: Sites of regulation and relationship to leaf injury. Oecologia. Vol. 53, 179-186.

(5) BLACK, V.J. and UNSWORTH, M.H. (1979). Resistance analysis of sulphur dioxide fluxes to Vicia faba. Nature. Vol. 282, 68-69.

(6) UNSWORTH, M.H. (1982). In: Effects of gaseous air pollution in agriculture and horticulture (Unsworth, M.H. and Ormrod, D.P., eds.), Butterworth Scientific, London.

(7) GOUDRIAAN, J. (1977). Crop micrometeorology: a simulation study. Ph.D. Thesis, Agricultural University of Wageningen, PUDOC Wageningen, The Netherlands.

SULFUR CONTAINING AIR POLLUTANTS AND THEIR EFFECTS ON PLANT METABOLISM

L.J. DE KOK, F.M. MAAS, I. STULEN and P.J.C. KUIPER
Department of Plant Physiology, University of Groningen
P.O. Box 14, 9750 AA Haren, The Netherlands.

Summary

H_2S and SO_2 affected plant yield at relative low levels without causing visible injury. In general H_2S was more toxic than SO_2. Short-term exposure (24 to 48 h) of plants to the pollutants affected sulfur, nitrogen and energy metabolism. Both exposure of plants to H_2S and SO_2 resulted in an accumulation of glutathione in the leaves; this accumulation was higher in H_2S than in SO_2 exposed plants. SO_2 exposure resulted in a rapid accumulation of sulfate, which was absent in H_2S exposed plants. Amino acid metabolism in spinach leaves was strongly affected by short-term H_2S exposure. Besides a strong increase in the cysteine content, there was a substantial decrease in the serine content. The content of most other amino acids and ammonia increased upon exposure, which indicated an overall effect of H_2S on amino acid/protein metabolism. *In vivo* nitrate reductase activity measured under aerobic conditions was increased in H_2S exposed spinach leaves. Leaf respiration was not affected by H_2S exposure. NADH oxidation by other cytosolic enzymes was inhibited in H_2S exposed plants, and the level of inhibition correlated with the reduction in relative growth rate by H_2S.

1. INTRODUCTION

Only little is known about the physiological background of the phytotoxicity of H_2S and SO_2. In our laboratory research is focused on the initial effects of H_2S and SO_2 on sulfur, nitrogen and energy metabolism of crop plants. Plants are exposed to pollutant levels which locally occur as peak levels in industrial areas in The Netherlands. The aim of our study is to get insight into the initial effect of the air pollutant on the physiology of the plant, in order to obtain physiological parameters which can be used as an early indication for the phytotoxicity of H_2S and SO_2.

2. EFFECT ON YIELD

Both H_2S and SO_2 are phytotoxic and affect crop yield at relative low levels. In some species H_2S exposure to levels lower than 0.1 $\mu l\ l^{-1}$

resulted in an increased yield (1,11). Even at 0.25 µl l^{-1} H_2S yield of Phaseolus vulgaris was increased after a two-week exposure (Table 1). In general, H_2S levels of 0.1 µl l^{-1} reduced plant yield (1,6,7,9,11; Table 1), though the level of inhibition depended on the growth temperature (7). In most species H_2S is more toxic than SO_2 (5,9; Table 1). Only high levels of H_2S resulted in visible injury (11).

Table 1. The effect of a two-week exposure of various plants to 0.25 µl l^{-1} H_2S or SO_2 on the yield of the shoot. The yield (g) represents the mean of at least 12 measurements (±SD). Statistically significant differences between control and fumigated plants are indicated by asterisks: *, $p<0.01$; **, $p<0.001$. Derived from Maas et al. (7,9)

	Control	H_2S	SO_2
Glycine max	1.40 ± 0.35	1.31 ± 0.34	1.09 ± 0.24*
Phaseolus vulgaris	4.86 ± 0.57	5.55 ± 0.70*	4.99 ± 1.12
Spinacia oleracea	0.88 ± 0.23	0.16 ± 0.06**	0.78 ± 0.24
Trifolium pratense	0.34 ± 0.11	0.21 ± 0.05**	0.35 ± 0.08

3. EFFECT ON SULFUR ASSIMILATION

Foliar uptake of H_2S and SO_2 disturbed regulation of sulfur assimilation and resulted in a rapid accumulation of water-soluble non-protein SH-compounds (1,2,3,6,7,8,9). The level of accumulation depended on the concentration of the pollutant and varied between species (1,8,9; Table 2) and was maximal after 24 to 48 hours of fumigation with H_2S (2,8). The SH-accumulation was for the greater part due to an enhanced content of the tripeptide glutathione (1,2,3,9; GSH contains the amino acids glutamate, cysteine and glycine), even though part of the accumulation of SH-compounds in some species may be due to homoglutathione (contains the amino acids glutamate, cysteine and alanine; 10). In general H_2S exposed plants accumulated more SH-compounds in their leaves/shoots than SO_2 exposed plants (9; Table 2). SO_2 exposure resulted in a rapid accumulation of sulfate in the leaves/shoots, which already was significant after one day exposure to 0.25 µl l^{-1} (9). A similar rapid sulfate accumulation was absent during H_2S exposure; only prolonged exposure resulted in enhanced sulfate levels in the shoots (9). The order of theoretical rates of deposition of H_2S and SO_2 in various species, which was derived from transpiration measurements, correlated with the observed increases in SH-content during the first 24 h of exposure in both H_2S and SO_2 fumigated plants and with the increase in the sulfate content in SO_2 fumigated plants. However, the increases in

Table 2. The effect of a 24 h exposure of various plants to 0.25 $\mu l\ l^{-1}$ H_2S or SO_2 on the water-soluble non-protein SH-content of the leaves. SH-content is expressed as $\mu mol\ g$ fresh weight^{-1} and represents the mean of four measurements ($\pm$SD). Statistically significant differences between control and fumigated plants are indicated by asterisks: **, p<0.001. Derived from Maas et al. (7,9).

	Control	H_2S	SO_2
Glycine max	0.50 ± 0.02	1.02 ± 0.07**	0.62 ± 0.02**
Phaseolus vulgaris	0.48 ± 0.03	1.03 ± 0.04**	0.54 ± 0.04
Spinacia oleracea	0.30 ± 0.01	1.36 ± 0.09**	0.58 ± 0.03**
Trifolium pratense	0.51 ± 0.05	1.26 ± 0.05**	1.15 ± 0.30**

SH-content were only 8 % of the theoretical H_2S and SO_2 deposition fluxes, whereas sulfate accumulation accounted for at least 57 % of the theoretical SO_2 deposition flux (2,9). When the H_2S fumigation was ceased the accumulated glutathione in spinach shoots rapidly decreased (2,3,8). The decrease in glutathione content could not be explained by oxidation to oxidized glutathione. Both in H_2S exposed and control plants glutathione was for more than 86% present in the reduced form, and this remained unaltered after the fumigation was ceased, (3). Glutathione reductase activity in spinach shoots was not substantially affected by short-term H_2S fumigation (3). No detectable H_2S emission was noted after the fumigation was terminated; thus the decrease of glutathione content was not due to a breakdown of glutathione and subsequent desulfhydration (3). Likewise there was no substantial accumulation of SH-compounds in the root (2). It is suggested that glutathione functions as a temporary storage compound of reduced sulfur in the presence of excess H_2S in the ambient air. Even though cyst(e)ine content was also elevated upon H_2S fumigation (12), the observation that glutathione rather than cysteine is used for storage may have the advantage that glutathione does not participate in the regulation of sulfur assimilation to such great extent as cysteine (10). It is also unlikely that the phytotoxicity of H_2S and SO_2 is directly due to elevated glutathione levels in the cell, since there existed no direct relation between the phytotoxicity of sulfur gasses and the accumulation of glutathione in the shoots/leaves (6,9). In addition it was observed that growth of spinach was less reduced by fumigation with 0.25 $\mu l\ l^{-1}$ H_2S at 15 °C than at 25 °C. However, SH/glutathione accumulation was higher at 15 °C than at 25 °C (7).

4. EFFECT ON NITROGEN METABOLISM

Exposure of spinach plants to H_2S strongly affected amino acid metabolism of the leaves (12). A two day fumigation period at 250 ppb resulted in a large increase of the cyst(e)ine content of the free amino acid pool; the content of the untreated and H_2S exposed plants was <0.1 and 1.5 µmol g dry weight^{-1}, respectively. The content of serine, which is a precursor for cysteine and thus glutathione, decreased strongly upon fumigation (from 22.7 to 7.3 µmol g dry weight^{-1}). The content of most other amino acids increased up to 100% (12). The ammonia content increased also upon H_2S fumigation. Besides a withdrawal of serine for the metabolization of sulfide, the increase of the content of almost all other amino acids suggested a general effect of H_2S fumigation on the amino acid metabolism of spinach leaves; e.g. an increased rate of nitrate reduction, an increased rate of protein degradation or an inhibited protein synthesis. Short-term H_2S fumigation did not affect the apparent kinetics of glutamine synthetase and glutamate dehydrogenase (12). Short-term fumigation of spinach plants with 0.25 µl l^{-1} H_2S at a relative low photon fluence rate of 35 µmol m^{-2}s^{-1} (within the 400-700 nm range) did not affect either _in vitro_ nitrate reductase activity or _in vivo_ activity measured under anaerobic conditions (4). _In vivo_ nitrate reductase of untreated plants was apparently inhibited in the presence of oxygen. However, short-term H_2S fumigation at concentrations of 0.04 µl l^{-1} and higher increased _in vivo_ nitrate reductase activity measured under aerobic conditions (4). Activity increased two to five-fold of that of untreated plants at 0.25 µl l^{-1}H_2S. Since, in an _in vivo_ nitrate

Table 3. Effect of a 24 and 48 h exposure of _Spinacia oleracea_ to 0.25 µl l^{-1} H_2S on _in vivo_ "aerobic" nitrate reductase activity (NRA) and on leaf respiration of leaves. NRA is expressed as µmol nitrite g fresh weight^{-1}h^{-1} and represents the mean of two experiments with leaves of eight (control) and four (exposed) shoots in each (±SD). Leaf respiration was measured under similar conditions as _in vivo_ NRA assay and is expressed as µmol O_2 g fresh weight^{-1}h^{-1} and represents the mean of six (control) and three (exposed) measurements with leaves of five shoots each (±SD). Derived from De Kok et al. (4).

	In vivo "aerobic" NRA	Leaf respiration
Control (average)	0.91 ± 0.34	7.0 ± 1.3
24 h H_2S	1.60 ± 0.58	5.0 ± 1.3
48 h H_2S	2.77 ± 0.60	5.9 ± 0.6

reductase assay no exogenous NADH is added, and nitrate is supplied in a saturating concentration, it is believed that the _in vivo_ nitrate reductase activity is the resultant of the amount of enzyme and the endogenous supply of NADH to the enzyme. H_2S did not affect the in vitro nitrate reductase activity, thus not the amount of the enzyme. It was proposed that the increase of in vivo nitrate reductase activity in the presence of oxygen was due to an enhanced supply of NADH to the enzyme (4). The latter could not be explained by an altered competition between mitochondrial (leaf) respiration and nitrate reductase for NADH, since leaf respiration was not significantly affected by H_2S exposure (4; Table 3). The suggestion that H_2S fumigation altered the NADH supply in the leaves/shoot is supported by the observations that NADH oxidation by 30.000 g supernatant of spinach leaf extracts from H_2S fumigated plants was inhibited. Preliminary experiments demonstrated that inhibition of oxidation was proportional with the inhibition of the relative growth rate of spinach shoots at various levels of H_2S (Table 4).

Table 4. Effect on H_2S of relative growth rate (RGR) and the capacity to oxidize NADH. Data are expressed as percentage of the control (Maas, F.M., De Kok, L.J. and Stulen, I., unpublished results).

H_2S concentration	Reduction RGR	Reduction NADH oxidation
0.075 µl l^{-1}	5	5
0.150 µl l^{-1}	22	25
0.300 µl l^{-1}	33	38

5. CONCLUSIONS

Our research demonstrated that short-term exposure of crop plants to sulfur-containing gasses resulted in an altered sulfur, nitrogen and energy metabolism. The accumulation of glutathione can be used as an early indication of H_2S and SO_2 stress on plants. In addition changes in energy metabolism by H_2S illustrated by an altered in vivo "aerobic" nitrate reductase activity and inhibition of NADH oxidation in 30.000 g supernatant extracts of shoots/leaves from fumigated plants most likely can be used to predict the rate of growth reduction by H_2S.

REFERENCES

(1) DE KOK, L.J., THOMPSON, C.R., MUDD, J.B. and KATS, G. (1983). Effect of H_2S fumigation on water-soluble sulfhydryl compounds in shoots of crop plants. Z. Pflanzenphysiol. 111: 85-89.

(2) DE KOK, L.J., BOSMA, W., MAAS, F.M. and KUIPER, P.J.C. (1985). The effect of short-term H_2S fumigation on water-soluble sulphydryl and glutathione levels in spinach. Plant, Cell and Environ. 8: 189-194.

(3) DE KOK, L.J., MAAS, F.M., GODEKE, J., HAAKSMA, A.B. and KUIPER, P.J.C. (1986). Glutathione, a tripeptide which may function as a temporary storage compound of excessive reduced sulphur in H_2S fumigated spinach plants. Plant and Soil 91: 349-352.

(4) DE KOK, L.J., STULEN, I., BOSMA, W. and HIBMA, J. (1986). The effect of short-term H_2S fumigation on nitrate reductase activity in spinach leaves. Plant Cell Physiol. 27: 1249-1254.

(5) KRAUSE, G.H.M. (1979). Relative Phytotoxicität von Schwefelwasserstoff. Staub-Reinhalt. Luft 39: 165-167.

(6) MAAS, F.M., DE KOK, L.J. and KUIPER, P.J.C. (1985). The effect of H_2S fumigation on various spinach (Spinacia oleracea L.) cultivars. Relation between growth inhibition and accumulation of sulphur compounds in the plant. J. Plant Physiol. 119: 219-226.

(7) MAAS, F.M., DE KOK, L.J., HOFFMANN, I. and KUIPER, P.J.C. (1987). Plant responses to H_2S and SO_2 fumigation. I. Effects on growth, transpiration, and sulfur content of spinach. Physiol. Plant. in press.

(8) MAAS, F.M., DE KOK, L.J., STRIK-TIMMER, W. and KUIPER, P.J.C. (1987). Plant responses to H_2S and SO_2 fumigation. II. Differences in metabolism of H_2S and SO_2 in spinach. Physiol. Plant. in press.

(9) MAAS, F.M., DE KOK, L.J., PETERS, J.P. and KUIPER, P.J.C. (1987). A comparative study on the effects of H_2S and SO_2 fumigation on growth and accumulation of sulfate and sulfhydryl compounds in Trifolium pratense L., Glycine max Merr. and Phaseolus vulgaris L. J. Exp. Bot. in press.

(10) RENNENBERG, H. (1982). Glutathione metabolism and possible biological roles in higher plants. Phytochemistry 21: 2771-2781.

(11) THOMPSON, C.R. and KATS, G. (1978) Effects of continuous H_2S fumigation on crop and forest plants. Environ. Sci. Technol. 12: 550-553.

(12) VAN DIJK, P.J., STULEN, I. and DE KOK, L.J. (1986) The effect of sulfide in the ambient air on amino acid metabolism of spinach leaves. In: Fundamental, Ecological and Applied Aspects of Nitrogen Metabolism in Higher Plants. Eds. H. Lambers, J.J. Neeteson and I. Stulen. Martinus Nijhoff/Dr. W. Junk Publishers, Dordrecht. pp. 207-209.

ANALYSE DES MECANISMES DE TRANSFERT DES PARTICULES A TRAVERS L'APPAREIL STOMATIQUE

JULIENNE G. et LOUGUET P.

Laboratoire de Physiologie Végétale et d'Ecophysiologie Végétale Appliquée.
UFR de Sciences et Technologie, Université Paris 12,
Avenue du Général de Gaulle 94010 CRETEIL CEDEX (FRANCE)

Résumé

Quatre plants de fèves sont placés dans une enceinte expérimentale et soumis à des aérosols d'uranine secs polydispersés (DMM=1,05 µm). On procède à deux séries de pulvérisation de sept heures par jour durant six jours : la première à la lumière, la seconde à l'obscurité. Tous les deux jours, des feuilles sont prélevées puis lavées après fermeture des stomates. L'absorption ainsi que le dépôt foliaires de l'uranine sont mesurés par dosage fluorimétrique lors des deux séries d'expériences (à la lumière et à l'obscurité). Les premiers résultats montrent que la pénétration d'uranine est plus forte dans les feuilles pulvérisées stomates ouverts par rapport à celles pulvérisées stomates fermés. Le rapport des absorptions foliaires stomates ouverts sur stomates fermés peut atteindre une valeur voisine de 9.

Summary

Four plants of Vicia faba cv Long Pod were placed in an experimental chamber and subjected to polydispersed dry aerosols of uranine (MMD=1.05 µm) under light (open stomata) or dark (closed stomata) conditions. Pulverizations were performed 7 hours per day during 6 days. Leaves were sampled every two days during the week long experiment and washed several times after stomatal closing. Both uranine deposition and absorption by leaves with open or closed stomata were compared. Foliar deposition and absorption of uranine were determined. The first results showed that leaves with open stomata (light) contained more uranine than those with closed stomata (darkness). A ratio of about 9 for light/dark absorption was found. Other experiments are currently available and will give statistical significance to the results.

1 - INTRODUCTION

La littérature est relativement pauvre en données sur les mécanismes de la pénétration des particules polluantes dans les végétaux. Pourtant, l'atmosphère contient des particules de granulométrie comprise entre (10^{-3} et 100 µm).

Comme on peut le constater, le diamètre des ostioles à stomates ouverts (de l'ordre de 10µm) laisse supposer que les particules microniques ou submicroniques peuvent pénétrer dans le végétal et agir selon leur toxicité. Ainsi, GMUR et al (2) ont pulvérisé pendant trois semaines des haricots avec des aérosols de sulfate d'ammonium en majorité submicroniques qui ont entraîné l'apparition de nécroses, chloroses et une diminution de la turgescence des feuilles. Le parenchyme lacuneux est le premier touché, l'épiderme adaxial ainsi que le parenchyme palissadique les derniers.

Après étude de ces symptômes, les auteurs suggèrent la probabilité d'une pénétration des aérosols par la voie stomatique sans toutefois le démontrer par une étude du comportement des stomates et des variations de la résistance à la diffusion des gaz des feuilles.

2 - METHODES ET TECHNIQUES

a - Matériel végétal

Nous avons choisi d'étudier la fève (*Vicia faba*) variété Aquadulce à très longues gousses de chez Vilmorin. Les plantules croissent en serre sur un mélange de tourbe vermiculite (2/3 - 1/3) sous éclairage artificiel (lampe OSRAM HQIE 400 W) dans les conditions suivantes : 400 µmol $m^{-2}s^{-1}$; cycle jour/nuit : 12h/12h ; HR de 60% (25°c) le jour et 80% (20°c) la nuit.

Lorsqu'elles sont âgées de 15 à 20 jours, les plantes peuvent être utilisées pour les expériences.

b - Dispositif de pulvérisation.

La chambre de pulvérisation (figure) est constituée par une boîte à gants de 0,26m^3 de volume dont quatre des faces sont en plexiglas transparent.

L'air filtré chargé de particules qui pénètre dans la chambre est obtenu par mélange de deux flux d'air : le premier est émis du dispositif de production d'aérosols, le second est un flux d'air filtré servant pour le séchage et la dilution des aérosols.

La filtration et l'assèchement se font grâce à la succession de plusieurs filtres. L'assèchement est obtenu par coalescence des molécules d'eau.

Le débit de renouvellement de l'atmosphère de la chambre est de 5,62 m^3 h^{-1} (à la pression atmosphérique) soit un renouvellement de 22 fois par heure.

Afin d'éviter une sédimentation rapide des particules microniques lors de leur entrée dans la chambre, 2 ventilateurs (marque Tosel) ont été mis en place.

La présence d'une grille (de maillage 0,6x0,6cm) dans la chambre entre les ventilateurs et les plantes permet de mieux canaliser l'arrivée d'air sur les feuilles (figure).

Quatre plants de fève sont disposés dans la chambre. Les plantes sont éclairées par une lampe OSRAM HQIE 400 W suivant un cycle jour/nuit 12h/12h. L'intensité lumineuse sous la lampe dans la chambre suit un gradient vertical comme suit : 1450 µmol m^{-2} s^{-1} au niveau du toit à 120 µmol m^{-2} s^{-1} au niveau du sol.

Afin d'éviter la contamination de la terre par l'uranine, les pots contenant les fèves sont enfermés dans un sac en polyéthylène resserré au niveau du collet par un lien.

Pendant les 7 heures de traitement quotidien, les pots sont tournés d'un huitième de tour (toutes les 52 mn) et permutés tous les soirs.

On peut donc estimer que les plantes ont été exposées de la même manière aux aérosols après deux jours de pulvérisation.

L'humidité relative de l'air de dilution est d'environ 15% à 20°C ce qui donne une hygrométrie d'entrée dans la chambre pour les deux températures utilisées : 20°C et 25°C (respectivement les températures lors des pulvérisations à l'obscurité et à la lumière) de 27 et 23%.

On procède à deux séries de traitements : la première se déroule à la lumière, les plantes sont alors exposées aux aérosols, stomates ouverts ; la seconde à l'obscurité, les stomates sont alors fermés pour cette valeur basse de HR chez *Vicia faba*.

c - Génération des aérosols

Une solution aqueuse d'uranine de 10^{-2}g cm^{-3} est pulvérisée à l'aide d'un générateur pneumatique de Gauchard type G1 (vendu par les établissements Doyer) sous une pression de 1,2 bar.

Les micro-gouttelettes ainsi générées sont ensuite séchées par un air de dilution sous un débit de $5{,}16 m^3 h^{-1}$ (à la pression atmosphérique).

Le débit de renouvellement d'air de la chambre étant de $5{,}62 m^3 h^{-1}$, la teneur en uranine dans la chambre est voisine de $2{,}7 \cdot 10^{-8}$ g cm^{-3} (valeur théorique).

d - Granulométrie des aérosols générés.

Elle a été déterminée à l'aide d'un impacteur Andersen II. Cet appareil effectue un classement physique des particules en fonction de leur diamètre. Les aérosols sont prélevés à l'entrée de la chambre à raison de 28,3 l mn^{-1}, à la température de 19°c et pour une humidité relative de 16%. Après traitement informatique des données, on obtient le Diamètre Médiant en Masse (DMM). Ce dernier est ici de 1,05 µm.

e - Interaction des contraintes physiques avec les contraintes biologiques

- Obtention de stomates ouverts ou fermés

Les stomates sont maintenus ouverts en éclairant la plante en présence d'une teneur en CO_2 de 350 VPM. De même, la fermeture des stomates et son maintien, à l'inverse, sont obtenus à l'obscurité.Pour les pulvérisations stomates fermés, les plantes sont maintenues à l'obscurité pendant la journée car il est nécessaire de surveiller en permanence le fonctionnement de l'installation. Le cycle photopériodique a alors été inversé.

f - Mesure des dépôts et de la pénétration d'uranine

- Mesure de la fluorescence de l'uranine

Les mesures ont été effectuées sur un spectrofluorimètre Jobin Yvon (type J.Y.3). 486 et 503 nm sont respectivement les longueurs d'onde d'excitation et d'émission de l'uranine optimales pour la mesure de la fluorescence. La largeur de fente de 10nm permet la détection de 10^{-10} g.cm^{-3} au minimum. La droite d'étalonnage est obtenue entre les concentrations de 10^{-10} et 10^{-7} g.cm^{-3}. Le maximum de fluorescence est obtenu à des pH supérieurs à 8,5. Nous utilisons de la potasse 0,01M pour alcaliniser les solutions à doser (pH aux environs de 10).

- Mesure des dépôts et pénétration dans les feuilles de fève.

Tous les deux jours, deux feuilles d'une même plante situées à un niveau voisin de la chambre sont prélevées puis lavées.

Les solutions récoltées sont alcalinisées puis leur fluorescence mesurée selon le même principe que pour les supports inertes. Après leur lavage, les feuilles sont découpées puis broyées. L'étude de la fluorescence du broyat moins le bruit de fond nous donne la quantité d'uranine absorbée.

. Afin de limiter pendant la phase d'extraction de l'uranine, les échanges éventuels d'uranine entre la feuille et le milieu de lavage, il est indispensable que les stomates des feuilles soient fermés avant le début des manipulations.

La section du pétiole des feuilles lors de leur prélèvement entraîne une ouverture rapide et temporaire des stomates suivie de leur fermeture : c'est l'effet IWANOFF (3), (mesures faites au poromètre à diffusion d'hydrogène (4)). La fermeture survenant 2 à 4 heures après le prélèvement des feuilles, nous avons choisi d'attendre 3 heures entre la section et les extractions d'uranine.

. Mesure des dépôts foliaires.

Les feuilles sont agitées manuellement successivement dans les bains suivants :

1 - 750 ml d'eau distillée 2 mn
2 - 250 ml " " 1 mn
3-4-5 - 1 litre de triton X100 ; 0,1% 1 mn
6-7 - 200 ml d'eau distillée 1 mn

Les deux premiers (1) et (2) après lavage des feuilles et adjonction de potasse servent à déterminer le dépôt foliaire par mesure de fluorescence.

. Mesure de la pénétration d'uranine dans la feuille.

Les pastilles de feuille sont ensuite broyées dans un potter dans l'acétone. L'homogénat est ensuite centrifugé durant 5 mn à 6000t mn^{-1}, puis épuisé 5 fois à l'acétone et centrifugé dans les mêmes conditions que précédemment.

On procède alors à une extraction lipidique par l'éther de pétrole suivie d'apport d'eau distillée. On recueille la phase aqueuse dépourvue de pigments chlorophylliens qui sont lipophiles contrairement à l'uranine.

L'acétone contenu dans la solution est évaporé à 35°C. L'extrait est ensuite amené à pH basique par adjonction de potasse et ajusté à 100ml avec de l'eau distillée.

L'ensemble est filtré sur membrane Millipore (ayant des pores de 0,45 µm de diamètre) afin d'éliminer les particules en suspension risquant de perturber la réception du rayonnement émis.

Chaque filtrat est divisé en deux lots :

1 - la solution à analyser :

Elle correspond à la solution obtenue après extraction

2 - la solution de référence :

On ajoute à un certain volume de solution à analyser une quantité connue d'uranine (5).

La détermination des intensités de fluorescence des deux lots permet de déduire la fluorescence (fluo) de l'uranine absorbée par ces feuilles.

A partir de 3 expériences, une valeur de rendement d'extraction moyenne égale à 94,4% a été obtenue.

Cette perte de rendement étant supposée équivalente en valeur relative pour les différents échantillons, il n'en sera pas tenu compte par la suite.

3 - RESULTATS

a - Dépôts

Les valeurs des dépôts foliaires en uranine sont présentées dans le tableau. On note, pour les 2 expériences, une augmentation parallèle des dépôts entre le 2ème et la fin du 4ème cycle bien que la quantité d'uranine déposée soit supérieure à l'obscurité par rapport à la lumière ; ce n'est plus le cas entre le 4ème et le 6ème cycle.

b - Pénétration de l'uranine dans les feuilles.

La fluorescence mesurée de chacun des broyats est la résultante de la fluorescence endogène de 2 feuilles augmentée de la fluorescence de l'uranine absorbée par les stomates et adsorbée dans la cuticule.

Les résultats des mesures sont présentés dans le tableau. Chaque valeur est obtenue à partir du broyat de 2 feuilles d'une même plante.

Après 2 cycles de pulvérisation, la pénétration est faible (environ 2.10^{-8} $g.cm^{-2}$) à la lumière et nulle à l'obscurité. Un pic d'absorption est observé après le 4ème cycle soit une valeur (14,4 g $.10^{-6}$ g cm^{-2} à la lumière contre 1,6 10^{-8} g cm^{-2} à l'obscurité). Le rapport de ces deux pics est alors d'environ 9.

La fin du 6ème cycle est marquée par une chute des absorptions tant à la lumière qu'à l'obscurité (respectivement 3,3 .10^{-8} contre 0,85 .10^{-8} g cm^{-2}).

4 - DISCUSSION ET CONCLUSION

a - Dépôts

Comme on peut le remarquer, les écarts de dépôts entre deux feuilles peuvent être importants. Cependant, une vitesse moyenne de dépôts à partir des mesures faites entre le 2ème et le 4ème cycle de pulvérisation peut être calculée; elle est alors d'environ 5,6 .10^{-6} g. cm^{-2} h^{-1}.

Les dépôts des aérosols sont fonction de leur concentration, de leur granulométrie et du brassage de l'air, ainsi que de la forme et de la rugosité des feuilles. On constate ici qu'un échantillonnage limité à deux feuilles pour l'étude des dépôts d'uranine est insuffisant.

b - Pénétration de l'uranine dans les feuilles.

Malgré un dépôt foliaire plus élevé à l'obscurité, on constate une absorption d'uranine par les feuilles plus importante à la lumière (c'est à dire stomates ouverts) après le 4ème cycle de pulvérisation.

Ces premiers résultats sont en faveur de l'hypothèse de la pénétration possible d'aérosols par voie stomatique.

Le 5ème cycle ayant été séparé du 4ème par 2 journées sans traitement, la chute observée à la fin du 6ème cycle est difficile à expliquer. Elle peut avoir plusieurs origines ; par exemple une dégradation de l'uranine par la feuille ou un transport dans la plante.

L'hypothèse de travail repose sur le fait que la fluorescence mesurée des broyats est la résultante de la fluorescence endogène de la feuille augmentée de la fluorescence de l'uranine absorbée par les stomates et adsorbée dans la cuticule. CHAMEL (1) étudiant cette dernière sur des épidermes foliaires de poirier à partir d'une solution contenant du cadmium, constate qu'elle peut être importante. Elle peut provenir d'une infiltration des solutés sous forme de solution ou d'un échange d'ions avec les parois. Le pH de la solution aqueuse d'uranine pulvérisée est voisin de 7 ; c'est à dire que les molécules de colorants peuvent être sous deux formes : dissociées et non dissociées. Les deux modes d'infiltration de la cuticule foliaire sont théoriquement possibles. Dans ce cas, la méthode employée peut être critiquée car on dose l'uranine ayant pénétré par la voie stomatique mais également celle ayant infiltré la cuticule. Cependant il est permis de penser que l'imprégnation des parois des cellules épidermiques est du même ordre de grandeur pour des feuilles pulvérisées à la lumière ou à l'obscurité. Dans ce cas la différence de fluorescence ne peut être due qu'à la pénétration par les stomates.

Les résultats obtenus jusqu'à présent ne sont pas suffisamment nombreux pour qu'ils puissent être confortés par l'analyse statistique. De nouvelles expériences sont en cours qui tenteront de répondre à ce problème par des mesures de trois lots de feuilles au lieu d'un seul.

REFERENCES

(1) CHAMEL A.R., GAMBONNET B., GENOVA C. et JOURDAIN A., 1984 - Cuticular behavior of cadmium studied using isolated plant cuticles. J. Environ. Qual., 13, n°3, 483-487.

(2) GMUR N.F., EVANS L.S. et CUNNINGHAM E.A., 1983 - Effects of ammonium sulfate aerosols on vegetation. II. Mode of entry and responses of vegetation. Atmospheric Environment, 17, n°4, 715-721.

(3) IWANOFF l., 1928 - Zur Methodik der Transpirationsbestimmung am Standort. Ber. deutsch. Bot. Ges., 46, 306-310.
(4) LOUGUET P., 1965 - Influence de la teneur en gaz carbonique de l'air sur le mouvement des stomates à l'obscurité. Physiol. Vég., 3, 345-353.
(5) WAGATSUMA T. et PAYLING WRIGHT H., 1964 - The estimation of uranin (fluorescein sodium) in blood. J. clin. Path., 17, 271-272.

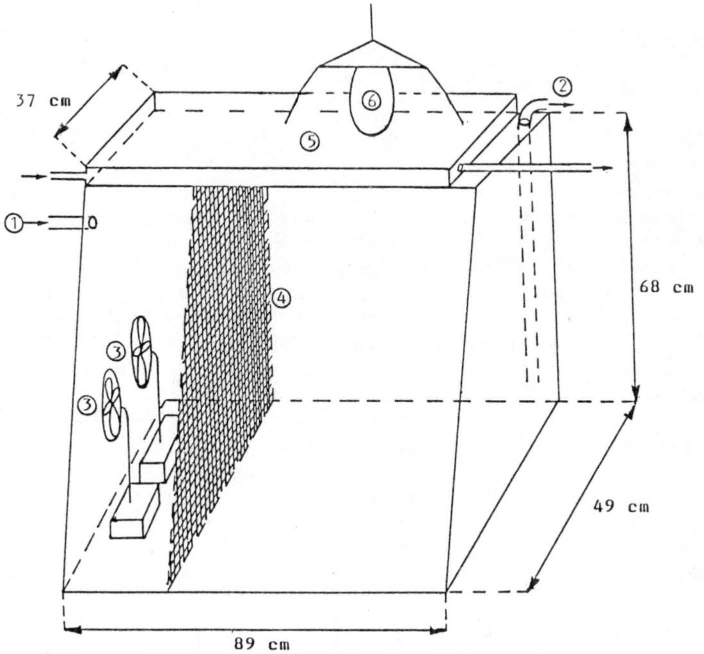

Figure : Schéma de l'enceinte de pulvérisation.
1 : arrivée des aérosols ; 2 : sortie des aérosols par aspiration d'une turbine ; 3 : ventilateur (2) ; 4 : grille ; 5 : circulation d'eau froide ; 6 : lampe.

Plantes éclairées (Bruit de fond)		Plantes à l'obscurité (Bruit de fond)	
Matières fluorescentes de la surface foliaire	Matières fluorescentes des tissus foliaires	Matières fluorescentes de la surface foliaire	Matières fluorescentes des tissus foliaires
2.10^{-10}	$3.9.10^{-9}$	2.1^{-10}	$5.9.10^{-9}$

PULVERISATIONS D'URANINE

Lumière				Obscurité				Lumière/Obscurité	
Cycles de 7 h	Dépôts foliaires $\times 10^6$	Uranine absorbée $\times 10^8$	Uranine absorbée /moyenne des dépôts $\times 10^2$	Cycles de 7 h	Dépôts foliaires $\times 10^6$	Uranine absorbée $\times 10^8$	Uranine absorbée /moyenne des dépôts $\times 10^2$	Dépôts foliaires	Uranine absorbée
2	1,2	1,7	1,5	2	9,7	0	0	0,1	-
4	9,3	14,4	1,5	4	17,1	1,6	0,09	0,5	8,9
6	20,3	3,3	0,1	6	14,7	0,9	0,06	1,3	3,8

TABLEAU : DEPOTS ET PENETRATION D'URANINE AU NIVEAU DES FEUILLES DE FEVES EN g cm^{-2}

OPEN-TOP CHAMBER STUDIES TO ASSESS THE EFFECTS OF SULPHUR DIOXIDE AND ACID RAIN, SINGLY OR IN COMBINATION, ON AGRICULTURAL PLANTS

H.J. WEIGEL, G. ADAROS and H.-J. JÄGER
Institut für Produktions- und Ökotoxikologie, Bundesforschungs-anstalt für Landwirtschaft, Bundesallee 50, D-3300 Braunschweig, Federal Republic of Germany

Summary

Some problems of investigating effects of low-level pollutant mixtures, as a cause for explaining possible damages to crop plants, are briefly discussed. An investigation with *Vicia faba* L. in 8 open-top chambers was carried out, as a contribution to clarify effects of low levels of SO_2, acid rain, or both in combination, on plant physiology, growth and yield. Plants were exposed to SO_2 concentrations within the range of 10 to 20 (control) or 55 to 90 $\mu g.m^{-3}$ and to rain events with solutions of pH 5.6 (control) or of pH 3.0/4.0. Events quoted a weekly precipitation of 1.6 mm, i.e. of 36.8 mm for the whole growing period. Plants were sown either in soil or in quartz gravel (applying nutrient solution). Physiological parameters were mostly determined in leaves. SO_2 stimulated growth of plants in soil but inhibited growth of plants in quartz. With regard to substrate sulphur content showed no differences. SO_2 caused an increase of S and sugar contents and a slight negative effect on content of K, Ca, Mg and total protein. On the other side, acid rain inhibited growth of plants in soil, and generally stimulated growth of plants in quartz. Partially, also content of S and sucrose was increased, whereas content of K, Ca and total protein was slightly reduced. Some interactive effects of both pollutants were observed.

1. INTRODUCTION

Investigations into the effects of single pollutants on plants are numerous, and responses especially of agricultural plants to gases like SO_2, O_3 and NO_x, and to acid rain, are well documented (1,2). However, nowadays, more and more attention is being paid to the effects of pollutant combinations on plant growth and productivity (3,4), a combination of two gaseous pollutants being the most investigated (5). Regarding the combination of acid rain with a gaseous pollutant, only a few results of investigations are known (6,7); this is particularly true for the combination with SO_2 (8).

Plant reaction to a specific pollutant is probably easy to determine when the selected dosis is high. In a concentration closer to reality, plant response is not always detectable. Therefore, a selected research method is the most important precondition for carrying out experiments, including not only the exposure system and its climatic performance, but also the system of feeding and controlling pollutant concentration. This is all the more important in the case of mixtures of two or more gases. The effect of a pollutant is heavily conditioned by climatic factors such as radiation, temperature, relative humidity, and wind. To facilitate interpretation of results they must be controlled continuously.

Considering also the daily course of natural light, with a well designed plant growth chamber it is possible to achieve reproducible climates (9). But most chambers are only suitable for monitoring constant climatic parameters. However, the tendency now is to investigate pollutant effects under natural conditions, which means under all possible combinations of climatic factors that may occur. In such a situation, results can be considered as realistic. However, in this case it is not possible to elucidate which particular climatic situation most significantly stimulated a plant reaction to a pollutant. Because up to now a growth chamber capable of simulating natural climatic conditions is non-existent, the alternative of open-top chambers appears to offer the most acceptable way of investigating plant reaction to gaseous air pollutants which may affect crop production directly (10).

In the present study, an investigation with broad beans in open-top chambers was carried out as a contribution to clarify if either SO_2 or acid rain, or both in combination, in levels frequently found in nature affect physiological and biochemical parameters, and also plant development and yield. Experiments took place at the Federal Agricultural Research Center, Braunschweig (FRG), in the growing season of 1985.

2. MATERIAL AND METHODS

Plants of *Vicia faba* L. cv. 'Con Amore' were directly sown in PE containers of 3 litre volume, filled with either a standard soil or quartz gravel of 0.7 to 1.2 mm mesh. Plants in soil were watered periodically with tap water, these in quartz gravel with a modified nutrient solution after Penningsfeld (11). A thin gravel layer of 2.0 to 3.2 mm mesh on the ground surface prevented by the latter the growth of algae. 20 days after germination containers were removed from the greenhouse to the open-top chambers and exposed to the different treatments. The position of the containers within the chamber was periodically changed to avoid marginal effects.

Eight open-top chambers were used allowing a 2 x 2 factorial design. The SO_2 treatments were: charcoal-filtered air alone (control) or with addition of SO_2. SO_2 concentration in the control chambers ranged between 1 to 30 $\mu g.m^{-3}$, mostly being within the range of 10 to 20 $\mu g.m^{-3}$. The SO_2 fumigated chambers had a concentration range of 55 to 90 $\mu g.m^{-3}$. Rain solution treatments were: deionized water (control, pH 5.6) and an artificial rain solution (pH 3.0/4.0). Rain events occurred three times a week for 8 to 9 min. Acid rain treatment started with the pH 3.0 solution for 3 min, and continued with the pH 4.0 solution. Rain events quoted a weekly precipitation of 1.6 mm and a total precipitation of 36.8 mm. Acid rain solutions were prepared by equilibrating deionized water with a known mixture of 95-98 % H_2SO_4 and 65 % HNO_3, resulting contents per litre of 6.4 mg N and 14.5 mg S (pH 3.0) or 0.6 mg N and 9.2 mg S (pH 4.0). The total deposition of H, N and S was 0.15, 0.89 and 2.07 kg per ha, respectively. Rain application resulted from pumping solutions through PVC nozzles installed in the middle of each chamber at a height of 2.6 m. This guaranteed a fully moistening of all leaves of each plant.

SO_2 concentrations were monitored continuously by a fluorescent analyser (Monitor Labs, Inc., San Diego, CA, USA), and according to this, SO_2 feeding was controlled by means of a thermic mass flowmeter. Photosynthetically active radiation (PAR), air temperature and relative humidity were also recorded continuously by means of a potentiometric recorder.

During plant development, growth analyses were carried out by measuring the following parameters: a) fresh and dry weight of leaves, stems, fruits, roots and whole plants; b) number of pods and seeds per plant, and seeds per pod, c) leaf area, and d) plant height. Additionally, analysis of the following chemical and biochemical criteria was carried out mostly in leaves: S, K, Ca, Mg, total and soluble protein, proline, glucose, fructose, and sucrose. 5 to 16 plants were sampled from each chamber for each determination. Plant material was dried at 95 °C in the course of two days. Such material was used for analysis of total S and N as well as for the ions. To determine sugar and proline contents, fresh material was immediately frozen by means of liquid nitrogen and later freeze-dried in the course of four days.

Analysis of variances were carried out by computing data after SPSS 9 (MANOVA).

3. RESULTS AND DISCUSSION

Different growth responses to the pollutants were found between plants grown in soil and those grown in quartz gravel.

For plants grown in soil SO_2 promoted during the periods of flowering and fruit filling fresh and dry weight of the whole plant and plant parts (+14 to +18 %), the number of blossoms and pods (+18 to +70 %), and leaf area (+16 %). Stimulating growth effects caused by low SO_2 concentrations have repeatedly been observed (12,13). Plants grown in quartz gravel could not confirm such a positive effect of SO_2 on the total fresh and dry weight of the whole plant, probably because the stimulating effect on leaves (+16 %) was weakened by a detrimental effect on pods (-20 %). The number of blossoms and pods, as well as leaf area did not show any reaction to SO_2. Although at first sight an interaction between SO_2 and soil substrate might be supposed, it is difficult to validate, if a different availability of S in substrate had influenced SO_2 uptake from air and, consequently, various effects on plant growth can be expected. In any case, sulphur content of leaves of plants, both in soil and in quartz gravel, did not show any difference related to substrate; it increased with the SO_2 treatment significantly (by plus 59 %). The content of micronutrients (K, Ca and Mg) in leaves was mostly reduced, as well as the content of total protein (-9 %). Determination of sugar contents of plants in gravel showed a considerable increase of glucose (+41 %), fructose (+19 %) and sucrose (+277 %). Controversy in literature regarding reactions of plants to long-term exposures to low levels of SO_2 is known (14,15). Roberts (15) noted that the variability in responses may be due to differences e.g. in in species or cultivar sensitivity, type of exposure chamber, plant age, and to interactions with other stress factors.

For plants grown in soil, acid rain had a detrimental effect on fresh and dry weight of the whole plant and on dry weight of pods (-6 to -17 %). For plants grown in quartz gravel, acid rain in some cases caused a positive effect on the growth parameters, but only in vegetative plants at the beginning of treatments: growth of stalks, leaves and even roots was stimulated. However, such effect was not shown by pods. Acid rain also stimulated sulphur content of leaves from plants grown in soil (by plus 24 %) but not of plants grown in gravel. Also a slight reduction of the micronutrients K and Ca, as well as total protein was ascertained for plants grown in gravel but at the same time acid rain incremented sucrose content of leaves (+40 %). Growth stimulations, caused by similar acid rain treatments, were also reported by other investigators (16,17). Yield losses observed in plants grown in soil are somewhat surprising as compared with numerous other investigations with artificial rain, which showed that, on the basis of H^+ deposition, a total H^+ deposition of 0.15 kg per ha, as used in our study, either did not affect crop growth or even resulted in growth stimulations (16,17).

Some interactive effects of SO_2 and acid rain were mostly observed in plants grown in soil. The presence of both pollutants appeared to stimulate the number of pods at the beginning of their development, but at the end no differences were established. On the contrary, a negative effect on the area of leaves was observed, which was partially reflected in their dry weight. Plants also showed an interaction of the two pollutants with respect to the content of S and of sucrose. To elucidate more of such interactions, it is probably indispensible to carry out experiments with more repetitions than was possible to make in the present work. In the only study about the effects of a combination of SO_2 and acid rain, Irving and Miller (8) did not find any interactions on soybeans. Interactive effects of acid rain with ozone were found by Jacobson et al. (18) and Troiano et al. (6) also on soybeans.

From the present results, it is obvious that the direction and intensity of pollutant effects are strongly dependent on factors like growing substrate and developmental stage of plants. General conclusions concerning pollutant impacts are difficult to give. However, apart from the effects of a single application of SO_2 or acid rain, the apparent interactions observed here indicate that their combination may be all the more detrimental to crop growth.

REFERENCES

(1) EVANS, L.S. (1984). Botanical aspects of acidic precipitation. The Botanical Review 50, 449-490.

(2) HEGGESTAD, H.E. and BENNETT, J.H. (1984). Impact of atmospheric pollution on agriculture. In: Air Pollution and Plant Life. M. Treshow (Ed.). J. Wiley and Sons Ltd., Chichester, New York, Brisbane, Toronto, Singapore. pp. 357-395.

(3) ENVIRONMENTAL PROTECTION AGENCY (1984). A review and assessment of the effects of pollutant mixtures on vegetation - Research recommendations. EPA-600/3-84-037. Environmental Research Laboratory, Corvallis, Oregon, 97333.

(4) REINERT, R.A. (1984). Plant responses to air pollutant mixtures. Annual Review of Phytopathology 22, 421-442.

(5) HECK, W.W., HEAGLE, A.S. and SHRINER, D.S. (1986). Effects on vegetation: native, crops, forest. In: Air Pollution. A.S. Stern (Ed.). Academic Press, New York. 6, 247-350.

(6) TROIANO, J., COLAVITO, L., HELLER, L., McCUNE, D.C. and JACOBSON, J.S. (1983). Effects of acidity of simulated acid rain and its joint action with ambient ozone on measures of biomass and yield in soybean. Environmental and Experimental Botany 23, 113-119.

(7) REBBECK, J. and BRENNAN, E. (1984). The effects of simulated acid rain and ozone on the field and quality of glasshouse-grown alfalfa. Environmental Pollution 36, 7-17.

(8) IRVING, P.M. and MILLER, J.E. (1981). Productivity of field-grown soybeans exposed to acid rain and sulphur doixide alone and in combination. Journal of Environmental Quality 10, 473-478.

(9) ADAROS, G. and DAUNICHT, H.-J. (1985). A movable, dewpoint-controlled daylight growth chamber, equipped for gas exchange measurements at high ventilation rates.Angewandte Botanik 59, 415-424.

(10) HEAGLE, A.S., PHILBECK, R.B., ROGERS, H.H. and LETCHWORTH, M.B. (1979). Dispensing and monitoring ozone in open-top field chambers for plant-effects studies. Phytopathology 69, 15-20.

(11) PENNINGSFELD, F. (1954). Hydrokultur. Forschungsberichte des Institutes für Bodenkunde und Pflanzenernährung. Weihenstcphan (FRG).

(12) SHIMIZU, H., FURUKAWA, A. and TOTSUKA, T. (1980). Effects of low concentrations of SO_2 on the growth of sunflower plants. Environmental Control Biology 18, 39-47.

(13) SAXE, H. (1983). Long-term effects of low levels of SO_2 on bean plants (Phaseolus vulgaris). II. Immission response effects on biomass production: quantity and quality. Physiologia Plantarum 57, 108-113.

(14) COWLING, D.W. and KOZIOL, M.J. (1982). Mineral nutrition and plant response to air pollutants. In: Effects of Gaseous Air Pollution in Agriculture and Horticulture. M.H. Unsworth and D.P. Ormrod (Eds.). Butterworth Scientific, London. pp. 349-375.

(15) ROBERTS, T.M. (1984). Long-term effects of sulphur dioxide on crops: An analysis of dose-response relations. Philosophical Transactions of the Royal Society of London, B 305. pp. 299-316.

(16) IRVING, P.M. (1983). Acidic precipitation effects on crops: A review and analysis of research. Journal of Environmental Quality 12, 442-453.

(17) VDI (1983). Säurehaltige Niederschläge - Entstehung und Wirkungen auf terrestrische Ökosysteme. Verein Deutscher Ingenieure, Kommission Reinhaltung der Luft, Düsseldorf (FRG).

(18) JACOBSON, J.S., TROIANO, J., COLAVITO; L., HELLER, L.I. and McCUNE, D.C. (1980). Polluted rain and plant growth. In: Polluted Rain. T.Y. Toribara, M.W. Miller and P.E. Morrow (Eds.). Plenum Press, New York. pp. 291-299.

OZONE SENSITIVITY OF OPEN-TOP CHAMBER GROWN CULTIVARS OF SPRING WHEAT AND SPRING RAPE

I. Johnsen, L. Mortensen, L. Moseholm and H. Ro-Poulsen.

University of Copenhagen, Institute of Plant Ecology, and National Agency of Environmental Protection, Air Pollution Laboratory.

Summary

Four cultivars of *Triticum aestivum* and five cultivars of *Brassica napus* were screened for ozone sensitivity in open-top chambers during summer 1986. Three different ozone levels were applied : 25, 50 and 75 ppb. The wheat cultivars alt reacted with negative effects at the 75 ppb level, and some growth parameters were negatively affected at the 50 ppb level. The rape cultivars showed promoted leaf senescense without affecting aboveground biomass. One cultivar even seemed to be stimulated by ozone.

INTRODUCTION

Ealier investigations at our laboratory have indicated that at least some important Danish cultivars of wheat and rape are adversly affected by relatively low levels of ozone pollution. In the European open-top chamber programme a comprehensive investigation throug several years of one single cultivar is planned. Therefore, it is essential to get an estimate of the relative differences in pollution response amongst cultivars. Four to fire cultivars of each species were exposed to three ozone levels in open-top chambers during May to July 1986.

PLANT MATERIAL

Brassica napus, cvs ('Petranova', 'Global', 'Concord', 'Rally' and 'Topas' and *Triticum aestivum*, cvs 'Wilham', 'Cornette', 'Vitus' and 'Nalle' were grown in 1 liter pots with a standard soil mixture ('K-soil'). It is a 3:2 v/v mixture of peat moss and clay, added nutrients. Sowing was done outside the chambers. One to two weeks after plant germination the pots, each with one plant, were placed at the bottoms of 6 open-top chambers, approximately 3o plants per cultivar per chamber.

Brassica napus and *Triticum aestivum* were grown together in the chambers from May 28 and were harvested in the period June 25-3o and July 2-1o, respectively.

During the experimental periods the pots were kept permanently moistened by hand watering, during hot periods twice a day. By harvest, above-ground growth and development parameters were measured and dry matter determination was done on the varius plant parts.

POLLUTANTS EXPOSURE

Three ozone levels were applied in the open-top chambers:. 1) charcoal filtered air, 2) ambient air added small amounts of ozone, and 3) ambient air added higher amounts of ozone. The addition took place from 9a.m. to 5.m. each day. Ozone was produced by a Sander Lab. Ozonizer, diluted by dry, pressurized air and distributed through teflon tubes to the chambers from a manifold with teflon needle valves at each outlet. Ozone were measured in the chambers with a Monitorlabs chemiluminesscence ozone analyser calibrated by means of NO gas titrattion.

Two replicate chambers were per ozone level. The 9a.m. to 5p.m. concentrations were kept in the range 2o-3o ppb, 4o-6o ppb and 7o-1oo ppb, respectively. SO_2 and NO_2 concentrations were for all chambers in the range o-1o ppb and 5-15 ppb, respectively (ambient levels).

RESULTS

Figure 1 presents the results from af preliminary analysis of the *Triticum aestivum* material. Effects of the 75 ppb treatment were seen on height, leaf area, stem, total shoot biomass and number of heads. Total leaf d.w. was not affected, but ozone caused the proportion of senescent leaves to raise. Also, the total leaf biomass was increased relatively to the whole shoot biomass. Some of the parameters are already affected by the 5o ppb treatment. This was observed on height and increasing dead leaf d.w. for some of the cultivars.

Results from the analysis of the *Brassica napus* cultivars, grown and exposed together with the wheat cultivars, are shown in Fig. 2.

The most through-going effect of ozone in spring rape was a premature senescense of leaves, and for most of the varieties this was at first seen at the 75 ppb level. Also the ratio dead leaves to green leaves is raised, as green leaf d.w. was decreased. One variety was stimulated by ozone, but with a changed dry matter partitioning as the leaves d.w. comprised a smaller percentage of shoot d.w.

CONCLUSIONS

The ozone effects on the four wheat cultivars seem to be very similar. Total above-ground dry matter production was reduced for all cultivars, probably because of the premature senescense observed, reducing the amount of photosynthetically active tissue.

No through-going differences in cultivar response were seen, with the exception of cv. "Vitus", which showed higher threshold level.

The plants were harvested at the time of head emergence. Reduced head formation with increasing ozone leves was observed. This reduction may be due to delay in head emergence, and permits no final conclusions on the seed yield.

The responce of rape showed no consistant trends between cultivars, except for the d.w. of dead leaves, which increased with ozone level, However, this responce cannot be related to yield. This species will not be considered for the future OTC-programme.

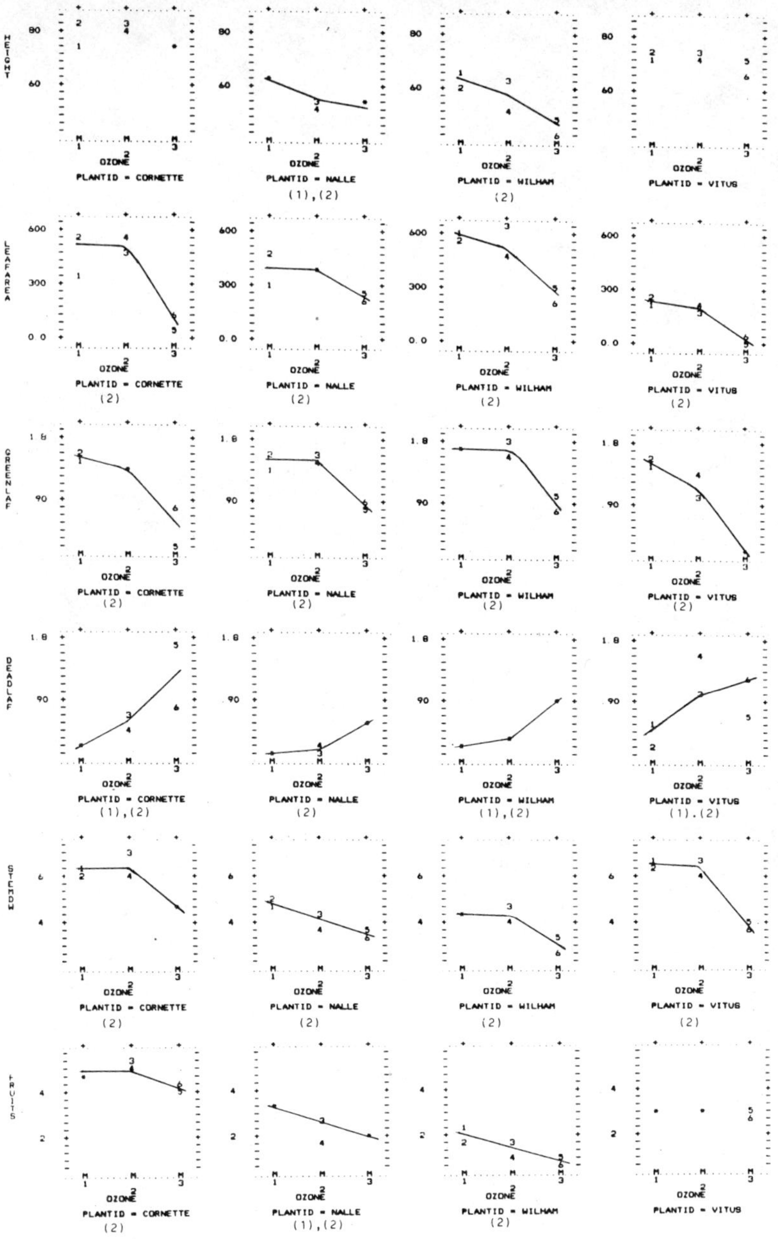

Fig. 1. Ozone effects measured on four cultivars of wheat, Triticum aestivum, shown on May 7th and grown in open-top chambers from May 28th to July 7th. Treatments were: ozone 1:25 ppb; ozone 2:5o ppb; ozone 3:75 ppb (approx. mean values p.am. to 5p.m.). Two replicate chambers with each 32 plants were used for each treatment. The diagrams show the mean value of each measured parameter marked with chamber number. Responsecurves are drawn when means are significantly different. (1): Significant (P 0.05) effect of ozone 2 treatment; (2): significant effect of ozone 3 treatment, both compared with the ozone 1 treatment.

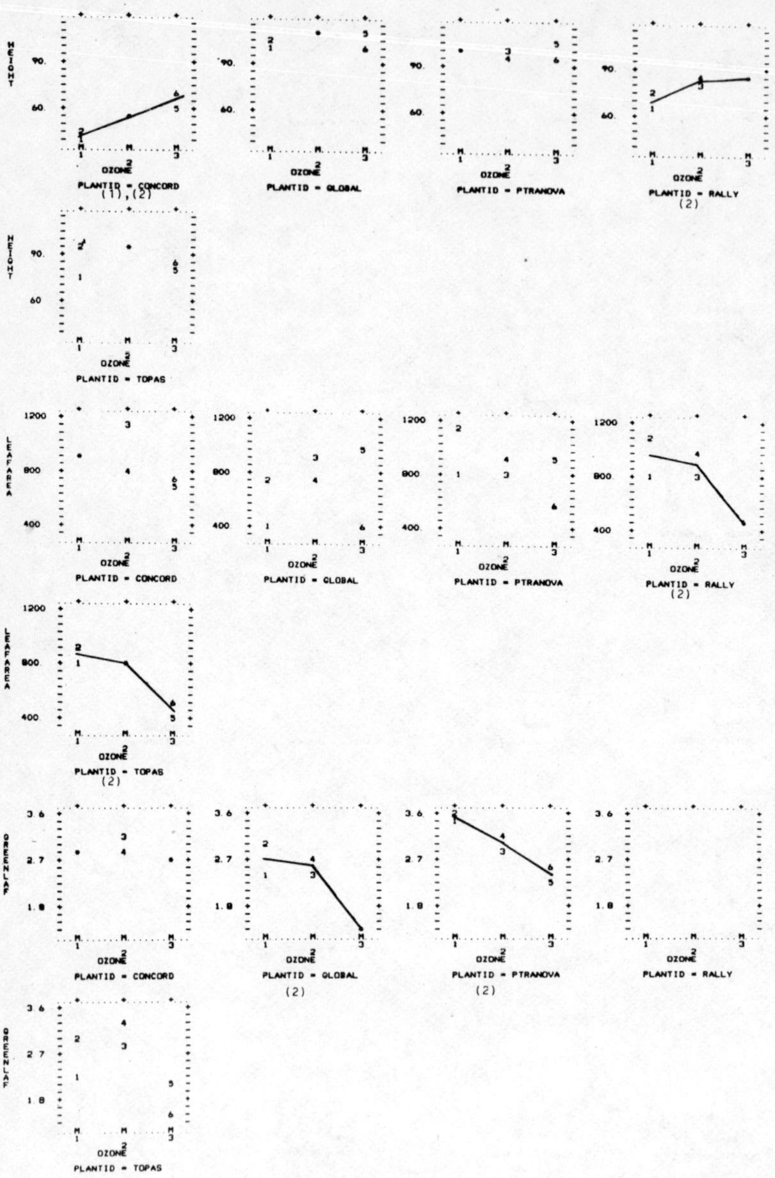

Fig. 2. Ozone effects measured on five cultivars of rape, Brassica napus, sown outdoors May 7th and grown in open-top chambers from May 28th to June 28th. Treatments were: ozone 1:25 ppb, ozone 2:5o ppb, ozone 3:75 ppb (approx. mean values 9a.m. to 5p.m.). Two replicate chambers with each 32 plants were used for each treatment. The diagrams show the mean value of each measured parameter marked with chamber number. Response-curves are drawn when means are significant different. (1): significant (P o.o5) effects of ozone 2 treatment· (2): significant effect of ozone 3 treatment, both compared with the ozone 1 treatment.

EFFECTS OF AMBIENT AIR POLLUTION ON CROP SPECIES IN AND AROUND LONDON

M.R. ASHMORE, A. MIMMACK and J.N.B. BELL
Department of Pure and Applied Biology
Imperial College of Science and Technology, London

Summary

This paper describes some results from a programme of research into the effects of ambient air pollution in and around London on the growth and yield of crop species. Three experimental approaches - a transect study, filtration studies at two locations, and a fumigation study - have been used in an integrated fashion in this programme. The experimental approach is illustrated by results obtained using *Pisum sativum* in 1983 and 1984. These showed significant beneficial effects of filtration on yield, a significant reduction in yield with increasing proximity to London, and a significant reduction in the yield of plants fumigated with low concentrations of SO_2 and NO_2. The responses of different cultivars and of different components of yield to these treatments were compared in order to understand further the role of individual pollutants in the three experiments. There was a high frequency of elevated O_3 concentrations during these experiments and the results of the fumigation experiment suggest that synergistic interactions between SO_2 and O_3 may have made a major contribution to the observed adverse effects of London's air.

1. INTRODUCTION

There is a growing realisation that in western Europe as a whole, the effects of ambient air pollution on crop growth can only be considered in terms of interactions between the principal phytotoxic components, notably SO_2, NO_x and O_3. It is well-established that interactions can occur between these three pollutants with respect to their effects on plants, with both synergism and antagonism having been demonstrated, but there is scarcely any information at gas concentrations which occur in rural or urban fringe areas (1). As such, it is currently not possible to make quantitative estimates of the relative contribution of SO_2, NO_x and O_3 to agricultural and horticultural crop yield reductions. However, such information is vital for the establishment of meaningful air pollution control policy, as the sources and control technology differ for the three pollutants. This paper describes part of a major research programme aimed at addressing this problem using three different techniques in an area to the west of London.

2. EXPERIMENTAL APPROACH

All experimental techniques for the study of air pollution effects on plants have their individual merits and disadvantages. The most realistic approach is to expose plants outdoors in standard cultures at sites with different air pollution regimes and compare their performance over a period during which the pollutants are monitored at each site: this has the advantage of the plants being exposed to ambient pollution under essentially natural conditions, but there can be severe problems of interpretation of responses because of differences in climatic factors between sites. A contrasting approach is the artificial fumigation of plants in chambers, using controlled concentrations of pollutants, either alone or in combination: in this case it is possible to define precisely the pollutant regimes, but it may be difficult to relate these to ambient regimes, and the chambers produce modifications of the ambient climate which may interact with pollutant effects. A third approach is intermediate in realism between those described above. This involves growing plants at a single location in chambers, which are ventilated with either ambient or filtered air: the effect of removal of the ambient pollutants is thus determined, but the relative importance of the individual gases is not readily understood and the chamber environment may modify responses compared with the ambient.

The present programme represents an attempt to avoid some of the limitations imposed by individual experimental techniques by utilising the three approaches described above in a coordinated fashion. By integrating the results obtained using the three techniques, it is hoped that advances can be made into understanding the relative importance of SO_2, NO_x and O_3 for plant performance and crop yield under ambient conditions. In the course of this programme the responses of four crop species - *Pisum sativum, Hordeum vulgare, Spinacia oleracea* and *Trifolium pratense* have been investigated in turn. In each study, four cultivars of the species concerned were investigated; these were selected on the basis of earlier screening studies, for their different sensitivities to SO_2, NO_2 and O_3. Collation of all the experimental data is currently in progress; in this paper, we present some important results obtained in the experiments with *Pisum sativum*.

3. TRANSECT STUDY

This study has exploited the various pollutant regimes occurring westwards from London into rural areas, based on the premise that there is a gradient of declining SO_2 and NO_2 concentrations with increasing distance from central London. In addition, under appropriate meteorological conditions during the summer, episodes of elevated O_3 concentrations will be superimposed on this gradient at concentrations which show a less consistent relationship with distance from London. Plants have been grown in containers with a standard soil at 15-18 sites along a 40 km transect from central London to our laboratory at Ascot, west of London. At each site, integrated measurements of SO_2 and NO_2 levels were obtained using simple chemical techniques, and of O_3 by means of biomonitoring with differentially sensitive cultivars of *Nicotiana tabacum*.

Results obtained in the summers of 1983 and 1984 using *Pisum sativum* showed a significant difference in yield along the transect. In 1984, all four cultivars showed a significant positive correlation between yield and distance from London, with the yield in central London being about 50% of that at the most rural sites (2). However, the components of yield affected in this manner showed some major differences between cultivars (Table 1). For one cultivar (Waverex) the effect of distance along the transect was on the number of pods produced per plant, rather than on the mean pod weight. Two cultivars (Banff and Douce Provence) showed no significant effects on the total number of pods, but significant effects on the total number of peas; therefore, the reductions in yield in these cases were primarily due to a reduction in the mean number of peas within each pod. The fourth cultivar (Progreta) showed a further distinct pattern; in this case, the mean number of peas within each pod was unaffected, and reductions in yield were due to a combination of a smaller number of pods per plant, and a reduced mean weight of individual peas.

Table 1. Effects of distance from London on components of yield of *Pisum sativum*.

Cultivar	Number of peas	Mean dry weight per pea	Total pea dry weight	Number of pods	Mean dry weight per pod	Total pod dry weight
Banff	.662	ns	.623	ns	.718	ns
Douce Provence	.587	ns	.618	ns	.781	.623
Progreta	.764	.695	.875	.753	.823	.860
Waverex	.738	ns	.798	.849	ns	.755

Tabulated values are the correlation coefficients between the values of the parameter at each site and its distance from London. A positive value indicates that the value of the parameter increases with increasing distance from London; ns indicates that the correlation coefficient was not significant at $P = 0.05$.

These correlations between distance from London and crop yield can only suggest an effect of ambient air pollutants. Other environmental variables are also related to distance along the transect, and, furthermore, it is impossible to identify the roles of individual pollutants. We are currently analysing the data from all the transect experiments using a step-up multiple regression procedure, to evaluate the relative importance of different pollutants, and of other environmental variables such as temperature and rainfall, in influencing plant growth.

Nevertheless, there is some direct evidence of the importance of one pollutant in contributing to the observed effects on *Pisum sativum*. The summers of both 1983 and 1984 were relatively hot, and there was a high frequency of elevated O_3 levels; a concentration of 80 ppb was exceeded at Ascot on 15 days in 1983 and 13 days in 1984 (3). Visible leaf injury, in the form of tip necroses on older leaves was observed after O_3 episodes in both summers, although it was more severe in 1983; these symptoms are similar to those found after controlled laboratory fumigations with O_3.

4. FILTRATION EXPERIMENTS

The effects of filtration have been determined in two sets of eight open-top chambers located at Ascot and at Kew Gardens, in a suburban area 12 km west of central London (4). Half of these chambers were ventilated with ambient air while the other half were ventilated with charcoal-filtered air. The filtration studies carried out during 1983 confirmed that the visible leaf injury observed on *Pisum sativum* plants in the transect experiment was due to ozone since significantly less injury was observed in the filtered air treatment on all four cultivars at both sites (5).

The filtration experiments carried out with *Pisum sativum* also demonstrated a significant adverse effect of ambient air pollution on yield, thus providing support for the hypothesis that effects on yield observed in the field were due to air pollutants. However, the components of yield affected in these experiments were not identical to those affected in the transect experiment. Table 2 summarises data obtained in the Ascot experiment, in 1984. Although three of four cultivars showed a significant reduction in yield, no significant effects on the number of peas or the number of pods were found. Instead, the most significant effect was on the mean dry weight per pea, a parameter which was unaffected in three of the four cultivars in the transect study (cf. Table 1).

Table 2. Summary of effects of filtration on components of yield of *Pisum sativum*.

Cultivar	Number of peas	Mean dry weight per pea	Total pea dry weight	Number of pods	Mean dry weight per pod	Total pod dry weight
Banff	ns	74	82	ns	89	ns
Douce Provence	ns	90	87	ns	ns	ns
Progreta	ns	86	ns	ns	ns	ns
Waverex	ns	77	73	ns	74	79

The tabulated values are those in the unfiltered air, expressed as a percentage of the value in filtered air; ns indicates that the difference between treatments was not significant at P = 0.05.

Mean ambient concentrations of SO_2 and NO_2 at Ascot during this experiment were 3 ppb and 6 ppb respectively, while O_3 concentrations above 60 ppb were recorded on 21 days. It is thus probable that O_3 was the major pollutant contributing to the observed effects of filtration. Along the transect the major change with increasing proximity to London will be an increased concentration of SO_2 and NO_2; for O_3 in contrast, there were fewer days with elevated O_3 in central London than at Ascot (3). Thus the differences between the experiments in the components of yield affected may be due to differences in the pollutants most responsible for the observed effects.

5. FUMIGATION EXPERIMENTS

An attempt has been made to determine the effects of individual pollutants, and of different pollutant combinations, by means of controlled fumigation experiments in eight closed outdoor chambers based at Ascot. Half of these chambers were ventilated with ambient air, while the remainder were ventilated with charcoal-filtered air; one chamber in each set of four also received either SO_2, NO_2 or SO_2 + NO_2. The SO_2 and NO_2 have been added at levels which simulate the pollution regime at Kew. Predictions were made, one day in advance, of the ambient SO_2 and NO_2 concentrations at Kew in each hour of the day, and these concentration patterns were then administered using a modified version of a computer controlled fumigation system designed for previous work (6). The predictions were made from the daily weather forecasts, on the basis of previous measurements of SO_2 and NO_2 levels at Kew in relation to meteorological conditions (7).

The fumigation experiment carried out using *Pisum sativum* in the summer of 1984 provided clear evidence that SO_2 and NO_2 at mean concentrations (8 ppb and 6 ppb respectively), which are typical of suburban London do have adverse effects on plant growth and yield. Significant interactions between SO_2 and NO_2 were only found with one cultivar, but a strong synergistic interaction between SO_2 and filtration in reducing pod dry weight was found with all four cultivars. The addition of 8 ppb SO_2 had little effect when added to filtered air, but caused substantial reductions in yield when added to unfiltered air. Our interpretation of this result is that there is an interaction between the added SO_2 and the high levels of O_3 in the ambient air at Ascot during this experiment.

This suggests that the large reductions in pea yield along the transect from Ascot to central London were to a large extent due to an effect of SO_2 acting synergistically with relatively high levels of ambient O_3 during the experiment. Some support for this hypothesis is given by an analysis of the effects of SO_2 plus unfiltered air on the components of yield (Table 3); the components of yield affected are rather similar to those affected in the transect experiment (Table 1). In both cases, there was a large effect on the total number of peas, but not on the mean dry weight per pea, thus indicating that the major component of yield affected was the number of peas per pod; one exception to this generalisation was found with cv. Progreta in the transect experiment. In both experiments, there was an additional effect on the number of pods produced on two cultivars, and in both experiments, it was the same cultivars (Progreta and Waverex) which were affected.

Table 3. Effects of addition of SO_2 to unfiltered air at Ascot on components of yield of *Pisum sativum*.

Cultivar	Number of peas	Mean dry weight per pea	Total pea dry weight	Number of pods	Mean dry weight per pod	Total pod dry weight
Banff	69	ns	73	ns	ns	74
Douce Provence	49	ns	51	ns	52	52
Progreta	38	ns	32	68	49	33
Waverex	47	ns	39	52	79	41

Tabulated figures are the value of each yield component in the SO_2 + unfiltered air treatment, expressed as a percentage of that in the filtered air treatment; ns indicates that there was no significant difference between these two treatments at $P = 0.05$.

6. CONCLUSIONS

The results reported in this paper provide strong evidence from transect, filtration and fumigation studies that ambient air pollution had an adverse effect on the growth and yield of *Pisum sativum* in summers in which relatively high levels of ambient O_3 were recorded. The results of the fumigation experiment and the transect experiment suggest that these are partially due to SO_2 and NO_2, although these gases were only present at concentrations below the generally accepted threshold for adverse effects. However, most controlled fumigation experiments with these gases have been carried out in the absence of O_3 and there are data reported from fumigation studies which demonstrate that exposures to SO_2 + NO_2, at concentrations of about 20 ppb, which had no effect alone, substantially reduced growth in the presence of only 30 ppb O_3 (8). Our data support the view that there are interactions between SO_2, NO_2 and O_3 at levels characteristic of substantial areas of rural western Europe which may be of considerable importance in terms of their effect on crop yield.

7. ACKNOWLEDGEMENTS

This work was supported financially by the C.E.C. 3rd Environment Programme and by the U.K. Natural Environment Research Council. We gratefully acknowledge the assistance of Cathy Dalpra, Dina Shah, Sally Power, Fiona McHugh and Sue Turner.

8. REFERENCES

(1) ROBERTS, T.M., DARRALL, N.M. and LANE, P.I. (1983). Effects of gaseous air pollutants on agriculture and forestry in the U.K. *Adv. Applied Biology* 9: 1-141.

(2) ASHMORE, M.R. and DALPRA, C. (1985). Effects of London's air on plant growth. *London Environmental Bulletin* 3: 4-5.

(3) UK PHOTOOXIDANT REVIEW GROUP (1987). Ozone in the United Kingdom. Department of the Environment, London.

(4) ASHMORE, M.R., MIMMACK, A., MEPSTED, R. and BELL, J.N.B. (1987). Research at Imperial College with open-top chambers 1976-1986. In: Proceedings of 2nd Open-Top Chambers Workshop, "Environmental Management in Open-Top Chambers". CEC, Brussels (in press).

(5) ASHMORE, M.R. (1984). Effects of ozone on vegetation in the United Kingdom. In: Ozone (P. Grennfelt, ed.), pp. 92-104. Published by IVL, Goteborg.

(6) LANE, P.I. and BELL, J.N.B. (1984). The effects of simulated urban air on grass yield. I. Description and simulation of ambient pollution. *Environmental Pollution, Series B* 8: 245-263.

(7) USHER, S.M. (1984). The effects of London air pollution on vegetation. Ph.D thesis, University of London.

(8) WOLTING, H.G. and von REMORTEL, E.A.M. (1985). Reacties von enkele plantesoorten op mengsels van NO_2, O_3 en SO_2 bij realistische concentratieniveaus. *Lucht en Omgeving* 2: 58-62.

EFFECTS OF OZONE AND CALCIUM NUTRITION ON NATIVE PLANT SPECIES

M.R. ASHMORE, C. DALPRA and A.K. TICKLE
Department of Pure and Applied Biology
Imperial College of Science and Technology, London

Summary

The sensitivity to ozone of about 150 native British herbaceous species has been assessed. Certain families, such as the Papilionaceae, contain a high proportion of sensitive species, while others, such as the Compositae, contain very few. There is a general trend for families which are more evolutionarily primitive to contain more sensitive species. Species characteristic of calcareous habitats tend to be more sensitive than species characteristic of acid habitats. A link between calcium nutrition and ozone sensitivity has been confirmed in two ways. Experimental studies have shown that more visible injury is produced by ozone on plants grown in higher levels of calcium, while field collections of plants of the same species collected from locations with higher levels of calcium were also more sensitive to ozone. However, preliminary experiments involving longer-term exposures to moderate ozone levels have shown a complex relationship between calcium nutrition and effects on plant growth.

1. INTRODUCTION

In most summers, the maximum O_3 concentration recorded in the U.K. exceeds 100 ppb, while under exceptionally hot summers, concentrations above 200 ppb have been measured (1). These levels are high enough to cause visible leaf injury to sensitive plant species, a fact which has been demonstrated experimentally in open-top chamber filtration experiments using *Pisum sativum* and *Trifolium repens* (2,3). These experiments have also shown decreased growth and yield of sensitive crop species caused by ambient air in rural southern England in summers with a significant number of days with O_3 concentrations above 60 ppb (3).

Although the responses of crop and forest trees to ozone have been extensively studied, there has been little work on the responses of wild herbaceous plants to this pollutant. However, since O_3 occurs in high concentrations over wide areas, it may have substantial effects on natural plant communities, and it is thus of considerable interest to identify the species or communities on which the impact of O_3 is likely to be greatest. This paper describes the results of a screening programme to identify sensitive native British species, and summarises the results of a series of experiments to investigate how one particular ecological factor may alter the responses of these species to ozone.

2. RELATIVE SENSITIVITY OF NATIVE SPECIES

During the past ten years, we have assessed the sensitivity to O_3 of a range of plant species. The plants are routinely exposed to 250 ppb O_3

for four hours in a controlled environment cabinet at 25°C and 80% relative humidity. The sensitivity to O_3 is based on the amount of visible leaf injury caused by this standard exposure to ozone. Comparisons of sensitivity between species are based on the average percentage areas injured on the most damaged leaf of each individual plant; this index is insensitive to differences in numbers and age of leaves between species.

To date, the sensitivity of approximately 150 native British herbaceous species has been assessed. On the basis of the assessment of mean visible injury to the most sensitive leaf, each species has been assigned to one of four sensitivity classes, defined as follows:- very sensitive, > 60%; moderately sensitive, 30-60%; slightly sensitive, 10-30%; and resistant, < 10%. Only a relatively small proportion of the species screened fall into the two most sensitive classes; a total of 35 species out of the 150 have been classed as very sensitive or moderately sensitive. These are listed in Table 1. However, this list must be treated with some caution; this assessment is inevitably only based on a very small sample of the entire gene pool of a particular species, and there will be many other sensitive species which we have not been able to identify. For these reasons we have examined the data to examine whether there is any consistent pattern as to which species are sensitive.

Table 1. Summary of British native herbaceous species sensitive to O_3.

Species with mean percentage injury of the most sensitive leaf		
> 70%	45-70%	30-45%
Anagallis arvensis	*Trifolium pratense*	*Prunella vulgaris*
Lotus uliginosus	*Avena fatua*	*Plantago maritima*
Lotus pedunculatus	*Hypericum perforatum*	*Erodium circutarium*
Trifolium repens	*Ulex europeus*	*Agrostis tenuis*
Reseda lutea	*Conium maculatum*	*Sherardia arvensis*
Catapodium rigidum	*Coriandrum sativum*	*Ranunculus acris*
Sanguisorba minor	*Vicia hirsuta*	*Agrostis canina*
Humulus lupulus	*Polemonium caeruleum*	*Dactylis glomerata*
Medicago lupulina	*Myostis arvensis*	*Iberis amara*
Lathyrus aphaca	*Papaver rhoeas*	*Pimpinella saxifraga*
Geum rivale	*Hyoscyamus niger*	
Agrostis stolonifera	*Lathyrus montanus*	
	Phleum pratense	

The species are tabulated in descending order of sensitivity, based on the average of the percentage of the area injured on the leaf of each plant showing the most injury.

The first observation was that the frequency of sensitive species within different families varied greatly. Of the families from which more than ten species were screened, the Papilionaceae contains the highest frequency of sensitive species, with 45% being classed as either moderately or very sensitive. In contrast, we have found no member of the Compositae which is included in either of these sensitivity classes. For many families it has only been possible to test a small number of species, and we therefore examined the whole data set in relation to the evolutionary status of the families to which the tested species belong. For this purpose, we have used Sporne's Index of Evolutionary Advancement (4). The results, summarised in Table 2a, show a large and statistically significant effect on O_3 sensitivity, with the more primitive families containing a higher proportion of sensitive species.

We have also compared the frequency of sensitivity among species characteristic of different habitats. The largest difference found has been between species characteristic of calcareous and acid habitats (Table 2b). Within the Graminae, for example, species such as *Catapodium rigidum*, which are characteristic of dry chalk grassland, are sensitive, while species such as *Nardus stricta* and *Deschampsia cespitosa*, which are characteristic of nutrient-poor acid grasslands, are highly resistant to O_3. This observation of increased sensitivity in more calcareous habitats led us to examine experimentally the relationship between calcium nutrition and O_3 sensitivity.

Table 2. Some factors influencing the sensitivity of native species.

(a) Effect of evolutionary status

Families with a value of Sporne's Advancement Index of	Percentage of species classed as		
	Resistant	Sensitive	Very sensitive
> 60	64.5	45.2	12.5
50-60	58.1	32.2	25.8
< 50	31.4	28.6	40.0

(b) Effect of habitat

Habitat	Percentage of species classed as		
	Resistant	Sensitive	Very sensitive
Acid	67.8	14.3	10.7
Calcareous	26.9	30.8	46.4

3. CALCIUM AND VISIBLE INJURY

The hypothesis that there is a link between calcium nutrition of a plant and its sensitivity to O_3, has been tested in two ways. The first experiment was to examine the sensitivity of plants of the same species growing in differing habitats. Plants of *Agrostis capillaris* were collected from a total of 16 sites covering a wide range of soil types. All sites were within 15 km of our laboratory, and thus should differ little in climatic conditions. The collected material was potted, allowed to grow on for one week and then exposed to 250 ppb O_3 for 4 hours. The results (Table 3) showed that the amount of visible leaf injury caused by the fumigation was much greater on material collected from sites on alkaline soils rich in calcium. Leaf calcium concentrations were determined for each of the fumigated plants; these showed a very high positive correlation ($r = 0.868$; $P < 0.001$) between the calcium contents of individual plants and the amount of visible injury on their leaves. A second species, *Agrostis setacea*, was also included in this study; this is an acidophilous species, and could only be collected from the more acidic sites. Nevertheless a significant positive correlation between leaf calcium content and visible injury was also found with this species.

Table 3. Effect of soil type on the O_3 sensitivity of *Agrostis capillaris*.

Soil	Exchangeable calcium content of soil (meq 100 g^{-1})	Mean leaf calcium content (mg g^{-1})	Mean percentage leaf injury
Bagshot sand	2.2	1.8	0.39
Barton sand	1.5	2.9	1.24
Reading beds	8.4	4.7	2.60
London clay	6.4	5.9	3.18
Upper chalk	8.4	5.9	3.32
Thames alluvium	8.6	12.3	7.20

Data of J. Husband. Plants were collected from 16 sites in East Berkshire; data have been grouped into the six major soil types encountered. Soil to a depth of 10 cm was collected adjacent to the sampled plants, and exchangeable bases were extracted with neutral N ammonium acetate. Plant material was digested in 2 M nitric acid. Calcium was determined by atomic absorption spectrophotometry.

These results strongly suggest a causal link between calcium levels and sensitivity to O_3; nevertheless the possibility of some other correlated ecological variable being responsible cannot be excluded. Therefore the effects of growing plants in nutrient solutions containing a range of calcium concentrations has been examined in controlled laboratory experiments. After growing the plants for three weeks in half-strength Rorison's solution (5) containing a range of concentrations of $CaCl_2$ the plants were then exposed to 250 ppb O_3 for 4 hours, and the amount of leaf injury produced was assessed. The results of experiments carried out with *Agrostis capillaris* and *Trifolium repens* are summarised in Table 4. A significant effect of treatment was found with *Trifolium repens*, but no effect of treatment was found with *Agrostis capillaris*. Correlation coefficients between leaf calcium concentration and visible injury of individual plants were also calculated; a highly significant positive coefficient was found for *Trifolium repens* ($r = 0.723$; $P < 0.001$), but not for *Agrostis capillaris*. A likely explanation of the difference in response of the two species is the higher rate of calcium accumulation by *Trifolium repens*. Leaf calcium concentrations for *Agrostis capillaris* were 10-15% of those of *Trifolium repens* grown at the same solution concentrations, and were also substantially less than those found in the field collections (cf. Table 3).

Most plants maintain extremely low (micromolar) cytsolic concentrations of calcium against a higher extracellular concentration by

Table 4. Effects of calcium concentration on O_3 sensitivity.

Calcium concentration in solution (mM)	*Agrostis capillaris*		*Trifolium repens*	
	Leaf calcium content (mg g^{-1})	Percentage leaf area injured	Leaf calcium content (mg g^{-1})	Percentage leaf area injured
0	0.49	27.7	3.17	38.3
0.1	0.56	33.4	4.23	42.6
0.5	0.94	29.5	8.35	50.6
1.0	0.98	26.5	9.67	49.2
5.0	1.87	30.2	17.75	61.2

Plants were grown in sand culture regularly watered with solutions containing the tabulated calcium concentrations. Plant material was digested in 2 M nitric acid and calcium was determined by atomic absorption spectrophotometry.

means of an active efflux mechanism (6). Plants from calcareous soils, or grown at high calcium concentrations, must maintain a steeper diffusion gradient into the cell by means of this active efflux. Ozone is known to have deleterious effects on cell membrane permeability, and there is some evidence that it specifically affects active transport sites (7). Any such effects could lead to a more rapid influx of calcium down the steeper diffusion gradient, and thus would lead to breakdown of cell processes, and result in cell death and visible injury.

4. CALCIUM AND GROWTH

The experimental work described above shows a strong relationship between calcium content and the sensitivity of plants to high concentrations of ozone. It is, however, clearly important to test whether calcium nutrition influences the responses of plants to longer-term exposure to more realistic O_3 levels, and we are currently investigating this in controlled laboratory experiments. Preliminary results with *Trifolium repens* and *Agrostis capillaris* suggest that the relationship is more complex than that found in our studies with visible injury. In general, the effect of O_3 on plant growth was less when calcium was deficient or supplied at luxury levels than at levels considered optimal for plant growth. However, our results have not all shown this response, and more work is needed to define the relationship between calcium nutrition and long-term effects of O_3 precisely.

5. CONCLUSIONS

Our studies have demonstrated a clear link between calcium nutrition and the sensitivity of plants to short-term exposure to high concentrations of O_3. Although other elements, such as nitrogen, phosphorus, potassium and sulphur, have been shown to influence plant responses to O_3 (8), we know of no previous report of the importance of calcium nutrition in this context. Current concern about the impact of pollutants on plant communities in Europe is focussed on those of poorly buffered acid soils, which are sensitive to increased acid deposition and nitrogen inputs. In contrast, the results of our screening and our experimental studies suggest that these are the communities least likely to be affected by O_3, and that changes in species composition are more likely to be caused by O_3 in calcareous habitats.

6. ACKNOWLEDGEMENTS

This work has been supported by a grant from the U.K. Natural Environment Research Council and a N.E.R.C. studentship to A.K. Tickle.

7. REFERENCES

(1) U.K. PHOTOCHEMICAL OXIDANTS REVIEW GROUP (1987). Ozone in the United Kingdom. Department of the Environment, London.

(2) ASHMORE, M.R., BELL, J.N.B., DALPRA, C. and RUNECKLES, V.C. (1980). Visible injury of crop species by ozone in the United Kingdom. *Environ. Pollut.*, 21, 209-215.

(3) ASHMORE, M.R. (1984). Effects of ozone on vegetation in the United Kingdom. In: *Ozone*. (P. Grennfelt, ed.), pp. 92-104. TVL, Goteborg.

(4) SPORNE, K.R. (1980). A re-investigation of character correlations among dicotyledons. *New Phytol.*, 85, 419-449.

(5) HEWITT, E.J. (1966). Sand and water culture methods used in the study of plant nutrition. Commonwealth Agricultural Bureaux, Farnham Royal, U.K.

(6) CLARKSON, D.T. (1984). Calcium transport between tissues and its distribution in the plant. *Plant, Cell and Environment*, 7, 449-456.

(7) TING, I.P. and HEATH, R.L. (1975). Responses of plants to air pollutant oxidants. *Advances in Agronomy*, 27, 89-121.

(8) COWLING, D.W. and KOZIOL, M.J. (1982). Mineral nutrition and plant response to air pollutants. In: *Effects of Gaseous Air Pollution in Agriculture and Horticulture*. (ed. by M.H. Unsworth and D.P. Ormrod), pp. 349-375. Published by Butterworth Scientific, London.

SECONDARY EFFECTS OF PHOTOCHEMICAL OXIDANTS ON CEREALS: ALTERATIONS IN SUSCEPTIBILITY TO FUNGAL LEAF DISEASES

A. v. TIEDEMANN and H. FEHRMANN
Institut für Pflanzenpathologie und Pflanzenschutz
der Georg-August-Universität, D-3400 Göttingen, W.-Germany

SUMMARY

In a nine-day fumigation experiment with 60-80 ppb ozone and 6-8 ppb PAN (peroxyacetyl-nitrate) predisposition of wheat and barley towards necrotrophic leaf pathogens was enhanced although leaves were not visibly injured by the treatment. Particularly, spot blotch (*Drechslera sorokiniana*) and tan spot (*D. tritici-repentis*) on wheat and spot blotch on barley were favoured by either of the two pollutants. The pollutant mixture exerted a highly synergistic action on plant susceptibility. In another experiment, wheat leaf attack by *Septoria nodorum* was highly correlated with increasing levels of ozone (0, 60, 90 and 120 ppb; 7 days; 7 h per day), applied before inoculation. Net blotch of barley (*D. teres*) was favoured especially at the lower ozone concentrations and on the older leaves. Results are discussed with respect to the increase of leaf diseases in cereal production since about two decades.

1. Introduction

During the last two decades fungal leaf diseases in German cereal production have steadily increased in importance. Except powdery mildew this applies mainly to weak pathogens on the leaves which had been without major economic importance in the preceding decades. Since a fundamental progress in crop production has taken place during this period, current explanations for the aggravated disease situation deal with the introduction of growth retardents, higher nitrogen fertilization, changes in crop rotation, side-effects of fungicides and others. Nevertheless, additional factors like stress due to phytotoxic air pollutants can no more be excluded, since forest decline in West Germany has indicated their harmful potential in natural ecosystems.

The few series of ozone measurements available for Western Europe support estimations of a doubling of mean concentrations during the past 30 years (2). Nowadays, at rural sites, 7-hour-mean concentrations of 50-125 ppb and peaks of above 300 ppb are currently recorded (3). Another photooxidant ocurring with photochemical smog is PAN (peroxyacetyl-nitrate) which has been first reported from the Los Angeles region in the early fifties. West European measurements are rare but indicate about 3-7 ppb PAN as diurnal maximum values (3, 7), exceptionally having climbed up to 16.6 ppb in London (9). Thus, in Europe subacute doses of photooxidants for most plants still are predominating. In this paper data about the influence of ozone and PAN in such low concentrations on several cereal leaf diseases resulting from controlled fumigation experiments are presented.

2. Experimental Methods

Winter wheat (cv.`Diplomat`) was sown in flower pots and vernalized at -1 to -6°C for 23 days after germination. Thereafter the plants were grown at max. 20°C in the greenhouse (16 h light period) until shooting stage (EC 30/31) together with spring barley (cv.`Apex`), cultivated under same conditions until ligula stage (EC 37-39). Fumigation was conducted in the chambers at the GSF, München-Neuherberg, using a unit of four gas cabins placed together in one phytotron, providing identical climate conditions. Climate was arranged to simulate an average day in early June: temperature (day) 22°C to 15°C (night), relative humidity between 7o and 90% respectively and light period 16 h (25.000 lx). After inoculation night temperature was raised to 18°C and r.h. to 95% for 72 hours.

The four fumigation variants were: (1) charcoal-filtered air, (2) 60-80 ppb ozone, (3) 6-8 ppb PAN and (4) the mixture of (2) and (3). Ozone was dosed for max. 7 h daily (from 10 a.m. to 5 p.m.) and PAN for max. 6 h (from 11 a.m. to 5 p.m.).

After 6 days of fumigation wheat was sprayed with spore suspensions of <u>Ascochyta</u> sp. (10^5 spores/ml), <u>Drechslera sorokiniana</u> (syn. Helminthosporium sativum)($5x10^3$) and <u>Drechslera tritici-repentis</u> (10^4 cfu/ml) and barley with <u>D. sorokiniana</u> ($3x10^2$). In the first set of experiments fumigation was continued immediately after inoculation for the 3 following days as described above. After that, plants were transfered to ambient air in the greenhouse for symptom development. Disease severity was assessed 6-10 dpi as percentage of attacked leaf area or number of spots per leaf.

In another set of experiments, performed in a similar fumigation unit at Göttingen, dose-response relations of ozone and disease development were investigated. Spring wheat (cv.`Schirokko`; EC 61-69) and spring barley (cv.`Apex`; EC 39-49), grown in flower pots under normal greenhouse conditions were simultaneously exposed to (1) filtered air, (2) max. 60 ppb, (3) max. 90 ppb and (4) max. 120 ppb ozone at the same diurnal doses as described above on 7 consecutive days prior to inoculation. Climatic conditions during exposure were such as described for the Munich experiments. After 7 days, ozone fumigation was stopped, wheat was inoculated with <u>Septoria nodorum</u> ($7x10^5$ spores/ml) and barley with <u>Drechslera teres</u> ($3x10^3$). During 48 hours following spore treatment plants were held under 100% r.h. (artificial fog). Disease severity was evaluated as described above. Spore formation of <u>S. nodorum</u> was assessed by incubating the leaves in humidity chambers under near UV light at max. 18°C for 4 days. After that, leaves were watered for another 24 h in distilled water (2 ml per leaf) to make the pycnidia release their spores. Spores in the yielded suspension were counted with a haemocytometer (Fuchs-Rosenthal). Number of spores was related to leaf area, measured by a computerized image processing video system.

3. Results and Conclusions

After 9 days of pollutant exposure, none of the noninoculated check plants - neither wheat, nor barley - showed any visible damage due to fumigation. Only with the inoculated plants, pollutant effects could be differentiated when evaluating the degree of leaf attack.

As the following two tables show, generally ozone alone exhibited only a slight to moderate increase in disease severity, mainly recognizable on the older leaves (f-1 and f-2, the two leaves below the flag leaf) . Thus, leaf attack of wheat by *D. sorokiniana* on f-2 was about to be doubled from 3.2 to 8.8% (table I), the same being true with spot blotch (*D. sorokiniana*) on barley (f-1 in table II). Contrary to these results, the pre- and postinoculative treatment with only ozone reduced significantly the amount of *Ascochyta* leaf spots on wheat, whereas tan spot disease (*D. tritici-repentis*) remained more or less unaffected (table I). Quite comparable, PAN alone exhibited only slight changes in disease development, although elevating number of tan spots on wheat (table I) and severity of spot blotch on barley (table II) significantly up to the twofold of the filtered air check.

Table I: Effects of ozone and PAN on *Ascochyta* leaf spots, tan spot disease (*D. tritici-repentis*) and spot blotch disease (*D. sorokiniana*) of wheat (cv.`Diplomat`)

Fumigation/ Pathogen	Leaf 1	Leaf 2	Leaf 3
Ascochyta (9 dpi)	Percentage	Diseased Leaf	Area
FA	0.7	3.0	3.7
ozone	0.3	1.2*	1.0**
PAN	0.8	4.2	4.4
ozone + PAN	1.3	5.6**	7.0**
D. sorokiniana (6 dpi)			
FA	1.0	3.3	3.2
ozone	1.6	3.7	8.8
PAN	2.2	7.0	6.6
ozone + PAN	5.7***	16.7***	46.7***
D. tritici-repentis (10 dpi)			
	Number of	Tan Spots per	Leaf
FA	0	1.0	2.0
ozone	0.2	1.0	1.4
PAN	0.2	2.6**	3.2*
ozone + PAN	0.8**	2.8***	2.3

FA: filtered air; (for pollutant doses see chapter `Methods`)
Leaf 1,2 and 3: youngest, fully developed leaf and the two leaves below
*, **, ***: significance levels for $p = 0.05$, 0.01, 0.001 following multiple T-test

Simultaneous influence of both pollutants in mixture changes the situation drastically. Ascochyta, being reduced by ozone alone, now causes twice as much diseased leaf area as the control plants show. Similarly the number of tan spots climbs up under the influence of the pollutant mixture. Much more striking however, severity of spot blotch disease is enhanced, this being true both on wheat and barley. As tables I and II show, leaf attack is enhanced between two- and fifteenfold on each of the three leaf levels. Thus, the two photooxidants in mixture are causing a highly synergistic (more-than-additive) action with respect to plant susceptibility.

Table II: Effects of ozone and PAN on spot blotch disease (*D. sorokiniana*) of barley (cv.`Apex`)

Fumigation	Percentage Diseased Leaf Area (7 dpi)		
	Leaf 1	Leaf 2	Leaf 3
FA	0.3	1.8	7.9
ozone	1.4	4.3	8.3
PAN	0.5	3.3	12.1*
ozone + PAN	2.6*	10.4***	17.2**

In the second set of experiments dose-response-relations of single ozone doses were tested with several host-pathogen systems. Simulating natural conditions during the main growth period in May, June and July, ozone exposure was stopped the day before inoculation. Thus, direct pollutant effects on the infecting pathogen were excluded and alterations in plant susceptibility could be clearer interpreted. Results as listed in table III and IV indicate a close correlation between ozone concentration and disease development, being most significant for *S. nodorum* on wheat. Here, even on the youngest leaf (flag leaf), disease severity increases considerably, reaching the highest levels generally at 120 ppb ozone. In contrast to this, the two youngest leaves of barley seem to be unaffected in relation to net blotch infection in any of the three pollutant concentrations (table IV). Looking on the third and fourth leaf, however, reveals similar reactions as described before for *S. nodorum* on wheat, except with the two highest pollutant doses where the curves tend to decline again.

Table III: Dose-response relationship of ozone and leaf attack and spore production by *S. nodorum* on wheat (cv.`Schirokko`)

Ozone Concentration (ppb)	Percentage Attacked Leaf Area			Spores/cm² on f and f-1
	f	f-1	f-2	
0	3.7	18.6	28.9	83000
60	5.3	29.6	-	133200
90	14.3*	30.5	38.8	174400
120	18.4**	38.3*	42.0	229800

f, f-1, f-2: flag leaf and the two leaves below

Results from spore counts confirm the leaf evaluation data. (table III). Total spore production per cm^2 on the two uppermost leaves of wheat increases from 83000 in filtered air up to 230000 spores in the highest ozone variant thus paralleling ozone doses and data of leaf attack.

Table IV: Dose-response relationships of ozone and net blotch disease (D. teres) of barley (cv.`Apex`)

Ozone Concentration (ppb)	Percentage Diseased Leaf Area (7 dpi) f	f-1	f-2	f-3
0	4.8	11.3	23.5	33.7
60	5.6	14.8	39.6**	60.6**
90	5.8	11.3	39.6**	54.7**
120	6.0	13.2	31.4	54.3**

The results demonstrate that air pollutants, particularly PAN, might change plant susceptibility at quite low and subacute concentrations. Recent investigations (reviewed by 4) give some examples of disease enhancement by mainly preinoculative ozone treatments in case of necrotrophic pathogens such as Botrytis cinerea on onions (10) or Alternaria solani on potato leaves (6). With biotrophic parasites like rusts and powdery mildews the effects generally turn to the opposite and reveal a certain inhibition of disease development (4, 1). Thus, in our experiments with facultative parasites on cereals results were not unexpected. Moreover, they demonstrate predisposing effects of photooxidants even when fumigation continues during the infection process. However, in this case reduction of even facultative parasites may occur. Nevertheless, preinoculative treatment with ozone refers much more to ambient conditions in Western Europe since alternating periods of dry, sunny (ozone concentration elevated) and cool, humid (ozone low) episodes are predominating the main growing season between May and July. Using a corresponding experimental design ozone at fairly low concentrations such as commonly found in rural areas of Germany caused a tremendous increase of disease development after inoculation with two major pathogens of wheat (S. nodorum) and barley (D. teres). Higher ozone concentrations far above average values from outdoor do not necessarily exhibit the major effects as results of the net blotch experiment indicate (table IV).

Not many efforts have been made to test the influence of ozone and PAN in mixture at `natural` mixing ratios. Several workers have found a pronounced less-than-additive effect of both compounds when evaluating direct injury to the leaves (8), or growth responses (11). This is explained by the different sites of action in different leaf tissues and related to different leaf ages. Surprisingly, our experiments resulted in completely different findings. Particularly with D. sorokiniana on barley and wheat but also with Ascochyta on the latter, drastic synergistic effects were enhancing plant susceptibility. This indicates that predisposing action of ozone as described in the dose-response experiments may be highly influenced in the presence of low PAN concentrations. In addition, it can be concluded that actual subacute concentrations of photooxidants like ozone and PAN may have contributed considerably to the elevated disease potential in cereals observed during the last decades.

REFERENCES

(1) DOHMEN, G.P., 1987. Secondary effects of air pollution: Ozone decreases brown rust disease potential in Environ. Poll. 43: 189-194

(2) FEISTER, U. and W. WARMBT, 1987. Long-term measurements of surface ozone in the German Democratic Republic. J.Atm. Chem. 5(1): 1-22

(3) GUDERIAN, R.,(ed.), 1985. Air pollution by photochemical oxidants. Springer Verlag.

(4) HEAGLE, A.S., 1975. Response of three obligate parasites to ozone. Environ. Poll. 9: 91-95

(5) HEAGLE, A.S., 1982. Interactions between air pollutants and parasitic plant diseases. In: Effects of gaseous air pollution in agriculture and horticulture. Butterwoths London, p. 333-348

(6) HOLLEY, J.D., G. HOFSTRA and R. HALL, 1985. Appearance and fine structure of lesions caused by the interaction of ozone and _Alternaria_ _solani_ in potato leaves. Can. J. Pl. Path. 7: 277-282

(7) LANDOLT, W., F. JOOS and H. MÄCHLER, 1985. Erste Messungen des PAN-Gehaltes der Luft im Raume Birmensdorf (ZH). Schweiz. Z. Forstw. 136(5): 421-426

(8) NOUCHI, I., H. MAYUMI and F. YAMAZOE, 1984. Foliar injury response of petunia and kidney bean to simultaneous and alternate exposures to ozone and PAN. Atm. Environ. 18(2): 453-460

(9) PENKETT, S.A., F.J. SANDALLS and B.M.R. JONES, 1977. PAN-measurements in England - analytical methods and results. VDI-Berichte Nr. 270: 47-54

(10) RIST, D.L. and J.W. LORBEER, 1984. Moderate dosage of ozone enhance infection of onion leaves by _Botrytis cinerea_ but not by _B._ _squamosa_. Phytopath. 74: 761-767

(11) TEMPLE, P.J. and O.C. TAYLOR, 1985. Combined effects of peroxyacetyl nitrate and ozone on growth of four tomato cultivars. J. Environ. Qual. 14(3): 420-424

COMBINED EFFECTS OF OZONE AND ACID MIST ON TREE SEEDLINGS

M.R. ASHMORE, C. GARRETY, F.M. MCHUGH and R. MEPSTED
Department of Pure and Applied Biology
Imperial College of Science and Technology, London

Summary

This paper describes results of a programme of research designed to test the hypothesis that O_3 and acid mists acting in combination are a major cause of the forest decline now widespread in central Europe. Initial experiments using *Picea abies* and *Picea sitchensis* seedlings showed that intermittent exposures to 120 ppb O_3 and pH 3.3 mist had little effect on growth or on visible appearance. Furthermore, needle concentrations of Ca and Mg were generally increased in seedlings exposed to high levels of O_3 and mist acidity, rather than decreased as would be suggested by this hypothesis. Subsequent experiments have involved longer-term continuous exposure of seedlings to levels of O_3 typical of high and low elevation sites in West Germany. Increased needle loss from *Pinus sylvestris*, and accelerated leaf fall from *Fagus sylvatica* been found in higher O_3 treatments; weekly treatments with pH 3.5 mist significantly increased needle loss and leaf fall in combination with high O_3 treatments, but not in combination with background O_3 levels. These data do show a synergistic effect of O_3 and acid mist at realistic concentrations, and thus strongly support the O_3/acid mist hypothesis.

1. INTRODUCTION

The causes of the extensive forest decline affecting West Germany and other countries of central Europe have been the subject of intensive scientific debate in recent years (1,2). One of the hypotheses which has received greatest attention is that the damage is primarily due to ozone (O_3) acting in combination with acid mists (3,4). Laboratory experiments have clearly demonstrated that exposure to high concentrations of O_3 increases the rate of leaching of elements such as magnesium and potassium, which are often found to be deficient in affected trees in the field (3, 5). This paper describes some important results from a programme of research designed to test this hypothesis.

2. EFFECTS OF INTERMITTENT O_3 EXPOSURE

Our initial experiments involved a study of the effects of intermittent exposure to high concentrations of O_3 followed by treatment with mist of differing acidities. In the summer of 1985, 3-year old seedlings of *Picea abies* and *Picea sitchensis* were exposed for 6 h to 120 ppb O_3 at two-weekly intervals in glasshouse fumigation chambers; during the intervening periods, the trees were maintained in closed outdoor chambers ventilated with filtered air. The seedlings were subjected to overnight mist treatments in controlled environment cabinets immediately after, or 24 h after, O_3 exposure. The mist solution applied was a standard mixture of ions adjusted to pH 5.3 or pH 3.3 using a 30/70 mixture of nitric and sulphuric acid. The mist, which had a mean droplet size of 70 μm, was applied for 3 h at a rate equivalent to 2 mm h^{-1}.

The experiment lasted for 20 weeks, after which the seedlings were harvested, their component parts were weighed, and the concentrations of Ca, Mg and K in needles of differing age classes were determined. Very few significant effects of treatment on plant growth were found, but there was evidence of significant effects of treatment on needle concentrations of calcium and magnesium. These results (Table 1) show an interactive effect of gas and mist treatment, which is frequently significant and which is consistent for element, species, and needle age. For seedlings treated with pH 5.3 mist, the effect of O_3, whenever it was significant, was to reduce needle concentrations; however, for seedlings treated with pH 3.3 mist, the effect of O_3, whenever it was significant, was to increase needle concentrations. Similarly, for trees treated with filtered air, the effect of increased mist acidity, whenever significant was to reduce needle concentrations; however, for seedlings treated with O_3, the effect of increased mist acidity, whenever significant, was to increase needle concentrations.

Other workers (6) have reported increased needle concentrations of Ca and Mg in tree seedlings exposed to high concentrations of O_3, but did not find any evidence of interactive effects with mist acidity. The effect of increased levels of mist acidity and O_3 in increasing needle concentrations in our experiment is the opposite of that predicted by the original O_3/acid mist hypothesis (3). Since leaching rates of Ca and Mg are increased by increasing acidity and since Ca is an element which is not mobile in the phloem, the most likely reason for these increased needle concentrations is an increased rate of uptake.

3. EFFECTS OF CONTINUOUS O_3 EXPOSURE

In subsequent experiments, we have attempted to examine the longer-term effects of continuous exposure to O_3 under more realistic environmental conditions. These experiments have been carried out in six outdoor semi-open-top chambers, which are 2.3 m high and 3.3 m in diameter. The experimental treatments consist of three O_3 fumigation regimes in combination with two mist acidities. The experimental treatments are regularly switched between chambers, to minimise chamber effects. Two-year old seedlings, of West German origin, of *Pinus sylvestris, Fagus sylvatica, Picea abies* and *Abies alba* have been used.

The three O_3 treatments were chosen to represent, in a simple manner, concentrations in unpolluted air, a concentration pattern typical of low elevation German sites, and a concentration pattern typical of high elevation German sites. Elevated O_3 concentrations (of 60, 80, 100 or 120 ppb) are given on 25% of days during the summer. Diurnal fluctuations in O_3 levels are not normally found at high elevation sites in West Germany; accordingly, the 'high elevation' O_3 treatment receives the same concentration in both day and night. No elevated O_3 concentrations are given during the winter. O_3 is generated from dry air by a high voltage discharge tube and its concentration determined using a UV photometer, with each chamber being sampled for 8 minutes per hour. Exposure statistics for the three O_3 treatments are summarised in Table 2.

Exposure to mist is carried out overnight, after switching off the air flow and sealing the chamber mouth. One exposure of 10 h is carried out per week. Mist is generated by spinning disc humidifiers, which produce droplets with diameters in the range 5-30 μm at a rate equivalent to 0.1 mm h^{-1}. The pH 4.8 mist solution contains: 60 μeq l^{-1} NH_4^+, 80 μeq l^{-1} Ca^{++}, 25 μeq l^{-1} Mg^{++}, 25 μeq l^{-1} Na^+, 10 μeq l^{-1} K^+, 10 μeq l^{-1}

Table 1. Summary of Treatment Effects on Needle Concentrations of *Picea sitchensis* and *Picea abies*.

Species and needle age		Treatment: pH 5.3 Control	pH 5.3 O_3	pH 3.3 Control	pH 3.3 O_3	Significance of effect of: Gas	Mist	Gas & Mist
Calcium (mg g^{-1})								
P. sitchensis	0 yr	9.7 a	9.9 ab	8.9 a	10.7 b	**	ns	*
P. sitchensis	1 yr	12.4 a	11.6 a	12.1 a	12.7 ab	ns	ns	*
P. abies (A)	0 yr	6.2 b	5.4 a	5.4 a	5.8 ab	ns	ns	*
P. abies (A)	1 yr	17.5 c	13.7 a	13.5 a	15.4 b	ns	ns	***
P. abies (B)	0 yr	6.6 a	6.9 a	6.9 a	8.4 b	ns	*	*
P. abies (B)	1 yr	14.4	15.4	15.0	18.2	**	**	ns
Magnesium (mg g^{-1})								
P. sitchensis	0 yr	2.36ab	2.46b	2.21a	2.68b	***	ns	*
P. sitchensis	1 yr	1.32	1.34	1.27	1.36	ns	ns	ns
P. abies (A)	0 yr	2.01b	1.61a	1.82b	1.97b	ns	ns	*
P. abies (A)	1 yr	1.80b	1.24a	1.32a	1.75b	ns	ns	***
P. abies (B)	0 yr	2.01	2.06	1.98	2.33	*	ns	ns
P. abies (B)	1 yr	1.41	1.49	1.63	1.92	ns	*	ns

Significance of treatments is indicated as follows:- ***, $P < 0.01$; **, $P < 0.01$; *, $P < 0.05$; ns, $P > 0.05$. Means followed by the different letters are significantly different at $P = 0.05$. Groups A and B represent two different soil types on which *Picea abies* was grown.

HCO_3^-, 85 µeq l^{-1} Cl^-, 60 µeq l^{-1} SO_4^{--}, 60 µeq l^{-1} NO_3^- and 15 µeq l^{-1} H^+, in the pH 3.5 mist the concentrations of H^+, SO_4^{--} and NO_3^- are increased to 330 µeq l^{-1}, 285 µeq l^{-1} and 150 µeq l^{-1} respectively.

Table 2. Summary of Summertime O_3 Exposure Statistics.

Treatment	Mean O_3 concentration (ppb)	Percentage of days with O_3 conc. > 50 ppb	Percentage of days with O_3 conc. > 75 ppb	Percentage of hours with O_3 conc. > 50 ppb	Percentage of hours with O_3 conc. > 75 ppb
Control	26.7	0.0	0.0	0.0	0.0
Low elevation	29.4	25.0	10.7	8.3	3.6
High elevation	48.2	25.0	10.7	25.0	10.7

Experimental treatments began on 2 May 1986. Visible effects of the experimental treatments were first noticed in early September in the form

Table 3. Summary of Effects on Needle Fall from *Pinus sylvestris*.

O_3 treatment	Mist acidity	
	pH 4.8	pH 3.5
Control	0.57 a	0.73 ab
Low elevation	1.37 b	1.09 ab
High elevation	2.15 c	3.10 d

Variance ratio for:-

Gas : 43.9 ***
Mist : 2.4 ns
Gas & mist : 4.2 *

Significance of treatment effects and individual comparisons are indicated as in Table 1. Tabulated values are dry weights of fallen needles in g $tree^{-1}$.

---

of chlorosis of older needles of *Pinus sylvestris*. The percentage of trees in each O_3 treatment showing chlorosis on 22 September was as follows:-

Control : 28%
'Low elevation' : 37%
'High elevation' : 74%

Differences between the frequency of chlorosis in the 'high elevation' O_3 treatment and the other two treatments were found to be significantly different ($\chi^2 = 14.8$; $P < 0.001$). No significant effect of mist acidity on the frequency of chlorosis was found.

By early October, there was substantial abscission of both 1 yr and 2 yr needles of *Pinus sylvestris*; however there was no visible injury or abscission of current years' needles. Fallen needles were collected, dried and weighed. There was a large and significant effect of O_3 treatment on the weight of fallen needles (Table 3). Needle loss was significantly greater in both O_3 treatments than in the control treatment, and also significantly greater in the 'high elevation' O_3 treatment than in the 'low elevation' O_3 treatment. Furthermore, there was a significant interaction between O_3 treatment and mist acidity; needle loss was significantly greater in the pH 3.5 mist than in the pH 4.8 mist only when accompanied by the 'high elevation' O_3 levels.

The progress of leaf fall from seedlings of *Fagus sylvatica* was monitored at regular intervals through the autumn period. The fallen leaves were collected, dried and weighed. The cumulative leaf loss through the autumn, expressed as a percentage of the total leaf complement of each tree, showed clear evidence of a synergistic interaction between O_3 and mist acidity. Leaf loss began earlier and proceeded more rapidly in the higher O_3 treatments; by the end of October, only 5% of leaves had fallen from trees in the control treatment, but trees exposed to the 'high elevation' O_3 treatment in combination with pH 3.5 mist had lost 50% of their leaves (Table 4). A significant effect of mist acidity was found only with the 'high elevation' treatment; mist acidity had no significant effect with the control or 'low elevation' treatments.

These results show a number of features which are consistent with the hypothesis that O_3 and acid mist acting in combination are a major cause of forest decline in central Europe:-

(1) There were significant adverse effects of O_3 treatments at concentrations which are representative of levels recorded in West Germany.

Table 4. Effects on Leaf Fall from *Fagus sylvatica*.

O_3 treatment	Mist acidity	
	pH 4.8	pH 3.5
Control	3.8 a	7.1 ab
Low elevation	15.6 abc	17.0 bc
High elevation	22.3 c	44.6 d

Variance ratio for:-

Gas	:	20.2 ***
Mist	:	6.2 *
Gas & mist	:	3.4 *

Tabulated values are the mean percentage of leaves per tree which had fallen by 27 October 1986. Significance of treatment effects and individual comparisons are indicated as in Table 1.

(2) The effect of O_3 was greater when given in concentration patterns typical of high elevation areas, where forest decline is generally more advanced.

(3) There were significant adverse effects of mist acidity at levels which are representative of those recorded in certain areas of West Germany.

(4) The effect of mist acidity was only significant in treatments with higher O_3 concentrations. This demonstrates the synergistic interaction between O_3 and mist acidity which is an essential feature of this hypothesis.

(5) The symptoms observed - chlorosis and premature loss of older needles of *Pinus sylvestris* and premature leaf fall of *Fagus sylvatica* - are symptoms found on mature trees in damaged forests in West Germany.

Our results do not support the recent contention that interactions between O_3 and cold stress are of major importance (7). Needle chlorosis was already visible by early September, while the lowest air temperature recorded before the *Pinus sylvestris* seedlings were harvested in early December was 1.0^{o}C.

To date, we have little evidence of adverse effects on *Picea abies* and *Abies alba*. We intend to continue the experimental treatments for a further summer in order to test the effects of longer-term exposures on these species. In addition, we are currently determining the effects of the O_3 and mist treatments on needle contents and leaching rates of calcium, magnesium and potassium in order to evaluate the importance of changes in nutrient cycling in the observed responses.

4. CONCLUSIONS

Previous experiments at other laboratories in the U.K. and in the U.S.A. in which tree seedlings have been exposed to combinations of O_3 and acid mist have not shown any major interactive effects between the two pollutants (6,8). Our results, in contrast, provide a clear experimental demonstration of interactions between these pollutants affecting nutrient concentrations, and leaf senescence and abscission. These results are not, however, consistent between our two experiments, and this suggests that the characteristics of the O_3 fumigation regime employed may be crucial in determining plant response. We believe that the adverse effects of realistic exposures found after only a few months in our second experiment indicate that there may be a substantial impact of the longer-term expsoures to these two pollutants experienced by mature trees in the field.

5. ACKNOWLEDGEMENTS

This work was supported financially by the Commission of the European Communities and the U.K. Department of the Environment. We thank Deborah Parsons and Anthony Fletcher for their excellent technical assistance, and Nigel Bell for his advice and encouragement.

6. REFERENCES

(1) KRAUSE, G.H.M., ARNDT, U., BRANDT, C.J., BUCHER, J., KENK, G. and MATZNER, E. (1986). Forest decline in Europe: possible causes and etiology. *Water, Air and Soil Pollution*. (in press)

(2) SCHUTT, P. and COWLING, E.B. (1985). Waldsterben, a general decline of forests in Central Europe: symptoms, development and possible causes. *Plant Disease*, 69, 548-558.

(3) PRINZ, B., KRAUSE, G.H.M. and STRATMANN, H. (1982). Waldschaden in der Bundesrepublik Deutschland. Landesanstalt fur Immisonsschutz des Landes Nordrhein-Westfalen, Essen.

(4) ASHMORE, M.R., BELL, J.N.B. and RUTTER, A.J. (1985). The role of ozone in forest damage in West Germany. *Ambio*, 14, 81-87.

(5) KRAUSE, G.H.M., JUNG, K.-D. and PRINZ, B. (1983). Neuere Untersuchungen zur Aufklarung immissionsbedingter Waldschäden. *VDI-Berichte*, 500, 257-266.

(6) SKEFFINGTON, R.A. and ROBERTS, T.M. (1985). The effects of ozone and acid mist on Scots pine saplings. *Oecologia*, 65, 201-206.

(7) BROWN, K.A., ROBERTS, T.M. and BLANK, L.W. (1987). Interaction between ozone and cold sensitivity in Norway spruce: a factor contributing to the forest decline in Central Europe? *New Phytologist*, 105, 149-155.

(8) TAYLOR, G.E.Jr., NORBY, R.J., MCLAUGHLIN, S.B., JOHNSON, A.H. and TURNER, R.S. (1986). Carbon dioxide assimilation and growth of red spruce (*Picea rubens* Sarg.) seedlings in response to ozone, precipitation chemistry and soil type. *Oecologia*, 70, 163-171.

DOES EXHAUST AIR FROM MOTORWAY TUNNELS AFFECT THE SURROUNDING VEGETATION?

SABINE BRAUN and WALTER FLUECKIGER
Institute for Applied Plant Biology
Sandgrubenstr. 25, CH-4124 Schönenbuch (Switzerland)

Summary

Fir trees in the surroundings of an exhaust funnel from a motorway were examined for lead contamination and growth. During a shut down period of two years the firs in the vicinity of the funnel showed a constant or slightly increased growth, whereas the growth of firs from adjacent sites was depressed at the same time. The changes in the lead content demonstrated that the emissions from the funnel were measurable even in a distance of up to 3 km. With air pollution measurements at a site 1.6 km in the west of the funnel significant increases of airborne dust and, with some limitations, also of NO_x could be detected during the operation time of the funnel.

1. INTRODUCTION

Although vehicular tunnels are common in alpine countries such as Switzerland, there is not very much knowledge about the effect of the corresponding funnels blowing their exhaust gases to mountaineous or alpine regions. There is not only a possibility of direct effects of auto exhaust components but also the formation of secondary pollutants such as ozone or acid mist might be enhanced especially in high regions with increased irradiation. Keller (1) examined pines (Pinus mugo Turra) in the surroundings of the ventilation port of the San-Bernardino vehicular tunnel (Switzerland) in an altitude of 1950-2050 m. He found elevetad levels of lead in the needles but neither nitrogen content nor buffering capacity of the needles were changed within some 300 m from the chimney compared to control needles from the same region. The tunnel had a frequency of about 2'300-10'000 vehicles per day, depending on the season.

In the context of examinations about forest decline, rather high damages were observed on beech trees in the vicinity of a funnel which ventilates a motorway tunnel in the Jura region of Switzerland (Oberbölchen). Thus, the question of the effect of this funnel on the surrounding forests was raised. Fortunately there was an opportunity to examine the vegetation during normal operation of the funnel and during a shut down period of two years.

2. MATERIALS AND METHODS

The motorway in examination has a frequency of about 30'000-40'000 vehicles per day. It runs at an altitude of about 600 m, whereas the site of the exhaust funnel is at 800 m above sea level.

Fir (Abies alba Mill.) was chosen for examination because of its widespread occurrence in this region. Before the funnel was shut down, branches of full grown firs (7th whirl) from 32 sites were collected by helicopter in spring 1984. These sites were arranged in concentric rings of 1,2,3 and 4 km radius around the funnel. The sampling was repeated in spring 1985. The collected branches were used on the one hand to measur growth backwards for five years and on the other hand to analyze the lead

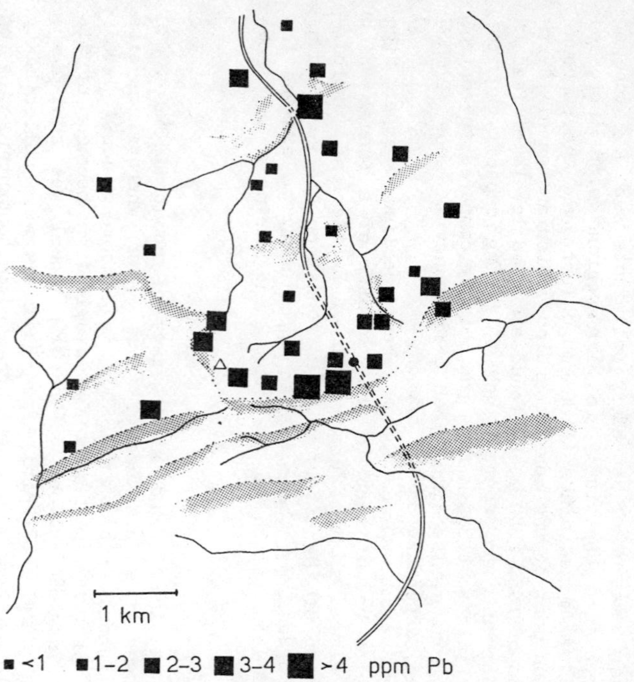

Fig. 1: Distribution of the lead content in one year old fir needles in spring 1984, before shut down of the funnel. Ridges are marked by dotted lines, south or southeast slopes are screened.
● Position of the funnel △ Site of air pollution measurements

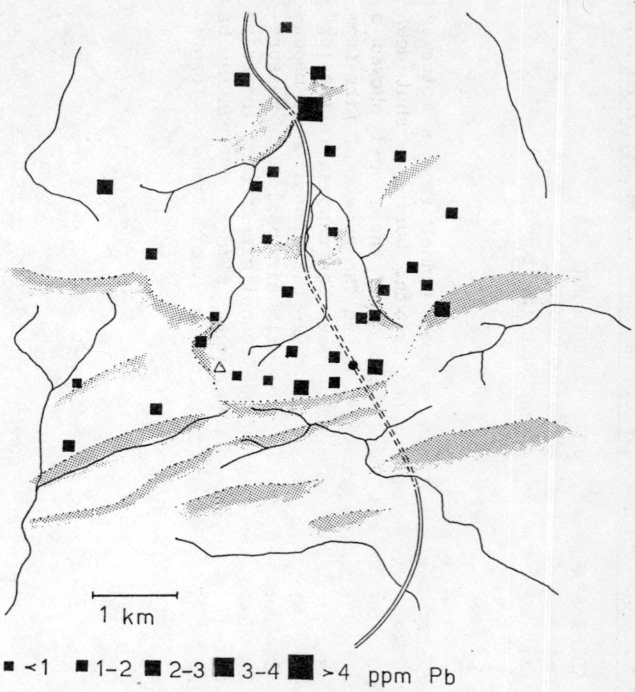

Fig. 2: Distribution of the lead content in one year old fir needles in spring 1985, one year after shut down of the funnel. Further explanations see Fig. 1.

content in the youngest needles. For lead analysis, the (unwashed) needles were dried, ground and digested in HNO_3/H_2O_2. Lead was determined by flame atomic absorption using a SP 9 Pye Unicam atomic absorption spectrophotometer equipped with a slotted tube atom trap accessory and background compensation.

Air pollution was measured with a mobile station which was situated 1,6 km in the west of the funnel, at a height of 970 m which is 170 m higher than the funnel. The monitors used were Horiba APNA 2000 for NO_x (chemiluminescence detector), Horiba APOA 2000 for O_3 (chemiluminescence detector) and Horiba APSA 2100 for SO_2 (flame photometric detector). Airborne dust was collected with a high volume sampler during 24 hours each time and analyzed gravimetrically. Wind direction, wind speed, sun radiation, air temperature and air humidity were recorded at the same time as air pollutants. A permanent, identically equipped station was located about 30 km northwest of the experimental site, at Schönenbuch.

A preliminary meteorological classification was done using the temperature courses of the two monitoring stations. Three classes were distinguished, depending on a temperature inversion being present only at night, during the whole day or lacking at all.

3. RESULTS

3.1. Lead content

The emissions from the funnel caused a wide spread increase of the lead content of fir needles (Fig. 1). They are in part above the 3 ppm given by Ewers and Schlipköter (2) as a limit for "non polluted" areas. Surprisingly enough, the highest values were not measured just beneath the funnel but on a ridge in the south, apart from the sites at the verge of the motorway itself. The shut down caused a clear reduction in the lead content of the needles in most of the examinated sites, even in a distance of 2-3 km, by up to 70% (Fig. 2).

3.2. Growth

In general, growth showed a more or less continuous decrease between 1980 and 1985 which is in accordance to results obtained in beech (3). At some of the sites near the funnel, however, this decrease was interrupted in 1984, when the funnel was shut down. Because the sites near the funnel had generally the highest altitude, and because growth shows a clear dependance from the altitude, a comparison of the absolute growth values was not possible. The relative growth change to the previous year was therefore chosen as a measure, especially the change between the years 1983 and 1984. It was compared to the changes in lead content during the same time course which was taken as a measure for funnel impact. As Fig. 3 shows, the growth change was most positive at sites with a strong decrease of lead, i.e. within the zone of high funnel impact.

There are some more indications that the shut down period of the funnel was beneficious for the vegetation. One of 34 beech observation sites of northwestern Switzerland is situated in the vicinity of the funnel. After having showed extraordinarily high damages in 1983, a clear reduction of damage could be observed in this site between 1984 and 1986, whereas in most other sites the damage increased or remained constant (4).

3.3. Air pollution

The air pollution was measured during five campaigns of about two weeks each. During part of this time, the funnel ventilation was working. Because of the high day to day variation of air pollution which is partly caused by different meteorological conditions, a clear influence of the

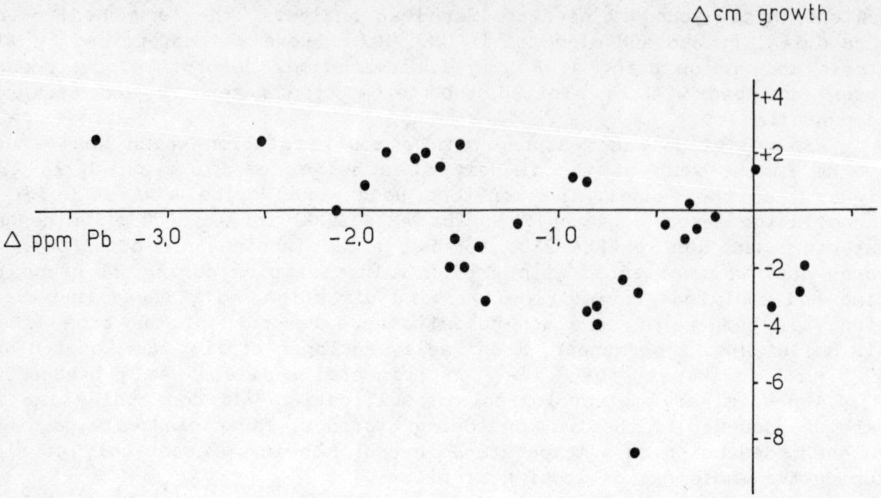

Fig. 3: Changes in lead content of one year old fir needles between spring 1984 and spring 1985 (Δ ppm Pb) in relation to changes in shoot growth (Δ cm growth) during the same time period. There is a significant negative correlation between the development of lead content and of shoot growth ($p<0,01$, r Spearman = -0,533)

funnel status on the measurements could be observed neither directly nor in the frequency distribution in relation to wind direction. However, part of this variation can be eliminated by correlating the daily means with the daily means of a permanent reference station (5). In the case of airborne dust, the days with operating funnel could now be separated clearly from those with the funnel shut down (Fig. 4). Some influence of the funnel could also be evaluated for NO_x when some meteorological conditions were taken into consideration. Confining the correlation with the reference station to days with a temperature inversion only at night, a significant influence of the funnel status was also observed (Fig. 5), although the variation was high and the distribution in time not quite homogeneous. It should be emphasized, however, that the site of air pollution measurements was not in the direct plume of the funnel; the main wind directions as measured by the mobile station were ENE and SW, whereas the funnel was situated in E direction. The ozone values were generally high in summer, with 98%-values of up to 220 ug/m^3 (113 ppb) and maximal half hourly means of more than 250 ug/m^3 (130 ppb), but no influence of the funnel status on the ozone level could be found.

4. DISCUSSION

The ample reduction in the lead content of fir needles as a consequence of shut down of the funnel indicates that the impact of the funnel reaches rather far, although the normal working time of the ventilation is only a few hours per day. This widespread distribution of lead may be explained by the size of the lead containing particles; according to (6) they have a mean size of 0,55 um. The lead results are in accordance to the finding that the total airborne dust is also significantly increased in a distance of 1.6 km when the funnel is operating.

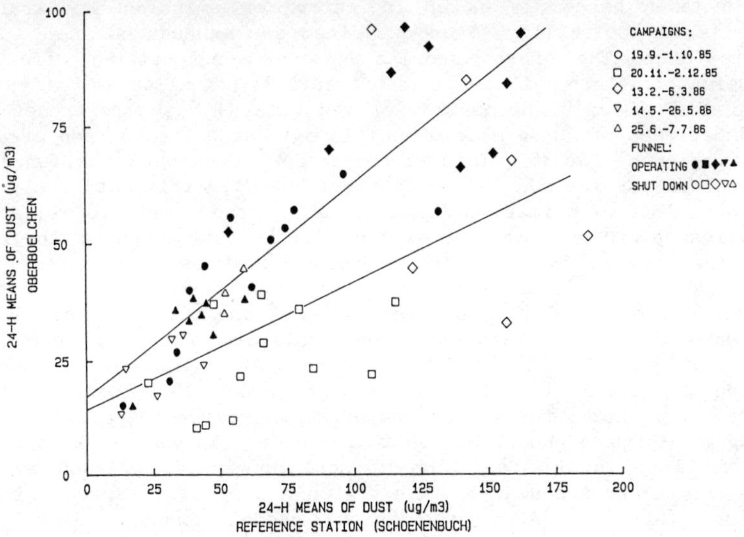

Fig. 4: Correlation between daily means of dust at Oberbölchen and at the reference station (Schönenbuch). The dust values at Oberbölchen are significantly increased when the funnel is working (upper line) compared to days with funnel shut down (lower line) ($p<0,05$; test for the difference between two regression lines after (9))

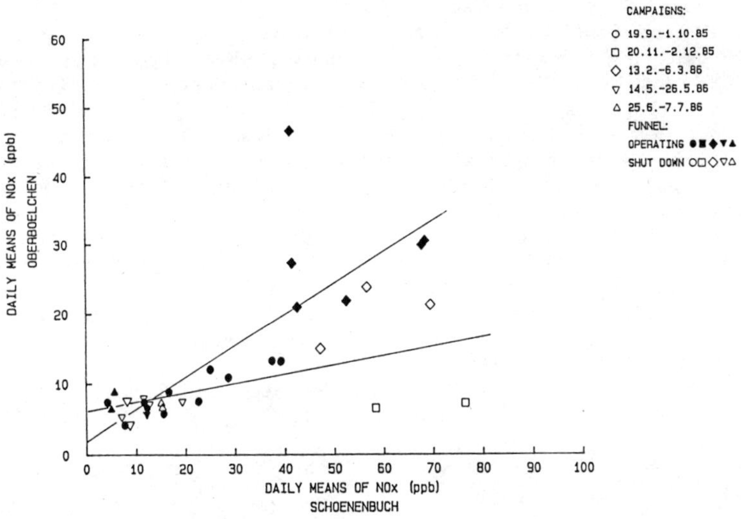

Fig. 5: Correlation between daily means of NO_x at Oberbölchen and at the reference station (Schönenbuch) for days with nightly temperature inversion. The difference between days with (upper line) and without funnel (lower line) is significant at $p<0,05$ (for the test, the logarithms of the daily means were used) (details see Fig. 4).

Lead is taken here only as an indicator for emissions by traffic. Because of its low solubility, lead dust from car exhaust has not a very high phytotoxicity. The growth results, however, suggest an effect of funnel emissions. The air pollution measurements do not allow a clear interpretation of this fact; by comparison of the data with the reference station, an annual mean of 20 ug/m^3 (10 ppb) is estimated for NO_2 and of about 1-2 ug/m^3 (0.5-1 ppb) for SO_2. This is rather low. Even when the funnel is suggested to increase the NO_2 values by about 30-50%, it is not likely that these concentrations of primary pollutants affect the vegetation directly. The ozone values are high, but they seem not to be influenced by the emissions from the funnel. Thus, other mechanisms must be involved for the growth response to the funnel emissions. One possibility are interactions of the emitted primary pollutants with the high ozone concentrations. In addition, peaks of NO_2 may be relevant for plants even when the annual mean is as low as 10 ppb (7), and short time measurements as in the present investigation may overlook such concentration peaks. High concentrations were measured e.g. during the winter campaign, where concentrations of SO_2 up to 233 ug/m^3 (90 ppb) and of NO_2 up to 136 ug/m^3 (73 ppb) were measured. Another possibility is the formation of secondary products such as acid mist; mists are rather frequent in this region and might interact with SO_2 and NO_x as well as with the high ozone concentrations (8).

5. ACKNOWLEDMENTS

The present investigation was financially supported by the civil engineering inspectorate of Basle-Country. We thank R. Lareida, motorway service in Sissach, M. Fischer, forest inspectorate in Liestal, and L. Förderer for their cooperation in installing the pollution monitoring station, harvesting the shoots and statistical advice respectively.

6. REFERENCES

(1) KELLER, Th. (1983). Beeinflussen alpine Tunnelentlüftungen Bergföhren der Umgebung? Strasse und Verkehr 69 (11)
(2) EWERS, U. and SCHLIPKöTER, H.W. (1984). Blei. In: E. Merian (ed.), Metalle in der Umwelt. Verlag Chemie, Weinheim, pp. 351-374
(3) BRAUN, S. and FLüCKIGER, W. (1987). Untersuchungen an Gipfeltrieben von Buche (Fagus sylvatica L.). Botanica Helvetica 97 (in press)
(4) FLüCKIGER, W. et al. (1986). Untersuchungen über Waldschäden in festen Buchenbeobachtungsflächen der Kantone Basel-Landschaft, Basel-Stadt, Aargau, Solothurn, Bern, Zürich und Zug. Schweiz. Zeitschr. Forstwes. 137, 917-1010
(5) BUNDESAMT FüR UMWELTSCHUTZ (1987). (in preparation)
(6) MILFORD, J.B., DAVIDSON, C.I. (1985). The sizes of particulate trace elements in the atmosphere - a review. JAPPA 35, 1249-1260
(7) FOWLER, D. (1987). Nitrogen oxides. COST 612 Workshop "Definition of European Pollution Climates and their Perception by Terrestrial Ecosystems" Bern, 27-30 April 1987
(8) SKEFFINGTON, R.A. et al. (1985). Schadsymptome an Fichte und Kiefer nach Belastung mit Ozon und saurem Nebel. Allg. Forstzeitschrift 40, 1359-1362
(9) CIBA-GEIGY AG (1980). Wissenschaftliche Tabellen Geigy, Teilband Statistik, 8. Auflage, Basel

ROLE DE LA CUTICULE DES PLANTES DANS LE TRANSFERT DES XENOBIOTIQUES DANS L'ENVIRONNEMENT

A. CHAMEL, B. GAMBONNET
DRF/LBIO/Biologie Végétale, CEN-G, 85 X, 38041 GRENOBLE CEDEX (France)

Résumé

Le sort des polluants aériens non gazeux, au sein des écosystèmes terrestres, dépend en partie de leur comportement au niveau de la barrière cuticulaire qui sépare l'atmosphère des tissus vivants de la plante.

Le rôle de la cuticule dans le transfert de ces produits peut être étudié à l'aide de "cuticules isolées". Les méthodes utilisées pour mesurer la rétention et la pénétration cuticulaires sont présentées après un bref rappel de la structure et de la composition chimique de la cuticule des plantes. Les résultats déjà acquis dans cette voie avec des métaux lourds et des pesticides montrent clairement que la cuticule représente un compartiment de l'environnement pouvant concentrer des xénobiotiques. Les études de perméabilité ont démontré qu'elle constitue aussi une voie d'accès aux tissus de la plante pour les différents types de polluants. Ces phénomènes de rétention et de pénétration au niveau de la cuticule dépendent de facteurs variés en relation avec le polluant, les conditions climatiques et l'espèce végétale.

Summary

The fate of aerial non gaseous pollutants, in the terrestrial ecosystems partly depends on their behaviour at the level of the cuticular barrier separating the atmosphere from living plant tissues. The function of the plant cuticle in the transfer of xenobiotics can be studied using isolated cuticles. The methods for measuring the cuticular sorption and permeability are described after a brief presentation of the structure and chemical composition of the plant cuticle. Results obtained in this way concerning heavy metals and pesticides clearly show that the cuticle represents an important sorption compartment in the environment. Permeability studies have shown that the cuticle also constitutes a means of access for different types of pollutants to the plant tissues. The cuticular sorption and permeability depend on various factors concerning the pollutant, the environmental factors and plant species.

La surface des organes aériens (feuilles, fruits) des plantes supérieures est recouverte d'un revêtement protecteur lipidique, extracellulaire, hétérogène au point de vue structural et chimique, et résistant à la biodégradation : la cuticule. Le sort des polluants aériens non gazeux, au sein de l'écosystème, dépend en partie de leur comportement au niveau de cette barrière séparant l'amosphère des tissus de la plante (figure 1).

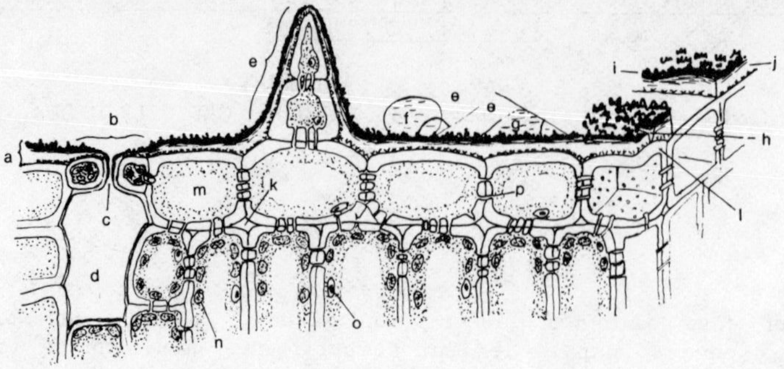

a : cuticule
b : stomate
c : ostiole
d : chambre sous stomatique
e : poil
f : sans agent mouillant
g : avec agent mouillant
e : angle de contact
h : craquelure
i : cires épicuticulaires
j : cutine
k : espace intercellulaire
l : paroi cellulaire
m : vacuole
n : chloroplaste
o : noyau
p : plasmodesme

Figure 1 : Diagramme représentant l'épiderme d'une feuille.

1. RAPPEL DE LA STRUCTURE ET DE LA COMPOSITION CHIMIQUE DE LA CUTICULE DES PLANTES (1)

La cuticule, dont l'épaisseur varie de 0,15 à 15 µm selon les espèces, est composée de plusieurs couches dont le nombre, l'épaisseur et la limite varient avec l'espèce végétale et le stade de développement. On distingue généralement, à partir de la surface (figure 2) :

. les cires épicuticulaires (figure 3),

. la cuticule *stricto sensu* ou cuticule proprement dite, formée de cutine et de cires intracuticulaires,

. une ou plusieurs couches cuticulaires pouvant être constituées, outre la cutine et les cires, de matériel dérivant de la paroi externe des cellules épidermiques,

. une couche de pectine ou lamelle de pectine.

Le constituant structural principal de la cuticule est la cutine, ce terme désignant des polyesters lipidiques insolubles dérivant d'acides gras en C_{16} ou C_{18}. Les cires épicuticulaires sont constituées de mélanges complexes de composés aliphatiques à longue chaîne et de constituants cycliques : alcools primaires, alcanes, alcools secondaires, dicétones, triterpénoïdes comme constituants majeurs.

La structure de la cuticule des plantes est si hétérogène qu'il n'existe pas de cuticule-type.

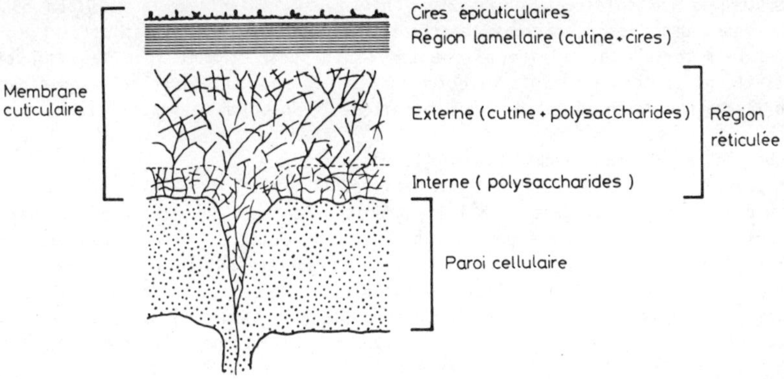

Figure 2 : Schéma simplifié de la structure d'un modèle de cuticule.

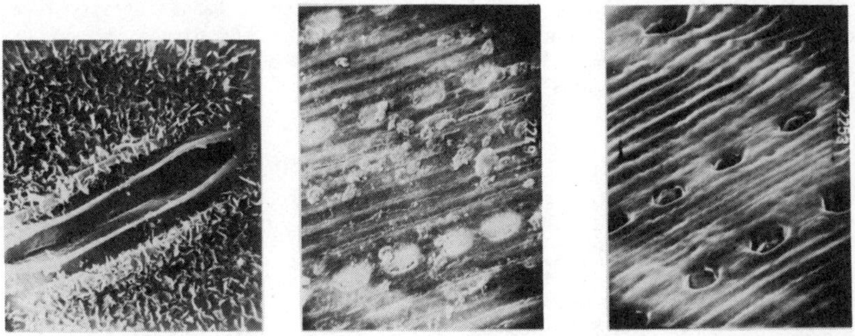

avec cires   sans cire (aiguille décirée)

Feuille de maïs   **Aiguille de pin**

Figure 3 : Aspect des cires épicuticulaires.

2. METHODES D'ETUDES

Le rôle de la cuticule dans le transfert des polluants peut être étudié à l'aide de "cuticules isolées" (figure 4) obtenues en séparant la cuticule des tissus sous-jacents par des méthodes enzymatiques (mélanges d'enzymes pectinase et cellulase) (2).

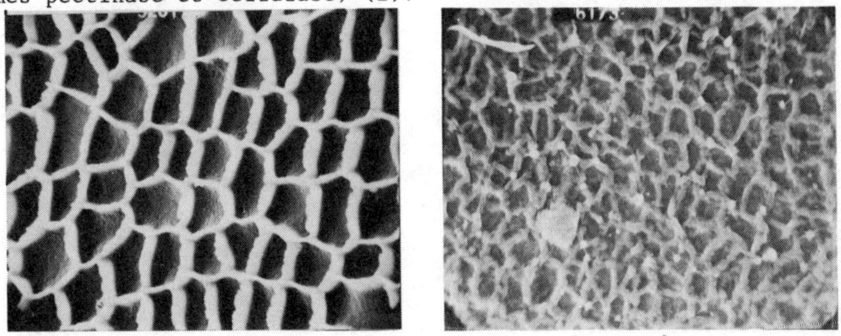

feuille (poirier)   fruit (tomate)

Figure 4 : Surface interne de cuticules isolées en microscopie électronique à balayage.

2.1. **Mesure de la fixation cuticulaire** (2):

Des disques cuticulaires (10 ou 25) sont mis au contact de la solution contenant le produit marqué, dans des conditions contrôlées. Lorsque l'équilibre est atteint, la quantité fixée par le matériel cuticulaire est mesurée soit indirectement (par la diminution de la radioactivité de la solution) soit directement (par la radioactivité des disques cuticulaires après un lavage rapide au moment de leur retrait de la solution radioactive).

2.2. **Mesure de la pénétration cuticulaire** (1, 2) :

Les deux dispositifs suivants ont été utilisés :

- un disque cuticulaire est fixé à l'extrémité d'un tube de verre à paroi épaisse lequel plonge dans un flacon de comptage pour scintillation liquide. La solution radioactive (0,5 ml) est déposée à l'intérieur du tube au contact de la surface cuticulaire externe. Les niveaux liquides du tube et du flacon sont égalisés. Le flacon receveur est échangé à différents temps et la radioactivité de la totalité de son contenu (15 ml) est mesurée chaque fois.

- un disque cuticulaire est placé entre les deux compartiments symétriques d'un dispositif de perméabilité en plexiglas installé dans un bain thermostaté à 25°C (figure 5). En début d'expérience, chaque compartiment reçoit le même volume (10 ml) de liquide, le produit marqué étant apporté du côté de la surface cuticulaire externe. Au temps zéro et à différents intervalles de temps, une fraction aliquote de 100 µl est prélevée dans le compartiment interne et sa radioactivité est mesurée. Après chaque retrait, le même volume de liquide est rajouté pour maintenir un volume constant. Ce dispositif est utilisé pour mesurer le coefficient de perméabilité ($P = \frac{dn}{dt} \cdot \frac{1}{C}$) où C représente la concentration de l'isotope radioactif dans la solution externe).

Figure 5 : Dispositif permettant de mesurer le passage de pesticides à travers la cuticule des plantes.

3. RESULTATS

3.1. La cuticule représente un compartiment de l'environnement pouvant concentrer les xénobiotiques : ce résultat est illustré par les données des tableaux 1 et 2 concernant des métaux lourds (Cd, Cu, Zn, Mn) et des molécules organiques à caractère herbicide.

Tableau 1 : valeurs de la rétention cuticulaire de quelques métaux lourds.

Les cuticules isolées ont été immergées durant 22 heures dans $^{115m}CdCl_2$, $^{64}CuCl_2$, $^{65}ZnCl_2$, ou $^{54}MnCl_2$ (concentration : 10^{-4}M) puis ont été lavées à l'eau durant 5 minutes avant d'être analysées. Chaque essai a été réalisé avec 10 cuticules.

Cuticules	Quantité d'élément retenu µg/cm²	mg/g cuticule	r
	Cadmium		
Poirier Passe Crassane (feuille) moyenne 8 essais	1.0 ±0.2	4.3 ±0.8	354
Pommier "Akane" (fruit) moyenne 4 essais	1.9 ±0.2	1.1 ±0.1	94
Tomate (fruit) moyenne 4 essais	1.1 ±0.2	0.4 ±0.06	31
	Cuivre		
Poirier Passe crassane (feuille)	4.1 ±0.6	9.5 ±1.4	1494
*	0.7 ±0.2	1.8 ±0.4	
	Zinc		
	0.7 ±0.1	1.6 ±0.3	250
*	0.2 ±0.04	0.5 ±0.09	
	Manganèse		
	0.02±0.002	0.08±0.009	14
*	0.02±0.02	0.1 ±0.07	

* dans ces essais, seule la face externe de la cuticule était mise en contact avec la solution radioactive.

r : $\frac{\text{concentration de l'élément dans la cuticule (µg/g)}}{\text{concentration de l'élément dans la solution (µg/ml)}}$

± : σ

Tableau 2 : valeurs de la rétention cuticulaire de 3 herbicides.

Les cuticules isolées ont été immergées durant 24 heures dans une des solutions herbicides (concentration : 10^{-3}M, pH = pKa) puis ont été lavées à l'eau durant 5 minutes avant d'être analysées. Chaque essai a été réalisé avec 25 cuticules.

Cuticules	Herbicide	Quantité d'herbicide retenu µg/cm²	mg/g cuticule	r
Gui	2.4 D	7.3±1.1	6.3±0.9	114
(feuilles)	Dicamba	4.1±0.8	3.6±0.7	59
	2.4 DB	18.7±2.9	15.8±2.5	216
Tomate	2.4 D	10.4±2.0	8.3±1.6	99
(fruits)	Dicamba	7.6±1.1	5.7±0.8	62
	2.4 DB	20.8±3.7	15.9±2.8	141

r : $\frac{\text{concentration de l'herbicide dans la cuticule µg/g}}{\text{concentration finale de l'herbicide dans la solution (µg/ml)}}$

± : σ

2.4 D : acide 2.4 dichlorophénoxyacétique

2.4 DB : acide 4 (2.4 dichlorophénoxy) butyrique

La récupération, par lavage à l'eau, des substances fixées par la cuticule n'est souvent que partielle ; elle dépend de l'élément considéré (3, 6).

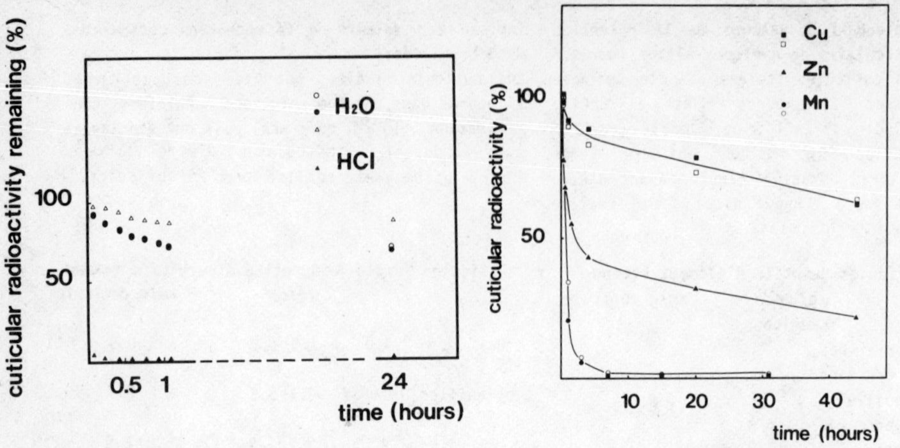

Cadmium (liquide de lavage : eau pure ou HCl dilué)

Cuivre - Zinc - Manganèse (liquide de lavage : eau pure)

Figure 6 : Effet du lavage sur la perte de Cadmium, Cuivre, Zinc et Manganèse par des cuticules isolées de feuilles de poirier.
Les disques cuticulaires ont été immergés avant le lavage dans 0,1 mM $^{115m}CdCl_2$ ou $^{64}CuCl_2$ ou $^{65}ZnCl_2$ ou $^{54}MnCl_2$ durant 22 heures (d'après 2, 3).

3.2. La cuticule constitue aussi une voie d'accès aux tissus vivants de la plante pour les différents types de polluants. Les expériences de perméabilité sur cuticules isolées dépourvues de stomate ont permis de démontrer que les ions minéraux et les molécules organiques peuvent diffuser à travers la cuticule des plantes (1, 2). La figure 7 illustre ce résultat dans le cas d'un pesticide. La diffusion de l'herbicide considéré, qui est un acide organique faible, est beaucoup plus importante à pH 3 par rapport à pH 8, ce qui suggère que le passage de cette molécule à travers la cuticule s'effectue principalement sous forme non dissociée.

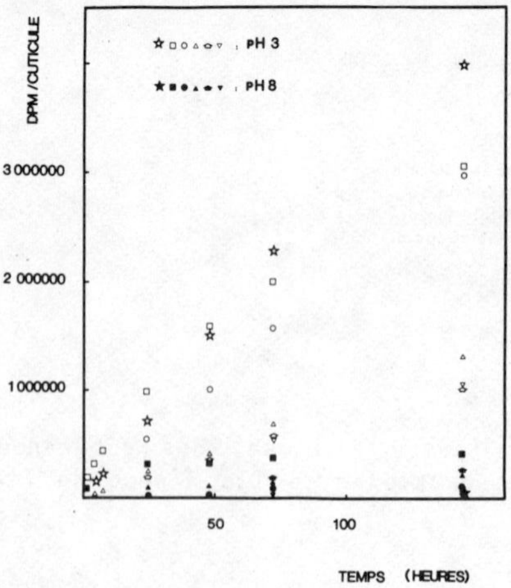

Figure 7 : Diffusion d'un pesticide (2,4 DB) à travers des cuticules isolées de feuille de poirier, à 2 valeurs de pH.
2,4 DB : acide 4-(2,4 dichlorophénoxy) butyrique.

Ces phénomènes de rétention et de pénétration au niveau de la cuticule dépendent de facteurs variés (1, 2) en relation avec :
- le polluant : nature, forme chimique, pH, concentration...
- les conditions climatiques : humidité, température...
- l'espèce végétale : structure et composition chimique de la cuticule, épaisseur...

Ces aspects sont importants à considérer pour les produits consommés par l'homme ou l'animal (légumes, fruits, fourrages...).

La cuticule des plantes peut être considérée comme un capteur de polluants dont la surface dans les communautés végétales, est considérable, avec une masse variant de 180 à 1500 kg/ha en zone tempérée. Elle constitue en particulier un important réservoir lipophile dans l'environnement pour les résidus agrochimiques.

REFERENCES

(1) CHAMEL, A. (1986). Foliar absorption of herbicides : study of the cuticular penetration using isolated cuticles. Physiol. Vég. **24**, 491-508.

(2) CHAMEL, A. (1986). Survey of different approaches to determine the behaviour of chemicals directly applied to aerial parts of plants. In Foliar Fertilization, Alexander (Ed.) Martinus Nijhoff Publishers, 66-86.

(3) CHAMEL, A., GAMBONNET, B., GENOVA, C., JOURDAIN, A. (1984). Cuticular behaviour of cadmium studied using isolated plant cuticles. J. Environ. Qual. **13**, 483-487.

(4) CHAMEL, A., BOUGIE, B. (1977). Absorption foliaire du cuivre : étude de la fixation et de la pénétration cuticulaires. Physiol. Vég. **15**, 679-693.

(5) CHAMEL, A., GAMBONNET, B. (1979). Etude, avec des feuilles *in situ* et des cuticules isolées, du comportement du cuivre et du zinc fournis par voie foliaire. Proc. Symp. Isotopes and Radiation in Research on Soil-Plant Relationship, Vienne, 373-391.

(6) CHAMEL, A., GAMBONNET, B. (1982). Study with isolated cuticles of the behaviour of zinc applied to leaves. J. Plant Nutrition **5**, 153-171.

(7) CHAMEL, A., NEUMANN, P. (1987). Foliar absorption of nickel : determination of its cuticular behaviour using isolated cuticles. J. Plant Nutrition **10**, 99-111.

POLLUTION ATMOSPHERIQUE ET METABOLISME RESPIRATOIRE DE L'EPICEA

A. CITERNE, J. BANVOY et P. DIZENGREMEL
Laboratoire de Physiologie Cellulaire Végétale
Université de Nancy I, B.P. 239
54506 Vandoeuvre-les-Nancy Cedex

Résumé

Une stimulation de l'intensité respiratoire des aiguilles apparaît caractéristique de l'Epicea (*Picea abies*) dépérissant (classe 3b) en situation naturelle. Des expériences de simulation réalisées en chambre à ciel ouvert sur trois clones d'Epicea montrent qu'une nette stimulation des activités enzymatiques mitochondriales (fumarase, enzyme malique à NAD) dans des extraits d'aiguilles est corrélée à la présence d'ozone. Le dioxyde de soufre n'entraîne qu'une faible augmentation des activités enzymatiques. L'effet de polluants potentiels a été testé sur des mitochondries isolées d'organes foliaires. Le sulfite, à faible concentration (0,1 mM), diminue l'efficacité de la phosphorylation oxydative et inhibe le transport d'électrons. La possibilité de relier les résultats obtenus *in vitro* aux modifications observées sur les arbres sera examinée.

Summary

A rise in respiration of needles appears to characterize the diseased (class 3b) spruce trees (*Picea abies*), in field conditions. Experiments performed in open-top chambers with three spruce clones showed that the rise in mitochondrial enzymic activities (fumarase and NAD-malic enzyme) could well be related to the presence of ozone as pollutant. SO_2 only caused a weak increase in enzymic activities. The effect of potential pollutants was measured on mitochondria isolated from foliar organs. Low concentrations of sulfite (0.1 mM) notably decreased both the efficiency of oxidative phosphorylation and oxygen uptake. A possible relationship between *in vitro* studies and the observed changes on whole trees will be discussed.

1. INTRODUCTION

Le phénomène du dépérissement des forêts résulte probablement d'un faisceau de causes parmi lesquelles la présence dans l'atmosphère de polluants (SO_2, O_3, NO_x, PAN), seuls ou associés, doit jouer un rôle important (1).

Les recherches menées jusqu'à présent traitaient principalement des effets des polluants sur les plantes herbacées. Les dégâts visibles affectant l'appareil foliaire, les mécanismes photosynthétiques furent étudiés de manière privilégiée (2). La photosynthèse est souvent inhibée et, en ce qui concerne les conifères, les atteintes à la composition pigmentaire et au fonctionnement de l'appareil photosynthétique peuvent être reliées à l'aspect maladif des arbres (3).

Les études portant sur la respiration sont moins nombreuses et concernent encore très peu les végétaux ligneux (4, 5). Une augmentation de l'intensité respiratoire découle généralement d'une pollution naturelle ou contrôlée (6). Les mécanismes biosynthétiques nécessaires à la croissance des arbres requièrent de l'énergie véhiculée par des molécules particulières (ATP). La respiration qui participe, tant à la lumière qu'à l'obscurité, au processus intracellulaire de production d'énergie, est donc une cible potentielle des polluants.

Les recherches entreprises concernent le métabolisme respiratoire des aiguilles d'Epicea. Les arbres testés proviennent soit de placettes forestières, soit de chambres à ciel ouvert (tests de simulation). Des mesures d'échanges gazeux ont été effectuées ainsi que des études sur différentes enzymes impliquées dans le processus respiratoire. Enfin, des tests *in vitro* sur des mitochondries isolées sont réalisés afin de mieux cerner les mécanismes d'action des polluants.

2. MATERIEL ET METHODES

Les arbres utilisés pour les expériences sont soit des perchis provenant de placettes situées dans le massif du Donon, soit de jeunes arbres (5 ans) de trois clones différents (Gérardmer, Istebna et Lac de Constance) soumis à des simulations de pollution dans des chambres à ciel ouvert à Pau-Montardon (Laboratoire de M. Bonte). Les arbres clonés sont en présence d'une atmosphère filtrée ou d'une atmosphère contenant SO_2, O_3 ou SO_2+O_3, à des concentrations identiques à celles enregistrées à la station de mesure de l'ASPA au Donon une semaine auparavant.

Les intensités respiratoires des aiguilles de perchis sont mesurées par consommation d'O_2 par la méthode manométrique de Warburg. Les extraits servant à la mesure des activités enzymatiques sont obtenus par homogénéisation des aiguilles dans un milieu contenant du tampon phosphate (0,2 M ; pH 7,6), du $MgCl_2$ (5 mM), du polyéthylèneglycol, (50 %, P/P) et du dithiothreitol (5 mM). Après centrifugation (100 000 g x 20 min), l'extrait est passé sur Sephadex G 25 avant utilisation. L'activité enzyme malique à NAD (EM) est mesurée selon la méthode de Davies et Patil (7) et l'activité fumarase selon la méthode de Hatch (8). Les mesures des effets *in vitro* des ions sulfite sur les propriétés oxydatives et phosphorylantes des mitochondries sont effectuées polarographiquement sur des préparations mitochondriales obtenues à partir de feuilles de Soja (*Glyxine max*) selon la technique de Tuquet et Dizengremel (9).

3. RESULTATS

Modifications du métabolisme respiratoire liées au dépérissement

Le tableau I regroupe les valeurs d'intensité respiratoire des aiguilles de l'année et de celles d'un an en fonction du dépérissement et de la période de l'année. La respiration des aiguilles de l'année est toujours plus intense que celle des aiguilles vieilles d'un an. D'autre part, c'est en Juin que l'intensité respiratoire des aiguilles est la plus forte. Enfin, le dépérissement se caractérise par une augmentation de respiration de 25 à 30 % chez les arbres de classe 3b par rapport aux arbres considérés comme sains (classe 1). Ce phénomène, particulièrement visible en Juin, se retrouve aussi bien chez les aiguilles de l'année que chez les aiguilles d'un an.

Résistance au cyanure

Chez les végétaux, des conditions particulières peuvent provoquer la mise en action ou intensifier le fonctionnement d'une chaîne particulière de transport d'électrons, mitochondriale mais non liée à la production d'énergie et caractérisée par sa résistance au cyanure et à l'azoture (10). Les expériences d'inhibition par le cyanure, menées sur les aiguilles

TABLEAU I

RELATION ENTRE LE DEPERISSEMENT ET L'ACTIVITE RESPIRATOIRE DES AIGUILLES D'EPICEA
RESPIRATORY RATE OF HEALTHY AND DISEASED SPRUCE NEEDLES.

Perte d'aiguilles (%)	Date de prélèvement							
	Juin 1985		Décembre 1985		Juin 1986		Octobre 1986	
	aiguilles 1984	aiguilles 1985	aiguilles 1984	aiguilles 1985	aiguilles 1985	aiguilles 1986	aiguilles 1985	aiguilles 1986
0-10 % (classe 1)	198	433^{b}	128	$255^{b_1,b_2}$	165	*221*	123^{a}	148^{a}
10-25 % (classe 2)	210	497	*220*	*285*	-	-	142	162
25-40 % (classe 3a)	-	-	-	$162^{b_1,a}$	164	255	152	187
40-60 % (classe 3b)	222	527^{b}	*220*	$205^{b_2,a}$	196	*432*	164^{a}	217^{a}

Valeurs obtenues à partir de 2 à 7 arbres pour la classe 3b, 2 à 4 arbres pour les classes 2 et 3a, 2 à 8 arbres pour la classe 1. Les valeurs en italique sont obtenues à partir d'un seul arbre. Les indices a, b, b_1, et b_2 montrent les moyennes significativement différentes 2 à 2 pour chaque prélèvement par analyse de variance et test de BONFERRONI.
a : au seuil de 5 % ; b : au seuil de 1 %. Valeurs exprimées en μl d'$O_2.h^{-1}.g^{-1}$ matière fraîche

d'Epicea sont difficiles à réaliser, la concentration de l'inhibiteur étant faible et la pénétration freinée par la structure des aiguilles. L'utilisation de l'azoture de sodium à 5 mM a malgré tout permis de rendre compte d'une tendance à une plus forte résistance aux inhibiteurs dans les aiguilles d'arbres classés 3b.

Evolution des capacités enzymatiques d'aiguilles d'Epicea en chambres de simulation

Les trois clones sont soumis de manière différée aux taux de pollution en SO_2 et O_3 enregistrés au Donon. Une chambre à air filtré sert de témoin. La figure 1 montre que ce sont les aiguilles du clone Lac de Constance qui présentent les plus faibles activités. D'autre part, la fumarase répond avec beaucoup plus d'amplitude à la présence des polluants. L'ozone provoque la plus grande stimulation de capacité quel que soit le clone et l'enzyme considérés (fumarase : 45 % pour Gérardmer ; 58 % pour Istebna ; 62 % pour Lac de Constance). Le dioxyde de soufre stimule d'environ 25 % la capacité de la fumarase chez Istebna et Lac de Constance mais n'affecte que très peu le clone Gérardmer. L'addition du SO_2 à l'ozone résulte, quelle que soit l'enzyme considérée, en une stimulation de la capacité enzymatique plus faible que celle enregistrée lors d'une pollution avec O_3 seul (Figure 1).

Effet des ions sulfite (SO_3^{2-}) sur les propriétés de mitochondries isolées d'organes foliaires

L'oxydation du malate par les mitochondries de feuilles de Soja est bien couplée à la phosphorylation, l'ADP/O représentant l'efficacité de

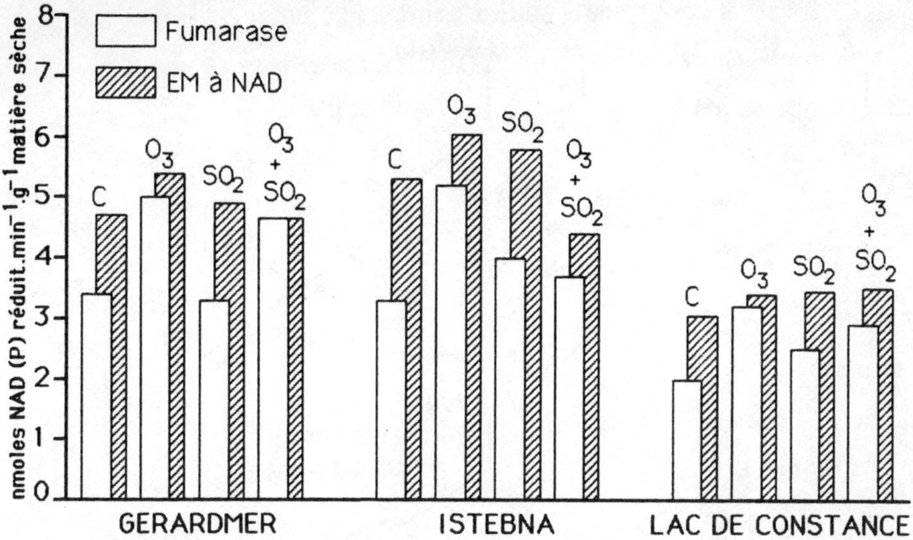

Figure 1 : Effet d'une pollution contrôlée en SO_2 et O_3 sur les capacités enzymatiques mitochondriales d'aiguilles de trois clones d'Epicea. C : contrôle, air filtré.

Effect of a controlled SO_2 and O_3 pollution on some mitochondrial enzyme capacities in needles from three spruce clones. C : control with filtered air.

la phosphorylation oxydative atteignant une valeur correcte (Figure 2A). Du sulfite de sodium 0,1 mM ajouté après le premier couplage (Figure 2A) ou avant le substrat (Figure 2B) provoque une inhibition tangible de la phosphorylation oxydative. Le rapport ADP/O diminue de 31 % tandis que le contrôle exercé par l'oxydation sur la phosphorylation (CR) baisse de 59 %. Cette baisse est due à l'inhibition de la vitesse d'oxydation en état 3 (en présence d'ADP). Une concentration 10 fois plus élevée en sulfite fait disparaître totalement la phosphorylation (Figure 2).

4. DISCUSSION

L'activité respiratoire des aiguilles des Epiceas est sans ambiguïté de 25 à 30 % plus élevée chez les arbres dépérissants (classe 3b) que chez les arbres sains (classe 1). L'hétérogénéité qui existe à l'intérieur de chacun des deux groupes est inférieure aux différences mesurées entre ces deux groupes. Chez les arbres dépérissants, les aiguilles de l'année sont vertes en Juin et jaunissantes en Octobre sans toutefois présenter de nécroses tandis que les aiguilles d'un an, dont la respiration est également plus intense, sont jaunes et parfois légèrement nécrosées. Les recherches menées sur l'activité respiratoire des aiguilles de conifères en conditions naturelles sont extrêmement rares. Par contre, toutes les études menées en laboratoire font état d'une stimulation de la respiration en présence d'ozone, de dioxyde de soufre ou de fluorures (4, 5).

Les tests de simulation en chambres à ciel ouvert ont permis de montrer un effet stimulateur du SO_2 et surtout de l'ozone sur les capacités enzymatiques mitochondriales de trois clones d'Epicea.

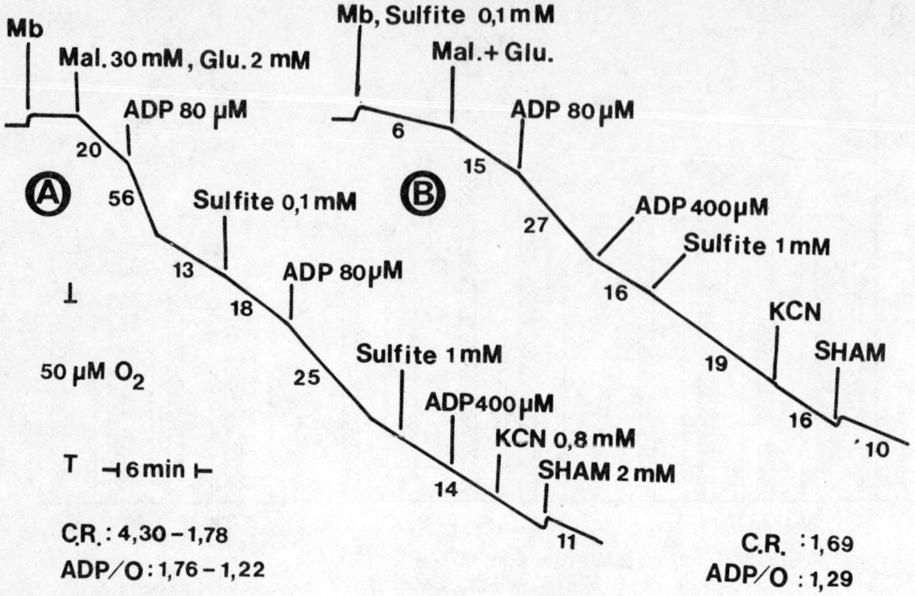

Figure 2 : Effet du sulfite de sodium sur les propriétés oxydatives et phosphorylantes de mitochondries de feuilles de Soja. Mal : malate ; Glu : glutamate. Les chiffres le long des traces sont en nmoles $O_2.min^{-1}.mg^{-1}$ protéines.

Effect of sodium sulfite on oxidative and phosphorylative properties of soybean leaf mitochondries.

Il est tentant de suggérer que l'augmentation de respiration des arbres dépérissants est reliée à un fonctionnement accéléré du système mitochondrial. A ce propos, il est à noter qu'une augmentation du rapport ATP/ADP dans les aiguilles d'arbres dépérissants a été récemment mise en évidence (11).

Afin de préciser les mécanismes d'action des polluants potentiels au niveau cellulaire, l'effet du sulfite de sodium a été testé sur des mitochondries isolées de feuilles de Soja, l'isolement de mitochondries d'aiguilles d'Epicea ne donnant, pour le moment, que des résultats insatisfaisants. Le sulfite à 1 mM inhibe fortement le pouvoir oxydatif des mitochondries et supprime la production d'ATP. Une concentration 10 fois moins élevée exerce déjà un effet inhibiteur sur les deux processus. Les ions sulfite, produits dans la cellule lors d'une pollution par le SO_2 sont connus pour leur toxicité au niveau du processus photosynthétique (12). Un effet toxique sur la production d'ATP avait déjà été signalé, mais pour des concentrations bien plus importantes (13). Les teneurs en sulfite utilisées dans ce travail ne semblent pas incompatibles avec des épisodes de pollution aigüe en SO_2 telles qu'il s'en produit régulièrement dans les Vosges chaque hiver (jusqu'à 300 $\mu g.m^{-3}$ en Janvier 1987).

La réponse des arbres à une pollution atmosphérique caractérisée par des pointes en concentration des deux polluants les plus étudiés (SO_2 et O_3) consiste en une respiration exacerbée, due, en partie au moins, à une stimulation d'activités mitochondriales. Cette situation a déjà été observée en conditions de fumigation contrôlée par le SO_2 seul. Il pourrait être inféré de ces résultats que la mitochondrie réagirait à la pollution en activant son fonctionnement, produisant l'énergie et le pouvoir réducteur nécessaires aux mécanismes de réparation ou de survie, en accord avec une récente proposition de Alscher et Amthor (Symposium de Münich, Avril 1987).

Remerciements. Ce travail, dans le cadre du programme DEFORPA, a bénéficié de subventions du Ministère de l'Environnement (N° 85 193 et 86 288) et de la CEE (N° ENV-857 F). Monsieur André Citerne bénéficie en outre d'une allocation MRT. Les auteurs tiennent aussi à remercier l'aide appréciable fournie par M. Queiroz et Mme Pierre de l'Institut de Physiologie Végétale à Gif-sur-Yvette.

BIBLIOGRAPHIE

(1) BONNEAU M.et FRICKER C. (1985). Le dépérissement des forêts dans le massif vosgien : relations possibles avec la pollution atmosphérique. Dans : Regards sur la santé de nos forêts. Rev. For. Fr., 37, 105-126.
(2) HEATH R.L. (1980). Initial events in injury to plants by air pollutants. Ann. Rev. Plant. Physiol., 31, 395-431.
(3) LICHTENTHALER H.K. et BUSCHMANN C. (1984). Photooxidative changes in pigment composition and photosynthetic activity of air-polluted spruce needles (*Picea abies* L.). Dans : Advances in photosynthesis Research, Vol. IV (3), C. Syberma ed., Martinus Nijhoff/W. Junk Pub., The Hague.
(4) BLACK V.J. (1984). The effect of air pollutants on apparent respiration. Dans : Gaseous air pollutants and plant metabolism, M.J. Koziol et F.R. Whatley ed., Butterworths.
(5) DIZENGREMEL P. et CITERNE A. (1987). Air pollutant effects on mitochondria and respiration. Proceedings of the 2nd int. Symposium on Air Pollution and Plant Metabolism, in press.
(6) REICH P.B. (1983). Effects of low concentration of O_3 on net photosynthesis, and chlorophyll contents in aging hybrid poplar leaves. Plant Physiol., 73, 291-296.
(7) DAVIES D.D. et PATIL K.D. (1975). The control of NAD specific Malic Enzyme from cauliflower bud mitochondria by metabolites. Planta, 126, 197-211.
(8) HATCH M.D. (1978). A simple spectrophotometric assay for fumarate hydratase in crude tissue extracts. Anal. Biochem., 85, 271-275.
(9) TUQUET C. et DIZENGREMEL P. (1984). Changes in respiratory processes in Soybean cotyledons during development and senescence. Z. Pflanzenphysiol., 114, 355-359.
(10) LANCE C., CHAUVEAU M. et DIZENGREMEL P. (1985). The cyanide-resistant pathway of plant mitochondria. In Encyclopedia of Plant Physiology, New Ser., R. Douce, D.A. Day eds., Springer-Verlag, Berlin, (18), pp 202-247.
(11) BENZ T. et HAMPP R. (1986). Levels of adenine nucleotides (ATP, ADP, AMP) and of inorganic phosphate in needles of *Picea abies*, representing different stages of development and of pollution dependence. 2. Statuskolloquium des PEF, Karlsruhe.
(12) MARQUES I.A. et ANDERSON L.E. (1986). Effects of arsenite, sulfite, and sulfate on photosynthetic carbon metabolism in isolated pea chloroplasts. Plant Physiol., 82, 488-493.
(13) BALLANTYNE D.J. (1984). Phytotoxic air pollutants and oxidative phosphorylation. Dans : Gaseous air pollutants and plant metabolism, M.J. Koziol et F.R. Whatley ed., Butterworths.

VERGLEICHENDE UNTERSUCHUNGEN DER WACHSMORPHOLOGIE AN NADELN VON KLONFICHTEN UND ALTFICHTEN

T. EUTENEUER UND R. DEBUS

Fraunhofer-Institut für Umweltchemie
und Ökotoxikologie
D-5948 Schmallenberg-Grafschaft

Summary

Results of a quantitative evaluation of the surface morphology of epicuticular waxes from spruces (Picea abies) are presented. Needles from spruce of varying exposition, age and genetic constitution show different degrees of melting and amount of epicuticular waxes. The effects of polluted air are investigated by means of open-top chamber experiments with clone spruces under controlled conditions. The influence of immissions on the surface morphology of the waxes is discussed.

1. EINLEITUNG

Im Rahmen eines vom European Open-Top Chamber Programme geförderten Projektes wurde auf dem Gelände des Fraunhofer-Instituts, Grafschaft eine Open-Top-Anlage installiert. Die Open-Top-Kammern sind mit 4-jährigen Klonfichten bepflanzt, die von gefilterter und ungefilterter Standortluft umströmt werden. Zusätzlich werden Altfichten unterschiedlich belasteter Standorte in die Untersuchungen u. a. der epikutikularen Wachse einbezogen. Es gilt als sicher, daß die morphologische und chemische Beschaffenheit der Wachse durch klimatische Faktoren beeinflußt wird (Baker 1974, Grill 1973, Hull 1979). Zusätzlich ist eine altersbedingte Degradation der Wachse gegeben (Hanover 1971, Jeffree 1971, Mueller 1954). Kontrovers diskutiert wird allerdings die Frage, ob Immissionen einen direkten Einfluß auf die epikutikularen Wachse haben (Huttunen 1984, Kim 1985, Trimble 1982).

Anhand eines neuerarbeiteten Bonitierungskataloges (Euteneuer, 1987) erfolgt die quantitative Bewertung der Morphologie epikutikularer Wachse in Abhängikeit von Nadelalter und Standort.

2. MATERIAL UND METHODEN

2.1 Standort - Open-Top Anlage

Der Aufbau der Open-Top Anlage erfolgte auf einem 10 m x 15 m großen Pflanzbeet, das zuvor mit Waldboden aufgefüllt wurde. Von den 12 mit Klonfichten bepflanzten Testflächen der Anlage sind je 4 zu einer Behandlungseinheit zusammengefaßt. Dadurch ergeben sich folgende Expositionen: Klonfichten in Open-Top Kammern in gefilterter oder ungefilter Standortluft und Klonfichten auf Freiflächen in Standortluft. Auf allen Testflächen werden neben dem Mikroklima kontinuierlich die Schadgase SO_2, O_3 und Stickoxide gemessen.

2.2 Standort - Hunau

An der Meßstation Hunau (770 m üNN) werden ebenfalls die Konzentrationen der oben genannten Gase erfaßt. Die auf einem Nord-West exponierten Hang stockenden Altfichten werden jeweils zu Beginn und am Ende der Vegetationsperiode beprobt. Dazu erfolgt die Astnahme aus dem 7. Wirtel (Knabe 1983). Die Untersuchungen werden an Nadeln der Jahrgänge 1 bis 3 und 5 durchgeführt.

2.3 Bonitierung der Wachsmorphologie

- Präparation

Für die REM Untersuchungen werden die eingefrorenen Nadeln nach Nadeljahrgängen getrennt zur Präparation aufgetaut. Die Nadeln werden in jeweils gleicher Position mit Leitsilber auf den Probenzylindern befestigt und anschließend mit Gold besputtert (Au, 75 sec., 1 KV, 90 mtorr, 5 mV). Im REM (Jeol JSM 35 bzw. T 20) wird die Analyse der Wachsmorphologie vorgenommen. Analysiert werden je Probebaum 5 Nadeln jedes Nadeljahrganges. Auf jeder Nadel gelangen 30 Stomata an definierten Stellen bei 2000 x und 7500 x Vergrößerung zur Bewertung.

- Bonitierungskatalog

Die Bonitierung der Wachsmorphologie erfaßt quantitativ die Merkmale Wachsverschmelzung und Wachsmenge. Beide Parameter werden unabhängig voneinander in einer 5-stufigen Skala unterteilt (Euteneuer, 1987). Die Differenzierung der Wachsverschmelzung erfolgt nach ansteigendem Verschmelzungsgrad (prozentualer Anteil unverschmolzener Wachsstrukturen und verschmolzener Wachsplatten an der Stomatafläche).

Die Wachsmenge wird in abnehmender Tendenz mit den Klassen a-e beurteilt. Bezugsparameter ist der prozentuale Wachsfüllungsgrad der Stomatavorhöhle. Die Sonderklasse 5 f erfaßt vollständig wachsfreie Stomatavorhöhlen mit sichtbarer Spaltöffnung. Bei der Bonitierung wird die Wachsstruktur jedes Stomata für beide Merkmale eingeordnet und erhält einen der 26 möglichen Klassenwerte (1 a bis 5 f) zugeteilt.

Bei der statistischen Auswertung werden Mittelwerte und Standardabweichungen für einzelne Nadeln, Nadeljahrgänge und Meßstationen berechnet.

3. ERGEBNISSE

3.1 Wachsmorphologie von Altfichtennadeln des Standortes Hunau

Die Bonitierung der Wachsmorphologie von 6 Probebäumen des Standortes Hunau zeigt Abb. 1. Dargestellt sind die Mittelwerte und Standardabweichungen für die Untersuchungsmerkmale, Wachsverschmelzung und Wachsmenge der Nadeljahrgänge 1985, 1984, 1983 und 1981. Die Höhe der Säule repräsentiert den Grad der Wachsverschmelzung, die räumliche Stellung im Diagramm die Wachsmenge.

Aus der Abbildung geht hervor, daß die epikutikularen Wachse im Bereich der Stomatavorhöhlen mit zunehmendem Nadelalter verschmelzen. Im Nadeljahrgang 1981 bedeutet ein mittlerer Verschmelzungsgrad von 3.1, daß 50 bis 90 % der Wachse in verplatteter Form die Stomatavorhöhlen verschließen. Im Nadeljahrgang 1985 traten weniger verplattete Wachse auf. Das Merkmal Wachsverschmelzung unterscheidet sich bei den beiden Nadeljahrgängen um eine halbe Bewertungseinheit. Die Wachsmenge nimmt mit zunehmendem Nadelalter sukzessive ab. Sind die Stomatavorhöhlen im Nadeljahrgang 1985 durchschnittlich noch zu 50 bis 75 % mit Wachs gefüllt, so ist die Wachsmenge im Nadeljahrgang 1981 um 75 bis 90 % reduziert. Der Unterschied zwischen den Nadeljahrgängen beträgt bei der Wachsmenge eine ganze Bewertungsstufe. Die Bonitierung ändert sich von 3 c auf 3 d.

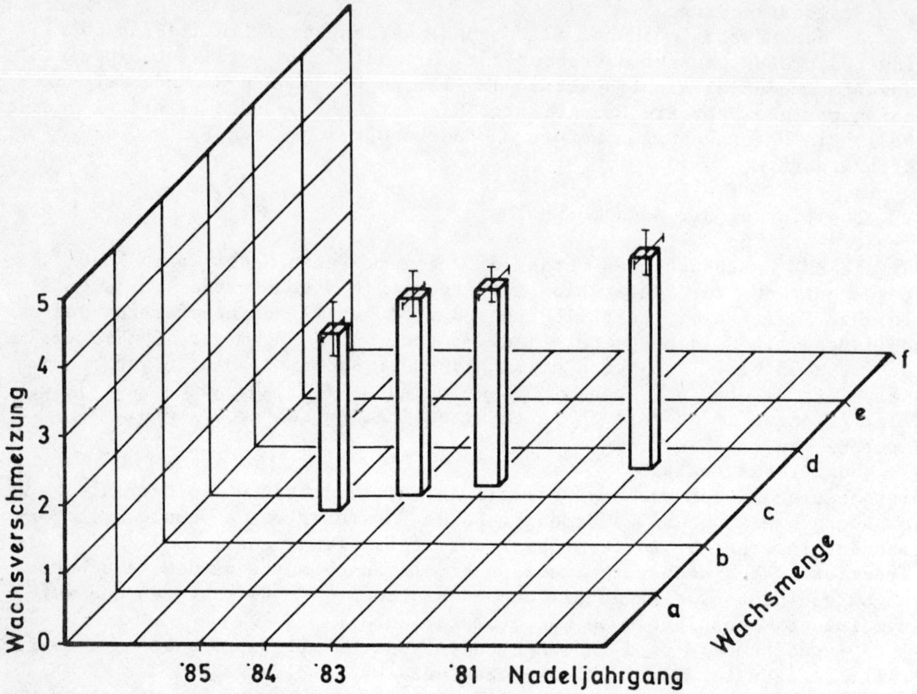

Abb.1: Bonitierung der Wachsmorphologie von Altfichtennadeln, Standort Hunau

3.2 Wachsmorphologie von Klonfichtennadeln der Open-Top-Anlage

Die Bewertung des Nadeljahrganges 1986 von 9 Klonfichten zu Expositionsbeginn ergibt einen Mittelwert von 2 b (1.5/2.1). Die Wachsverschmelzung ist sehr gering; es dominierten überwiegend röhrenförmige Wachsstrukturen, die geringe Verschmelzungstendenzen aufweisen. Verplattete Wachse verschließen durchschnittlich unter 25 % der Stomatavorhöhlen. Die Wachsmenge ist im Mittel um weniger als 25 % reduziert. Ein wesentlich differenzierteres Bild der Wachsmorphologie zeigt Abb. 2. Dargestellt ist die Häufigkeitsverteilung der Bonitierungsklassen, wobei zusätzlich nach drei Nadelbezirken differenziert wird. Jedes Feld auf der Abbildung repräsentiert eine der 26 Bonitierungsklassen des Kataloges. Die Höhe der Säulen gibt die Häufigkeit in % an, in der die jeweilige Bonitierungsklasse auftritt. Aus der Abb. 2 geht hervor, daß die Klonfichtennadeln am häufigsten mit der Bonitierunsklasse 1 b bewertet wurden und den errechneten Mittelwert von 1.5/2.1 gut repräsentieren. Es wurde ferner beobachtet, daß hauptsächlich die Bonitierungsklassen mit geringem Verschmelzungsgrad 0 bis 2 und geringer Wachsmengenreduktion a bis c überwiegen. Differenziert nach Nadelbezirken wird deutlich, daß an Nadelbasis und Nadelmitte die Bonitierungsklassen mit geringer Wachsverschmelzung und hoher Wachsmenge (0a, 0b, 1a, 1b) dominieren. An der Nadelspitze dagegen findet man Stomata mit hohem Wachsverschmelzungsgrad (4a, 4b, 4c) und reduzierter Wachsmenge (0c 1c, 1d, 2c).

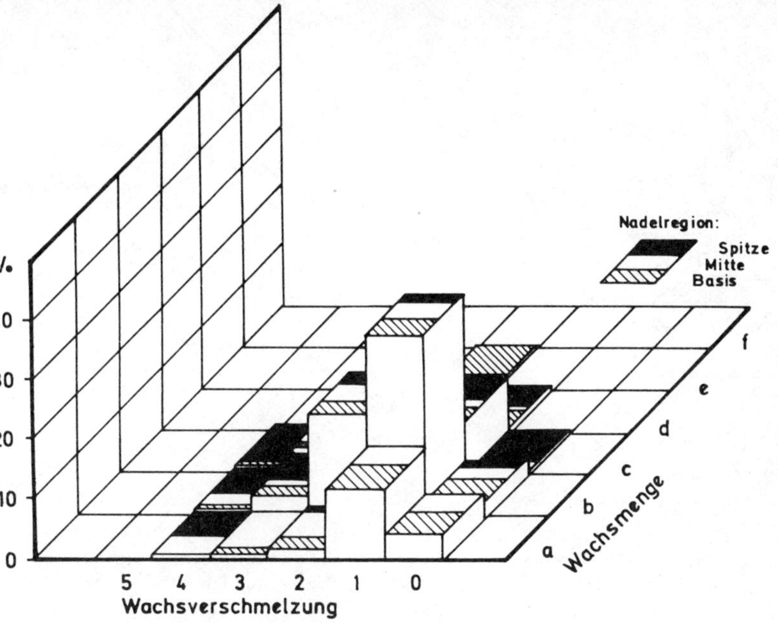

Abb. 2: Häufigkeitsverteilung der Bonitierungsklassen (in %) differenziert nach Nadelbezirken, Klonfichten, Nadeljahrgang 1986

3.3 Vergleich der Wachsmorphologie von Klonfichten- und Altfichtennadeln

Die Gegenüberstellung der Häufigkeitsverteilungen der Bonitierungsklassen von Altfichtennadeln der Jahrgänge 1985 und 1981 zeigen Abb. 3 und 4. Im Vergleich zu den Klonfichtennadeln (1986) ist bei den Altfichtennadeln von 1985 der Bonitierungswert auf 3 c (2.6/2.8) gegenüber 2b (1.5/2.1) erhöht. Die Wachsverschmelzung ist auf den Nadeln der Altfichten weiter fortgeschritten und die Wachsmenge stärker reduziert. Die Bonitierungsklassen mit fast unverschmolzenen Wachsen 0a, 0b, 0d, 0e und 1a treten hier nicht auf. Bei den Altfichten ist eine deutliche Verschiebung zu den Bonitierungsklassen mit höherer Wachsverschmelzung und geringerer Wachsmenge festzustellen. Die Differenzierung nach Nadelbezirken zeigt, daß an der Nadelbasis und Nadelmitte geringere Wachsverschmelzungen und größere Wachsmengen auftreten, während die Wachse an der Nadelspitze schon stark reduziert sind.

Der Mittelwert für die Nadeln der Altfichten des Jahrganges 1981 liegt bei 3 d (3.1/3.7), d. h. hohe Wachsverschmelzung (50 bis 90 % verplattete Wachse) und stark reduzierte Wachsmenge (75 bis 90 %). Es dominieren die Bonitierungsklassen 3 c und 3 d, während die Klassen 0 a-e, 1 a b und 2 a nicht auftreten. Bezogen auf Nadelbezirke ist das Ergebnis bei den Altfichtennadeln des Jahrganges 1981 ähnlich wie beim Nadeljahrgang 1985: Nadelmitte und -basis mit relativ geringer Verschmelzung und höherer Wachsmenge als an der Nadelspitze. Es ist jedoch eine weitere Verschiebung zu den Bonitierungsklassen festzustellen, die eine höhere Verschmelzung und Reduktion der Wachse anzeigen.

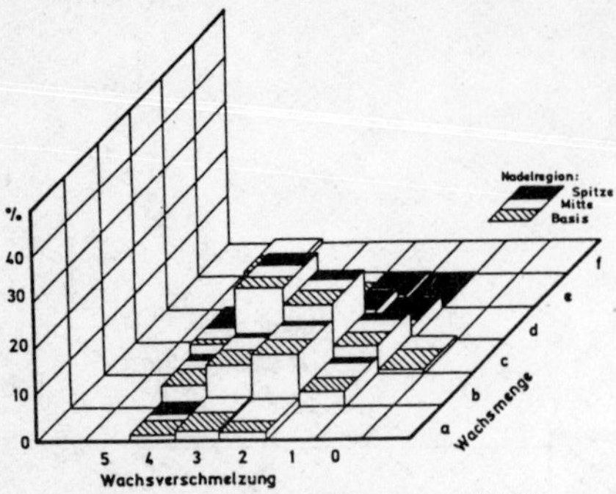

Abb. 3: Häufigkeitsverteilung der Bonitierungsklassen (in %) differenziert nach Nadelbezirken, Altfichten Hunau, Nadeljahrgang 1985

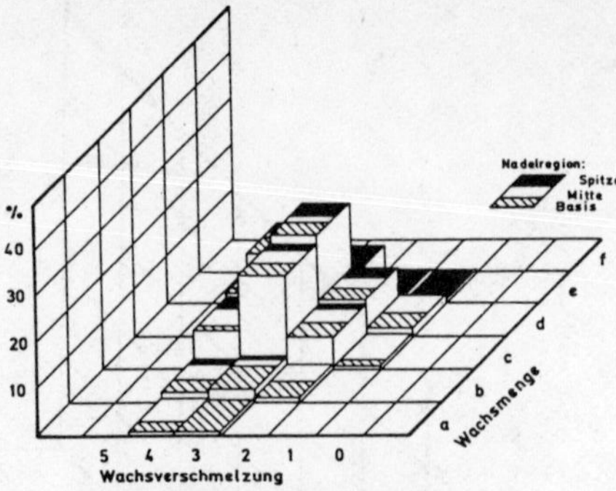

Abb. 4: Häufigkeitsverteilung der Bonitierungsklassen (in %) differenziert nach Nadelbezirken, Altfichten Hunau, Nadeljahrgang 1981

4. DISKUSSION

In früheren Arbeiten beschrieben Mueller (1954), Grill (1973a), Jeffree (1971) und Hanover (1971) eine altersbedingte Degradation der epikutikularen Nadelwachse. An jüngeren Fichtennadeln fanden sie durchweg röhrenförmige Wachsstrukturen. Mit zunehmendem Nadelalter kommt es zur Verminderung der Wachsbedeckung und zum Aggregieren der Wachsstrukturen (Grill, 1973a). Diese Ergebnisse konnten vom Autor mit Hilfe eines Bonitierungskataloges quantifiziert werden (Euteneuer, 1987). Fowler (1980) und Sauter (1986) entwickelten eine 5 bzw. 4 stufige Arbeitsskala zur Quantifizierung der Wachsdegradation. In dieser Untersuchung werden die Merkmale Wachsverschmelzung und Wachsmenge kombiniert mit einem 26stufigen Bewertungskatalog erfaßt. Es wird nach drei verschiedenen Nadelbezirken differenziert bewertet. Die Ergebnisse zeigen deutliche quantitative Unterschiede in Wachsverschmelzung und Wachsmenge bei Nadeln unterschiedlichen Alters. Die Bonitierung ergab ferner, daß an Nadelmitte und -basis die epikutikularen Wachse deutlich besser bewertet werden konnten als an der Nadelspitze. Als Einflußfaktoren auf die unterschiedlich ausgeprägte Wachsmorphologie werden das Mikroklima (Hull, 1979) und die genetische Konstitution (Wettstein-Knowles, 1986) angesehen. In neueren Untersuchungen wird der Einfluß unterschiedlicher Immissionsbelastungen auf die Ausprägung der Wachsmorphologie diskutiert (Cape 1981, Fowler 1980, Grill 1973, Hafner 1986, Huttunen 1984, 1983, Kim 1985, Riding 1985, Trimble 1982).

An belasteten Standorten konnte eine forcierte Degradation der Wachse registriert werden (Fowler 1980, Hafner 1986, Huttunen 1983, Kim 1985, Riding 1985, Sauter 1986). Bei der direkten Begasung mit Ozon wies Trimble (1982) keine signifikante Auswirkung auf die Wachsmorphologie nach. Bei der Beurteilung von Immissionseinwirkungen auf Strukturveränderungen der epikutikularen Nadelwachse müssen die anderen Einflußparameter (Mikroklima, genetische Konstitution und Altersdifferenzierung) mitberücksichtigt werden.

Aus den Untersuchungen in den Open-Top-Kammern werden Aussagen über die Wirkung unterschiedlicher Immissionsbelastungen bei weitgehend standardisierten und kontrollierten Einflußfaktoren erwartet.

LITERATUR

BAKER, E. A. (1974). The influence of environment on leaf wax development in Brassica oleracea. New Phytol. 73, 955-966.

CAPE, J. N.; FOWLER, D. (1981). Changes in epicuticular wax of Pinus sylvestris exposed to polluted air. Silva Fennica 15, 457-458

EUTENEUER, T.; STEUBING, L. (1987). Quantitative Bewertung der Morphologie epikutikularer Wachse von Nadeln von Picea abies. In Vorbereitung

FOWLER, D. et al. (1980). The influence of a polluted atmosphere on cuticle degradation in Scots pine (Pinus sylvestris). Proc. Int. Conf. Ecol. Impact Acid Precip. Norway 146

GRILL, D. (1973). Rasterelektronenmikroskopische Untersuchungen an Wachsstrukturen der Nadeln von Picea abies (L.) Karst.. Micron 4, 146-154

GRILL, D. (1973). Rasterelektronenmikroskopische Untersuchungen an SO_2 belasteten Fichtennadeln. Phytopath. Z. 78, 75-80

HAFNER, L. (1986). Zur Feinstruktur der geschädigten Kiefernnadel. AFZ 45, 1119-1121

HANOVER, J. W.; REICOSKY, D. A. (1971). Surface wax deposits on foliage of Picea pungens and other conifers. Amer. J. Bot. 58, 681-687

HULL, H. M. et al. (1979). Environmental modification of epicuticular wax structure of Prosopis leaves. J. Arizona-Nev. Acad. Sci. 14, 39-42

HUTTUNEN, S. et al. (1984). Effects of acid deposition on needle surfaces. Proc. 2nd Eur. Conf. Chem. Environm., Lindau 219-220

HUTTUNEN, S.; LAINE, K. (1983). Effects of air-borne pollutants on the surface wax structure of Pinus sylvestris needles. Ann. Bot. Fennici 20, 79-86

JEFFREE, C. E. et al. (1971). Epicuticular wax in the stomatal antechamber of Sitka spruce and its effects on the diffusion of water vapour and carbon dioxide. Planta 98, 1-10

KIM, Y. S. (1985). REM-Beobachtungen immissionsbeschädigter Fichtennadeln Cbl. ges. Forstwesen 102, 96-105

KNABE, W. (1983). Immissionsökologische Waldzustandserfassung in NRW (IWE 1979) Fichten und Flechten als Zeiger der Waldgefährdung durch Luftverunreinigung. Forschung und Beratung, Reihe C, Heft 37

MUELLER, L. E. et al. (1954). The submicroscopic structure of plant surfaces. Amer. J. Bot. 41, 593-600

RIDING, R. T.; PERCY, K. E. (1985). Effects of SO_2 and other air pollutants on the morphology of epicuticular waxes on needles of Pinus strobus and Pinus banksiana. New Phytol. 99, 555-563

SAUTER, J. J. (1986). SEM-observations of the structural degradation of epistomatal waxes in Picea abies (L.) Karst. - and its possible role in the "Fichtensterben". Eur. J. Forest Pathol. 16, 408-423

TRIMBLE, J. L. et al. (1982). Chemical and structural characterization of the needle epicuticular wax of two clones of Pinus strobus differing in sensitivity to ozone. Phytopathol. 72, 652-656

WETTSTEIN-KNOWLES, P. (1986). Role of cer-cqu in epicuticular wax biosynthesis. Biochem. Soc. Transactions 14, 576-579

DEPERISSEMENT FORESTIER ET PRECIPITATIONS ACIDES DANS LES PYRENEES CENTRALES

V. CHERET*, J. DAGNAC*, F. FROMARD* et D. GUILLEMYN**
*Laboratoire de Botanique et Biogéographie, U.A. 700 CNRS
**Institut de la Carte Internationale de la Végétation U.A. 688 CNRS

Résumé
Le dépérissement affectant le sapin (*Abies alba* M.) dans les Pyrénées centrales se caractérise par des perturbations morphologiques (jaunissement important du feuillage, perte d'aiguilles) et des troubles nutritionnels (carence en potassium, déficit en calcium). Les précipitations acides mises en évidence dans cette région contribuent à ce phénomène, agissant en synergie avec d'autres facteurs favorisant le dépérissement constaté (sols oligotrophes, déséquilibres structuraux, accidents climatiques).
Summary
In central Pyrenees, morphological perturbations (yellowing of the foliage, early leaf-fall) and deficiencies in mineral nutrition (lack of potassium and calcium) are the main symptoms of the fir-stand decay. Acid rains occurring in the same area contribute to this phenomenon with others constraints (oligotrophic soils, unbalanced forest structures, particularly dry years).

1. INTRODUCTION

Le dépérissement observé dans les Pyrénées centrales concerne essentiellement le sapin pectiné (*Abies alba* M.), essence spontanée de l'étage montagnard (1000 à 1700 mètres d'altitude) tout au long de la chaîne (55000 ha environ). L'épicéa utilisé dans quelques reboisements est également atteint par place. Nous n'évoquerons pas ici le cas du hêtre, les symptômes qui l'affectent étant, d'après nos observations, en relation directe avec les sécheresses successives des années 1983, 1984 et 1985. C'est dans la région de Bagnères-de-Luchon (Pyrénées hautes-garonnaises, fig.n°1) que le phénomène est le plus développé. Des études intensives y ont été menées (1, 2) et sont actuellement étendues à d'autres secteurs des Pyrénées.

2. LES SYMPTOMES

Le dépérissement se caractérise par un jaunissement des aiguilles âgées de 2 ans et plus, une chute prématurée de ces aiguilles, une allure en "plateau" de la cime des arbres atteints. On note aussi fréquemment l'apparition de pousses de détresse et une production de cônes particulièrement abondante. Les sapins dépérissants sont le plus souvent des adultes dominants, mais aussi quelquefois de jeunes sujets, ils sont distribués en petits bouquets ou isolés dans des peuplements apparemment sains. Outre ces symptômes qualitativement comparables à ceux décrits dans d'autres régions d'Europe, les sapinières étudiées sont caractérisées par un déséquilibre structural certain (vieillissement et régularisation excessive des futaies), une emprise importante du gui (Viscum album L.) et des lichens épiphytes (Usnea barbata L.). Elles sont installées sur pentes souvent fortes, sur des sols de type brun acide à moder ou mull-moder, peu profonds et à faibles potentialités nutritives, mais aucune corrélation n'a cependant pu être établie entre caractéristiques édaphiques et dépérissement (1).

Le réseau de placettes d'observations mis en place dans les Pyrénées centrales en 1985 (O.N.F., C.E.M.A.G.R.E.F.) permet d'évaluer l'ampleur et d'apprécier l'évolution du dépérissement sur 2 années consécutives (85 et 86). Si pour l'ensemble du réseau français la situation du sapin s'est améliorée (passant de 18,6 à 15,3 % pour les arbres ayant perdu au moins 25 % de leur feuillage (3)), elle s'est aggravée dans les Pyrénées (2,3 % à 8 %), avec des nuances notables selon les secteurs (13,7 % pour la haute-chaîne). Le critère coloration du feuillage montre qu'en 1986 31 à 60 % des sapins, selon les régions pyrénéennes considérées, étaient jaunissants. L'importance de ce critère reste donc essentielle pour caractériser le phénomène, ce symptôme étant moins étendu dans les autres massifs forestiers (23 % de coloration anormale en Alsace, 21 % en Lorraine).

Outre le suivi annuel sur le terrain de placettes d'observations, une étude en télédétection rapprochée (4,5) a été réalisée. Les photographies aériennes obtenues (films couleurs, échelles 1/5000ème au 1/8000ème) permettent de distinguer plusieurs classes d'arbres (sains, dépérissants par jaunissement, dépérissants par cime sèche, arbres recouverts de lichens, arbres morts sur pied). Les deux procédés d'investigation menés en parallèle sur plusieurs transects (terrain, missions avion) ont amené à une cartographie de l'état sanitaire des peuplements (fig. n°2).

3. LES PROBLEMES NUTRITIONNELS

Un diagnostic foliaire réalisé sur sapins sains et sapins dépérissants (aiguilles de 1 an et 4 ans) a permis de caractériser le phénomène sur le plan de la nutrition minérale (tab. n°1). Les concentrations en phosphore, magnésium, fer, manganèse, cuivre et zinc se situent au niveau optimal et ne sont pas significativement différentes entre les deux classes d'individus ; les teneurs en azote et calcium sont seulement moyennes par rapport aux valeurs généralement admises et une déficience significative apparaît chez les sapins dépérissants pour ce dernier élément. Les teneurs en potassium chez les sapins dépérissants sont inférieures au seuil de carence et une différence significative (au seuil de 1 %) s'observe entre les sapins malades et les sapins sains ; pour ces derniers la nutrition en potassium est peu élevée mais non carencée.

L'existence d'une carence en potassium et d'une déficience en calcium associée au phénomène de dépérissement a été confirmée par de nouvelles analyses réalisées 2 ans plus tard sur les mêmes arbres, les valeurs obtenues pour le potassium étant les suivantes :

	Age Aiguilles	Sapins sains		Sapins dépérissants	
		moy.	min.-max.	moy.	min.-max.
Teneurs en potassium en % de m.s.	1 an (**)	0,40	0,31-0,48	0,27	0,18-0,37
	4 ans (**)	0,25	0,16-0,35	0,17	0,10-0,25

(**) différence significative au seuil de 1 %

Contrairement aux résultats acquis dans les Vosges (6), aucun déficit en magnésium n'a été décelé chez les sapins dépérissants, ces derniers présentant au contraire des teneurs sensiblement plus élevées que les sapins sains. Par contre une baisse du taux de chlorophylle totale a été observée chez les aiguilles atteintes, baisse s'accentuant avec l'âge des aiguilles (2). Enfin, les teneurs en soufre mesurées sont relativement élevées et dépassent le seuil naturel dans les aiguilles de 4 ans des individus sains et des individus dépérissants. Il a donc été établi que des pertes en éléments nutritifs (K, Ca) sont liées au dépérissement. Ces troubles nutritionnels peuvent avoir de multiples causes. Outre les déséquilibres écologiques déjà évoqués ainsi que les accidents climatiques de ces dernières années (sécheresses notamment), les précipitations acides récemment mises en évidence (1) favorisent, par un phénomène de lixiviation des cations au niveau des aiguilles, le dépérissement constaté.

4. MISE EN EVIDENCE ET ANALYSE DES PRECIPITATIONS ACIDES

Un collecteur automatique d'eau de pluie placé dans le Luchonnais à proximité des peuplements forestiers les plus atteints a permis de recueillir durant 3 mois d'été plusieurs événements pluvieux ; le pH mesuré sur les échantillons varie entre 3,5 et 4,7 pour une moyenne de 4,04, valeur très faible pour une région pourtant éloignée de toute source polluante importante (fig. n°3). Aucune pluie alcaline n'a été enregistrée. Le dosage des principaux anions et cations dans ces eaux (tab. n°2) met en évidence de fortes concentrations en sulfates, nitrates, ammonium et calcium. La part respective des anions majeurs est de 65 % pour SO_4^{--}, 27 % pour NO_3^-, 8 % pour Cl^-, proche des pourcentages ordinairement observés BEILKE in (7)). Comparativement aux valeurs données pour la France (8) les teneurs en sulfates, nitrates et ammonium sont nettement plus élevées, suggérant ainsi une forte contamination de l'atmosphère dans le secteur d'étude par des sources anthropogéniques.

L'analyse des corrélations entre les constituants chimiques montre que :

- seuls les sulfates et les nitrates sont corrélés significativement aux ions H^+, confirmant ainsi que l'acidification des pluies est essentiellement due aux acides forts H_2SO_4 et HNO_3,
- les composés alcalins participant à la neutralisation d'une partie des composés acides sont surtout NH_4^+ et dans une moindre mesure Ca^{++},
- de fortes corrélations existent entre SO_4^{--}, NO_3^- et NH_4^+, indiquant l'origine commune principalement anthropogène de ces 3 éléments. Le calcul des rapports SO_4^{--*}/H^+ et $SO_4^{--*}+NO_3^-/H^+$ pour chaque événement pluvieux (fig. n°4) montre que, dans la majorité des cas, les sulfates sont en quantité largement suffisante pour expliquer à eux seuls l'acidité de l'eau de pluie. Lorsque ce dernier rapport est inférieur à 1, comme c'est le cas pour 3 événements pluvieux, on peut cependant envisager la participation à la balance ionique d'autres composés acides, minéraux et/ou organiques.

5. CONCLUSIONS

L'incidence directe des précipitations acides sur les parties aériennes des sapins serait de provoquer ou d'accentuer les déficiences minérales observées dans les aiguilles (K, Ca) en lixiviant une partie des cations échangeables, perte difficilement compensable en raison des faibles réserves nutritives des sols. Cette hypothèse a été confirmée par l'analyse de la composition chimique des pluviolessivats collectés sous la frondaison des sapins (1).

Cette première expérimentation réalisée dans le Luchonnais a été étendue durant l'été 86 à d'autres régions pyrénéennes et se poursuivra dans le cadre d'un programme de recherche franco-espagnol. D'autre part, des dosages d'oxydes d'azote et de soufre seront réalisés. Les mesures ponctuelles d'ozone effectuées en divers secteurs des Pyrénées (Luchon, vallée d'Ossau) n'ont pas révélé de fortes concentrations (maximums entre 40 et 50 ppb). Par contre les nécroses apparues sur les cultivars BELW3 du tabac Nicotiana tabacum L., marqueurs biologiques réagissant spécifiquement aux attaques de l'ozone et installés dans les mêmes stations, suggèrent l'existence de périodes pendant lesquelles cet élément serait en concentration plus importante. Cette expérimentation sera également reconduite au cours de l'année 87, parallèlement à l'extension des études symptomatologiques à d'autres secteurs des Pyrénées.

REFERENCES

(1) CHERET, V. (1987). La sapinière du Luchonnais (Pyrénées hautes-garonnaises) : étude phytoécologique, recherches sur le phénomène de dépérissement forestier. Thèse de Doct. de l'Univ. Paul Sabatier, spécialité Phytoécologie, Toulouse : 287.
(2) CHERET, V., DAGNAC, J. et FROMARD, F. (1987). Le dépérissement du sapin dans les Pyrénées luchonnaises. Rev. For. Franç., XXXIX, 1 : 12-24.
(3) Inventaire forestier national (1987). Observations en 1986 de l'état sanitaire des arbres forestiers. Document interne.
(4) GUILLEMYN, D.R. (1985). Contribution de la télédétection par photographie aérienne à la symptomatologie de la sapinière luchonnaise. D.E.A. option Ecologie, Univ. Paul Sabatier, Toulouse : 57.
(5) GREPIN, G. (1985). Télédétection par avion léger appliquée au dépérissement de la sapinière luchonnaise. D.E.A. option Ecologie, Univ. Paul Sabatier, Toulouse : 85.
(6) LANDMANN, G., BONNEAU, M. et ADRIAN, M. (1987). Le dépérissement du Sapin pectiné et de l'Epicéa commun dans le massif vosgien est-il en relation avec l'état nutritionnel des peuplements ? Rev. For. Franç., XXXIX, 1 : 5-11.
(7) MARTIN, M. (1984). L'acidité atmosphérique. Livre blanc sur les pluies acides, Secrétariat d'Etat à l'Environnement et à la Qualité de la vie : 103-117.
(8) GRANAT, L. (1972). On the relation between pH and the chemical composition in atmospheric precipitation. Tellus, 24 : 551-560.

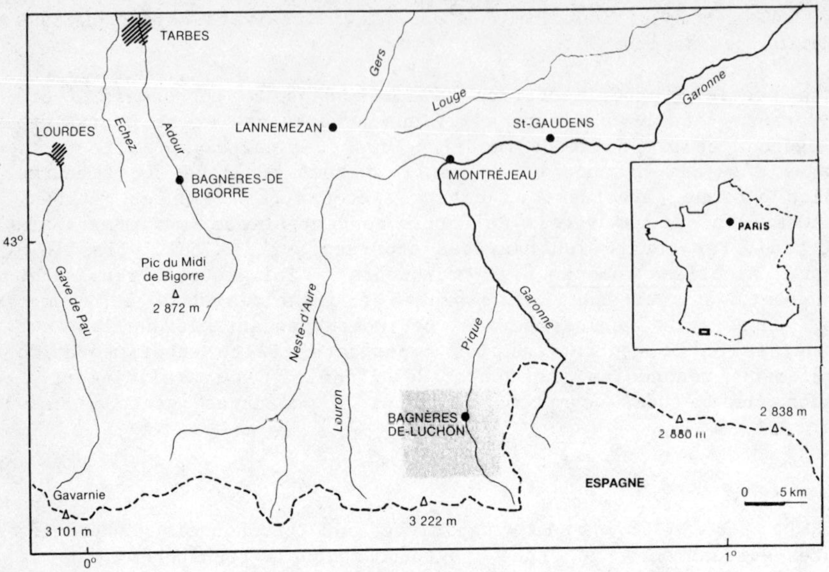

Fig. n°1
Localisation du secteur d'étude

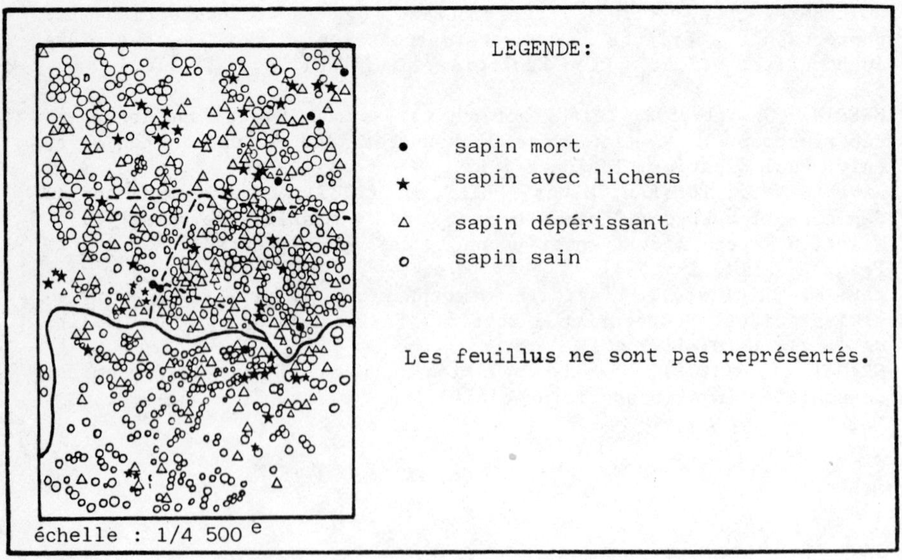

Fig. n°2
Exemple de cartographie réalisée par photo-interprétation et contrôle sur le terrain

Elément	Age Aiguilles		Sapins sains (n=7) moy.	Sapins sains (n=7) min.-max.	Sapins dépérissants (n=13) moy.	Sapins dépérissants (n=13) min.-max.
N%	1 an		1,23	1,07-1,43	1,31	1,08-1,68
	4 ans		1,19	1,15-1,31	1,13	0,91-1,32
P%	1 an		0,21	0,19-0,24	0,22	0,18-0,25
	4 ans		0,19	0,18-0,21	0,20	0,15-0,22
K%	1 an	(**)	0,38	0,31-0,48	0,20	0,14-0,31
	4 ans	(**)	0,33	0,25-0,48	0,16	0,11-0,31
Ca%	1 an	(**)	0,73	0,61-0,95	0,51	0,24-0,71
	4 ans	(*)	1,43	0,84-2,15	1,04	0,56-1,58
Mg%	1 an		0,12	0,09-0,17	0,15	0,08-0,19
	4 ans	(**)	0,12	0,07-0,16	0,16	0,11-0,21
Fe%	1 an		0,18	0,14-0,21	0,18	0,15-0,24
	4 ans		0,19	0,15-0,24	0,20	0,15-0,30
S‰	1 an		1,56	1,22-2,38	1,38	1,12-1,58
	4 ans		1,80	1,44-2,24	1,72	1,46-1,92
Mn‰	1 an		0,83	0,31-1,19	0,73	0,33-1,07
	4 ans		1,43	0,39-2,07	1,40	0,79-2,19
Cu‰	1 an		0,007	0,006-0,008	0,007	0,004-0,008
	4 ans	(*)	0,007	0,006-0,008	0,006	0,006-0,006
Zn‰	1 an		0,051	0,042-0,072	0,041	0,027-0,054
	4 ans		0,052	0,030-0,084	0,054	0,029-0,093

Tableau n°1
Teneurs en éléments minéraux des aiguilles de sapins (% ou ‰ de m.s.)
Comparaison entre sapins sains et sapins dépérissants
(**) différence significative au seuil de 1%
(*) différence significative au seuil de 5%

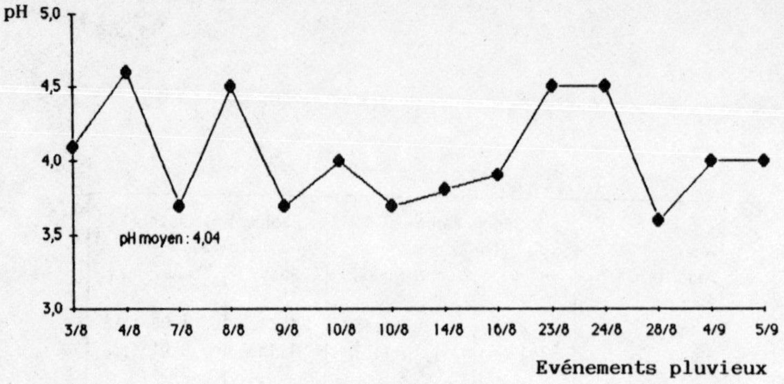

Figure n°3
Evolution du pH en fonction des événements pluvieux

Eléments ppm	Luchon	France (L. Granat) (8)
SO4=	7.34	2.79
NO3-	3.09	1.98
Cl-	0.94	2.13
F-	0.17	–
NH4+	1.56	0.27
Ca++	0.99	0.68
Mg++	0.16	0.39
K+	0.61	0.16
Na+	0.34	0.92

Tableau n°2
Composition chimique des eaux de précipitations

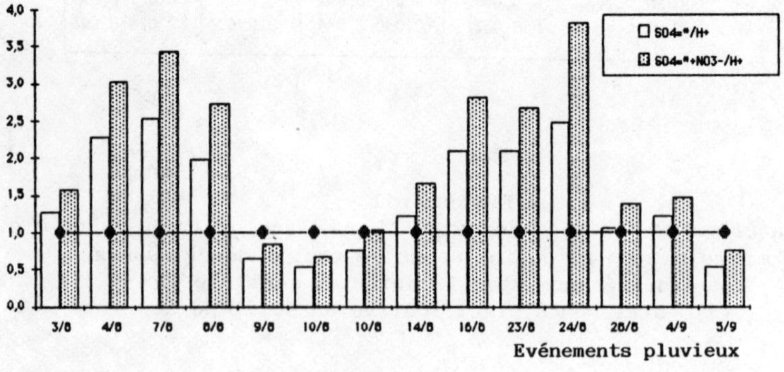

Figure n°4
Variation des rapports de concentration en fonction des événements pluvieux
$SO_4{}^{=*}$: fraction des sulfates d'origine continentale et anthropique calculée selon la formule de W.C. KEENE et al. (in (1))

INDIRECT EFFECTS OF ACID MIST UPON THE RHIZOSPHERE AND THE LEAVES' BUFFERING CAPACITY OF BEECH SEEDLINGS

S. Leonardi and W. Flückiger
Institute for Applied Plant Biology
CH-4124 Schönenbuch

Summary

Beech seedlings (Fagus sylvatica L.) in water culture were treated with an artificial acidic mist (pH 4.6, 3.6, 2.6). The leaching of particularly Ca and Mg was positively correlated to the acid load (126 - 12560 Mol $\mu H^{+}/m^{2}$). The increased leaching from the leaves was followed by an increased uptake of cations. As a consequence, more protons were released from the roots. This acidification of the rhizosphere was made visible with a pH-indicator in an agar medium. - The buffering capacity of the leaves decreased with the more acidic mist treatments. The chemical analysis of leaf fractions revealed a close relationship between the contents of exchangeable cations and the buffering capacity of the extracellular space. - The transpiration of the seedlings was increased during this short-term experiment (8 days).

1. INTRODUCTION

The leaching of mainly cations due to the impact of acid depositions is a well-known phenomenon. It could be demonstrated that leached ions may be reabsorbed by the roots (1). As long as the mineral supply is high enough, increased amounts of leached ions needn't lead to decreased leaf contents (2,3). This biogeochemical cycling induced by leaching is thought to ba accelerated by acid deposition. Eventually, combined effects with other stress factors (e.g., ozone, frost, aphids) may even increase the leachability.

The reabsorption of the lost cations is connected with a proton-efflux, meaning that cation leaching from leaves entails acidification of the rhizosphere (4). The acidification is strongly dependent on the buffering capacity of the soil (5) and on intact uptake mechanisms of the roots. However, altered pH-gradients across the rhizosphere may lead to distinct impacts on the availability of heavy and toxic metals as well as on the acquisition of nutrients.

The processes of leaching and reabsorption involve control systems for the allocation of the respective minerals. During the course of allocation short-term inbalances and destabilizations may impede the normal metabolism. The buffering capacity (B.C.) of leaves, used as an indicator of pollution impact on trees (6), was measured in the present study to observe physiological alterations due to an acid mist treatment. Since B.C. decreases with decreasing cation-excess of the leaf (7), a decrease in B.C. following cation leaching should be expected.

2. MATERIALS AND METHODS

Beech seedlings (Fagus sylvatica L.) were grown in water culture. In August, the leaves were sprayed with an acidic solution (H_2SO_4 : HNO_3 : HCl = 1 : 1 : 1) of pH 4.6, 3.6, and 2.6. Five mist treatments within seven days led to a total application of 5 mm mist, corresponding to proton depositions of 126, 1256, and 12560 µMol H^+/m^2. After the mist treatments the leaves were rinsed with H_2O to get the leachates.

Samples of the culture solution were taken at the beginning and at the end of the experiment to calculate uptake and release of nutrients and protons.

To visualize effects of pH-changes in the rhizosphere following the application of the acid mist to the foliage the method of (9) was adopted; the concentrated agar was mixed with the culture solution and bromocresol green as pH-indicator. As a consequence of misting the embedded plants with pH 2.6 colour changes from blue to yellow around the fine roots could be observed after approx. 15 hours. Distinct differences in the discolour of the medium between the treatments with either pH 2.6 and pH 4.6 were obtained after 20 hours.

For chemical analysis the leaves were harvested at the end of the mist treatments. Each leaf was "fractionated" as follows: after homogenizing the leaf with 25 ml H_2O the homogenate was filtered (0.45 µm). This leaf fraction was ascribed FH, representing the water-soluble extract of the homogenate. The residues were resuspended with 25 ml H_2O. The latter fraction, termed SH, represents the non-filterable (residue-bounded) fraction of the homogenate. The buffering capacity (B.C.) was determined in both FH and SH, following the proposal of (8). The B.C. was expressed as µeq H^+ required to change the pH of FH or SH equivalent to 1 g oven-dried foliage by one pH unit (BCd1) or to pH 3 (BC-3) (7). One major correction of the calculation of (8) was made to compensate the volume of titrant used for changing the pH of the added water (9).

3. RESULTS AND DISCUSSION

The leaching of Ca^{2+} and Mg^{2+} was significantly increased if the foliage had been sprayed with acidid solutions of pH 3.6 and pH 2.6, compared to the pH 4.6 treatment. K^+ wasn't leached in greater amounts following the more acidic mistings. Maximum leaching was found for Ca^{2+}: 20.3 % of the initial leaf content were removed by the application of 5 mm of the pH 2.6 spraying solution.

The absorption of the studied cations and the concomittant release of protons are listed in Table 1. The uptake of all cations was increased markedly; however, only NH_4^+ showed a significant increase. But ammonium was leached in small amounts just from a few leaves; hence, it is suggested that the reabsorption of leached cations is not very specific. Both the NH_4^+-uptake and the H^+-efflux were at their maximum in the pH 3.6 treatment in spite of the greater leaching found in the pH 2.6 treatment. It is thought, that the absorption mechanisms were over-stressed following the highest proton-load of 12.6 mMol H^+/m^2*8 days. This conclusion is further supported by linear regression of the paired data of cation uptake and proton release: whilst there were found significant correlation coefficients for the studied cations in the pH 4.6 and pH 3.6 treatments, this was true only for ammonium in the pH 2.6 treatment.

A signifacant increase in transpiration due to the acid mist application could be observed. The possible role of the high Ca^{2+}-loss - leading to stomatal opening - must be further examined.

Table 1 Uptake of K^+, Ca^{2+}, Mg^{2+} and NH_4^+, release of H^+ (all in meq/m^2 leaf-area) and transpiration (l/m^2 leaf-area) of the beech seedlings

Treatment	H^+	K^+	Ca^{2+}	Mg^{2+}	NH_4^+	H_2O
pH 4.6	3.47^a	5.74	5.96	1.62	27.00^a	3.84^a
pH 3.6	22.03^b	11.59	11.06	1.30	48.00^b	8.43^b
pH 2.6	13.68^{ab}	10.10	10.04	1.35	38.11^{ab}	7.99^b
p(F)	0.01	0.67	0.61	0.88	0.03	0.05

p(F): significancy level of different means (n=8)
means within columns with no or same letters do not differ significantly

The dependence of the leaves' buffering capacity (B.C.) on the cation/anion-relation has been mentioned by (6) and (7). As a consequence of the exchange of mist-derived protons against cations of the leaf the B.C. of the examined leaf-homogenates decreased: Table 2. This was true particularly in the SH-fraction of the homogenate, meaning that the buffering capacity of the apoplast is stressed at first. Note the decreases of the pH in both fractions.

Table 2 Buffering capacities (µeq H^+/g dry weight) of beech leaves

Treatment	FH: water-soluble filtrate of the homogenate			SH: suspension of the residues		
	pH_o	BCd1	BC-3	pH_o	BCd1	BC-3
pH 4.6	5.75	67.29	240.60	5.90	57.83	232.46
pH 3.6	5.57	55.26	237.63	5.70	55.11	216.13
pH 2.6	5.51	66.32	232.39	5.70	52.03	168.03

The most interesting results were obtained in the SH-fraction: there were found close interdependences between the contents of exchangeable cations (measured in the filtrate of SH after its titration with HCl) and the respective buffering capacities (titrated to pH 3). This indicates exchange reactions on the cell walls. An example of this relationship is given for Ca^{2+} in Figure 1. The correlation coefficient of the linear regression of the paired data was r = 0.73.

Figure 1 Buffering capacity and exchangeable Ca^{2+}-content of a suspension of homogenate residues of beech leaves

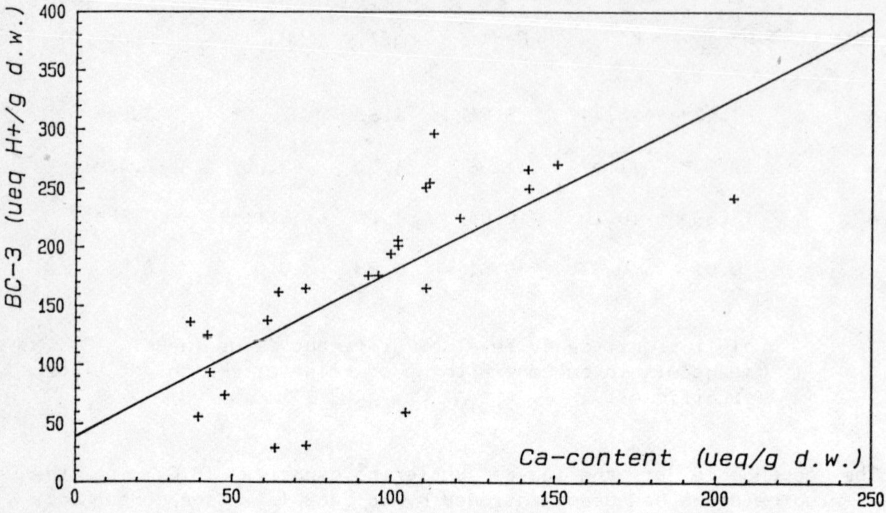

4. CONCLUSIONS

The wetting of the foliage by acid solutions such as rain and mist leads to a H^{+}-input into the plant, followed directly by cation-leaching from the leaves and indirectly by a H^{+}-efflux from the roots. The two processes stress both the buffering capacities of the leaf and of the rhizosphere. Metabolic inbalances and altered ionic environments of the roots may be the consequences.

The biogeochemical cycling of nutrients is linked through the leaching and reabsorption of the respective ions. This cycle is accelerated by acid deposition. Possible degradations of leaves and/or fine roots would stress the cycle even more.

REFERENCES

1) Tukey, H.B.Jr., R.A. Mecklenburg, & J.V. Morgan (1965). In: Isotopes and radiation in soil-plant nutrition studies. IAEA, Vienna, 371-385
2) Skeffington, R.A., & T.M. Roberts (1983). Oecologia 65: 201-206
3) Glavac, V. (1987). Allg. Forstz. 42: 303-305
4) Miller, H.G. (1984). Phil. Trans. R. Soc. Lond. B 305: 339-352
5) Röhmheld, V. (1986). Kali-Briefe (Büntehof) 18: 13-30
6) Jäger, H.-J., & H. Klein (1976). Eur. J. For. Path. 6: 347-354
7) Beese, F., & A. Waraghai (1985). Allg. Forstz. 40: 1164-1166
8) Sidhu, S.S., & J.G. Zakrevski (1982). Plant and Soil 66: 173-179
9) Marschner, H., V. Röhmheld, & H. Ossenburg-Neuhaus (1982): Z. Pflanzenphysiol. 105: 407-416
10) Leonardi, S., & W. Flückiger. Forstw. Cbl., submitted

RECHERCHES SUR LES CAUSES DU DEPERISSEMENT DES FORETS. MISE EN PLACE DE TESTS D'EXCLUSION AU COL DU DONON (VOSGES - FRANCE)

J.P. GARREC et M. BERTEIGNE

I.N.R.A. - Centre de Recherches Forestières
Laboratoire d'Etude de la Pollution Atmosphérique
CHAMPENOUX - 54280 SEICHAMPS - FRANCE

Investigations of the causes of Forest Dieback. Development of exclusion experiments in Open-Top chambers at Donon Pass (Vosges-France).

Les origines exactes du dépérissement des forêts sont encore inconnues. Si la pollution de l'air est fortement suspectée (en particulier l' ozone), l'influence des précipitations acides ne doit pas être négligée (en particulier les brouillards acides).

Pour tester la validité de ces hypothèses et rechercher l'importance relative de chacun de ces deux paramètres, de jeunes clones d'épicéas de 4-5 ans et de trois origines différentes (ISTEBNA (Pologne); LAC DE CONSTANCE (Allemagne) et GERARDMER (France)) ont été placés dans deux chambres à ciel ouvert (Open-Top). Ces chambres sont situées au Col du Donon dans les Vosges, où un dépérissement a été constaté.

Dans la première chambre, l'air amené par un ventilateur est parfaitement filtré et dépollué au moyen d'un filtre en charbon actif (les polluants SO_2, NO, NO_2, O_3 et les hydrocarbures sont éliminés). Parallélement, la pluie est éliminée au moyen d'un toit mobile automatique, et remplacée au même moment par un arrosage équivalent au moyen d'eau de source.

La deuxième chambre où l'air n'est pas filtré et qui ne possède pas de toit sert à mesurer l'effet de la chambre sur les arbres témoins recevant la pollution locale.

Parallélement, une troisième population d'arbres placée hors chambre, sert à mesurer "l'effet serre des chambres sur l'évolution des lots d'arbres.

Ces expériences d'exclusion qui ont débuté à l'automne 1985, n'ont pas permis jusqu'à présent (1 an de fonctionnement) de mettre en évidence des différences significatives au niveau des paramètres de croissance des épicéas entre la chambre dépolluée et la chambre témoin. (mesure de la date de débourrement, de la hauteur de la pousse terminale).

Il semble cependant qu'un léger jaunissement des aiguilles apparaisse dans la chambre témoin, comparativement à la chambre dépolluée, et ceci avec une réponse différente suivant l'origine des clones (le clone polonais se montrant particulièrement résistant). De même, des mesures physiologiques plus fines effectuées dans le cadre du programme français DEFORPA sur ces arbres, semblent indiquer des réponses différentes au bout d'un an, des arbres poussant dans le chambre dépolluée comparativement à la chambre témoin.

Si pour l'instant, nous éliminons dans ces chambres les deux facteurs supposés être à l'origine du dépérissement des forêts, il est envisagé par la suite de ne faire fonctionner que la filtration de l'air, ou que le toit mobile, ceci afin d'estimer la part respective de chacun de ces facteurs sur le dépérissement.

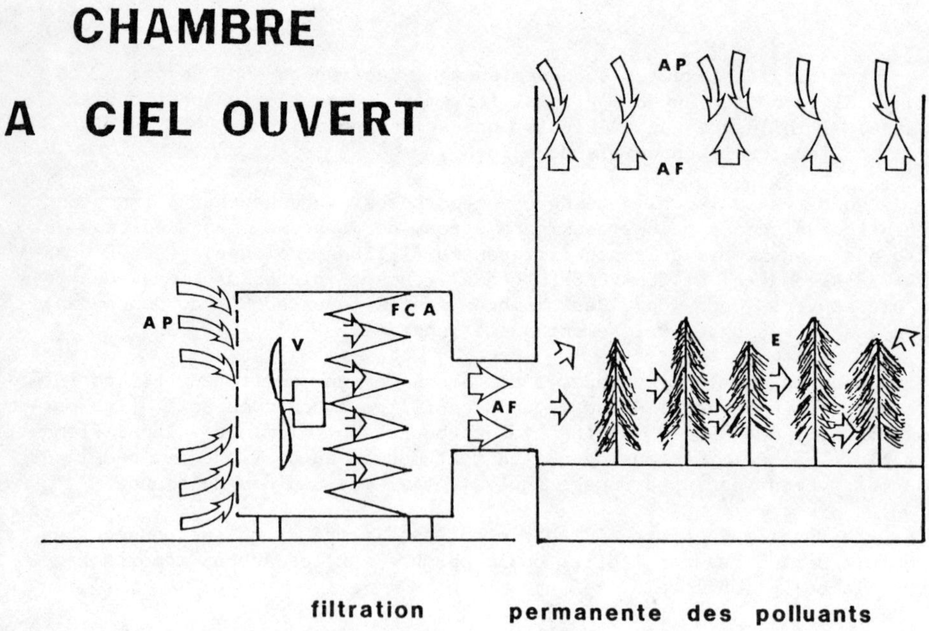

Figure 1 : Fonctionnement d'une chambre à ciel ouvert ("open-top").

EXCLUSION TOTALE

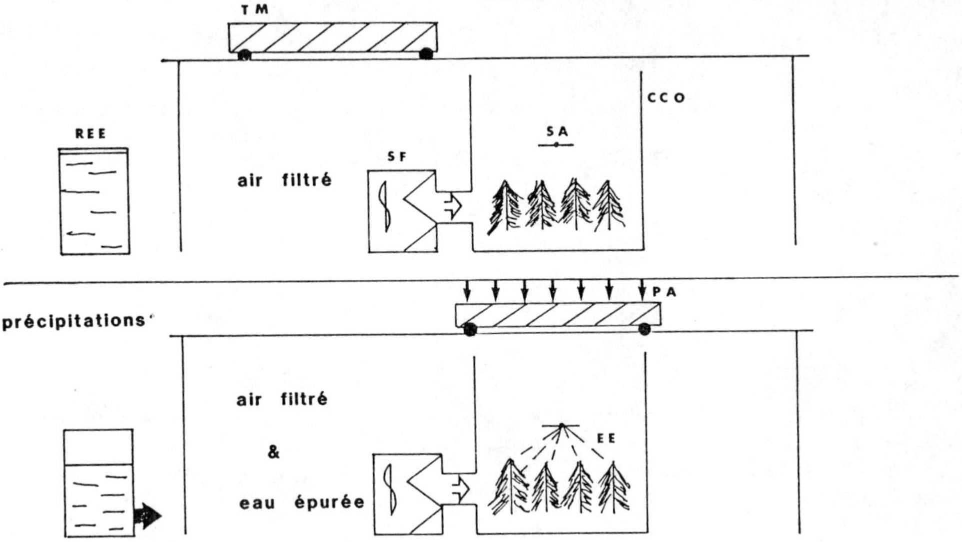

AF	*: air filtré*	*PA*	*: pluie acide*
AP	*: air pollué*	*REE*	*: réservoir d'eau épurée*
CCO	*: chambre à ciel ouvert*	*SA*	*: système d'arrosage*
E	*: épicéa*	*SF*	*: système de filtration*
EE	*: eau épurée*	*TM*	*: toit mobile*
FCA	*: filtre en charbon actif*	*V*	*: ventilateur*

Figure 2 : Principe de l'exclusion totale. En plus d'une filtration permanente, un toit mobile vient recouvrir la chambre à ciel ouvert au cours des précipitations. Au même moment, un arrosage d'eau épurée, de même intensité que les précipitations, se déclenche dans la chambre, au-dessus des épicéas.

EFFETS DES "PLUIES ACIDES" SUR LES CUTICULES ET LES SURFACES FOLIAIRES

J.P. GARREC et C. LAEBENS

I.N.R.A. - Centre de Recherches Forestières
Laboratoire d'Etude de la Pollution Atmosphérique
CHAMPENOUX - 54280 SEICHAMPS - FRANCE

Effects of acid rains on cuticles and foliar surfaces.

A partir de prélévements effectués dans les forêts vosgiennes où des signes de dépérissement des arbres apparaissent, l'observation au microscope électronique à balayage d'aiguilles d'épicéas (Picea abies) plus ou moins dépérissants, a permis de mettre en évidence de nettes perturbations des cires épicuticulaires.

Ces perturbations sont, dans la majorité des cas de deux sortes :

- une modification de la structure des cires remplissant les chambres épistomatiques (cires épistomatiques). En effet, nous constatons une agrégation beaucoup plus rapide au cours du temps de la fine structure en aiguilles de ces cires sur les arbres dépérissants comparativement aux sains.

- une érosion des cires peri-stomatiques. En effet, nous constatons une disparition beaucoup plus rapide au cours du temps des "bouquets" de cires épicuticulaires entourant les stomates sur les arbres dépérissants comparativement aux sains.

Ces types d'observation ont déjà été reportés par des chercheurs étudiant le problème du dépérissement des arbres forestiers.

Cette agrégation des cires au niveau de l'ostiole est susceptible de modifier les résistances stomatiques dans les aiguilles d'arbres dépérissants avec les conséquences physiologiques que celà implique. Paralléle-ment l'érosion des cires péri-stomatiques est susceptible de modifier la perméabilité cuticulaire à l'eau et aux ions avec également de graves répercussions physiologiques.

En particulier ces perturbations des cires épicuticulaires pourraient être interprétées comme les signes d'un vieillissement précoce des aiguilles induit par le phénomène à l'origine du dépérissement des arbres.

Pour expliquer ce dépérissement des forêts attribué aux pluies acides, de nombreuses hypothèses ont été avancées. Parmi celles-ci, la carence minérale est souvent avancée (en Mg, mais également en Ca et K). Cette hypothèse est basée sur des observations de terrain (jaunissement des aiguilles) et sur de nombreuses analyses d'aiguilles de résineux.

Ces dépérissements liés à une carence en Mg ou K principalement, auraient pour origine le lessivage des aiguilles par les dépôts acides (pluies, brouillards), lessivage accéléré par des effets de l'ozone sur les perméabilités cuticulaire et membranaire.

Pour "tester" cette hypothèse et pour comprendre les mécanismes de dégradation des arbres forestiers en rapport avec les perturbations physiologiques au niveau de la nutrition minérale, nous avons étudié en laboratoire les modifications de la perméabilité cuticulaire sous l'action des "pluies acides" et des polluants atmosphériques.

Pour ce faire, des cuticules isolées enzymatiquement ont été soumises à une pollution soit à l'ozone (1000 ppb pendant une semaine), soit à une solution acide (mélange acide sulfurique/acide nitrique, rapport 2/1, pH de 2 à 5 pendant une semaine), soit aux deux simultanément.

Les premiers résultats obtenus ne semblent pas indiquer qu'il y ait un effet important de ces deux agents sur les propriétés des cuticules, en particulier sur la perméabilité de l'eau au travers de ces cuticules isolées ($P_{H_2O}^{témoin}$: 0,8. 10^{-5} $m.s^{-1}$; $P_{H_2O}^{O_3}$: 1,8. 10^{-5} $m.s^{-1}$; $P_{H_2O}^{acide}$: 1,6 10^{-5} $m.s^{-1}$)

Les expériences similaires sur les modifications de la perméabilité de la cuticule aux ions sont en cours de réalisation.

Ces premiers résultats indiqueraient que :

1) Les carences observées au niveau des feuilles dans les zones de dépérissement des forêts proviendraient beaucoup plus d'une mauvaise nutrition minérale au niveau des racines qu'un lessivage anormal des ions par les dépôts acides au niveau des feuilles.

2) Le phénomène de lessivage qui existe au niveau des rameaux d'arbres dépérissants proviendrait avant tout, d'un départ des ions au travers des tiges et des petioles qu'au travers des cuticules des feuilles. Le rôle des stomates dans ce lessivage reste encore inconnu.

PHOT 1

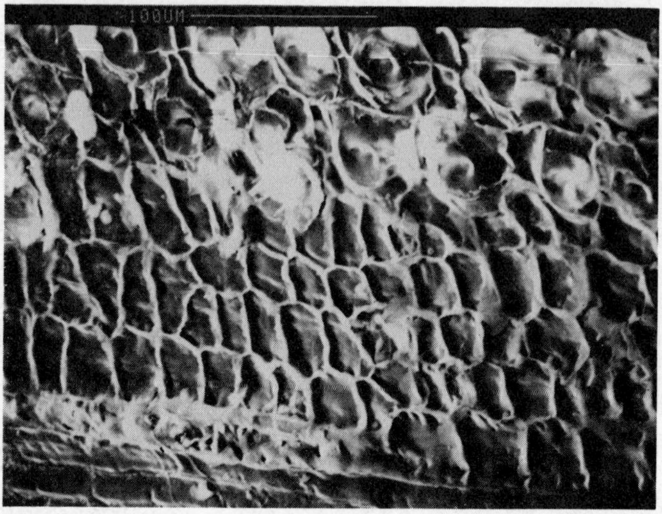

Face interne d'une cuticule isolée inférieure d'aiguille de sapin (Abies alba). On note les orifices de deux rangées parallèles de stomates et les saillies de la cuticule entre les cellules épidermiques.

PHOT 2

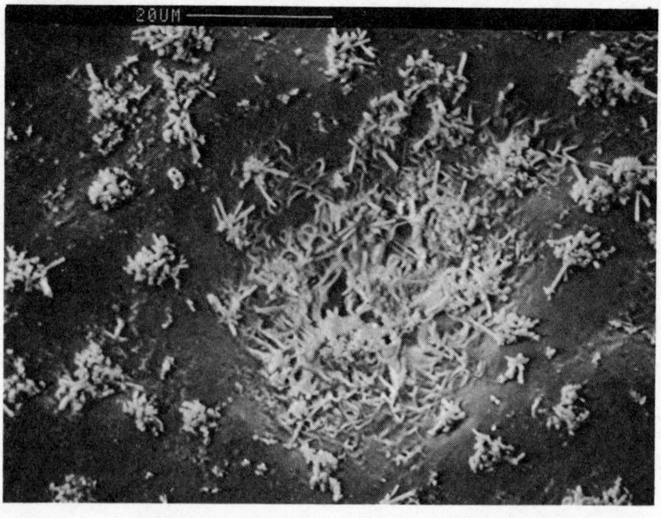

Aspect normal des cires épicuticulaires d'un stomate d'aiguilles d'épicéa sain : les cires épistomatiques forment de très fines aiguilles et les cires péristomatiques sont très abondantes.

PHOT 3

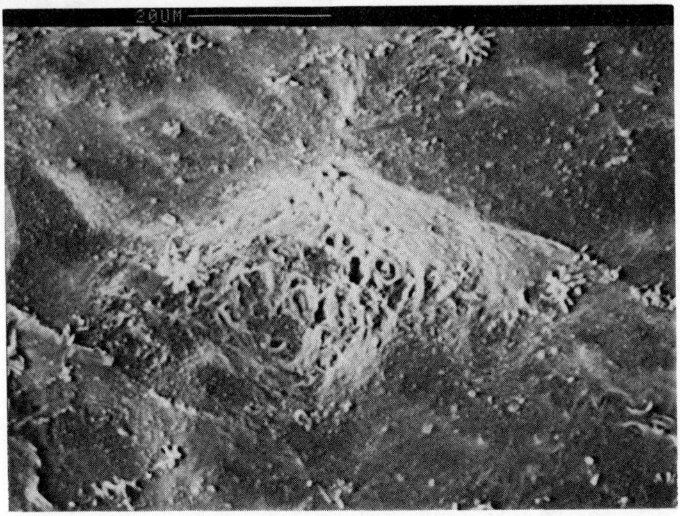

Un aspect des cires épicuticulaires d'un stomate d'aiguilles d'épicéa dépérissant : les cires épistomatiques sont fortement agglomérées et les cires péristomatiques sont peu abondantes (érosion ?).

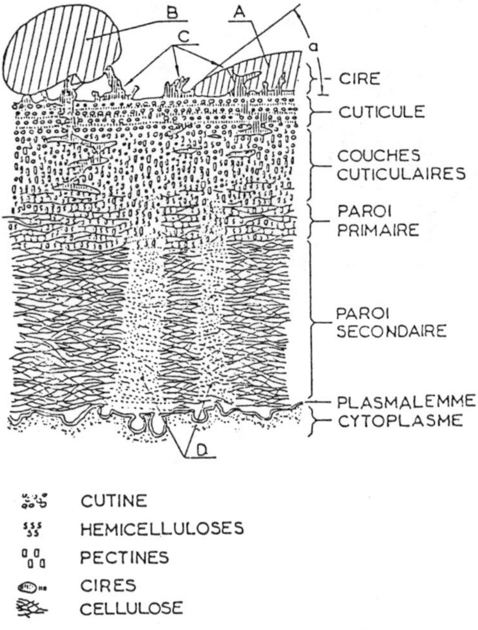

Figure 1 : Schéma simplifié de la structure d'une surface foliaire d'après FRANKE, 1967 et CHAMEL (cuticule + paroi des cellules épidermiques).

A : gouttelette d'eau avec tensioactif - B : sans tensioactif
C : baguettes de cire - D : ectodesme - α : angle de contact

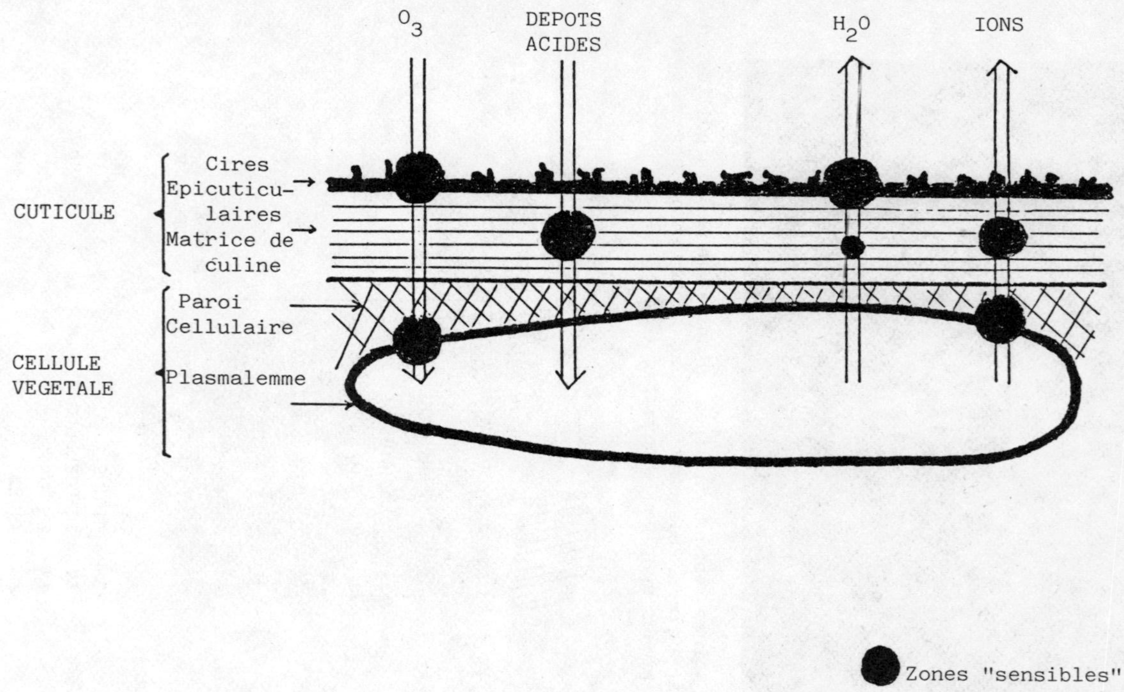

Figure 2 : Représentation schématique de nos hypothèses sur l'action des "pluies acides" au niveau de la cuticule et des cellules adjacentes, et sur les conséquences sur le départ d'eau et d'ions (avec l'importance présumée des zones sensibles).

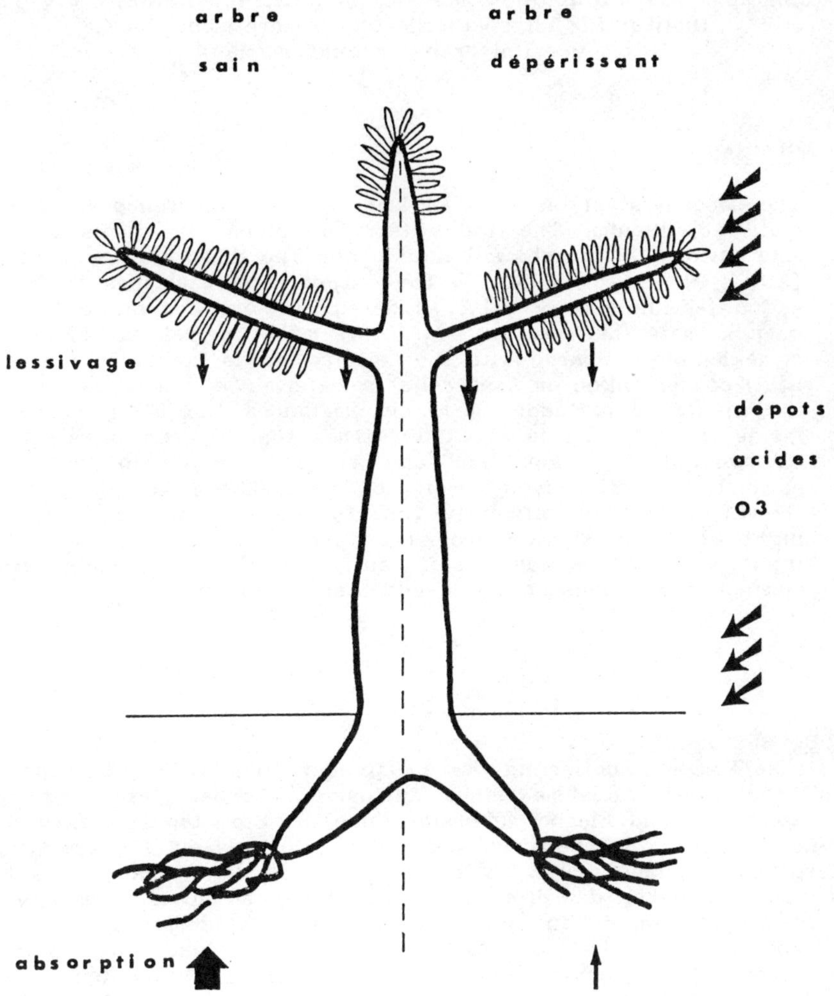

Figure 3 : Représentation schématique de nos hypothèses sur l'origine des carences minérales foliaires observées chez les conifères dépérissants comparativement aux conifères sains (avec l'importance présumée des variations au niveau de l'absorption radiculaire et du lessivage par les feuilles et par les tiges).

DIFFUSIONS/REAKTIONSMODELL FüR SO_2 AM BLATTPFAD

G.H.KOHLMAIER, F.W.BADECK, M.PLÖCHL, K.SIEBKE, O.SIRE, C.WIENTZEK
Institut für Physikalische und Theoretische Chemie
der Universität Frankfurt/Main

SUMMARY

The concentration of S(IV) in leaf cells is determined by a steady-state equilibrium. The input term is given by diffusion of SO_2, solution of SO_2 in cell wall liquids and the dissociation-equilibrium of $[SO_2 \cdot H_2O]$ $[SO_3^{2-}]$ $[HSO_3^-]$. The output term describes the metabolisation (elimination) of S(IV). Under physiological conditions and with open stomata (i.e. resistance « $10^4 \sec \cdot m^{-1}$) the velocity of elimination is neglegible compared with the velocity of the input. In this case the S(IV)-concentration in leaf cells is determined almost exclusively by the input term. Under these circumstances the S(IV) concentration inside the cell (c_5) is about 10^5 times the SO_2 concentration in the environment (c_1). For the same reason changes in vmax of the elimination reaction over a range of 10^2 (0.000054-0.0054) don't change the S(IV) level significantly (differences < 0.1%). Only during the night when the stomata are closed (resistance: $1.3x10^4$ $\sec \cdot m^{-1}$) the input velocity decreases sufficiently to show a visible effect of elimination reactions on the steady-state equilibrium.

MODELLBESCHREIBUNG

Die Kompartimentierung des Blattpfades für den SO_2-Transport wird hinsichtlich der Annahme eines diffusiven Transportes vorgenommen. Eintrittsstellen sind hierbei in erster Linie die Stomata, da aufgrund eines 10^6-fach höheren Diffusionswiderstandes die Cuticula als Transportweg eine untergeordnete Rolle spielt. Mitentscheidend für die Aufnahme des SO_2 in das Blatt sind neben der Stomata-Apertur die turbulenten Strömungen über der Blattoberfläche, die in Abb. 3 als Grenzschichtwiderstand r_{12} eingehen. Die Formulierung des Grenzschichtwiderstandes und des stomatären Widerstandes (r_{23}) sind (2) entnommen. Saisonale Einflüsse und Tagesgang des Lichtangebots werden in einer Formulierung nach (3) berücksichtigt. Der Einfluß dieser Größen auf die Stomata-Apertur ist somit im Modell berücksichtigt.

Der Übertritt des SO_2 aus den Interzellularen in die Flüssigphase der Zellwand wird nicht nur durch dessen Löslichkeit, ausgedrückt durch die Henry'sche Konstante, bestimmt, sondern auch durch die pH-Wert-abhängigen Dissoziationsverhältnisse. Unter der Annahme, daß im Zellwandwasser ein Puffersystem mit dem pH-Wert 5 und im Zellinneren ein solches mit dem pH-Wert 7,5 aufrechterhalten werden können, liegt S(IV) in beiden Kompartimenten hauptsächlich als Hydrogensulfit und Sulfit vor (χ_4^1=0.0005, χ_4^2=0.9995, χ_5^2=0.72, χ_5^3=0.28) (vgl. Abb.2 u. 3). Die Geschwindigkeit der Anreicherung von S(IV) in den Chloroplasten wird somit bestimmt durch

1. das Verhältnis der Konzentrationsdifferenz zwischen c_1 und c_5, wobei diese um den Faktor γ modifiziert ist, zu dem effektiven Gesamtwiderstand r_{15}^{eff}
2. die Geschwindigkeit der S(IV)-abbauenden Schritte.
   Diese sind die Oxidation zu S(VI) und Reduktion zu S^{-2}, welches als H_2S exhaliert oder zu Cystein metabolisiert wird (siehe auch Abb.2). Die Eleminationsgeschwindigkeiten und Michaelis-Konstanten dieser Reaktionen werden in einem Term zusammengefaßt.

ERGEBNISSE UND DISKUSSION

Ein Vergleich der Abbildungen 4 und 5 zeigt, daß für verschiedene Szenarien (c_1: niedrig, hoch, sehr hohe Kurzzeitbelastung) in wenigen (<10 Stunden) ein Gleichgewichtszustand für c_5 erreicht wird, der etwa 10^5 mal größer ist als c_1. In den Abbildungen 6 und 7 ist zu erkennen, daß der Einfluß der Elimination auf den Gleichgewichtszustand nur in den Nachtstunden relevant wird. Dieser beruht in erster Linie auf dem Verhältnis der Konzentrationsdifferenz zum Gesamtwiderstand und gibt den Einfluß von γ wieder. Nur in den Nachtstunden wird die Eintrittsgeschwindigkeit klein genug, um den Einfluß der Elimination auf den Gleichgewichtszustand deutlich werden zu lassen. Die Unterschiede in den Gesamtwiderständen von Nadel- und Laubblatt (etwa 2:1) bewirken eine Änderung der Gleichgewichtskonzentration um weniger als 2% (Abb.8). Nach Messungen aus (1) akkumuliert Picea abies eine Gesamtschwefelmenge von 3 $mol\cdot m^{-3}$ in 250 Tagen. Ein Vergleich der Integration der eliminierten Schadstoffmengen (Abb. 9) mit diesen experimentell gewonnenen Daten zeigt, daß wir die maximalen Eliminationsgeschwindigkeiten vmax und die zugehörigen Konstanten Km in den richtigen Größenordnungsbereichen gewählt haben.

LITERATUR

(1) Jaeger, H.-J., H.-J. Weigel & L. Grünhage, 1986. Physiologische und biochemische Aspekte der Wirkung von Immissionen auf Waldbäume. Eur.J.For.Path. 16, 98-109

(2) Lommen, P.W., C.R. Schwintzer, C.S. Yoccum & D.M. Gates, 1971. A Model Describing Photosynthesis in Terms of Gas Diffusion and Enzyme Kinetics. Planta 98, 195-220.

(3) Richter, O., 1985. Simulation des Verhaltens ökologischer Systeme. Verlag Chemie, Weinheim, 219 S.

Die Arbeit wurde vom BMFT gefördert im Rahmen des Projektes Computersimulation des Stoff- und Energiehaushaltes gesunder und geschädigter Waldökosysteme.

Abb. 1

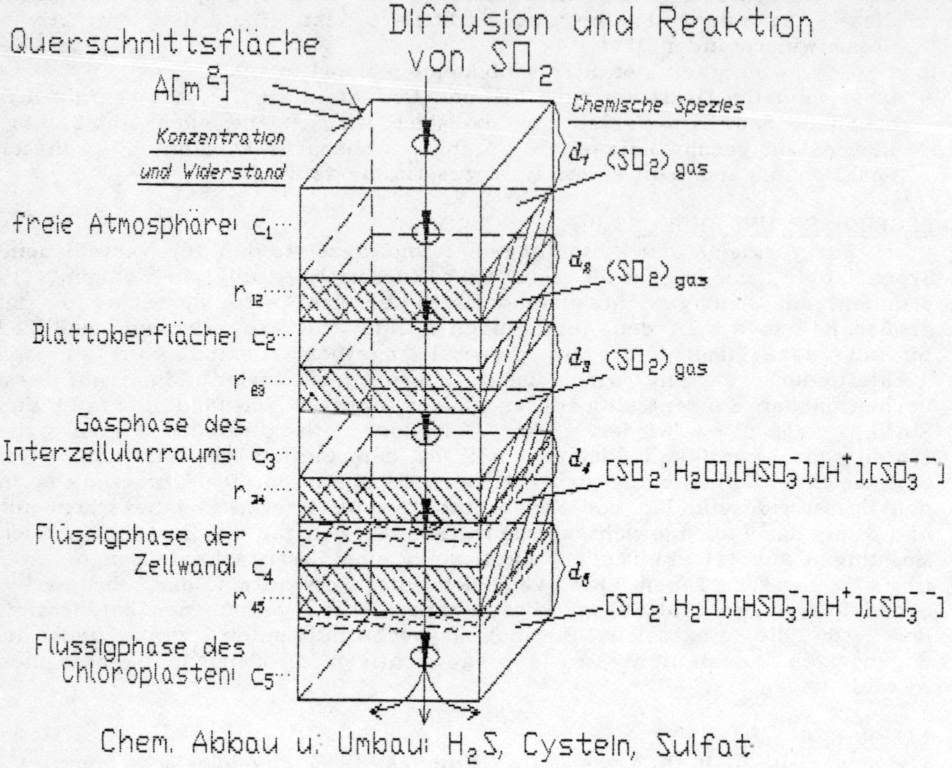
Diffusion und Reaktion von SO_2
Querschnittsfläche A[m^2]
Konzentration und Widerstand
Chemische Spezies
freie Atmosphäre: c_1
r_{12}
Blattoberfläche: c_2
r_{23}
Gasphase des Interzellularraums: c_3
r_{34}
Flüssigphase der Zellwand: c_4
r_{45}
Flüssigphase des Chloroplasten: c_5
d_1 $(SO_2)_{gas}$
d_2 $(SO_2)_{gas}$
d_3 $(SO_2)_{gas}$
d_4 $[SO_2 \cdot H_2O],[HSO_3^-],[H^+],[SO_3^{--}]$
d_5 $[SO_2 \cdot H_2O],[HSO_3^-],[H^+],[SO_3^{--}]$
Chem. Abbau u. Umbau: H_2S, Cystein, Sulfat

Abb.2

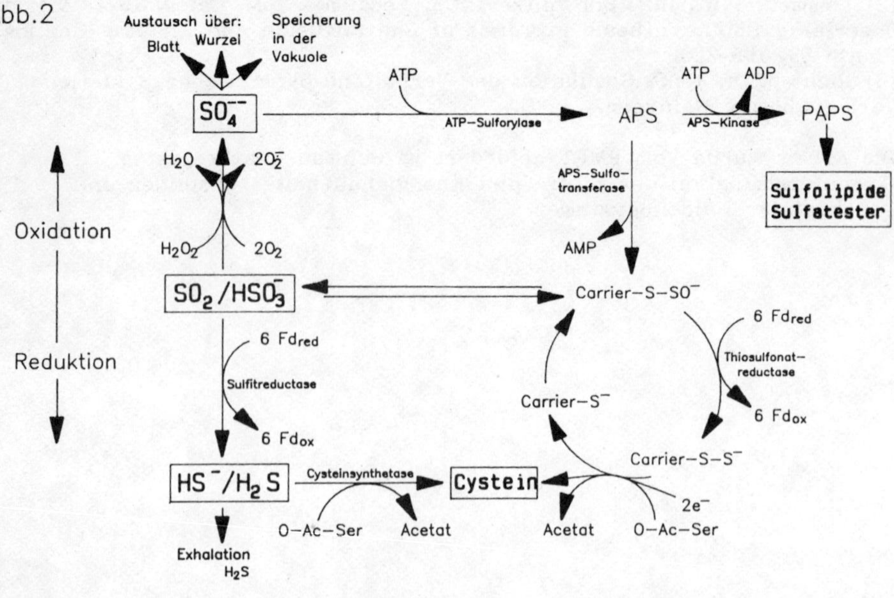
Austausch über: Blatt Wurzel
Speicherung in der Vakuole
SO_4^{--}
ATP
ATP-Sulforylase
APS
ATP ADP
APS-Kinase
PAPS
Sulfolipide Sulfatester
APS-Sulfotransferase
AMP
H_2O $2O_2^-$
H_2O_2 $2O_2$
Oxidation
Reduktion
SO_2/HSO_3^-
Carrier-S-SO$^-$
6 Fd_{red}
Thiosulfonat-reductase
6 Fd_{ox}
Carrier-S$^-$
Sulfitreductase
6 Fd_{red}
6 Fd_{ox}
Carrier-S-S$^-$
HS^-/H_2S
Cysteinsynthetase
Cystein
2e$^-$
O-Ac-Ser
Acetat
Acetat
O-Ac-Ser
Exhalation H_2S

Abb. 3

Gleichungssystem für Transport und Reaktion von SO_2 am Blattpfad

1. c_1: Konzentration von SO_2 in der Außenluft

$$c_1 = c_1(t)$$

2. c_2: Konzentration von SO_2 an der Blattoberfläche

$$\frac{dc_2}{dt} = \frac{1}{d_2}\left[\frac{c_1-c_2}{r_{12}} - \frac{c_2-c_3}{r_{23}}\right]$$

3. c_3: Konzentration von SO_2 im Gas-Interzellularraum

$$\frac{dc_3}{dt} = \frac{1}{d_3}\left[\frac{c_2-c_3}{r_{23}} - \frac{c_3-c_4\, X_4^{SO_2}\, H(SO_2)^{-1}\, V_L^{-1}}{r_{34}}\right]$$

$X_4^{SO_2}$ = *Molenbruch* SO_2 $H(SO_2)$ =*Henry Konst.* V_L =*Molvolumen*

4. c_4: Konzentration von $[S(IV)]$ im Zellwandwasser

$$\frac{dc_4}{dt} = \frac{1}{d_4}\left[\frac{c_3-c_4\, X_4^{SO_2}\, H(SO_2)^{-1}\, V_L^{-1}}{r_{34}} - \sum_{i=1}^{3}\frac{c_4 X_4^i - c_5 X_5^i}{r_{45}^i}\right]$$

Die Moleküle $X_4^{SO_2/HSO_3^-/SO_3^{--}}$ *und* $X_5^{SO_2/HSO_3^-/SO_3^{--}}$ *sind* p_H *abhängig*

5. c_5: Konzentration von $[S(IV)]$ im Chloroplast

$$\frac{dc_5}{dt} = \frac{1}{d_5}\left[\sum_{i=1}^{3}\frac{c_4 X_4^i - c_5 X_5^i}{r_{45}^i}\right] - k^{SO_4^{--}} c_5 - \frac{V_{max}^{H_2S}\, c_5}{k^{H_2S}+c_5} - \frac{V_{max}^{Cystein}\, c_5}{k^{Cystein}+c_5}$$

i: $1=[SO_2\cdot H_2O]$; $2=[HSO_3^-]$; $3=[SO_3^{--}]$

6. Effective Gleichung für $[S(IV)]$ im Chloroplast

$$\frac{dc_5}{dt} = \frac{1}{d_5}\cdot\frac{1}{r_{15}^{eff}}\{c_1 - \gamma c_5\} - R(c_5)$$

$$\gamma = X_4^{SO_2}\, H(SO_2)^{-1}\, V_L^{-1}\, \frac{\sum_{i=1}^{3} X_5^i/r_{45}^i}{\sum_{i=1}^{3} X_4^i/r_{45}^i}$$

$$r_{15}^{eff} = \left(\left(X_4^{SO_2}\, H(SO_2)^{-1}\, V_L^{-1} + r_{14}\sum_{i=1}^{3} X_4^i/r_{45}^i\right)\Big/\left(\sum_{i=1}^{3} X_4^i/r_{45}^i\right)\right.$$

Abb. 4

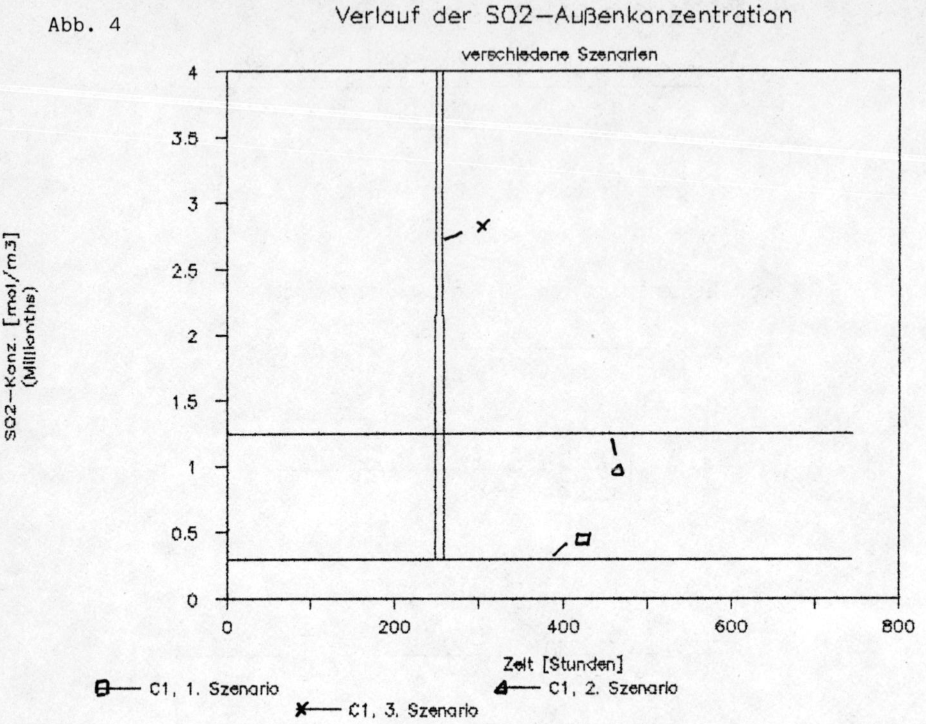
Verlauf der SO2–Außenkonzentration
verschiedene Szenarien
SO2–Konz. [mol/m3] (Millionths)
4
3.5
3
2.5
2
1.5
1
0.5
0
0
200
400
600
800
Zeit [Stunden]
C1, 1. Szenario
C1, 2. Szenario
C1, 3. Szenario

Abb. 5

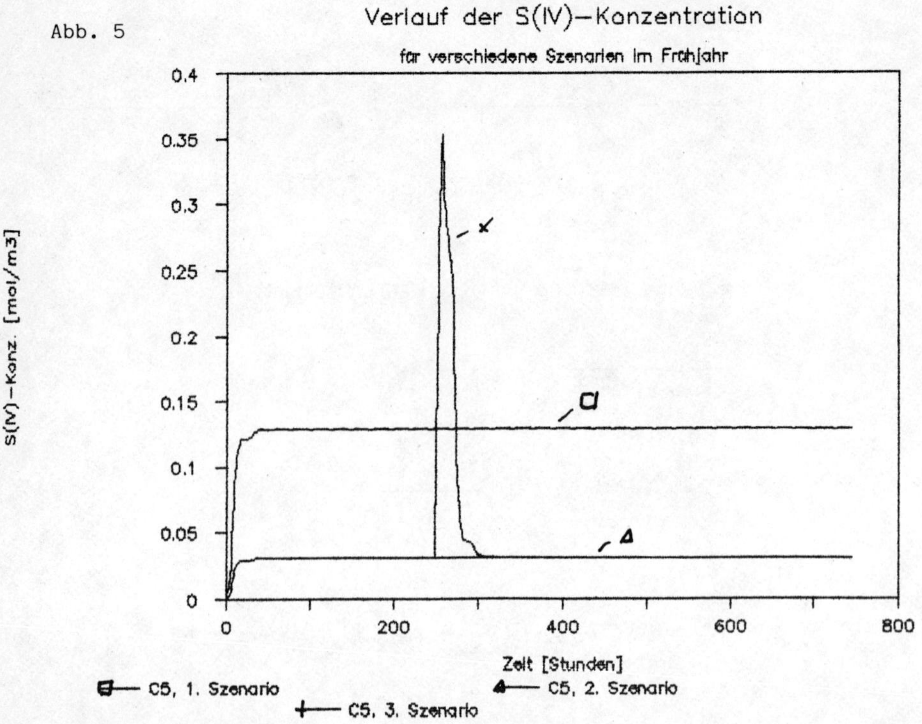
Verlauf der S(IV)–Konzentration
für verschiedene Szenarien im Frühjahr
S(IV)–Konz. [mol/m3]
0.4
0.35
0.3
0.25
0.2
0.15
0.1
0.05
0
0
200
400
600
800
Zeit [Stunden]
C5, 1. Szenario
C5, 2. Szenario
C5, 3. Szenario

Abb. 6

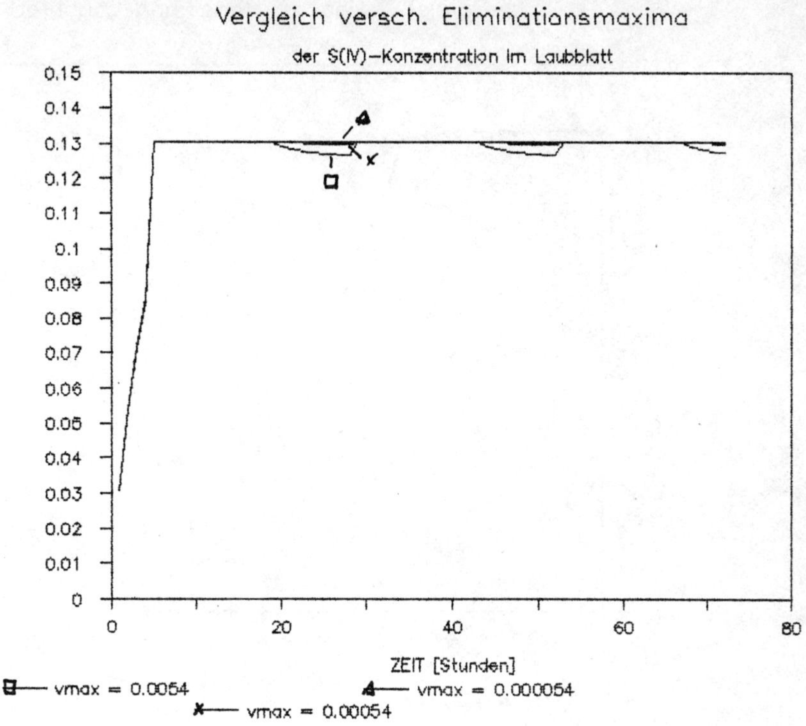
Vergleich versch. Eliminationsmaxima
der S(IV)-Konzentration im Laubblatt
S(IV) [mol/m3]
0.15
0.14
0.13
0.12
0.11
0.1
0.09
0.08
0.07
0.06
0.05
0.04
0.03
0.02
0.01
0
0
20
40
60
80
ZEIT [Stunden]
vmax = 0.0054
vmax = 0.000054
vmax = 0.00054

Abb. 7

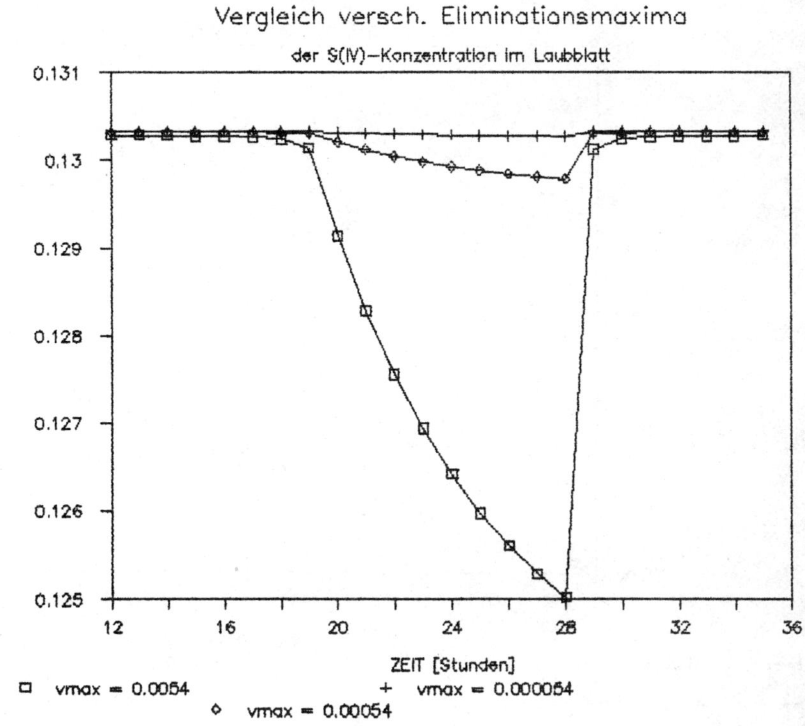
Vergleich versch. Eliminationsmaxima
der S(IV)-Konzentration im Laubblatt
S(IV) [mol/m3]
0.131
0.13
0.129
0.128
0.127
0.126
0.125
12
16
20
24
28
32
36
ZEIT [Stunden]
vmax = 0.0054
vmax = 0.000054
vmax = 0.00054

Abb. 8

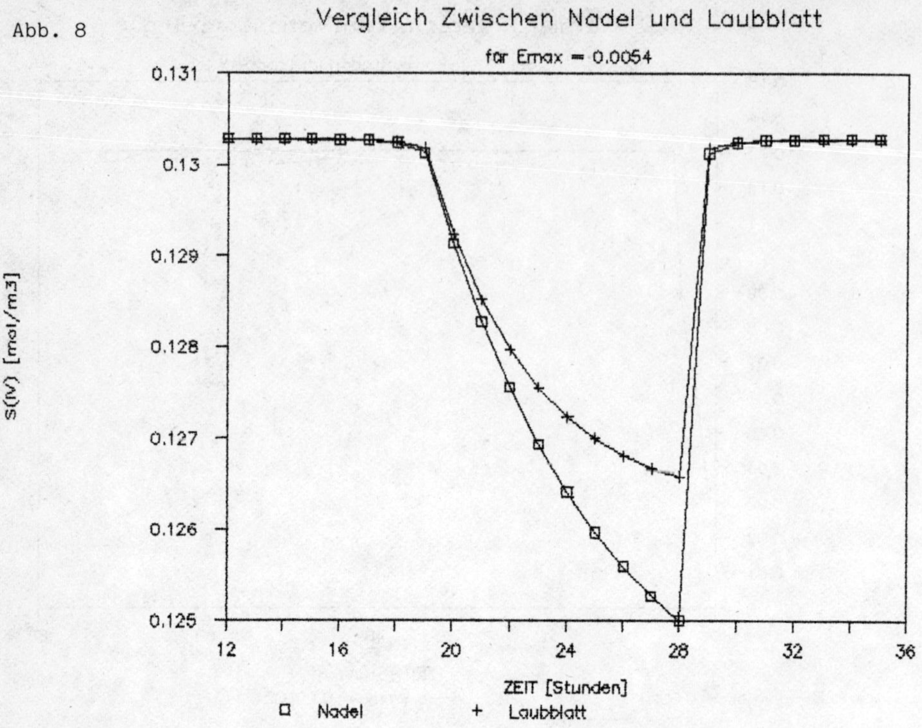

Vergleich Zwischen Nadel und Laubblatt
für Emax = 0.0054
0.131
0.13
0.129
0.128
0.127
0.126
0.125
S(IV) [mol/m3]
12
16
20
24
28
32
36
ZEIT [Stunden]
Nadel
Laubblatt

Abb. 9

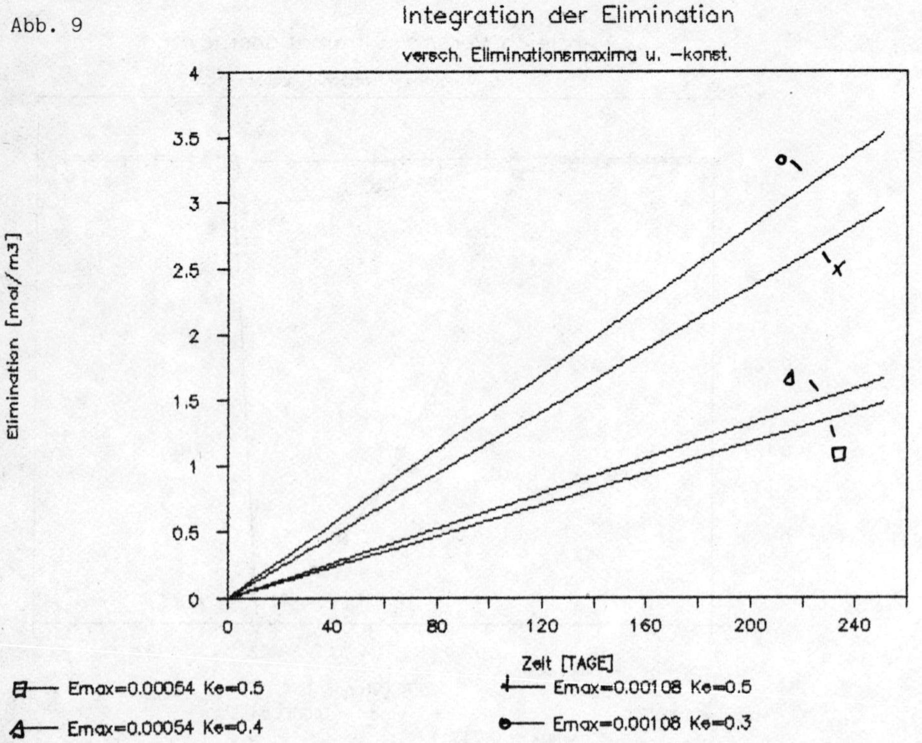

Integration der Elimination
versch. Eliminationsmaxima u. -konst.
4
3.5
3
2.5
2
1.5
1
0.5
0
Elimination [mol/m3]
0
40
80
120
160
200
240
Zeit [TAGE]
Emax=0.00054 Ke=0.5
Emax=0.00054 Ke=0.4
Emax=0.00108 Ke=0.5
Emax=0.00108 Ke=0.3

Effets d'une pollution de l'atmosphère par l'Ozone et le Dioxyde de Soufre sur la croissance, la conductance stomatique et la teneur en composés phénoliques de 3 clones d'épicéas en chambre à ciel ouvert.

P. MALKA*, CONTOUR-ANSEL D.**, P. LOUGUET** ET J. BONTE***

*ADERA/CDERE.Contrat ELF. **Laboratoire de Physiologie Végétale et Ecophysiologie Végétale Appliquée. Université PARIS XII. ***Centre Départemental d'Etudes et de Recherches sur l'Environnement.

SUMMARY

To evaluate the influence of atmospheric SO_2 and O_3 suspected to be involved in forest dieback, three Norway spruce clones (Picea abies L.) growing in Open-Top-Chambers have been exposed since June 1986 to a Vosgien forest typical SO_2 and O_3 concentrations profile. Growing parameters, needle polyphenolic contents and stomatal conductance have been measured. No difference between exposed or control trees have been detected yet, but results show significative differences between clones and also between trees growing outside or inside Open-Top-Chambers.

RESUME

Afin d'évaluer la part de responsabilité du SO_2 et de l'O_3 atmosphériques mis en cause dans le problème de dépérissement des forêts, trois clones d'épicéas (Picea abies L.) placés en Chambre à Ciel Ouvert sont exposés depuis Juin 1986 à une pollution en O_3 et SO_2 caractéristique de l'atmosphère du Donon (zone dépérissante des Vosges). Les critères symptomatologiques choisis sont: paramètres de croissance des arbres (longueur moyenne des pousses, diamètre du tronc, hauteur), teneurs en phénols totaux des aiguilles et conductivité stomatique. Au terme de plusieurs mois d'expérimentation, aucune différence significative entre témoins et pollués n'a pu être mise en évidence. On a pu montrer, à travers les paramètres écophysiologiques mesurés, une nette variabilité génétique, ainsi qu'un "effet de chambre" .

I. INTRODUCTION

Le fonctionnement de l'appareil stomatique est susceptible d'être directement altéré par la présence de polluants atmosphériques ou indirectement en reflétant l'existence d'un dérèglement physiologique profond de l'arbre soumis à l'influence d'un environnement défavorable. Le but des recherches entreprises est d'évaluer dynamiquement, c'est à dire, au cours du temps et in situ, l'influence de doses subnécrotiques d'OZONE (O_3) et de DIOXYDE DE SOUFRE (SO_2) sur le degré d'ouverture des stomates de 3 clones d'épicéas (*Picea abies* L.) dans des conditions semi-contrôlées.

II. MATERIEL ET METHODES

* Chambres a ciel ouvert (CCO)

Les trois clones d'épicéas ont été soumis en chambre à ciel ouvert, depuis le 6 juin 1986, à différents traitements :

- une pollution par O_3 et SO_2, dont le niveau et la fréquence des concentrations moyennes horaires ont été fidèlement reproduits 15 jours après leur occurrence réelle sur le site du DONON (Vosges) où le dépérissement des arbres est nettement observé.
- une pollution en O_3 et SO_2 de type rural, qui correspond à la pollution de fond locale du site expérimental de MONTARDON (Pyrénées-Atlantiques).
- l'absence pratiquement de pollution, grâce à la filtration de l'air sur du charbon actif.

* Technique de mesure de la Résistance stomatique (RS) des aiguilles d'Epicea à l'aide du poromètre ΔT Devices

L'aiguille d'épicéa, détachée du rameau, est introduite dans la chambre porométrique contenant des sondes qui mesurent la température ambiante et l'humidité relative de l'air. L'augmentation de la teneur en vapeur d'eau dans la chambre se fait d'autant plus rapidement que l'aiguille transpire activement; l'appareil donne le temps nécessaire pour faire passer une humidité relative HR donnée de la chambre porométrique à la valeur HR + 5%. La mesure exprimée en temps (1 unité de mesure = 5m sec) est ensuite convertie en résistance stomatique (RS en s cm^{-1}) à l'aide de courbes d'étalonnage, établies dans les mêmes conditions de température et d'humidité de l'atmosphère, à partir d'une plaque de calibration évaporante. La conductance stomatique (g_s) se déduit aisement de RS, sachant que $g_s = 1/RS$ ($cm\ s^{-1}$)
L'adoption de cette technique habituellement utilisée pour des feuilles planes, nous a conduit à effectuer un certain nombre de travaux préliminaires destinés à mieux cerner les problèmes pouvant compromettre la validité des mesures effectuées sur des aiguilles isolées et en conditions naturelles, tels que: conditions de calibration, choix du cycle de mesure, influence de la section du pétiole sur la g_s et évaporation possible d'un exsudat à ce niveau, influence du déficit hydrique sur la g_s, mise en évidence d'une corrélation linéaire entre la longueur et la surface des aiguilles (1). Les variations journalières de g_s d'épicéas ont conduit à rejeter le principe de mesures ponctuelles, et à adopter un protocole de mesures se déroulant sur une journée complète. Les mesures de g_s ont été effectuées sur des aiguilles de l'année 1986, de même rang, et situées sur le tiers supérieur de la couronne.

* Dosage des composés phénoliques

Des prélèvements ont été effectués sur les 3 clones à raison de 5 individus par chambre : au total, 20 pousses par chambre et par clone ont été récoltées. Après lyophilisation, le dosage des composés phénoliques est effectué par la technique du Folin, en mesurant la D.O. à 760 nm (2)

* Mesures de croissance et de développement:
Durant la période de végétation active, les stades phénologiques ainsi que la taille moyenne des pousses des 2 premiers verticilles ont été relevés

III. RESULTATS

*Caractéristiques des g_s en conditions naturelles et en chambre à ciel ouvert

On constate une diminution des g_s (fermeture des stomates) du clone CONSTANCE (CST) en fonction du déficit hydrique de l'air (δe en g cm^{-3}), puis une légère réouverture stomatique en milieu d'après-midi (16h-17h) suivie de l'amorce d'une fermeture coïncidant avec le coucher du soleil. Cette évolution confirme l'influence de δe sur le degré d'ouverture des stomates déjà mise en évidence dans une chambre à transpiration en conditions contrôlées sur un individu CST au cours des phases préliminaires de mise au point (1). Ce phénomène, classiquement décrit sous le nom de "dépression de midi" permet la régulation de la transpiration par un mécanisme probablement du type "feedforward" (3). Les g_s les plus élevées, pour les 3 clones, se situent le matin quand l'humidité relative de l'air est encore élevée et δ e faible. On a pu noter une diminution générale des g_s au cours du temps : les g_s mesurées en octobre sont plus faibles qu'en août (voir figure). Il est possible d'expliquer cette évolution par une activité décroissante du métabolisme de la plante et des cellules stomatiques à la fin de la période de vie active de l'arbre. Pendant la période hivernale, la conductivité stomatique est très faible. SMITH and al. (4) ont également constaté la "fermeture automnale" des stomates sur 6 espèces de conifères dont Picea engelmannii. Aucun effet particulier des diverses conditions de pollution dans les CCO n'est encore visible actuellement. La figure résume les résultats principaux obtenus au cours de 6 journées de mesures pendant la période de végétation , après la différenciation stomatique, et pendant la période hivernale, dans les 5 CCO et l'AA. Nous avons tenté de rechercher et de quantifier des critères utiles permettant l'exploitation des résultats. Nous avons retenu : les valeurs de g_s maximales,minimales mesurées au cours de la journée ou obtenues en fin de journée, la moyenne journalière de g_s et les g_s minimales obtenues à l'obscurité.Le critère le plus représentatif est la g_s maximale: en effet, pour un clone donné, la variabilité entre les aiguilles d'individus distincts est la plus faible au moment de l'ouverture stomatique maximale. On observe des g_s plus faibles pour CST : 90% des g_s de CST, pendant la période de végétation, tous critères confondus, c'est à dire moyenne journalière, minima, maxima, sont inférieures aux g_s de IST et GER. Les g_s des aiguilles d'épicéa en AA sont nettement plus faibles que pour les épicéas en chambre à ciel ouvert. Toutefois les mesures effectuées en Février 1987 révèlent une inversion de cette tendance. Un "effet de chambre" sur un certain nombre de paramètres écophysiologiques a été observé par ARNDT et SEUFERT (5). De même, la croissance des épicéas en air ambiant est nettement plus élevée que celle des arbres placés en CCO (6). Il faut rappeler à cet égard que des différences parfois importantes de température, d'humidité relative, et de rayonnement solaire ont été enregistrées entre l'air ambiant et les CCO.

Compte-tenu de la durée trop brève de fumigation , il est actuellement prématuré de proposer une interprétation fiable des différences entre les traitements. Sur la plupart des espèces soumises à une pollution aigue par l'ozone, on admet un mécanisme de fermeture des stomates, modulé selon les conditions de saturation de l'air (7). Les premiers résultats obtenus sur des pins (Pinus sylvestris) en CCO (fumigation à l'ozone à 25 ppb au-dessus du niveau ambiant de 11h à 18h pendant 4 mois) ont mis en évidence une chute de 40% de la transpiration (8).

*Teneur en composés phénoliques:

Les résultats des dosages de polyphénols dans les aiguilles d'épicéas prélevées dans les 6 chambres à ciel ouvert, après 132 jours d'exposition à l'ozone, sont consignés dans le tableau 1 : les teneurs moyennes pour chaque clone, calculées à partir des valeurs des 6 chambres (tab. 1A) ne sont pas significativement différentes entre les clones C et G (test t à p = 0,05) ; par contre, la différence est significative pour le clone I, aussi bien vis à vis de C que vis à vis de G. Si l'on détaille les résultats en tenant compte des différences entre chambres (tab. 1B), on constate que le clone I se différencie à nouveau pour toutes les chambres, les valeurs moyennes calculées pour ce clone étant toujours inférieures à celles des 2 autres.

Enfin, si l'on fait les moyennes des résultats par type de traitement, tous clones confondus (tab. 1C), il ressort nettement que l'exposition à l'ozone à dose sub-nécrotique, n'a pas eu d'incidence pour l'instant sur le taux de polyphénols (les différences entre traitements considérés 2 à 2, ne sont pas significatives sauf entre AA et AF).

* Effets visibles:

Le tableau 2 montre que comme pour les autres paramètres aucun effet ne s'est manifesté ni sur l'aspect général des arbres ni sur leur croissance. L'influence de la présence de la CCO semble prévaloir sur l'effet des différents traitements (7).

IV.CONCLUSIONS

Malgré les limites d'utilisation du poromètre Δ T, la détermination des conductances stomatiques d'aiguilles d'épicéas soumis à une pollution contrôlée en chambre à ciel ouvert, est réalisable à condition de respecter les critères d'échantillonnage et les impératifs techniques définis au cours de la phase de mise au point . Nous avons pu préciser les causes principales d'erreur et chiffrer la variabilité individuelle (1). Nous avons en outre mis clairement en évidence la sensibilité des stomates de Picea abies au déficit de saturation atmosphérique, la plus grande résistance des aiguilles du clone CONSTANCE à la diffusion de la vapeur d'eau et la diminution globale de l'ouverture stomatique dans la période préhivernale dans les clones étudiés. A l'heure actuelle, même si certaines tendances apparaissent, aucune différence significative entre les diverses CCO n'a pu être mise en évidence. La phase de mise au point technique de mesure de g_s ayant coïncidé avec la mise en place des CCO sur le site expérimental de MONTARDON, la prochaine campagne de mesures en 1987 devrait permettre de progresser dans l'évaluation des effets des polluants. L'utilisation de 3 clones offre la possibilité de mettre en évidence un éventuel mécanisme de résistance à la pollution et d'étudier la sensibilité clonale aux polluants avant l'apparition d'effets visibles. La traduction de ces divers processus devrait se manifester par une modification symptomatique du comportement stomatique.

REFERENCES

(1) LOUGUET P., MALKA P., CONTOUR-ANSEL D. Variations de la résistance stomatique et de la teneur en composés phénoliques de feuilles d'épicéas soumis à une pollution contrôlée par le dioxyde de soufre et l'ozone en chambre à ciel ouvert. Rapport DEFORPA. A paraître.

(2) CONTOUR-ANSEL D., LOUGUET P. 1986. Variations du taux de polyphénols dans les aiguilles d'épicéas (Picea abies) présentant différents degrés de dépérissement. Pollution Atmosphérique. n°112.
(3) FARQUHAR G.D, SHARKEY T.D. 1982.Stomatal conductance and photosynthesis. Ann. Rev. Plant. Physiol.233,317-45.
(4) SMITH W.K and al.1984. Autumn stomatal closure of six conifers species of the Central Rocky Mountains.Oecologia (Berlin). 63:237-242.
(5) ARNDT, SEUFERT G. 1986. Excursion to Hohenheim. 2nd European Open-Top-Chamber Worshop. 17-19 sept 1986.
(6) BONTE J., CANTUEL J., DUBOIS C., MALKA P. 1986. Simulation de l'action du dioxyde de soufre et de l'ozone seuls ou en mélange sur l'épicéa en chambre à ciel ouvert. DEFORPA.Rapport Final. A paraître.
(7) UNSWORTH M.H., BLACK V.J, 1981. Stomatal response to pollutants. Stomatal Physiology. Seminar Series 8. Ed. PJ.JARVIS & TA. MANSFIELD. 188-203.
(8) SKARBY L. and al. 1986. Effects of ozone on Scott pine (Pinus sylvestris) and Norway spruce (Picea abies) grown in open-top-chambers at Rorvik, Sweden. COST WORKSHOP. 19-23 oct.1986.

FIGURE : EVOLUTION DES CONDUCTANCES STOMATIQUES MAXIMALES ($g_{s\ max}$) DES AIGUILLES D'EPICEAS EN FONCTION DE LA DUREE D'EXPOSITION .

$g_{s\ max}$: valeur maximale de conductance d'aiguille obtenue sur une journée complète de mesures. Chaque valeur représente la moyenne de 3 arbres (1 individu par clone et par chambre).

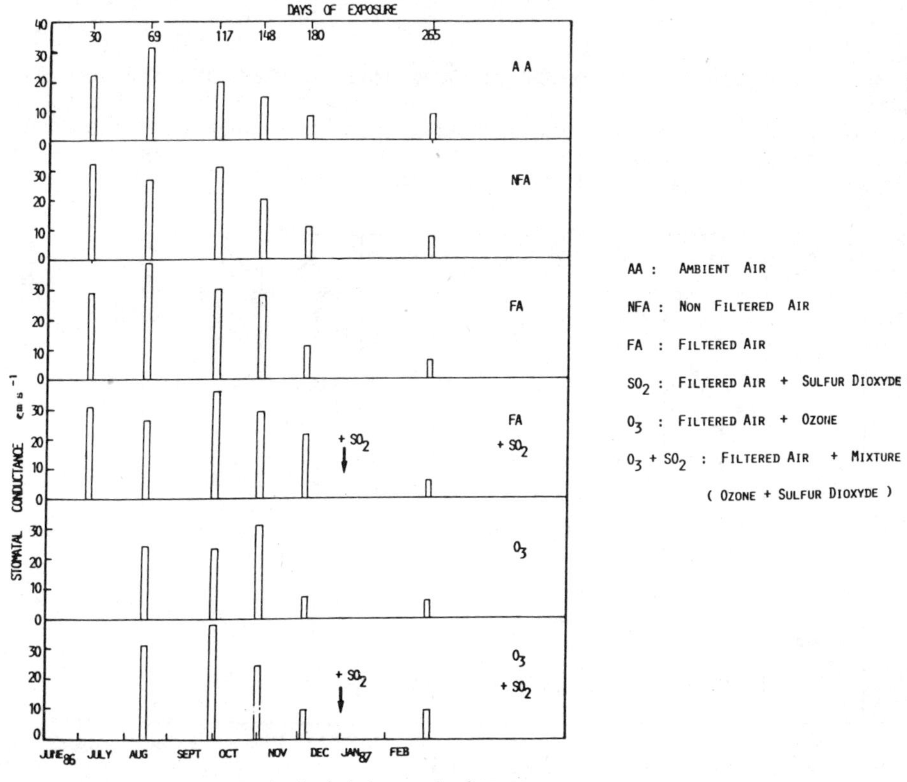

$g_{s\ max}$ is the maximum needle conductance value obtained over one day of measures
each value represents the mean $g_{s\ max}$ for 3 trees (one per clone)

TABLEAU 1 : ANALYSE QUANTITATIVE DES COMPOSES PHENOLIQUES DANS LES AIGUILLES DE PICEA ABIES, PRELEVEES DANS LES 6 CHAMBRES, APRES 132 JOURS D'EXPOSITION A L'OZONE

CLONE	$\bar{M}$	σ	SIGNIFICANCE	
C	79.6	19.6	I/C	(S)
I	49.3	9.3	I/G	(S)
G	68.5	15.9	C/G	(N.S)

1 A : MEAN VALUES FOR EACH CLONE (AVERAGE ON THE 6 CHAMBERS)

TREATMENT \ CLONE	C	I	G
AA	97.6	61.8	81.9
NFA	58.2	**	94.2
FA*	70.3	46.8	67.9
FA+O_3*	90.4	44.9	55.7

1 B : MEAN VALUES FOR EACH CLONE AND EACH TREATMENT

TREATMENT	$\bar{M}$	σ	SIGNIFICANCE	
AA	77.0	15.3	AA/NFA	(N.S)
NFA	70.2	17.5	AA/FA	(S)
FA*	60.6	15.1	AA/FA+O_3	(N.S)
FA+O_3*	63.7	22.9	NFA/FA	(N.S)
			NFA/FA+O_3	(N.S)
			FA/FA+O_3	(N.S)

1 C : MEAN VALUES FOR EACH TREATMENT (AVERAGE ON THE 3 CLONES)

LEGENDS

$\bar{M}$: MEAN VALUES, IN MG G^{-1} DRY WEIGHT (TAKING GALLIC ACID AS REFERENCE)

σ : STANDARD DEVIATION

(S) : SIGNIFICANT DIFFERENCE BETWEEN 2 CLONES (T TEST, P=0.05)

(N.S) : NON SIGNIFICANT (T TEST, P=0.05)

* : MEAN VALUES BETWEEN THE 2 CHAMBERS

** : NOT AVAILABLE

TABLEAU 2 : CROISSANCE DE LA POUSSE APICALE APRES 70 JOURS D'EXPOSITION A L'OZONE.

Chaque valeur représente la longueur moyenne et l'intervalle de confiance de la pousse apicale des 5 arbres par clone et par chambre. Les mesures ont été effectuées en fin de période de croissance, le 14/8/86.

CLONES \ O.T.C.	A A	N F A	F A	F A*	O_3	O_3*
CONSTANCE	 15.8 ± 6.2	 14.3 ± 5.5	 14.0 ± 6.0	 7.5 ± 7.6	 12.6 ± 4.6	 16.4 ± 4.2
ISTEBNA	 23.3 ± 6.7	 15.8 ± 15.4	 17.5 ± 4.5	 18.6 ± 10.8	 20.2 ± 5.3	 17.8 ± 4.6
GERARDMER	 15.6 ± 12.9	 10.9 ± 6.7	 11.0 ± 8.3	 9.1 ± 4.0	 11.3 ± 3.4	 11.5 ± 6.3

* : SO_2 ADDED IN THESE CHAMBERS DURING WINTER (SINCE JANUARY 87)

Each value is the mean apical shoot lenght, averaged over the 5 trees per clone located in each O.T.C. (Open-Top-Chamber) and their respective Standard Error. Measurements were made at the end of growing stage (14/08/87)

EFFETS DE LA POLLUTION ATMOSPHERIQUE SUR LE SYSTEME SECRETEUR ET LA COMPOSITION TERPENIQUE DES AIGUILLES DE PICEA ABIES

A. SAINT-GUILY

Laboratoire de Physiologie Cellulaire Végétale
Unité associée au C.N.R.S. n°568
Université de Bordeaux I-33405 TALENCE Cedex

Résumé

Sur des aiguilles de Picea abies à différents stades de dépérissement, on a réalisé une analyse qualitative et quantitative de la composition en mono- et sesquiterpènes de l'essence, ainsi qu'une étude ultrastructurale du système sécréteur. Les prélèvements, en conditions naturelles (Vosges, Massif du Donon), sur des arbres de même âge nous ont permis de mettre en évidence une grande variabilité individuelle de la composition en terpènes des essences. Il faut donc rester prudent dans les interprétations des analyses comparatives réalisées sur les arbres à différents stades de dépérissement. Toutefois, par le biais de la morphométrie, nous avons pu mettre en évidence une différence significative du volume leucoplastidial, celui-ci étant plus important pour les arbres dépérissants que pour les arbres ne présentant pas de symptômes de dépérissement.

Summary

Qualitative and quantitative analysis was determined on the mono- and sesquiterpenes composition of the essential oil from Picea abies needles. An ultrastructural study of the secretory system was performed. The collections, made in the Donon Forest (Vosges), from trees of the same age, allow us to display individual variability on terpenes composition of the essential oil. Therefore, comparative analysis interpretations, achieved on trees at different decline steps, must be taken with care. However, with morphometry method, we were able to show a significant difference of the leucoplastidial volume which is larger for wasting trees than for trees without decline symptoms.

1. INTRODUCTION

Dans les conditions physiologiques normales, la production des essences et des résines chez les Gymnospermes est caractérisée par une compartimentation rigoureuse des étapes de biosynthèse et des sites d'accumulation (Bernard-Dagan et al., 1982). La synthèse des monoterpènes est un phénomène de courte durée qui se produit au sein de cellules spécialisées riches en organites particuliers : les leucoplastes. D'autre part, on sait que la sécrétion des résines peut être fortement accrue par l'effet d'agents extérieurs variés : blessures (Vassilyev et Carde, 1976), attaque d'insectes, infestation par des champignons (Cheniclet, 1987) ou application d'herbicides (Clason, 1976). Nous pouvons donc raisonnablement penser que les pollutions affectant les peuplements de Picea abies peuvent induire des perturbations du métabolisme terpénique, en relation avec les phénomènes de dépérissement décrits sur les arbres. Ces perturbations se traduiraient par un déséquilibre qualitatif et quantitatif des composés isopréniques élaborés, ainsi qu'éventuellement par des modifications structurales.

2. MATERIEL ET METHODES

Les prélèvements ont été effectués en juin et novembre 1985 et juillet 1986 (Massif du Donon). Les arbres prélevés, âgés de 15 à 20 ans, sont classés de 1 à 3b selon leur degré d'atteinte (jaunissement et chute du feuillage) : les arbres de la classe 1 sont considérés comme résistants ; ceux de la classe 2 ont de 0 à 10% d'aiguilles tombées ; dans la classe 3a, les arbres ont perdu 10 à 20% de leurs aiguilles ; les arbres classés 3b ont 35 à 50% d'aiguilles tombées et présentent un jaunissement du feuillage. Sur chaque arbre a été prélevée une branche du quatrième verticille, porteuse de pousses et d'aiguilles de trois années consécutives. Les prélèvements de novembre 1985 ont porté sur 20 individus de la classe 1 ainsi que sur un individu de chacune des classes décrites. En juillet 1986, 3 ou 4 arbres par classe ont été étudiés.

Une double fixation séquentielle a été réalisée : par le glutaraldéhyde à 2,5% dans un tampon phosphate 0,1 M pH 7,2 contenant 3% de saccharose ; par le tétroxyde d'osmium à 1% dans le même tampon. Les objets sont ensuite déshydratés dans l'éthanol (de 30° à 100°) et inclus dans une résine epoxy (epon). Les observations sont faites à l'aide d'un microscope électronique à transmission PHILIPS 301 sur des coupes fines (60 à 90 nm d'épaisseur).

Les extraits pentaniques de matériel frais (4 à 8 grammes d'aiguilles) sont déshydratés sur sulfate de sodium anhydre. Les hydrocarbures (HC) et les composés oxygénés (CO) terpéniques sont séparés sur une colonne de silice. Les mélanges 98/2 et 90/10 éluent respectivement les HC et les CO. Les extraits récupérés sont analysés par chromatographie en phase gazeuse sur une colonne capillaire apolaire C.P.Sil-5 de 25 mètres. La détermination de la nature des pics principaux a été faite par chromatographie en phase gazeuse (colonne C.P.Sil-5, 50 mètres) couplée à la spectrométrie de masse (U.G. Micromass 16F).

3. RESULTATS

3.1. Etude analytique

La variabilité de la composition terpénique de différents résineux, entre provenances et individus d'une même provenance, a été démontrée (Marpeau-Bézard et Baradat, 1979). Il était donc important de connaître cette variabilité chez Picea abies avant d'étudier l'effet de la pollution sur le métabolisme terpénique.

L'étude qualitative des hydrocarbures mono- et sesquiterpéniques réalisée sur une vingtaine d'arbres de la classe 1 a permis de mettre en évidence une grande variation des teneurs en camphène, ß-pinène, myrcène et limonène (figure 1). Pour ces composés, les arbres semblent répartis en trois classes correspondant à une richesse différente en terpènes.

Sur le plan quantitatif, les analyses faites sur les prélèvements de juillet 1986 (figure 2) montrent qu'il existe également une importante variabilité individuelle pour les rendements en terpènes. L'augmentation apparente du rendement en terpènes avec l'âge des feuilles et le dépérissement de l'arbre (prélèvement de novembre 1985) ne peut donc pas, dans l'immédiat, être interprétée comme une accumulation de terpènes correspondant à une synthèse plus importante chez les arbres dépérissants.

3.2. Etude ultrastructurale

L'étude structurale des aiguilles de Picea abies nous a permis de mettre en évidence deux canaux sécréteurs sous-épidermiques discontinus (figure 3). Les observations cytologiques mettent en évidence un fonctionnement de courte durée des cellules épithéliales des canaux sécréteurs foliaires (responsables de la synthèse des terpènes) : seules, les aiguilles de l'année, pendant leur période de croissance (mai, juin, début juillet), comportent des canaux à cellules en phase sécrétrice active.

Sur les fixations de juin 1985, a été estimée la densité de volume du leucoplastidome à l'aide de techniques morphométriques (Weibel, 1969). Sur un échantillonnage de 20 à 30 cellules observées par classe, une moyenne de l'estimation de la densité de volume plastidial et son écart-type ont été calculés :

Classe	1	2	3a	3b
Nombre de cellules observées	21	36	31	28
Densité de volume moyen	110,6	188,6	160,1	242,7
Ecart à la moyenne (±)	35	26,1	23,2	42

La densité de volume plastidial paraît être plus faible pour la classe 1 que pour les autres classes, ce qui peut correspondre à un développement plus important du leucoplastidome des arbres considérés dépérissants.

Dans une première approche, seuls des arbres atteints en conditions naturelles ont été étudiés. A présent, il est nécessaire de travailler sur des arbres appartenant à un même clone et soumis à des pollutions contrôlées.

REFERENCES

BERNARD-DAGAN, C., PAULY, G., MARPEAU, A., GLEIZES, M., CARDE, J.P. et BARADAT, P. (1982). Control and compartmentation of terpene biosynthesis in leaves of Pinus pinaster. Physiol. Vég., 20, 775-795.

CHENICLET, C. (1987). Effects of wounding and fungus inoculation on terpene producing systems of maritime Pine. J. Exp. Bot., sous presse.

CLASON, T.R. (1976). Observations on paraquat induced oleoresin formation in slash Pine (Pinus elliotti Engelm.). Ph.D., University of Georgia, Agriculture, general.

MARPEAU-BEZARD, A. et BARADAT, P. (1979). Etat actuel des recherches sur la génétique des terpènes chez le Pin maritime. Applications diverses. 104ème Congrès des Sociétés savantes, Bordeaux, sciences, fasc. II, 287-297.

VASSILYEV, A.E. et CARDE, J.P. (1976). Effets du gemmage sur l'ultrastructure des cellules sécrétrices des canaux de l'écorce des tiges de Pinus sylvestris (L.) et Picea abies (L.). Protoplasma, 89, 41-48.

WEIBEL, E.R. (1969). Stereological principles for morphometry in electron microscopic cytology. Int. Rev. Cytol., 26, 235-302.

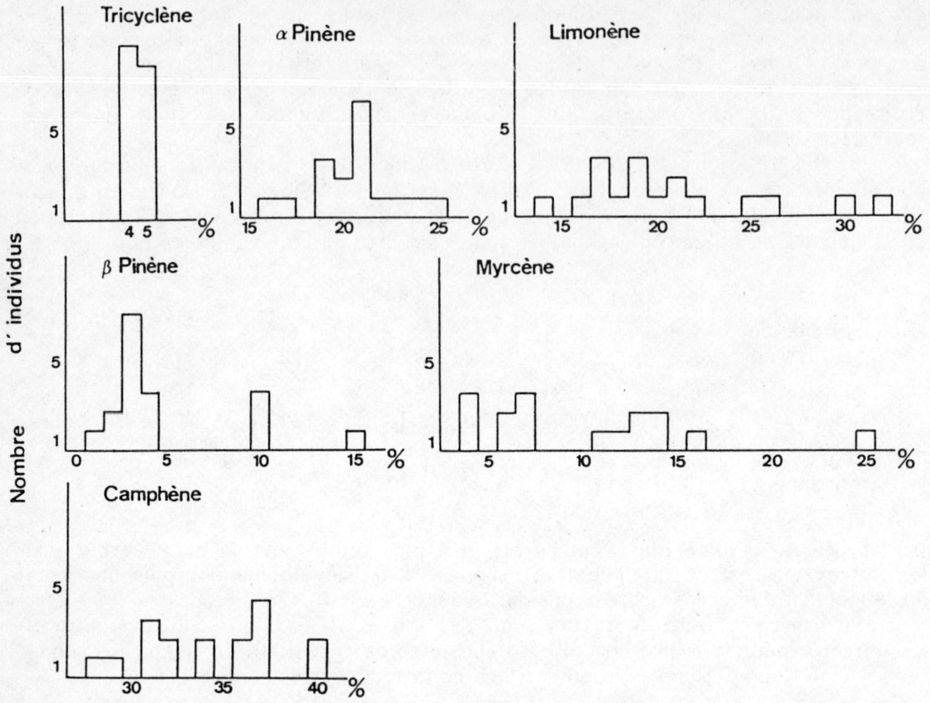

Figure 1 : Répartition d'une vingtaine d'individus en fonction des teneurs (%) en terpènes des aiguilles de <u>Picea abies</u> (tricyclène, α pinène, limonène, ß pinène, myrcène, et camphène).

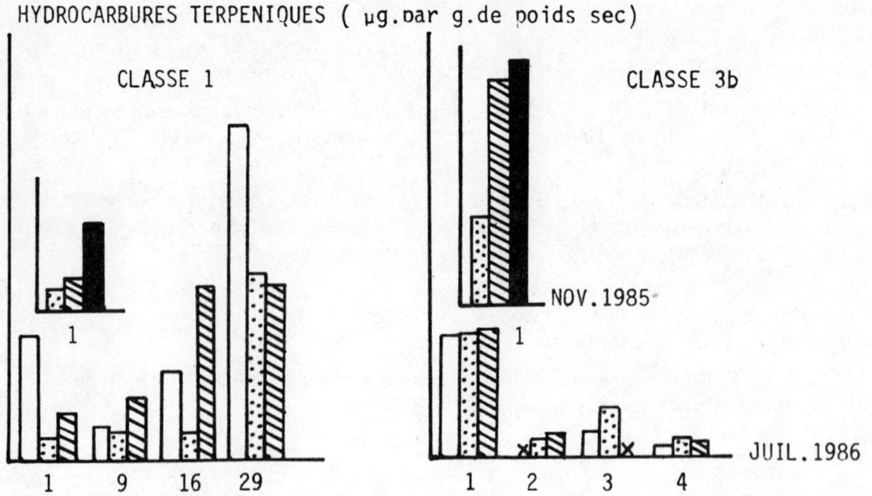

Figure 2 : Rendements en hydrocarbures terpéniques obtenus en novembre 1985 et juillet 1986 sur les feuilles 1983 ; 1984 ; 1985 ; 1986 .
Les arbres choisis appartiennent aux classes 1 (1, 9, 16, 29) et 3 b (1, 2, 3, 4).

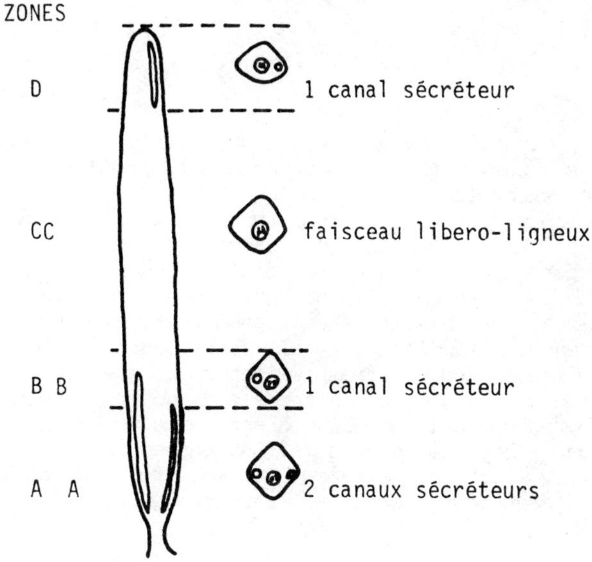

Figure 3 : Reconstitution schématique de la répartition des canaux sécréteurs d'une aiguille de <u>Picea abies</u>, à partir de coupes sériées à main levée.

DIRECT EFFECTS OF ACID WET DEPOSITION ON PHOTOSYNTHESIS, STOMATAL CONDUCTANCE AND GROWTH OF POPULUS CV. BEAUPRE

P. VAN ELSACKER, C. MARTENS and I. IMPENS
Department of Biology, University of Antwerp (UIA), Belgium

Summary

Poplar cuttings (Populus cv. Beaupré) were grown in a greenhouse and sprayed three times a week with simulated acid rain of pH 4.3 and 3.5 or with de-ionized water (pH 5.6). In August the "old" shoots were cut and "new" shoots developed from two axillary buds. Net photosynthetic rate and stomatal conductance of both leaf sides decreased with decreasing pH both for old and new shoots. The reduction of net photosynthesis is not only due to a decreased stomatal conductance, but also to a decreased residual conductance. Water use efficiency is similar for all treatments. Chlorophyll content of the new shoots and dry/fresh weight ratio are decreased at pH 3.5 and 4.3, specific leaf area is increased and chl a/b ratio is not changed. The reduced chlorophyll content seems to be important for the reduction of net photosynthesis. At the end of the season shoot lenght and dry weight of stem and root were reduced at pH 4.3 and 3.5. No significant change of total leaf area or leaf dry weight is found.

1. INTRODUCTION

Poplar represents more than 15% of the Belgian wood production on 5% of the forested area and is an important element in the Flemish landscape (Northern Belgium). Exposition to air pollution and acid precipitation may have important ecological and economical consequences. Poplar is known to be sensitive to gaseous air pollutants (2, reviews in 3, 4). Its leaf surface was injured by simulated acid rain of pH $\leq$ 3.1 and galls were formed (1), the sensitivity differed between clones. The objective of this study was to determine direct effects of acid precipitation on the gas exchange (carbon dioxide, water vapour) and biomass production of a high yielding poplar clone.

2. MATERIALS AND METHODS

Cuttings of Populus cv. Beaupré were potted in April and grown in a greenhouse. In the beginning of August half of the shoots were cut, leaving two axillary buds from which new shoots developed (named "old" and "new" resp.). Throughout the growing season three groups of 16 plants each were treated three times a week with de-ionised water of pH 5.6 (control) or with simulated acid rain of pH 4.3 or 3.5. The solution was applied with a plastic hand pumped sprayer (Birchmeier) until the leaves were thoroughly wetted and water began to run off (0.06 ml/cm^2 leaf area was intercepted). The ion concentrations in mg/l are assumed to represent an average composition of acid rain (7) : 4.50 SO_4^{2-}, 0.40 HSO_3^-(dissolved

SO_2), 2.00 NO_3^-, 1.20 NH_4^+, 0.99 Na^+, 1.45 Cl^-, 0.30 K^+, 0.60 Ca^{2+}, 0.30 Mg^{2+} (7). A mixture of sulphuric and nitric acid (weight ratio sulphate/nitrate = 2.4) was used to acidify the solution to pH 4.3 (approx. European average) or 3.5 (realistic minimal value).

Net photosynthetic rate (PN) was measured in the greenhouse at PPFD saturation (800 µmol m^{-2} s^{-1}) by placing the plants under sodium lamps (Philips, SON/T). A small perspex clamp-on-leaf cuvette was connected to an IRGA (Leybold-Heraeus, type Binos 1) in a closed system. An electronic timer-differentiator measured the time (<20s) of a fixed CO2 depletion (5 or 10 µl/l). Stomatal conductance was measured with an Automatic Porometer MkII (Delta-T Devices) before the CO2 measurements. Measurements were done on mature leaves during the summer months from July until the beginning of September.

At the end of the experiment shoot length, leaf area and dry weight of root, stems and leaves were determined. Chlorophyll was extracted from leaf samples in N,N-dimethylformamide (5). The absorbance was measured with a Shimadzu UV-160 spectrofotometer.

3. RESULTS

3.1. Gas exchange

The daily means of net photosynthesis at saturating PPFD (PN) and stomatal conductance (Gs) showed no clear seasonal trend, so the data of each pH-group were pooled for old and new shoots separately.

PN decreases with decreasing pH for both shoots (Fig. 1.), this decrease was significant at $p<0.001$ (Kruskal-Wallis test). Differences between the three pH-groups of the old shoots were significant at $p<0.05$ (Dunn test): 3.5 < 4.3 < 5.6. In comparison with the control group PN was reduced by 12 and 20% for pH 4.3 and 3.5 respectively. The Dunn test for PN of the new shoots showed that pH 3.5 = 4.3 < 5.6 at $p< 0.05$; PN of both acid rain group was reduced by 33%.

Gs is also decrease by the acid precipitation treatment, with a similar response of upper and lower leaf sides of old and new shoots (Fig. 2.). This effect was significant at $p<0.001$ (Kruskal-Wallis) except for the lower leaf side of new shoots due to a large variation ($p = 0.060$). The Dunn test showed following significant differences ($p < 0.05$) for the old shoots: pH 3.5 < 4.3 < 5.6 (lower side) and 3.5 = 4.3 < 5.6 (upper side); for the new shoots was pH 3.5 < 5.6 with an intermediate Gs value at pH 4.3 (upper side). On the whole reduction of the stomatal conductance (upper and lower leaf side) is approximately 20% (pH 4.3) to 30% (pH 3.5).

3.2. Chlorophyll content - SLA

Chlorophyll content, specific leaf area (SLA) and dry/fresh weight ratio were determined on new shoot leaves (Fig. 3.). Obviously, differences between pH 4.3 and 3.5 were not significant. Total chlorophyll content per unit leaf area and per unit dry weight were significantly lower at both acid treatments in comparison with the control, whereas chl a/b ratio was the same. Leaves were visibly paler. Acid precipitation appeared to affect leaf structure : the dry weight fraction was lower and the SLA was larger. This is an indication for lighter and less robust (thinner?) leaves.

3.3. Growth and production

Mean total shoot length, leaf area and dry weights of shoot, root and leaves of new shoots are shown in Fig.4. Total shoot length (sum of two shoots per cutting) was obviously lower at pH 4.3 and 3.5, with no significant differences between them. The same was observed for shoot dry weight. The largest reduction was measured for root biomass : 30 % (pH 4.3) to 50 % (pH 3.5). Leaf dry weight did not show a clear trend. Total plant leaf area was largest for pH 3.5 (large SLA), but differences were not significant. The estimated reduction of total plant biomass is 30 to 38 % for pH 4.3 and 3.5 respectively.

4. DISCUSSION

Spraying with simulated acid rain obviously affected the water vapour and carbon dioxide exchange : stomatal conductance and net photosynthesis were lower at low pH. This reduction was not only caused by stomatal closure (lower Ga). Calculation of the residual conductance G'r for CO2 transport and fixation inside the leaf (6) show that photosynthetic capacity was also internally affected : G'r of old and new shoots decreased with decreasing pH (Table I). The water use efficiency, calculated as the ratio of resistances for water vapour to those for carbon dioxide (neglecting the boundary layer), was similar for all treatments.

The reduced PN and G'r of the new shoots correlated with a lower chlorophyll content. When calculating PN on a chlorophyll weight basis (Table I) there were no clear differences between the treatments. So, the reduction of the chlorophyll seems to be an important factor in the CO2 exchange response to acid precipitation.

The reduction of the photosynthetic capacity by acid rain appears to be linked to a reduction of shoot lenght and dry matter accumulation (esp. stem and root). Such an adverse effect of acid wet deposition might have important consequences for the biomass production of poplar.

REFERENCES

(1) EVANS, L.S., GMUR, N.F. and DA COSTA, F. (1978). Foliar responses of six poplar clones of hybrid poplar. Phytopathology 68, 847-856.
(2) KIMMERER, T.W. and KOZLOWSKI, T.T. (1981). Stomatal conductance and sulfur uptake of five clones of Populus tremuloides exposed to sulfur dioxide. Plant Physiol. 67, 990-995.
(3) KOZIOL, M.J. and WHATLEY, F.R. (1984). Gaseous air pollution and plant metabolism. Butterworths, London.
(4) MOOI,J. (1983). Responses of some poplar species to mixtures of SO2, NO2, O3. Aquilo Ser. Bot 19, 189-196.
(5) MORAN, R. (1982). Formulae for determination of chlorophyllous pigments extracted with N,N-dimethylformamide. Plant Physiol. 69, 1376-1381.
(6) VAN ELSACKER, P. and IMPENS, I. (1984). Photosynthetic performance of poplar leaves at different levels in the canopy. In: Advances in photosynthesis research. Edited by SYBESMA, C. Martinus Nijhoff/Dr W. Junk Publishers, The Hague-Boston-Lancaster. Vol. IV, pp. IV.2.129-132.
(7) VAN ELSACKER, P. and IMPENS, I. (1986). Direct effects of simulated acid wet deposition on gas exchange of Norway spruce. EEC (COST) Workshop on Direct Effects of Dry and Wet Deposition on Forest Ecosystems, Lokeberg, Kungalv (Sweden), 19-23 Oct. 1986.

Table I. Residual conductance (G'r, mm/s), water use efficiency (WUE) and net photosynthesis (PN/Chl, μg CO2 s-1 (mg Chl)-1) for old and new shoots at various pH treatments.

		pH 3.5	pH 4.3	pH 5.6
old shoots	G'r	1.48	1.63	1.83
	WUE	0.15	0.15	0.14
new shoots	G'r	0.93	0.99	1.57
	WUE	0.08	0.08	0.09
	PN/Chl	2.3 (0.4)	2.5 (0.4)	2.2 (0.2)

(): 95% confidence limits

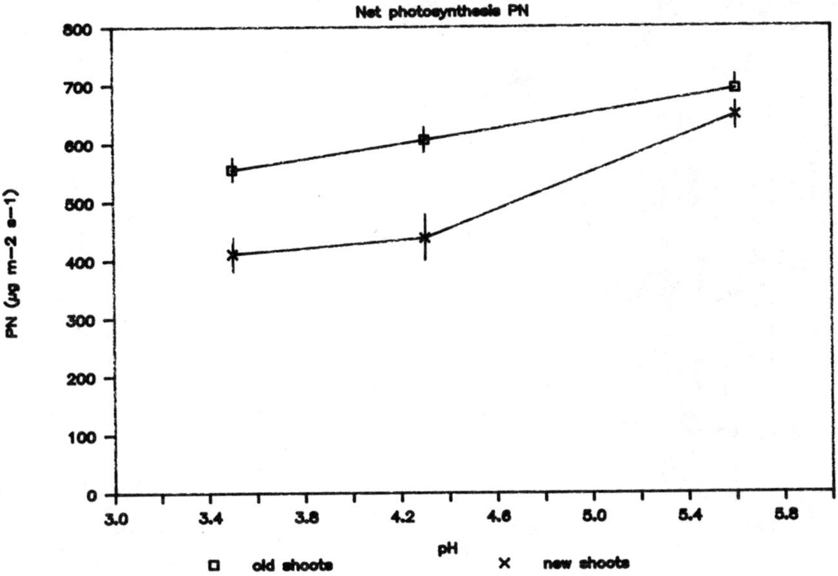

Fig. 1. Net photosynthesis PN of old and new shoots exposed to simulated acid precipitation of various pH. Bar: 95% confidence limit.

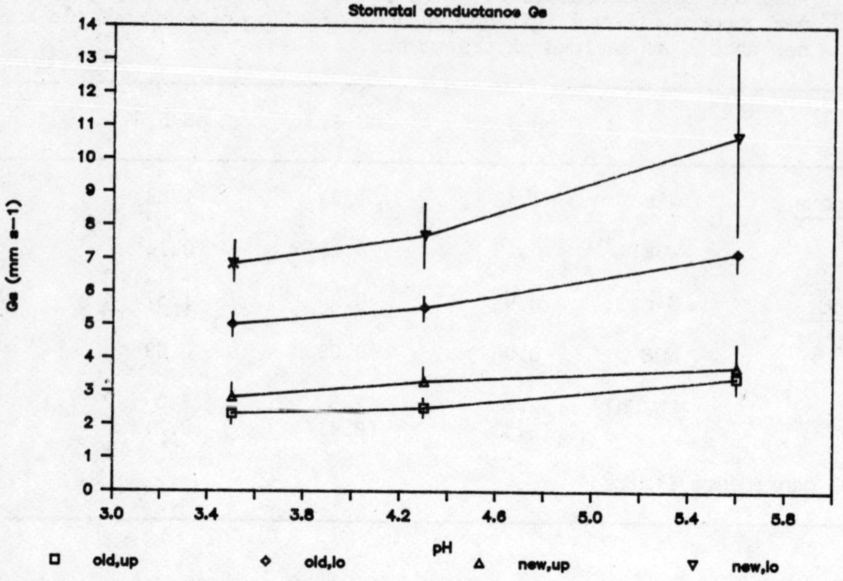

Fig. 2. Stomatal conductance Gs of upper (up) and lower (lo) leaf surface of old and new shoots at various pH.

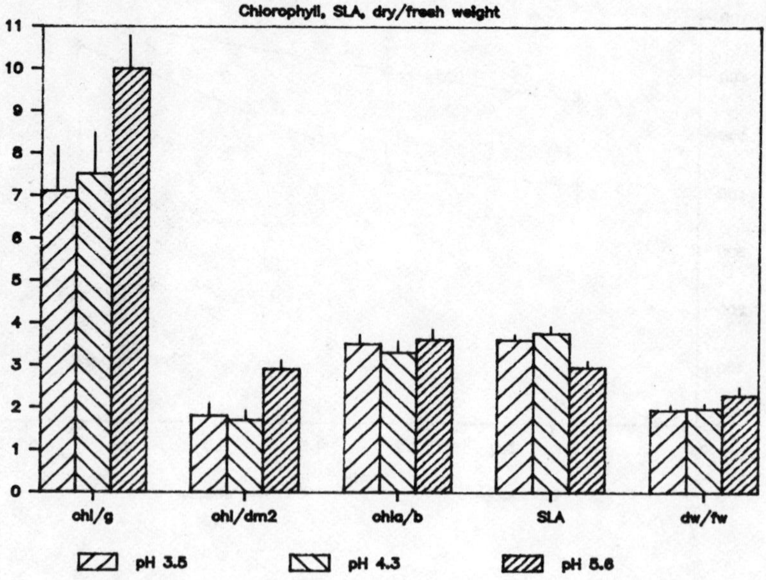

Fig. 3. Total chlorophyll content (Chl/g: mg/gdw, Chl/dm2: mg/dm2), Chl a/b ratio, specific leaf area (SLA, dm2/gdw) and dry/fresh weight ratio (dw/fw: x10) at various pH.

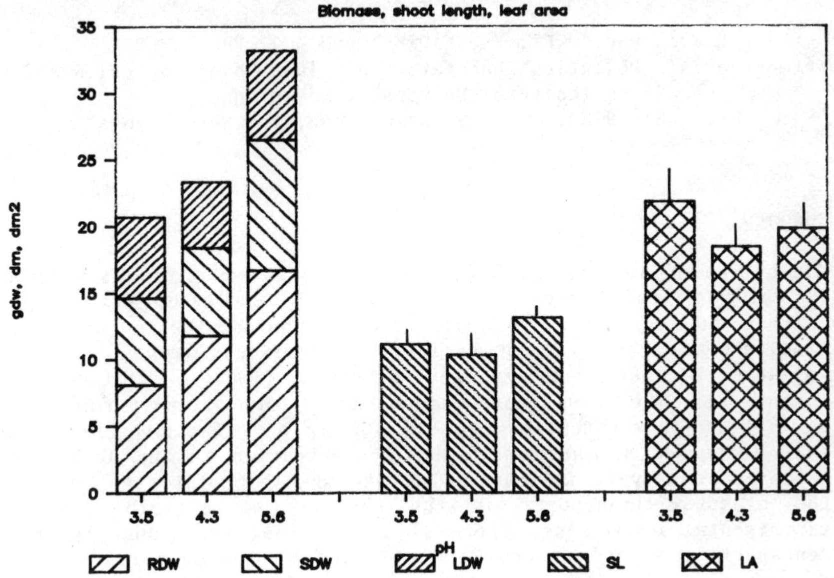

Fig. 4. Mean root (RDW), shoot (SDW) and leaf (LDW) dry weight (g), total shoot lenght (LS : dm) and leaf area (A : dm2) per plant at various pH.

THE UPTAKE OF ATMOSPHERIC AMMONIA BY LEAVES

L.W.A. van HOVE[1], E.H. ADEMA[1] and W.J. VREDENBERG[2]
Department of Air Pollution[1]/Laboratory of Plant Physiological Research[2]
Agricultural University Wageningen
P.O. Box 8129, 6700 EV Wageningen, The Netherlands

Summary

The uptake of ammonia (NH_3) by leaves is studied using a leaf chamber, in which a leaf attached to the plant is enclosed. With this leaf chamber NH_3 uptake, transpiration and photosynthesis are measured simultaneously. In a first series of experiments leaves of Phaseolus vulgaris L. were exposed for a short time to different light intensities. The leaves appeared to have a high affinity for NH_3. Even at high NH_3 concentrations (up to 400 $\mu g\ m^{-3}$) the NH_3 flux into the leaf increased linearly with NH_3 concentration in the leaf chamber. Resistance analysis indicated that NH_3 transport into the leaf is via the stomata: transport via the cuticle is negligible under the experimental conditions. There is no internal resistance against NH_3 transport. Similar results were obtained with poplar leaves (Populus euramericana L.). Also the adsorption of NH_3 on the leaf surface is studied. Preliminary measurements showed that the adsorbed quantity of NH_3 is strongly dependent on relative humidity of the air. More over the adsorbed quantity appeared to be proportional with NH_3 concentration.

1. INTRODUCTION

Serious dieback of forests and strong decline of plant species is observed in areas in the Netherlands, where intensive livestock breeding is concentrated. There is increasing evidence, that the high emission of ammonia (NH_3), caused by NH_3 volatilization of animal manure, is to a large extent responsible for these effects (3,4,7).

It is estimated that about two third of the total deposition of NH_3 is via dry deposition (1). Many aspects of the dry deposition of NH_3 on vegetation surfaces are uncertain. In particular more information is needed about transport processes within plant canopies and quantities ad- and absorbed by foliar surfaces under different environmental conditions. Information about these processes is required for a better understanding of the mechanism and effects of NH_3 deposition on vegetation.

In the present study the uptake of atmospheric NH_3 by leaves is analysed. The objective is to develop a mechanistic model describing this process. The study contains the following subjects: 1) the relationship between concentration, transport resistances and fluxdensities at a short and long term exposure of NH_3, 2) the influence of environmental variables and plant properties on this relationship and 3) the interaction with other gaseous pollutants such as sulphur dioxide.

A simulation model for the dry deposition of NH_3 on vegetation could be developed by incorporating this model in already existing micrometeorological models such as 'microweather' (5).

2. THE MODEL

In analogy with Ohm's electricity law the uptake of NH_3 can be described as a flux (current) driven by a difference in gas concentration (potential difference) and restricted by diffusion resistances of the leaf boundary layer, stomata and cuticle. Figure 1 shows a transverse section through a typical leaf and gives the resistance analogue describing the NH_3 uptake by the leaf.

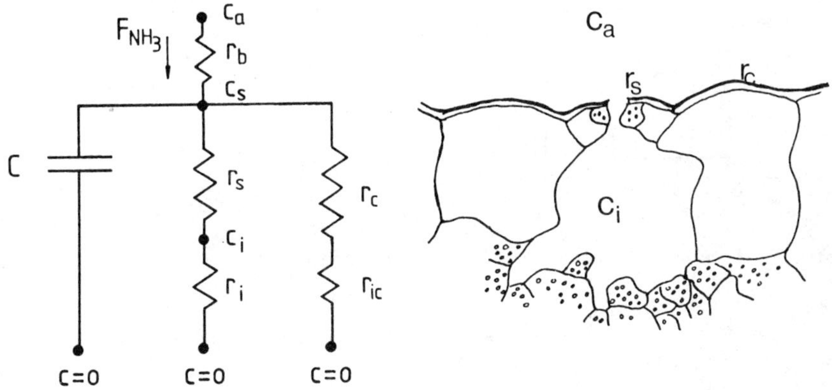

Figure 1. A resistance analogue for the uptake of NH_3 by one side of a leaf. F_{NH_3} is the total flux density towards the leaf. C_a, C_s and C_1 are NH_3 concentrations in the leaf chamber, at the leaf surface and the substomatal cavity of the leaf respectively. r_b is the boundary layer resistance of the leaf, r_s the stomatal resistance and r_c the cuticular resistance. r_i and r_{ic} are internal resistances. The capacity (C) represents surface sorption of NH_3.

3. EXPERIMENTAL PROCEDURE

The uptake of NH_3 is measured using a leaf chamber, in which a leaf attached to the plant is enclosed. Both the plant and the leaf chamber are housed in a controlled environment room and illuminated by two high pressure iodine vapour lamps (Philips HPLI 400 W). Light intensity of the lamps can be varied. Figure 2 gives a schematic overview of the fumigation system.

A continuous flow of conditioned air, containing a known amount of NH_3, is led through the leaf chamber. At the inlet and outlet of the chamber the NH_3 concentration is measured. The fluxdensity is calculated from the difference between inlet and outlet concentration (ΔC in g m^{-3}), the gas flow rate (f in m^3 s^{-1}) and leaf area (A in m^2). In formula:

$$F = (f \times \Delta C) / A \ (g\ m^{-2}\ s^{-1}) \qquad (1)$$

Simultanuously, transpiration and photosynthesis of the leaf are measured. In this way the fluxdensity of NH_3 can be directly related to stomatal behaviour and photosynthesis of the leaves.

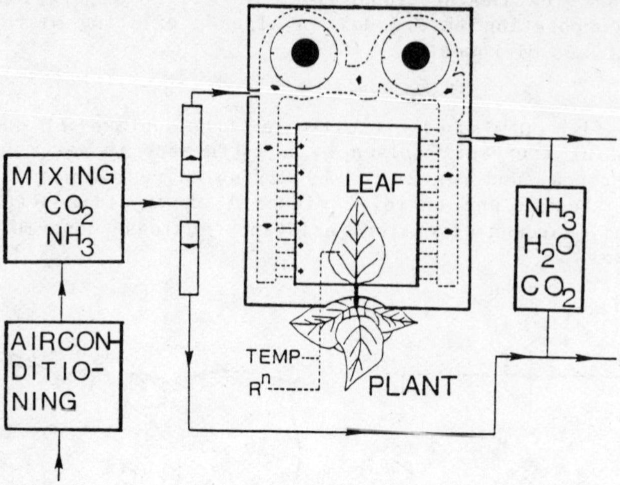

Figure 2. Schematic overview of the fumigation system.

4. MATHEMATICAL ANALYSIS

In case of a steady state situation in the leaf chamber the total flux-density of NH_3 towards the leaf may be written as:

$$F = C_a \,/\, r_t \qquad (2)$$

$$r_t = r_b + ((r_s + r_i)^{-1} + (r_c + r_{ic})^{-1})^{-1} \qquad (3)$$

r_t is obtained from equation (1). From the rate of transpiration the stomatal resistance for NH_3 can be derived (6,8). r_b is obtained by measuring the evaporation of a water saturated piece of filter paper of the same shape and size as the leaf. The resistances r_i and $(r_c + r_{ic})$ are determined indirectly by using a graphical method applied by Black and Unsworth (2).

5. RESULTS AND DISCUSSION

The NH_3 uptake by leaves exposed for a short time to NH_3.

The experiments are carried out with bean plants (Phaseolus vulgaris L.) and with poplar shoots (Populus euramericana L.) obtained from cuttings. In a first series of experiments bean leaves were exposed for 9 hours to different NH_3 concentrations, at different light intensities. The results are shown in figure 3.

The leaves have a high affinity for NH_3 in the light. Even at high NH_3 concentrations the NH_3 flux into the leaf increased linearly with NH_3 concentration in the leaf chamber. Resistance analysis indicated that at a short term exposure NH_3 uptake is via the stomata: transport through the cuticle is negligible under the experimental conditions. There is no internal resistance against NH_3 transport (6). Similar results were obtained with poplar leaves. So NH_3 flux into the leaf can be predicted by data on the boundary layer resistance, stomatal conductance of H_2O and ambient NH_3 concentration.

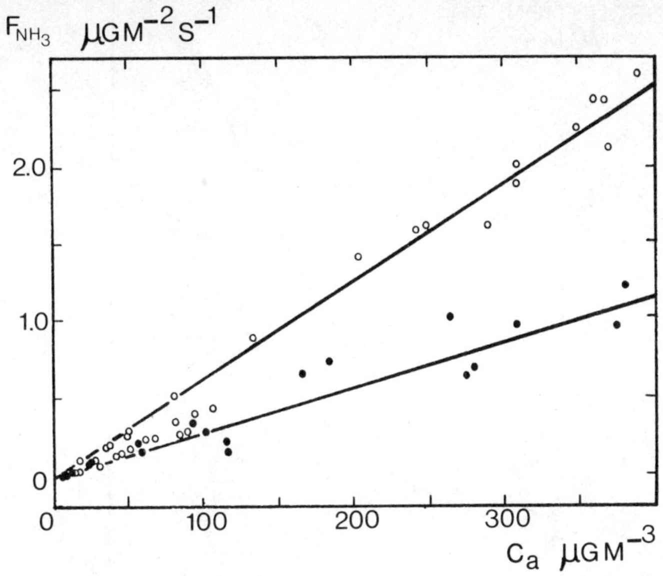

Figure 3. The rate of NH_3 uptake by leaves of Phaseolus vulgaris L. at a light intensity of 11 (.) and 70 $W.m^{-2}$ ( ) (PAR) as a function of ambient concentration (c_a) (temperature: 22 - 25 °C); relative humidity: 60% ± 5%.

The adsorption of NH_3 on the leaf surface.

In determining NH_3 adsorption on the leaf surface uptake of NH_3 via the stomata have to be avoided. This is achieved by cutting off the leaf from the plant, after which the petiole is held in an ABA solution. After enclosure in the leaf chamber the leaf is kept in the dark to ensure complete stomatal closure. When NH_3 is injected into the gas flow passing the leaf chamber, the outlet concentration (C_{out}) as a function of time (t) can be described by the expression:

$$C_{out} = C_{in} \ (1 - e^{-kt}), \qquad (4)$$

where C_{in} is the inlet concentration and k a time constant (sec^{-1}). The obtained curve contains however NH_3 adsorption on the internal surfaces of the leaf chamber as well. Therefore, a similar curve is determined for the empty chamber. From the difference between the two curves the quantity of adsorbed NH_3 can be calculated. The first results obtained with bean leaves are shown in figure 4.

The adsorbed quantity of NH_3 is strongly dependent on relative humidity of the air. Also the adsorbed quantity appeared to be proportional with NH_3 concentration. Evidently, an equilibrium between the adsorbed quantity of NH_3 and its concentration in the atmosphere exist. This situation can be compared with the dissolving of NH_3 in water. It is therefore postulated that the leaf surface is covered or partly covered with a waterfilm. This waterfilm may play an important role in the deposition of other air pollutant compounds like SO_2 on vegetation.

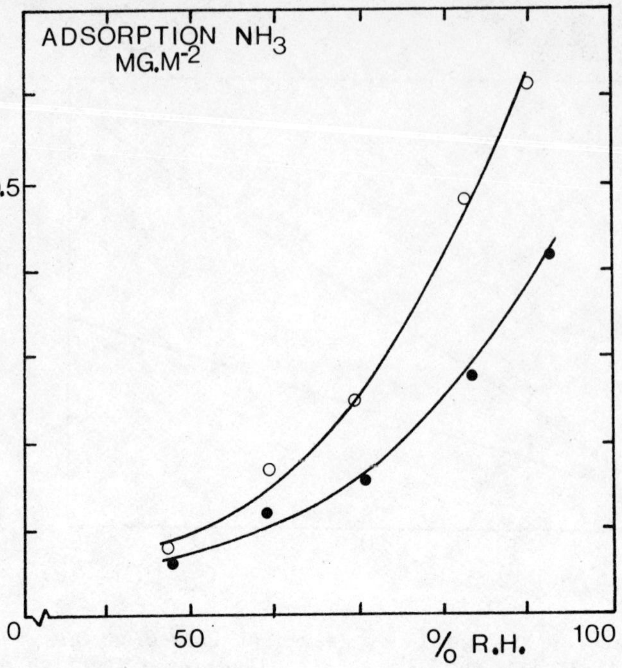

Figure 4. NH_3 adsorption on the leaf surface of Phaseolus vulgaris L. as a function of relative humidity of the air. NH_3 concentrations were 56 $\mu g\ m^{-3}$ (.) and 101 $\mu g\ m^{-3}$ (°).

REFERENCES

(1) AALST, R.M. van (1984). Depositie van verzurende stoffen in Nederland. In: Zuren regen, oorzaken, effecten en beleid, pp. 66-70. Proc. Symp. Zure Regen, 's Hertogenbosch, 17-18 November 1983. Eds. E.H. Adema and J. van Ham. Pudoc, Wageningen.

(2) BLACK, V.J. and UNSWORTH, W.H. (1979). Resistance analysis of sulphur dioxide fluxes to Vicia faba. Nature 282: 68-69.

(3) BREEMEN, N. van, BURROUGH, P.A., VELTHORST, E.J., DOBBEN, H.F. van, WIT, T. de, RIDDER, T.B. and REIJNDERS, H.F.R. (1983). Soil acidification from atmospheric ammonium sulphate in forest canopy throughfall. Nature 299: 548-550.

(4) EERDEN, L.J.M. van der (1982). Toxicity of ammonia to plants. Agriculture and Environment 7: 223-235.

(5) GOUDRIAAN, J. (1979). 'Microweather', simulation model, applied to a forest. In: Comparison of forest water and energy exchange models. S. Halldin (ed.): 47-57.

(6) HOVE, L.W.A. van, KOOPS, A.J., ADEMA, E.H., VREDENBERG, W.J. PIETERS, G.A. (1987). Analysis of the uptake of atmospheric ammonia by leaves of Phaseolus vulgaris L. (submitted).

(7) ROELOFS, J.G.M., KEMPERS, A.J., HOUDIJK, A.L.F.M. and JANSEN, J. (1985). The effect of air-borne ammoniumsulphate on Pinus nigra var. maritima in the Netherlands. Plant and Soil, 84: 45-56.

(8) UNSWORTH, M.H., BISCOE, P.V. and BLACK, V. (1976). Analysis of gas exchange between plants and polluted atmospheres. In: Effects of air pollutants on plants, pp. 5-16, T.A. Mansfield (ed.). Cambridge University Press.

INVESTIGATING EFFECTS OF AIR POLLUTION BY MEASURING THE WATERFLOW VELOCITY IN TREES

J.H.A. van WAKEREN, H. VISSER and F.B.J. KOOPS
Environmental Research Department, N.V. KEMA
P.O. Box 9035, 6800 ET Arnhem, The Netherlands

Summary

During two years the waterflow velocity has been monitored in trees with an automized measurement grid. Time series analysis shows significant correlation with meteorological parameters. The correlation with air pollution is statistically not significant.

INTRODUCTION

Within the framework of research on effects of acid rain and air pollution on trees, we have measured the waterflow velocity in trees like oak, Douglas-fir and beech. The waterflow in a tree can be considered as an activity of this tree and it is to be expected that this activity may depend on parameters like the vitality of the tree itself (the condition of fibrils, stem and leaves) and the environment e.g. the meteorological parameters and air pollutant components like SO_2, NO_x and O_3.

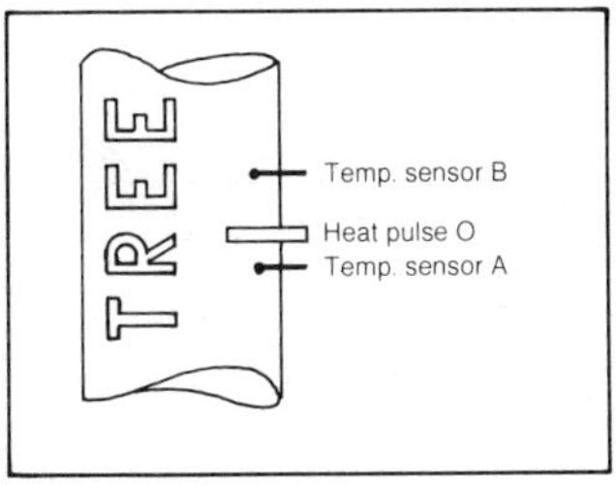

Fig. 1 A small amount of heat is generated in the tree bij sending an electric pulse through the resistor mounted in the xylem of the stem of the tree at position O. One temperature sensor is mounted at position A below O, another at position B above O

PRINCIPLE OF MEASUREMENT

The principle of measurement is based on the Heat Pulse Velocity method (HPV). This HPV method functions as follows. A small pulse of heat (10 Joule) is injected in the stem of a tree by means of an electric current through a resistor which is implanted in the xylem. Two sensitive temperature sensors are mounted a-symmetrically above and below the place of heat input. The lower sensor has a distance of 8 mm from the heater and the upper sensor has a distance of 10 mm from the heater, see Fig. 1. The injected heat will move through the xylem by two different mechanisms: diffusion and waterflow. Because the waterflow velocity in a tree is relatively slow, say in the order of 10 cm per hour, the diffusion term is much larger than the waterflow term. A sketch of the temperature increase in the xylem at a time τ_1 after the heat pulse is visible in Fig. 2.

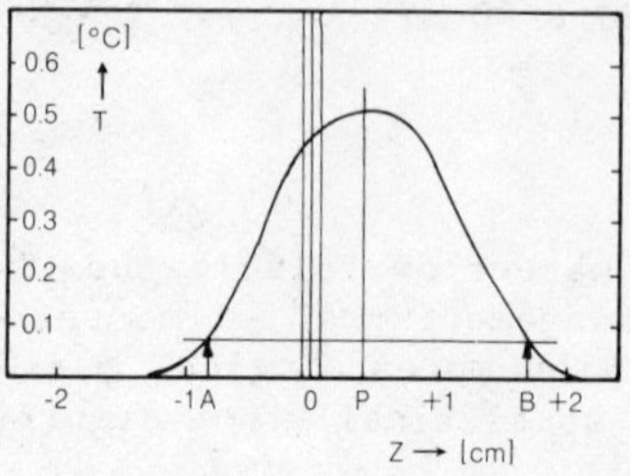

Fig. 2 Propagation of the heat pulse in the tree. At this moment the temperature increase at sensor positions A and B is equal. The water in the tree has travelled a distance OP from the moment of injecting the heat pulse

The bell shaped curve shows the temperature increase which has spreadout by diffusion. The maximum temperature is found at point P at the centre of the bell shape. The length OP is the distance the water has run during time τ_1, this distance is in the order of only 1 mm. At a certain time τ after the heat pulse the two temperature sensors placed e.g. in A and B will both measure an equal increase of temperature. Noting that $OP = (OB - OA)/2$, the heat pulse velocity v can be calculated as $v = (OB - OA)/2\tau$.

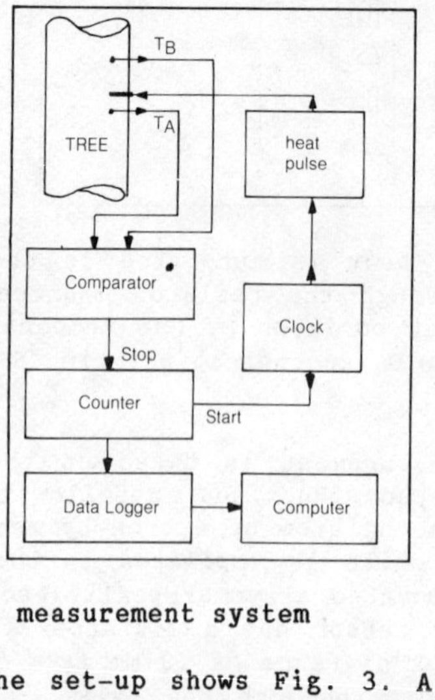

Fig. 3 Flow chart of the measurement system

A flow chart of the set-up shows Fig. 3. A clock starts the heat pulse and a counter. The signals of the temperature sensors are continuously compared and at the moment that both signals are equal, the counter stops. The value of this counter is stored on magnetic tape. The measurement range is $0.5 < v < 100$ cm/h. Each hour one heat pulse of one second is sent to each tree. An automized measuring grid is set up, which monitors 32 trees.

CALIBRATION

The system has been calibrated in the laboratory by means of an artificial tree, consisting of a plastic pipe with heater and sensors. A precision peristaltic pump controlled an adjustable flow of water. The calibration showed that the real waterflow velocity is systematically a factor 2,5 higher than the velocity determined by the heat pulse method. This factor shows good agreement with a computer simulation of Swanson who explained these differences by the geometry of heater and sensors in his study.

The length of the heater is 10 mm and the lengths of the sensors are 1 mm; all diameters are 2 mm. The mounting of heater and sensors in the xylem in the tree is critical. It is important that the holes in the tree must be drilled precisely to avoid errors larger than about 30%.

MEASUREMENTS

During two years measurements have been performed with 32 waterflow velocity measuring units. A characteristic picture during a summer period of the heat pulse velocity as a function of time is given in Fig. 4.

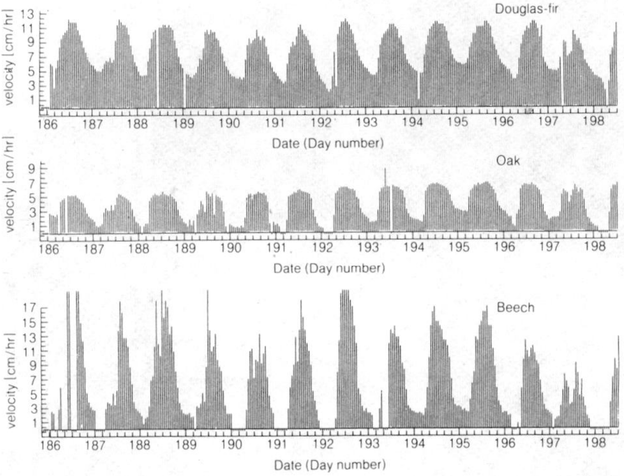

Fig. 4 Measurements of the waterflow velocity in douglas-fir, oak and beech

A day-night rhythm is clearly visible. During periods of rain the velocity slows down or even stops. During winter periods the waterflow velocity falls down below 0.5 cm/h, which is the minimum in the measuring range. During summer nights, when it is not raining, the waterflow does not stop, but maintains a value of about 2 cm/h. We also have connected 8 waterflow velocity measuring units on one single tree to detect inhomogenities in height and azimuth. The mutual differences appeared to be within ± 30%, which can be attributed to inaccuracy due to errors in the accuracy of drilling the holes in the xylem to mount the sensors.

STATISTICAL RESULTS

Time series analysis is used to detect statistical correlation between waterflow velocity, air temperature, relative humidity, vapour pressure deficit, photosynthetic active radiation, and concentration of the air pollutants SO_2, NO_2 and O_3.

Statistical correlation ρ^2 of the waterflow velocity with:
NO_2, negative $\rho^2 = 4\%$
O_3, positive $\rho^2 = 30\%$
SO_2, positive $\rho^2 = 2\%$

However, if we take the meteorological parameters into account, none of the calculated correlations are significant.

Statistical correlation ρ^2 of the waterflow velocity with:
vapour pressure deficit, positive $\rho^2 = 53\%$
photosynthetic active radiation, positive $\rho^2 = 46\%$
air temperature, positive $\rho^2 = 37\%$

Vapour pressure deficit and photosynthetic active radiation explain 60% of the variation in the waterflow velocity.

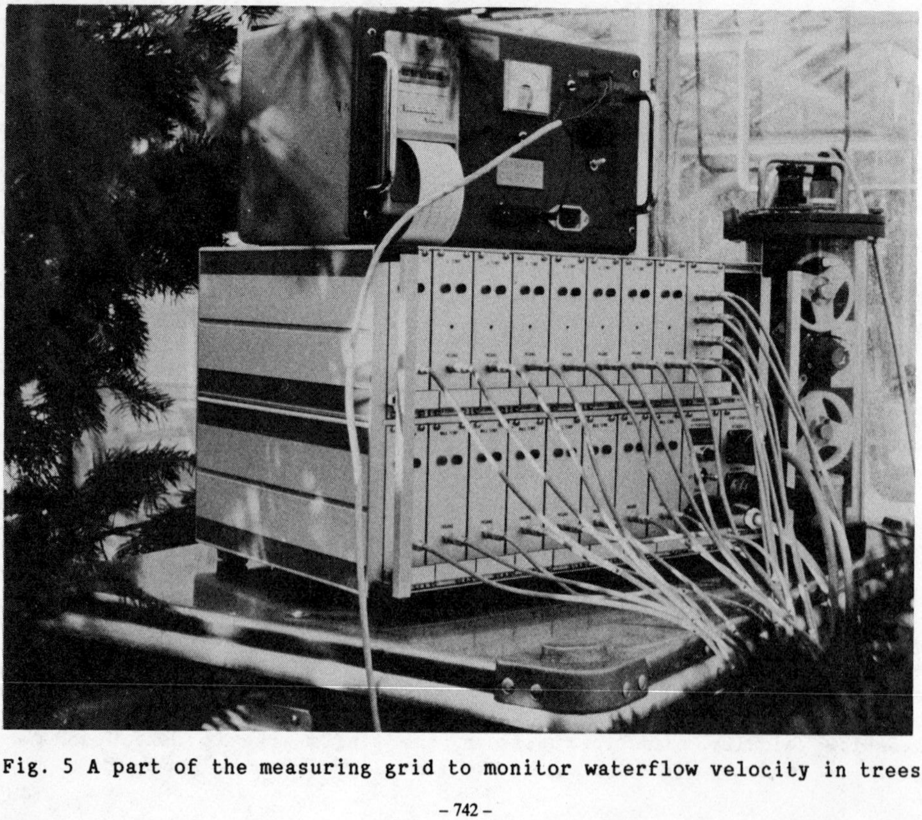

Fig. 5 A part of the measuring grid to monitor waterflow velocity in trees

UNTERSUCHUNGEN ZUM KOHLENHYDRATSTOFFWECHSEL AN FICHTE (PICEA ABIES KARST.) AUS KLIMAKAMMEREXPERIMENTEN UND AUS GESCHÄDIGTEN WALDBESTÄNDEN

K. Vogels und T. Lambrecht
Universität GHS Essen
Institut für Angewandte Botanik

Abstract

In different seasons the carbohydrate metabolisms of slightly and severely damaged spruce trees taken from the mainly SO_2 polluted Fichtel Gebirge and the predominantly O_3 damaged Black Forest respectively were analysed and the results compared with those of SO_2 and O_3 fumigation experiments. The physiological capacities of the test trees were determined by measuring the photosynthetic rates. The carbohydrate metabolisms were analysed by measuring the contents of selected mono-, di- and trisaccharides and of total sugar and starch in the needles. The amounts of amyloplasts in the thin roots were to indicate the changes in partitioning. The results of the encymatic determination of starch in the needles corresponded to histological investigations of the extension of starch granules in the chloroplasts.

1. EINLEITUNG

Der Gesundheitszustand des Waldes in der Bundesrepublik Deutschland hat sich auch in den letzten Jahren weiter verschlechtert. Trotz günstiger Witterungsbedingungen in den Jahren von 1984 bis 1986 stieg das Ausmaß der Schädigung auf 54% der gesamten Waldfläche an (1). Besonders betroffen sind der Schwarzwald, die Bayerischen Alpen sowie der Bayerische Wald, das Fichtelgebirge und der Fränkische Wald. Hier beträgt der Anteil der geschädigten Wälder über 60%; 30% weisen mittlere bis starke Schädigungen auf. Überdurchschnittlich zugenommen haben die Schäden im mittleren Teil der Bundesrepublik (geschädigte Waldfläche 50-60%), vornehmlich im westfälischen, niedersächsischen und hessischen Bergland.

Als Ursache für die großräumigen Waldschäden kommen nach derzeitigem Wissen vornehmlich gasförmige Immissionen in Betracht, die entweder direkt schadauslösend oder praedisponierend gegenüber biotischen und abiotischen Faktoren wirken (2, 3, 4, 5, 6, 7). Die gegenwärtige Immissionssituation in der Bundesrepublik Deutschland ist durch das Auftreten mehrerer ökotoxikologisch relevanter Stoffe in relativ niedrigen Konzentrationen gekennzeichnet; dennoch lassen sich in den genannten Waldschadensgebieten unterschiedlich dominierende Immissionskomponenten feststellen. Während die Vegetation im Südwesten der Bundesrepublik (Schwarzwald) vornehmlich durch hohe Ozon-Konzentrationen gefährdet ist, charakterisieren erhöhte Schwefeldioxid-Konzentrationen die Belastungen in den südöstlichen Mittelgebirgen wie Fichtelgebirge und Bayerischer Wald. Im mittleren Teil des Landes (Weserbergland, Hessisches Bergland) sind dagegen beide Schadkomponenten annähernd gleich stark vertreten.

Im Rahmen eines von der Europäischen Kommission geförderten Forschungsvorhabens (Projekt-Nr. ENV 846 D) wurden unter Beachtung der o.g. Immissionstypen an ausgewählten Standorten im Egge-Gebirge (Weserbergland), im Kaufunger Wald (Hessisches Bergland), im Fichtelgebirge und im Südschwarzwald physiologische und histologische Untersuchungen an gering (Schadensklasse 0-1) und stark geschädigten Fichten (Schadensklasse 2-3) vorgenommen. Die gewonnenen Ergebnisse wurden mit Untersuchungsbefunden an Pflanzen aus parallel verlaufenden Klimakammerexperimenten mit chronischen SO_2 und O_3 Belastungen verglichen.

Ziel der hier vorgestellten Untersuchungen ist es, die Wirkungen gasförmiger Immissionen auf den Kohlenhydratstoffwechsel aufzuzeigen. Im Klimakammerexperiment soll dabei nachgewiesen werden, wie sich eine schadgasbedingt veränderte CO_2-Assimilation auf den Zucker- und Stärkehaushalt der Nadeln und als Folge von Veränderungen im "partitioning", d.h. in der Aufteilung der Assimilate, auf den Amyloplastenanteil in den Wurzeln auswirkt. In Freilanderhebungen wird darüber hinaus der Einfluß der unterschiedlichen jahreszeitlichen Bedingungen auf den Kohlenhydratstoffwechsel aufgezeigt.

2. MATERIAL UND METHODEN

Dreijährige Fichtenklone der Provenienz Kartuzy wurden in der Klimakammer 20 Wochen mit 258 ug SO_2/m^3 (kontinuierlich) und 176 ug O_3/m^3 (14 Stunden/Tag) einzeln und in Kombination begast.
Die geobiologischen Erhebungen, die im Rahmen dieser Arbeit auf die Standorte im Fichtelgebirge (Schneeberg, 1000 m über NN) und Südschwarzwald (Belchen, 1050 m über NN) beschränkt bleiben, wurden an 40-60-jährigen Fichten durchgeführt. Die Probenahme erfolgte mit einer fahrbaren Zweiradleiter im oberen Kronenbereich (10-14 m über Grund) zu unterschiedlichen Jahreszeiten.
Zur Durchführung der physiologischen Untersuchungen wurden die geernteten Nadeln in flüssigem Stickstoff fixiert, gefriergetrocknet und gemahlen. Die Zucker wurden mit Äthanol extrahiert und im Rotationsverdampfer eingeengt. Die Bestimmung der Glucose, Fructose, Saccharose und Raffinose erfolgte mit der Hochdruckflüssigkeitschromatographie; enzymatische Tests sicherten die Ergebnisse ab. Der Gesamtzuckergehalt wurde mit Anthron/Schwefelsäure nach einem von Ebell (9) beschriebenen Verfahren photometrisch ermittelt. Die nach der Alkoholextraktion im Rückstand verbliebene Stärke wurde mit Dimethylsulfoxid gelöst, hydrolysiert und enzymatisch als Glucose nach Keppler und Decker (10) bestimmt.
Die Erfassung der Amyloplastenhäufigkeit in den Feinwurzeln erfolgte durch eine rechnergesteuerte Bildauswertung. Hierzu wurden mit dem Gefriermikrotom Semidünnschnitte angefertigt und

mit Jod-Jod-Kalium angefärbt. Eine Videokamera übertrug das mikroskopische Bild auf einen Kleincomputer, der durch ein Zusatzgerät die einzelnen Bildpunkte vier verschiedenen Graustufen zuordnen und deren Flächenanteil berechnen konnte. Ein angeschlossenes Graphiktablett ermöglichte es, die Zellumina, Zellwände und Amyloplasten manuell voneinander abzugrenzen, unterschiedlich anzufärben und somit die Genauigkeit der Auswertung zu verbessern.

3. ERGEBNISSE

3.1 Klimakammeruntersuchungen

Eine 20-wöchige Begasung mit 258 $\mu g\ SO_2/m^3$ und 176 $\mu g\ O_3/m^3$ bewirkte eine Abnahme der Photosyntheserate um 38% bzw. 10%. Nach Kombinationsbegasung war die Photosyntheseleistung um 52% vermindert; eine, wenn auch nur geringe, synergistische Wirkung (Abb. 1A). Als Folge der Begasung und der daraus resultierenden Minderung der Kohlenstoff-Assimilation konnte eine deutliche Abnahme der Amyloplastenhäufigkeit in den Wurzeln beobachtet werden (Abb. 1B). Ebenfalls vermindert war der Stärkegehalt in den Nadeln nach Einzelbegasung (Abb. 1E, F). Nach Kombinationsbegasung konnte im Vergleich zu den Einzelbegasungen ein Anstieg im Stärkegehalt verzeichnet werden; ein tendentiell auch nach Klimakammerexperimenten mit verstärkt akut wirkenden Belastungen beobachtetes Ergebnis. Die reduzierte Photosyntheseleistung hätte einen eingeschränkten Zuckergehalt erwarten lassen. Tatsächlich waren die Gehalte in den Alt- und Neutrieben jedoch erhöht (Abb. 1C, D). Maximalwerte wurden jeweils nach O_3-Begasung beobachtet, bei jenem Versuchsglied, bei dem die Photosyntheseleistung nur geringfügig vermindert war. Begasungsversuche mit höheren Konzentrationen als oben angegeben führten in den belasteten Pflanzen durchweg zu niedrigeren Zuckergehalten als in den Kontrollen.

Faßt man die gewonnenen Ergebnisse aus den Klimakammerexperimenten zusammen, so steht nach einer eher chronisch wirkenden Begasung den Minderungen der Photosyntheseleistung und Amyloplastenhäufigkeit ein erhöhter Gesamtzuckergehalt gegenüber. Dies läßt darauf schließen, daß der Transport der Assimilate vom Ort der Bildung zum Ort der Speicherung gestört ist; eine Schädigung oder Beeinträchtigung des Phloems muß in Betracht gezogen werden. Nach Begasungsversuchen mit verstärkt akut wirkenden Belastungen wurde dagegen ausschließlich ein geringerer Zuckergehalt als in den Kontrollen ermittelt. Hier war die Photosyntheseleistung derart eingeschränkt, daß die Assimilatproduktion nicht einmal mehr zur Aufrechterhaltung des Stoffwechselmetabolismus in den Nadeln ausreichte.

3.2 Freilanderhebungen

Bereits in früheren Veröffentlichungen konnte nachgewiesen werden (11, 12), daß sowohl im Fichtelgebirge als auch im Schwarzwald die gering geschädigten Referenzbäume eine deutlich höhere Photosyntheseleistung aufwiesen als die mittel- bis stark geschädigten. Obwohl auch bei den Referenzbäumen die Photosyntheseraten mit zunehmendem Nadelalter abnahmen, blieben sie dennoch auf einem hohen Niveau. Da die entsprechenden Atmungsraten ebenfalls leicht vermindert waren, verblieb ins-

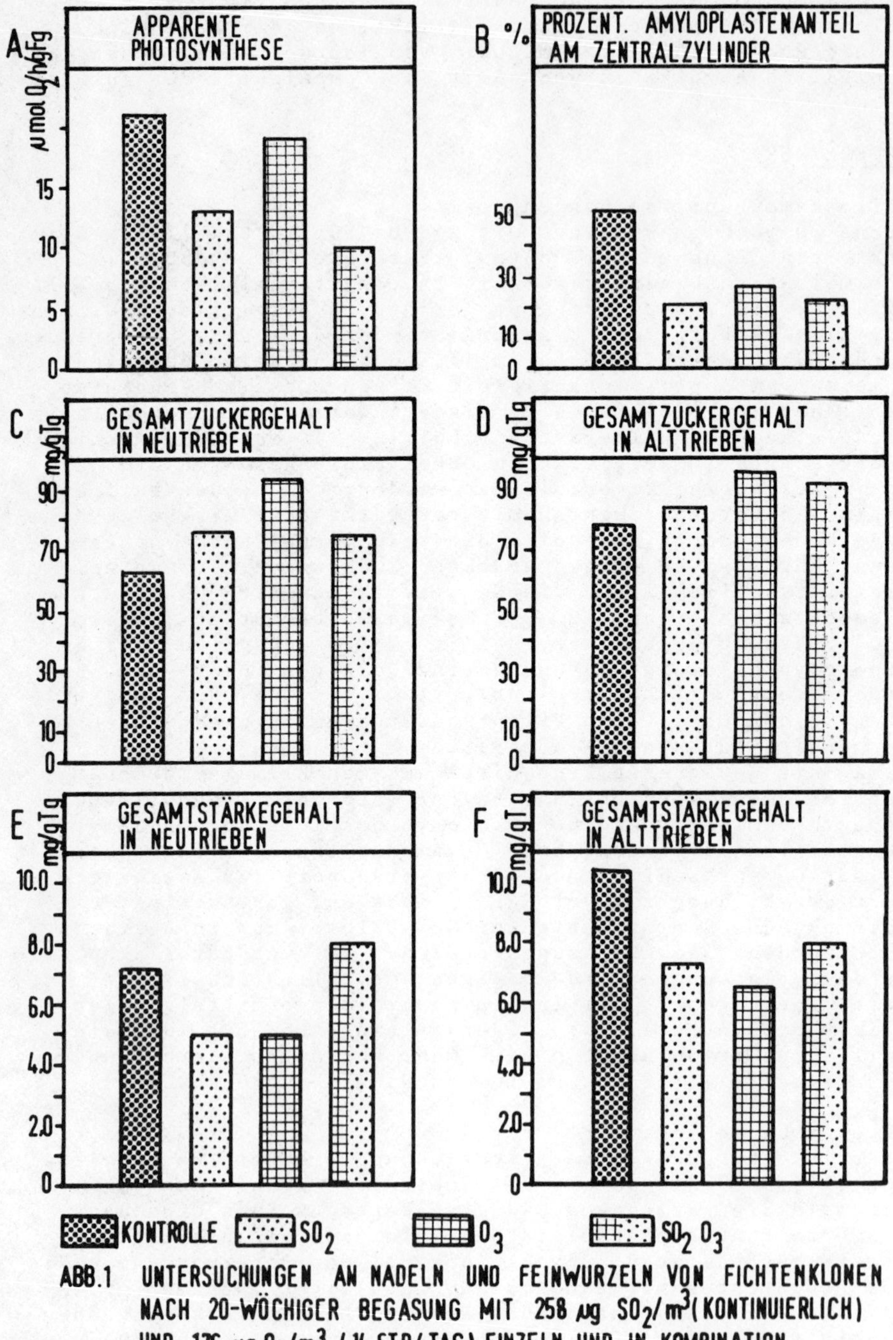

ABB.1 UNTERSUCHUNGEN AN NADELN UND FEINWURZELN VON FICHTENKLONEN NACH 20-WÖCHIGER BEGASUNG MIT 258 µg SO_2/m^3 (KONTINUIERLICH) UND 176 µg O_3/m^3 (14 STD/TAG) EINZELN UND IN KOMBINATION

gesamt bei allen untersuchten Nadeljahrgängen eine positive Stoffbilanz. Dagegen war in den stark geschädigten Versuchsbäumen die Photosyntheseleistung mit ansteigendem Nadelalter zunehmend eingeschränkt, während die entsprechenden Dunkelatmungsraten anstiegen, so daß die Ernährung der Fichten auf 2-3 Nadeljahrgänge beschränkt blieb.
Vor diesem Hintergrund müssen die Ergebnisse der Untersuchungen über die Amyloplastenhäufigkeit in den Feinwurzeln gesehen werden. Offensichtlich durch die hohe Photosynthesedepression bedingt, zeichneten sich die stark geschädigten Fichten durch erhebliche Einbußen in der Häufigkeit der stärkespeichernden Plastiden aus (Abb. 2). Diese Minderungen beliefen sich im Schwarzwald auf 50%, im Fichtelgebirge sogar auf 63%.

In weit geringerem Maße als bei den Amyloplasten waren bei den Zuckern Unterschiede zwischen gering und stark geschädigten Bäumen zu beobachten. Die graphische Darstellung blieb daher auf den Glucose-, Fructose- und Saccharosegehalt der Fichtennadeln im Schwarzwald beschränkt (Abb. 3). Eine Analyse der einzelnen Zuckerkomponenten läßt erkennen, daß vornehmlich jahresperiodische und nur geringe immissionsbedingte Schwankungen auftraten. So konnte die Raffinose (o. Abb.) mit dem in Kap. 2 beschriebenen Verfahren nur im Dezember nachgewiesen werden, wobei die Referenzbäume generell um 10-20% höhere Werte als die Schadbäume aufwiesen. Während bei der Fructose insgesamt nur geringfügige Veränderungen beobachtet werden konnten, ereichte die Glucose am Ende des Neuaustriebes und im Dezember hohe, im Sommer dagegen niedrige Gehalte. Die Saccharose war am Ende des Neuaustriebes nur in geringen Mengen nachweisbar, der Gehalt stieg aber im Laufe des Jahres weitgehend kontinuierlich an und erreichte im Dezember einen maximalen Wert.

Der Vergleich zwischen dem weitgehend gesunden und dem kranken Baum läßt erkennen, daß die geschädigte Fichte im Dezember niedrigere, im Juli und September dagegen höhere Zuckergehalte als der Referenzbaum aufwies. Im Sommer, zur Zeit der höchsten physiologischen Aktivität, war somit trotz eingeschränkter Photosyntheseleistung ein Anstau der gelösten Zucker in den geschädigten Fichten zu verzeichnen. Analog zu den Klimakammerexperimenten deutet auch dieses Ergebnis auf Störungen im Assimilattransport hin.
Den Zuckergehalten entgegengesetzt verhielten sich die jahresperiodischen Schwankungen im Stärkegehalt (Abb. 4). Die höchsten Werte wurden im Mai und zum Teil im Juni vor und am Ende der Ausbildung von Neutrieben beobachtet; im September war nur wenig, im Dezember überhaupt keine Stärke mehr nachweisbar. Während die Zucker zum Winter hin anstiegen, nahm der Gehalt an Stärke ab. Ein Vergleich zwischen gering und stark geschädigten Fichten läßt teilweise erhebliche Unterschiede in der Stärkeakkumulation erkennen, die im Schwarzwald durchweg deutlicher ausgeprägt war als im Fichtelgebirge. Die höchsten Einbußen waren bei den Schadbäumen im Mai zu verzeichnen. Bezogen auf den jeweiligen Referenzbaum konnte im Fichtelgebirge ein um durchschnittlich 36%, im Schwarzwald sogar ein um 68% verminderter Stärkegehalt beobachtet werden. Den geschädigten Fichten standen somit in weit geringerem Maße als den gesunden

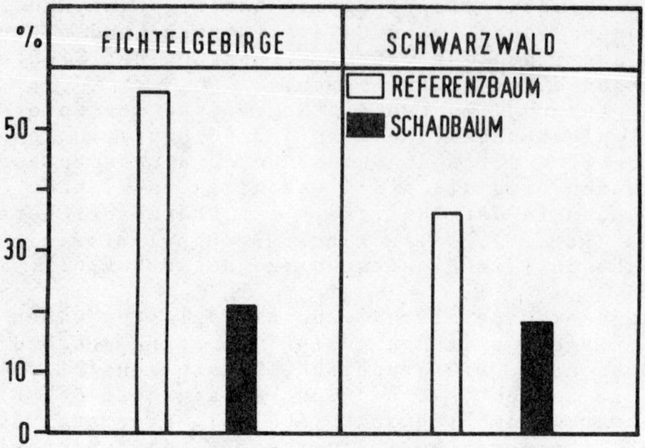

ABB.2 PROZENTUALER AMYLOPLASTENANTEIL AM ZENTRALZYLINDER IN WURZELN DES REFERENZ- UND SCHADBAUMES IM SCHWARZWALD UND IM FICHTELGEBIRGE

Reservesubstanzen zur Ausbildung der Neutriebe zur Verfügung. Die Möglichkeit, im Frühjahr neue und somit gesunde photosynthetisch aktive Nadelmasse auszubilden, war erheblich eingeschränkt.
Im Gegensatz zu den teilweise erheblichen Minderungen der Stärkegehalte im Frühjahr konnte im Sommer ein Stärkestau beobachtet werden. In Übereinstimmung mit den bisher vorgestellten Ergebnissen muß davon ausgegangen werden, daß als Folge von Beeinträchtigungen im Transportsystem Störungen im "partitioning" auftraten.

Die erheblichen jahresperiodischen Schwankungen im Stärkegehalt der Nadeln wurden auch im elektronenmikroskopischen Bild sichtbar (Publikation in Vorbereitung). Während die Stärkegranula vor dem Neuaustrieb im Mai einen erheblichen Anteil am Chloroplasten einnahmen, überwogen im September Stroma- und Granathylakoide. Die geschädigten Fichten wiesen im Sommer einerseits eine erhebliche Degeneration der Chloroplasten auf, andererseits waren die Stärkegranula durchschnittlich größer als in den Referenzbäumen ausgebildet.

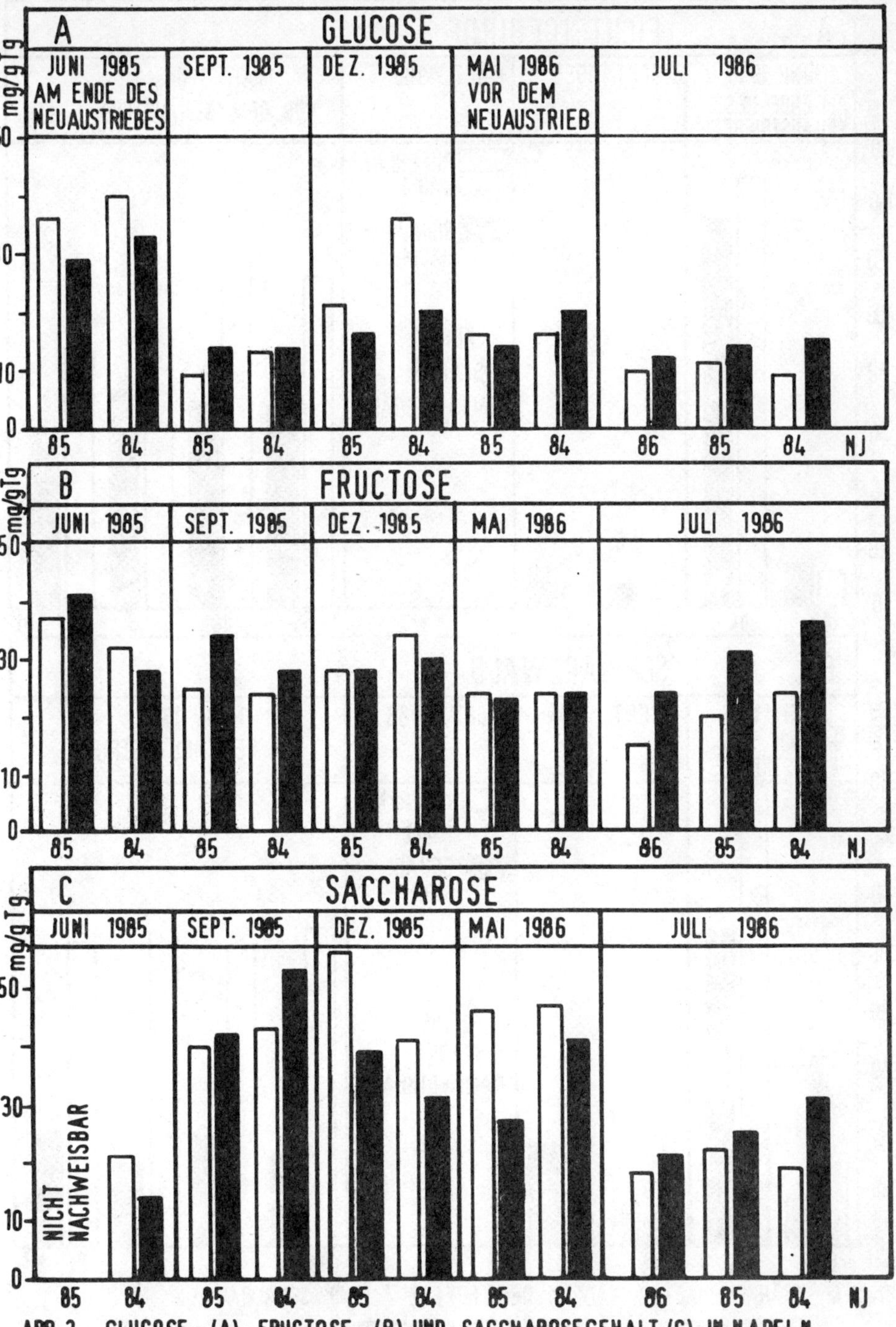

ABB. 3 GLUCOSE-(A), FRUCTOSE-(B) UND SACCHAROSEGEHALT (C) IN NADELN DES REFERENZ- UND SCHADBAUMES IM SCHWARZWALD (NJ = NADELJAHRGANG ▭ REFERENZBAUM ■ SCHADBAUM)

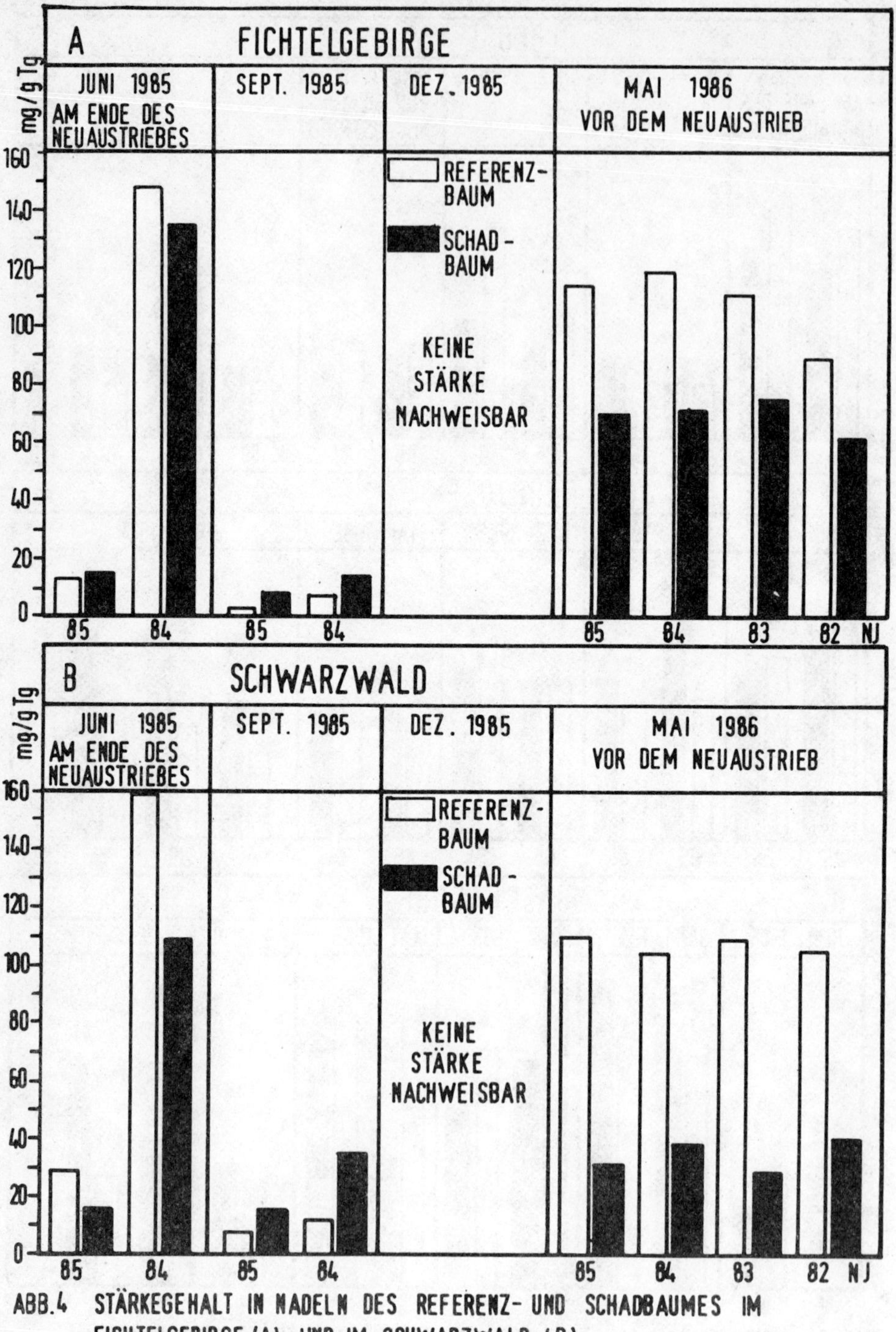

ABB. 4 STÄRKEGEHALT IN NADELN DES REFERENZ- UND SCHADBAUMES IM FICHTELGEBIRGE (A) UND IM SCHWARZWALD (B) (NJ = NADELJAHRGANG)

4. DISKUSSION

Physiologische Untersuchungen zum Kohlenhydratstoffwechsel an Fichten in Abhängigkeit von der Jahreszeit liegen von Jeremias (13) vor. Danach treten im Sommer nur geringfügige Konzentrationsschwankungen bei den beiden Monosacchariden Glucose und Fructose auf, während die Saccharose im Hochsommer ein Maximum erfährt und die Raffinose nicht nachweisbar ist. Im Spätherbst konnte ein Anstieg aller Zucker, vornehmlich der Raffinose, beobachtet werden, der zum Frühjahr hin wieder abnahm. Der Zuckerrückgang von der Winter- zur Sommerphase war mit einer wesentlichen Vermehrung des Stärkegehaltes verbunden.

Auf die Bedeutung der Zucker, vornehmlich der Di- und Trisaccharide für die Frosthärtung der Pflanzen wird von Santarius (14) verwiesen.

Feinstrukturelle Untersuchungen über die jahreszeitlichen Veränderungen des Stärkegehaltes in den Chloroplasten von Fichtennadeln wurden von Senser und Beck (15) vorgenommen. Danach werden im Frühjahr, vor dem Neuaustrieb, erhebliche Mengen an Stärke in die Chloroplasten eingelagert, die Plastiden erreichen ihr größtes Volumen und das Thylakoidsystem wird durch die Stärke an den Rand gedrückt. Im Sommer bildet sich ein dichtes Membransystem aus Grana- und Stromathylakoiden aus, in dem tagsüber Assimilationsstärke auf- und nachts wieder abgebaut wird. Im Spätherbst, mit Einsetzen des Wechsels von der Stärke- zur Oligosaccharidproduktion, ist auch nach Belichtung in den Plastiden keine Assimilationsstärke mehr vorhanden.

Die zitierten Literaturergebnisse über den jahreszeitlichen Verlauf des Kohlenhydratstoffwechsels in unbelasteten Fichten stimmen somit weitgehend mit den eigenen Befunden überein.

Eine von Koziol (16) vorgenommene Zusammenstellung von Arbeiten über den Einfluß von Luftverunreinigungen auf den Kohlenhydratstoffwechsel läßt erkennen, daß unabhängig von der Immissionskomponente alle drei möglichen Wirkungsformen wie Stagnation, Depression und Stimulation der Zucker- und Stärkegehalte auftreten. Allerdings konnten Koziol und Jordan (17) in Begasungsexperimenten an Buschbohne (Phaseolus vulgaris) nachweisen, daß vornehmlich chronisch wirkende Belastungen Steigerungen, akut wirkende Begasungen dagegen Minderungen der Zucker- und Stärkegehalte nach sich ziehen.

Nach Begasungsversuchen mit SO_2 an Buschbohne beobachteten Noyes (18) sowie Teh und Swanson (19) erhebliche Störungen im "partitioning". Als mögliche Ursache vermuteten die Autoren Störungen im Phloemtransport. Ein vorzeitiger Phloemkollaps, verbunden mit einer Hypertrophierung der Strasburger-Zellen, wurde in licht- und elektronenmikroskopischen Untersuchungen an immissionsgeschädigten Fichten und Tannen aus dem Schwarzwald nachgewiesen (20).

Faßt man die Ergebnisse der genannten Autoren und die eigenen Resultate aus Klimakammerbegasungen und Freilanderhebungen zusammen, so zeichnet sich durch Störungen des Kohlenhydratstoffwechsels für die immissionsgeschädigten Fichten im Fichtelgebirge und Schwarzwald folgendes Bild ab:

- Im Sommer, zur Zeit der höchsten photosynthetischen Aktivität, entwickelt sich aufgrund von Schädigungen am Phloem ein Stärkestau im Mesophyll mit der Folge unzureichender Speicherung von Assimilaten in Wurzel und Stamm
- Im Winter vermindern niedrigere Zuckergehalte die Frosthärte
- Bedingt durch die eingeschränkte Stärkespeicherung in den Amyloplasten steht im Frühjahr weniger Reservesubstanz zur Ausbildung neuer Triebe und damit zur Sicherung der künftigen Entwicklung zur Verfügung

LITERATUR

(1) BUNDESMINISTERIUM FÜR ERNÄHRUNG, LANDWIRTSCHAFT UND FORSTEN (1986). Waldschadenserhebung.

(2) PRINZ, B., KRAUSE, G.H.M., STRATMANN, H.(1982). Waldschäden in der Bundesrepublik Deutschland. LIS-Bericht Nr. 28, Essen.

(3) ARNDT, U., SEUFFERT, G., NOBEL, W. (1982). Dic Beteiligung von Ozon an der Komplexkrankheit der Tanne (Abies alba Mill.) - eine prüfenswerte Hypothese. Staub Reinhalt. Luft 42: 243-246.

(4) GUDERIAN, R., TINGEY, D.T., RABE, R. (1985a). Effects of photochemical oxidants on plants, 126-346. In: GUDERIAN R. (ed): Air Pollution by Photochemical Oxidants. Ecological Studies 52, Berlin/Heidelberg/New York/Tokyo (Springer).

(5) GUDERIAN, R., KÜPPERS, K., SIX, R. (1985b). Wirkungen von Ozon, Schwefeldioxid und Stickstoffdioxid auf Fichte und Pappel bei unterschiedlicher Versorgung mit Magnesium und Kalzium sowie auf die Blattflechte Hypogymnia physodes. VDI-Berichte 560: 657-701.

(6) KRAUSE, G.H.M., JUNG, K.D., PRINZ, B. (1985). Experimentelle Untersuchungen zur Aufklärung der neuartigen Waldschäden in der Bundesrepublik Deutschland. VDI-Berichte 560: 627-656.

(7) REHFUESS, K.E., BOSCH, C. (1986). Experimentelle Untersuchungen zur Erkrankung der Fichte (Picea abies (L.) Karst) auf sauren Böden der Hochlagen: Arbeitshypothese und Versuchsplan. Forstw. Cbl. 105: 201-206.

(8) GUDERIAN, R. (1986). Möglichkeiten zur Erfassung und Bewertung von Immissionswirkungen auf Pflanzen. Umweltbundesamt, Berichte 7/86, Erich Schmidt Verlag Berlin.

(9) EBELL, L.F. (1969). Variation in total soluble sugars of conifer tissues with method of analysis. Phytochemistry, Vol 8: 227-233.

(10) KEPPLER, D. und DECKER, K. (1974). In Methoden der enzymatischen Analyse (Bergmeyer, H.U., Hrsg.) 3. Aufl., Bd. 2, 1171-1176. Verlag Chemie, Weinheim und (1974) in Methods of Enzymatic Analysis (Bergmeyer, H.U., ed.) 2nd ed., Vol 3: 1127-1131. Verlag Chemie, Weinheim, Academic Press, Inc. New York and London.

(11) GUDERIAN, R., VOGELS, K., MASUCH, G. (1986). Comparative physiological and histological studies on Norway Spruce (Picea abies Karst.) by climatic chambers experiments and field studies in damaged forest stands. Proc. of the Seventh World Clean Air Congress, Syndney, Vol. III: 148-156.

(12) VOGELS, K., GUDERIAN, R., MASUCH, G. (1986). Studies on Norway spruce (Picea abies Karst.) in damaged forest stands and in climatic chambers experiments. Intern. Conf. on Acidification and its Policy Implications, organized by the Government of the Netherlands in cooperation with the United Economic Comission for Europe (ECE). Amsterdam, 5.-9. May 1986.

(13) JEREMIAS, K (1969). Zur winterlichen Zuckeranhäufung in vegetativen Pflanzenteilen. Ber. Dtsch. Bot. Ges. Bd 82, H. 1/2: 87-97.

(14) SANTARIUS, K.A. (1971). Ursachen der Frostschäden und Frostadaptation bei Pflanzen. Ber. Dtsch. Bot. Ges., Bd 84 H. 7/8: 425-436.

(15) SENSER, M., BECK, E. (1979). Kälteresistenz der Fichte. Ber. Dtsch. Bot. Ges., Bd 92: 243-259.

(16) KOZIOL, M.J. (1984). Interactions of gaseous pollutants with carbohydrate metabolism. In: Koziol M.M., Whatley F.R. (eds.) Gaseous Air Pollutants and Plant Metabolism. Butterworths, London.

(17) KOZIOL, M.J., JORDAN, C.F. (1978). Changes in carbohydrate levels in red kidney bean (Phaseolus vulgaris L.) exposed to sulphur dioxide. Journal of Experimental Botany 29: 1037-1043.

(18) NOYES, R.D. (1980). The comparative effects of sulfur dioxide on photosynthesis and translocation in bean, Physiological Plant Pathology 16: 73-79.

(19) TEH, K.H. and SWANSON, C.A. (1982). Sulfur dioxide inhibition of translocation in bean plants, Plant Physiology 69: 88-92.

(20) PARAMESWARAN, N., FINK, S., LIESE, W. (1985). Feinstrukturelle Untersuchungen an Nadeln geschädigter Tannen und Fichten aus Waldschadensgebieten im Schwarzwald. Eur. J. For. Path. 15: 168-182.

COMPARATIVE INVESTIGATIONS ON THE PHOTOSYNTHETIC ELECTRON TRANSPORT CHAIN OF SPRUCE (PICEA ABIES) WITH DIFFERENT DEGREES OF DAMAGE IN THE OPEN AIR

A. WILD, B. DIETZ, U. FLAMMERSFELD, and I. MOORS
Institute for General Botany of the Johannes Gutenberg-University, D-6500 Mainz, Federal Republic of Germany

Summary

Several components of the photosynthetic electron transport chain (P-700, cytochrome f, Q_B-protein) as well as the rate of electron transport and the chlorophyll content of the needles of spruce trees with different degrees of damage were investigated. The investigations were carried out in pair comparison at a location in the Hunsrück (West Germany). The rate of electron transport was determined as photoreduction of 2,6-dichlorophenolindophenol. Significant damage to the electron transport system is shown in the thylakoids of the damaged trees compared to the less severely damaged trees. In the spruce trees with more damage, the rates of electron transport are significantly lower. The investigation of the water-splitting enzyme system by feeding in electrons by means of diphenylcarbazide indicates that the electron transport on the oxidizing side of photosystem II is impaired. In the more damaged trees the content of Q_B-protein, cytochrome f as well as P-700 referred to dry weight is markedly lowered. On the other hand, in relation to chlorophyll only the content of P-700 and Q_B-protein is significantly decreased. These investigations imply that the thylakoid membranes are sites of early damage from air pollution.

1. INTRODUCTION

For several years, "new forest injury" has been occuring over large areas in Europe. According to the state of research, this complex disease appears to be of multifactorial origin, i.e. the causes are very diverse and one stressor alone rarely plays the major role. Today, the role of climatic, edaphic and microbial aspects is undisputed at limited locations and times, and may indeed be of crucial significance. However, in superordinate terms, anthropogenic air pollution is likely to be responsible for spreading over large regions and the exceptional intensity of damage.

The investigations described here were carried out in the context of our research project on the physiological, biochemical and cytomorphological characterization of spruce with different degrees of damage in the open air (1-3). Besides the photosynthesis (4), the water balance, the nitrogen assimilation (5), the content of chemical elements as well as light and electron microscopic investigations (6, 7), the photosynthetic electron transport was explored, too.

2. MATERIAL AND METHODS

The measurements were carried out in 1986 on a 20-25 year old spruce (Picea abies) plantation near Idar-Oberstein (Hattgenstein Forest District) and were set up as a pair comparison i.e. the measurement data of a tree with symptoms of damage were always compared with a tree in the immediate vicinity which phenotypically showed less damage. The trial area was on a moderately severe south slope on the west side of the Hunsrück mountains at about 660 m above sea level. We only used needles of 1984 and 1985 shoots from the seventh whorl and the same crown region of the trial trees.

The thylakoid suspension was obtained according to (8) with several modifications.

The rate of electron transport (ETR) was determined as photoreduction of 2,6-dichlorophenolindophenol (DCPIP) at pH 6.5 according to (9). To investigate the impairment of the water-splitting system, diphenylcarbazide (DPC) was added to the test system.

The values for Q_B-protein were determined by the data obtained from binding of ^{14}C-atrazine (atrazine titration).

Cytochrome f and P-700 were measured by the oxidized-minus-reduced difference spectrum (10). For the calculation of the concentration the extinction coefficients 17.7 cm^2 μmol^{-1} (cytochrome f) and 64 cm^2 μmol^{-1} (P-700) were used.

3. RESULTS AND DISCUSSION

The chlorophyll (chl) content in the needles of the more severely damaged spruce trees is significantly reduced (Fig. 11).

On average, the Q_B-protein is lowered in the more damaged spruce trees, i.e. inter alia the reducing side of the photosystem II (PS II) is also affected by damage (Fig. 1+2).

The more damaged trees do not show any difference in the cytochrome f content in relation to chl (Fig. 3). This means that the content of cytochrome f and chl are reduced to the same extent in damage. The concentration of cytochrome f per chl only decreases in case of severe damage. However, if the cytochrome f content is related to the dry weight, a marked difference between the tree partners is shown (Fig. 4).

The concentration of P-700 (Fig. 5+6) decreases significantly in the more damaged trees (to a greater extent in the 1984 needle generation than in the 1985 needle generation).

The electron transport rates (ETR) are clearly lowered in both needle generations in the more damaged trees (Fig. 8+9).

An increase of ETR can be detected after addition of DPC to the measurement system both in the more damaged and in the relatively asymptomatic spruce trees. The increase of ETR is very much more pronounced in the more damaged partners of tree pairs (Fig. 10). This indicates that the water-splitting system is impaired on the oxidizing side of PS II.

The results show that the photosynthetic electron transport chains are sites of injurious effects occuring at an early stage. This is in agreement with our measurement on CO_2 exchange (4) and the results of our electron microscopic investigations (7), which clearly show that the thylakoid membranes of damaged trees show severe impairments even before the occurrence of outwardly visible symptoms of damage on the needles.

The results of our physiological and cytomorphological investigations on spruce from different locations consistently show the presence of early damage to the cell membranes (11, 12). This membrane damage already occurs to a substantial extent in the green needles of damaged trees. On the cellular level, this membrane damage - in the genesis of which anthropogenic air pollutants are essentially involved - evidently plays a central role in the pathogenesis of the new forest injury.

REFERENCES

(1) WILD,A., BODE, J. (1986). Physiologische, biochemische und anatomische Untersuchungen von immissionsbelasteten Fichten verschiedener Standorte. In: Wirkungen von Luftverunreinigungen auf Waldbäume und Waldböden, Spezielle Berichte der Kernforschungsanlage Jülich-Nr. 369, 166-181, ISSN 0343-7639.

(2) WILD, A., (1987). Physiological and cytomorphological investigation of spruce-needles exposed to air pollution. In Sonderheft: Pflanzenphysiologische Untersuchungsergebnisse als Beiträge zur Waldschadensforschung. Allg. Forstz. (AFZ), in press.

(3) BODF, J., KÜHN, H.-P., WILD, A. (1985). Die Akkumulation von Prolin in Nadeln geschädigter Fichten (Picea abies (L.) KARST.). Forstw. Cbl. 104, 353-360.

(4) BENNER, P., WILD, A. (1987). Measurement of photosynthesis and transpiration in spruce trees with various degrees of damage. J. Plant Physiol., in press.

(5) DÜBALL, S., WILD, A. (1987). Investigations on the activity of glutamine synthetase, the content of free ammonium, soluble protein and chlorophyll in spruce trees with different degrees of damage in the open air. Planta, in press.

(6) HASEMANN, G., JUNG, G., WILD, A. (1987). Losses of structural resistance in damaged spruce needles of locations exposed to air pollution. I. Mesophyll and central cylinder. Eur. J. For. Path., submitted.

(7) JUNG, G., WILD, A. (1987). Electron microscopic studies of spruce needles in connection with the occurrence of new kinds of tree damage. I. Investigations of the mesophyll. Phytopath. Z., in press.

(8) SENSER, M., BECK, E. (1978). Photochemically active chloroplasts from spruce (Picea abies (L.) KARST.). Photosynthetica 12, 323-327.

(9) ÖQUIST, G., MARTIN, B., MARTENSSON, O. (1974). Photoreduction of 2,6-dichlorophenolindophenol in chloroplasts isolated from Pinus silvestris and Picea abies. Photosynthetica 8, 263-271.

(10) WILD, A., KE, B., SHAW, E.R. (1973). The effect of light intensity during growth of Sinapis alba on the electron-transport components. Z. Pflanzenphysiol. 69, 344-350.

(11) WILD, A. (1987). Licht als Streßfaktor bei Waldbäumen. Naturwissenschaftliche Rundschau, in press.

(12) WILD, A. (1987). Licht als Streßfaktor bei Holzgewächsen. In: Klima und Witterung im Zusammenhang mit den neuartigen Waldschäden, Symposium Gesellschaft für Strahlen- und Umweltforschung München, in press.

Figures: **1-11**.
Pair comparison with three tree pairs,
□ = slightly damaged partner,
■ = more severely damaged partner,
∅ = mean value over the investigation period 1986,
'85 = the 1985 needle generation,
'84 = the 1984 needle generation.

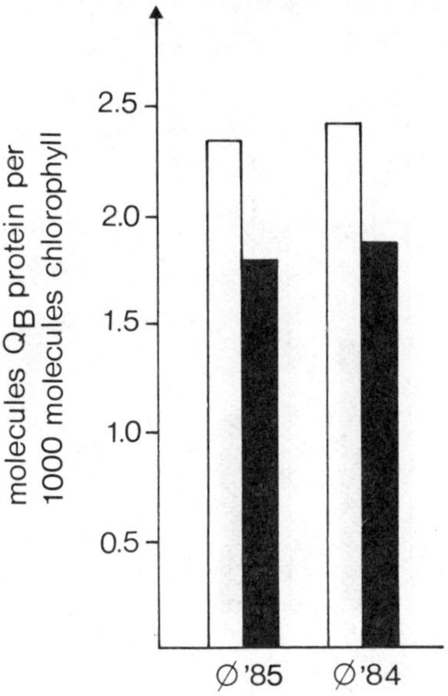

Figure 1

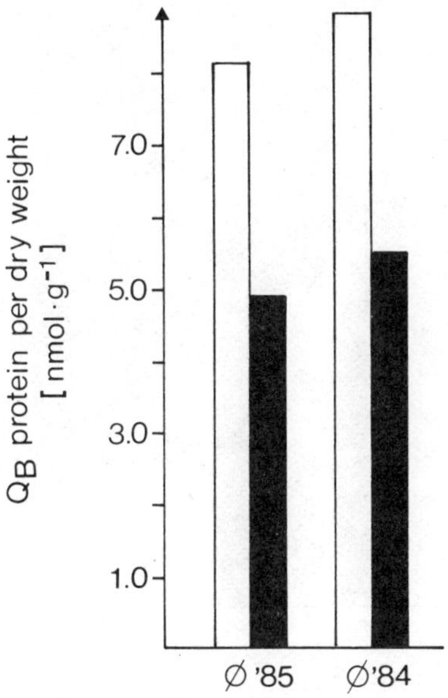

Figure 2

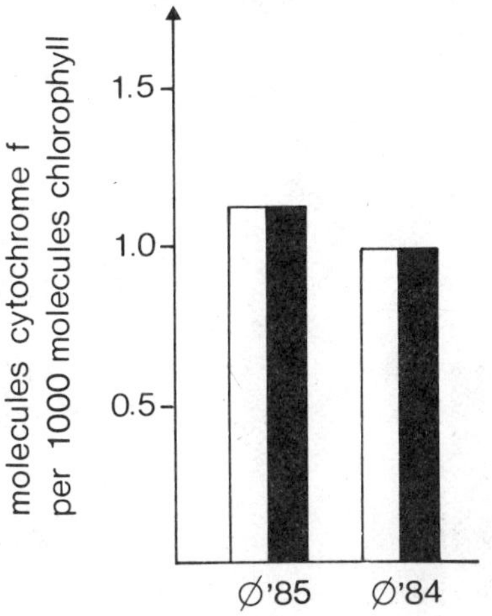

Figure 3

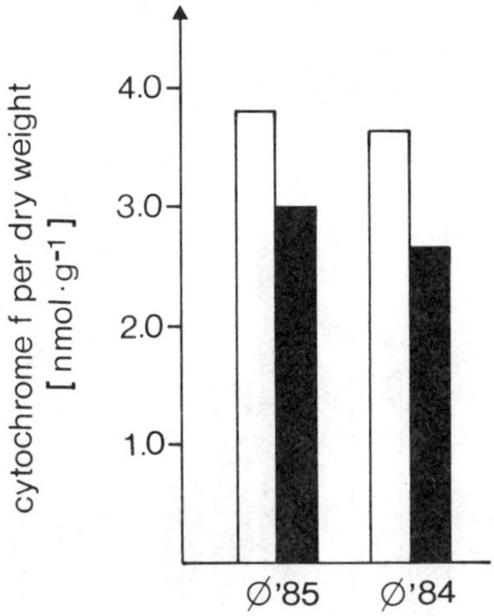

Figure 4

Figures : 1-11.
Pair comparison with three tree pairs,
□ = slightly damaged partner,
■ = more severely damaged partner,
Ø = mean value over the investigation period 1986,
'85 = the 1985 needle generation,
'84 = the 1984 needle generation.

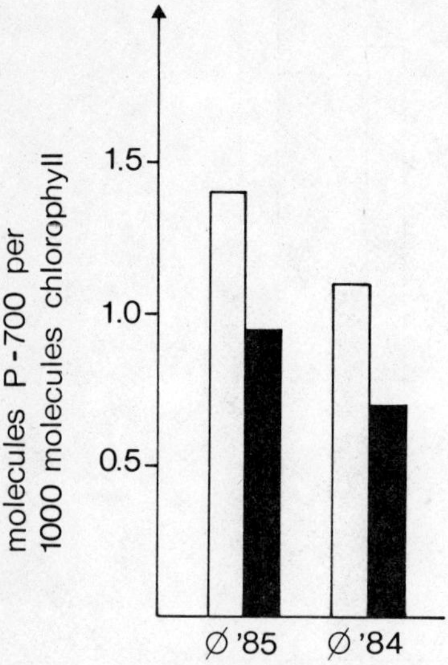

Figure 5

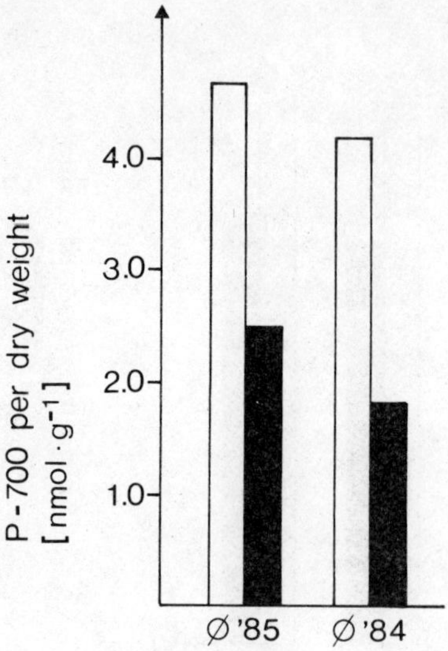

Figure 6

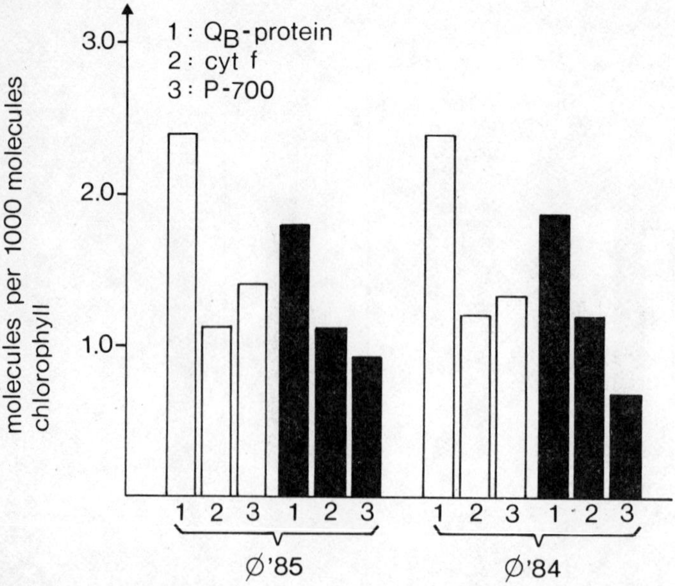

Figure 7

Figures: 1-11.
Pair comparison with three tree pairs,
□ = slightly damaged partner,
■ = more severely damaged partner,
Ø = mean value over the investigation period 1986,
'85 = the 1985 needle generation,
'84 = the 1984 needle generation.

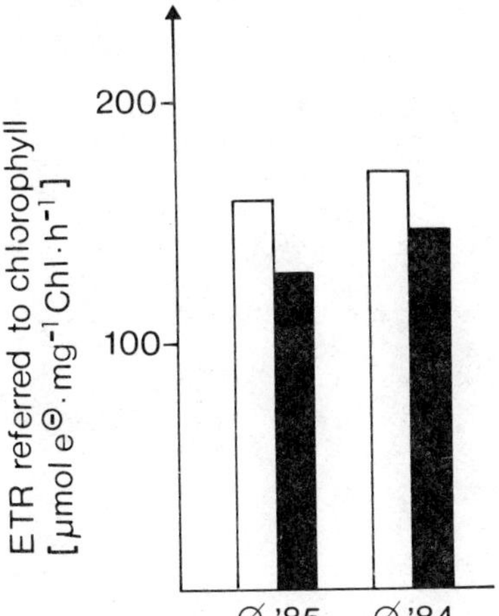

Figure 8

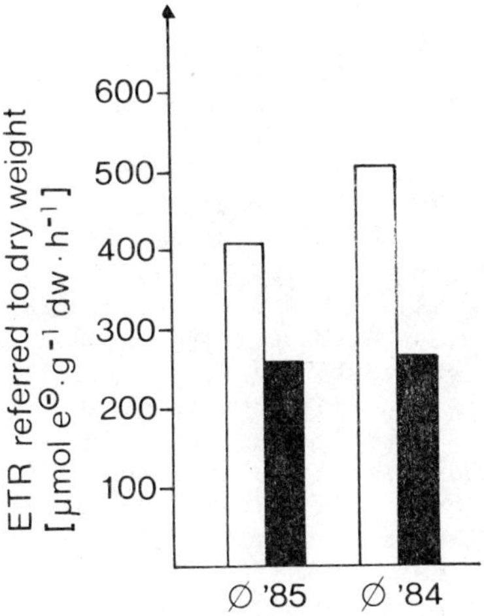

Figure 9

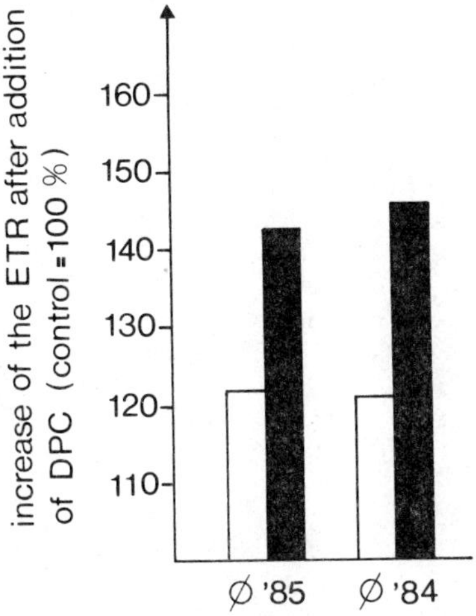

Figure 10

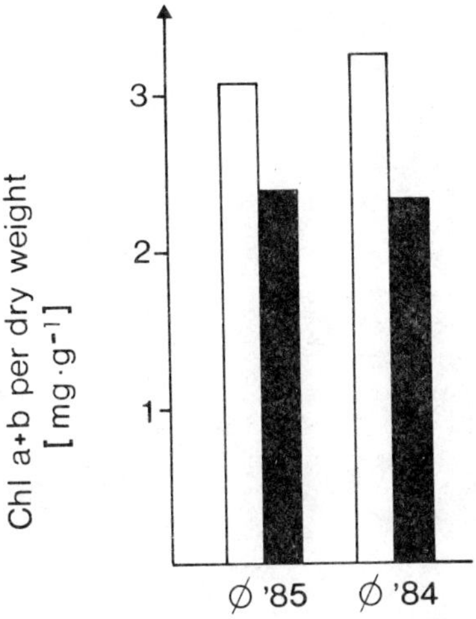

Figure 11

SOME EFFECTS OF LOW LEVELS OF SULPHUR DIOXIDE AND NITROGEN DIOXIDE ON THE CONTROL OF WATER-LOSS BY *Betula* spp.

E.A. WRIGHT
Department of Biological Sciences, University of Lancaster,
Bailrigg, Lancaster, Lancashire. LA1 4YQ
United Kingdom

SUMMARY

Leaves of clones of both *Betula pendula* and *B. pubescens* were found to lose water more quickly if they had been previously exposed to concentrations of sulphur dioxide and nitrogen dioxide between 20 and 40 parts per billion (ppb = nl l^{-1}) for a month. Further experiments revealed that such trees suffered more acutely in the event of water-stress, shedding their older leaves prematurely.

The damage was found to be confined to the lower surface of the leaves only, where all the stomata are situated. Scanning electron micrographs of this surface of polluted leaves revealed evidence of damage to subsidiary cells of some of the stomata, resulting in the guard cells being unable to close.

A study of the gas exchange of whole plants of *Betula pubescens* exposed to high and low concentrations of carbon dioxide suggested that the stomata of polluted birch still had a normal physiological response to CO_2, although there were signs that it was reduced and was more variable after exposure to 40 or 60 ppb SO_2+ NO_2. This study also showed that their mean basal transpiration rates compared with the controls were almost doubled and trebled respectively.

These results indicate that the enhanced water-loss by polluted birch leaves is produced partly by incomplete stomatal closure and partly by increased cuticular transpiration.

1. INTRODUCTION

Sulphur dioxide and nitrogen dioxide have been found to reduce the growth of birch trees when present together at levels of about 65 ppb (1). A symptom of the damage is premature leaf-loss which generally occurs after the development of visible lesions. The following experiments were conducted to attempt to assess the effect of leaf damage not necessarily involving visible lesions on the control of water-loss by birch trees.

All the plants used in this study were clonal in origin. They were grown in pots for 1-2 months in growth cabinets in a controlled environment. They were exposed there to clean air, 20 ppb, 40 ppb, or 60 ppb SO_2 + NO_2 continuously for that period.

2. WATER-LOSS BY EXCISED LEAVES OF *Betula pubescens*

Figure I shows the cumulative percentage weight loss of leaves of equivalent age excised from clonal plants polluted as described in the introduction. They were weighed immediately after excision and then at intervals over the next 2 days. The weight loss in these circumstances can all be ascribed to loss of water. The figure shows that the greater the concentration of pollutants the greater the rate of weight loss, implying that exposure to these pollutants causes a reduction in the capacity of the leaves to limit their water-loss.

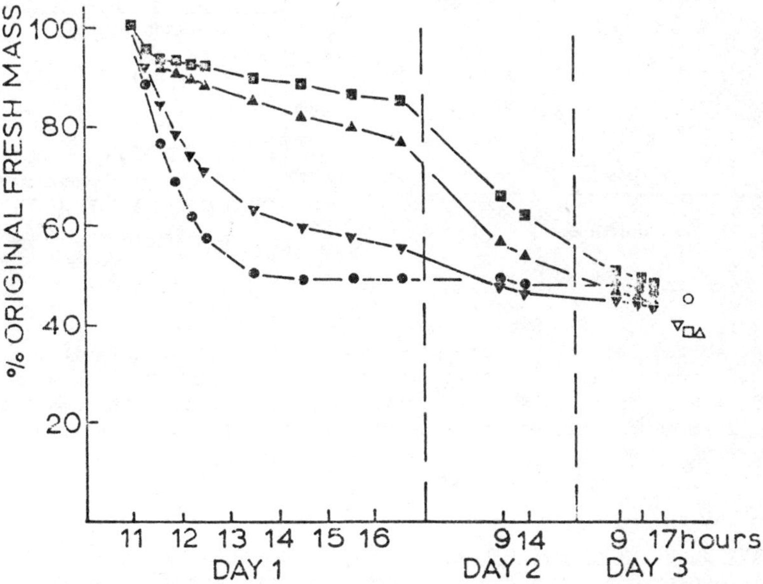

Figure I

Fresh weight of excised leaves measured over time from one clone of *Betula pubescens grown* for 30 days in clean air (■), 20 ppb SO_2 + NO_2 (▲), 40 ppb SO_2 + NO_2 (▼), 60 ppb SO_2 + NO_2 (●). Each point is a mean of nine replicates consisting of one leaf from one tree. The open symbols show the final dry mass, after oven drying at 80 C.

3. RESPONSES OF POLLUTED BIRCH TO WATER-STRESS

Three replicates of two clones of *Betula* from both the control and 40 ppb SO_2 + NO_2 treatments were subjected to a period of drought simply by witholding water from Day 1. Figure II shows the transpiration rates and effective leaf areas of whole *Betula pendula* shoots as they became water-stressed. The figure reveals that the polluted plants transpired more at the beginning of the drying period and were not able to maintain a low transpiration rate when water-shortage was acute without shedding leaves. The polluted plants all died before the controls, despite having a smaller initial leaf area. An identical experiment conducted simultaneously on the clone of *B. pubescens* produced similar results.

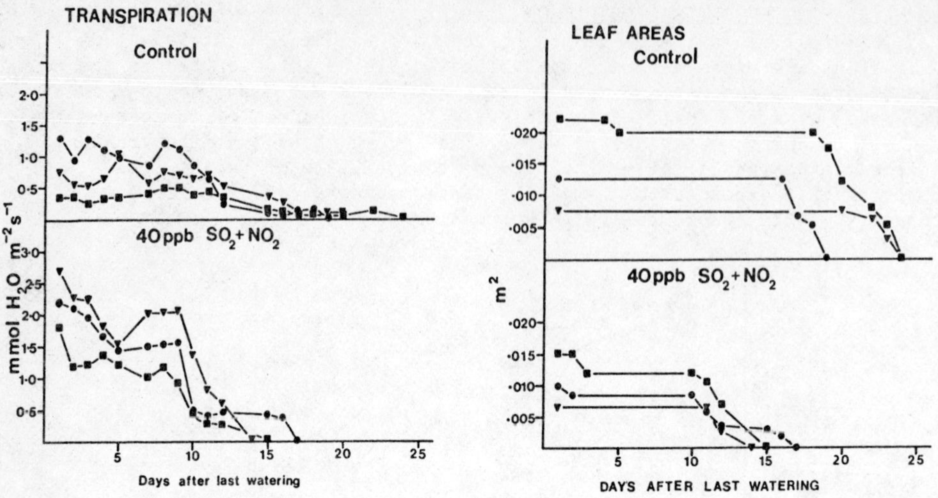

Figure II

Transpiration rates (left) and effective leaf areas (right) in one clone of *Betula pendula* after being grown in clean air or air containing 40 ppb SO_2 + NO_2 for 35 days. The plants were watered for the last time on Day 0. Each symbol represents a measurement of one plant only.

4. SCANNING ELECTRON MICROSCOPY OF THE UNDER-SURFACE OF BIRCH LEAVES

An experiment not shown revealed that coating the lower surface of the leaves with vaseline prevented the polluted leaves losing more water than the control leaves, strongly implying that the leaf under-surface is the site of the damage. The scanning electron micrographs show the under-surface of control leaves (figure III) and leaves exposed to 40 ppb SO_2 + NO_2 (figures IV and V) for over a month.

The leaves were excised from the plant and left to transpire freely for 15 minutes to allow the stomata to close. They were than mounted on aluminium stubs and frozen in liquid nitrogen before examination in a scanning electron microscope adapted for examination of specimens at very low temperatures (2). The photomicrographs reveal that the stomata on the control leaves were almost all shut whereas many on the polluted leaves were wide open. In the higher magnification print the subsidiary cells of the guard cells appear collapsed, and this might have affected the ability of the guard cells to close the pore. The polluted leaves used for this study displayed visible lesions at the time the samples were taken.

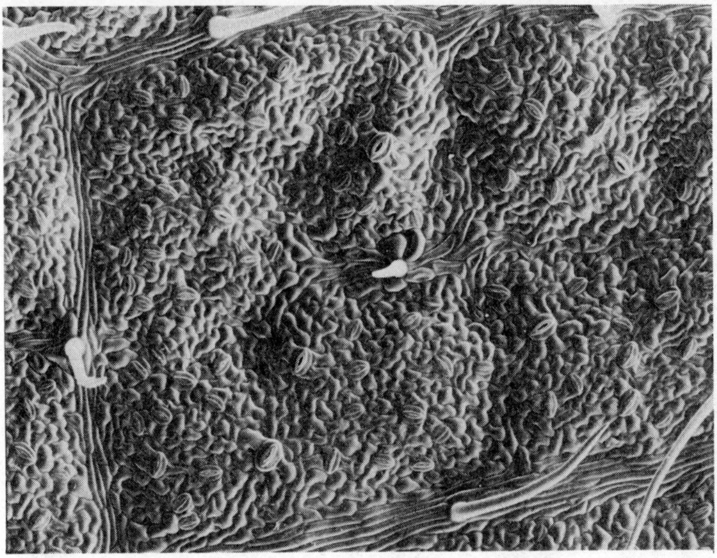

Figure III
The under-surface of a *Betula pubescens* leaf exposed to clean air for 40 days. Magnification = x 150.

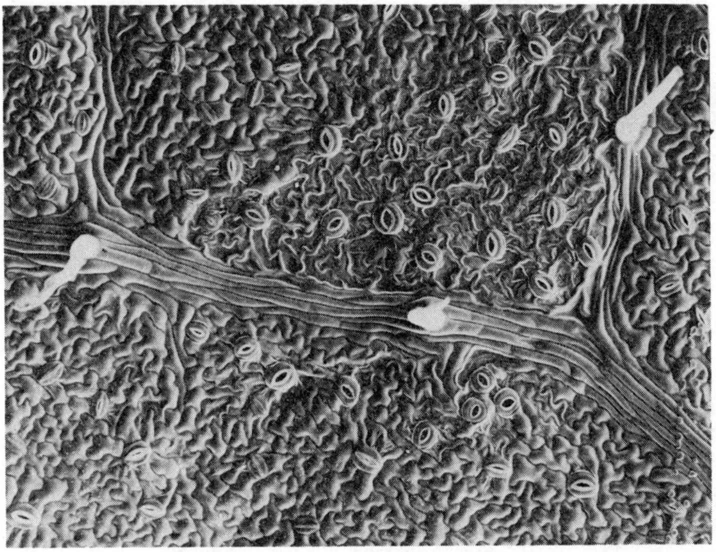

Figure IV
The under-surface of a *Betula pubescens* leaf exposed to 40 ppb SO_2 + NO_2 for 40 days. Magnification = x 150.

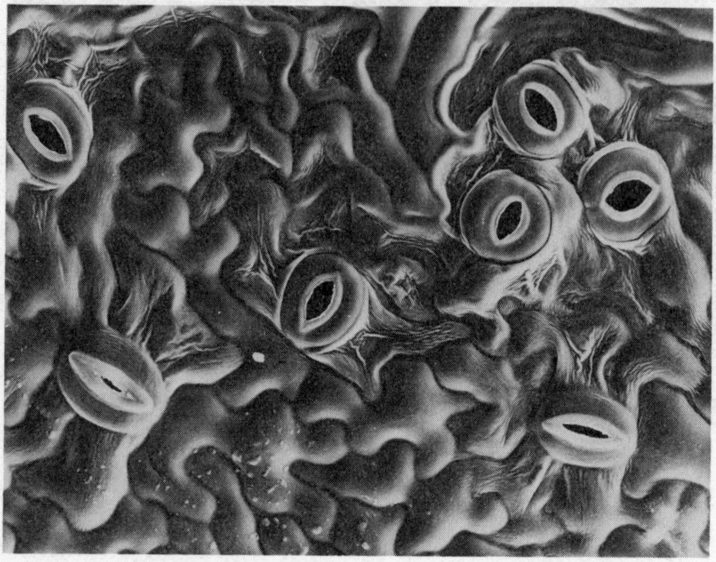

Figure V
The under-surface of a *Betula pubescens* leaf exposed to 40 ppb SO_2 + NO_2 for 40 days. Magnification = x 450.

5. WHOLE PLANT GAS EXCHANGE AT HIGH AND LOW CARBON DIOXIDE CONCENTRATIONS

Figure VI shows the results of transpiration measurements made on plants of *Betula pubescens* exposed to high (≃ 1000 ppm) and low (≃ 150 ppm) concentrations of CO_2. High CO_2 induces stomatal closure in birch trees, while low CO_2 induces stomatal opening. The figure reveals that birch exposed to 40 and 60 ppb SO_2 + NO_2 showed a slightly reduced and more variable stomatal response, and more conclusively, a greatly increased basal transpiration rate.

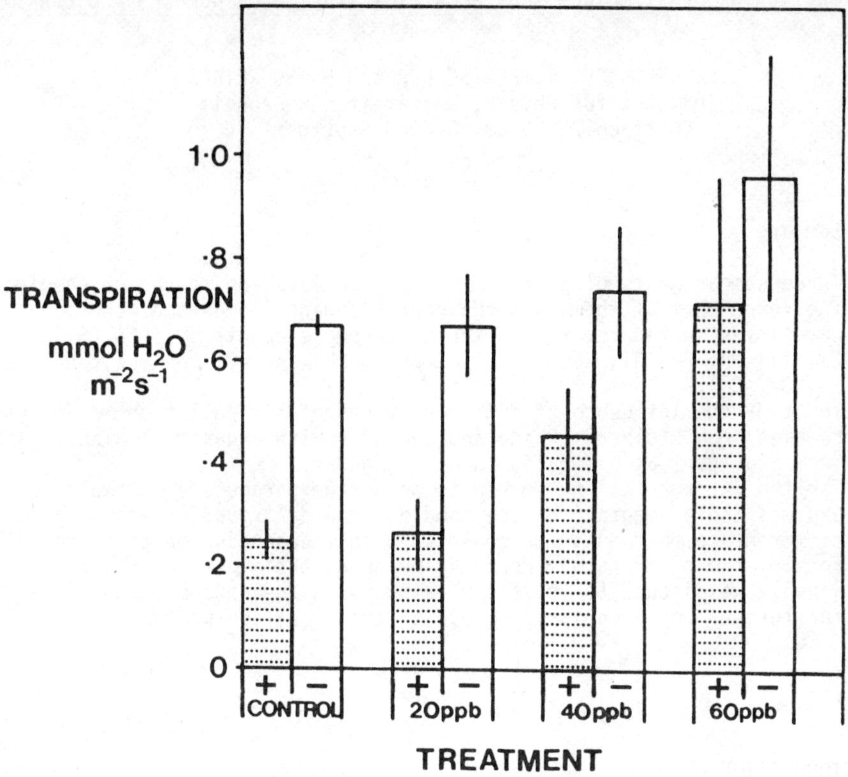

Figure VI
Transpiration rates of *Betula pubescens* clones after exposure for over two months to clean air, 20, 40 or 60 ppb SO_2 + NO_2. Rates were measured either at high CO_2 (+) or low CO_2 (-). Vertical bars are 2 x standard error. There are three replicates per bar

6. CONCLUSIONS

Sulphur dioxide and nitrogen dioxide together, particularly at levels above 20 ppb, cause birch to be less efficient at controlling water loss. From the evidence presented it appears that a combination of stomatal and epidermal damage is responsible.

REFERENCES

(1) GARDNER,D.L., O'CONNOR,P. & OATES,K (1981). Low temperature scanning electron microscopy of dog and guinea-pig hyaline articular cartilage. **Journal of Anatomy**, 132: 267-282.

(2) WRIGHT,E.A. (1987). Some effects of sulphur dioxide and nitrogen dioxide, singly and in mixture on the macroscopic growth of three birch clones. **Environmental Pollution**, in press.

DEPOSITION OF HEAVY METALS AND THEIR DISTRIBUTION ON NEEDLES AND BARK

V. CERCASOV, I. RENTSCHLER and H. SCHREIBER
Institut für Physik, Universität Hohenheim
Postfach 70 05 62, D-7000 Stuttgart 70

Summary

Three compartments of a tree: the surface layer on leaves or needles, the inner part of them, and the corresponding segments of the bark were investigated for the contents of some elements by XRFA (S, K, Ca, Ti, Mn, Fe, Ni, Cu, Zn, Br, Sr, Rb, Pb) and by INAA (Fe, Zn, Cr, Sb, Se, Co).
To study the influence of different types of air pollution we analysed samples from Stolberg (Rhineland), a site with industrial plants, and from Stuttgart at a location near a highway.
For the interpretation of results we define parameters suitable to express the elemental surface load and the distribution among the different compartments. We could show that depending on the chemical form of the elements the bark may be an essential store in a tree. Finally we discuss the possible pathways in the canopy and the need for further research in the field of translocation processes.

1. INTRODUCTION

Among the different compounds of atmospheric pollutants the particulate matter, containing heavy metals, should not be neglected. Besides the catalytic properties important in the field of gas to particle conversion the toxicity of some elements as Cu, Cd, Se, As, Pb, Zn, Hg have to be considered. The understanding of a direct influence upon plants, animals and human beings must be extended to the long term changes in the biosphere especially by contamination of the soil. This point of view requires the knowledge of the amounts of influxes direct into the soil and into the vegetation.

We tried to get informations on the first step of these processes investigating the deposition of some elements and their transport along different pathways on and into plants. We studied the distribution mainly in three compartments of trees: the surface layer on leaves and needles, the inner part of these and the bark. In this work we present some results of such distributions depending on the age of plant segments as well as on the chemical and physico-chemical form of the contaminants originating from different sources. We interpret the age-depending changes partly as a consequence of translocation processes between the compartments.

2. METHODS

A. Plant material

After gathering the branches of Picea abies and Picea omorica containing usually four age classes the geometrical properties as areas of needles and bark, needle number density and mass per cm^2 were determined. On a definite number of needles we applied the film stripping method in order to remove the surface contamination (1), (2). We prepared six kinds of samples: untreated and stripped needles for individual measurements, pellets or collections of them and of the films containing the surface contamination, and finally segments of the bark.

B. Instrumental analyses

For XRFA the samples supported by Hostaphan® foils were measured with a tube excited (Rh) energy dispersive X-ray fluorescence analysis device (Finnigan Inc.) under following conditions: vacuum, 40 kV, 0.10 mA, 4000 s. Absorption and enhancement effects were corrected (3). Reliable values can be obtained, in general, for S, K, Ca, Ti, Mn, Fe, Ni, Cu, Zn, Br, Sr, Rb, Pb. The typical detection limits range from 1 to 20 µg/g (4).

For INAA the samples were irradiated in quartz vials together with multielement standards at the nuclear reactor of the GKSS Research Center, Geesthacht for 3 days at a flux of about 10^{14} thermal neutrons$\cdot cm^{-2}\cdot s^{-1}$. Gamma spectrometrical measurements were performed 1 month after the irradiation using a high purity Ge detector. The elements Fe, Zn, Cr, Sb, Se and Co were determined. On account of missing reference values we present for Ag and Hg only data refering to ratios of counting rates.

SEM micrographs of the surface of needles have been documented for further interpretation.

3. RESULTS AND DISCUSSION

A. Age-depending concentrations in needles and bark

Without regard to the possible variations depending on species and locations we present some results of measurements of the elements K and Mn in the needles of a spruce (fig. 1). This demonstrates the variability in one single plant and the development in the course of time showing for these elements a decrease with age. This trend is typical for numerous results on other trees and locations.

The contents of other elements increase with time, examples are Pb, Fe, Zn and Cu. In fig. 2 we represent additionally the mean concentrations of the bark segments belonging to the different stages of age.

We find that at this location the concentrations in the bark exceed those of the needles by a factor of nearly 10. For other locations this ratio can be considerably higher or lower. On the contrary the contents of K and Ca are generally higher in the needles than in the bark. This indicates that particularly with increasing age the bark contains less living substance in relation to the needles. Of all samples we ever measured the elements S, Fe, Co, Cu, Zn, Br, Sb, Ag and Pb show variations that seem to depend on locations i.e. source influences. In order to study the distribution of elements in the two compartments needles and bark we need also the amounts of dry-weight of needles and bark segments.

B. Distribution of elements in needles and bark

Especially at conifers we could gain numerous data of the age depending values of bark area and mass, needle geometry and weights and of the corresponding values of their number density on the bark surface. With these parameters it is possible to calculate the burden of elements in the age-classified compartments needles and bark. The ratio of the amounts in the bark segments related to the associated needles we call coefficient of distribution - C_{BN}. It expresses the relative significance of the bark as a store for elements particularly if we want to discuss elemental input and output as well as phenomena of translocation in the tree.

To give evidence of these questions we present in table 1 values from spruce trees at Stolberg (Rhineland), a site with relatively high emissions from industrial plants (lead ore processing, re-entrainment of dumped material).

It is obvious that the values in table I are characteristic of this special source configuration and of the given composition of pollutants. The analyses of airborne particles, and deposition measurements (5) show a high concentration of the elements Pb, Zn and Cu. The data in table I reflect this fact by high values in needles and bark. The low values of Br proof that the main part of the lead contamination does not originate from traffic. In the table I we arranged the elements in the order of decreasing coefficients of distribution (C_{BN}) for the older segment. The development with age of these coefficients can be seen in greater detail in fig. 3 for two different species of spruce and other locations.

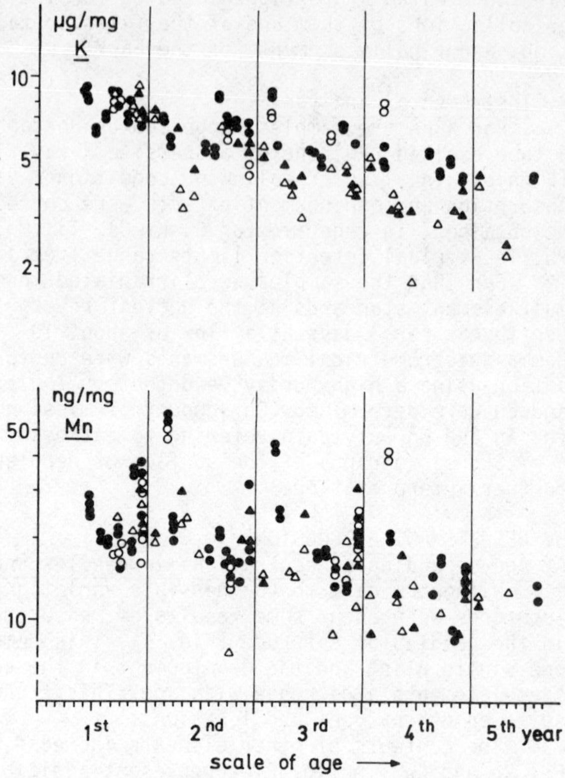

Fig. 1. K and Mn contents in the needles of a spruce near a highway (▲ single needles, ○ pellets)

We interpret the coefficients of distribution in a sense of enrichment factors ranging in these samples from values below 1 up to more than 10. The data in table I for Pb and Cu are not extremely high; we found in the Stolberg region values up to 30 combined with concentrations in the bark of 0.7 %.

For a first attempt it seems reasonable to classify the investigated elements into three groups:

$$C_{BN} < 1 \;:\; \text{K, Ca}$$

$$1 < C_{BN} < 5 \;:\; \text{S, Br, Zn, Mn, Hg, Sb, Cr, Fe, Se}$$

$$5 < C_{BN} \;:\; \text{Co, Cu, Ag, Pb}$$

In the case of the tree represented in table I the decrease of K and Ca in the course of 3 years is partly caused by the loss of needles and by yearly variation of needle growth, an effect that should not be neglected and must be considered in the elemental balance of a tree in detail. For the relative classification of different elements the needle loss has no effect. This influence can be judged by considering the development of the concentration values in the needles.

The concentrations itselves, the coefficients of distribution and their development with age give reason for question on their causes. One of the

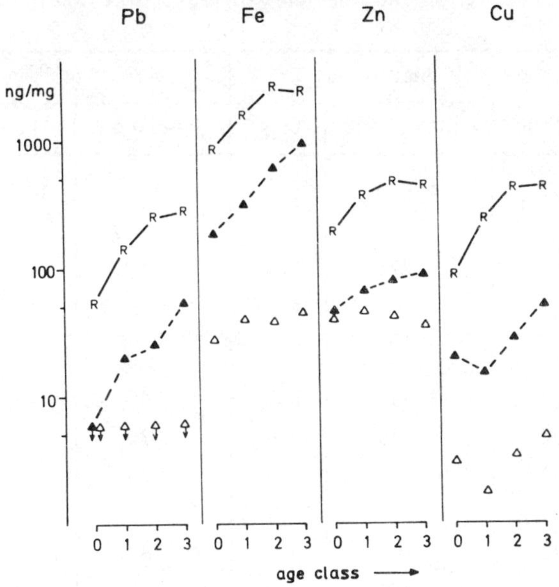

Fig. 2. Elemental contents in needles (▲) and bark segments (R) of a spruce (Picea omorica) at Stolberg (△ needles, surface contamination removed)

causes is obviously related to the input along the aerial pathway - eventually giving rise to different degrees of surface contamination.

C. Discrimination between surface contamination and inner content of leaves and needles

A method for the separation of the dust deposited on the surface of leaves was developed and published in 1982 (1). It consists in applying a plastic solution and after drying stripping off the thin film from the leaves or needles. The blank values of the plastic material are very low for most of the elements (2). The stripped film material containing the removed surface contamination was analysed by XRFA and INAA.

For the same tree as reported in table I we present in the following table the element contents of two compartments: the surface layer ("film") and the inner part of the needles ("stripped needles"). If one intends to discuss these contents as a result of the input originating from the deposition of airborne pollutants, it is useful to express the values in relation to the area of the acceptors, that is the area of needles. In this intention the values in table II can be understood as a kind of surface density or elemental density thickness (6).

The elements in table II are arranged in the same order as in table I, that is with decreasing coefficients of distribution bark:needles. It is obvious that generally high values of the bark enrichment factor (C_{BN}) correspond with high degrees of the surface contamination of needles (F_i). This regularity supports the hypothesis that the high contents in the bark

Table I. Amounts of elements in needles and bark corresponding to 1 cm^2 bark area (ng/cm^2) and the coefficients of distribution (C_{BN})

Element	1st year's			4th year's		
	Needles	Bark	C_{BN}	Needles	Bark	C_{BN}
Pb	14 500	11 600	0.8	16 000	148 000	9.3
Ag	+	+	5.1	+	+	8.8
Cu	1 290	771	0.6	513	4 330	8.4
Co	3.2	4.7	1.5	5.5	35	6.4
Se	6.0	3.5	0.6	9.7	46	4.7
Fe*	15 500	11 600	0.7	12 500	45 700	3.7
Cr	38	75	2.0	70	245	3.5
Sb	139	123	0.9	879	2 310	2.6
Hg	+	+	1.4	+	+	2.3
Mn	2 590	704	0.3	1 180	2 040	1.7
Zn*	7 390	4 750	0.6	14 300	22 600	1.6
Br	412	222	0.5	356	505	1.4
S	209 000	25 600	0.1	108 000	126 000	1.2
Ca	635 000	91 300	0.1	401 000	320 000	0.8
K	677 000	226 000	0.3	217 000	151 000	0.7

+) without reference values
*) mean values of XRFA and INAA

compartment are likewise results of airborne contamination as the high burden on the surface of needles. This interpretation is also based on the fact that the concentration in the outside part of the bark for some elements, particularly evident at Stolberg for Fe, Cu, Zn and Pb, is considerably higher than in the inner part. The distribution is opposite for the nutrient elements, especially in the young bark, and in all age classes for K. These results have been gained by XRFA measurements.

A similar classification of elements concerning deposition and uptake is reported in the literature (7). Indeed, high values of surface contamination on needles or leaves and on the outer part of the bark give a strong hint, if not a proof, for a considerable input of air pollutants into trees, but the reverse conclusion is not correct: low values of surface contamination don't indicate a negligible input of such elements. For example, it is sure that in traffic influenced locations the influx of airborne Br can be extremely high, however connected with very low surface contamination on the leaves and needles (1). The cause is that the deposited lead halide particles are leached, followed by an uptake of the soluble Br and Cl.

D. Transfer of deposited material in the canopy

Finally we will try to discuss consequences to the balance of some elements in the branches of a tree. If we compare the surface density of bark (table I) with the values of the needles (table II), we recognize that these compartments contain partly very different loads on the same area of acceptor. At least for the elements that contribute normally relatively little to the content of organic materials and also showing negligible uptake via roots, the enrichment of the bark should be explained. Considering the different surface structures of needles and bark, it seems reasonable that the loss of the deposited contamination, that is the carry off by rain, dew and fog precipitation, should be much higher for the needles. The question is, if this flux will appear completely below the canopy or if a considerable part can be transfered to the bark and be retained in this compartment. To decide this question it is necessary to measure the total input above the canopy and the resulting throughfall including the stem flow. But also on the basis of the presented values we can offer an answer in a qualitative manner: it is sure that the element Mn is subject to strong leaching effects, and therefore the observed contents in the throughfall (7) are in some locations rather determined by leaching than by the transport of the air pollutants deposited on leaves and needles. In our measurements at Stolberg we found a considerably enrichment of Mn in the bark. If we draw from table I the Mn load refered to 1 cm^2 bark area of the 4th age class, we recognize that the major part of the Mn content can be found in the bark. Also the observed concentrations in the bark were twofold higher than in the needles. While the Mn contents of the needles did not change essentially with age, the concentration and the surface density of the bark redoubled in the course of 3 years. These facts can serve as a preliminary indicator for the possibility that the transfer processes from the leaf and needle compartment contribute to the relative enrichment of some elements in the bark. Measurements of the input and output fluxes are necessary to balance this transfer for the different elements.

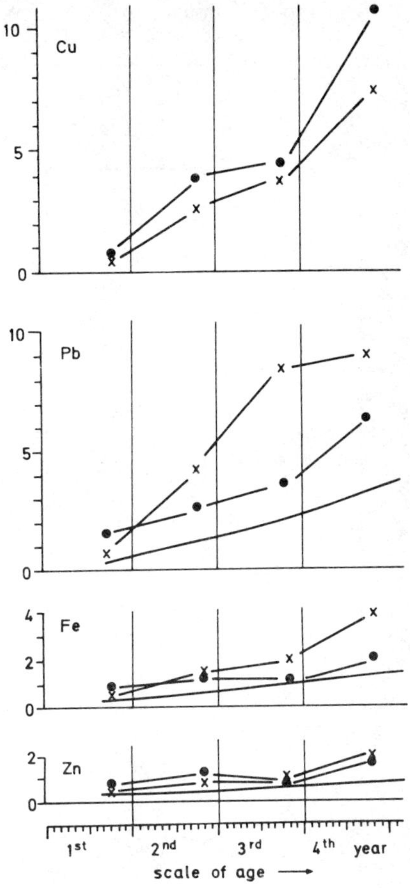

Fig. 3. Coefficients of distribution (C_{BN})
(● Picea omorica Stolberg,
× Picea abies Stolberg,
— Picea abies Stuttgart near a highway)

However the chemical state of the elements may play an essential role. For example, we found for Zn emitted in a vitrious form from tailings at Stolberg a high degree of surface contamination on the needles joined with high concentration in the bark (see fig. 2), while in the neighbourhood of the sintering plant, emitting Zn vapors, we found low values of surface contamination of the needles (see table II) and moderate values of the con-

Table II. Elemental surface density (ng/cm^2) refered to 1 cm^2 acceptor area and the degree of surface contamination (F_i)

Element	1st year's			4th year's		
	Film	Stripped Needles	F_i	Film	Stripped Needles	F_i
Pb	1 040	111	9.4	4 630	891	5.2
Ag	+	+	4.2	+	+	5.5
Cu	38.3	39.8	1.0	78.7	63.5	1.2
Co	0.563	0.341	1.7	0.705	0.169	4.2
Se	0.658	-	-	0.184	0.401	0.5
Fe*	1 060	383	2.8	2 080	402	5.2
Cr	23.2	1.05	22.1	21.8	< 0.7	> 30
Sb	19.2	2.58	7.4	119	12.6	9.4
Hg	+	+	0.7	+	+	0.1
Mn	39.7	137	0.3	34.5	156	0.2
Zn*	96.2	580	0.2	179	2 120	0.08
Br	-	45.7	< 0.2	-	67.9	< 0.2
S	606	10 700	0.06	1 850	21 200	0.09
Ca	435	30 900	0.01	918	77 400	0.01
K	457	37 200	0.01	623	44 300	0.01

+) without reference values
*) mean values of XRFA and INAA

centration in the bark. As a third case, near a highway the surface contamination of the needles is very low and the enrichment of Zn in the bark is small (see fig. 3).

4. CONCLUSIONS

Our results presented in form of coefficients of distribution at Stolberg and at other locations near a highway let derive the requirement that deposition measurements of air pollutants on trees, especially conifers, must include the bark compartment and cannot be restricted only to needles or leaves.

The transport processes in a tree have to be investigated in further details taking into account the bark compartment.

Acknowledgements

The authors gratefully acknowledge the Projekt Europäisches Forschungszentrum für Maßnahmen zur Luftreinhaltung (PEF) for financial support under No. 83/008/2.

The neutron irradiation was kindly provided by the GKSS Forschungszentrum, Geesthacht.

For technical assistance in the field and in the laboratory we thank Mrs. M. Hepp. For preparing the plant material for INAA we are indebted to

Mrs. I. von Nickisch-Rosenegk.

REFERENCES

(1) RENTSCHLER, I., Naturwiss. 69 (1982) 240.
(2) CERCASOV, V., Atmos. Environ. 19 (1985) 681.
(3) RENTSCHLER, I. and SCHREIBER, H., in Projekt Europäisches Forschungszentrum für Maßnahmen zur Luftreinhaltung (PEF), KfK-PEF 2, 1985, 413.
(4) RENTSCHLER, I. and SCHREIBER, H., in Projekt Europäisches Forschungszentrum für Maßnahmen zur Luftreinhaltung (PEF), KfK-PEF 4, 1986, 703.
(5) NÜRNBERG, H.W. et al., in Anhang III von Umweltprobleme durch Schwermetalle im Raum Stolberg 1983, Ministerium für Arbeit, Gesundheit und Soziales, NRW, 1983.
(6) CERCASOV, V. and SCHREIBER, H., in 7th International Conference Modern Trends in Activation Analysis, Isotope Division Risø National Laboratory, 1986, 973.
(7) MAYER, R. and ULRICH, B., in Deposition of Atmospheric Pollutants, D. Reidel Publ. Comp., 1982, 195.

AMMONIA AND PINE TREE DIEBACK IN BELGIUM

L. DE TEMMERMAN, A. RONSE, K. VAN DEN CRUYS, K. MEEUS-VERDINNE
Institute for Chemical Research,
Ministry of Agriculture, BELGIUM

Summary

Ammonia and ammonium depositions are supposed to be involved in pine dieback in the north of Belgium. Dieback symptoms appear mainly on Corsican and Austrian pines. In some cases damage has also been found on Scots pines.
Frost damage occurred on Corsican pines, mainly after the severe winter of 1984-1985.
Outbreaks of fungal infections, mainly *Sphaeropsis sapinea* (Fr) Dyko & Sutton, has been observed mostly on Corsican, Austrian and sometimes on Scots pines. In the most severe cases, sports consisting of 5 to 10 trees were dying at several places, even in young pine stands.
As most severe effects occur in areas with very intensive animal husbandry, high ammonia and ammonium input are likely to be involved. It has been shown that Corsican pine trees with frost damage and severe fungal diseases contain in general much more nitrogen than undamaged trees.
Dieback due to ammonia emissions can occur in areas where the estimated nitrogen production from animal wastes amounts to more than 350 kg N $ha^{-1}.year^{-1}$(arable land).
The high ammonium input in the forest ecosystems on poor sandy soils is also disturbing the nutrient supply of the trees. Relative deficiency of mineral elements such as magnesium and phosphorus occur,which makesthe trees more sensitive to pests and climatic stresses.

1. INTRODUCTION

A very intensive bioindustry has developed in the northern part of Belgium especially in areas with very poor sandy soils. Calculated on the municipality level the nitrogen production reaches 300 to more than 700 kg N per hectare arable land and per year (1).

Since 1985 problems of pine dieback have been observed in the areas with the most intensive bioindustrial activities namely in the north of Antwerp and West Flanders. The observed effects were frost damage on Corsican pines and also rather large infections of fungal diseases, mostly *Sphaeropsis sapinea* (Fr) Dyko & Sutton (2).

In the case of *Sphaeropsis* infection, the shoots of the current year are not growing out completely which gives them a bushy appearance. These problems are likely to be related to the high ammonia emissions as is shown in the schematic overview in fig. 1. Indeed, ammonia and ammonium are involved in several processes in air, soil and plant systems as well by direct depositions on the leaves as by input in the soil. In addition, ammonium is causing indirect effects by enhancing acid deposition, soil acidification, etc.

Since ammonia disturbs the nutrient balance in poor soils (relative) deficiency of mineral elements occurs, which is reflected by the ratios of total nitrogen contents to contents of these elements. Especially magnesium deficiency has been found, together with relative phosphorus and calcium deficiency (2). Frost damage on Corsican pines occurred, mainly after the severe winter of 1984-1985. The symptoms observed were : discolouration of needles, precocious loss of the oldest needles and black discolouration of the inner part of the bast (phloem) of the stem at 1.50 m above the ground level.

Ammonia emissions have not been measured yet but nitrogen productions by animal wastes per hectare arable land have been calculated at the municipal scale.

2. OBSERVATIONS IN FEBRUARY AND APRIL 1987

In February and April, the research has been focused on the most western province of the north of Belgium, West Flanders. 13 localities have been observed. Special attention has been paid to the occurrence of fungal attacks. The observed tree species were *Pinus nigra* subsp. *laricio* and *austriaca, Pinus sylvestris* and *Pseudotsuga menziessi*.

Sphaeropsis sapinea has been found to be a widely spread disease occuring mostly on *Pinus nigra*. Corsican pines as well as Austrian pines were infected. In some cases Scots pines were infected as well. The most severe effects were found in areas with a very high nitrogen production (from livestock) per hectare arable land. In four localities dieback of groups of 5 to 15 trees has been found at several places in pine stands. Even young stands of Corsican and Austrian pines were severely damaged by the fungus.
On 5 other plots, the disease occured but no dieback of trees or only dieback of a single specimen has been observed.

3. RESULTS OF NEEDLE ANALYSIS AND DISCUSSION

Needle samples have been taken in February as well as in April 1987. In addition to samples of Corsican pines also needles of Scots pine and Douglas fir have been taken. As there was only one sampling place of Austrian pine, it has been included in the group of Corsican pines.

The parameters, which have been found to be directly related to the high ammonia/ammonium input in the soils and on the trees, are the total nitrogen content and probably the extractable ammonium content of the needles.

The determinations were carried out on dried (75°C) and ground needles. The total nitrogen content was determined with the Kjeldahl method while the extractable ammonium content was determined by means of a specific ammonia electrode in a period of maximum 1 hour after an extraction of the samples with water. The extractable ammonium content consists of the free ammonium fraction in the needles and probably also the weakly bound ammonium in cell compounds.

TABLE I. TOTAL NITROGEN AND EXTRACTABLE AMMONIUM CONTENT OF NEEDLES

		Total nitrogen content %		Ammonium content mg kg^{-1}	
		1st year needles	2nd year needles	1st year needles	2nd year needles
Pinus nigra	$\bar{x}$	1.45	1.50	89	81
subsp. *laricio* n = 40	min - max	1.1-2.0	1.15-3.0	16-379	14-303
Pinus sylvestris	$\bar{x}$	1.83	1.88	48	65
n = 20	min - max	1.62-2.08	1.61-2.51	20-135	25-179
Pseudotsuga	$\bar{x}$	1.64	1.93	56	99*
n = 7 *menziessi*	min - max	1.46-1.81	1.61-2.26	23-95	44-267

* 2 nd.-5th year classes

Normal amounts of total nitrogen in needles are estimated to be between 1 and 1.5 % for trees growing on poor sandy soils (2).
As shown in table I, the average nitrogen contents of the needles are equal or higher than 1.5 %. It is impossible to make a clear distinction between the total nitrogen contents of Scots pine, Corsican pine and Douglas fir because the exposure of the different plots to ammonia/ammonium in ambient air is different. Indeed the ambient concentrations may differ a lot on relatively short distances because of the large spread of permanent (stables) and temporary (spraying manure) emission sources.

The extractable ammonium content of pine needles is normally 10 to 20 mg kg^{-1}. Again it is rather difficult to make a distinction in accumulation rate for the different tree species but far the highest concentrations have been found in the needles of *Pinus nigra*.

It is not yet clear if there is a relation between the extractable ammonium content of the needles and outbreak of diseases. Although Corsican pines are much more subject to *Sphaeropsis* infection than Scots pines in the studied area.

As has been described previously, the high nitrogen input in the forest ecosystem causes a general phosphorus deficiency and in some cases also a magnesium deficiency (2). The N/K, N/Mg and N/P ratios show also a relative potassium, magnesium and phosphorus deficiency (2). The data obtained during the most recent sampling periods in West Flanders are confirming those obtained earlier (tabel II). A relative calcium and potassium deficiency rarely occurs in the studied area.

TABLE II. NUTRIENT STATUS OF CORSICAN PINE NEEDLES IN WEST FLANDERS (1987)

		undamaged n = 4	moderately damaged n = 13	severely damaged n = 9	relative deficiency
Total N %	mean	1.32	1.34	1.68	-
	min-max	1.19-1.50	1.16-1.59	1.45-2.0	
N/Ca ratio	mean	5	5	8	> 10
	min-max	3-8	2-10	3-16	
N/K ratio	mean	2.7	2.6	2.8	> 4.3
	min-max	2.3-3.1	2.0-3.9	1.8-4.8	
N/Mg ratio	mean	19	18	26	> 25
	min-max	13-21	12-31	17-40	
N/P ratio	mean	17	17	17	> 15
	min-max	14-18	12-23	13-22	
Distance to emission source (m)	average	> 3000	1200	600	
	min-max		200-3000	500-1000	

Up to now however no clear relationships could be found between the nitrogen production per ha arable land and the ammonium contents of the needles. Only a weak relation between the nitrogen production of livestock and the total nitrogen in the needles has been found (2). It must be stressed however that the nitrogen production per ha arable land is a very rough estimation of the ammonia/ammonium load in the neighbourhood of pine stands. Moreover the most severe outbreaks of *Sphaeropsis* disease are found in areas with a very high nitrogen production but not every pine stand is damaged to the same extent. Indeed, the epidemiology of fungal diseases is dependent on more parameters than only high ammonia/ammonium deposition on the needles. It has been found however that dieback of Corsican pines will occur mostly in areas where the estimated nitrogen production from animal wastes amounts to more than 350 kg $N.ha^{-1}.year^{-1}$ (2).

A more detailed study of the geographical distribution of the permanent sources in the studied area revealed that the most severe effects on pine stands occur always in a zone of maximum 1000 meter from bio-industrial emissions, while undamaged pine stands are more than 3000 meter away from important permanent emission sources.

Relative deficiency of magnesium and to a lesser extent calcium and potassium is occuring more frequently in severely damaged Corsican pine stands. Relative phosphorus deficiency is a more general fenomenon in the studied area.

Acknowledgement

The authors are much indebted to P. Coosemans and R. Van Cauter for the analytical assistance.

REFERENCES

(1) MEEUS-VERDINNE, K., SCOKART, P.O., GUNS, M., (1985). L'ammoniac émis par les déchets animaux et la pollution atmosphérique. Revue de l'agriculture, 38, 2, 239-51.

(2) RONSE, A., DE TEMMERMAN, L.O. and MEEUS-VERDINNE, K., (1987). Possible effects of ammonia and ammonium input in forest ecosystems. To be published in : Proceedings of the SCOPE meeting in Brussels on 28 November 1986 : "Agriculture and Environment".

Fig. 1. THE POSSIBLE ROLE OF AMMONIA IN THE PINE TREE DIEBACK

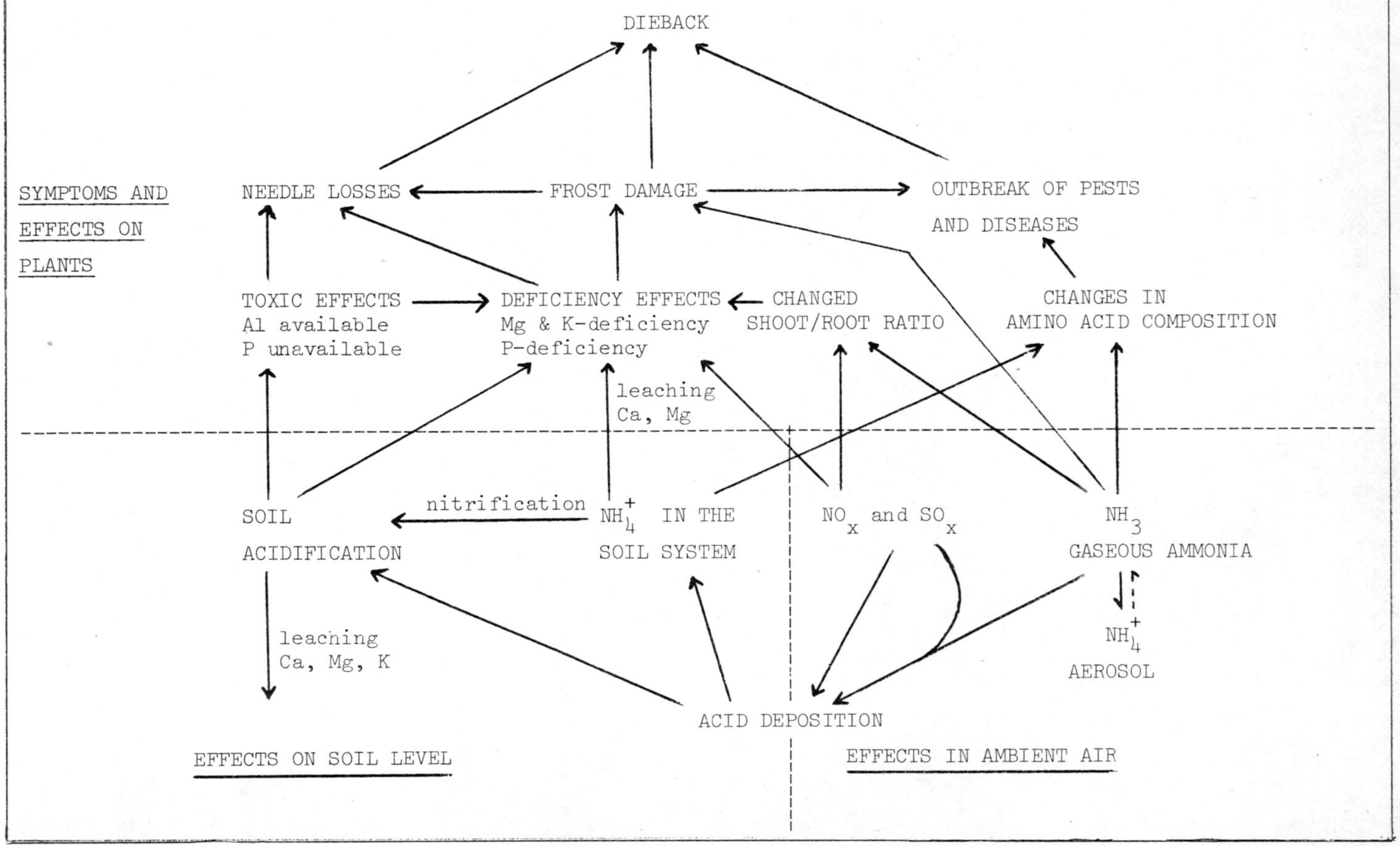

THE EFFECTS OF ACID MIST ON CONIFER APHIDS AND THEIR IMPLICATIONS FOR TREE HEALTH

N.A.C.KIDD and M.B. THOMAS
Department of Zoology, University College, Cardiff

Summary

In this study we investigated the responses of the aphid *Schizolachnus pineti* to both acid mist and "water stress" treatment of its host tree, *Pinus sylvestris*. Aphids on acid mist treated plants grew faster and attained higher population numbers than those on control plants. Water-stressing the trees, however, tended to reduce aphid growth performance. The proportion of winged (migratory) forms amongst the adults was also found to be higher on the acid-treated plants, suggesting that the presence of acid mist or acid rain may promote the spread of aphid infestation between trees. As high infestations of the trees can have damaging effects on the pine trees, the interaction between aphids and air pollutants may have a role to play in the current forest decline.

1. INTRODUCTION

A widespread decline in the forests of some central European countries has become apparent in recent years, affecting a number of tree species, particularly Silver Fir, Norway Spruce, Scots Pine and Beech (1). The symptoms of this "forest die-back syndrome" vary with species, but generally include needle-loss leading to thinning of crowns, colour changes in the needles (e.g. yellowing similar to that produced by Mg deficiency), death of roots and growth reduction (6).

So far, air pollutants such as sulphur dioxide and ozone have been considered as the most likely cause. However, some of the symptoms associated with the European forest decline are also commonly associated with attack by certain insect species, notably aphids. These parasites can infest needles, bark and roots, extracting sap from the phloem tissues as a food source. On Scots pine, for example, high numbers of the green aphid, *Eulachnus agilis*, induce yellowing, early senescence and dropping of the needles (3).

As these insects generally occur throughout the geographical range of their host plants, their possible involvement in the development of the forest die-back syndrome cannot be ignored. Aphids have a very intimate relationship with their hosts, influencing plant physiology through feeding and salivary secretions and, in turn, responding to subtle changes in plant chemistry. Changes in plant physiology, for example, can result in significant alterations in aphid growth rates, reproduction and survival (4). Pollution damage to plants has been shown to cause these effects in aphids infesting beans and peas (2,7). Were this also to be true of conifer aphids, it could, in turn, affect the susceptibility of the tree to further damage. Under certain circumstances, therefore, aphids could be an important link in the chain of events leading to forest decline.

As a first step in testing this hypothesis, we investigated in the laboratory the responses of the needle-feeding aphid *Schizolachnus pineti* to both acid mist and "water stress" treatment of its host tree, Scots Pine (*Pinus sylvestris* L.). This aphid was chosen from a number of species commonly infesting Scots Pine, as its effects on tree health have been examined in detail (5).

2. METHODS

The growth performance of the aphids was measured on 2 year old potted Scots pine saplings maintained at 15°C with an 18 hour daylength. At the start of the experiment the trees were in spring condition with buds expanding. To vary the "water stress" conditions, trees were either left unwatered throughout the experiment (drought), given a standard amount each day (normal), or left with the roots totally immersed during the experiment (waterlogged). In each water treatment, the foliage of half the trees were sprayed twice daily with 9 ml. sulphurous acid solution (857 µeq/l.) per tree, and the other half were sprayed with an equal quantity of distilled water.

Twenty newly-born aphids were first weighed and then introduced on to each tree. At the fourth instar stage, the presumptive form of each aphid was noted - apterous (wingless) or alate (winged). The presumptive apterae were then weighed again, and their mean relative growth rates (4) calculated. Ten of the presumptive apterae from each tree were then returned to grow to adulthood, reproduce and initiate a colony. The population numbers of these tree colonies were subsequently monitored over a period of thirty six days. At the end of the experiment the growth of the leader shoot of each tree was measured.

3. RESULTS

While no significant differences were found in individual aphid growth rates between water treatments, those aphids on acid-mist treated trees grew at a faster rate (Figure 1), with a greater proportion becoming winged adults (Figure 2). However, the watering regime also affected the proportion of winged forms, with fewer resulting from the waterlogged and unwatered trees.

The subsequent performance of the tree colonies was also very much influenced by water treatment. After 35 days population numbers on the normally-watered trees were significantly higher than those on the water-stressed trees (Figure 3). In fact, aphid numbers remained very low on the unwatered trees. In all water treatments, however, the acid-treated trees produced significantly higher aphid population levels (about 30% higher) than the controls (Figure 3).

At the end of the experiment, after 36 days, the growth of the leader shoots was slightly greater in the normally watered trees, compared with the waterlogged trees. In the unwatered trees there was no leader growth. Acid mist treatment had no significant effect on leader extension in any of the watering regimes.

4. CONCLUSIONS

The aphids on the acid mist treated trees grew faster and achieved higher population levels than those on control plants. Water-stressing the trees, however, tended to reduce the aphid growth performance. The proportion of alatae (migratory forms) amongst the adults was also found to be higher on the acid-treated plants, this being independent of aphid density. This result suggests that the presence of acid mist or acid rain may promote the spread of aphid infestation between trees.

Heavy infestations of *S. pineti* are also known to retard the growth rate of P. sylvestris and reduce its survival (5). Our results, therefore, support the view that aphid populations may have a role to play in the interactions between air pollutants and tree health and so should not be ignored in future investigations into the causes of forest decline.

REFERENCES

(1) BINNS, W.O. and REDFERN, D.B. (1983). Acid rain and the forest decline in West Germany. Forestry Commission Research and Development Paper 131.

(2) DOHMEN, G.P., McNEILL, S. and BELL, J.N.B. (1984). Air pollution increases *Aphis fabae* pest potential. Nature, 307, 52-53.

(3) KEARBY, W.H. and BLISS, M. (1969). Field evaluation of three granular systemic insecticides for control of the aphids, Eulachnus agilis and Cinara pinea. Journal of Economic Entomology, 62, 60-62.

(4) KIDD, N.A.C. (1985) The role of the host plant in the population dynamics of the large pine aphid, Cinara pinea. Oikos, 44, 114-122.

(5) LEWIS, G.B. (1987) Regulating interactions between pine aphid colonies and host plant growth. Ph.D. Thesis, University of Wales.

(6) SKEFFINGTON, R.A. and ROBERTS, T.M. (1985) The effects of ozone and acid mist on Scots pine saplings. Oecologia, 65, 201-206.

(7) WARRINGTON, S. (1987) Relationship between SO_2 dose and growth of the pea aphid, Acyrthosiphon pisum, on peas. Environmental Pollution (Series A). (in press).

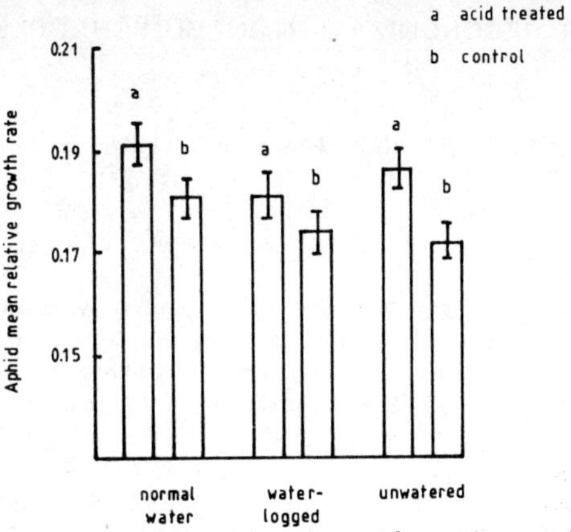

FIGURE 1 The mean relative growth rates of <u>S. pineti</u> in relation to acid mist treatment and water stress (with Standard Errors).

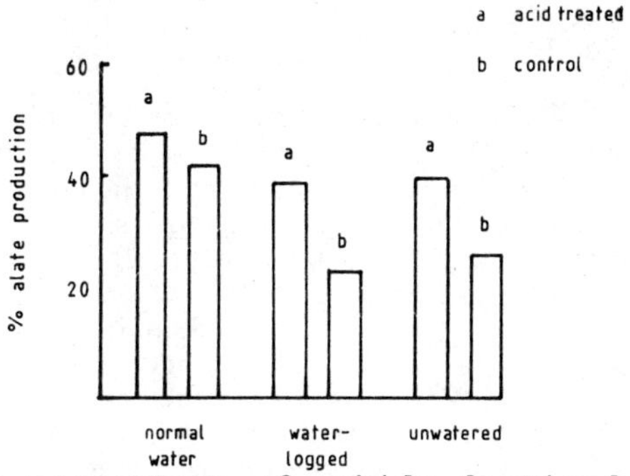

FIGURE 2 The percentage of aphids destined to be winged alatae in relation to acid mist treatment and water stress.

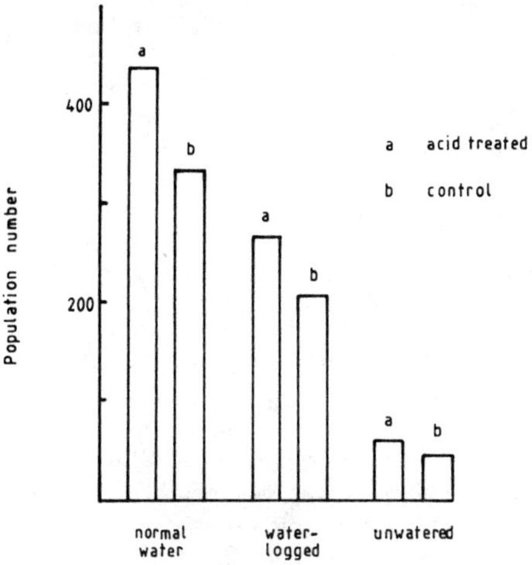

FIGURE 3 The average population densities of aphids per tree after 36 days in relation to acid mist treatment and water stress.

LICHEN AND CONIFER RECOLONIZATION IN MUNICH'S CLEANER AIR

O. KANDLER
Botanical Institute, University of Munich, Menzingerstr.67, D-8000 München 19

Summary

At the end of the 19th century the center of the city of Munich became deprived of lichens and spruce because of high concentrations of SO_2.However, since the end of the 1960ies, when an increasing portion of coal and oil was replaced by natural gas and electricity from atomic power stations, the air became cleaner (reduction of dust and SO_2 pollution), and even the city center was recolonized by spruce (Picea abies (L.) Karst.) and moderately sensitive lichens within a few years. Although air pollution caused by traffic has increased during the last few decades, no significant air pollution-dependent damage or growth inhibition of spruce or other trees can be observed even along roads with heavy traffic.

1. INTRODUCTION

Lichens and conifers are well known indicators of air pollutants (e.g. dust, SO_2, HF, NO_x, NH_3, etc.). Thus the mapping of lichens belonging to different sensitivity groups (1) allows the monitoring of air pollution even when quantitative analytical data on particular pollutants are missing.Thanks to the detailed description of the lichens in the Munich area by Arnold (2) at the end of the 19th century and similar work by later authors (3,4,5,6,7), Munich possesses the probably most complete record on the change of the lichen flora caused by the rise and fall of air pollution over the last 100 years. In addition, reports of Munich botanists on the damage of conifers (8,9) and chemists on the sulfuric acid content of snow (10) at the end of the 19th century are available. In the following paragraphs these historical data will be correlated with recent biological findings and chemical data on air pollutants.

2. LICHENS

"No lichens can be found anymore in the center of Munich, i.e. the part of town originally surrounded by a wall", reports Arnold (1) in 1891, although the occurrence of lichens in the very center of Munich had been reported in the 3rd quarter of the 19th century. In accordance with Nylander's earlier observations (11) in Paris, Arnold suggested that sulfurous acid, originating from the increasing use of coal instead of wood, was the causing agent. Based on Arnold's papers and collection of more than 1000 lichen specimen, still kept at the Herbarium of the Bavarian State, Schmid (3) and Mägdefrau (4) drew a map of the distribution of lichens in 1891 and compared it with the map derived from their own lichen monitoring in 1955 (Fig.1). The area of the so-called lichen desert grew from 8 km² in 1891 to 58 km² in 1955, thus reflecting the increase in population (from 300 000 to 1 million) and in industrialization. The zone exhibiting a damaged and reduced lichen flora (not shown in Fig.1) covered the whole built city area.

Although the population of Munich increased further to 1.3 million and industrialization continued, the area of the lichen desert was found to have decreased by 30 % when the next mapping was performed by Jürging in 1967 (Fig.1). Before Jünging published his study (5), he reinvestigated some of the lichen-free locations in 1973 and found now small thalli of 3 species of foliose lichens on trees in the Old Botanical Garden and the Hofgarten (Tab.1). At the Theresienwiese (Fig.1), an enclave in the lichen desert where only the crustacous lichen

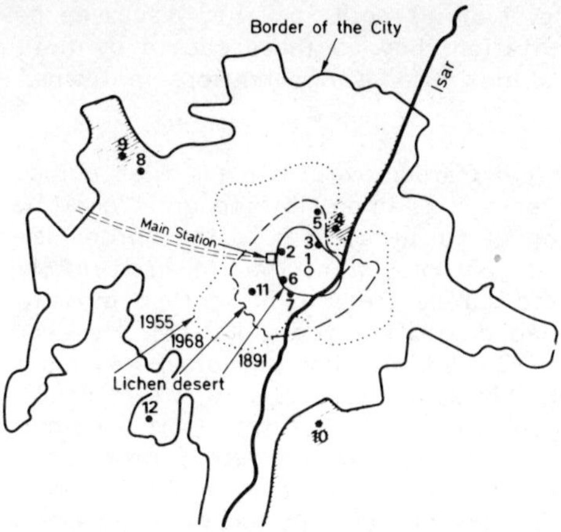

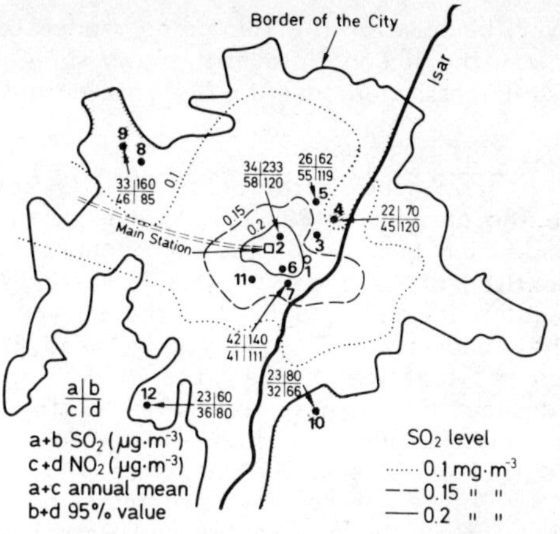

Fig.1: Extension of the lichen-free area (lichen desert) in Munich from 1891 to 1968.

Fig.2: Isopleths of SO_2 concentrations in 1967, annual mean and 95 % values of SO_2 and NO_2 levels in 1985.

1 = City center (Cathedral); 2 = Old Botanical Garden; 3 = Hofgarten; 4 = English Garden; 5 = Academy of Fine Arts; 6 = Sendlingertorplatz (near Institute of Hygiene); 7 = Southern Cemetery; 8 = Botanical Garden Nymphenburg; 9 = Kapuzinerhölzl; 10 = Perlacher Forest; 11 = Theresienwiese (a fair ground); 12 = Forstenried.

Lecanora varia had been found in 1968, several foliose species where present in 1973, indicating the onset of recolonization. A reinvestigation of the lichen flora in the city center by Kandler and Poelt (6) in 1983, revealed a set of 12 mainly foliose species growing abundantly not only on the trunks but also up in the branches of trees in the Old Botanical Garden and in the Hofgarten (Table 1). Also in the outer regions of the town, an increase in the more sensitive species was found. Thus, a distinct progress of recolonization was obvious. These findings were corroborated by a thorough quantitative monitoring of lichens commissioned by the municipial council in 1985 and carried out by Macher and Steubing who state (7) that no lichen desert exists anymore in the city area of Munich. How-

Table 1: Migration progress of lichens during the recolonization of the city center of Munich.

Species	present in 1983 Hof-garten	present in 1983 Old Bot. Garden	Distance of find closest to Old Bot. Garden in 1968 (km) 1.5	2	3.5	4-5	Sensitivity (1) Group	Growth limit $\mu gSO_2/m^3$
Candelariella reflexa	-	x	-	-	-	-	-	-
Hypogymnia physodes	x b)	-	-	-	x	x	5-4	60-70
Lecanora hageni	x	-	x	x	x	-	3	125-150
Parmelia exasperatula	x	- a)	-	-	-	x	6-7	40-50
Parmelia sulcata	x b)	x b)	-	x	x	x	4-5	50-60
Physcia adscendens	x b)	x b)	-	x	x	x	4	70
Physcia dubia	x	x					-	-
Physcia insignis	-	x					6-5	50-60
Physcia orbicularis	-	x	-	-	-	x	6-5	50-60
Physconia grisea	-	x					6-5	50-60
Xanthoria fallax	-	x					-	-
Xanthoria parietina	x b)	x	-	x	x	x	5-4	60-70

a) present in 1985 (7), b) first thalli observed in 1973 (5) ever, because of the remaining moderate level of air pollution - as discussed below - the lichen flora still shows some limitations beyond those caused by the usual restrictions due to the unfavourable climatic and soil conditions in towns.

3. CONIFERS

K.v.Goebel, director of the Old Botanical Garden, wrote in his first guide to the garden in 1899 (9): "Since several years it is impossible to grow conifers in Munich because of the high concentration of sulfurous acid ... the garden is deprived of its beauty". Also R.Hartig, forest pathologist at the University of Munich, reported in 1896 that the 60 years old spruce trees in his garden close to the Academy of Fine Arts (Fig.1 and 2) had died off 8 years ago (8). He also noticed that the spruce trees in the English Garden and the Leopold Park, located about 500 m east and 200 m north from his home, respectively, were also heavely damaged or dying. Thus even a larger area than that found by Arnold to be free of lichens, was also free of spruce at the end of the 19th century.

Today, spruce trees are found not only in the garden of the Chemical Institute, originally part of the Old Botanical Garden, but also in the immediate neighbourhood of Hartig's former home, of the Leopold Park and of the English Garden. The parks themselves are free of spruce since the park administration is still reluctant to plant conifers again. However, thanks to private people who never stopped planting conifers in their gardens, several hundred spruce trees are presently found within the original lichen and spruce desert. None of these trees is older than about 25 years indicating that planting of spruce was only successful after the early 1960ies, the time when also the recolonization of the lichen desert started. Almost all these spruce trees look very healthy, with up to 10 green needle seasons and an annual hight increment of up to 80 cm.

4. AIR POLLUTANTS

As early as in 1885, the alarming weathering of marble monuments in Munich led Sendtner (10) to analyse the sulfuric acid content of snow. Samples collected in winter 1885/86 in the Southern Cemetery and in the yard of the Institute of Hygiene showed an increase in sulfuric acid from 8 mg/kg on the day of snow fall to about 30 and 75 mg/kg in 2 and 10 d old snow. On the other hand, 2 and 10 d old snow collected in January 1987 at about the same locations (12), contained only 2.5 and 20 mg sulfuric acid/kg, respectively. The comparison of the old and new data reveals a several times higher accumulation of sulfuric acid in snow in 1885/86.

The first reliable measurements of SO_2 in Munich in the 1950ies revealed SO_2 levels of 50 to several hundred µg/m³ and maximum values of >2500 µg/m³ for several hours (14), depending on the season and location (13). The air pollution culminated in the early 1960ies, when the "Raumordnungsbericht" of the Ba-

Table 2: Sulfate content of snow at different locations in Munich in winters of 1885/86 and of 1986/87. (mg SO_4^{2-}/kg snow)

Location	Winter 1986/87 Age of snow (d)			Winter 1885/86 Age of snow (d)			
	1-2	4-5	9-10	0	2	10	14
Kapuzinerhölzl	1.1	2.3	10.7	-	-	-	9.3 a)
Sendlingertorplatz	2.5	-	19.5	8.4 b)	34.7	74.6	73.2
Southern Cemetery	2.2	-	21.0	8.6	29.2	-	-

a) Forstenried; distance to city center similar Kapuzinerhölzl (Fig.1).
b) Yard of the Institute of Hygiene ca. 300 m from Sendlingertorplatz.

Table 3: Annual increment in spruce and annual mean concentration of pollutants in Kapuzinerhölzl at various distances from Menzingerstraße (48 000 vehicels/d).

Group (Fig.3)	Age of trees	Distance m	n	Mean annual ring width (mm) 1940-'49	'50-'59	'60-'69	'70-'74	'75-'79	'80-'86	S-content of needles g/kg dw.	Pollutants[d)] in µg/m³ CO	SO_2	NO	NO_2	Pb	ng/m³ Cd
1	25-30	0.0[a)]	10	-	-	3.26[b)]	4.05	4.39	4.07	1.203	4010	57	198	122	1.5	3.4
2	30-35	3-30	9	-	-	4.16	4.30	4.60	3.89	1.202	3590	52	160	116	1.2	3.0
3	60-100	70-100	6	1.69	1.81	1.75	2.24	2.39	2.42	1.090	2430	37	54	76	0.5	1.8
4	80-100	ca. 150	5	2.51	2.21	1.96	2.47	3.18	3.07	1.077	2300	39	42	66	0.4	1.7
5	80-100	250-300	4	1.1	1.05	0.70	1.02	1.31	1.22	1.045	2210	34	40	51	0.4	1.6
										Basal load without traffic	2100 (4500)[c)]	33 (90)	30 (60)	44 (80)	0.3 (1.2)	1.5 (3.0)

a) traffic island; b) period 1965-'69; c) 95 % value;
d) Pollution gradient kindly calculated by Dr. Rabel, Bayerisches Landesamt für Umweltschutz.

varian government 1963/64 reported an annual mean of 260 µg SO_2/m³ and a maximum daily mean of 2500 µg/m³. From then on, the progressing replacement of coal and oil by natural gas and by electricity derived from atomic power led to a dramatic drop of dust and SO_2 output. The SO_2 concentration decreased to an annual mean <40 µg/m³ in the early 1970ies and became stabilized at this level (15, Fig.3). The originally radial pollution gradient in the city area with its center around the main station (Fig.2), where also several breweries are located, changed to an almost uniform distribution of pollutants with a slight gradient from NW to SE as indicated by the SO_2 and NO_2 levels given for a few locations in Fig.2 (data taken from (16)). The highest pollution levels, especially with respect to NO_x, are found along the feeder roads of the autobahn. Here, the basic pollution load is overlaid by the steep gradient of pollutants generated by traffic. The Botanical Garden Nymphenburg and the Kapuzinerhölzl, an oak and beech forest of about 15 ha with some scattered spruce, are such areas. Both are located along Menzingerstraße (Fig.4), the feeder road of the autobahn to Stuttgart (48 000 vehicels/24 h).

In spite of the distinct gradient of various pollutants including heavy metals (Table 3), also reflected by the lichen flora (Wiegel, unpubl.) and a slightly increased sulfur content in the youngest needle season of trees close to the road (group 1,2, Table 3), the trees look healthy throughout the forest and the annual

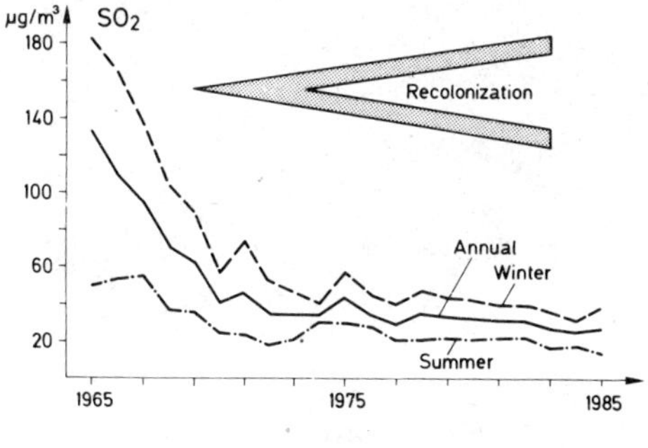

Fig.3: Decrease in SO_2 concentrations in the city of Munich from 1965 to 1985 (mean values from 10 monitoring stations).

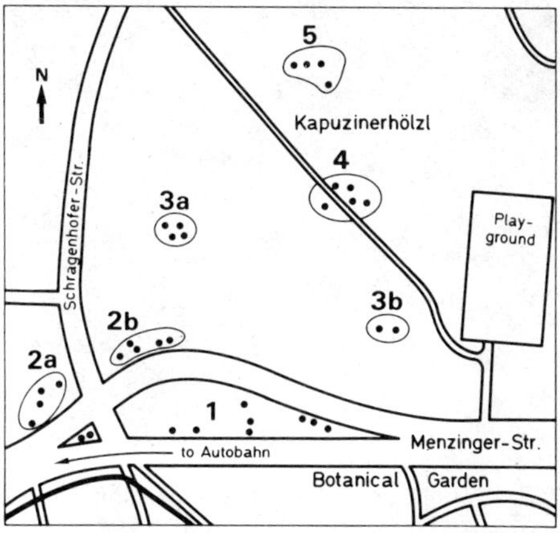

Fig.4: Map of Kapuzinerhölzl opposite the Botanical Garden Nymphenburg. Numbers indicate location of the groups of spruce listed in Table 3.

Fig.5: Condition of a group of conifers in the Botanical Garden, 5-30 m from Menzingerstraße, in 1974 and 1986. Spruce trees showing crown transparency are indicated by an arrow (Foto K.Liedl). 1) Picea abies; 2) P.abies "Finedonensis"; 3) P.abies "Aurea"; 4) P.abies "Argenteo-spica"; 5) P.abies "Cupressina"; 6) Picea glauca; 7) Picea omorica; 8) Thuja occidentale "Globosa"; 9) Thuja occidentale var.; 10) Thujopsis dolabrata.

increment of spruce shows no correlation with the pollution gradient (Table 3). Since annual increment depends on the age of the tree and soil conditions, the course of increment rather than its absolute level is relevant for the present consideration. Actually, a decrease in annual increment was to be expected during the last few decades as traffic has increased >2-fold since 1960 in Menzingerstraße. However, no decrease in annual increment of spruce was observed (Table 3) indicating that air pollution has not yet reached a level at which the inhibitory potential of the pollutants leads to apparent inhibition. Correspondingly, comparison of 2 photographs taken in 1974 and 1986 of a group of conifers growing in the Botanical Garden within a zone of 5-30 m from Menzingerstraße (Fig. 5) shows no recent damage. The transparency of crowns in some of the spruces nowadays often considered as one of the new symptoms subsumed under the term "Waldsterben", and as an indicator of decay within 3 years (17), was at least in 1974 as distinct as today and may also be seen on much older photographs (18). Thus, crown transparency, neither a new symptom nor an indicator of fast decay, does not show a correlation with the course of air pollution, but may be caused by various conditions not fully understood as yet. Moreover, no distinctly pollution-related symptoms or damages have been observed in any of the 15.000 plant species grown in the Botanical Garden - with one exception: the flowers of some of the tropical orchids kept in the green-house close to Menzingerstraße exhibit distinctly enhanced senescence during inversion episodes when air pollution culminates. Furthermore, trees in other areas of the city monitored by aerial photographs, in general, showed good health. Occasional damage was considered to be caused by unfavourable climatic and soil conditions (19). In contrast, so-called "Waldsterben" has been reported from the forests surrounding Munich (17) since the early 1980ies (e.g. Perlacher Forst, Forstenried (20)), although all the known air pollutants, with the exception of ozone, show a significantly lower level in those areas than in the city. This discrepancy has also been observed in other cities as recently pointed out by Benarie (21) thus questioning the assumed role of air pollutants in the so-called "Waldsterben".

5. FINAL REMARKS

The history of the rise and fall of SO_2 levels and the devastation and recolonization of the city center of Munich shows once again that SO_2 is the dominating vegetation-relevant component of air pollution in large cities. History also shows that poisoning air pollution is not a one-way street unequivocally coupled with industrialization and growing cities. Technical development and rational measures have reversed the increase of the most destructive components, SO_2, Pb and dust, not only in Munich but also in most other cities, even in highly industrialized areas, e.g. the Ruhr area (22), at a time when ENVIRONMENT was not a main subject of public opinion and journalism. This experience makes us hope that also remaining and new problems arising from new types of pollutants will be kept under control by the combined effort the scientific community and administration, and the good will of people.

6. REFERENCES

(1) HAWKSWORTH, D.L. and ROSE, F. (1970). Qualitative scale for estimating sulphur dioxide air pollution in England and Wales using epiphytic lichens. Nature **227**, 145-148.
(2) ARNOLD, F. (1891 to 1901). Several papers in: Ber.d.Bayer.Botan.Ges. Vol. **1** to **8**.
(3) SCHMID, A. (1957). Die epixyle Flechtenvegetation von München. Thesis

Universität München.
(4) MÄGDEFRAU, K. (1960). Flechtenvegetation und Stadtklima. Naturw. Rdsch. **6**, 210-214.
(5) JÜRGING, P. (1975). Epiphytische Flechten als Bioindikatoren der Luftverunreinigung - dargestellt an Untersuchungen und Beobachtungen in Bayern. Bibliotheca Lichenologica **4**, 1-164.
(6) KANDLER, O. and POELT, J. (1984). Wiederbesiedlung der Innenstadt von München durch Flechten. Naturw. Rdsch. **37**, 90-95.
(7) MACHER, M. and STEUBLING, L. (1985). Flechtenzonierung des bebauten Stadtgebietes von München. Report of the Municipial Council Munich.
(8) HARTIG, R.(1986). Über die Einwirkungen des Hütten- und Steinkohlenrauches auf die Gesundheit der Nadelwaldbäume. Forstl.-naturwiss.Zeitschr.**V**, 245-290.
(9) GOEBEL, K. (1899). Führer durch den Kgl. botanischen Garten in München. Val. 'Höfling, München.
(10) SENDTNER, R. (1887). Schweflige Säure und Schwefelsäure im Schnee. Bayer. Industrie- & Gewerbeblatt **1887**, 67-81.
(11) NYLANDER, W. (1866). Les Lichens du Jardin du Luxembourg. Bull. Soc. Bot. France **13**, 364-372.
(12) KANDLER, O., NÖTH, H. and SCHÄFER, R. (1987). Sulfuric acid of snow in the city area of Munich in the winter 1986/87 (paper in preparation).
(13) STRATMANN, H. (1959). Schwefeldioxyd-Immissionen eines Heizkraftwerkes in München. Staub-Reinhaltung der Luft **19**, 352-360.
(14) KELLER, H. (1958). Beiträge zur Erfassung der durch schweflige Säure hervorgerufenen Rauchschäden an Nadelhölzern. Veröffentl.aus dem Forstbot. Inst.d.Bayer.Forstl.Forschungsanstalt München. Paul Parey, Hamburg-Berlin.
(15) ANONYMUS (1986). Lufthygienischer Jahresbericht 1985. Schriftenreihe Bayerisches Landesamt f.Umweltschutz, Rosenkavalierplatz 3, D-8000 München 81.
(16) HERMANN, J. (1986). Immissions-Vorbelastung im Gebiet um die Heiz- und Wärmekraftwerke der Stadtwerke München. Report of the Stadtwerke München.
(17) SCHÜTT, P. and COWLING, E.B. (1985). Waldsterben, a General Decline of Forests in Central Europe: Symptoms, Development, and Possible Causes. Plant diesease **69**, 548-558.
(18) KANDLER, O. (1987). Klima und Baumkrankheiten. p.269-275. In: Proceedings of the Symposium: Klima und Witterung in Zusammenhang mit den neuartigen Waldschäden. GSF-Bericht 10/87, GSF, Ingolstädter Landstraße, D-8042 München-Neuherberg.
(19) HAYDN, R. et al. (1985). Interpretation des Vitalitätszustandes von Bäumen aus digitalisierten Farbinfrarot-Luftaufnahmen. Das Gartenamt **34**, 687-695.
(20) SCHMIDT, A. and HARTMANN, R. (1984). Die Abhängigkeit der neuartigen Waldschäden von Standort und Wasserversorgung im Forstamt München. Allg. Forstzeitschrift, 552-553.
(21) BENARIE, M. (1985). Lack of relationship between acidic rain and tree damage in urban areas of Europe. The Science of the Total Environment **43**, 185-186.
(22) WIEGEL, H. and RABE, R. (1985). Wiederbesiedlung des Ruhrgebiets durch Flechten zeigt Verbesserung der Luftqualität an. Staub-Reinhaltung der Luft **45**, 124-126.

ACKNOWLEDGEMENT:

The data compiled in Table 3 are taken from a project commissioned to the author and financed by the Bayerisches Staatsministerium für Landesentwicklung und Umweltfragen.

THEME IV : FATE OF AIR POLLUTANTS IN SOILS. EFFECTS ON SOILS AND CONSEQUENCES FOR PLANTS

THEME IV : SORT DES POLLUANTS DANS LES SOLS. EFFETS SUR LES SOLS ET CONSEQUENCES POUR LES PLANTES

THEMA IV : SCHADSTOFFVERLAUFE IN DEN BODEN. WIRKUNGEN AUF DEN BODEN UND ETWAIGE KONSEQUENZEN FUR PFLANZEN

LONG-TERM CHANGES IN THE POLYNUCLEAR AROMATIC HYDROCARBON CONTENT OF AGRICULTURAL SOILS

K.C.JONES[1], J.A. STRATFORD[1],
K.S. WATERHOUSE[1] and A.E. JOHNSTON[2]

[1]Department of Environmental Science, University of Lancaster, Lancaster LA1 4YQ, U.K.

[2]Soils and Plant Nutrition Department, Rothamsted Experimental Station Harpenden, Hertfordshire AL5 2JQ, U.K.

Summary

Soil samples collected from Rothamsted Experimental Station in south east England at various times since the mid-1800's and up to the present day have been analysed recently for polynuclear aromatic hydrocarbons (PAHs). All the soils were collected from the plough layer (0-23 cm) of the Broadbalk experimental plot, for which atmospheric deposition will have been the only source of PAH input. The total PAH burden of the plough layer has increased approximately five-fold since the 1880/90's, with some compounds (notably benzo(b)fluoranthene, benzo(k)fluoranthene, benzo(a)pyrene, benzo(e)-pyrene, pyrene, benzo(a)anthracene and indeno (1,2,3-cd)pyrene) showing substantially greater increases. Average rates of increase for individual PAHs in the Rothamsted plots over the century since 1880/90 vary between 0.05 - 0.68 (mean = 0.3) mg m^{-2} $year^{-1}$. These fluxes are similar to contemporary atmospheric deposition rates to semi-rural locations. Regional fallout of anthropogenically-generated PAHs derived from the combustion of fossil fuels will be the principal source of PAHs to the Rothamsted soils. It is suggested that the increase in soil PAHs observed this century at Rothamsted are representative of those likely for agricultural soils in many industrialised countries or regions.

1. INTRODUCTION

Polynuclear aromatic hydrocarbons (PAHs) are a group of environmental contaminants, many of which are known to be mutagenic and carcinogenic. Many previous studies have examined temporal trends in their environmental levels using dated marine and freshwater sediment cores as the sampling medium, with the majority of work undertaken in the USA and mainland Europe (1). In general these studies have found: (i) an increasing sediment PAH burden since the mid-1800's, with a peak in the 1950/60's, and (ii) a constant qualitative PAH pattern for most of the locations studied, with an increase in PAH abundance near urban centres. These observations, and other studies, have pointed to anthropogenic combustion of fossil fuels as the major source of environmental PAHs and to the significance of long range atmospheric processes in dispersing PAHs through the environment.

This is the first study to investigate long-term changes in the atmospheric fallout of PAHs by using archived soils as the sampling medium, but there are several advantages to this approach as opposed to using sediment cores. Firstly, the sampling dates are known with certainty and the samples have been undisturbed since collection and preparation for storage. Secondly, the control plots at Broadbalk will only have received inputs via the atmosphere, whereas the interpretation of some sediment cores may be confused by additional inputs of runoff from the surrounding catchment. Thirdly, the soils will have received inputs directly from the atmosphere, whilst material deposited in sediments may have undergone chemical/physical changes during passage through the water column, and received additional inputs of biogenic PAHs. Two further differences should be acknowledged - the potential for direct atmosphere-soil exchange of PAH via the vapour phase prior to sampling, and photolysis of PAH adsorbed to particulates at the soil surface. Once in storage, however, photolytic and microbial degradation is likely to have been negligible as the samples were kept in the dark in sealed glass containers.

In addition to using these samples as an historical monitoring tool, there is another purpose to the investigation of PAH in soils. Dietary intake has been identified as the principal route of human exposure to PAH for non-smokers, with plant-based foodstuffs constituting roughly 50 per cent of the total PAH intake in a typical UK diet. Whilst it is likely that the majority of PAH associated with crop plants result from direct atmospheric deposition and soil splash onto the leaves and shoots, or are introduced in the preparation and cooking of foods, uptake of PAH via the root system has also been reported.

The major long-term toxic effects of PAHs is considered to be carcinogenesis, primarily at the point of entry to the body. While there does appear to be some evidence linking atmospheric PAH pollution and cigarette smoking in particular, with increased lung cancer incidence there has been no evidence reported as yet implicating ingested PAHs with increased cancer incidence. The effect of diet generally has not been considered in studies to elucidate the possible carcinogenicity of PAHs to human populations.

2. MATERIALS AND METHODS

Rothamsted (grid reference TL 120137) is a 'semi-rural' location in western Europe, 42 km north of central London (2). Soil collection, processing and analysis for PAHs are described in detail elsewhere (3,4). In brief, soil samples collected from the Broadbalk continuous wheat experiment (2) and stored in 1846, 1881, 1914, 1956 and 1980 were analysed by gas chromatography-mass spectrometry. Additional samples were analysed by another procedure, but these are referred to elsewhere (3).

3. RESULTS AND DISCUSSION

Data for PAH compounds and the 'total PAH' content of each sample are presented in Table I, together with calculated flux rates to the plots. Total PAH data are also plotted in Figure 1.

Temporal trends in soil PAH

The soil plough layer PAH burden at Broadbalk has increased roughly five-fold between 1880/90 and the present. Samples collected prior to 1865 at Rothamsted were not processed and stored in the same way

TABLE I. PAH CONTENT OF BROADBALK SOILS, ROTHAMSTED ($\mu g\ kg^{-1}$ dry weight) AND THE CALCULATED NET CHANGE IN SOIL PLOUGH LAYER PAH BURDEN ($mg\ m^{-2}\ year^{-1}$) SINCE 1880/90.

YEAR	PHENAN.	ANTHRAC.	FLUORAN.	PYRENE	B(a)ANTH	CHRYSENE	B(e)P	B(a)P	PERY
1846	53	3.6	34	23	29	27	27	18	17
1881	68	13	45	14	9	16	13	6.7	0.86
1914	89	ND	37	11	5.9	18	11	19	ND
1956	121	10	190	124	69	87	65	73	15
1980	137	13	206	154	114	122	144	133	18
Net change $mg\ m^{-2}\ y^{-1}$	0.20	-	0.46	0.40	0.29	0.30	0.38	0.36	0.05

YEAR	ANTHAN.	B(ghi)PY	B(b)Fln	B(k)Fln	NAPTH.	ACENAPTHYL	CORO	ΣPAH
1846	0.28	22	18	17	39	1.6	10	339
1881	0.12	8.4	12	8.4	38	0.73	5.4	259
1914	ND	6.3	86	6.2	53	ND	5.4	348
1956	1.2	55	76	73	28	3.4	18	1009
1980	3.6	66	244	247	23	5.0	22	1652
Net change $mg\ m^{-2}\ y^{-1}$	0.10	0.16	0.67	0.68	-	0.12	0.05	4.22

Compounds in full: Phenanthracene; anthracene; fluoranthene; pyrene; benzo(a)anthracene; chrysene; benzo(e)pyrene; benzo(a)pyrene; perylene; anthanthrene; benzo(ghi)perylene; benzo(b)fluoranthene; benzo(k)fluoranthene; napthalene; acenapthylene; coronene.

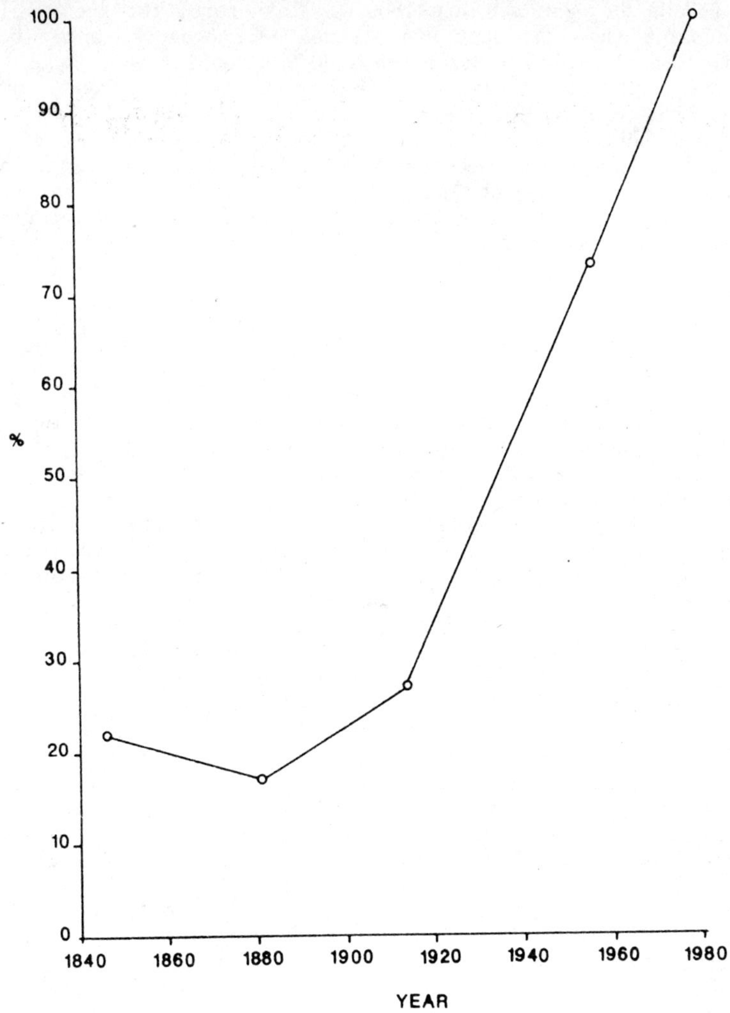

Figure 1. Temporal trends in soil plough layer PAH content at Broadbalk, Rothamsted. Data normalised against the most recent sample.

as those taken subsequently; i.e. pre-1865 samples were not air-dried and sieved, thereby allowing microbial activity to continue during sample storage. Consequently, samples taken earlier than this date have often yielded anomalous results in previous studies on the carbon and nitrogen content of the Broadbalk soils, despite the pre-1865 samples being re-processed (i.e. dried and sieved) in the late 1860's. Our data also yield rather surprising results for the 1846 sample, in that its total PAH content is higher than that measured in the later sample from 1881. Based on the evidence of dated lake sediment cores collected in mainland Europe and the USA one would expect an increase in the atmospheric deposition of PAHs through the late 1800's. Therefore, for the remainder of this discussion the 1846 data are treated as anomalous, and only the trends revealed in the century since the 1880/90 considered.

The total PAH data indicate a continued steady increase in the soil plough layer PAH burden at Broadbalk. This is of interest, in that the marked post-1950's decreases in the deposition flux of PAHs to both lacustrine and marine sediments which have been observed in North America and mainland Europe are not observed in this soil time series. However, without corroborating evidence, this should not be interpreted as indicative of a significantly different atmospheric PAH history for the U.K.

Deposition fluxes

Average annual rates of increase in soil PAHs at Broadbalk are presented in Table I. Since the plough layer at Broadbalk contains 2870 t soil ha^{-1}, changes in soil PAH concentrations with time may readily be converted to a net deposition flux (mg PAH m^{-2} $year^{-1}$). Fluxes for individual PAH compounds vary between 0.05 - 0.68 (mean = 0.30) mg m^{-2} $year^{-1}$, the most abundant of which are B(b)fln, B(k)fln, fluoran, pyrene, B(e)P, B(a)P, B(a)anth and phenan.

Losses of PAHs from the soil system are possible through microbial breakdown, photo-oxidation, vapourisation, crop offtake and leaching. The rates of increase in soil PAH at Broadbalk are therefore not necessarily equivalent to deposition fluxes to the soil. However, it is worth comparing the rates of increase with deposition flux data reported for other locations. Remote sites in the USA consistently have present-day deposition rates for individual PAHs of $\sim$0.01 mg m^{-2} $year^{-1}$; sites located nearer to urban centres have much greater current inputs (average of 0.35 mg m^{-2} $year^{-1}$)(5). This suggests that Rothamsted now receives substantial inputs of anthropogenically-derived PAHs from regional fallout from surrounding urban/industrial conurbations. Another possible source of PAHs to Rothamsted soils is localised fallout from stubble burning, a practice which is commonly adopted in the UK, but not on the Station's experimental plots. The importance of PAH inputs to soils from this source is unknown.

4. GENERAL COMMENTS AND CONCLUSIONS

The total PAH burden at Rothamsted in 1880/90 was similar to that observed in contemporary remote locations in the UK (4). Data on the PAH content of contemporary soils for numerous locations varying between 'remote' and 'urban' have been published elsewhere (see 3). With respect to these data it is apparent that the contemporary total PAH content at Rothamsted lies well within the range between the two, indicating the 'semi-rural' character of this site. It therefore seems reasonable to postulate that the increases in plough layer PAH levels observed at Rothamsted are not site specific and agricultural soils in many parts of the industrialized world will have been subject to similar dramatic increases over the last century. Soils nearer to major conurbations may be expected to contain levels considerably greater than those at Rothamsted, as will soils near to point sources and those receiving PAH inputs in other forms (e.g. sewage sludge). Furthermore, atmospheric inputs are likely to remain fixed in the top few centimetres of undisturbed (i.e. unploughed) soils, as opposed to the situation at Broadbalk where surface layer increases have been 'diluted' through the 0-23 cm depth.

The rates of PAH input (i.e. atmospheric deposition) clearly exceed rates of output (microbial breakdown, photo-oxidation, vapourisation, crop offtake and leaching) at Broadbalk. The similarities between the average annual rates of increase in the soil PAH burden and the likely average atmospheric deposition flux at Rothamsted suggest that losses via these five possible mechanisms probably effectively remove only a relatively small proportion of the total annual input. This implies a long residence time for PAHs in soils.

Intake of plant-based foodstuffs constitute a substantial proportion of ingested PAHs in the human diet. It is therefore also of importance to establish the significance of increases in soil PAH on the composition of major foodstuffs. In this connection we will subsequently be presenting data on PAHs in grain samples harvested over the last century from the Broadbalk plots.

ACKNOWLEDGEMENT

The authors are grateful to Prof. R.A. Hites and Dr. E.T. Furlong (Indiana University) for their considerable assistance with the analysis.

REFERENCES

(1) MONITORING AND ASSESSMENT RESEARCH CENTRE (1985). Historical Monitoring. MARC Technical Report No. 31, University of London.

(2) JONES, K.C., SYMON, C.J. and JOHNSTON, A.E. (1987). Retrospective analysis of an archived soil collection I. Metals. Sci. Total Environ. 61: 131-144.

(3) JONES, K.C., STRATFORD, J.A., WATERHOUSE, K.S., FURLONG, E.T., GIGER, W., HITES, R.A., SCHAFFNER, C. and JOHNSTON, A.E. Retrospective analysis of an archived soil collection III. Polynuclear aromatic hydrocarbons. For submission to Environ. Sci. Technol.

(4) JONES, K.C., STRATFORD, J.A., WATERHOUSE, K.S. and JOHNSTON, A.E. (1987). Polynuclear aromatic hydrocarbons in U.K. soils: long-term temporal trends and current levels. In: Proc. XXI Annual Conf. Trace Substances and Environmental Health, St. Louis, Missouri, USA.

(5) GSCHWEND, P.M. and HITES, R.A. (1981). Fluxes of polycyclic aromatic hydrocarbons to marine and lacustrine sediments in the northeastern United Stated. Geochim. Cosmochim. Acta 45: 2359-2367.

MOBILISATION DU FLUOR PAR ACIDOLYSE DES SOLS

J.A. BARDY et C. PERE
A.P.P.A. Comité Bordeaux-Aquitaine
Laboratoire Municipal de Bordeaux

Summary

Natural and anthropogenic soil-borne fluorine is leached when soil columns are continuously percolated with simulated acid rain (H_2SO_4 : 0,01 N). The peak concentration of fluorine which appears in the outflow samples is proportional to the height of the column. This is due to the acidolysis of clays in the upper layers which first liberate Al^{3+} and F^- . Then Al^{3+} is precipitated as $Al(OH)_3$ and F^- coprecipitated in deeper and deeper layers of the column. So, $Al(OH)_3$ and F^- are accumulated at the bottom. The adsorption of fluorine by $Al(OH)_3$ and clays is favoured by moderate acid pH conditions, but when pH values become sufficiently low $Al(OH)_3$ and F^- are released, giving soluble aluminium fluoride species which are less acid than Al^{3+}. The increase of the F/Al ratio in the leachates with the increase of the height of the peak concentration is not well elucidated. In sandy soils, poor in clays and treated with raw calcium phosphate, fluorine is more easily leached and appears in the leachates mainly in a free form aF^-, whereas in leachates of clay soils it is combined with Al. So, it may be thought that fluorine, of natural or anthropogenic origin, and mobilized by acid rain, is detrimental to forests.

1. PARTIE EXPERIMENTALE : ACIDOLYSE DE COLONNES DE TERRE

Résultats des essais précédents -

Dans une précédente étude (1), nous avions traité en continu des colonnes de terre de différentes hauteurs (h = 20 cm, 40 cm, 50 cm, Ø = 6,5 cm) par une solution sulfurique 0,01 N. Un test préliminaire consistait à traiter 50 g de terre par 100 ml de différentes solutions d'attaque : eau distillée, solutions sulfuriques 0,0001 N, 0,001 N, 0,01 N et 0,1 N, pour vérifier l'aptitude de chaque terre à libérer du fluor.

Les deux terres étudiées provenaient l'une du Jura et la seconde du Massif Central (PUY-MARY). La hauteur du pic de concentration en fluor qui apparaissait après l'addition d'un certain volume de la solution sulfurique était proportionnelle à la hauteur de la colonne : la figure I (lixiviation de la terre du Massif Central) est significative.

Avec la terre du Jura, le pic de concentration en fluor était atténué. Il s'agissait d'une terre pauvre, humifère et le fluor apparaissait dans les premières fractions de lessivats, ce qui ne permettait pas son accumulation. Une autre cause de la diminution du pic était due à la cheluviation des acides humiques qui complexaient le fluor à l'état organique non dosable à l'électrode spécifique.

Lors de l'acidolyse des colonnes de terre du Massif Central, les phénomènes suivants avaient ainsi été observés :

- Le pic de concentration en fluor apparaît d'autant plus tardivement dans les lixiviats et il est d'autant plus élevé que la colonne est haute.

- Les divers pics correspondent à une teneur maximale en potassium.
- Le pH des lixiviats, lors des maxima fluorés, est d'autant plus élevé que la teneur en fluor est importante.

Nous en avions déduit que le fluor était engagé dans des molécules solubles du type K_3AlF_6 dans lesquelles l'aluminium n'est pas hydrolysé. En fait, l'étude bibliographique sur le fluor dans le sol et sa spéciation dans les solutions permet de rectifier et d'affiner notre précédente interprétation :

- l'acidolyse décompose les argiles et solubilise le fluor et l'aluminium qui migrent et s'accumulent progressivement dans les parties de plus en plus profondes de la colonne.
- L'accumulation du fluor en bas de la colonne est donc proportionnelle à la hauteur de celles-ci. Cependant, il existe un facteur limitant à cette accumulation, qui correspond à la capacité maximale d'adsorption des argiles et de $Al(OH)_3$ vis-à-vis du fluor. Cette capacité maximale est d'ailleurs favorisée par des pH acides tant qu'il n'y a pas d'acidolyse notable des argiles.
- Lorsque la capacité maximale d'adsorption des argiles, non encore acidolysées en bas de la colonne, est dépassée, le fluor apparaît, dans les lixiviats, complexé à l'aluminium. Pour chaque terre, il existe donc une hauteur maximale de colonne à partir de laquelle le pic de concentration en fluor a atteint son maximum. L'importance du pic dépend de la teneur en fluor de la terre, de son taux en argiles et de la nature et du pouvoir tampon de celles-ci.
- Les valeurs des pH des lixiviats correspondant aux pics de concentration sont d'autant moins acides que le rapport F/Al est élevé dans ceux-ci, car F^- complexe Al^{3+} sous forme de AlF^{2+} moins acide que Al^{3+}.
- L'accroissement du rapport F/Al dans les lessivats avec l'accroissement de la hauteur du pic de concentration en fluor n'est pas bien élucidé. Plusieurs causes sont possibles : la grande aptitude de $Al(OH)_3$ à adsorber F ; la formation au moment de la dissolution de $Al(OH)_3$ de complexes fluorés dont la solubilité n'est pas dépendante du coefficient de solubilité de CaF_2 ; enfin peut-être la moindre adsorption de ces complexes en milieu acide par les argiles non encore acidolysées en bas de la colonne.
- Après l'apparition du pic, le fluor diminue car son accumulation est freinée par son exportation dans les lessivats.
- Les variations concomitantes de F et K résultent des coprécipitations simultanées de ces éléments avec $Al(OH)_3$. Mais, elles n'ont pas lieu pour toutes les terres étudiées.

Remarques - Les pics de concentration en fluor sont à comparer avec les pics de concentration en aluminium obtenus par JANINA POLOMSKI & al (2) en lessivant une terre par une solution de sel fluoré. Mais, dans leurs essais c'est le fluor de la solution de lessivage qui réagit avec l'aluminium mobilisé et non le fluor libéré in situ par acidolyse.

- Le fluor des lixiviats est susceptible d'être concentré par évaporation lente de ceux-ci.

Résultats des nouveaux essais -

Les études précédentes ont été poursuivies sur deux autres terres :
- La terre N°4 a été prélevée sur la commune d'AUDENGE (Gironde) dans un semis de jeunes pins. Il s'agit d'un podzol atlantique humifère, constitué essentiellement de sable siliceux à grains ronds, usés, de taille inférieure ou égale à 0,4 mm. - Analyse chimique : F = 0,02 % -
- La terre N°5 provient de la forêt d'IRATY dans les Pyrénées. Elle a été prélevée à 1.130 m d'altitude dans une hêtraie sapinière à humus doux sur sol argileux gréseux. Le soubassement est à base de marnes et grés.
- Analyse chimique : F = 0,001 % -

Les résultats des tests préliminaires par acidification directe sont rassemblés dans le tableau I. Les résultats analytiques des lixiviats des terres 4 et 5 sont reportés respectivement dans les tableaux II et III.
L'examen du tableau II fait apparaître :
- Le fluor est présent dans les premières fractions de lessivats et sa concentration augmente progressivement. Cependant, le pic de concentration est beaucoup plus élargi que lors de l'acidolyse de la terre N°2.
- La teneur importante en silice et surtout en phosphate.
C'est la présence de ce dernier qui nous a permis de supposer que la terre, prélevée dans un semis de jeunes pins, avait été amendée quelques années auparavant avec du phosphate de calcium. Après enquête, notre hypothèse s'est révélée exacte. Cependant, pour s'assurer que le fluor provient des engrais la terre a été lavée, le sable grossier a été décanté et broyé finement. Par acidolyse directe, il ne libère pas de phosphates et seulement des traces de fluor (tableau I : terre 4 bis).
- En raison de la faible teneur en argile et de l'inertie presque totale du sable siliceux vis-à-vis des acides, les dernières fractions de lixiviat sont essentiellement constituées d'acide sulfurique. En revanche, lors de l'acidolyse de la terre du Massif Central, l'aluminium des argiles tamponnait les percolats, en fin d'essai, à des valeurs de pH de 3,5 - 3,6.
- La présence d'activités fluor. Celles-ci ont été dosées à l'électrode spécifique en l'absence du tampon classique qui libère le fluor complexé. La courbe d'étalonnage a été établie dans les mêmes conditions.

La présence d'activités fluor nécessiterait une étude théorique et expérimentale qui ne peut entrer dans le cadre de cet article. En première approximation, on peut cependant affirmer qu'elle résulte des teneurs élevées en fluor par rapport à l'aluminium. Ce dernier est en quantité insuffisante pour complexer tout le fluor dans les conditions existantes de pH.

En résumé, l'acidolyse en continu de la terre N°4 mobilise très rapidement le fluor dont les activités apparaissent dans les premiers lixiviats et augmentent rapidement. Ceci est dû au caractère labile du fluor contenu dans le phosphate de calcium et à l'absence de quantités notables d'argiles.
- L'examen du tableau III montre que la terre d'IRATY est très pauvre en calcium et magnésium et le pic de concentration en fluor n'est pas très élevé.

2. CONCLUSIONS

Nos essais d'acidolyse ne sont pas rigoureusement extrapolables aux sols forestiers. En effet, les précipitations acides n'ont jamais un pH de 2 et nous n'avons pas simulé de périodes de sécheresse prolongée. Or, celles-ci peuvent provoquer, par vieillissement de $Al(OH)_3$, sa polycondensation et finalement la formation de gibbsite qui a, vis-à-vis du fluor, un pouvoir adsorbant bien inférieur à celui de l'hydrate d'aluminium fraîchement précipité.

Cependant, la libération du fluor par acidolyse des sols existe certainement dans les sols en place. En effet, dans les eaux de certains lacs acides de l'Amérique du Nord, alimentés par draînage, on trouve du fluor entièrement complexé par l'aluminium (3) (4) (5). Celui-ci ne peut provenir que de l'acidolyse ménagée, en surface, des sols des bassins versants car les lacs uniquement alimentés par les pluies n'en contiennent pas, mais les teneurs en fluor et aluminium sont toutefois très faibles.

En revanche, dans le sol des forêts soumises aux pluies acides depuis des décennies, les solutions peuvent contenir plusieurs dizaines de mg/l d'aluminium. Il serait donc intéressant de savoir si ces teneurs massives en aluminium ne sont pas partiellement complexées pour des quantités non négligeables de fluor.

Les données manquent à ce sujet. LINDSAY (6) mentionne que les activités fluor aF^- dans les sols peuvent varier de 10^{-10} à 10^{-4} M, soit au plus 1,9 mg/l d'aF^-. Mais ces activités fluor correspondent au fluor réellement libre et non au fluor total en solution : $aF^- + AlF^{2+} + AlF_2^+$ etc...

Si des quantités notables de fluor étaient trouvées dans les solutions du sol des forêts dépérissantes, la chronologie des effets des pluies acides pourrait être la suivante :

- Pendant des décades, les pluies acides auraient favorisé la croissance des arbres par solubilisation des éléments nutritifs. En effet, les expériences réalisées sur le terrain par ABRAHAMSEN et TVEITE (7) ont mis en évidence que des pluies acides simulées provoquent d'abord une croissance accélérée de jeunes pins. Mais après plusieurs années, ceux-ci déclinent.
- Ensuite, le fluor serait apparu dans les solutions du sol dans la zone d'accumulation, en même temps que l'aluminium. Ceci expliquerait que des forêts récemment encore florissantes ont rapidement périclité à la suite d'étés chauds, secs et ensoleillés. Ceux-ci induisent en effet des "poussées d'acidification" par accélération de la nitrification (8). Ces poussées d'acidification provoquent dans le voisinage immédiat des racines fines une mobilisation de l'aluminium, donc vraisemblablement du fluor accumulé dans le même horizon.
- Finalement, en adoptant le point de vue de PRENZEL (9), selon lequel la formation de $AlOHSO_4^\circ$ constitue une réserve d'acidité dans le sol, on peut supposer que la formation de $AlFSO_4^\circ$: $AlF^{2+} + SO_4^{2-} \rightleftharpoons AlFSO_4^\circ$ pourrait constituer une réserve en fluor disponible dans le sol, même si le phénomène des pluies acides cessait rapidement.

REFERENCES

(1) BARDY J.A. et PERE C. (1986). Solubilisation du fluor par acidolyse des argiles. Pollution Atmopshérique N°110, Avril-Juin 86, 105-111.

(2) JANINA POLOMSKI, FLUHLER H., BLASER P; (1982). Fluoride-induced mobilization and leaching of organic matter, iron, and aluminium. J. Environ. Qual. 11, 452-456.

(3) DRISCOLL C.T. Jr, BAKER J.P., BISOGNI J.J. Jr, SCHOFIELD C.L. (1980). Nature (London), 284, 161-164.

(4) LAZERTE Bruce D. (1984). Can J. Fish Aquat. Sci., 41, 766-776.

(5) JOHNSON N.M., DRISCOLL C.T., EATON J.S., LINKENS G.E., Mc DONEL W.H. (1981). Geochem. Cosmochim. Acta, 45, 1421-1437.

(6) LINDSAY W.L. (1979). Chemical Equilibria in soils. WILEY, NEW-YORK.

(7) ABRAHAMSEN G. and TVEITE B. (1983). Effects of air pollutants on forest growth. In Ecological effects of acid deposition. pp.199-210, National Swedish Environment Protection Board Report, PM 1636, cité dans Effect of acidic deposition on forest soil and vegetation. ABRAHAMSEN G., Phil. Trons. R. Soc. Lond. B 305, 369-382 (1984).

(8) HUTTERMANN A. and ULRICH B. (1984). Solid phase-solution-root interactions in soils subjected to acid deposition. Phil. Trans. R.Soc.Lond. B 305, 353-369.

(9) PRENZEL J. (1983). A mechanism for storage and retrieval of acid in acid soils. B. ULRICH and J. PANKRATH Eds. Effects of accumulation of Air pollutants in forest ecosystems, 157-170. Copyright by D. REIDEL Publishing Company.

		EAU DISTILLEE	H_2SO_4 1 x 10^{-5}	H_2SO_4 1 x 10^{-4}	H_2SO_4 1 x 10^{-3}	H_2SO_4 1 x 10^{-2}	H_2SO_4 1 x 10^{-1}
TERRE N° 4	pH	5,78	5,51	5,24	4,36	2,72	1,30
	Ca^{++} mg/l	3,00	3,00	3,00	15,00	47,00	21,00
	Mg^{++} mg/l	0,30	0,30	0,30	0,80	1,20	2,50
	Na^{+} mg/l	1,00	0,80	1,00	1,00	2,00	2,50
	K^{+} mg/l	0,40	0,20	0,20	0,40	0,90	1,60
	F^{-} mg/l	0,08	0,07	0,07	0,50	4,00	4,60
TERRE N° 4Bis (Sable broyé de la Terre N° 4)	pH					2,42	1,41
	Ca^{++} mg/l					7,00	5,00
	Mg^{++} mg/l					0,20	1,00
	Na^{+} mg/l					0,80	8,50
	K^{+} mg/l					0,80	35,20
	F^{-} mg/l					0,36	0,40
TERRE N° 5	pH	4,70	4,66	4,55	4,07	3,33	1,97
	Ca^{++} mg/l	1,00	1,00	1,00	2,00	5,00	2,00
	Mg^{++} mg/l	1,00	0,70	0,70	0,60	1,40	3,10
	Na^{+} mg/l	1,80	1,80	1,80	1,80	1,80	1,80
	K^{+} mg/l	7,60	5,80	5,60	4,00	7,80	11,60
	F^{-} mg/l	0,04	0,03	0,03	0,03	0,11	0,09

TABLEAU I

Volume total	pH	F^-	aF	Ca^{2+}	Mg^{2+}	Na^+	K^+	SO_4^{2-}	Fe	P_2O_5	SiO_2	Al^{3+}
ml						-	mg/l	-				
100	4,93	1,80	1,22	160	37,6	18	17,6	513	Néant	88	60	0,52
200	4,81	3,20	1,60	188	34,8	11	11,4	505	Néant	126	72,6	--
300	4,70	5,00	2,00	240	22,8	8	5,2	505	Néant	192	96,2	--
400	4,63	7,00	1,42	260	14,8	8	4,0	505	0,05	214	102	--
700	4,65	9,93	1,72	268	8,5	8	2,4	480	0,09	242	111	5,6
1100	4,68	14,70	2,18	284	3,4	7	1,0	480	0,16	242	112,4	8,0
1500	4,80	12,30	1,94	276	3,7	6	1,0	460	0,27	231	109,4	6,3
1900	4,85	16,62	3,31	272	2,4	6	0,6	470	0,35	225	112	7,3
2300	4,95	22,80	8,62	292	2,2	6	0,6	480	0,80	231	109	6,5
3300	5	26,00	13,50	280	2,0	6	0,4	460	0,56	239	99	5,0
3650	5,20	20,00	8,00	272	2,6	8	1,6	480	1,15	220	113	--
4150	4,73	22,00	10,00	268	2,5	4	1,0	480	4,71	200	108	--
4650	4,93	28,00	12,55	272	2,2	5	0,6	480	2,75	200	96	--
5150	5,02	28,00	12,80	284	2,3	6	0,6	480	2,73	228	116	--
5650	4,92	30,00	14,10	280	2,3	6	0,6	480	4,45	220	107	--
6150	3,81	38,00	18,60	280	1,7	8	0,5	500	5,10	212	102	4,5
6650	3,52	38,00	18,50	276	1,8	8	0,4	500	9,35	228	105	--
10650	2,94	30,00	9,10	232	1,5	7	0,4	480	3,28	185	92	--
12250	2,36	12,00	0,93	60	0,7	2	0,8	460	2,64	44	40	3,1
12750	2,24	0,51	0,015	2	0,1	0	0,2	460	2,50	19	3,6	--
14750	2,20	0,30	0,007	1	0,1	0	0,2	455	0,35	10	2,8	--
18550	2,19	0,14	0,006	1	0,1	0	0	464	0,30	5	1,8	--
22350	2,20	0,10	0,003	1	0,05	0	0	460	0,10	2	2	0,8

TERRE d'AUDENGE 1500 g

Hauteur colonne 40 cm H_2SO_4 0,01 N

TABLEAU II

Volume total en ml	pH	F^- mg/l	Ca^{2+} mg/l	Mg^{2+} mg/l	Na^+ mg/l	K^+ mg/l
200	4,27	0,06	3,00	0,60	5,00	3,40
400	4,48	0,06	2,00	0,30	5,00	5,40
600	4,30	0,06	5,00	2,60	4,00	14,40
800	4,03	0,14	19,00	7,40	0,80	25,60
1000	4,01	0,27	20,00	5,60	0,80	27,20
1250	3,87	0,65	14,00	2,80	0,80	24,80
1500	3,73	0,58	5,00	1,20	0,40	15,40
1750	3,70	0,50	3,00	0,60	0	11,80
2000		0,30				

TERRE IRATY 1240 g

Hauteur colonne 40 cm $H_2 SO_4$ 0,01 N

TABLEAU III

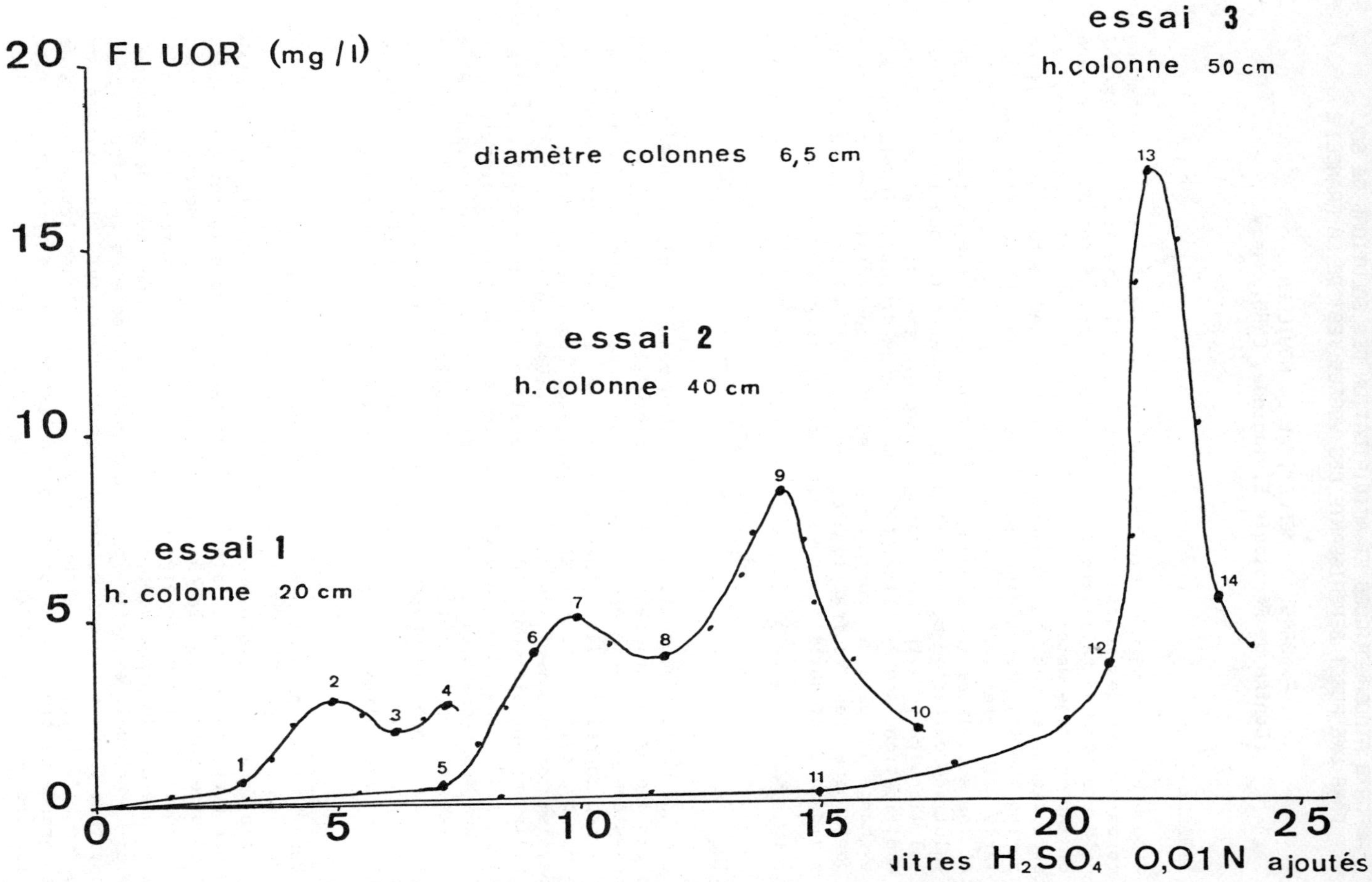

FIGURE 1 LIXIVIATION DE LA TERRE N°2

APPORTS ATMOSPHERIQUES ET ACIDIFICATION DES SOLUTIONS DE SOL DANS UNE FORET DEPERISSANTE DES VOSGES (EST DE LA FRANCE)

F. GRAS, D. MERLET, et J. ROUILLER
Centre de Pédologie Biologique, CNRS, Nancy

Summary

Atmospheric inputs and acidification of soil solutions in a withering forest of Vosges mountains (Eastern France). Chemical contents of rain and through fall waters at several levels of the biosphere and of the drainage waters in a forest ecosystem are studied during an exceptional season. The amounts of cations and sulfurs are modified during the transfer of the solutions through the ecosystem particularly with retention in soil.
At the soil level these changes are confirmed by the analysis of soil solutions extracted by centrifugation. However the quality of these solutions vary with the storage time of the soil samples. These modifications are explained by simultaneous diffusion in fluids and exchange on colloïds surfaces processes. Appropriate chemical reagents were used to fractionate the pool of sulfates bound to the solid phase of soil into several compartments.

SOMMAIRE

L'étude de la composition chimique des eaux de pluie, des pluviolessivats à plusieurs niveaux de la biosphère et des eaux de drainage dans un écosystème forestier a été effectuée au cours d'une saison climatique assez exceptionnelle. Elle a permis de mettre en évidence les modifications des quantités de soufre et de cations au cours du transit à travers l'écosystème et notamment les rétentions au niveau du sol.

Celles-ci ont été confirmées par l'analyse de solutions extraites par centrifugation. Cependant, la qualité de ces solutions varie avec le temps de stockage des échantillons de sol. Ces modifications ne peuvent être interprétées qu'en faisant intervenir simultanément les processus de diffusion au sein de la phase liquide et d'échange à l'interface avec les colloïdes. L'utilisation de réactifs chimiques appropriés permet de fractionner les sulfates retenus dans le sol en plusieurs compartiments.

1. INTRODUCTION

Dans le cadre du programme français de recherches sur le dépérissement des forêts (DEFORPA), une étude a été entreprise en 1985 sur un écosystème forestier (sapinière à hêtres) représentatif de l'étage montagnard sur grès, fortement atteint de dépérissement, dans le but de déterminer dans ce cas précis, le rôle du sol. Etant donnés les désaccords des scientifiques sur les propriétés du sol susceptibles d'intervenir [3] nous avons adopté une démarche n'éliminant, a priori, aucune cause de dommages. Celle-ci a conduit à mettre en avant quatre facteurs de stress pour l'écosystème [7,8] : le dessèchement périodique du sol, la toxicité du manganèse, l'accumulation saisonnière de l'aluminium dans les solutions de sol en grande partie liée à un dysfontionnement dans le cycle de l'azote et enfin les dépôts de sulfates d'origine atmosphérique. Ce dernier aspect du problème a retenu ici plus particulièrement notre attention car si de nombreux scientifiques soupçonnent les sulfates d'être en grande partie responsables des dégâts observés [4,5,6,7], il n'existe, par contre, peu de travaux concernant les mécanismes physico-chimiques induits par les augmentations de sulfates dans les eaux du sol [4,6].

2. DESCRIPTION DE L'ECOSYSTEME

Située dans la région de Saint-Dié, à 600 mètres d'altitude, la forêt est une futaie mixte (sapins-hêtres) d'une centaine d'années. L'état sanitaire des arbres est déplorable : 95 % des arbres étaient considérés en 1986 comme dépérissants, l'état de la cime étant noté entre 3 et 5 (cote ONF). En outre, les analyses foliaires effectuées en 1984 et 1986 indiquent des déficiences élevées en potassium, calcium, magnésium, phosphore ; en revanche, les teneurs en manganèse sont considérables (2 à 5,8 ‰) et les teneurs en aluminium sont identiques à celles des résineux malades de la région de Göttingen [1].

Le sol appartient à la catégorie des sols bruns acides avec un humus de type mull (C=5,3 %) ; C/N=18) à forte activité biologique malgré un pH voisin de 4. La texture est sableuse sur l'ensemble du profil L'horizon humifère A_1 de faible épaisseur présente une porosité très importante, une réserve en eau utile assez élevée (21 mm par dm), une forte acidité d'échange due à la présence d'aluminium échangeable qui représente plus de 50 % de la CEC effective et des quantités notables de manganèse échangeable.

Les horizons minéraux du sol (B_1, B_2, C) dont l'épaisseur peut varier entre 40 et 80 cm, sont beaucoup plus tassés et ne disposent plus que d'une faible réserve en eau utile (15 mm par dm de sol). Du point de vue chimique, ils s'appauvrissent encore en "bases échangeables" tandis que l'acidité d'échange augmente encore, pour dépasser 70 % de la CEC effective avec prédominance de l'aluminium.

3. BILAN DES IONS AU DESSUS ET EN DESSOUS DE LA SURFACE DU SOL EN AUTOMNE 1985

Nous avons mis à profit les conditions climatiques exceptionnelles de l'Est de la France en automne 1985 - avec notamment une longue période de 6 semaines privée de pluies, suivie de deux épisodes pluvieux de 14 jours chacun pour étudier les effets des apports atmosphériques sur le sol.

3.1. Modifications des dépôts dans les pluies et les pluviolessivats

La figure 1 représente les quantités de H^+ Ca^{++} Na^+ et SO_4^{--} apportées par les eaux de pluies avant l'entrée dans la biosphère, les modifications (gains : + ou pertes : -) subies au cours de l'interception par le houppier des sapins et des hêtres ainsi que par la strate très dense de *Festuca sylvatica* et enfin les quantités réellement introduites dans le sol par infiltration pendant la période considérée (1er novembre-12 décembre).

3.1.1. Le pH des pluies est en moyenne de 4,4 avec un minimum de 3,8 dans les premières précipitations, en association avec des teneurs relativement élevées en Ca, SO_4 (ainsi que K).

3.1.2. Sous les arbres, on observe un gain très important de H^+ (pH moyen 4,0), de Na, et surtout de Ca et de SO_4 ; signalons également des accroissements très nets en NO_3, beaucoup moins élevés en Mg et K et insignifiants en Al et Mn.

Les concentrations les plus importantes ont été observées pour tous les éléments au cours des premières retombées de novembre, notamment pour les sulfates (750 µe/l), les nitrates (714 µe/l). Des quantités appréciables de plomb (100 ppb) et de zinc (600 ppb) ont également été mises en évidence.

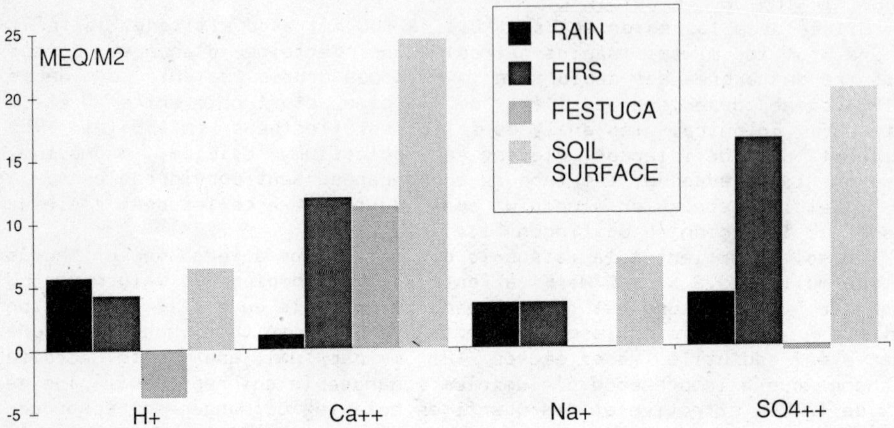

Figure 1 - Apports par les pluies et modifications subies par les pluviolessivats durant l'automne 1985.

Il est incontestable que les apports par la pluie ne représentent qu'une partie des pollutions d'origine atmosphérique. En se fondant sur les travaux réalisés dans de nombreux pays, on peut envisager plusieurs origines aux éléments contenus dans les pluviolessivats recueillis sous les arbres :

- Apports sous forme d'aérosols (brumes et brouillards) dont les gouttelettes se fixent à l'extérieur et à l'intérieur des surfaces foliaires. Ces éléments (SO_2, Na, Ca) sont lavés et entraînés avec les premières pluies. SO_2 et Na ne semblent pas être adsorbés par les tissus car ils sont absents des aiguilles. Mais le premier est susceptible de se combiner à l'oxygène de l'air et de s'hydrater pour former de l'acide sulfurique H_2SO_4.

- Apports de poussières très fines en suspension dans l'air et se déposant sur les feuilles et surtout les aiguilles qui présentent une forte capacité de capture.

- Cations exsudés par les tissus foliaires en échange des protons, notamment ceux qui accompagnent les sulfates. Telle serait l'origine d'une grande partie de Ca et K qui, du 15 septembre au 30 octobre, ont été désorbés et ont ensuite été lixiviés par les pluies.

- Anions tels que NO_3 faisant partie du cycle interne de l'écosystème et susceptibles d'être également désorbés en échange de OH^- ou de CO_3H^-.

En ce qui concerne les métaux, les expériences OTC réalisées en R.F.A., semblent montrer que si Mn et Zn peuvent être lixiviés (leaching), par contre, Al et Pb ne sont pas mentionnés et seraient donc localisés dans les aérosols ou les poussières.

3.1.3. La strate herbacée joue, à cette époque de l'année, un rôle régulateur très important qui se manifeste par une élévation du pH de plus d'une unité, par une désorption très importante de Ca et modérée de K. En revanche, la majeure partie des sulfates et des nitrates et la totalité de Na traversent le tapis de fétuque sans subir de modifications.

3.2. Bilan entrée/sortie dans le sol

L'étude des flux a été réalisée à l'aide de lysimètres fermés de $78cm^2$ de diamètre. A la base du profil (1,20 m), les bilans sont positifs pour H^+, Ca, K, Na, SO_4 et légèrement négatifs pour Al, Mn et surtout NO_3 (tableau I).

Tableau I - Gains et pertes d'éléments entre la surface du sol et les niveaux 60 cm et 120 cm (en me/m^2)

Elements	H^+	Ca	K	Na	Al	Mn	NO_3	SO_4
Apports atmosphériques au sol	5,8	1,2	1,0	3,5	0,25	0,8	0,2	4,3
Bilan à 60 cm	-3,7	-12,9	-2,5	-0,5	-113	-53	-276	0,7
Bilan à 120 cm	-2,8	13,3	6,1	5,7	- 8,5	- 1,4	- 15	11,3

Le sol s'enrichit donc globalement en Ca, SO_4 et Na et s'appauvrit en NO_3, Al et Mn. En réalité, intervient un deuxième processus qui joue un rôle considérable dans le fonctionnement de ce sol, c'est la redistribution des éléments dans celui-ci. La comparaison entrée/sortie entre 60 et 120 cm montre qu'il y a une accumulation (au moins provisoire) de tous les éléments entre ces deux niveaux avec, par ordre d'importance, NO_3, Al, Mn, SO_4, Ca, K et Na. Bien que les lysimètres donnent une image déformée de la réalité (notamment en éliminant l'effet des racines des arbres), il n'est pas interdit de penser que les éléments qui ne subissent aucun prélèvement biologique (Na) ou des prélèvements peu importants (SO_4) se comportent de la même façon dans le sol en place en se redistribuant entre l'horizon de surface et les horizons profonds.

4. TRANSFORMATIONS AU COURS DU TEMPS DES SOLUTIONS DE SOL

4.1. La compréhension des mécanismes d'évolution des apports au cours de leur séjour dans le sol a été abordée par l'étude des solutions du sol. Celles-ci ont été extraites à l'aide d'une centrifugeuse équipée de godets constitués de deux éléments emboîtés séparés par une grille en acier.

Les prélèvements ont été effectués le 12 décembre 1985, de la façon suivante : 23 sondages à la tarière porte-cylindres (volume : 250 cm^3) ont été effectués, tous les mètres, le long d'un transect rectiligne situé à 2,50 m de 3 sapins, très dépérissants. Trois niveaux ont été prélevés : 0-10 cm (horizon A_1), 20-30 cm (horizon B_1), 40-50 cm (horizon B_2).

Les échantillons provenant des sondages 1-13 ont été traités dans le mois qui a suivi le prélèvement, les échantillons des sondages 13-23 ont subi les mêmes manipulations mais après un temps de stockage (à 4°C) plus long. Les analyses réalisées de façon parfaitement identiques ont donné, pour un même niveau, des résultats très divergents en ce qui concerne les concentrations des éléments suivants : H, K, Na, Al, Mn, SO_4 (figure 2).

Le premier lot (1-13) se caractérise par des pH voisins de 5 en surface et de 6 en profondeur, des teneurs en sulfates (400 à 900 µe/l) et Na (1000 à 2000 µe/l) très élevées et, inversement, des teneurs en Al (24 à 140 µe/l) et K (30 à 200 µe/l) très faibles. Ces solutions présentent des caractéristiques physico-chimiques voisines de celles des eaux qui ont pénétré dans le sol au cours des épisodes pluvieux antérieurs aux prélèvements (voir figure 1, soil surface).

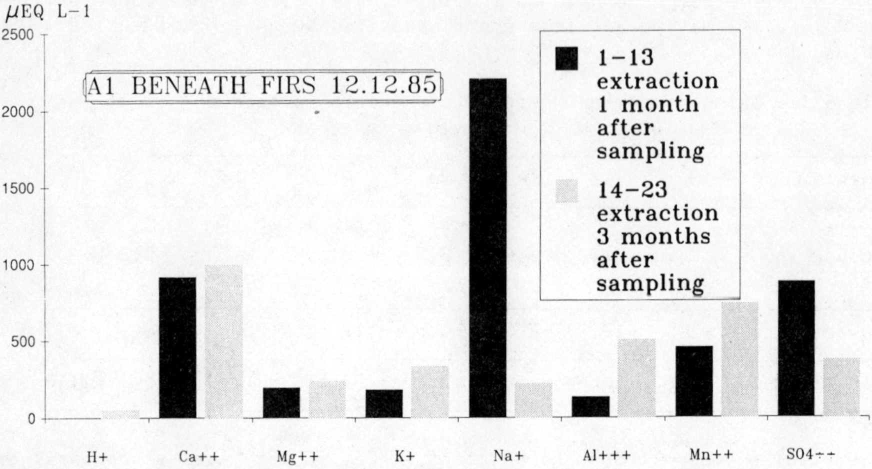

Figure 2 - Evolution, au cours du temps, des solutions de sol sous les sapins dépérissants.

A l'opposé, le second lot contient deux fois moins de sulfates en A_1 et 4 à 8 fois moins en-dessous, les pH sont inférieurs d'une unité aux pH précédents et les teneurs en aluminium sont 4 à 10 fois plus élevées.

Cette modification des solutions de sol paraît, à première vue, surprenante. Nous avons tenté de l'expliquer en nous fondant sur un certain nombre de travaux récents [4,6]. Soulignons d'abord que les résultats de la deuxième série d'analyses sont plus proches des caractéristiques chimiques des eaux lysimétriques que ceux de la 1ère série. Tout se passe comme si, dans les lysimètres, les eaux de percolation n'étaient pas, à un moment donné, en équilibre avec les solutions plus intimement liées au sol.

4.2. INTERPRETATIONS

REUSS (1986) a déterminé l'influence des apports de sulfates sur l'alcalinité (ANC) et sur le pH des eaux percolant à travers des colonnes de sols à taux de saturation en cations variant entre 5 et 20 % (figure 3). La comparaison entre les points représentatifs des deux horizons A_1 et B_1 pour la 1ère série (cadre blanc) et pour la 2ème série (cadre noir) et les courbes, 10 % pour A_1 et 5 % pour B_1, indique : (1) un déplacement d'équilibre dans le sens d'une diminution du pH et des teneurs en sulfates, dans les solutions du sol ; (2) au départ, des valeurs de pH plus élevées d'une unité (compte-tenu des teneurs en SO_4) comparativement à celles indiquées par les courbes d'équilibres ; (3) après 3 mois, des valeurs de pH plus faibles (3 à 8 dixièmes d'unité pH) que les données de REUSS.

L'interprétation que nous donnons de ce comportement repose sur la nature des apports de sulfates que nous avons analysée plus haut. Rappelons, en effet, que les solutions qui pénètrent dans le sol, sont peu acides et riches en Na, Ca et SO_4, et que les solutions capillaires extraites par centrifugation dans les semaines qui suivent les prélèvements ont des caractéristiques de pH et de concentration très voisines. Mais progressivement se développent, dans l'eau du sol, des processus de <u>diffusion</u> qui affectent les éléments dont les gradients de concentration sont les plus élevés :

- <u>diffusion centripète</u> en direction des surfaces électriquement chargées pour Na^+ et SO_4^{--},
- <u>contre-diffusion</u> centrifuge de H^+, Al^{+++} et K^+ et sans doute de NO_3.

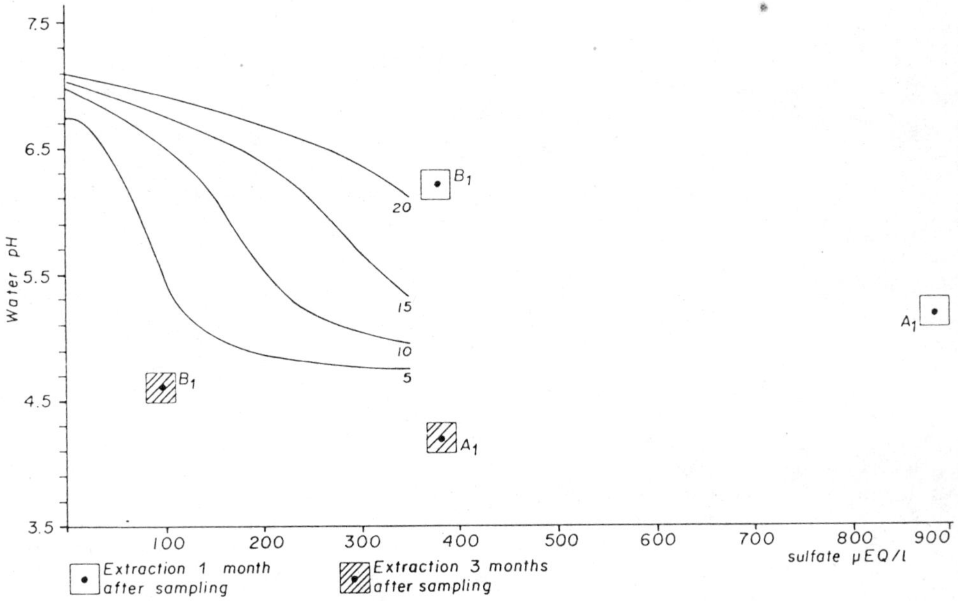

Figure 3 - Effet de l'accroissement des sulfates sur le pH des eaux de drainage pour différents taux de saturation en bases du sol (REUSS, 1986).

Force est de constater que les flux de diffusion sont très lents et aboutissent, au bout de 3 mois seulement, à une égalisation des activités chimiques des différents constituants. Il est intéressant d'examiner à l'équilibre les pH et les concentrations en Al^{+++} et en SO_4 des solutions extraites. Dans B_1 (pH proche de 4,5, [Al]=0,1 mmole et SO_4 = 0,1 mmole). Les conditions sont favorables à la formation de produits mal cristallisés, à base de sulfate d'aluminium [6] : jurbanite de formule $Al(OH)SO_4$ $5H_2O$ ou basaluminite de formule $Al_4(OH)$, $10SO_4$, $5H_2O$.

4.3. <u>Quantités et formes du soufre accumulés dans le sol</u>

Des échantillons du même sol prélevés en décembre 1985 en profondeur, et en mai 1986 en surface et en profondeur, ont été mis en contact d'une part avec NH_4Cl 0,5N et, d'autre part, avec NaOH 0,1N (Tableau II).

Les quantités contenues dans les solutions de sol ne représentent donc qu'une très faible fraction des quantités susceptibles d'être extraites par les deux réactifs.

Tableau II - Quantités de SO_4 extraites par NH_4Cl 0,5N et NaOH 0,1N (µe/100 g sol sec) comparées aux quantités dans les solutions de sol.

	Eaux capillaires SO_4	NH_4Cl SO_4	NaOH SO_4
Décembre 1985 (P)	8,9	291	708
Mai 1986 (S)	13,1	68	652
Mai 1986 (P)	4,4	131	479

Par ailleurs, la soude (pH 13) extrait 3 à 10 fois plus de SO_4 que le réactif NH_4Cl non tamponné qui, en contact avec le sol, acquiert une forte acidité ($pH < 4$).

Sous réserve d'une étude plus approfondie de l'action respective de ces deux électrolytes, on peut fournir les explications suivantes : NH_4Cl désorbe, par échange avec Cl^- les sulfates fixés sur les sites à charges + entourant notamment les hydroxydes de fer et d'aluminium amorphes, tandis que NaOH dissout également les minéraux à base de sulfate d'aluminium et le soufre organique.

Deux résultats importants ressortent également du tableau II :

1) Les quantités de SO_4 extractible par NH_4Cl, semblent plus élevées en profondeur qu'en surface, sans doute en raison d'une augmentation des produits alumineux amorphes et d'un ZPC plus élevé, tandis que SO_4 extrait à la soude est plus abondant en surface à cause de la présence de soufre dans la matière organique.

2) Il y a, semble-t-il, moins de sulfates après la période hivernale qu'avant celle-ci. Les réserves en sulfates du sol pourraient donc subir des variations saisonnières.

CONCLUSION

Les apports de soufre d'origine atmosphérique au cours des semaines de sécheresse, de septembre-octobre 1985 sont incontestables. La presque totalité de ce soufre pénètre dans le sol dans lequel il est retenu. Il se transforme totalement en sulfates. Ceux-ci subissent progressivement, au sein des solutions de sol, des processus chimiques de diffusion, d'adsorption sur les surfaces des colloïdes et sans doute de précipitation sous forme de sulfates d'aluminium hydratés susceptibles de se redissoudre lorsque les équilibres géochimiques sont modifiés. Ces produits, encore mal définis, pourraient exercer des effets sur la rhizosphère qui s'additionneraient aux autres facteurs de stress que nous avons mentionnés en introduction : stress hydrique, acidification liée au dysfonctionnement du cycle de l'azote, toxicité du manganèse et de l'aluminium.

REFERENCES

[1] GEHRMANN, J., GERRIETS, M., PUHE, J. and ULRICH, B. (1984). Untersuchwingen an Boden, Wurzeln, Nadeln und erste Ergelnisse von depositions-messungen in Hills. Im Berichte des Forschungszentrons Waldökosystem/Waldsterben, Göttingen.

[2] GILLMAN, G.P. (1984). Using variable charge characteristics to understand the exchangeable cation status of oxic soils. Aust. J. Soil Sci. Res., 22, 71-80.

[3] HINRICHSEN, D., (1986). Multiple pollutants and forest decline. Ambio, XV, 5, 258-265.

[4] JOHNSON, D.W. (1985). Sulfur cycling in forests. Biogeochemistry, 1, 29-43.

[5] MILLER, H.G., MILLER, J.D. and COOPER, J.M. (1979). Changes in sulphur content and acidity of rainwater on passing through the canopy of Sitka spruce forest. International Symposium Sulphur Emissions and the Environment. London, 8-10 mai 1979.

[6] REUSS, J.O. and JOHNSON, D.W. (1986). Acid deposition and the acidification of soils and waters. Springer-Verlag. Ecological Studies, 59, 119 p.

[7] ULRICH, B. (1983). Stabilität Waldökosystemen unter dem Einfluss des "Sauren Regens".

[8] ULRICH, B. (1984). Effects of air pollution on forest ecosystems and waters. The principles demonstrated at a case study in Central Europe. Atmospheric Environnment. Vol. 18, 3, 621-628.

[9] VAN GRINSVEN, J.J.M., MULDER, J. and VAN BREEMEN, N. (1984). Hydrochemical budgets of some dutch woodland soils with high inputs of atmospheric acid deposition : mobilisation of aluminium. Hydrochemical balances of freshwater systems. Proceedings, Uppsala Symposium, September 1984.

A SIX YEAR NUTRIENT BUDGET IN A CONIFEROUS WATERSHED UNDER ATMOSPHERIC POLLUTION

A. HAMBUCKERS and J. REMACLE
Ecologie Microbienne - Departement de Botanique
Université de Liège
Sart Tilman - 4000 Liège
Belgique

SUMMARY

The biogeochemical fluxes were studied since 1981 in a coniferuus watershed in which dieback evidences were observed since 1983. We present here some results concerning the throughfall, (i) the evolution of their quality since the first years and (ii) comparisons between stands with different health states. These observations confrontated with other ones concerning the brook chemistry and the soil solution composition lead to assume misfunctionnal root absorption of the trees in the watershed.

1. INTRODUCTION

The nutrient budget of a coniferous (Picea abies (L.) Karst.) watershed (81 ha.) was followed since 1981. It is located in the "Haute Ardenne" region (Belgium) at about 500 m a.s.l..

The following compartments were examined : the precipitation in open field, the throughfall, the seepage under organic layers and the streamflow at the output. The analysis of the transfers of water through the forest up to the brook put in light the influence of the biotic and abiotic compartments (1,2,3,4,5).

However, dieback evidence was observed since 1983 and is still more obvious in 1987. So it appears interesting to compare the fluxes in the watershed along the years with the aim of detecting evolution and putting in light the mechanism of dieback process (cf. fig. 1).

In addition, as it concerns the throughfall collection, the current experimental design leads to compare spruce stands showing unequal proportion of healthy trees (cf. fig. 2).

2. RESULTS

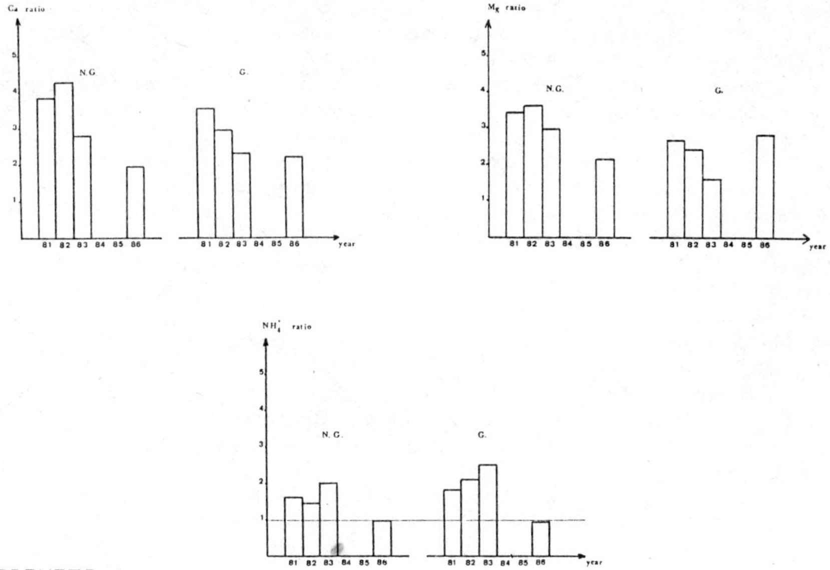

FIGURES 1 :

Averages of the ratios throughfall input / open rain deposition for the october to march periods (N.G. : non-growing period) and for the april to september periods (G. : growing period).

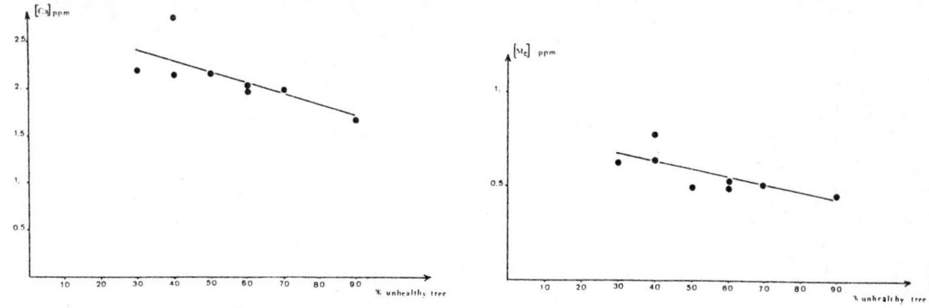

FIGURES 2 :

Plots of the average throughfall concentrations vs the proportions of unhealthy trees under which the throughfalls were collected. The correlations were significant at level 0.05 only for the cations calcium and magnesium. The differences in concentration between the stands were not significant for most of the others nutrients.

3. PRELIMINARY CONCLUSIONS

The experimental design permits to confrontate comparisons in time (fig. 1) and comparisons in space (fig. 2). Both approachs show decreasing trends for the concentrations of calcium and magnesium in the throughfall under spruces showing dieback symptoms.

This could result from decreased needle concentrations induced by misfunctionnal root absorption. This hypothesis is supported by increased calcium and magnesium concentrations in the brook at the exit of the watershed for the april to september periods in the last years (in press).

Moreover, first results concerning soil solution of "O" horizons show high heavy metal concentrations (more than 0.050 ppm in cadmium and 0.50 ppm in lead) that could induce detrimental effects on higher plants, particularly on the root system (6,7).

As it concerns ammonium, we observed since 1986 very low values for the ratio of deposition rates (below 1) indicating high retention rate by the needles. Besides, ZEDLER *et al.* (9) and VAN PRAAG and WEISSEN (8) have stressed perturbations in the nitrogen metabolism that ZEDLER *et al.* (9) attribute to high deposition rates of ammonium combined with aluminium concentrations in the soil inhibiting nitrate absorption.

4. REFERENCES

(1) BULDGEN, P. (1982).Oïkos, 38 : pp. 99-107.

(2) BULDGEN, P., CAJOT, O., MONTJOIE, A. and REMACLE, J. (1984).Physio-Geo, 9 : pp. 47-45.

(3) BULDGEN, P., DUBOIS, D. and REMACLE, J. (1983). Soil Biol. Biochem., 15 : pp. 511-518.

(4) BULDGEN, P. and REMACLE, J. (1981).Soil. Biol. Biochem., 13 : pp. 143-147.

(5) BULDGEN, P. and REMACLE, J. (1984).pp.163-172 *in* SCOPE BELGIUM-PROCEEDINGS "Acid deposition and sulfur cycle". BRUSSELS.

(6) GODBOLD, D.J. and HUETTERMANN, A. (1986).Water Air Soil Pollut., 31 : pp. 509-515.

(7) MITCHELL, C.D. and FRETZ, T.A. (1977). J. Am. Soc. Hortic. Sci., 67 : pp. 81-85.

(8) VAN PRAAG, H.J. and WEISSEN, F. (1986).Tree Physiol.,1 : pp. 169-176.

(9) ZEDLER, B., PLARRE, R and ROTHE, G.M. (1986). Environ. Pollut. (series A), 40 : pp. 193-212.

A MATHEMATICAL MODEL FOR ACIDIFICATION AND NEUTRALIZATION OF SOIL PROFILES EXPOSED TO ACID DEPOSITION.

Harald U Sverdrup & Per G Warfvinge
Departement of Chemical Engineering, Lund Institute of Technology
Box 124, Chemical Center, S-221 00 Lund, Sweden

Ulf von Brömssen
Swedish Environmental Protection Board (SNV)
Box 1302, S-171 25, Solna, Sweden.

ABSTRACT

A simple model for soil acidification describes the observed pH and base saturation profiles in a Swedish podzol affected by acid deposition. The model is based on a mechanistic respresentation of ion exchange and the weathering rate and is based on a mass balance for calcium and alkalinity. The model has been extended to include a dynamic model for dissolution of limestone in soils. This model has a potential use as a tool for planning and evaluation of different terrestial liming operations.

1 INTRODUCTION

The acidification of lakes and streams and the impact of acidification on aquatic ecosystems is one of the major environmental concerns in Scandinavia at present. The acidification observed in our lakes and streams may to a large extent only be the symptoms of large scale chemical changes in the soils of the watershed. The chemical composition will be influenced by the soil processes such as weathering and ion exchange, as all runoff have percolated through some part of the soil.

This paper is focused on the modeling of the impact on weathering kinetics on steady state soil chemical profiles, and neutralization of acidified soil systems. The work has been funded by the Swedish Environmental Protection Board (SNV), as a part of research directed to explore cause of and mitigations strategies for acid groundwater.

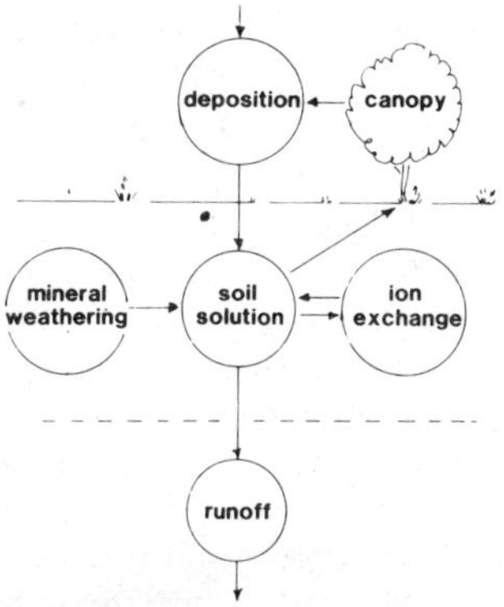

Figure 1: A schematic representation of the interaction of the most important soil processes for acidifying soils. It is important to recognize that the soil processes affect each other through the soil solution only.

2 KINETICS OF WEATHERING IN THE SOIL

The minerals in Swedish podzols have been exposed to weathering since the last glaciation. Accordingly we have reason to assume that the production of cations in the soil will follow long term weathering kinetics. We may thus neglect the initial and intermediate processes, as these probably are long past.

An expression for the production of cations by weathering of minerals is crucial to any acidification model. In earlier acidification simulation models the weathering rate has been modeled by postulating a very simple model based on one or two coefficients, determined by calibration of the model to a large data set. As a result the general applicability of these models is severly limited.

Several studies ([1], [2], [3]) have shown that several parallel elementary reactions contribute to the total dissolution rate, each forming its own activated surface complex with the mineral surface reaction sites. The unit dissolution reactions are:

- The reaction with the hydrogen ion
- The reaction with water
- The reaction with the hydroxyl ion
- The reaction with carbon dioxide

For silicate minerals it has been shown that the kinetic equation under certain conditions may be reduced to a simple expression:

$$R = k_{H^+} \cdot a^n_{H^+} + k_{CO_2} \cdot P_{CO_2}{}^m + k_W + k_{OH} \cdot a^w_{OH^-} - k_B \cdot a^x_{cation} \cdot a^y_{Al^{3+}} \cdot a^z_{Si(OH)_4}$$

k_{H^+} is the rate coefficient for the reaction with H^+, k_{CO_2} for the reaction with CO_2 and k_W the rate coefficient for hydrolysis. The last term would include true backward reactions if they occur, precipitation of secondary minerals as well as the deactivation of the activated surface complexes. In soils the availability of moisture may be limited, and the activity of the exposed mineral surface may be less than unity.

Weathering rate coefficients have been determined for a number of minerals important in acid sensitive areas. In Table I, coefficients are given for a number of these.

The rate equation for alkalinity production from weathering reactions thus becomes:

$$\frac{dAlk_W}{dt} = R \cdot A_W \cdot \Theta \cdot X_W \cdot \frac{\rho(rock)}{\rho(mineral)}$$

A_W is the exposed surface area of the mineral to the solution, X_M is the fraction of the rock being the mineral considered, and R the chemical kinetic expression. Θ is the soil moisture saturation, taken to be equal to the activity of the mineral surface. It is assumed that the reactions only will take place on wetted surfaces. X_W is the weight fraction of the mineral to be considered in the soil matrix, while ρ denotes density.

The importance of carbon dioxide is often stressed in the debate concerning acidification and weathering. Table II serves to demonstrate the relative importance of carbon dioxide weathering, in relation to the other weathering unit reaction.

In modest or dilute solution neither sulfate, nitrate or chloride seem to form any surface complexes with the minerals investigated. There is qualitative information however that indicates that certain organic acids may form surface complexes and react with the mineral ([4], [5]), but net effect of organic acid ligands for the weathering rate in natural soil conditions seem to be small [3].

Mineral	pk_{H^+}	n	pk_W	k_{CO_2}	m	pk_{OH}	w
Albite	12.8	0.55	15.4	14.3	0.7	13.3	0.3
Wollastonite	8.0	0.65	13.2	12.4	0.7	-	-
Biotite	12.8	0.55	15.6	-	-	-	-

Table I: Rate coefficients for the chemical weathering rate of some minerals in the soil. The weathering rate is expressed as keq/m^2s

Mineral	Reaction	CO_2 partial pressure			
		0.03%	0.3%	3%	30%
Albite	H$^+$-reaction	56%	52%	38%	17%
	Hydrolysis	42%	39%	29%	12%
	CO_2 reaction	2%	9%	33%	71%
Enstatite	H$^+$-reaction	47%	46%	44%	33%
	Hydrolysis	53%	53%	50%	38%
	CO_2 reaction	0	1%	6%	29%

Table II: The relative importance of the reaction of a mineral and carbon dioxide versus the reaction with water at pH 4.5.

3 A MASS BALANCE FOR CATIONS IN THE SOIL

In a first model approach, only calcium ions and acidity has been considered. We define acidification of the soil profile to be a change in the base saturation towards a lower value. A equilibrium the change will be zero, and the net supply of cations from weathering and deposition will balance the amount removed by leaching. Considering only calcium cation exchange,we may form two mass balances for each soil horizon element, assuming each strata to be mixed:

$$\frac{d[Alk]}{dt} = \frac{1}{\tau} \cdot ([Alk_{in}] - [Alk]) + \frac{dAlk_d}{dt} + \frac{dAlk_x}{dt} + \frac{dAlk_w}{dt}$$

and

$$\frac{d[Ca^{2+}]}{dt} = \frac{1}{\tau} \cdot ([Ca^{2+}_{in}] - [Ca^{2+}]) + (\frac{dAlk_d}{dt} + \frac{dAlk_x}{dt} + f \cdot \frac{dAlk_w}{dt})/2$$

where Alk_x is the contribution from buffering cation exchange reactions, Alk_w is the alkalinity from weathering reactions. Alk_d is a general sink/source term. τ is the retention time in the soil element at each point in time, while f is a stoichiometric factor.

The model was applied to a soil profile in a reseach watershed near Laxå, Sweden, to calculate the pH-values and the base saturation in such a soil profile in equilibrium with the present deposition level. The model need input data to calculate from, all which may be measured in some way from the conditions at the site.

For the calculation of the Laxå profiles the values for exposed mineral surface area was guessed, and the mineral composition was estimated using data from similar watersheds. The result of the calculations are shown in figure 2. The results seem to indicate that the soil is in equilibrium with the precipitation at least down to 70 cm below the surface.

At present the hydrological condititoins are fed to the model as input data. Work is being carried out however to integrate the model with a hydrological watershed model [6] called 'PULSE" used for a calibrated simulation model.

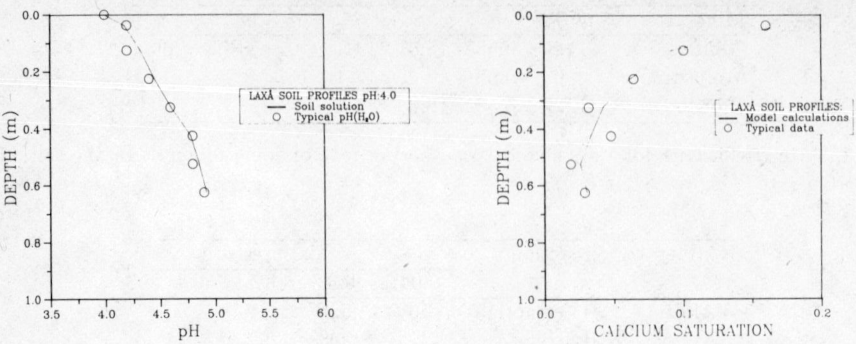

Figure 2: Calculated and observed values for pH and base saturation in a soil profile near Laxå, Sweden. Some of the physical data needed as input has been estimated as they are yet not available at this point in time.

4 DISSOLUTION OF LIMESTONE IN SOIL

As discussed earlier, the adverse effects of acid deposition in terms of reduced alkalinity and pH–values of groundwater and surface waters, can be deduces to insufficient capability of the top soils to neutralize the acid input. By adding a mineral to the top soil that dissolves more easily such as limestone, it would be possible to neutralize acid water in the soil profile as well accumulated acidity in the soil material.

Figure 3 serves to illustrate some situations where soil liming is a potential measure to mitigate some adverse effects of acid deposition:

- Liming of acidified forest land
- Liming of discharge areas to provide alkalinity to acidified surface waters.
- Liming of recharge areas in order to affect groundwater aquifers

Calcite in soil reacts with dissolved species in the soil solution producing two moles of alkalinity and one mole of calcium-ions for each mole of calcium carbonate that dissolves. These calcium-ions may then react further with exchangeable cations on a solid phase. In a severely acidified, humic profile, the most important exchange reaction will be with the H^+–ion. The chemical reactions thus proceeding in parallel in an acid soil solution are:

$$CaCO_3 + H^+ \longrightarrow Ca^{2+} + HCO_3^-$$

and

$$2R - H + Ca^{2+} \longrightarrow 2R - Ca_{1/2} + 2H^+$$

where $R - H$ and $R - Ca$ represent exchangeable H and Ca.

Thus, the dissolution and the $H - Ca$ –cation exchange reactions are coupled both by Ca^{2+} and, through the carbonate system, the H^+–ion.

The overall reaction may be written:

$$CaCO_3 + 2R - H \longrightarrow 2R - Ca_{1/2} + HCO_3^- + H^+$$

The hydrogen ions may then react further with undissolved limestone. This concepual model is illustrated in Figure 4.

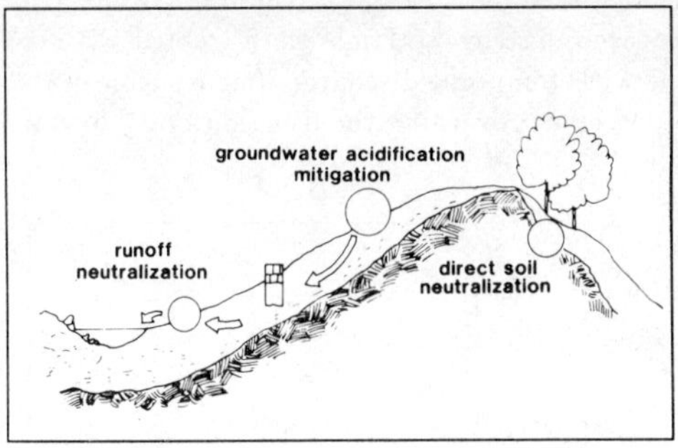

Figure 3. Soil liming treatment locations

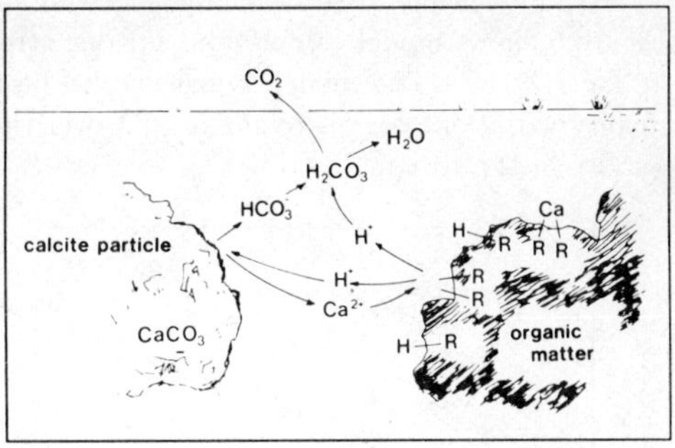

Figure 4. Conceptual soil liming model

Based upon this conceptual model, a mathematical model for dissolution of limestone in soils was developed. This includes four fundamental chemical reaction systems:

- Dissolution of calcite in a stagnant aqueous system
- $H - Ca$ cation exchange reactions
- Leaching and accumulation of dissolved components
- $H_2O - CO_2$ equilibrium reactions

The first three processes are dynamic while the latter is modelled as equilibrated with a given CO_2 partial pressure. The dynamics of these systems has been analysed theoretically and experimentally ([7], [8], [9]). The model is based on physically processes, and input requirements include pre-liming soil pH and base saturation, precipitation chemistry, as well as design parameters such as limestone dose and particle size distribution.

The model has been applied to different soil liming situations with satisfactory results. Figure 4 illustrates the change in key soil chemistry parameters at an experimental area in Laxå in southern Sweden, following the application of 30 kg $CaCO_3/\mathrm{m}^2$.

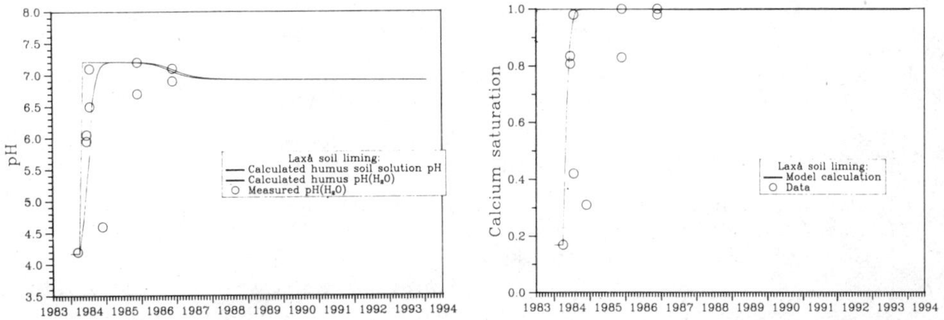

Figure 5: Model calculations and experimental data for the limed areas in Laxå. The data must be viewed in the light of the natural spatial variability and inherent difficulties in evaluating soil chemistry of soils containing undissolved limestone.

The model has also been applied to liming of discharge areas near running waters [10]. Figure 5 shows model calculations for one stream were nearby wetlands were treated with 0.5 kg $CaCO_3/\mathrm{m}^2$. The model, which has to be fed with temporal discharge and baseline water quality variations, serves to shred light over the processes governing the dynamics of the water quality in the stream.

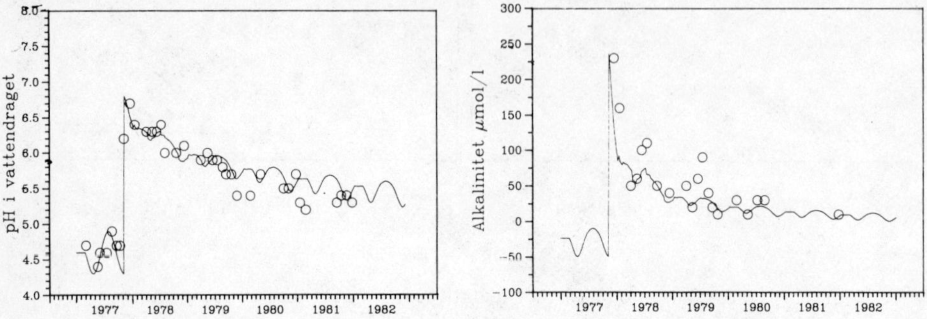

Figure 6: Model calculations for the water quality in River Västra Skälsjöbäcken, following liming of the discharge areas. The peak in stream alkalinity shortly after liming can not be modeled with a model assuming equilibrium cation exchange

References

(1) M. Lagache. Contribution a l'etude de la alteration des feldspats, dans l'eau, entre 100 et 200 °C, sous diverse pressions de CO_2, et application a la synthese des minerals argileux. *Bull. Soc. Franc. Miner. Christ.*, 223–253, 1965.

(2) L. Chou and R. Wollast. Steady state kinetics and dissolution mechanisms of albite. *American Journal of Science*, 285:963–993, 1985.

(3) H. Sverdrup and P. Warfvinge. *The Kinetics of Mineral Weathering.* Technical Report, Department of Chemical Engineering II, Lund Institute of Technology, 1987.

(4) E.P. Manley and L.J. Evans. Dissolution of feldspars by low molecular weight aliphatic and aromatic acids. *Soil Science*, 141:106–112, 1986.

(5) D.E. Grandstaff. The dissolution rate of forsteritic olivine from hawaiian beech sand. In *3rd International water-rock interaction symposium*, Geochem. cosmochim. soc, Alberta research council, 1980.

(6) S. Bergström, B. Carlsson, G. Sandberg, and L. Maxe. Integrated modelling of runoff, alkalinity and ph on a daily basis. *Nordic Hydrology*, 16:89–104, 1985.

(7) H. Sverdrup and P. Warfvinge. *Upplösning av kalksten och andra neutralisationsmedel i mark.* SNV Report 3311, Statens Naturvårdsverk, Solna, Sweden, 1987.

(8) P. Warfvinge and H. Sverdrup. *Upplösning av kalksten i sura, stagnanta vattenlösningar (Dissolution of calcite in an acid, stagnant aqueous system).* Technical report, Department of Chemical Engineering II, Lund Institute of Technology, 1987.

(9) P. Warfvinge. *Neutralization of soil sytems. CODEN LUTKDH/TKKT/1002/1-220/1986*, Department of Chemical Engineering II, Lund Institute of Technology, 1986. Licentiate Thesis.

(10) P. Warfvinge and H. Sverdrup. Upplösning av kalksten i mark – våtmarkskalkning (dissolution of calcite in soils – wetland liming). *Vatten*, 43:59–64, 1987.

LE BASSIN VERSANT DU STRENGBACH A AUBURE (HAUT-RHIN, FRANCE) POUR L'ETUDE DU DEPERISSEMENT FORESTIER DANS LES VOSGES (PROGRAMME DEFORPA): I EQUIPEMENT CLIMATIQUE, HYDROLOGIQUE, HYDROCHIMIQUE.

D. VIVILLE[1], B. AMBROISE[1], A. PROBST[2], B. FRITZ[3], E. DAMBRINE[4], D. GELHAYE[4], C. DELOZE[4]

[1] Centre d'Etudes et de Recherches Eco-Géographiques (CEREG, UA 95 CNRS) ULP, 3 rue de l'Argonne, F 67083 Strasbourg Cedex.
[2] Institut de Géologie, ULP, 1 rue Blessig, F 67084 Strasbourg Cedex.
[3] Centre de Sédimentologie et de Géochimie de Surface (CSGS, LP 6251, CNRS), 1 rue Blessig, F 67084 Strasbourg Cedex.
[4] Laboratoire "Sols et Nutrition des Arbres Forestiers", Institut National de la Recherche Agronomique - Centre National de Recherches Forestières (INRA-CNRF), Champenoux, F 54280 Seichamps.

Résumé

L'étude de la dynamique des éléments minéraux dans les écosystèmes forestiers, la connaissance de la participation de chaque élément chimique dans les apports atmosphériques et dans les exports par le ruisseau nécessitent une double approche au niveau de la station et du bassin versant élémentaire. La mesure quantitative et qualitative des apports atmosphériques, de leur transit à travers le couvert végétal et le sol à leur sortie par le ruisseau demande un important équipement climatique, hydrologique, hydrochimique et biogéochimique. C'est ce qui a été entrepris sur le petit bassin versant granitique (80 ha) du Strengbach.

Summary

In front of the general pollution climate in Western Europe and with the new problem of forest decline, it appeared obvious to study the influence of acid rain water on a forested ecosystem. A part of the DEFORPA program is to study hydrochemical balance in a small forested basin of the granitic Vosges massif (East of France): the Strengbach basin at Aubure (Haut-Rhin). To understand the chemical behaviour of atmospheric inputs in the basin, one need an important hydrochemical equipment. This basin has been equiped since november 1985. Climatical parameters (temperature, air moisture content, wind speed and direction, global radiation) and air·quality parameters (SO_2, NO_x, O_3) are measured in an automatical station. Total precipitations are measured in 4 sites with open collectors; one station is equiped by 2 automatic samplers for the collection of wet precipitation and estimation of dry deposits. Throughfall waters and soils solutions are measured in an experimental station: automatic samplers of total and wet precipitations, litter collectors, lysimeter plates for soil solutions. At basin outlet, a H-flume and a water level recorder give discharge data. A programable water sampler is connected to the water level recorder. First chemical results are presented in (9).

1. POSITION DU PROBLEME ET OBJECTIFS

Le dépérissement forestier, constaté en Europe Occidentale et depuis peu dans les Vosges, serait attribué à la pollution atmosphérique qui interviendrait à différents niveaux:

- par ses effets toxiques directs sur le feuillage,
- par ses effets indirects sur l'acidification progressive des sols: les apports de polluants atmosphériques développeraient certaines carences (Mg) ou toxicités (Al) néfastes pour les végétaux.

Tester cette seconde hypothèse peut se faire par des mesures in situ dans une station. Mais, pour tenir compte de l'hétérogénéité spatiale du sol et de la végétation, très grande en milieu forestier, il est nécessaire de coupler cette étude stationnelle des mécanismes à une étude plus globale faite à l'échelle d'un petit bassin versant: ceci permet d'intégrer les effets de cette variabilité spatiale et de contrôler la validité des mesures des flux stationnels.

Cette double approche associant des études aux deux échelles complémentaires de la station et du bassin versant élémentaire a été retenue dans le cadre du programme DEFORPA (Dépérissement Forestier Attribué à la Pollution Atmosphérique) pour étudier le chimisme des apports atmosphériques, leur devenir et leur influence sur la dynamique géochimique actuelle en relation avec le dépérissement forestier.

Un petit bassin versant forestier dépérissant est étudié dans les Vosges depuis septembre 1985 afin:

- d'approfondir les connaissances sur la dynamique des éléments minéraux dans les écosystèmes forestiers;
- de connaître pour chaque élément chimique sa participation dans les apports atmosphériques et dans l'exportation à l'exutoire du bassin;
- d'établir des bilans annuels "entrées- sorties" pour les principaux éléments chimiques mobilisés par l'altération naturelle et par les transferts de polluants acides et azotés dans le bassin.

Ces objectifs nécessitent un dispositif expérimental important décrit dans cet article. Les premiers résultats géochimiques sont présentés dans (9).

2. LE BASSIN DU STRENGBACH (Fig.1)

2.1 Raisons du choix

Atteint par le dépérissement forestier, le bassin amont du Strengbach à Aubure (Haut-Rhin), représentatif des stations forestières vosgiennes d'altitude sur roches acides, répond aux critères essentiels pour une telle étude:

- les contours hydrographiques sont bien délimités;
- le substrat géologique est homogène;
- la couverture forestière est quasiment complète (97%) et constituée d'essences présentant des symptômes de dépérissement (jaunissement et défoliation prononcés);
- il n'existe pas de source de pollution locale.

D'autre part, sa taille limitée permet la comparaison avec d'autres bassins versants étudiés ailleurs: Ringelbach, Mont Lozère, Ardennes belges (4, 8, 2, 3,1).

2.2 Description

Ce petit bassin (80 Ha), situé entre 883m et 1146m d'altitude, d'axe W-E, de pente moyenne 14°6, présente une dissymétrie dans son profil en long. En effet, les pentes sont faibles dans la partie supérieure du bassin et très fortes (jusqu'à 40°) dans la partie aval. Cette différence est due à l'englacement ,ou tout au moins à la présence d'un important névé, lors des dernières périodes froides comme en témoignent les blocs morainiques du vallon.

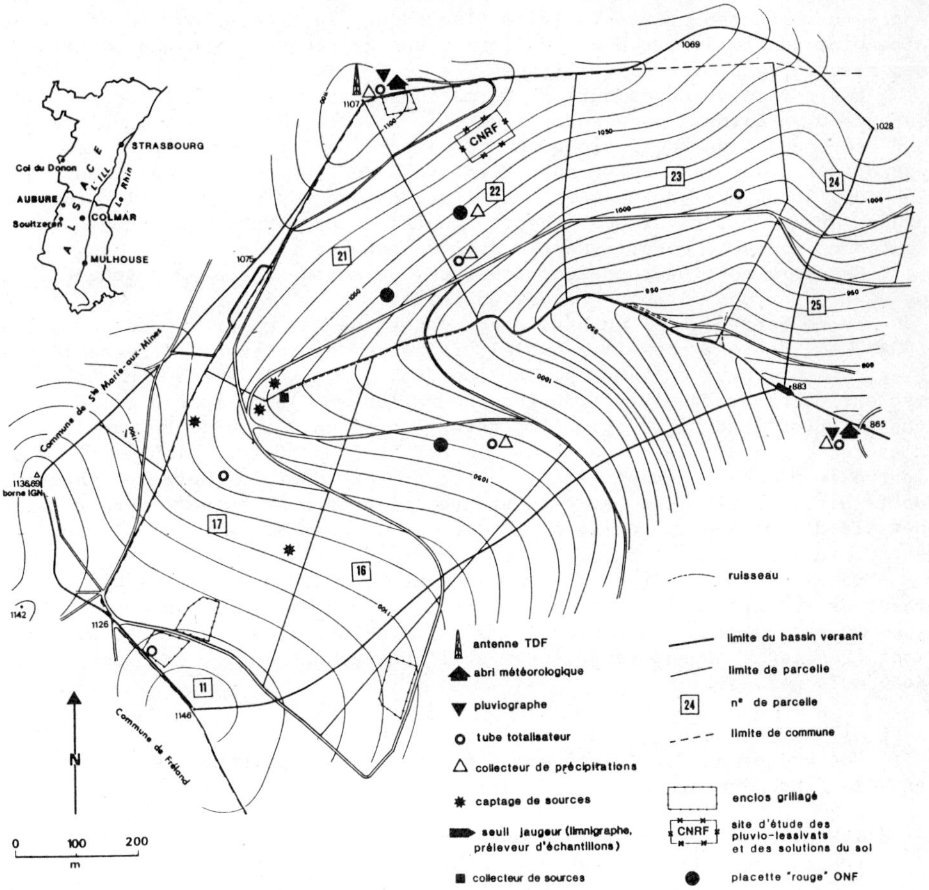

Figure 1: Le bassin versant du Strengbach à Aubure: localisation des différents sites de mesure.

Le substrat est composé d'un granite grossier à deux micas (biotite assez rare et souvent rubéfiée, muscovite difficile à identifier). Le faciès rencontré sur le versant Sud est un faciès microgrenu assurant le passage à un gneiss rubané présent de façon peu importante dans la partie Nord-Est du bassin.

Les formations superficielles sont le produit de l'altération du granite en une arène grossière comportant de nombreux cailloux. Leur épaisseur, mesurée par résistivité électrique, est généralement de 3m et n'atteint que 7m au maximum. Ces formations ne peuvent donc pas constituer des réservoirs aquifères importants.

Les sols développés sur ces formations sont relativement épais (1m), de type brun acide à ocre podzolique (6). La partie supérieure du versant Sud est occupée par des sols bruns et la partie inférieure par des sols podzoliques. Sur le versant Nord, les sols ocres podzoliques sont très étendus. Quant aux bas de versants, ils sont occupés par des sols bruns ou bruns ocreux colluviaux. Une petite unité hydromorphe (stagnogley) est également présente dans la partie aval du fond de vallon.

Ces sols sont profondément désaturés de sorte qu'ils sont susceptibles de développer des carences ou des toxicités.

Le couvert végétal du bassin est constitué en majorité de conifères (80%), de l'épicéa pour l'essentiel, le hêtre constituant le complément. L'épicéa se trouve en jeune plantation de 5 à 30 ans -notamment dans le secteur de propriété privée du fond de vallon -en jeune futaie de 30 à 45 ans d'aspect irrégulier selon les parcelles et en futaie de 80 à 130 ans d'aspect régulier, le plus souvent pur, et parfois mélangé au sapin (parcelle n° 23). Le sapin est en outre présent dans la hêtraie sapinière de faible extension. Le hêtre se trouve à l'état pratiquement pur dans la hêtraie d'altitude (parcelle n° 16) et en régénération dans la hêtraie sapinière.

Le climat du bassin est typiquement océanique. Les précipitations mesurées au village distant de 2 km et à 800 m d'altitude sont de 1200 à 1300 mm par an dont environ 15% sous forme de neige. Sur le bassin, elles sont très certainement supérieures et l'enneigement au sol est de l'ordre de 4 mois par an.

3. EQUIPEMENT DU BASSIN

Cet équipement progressivement mis en place depuis novembre 1985 est en voie d'achèvement.

3.1 Mesure des paramètres climatiques

Les mesures climatiques sont faites en deux points du bassin:

- en un site aval, à 868 m d'altitude, proche de l'exutoire, dans un abri météorologique standard (thermo-hygrographe, thermomètre maxima et minima, évaporimètre Piche);

- et surtout en un site amont sur la crête, à 1101 m d'altitude, par une station météorologique automatique (température, humidité de l'air, vitesse et direction du vent, rayonnement global), interrogeable par téléphone et gérée conjointement avec le Service Météorologique Inter Régional du Nord Est (SMIRNE) de Strasbourg-Entzheim; cette station est couplée à une station de mesure de la qualité de l'air (SO_2, NO_x, O_3) gérée par l'Association de Surveillance de la Pollution Atmosphérique d'Alsace (ASPA).

3.2 Mesure des apports atmosphériques secs et humides hors couvert forestier

La mesure quantitative des apports se fait d'une part par des pluviographes installés à proximité des abris météorologiques et d'autre part par 7 tubes totalisateurs en plexiglas, gradués et répartis sur tout le bassin permettant de mesurer en toute saison les précipitations (saumure antigel, huile anti-évaporation).

La mesure qualitative est faite par 4 collecteurs à entonnoir (prélèvement hebdomadaire) et par 2 pluviomètres automatiques (prélèvement quotidien), dont l'un est à obturateur, pour estimer les dépôts secs et humides.

3.3 **Mesure des flux sous couvert**

Les bilans à l'échelle de la station se font dans une parcelle de 0.4 ha de superficie, installée à 1080 m d'altitude sur un versant Sud, constituée d'un peuplement homogène d'épicéas dépérissants agés ,d'environ 80 ans.

Dans cette station, les litières (aiguilles, mousses) sont recueillies dans une vingtaine de bacs rectangulaires disposés à 20 cm du sol.

Les pluviolessivats au sens large (pluviolessvats humides, dépôts secs, litières) sont prélevés dans 6 gouttières de 2 m de long et 10 cm de large disposées en transect entre les arbres; en hiver, le prélèvement se fait dans 2 bacs à fond chauffant de 2 m de long.

Quant aux pluviolessivats au sens strict, ils sont prélevés par 3 pluviomètres (2 bacs de 2 m de long et un entonnoir de 30 cm de diamètre) à système d'ouverture automatique lors des épisodes pluvieux (5). Ces deux appareils sont par ailleurs équipés de collecteurs quotidiens pour l'estimation du dépôt sec (7).

Les solutions du sol sont recueillies par 32 plaques lysimétriques installées à différentes profondeurs (5, 10, 30, 60cm) correspondant aux limites des différents horizons pédologiques ainsi que par 10 bougies poreuses installées 2 par 2 à 30, 60, 80 et 150cm.

3.4 **Mesure et échantillonnage des écoulements**

Le site retenu comme exutoire permet de fixer nettement les limites hydrographiques du bassin et se trouve sur des colluvions peu épaisses, en aval d'une petite zone saturée; de plus, un film en polyéthylène barre le vallon pour limiter les écoulements souterrains.

L'équipement du seuil hydrologique comprend:

- un canal jaugeur en fibre de verre de type "H-flume" à déversoir trapézoidal calibré permettant de mesurer avec une bonne précision les basses et les hautes eaux: les débits déjà observés varient entre 3 l/s et plus de 200l/s;
- un limnigraphe à retournement;
- un préleveur d'échantillons, programmable, couplé au limnigraphe.

Il existe dans ce bassin, en amont, un captage de sources et deux canalisations en sortent; elles ont été équipées de compteurs d'eau afin de connaître les volumes exportés.

Les prélèvements d'échantillons se font de façon systématique chaque semaine dans le collecteur et à l'exutoire; de plus, le préleveur peut échantillonner de façon fine les événements hydrologiques.

4. <u>CONCLUSION</u>

L'équipement de ce bassin est en voie d'achèvement, un dispositif de mesure des flux sous couvert devant encore être installé au printemps 1987 à des fins comparatives sous un jeune peuplement d'épicéas.

L'ensemble du dispositif, trop récent, ne permet de présenter que des premiers résultats indicatifs du comportement des différents éléments chimiques (8,9).Le premier bilan hydrogéochimique ne pourra être établi qu'à la fin de cette année hydrologique (automne 1987) avant de l'être en routine.

Par la suite, les études porteront davantage sur la variabilité spatiale et temporelle (en fonction des types de temps) des précipitations et sur une analyse plus détaillée des événements hydrologiques.

REMERCIEMENTS

Ces recherches sont rendues possibles grâce au Programme DEFORPA (CEE-DG XII, Ministère de l'Environnement), au Ministère de la Recherche et de l'Enseignement Supérieur, à la Région Alsace.

REFERENCES

(1) BULGEN P. (1984): Etude écosystémique de deux bassins versants boisés en Haute-Ardenne. Thèse Doc. Sci. Botaniques, Univ. Liège, 218 p..

(2) DUMAZET B. (1983): Modification de la charge chimique des eaux au cours du transit à travers trois écosystèmes distincts du Mont Lozère. Thèse 3ième cycle, Univ. Orléans, 147 p..

(3) DUPRAZ C. (1984): Bilans des transferts d'eau et d'éléments minéraux dans trois bassins versants comparatifs à végétations contrastées (Mont Lozère, France). Thèse Doc. Ing., Géologie appliquée (hydrologie), Univ. Orléans, 363 p. et annexes.

(4) FRITZ B., MASSABUAU J.C., AMBROISE B. (1984): Physico-chemical characteritics of surface waters and hydrological behaviour of a small granitic basin (Vosges massif, France): annual and daily variations. In Hydrochemical balances of freshwater systems, (Uppsala symp., September 1984), IAHS Publ. n° 150, 249-261.

(5) GELHAYE D., HURPEAU A., PERRIN J.R. (1987): Un pluviomètre original pour la collecte quantitative des pluviolessivats.Revue Forestière Française n° 3, (à paraître).

(6) LEFEVRE Y. (1987): Répartition et caractérisation analytique des sols du bassin versant d'Aubure. Relation avec le dépérissement. Revue Forestière Française (à paraître).

(7) LINDBERG S.E., LOVETT G.M. (1985): Field measurements of particle dry deposition rates to foliage and inert surfaces in a forest canopy. Envir. Sci. Technol., Vol. 19 n° 3, 238-243.

(8) PROBST A., FRITZ B., AMBROISE B., VIVILLE D. (1987): Forest influence on the surface water chemistry of granitic basins receiving acid precipitation in the Vosges massif, France. (Vancouver symp., August 1987), IAHS, (accepté pour publication).

(9) PROBST A., FRITZ B., AMBROISE B., VIVILLE D. (1987): Le bassin versant forestier du Strengbach à Aubure (Haut-Rhin, France) pour l'étude du dépérissement dans les Vosges (Programme DEFORPA): II- Influence des précipitations acides sur la chimie des eaux de surface. Symposium sur les effets de la pollution de l'air sur les écosystèmes terrestres et aquatiques, Grenoble 18 - 22 mai 1987, (accepté pour publication).

LE BASSIN VERSANT DU STRENGBACH A AUBURE (HAUT-RHIN, FRANCE) POUR L'ETUDE DU DEPERISSEMENT FORESTIER DANS LES VOSGES (PROGRAMME DEFORPA)
II- INFLUENCE DES PRECIPITATIONS ACIDES SUR LA CHIMIE DES EAUX DE SURFACE

A. PROBST*, B. FRITZ**, B. AMBROISE***, D. VIVILLE***

* Institut de Géologie, ULP, 1 rue Blessig, F 67084 Strasbourg Cedex
** Centre de Sédimentologie et de Géochimie de la Surface (C.S.G.S., LP 6251 CNRS), 1 rue Blessig, F 67084 Strasbourg Cedex
***Centre d'Etudes et de Recherches Eco-Géographiques (C.E.R.E.G., UA 95 CNRS), ULP, 3 rue de l'Argonne, F 67083 Strasbourg Cedex

Résumé

Le petit bassin versant du Strengbach (Haut-Rhin, Vosges) a été choisi pour établir des bilans hydrogéochimiques en relation avec le dépérissement forestier (Programme DEFORPA). La description du bassin et le but des recherches sont donnés dans (14). Les premiers résultats géochimiques concernent les eaux de pluie, de source et du ruisseau. L'acidité des eaux de pluie (pH moyen= 4.5) est déjà neutralisée dans les eaux de sources et du ruisseau mais les anions d'acides forts dominent la charge anionique des eaux de surface. Le pouvoir tampon, déjà très faible, est affecté pendant la fonte des neiges et à la reprise des écoulements à l'automne.Les concentrations élevées en sulfate dans le ruisseau peuvent être expliquées par d'importants dépôts secs de soufre. Ce que confirment les premières analyses de pluviolessivats.Le contraire s'observe dans le bassin non forestier du Ringelbach où l'alcalinité domine la charge anionique des eaux de surface drainant aussi une roche mère granitique.

Summary

The Strengbach basin (Haut-Rhin, Vosges massif) has been chosen to study hydrochemical balances in surface waters in relation with forest decline (DEFORPA Program). The caracteristics and the equipment of the basin are described in part I. The first results concern rain,spring and river waters. The acidity of rain waters (mean pH=4.5) is already neutralized in spring and stream waters but the strong acid anions always dominate the anionic charge of surface waters: sulfate represents 60 % of the stream anionic charge, while the alkalinity remains low (44 μeq/l). The buffering capacity is then very weak and is affected during snowmelt and with the first increase in automn runoff. The high concentration of sulfur in stream water can be explained by important dry deposits of sulfur: the first analyses of chemical composition of throughfall confirm this hypothesis. The contrary is observed in the non forested Ringelbach basin where alkalinity dominates the anionic charge of the stream waters also draining granitic bedrock.

1. INTRODUCTION

Il y a quelques années, des études régionales ont montré le caractère acide des eaux de pluie sur la plaine d'Alsace et les Vosges (2, 6) et leur influence sur l'altération des grès des monuments (5). Aujourd'hui, dans le cadre du programme DEFORPA (Dépérissement des Forêts attribué à la

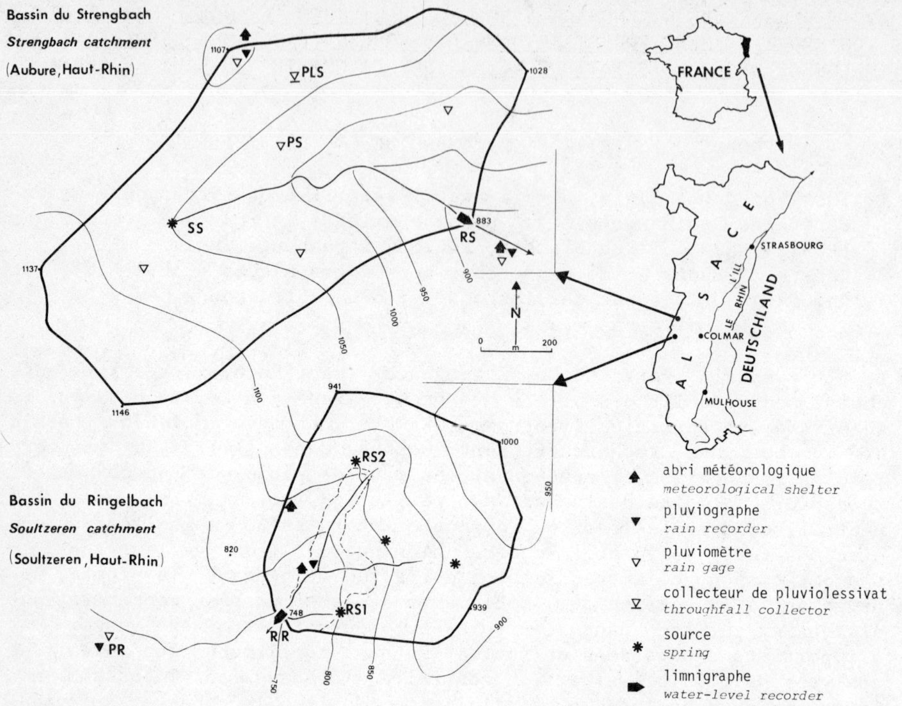

Fig. 1 Carte de localisation générale des bassins versants du Strengbach et du Ringelbach en Alsace (Est de la France).

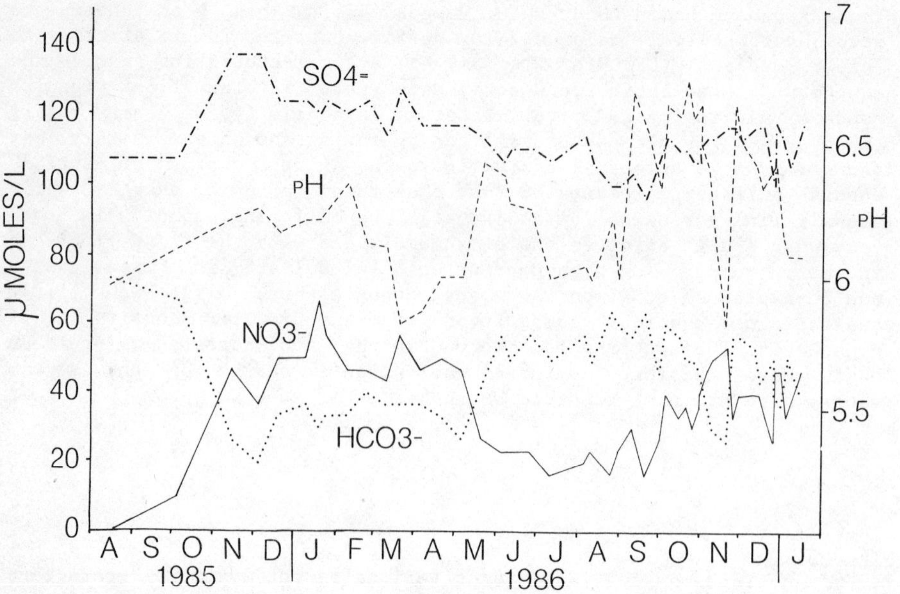

Fig. 2 Variations temporelles des concentrations (sulfate, nitrate, alcalinité) et du pH dans les eaux du Strengbach.

Pollution Atmosphérique), le bassin versant du Strengbach à Aubure (Haut-Rhin) a été choisi pour établir des bilans hydrogéochimiques en relation avec le dépérissement forestier. L'objectif des recherches, les caractéristiques de ce bassin versant et de son équipement climatique et hydrochimique sont données par ailleurs (14). Les résultats que nous présentons ici concernent les premières données de qualité chimique des eaux de pluie, de source et de ruisseau depuis septembre 1985. Le premier bilan hydrogéochimique sera établi pour l'année hydrologique octobre 1986 - octobre 1987. Une comparaison des ces données sera établie avec un autre petit bassin (36 ha) granitique non forestier des Vosges (fig. 1, Ringelbach, Haut-Rhin) soumis à l'influence de pluies de composition chimique analogue à celles du bassin du Strengbach.

2. METHODOLOGIE

Dans le bassin du Strengbach, les eaux de pluie, de source et de ruisseau ont été collectées initialement à un pas de temps bimensuel; depuis août 1986, la collecte des eaux de ruisseau et de source s'effectue une fois par semaine et plus fréquemment pendant les évènements hydrologiques. D'autre part, quelques échantillonnages de pluviolessivats ont été effectués avant que la station du CNRF de Nancy (mesure de pluviolessivats et des solutions de sol) soit opérationnelle. Pour les résultats présentés sur le bassin du Ringelbach (non foresté), l'échantillonnage s'est effectué une fois par mois pendant un an (mi 81-mi 82), puis tous les 3 mois jusqu'à aujourd'hui pour confirmer le domaine de variation de la composition chimique de ses eaux (6). Les précautions d'échantillonnage et les méthodes d'analyses sont détaillées dans (8).

3. RESULTATS ANALYTIQUES

3.1 Caractéristiques et variation temporelle du chimisme des apports

Pour l'ensemble des prélèvements, les eaux de pluie sont des solutions très diluées. La minéralisation est en moyenne d'environ 7 mg/l et s'étale de 2 mg/l (neige) à exceptionnellement 49 mg/l (très faibles pluies).

Ces solutions sont acides (pH moyen= 4.5) avec une variation de pH allant généralement de 4 à 5.3 et exceptionnellement à 6.4 (13). Les éléments dominants parmi les anions sont les anions d'acides forts ($SO_4^{=}$, NO_3^-, Cl^-), respectivement 51 %, 28 %, 14 % de la charge anionique avec une alcalinité nulle ou négative (13). Parmi les cations le proton est dominant (36 %), suivi de l'ammonium (25 %) et du calcium (19 %).

Les nitrates sont dominants par rapport aux chlorures dans les eaux de pluie d'automne et de printemps et c'est l'inverse dans les eaux de neige et pluies d'hiver (décembre à mars). Le sulfate est toujours l'anion majeur à l'exeption des chutes de neige de janvier où la proportion de chlorures est égale ou légèrement supérieure.

Parmi les cations, le sodium devient prépondérant par rapport au calcium dans les précipitations hivernales. Ceci correspond à une origine essentiellement océanique pour le sodium (masses d'air maritime en période hivernale) et une origine plus continentale (poussières) dans le cas du calcium (3, 9).

Pour la période considérée, ce sont les pluies de printemps (avril 86 -juillet 86) qui sont les plus acides, avec un pH moyen de 4.3, et les plus chargées en polluants anthropogènes. Le proton atteint 50 % de la charge en cations, sulfates et nitrates respectivement 62 % et 38 % des anions. On peut également noter que NH_4^+, $SO_4^{=}$, NO_3^- ont des variations temporelles semblables (très bonnes corrélations entre ces éléments), traduisant une origine commune essentiellement dérivée de polluants atmosphériques.

3.2 Caractéristiques et variations temporelles du chimisme des eaux de source et de ruisseau

Ces eaux sont légèrement plus concentrées que les eaux de pluie: de 31 à 43 mg/l pour les sources et de 33 à 45 mg/l pour le ruisseau.

Le pH varie de 5.5 à 6.7 pour les sources et de 5.8 à 6.7 dans le cas du ruisseau. L'acidité des eaux de pluie ayant transité dans le bassin est donc au moins neutralisée à ce niveau. Cependant la charge anionique reste dominée par les anions d'acides forts: $SO_4^{=}$ (60%), NO_3^-, Cl^- (13).

L'alcalinité moyenne des eaux du ruisseau et des sources est très faible pour des eaux sur granite (respectivement 45 µeq/l et 49 µeq/l). Les plus fortes valeurs correspondent aux périodes d'étiages (août -octobre). La valeur maximale enregistrée pour le ruisseau (72 µeq/l) ne représente que 22 % de la charge anionique, la plus faible (20 µeq/l) seulement 9 % . Lors des premières pluies d'automne (reprise des écoulements) en novembre 85 et 86, une chute de l'alcalinité se produit (fig. 2) . En novembre 86, cette baisse d'alcalinité s'accompagne d'une chute progressive du pH et d'une augmentation des sulfates. Ce phénomène se produit avec un décalage par rapport aux fortes pluies de fin octobre (120 mm) qui ont entraîné une crue importante. Le départ du soufre stocké dans les bassins avec les premières pluies qui suivent une période estivale sèche a déjà été mis en évidence sur de nombreux bassins versants (15, 7).

Il faut noter que le pouvoir tampon des eaux de surface est également affecté lors de la fonte du manteau neigeux en mars -avril (fig. 2). L'apport de sulfate des eaux de neige entraîne également une chute de l'alcalinité et du pH.

Nitrate et **potassium** sont influencés par le cycle végétatif forestier (maxima d'hiver, chute au printemps).

Les cations dominants sont calcium > sodium > magnésium > potassium. **Calcium** et **magnésium** sont bien corrélés par leur origine liés à l'altération sur le bassin. Ils sont également recyclés partiellement par l'écosystème forestier comme l'ont montré d'autres auteurs (1, 9).

4. DISCUSSION

La comparaison des chimismes des eaux du bassin forestier du Strengbach et du bassin non forestier du Ringelbach, recevant des précipitations acides, montre des différences essentielles (fig. 3):

a) dans le bassin non forestier du Ringelbach, l'acidité incidente de l'eau de pluie a été efficacement neutralisée durant son transit dans le bassin. L'influence des processus d'altération est nette (RS1, RS2 et RS) avec une augmentation de l'alcalinité carbonatée dans la charge anionique et une disparition des H^+ parmi les cations (6).

b) dans le bassin forestier du Strengbach (RS et SS), l'acidité a aussi été neutralisée mais la charge anionique reste dominée par les anions d'acides forts (13). Dans ce bassin le pouvoir tampon des processus d'altération reste faible, les cations bivalents sont remplacés par les protons et des ions ammonium comme sur d'autres basins acidifiés (12).

Dans le bassin du Strengbach, la concentration en sulfate du ruisseau est beaucoup plus forte que celle de l'eau de pluie (4 fois plus concentrée). Ce rapport élevé de concentrations, inexpliquable par la seule évaporation, ne semble pas pouvoir être attribué à un apport de soufre du bassin dans cet environnement granitique. L'importance des sulfates quittant le bassin laisse supposer qu'une part importante du SO_2 atmosphérique est filtrée sous forme de dépôts secs par les conifères et non prise en compte par les collecteurs de précipitations à découvert (9, 10, 7). Les pluviolessivats récoltés sur le bassin en automne 1986 ont des concentrations sous pessière 5 à 12 fois plus élevées que les pluies hors couvert. Ces résultats sont comparables à ceux d'autres auteurs (3, 11).

La station du CNRF devrait donner des résultats essentiels pour comprendre le transit de certains éléments et notamment des sulfates à travers le couvert végétal et les sols du bassin avant d'atteindre l'exutoire.

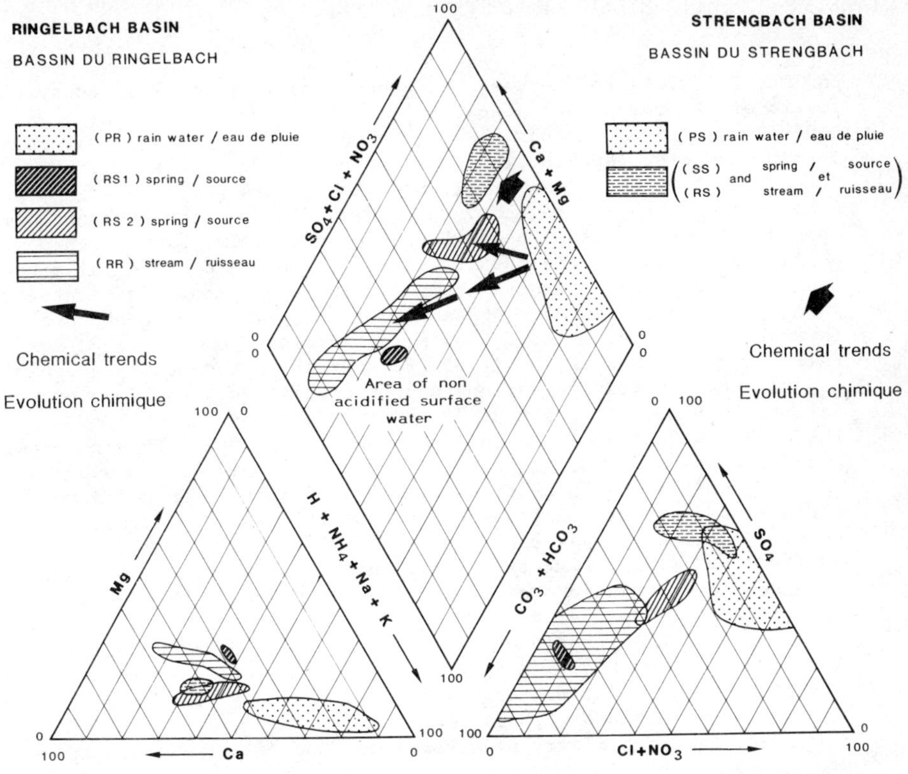

Fig. 3 Diagramme de Piper pour les eaux des deux bassins.

5. CONCLUSION

Le bassin versant du Strengbach à Aubure reçoit des pluies acides confirmant les analyses effectuées par (2). Comme dans le bassin non forestier du Ringelbach, cette acidité est neutralisée lors du transit à travers l'écosystème. Cependant, dans le bassin forestier le chimisme des eaux de sources et de ruisseau reste dominé par les anions d'acides forts, traceurs de pollution atmosphérique (sulfates et nitrates).

Le pouvoir tampon des processus d'altération reste faible et l'alcalinité carbonatée chute ainsi que le pH à la reprise des écoulements (automne) et lors de la fonte du manteau neigeux au printemps.

L'importance des sulfates dans les eaux du Strengbach semble être imputable aux dépôts secs de SO_2 filtrés par les conifères. Le chimisme des pluviolessivats d'automne montre un enrichissement considérable en sulfates par rapport au pluies hors couvert.

Ces premiers résultats seront confirmés après une année hydrologique complète et une année de mesure stationnelle des pluviolesssivats et des solutions de sol qui permettra de boucler le premier bilan hydrochimique.

REMERCIEMENTS

Ces recherches sont rendues possibles grâce au Programme DEFORPA (CEE

-DG XII, Ministère de l'Environnement), au Ministère de la Recherche et de l'Enseignement supérieur, à la Région Alsace. Les analyses chimiques sont effectuées au C.S.G.S.. Nous remercions tout particulièrement Mme Y. HARTMEIER et MM. D. MILLION et G. KREMPP.

REFERENCES

(1) BARTOLI F. (1986) Les cycles biogéochimiques dans les écosystèmes forestiers tempérés. Sci. Géol. Bull., 39, 2, pp. 195-209.

(2) BOURRIE, G. (1978). Acquisition de la composition chimique des eaux en climat tempéré. Application aux granites des Vosges et de la Margeride. Sci. Géol., Mém., 52, Strasbourg, 174 p..

(3) BULDGEN, P. (1984) Etude écosystémique de deux bassins versants boisés en Haute-Ardenne. Thèse Doc. Sci. Botaniques, Univ. Liège, 218 p..

(4) DUPRAZ, C. (1984) Bilans des transferts d'eau et d'éléments minéraux dans trois bassins versants comparatifs à végétations contrastées (Mont -Lozère, France). Thèse Doc. Ing., Géologie appliquée (hydrologie), Univ. Orléans, 363 p. et annexes .

(5) FRITZ, B. et JEANNETTE, D. (1981) Pétrographie et contrôle géochimique expérimental de transformations superficielles de grès de monuments. Sci. Géol. Bull., Strasbourg, 34, 4, 193-208.

(6) FRITZ, B., MASSABUAU, J.C. and AMBROISE, B. (1984) Physico-chemical characteristics of surface waters and hydrological behaviour of a small granitic basin (Vosges massif, France): annual and daily variations. In Hydrochemical balances of freshwater systems, (Uppsala symp., September 1984), IAHS Publ. n°. 150, 249-261.

(7) HULTBERG, H. (1985) Budgets of base cations, chloride, nitrogen and sulphur in the acid Lake Gardsjön catchment, SW Sweden. Ecol. Bulletins, 37, 133-157.

(8) KREMPP, G. (1987) Techniques de prélèvement des eaux naturelles et des gaz associés. Méthodes d'analyse des eaux et des roches. Notes techniques n°19, Institut de Géologie, Strasbourg, France.

(9) LIKENS, G.E., BORMANN, F.H., PIERCE, R.C., EATON, J.S. and JOHNSON W. M. (1977) Biogeochemistry of a Forested Ecosystem. Springer verlag, New York, 147 p..

(10) LOVETT, G.M. and LINDBERG, S.E. (1984) Dry deposition and canopy exchange in a mixed oak forest as determined by analysis of throughfall. Journal of Applied Ecology, 21, 1013-1027.

(11) MAYER R. and ULRICH B. (1980) Input to soil, especially the influence of vegetation in intercepting and modifying inputs - a review. In Effects of acid precipitation on terrestrial ecosystems, Hutchinson and Havas, NATO conference series, Series I: Ecology, 4, 173-182.

(12) PACES T. (1985) Sources of acidification in Central Europe estimated from elemental budgets in small basins. Nature, v.315, n° 6014, 31-36.

(13) PROBST A., FRITZ B., AMBROISE B., VIVILLE D. (1987) Forest influence on the surface water chemistry of granitic basins receiving acid precipitation in the Vosges massif. AIHS Symposium, Vancouver, 9-22 Août 1987 (accepté pour publication).

(14) VIVILLE D., AMBROISE B., PROBST A., FRITZ B. (1987) le bassin versant du Strengbach à Aubure (Haut-Rhin, France) pour l'étude du dépérissement forestier dans les Vosges (Programme DEFORPA). I-Equipement climatique, hydrologique, hydrochimique. Symposium sur les effets de la pollution de l'air sur les écosystèmes terrestres et aquatiques, Grenoble, 18-22 mai 1987 (accepté pour publication).

(15) WRIGHT R.F. et JOHANNESEN M. (1980) Input and output budgets of major ions at gauged catchments in Norway. In: Drablos , D. and Tollan, A. (eds), Ecological Impact of acid precipitaion. SNSF project, As, Norway, pp. 250-251.

FATE OF MINERAL NITROGEN IN ACID HEATHLAND AND FOREST SOILS

J.W. VONK, D. BARUG, and T.N.P. BOSMA
TNO Technology for Society, P.O. Box 108,
3700 AC Zeist (NL)

Summary

Ammonium added to heathland soil (pH 3.7) or to Douglas forest soil (pH 3.7) appeared to be rather stable. This contrasted with a sandy agricultural soil (pH 5.5) from which ammonium disappeared rapidly.
The numbers of autotrophic ammonium-oxidizing bacteria in heathland and forest soil were very low, and tests for potential ammonium-oxidizing capacity were negative.

Nitrate formation from added ammonium was irregular: it appeared and disappeared again after some period. However, in incubation experiments with forest soil samples without addition, nitrate was slowly formed. The isolation of several heterotrophic nitrifying bacteria suggests that nitrate formation in these acid soils is mainly heterotrophic.

Ammonium concentration increase inhibited the net mineralisation of nitrogen in Douglas forest soil.

1. INTRODUCTION

Air pollution by ammonia from intensive indoor cattle-breeding and soil manuring is a problem in the Netherlands (1). It is currently recognized as a problem in other countries as well. Dry and wet deposition of NH_3 /NH_4^+ occur at relatively short distances. Ammonium may contribute to soil acidification by its conversion by soil microorganisms into nitrate (nitrification). This process releases protons into the soil environment.

It is well known that autotrophic nitrification is inhibited at low pH (2), but it is still controversial whether nitrification in acid soils is mainly heterotrophic or autotrophic (3, 4, 5). Also, not much information is available on the rate of nitrification in weakly buffered, acid soil.

Since acidification and nitrogen eutrophication may affect terrestrial soil ecosystems, we investigated the fate of ammonium and other nitrogen conversions in acid (pH 3-4) heathland and acid Douglas forest soil.

2. MATERIALS AND METHODS

2.1 Soils

Two locations with weakly buffered, acid soils were selected. One location was a sandy podzolized heathland at Terlet (NL). The dominant vegetation was heather (Calluna vulgaris), but a part was covered with various grasses. The other location near Garderen (NL) had a sandy soil with Douglas (Pseudotsuga menziessii) forest. It was part of the monitoring location of the Dutch Priority Programme on Acidification.

Sampling of the heathland soil was carried out at 10 different sites. The mineral soil layer was taken (pH 3.8-4.4). Soil with mature heather plants was studied in more detail: soil was taken from the decomposed litter layer (pH 3.6), the mineral layer (0-20 cm, pH 3.9) and from around the roots (rhizosphere, pH 3.9).

The soil of the Garderen location was sampled in two layers: the decomposed litter layer (0-10 cm, pH 3.7) and the upper humic mineral layer (pH 3.7). A sandy agricultural soil (pH 5.5) was taken from Wageningen (NL).

When not used immediately, the soils were stored at + 2°C. Before use the soils were partly air-dried, sieved (2 mm) and restored to their original moisture content.

2.2 Incubation of soil samples

Conversion of inorganic and organic nitrogen was determined in incubation experiments after addition of ammonium sulphate or lucerne meal to soil samples (50 g). Ammonium sulphate (589 mg/kg) was added as a solution to partly dried soil. The original moisture content of the soil was restored.

Lucerne meal (0.5%) was added as a dry powder. After addition the soil samples were thoroughly mixed and incubated in a moist atmosphere at 20°C.

2.3 Tests for potential autotrophic ammonium-oxidizing capacity

This test was carried out according to Belser and Mays (6). Soil samples (10 g) were incubated in 50 ml of water or phosphate buffer (1 mM) adjusted to pH 7. To this suspension 4.0 mg ammonium sulphate and 61 mg potassium chlorate were added. After shaking at 24°C for periods up to 8 days samples of the suspension were centrifuged after addition of $CaCl_2$. Nitrite was determined in the clear supernatant.

2.4 Determination of ammonium, nitrate and nitrite

Soil samples (50 g) were extracted with 2 M KCl (100 ml). The extract was filtered and the concentrations of ammonium, nitrate and nitrite were determined with the aid of an Autoanalyser.

2.5 Most probable numbers (MPN) of autotrophic nitrifyers

MPN counting of autotrophic nitrifying bacteria was carried out according to Schmidt and Belser (7). Plates with microcups (Costar) were used. Every cup was filled with 100 μl medium. Counts were made at pH 4 and pH 7.2. Cups were tested for the presence of nitrite and nitrate to detect ammonium oxidizers and in a separate experiment for the absence of nitrite to detect nitrite oxidizers.

2.6 Isolation of heterotrophic nitrifying bacteria

Heterotrophic nitrifying bacteria were isolated from heathland soil by serial dilution on agar plates containing a medium of sodium acetate (5 g/l), Mg SO_4 (0.6 g/l), NH_4 Cl (0.3 g/l), $NaHCO_3$ (0.1 g/l) and $CaCl_2$ (0.01 g/l) in phosphate buffer (0.5 mM) containing

Hoaglandt trace elements. Isolations were carried out at pH 4,5 and 7.5. Isolates were tested for their ability to form nitrate or nitrite in a glucose-peptone-yeast extract medium.

3. RESULTS

3.1 Rate of ammonium conversion

Fig. 1 shows the rate of conversion of added ammonium in the mineral layers of Douglas forest soil, heathland soil and agricultural soil. The rate of disappearance of ammonium in the acid soils was much lower than in the less acid agricultural soil.

The numbers of autotrophic ammonium oxidizing bacteria in the heathland and forest soils were low compared with the agricultural soil (Table I). However, relatively high numbers of nitrite oxidizing bacteria were found, especially in the rhizosphere. Tests for potential ammonium-oxidizing capacity were negative for all soils from the 10 heathland sites sampled and for the organic and mineral Douglas forest soils.

3.2 Formation of nitrate

Results of experiments on the formation of nitrate and nitrite from ammonium and lucerne meal in heathland soil are given in Fig. 2 and 3. Small amounts of nitrate and nitrite were formed, but these disappeared after a certain period. This phenomenon was observed with all soil samples of the 10 heathland sites sampled.

In the organic and mineral layers of the acid Douglas forest soil, ammonium added decreased in concentration after about 30 days. Shortly thereafter the concentration increased again. Nitrate formation increased at first but decreased later on (Fig. 4) After addition of ammonium the net total mineralization (NH_4^+ + NO_3^-) after about 100 days was inhibited (Table II).

Addition of N-rich substrate to forest soil in the form of lucerne meal stimulated the formation of nitrate in mineral soil, but not in the organic soil (Table II). A strong accumulation of ammonium from mineralization of lucerne meal was observed in the organic layer.

3.3 Isolation of heterotrophic nitrifying bacteria

Seven cultures of heterotrophic nitrifying bacteria were obtained. Of these, six formed nitrite during incubation in peptone medium, whereas one formed nitrate. The maximal concentration of nitrite and nitrate formed were 1.3 and 1.7 mgN/l, respectively.

4. DISCUSSION

The rate of conversion of added ammonium into nitrate (potential nitrification) in weakly buffered, acid heathland and Douglas forest soil was low. Moreover the process proceeded very irregularly. This result agrees with observations of Martikainen (8) who found almost no nitrification in a coniferous forest soil. The low rates are understandable in view of the low counts of autotrophic nitrifyers. Several authors reported the isolation of these organisms from acid soils, but did not indicate their actual importance.

In our study we found a slow formation of nitrate from native organic matter upon incubation of forest soil. Since the formation of nitrate was stimulated by the addition of lucerne meal and since heterotrophic nitrifying organisms could readily be isolated from weakly buffered acid soil, heterotrophic nitrification is probably responsible for nitrate formation. Lang and Jagnow (3) isolated a large number of fungi and heterotrophic bacteria from acid soil which were capable to form nitrite and nitrate in peptone solution.

The disappearance of nitrate after addition of ammonium has been observed several times in our experiments. This could be explained by an increased microbial biomass due to the higher amounts of readily available ammonium.

The inhibition of total net mineralization by addition of extra ammonium is remarkable. It suggests that increase of ammonium concentrations by deposition might strongly affect the microbial soil ecology. Further research to elucidate this phenomenon is needed.

Acknowledgements

This work was supported by the Dutch Ministry of Housing, Physical Planning and Environment. We thank Mrs. M.M.J. Budding for technical assistance and Dr. F. Stams for making available bacterial counts in forest soil.

REFERENCES

(1) VAN BREEMEN, N. et al. (1982). Soil acidification from atmospheric ammonium sulphate in canopy throughfall. Nature. 299, 548-550.

(2) FOCHT, D.D. and VERSTRAETE, W. (1977). Biochemical ecology of nitrification and denitrification. Advances in Microb. Ecology 1, 135-214.

(3) LANG, E. and JAGNOW, G. (1986). Fungi of a forest soil nitrifying at low pH values. FEMS Micr. Ecol. 38, 257-265.

(4) HANKINSON, T.R. and SCHMIDT, E.L. (1984). Examination of an acid forest soil for ammonia- and nitrite-oxidizing autotrophic bacteria. Can. J. Microbiol. 30, 1125-1132.

(5) WALKER, N. and WICKRAMASINGHE, K.N. (1979). Nitrification and autotrophic nitrifying bacteria in acid tea soils. Soil. Biol. Biochem. 11, 231-236.

(6) BELSER, L.W. and MAYS, E.L. (1980). Specific inhibition of nitrite oxidation by chlorate and its use in assessing nitrification in soils and sediments. Appl. Env. Microb. 39, 505-510.

(7) SCHMIDT, E.L. and BELSER, L.W. (1982). Nitrifying bacteria. In: Methods of Soil Analysis, Part 2, A.L. Page et al. (Eds.). Am. Soc. Agron., Madison, pg 1027-1042.

(8) MARTIKAINEN, P.J. (1984). Nitrification in two coniferous forest soils after different fertilization treatments. Soil Biol. Biochem. 16, 577-582.

Table I Numbers of autotrophic nitrifying bacteria

	Numbers of bacteria (g^{-1})			
	NH_4^+-oxidizing		NO_2^--oxidizing	
Soil type	pH 4	pH 7.2	pH 4	pH 7.2
Heathland, mineral	10	10	5×10^3	6×10^3
organic	10	60	2×10^3	2×10^3
rhizosphere	10	10	2×10^5	7×10^3
Forest, mineral	-	< 30	-	3×10^4*
organic	-	< 30	-	18×10^4*
Agricultural	10	3×10^2	5×10^5	3×10^5

* pH 6.0

Table II Rates of N-mineralization and nitrification in Douglas forest soil at 20°C.

Addition	Soil layer	Rate of formation (mgN/kg/day)	
		NH_4^+	NO_3^-
None	Organic	0.8	0.3
"	Mineral	0.10	0.08
$(NH_4)_2SO_4$ (589 mg/kg)	Organic	0.64	0.0
"	Mineral	0.0	0.0
Lucerne meal (5 g/kg)	Organic	2.65	0.25*
"	Mineral	0.08	0.13

* Strong increased formation of NO_3^- during the period 40-80 days.

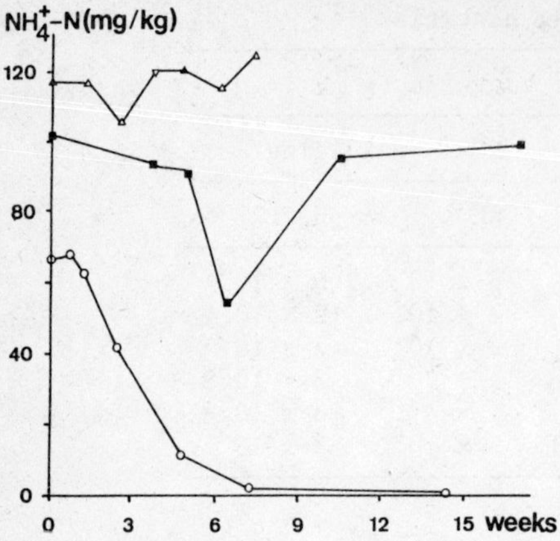

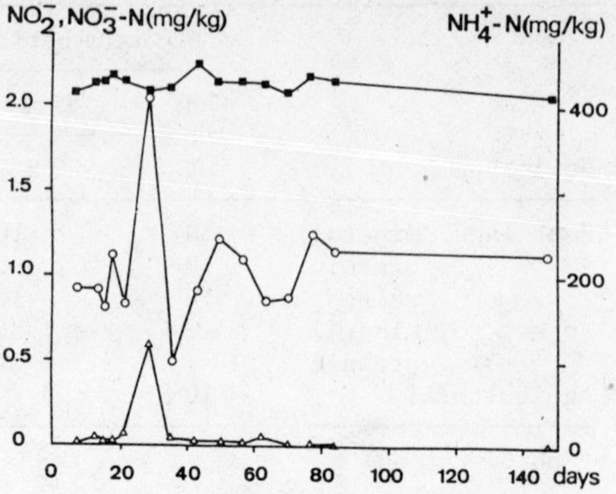

Fig. 1 Rate of ammonium disappearance in three soil types. Δ-Δ heathland mineral soil, pH 3.9; ■-■ Douglas forest, mineral layer, pH 3.7; o-o agricultural humic sandy soil, pH 5.5. Incubation at 20°C.

Fig. 2 Formation of nitrate (o-o) and nitrite (Δ-Δ) from ammonium (■-■) after incubation in heathland mineral soil at 20°C.

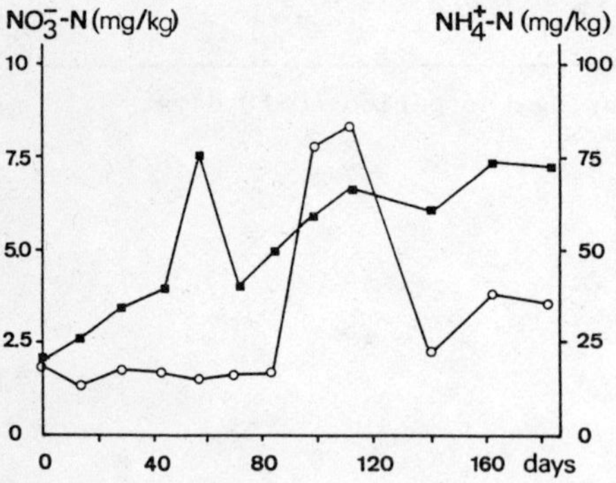

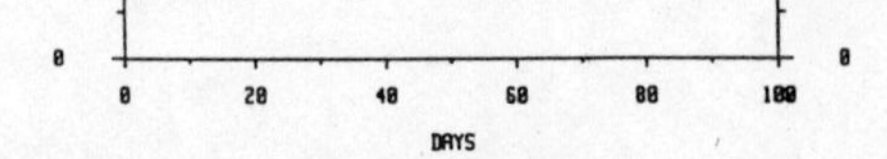

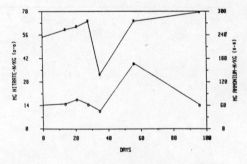

Fig. 3 Formation of nitrate (o-o) and ammonium (■-■) from lucerne meal (0.5%) in heathland mineral soil at 20°C.

Fig. 4 Formation of nitrate (o-o) in the organic layer of Douglas forest soil after addition of ammonium (*-*) (125 mg N/kg) and incubation at 20°C.

THE EFFECT OF ATMOSPHERIC DEPOSITION, ESPECIALLY OF NITROGEN, ON GRASSLAND AND WOODLAND AT ROTHAMSTED EXPERIMENTAL STATION, ENGLAND, MEASURED OVER MORE THAN 100 YEARS

K.W.T. GOULDING, A.E. JOHNSTON and P.R. POULTON
Soils and Plant Nutrition Department, Rothamsted Experimental Station, Harpenden, Herts, AL5 2JQ, UK.

Summary

Analyses of precipitation at Rothamsted began in 1853 and have continued periodically. Soil and plant samples have been taken at intervals from the permanent grassland of the Park Grass Experiment (begun in 1856) and from the deciduous woodland of the Broadbalk and Geescroft Wilderness Experiments (begun in the 1880s). By combining precipitation chemistry with estimates of other inputs, the relative importance of atmospheric deposition to acid inputs can be estimated, as well as the effect of these on the composition and yield of grassland and woodland. Acid rain (wet deposited H^+) has been a negligible source of acidity throughout the 100 year period of the experiments. Total atmospheric deposition, including dry deposited SO_2 and NO_X, and NH_4^+ which may be nitrified, may comprise up to 30% of the total acidifying inputs when soils are at or near pH 7 and up to 70% in more acid soil. Acidification slightly reduces the yield of herbage and the number of species present in grassland and the stand weight of woodland; different tree species dominate acid woodland soils. There has been greater acidification under deciduous woodland than grassland which is not grazed. This has important implications for the possible movement of land from agriculture into forestry or agroforestry.

1. INTRODUCTION

Rothamsted Experimental Station is at a semi-rural location in western Europe, 42 km north of central London. Analyses of precipitation there began in 1853 and have continued periodically since (2). Soil and plant samples have been taken at intervals from Park Grass, the permanent grassland experiment which began in 1856 (6, 7), and from the naturally regenerating deciduous woodland of the Broadbalk and Geescroft Wilderness Experiments, which began in the 1880s (3). By combining precipitation chemistry with measurements or estimates of other acidifying inputs - dry deposition, natural soil processes, eg respiration and mineralization, the effect of manures, and nutrient balances - the relative importance of these inputs can be assessed. The data also show the effect of acidification on both plant yield and composition of grassland and woodland, and provide some indication of the effects of moving land from agriculture into forestry or agroforestry or of the abandonment of crop production.

2. ATMOSPHERIC DEPOSITION

The first rain gauge at Rothamsted, built during 1852/3, provided precipitation inputs to improve nutrient balance sheets. Present-day interest in precipitation chemistry focuses on pollutants rather than nutrients, and the data has been examined in the context of Acid Rain (2). The measurement of pH was not possible before the 1920s and no precipitation samples from before this have been kept. Therefore the acidity of pre-1920s precipitation has been estimated by extrapolation. The precipitation at Rothamsted has never been very acid (Table I), although acidity has been increasing recently.

Table I. Estimated amounts of wet and dry deposited H^+ at Rothamsted

Input	Amounts ($keq\ ha^{-1}\ a^{-1}$) for the periods: 1860s	1920s	1980s
Wet deposited			
H^+	0.05	0.07	0.10
NH_4^+	0.4	0.4	2.0
Dry deposited			
SO_2 plus NO_X	0.6	1.2	1.8
Total	1.1	1.7	3.9

Dry deposition of S was measured in 1974 (1), and of NH_3 in 1976/7 (5). These data were used (4), together with those from other monitoring programmes and from models, to estimate dry deposition over the course of the experiment (Table I). Dry deposited S and N have always made up >90% of the atmospheric input of acidity. Wet deposition of H^+, Acid Rain, is and always has been a minor source, comprising <10% of total atmospheric inputs throughout the period of monitoring. However, becoming increasingly important is the potential acidity from the nitrification of NH_4^+ in precipitation. This may be adding 2.0 keq H^+ $ha^{-1}\ a^{-1}$, which is more than the H^+ derived from the dry deposition of SO_2 and NO_X (1.8 keq H^+ $ha^{-1}\ a^{-1}$). The probable source of this N is predominantly agriculture. Much NH_4^+ may come from animal excreta, some by volatilization from fertilizers. The rest comes from the combustion of fuels. There is a close association of NH_4^+ and NO_3^- in Rothamsted precipitation (2).

3. THE PARK GRASS EXPERIMENT

The Park Grass Experiment, begun in 1856, examined the effect of various manurial treatments on 300 year-old permanent grassland. Sited on a flat plateau at an altitude of 130 m, the soil is a non-calcareous silty clay loam. Data from unmanured and N only plots are considered here. The experiment is described more fully elsewhere (4, 6, 7).

The pH of the unmanured soil (0-23 cm) has declined steadily from 5.6-5.8 in 1856 to about 5.1 at present, except from 1876 to 1923 when small amounts of lime were added (Figure 1). Initially atmospheric inputs contributed only a little to this acidification (Table II) but they now

Table II. Estimated amounts and relative contributions of the acidifying inputs to Park Grass soils in the 1860s, 1920s and 1980s

Acidifying input	Amounts, keq H^+ $ha^{-1}\ a^{-1}$, and percentage contributions in parentheses — Period: 1860s	1920s	1980s
1. Minimum H^+ input	6.0	7.0	5.5
2. Atmospheric deposition	1.1(18)	1.7(24)	3.9(71)
3. Nutrient uptake	0.5(8)	0.5(7)	0.5(9)
4. Soil-derived natural sources	? (74)	? (69)	? (20)

comprise over 70% of inputs. However, the soil is apparently not acidifying further, presumably because the lime potential of the precipitation is greater than that of the soil solution (2).

The pH of soils receiving N fertilizer are included in Fig. 1 to show how relatively small atmospheric effects are compared with those of nitrogen fertilizers. Plots receiving 96 kg ha^{-1} N as ammonium sulphate showed a rapid decrease in soil pH to an equilibrium value of 3.6 ± 0.2 (Fig. 1). Where 48 or 144 kg ha^{-1} N as ammonium sulphate was applied the rate of decrease was in direct proportion to the amount of N added. This acidification was caused primarily by nitrification of the NH_4^+, which, with the largest amount of N, may have contributed 20 keq H^+ ha^{-1} a^{-1} at the beginning of the experiment, but much less as soil pH has fallen and nitrification has been reduced (4). By contrast, against a background of acidification from natural sources, N added as sodium nitrate has decreased the rate of soil acidification. The reason is not clear, but NO_3-N may have been lost by microbial fixation of N and denitrification, and sodium rather than calcium may have been leached.

No experiments have been done to show direct effects of atmospheric deposition on the herbage. Indirect effects of acidification, primarily via the soils, can be assessed by comparing plant composition and yield of the various plots.

Tables III and IV show the effects of soil acidification on the yield and botanical composition of some of the plots of Park Grass, and the effects of ameliorating the acidity with dressings of chalk since 1903 (for details see ref. 7). The change in method of harvest has had a major effect on yield. On the unmanured ('None') plots the small increase in yield on the limed half plots suggests that lack of nutrients, mainly N, rather than acidification to pH 5.1 had the major effect on yield. On the NH_4-N plots, there was a large benefit from liming and the fall in pH to 3.6 in the absence of lime has diminished yields. The same comparisons for botanical composition (Table IV) suggest that there is a complex interaction between pH and nutrition (for further details see 6). Supplying N greatly increased yield (Table III), but also reduced enormously the number and type of species (Table IV).

Table III. Effect of acidity on the yield of herbage from the Park Grass Experiment

Date	Yield of dry matter in t ha^{-1} from the treatments:				
1856	2.8 over whole site				
	None	None+lime	NH_4-N	NH_4-N+lime	NO_3-N
1886-1895	2.1	-	4.9	-	5.7
1920-1959	1.5	1.6	4.7	5.6	6.2
1965-1973*	3.0	3.3	6.7	9.2	9.2
1974-1982*	2.5	3.3	5.3	7.6	7.3

* Yields of hay estimated from total herbage yields, ie no hay-making losses

4. BROADBALK AND GEESCROFT WILDERNESS

Broadbalk and Geescroft Wildernesses are on the same gently undulating plateau as Park Grass, the same soil series, and at the same elevation. Broadbalk Wilderness is about 600 m north of Park Grass and Geescroft Wilderness about the same distance to the east. Both experiments are on parts of fields which were shown to be in arable cultivation on an estate map of 1623. They were fenced off and left untended in 1882, Broadbalk, and 1886, Geescroft. Subsequently both have reverted to deciduous to woodland (3).

Table IV. Effect of acidity on the botanical composition of the permanent pasture of Park Grass.

Date	Treatment	Soil pH	Plant species			
			Number	% by weight		
				Grasses	Legumes	Other
1856	-	5.7	23*	76	5	19
1877	None	5.3	52	71	8	21
	NH_4-N	5.0	28	95	0	5
	NO_3-N	5.9	28	88	1	11
1948/9	None	5.3	36	53	7	40
	None+lime	7.1	32	36	16	48
	NH_4-N	3.8	6	99	0	1
	NO_3-N	6.1	16	94	2	4

* Not known accurately. Certainly an underestimate.

Broadbalk soils received large dressings of chalk in the 18th or early 19th century which have buffered its pH at 7.9-8.0 for nearly 200 years. By contrast, Geescroft had no reserves of lime and the pH of its soil has decreased throughout the profile (Fig. 2). Its surface (0-23 cm) soil is approaching an equilibrium pH of 3.6. This value is much lower than that on the unmanured grassland of Park Grass; it is much closer to that of the soils which have received the additional acidifying inputs from nitrogen fertilizers.

Table V. Estimated amounts and relative contributions of the acidifying inputs to unmanured woodland (mainly deciduous species) soils in the 1890s, 1930s and 1980s

Acidifying input	Amounts, keq H^+ ha^{-1} a^{-1}, and percentage contributions in parentheses		
	1890s	1930s	1980s
1. Minimum H^+ input	14	9	5.5
2. Atmospheric deposition	1.1(8)	1.7(19)	3.9(71)
3. Nutrient uptake	0.1(1)	0.2(2)	0.3(5)
4. Soil-derived natural sources	? (91)	? (79)	? (24)

The causes of the acidification on Geescroft are listed in Table V (see also ref. 4). The soil-derived natural processes of respiration and mineralization were predominant in the early part of the experiment when soil pH values were near-neutral. However, as soil pH declined and mineralization and respiration decreased, atmospheric deposition has become a large proportion of the acidifying input. The estimate of total atmospheric deposition used here may be too small because trees are much more efficient than grass at capturing wet and dry deposition.

The effects of acidification can be seen by comparing the two Wilderness sites (Table VI and VII). Stand weight is smaller on Geescroft than on Broadbalk, and stand composition is not the same (Table VI). Geescroft is almost entirely comprised of oak (58%) and ash (26%) with some hawthorn (8%) and elm (5%). Broadbalk is a mixed ash (38%), hawthorn (32%) and sycamore (18%) wood, with some maple (8%). Both ash and maple dislike acid soils, and the long term decline in pH on Geescroft could affect their ability to compete and to regenerate, making oak even more dominant. But there is evidence that low pH may be preventing the regeneration of oak.

On Geescroft there has been less accumulation of organic C, N, S and P, than on Broadbalk because of reduced inputs and/or decomposition (Table

Table VI. Stand measurements (dry weight in t ha^{-1}) of Broadbalk and Geescroft Wilderness sites

Species	Broadbalk	Geescroft
Ash (*Fraxinus excelsior*)	104.1	46.7
Elder (*Sambucus nigra*)	0.1	1.1
Elm (*Ulmus spp.*)	0	8.4
Hawthorn (*Craetagus monogyna*)	88.6	14.5
Hazel (*Corylus avellana*)	6.1	0.5
Maple (*Acer campestre*)	21.5	1.5
Oak (*Quercus robur*)	3.6	105.1
Sycamore (*Acer pseudoplatanus*)	50.3	1.6
Silver birch (*Betula pendula*)	0	1.4
All species	274.3	180.8

Table VII. Mean annual gains (kg ha^{-1} a^{-1}) in organic C,N,S and P, and in inorganic S in the topsoil (0-23 cm) of Broadbalk and Geescroft

	Broadbalk	Geescroft
Organic C	530	250
N	45	13
S	6.9	3.7
P	5.6	2.3
Inorganic S	0.2	1.6

VII and reference 3). The greater gain of inorganic S is because of the greater number of positive, pH-dependent adsorption sites on soils of low pH.

5. CONCLUSIONS

At Rothamsted the equilibrium pH of non-calcareous surface (0-23 cm) soil under unmanured grassland is approximately 5.1 compared with that under deciduous woodland of 3.6. This greater acidification under woodland arises from its greater interception of atmospheric deposition and/or the greater soil respiration and mineralization. Such an effect has considerable implications for any future conversion of land from agriculture into forestry of agroforestry.

REFERENCES

(1) BROMFIELD, A.R. and WILLIAMS, R.J.B. (1974). The direct measurement of sulphur deposition on bare soil. Nature Vol. 252, No. 5483, 470-471.

(2) GOULDING, K.W.T., POULTON, P.R., THOMAS, V.H. and WILLIAMS, R.J.B. (1986). Atmospheric deposition at Rothamsted, Saxmundham and Woburn Experimental Stations, England, 1969-1984. Water, Air and Soil Polution. Vol. 29, 27-49.

(3) JENKINSON, D.S. (1971). The accumulation of organic matter in soil left uncultivated. Rothamsted Experimental Station, Report for 1970, Part 2, 113-137.

(4) JOHNSTON, A.E., GOULDING, K.W.T. and POULTON, P.R. (1986). Soil acidification during more than 100 years under permanent grassland and woodland at Rothamsted. Soil Use and Management, Vol. 2, 3-10.

(5) RODGERS, G.A. (1978). Dry deposition of atmospheric ammonia at Rothamsted in 1976 and 1977. Journal of Agricultural Science, Cambridge. Vol. 90, 537-542.

(6) THURSTON, JOAN M., WILLIAMS, E.D. and JOHNSTON, A.E. (1976). Modern developments in an experiment on permanent grassland started in 1856: Effects of fertilizers and lime on botanical composition and crop and soil analyses. Annales agronomiques. Vol. 27, 1043-1082.
(7) WARREN, R.G. and JOHNSTON, A.E. (1964). The Park Grass Experiment. Rothamsted Experimental Station, Report for 1963, 240-262.

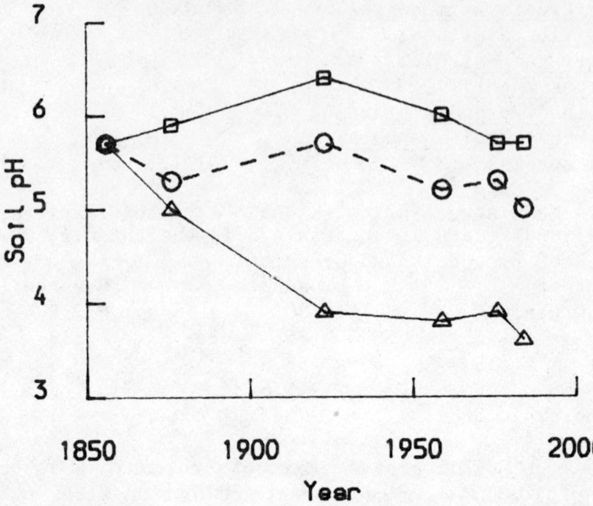

Figure 1. The pH values of soil samples taken from the unmanured (O), ammonium-N (Δ), and nitrate-N (□) treated plots of the Park Grass Experiment.

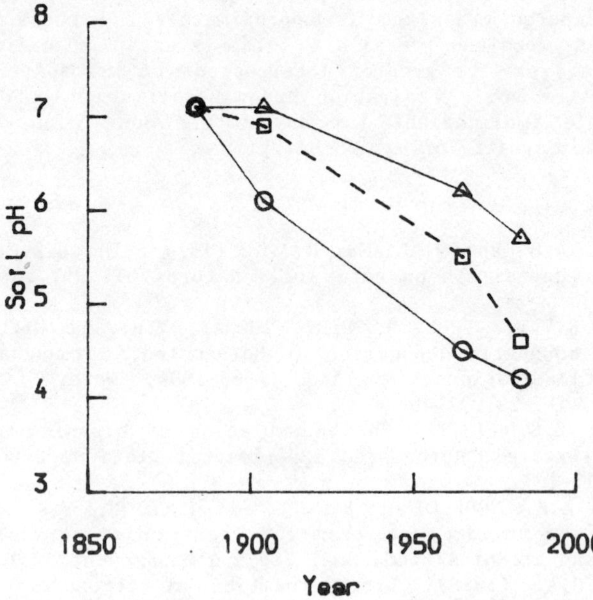

Figure 2. The pH values of soil samples taken from the 0-23 cm (O), 23-46 cm (□), and 46-69 cm (Δ) horizons of Geescroft Wilderness.

EFFECTS OF AMMONIA DEPOSITION ON ANIMAL-MEDIATED NITROGEN MINERALIZATION AND ACIDITY IN CONIFEROUS FOREST SOILS IN THE NETHERLANDS

H.A. VERHOEF and F.G. DOREL
Department of Ecology & Ecotoxicology, Free University,
De Boelelaan 1087, 1081 HV Amsterdam, the Netherlands

Summary
In forests in the south-east of the Netherlands, the available nitrogen in air and precipitation has greatly increased, mainly due to the evaporation of ammonia from, mainly liquid, farmyard manure. Fresh Pinus-needles from that area have on the tree a much higher nitrogen content than that of needles from a "clean" Pinus-stand. In microcosms the nitrogen mineralization of the litter and fermentation layer from both stands has been compared. Ammonification dominated only in the litter layer of the "clean" stand. In its fermentation layer and in both layers of the "saturated" stand, however, nitrification dominated. The collembolan Tomocerus minor increased mineralization in all layers but the "saturated" fermentation layer; there it decreased nitrogen mineralization. Decomposition, measured as dry weight loss, of the "saturated" layers was lower than that of the "clean" layers. Fumigation with NH_3 on "clean" layers neutralizes the acidifying effect of SO_2 treatment.

1. INTRODUCTION

The increase of available nitrogen in air and precipitation during the last decades has changed the nitrogen conditions in the temperate forests of northern Europe, so that the trees have become "over-saturated" with nitrogen. Healthy forests should get their main amounts of nitrogen from biological turnover of organic matter in the soil. In acid coniferous forests ammonification, at which ammonia is created by different bacteria and fungi, is supposed to surpass nitrification, at which ammonium is oxidized to nitrite and nitrate. The uptake of ammonium by the roots is accompanied by the release of protons from the plant roots into the soil solution. This causes leaching of magnesium, calcium and potassium from the upper soil horizon, and may lead to relative potassium and magnesium deficiences.

This is enhanced by the uptake of gaseous ammonia and ammonium particles (as ammonium sulfate) through the stomata of the needles, which is accompanied by cation losses (potassium and magnesium), too. Furthermore, increased amounts of needle-nitrogen cause a decrease in frost hardiness and increased susceptibility to attacks by insects, fungi, bacteria and viruses (3, 4, 6).

2. RESULTS

In the Netherlands in 1986 the percentage of forests which are vital further decreased to 46,9%, according to the monitoring program of the Netherlands State Forest Management (5). Forest decline is obvious in areas with a high ammonia emission from livestock wastes, which can be as high as 320 t of NH_3/5x5 km/year (see Fig. 1) (2).

In spring 1985 six stands of Pinus sylvestris L. were studied, one in an area with low emission (stand 1) and five in areas with relatively high emission (stands 2 to 6) (See Fig. 1). In each stand fresh needles were collected and samples were taken from the organic

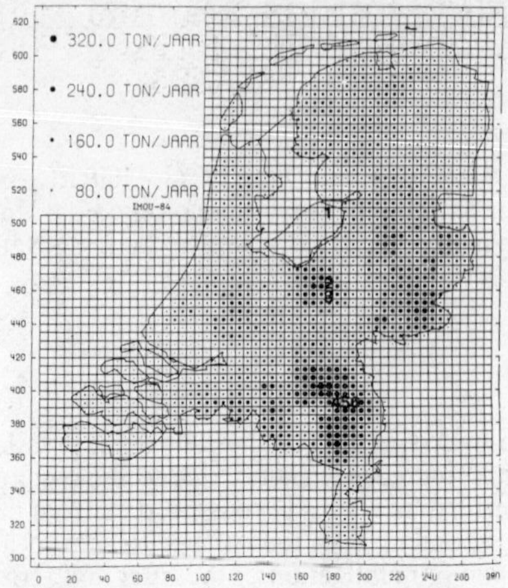

Fig. 1. Ammonia emission from farmyard manure. (2)

layer: the litter and the fermentation layer (see Table 1).

Table 1. Total N content (%) of fresh needles and litter and fermentation layers of 6 different *Pinus sylvestris* stands.

Location	Fresh needles	Litter layer	Fermentation layer
1. Roggebotzand	1.33	1.61	1.78
2. Loobos	2.24	1.61	1.66
3. De Valouwe/ Wekerom	4.27	1.70	2.13
4. Stippelberg	2.49	2.17	2.00
5. Klotterpeel	2.62	1.37	1.70
6. Rouwkuilen	4.51	1.85	2.38

Sampling date: 28/29 May 1985.

These data show that there is a clear increase of the nitrogen content of the needles of stands 2 to 6 in the areas with a high emission, compared with that in the area with a low emission. Concerning the soil layers both litter and fermentation layers of stand 3, 4 and 6 have higher values than those of stand 1. Thus, it can be concluded that not only in the trees, but also in, deeper, soil layers there is an effect of the deposition of ammonia.

From these six stands two were selected for microcosm studies; stand 1: Roggebotzand and stand 3: De Valouwe/Wekerom.

For these studies perspex cylindrical jars (height, 7.4 cm; diameter, 5.6 cm) were used, with gauze floors and gauze lids. The jars were filled with aliquots of defaunated soil layer, either litter or fermentation layer. To half of the jars 15 specimens of the collembolan Tomocerus minor (Lubbock) were added. The jars were placed on oasis, which was kept wet permanently. Every two weeks the microcosms were irrigated with distilled water and the leachates were analysed for mineral-N as nitrate and ammonium. The results, given in Fig. 2A show that in stand 1 (Roggebotzand) in the litter layer ammonification dominates over nitrification (which is hardly measurable), whereas the reverse is true in the fermentation layer.

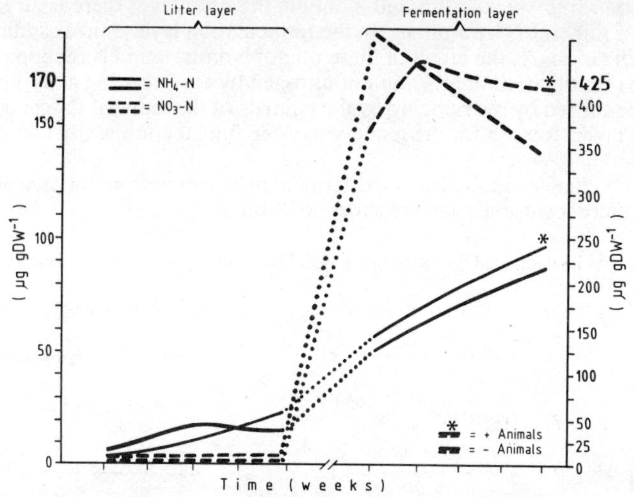

Fig. 2.A. NH_4-N (left axis) and NO_3-N (right axis) leaching from litter and fermentation layers from stand 1 with and without animals.

In the fermentation layer surprisingly high numbers of nitrifying bacteria have been determined (Verhoef, unpublished data). In stand 3 (De Valouwe/Wekerom) in the litter as well as in the fermentation layer nitrification dominates over ammonification (Fig. 2B). The high amounts of ammonium and nitrate are considered to be products of the ammonia deposition and not originating from biodegradation. The dry weight loss in the soil layers from stand 3 is very low compared with that of stand 1.

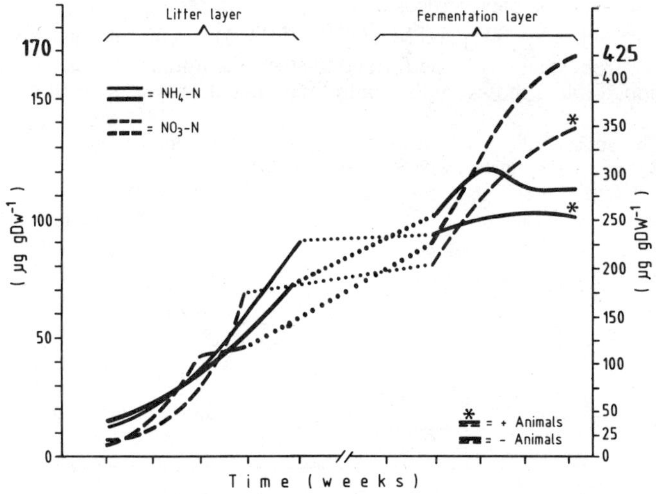

Fig. 2.B. NH_4-N (left axis) and NO_3-N (right axis) leaching from litter and fermentation layers from stand 3 with and without animals.

Animal treatment causes in stand 1 an increase in mobilization of mineral-N. In the litter layer this effect can be seen after eight weeks. In the fermentation layer this already happens after two to four weeks. In stand 3 only in the litter layer there is an enhancement of mobilization of mineral-N by animals. In the fermentation layer animal addition causes a decrease in mobilization. As the effect of these microbivorous animals is supposed to be the disruption of the microbial immobilization of nitrogen by their feeding activities (1, 7), this decrease might be caused by overgrazing by the animals of the reduced fungal community in the fermentation layer. Reason for the reduction of the fungal community could be the high input of ammonia.

The acidity changes in the litter and fermentation layers from the two stands during the 3.5 months' microcosm study are presented in Table 2.

Table 2. Changes in acidity (pH) of the organic soil layers during a 3.5 months' microcosm study.

Soil layer	Roggebotzand (1) initial		Roggebotzand (1) final		Wekerom (3) initial		Wekerom (3) final	
Litter layer	4.70 →	a	5.79 / 5.67	b	3.95 →	a	5.84 / 4.57	a
Fermentation layer	4.80 →	a	4.12 / 4.27	b	4.02 →	a	4.93 / 4.46	a

a = significant; b = non-significant difference.

Stand 1 has a higher pH than stand 3, concerning both litter and fermentation layer. During the stay in the microcosm the pH increased in the litter layer and decreased in the fermentation layer. This is in accordance with the prevailing ammonification in the litter and the nitrification in the fermentation layer. In stand 3 both in the litter and in the fermentation layer an increase has been found. It appears that this is greater in the upper part than in the lower part of the soil layer in the microcosm.

Fumigation with SO_2, NH_3 and SO_2/NH_3 (100 µg m^{-3}) on microcosms filled with fermentation layer from stand 1 was performed to study the initial reaction in a "clean" soil layer on deposition. Table 3 shows for the control the same decrease as given in Table 2.

Table 3. Changes in acidity (pH) of the fermentation layer from Roggebotzand (1) after 2 months of fumigation with SO_2, NH_3 and SO_2/NH_3.

	Control		$+SO_2$		$+NH_3$		$+SO_2/NH_3$	
Initial value	4.55		4.55		4.55		4.55	
Final value	4.08 / 4.39	a	3.96 / 4.25	a	4.37 / 4.33	b	4.27 / 4.26	b
Control vs. treatment			a		a		b	

SO_2 causes a greater decrease, NH_3 an increase, compared with the control, whereas the combination of NH_3 and SO_2 results in a change similar to that of the control. Thus, it seems that NH_3 treatment neutralizes the acidifying effect of the SO_2 treatment in this "clean" soil layer. The effect of animals on these processes is the subject of future research.

Acknowledgements - Ms D. Hoonhout is acknowledged for typing the manuscript.

REFERENCES

(1) Anderson, J.M., Huish, S.A., Ineson, P., Leonard, M.A. and Splatt, P.R. (1985). Interactions of invertebrates, microorganisms and tree roots in nitrogen and mineral element fluxes in deciduous woodland soils. In: "Ecological interactions in soil" (Ed. by A.H. Fitter et al.) Blackwell Scientific Publications, 377-392.

(2) Buijsman, W. (1984). Een gedetailleerde ammoniak emissiekaart van Nederland.

(3) Nihlgård, B. (1985). The ammonium hypothesis; an additional explanation to the forest dieback in Europe. Ambio, 14, 1, 1-8.

(4) Roelofs, J.G.M., Kempers, A.J., Houdijk, A.L.F.M. and Jansen, J. (1985). The effect of air-borne ammonium sulphate on Pinus nigra var. maritima in the Netherlands. Plant and Soil 84, 45-51.

(5) Staatsbosbeheer (1987) De vitaliteit van het Nederlandse bos in 1986. Staatsbosbeheer, Utrecht.

(6) Van Breemen, N., Mulder, J. and Driscoll, C.T. (1983). Acidification and alkalinization in soils. Plant and Soil 75, 283-308.

(7) Verhoef, H.A. and de Goede, R.G.M. (1985). Effects of collembolan grazing on nitrogen dynamics in a coniferous forest. In: "Ecological interactions in soil". (Ed. by A.H. Fitter et al.) Blackwell Scientific Publications, 367-376.

INVESTIGATIONS ON THE TELLURIC MICROBIOLOGICAL FEATURES OF FOREST DIEBACK IN VOSGES

D.ESTIVALET*,R.PERRIN* AND F.LE TACON**
*Station de recherches sur la flore pathogène dans le sol 17 rue Sully 21034 DIJON cedex FRANCE
** Laboratoire de microbiologie forestière CRF Champenoux B.P.35 54280 SEICHAMPS

Summary

The microbiological aspects of forest dieback has been investigated from an inventory drawn up in Norway spruce stands in Vosges area and from an experimental approach in greenhouse. Declining trees have generally less living mycorrhizae than healthy ones.But because of the influence of various factors it's not possible to reveal a clear relationship between rootlets decline and forest dieback.None of the notorious root pathogens is involved in the alteration of rootlets.A great number of morphological types of mycorrhizae and a great variability among the trees have been observed without any relationship with forest dieback.The principle of experimental design is to subject the soil to various treatements in order to change it's microbiological balance.The effect of each treatement on development of Norway spruce saplings has been followed.After one growth season in greenhouse the sapling growth is highly improved by some soil treatment especially pasteurization(70°C) indicating that a part of the microflora has a depressive influence on plant growth.These deleterious microorganisms and mycorrhizal fungi play a prominent role in the striking root of sapling on soils coming from dieback forest stands in the Vosges area.

I. INTRODUCTION

Could forest decline be associated with disturbances of the soil microbiological balance, studied through two of it main components : mycorrhizae and pathogens ?

Investigations include an inventory drawn up in Norway spruce stands in the Vosges area and a experimental approach in greenhouse.

II.INVENTORY

The following observations have been made from roots samples collected by digging at various time throughout the year:

-Distribution of short roots in various classes (7), according to the visual state of apice, cortex, and central cylinder under stereoscopic microscop.(table 1)

-Percentage of different morphological types of mycorrhizae (1)

-Attempts to isolate pathogenic fungi from roots and rhizospheric soil.

Trees of declining stands (figure 1 A1,B3) have generally less living mycorrhizae than healthy ones (B1,B2).These differences are especially evident for roots belonging to class 1.

Nevertheless distribution of short roots is not always in relation with the declining state of trees within the stand.(figure 2) We observed great differences in the distribution of short roots between two healthy trees (III and IV), and the percentage of living mycorrhizae of the healthy tree is much less (IV) than those of the early declining tree (I) and equivalent to those of late declining tree (II).

A great number of morphological types of mycorrhizae and a high variability among trees have been observed (figure 3) without any relationship with forest dieback.The three dominant types of mycorrhizae is equally developped on the root system of the trees studied.

None of the notorious root pathogens (Pythium spp., Fusarium spp., Rhizoctonia spp.,...) has been isolated from the root system.Nevertheless, some species belonging to Trichoderma, Penicillium, Cylindrocarpon, Acremonium,....can be regularly recovered from the altered cortex of short roots.The role of these fungi is not clearly understood.(2)

A clear relationship between rootlets decline and forest dieback cannot be established from inventory results.

III.EXPERIMENTAL APPROACH

Principle:Various soils collected under declining or healthy stands are subjected to different treatments in order to change microbiological balance or nutrient status.Single treatments are :steam desinfectation, (100°C-1 hour), pasteurization (70°C-1/2 hour),nutrient solution applied each month, monthly fongicidal drenches (benomyl, thiabendazole, propamocarb) and control.In addition desinfectation and pasteurization have been combined with inoculation of Laccaria laccata, or nutrient solution, or mixing with 10% of soil from healthy or declining stand.

The effect of each treatment on the development of Norway spruce saplings (2+0) has been followed by assessing the shoot and root dry weight, the level of mycorrhizal association, and the state of health of roots.

After one growing season in greenhouse the saplings growth is hightly improved after desinfectation, pasteurization (70°C), and mycorrhizal inoculation with Laccaria laccata.(Figures 4 and 5).Pasteurization alone or associated with nutrient solution (healthy stand figure 4) or with mycorrhizal inoculation (declining stand figure 5) provides the best growth increment.

These results reveal that the part of the soil microflora supressed by the 70°C soil treatment (deleterious microorganisms) has a depressive influence on plant growth without any sign of root alteration.This fact is supported by the negative effect of reintroducing soil from declining stand after pasteurization (figure 5)

Growth stimulation is not in relation with the level of mycorrhizal association.

We can provisionally conclude that deleterious microorganisms and mycorrhizae (from a qualitative point of view) play a prominent role in the striking root of saplings on soil collected under dieback forest stands in the Vosges area.

REFERENCES
(1) DOMINIK, T., (1969). Key to ectotrophic mycorrhizae.Folia Forestalia polonica. Ser.A,15,309-321.
(2) SCHONHAR, S., (1985).Infektionsversuche an Fichten-und Kiefernkeimlingen mit auskranken Fichtenfeinwurzeln isolierten Pilzen.Allg.Forst-u.J. -Ztg.,157,5,97-98.

Table 1:Signification of different classes

+ apparently heathly
- altered

class	apice	cortex	central cylinder
1	+	+	+
2	–	+	+
3	+	–	+
4	–	–	+
5	+	–	–
6	–	–	–
7	--------------broken roots-------------		

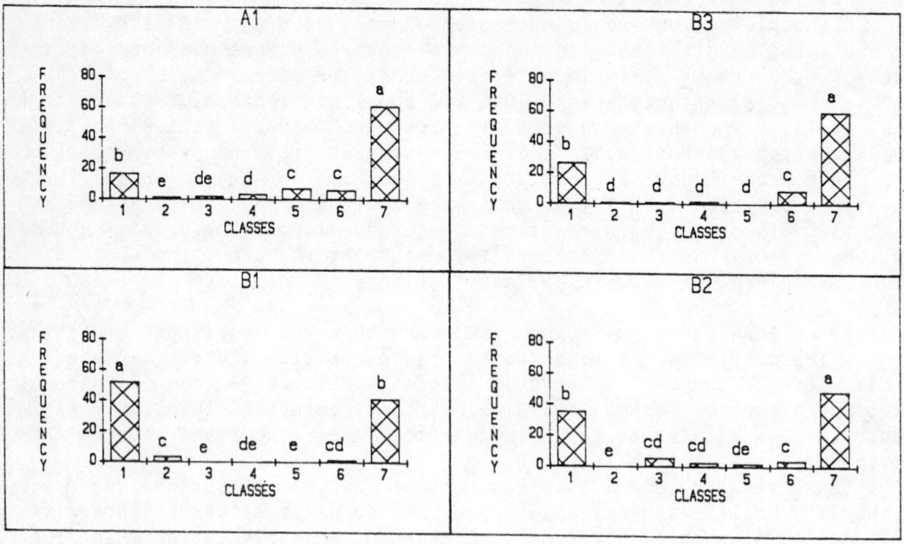

Figure 1:Distribution of short roots according to the state of their apice, cortex, and central cylinder from samples collected on different 15-20 years old trees within stands:

-A1:declining natural regeneration of Norway spruce.

-B3:declining natural regeneration of Norway spruce under high stands of mixed declining Silver fir and Norway spruce.

-B1:Healthy Norway spruce plantation.

-B2:Healthy Norway spruce natural regeneration.

Columns surmounted by same letter are not significantly different according to the U test (P=0,05).

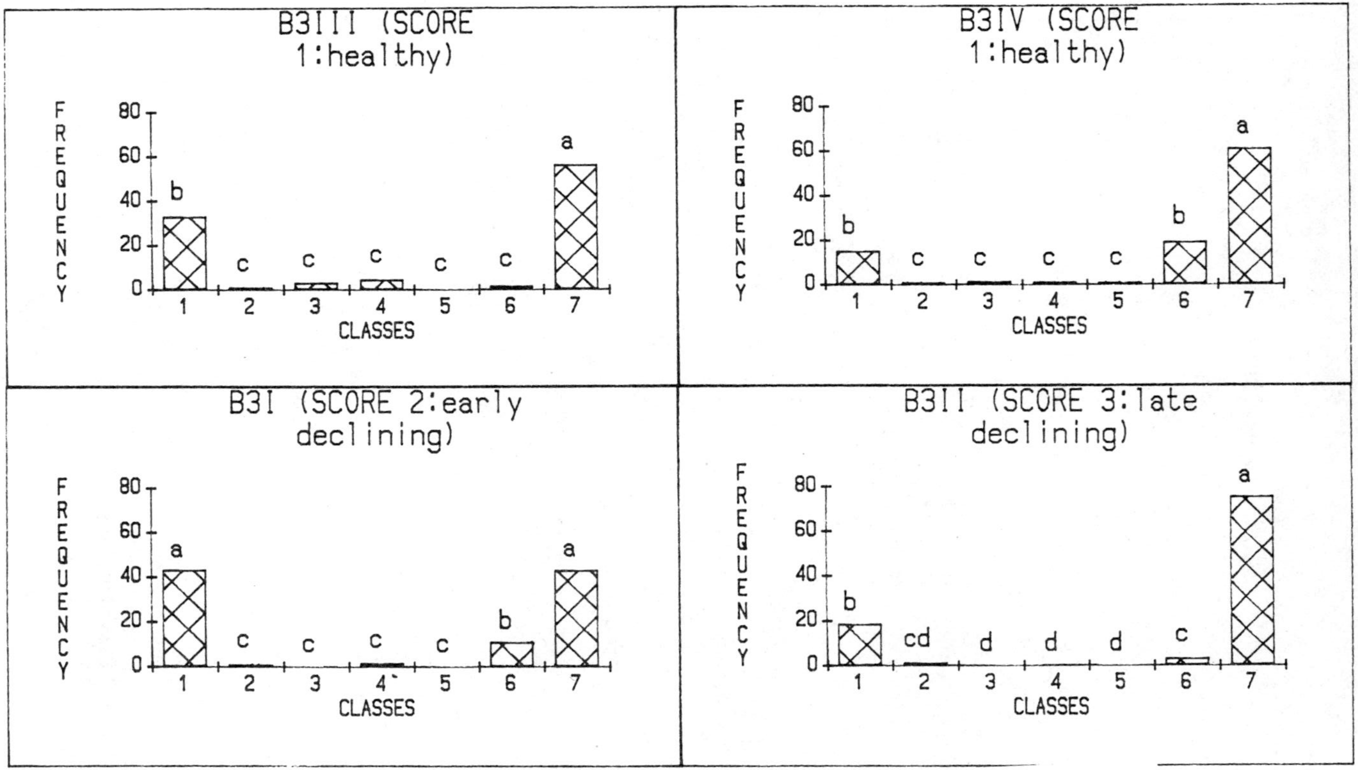

Figure 2:Distribution of short roots according to the state of their apice, cortex, and central cylinder from samples collected on declining trees (I,II) and healthy trees (III,IV) of the same stand B3. Dieback state is related to the french scoring system.

Columns surmounted by same letter are not significantly different according to the U test (P=0,05)

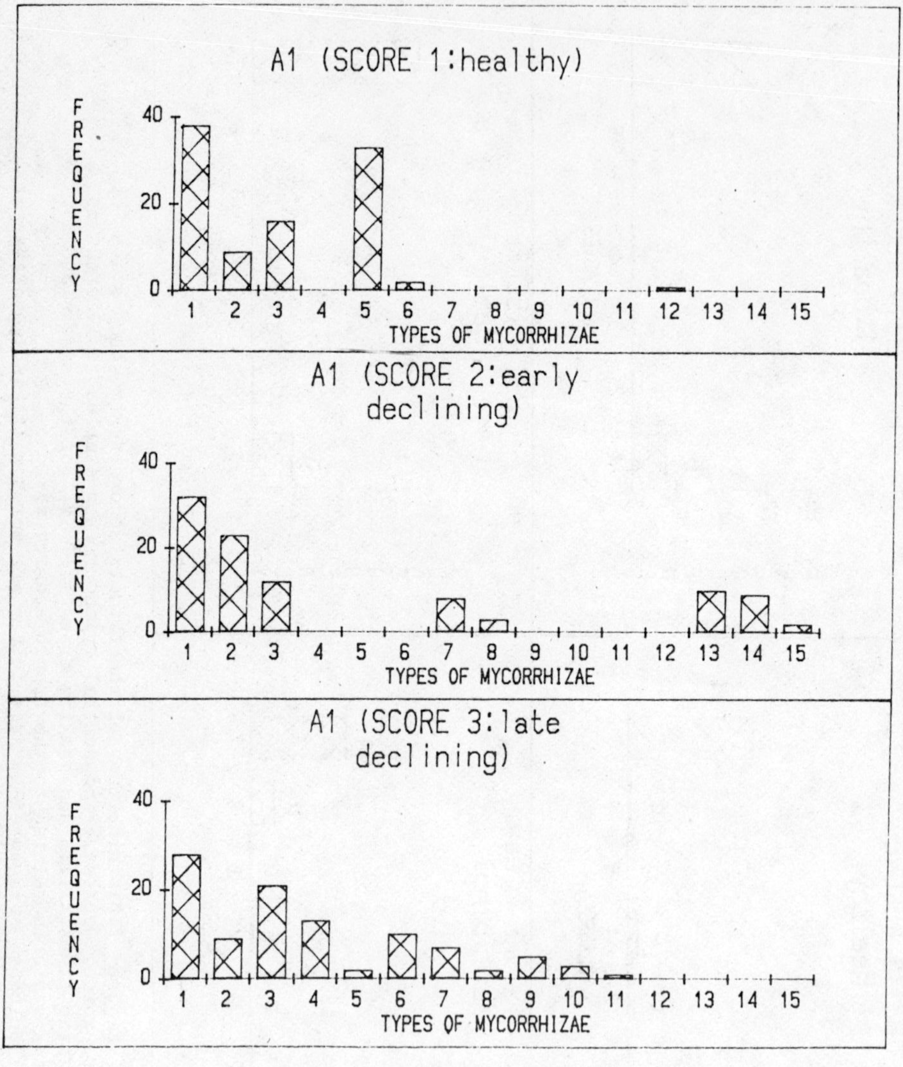

Figure 3:Distribution of different morphological types of mycorrhizae related to 3 trees from a natural regeneration (A1).Types of mycorrhizae has been separated according to the DOMINIK classification.

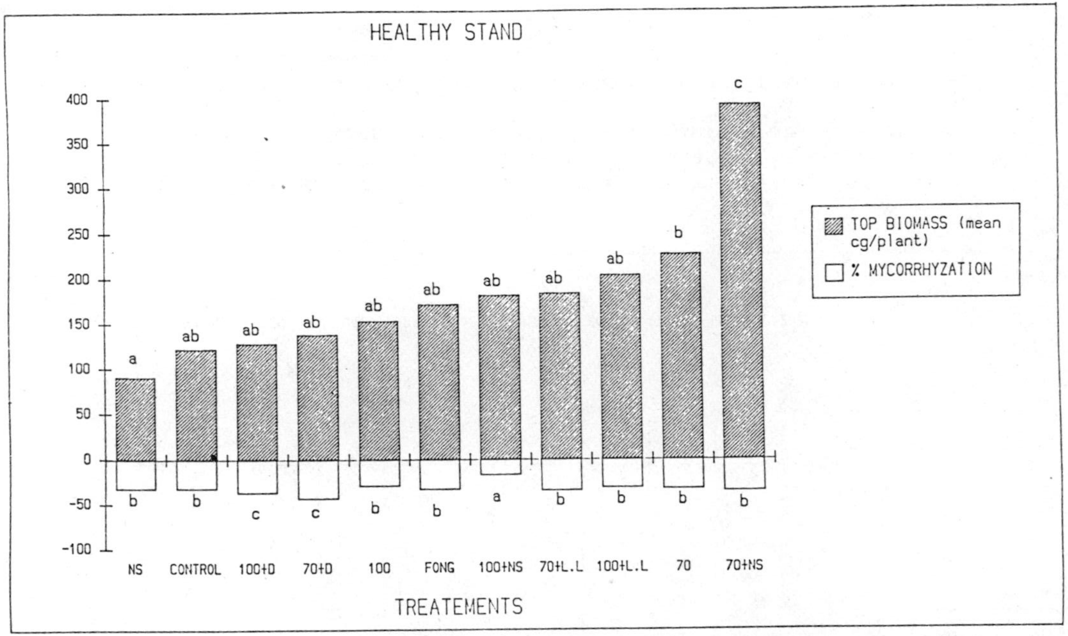

<u>Figure 4</u>:Average top biomass produced after one growing season in greenhouse by saplings growing on soil collected under healthy stand according soil treatment in relation with level of mycorrhizal association.NS:nutrient solution,100+D:desinfectation + 10% of soil collected under declining stand,70+D:pasteurization + 10% of soil collected under declining stand,100:desinfectation,Fong:fongicidal drenches,100+NS:desinfectation + nutrient solution,70+LL:pasteurization + inoculation with <u>Laccaria laccata</u>,100+LL:desinfectation + inoculation with <u>Laccaria laccata</u>,70:pasteurization,70+NS:pasteurisation + nutrient solution.

Means associated with same letter are not significantly different according Newman and Keuls test (P=0,05)

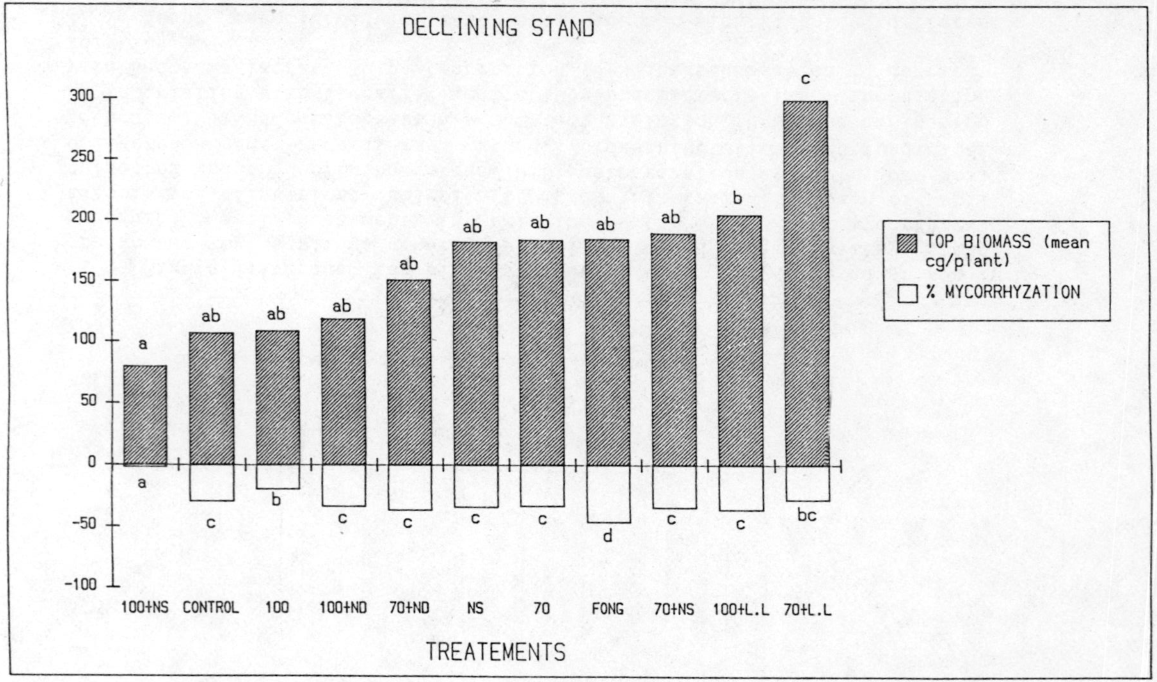

Figure 5:Average top biomass produced after one growing seasonin greenhouse by saplings growing on soil collected under declining stand according soil treatment in relation with level of mycorrhizal association.

Means associated with same letter are not significantly different according Newman and Keuls test (P=0,05).

THE INFLUENCE OF ACID RAIN ON MYCORRHIZAL FUNGI AND MYCORRHIZAS OF DOUGLAS FIR (PSEUDOTSUGA MENZIESII) IN THE NETHERLANDS

A.E. JANSEN
Agricultural University, Department of Phytopathology,
Binnenhaven 9, 6709 PD WAGENINGEN, The Netherlands

Summary

The occurrence of mycorrhizal fungi and mycorrhizas in Douglas fir stands was studied in more and less polluted regions in The Netherlands. The number of species of mycorrhizal fungi and the mycorrhizal frequency were negatively correlated with the level of air pollution. In less polluted regions only old stands (> 40 years) were affected, in more polluted areas also medium old stands (> 20 years) and there the number of species was about 60% lower compared to the less polluted regions. In a pot experiment, mycorrhizal frequency was significantly lowered by fumigation. Both a direct effect (via the above ground parts) as an indirect effect (via the soil) seem likely. In studying the mycorrhizas occurring in the plots several types were distinguished and about 10 types could be identified as belonging to a fungal genus or species.

1. INTRODUCTION

In The Netherlands a strong decline in number of fruitbodies of macrofungi was reported (1). Decrease was strongest in mycorrhizal fungi, especially those associated with conifer trees on dry, acid, nutrient poor and humus poor sandy soils. Decreased are e.g. species belonging to the Hydnaceous fungi, Clavarioid fungi, and from the genera *Cantharellus*, *Tricholoma*, *Suillus*, *Cortinarius* (Fig. 1).

Patterns of decline or disappearance show a strong correlation with patterns of air pollution: decline started in the end of the fifties in the southern part of the country (where air pollution is relatively high) and reached the north-eastern part (where air pollution is relatively low) in the midst of the seventies.

From the decrease of mycorrhizal fungi and the (well known) decrease of tree vitality, questions rose like:

- Is decrease of fruitbody production correlated with a decrease of mycorrhiza frequency.
- Does decrease of mycorrhiza frequency cause the decrease in tree vitality or (alternatively) does decrease of tree vitality cause the decline in mycorrhizal infection.
- Will air pollution affect mycorrhiza directly or is there an indirect effect by e.g. lower pH, changed aluminium concentrations.

2. METHODS

Twenty five plots were established for observations on the occurrence of fruitbodies of mycorrhizal fungi, mycorrhizal infection and tree vitality. Plots were chosen in the north, centre and south of The Netherlands

(Fig. 2) in the gradient of levels of SO_2 and NH_3. Fruitbodies were observed in the autumns of 1985 and 1986.

Mycorrhizas were sampled at 0 - 5 cm depth in fixed volumes of 110 cm^3; number and proportion of mycorrhizal and non-mycorrhizal roottips were determined.

In a fumigation chamber, 2 years old Douglas fir plants were treated with

- air artificially polluted with NH_3, concentration 200 $\mu g/m^3$, sprayed with clean water
- water artificially polluted with $(NH_4)_2SO_4$, concentration 2500 μ Mol/l, precipitation was 15 mm/week, air was filtered
- both artificial polluted air and rain
- filtered, so unpolluted, air and sprayed with clean water.

After 3 months, samples of the root system were studied for the proportion of new and old mycorrhizal root tips.

3. RESULTS AND DISCUSSION

a. Observation on fruitbodies and mycorrhizas

Results have to be regarded as provisional, as for reliable results observations on fungi should cover 4 years.

Plots in the northern part had more species of mycorrhizal fungi (average 11) than in the central (av. 9) and in the southern part (av. 4) (Fig. 2). The average number of fruitbodies was 66 in plots in the northern part, 86 in the central and 22 in the southern part. In all three parts of the country, the number of species in young plots was higher than in the medium old and old plots (Table I). The effect of age was statistically significant ($p < 0.001$), that of geographical area almost significant ($0.10 > p > 0.05$). Old stands in the central part have the same number of species as old stands in the northern part, but the decrease starts about 20 years earlier in stand development. In the southern part, where concentrations of both SO_2 and NH_3 are highest, the decrease starts early in stand development and in medium old and old stands only 60% of the expected number of species was found if the same reduction as in the northern part is assumed.

In mycorrhiza frequency the same trends held. Although not yet all samples were counted, for the present it can be concluded that in the northern part mycorrhiza frequency was highest, average 35%, in central and southern part much lower, average 12% and 10% respectively. There is a significant negative correlation between mycorrhiza frequency and age of a stand (Fig. 3).

In studying the mycorrhizas, several types could be distinguished on macroscopical and microscopical characters, as surface structure, absence or presence of strands and colour. About 10 types could be identified as belonging to fungal genera or species, as *Lactarius, Laccaria, Dermocybe crocea, Russula ochroleuca, Scleroderma citrinum* and *Thelephora terrestris*.

Table I. Average number of species in the plots divided after age of the stand and geographic area.

geographic area	age of stand: young	medium	old
north	16	13	5
centre	16	7	6
south	13	2.5	2.3

As far as possible, identifications will be checked in near future by synthesis trials.

The reduction in number of fungal species was not expected. On the contrary, it was expected that old stands should have more species than younger ones, as niches should be more differentiated. In Picea stands in Norway, a much less polluted area, this was indeed found by Bendiksen (2).

In a study on Pinus in The Netherlands, Termorshuizen & Schaffers (3) found a significant correlation between concentrations of SO_2 and NH_3 and the mycoflora (number of species and number of fruitbodies) of old stands. This seems likely for Douglas fir stands too.

Number of mycorrhizal species in young Douglas fir stands and young Pinus stands were relatively unaffected. This might be caused by ploughing the soil before planting.

b. Fumigation experiment

An experiment on the influence of air polluted with NH_3 and of rain polluted with $(NH_4)_2SO_4$ on 2 years old Douglas fir plants during 3 months, revealed a strong, statistically significant, reduction of mycorrhiza frequency: NH_3 polluted air reduced mycorrhiza frequency by 94% $(NH_4)_2SO_4$ polluted rain by 42% (Table II). So, mycorrhiza frequency was mainly affected directly by polluted air, but there might be an indirect effect by changes in the soil too.

Table II. Influence of NH_3 and $(NH_4)_2SO_4$ on mycorrhiza of Douglas fir plants.

A = air, artificially polluted with NH_3, concentration 200 $\mu g/m^3$; R = water, artificially polluted with $(NH_4)_2SO_4$, conc. 2500 $\mu M/1$, 15 mm rain per week; C = control, filtered air and clean rain.
Mean values are given for the 4 treatments. Each treatment had 6 plants, except for treatment AR where there were only 5 because one plant died during the experiment.

Treatment	counted number of roottips		mycorrhiza frequency (%)	
	old roots	young roots	old roots	young roots
A	54	132	35	1
R	50	103	43	11
AR	70	116	44	1
C	102	162	53	24

4. CONCLUSIONS

For the present it can be concluded that

- in areas where air pollution is relatively low the number of fungal species is significantly lowered
- in areas where air pollution is higher, this reduction is accelerated with about 20 years and the number of species is again reduced by 60%
- young stands are not or less affected
- mycorrhiza frequency is mainly affected directly by polluted air, but there might be an indirect effect by changes in the soil too.

5. REFERENCES

1. Arnolds, E. (ed.), (1985). Veranderingen in de paddestoelenflora (Changes in the mycoflora; in Dutch). Wet. Meded. K.N.N.V. 167.
2. Bendiksen, E. (1981). Mykorrhizasopp i forskjellige suksesjonsstadier av granskogssamfunn i Lunner, Oppland. K. Norske Vidensk. Selsk. Mus. Rapp. Bot. Ser. 1981-5; 246-258.

3. Termorshuizen, A.J. & Schaffers, A.P. (1987). Mycorrhizal fungi of Pinus sylvestris L. related to air pollution and tree vitality. Proceedings 7th NACOM (in press).

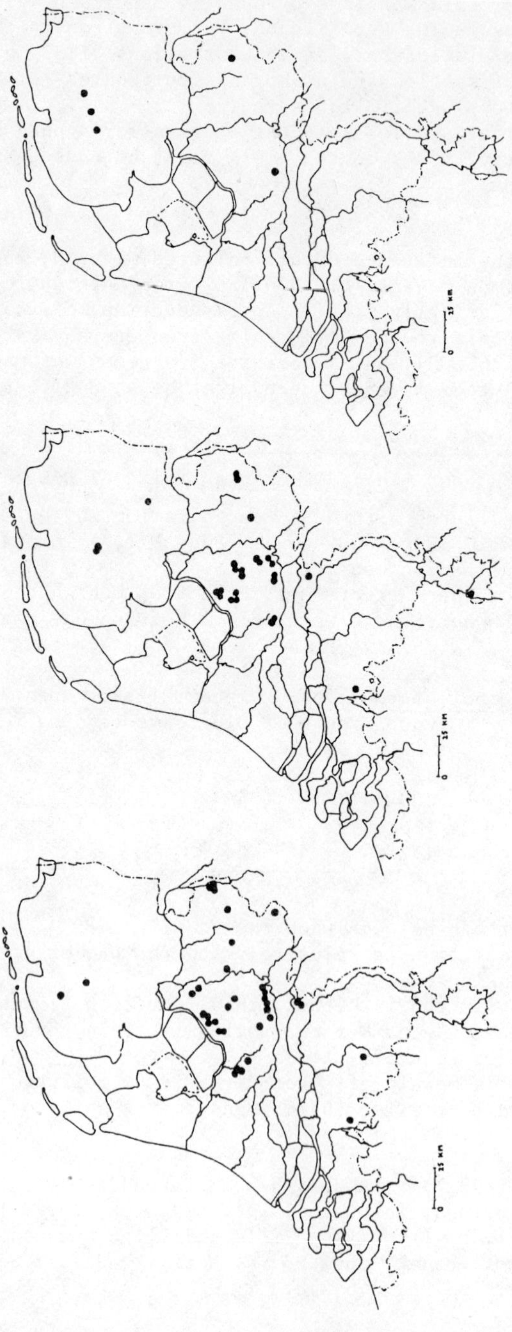

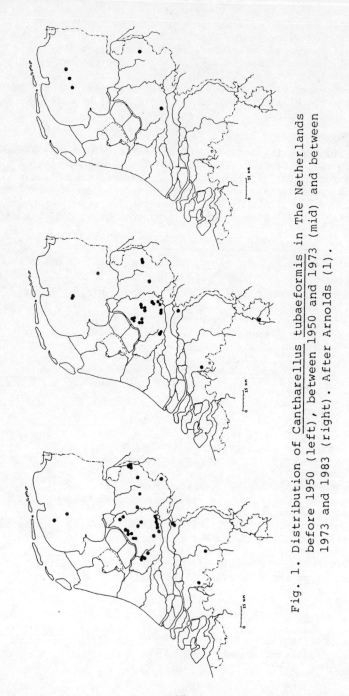

Fig. 1. Distribution of Cantharellus tubaeformis in The Netherlands before 1950 (left), between 1950 and 1973 (mid) and between 1973 and 1983 (right). After Arnolds (1).

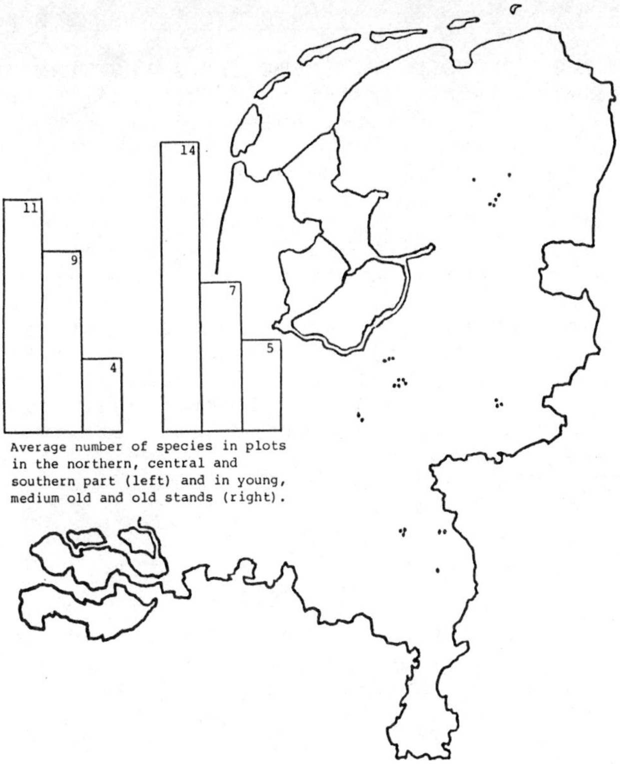

Fig. 2. Distribution of plots of Douglas fir stands and the average number of species of mycorrhizal fungi in plots of 3 regions and of 3 age classes.

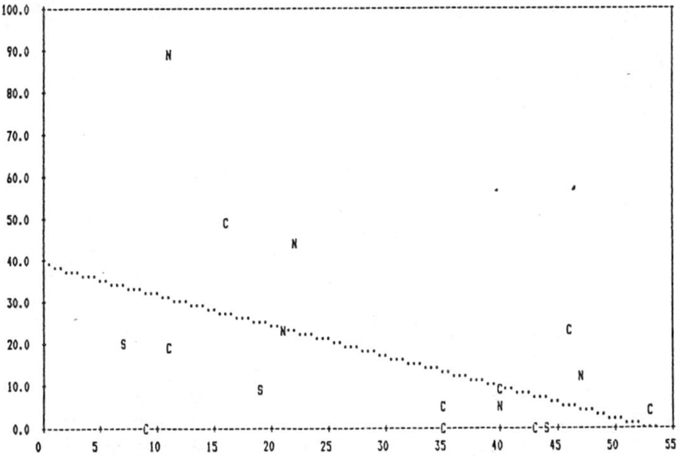

Fig. 3. Relation between age of stands and mycorrhiza frequency. The dotted line is the line Y= −0.74X + 39.9; correlation is −0.49. N, C, S refer to plots in the northern, central and southern part of The Netherlands.

LEAD CONTENTS AND DISTRIBUTION IN SPRUCE ROOTS

D.L. GODBOLD, E. FRITZ and A. HÜTTERMANN
Forstbotanisches Institut, Universität Göttingen,
3400 Göttingen, F.R.G.

Summary
Root elongation of *Picea abies* seedlings was inhibited by Pb in nutrient solution. Using X-ray microanalysis the distribution of Pb in root tips of Pb treated seedlings, and root and root tips of spruce roots taken from declining tree stands in the Harz mountains was estimated. In Pb treated seedlings Pb was found in mainly in cell walls and electron dense deposits. In field samples Pb was found in cell walls and in vacuolar deposits.

1. INTRODUCTION

Investigation of metal cycling in forest ecosystems has shown that Pb is accumulating in forest soils in Europe and North America (1,2). Lead inputs of 400 and 700 g ha^{-1} yr^{-1} have been found in forest in Northern Germany and Vermont respectively (2,3). These studys have also shown that the organic horizon is an effective sink for Pb and retains almost all of the Pb deposited.

The aim of this work was to assess the influence of Pb on roots of spruce. Root growth of seedlings maintained in Pb containing nutrient solutions was determined, and the distribution of Pb in the roots was compared to that of spruce roots from declining tree stands.

2. MATERIALS AND METHODS

2.1 Culture of plants

Plants of *Picea abies* L. Karst. were grown from seeds under aseptic conditions as previously described (4). The seedlings were then transferred to sterile nutrient solutions based on long term measurements of soil solutions at forest sites at Solling, West Germany. The nutrient solution had the following composition: (uM) $Ca(NO_3)_2$ 68, $CaCl_2$ 62, $MgSO_4$ 82, KH_2PO_4 16, KNO_3 334, $NaNO_3$ 174, $FeCl_2$ 5, $MnSO_4$ 3, H_3BO_3 5, Na_2MoO_4 0.1, $ZnSO_4$ 0.1, $CuSO_4$ 0.1, pH 4.0. Solutions were constantly aerated with sterile air. After 1 week the seedlings were transferred to non-sterile solutions and grown for a further 4 weeks. The seedlings were then transferred to nutrient solutions containing 2 uM (414 ug l^{-1}) Pb as $PbCl_2$ for 7 weeks. Growth conditions were 23/21°C day/night temperatures, 100 umol m^{-2} s^{-1} photon flux density (Osram L18w/25 lamps) and a 16h photoperiod.

2.2 Estmation of root elongation.

After growth under aseptic condition, seedlings were transferred to non-sterile nutrient solutions and allowed to equilibrate for 2 days, then transferred to nutrient solutions containing a range of Pb concentrations (0.1-2 uM) for 7 days. Root elongation was measured as previously described (5).

2.3. Collection of field samples.

Root samples were collected from 30 year old declining spruce stands at Altenau in the Harz mountains, West Germany, using a soil core borer. Roots in the mineral soil (0-5 cm) were carefully freed from the soil, cut into 2 mm segments and immediately prepared for X-ray microanalysis.

2.4. X-ray microanalysis

Samples were prepared and X-ray microanalysis carried out using the method of Fritz (6). This method is suitable for comparision of element contents between different sections, but does not permit a statement about absolute concentration of the elements.

3. <u>RESULTS</u>

The rate of root elongation was significantly inhibited by Pb after exposure for 1 day (Fig 1). The degree of inhibition was dependent upon the duration of exposure and the Pb concentration in the nutrient solution.

Using X-ray microanalysis, in root tips of Pb treated seedlings Pb was found primarily in cell walls and electron dense material in cortex cells and xylem vessels (Fig 2) Cortex cell walls had higher Pb contents than those of the stelar tissues. For comparison the contents of Ca and K in the cell walls are also shown. In both cortex and stele large amounts of Pb were found in electron dense deposits, at the inner edge of the cytoplasm (Fig 3), and were associated with very high amounts of K (tab 1). However in Pb treated roots, K containing electron dense deposits could also be found which did not contain Pb (Fig 3). In roots of control seedlings similar electron dense deposits could also be found in the stelar tissues (Fig. 4). X-ray microanalysis showed these deposits also contained large amounts of K (tab 1). The electron dense deposits appeared to be more common in Pb treated roots than control roots.

Table 1. Contents of K (cps) in electron dense deposits in root tips of spruce roots grown for 7 weeks in nutrient solutions with or without 2 uM Pb. Standard error in parentheses.

Treatment	Tissue	K contents
Control	Stele	7916 (772)
2 uM Pb	cortex	3465 (738)
2 uM Pb	Stele	5952 (865)

In the tips of root collected from the Harz mountains Pb could be located in the cell wall of the cortex and stele, and in the Hartig net of the associated mycorrhiza (Fig. 5). The counts for Pb in these tissues were found to be greater than those for K and Ca. In older root regions with secondary xylem tissue, Pb was found in the outer hyphal layer, Hartig net and cortex cell walls (Fig 6). Large amounts of Pb were found in electron dense deposits in vacuoles of endodermis cells, and in cell walls of phloem cells. In the cortex the counts for Pb again exceeded those of K and Ca.

4. DISCUSSION

In roots of both Pb treated seedlings and field samples, Pb could be found in cell walls and electron dense deposits. Similar electron dense deposits have been found on beans exposed to high levels of Zn (7). However, in this study the deposits were found in both Pb treated and control roots. In cell walls of both root tips and older root sections of roots collected from declining spruce stands in the Harz mountains, higher Pb contents were found than in those of Pb treated seedlings. The presence of a mycorrhizal sheath does not appear to prevent entry of Pb into spruce roots, as has often been speculated.

In the soil solution of the humic layer under declining spruce stands at Solling, West Germany the average Pb concentration is about 0.2 uM (1). This concentration is greater than that shown to inhibit root elongation. Furthermore, ion budget studies suggest that this Pb concentration is a considerable under estimation of the actual Pb that is plant available, and may only reflect the concentation of Pb which has not been taken up (1).

In forest soils Pb is often found in a chelated form, however the similar distribution patterns found in Pb treated seedlings and field samples suggest the roots must also be exposed to ionic Pb. This may then have drastic effects on root growth and turnover.

5. REFERENCES

(1) LAMERSDORF, N. (1987). Verteilung und Akkumulation von Spurenstoffen in Waldökosystemen. Ph.D thesis, University of Göttingen, in preparation.

(2) FRIEDLAND, A.J. and JOHNSON, A.H. (1985). Lead distribution in high elevation forest in Vermont. J. Environ. Qual. 14, 332-336.

(3) SCHULTZ, R. et al. (1987). Raten der Deposition, der Vorrats änderungen und des Austrages einiger Spurenstoffe in Waldökosytemen. Abschluß Bericht der DFG / Forschungs-antrages NUL 35/46/1.

(4) SCHLEGEL, H., GODBOLD, D.L. and HÜTTERMANN, A. (1987). Whole plants aspects of heavy metla induced changes in CO_2 uptake and water relations in spruce (Picea abies) seedlings. Physiol. Plantarum, 69, 265-270.

(5) GODBOLD, D.L. and HÜTTERMANN, A. (1986). The uptake and toxicity of lead and mercury to spruce (Picea abies) seedlings. Water, Air Soil Pollut. 31, 509-515.

(6) FRITZ, E. (1987). Gefriertrocknung und Vakuum-Druck

Infiltration mit Aether und Kunststoff für die Röntgen-microanalyse wasserlöslicher Ionen in Pflanzenzellen. In preparation.

(7) ROBB, J., BUSCH, L.and RAUSER, W.E. (1980). Zinc toxicity and xylem vessel alterations in white beans. Ann. Bot. 46, 43-50.

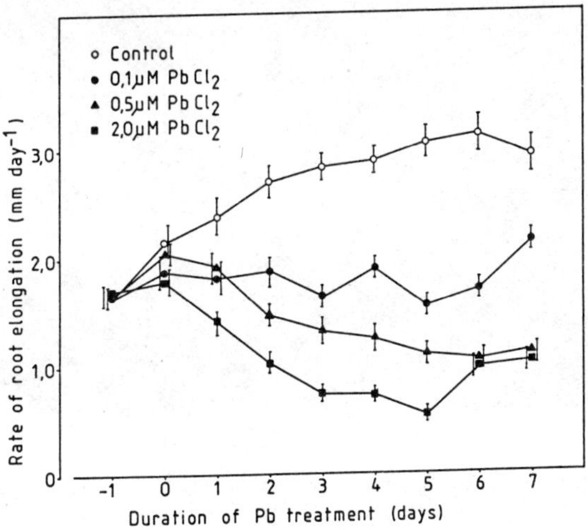

Fig. 1. Root elongation rate of <u>Picea</u> <u>abies</u> seedlings grown for 7 days in nutrient solutions containing a range of Pb concentrations. I=S.E.

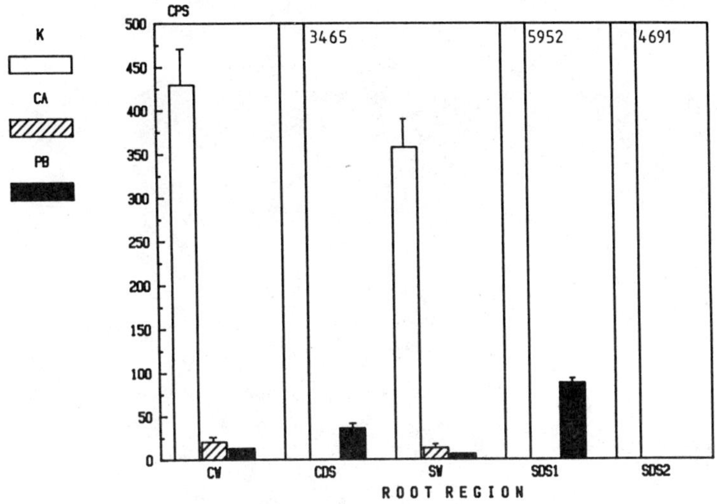

Fig. 2. Contents of Pb, K and Ca in root tips of <u>Picea</u> <u>abies</u> seedlings grown for 7 weeks in nutrient solutions containing 2uM Pb. Contents expressed in counts per second. CW = Cortex cell wall, CDS = Cortex electron dense layer, SW = Stele cell wall, SDS1/2 = Stele electron dense layer. I=S.E.

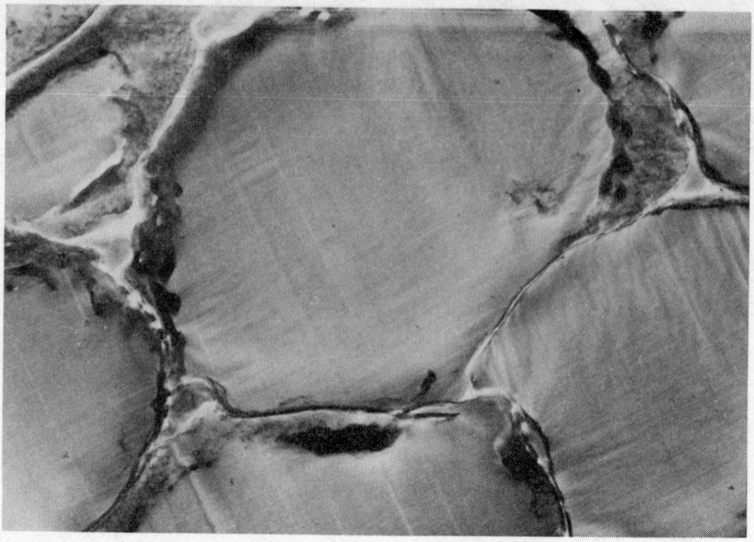

Fig. 3. Electron dense deposits in stelar tissue of Pb treated spruce seedlings (grown as in Fig 2).

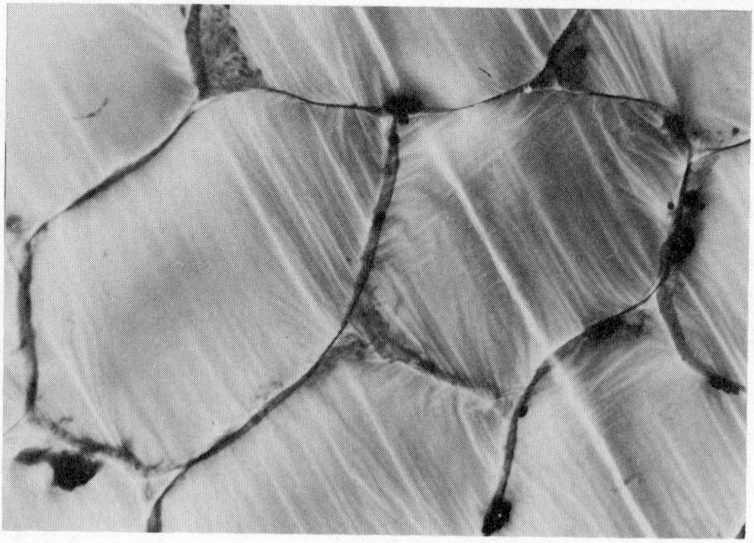

Fig. 4. Electron dense deposits in stelar tissue of control seedlings (grown as in Fig 2).

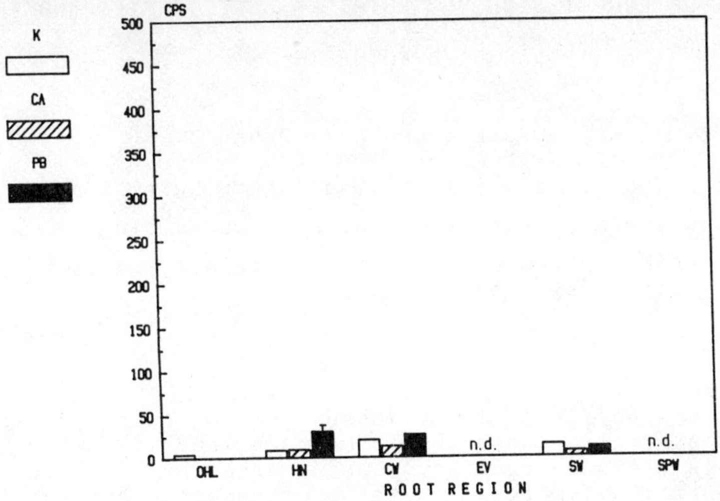

Fig. 5. Contents of Pb, K and Ca in root tips fine roots of Picea abies taken from the upper mineral soil at Altenau, Harz mountains, West Germany. Contents in counts per second. OHL = Outer hyphae layer of mycorrhiza, HN = Hartig net, CW = Cortex cell wall, EV = Endodermis vacuole, SW = Stele cell wall, SPW = stele phloem cell wall. I=S.E.

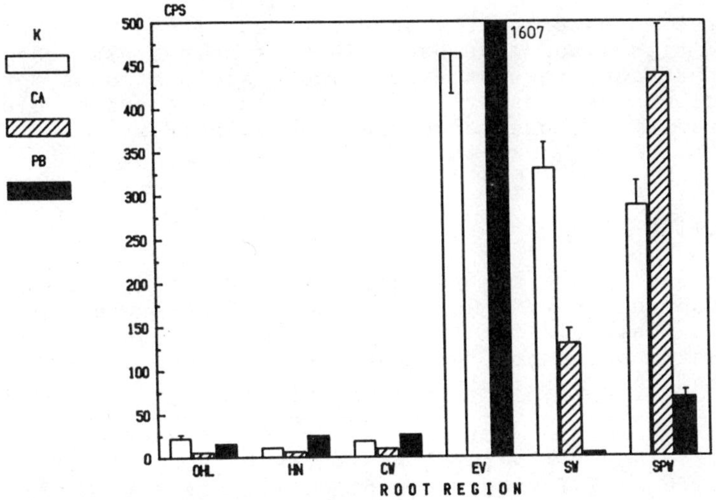

Fig. 6. Older root regions (see Legend Fig. 5)

"NEW TYPE" OF FOREST DECLINE, NUTRIENT DEFICIENCIES AND THE "VIRUS"-HYPOTHESIS

R.F. Huettl* and B.M. Mehne**

*Institute of Soil Science and Forest Nutrition

**Institute of Forest Botany and Wood Biology

Albert-Ludwigs-University, Freiburg
Bertoldstrasse 17, D-7800 Freiburg

SUMMARY

Beside unspecific foliage losses the so called "new type" of forest decline observed in West Germany is particularly in coniferous trees and stands marked by sitespecific discoloration symptoms. It was hypothesized that the foliar discoloration phenomena are caused by nutrient deficiencies - probably induced by air pollutants - or by virus infections. To test these hypotheses diagnostic fertilization trials and grafting experiments were carried out. From these two comprehensive research projects typical examples are presented.
In a K deficient Norway spruce stand (Picea abies Karst.) the application of a fast soluble K fertilizer led to the disappearance of the discoloration symptoms within only one growth period indicating that nutrient deficiencies are present and can be removed in declining stands.
Scions taken from the control trees of this trial stand and grafted onto healthy spruce trees turned green after mutual tissue had developed. These findings also underline that the frequently observed discoloration symptoms in declining stands are most likely not caused by biotic factors but are due to nutrient deficiencies.

1. INTRODUCTION

Since about one decade the so called "new type" of forest decline is observed in West Germany. In coniferous trees and stands beside unspecific foliage losses the damages are marked by discoloration symptoms. Amongst many other hypotheses it was suggested that particularly the discoloration symptoms might be caused by nutrient deficiencies. Needle analyses carried out by many investigators indicated for stands revealing yellowing of older foliage sitespecific Mg or K deficiencies. To explain these recent deficiency situations increased nutrient element leaching (e.g. Mg, K) from the canopy as well as from the soil due to various mechanisms provoked by higher atmospheric deposition of pollutants is assumed (cf. 5).

Another hypothesis relates the observed discoloration phenomena to the impacts of viruses or microorganisms (cf. 7, 3, 9). It was suggested that the phloem damages detected in yellowed needles via histological investigations (cf. 2, 1, 10) could be caused by the infection of phloemspecific biotic factors.

2. OBJECTIVES

In order to test the hypothesis of nutrient deficiencies diagnostic fertilization trials were established in declining forest stands (cf. 4). To investigate the virus hypothesis grafting experiments were carried out as grafting is a classical method to test whether disease symptoms are caused by biotic factors i.e. viruses or microorganisms. In case of virus diseases transference from the symptomatic tissue of the scion to the healthy tissue of the stock should occur after the development of mutual tissue. If the symptoms of the scions are primarily caused by infectious agents, the healthy trees should develop the same symptoms.

3. RESULTS

Following one typical diagnostic fertilization experiment is described. Further results of this comprehensive research project are presented by Huettl (4), Zoettl and Huettl (11), Zoettl et al. (12) and Huettl (6).

In fall 1983 a 15-yr-old Norway spruce stand (Picea abies Karst.) growing on a peatland soil in Southwestern Germany was marked by a pronounced lemon green discoloration and occasional tip-necroses of all needle year classes (s. Figure 1). The chemical soil analysis data indicate an insufficient K supply related to an extremely low K/Ca ratio (Table I). Needle analysis data are well related to substrate chemistry and reveal a strong K deficiency (fall 1983: 2.6 mg g^{-1} d.m.; cf. Table II).

Table I: Diagnostic fertilization trial Saulgau 2. Chemical soil status in spring 1984 (NH_4Cl extraction; exchangeable cations at soil pH); 20-30 cm soil depth, before fertilization.

Mg^{2+}	Ca^{2+}	K^+	Al^{3+}	H^+	Mg/Al	Ca/Al	K/Ca	base sat.	pH
µval g^{-1}					mol			%	($CaCl_2$)
45.5	1229	1.4	0	8.7	-	-	0.002	99	5.0
(preliminary threshold values; Huettl in prep.)					(0.05-) 0.1	0.1 (-1)	0.5 (-1)	(10-20?)	(4-5?)

To improve K supply a fast soluble K fertilizer was supplied in spring 1984. Only five months later the fertilized trees had greened up remarkably (Figure 1 and 2) coinciding with much improved K needle content (Table II). Re-investigations in fall 1985 and 1986 underline the positive and sustained effect of the K fertilizer application. Also three growth periods after fertilization the trees were characterized by vigorous growth, green foliage, and a good nutritional status. Interesting is the optimal supply of P which would not necessarily be expected on organic soils. On the control plot the trees were still marked by severe K deficiency symptoms, increased needle losses, and reduced height growth. Due to frost impacts a considerable number of unfertilized trees had died since 1983.

Table II: Diagnostic fertilization trial Saulgau 2. Element contents in current needles (top whorl) of Norway spruce; fall sampling (n=10); Kalimagnesia (30% K_2O; 10% MgO), 800 kg ha^{-1} in spring 1984.

plot	sampling year	N	P	K	Ca	Mg	Mn	Zn
		mg.g^{-1} d.m.					µg g^{-1} d.m.	
(deficiency threshold values)		12-13	1.1-1.2	4.0-4.5	1.0-2.0	0.7-0.8	20 (-80)	13
control	1984	18	3.9	2.4	3.2	0.7	70	12
control	1986	16	3.1	2.2	3.7	0.9	66	12
fertilized	1984	18	3.2	6.9	3.4	0.6	100	24
fertilized	1985	21	3.0	7.3	5.4	0.8	130	31
fertilized	1986	19	3.7	6.5	5.4	0.8	130	31

In spring 1986 2-yr-old scions characterized by decline symptoms were sampled from the control trees at this research area and grafted onto healthy 4-yr-old Norway spruce plants (Figure 3). In contrast to the expected transference of symptoms from the scions to the healthy trees the yellowed scions turned green (Figure 4) after mutual tissue had developed between grafts and healthy plants (Figure 5) which were well supplied with nutrients.

Since spring 1984 about 3000 graftings were carried out in order to test the "virus"-hypothesis for the different decline symptoms by sampling scions from all major decline areas in West-Germany. In not a single case transference of symptoms from the scions to the stocks could be observed. But whenever graftings with scions marked by discoloration symptoms were undertaken successfully the damaged scions regreened and stayed healthy for all subsequent growth periods (cf. 8).

4. CONCLUSIONS

From the presented data it can be concluded that foliar analyses and fertilization trials are solid tools to characterize and respectively remove nutrient deficiencies in declining forest trees and stands. These trials as well as classical grafting experiments indicate that the frequently observed discoloration symptoms in stands revealing "new type" forest damages are not (primarily) caused by biotic factors (e.g. viruses) but are due to nutritional disturbances (e.g. nutrient deficiencies).

5. ACKNOWLEDGEMENT

We would like to thank Prof.Dr. H.W. Zoettl and Prof.Dr. H.J. Braun for excellent advice and the Federal Ministry of Science and Technology, Bonn, FRG for financial support.

6. REFERENCES

(1) Fink, S.: 1983, Allg. Forstz. 38, 660-663.
(2) Fink, S. and Braun, H.J.: 1978, Allg. Forst- u. Jagdz. 149, 145-150.
(3) Frenzel, B.: 1985, in: Kortzfleisch, G. von (Ed.): Waldschaeden - Theorie und Praxis auf der Suche nach Antworten, 61-80.
(4) Huettl, R.F.: 1985, Freiburger Bodenkundl. Abh. 16, 195 p.
(5) Huettl, R.F.:1987a, European Symposium "Effects of Air Pollution on Terrestrial and Aquatic Ecosystems", Grenoble, 18-22 May 1987.
(6) Huettl, R.F.: 1987b, Allg. Forstz. 42, 289-299.
(7) Kandler,O.: 1983, Naturwiss. Rundschau 36, 488-490.
(8) Mehne, B.M. and Braun, H.J.: 1987, Pfropfversuche zur Pruefung der Epidemiehypothese als Erklaerungsansatz für die "neuartigen" Waldschaeden,BMFT (Federal Ministry of Science and Technology) Statusseminar "Ursachenforschung zu Waldschaeden",Jülich, 30.3.-3.4.1987.
(9) Nienhaus, F.: 1985, Allg. Forstz. 40, 119-124.
(10) Parameswaran, N., Fink, S., and Liese, W.: 1985, Eur.J.For.Path. 15, 168-182.
(11) Zoettl, H.W. and Huettl, R.F.: 1986, Water, Air, and Soil Pollution 31, 449-462.
(12) Zoettl, H.W., Huettl, R.F., Liu, J., and Ende, H.P.: 1987, Diagnostische Duengungsversuche in immissionsgeschaedigten Waldgebieten, BMFT (Federal Ministry of Science and Technology) Statusseminar "Ursachenforschung zu Waldschaeden", Jülich, 30.3.-3.4.1987.

Figure 1: Diagnostic fertilization trial Saulgau 2. Norway spruce exhibiting K deficiency symptoms in spring 1984 before fertilization.

Figure 2: Diagnostic fertilization trial Saulgau 2. The same tree five months after fertilization in fall 1984.

Figure 3: Grafting experiment to test the "virus"-hypothesis. 2-yr-old Norway spruce scion marked by typical decline symptoms sampled at site Saulgau 2 grafted onto healthy Norway spruce in April 1986.

Figure 4: Grafting experiment. The same scion five months after grafting in September 1986.

Figure 5: Grafting experiment. Three weeks after grafting mutual parenchymatous tissue was developed between scion and healthy plant permitting water and nutrient flow to the scion.

EFFECTS OF AIRBORNE AMMONIUM ON NATURAL VEGETATION AND FORESTS

J.G.M. Roelofs, A.W. Boxman and H.F.G. van Dijk
Laboratory of Aquatic Ecology
Catholic University
Toernooiveld
Nijmegen, The Netherlands

Summary

In weakly buffered ecosystems a high deposition of ammonium leads to acidification and nitrogen-enrichment of the soil. As a consequence many plant species characteristic of poorly buffered environments disappear. Among the acid tolerant species there will be a competition between slow-growing plant species and fast-growing nitrophilous grass or grass-like species. This process contributes to the often observed change from heath- and peatlands into grasslands.
In forest ecosystems a high input of ammonium leads to leaching of K^+, Mg^{2+} and Ca^{2+} from the soil, often resulting in increased ratios of NH_4^+ to K^+ and Mg^{2+} and/or Al^{3+} to Ca^{2+} in the soil solution. Field investigations show a clear correlation between these increased ratios and the condition of *Pinus nigra* var. *maritima* (Ait.) Melville, *Pseudotsuga menziesii* (Mirb.) Franco and *Quercus robur* L..
Ecophysiological experiments proved that coniferous trees take up NH_4^+ by the needles and compensate for this by excreting K^+ and Mg^{2+}.
This combination of effects often results in potassium and/or magnesium deficiencies, severe nitrogen stress, and as a consequence premature shedding of leaves or needles.
Furthermore the trees become more susceptible to other stress factors such as ozone, drought, frost and fungal diseases.

1. THE CHANGE OF HEATHLANDS INTO GRASSLANDS

The most obvious phenomenon in many heathlands during the last decades is the changing from heathland into grassland (Heil, 1984; Heil and Diemont, 1983; Roelofs et al., 1984; Roelofs, 1986). Particularly *Molinia caerulea* (L.) Moench and *Deschampsia flexuosa* (L.) Trin. expand strongly, at the expense of *Calluna vulgaris* (L.) Hull and other heathland species.

In order to estimate whether this phenomenon is related to changes in the physical-chemical environment, 70 grass-dominated and heather-dominated heathlands have been investigated (Roelofs, 1986). Many parameters such as the pH, showed hardly any differences. However, the nitrogen levels in grass-dominated heathlands appeared to be much higher (Table 1).

Both in grass-dominated and heather-dominated heathlands the ammonium levels were 10 - 20 times higher than the nitrate levels. Investigations clearly show that a major part of the nitrogen orginates from atmospheric deposition. Under natural conditions this atmospheric nitrogen deposition is only a few $kg.ha^{-1}yr^{-1}$. At the present time in The Netherlands the deposition on heathlands often varies between 20 and 60 $kg.ha^{-1}yr^{-1}$; 60 - 90% as ammoniumsulphate.

Table 1. The pH (H_2O) and average nutrient concentrations in the soil-solution of 70 investigated heathlands.

species	coverage	pH (H_2O)	NH_4^+	NO_3^-	PO_4^{3-}	K^+
			µmoles kg^{-1}			
Erica tetralix L.	> 60%	4.1	55	0.0	4.0	37
Calluna vulgaris (L.) Hull	> 60%	4.1	84	1.4	4.4	46
Molinia caerulea (L.) Moench	> 60%	4.2	248	17.2	4.7	88
Deschampsia flexuosa (L.) Trin.	> 60%	4.1	429	29.0	6.0	182

2. IMPACT OF AMMONIUM ON HEATHLAND VEGETATIONS

Soil acidification

Although heathland soils are often acidic by nature, there are often certain spots or areas where, due to natural causes (loamy places, a calcareous underground, upwelling deeper groundwater) or to human activities (digging, cattle drinking-places) the soil has become slightly buffered and thus less acidic (Roelofs et al., 1984). Here plant species occur which are restricted to these slightly buffered, less acidic sediments (Table 2).

Table 2. The distribution of some plant species from heathlands in relation to the soil pH.

species	n	pH (H_2O)		
		mean	min.	max.
Erica tetralix L.	> 10	4.1		
Calluna vulgaris (L.) Hull	> 10	4.1	4.0	4.3
Molinia caerulea (L.) Moench	> 10	4.2	3.8	4.7
Polygala serpyllifolia Hose	> 10	4.5	4.1	5.7
Lycopodium inundatum L.	> 10	4.6	4.4	4.9
Pedicularis sylvatica L.	> 10	4.7	4.2	5.9
Thymus serpyllum L.	> 10	5.1	4.7	5.6

These plant species like *Thymus serpyllum* L. and *Pedicularis sylvatica* L. never occur on sediments with a pH value as low as 4.1. The deposited ammonium at these slightly buffered locations is transformed into nitrate very quickly by nitrification, which causes acidification of the soil (Van Breemen et al., 1983; Roelofs et al., 1984).

Laboratory experiments with artificially buffered heathlands soils show that nitrification stops or is strongly inhibited in this type of soil at pH 4.1 (Roelofs et al., 1985). This appeared also to be the case for the average pH-value in both grass-dominated and heather-dominated heathlands, which indicates that the pH in heathlands is probably determined by the nitrification limit. The final result of high NH_4^+ deposition levels is that the differences in pH disappear and thus also the plant species of slightly buffered locations. A poor plant community remains, consisting of only a few acid resistant species.

Nitrogen enrichment

If the soil on which ammonium is deposited acidic, a strong accumulation of nitrogen occurs in the upper soil layer, because ammonium is bound much more strongly to the soil absorption complex than nitrate. When there is competition between heather species such as *Erica tetralix* L. and *Calluna vulgaris* (L.) Hull and grasses such as *Molinia caerulea* (L.) Moench. the grasses profit from these higher nitrogen levels (Scheikh, 1969; Heil and Diemont, 1983; Berendse and Aerts, 1984; Heil, 1984; Roelofs et al., 1984; Roelofs, 1986). Field fertilisation experiments have shown that nitrogen enrichment indeed stimulates the development of grasses in heathlands (Heil and Diemont, 1983).

However, the problem with these field fertilisation experiments is that the high atmospheric nitrogen deposition was not taken into account. For this reason, experiments were carried out in a greenhouse. A number of small heathlands were created, using undisturbed, natural heathland soils. Precipitation experiments during one year showed that the biomass development of the grasses *Agrostis canina* L. and *Molinia caerulea* is not influenced by the acidity of the precipitation (Fig. 1). If the precipitation contained ammonium sulphate, a strong increase in biomass with increasing NH_4^+ deposition was observed. The chosen annual ammonium deposition was comparable with the real field deposition. The increase in biomass of *Molinia* was the strongest between 1.4 and 2.8 $kmol.ha^{-1}yr^{-1}$ (= 20 and 40 $kg.ha^{-1}yr^{-1}$).

The results of these experiments show that the NH_4^+ deposition level in The Netherlands (20 - 60 $kg.ha^{-1}yr^{-1}$) cause a marked increase in biomass of the two investigated grass species. For this reason it can be concluded that the high atmospheric nitrogen enrichment is a main cause for changes from heather-dominated into grass-dominated heathlands.

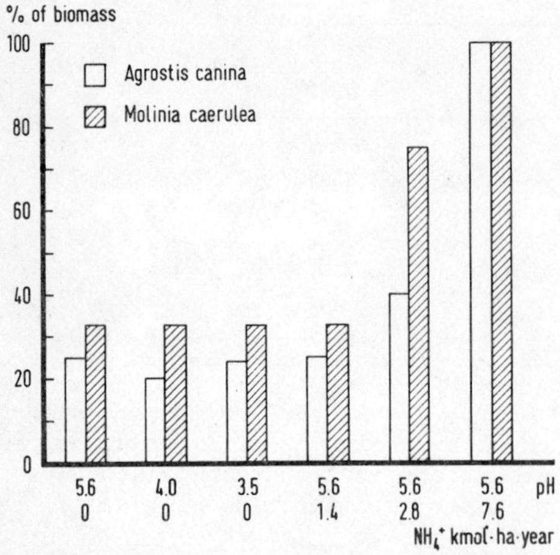

Fig. 1. The relative biomass development of *Agrostis canina* and *Molinia caerulea* on natural heathland soil during a one year treatment with precipitation with different pH and ammonium concentrations in a greenhouse.

3. AMMONIUM DEPOSITION AND THE CONDITION OF FORESTS

The condition of the Dutch forests is alarming. A recent investigation by the Dutch State Forest Service reveals that 50.1% of the forest stands show a decreased vitality (Anonymus, 1985). The geographical pattern of the damage does not fit in very well with the occurrence of well-known pollutants as SO_2, NO_X and O_3 (den Boer, 1986). The situation is most critical in the southeastern part of the country. Here nitrogen deposition in forest stands is very high and about 10 - 20 times the natural supply of 5 - 10 $kg.N.ha^{-} yr^{-}$. Due to the filtering action of the tree canopies deposition of gaseous ammonia, sulphur dioxide and ammonium sulphate is considerable higher in the forests than in the surrounding meadows (van Breemen et al., 1982; Nihlgard, 1985; Roelofs et al., 1985; see also table 3).

Several authors mention a relation between agricultural activities and the condition of Pine trees (Hunger, 1978; Janssen, 1982; Roelofs et al., 1985). Mainly four types of damage can be observed:

a. red or brown colouring of the needles of all year classes;
b. yellowing of the needles; the older needles more frequently;
c. yellowing of the youngest needles which is most pronounced at the base of the needles;
d. the occurrence of fungal or insect diseases.

All these damages can be related to high or disturbed nitrogen budgets (Roelofs et al., 1985; Roelofs, 1986; van Dijk and Roelofs, 1986). The first type of damage mainly occurs in the neighbourhood of ammonia sources like farms or fields dressed with animal slurry (Janssen, 1984).

Table 3. The average chemical composition of precipitation in a) open plots and b) throughfall in *Pinus nigra* forests in three different regions of The Netherlands during 1984 (μM/l).

	H^+	NH_4^+	K^+	Na^+	Ca^{2+}	Mg^{2+}	NO_3^-	Cl^-	SO_4^{2-}
North-West (Terschelling)									
open	100	65	25	308	47	62	48	370	72
throughfall	400	59	345	6700	460	1120	110	8000	860
South (Heeze)									
open	45	130	19	60	43	17	54	73	70
throughfall	6	1060	170	310	200	113	216	350	760
South-East (Venray)									
open	2	200	20	40	30	15	49	50	90
throughfall	1	2421	216	175	278	100	147	462	1400

It is caused by a combination of low temperature (frost) and high ammonia concentration in the air, probably as a result of a too low ammonia detoxifying capacity of the trees at low temperature (van der Eerden, 1982). The second type of damage, the yellowing of the needles, is related to potassium and/or magnesium deficiencies in the needles. These deficiencies are very significantly correlated with disturbed nitrogen budgets in both air and forest soil (Roelofs et al., 1985; Roelofs, 1986). Ecophysiological experiments proved that increased ratios of NH_4^+ to K^+ inhibit the growth of symbiotic fungi and the uptake of potassium and magnesium by the root system. At high NH_4^+/K^+ and Al^{3+}/Ca^{2+} ratios there is a net flux of Mg^{2+} and Ca^{2+} from the root system to the soil solution.
The third type of damage, the yellowing of the youngest needles is strongly correlated with extremely high arginin levels in the needles, high ammonium concentrations in the precipitation and disturbed nitrogen budgets in the soil solution (van Dijk and Roelofs, 1987).

The last mentioned type of damage, the fungal and insect diseases, may be related to the disturbed nutrient balance in the plant tissue. Investigations in *P. nigra* forests have shown that all trees infected with the fungus *Sphaeropsis sapinea* (Fr.) Dyko and Sutton had significantly higher nitrogen levels in the needles compared to non-infected healthy trees (Roelofs et al., 1985).

REFERENCES

(1) ANONYMUS (1985). De vitaliteit van het Nederlandse bos. Dutch State Forest Service, Utrecht.
(2) BERENDSE, F. and AERTS, R. Competition between *Erica tetralix* L. and *Molinia caerulea* (L.) Moench as affected by the availability of

nutrients. Oecologia Pl. 5 (1984) 1-13.
(3) BOER, W.M.J. DEN (1986). Ammonia, not only a nutrient but also a cause of forest damages. In: Int. Symp. "Neue Ursachenhypothesen", December 16-17, 1985, U.B.A. Berlin.
(4) BREEMEN, N. VAN, BURROUGH, P.A., VELTHORST, E.J., DOBBEN, H.F. VAN, WIT, T. DE, RIDDER, T.B. and REYNDERS, H.F.R. (1982). Soil acidification from atmospheric ammonium sulphate in forest canopy throughfall. Nature 229: 548-550.
(5) BREEMEN, N. VAN and JORDENS, E.R. (1983). Effects of atmospheric ammonium sulphate on calcareous and non-calcareous soils of woodlands in The Netherlands. In: Effects of Accumulation of Air Pollutants in Forest Eco-systems. Reidel Publ. Co, Dordrecht.
(6) DIJK, H.F.G. VAN and ROELOFS, J.G.M. (1987). Effects of airborne ammonium on the nutritional status and condition of Pine needles. Proc. workshop on direct effects of dry and wet deposition on forest ecosystems. In particular canopy interactions, Lökeberg (Sweden) okt. 19-23 1986. E.E.C. in press.
(7) EERDEN, L.J.M. VAN (1982). Toxicity of ammonia to plants. Agriculture and Environment, 7: 223-235.
(8) HEIL, G.W. (1984). Nutrients and the species composition of heathlands. Ph. D. Thesis, Univ. of Utrecht.
(9) HEIL, G.W. and DIEMONT, W.M. (1983). Raised nutrient levels change heathland into grassland. Vegetatio 53: 113-120.
(10) HUNGER, W. (1978). Über Absterbeerscheinungen an ältern Fichtenbeständen in der Nähe einer Schweinemastanlage. Beitr. Forstwirtsch. 4: 188-189.
(11) JANSEN, Th. W. (1982). Intensieve veehouderij in relatie tot ruimte en milieu. Dutch State Forest Service, Utrecht.
(12) NIHLGÅRD, B. (1985). The ammonium hypothesis - An additional explanation to the forest dieback in Europe. Ambio, 14: 2-8.
(13) ROELOFS, J.G.M. (1986). The effect of airborne sulphur and nitrogen deposition on aquatic and terrestrial heathland vegetation. Experientia, 42: 372-377.
(14) ROELOFS, J.G.M., CLASQUIN, L.G.M., DRIESSEN, I.M.C. and KEMPERS, A.J. (1984). De gevolgen van zwavel en stikstofhoudende neerslag op de vegetatie in heide- en heidevenmilieus. In: Zure regen, oorzaken, effecten en beleid, pp. 134-140. Proc. Symp. Zure regen, 's Hertogenbosch, 17-18 November 1983. Eds. E.H. Adema and J. van Ham. Pudoc, Wageningen.
(15) ROELOFS, J.G.M., KEMPERS, A.J., HOUDIJK, A.L.F.M. and JANSEN, J. (1985). The effect of air-borne ammonium sulphate on *Pinus nigra* var. *maritima* in The Netherlands. Pl. Soil 84: 45-56.
(16) SCHEIKH , K.H. (1969). The effects of competition and nutrition on the interrelations of some wet-heath plants. J. Ecol. 57: 87-99.

ACID RAIN STUDIES IN THE FICHTELGEBIRGE (NE-BAVARIA)

R. HANTSCHEL, M. KAUPENJOHANN, R. HORN and W. ZECH
Institute of Soil Science and Soil Geography,
University of Bayreuth FRG

Summary

An ecosystem balance of a damaged and a healthy spruce stand in a highly SO_2-polluted mountain area in NE-Bavaria is presented. Applying a canopy model shows the important factor of H^+-input with the SO_2. The canopy buffering of these protons causes needle leaching of base cations and weak organic acids. Besides this model proves the higher pollution of the damaged spruce stand. Consequently different chemical composition of soil solutions is measured in dependency to the amount of acidic input and the different soil types.
Buffering of acid precipitation by spruce canopies induces significant acidification of the rhizosphere. Mg-equivalents in H_2SO_4-extracts of undisturbed soil samples are well correlated with Mg-content of spruce needles, if the pH-value of the H_2SO_4 corresponds to the H^+-buffering of the canopy. These results indicate strong relations between acid deposition, canopy processes, acidification of the rhizosphere, and tree nutrition. Moreover acid deposition can reduce the Mg-content of needles directly at the end of vegetation period as well as P-uptake seems to be increased by proton input. These laboratory results prove field observations in the Fichtelgebirge. Therefore acid depositions seem to be the main factor causing forest decline in the Fichtelgebirge.

Water-, nutrient-, and pollutant transport in spruce stands

1. INTRODUCTION

Air pollutants are supposed to be the main factor causing forest decline, but the mechanisms are still being discussed. The significance of direct needle or leaf damages and indirect influences via soil has not been finally explained. Therefore we examined water and element fluxes in spruce stands as well as detailled ecosystem processes, like canopy buffering, to get more informations on the different influences of acid deposition on spruce ecosystems.

2. MATERIAL AND METHODS

The ecosystem balance was measured on one severely damaged (Oberwarmensteinach, podzol of phyllite) and one not damaged spruce stand (Wülfersreuth, podzolic cambisol of phyllite) in the Fichtelgebirge, Central Europe (NE-Bavaria), for two years. Experimental design in the field and kind of equipment was choosen according to MEIWES et al. (1). To differentiate element fluxes in the canopy we used the model of ULRICH (2), which produces gaseous SO_2-input, gaseous H^+-input, needle leaching of K, Ca, Mg, dry deposition and total H^+-deposition with the throughfall. Solutions were sampled weekly and the fluxes calculated on a two week base. To get the amount of soil water in different depths a one-dimensional, numerical water-balance-model was used (3).

3. RESULTS AND DISCUSSION

The figure shows the results of the ecosystem balance for 1985. No significant difference was analyzed for the SO_4-S- and the proton input with the precipitation between both stands. The N-input was the dominant factor for the ion balance in Oberwarmensteinach and Wülfersreuth, while the base cations are unimportant. Chemical composition of the precipitation water is changed in the canopy by a lot of interactions between the liquid phase and the needle surfaces (4). In the investigated stands the main canopy processes are the high SO_2-interception and the proton buffering by cation and weak acid leaching. A significant higher SO_2-input in the damaged stand (O) can be proved. Consequently the gaseous H^+-input is also significantly higher in Oberwarmensteinach than in Wülfersreuth and the damaged canopy has to buffer more protons than the healthy one. Therefore K- and Ca-leaching in Oberwarmensteinach is higher than in Wülfersreuth. For Mg no distinct differences can be tested, but it is important to remember, that Oberwarmensteinach, characterized by very low Mg needle contents (5), shows the same leaching rates than the much better Mg-supported spruces in Wülfersreuth (W). HANTSCHEL and KLEMM (6) showed the importance of weak acidity in precipitation and throughfall samples, probably caused by needle leaching, for the proton input. Result of these processes is a significant higher H^+- and SO_4-S-input with the throughfall in the damaged stand than in the healthy one. Comparing measurements in an old spruce stand in Oberwarmensteinach showed the much higher H^+- and SO_4-S-input with the throughfall due to the higher filter capacity of the old trees.

Calculating the proton input using only the pH-value in the samples underestimates the proton fluxes considerably (3), because the canopy-buffered protons induce a equivalent rhizosphere acidification by increased base cation uptake of needle-leached ions (7).

These input data reveal the different acidic input on the same sites, only caused by different stand structure, which causes different soil chemical processes on the same soil type (8).

Element input with the litterfall is only for nitrogen in the range of throughfall input. The mineral soil input, sampled with no-suction lysimeters, was influenced by additional mineralization of roots, which were cut during installation. Nevertheless the higher proton input into the mineral horizon in Oberwarmensteinach is obvious. The chemical composition of the output is a result of soil chemical processes. On the damaged stand more H^+ and Al-ions percolate through the soil profile than on the healthy stand, where H^+ and cation acids were buffered by base cation release. The high N-output of both stands, mainly NO_3-N, causes additional seepage acidification and base cation leaching, respetively (9). Calculating the input-output balance the damaged podzol stand is characterized by a net loss of aluminum, a small storage of protons and no significant flux changes for the base cations, SO_4-S and N. Contrary the healthy stand shows a very low Al-output, a higher proton storage causing higher base net losses of the podzolic cambisol and no differences for SO_4-S and N.

Results for 1985 in the 'Fichtelgebirge' -NE-Bavaria

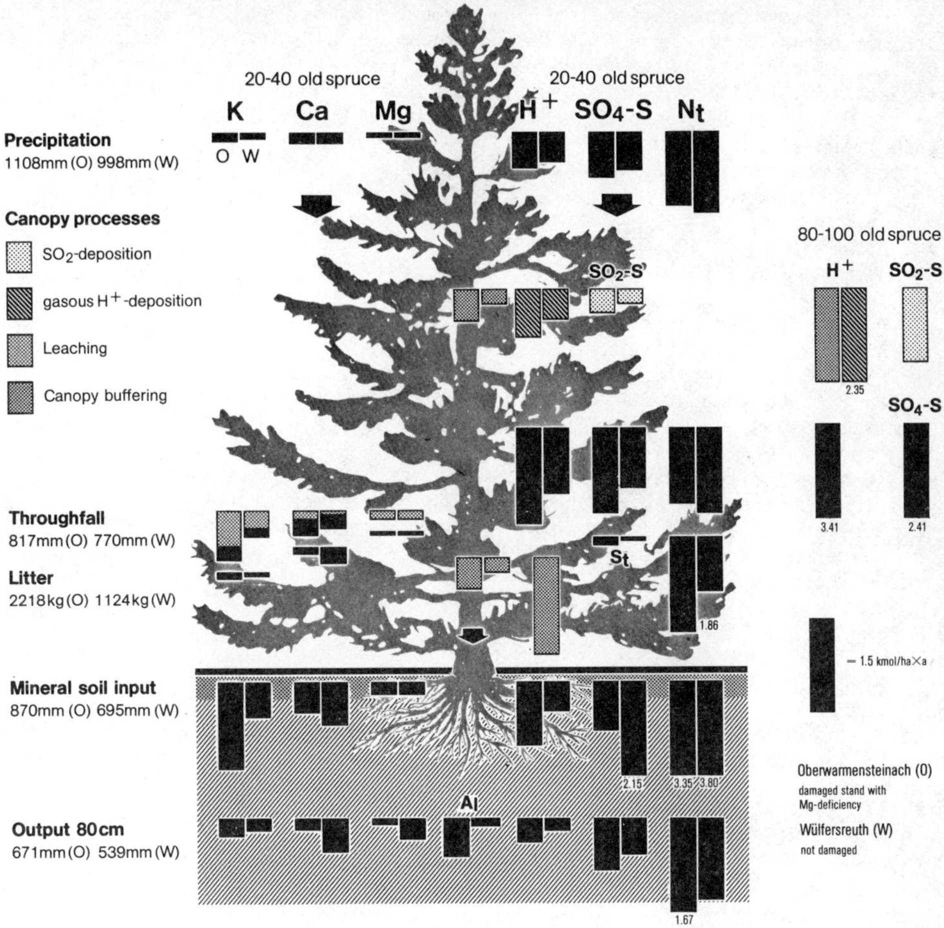

Effects of acidic deposition on forest nutrient status

1. INTRODUCTION

According to the model of ULRICH (2) precipitation acidity buffering induces K-, Ca-, and Mg-leaching from forest canopies. Evidence for this process is also given by greenhouse experiments (10). Although these authors measured significantly increasing base cation leaching from the canopy of spruce as a function of precipitation acidity, they could not prove base cation deficiency of those trees during the growing season. Nutrient deficiency, however, especially Mg-deficiency is a very common phenomenon of declining spruces in the highly SO_2-polluted Fichtelgebirge (11,3). The deficiency symptoms periodically intensify in autumn and winter, when the highest acid concentrations in fog were measured (12). To test acid rain effects on spruce nutrient status after the growing season (in contrast to

MENGEL et al. 10) the first experiment was carried out.

Furthermore base cation leaching from the canopy should increase nutrient uptake from the soil solution, which should induce additional acidification of the rhizosphere (2). The first experiment studies this process, too. Additional acidification of the rhizosphere should effect the chemical composition of the soil solution. Effects on spruce nutrient status via soil should be expected. Significance of this process was tested by means of the percolation soil solution (PBL-method 13,14) in the second experiment.

2. H^+-BUFFERING OF THE FOREST CANOPY

2.1 MATERIAL AND METHODS

Picea abies trees, three years old, were grown in nutrient solution. Leaf area was varied by removal of needles (pretreatment I = control, II = removal of 3 years old needles, III = removal of 2 and 3 years old needles). At the end of the vegetation period the trees were sprayed with 200 ml/day deionized water and sulfuric acid during a period of 10 days (A = control, B = deionized water, C = H_2SO_4 pH 2.7, D = H_2SO_4 pH 2.4, each treatment with 5 replications). K-, Ca-, and Mg-contents were analyzed in the throughfall. At the end of the experiment total acidity of the nutrient solution (titration up to pH 8.4 with 0.01n NaOH) and P-contents were analyzed as well as N-, P-, K-, Ca-, and Mg-contents of the needles (detailed information in 15).

2.2 RESULTS AND DISCUSSION

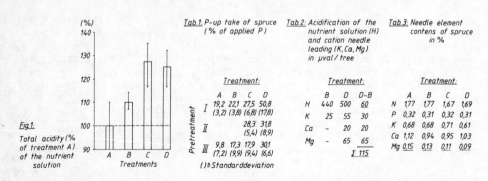

Figure 1 shows the effects of the treatments on total acidity of the nutrient solutions. The acidic treated trees C and D have 25-28% (0.29-0.32 mval H^+/l) higher total acidity than treatment A and B. These results confirm the assumptions of ULRICH (2). Comparable results on the acidification of soils by SO_2-fumigation were shown by MADHOOLIKAAGRAWAL and RAO (16). Acidic treatments significantly increased the P-uptake of spruce (table 1). This result reveals good agreement with field investigations. Earlier studies have proved, that declining spruce stands in South Germany often show high P-values in the needles (5). This experiment could not increase P-contents of the needles (table 3), probably because of the short experimental time. Comparison of treatment B and D indicates, that acidic spraying increased the leaching of K, Ca, and Mg out of the canopy (table 2). The difference between the treatments is 115 uval/tree. But additional acidification is only 60 uval/tree. The difference (115 µval M^+/tree - 60 µval/tree = 55 µval/tree) should result in a net base loss of the aci-

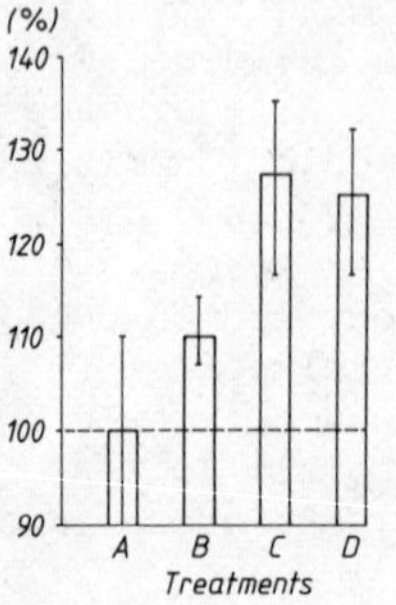

Fig.1: Total acidity (% of treatment A) of the nutrient solution

Tab.1: P-up take of spruce (% of applied P)

Pretreatment	Treatment: A	B	C	D
I	19,2 (3,2)	22,1 (3,8)	27,5 (6,8)	50,8 (17,8)
II			28,3 (5,4)	31,8 (8,9)
III	9,8 (7,2)	17,3 (9,9)	17,9 (9,4)	30,1 (6,6)

() ≙ Standarddeviation

Tab.2: Acidification of the nutrient solution (H) and cation needle leading (K, Ca, Mg) in µval / tree

	Treatment: B	D	D-B
H	440	500	60
K	25	55	30
Ca	-	20	20
Mg	-	65	65
			Σ 115

Tab.3: Needle element contens of spruce in %

	Treatment: A	B	C	D
N	1,77	1,77	1,67	1,69
P	0,32	0,31	0,32	0,31
K	0,68	0,68	0,71	0,61
Ca	1,12	0,94	0,95	1,03
Mg	0,15	0,13	0,11	0,09

dic treated trees. As shown in table 3 there is a sharp decrease of Mg-contents as a function of acidic treatments. Totally, the differences between treatment B and D are 45 μval Mg/tree, which is about the level expected by substracting acidification of the nutrient solution from base leaching from the canopy. Decrease of Mg-contents corresponds with field investigations (5). Results of the base leaching support the data of MENGEL et al. (10), too. But the authors could not decrease base contents of spruce needles with acidic fog treatments. The difference to the presented results may be explained by different experimental conditions. MENGEL et al. realize their study during the vegetation period, while the shown experiment took place at the end of the growing season.

3. EFFECTS OF ACIDIC DEPOSITION ON SOIL SOLUTION

3.1 MATERIAL AND METHODS

Soil solution was gained from undisturbed soil samples of a podzol (Oberwarmensteinach, pH_{KCl} 2.9), podzolic cambisol (Wülfersreuth, pH_{KCl} 2.8), eutrophic cambisol (Selb, pH_{KCl} 3.4), and a rendzina (Kloaschau, pH_{KCl} 7.0) by application of the percolation method (PBL, 13). Soil cores ($100\ cm^3$) of these top soils were percolated with H_2SO_4 (pH 5.6 - 2.5). Nutrient contents of the extracts were correlated with element contents of spruce needles. Moreover equilibrium soil solutions (GL) were extracted in batch experiments with H_2SO_4, too (14).

3.2 RESULTS AND DISCUSSION

Figure 2 shows the cation release of undisturbed soil samples as a function of pH of the extraction solution. With the exception of K, generally cation concentrations increased with increasing acidity of the percolation solution. Samples of Oberwarmensteinach reached maximum cation concentrations at about pH 3.0, i.e. the pH-buffer rate of this soil may be exceeded by acid rain events in the Fichtelgebirge (14). The correlation of Mg-equivalents in the PBL, estimated at pH of the percolation solution corresponding to the average pH-buffering of spruce canopies, with Mg-contents of spruce needles are very good (figure 3). The percolation results are an indication for close relations between acidic deposition, nutrient composition of the soil solution in the rhizosphere, and Mg-supply of spruce.

Equilibrium extracts, however, showed worse correlations. Preparation of the GL destroys soil aggregates, which often can not be penetrated by plant roots. Therefore these results support the work of HORN (17), who could show, that there are important differences in base saturation between the surface of aggregates and the inner part of them.

Fig.2: Cation release of undisturbed soil cores as a funktion of pH of the extruction solution (PBL)

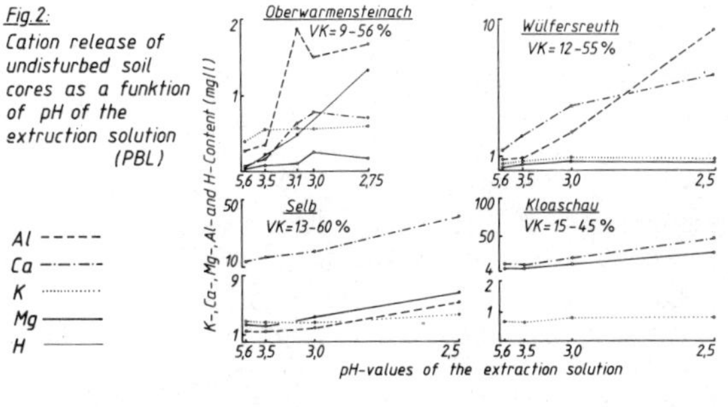

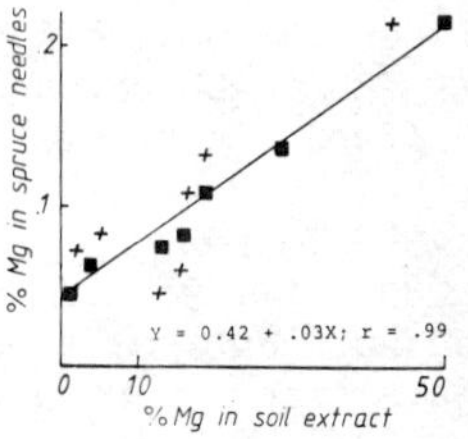

Fig.3: Mg-needle content (%) of spruce as a function of Mg (%) in soil extracts (PBL=■, GL= x)

REFERENCES

(1) MEIWES, K.J., KÖNIG, N., KHANNA, P.K., PRENZEL, J. and ULRICH, B. (1984). Chemische Untersuchungsverfahren für Mineralböden, Auflagehumus und Wurzeln zur Charakterisierung und Bewertung der Versauerung in Waldböden. Berichte des Forschungszentrums Waldökosysteme/Waldsterben. Bd. 7 1-67

(2) ULRICH, B. (1983). Interaction of forest canopies with atmospheric constituents: SO_2, alkali and earth alkali cations and chloride. IN ULRICH, B. and PANKRATH, J. (eds.). Effects of accumulation of air pollutants in forest ecosystems. Reidel Publishing Company

(3) HANTSCHEL, R. (1987). Wasser- und Elementbilanz geschädigter, gedüngter Fichtenökosysteme Im Fichtelgebirge unter Berücksichtigung physikalischer und chemischer Bodenheterogenität. Bayreuther Bodenkundl. Ber. 3 (in press)

(4) CRONAN, C.S. and REINERS, W.A. (1983). Canopy processing of acidic precipitation by coniferous and hardwood forests in New England. Oecologia (Berlin) 59 216-223

(5) KAUPENJOHANN, M., HANTSCHEL, R., HORN, R. and ZECH, W. (1987). Ergebnisse von Düngungsversuchen mit Magnesium an vermutlich immissionsgeschädigten Fichten (Picea abies (L.) Karts.) im Fichtelgebirge. Forstw. Cbl. 106 (in press)

(6) HANTSCHEL, R. and KLEMM, O. (1987). Characterization of weak acidity in selected precipitation samples from a forest ecosystem. Tellus (in press)

(7) REUSS, J.O. and JOHNSON, D.W. (1986). Acid deposition and the acidification of soils and waters. Ecological Studies 59. Springer-Verlag

(8) HORN, R., HANTSCHEL, R., KAUPENJOHANN, M. and ZECH, W. (1987). Influence of soil aggregation on forest nutrition and decline phenomena. Oecologia (reviewed)

(9) BEESE, F. und MATZNER, E. (1986). Langzeitperspektiven vermehrten Stickstoffeintrags in Waldökosysteme: Droht Eutrophierung ? Berichte des Forschungszentrum Waldökosysteme/Waldsterben Reihe B Bd. 3 182-204

(10) MENGEL, K., LUTZ, H.-J. und BREININGER, M.TH. (1987). Auswaschung von Nährstoffen durch sauren Nebel aus jungen intakten Fichten (Picea abies). Z. Pflanzenernähr. Bodenk. 150 61-128

(11) ZECH, W., SUTTNER, TH. and POPP, E. (1985). Elemental analyses and physiological response of forest trees in SO_2-polluted areas of NE-Bavaria. Water, Air and Soil Pollution 25 111-113

(12) SCHRIMPF, E., KLEMM, O., EIDEN, R.,FREVERT, T. and HERRMANN, R. (1984) Anwendung eines GRUNOW-Nebelfängers zur Bestimmung von Schadstoffen in Nebelniederschlägen. Staub-Reinhalt. Luft 44 72-75

(13) HANTSCHEL, R., KAUPENJOHANN, M., HORN, R. und ZECH, W. (1986). Kationenkonzentrationen in der Gleichgewichts- und Perkolationsbodenlösung (GBL und PBL) - ein Methodenvergleich. Z. Pflanzenernähr. Bodenk. 149 136-139

(14) KAUPENJOHANN, M. und HANTSCHEL, R. (1987). Die kurzfristige pH-Pufferung gestörter und ungestörter Waldbodenproben. Z. Pflanzenernähr. Bodenk. 150 (in press)

(15) KAUPENJOHANN, M., SCHNEIDER, B.U., HANTSCHEL, R., ZECH, w. und HORN, R. (1987). Sulfuric acid rain treatment of Picea abies: Effects on nutrient solution and throughfall chemistry as well as on spruce nutrition. Z. Pflanzenernähr. Bodenk. (submitted)

(16) MADHOOLIKAAGRAWAL, P.K. and RAO, D.N. (1985). Effects of sulfur dioxide fumigation on soil system and growth behaviour of Vicia faba plants. Plant and Soil 86 69-78

(17) HORN, R. (1987). Die Bedeutung der Aggregierung für die Nährstoffsorption in Böden. Z. Pflanzenernähr. Bodenk. 150 13-16

ACIDIFICATION RESEARCH ON DOUGLAS FIR FORESTS IN THE NETHERLANDS (ACIFORN PROJECT)

RECHERCHE D'ACIDIFICATION DES FORETS DU DOUGLAS-VERT AUX PAYS-BAS (PROJET "ACIFORN")

Peter Evers, Carla J.M. Konsten, Aart W.M. Vermetten

Postal adress ACIFORN group: Agricultural University -Department of Air Pollution
P.O.B. 8129 - 6700 EV Wageningen
The Netherlands

In 1985 the Dutch Priority Programme on Acidification was established. It will provide lacking scientific information on acid deposition to support the government in taking counter measures. Main topics of the programme are: research on exposure effect relationships, research on NH_3 emission control techniques and study of the effectiviness of control measures. The exposure effect studies focus on forests, natural vegetation and agricultural crops and compromise experimental field studies, comparitative field studies and studies under controlled conditions.

In one of the experimental field studies, the ACIFORN project, two stands of Douglas fir of different growth rate are monitored. On two field stations five research institutes perform an integrated study on the following topics (see figure):
-root dynamics (poster ACIFORN-2)
-hydrology (poster ACIFORN-3)
-soil chemistry (poster ACIFORN-4)
-air pollution (poster ACIFORN-5 + 6)
-tree physiology (poster-7 + 8)

Final aim of the ACIFORN project is to establish the influence of air pollution, among other stress factors, on the growth of Douglas fir.

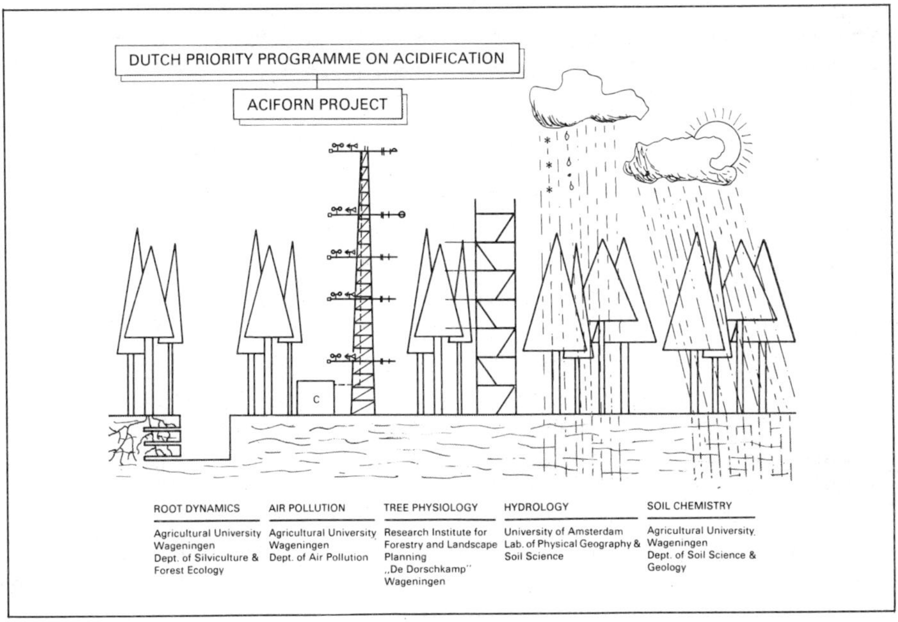

MONITORING OF ROOT GROWTH
(observation continue de la croissance racinaire)

A.F.M. Olsthoorn
Dept of Silviculture and Forest Ecology
Wageningen Agricultural University
P.O.Box 342, Wageningen, The Netherlands

AIMS

As part of the ACIFORN project on the impact of air pollution on the ecophysiology of Douglas-fir root growth is monitored to assess:
- total root growth;
- periodicity of root growth;
- turnover of fine roots;
- amount of biomass and necromass of roots;
- root intensity in different soil horizons;
- effects of air pollution on root growth.

Air pollution can affect root growth in two ways: through the soil and through the aboveground parts. If there is a distinct effect of air pollution on root growth it is of great importance to know if this is caused by growth abberations aboveground, e.g. resulting in less assimilates for root growth, or toxic substances in the soil solution. Data collected by other project participants on these aspects (e.g. assimilation, soil chemistry, water availability) will be combined with root growth data in simulation models for tree growth, hydrological and soil chemical processes, to determine the influence of air pollution on these processes.

METHODS

Root growth is monitored with different methods, as each method has its advantages and problems. Methods in this project are not to be destructive, as this would make monitoring impossible. Most of the work is done with recently developed methods. Information on periodicity of root growth is collected with the use of endoscopes in Perforons and Minirhizotrons. Root intensity can be calculated from results in Minirhizotrons and from root auger campaigns. Information on the turnover of roots will be collected with Ingrowth Cores. The methods are described below. Field measurements have started in 1987.

In addition to the field work some acidification processes are simulated in greenhouse trials with seedlings of Douglas-fir. This makes it possible to study one factor for its influence on root growth. In the Netherlands deposition of ammonium sulphate is a major air pollution component, therefore this is simulated in greenhouse trials.

PERFORON METHOD

Horizontal perforations (12 mm diam.) are made with a special auger in the soil, from the side of a container or a pit in the field. The method is developed in greenhouse containers. For the ACIFORN project, a special rootcellar has been constructed with one stainless steel wall, which has perforations 4 cm apart. The length of the perforations in the soil is approx. 30 cm. In this way a small part of the root system of a living tree can be observed, as roots of most tree species accept the perforations as natural holes. Root growth, branching and development of mycorrhiza can be monitored by direct observation with an endoscope and recording on video. Parts of the root system can be harvested with biopsy scissors for further analysis. Also, other aspects can be studied like root architecture, root competition, soil and rhizosphere fauna. Sometimes, the perforations induce abnormal root growth, resulting in curling of roots inside the perforations. With Douglas-fir these problems hardly ever occur (Bosch,1984).

INGROWTH CORES

Holes made with a root auger are filled again with soil. Roots can penetrate freely into these soil columns. The ingrowth cores are installed in spring 1987 and are sampled at regular intervals (½ year). The ingrowth cores will remain in the soil up to two years. Growth and death of fine roots in these cores can be retraced to assess the temporal variation in the amounts of living fine roots. The turnover of fine roots can be estimated (Persson, 1979).

MINIRHIZOTRONS

At an angle of 45 degrees a 8 x 8 cm square lexan tube is brought into an undisturbed soil profile up to a depth of 1 m. By means of a long endoscope photographs can be taken with regular intervals of roots in contact with the lexan tube. These photographs are interpreted to estimate root intensity, periodicity of root growth, turnover and decomposition of roots. Recently instead of the lexan tube a rubber tube (e.g. from a motorbicycle) is used to avoid an air space between soil and lexan surface. This tube is removed before photographical observation and re-installed and inflated. (van Noordwijk et al.,1985).

LITERATURE

Bosch, A.L.-1984- A new root observation method: the perforated soil system. Acta Oecologia/Oecologia Plantarum 5(9), nr 1: 61-74.

Noordwijk, M. van, A. de Jager & J. Floris -1985- A new dimension to observations in minirhizotrons: a stereoscopic view on root photographs. Plant and Soil 86: 447-453.

Persson, H. -1979- Fine root production, mortality and decomposition in forest ecosystems. Vegetatio 41: 101-109.

(This is the text of one of the posters of the ACIFORN-project, see also other contributions)

MONITORING OF HYDROLOGICAL PROCESSES UNDER DOUGLAS FIR

A. TIKTAK and W. BOUTEN
Dep. Physical Geography and Soil Science
University of Amsterdam
Dapperstraat 115
1093 BS AMSTERDAM
The Netherlands.

Summary.
Within the framework of the Dutch Priority Programme on Acidification an interdisciplinary study is carried out in two Douglas fir stands of different vitality. The aim is to obtain a better knowledge of the effects of atmospheric pollution on the behaviour of a forest ecosystem. As the hydrology strongly influences biogeochemical processes, a hydrological study supports the various topics in this program. Deterministic simulation models are used to calculate hydrological state variables and fluxes in the ecosystem. Precipitation, throughfall and soil water pressure heads are measured automatically; soil water contents are measured using a neutron depth probe and litter water contents are determined gravimetrically. Litter evaporation, plant evaporation, interception, plant water uptake, soil water fluxes and percolation are calculated.

1. INTRODUCTION.

Within the framework of the Dutch Priority Programme on Acidification an interdisciplinary study is carried out (Evers et al., 1987a). The aim is to obtain a better knowledge of the effects of atmospheric pollution on the behaviour of a forest ecosystem. Therefore various biotic as well as abiotic parameters and processes are monitored at the same field location. The project includes measurements of effects of air pollutants on the physiology of trees (Evers et al., 1987b; Olsthoorn, 1987), the quantification of gaseous pollution levels and deposition (Vermetten et al., 1987) and the monitoring of the soil chemical processes in the root environment (Konsten and v. Breemen, 1987). As the hydrology strongly influences biogeochemical processes, a hydrological study supports the various topics in the program.

Interception of water by the vegetation has a great impact on other processes. It reduces transpiration as long as the leaves are wet and, at the same time, it augments the filtering efficiency for certain air pollutants. It reduces the net downward soil moisture flow because evaporation rates are higher than transpiration rates. It also severely increases the spatial variability of fluxes. As water acts as a transporting agent for most chemical elements, the quantification of soil water fluxes is essential for the calculation of chemical budgets, often used for the research of element uptake, ion leaching and soil weathering (Bringmark, 1980; v. Grinsven et al., 1984). Also microbial processes depend upon the hydrology. Decomposition of litter and mineralization of soil organic matter are retarded by low water contents and low water fluxes. Under dry conditions the mobility of certain microorganisms is limited, because the waterfilled pores are too small to permit their passage (Griffin and Quail, 1968). However, nutrient diffusion and mass flow at higher water contents strongly augment microbiological activity (Papendick and Campbell, 1981). With respect to vegetation and its relationships to hydrological conditions Kozlowski (1982) deals thoroughly with the complex mechanism of responses of trees to the development of water stress.

2. HYDROLOGICAL MODELLING.

The research is carried out in two Douglas fir forests of different vitality, planted about 25 years ago. There is no undercover. The soil is composed of a litter layer ranging in thickness from 1 to 4 cm, underlain by a very sandy mineral soil with a groundwater table at a depth of more than 30 meters. As there is no catchment the soil water balance is to be simulated.

The most important hydrological fluxes and state variables in the forest ecosystem are shown in figure 1. Some can be monitored directly, others can only be calculated by means of a deterministic simulation model. Precipitation and throughfall quantities are measured with accurate time control. This allows a good estimation of the evaporative water losses, assuming above-ground absorption to be negligible. Moreover, part of the throughfall will not infiltrate the soil, but will evaporate from the litter layer. The losses cannot be measured directly. They are calculated by means of a combined litter and vegetation interception model, for which micrometeorological variables (provided by weather stations on the sites) are essential inputs.

Soil water fluxes are calculated with a deterministic model for unsaturated, non stationary flow, accounting also for water uptake by the vegetation (model SWIF; Soil Water In Forests). Reduction of transpiration during periods of low soil water pressure heads is accounted for too. The soil water model forms part of the FORHYD-package, which is developed within the framework of the Dutch Priority Programme on Acidification (FORest HYDrological model; Bouten and Tiktak, 1987). Important data for this model are soil physical properties and root distribution functions. The most important state variables are the soil water content and the soil water pressure heads, which are used for calibration and validation of the model. The flux, which is derived from the combined litter and interception model is used as the model's upper boundary condition. As a lower boundary condition, measured pressure heads at a depth of 200 cm. are used.

3. INSTRUMENTATION AND EXPERIMENTAL LAYOUT.

Quantities of precipitation and throughfall are measured automatically every 3 minutes by means of funnels, connected to capacitive water level recorders. This method is based on the work of Keizer (1984). In the forest, throughfall measurements are replicated 16 times to account for variability in the crown density (D in figure 2). The litter moisture contents are determined gravimetrically each fortnight from fresh litter samples.

Soil water pressure heads are read automatically twice a day at depths ranging from 5 to 200 cm (A in figure 2). A 24 port Scanivalve fluid switch is stepped every 3 minutes to connect 22 tensiometers and 2 reference heads to a single pressure transducer, converting pressure into voltage (Burt, 1978). Under dry conditions, soil pressure heads are estimated by gypsum blocks (B in figure 2). Soil water contents by neutron depth probe measurements are determined fortnightly for 3 profiles at each site. These measurements are calibrated with water contents of volumetric soil samples.

Soil temperatures are measured every 15 minutes for use in heat-flow models. Heat fluxes are used for energy balances in evaporation and transpiration studies (C in figure 2).

For the control and temporary storage of the automated measurements an ARCOM microcomputer is used. For long range storage the data are transmitted to a PDP-11 mini-computer on the site and stored on tape. The tapes are loaded on the University of Amsterdam mainframe computer for further data processing.

4. REFERENCES.

BOUTEN, W. and A. TIKTAK (1987). Documentation of FORHYD, a FORest HYDrological simulation model. In preparation.

BRINGMARK, L. (1980). Ion leaching through a podsol in a Scots pine stand. In: Structure and function of northern coniferous forests - an ecosystem study. T. Persson (Ed.). Ecol. Bull. 32, pp. 341-361.

BURT, T.P. (1978). An automatic fluid-scanning switch tensiometer system. Techn. Bull. 21. Brit. Geom. Res. Group, 29 pp.

EVERS, P., C.J.M. KONSTEN and A. VERMETTEN (1987a). Acidification Research in Douglas fir forests in the Netherlands (ACIFORN project). Symp. effects of air pollution on terrestrial and aquatic ecosystems. Grenoble, may 1987.

EVERS, P., P. CORTES, H. v.d. BEEK, W. JANS, J. DONKERS, J. BELDE, H. RELOU and W. SWART (1987b). Influence of air pollution on tree physiology. Symp. effects of air pollution on terrestrial and aquatic ecosystems. Grenoble, may 1987.

GRIFFIN, D.M. and G. QUAIL (1968). Movement of bacteria in moist particulate systems. Aust. J. Biol. Sci. (21), pp. 579-582.

GRINSVEN, J.J.M., van, J. MULDER and N. v. BREEMEN (1984). Hydrochemical budgets of some woodland soils with high inputs of atmospheric deposition: Mobilization of aluminium. In: Hydrochemical balances of freshwater systems. Erikkson, E. (Ed.). I.A.H.S. publ. 150, pp. 237-247.

KEIZER, G. (1984). Capacitive sensors for the measurement of level and dielectric constant. In: Sensors and actuators, Kluwer Technical Books, Deventer, Holland. pp. 73-78.

KONSTEN, C.J.M. and N. van BREEMEN (1987). Monitoring of soil chemical parameters under Douglas fir. Symposium effects of air pollution on terrestrial and aquatic ecosystems. Grenoble, may 1987.

KOZLOWSKI, T.T. (1982). Water suply and tree growth. Part I: Water deficits. For. Abstr. 43. pp. 57-95.

OLSTHOORN, A. (1987). Monitoring of root growth. Symp. effects of air pollution on terrestrial and aquatic ecosystems. Grenoble, may 1987.

PAPENDICK, R.I. and G.S. CAMPBELL (1981). Theory and measurement of water potential. In: Water potential relations in soil microbiology. Parr, J.F. (Ed.). S.S.S.A. Spec. Publ. 9, pp. 1-22.

VERMETTEN, A., P. HOFSCHREUDER and H. HARSSEMA (1987). Air pollution monitoring in a Douglas fir forest. Symp. effects of air pollution on terrestrial and aquatic ecosystems. Grenoble, may 1987.

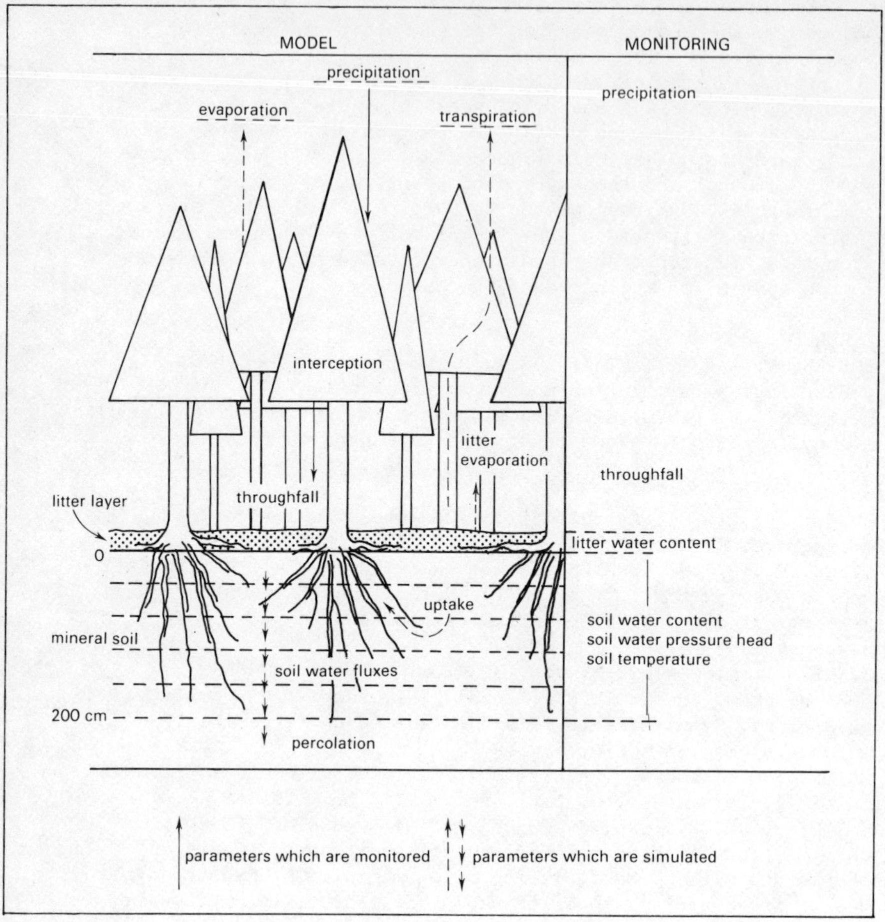

Figure 1. Hydrological parameters which are monitored or calculated

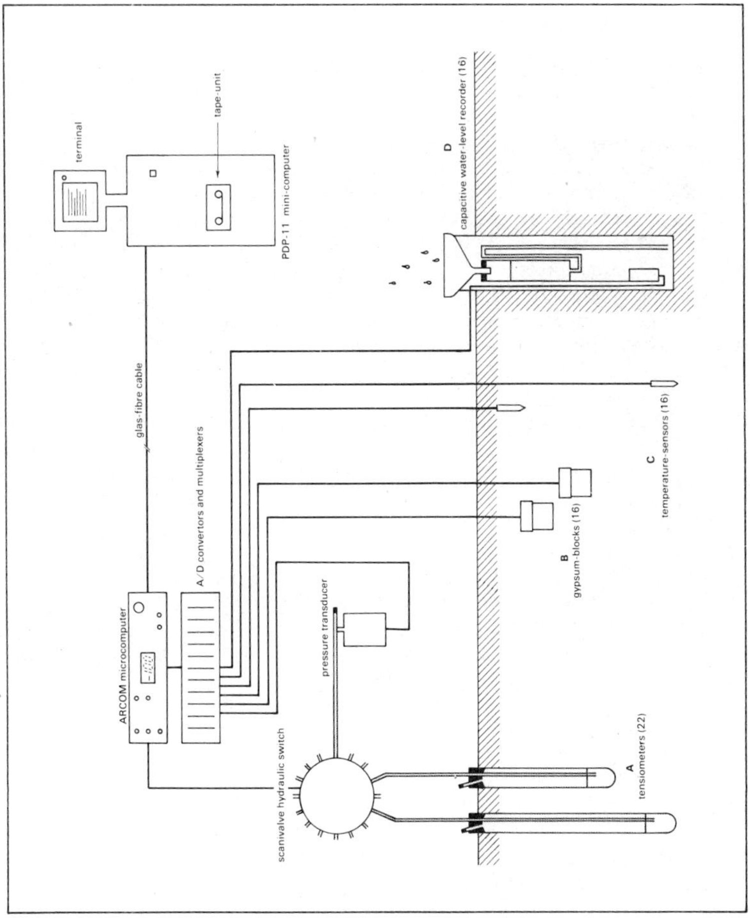

Figure 2. Automated measurements.
The micro-computer (ARCOM) is used for measuremts control; the mini computer (PDP-11) is used for central data storage.

MONITORING SOIL CHEMICAL PARAMETERS UNDER DOUGLAS FIR

Observation continue des parametres chimiques du sol sous Douglas vert (Pseudotsuga menziessi)

Carla J.M. Konsten and Nico van Breemen

Agricultural University, Dept. Soil Science & Geology,
P.O. Box 37, 6700 AA Wageningen, the Netherlands

Abstract

In March 1987 a multidiciplinary monitoring project started in two Douglas fir stands of different vitality. Main goal of this project (the ACIFORN project) is to provide insight in the causes of decline of vitality of forests in the Netherlands.

The present paper will discuss the collection of precipitation (bulk and wet-only), throughfall, litter fall and soil solution within this project. The chemical composition of the soil solution gives an impression of the chemical environment of the root system and of the availability of some nutrients. Combination of our chemical data and forest hydrological models, also developed within this project, result in chemical fluxes through various compartments of the forest ecosystems. These parameters form essential input for our chemical model, with which we evaluate the impact of acid precipitation on the forest soil and the internal soil processes.

The project further includes monitoring of hydrological parameters of the forests; concentration profiles of main air pollutants; the nutritional and hormonal status of the trees; photosynthesis in ambient, filtered and polluted air; and the dynamics of root growth.

Introduction

In 1985 the Dutch government launched the Dutch Priority Program on Acidification aimed at gaining a better insight in causes, effects and possible cures of vitality decline of Dutch forests. A nation-wide inventory of forest vitalitiy in 1986 showed a threefold increase in non-vital forest since 1984 (Staatsbosbeheer, 1986). Part of this program forms a multidisciplinary monitoring project (the ACIFORN project: ACIdification of FORests in the Netherlands) on two Douglas fir sites of different vitality. The monitoring project provides detailed, integrated information on the functioning of Douglas fir ecosystems. The data collected will serve to calibrate and validate dynamic models of forest ecosystems, which are developed within the Dutch Priority Program as well.

Our part of the program includes the monitoring of precipitation (bulk and wet-only), throughfall, litter fall and soil chemical parameters. First goal of our research is to study the chemical quality of the rooting zone. Second goal is to evaluate the impact of acid precipitation on the soil and the internal soil processes.

The program further includes the monitoring of soil hydrological parameters and the development of a hydrological model of Douglas fir forests (Tiktak & Bouten, 1987); the monitoring of the concentration of main gaseous pollutants and the development of a model for dry and wet deposition (Vermetten et al., 1987); the monitoring of nutritional and hormonal status and the growth of trees; measurements of photosynthesis in assimilation chambers, in which single twigs are exposed to **ambient air**, filtered air and air with various levels of one or more pollutants (Evers et al., 1987); and the monitoring of root growth by endoscopy in perforated soil chambers (Bosch, 1984; Olsthoorn, 1987).

Research sites

Both research sites are located in the Veluwe area, a large flat to undulating area with forests and moorlands in the central part of the Netherlands. The mean annual precipitation amounts to 800 mm. The soils are excessively well drained. The drainage is assumed to be vertical. The groundwater level is several to tens of meters below the soil surface.

One Douglas fir research plot is 2.5 ha in size and has a 27 year old stand of *Pseudotsuga menziesii*, of poor vitality. The soil is a Typic Dystochrept on sandy loam amd loamy sand textured Rhine sediments of Early and Middle Pleistocene age, which were pushed into ridges during the Saalian glaciation.

The second Douglas fir plot, which was planted in 1953, is 1.5 ha in size and also has a stand of *Pseudotsuga menziesii*, but of a better vitality. The soil is a Dystric Eutrochrept on loamy fine sand textured coversand deposits of Weichselian age.

Methodology

The availability of nutrients in de rooting zone is estimated from the chemical composition of the soil solution and the soil water contents. The soil water contents are derived from direct measurements (by neutrone probe) or calculated from soil water pressure heads and provided by the hydrological section of the ACIFORN project (Tiktak & Bouten, 1987).

The impact of acid deposition on the soil and soil processes is generally evaluated by calculations of budgets of main nutrients and of chemical fluxes in different compartments of the ecosystems (Likens et al., 1977; Matzner & Ulrich, 1981; Mollitor & Raynal, 1982; Van Breemen et al., 1983 and 1984; Van Grinsven et al., 1984). In fig. 1 the fluxes which are distinguished in our chemical model are depicted.

Some of the chemical fluxes are measured directly: <u>bulk and wet-only precipitation</u>, <u>throughfall</u> and <u>litter fall</u>. In part, also <u>chemical fluxes through the soil</u> are sampled instantaneously, as will be described below. Partly, the chemical fluxes through the soil are estimated by multiplying concentrations of nutrients by the water fluxes moving trough the soil. The water fluxes are simulated, using the Forest Hydrological Model (FORHYD; Tiktak & Bouten, 1987). The <u>nutrient uptake</u> by the vegetation is estimated from the increase of living biomass above the soil (derived from the diameter growth of the trees) and from root growth (Olsthoorn, 1987) plus the litter fall. <u>Decomposition of litter</u> and <u>mineralization</u> is studied in an additional project, using litter bags. Acidification due to <u>nutrification</u> is inferred from calculations of the nitrogen cycle. From the budget calculations <u>weathering of minerals</u> is estimated, as well as <u>ion exchange</u>.

Instrumentation

<u>Bulk precipitation and throughfall</u> are collected by means of 400 cm^2 black polythene funnels into 5 l polythene bottles, which are protected from light. At both forest sites twelve of these collectors are placed in a regular pattern, but at different distances from the nearest tree. With

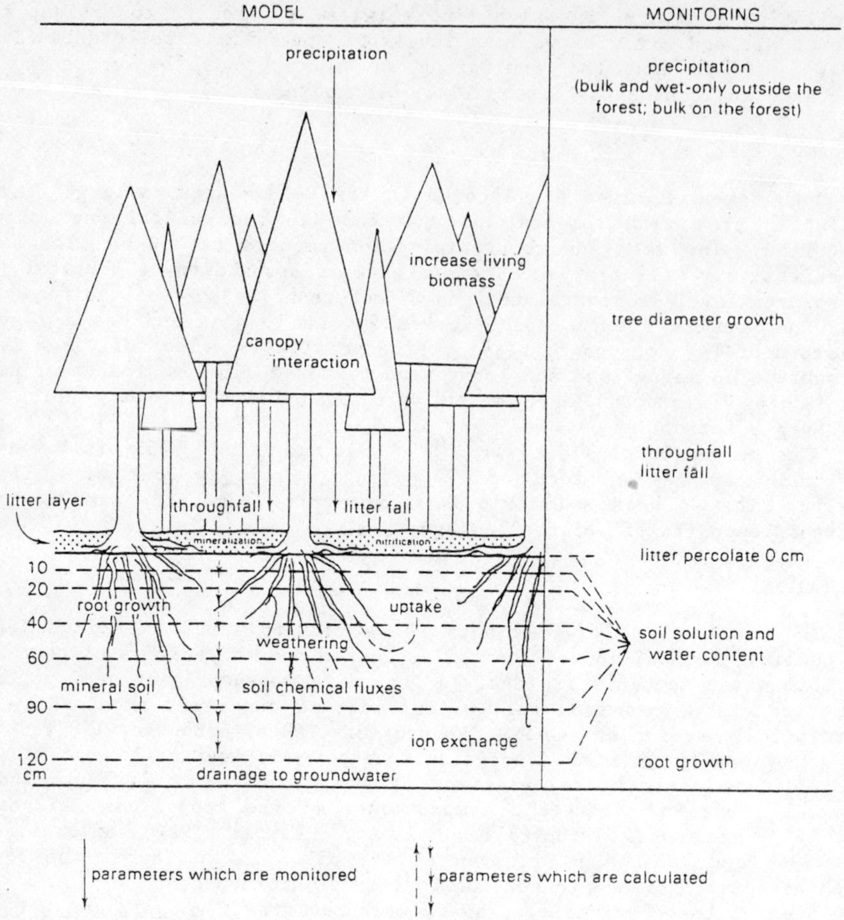

Figure 1. Schematic presentation of the chemical model and the parameters which are monitored on the Douglas-fir sites.

this amount of devices and lay-out we expect to account optimally for spatial variability, as examined by Duijsings et al. (1986), Rasmussen and Beier (1986) and Von Freiesleben et al. (1987). Bulk precipitation (in threefold) and wet-only precipitation are collected from non-forested areas near the forest sites. Throughfall and precipitation from each site are pooled for weekly analysis. Four weekly subsamples of each separate device are analyzed individually.

Litter fall is collected into 1 m^2 trays with plastic netting (pore size 1.5 mm). On each plot 6 collectors are used, which are evenly distributed over the plot.

Soil solution is collected in two ways. Firstly, from porous ceramic cups (Soil Moisture Equipment Co., type 655X1-B1M3), soil solution is sampled fortnightly into pre-evacuated glass bottles at depths of 10, 20, 40, 60, and 90 cm. Secondly, soil solution is collected quantitatively by an automated continuous sampling procedure, using filter plates and a suction which equals the pressure head in the soil (see fig. 2). The latter procedure is used to collect chemical fluxes from underneath the litter layer (in duplicate) and at a depth of 10 and 120 cm. The samples are collected fortnightly. The filter plates are constructed from porous polythene plate, covered by a Versapor 200 filter (Ø 0.2 μm), according to Driscoll et al. (1985) . The soil solution is sampled by a suction which is regulated electronically to equal the soil pressure head at the depth of the porous plate (measured with a tensiometer) and into a polythene bottle. We developed this sampling technique from the method of direct measurement of unsaturated soil water fluxes of Booltink et al. (1987). At both forest sites these sampling devices are replicated three times.

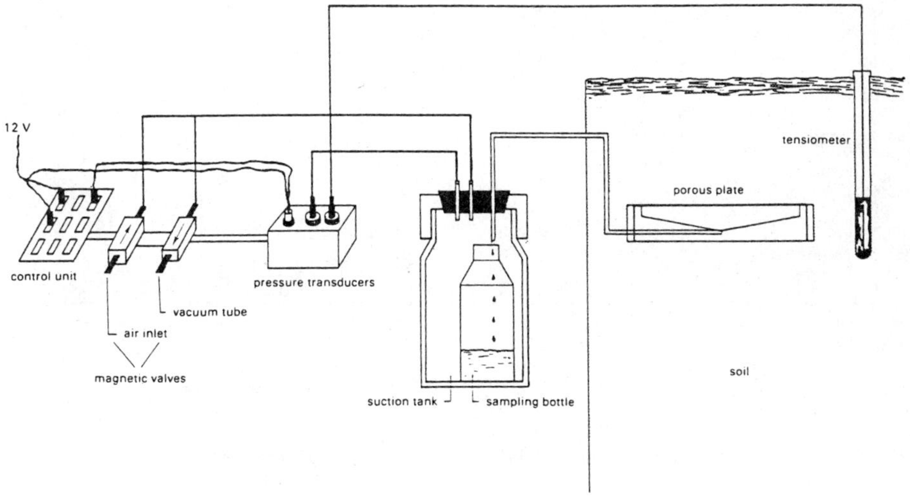

Figure 2. Automated soil solution sampling device (for explanation see text)

All water samples are analyzed for pH and electric conductivity (within 24 h from sampling), total and inorganic carbon (by TOC Analyzer), chloride, fluoride, sulphate and nitrate (by ion chromatography). Bicarbonate is calculated from pH and inorganic carbon content. Phosphate, silica and major cations (K, Na, NH_4, Ca, Mg, Al, Fe, Mn) are determined by spectrophotometry. Analysis quality control is carried out by the method of Stuyfzand (1983) for electric conductivity, and by comparing the sums of total cations and anions (whereby the amount of organic anions is calculated by the formula of Oliver et al. (1983)).

Monitoring of the <u>diameter growth</u> of the Douglas firs is carried out by monthly readings of the diameter ribbons on selected trees.

<u>Preliminary results</u>

In fig. 3 some of the first results are shown. The results show a predominance of sodium, ammonium, chloride and sulphate in throughfall water. In the soil solution aluminum is the principle cation, whereas nitrate and, in the lower part, sulphate are the major anions. Ammonium probably originates from ammonia, which volatilizes from manure, whereas sulphate is formed from sulphur dioxide from fossil fuels (Van Breemen et al., 1982). Ammonium has almost completely disappeared at the top of the mineral soil, with simultaneous lowering of pH and increases in the concentrations of aluminum and nitrate. These results do agree with

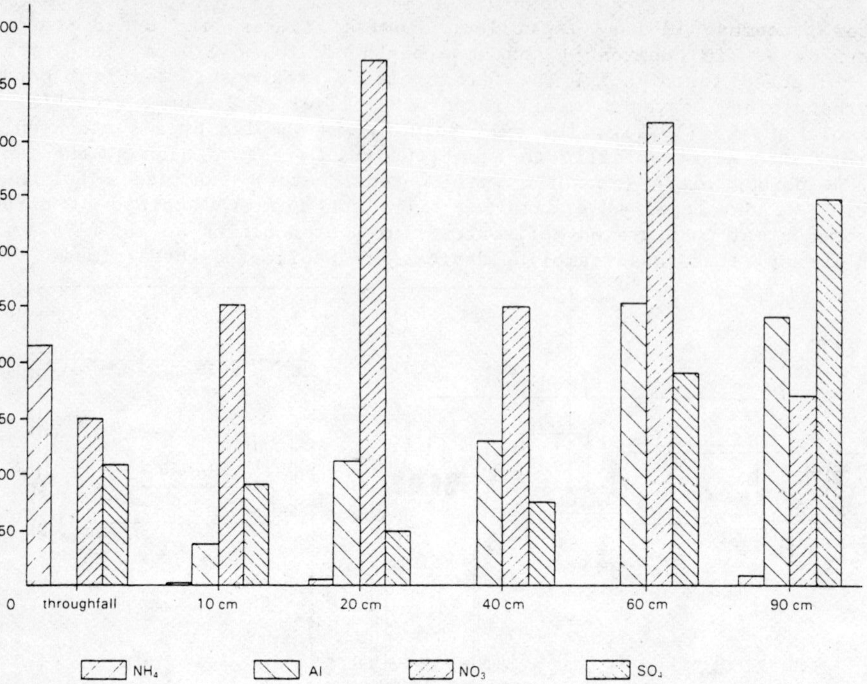

Figure 3. Composition of soil solution on the low-vital Douglas-fir site (January, 1987; mmol.m-3)

concentrations of aluminum and nitrate. These results do agree with results from other, nearby monitoring sites and may be explained from acidification due to nitrification of atmospheric ammonium and from mobilization of aluminum (Van Breemen et al., 1983 and Van Grinsven et al., 1984).

Concluding remarks

The integrated character of the research program provides for the monitoring of a large number of biotic and abiotic parameters at the same time and in the same place. This will result in a thorough knowledge of the functioning of the ecosystems under various environmental conditions. The joint data set of all participants in the field monitoring program will be used to validate and calibrate a dynamic model, which will permit the prediction of future developments. Our first step towards this model will be the calculation of yearly budgets of water and nutrients. During the first year of monitoring most attention will be paid to the calibration of the hydrological and chemical models.

References

Booltink, H.W.G., Van Grinsven, J.J.M., Bongers, N., Waringa, N., and Van Breemen, N., 1987. A new method to measure the unsaturated soil water flux. In: Proc. Int. Symp. Field Methods in Terrestrial Ecosystem Nutrient Cycling, Grange over Sands, 1986 (in press).

Bosch, A.L., 1984. A new root observation method: the perforated soil system. Acta Oecol./Oecol. Plant. 5: 61-74.

Driscoll, C.T., Van Breemen, N., and Mulder, J., 1985. Aluminum chemistry in a forested Spodosol. Soil Sc. Soc. Am. J. 49: 437-444.

Duijsings, J.J.H.M., Verstraten, J.M., and Bouten, W., 1986. Spatial variability in nutrient deposition under an oak/beech canopy. Zeitschr. Pflanzernähr. Bodenk. 149: 718-727.

Evers, P., Cortes, P., van de Beek, H., Jans, W., Donkers, J., Belde, J., Relou, H., and Swart, W., 1987. Influence of air pollution on tree physiology. Proc. Symp. Effects of air pollution on terrestrial and aquatic ecosystems, Grenoble, May 1987.

Likens, G.E., Bormann, F.H., Pierce, R.S., Eaton, J.S., Johnson, N.M., 1977. Biogeochemistry of a forested ecosystem. Springer Verlag, New York, 146 p.

Matzner, E., and Ulrich, B., 1981. Bilanzierung jährlicher Elementflüsse in Waldökosystemen im Solling. Z. Pflanzenernähr. Bodenkd. 144: 660-681.

Mollitor, A.V., and Raynal, D.J., 1982. Acid precipitation and ionic movement in Adirondack forest soils. Soil Sci. Soc. Am. J. 46: 137-141.

Oliver, B.G., Thurman, E.M., and Malcolm, R.L., 1983. The contribution of humic substances to the acidity of colored natural waters. Geochim. Cosmochim. Acta 47: 2031-2035.

Olsthoorn, A., 1987. Monitoring of root growth. Proc. Symp. Effects of air pollution on terrestrial and aquatic ecosystems, Grenoble, May 1987.

Rasmussen, L., and Beier, C., 1986. Improved methods for throughfall and stemflow collection. In: Andersson, F. (ed.): EEC-Proceedings Workshop on Direct Effects of Dry and Wet Deposition on Forest Ecosystems - in particular Canopy Interactions. Lökeberg, Kungälv.

Staatsbosbeheer, 1986. Verslag van het landelijke vitaliteitsonderzoek 1986. De vitaliteit van het Nedelandse bos 4, Staatsbosbeheer 1986-21, Utrecht.

Stuyfzand, P.J., 1983. De berekening van het elektrische geleidingsvermogen van natuurlijke wateren: een zeer nauwkeurige methode met voorbeelden van toepassing. SWE-83.001, Keuringsinstituut voor Waterleidingartikelen, Rijswijk.

Tiktak, A., and Bouten, W., 1987. Monitoring of hydrological processes under Douglas fir. Proc. Symp. Effects of air pollution on terrestrial and aquatic ecosystems, Grenoble, May 1987.

Van Breemen, N., Burrough, P.A., Velthorst, E.J., Van Dobben, H.F., De Wit, T., Ridder, T.B., Reijnders, H.F.R., 1982. Soil acidification from atmospheric ammonium sulphate in forest canopy throughfall.
Nature 299: 548-550.

Van Breemen, N., Van Grinsven, J.J.M., and Jordens, E.R., 1983. H^+-budgets and nitrogen transformations in woodland soils in the Netherlands influenced by high inputs of atmospheric ammonium sulphate. VDI-Berichte 500: 345-348.

Van Breemen, N., Mulder, J., and Driscoll, C.T., 1983. Acidification and alkalinization of soils. Plant and Soil 75: 283-308.

Van Breemen, N., Driscoll, C.T., and Mulder, J., 1984. Acidic deposition and internal proton sources in acidification of soils and waters. Nature 307: 599-604.

Van Grinsven, J.J.M., Mulder, J., and Van Breemen, N., 1984. Hydrochemical budgets of some woodland soils with high inputs of atmospheric acid deposition: mobilization of aluminium. In: Erikkson, E. (ed.): Hydrochemical balances of freshwater systems. I.A.H.S. Publ. 150: 237-247.

Vermetten, A.W.M., Hofschreuder, P., and Harssema, H., 1987. Air pollution monitoring in a Douglas fir forest. Proc. Symp. Effects of air pollution on terrestrial and aquatic ecosystems, Grenoble, May 1987.

Von Freiesleben, N.E., Ridder, C., and Rasmussen, L., 1987. Patterns of acid deposition to a danish spruce forest, Water Soil Air Poll. (in press).

AIR POLLUTION MONITORING IN A DOUGLAS FIR FOREST

LA POLLUTION DE L'AIR EN FORET DE DOUGLAS-VERT

Aart Vermetten, Peter Hofschreuder, Hendrik Harssema

Agricultural University Wageningen - Department of Air Pollution
P.O.B 8129 - 6700 EV Wageningen
The Netherlands

INTRODUCTION

As a part of the ACIFORN - project (Acidification of Forest in the Netherlands: monitoring of two stands of Douglas-fir) air pollution and micrometeorological variables are monitored..
The continuous measuring programme will probably last for two years, starting from April 1987. During the growing season of 1987 one of the field stations will already be fully operational.

AIMS OF THE PROJECT

Final aim of the ACIFORN - project is to evaluate the influence of air pollution, among other stress factors, on the vitality and growth rate of the Douglas-fir (Pseudotsuga menziesii).

The goals of the air pollution monitoring programme are:

-continuous determination of concentration levels of the main pollutants SO_2, NO_2, O_3 and NH_3 above, within and below the forest canopy.

-estimating the dry and wet deposition of these components to the forest canopy and soil.

-a study of the effects of air pollution on the physiology of the Douglas-fir, paying special attention to periods with high concentration levels

DESCRIPTION OF THE FIELD LOCATIONS

Both field sites are situated in the central part of the Netherlands. Air pollution in this region can be characterized by moderate levels of SO_2, NOx, low O_3 and rather high levels of NH_3.
The sites were selected on soil type and homogenity, age and origin of the trees and the aerodynamic structure of the stand and its surroundings. Both stands are approximately 30 years old. Average tree height is 15-20 meter; there is a large variability in height and vitality within the stands.

EXPERIMENTAL SET-UP (fig 1.)

In both stands the following equipment is installed:
-a 30 m tower at the center of the plot with arms bearing meteorological sensors

-teflon tubing with filter inlets for gaseous pollutants at five heights in the tower

-a small air-conditioned housing at the foot of the tower, containing gas analysers and a central computer

METEOROLOGICAL INSTRUMENTS

The position of the meteorological sensors is shown in figure 1. Adjustments to other

heights can be made. Sampling intervals vary from 10 seconds (wind sensors) upto 60 seconds for temperature and humidity.
Every 15 minutes averages are calculated and sent to the central computer for storage.

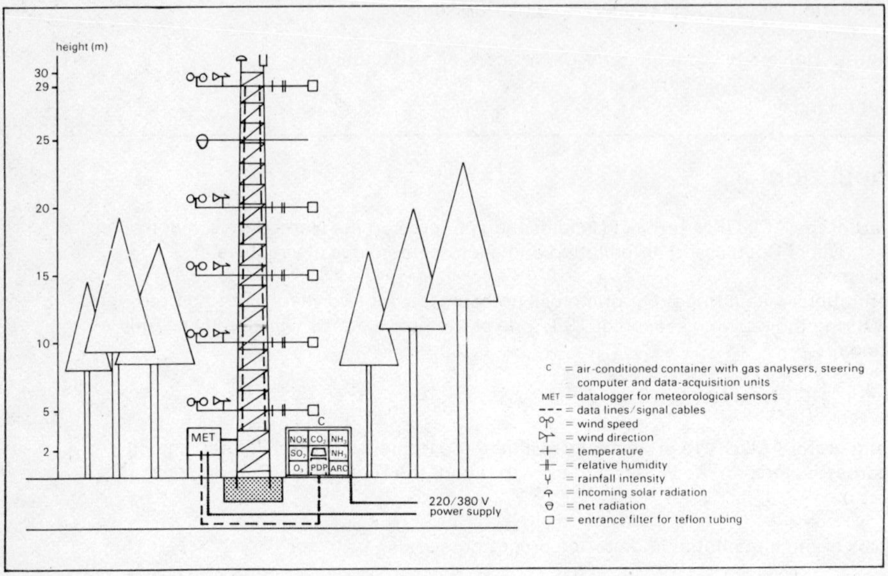

Figure 1. Experimental lay-out of the measuring sites.

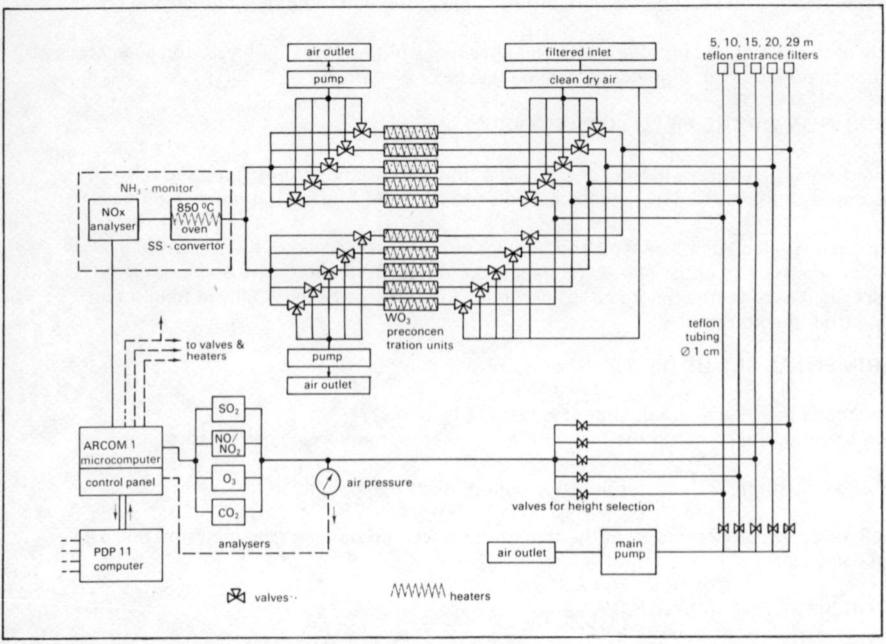

Figure 2. Data-acquisition and processing.

AIR SAMPLING LINES

A large pump pulls air from five heights down to the container through teflon tubing, which is slightly heated to avoid condensation of water vapour.
At the inlet of each sampling line a teflon filter is mounted to remove the bulk of the atmospheric aerosol. Residence times are kept at a few seconds to avoid chemical reactions within the tubing.

GAS ANALYSIS EQUIPMENT (fig. 2.)

At the end of the sampling tubes gas analysers switch between the five measuring heights with a cyclus time of 15 - 60 minutes. The use of only one analyser for each component prevents systematic differences between analysers.
NH_3 is preconcentrated for each height on tungsten (WO_3) denuders, after removal of HNO_3. The denuders are subsequently heated and flushed with clean air. The released NH_3 is passing through a stainless steel convertor at 850 °C and measured as NOx.

Table I. Gas analysers

Component	Analyser	Remarks
SO2	Thermo Electron 43W	Fluorescence
NO +NO2	Monitor Labs 8840	Chmemoluminescence
O3	Bendix 8002	Chmemoluminescence + ethylene
NH3	Monitor Labs 8840 + SS-convertor	Preconcentration on WO3
CO2	ADC 225-MK-3	Infrared Gas Analysis

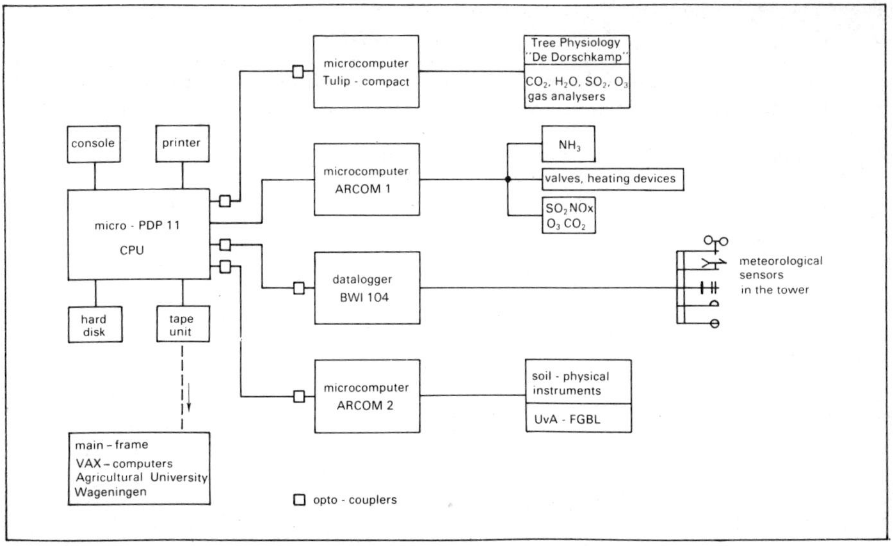

Figure 3. Steering processes for the air pollution measurements. All valves and heaters are controlled by the ARCOM microcomputer.

QUALITY CONTROL

Gas analysers are calibrated in the field on a regular base. During selected periods there will be comparisons with other gas measuring equipment, e.g. denuder systems for NH_3.

DATA-ACQUISITION, PROCESSING AND STORAGE (fig.3.)

The central PDP11-computer collects data from all the participants in the ACIFORN - project. For the connection with the other computer systems at the field location opto-couplers are used to guard the equipment against lightning.
Raw data are checked for validity range, converted into real units and stored on disk; backups are made on magnetic tape and transported to Wageningen for further processing on a mainframe computer system.

ADDITIONAL MEASUREMENTS

During a few selected periods there will be intensive measuring campaigns to evaluate:
-the aerosol size distribution and composition
-differences between the center and the edge of a forest stand
-the representativity of the measurements in one forest stand for a larger forested area
-concentrations of other gases, e.g. HNO_3, hydrocarbons
-chemical composition of fog and dew.

INTERPRETATION

-Concentration levels and frequency distributions.

-Deposition fluxes to the upper canopy will be calculated using flux-gradient relationships: $F=K*dc/dz$. Additional experiments with simultaneous use of fast sensors for momentum and heat exchange are needed to prove the validity of this approach.

-Throughfall and wet deposition data can be used to calculate dry deposition as a residu in the deposition balance of the entire forest ecosystem.

-A resistance model will be used to calculate the distribution and uptake of pollutants within the canopy.

-Stomatal resistances for uptake of pollutants have to be deduced from the exposure of single branches to gaseous pollutants in cuvettes (in coöperation with "de Dorschkamp").

-Direct effects during the periods with high levels of air pollutants.

INFLUENCE OF AIR POLLUTION ON TREE PHYSIOLOGY

INFLUENCE DE POLLUTION DE L'AIR AU PHYSIOLOGIE DE L'ARBRE

P. Evers, P. Cortes, H. v.d. Beek, W. Jans, J.Donkers, J. Belde, H. Relou, W. Swart

Dorschkamp Research Institute for Forestry and Landscape Planning - P.O.B. 23 - 6700 Wageningen - The Netherlands

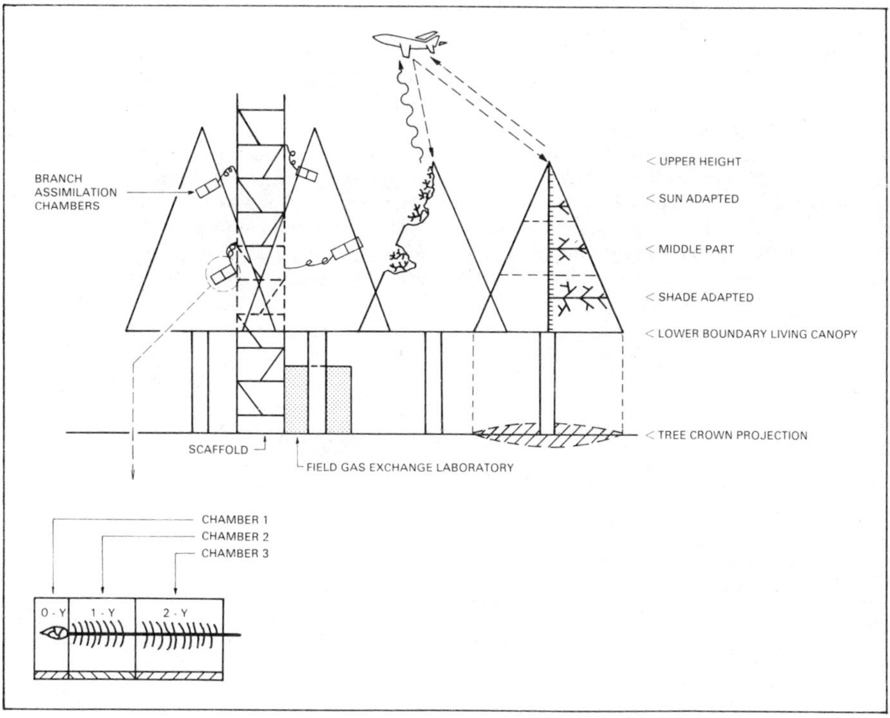

Figure 1. Diagram of the field gas exchange laboratory and growth analysis measurements.

PHYSIOLOGICAL PRE-VISUAL PARAMETERS OF THE INFLUENCE OF AIR POLLUTION OF DOUGLAS FIR

1- Photosynthetis, transparation and respiration (Figs. 1,2)
   in 2 monoculture ecosystems in contrasting stage of "vitality", on sun and shade adapted 0-, 1- and 2-years old needles (Kootwijk: moderate, Speuld: poor)
   light series on detached branches in a controlled environment to test the photochemical efficiency
   application of filtered and controlled polluted air next to the ambient supply in the chamers on all needle types.
   determination of chlorophyll a + b - protein complex in needles
2- Status of growth regulators in needles
   production of ethylene: analysis of precursor ACC and conjugate MACC, seasonal fluctuations and determination of the increased production after peak periods. Higher ACC levels were found in damaged test trees and in trees that received $(NH_4)2SO_4$ in rain experiments.

selection of sample trees through infrared reflection patterns on aireal photographs. Classification of the variability of the conditions of the trees is done by combining colour reflection with an assessment of the architecture of the top of the crowns. Only average condition trees are used.

3- Needle nutritional status

determination of N, P, K, Ca, Mg, Fe, and S in the same needle types as in (1). In the poor stand, needles accumulate more N and P in the sun adapted needles compared with shade adapted needles. Just after flushing, the new needles contain high levels of N and P which dilutes during the development. This dilution is slower in the poor stand.

determination of starch seasonal fluctuation, accumulation and disturbed translocation

4- The central, pre-visual parameter is growth. The other parameters, especially photosynthesis are tested for their influence on growth. Biomass modelling is the central tool to define the difference between the moderate and the poor stand as influenced by air pollution. Growth analysis is done in situ (Fig. 3), by sampling and by aerial and telephotography. In the poor stand, sun adapted needles have a lower dry weight and a smaller leaf area (Fig. 4). Extension growth in this forest starts later but ends at about the same absolute value (Fig. 5). However, avarage extension growth one year earlier (1985) was slower in Speuld as compared with Kootwijk. The reduction in growth in 1986 as compared with 1985 is dramatic (fig. 6). The leaf areas of the different needle types are simplified for model purposes using the volumes of parts of the crown (Fig. 7).

GROWTH ANALYSIS

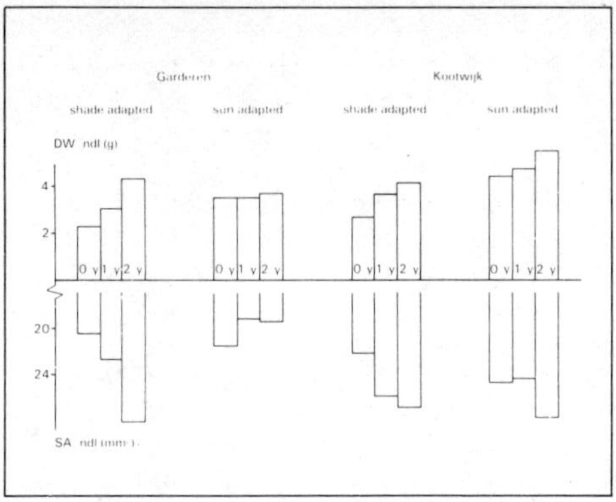

Figure 4. Mean dry weight (DW) per needle (g) and mean surface area (SA) per needle (mm²) of current year (1986), 1-year-old and 2-year-old shade adated and sun adapted needles at the Garderen (Speuld) and Kootwijk locations.

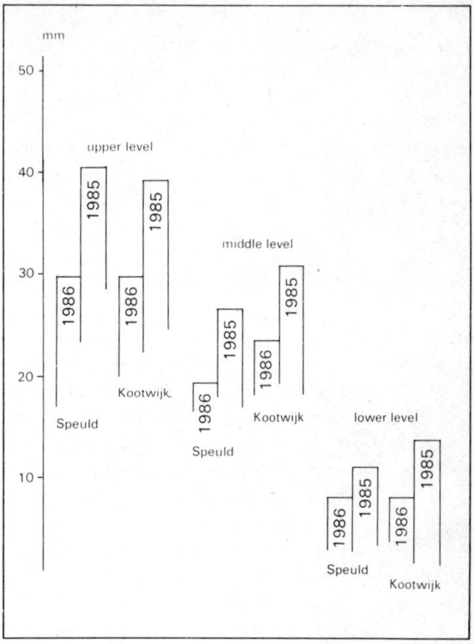

Figure 6. Mean extension growth of current year shoots at three levels in the canopy in 1985 and 1986

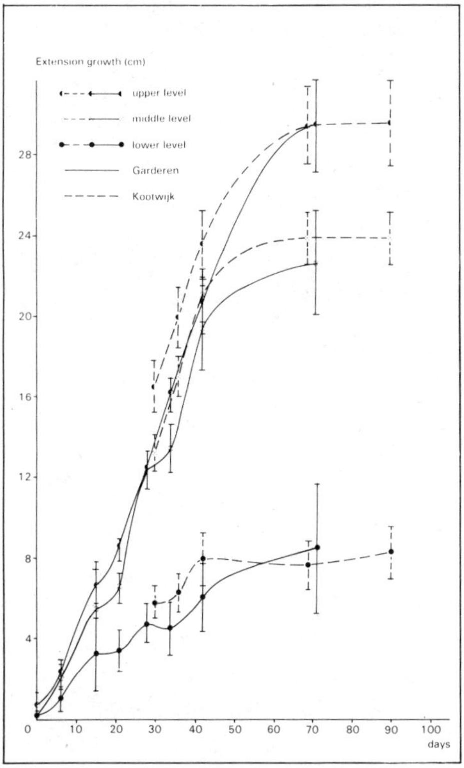

Figure 5. Mean extension growth of current year shoots at the Garderen and Kootwijk locations at 3 levels in the canopy (day 0=860520)

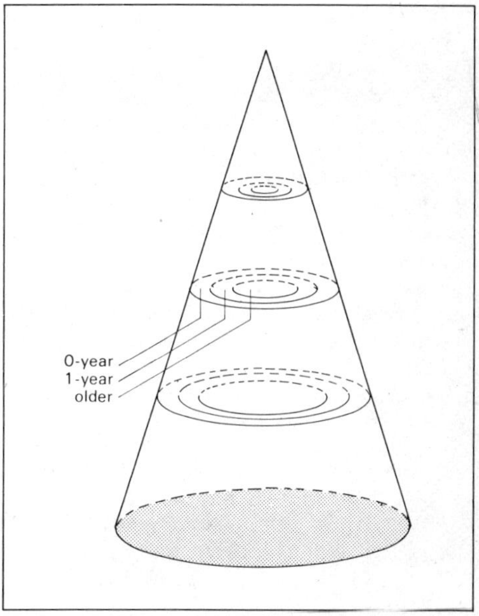

Figure 7. Idealised tree: modelled and simplified total leaf area per height and per age.

THE INFLUENCE OF ACID RAIN ON FINE ROOT DISTRIBUTION AND NUTRITION

B.U. Schneider and W. Zech
Institute of Soil Science, University of Bayreuth (FRG), B.O. Box 10 12 51

Summary

A comparison of fine root distribution and nutrition between two sites differing in the degree of proton input is presented. Additionally relations between the fine root biomass and plant nutrition and the reaction of root growth to Mg-fertilization are shown. High proton input seems to be responsible for intensified root growth in the upper part of the organic layer ($O_{L,F}$) as well as for a stronger decrease of the fine root biomass in the mineral soil. Root nutrition indicates that the release of protons from the roots especially restricts the uptake of Calcium and reinforces that of Phosphorus. At well supported sites more Magnesium is stored in fine roots of the mineral soil affecting a more lasting availability during drought. Correlations between fine root biomass and Magnesium in roots and in needles are restricted to the humus layer suggesting that tree nutrition mainly depends on nutrient uptake from the organic material. The negative correlation between fine root biomass and Phosphorus in the roots reflects the influence of high proton input causing reinforced P-uptake. In case of Mg-deficiency, fertilization with $MgSO_4$ and $CaCO_3$+ MgO causes an increase of the fine root biomass between 200-400% in the organic layer. In a soil with high amounts of exchangeable Ca and Mg fertilization of both elements may reinforce the antagonism towards K-availability and induce increasing root growth.

1. INTRODUCTION

Acid deposition in forest ecosystems may cause both direct damages of foliage (1) and changes of chemical conditions in the soil solution (2). According to the Ulrich model (3) the loss of cations from damaged foliage during the canopy buffering process of protons leads to an increasing acidification of the soil-root intersphere when reinforced cation uptake is followed by a release of protons from the root. Recent results of laboratory experiments done by Kaupenjohann et al. (4) and Flückinger (5) confirm this hypothesis. Exceeding the buffer rate of the soil (2) free protons may negatively influence root growth and nutrition (6). Increasing concentrations of acid cations may additionally change the relative nutrient availability in the soil. The disturbance of both root growth and nutrition should lead to specific deficiencies. At high altitudes in the Fichtelgebirge (North East Bavaria) Zech and Popp (7) could show that damaged stands are mainly suffering from Mg-deficiency. This study wants to show the effects of acid deposition on fine root biomass, correlations between concentrations of living fine roots and the plant nutrition and informs about the reaction of root growth to Mg-fertilization.

2. MATERIAL AND METHODS

In 1985 root samples were taken from two spruce ecosystems (40 years old) at Oberwarmensteinach (760 a.s.l.) and Wülfersreuth (680 a.s.l.) located 15 km from each other in the 'Fichtelgebirge' (North East Bavaria). In Oberwarmensteinach spruce is stocking on a phyllite-derived Podzol (pH KCl 2.9), trees in Wülfersreuth grow on a phyllite-derived podzolic Cambisol (pH KCl 2.8). The soil of the damaged site Oberwarmensteinach (42 kg SO_4-S/ha•a) shows a low base saturation ($<$ 3%) compared to Wülfersreuth (29 kg/ha•a) with 5-8%. In case of the fertilizer experiment a third site at Selb (560 a.s.l.) with high base saturation of the soil was investigated. There spruce is growing on an euthrophic Cambisol derived from basalt.

The sampling times coincided with specific phenological states: in April when soil was still frozen, in May at bud-break, in July when leaf growth was finished and in November at the end of the vegetation period. Five soil cores were taken from randomly chosen plots (five at each site) 1.2 m away from the trees. The sampling technique, the preparation of the fine roots ($\varnothing <$ 2 mm) from the soil and the criteria for the separation between living (fine root biomass) and dead roots (fine root necromass) were done as described by Murach (8).

Before weighing the roots were dried for two days at 70 ^{o}C. After pulverization 100 mg of each sample were digested 6h at 170 ^{o}C in 100 ml of HNO_3(65%) under pressure. The solution was measured by means of the ICP-method (IJ 48 ICP OE5) (9).

3. RESULTS AND DISCUSSION

3.1 Morphological Characteristics

Studies on forest dieback require consideration of each compartment of the ecosystem to obtain an impression of causal relationships as complete as possible. The rhizosphere, especially the soil-root intersphere is one part of the system we still need more information about. Root morphology and nutrition should reflect the growth relations depending from chemical and physical soil properties. The most efficient way to characterize root morphology is the measurement of the fine root mass. In comparison with other morphological characteristics this parameter is scarcely burdened with methodological problems and shows good correlations for example with both the total number of root tips (Fig. 1) and the ectomychorrizals (10).

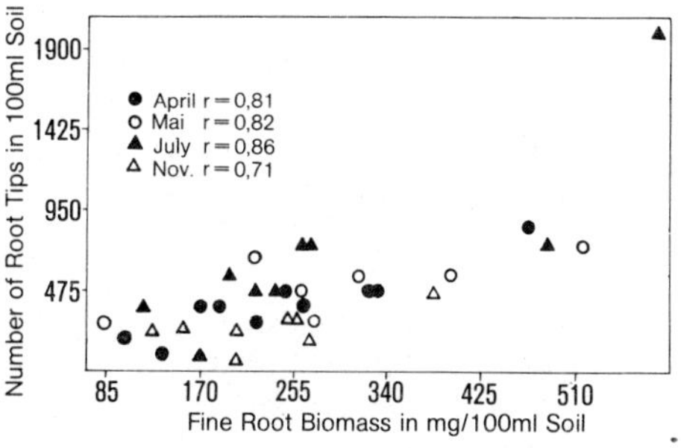

Fig. 1

The distribution of the fine root biomass (Fig. 2) in the humus layer and the upper mineral soil differs between both sites.

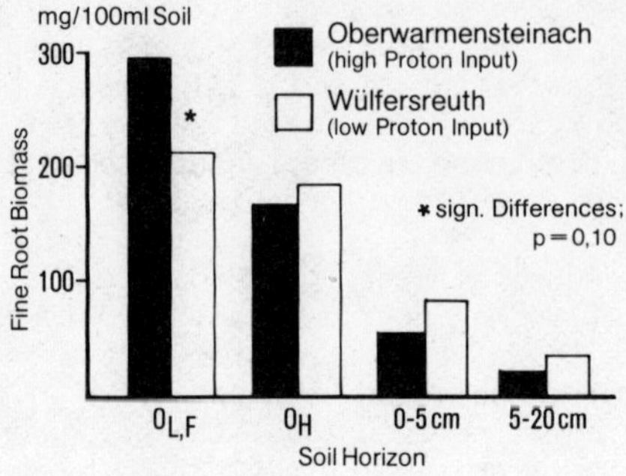

Fig. 2

There are significantly higher concentrations of living fine roots in the $O_{L,F}$ at Oberwarmensteinach (high proton input) compared to Wülfersreuth (low proton input). More roots tend to grow in the humified organic horizon (O_H) and the upper mineral soil at Wülfersreuth. The smaller decrease of the fine root biomass at this site possibly diminishes the danger of drought. The results suggest that the support of trees is mainly based on nutrient uptake from the upper parts of the humus layer ($O_{L,F}$).

3.2 Differences In Root Nutrition

The element contents in the fine root biomass taken from 0-5 cm depth (Fig. 3) may give an explanation for the lower concentrations of fine roots in the mineral soil at Oberwarmensteinach.

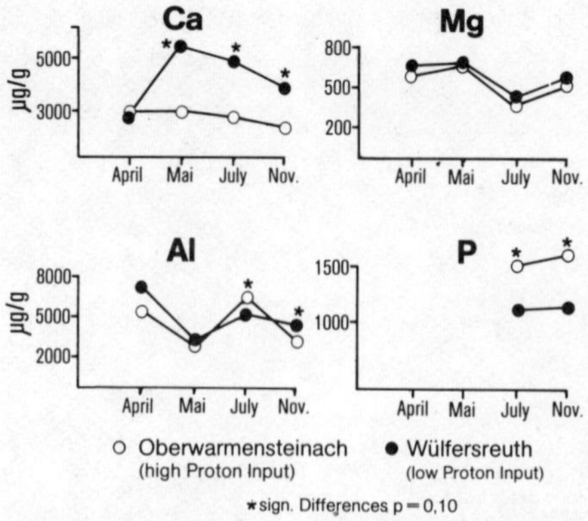

Fig 3

Due to the high mobility of Magnesium in plants (11) the contents in fine roots of both sites does not differ. In the course of full leaf expansion in July the requirement for Magnesium seems to cause a mobilization of this element from the roots since the Mg-concentrations on both sites drop down. At the end of the vegetation period the Mg-contents again increase.

Up from the second sampling in May the concentration of Calcium in living fine roots taken from the site Wülfersreuth was significantly higher compared to those taken from Oberwarmensteinach. Since the Al-contents show no clear differences between both sites the uptake of Calcium and Magnesium does not seem to be restricted by the presence of that element. This indicates other influencing factors causing Mg-deficiency as shown for spruce at Oberwarmensteinach (12). By means of the PBL-method Kaupenjohann (13) measured extremely low Calcium and Magnesium concentrations in the extracts he got when percolating sulfuric acid of pH 3.5 (due to the average proton input) through undisturbed soil samples from Oberwarmensteinach. A short-term increase of the proton load up to pH 3.0 would exceed the buffer rate of the soil (2) which may leave free protons in the soil solution. Especially protons restrict the uptake of Calcium as laboratory experiments of Stienen (6) could show, whereas Mg-uptake was not strongly affected.

Recent studies on forest decline generally show high P-contents in the foliage of damaged stands (14). Although the amounts of lactat-soluble Phosphorus in the soil is the same on both sites roots of the damaged site Oberwarmensteinach show higher P-concentrations compared to Wülfersreuth. Laboratory experiments of Kaupenjohann et al. (4) could show that there is a causal relationship between acid treatment of the shoot and reinforced uptake of Phosphorus from a nutrient solution. Both results support the assumption of a relation between proton input into forest ecosystems and high P-contents in the plant.

Figure 4 shows the store of Magnesium in the fine root biomass.

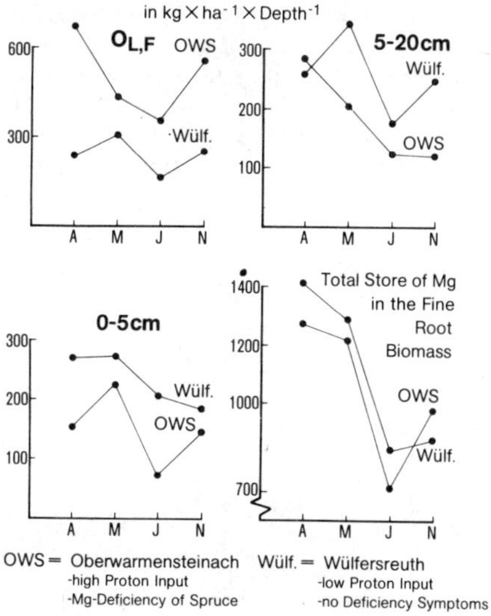

Fig. 4

Although the total amount is nearly the same on both sites , more Magnesium is stored in the fine root biomass of the mineral soil at Wülfersreuth. According to figure 3 the mobilization of Magnesium seems to depend on the requirement of the shoot. Since the distribution of fine roots in Oberwarmensteinach is comparatively worse, drought events may disturb the maintainance of Mg-mobilization from the roots much earlier than in Wülfersreuth.

3.3 Correlations between Fine Root Biomass and Plant Nutrition

Positive correlations between the fine root biomass and the Mg-contents (Fig. 5) in living fine roots can be calculated from fine root material of the humus layer.

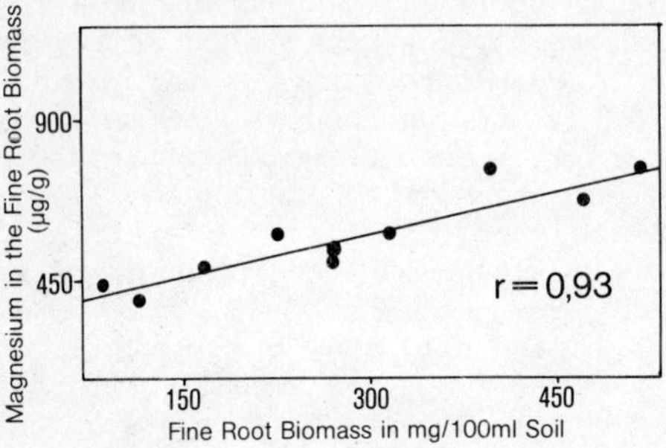

Fig. 5

In contrast no such relations can be shown for roots from the mineral soil suggesting that only in the O-layer intensified root growth leads to a corresponding increase of Mg-uptake. Recently laboratory experiments of Nätscher and Süsser (15) testing the buffering reactions of the organic layer at Oberwarmensteinach give evidence that protons at a low pH mainly exchange Magnesium and Calcium. If cation leaching from the canopy leads to a release of protons from the roots, as it was said before, the availability of both elements should increase where root growth in the organic layer is the highest, like in Oberwarmensteinach. The negative correlation between fine root biomass and P-contents in the roots (Fig. 6) coming from the mineral soil again emphasizes the role of this element as an indicator for the proton input and its influence on root growth in soils with low buffering rate.

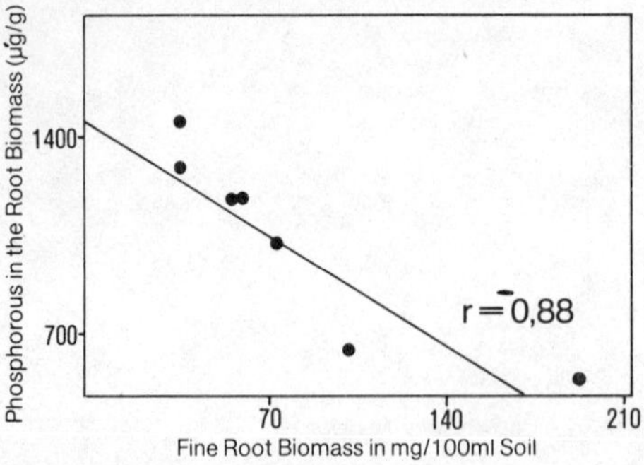

Fig. 6

The positive correlation between fine root biomass of the organic layer and the Mg-contents in the needles in early spring (Fig. 7) supports the assumption that nutrient support of spruce mainly depends on nutrient uptake from the humus layer.

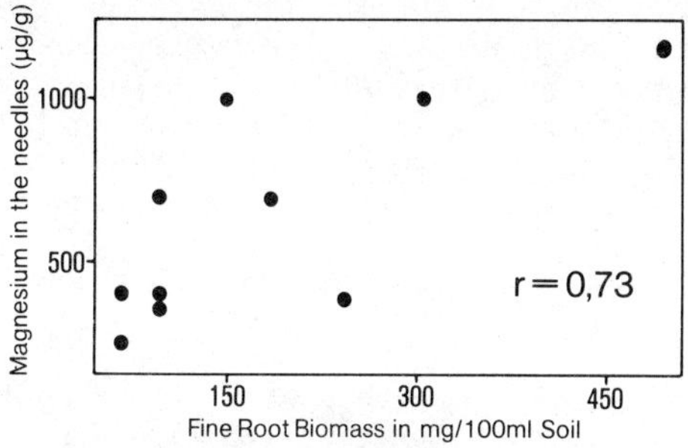

Fig. 7

3.4 Effects of Fertilization on Root Growth

In May 1983 at Oberwarmensteinach, Wülfersreuth, and Selb a fertilizer experiment was installed testing the effect of several Mg-fertilizers on revitalization of spruce. In April 1985 root samples were taken from the non-fertilized plot and those fertilized with $MgSO_4$ (1000 kg/ha) and $CaCO_3$ + MgO (10000 kg/ha) to study the influence of fertilization on root growth (Fig. 8).

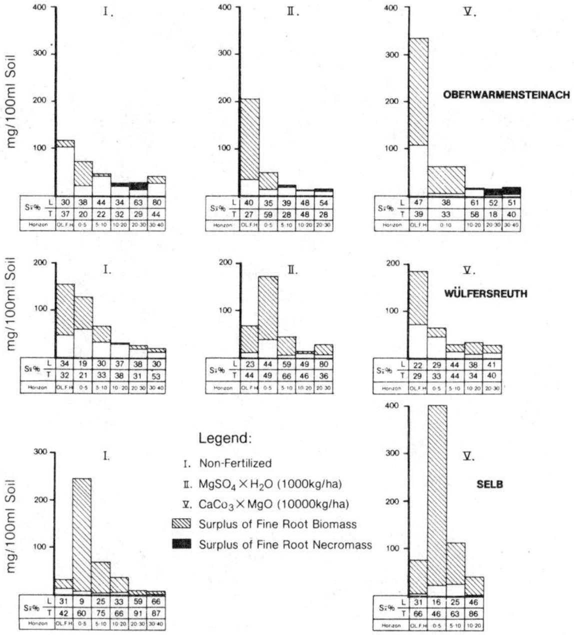

Fig. 8

At Oberwarmensteinach the fine root biomass shows an increase of 200-400% after fertilization. This effect was restricted to the humus layer suggesting that the chemical soil properties could not affect a significant ingrowth of roots into the soil. The tendency of improved root growth in the soil at the limed plot has still to be examined by means of a second sampling.

In Wülfersreuth, with respect to high variations no differences between treated and non-treated plots can be observed. Trees at this site are sufficiently supported with Magnesium as well as with Potassium and Calcium. In contrast to Oberwarmensteinach the better nutrient availability at this site does probably not necessitate the growth of more roots.

In Selb the high base saturation mainly consisting of Magnesium and Calcium seems not to limit root growth because of chemical soil properties. However, at this site the nutrient analysis of needles indicates an insufficient support with Potassium which may be due to an antagonism between this element and Magnesium/Calcium. After fertilization with $CaCO_3$ + MgO a sharp increase of fine root biomass can be measured. The reinforcement of the antagonism may explain this effect. In contrast to Oberwarmensteinach the potential soil properties enable the tree to grow more roots maintaining sufficient K-support. At both sites the state of deficiency seems to cause root growth. In case of soil chemical restrictions, growth can only be intensified in the organic layer.

REFERENCES

(1) MENGEL, K., LUTZ, H.-J. und BREININGER, M.TH. (1987). Auswaschung von Nährstoffen durch sauren Nebel aus jungen intakten Fichten (Picea abies). Z. Pflanzenernähr. Bodenk. 150 61-128

(2) KAUPENJOHANN, M., HANTSCHEL, R. (1987). Die kurzfristige pH-Pufferung gestörter und ungestörter Waldbodenproben. Z. Pflanzenernähr. Bodenk. 150 (in press)

(3) ULRICH, B. (1983). Interaction of forest canopies with atmospheric constituents: SO , alkali and earth alkali cations and chloride. In ULRICH, B. and PANKRATH, J. (eds.). Effects of accumulation of air pollutants in forest ecosystems. Reidel Publishing Company

(4) KAUPENJOHANN, M., SCHNEIDER, B.U., HANTSCHEL, R., ZECH, W. und HORN, R. (1987). Sulfuric acid rain treatment of Picea abies: Effects on nutrient solution and throughfall chemistry as well as on spruce nutrition. Z. Pflanzenernähr. Bodenk. (submitted)

(5) FLÜCKINGER, W., (1987). Untersuchungen über den Ernährungszustand und Ernährungsmaßnahmen in festen Beobachtungsflächen der Schweiz. In: Symposiumsberichte Oktober 1986, Wien, Univ. f. Bodenkultur(in press)

(6) STIENEN, H., (1985). Struktur und Funktion von Feinwurzeln gesunder und erkrankter Fichten (Picea abies(L.)Karst.) unter Wald und Kulturbedingungen. Dissertation Fachbereich Biologie, Univ. Hamburg

(7) ZECH, W., POPP, E. (1983). Magnesiummangel, einer der Gründe für das Fichten- und Tannensterben in NO-Bayern. Forstw. Cbl. 102, 50-55

(8) MURACH, D. (1983). Die Reaktion von Fichtenwurzeln auf zunehmende Bodenversauerung. AFZ 38 (26/27) , 683-686

(9) SCHRAMEL, P., WOLF, A., SEIF, R., KLOSE, B.-J. (1980). Eine neue Apparatur zur Druckveraschung von biologischem Material. Fresenius Z. Anal. Chem. 302, 62-64

(10) MEYER, J., SCHNEIDER, B.U., WERK, K., OREN, R., SCHULZE, E.-D.(1987) Root tip and ectomycorrhiza development and their relation to above ground and soil nutrients. (sumitted)
(11) MENGEL, K., (1979). Ernährung und Stoffwechsel der Pflanze. 5. Auflage, S. 366
(12) KAUPENJOHANN, M., HANTSCHEL, R., HORN, R., ZECH, W. (1987). Ergebnisse von Düngungsversuchen mit Magnesium an vermutlich immissionsgeschädigten Fichten (Picea abies(L.)Karst.) im Fichtelgebirge. Forstw. Cbl. 106 (in press)
(13) HANTSCHEL, R., KAUPENJOHANN, M., HORN, R., ZECH, W. (1986). Kationenkonzentrationen in der Gleichgewichts- und Perkolationsbodenlösung (GBL und PBL) - ein Methodenvergleich. Z. Pflanzenernähr. Bodenk. 149, 136-139
(14) ZECH; W., SUTTNER, T., KOTSCHENREUTHER, R. (1983). Mineralstoffversorgung vermutlich immissionsgeschädigter Bäume in NO-Bayern. Kali-Briefe (Büntehof) 16, 565-571
(15) NÄTSCHER, L., SÜSSER, P., SCHWERTMANN, U. (1987). Art, Menge und und Wirkungsweise von Puffersystemen in Böden a) in Auflagehorizonten und b) im Mineralboden. In: Ursachenforschung zu Waldschäden; Kurzfassung der Poster. S. 93-96

FUNCTION OF FOREST ECOSYSTEMS UNDER THE STRESS OF AIR POLLUTION

W. GRODZIŃSKI
Department of Ecosystem Studies, Jagiellonian University, and Institute of Fresh Water Biology, Polish Academy of Sciences
Karasia 6, 30-060 Kraków, Poland

Summary

A team project was developed to study the function of large forest ecosystems under the stress of air pollution in southern Poland, and to propose general methods of forest management in industrial regions. The study area includes a complex of lowland pine and deciduous forest /Niepołomice Forest - 11,000 ha/ situated 20-40 km east of Cracow. For the last 30 years the forest has been exposed to long-term, moderate level air pollution, chiefly SO_2, heavy metals and fluorine, emitted by the coal combustion industry in the region. The woodland habitat was already heavily contaminated by sulfur, and trace metals, and was also acidified.

The forest was studied as an input-output system; the input was measured as wet and dry atmospheric deposition, whereas the output was measured through the entire woodland watershed. Primary and secondary production and decomposition were evaluated in detail. Different behaviour of energy, nutrients /N, P, K, Ca and S/ and pollutants /S, Cd, Pb, Zn, Ni, Cr and Cu/ was observed and described. The field studies and a simulation model revealed a reduction of the potential photosynthetic activity of pine needles by 13-18%, and of the timber increment by at least 20%, and also an inhibition of litter decomposition by about 20-25% due to the possible inhibition of microbial activity. Consequently, the main disturbances of forest ecosystems in industrial regions concern such basic functions as: cycling of nutrients, primary production, and changes in top soil.

The following practical recommendations were established and are currently in use: /1/ to gradually rebuild the pollution-sensitive pine into a more resistant mixed forest, /2/ to fertilize the pine carefully from the air, /3/ to raise the water-table by amelliorations in order to provide more available nutrients for the roots, /4/ to optimize the wildlife management, and to reduce tourist pressure. However, without controlling industrial emissions, the above ecological recommendations are less useful.

A book originated from the project was published recently /Forest Ecosystems in Industrial Regions, Eds W. Grodziński, J. Weiner, P.F. Maycock, Springer-Verlag, Berlin-Heidelberg-New York-Tokyo, XVII+1-277 pp., 116 figs, 98 tab./.

CARBON ASSIMILATION AND NUTRITION OF TREES IN TWO PICEA ABIES (L.) KARST. STANDS IN BAVARIA EXHIBITING APPARENT DECLINE DIFFERENCES

R. OREN[1], E.-D. SCHULZE, U. SCHNEIDER, K. S. WERK and J. MEYER
[1]Lehrstuhl für Pflanzenökologie, Universität Bayreuth,
Postfach 10 12 51, 8580 Bayreuth, FRG

Summary

A significantly lower growth was found in a Norway spruce stand where some of the trees are becoming yellow, in comparison to a similar stand composed of green trees. The growth loss can only be explained by lower photosynthesis of yellow foliage because the green foliage in both stands photosynthesized similarly, and no difference in production below ground was detected. The yellowing of the foliage is a result of lower Mg^{++} uptake and, thus, lower Mg^{++} concentration in the foliage of the trees in the declining stand, including the foliage green trees. Correction of the leaf area index of the declining site, to account for the reduced productivity of the canopy, will result in a production which is appropriate for this effective leaf area index.

Norway spruce (Picea abies (L.) Karst.) trees in the Fichtelgebirge, North-East Bavaria, become yellow, loose needles and subsequently die. This phenomenon, first reported in 1978 (1), continues to spread. In order to gain some clues as to what are the causes of decline, we randomly positioned five plots in each of the following two populations of trees; one population showing clear signs of decline while the other appears healthy (see Table 1 for site description).

The stands are located such that climatic differences are minimal and their soil was developed from the same parent material, although the soil on the site with the declining stand is more podsolized (2). Based on the above similarities and the similar mean leaf area index in the two stands, we expected that growth and the production efficiency of the foliage would be similar (3). Our study results, however, do not support this hypothesis; the declining site had significantly lower stem wood growth than the healthy site (693 vs. 1073 g m^{-2} ground area yr^{-1}, $p=0.043$), because the production efficiency of its foliage was lower (58 vs 101 g wood m^{-2} leaf area yr^{-1}, $p=0.010$).

At a given competition level, as indicated by leaf area index, reduction in growth may be caused by lower photosynthesis or higher allocation to root growth. We find no differences between the stands in allocating carbohydrates to root growth (4). Lower photosynthesis may result directly from disruption of the structure and function of the photosynthetic apparatus, or indirectly, for example, from reduced availability of nutrients to maintain the apparatus. If either one or both processes are continuous (i.e. not a threshold type of response), we would expect apparently healthy (i.e. green) trees in the two stands to be different. We decided to examine only apparently healthy individuals in the two stands because dying trees are different from healthy trees in many ways which are not related to the causes of mortality. Our aim in this study was to identify a few key variables to separate healthy from declining trees before differences become apparent, and to test those variables in subsequent studies.

Tabel 1: Site and stand descriptions of Picea abies (L.) Karst. stands located near the villages of Oberwarmensteinach (apparently declining) and Wülfersreuth (apparently healthy) in the Fichtelgebirge, Bavaria, FRG (1 SE is given when appropriate, n=5).

site parameter	Oberwarmensteinach	Wülfersreuth
elevation (m)	755	675
aspect	SE	NW
slope (degree)	5	1
soil type	podsol	podsol brown earth
soil pH range	2.9 - 4.4	2.8 - 4.4
throughfall (mm) 7/854-6/85	1,424	1,106
input 1/7/84-30/6/85		
SO_4 (kmol ha^{-1} yr^{-1})	1.43	0.90
H^+ (kmol ha^{-1} yr^{-1})	1.24	0.75
stand parameter		
stand age (years)	30 (2)	30 (1)
stand density (trees m^{-2})	0.44 (0.14)	0.31 (0.02)
stand height (m)	7.93 (0.90)	10.04
(0.51)		
leaf area index ($m^2 m^{-2}$ ground area)	11.9 (1.1)	10.6 (0.6)
quadratic mean diameter (mm at 1.3 m above ground)	97 (11)	115 (4)

Assimilation rates under standard conditions of needles of six age classes did not differ between the stands, nor had stomatal behaviour of trees in the declining stand become less controlled with age, as could be expected of trees that are continuously exposed to directly damaging acid precipitation or air pollution (5).

The concentrations of Mg^{++} and Ca^{++} in the foliage presented the most apparent nutritional differences between the stands (Mg^{++}: 17 vs. 39, $p<0.01$, Ca^{++}: 106 vs. 212, $p<0.05$, µmol g^{-1} dry weight for the declining and healthy stand respectively). The foliage in the healthy stand had a Mg^{++} concentration above the level where growth is limited substantialy (37 µmol g^{-1} dry weight, see 6), while that in the declining site had a concentration just above the level of strong deficiency (8.2 µmol g^{-1} dry weight, see 6). The limitation appears to result at least in part from differences in the uptake as can be seen from the correlation (on a plot level) between concentration in xylem sap and concentration in one-year old foliage (Fig. 1). Differences in the uptake between the stands could not be attributed to differences in transpiration (2). The differences probably resulted from both higher concentrations of Mg^{++} and Ca^{++} in the soil solution, and from better distribution of roots throughout the soil profile and higher degree of mycorrhization in the healthy site (7).

Fig. 1: Mg^{++} and Ca^{++} concentration in one-year-old needles appear to be associated with the concentration of these elements in the xylem sap of Picea abies (L.) Karst. trees in 10 plots in the Fichtelgebirge (circles: Mg^{++}, triangles: Ca^{++}, filled symbols - Wülfersreuth, open symbols - Oberwarmensteinach).

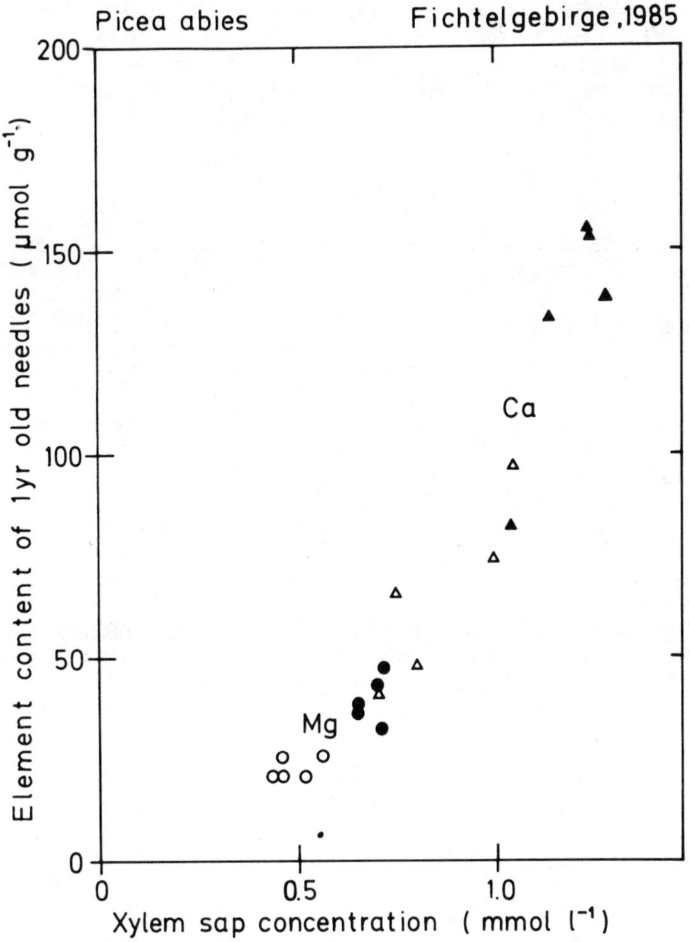

As a result of low Mg^{++} supply to trees in the declining stand, the concentration of Mg^{++} in the needles of certain proportion of the trees is below the level which has been shown to cause yellowing (6). Yellow needles have lower photosynthetic rates (8). Therefore, although the stands have similar canopy leaf area and, thus, similar competition level for light, the production efficiency of that leaf area is not the same in the two sites. The leaf area of the declining stand should be reduced by a factor combining the proportion of yellow leaf area and its reduction in photosynthetic rate. This adjustment would explain the differences in growth between the sites.

As some of the trees in the stand will die, competition for the limited Mg^{++} will ease and trees which are currently green and have Mg^{++} concentrations just above deficiency may escape mortality for a while. The result will be a stand of apparently healthy trees which is more open than the present stand. However, if Mg^{++} availability will become lower through the process of soil leaching, the stand will repeat the cycle, ending at a still lower leaf area and production. The decline caused by Mg^{++} limitation could also be enhanced by the ample nitrogen supply which may stimulate trees to grow to the verge of deficieny and permit no buffer against a further reduction of Mg^{++} availability (9).

REFERENCES

(1) ZECH, W., and POPP, E. (1983). Magnesiummangel, einer der Gründe für das Fichten- und Tannensterben in NO-Bayern. Forstwiss. Cbl. 102, 50-55.

(2) SCHULZE, E.-D., OREN, R., WERK, K. S., MEYER, J. and ZIMMERMANN, R. (1986). Kohlenstoff-, Wasser- und Nährstoffhaushalt von Fichten stark belasteter Hochlagenstandorte auf Phyllit in NO-Bayern. In: F. Führ, S. Ganser, G. Kloster, B. Prinz, E. Stüttgen (eds.) Wirkungen von Luftverunreinigungen auf Waldbäume und Waldböden. BMFT Statusseminar 1985, Jül-Spez-369, 106-117.

(3) OREN, R., WARING, R.H. STAFFORD, S. and BARRATT, J. (1987). Twenty-four years of Ponderosa pine stand growth in relation to canopy leaf area and understory competition. Forest Science (in press).

(4) OREN, R., WERK, K.S., MEYER, J. and SCHULZE, E.-D. (1987). Performance of _Picea abies_ (L.) Karst. at different stages of decline: I. Carbon relations. Oecologia (in press).

(5) ZIMMERMANN, R., SCHULZE, E.-D., OREN, R. and WERK, K.S. (1987). Photosynthesis and transpiration of green trees with different sulphate and proton input in Northern Bavaria, FRG. Proceeding of the 1986 COST-Workshop in Sweden on "Direct effects of air pollution on forest trees". Commission of European Communities, (in press).

(6) Ingestad, T. (1959). Studies on the nutrition of forest tree seedlings: II. Mineral nutrition of spruce. Physiologia Plantarum 12, 568-593.

(7) MEYER. J., OREN, R., WERK, K.S. and SCHULZE, E.-D. (1986). The effect of acid rain on forest tree roots. Proceedings of the Jülich COST Workshop. Comission of the European Communities 95-107.

(8) LANGE, O.L., GEBEL, J., ZELLNER, H. and SCHRAMEL, P. (1985). Photosynthesekapazität und Magnesiumgehalte verschiedener Nadeljahrgänge bei der Fichte in Waldschadensgebieten des Fichtelgebirges. In: Berichtband des Statusseminars "Wirkungen von Luftverunreinigungen auf Waldbäume und Waldböden". BMFT Statusseminar 1985, Jül-Spez-369, 106-117.

(9) SCHULZE, E.-D. (1987). Proceedings of the European Symposium on the "Effects of air pollution on terrestrial and aquatic ecosystems", Grenoble, 18-22 May, 1987 (this issue).

FOREST ECOSYSTEMS RESEARCH NETWORK

FERN

Anne Teller

European Science Foundation

Summary

FERN was launched in February 1986 by the European Science Foundation (ESF). The project is to be seen essentially as a coordination and concertation of national research activities already in hand or planned for the near future. It will establish a network of European research groups in the field of forest ecology. An inventory of all participating research teams is in press. The organization of topical workshops results in recommendations aimed at improving the coherence of the overall research carried out in Europe as well as harmonizing the techniques and at establishing common objectives for future investigations.

1. INTRODUCTION

ESF is carrying out a European-wide 5-year research project in the field of ecology called Forest Ecosystems Research Network (FERN) as an Additional Activity. The project addresses the basic scientific problems associated with the stability and destabilization of forest ecosystems with special emphasis on Europe. It is hoped that the project will also attract the cooperation of research groups in Eastern European countries, not affiliated with ESF, in view of the common nature of the problems to be investigated.

2. OBJECTIVES

The ultimate objective of FERN is to strengthen the scientific understanding of forest ecosystems in Europe in order to assess the true meaning of current changes and to predict the fate of these ecosystems on a mid- to long-term basis.

The immediate aim of FERN is to contribute to the elucidation of some of the key patterns and processes of forest ecosystems and to identify reliable indicators of forest ecosystem change (particularly in the context of forest decline).

Its aim is also to introduce an element of coherence into relevant scientific community in Europe by facilitating communication.

3. MANAGEMENT

Five Working Groups have been selected to launch coordinated activities. For each topic, a group of three persons (i.e., a pilot, co-pilot and member of the Steering Committee) was appointed.

1. Retrospective study of man-induced changes in European forest ecosystems - P. Piussi (Italy), E. Dahl (Norway), C.O. Tamm (Sweden).

2. Influence of fire and grazing on stability of Mediterranean forest ecosystems - P. Quézel (France), A. Escarre (Spain), Ph. Bourdeau (Belgium).

3. Changes in nitrogen status of European forests - H. Miller (Great Britain), F. Beese (Germany), H.W. Zöttl (Germany).

4. Soil-litter compartment - P. Ineson (Great Britain), I. Kottke (Germany), M. Ciric (Yugoslavia).

5. Architectural patterns in European forest ecosystems - R.A.A. Oldeman (Netherlands), M. Leikola (Finland), W.E.S. Mutch (Great Britain).

4. IMPLEMENTATION

FERN works through the organization of topical workshops, state-of-the-art reviews and the setting-up of collaborative research projects. The exchange of scientists for limited periods of time will be another important item in carrying out the project.

5. CALENDAR OF ACTIVITES

1. Workshop on "Field Methods in Terrestrial Ecosystem Nutrient Cycling" in collaboration with NERC (UK), Grange-over-Sands, 3-5 December 1986. The proceedings will be published by Elsevier. Among the principal conclusions of the Workshop, participants recommended the establishment by FERN of an international network of sites with basic variations in nutrient input and consitent management. As a first stage, a register of existing long term experiments would be computerized. Dr. Ineson, from ITE, Cumbria, is responsible for this inventory.

2. Workshop on the "Influence of fire on the stability of Mediterranean forest ecosystems", Giens, 23-26 March 1987. The proceedings of the Workshop are in press. In conclusion of the Workshop, it was recommended to initiate a European research project on experimental fires and ecological implications. Prof. Quézel, from the University of Marseille, is responsible for this project.

3. Review of forest modelling facilities in Europe, Edinburgh, autumn 1987, in order to set up a network with electronic communication to serve forest modelling in Europe. Dr. Mutch, from the University of Edinburgh, will coordinate the activity.

4. Prof. Oldeman, from the University of Wageningen, together with Dr. Walter, from the University of Strasbourg is organizing a Workshop in Strasbourg at the end of 1987, or early 1988 on the "Unification of methods for European forest pattern research".

5. Workshop on the "Use of ^{15}N methods in nutrient cycling experiments" is scheduled for early 1988, at I.A.E.A. in Vienna. The Workshop would include practical demonstrations for the research workers interested in using these techniques in ecological studies. Dr. Ineson is responsible for its organization.

6. State-of-the-art review on "Changes in nitrogen status in European forest ecosystems", Aberdeen, April 1988. The emphasis would be on the comparison and evaluation of methodologies to seek a possible standardization of methods at European level. Prof. Miller, from the University of Aberdeen, is organizing the meeting.

7. Workshop on "Ecological history of European forest ecosystems", Italy, Autumn 1988. Research workers who are willing to participate or to present a local or a regional synthesis of a case study as a communication or a poster are kindly requested to contact the Scientific Secretary or Professor Piussi of the University of Florence as soon as possible.

6. PARTICIPATION

About 200 research teams located in 17 European countries (Austria, Belgium, Denmark, Finland, France, Germany, Great Britain, Italy, Ireland, Netherlands, Norway, Portugal, Spain, Sweden, Switzerland, Turkey, Yugoslavia) have officially joined the project. An inventory of all participating organizations is in press. It is planned to publish FERN NEWSLETTER to keep all participants rapidly informed of the progress of the project.

For more information about FERN or related activities, please contact Anne Teller, Scientific Secretary, Commission of the European Communities, 200 rue de la Loi, B-1049 Brussels.

THEME V : EFFECTS OF AIR POLLUTION ON AQUATIC ECOSYSTEMS

THEME V : EFFETS DE LA POLLUTION DE L'AIR SUR LES ECOSYSTEMES AQUATIQUES

THEMA V : WIRKUNGEN DER LUFTVERSCHMUTZUNG AUF AQUATISCHE ÖKOSYSTEME

HISTORICAL DEVELOPMENT AND EXTENT OF ACIDIFICATION OF SHALLOW SOFT WATERS IN THE NETHERLANDS

G.H.P. ARTS
Laboratory of Aquatic Ecology, Catholic University, Toernooiveld, 6525 ED Nijmegen (The Netherlands)

SUMMARY

On the pleistocene sandy soils in the eastern and southern part of The Netherlands and in the coastal dune area 563 sites are known to be originally characterized by the occurrence of the isoetids *Littorella uniflora* (L.)Aschers., *Lobelia dortmanna* L., *Isoetes lacustris* L. and/or *Isoetes echinospora* Durieu. In the period 1983 - 1985 a sample survey of 147 still existing waters indicated that 55 % of them now were either characterized by the submerged peat mosses *Sphagnum cuspidatum* Hoffm. and/or *Sphagnum denticulatum* Bridel Brideri or by the complete absence of submerged macrophytes. Taking into account the autecology of aquatic macrophytes as well as old and recent pH and alkalinity data, the observed changes must be attributed to the effects of acidification.
Depending on the criteria applied (decrease of alkalinity or deterioration of Littorellion - communities) acidification of poorly buffered waters in The Netherlands has started around 1900 or 1925.
On average the pH in acidified waters decreased from 5.5 in the period 1900 - 1950 to 4.4 in the period 1981 - 1986. Recently, pH-values down to 3.5 were recorded.

1. INTRODUCTION

Small shallow pools on an originally sandy substrate are very characteristic of the pleistocene regions in The Netherlands. These regions enclose the eastern, central and southern parts of the country. Situated on sandy soils poor in lime and nutrients, the waters originally reflected low alkalinity and low ionic contents. Being largely dependent on rain water they have fluctuating water levels and can run dry during a shorter or longer period of the year. The original characteristic aquatic vegetation, which can cover a substantial part of the sediment, is geographically restricted to the atlantic and subatlantic areas of Western Europe. Besides a distinctive macrophyte species composition, *Littorella uniflora* (L.)Aschers., *Lobelia dortmanna* L., *Isoetes lacustris* L. and *Isoetes echinospora* Durieu ("Littorellion"-species) often dominated the vegetation. These isoetids are also dominant in other temperate soft-water lakes (5).

In view of the described properties soft waters are very sensitive to water and air pollution. Many of them are becoming increasingly acidic as a result of atmospheric deposition of acidifying components (2,7,10,13).

Until now, the quantitative importance of acidification in the observed decline of macrophytes communities in atlantic soft waters was not fully known. This gap may only be filled by historical studies, which offer a frame of reference.

This paper firstly presents the extent of acidification of soft waters in The Netherlands. These waters are representative of the West European atlantic type of soft water. Secondly it describes a history of acidification of these soft waters. The extent as well as the described timetrend of acidification are based on data concerning pH, alkalinity and the presence of aquatic macrophytes.

2. MATERIALS AND METHODS

In order to trace waters, in which the isoetids *Littorella uniflora*, *Lobelia dortmanna*, *Isoetes lacustris* and/or *Isoetes echinospora* grow or have been growing in this century, distribution maps, herbaria, literature and archives were studied. In the period 1983 - 1985 147 still existing waters, which could be exactly localized, were visited in summer. The aquatic vegetation was thoroughly investigated from a boat with the help of a rake and by exploring the shore. On each location the pH of the water layer was measured with a Metrohm model E 488 pH-meter and a model EA 152 combined electrode. Alkalinity was estimated by titration of 100 ml water with 0.01 N HCl solution down to pH 4.2. Besides this field work, physico-chemical data and data concerning the macrophyte species composition were collected from literature and the archive of the Research Institute for Nature Management, the archives of the State Forest Service, the archive of the Dutch Society for Conservation of Nature, the Rijksherbarium and personal archives of botanists.

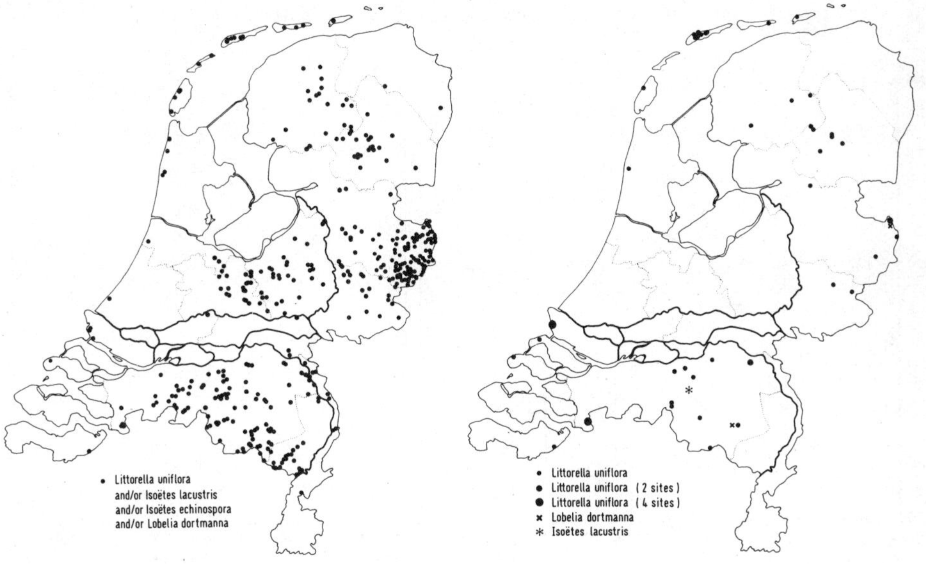

Fig. 1. Sites where isoetids have been recorded ever since 1834 (left) and in the period 1983-1986 (right).

3. RESULTS

Fig. 1 presents the sites, where the isoetids *Littorella uniflora*, *Lobelia dortmanna*, *Isoetes lacustris* and/or *Isoetes echinospora* have been recorded ever since 1834. Most records go back to the period 1900 - 1960. In the period 1983 - 1986 isoetids have only been found in 52 localities, most of them being sites with *Littorella uniflora* (Fig. 1). As compared with the original number of 563, this is less than 10 %.

The sample survey of 147 still existing waters in the period 1983 - 1985 indicated that in about half of them the submerged peat mosses Sphagnum cuspidatum Hoffm. and/or Sphagnum denticulatum Bridel Brideri were very characteristic. Being often present in submerged mats covering sediment and filling the water layer, they are frequently accompanied by submerged growing Juncus bulbosus L. (predominantly forma fluitans). Field observations have shown that the considered peat mosses seem to be restraint in their submerged distribution to pH values below 6.0 (Sphagnum denticulatum) and 5.0 (Sphagnum cuspidatum) and alkalinity-values below 0.09 (n=53 and n=56, respectively). In some waters submerged macrophytes were completely absent, while pH-values were estimated below 5.4.

Table I. Classification of aquatic macrophytes, characteristic of poorly buffered waters, with respect to pH and alkalinity of the water layer based on distribution data (1).

I	n=9	Species occurring in soft, slightly acid waters. The optimum alkalinity is below 1 meq/l and the optimum pH between 5.0 and 6.0. During acidification these aquatic plants can temporarily stand pH-values around 4.0. This group includes most isoetid water plants.
II	n=17	Species growing in soft or moderately hard, slightly acid or circumneutral waters. Usually they do not occur in waters with a pH below 5.0. The optimum alkalinity is below 2 meq/l.

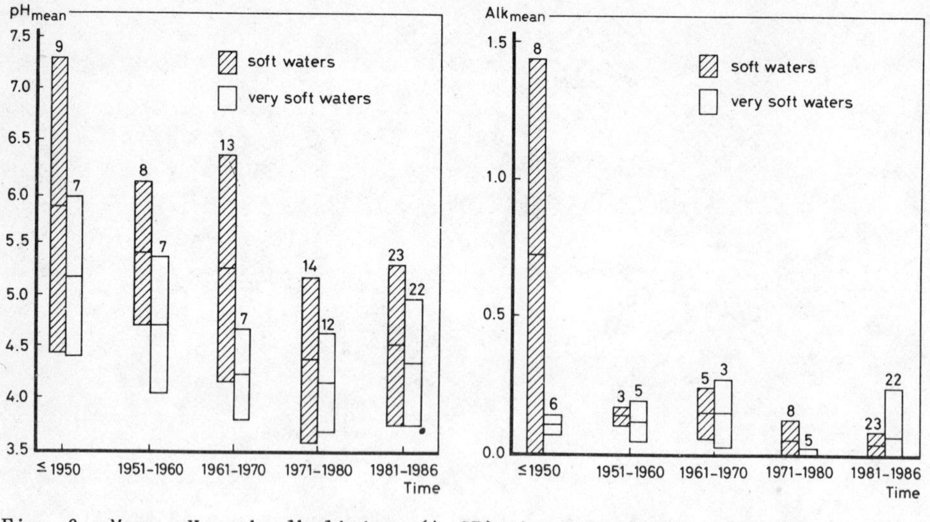

Fig. 2. Mean pH and alkalinity (± SD) in soft and very soft waters in different periods. The number of waters is indicated above the bars.

The waters as defined above (55 % of the sampling sites) can be divided into two types on the basis of historical data of aquatic macrophytes (i.e. submerged growing plants and plants with floating leaves), characteristic of poorly buffered waters. One type of water was inhabited by species belonging to group I (Table I). These waters could be characterized as species-poor communities of isoetid water plants. In a second group of waters species classified in group I as well as species belonging to group II were growing. Originally these waters were richer in species. Both types of water also differed in pH and alkalinity (Fig. 2). Based on the oldest data of pH and alkalinity, the former type of water can be characterized as "very soft", while the second type of water can be considered as "soft" water.

Species belonging to group I disappeared earlier from the very soft waters than from the soft waters. In the soft waters this has been preceded by the disappearance of species classified in group II.

In soft waters the sharpest fall in pH has occurred in the period 1971 - 1980, when compared with the preceding period (Fig. 2). Alkalinity has decreased earlier and was already as low as 0.15 in the period 1951 - 1960. In very soft waters pH constantly decreased until the period 1971 - 1980. A sharp decrease in alkalinity could not be demonstrated but may have occurred in the beginning of this century. On average pH in acidified waters decreased from 5.5 in the period 1900 - 1950 to 4.4 in the period 1981 - 1986.

Deterioration of communities of aquatic macrophytes in now acidified waters already started in 1925 (Fig. 3).

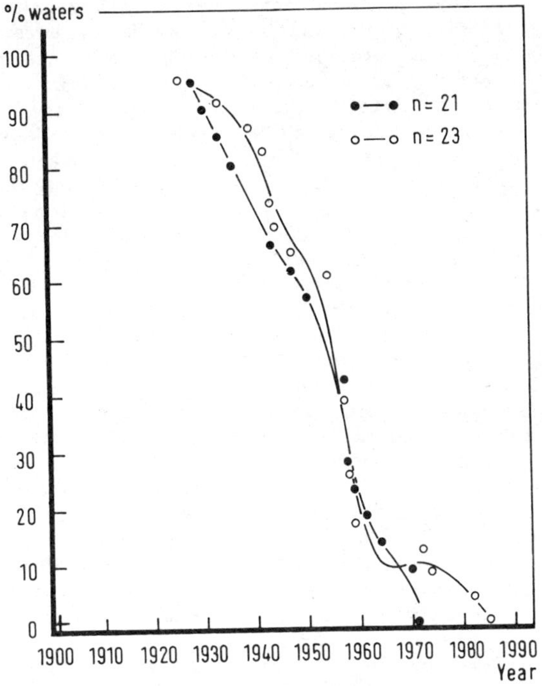

Fig. 3. Percentage of now acidified waters, in which communities of "Littorellion"-species are well developed, in relation to time.
soft waters o—o very soft waters •—•

4. DISCUSSION

The conditions in waters in former years have been described by means of old chemical data (Fig. 2) and macrophyte species indicating water quality (Table I). Waters, once characterized by species classified in group II (Table I), were originally slightly acid to circumneutral on the basis of the autecology of aquatic macrophytes (Table I) as well as old pH and alkalinity data (Fig. 2). On the other hand, characterization of waters with pure isoetid communities meets with difficulties. Isoetids, having their optimum in poorly buffered waters, are nowadays not always indicative of this type of water, because they can temporarily remain in acidifying and acidified waters. They are also recorded for acidic lakes in areas outside The Netherlands, e.g. the Adirondack region (U.S.A.) (9). If other studies are taken into account, waters in which isoetids (Littorella uniflora, Lobelia dortmanna, Isoetes lacustris, Isoetes echinospora) grow vary from slightly acid (1,4,6, this study) to circumneutral (8). Mean alkalinity in these waters is ranging from 0.2 to 0.7 meq/l with maximum values over 1 meq/l (1,4,12, this study). As distinct from the other references the data reported by Pietsch (8) are not limited to atlantic West European waters, but include all types of waters in Europe inhabited by the isoetids concerned. The actual occurrence of isoetids in acidified waters may have lowered the presented mean values (1).

This study demonstrates that the recorded submerged growing peat mosses are found in slightly acid and acid waters with low alkalinity (maximum alkalinity 0.08 meq/l). Consequently, the considered peat mosses grow in waters in which the chemical conditions are very distinct from those in waters inhabited by the "Littorellion"-species. They prove te be very good indicators for acidified originally low alkaline waters. The presence of large beds of Sphagnum in acidified waters has also been reported elsewhere (2,3). Taking into account the autecology of aquatic macrophytes and old and recent pH and alkalinity data, the observed changes in low alkaline waters must be attributed to the effects of acidification. In conclusion it can be stated that in The Netherlands 55 % of these waters have been acidified. As most acidified waters are hydrologically isolated and mainly fed by rain water, acidification may be attributed to the impact of atmospheric pollutants. Besides SOx and NOx, NHx as airborne acidifying agent plays an important role in The Netherlands (11).

This paper demonstrates that in atlantic to subatlantic West European soft and very soft waters historical changes in aquatic macrophyte flora are drastic and can be described in terms of species disappearing from waters in different phases of acidification (Table I). These considerable changes due to acidification may be closely connected with the extreme low pH-values recorded for these waters (mean pH 4.4 in the period 1981 - 1986; minimum value 3.5). In lakes in northern Europe and North America pH-values below 4.5 are rarely measured (14).

Acidification of atlantic soft and very soft waters has already been going on since the beginning of this century. Alkalinity already decreased in the first half of this century. Deterioration of aquatic macrophytes communities in now acidified waters has been observed since 1925. Very soft waters have been acidified earlier than soft waters. Apart from labour-intensive palaeolimnological studies an analysis of historical data originating from literature and archives has been proved a workable method for studying acidification in atlantic low alkaline waters.

ACKNOWLEDGEMENTS

A part of this study has been financed by the Ministry of Housing, Physical Planning and Environment, Directorate Air (project L1814054). The author is greatly indebted to Prof. Dr. C. den Hartog and Drs. J.A.A.R. Schuurkes for critically reading the manuscript. Mr. J.G.M. Roelofs stimulated discussions. Thanks are extended to the Graphics Department of the Faculty of Sciences for preparing the illustrations and to all other persons and authorities who have contributed in some way to this study.

REFERENCES

(1) De Lyon, M.J.H. and Roelofs, J.G.M. (1986). Aquatic plants in relation to water quality and sediment characteristics. Report Laboratory of

Aquatic Ecology, Catholic University Nijmegen to the Ministry of Agriculture and Fisheries. 106 pp. (part 1) and appendices (part 2). (in Dutch).

(2) Grahn, O. (1977). Macrophyte succession in Swedish lakes caused by deposition of airborne acid substances. Water, Air Soil Pollut. 7: 295-305.

(3) Hendrey, G.R. and Vertucci, R. (1980). Benthic plant communities in acid Lake Colden, New York. Sphagnum and the algal mat. In: D. Drabløs and A. Tollan (Editors), Proc. Int. Conf. Ecol. Impact Acid Precip. Norway. SNSF-Project 314-315.

(4) Heusden, G.H.P. van and Meijer, W. (1949). A chemical-botanical investigation of pools and moorland pools. Report Hugo de Vries Laboratory Amsterdam/Municipal Water Company Amsterdam. 50 pp. + appendices. (in Dutch).

(5) Hutchinson, G.E. (1975). A Treatise on Limnology. Limnological Botany Vol.3. Wiley, New York. 660 pp.

(6) Iversen, J. (1929). Studien über die pH-Verhältnisse dänischer Gewässer und ihren Einfluss auf die Hydrophytenvegetation. Bot. Tidskr. 40: 277-333.

(7) Last, F.T., Likens, G.E., Ulrich, B. and Walløe, L. (1980). Acid precipitation- progress and problems. Conference summary. In: D. Drabløs and A. Tollan (Editors), Proc. Int. Conf. Ecol. Impact Acid Precip. Norway. SNSF-Project 10-12.

(8) Pietsch, W. (1977). Beitrag zur Soziologie and Ökologie der europäischen Littorelletea- und Utricularietea-Gesellschaften. Feddes repert. Bot. Taxon. Geobot. 88: 141-245.

(9) Roberts, D.A., Singer, R. and Boylen, C.W. (1985). The submersed macrophyte communities of Adirondack Lakes (New York, U.S.A.) of varying degrees of pH. Aquat. Bot. 21: 219-235.

(10) Roelofs, J.G.M. (1983). Impact of acidification and eutrophication on macrophyte communities in soft waters in The Netherlands. I. Field observations. Aquat. Bot. 17: 139-155.

(11) Schuurkes, J.A.A.R. (1986). Atmospheric ammonium sulphate deposition and its role in the acidification and nitrogen enrichment of poorly buffered aquatic systems. Experientia 42: 351-357.

(12) Spence, D.H.N. (1967). Factors controlling the distribution of freshwater macrophytes with particular reference to the lochs of Scotland. J. Ecol. 55: 147-170.

(13) Wetzel, R.G. (1983). Limnology. 2nd Edn. Saunders College Publishing, Philadelphia. 860 pp.

(14) Wright, R.F., Conroy, N., Dickson, W.T., Harriman, R., Henriksen, A. and Schofield, C.L. (1980). Acidified lake districts of the world: a comparison of water chemistry of lakes in southern Norway, southern Sweden, southwestern Scotland, the Adirondack Mountains of New York and southeastern Ontario. In: D. Drabløs and A. Tollan (Editors), Proc. Int. Conf. Ecol. Impact Acid Precip. Norway. SNSF-Project 377-379.

ACIDIFICATION OF FRESHWATERS IN THE VOSGES MOUNTAINS (EASTERN FRANCE)

J.-C. MASSABUAU*, B. FRITZ** and B. BURTIN*
*Laboratoire d'Etude des Régulations Physiologiques, CNRS,
23 rue Becquerel, 67087 Strasbourg, France
**Centre de Sédimentologie et Géochimie de la Surface, CNRS,
1 rue Blessig, 67084 Strasbourg, France

Summary

The chemistry of surface waters has been studied in the Vosges mountains in relation with ecophysiology of aquatic animals. We demonstrate the existence of four acidified rivers and two sensitive lakes in the area of Cornimont-La Bresse. The chemistry of these waters is characterized by very low alkalinities and presence of strong acid anions (SO_4, NO_3, Cl) that dominate in the anionic charge. All the biotopes exhibit annual pH rhythms well correlated with the amount of rainfall. Following snow melt, minimum pH values of 4.7 were measured whereas towards the end of summer, pH can recover up to 6-7. In the four studied rivers, the trout *Salmo trutta fario* began disappearing 15-20 years ago. Today the fish population is completely lost.

1. INTRODUCTION

Acidification of freshwater is a well-known problem in Scandinavia, Canada, northeastern USA, southeastern Scotland and Central Europe (1, 2). Also in some restricted areas in Belgium, Denmark, Italy and the Netherlands some sites have been found (3). In France, the most sensitive area is the Vosges mountains where there are poorly buffered surface waters receiving acid rainfall with mean pH values around 4.5 (4). By the end of 1985, we had already reported the existence of a few river waters in the area of Cornimont that reached very low pH values following a rainy period (5). The present report confirm our observations by showing an atmospheric origin of the acidification in these rivers where the fish population is completely lost.

2. STUDIED AREA AND METHODS

The studied area is located at about 90 km south-west of Strasbourg on the western side of the Vosges mountains (Fig. 1). The elevation range is 600-1200 m o.m.s.l. and the climate is temperate oceanic mountainous. The mean annual precipitation is 1664 $\pm$ 342 mm at the weather center of Cornimont ($\bar{x}$ $\pm$ 1 standard deviation; minimum value 1030 mm, maximum value 2193 mm, Office National Météorologique). The area is generally covered by snow from December 15 to the end of April. It is nearly completely forested with conifers (spruce and white fir) and beeches that are affected by the recent forest decline. We have been studying rivers and lakes in this area since October 1985 (5). Values of pH were measured *in situ* and in the laboratory within 24 h. Water was sampled in polyethylene bottles and filtered either in the field or in the laboratory (Millipore 0.45 µm filter for all elements except aluminium, 0.02 µm filter). For technical reasons direct rain water collection for analysis was not feasible at Cornimont.

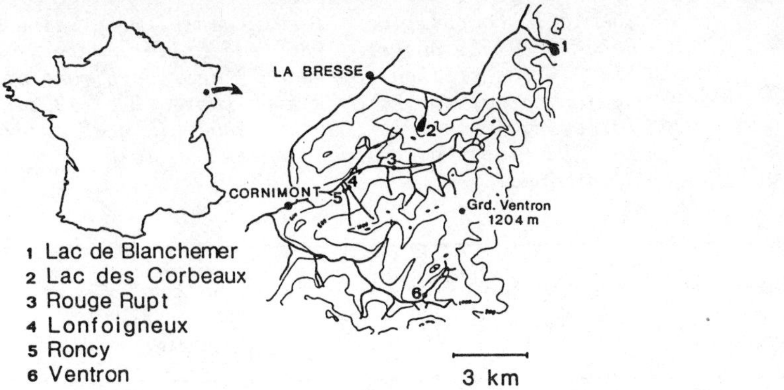

Fig. 1: Localization map of the studied sites in the Vosges mountains.

Rain water analysis corresponds to rainfall collected 35 km north-west in the Strengbach basin (7) for the period December 1985-August 1986. As results are not significantly different from those obtained throughout the Vosges mountains by Bourrié in 1973 (4) and in the Valley of Munster by Fritz *et al.* (6) in 1981-82, we think they must be reasonably representative of the rain at Cornimont. The major inorganic species are analyzed using mostly the classical techniques. Calcium, magnesium, sodium and potassium are determined by atomic absorption spectrophotometry; ammonium, chloride, nitrate, nitrite, sulfate by automatic colorimetry (Technicon apparatus). Alkalinity or acidity is determined by careful titration (near neutral solutions). Sampling cautions and particular schemes are detailed in Kremmp (8).

3. RESULTS

The mean chemical composition of acid rain water in the Strengbach basin for the period December 1985-August 1986 is presented in Table I. The solution is diluted (total dissolved salt, TDS = 6.1 $g \cdot L^{-1}$) and slightly acid by comparison to a meteoric water equilibrated with atmospheric CO_2 (pH 5.6). The mean pH value is 4.5 and the minimum 4.1. The anionic charge is dominated by strong acid anions (SO_4, NO_3 and Cl). Protons and NH_4 are

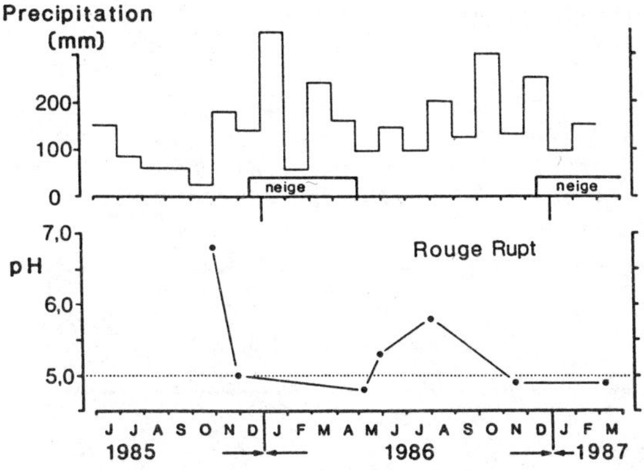

Fig. 2: Seasonal changes of precipitation at Cornimont and water pH in Le Rouge-Rupt river.

Table I

Chemical composition of rain water (mean value for December 1985-August 1986) and river waters (see Fig. 1 for station references) near Cornimont at the beginning of the rainy season (Nov. 29, 1985) and following snow melt (May 5, 1986). In µeq·L⁻¹ except for TDS (total dissolved salts) in mg·L⁻¹. nd, not determined.

Date	November 29, 1985			Dec.85-Aug. 86	May 9, 1986		
Station	3	4	5	Rain	3	4	5
pH	5.3	5.7	5.5	4.5	4.7	4.8	4.8
NH_4^+	< 1	< 1	< 1	23	< 1	< 1	< 1
Na^+	63	77	66	10	nd	nd	nd
K^+	4	5	5	7	nd	nd	nd
Mg^{++}	32	38	34	6	nd	nd	nd
Ca^{++}	68	76	76	22	52	nd	nd
H^+	6	4	3	31	19	17	15
Alkalinity + H^+	2	6	5	7	0	0	2
Cl^-	37	43	41	16	29	38	35
NO_3^-	28	43	29	22	26	37	14
SO_4^{--}	104	112	106	48	50	104	100
Σ cations	173	200	184	80	nd	nd	nd
Σ anions	172	204	181	68	nd	nd	nd
TDS	11.6	13.3	11.9	6.1	nd	nd	nd

the major cations. Figure 2 shows the effect of such acid rainfalls on the pH of Le Rouge-Rupt river. Precipitation changes are shown in parallel. Minimum pH values are clearly correlated with rainy season and snow melt. Conversely, pH goes up during summer when precipitation is minimum. It was following the exceptionally dry period of summer and autumn 1985 that we recorded the highest pH values (pH 6.8-7 on November 1, 1985, (5)). Table I presents the chemical composition of three rivers (Le Rouge-Rupt,

Table II

pH and aluminium concentration (µg·L⁻¹) in 4 streams and 1 lake at the beginning of snow melt (March 11, 1987, see Fig. 1 for station references).

Station	3	4	5	6	2 (surface)	2 (-9 m)
pH	4,9	5,2	5,3	5,3	6,0	6,3
Al	308	346	262	254	132	56

Lonfoigneux and Roncy) where the fish population is completely lost. Samples were taken three weeks after the onset of the rainy period in autumn 1985 (November 29), and one week after the main snow melt of May 1986 (May 9). These stream waters are slightly more concentrated than rain water but their alkalinity is extremely low. In particular, alkalinity is undetectable following snow melt when stream discharge is maximum. At that time pH reaches the minimum value of 4.7-4.8. Similarly, we also found low pH values in two other stations. In Le Ventron river where there are no fish anymore, pH reached 4.9 and in the lake Blanchemer it decreased to 5.5. The aluminium concentration in these waters is shown in Table II for the period of March 1987 (snow melt). In four rivers, the concentration is higher than 200 $\mu g \cdot L^{-1}$, which is the maximum acceptable value for drinking water (see below).

4. DISCUSSION

The environmental effect of acid precipitation has been clearly implicated in many cases of forest decline and in the acidification of freshwaters in North America and in Europe. In France the problem of forest decline was discovered some years ago. The present report demonstrates that the problem of acidified water also exists in the Vosges mountains.

In four rivers near Cornimont, Le Rouge-Rupt, Lonfoigneux, Roncy and Le Ventron the waters are acid and the fish population is lost. The analysis reveals the presence of the strong acid anions that are already in the acid rain (SO_4, NO_3,Cl). Three other observations confirm the atmospheric origin of the acidity. First, taking Le Rouge-Rupt river as an example, there is a good relationship between pluviometry and stream pH: pH is low during rainy season and minimum following snow melt. Second, a weathering origin for SO_4 is unlikely in this granitic mountain where we have no evidence of sulphide oxidation (pyrites). Third, the fish lost is recent (see below).

Our measurements in the area of Cornimont must be compared with what is observed in the Strengbach basin near Aubure (35 km in the north east, (7)) and in the Ringelbach basin (20 km in the north east, (6)). In the Strengbach water pH is near neutrality (mean 6.1, minimum 5.8, mean annual precipitation 1200-1300 mm, trouts are present) while in the Ringelbach water pH is slightly alkaline (mean 7.05, minimum 6.3, mean annual precipitation 1230 mm, trouts are also present) by comparison to equilibration with atmospheric CO_2. One of the major chemical differences between these three types of water lies in their different alkalinities that corresponds to the bicarbonate set free by the weathering of primary rocks (3). In the waters of Le Rouge-Rupt, Lonfoigneux and Roncy, the alkalinity is virtually zero, when in the Strengbach it is equal to 40 $\mu eq \cdot L^{-1}$ and in the Ringelbach it reaches 300 $\mu eq \cdot L^{-1}$. From this, one can derive the history of acidification in these three sites. Indeed one can estimate the preacidification alkalinity value - that is an index of the original buffer capacity - as (i) in unacidified oligotrophic waters bicarbonate alkalinity is present in approximatively equivalent amount as the sum of calcium and magnesium, and (ii) the acidification of freshwater can be assimilated to a large scale titration of bicarbonate solution with acids that are deposited from the atmosphere (3). In Le Rouge-Rupt, Lonfoigneux and Roncy, as the sum Ca + Mg is 100-115 $\mu eq \cdot L^{-1}$ (Table I) the initial alkalinity should have been the same whereas today it is close to zero. Similarly, one can conclude that in the Strengbach basin it is partially lost (Ca + Mg = 260 $\mu eq \cdot L^{-1}$ and alkalinity = 40 $\mu eq \cdot L^{-1}$, (7)) and that in the Ringelbach basin it has not significantly changed (Ca + Mg = 325 $\mu eq \cdot L^{-1}$ and alkalinity = 300 $\mu eq \cdot L^{-1}$, (6)). Considering that the three sites are receiving roughly comparable amounts of acid rainfall, in a relatively restricted area of the Vosges mountains, we have three examples of streams at three different stages of acidification. The rivers with the smaller buffer capacity are today acid. Such spatial variation is classical in the literature (9). But one must notice that the annual pH rhythm we observed in Le Rouge-Rupt river shows that there is still some residual buffer capacity in this area. Indeed, the pH changes we observed can be interpreted as a seasonal change in the ratio between the supply of small amount of well buffered groundwater and the supply of large amount of spring water highly influenced by unbuffered rain water.

The anglers of Cornimont claim that fish population, especially brown trouts *S. t. fario*, started disappearing in the late 1960s in many rivers around the city. They also say that their reproduction appears impaired in lake Blanchemer and in the lake des Corbeaux. It is in the latter that, to our knowledge, the first acidification measurement was done in 1981. On December 22, when the amount of precipitation was 421 mm from the beginning of the month (mean regular value, 100 mm), Schoen *et al.* (10) recorded pH 4.9 in this lake where alkalinity is also close to zero. From that time, our own measurements show that pH oscillates between 6.8 (November 31, 1985) and 5.6 (May 9, 1986) at the surface, when it is between 5.6 and 6.0 in the lower half.

Aluminium is a very important factor in the toxicity of acid waters in humans (11) and in aquatic animals (5). For drinking water, the directive erected by the European Community (CEE) on July 15, 1980 proposed that a regular value of aluminium concentration must be 50 $\mu g \cdot L^{-1}$ and that the maximum acceptable value is 200 $\mu g \cdot L^{-1}$. But in all the water rivers where we measured it, concentrations were above this norm. This will be a difficult problem to solve in the Vosges mountains where collecting of drinking water often depends on surface waters.

Acknowledgments: The authors thanks M. Gehin, F. Valentin and G. Gehin from the Association de Pêche de Cornimont for their field assistance. Project partially supported by programs PIREN-Eau/Alsace (CNRS, Ministère de l'Environnement, Région Alsace) and DEFORPA (CEE-DG XII, Ministère de l'Environnement).

REFERENCES

(1) WRIGHT, R.F. (1983). Acidification of freshwaters in Europe. Water Qual. Bull. 8: 137-142.

(2) PACES, T. (1985). Sources of acidification in Central Europe estimated from elemental budgets in small basins. Nature 315, No. 6014, 31-36.

(3) HENRIKSEN, A. (1980). Acidification of freshwater - a large scale titration. Proc. Int. conf. ecol. impact acid precip., Norway, SNSF project 68-74.

(4) BOURRIE, G. (1978). Acquisition de la composition chimique des eaux en climat tempéré. Application aux granites des Vosges et de la Margeride. Sci. Géol. Mém. 52: 174 p.

(5) MASSABUAU, J.-C. (1985). Pluies acides et physiologie des animaux aquatiques. Bull. Soc. Ecophysiol. 10 (fasc. 2): 59-74.

(6) FRITZ, F., MASSABUAU, J.-C. and AMBROISE, B. (1984). Physico-chemical characteristics of surface waters and hydrological behaviour of a small granitic basin (Vosges massif, France): Annual and daily variations. IAHS Publ. 150: 249-261.

(7) PROBST, A., FRITZ, B., AMBROISE, B. and VIVILLE, D. (1987). Forest influence on the surface water chemistry of small granitic basins receiving acid precipitation in the Vosges massif, France. IAHS Publ. (In press).

(8) KREMPP, G. (1982). Techniques de prélèvement des eaux naturelles et des gaz associés. Méthode d'analyse des eaux et des roches. Notes techniques n° 11. Institut de Géologie, Strasbourg, France, 59 p.

(9) GOLDSTEIN, R.A., GHERINI, S.A., CHEN, C.W., MOK, L. and HUDSON, R.J.M. (1984). Integrated acidification study (ILWAS): A mechanistic ecosystem analysis. Phil. Trans. R. Soc. Lond. B 305: 409-424.

(10) SCHOEN, R., WRIGHT, R.F. and KRIETER, M. (1983). Regional survey of freshwater acidification in West Germany (FRG). NIVA-Report, 5: 15 p.

(11) GALLE, P. (1986). La toxicité de l'aluminium. La Recherche, 179: 766-775.

MERCURY POLLUTION IN SWEDISH WOODLAND LAKES*

M. Meili
University of Uppsala, Institute of Limnology, Box 557, 75122 Uppsala, Sweden

SUMMARY

Airborne mercury has caused severe pollution of Swedish woodland lakes. In about 40 000 lakes, the mercury concentration in fish (standard: pike of 1 kg body weight) is estimated to be >0.5 mg/kg, which is the limit for fish products in most countries. Although mercury use in agriculture and industry has been significantly reduced during the last 20 years, the concentrations of mercury in fish are not decreasing. In addition to Swedish emissions, significant amounts of airborne mercury originating in continental Europe are deposited in Sweden. Origin, transport, fate and bioaccumulation of mercury are discussed, based on current results from a coordinated research project and on earlier Swedish studies.

1 INTRODUCTION

Mercury became well-known as an environmental toxin in the 1950's when many inhabitants of Minamata, Japan, became seriously ill and even died from mercury poisoning. The cause of the disaster was traced back to the consumption of fish and seafood which contained very high concentrations of methylmercury (up to 100 mg/kg), the source of which was an industrial effluent. In several other cases, mental retardation could be related to prenatal mercury exposure at much lower levels (21).

In most comparative toxicological studies, mercury is recognised as the most toxic of the heavy metals. In contrast to other metals it is able to pass through the biological barriers to the brain and to the foetus, and it can accumulate in most body tissues. This is due to its ability to form stable organomercuric compounds, even under natural conditions (14,15).

In Sweden, mercury pollution was first discovered in 1956 when several bird species were heavily decimated after consuming mercury-treated grain, or animals living on it. In 1964, high mercury concentrations were also observed in pike (*Esox lucius* L.), a predatory fish which is common in most Swedish lakes. In many cases, the sources of mercury were found to be paper and pulp mills which were discharging large amounts of mercury directly into rivers and lakes. Consequently, the use of mercury in agriculture and certain industrial processes was banned in 1967. In the same year, the government banned the sale of food that had a mercury concentration exceeding 1 mg/kg. In addition, lakes were blacklisted, if pike at a standard body weight of 1 kg had a mercury concentration higher than 1 mg/kg.

---

* This contribution is based mainly on work within a coordinated Swedish research project, Occurrence and turnover of mercury in the environment (project leader O. Lindqvist, references 24 and 25), in which the author participates with studies on the bioaccumulation of mercury in lake ecosystems.

In the 1960's, high mercury concentrations in freshwater fish were revealed even in remote woodland lakes that were free from direct discharge of mercury or other pollutants to the lake or the drainage water (20). Most of the accumulated mercury was identified as methylmercury (28), even in areas where inorganic mercury was emitted. Microbial methylation of mercury in the environment was suggested as being responsible for the transformation (14,15).

The background level of mercury in pike is estimated to be less than 0.2 mg/kg (20). A recent compilation of all available Swedish data on mercury concentrations in pike with a body weight of 1 kg indicated that at least 9000 Swedish lakes are exceeding the blacklisting limit of 1 mg/kg (2). In most other countries the limit for mercury in fish products is 0.5 mg/kg, derived from the provisional tolerable weekly intake of 0.2 mg methylated mercury established by FAO/WHO. An application of this limit in Sweden would imply that about 40 000 of a total of 83 000 Swedish lakes (surface area >0.01 km2), and the majority of lakes in southern and central Sweden, would have to be blacklisted (23,Fig.1). Moreover, the evaluation revealed that 20 years after the implementation of the mercury ban and the subsequent dramatic reduction of mercury emissions, the average mercury concentrations in pike have not decreased.

As a consequence, a coordinated research project on the occurrence and turnover of mercury in the environment was initiated by the Swedish National Environmental Protection Board in 1984 and will continue until 1989 (24). The research programme is divided into several subareas (emission, athmosphere, terrestrial systems, aquatic systems, analytical methods). This paper summarises some results of the current research as well as previous investigations on the mercury pollution of lake ecosystems in Sweden.

2 ORIGIN AND LOADING OF MERCURY IN THE ENVIRONMENT

Analyses of dated sediment profiles in a number of Swedish forest lakes revealed a dramatic increase in mercury loading to lakes during the 20th century (5). The mercury concentration in sediments increased most in southern Sweden (about fivefold), while they remained rather unchanged in the northernmost part of the country (18,19). Preliminary results from a Nordic station network for the deposition of mercury with rainwater (Å. Iverfeldt in 25) and an inventory of the mean amount of mercury accumulated in forest soil (A. Andersson, pers. comm.) indicate a similar pattern, both being at least five times higher in southern Sweden than in northern Sweden.

These findings may suggest that most of the mercury found in Swedish lakes originates in other European countries and is imported with southerly winds. This would correspond very well to the regional pattern of both the deposition of sulphuric acid (Fig. 3), soot, cadmium and other anthropogenic contaminants over Sweden, and the acidification of soil, groundwater and 15000 lakes. It has been estimated that only 10 % of the sulphur deposited in Sweden have their origin in Swedish emission sources, whereas other countries contribute at least 65 % (27).

An inventory of the mercury levels in the mor layer of forest soils revealed a more detailed regional pattern, apart from the south-north gradient mentioned above (A. Andersson in 25). The highest concentrations were found in specific areas scattered throughout the country (Fig. 2). A corresponding inventory of the mercury concentrations in pike resulted in a similar pattern (Fig. 1). Most of these areas are situated within a distance of about 100 km from previous point sources of mercury emissions to the atmosphere; namely from various branches of the paper and metal industries. Between 1935 and 1970, large amounts of mercury were emitted from a few point sources, mainly chlorine-alkali plants and a metal smelter (Fig. 4). A large part of these local emissions was deposited in

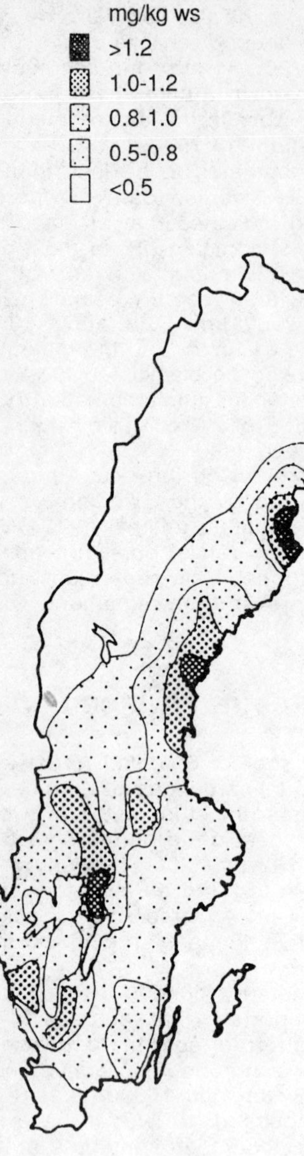

Figure 1:
Mercury concentration
in muscle tissue
of 1kg-pike 1981-1985
(2)

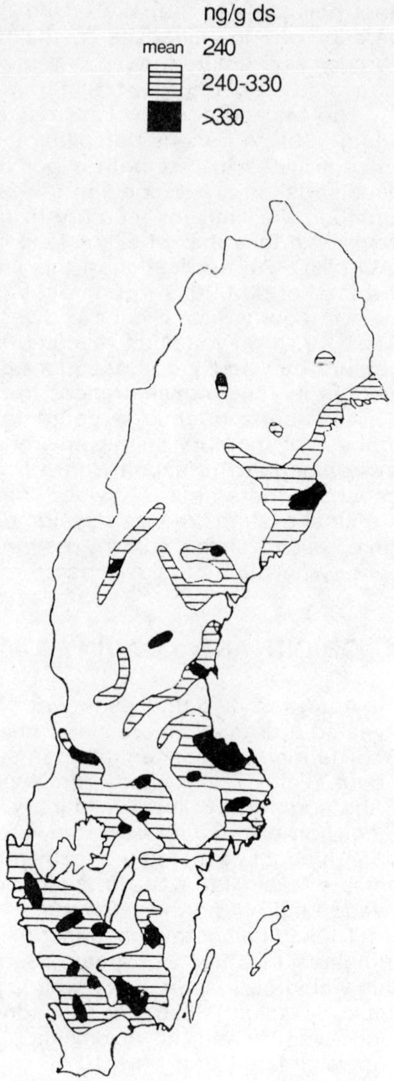

Figure 2:
Mercury concentration
in the mor layer
of forest soil
(25)

Figure 3:
Wet deposition
of acid (meq/m2/y)
1983-1985
(27)

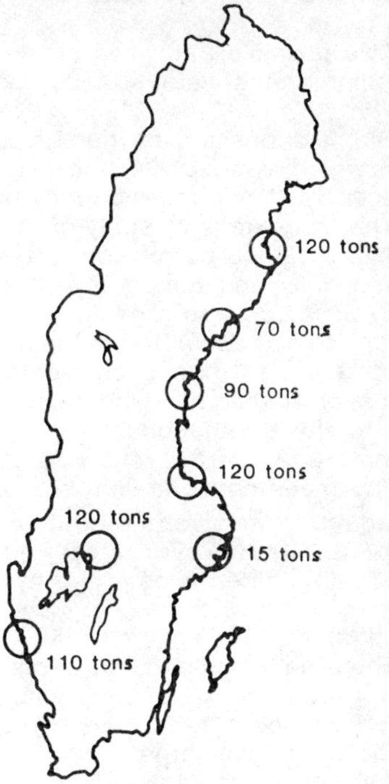

Figure 4:
Total emissions of mercury
from point sources
to the air 1940-1980
(23)

the vicinity of the point source (23), where it was added to the mercury from long-range transport.

The total emission from these point sources was estimated to be about 700 tons of mercury (23). Most of them are situated close to the east coast, where westerly winds exported a large part of the emitted mercury from the mainland. The total amount of mercury deposited in Sweden can be estimated to be 700-800 tons, considering the total mercury content in the soil and assuming the natural background to be negligible. Thus the local emissions probably only account for a part of the mercury deposited in the country.

Today, Swedish emissions from large point sources are estimated at 6 tons/year, with the largest part being contributed by combustion of household waste (mainly from batteries) (25). Rates and patterns of mercury deposition indicate that the mercury imported from other countries may be of the same order of magnitude.

3 TRANSPORT OF MERCURY TO LAKES

Two possible pathways of mercury transport to lakes are discussed: Direct deposition on the water surface and transport with runoff water from the drainage area.

Direct deposition of mercury can be estimated from the wet deposition of Hg(II) which is about 20 and 2 g/km2/year in southern and northern Sweden, respectively. Dry deposition may be of similar magnitude (23).

The first extensive study of mercury concentrations in drainage water started in 1985. A strong correlation with the concentration of humic matter was found, both spatially and temporally. This confirmed previous findings about the high affinity of mercury to organic material. High concentrations of both humus and mercury coincided with high water flows. Much of the annual export of mercury from forest soil thus occurred in short periods of time. In the north, these periods of significant export occurred mainly during the snow melt in spring, while in the south considerable amounts of mercury also were exported in autumn (12, A. Iverfoldt and K. Johansson in 25).

The mean concentration of total mercury in the drainage water was about 4 to 7 μg/m3 in most of the southern and central areas and 1 to 2 μg/m3 in the north (12,25). However, most of the regional differences can be attributed to considerable differences in the hydrological regimes and in the concentration of humic matter, which was very low in the runoff waters from all northern stations investigated.

The area specific export of mercury from the drainage area varied between 1 and 6 g/km2/year, with higher values in the southern stations as compared to the northern stations (12,25). Export appeared to be positively correlated with area specific water discharge. Moreover it was found that pH and conductivity were low during periods when water flux and humus and mercury concentration were high (12,25). In addition, mercury concentrations in groundwater appear to be low (M. Åstrup in 25). This indicates that mercury is leached from the top soil layer (12,25).

Using the data above, the relative importance of the two pathways discussed can be estimated. A rough estimation can be made of how much mercury loading is accounted for by direct deposition as compared to surface runoff. The calculations are based on normal values for some characteristics of typical headwater lakes in forested bedrock areas with different location and different water colour. Different approaches lead to similar results.

The estimates for the total mercury load to a lake with a surface area of 1 km2 range from 5 to 200 g/year, depending on location and hydrology of the lake. The calculations suggest that in clearwater lakes, the direct deposition may exceed the runoff transport several times, with the probability increasing from the north to the south and from the west to the east of Sweden. In brownwater lakes, however, the runoff transport may exceed the direct deposition. Mercury loading to lakes from their drainage areas varies considerably and seems to depend primarily on the hydrology of the lakes and the humus concentration in the inflow water, and only to a minor extent on the mercury deposition rate in the area.

In southern and central Sweden, the mercury content in surface sediments increases with increasing content of organic material in the sediment (23). The data generally show that the carbon/mercury ratio in the sediment was several times higher than in the runoff water. In the north, however, there was no evident correlation between mercury and organic material in the sediment, although the carbon/mercury ratio in the runoff water seems not to differ much between the two regions. This supports the hypothesis that the direct deposition of mercury on the lake surface and subsequent scavenging by organic matter can be quantitatively important.

The estimations above clearly show that only very small amounts of mercury are required to cause severe pollution problems. In addition, mercury has a very long residence time in the environment. The data used above allow to roughly quantify the mercury turnover in Swedish forest soils:

A comparison between the amounts of mercury deposited and mercury exported from forest soil indicates that net deposition still occurs in most locations (13).

The ratio between the mercury export and the mercury pool accumulated in Swedish forest soils indicates that less than 1% is exported every year. This implies that the theoretical half-time of mercury in forest soils may be several hundred years. Similarly, soil carbon, to which mercury probably is strongly bound, has a mean age of 200 to 300 years (A. Andersson, pers.comm.).

It is important to point out that most of these calculations are based on preliminary results, and in order to outline general aspects the wide local variations of mercury concentration, which have been found both in soils, runoff water and lake sediments, were not taken into consideration. Neither local deposition of mercury from point sources nor reemission of mercury from land and water have been considered. Moreover, the chemical composition and behaviour of locally deposited mercury may be different from mercury which is subject to long-range transport.

4 FATE OF MERCURY LOADING

Pollutants added to lakes can either be sedimented, transported downstream or taken up by biota.

The sedimentation rates of mercury in woodland lakes can be calculated from sediment characteristics which are typical for these lakes (23). A comparison between sedimentation rate and mercury loading indicates that a considerable part of the total mercury load to lakes is deposited in the sediment. However, sedimented mercury can still enter the food chains, either via benthic animals or after transformation into methylmercury and subsequent release into the water body.

The mercury uptake in fish communities can be calculated based on their productivity in woodland lakes (500-5000 kg/km2/year) and their average mercury content (0 to 1 mg/kg). It indicates that at most a few percent of the total annual mercury load are taken up by fish, despite a very effective retention of mercury in body tissues. It also shows that small amounts of mercury are sufficient to contaminate a fish community to the blacklisting limit.

The estimates of sedimentation rates and uptake rates in biota indicate that in most cases a major part of the total mercury load to a given lake is transported downstream. This contributes to the spreading of mercury pollution.

5 MERCURY IN BIOTA

Mercury contents in fish from different lakes vary widely, even between lakes within the same area. This indicates that other factors than the total mercury load to the lakes are governing bioaccumulation.

Most Swedish woodland lakes are situated in bedrock areas where weathering is very slow. Therefore, most lakes are poor in nutrients and buffering capacity, rich in humic matter and slightly acidic. They are also very sensitive to acid rain and contaminants.

A compilation of data from about 1500 Swedish lakes provided a map on the mercury content in pike of 1 kg body weight (Fig. 2). This map shows that mean mercury concentrations in pike with a body weight of 1 kg are above 0.5 mg/kg in most lowland areas of Sweden, except in the northernmost part and in some southern areas rich in calcareous soils where lakes are not acidic despite high deposition of acid. Concentrations are even higher in large areas around former large point sources and in southern Sweden (Fig.1 + 3).

In acid lakes, the mercury concentrations in fish are generally considerably higher than in pH-neutral lakes (1,2,4,10,11,23). The reasons for this are not known, but a number of chemical, physiological and ecological hypotheses to explain this have been discussed (3,4,23,26), and research is going on (26).

Another important abiotic factor is the concentration of humic matter. In humic lakes, mercury concentrations in fish and zooplankton seem to be higher than in clearwater lakes within the same area (1,26). This may be a consequence of increased mercury loading (see above). On a more general basis, and taking mercury loading into account by means of sediment data, no significant correlation was found between water colour and mercury concentration in pike (2). The role of humic substances other than as a mercury carrier has been discussed as well (23).

The previously mentioned comparison of mercury loading to a lake and mercury accumulated in fish indicates that not only the amount but also the availability of mercury for biota is of crucial importance. Thus, the quantitative relationship between the mercury directly deposited on the lake surface and the mercury input from the drainage area does not provide any information about their relative importance for the biotic compartments in the lake. Both fluxes are large enough to dominate the transport to the biota by themselves. More knowledge is required about the chemical character of the mercury compounds and their quantitative importance in the two fluxes. Moreover, little is known about the bioavailability of the different chemical species occurring in natural waters.

The most important compound in terms of biomagnification is methylmercury. It accounts for 90 % or more of the total mercury in most predatory fish (28). Methylmercury does not have to be supplied to the environment but can be formed under natural conditions from inorganic forms of mercury by bacteria (14,15). Even abiotic formation triggered by humic substances has been discussed. The concentration of free methylmercury in natural waters is, however, extremely low and difficult to analyse (22). This may be a consequence of a very efficient uptake by biota (9).

Due to its efficient retention in most biotic tissues, methylmercury readily accumulates in organisms. The highest concentrations of mercury are often found in old predatory fish. The concentrations in invertebrates are considerably lower. A comparison between pike with a body weight of 1 kg and zooplankton organisms from the same lake showed that the total mercury concentration in the fish is about 25 times higher. This ratio was about the same in eight different headwater lakes, regardless of mercury concentration in the animals from different lakes. Moreover, the ratio of methylated to total mercury was only about 60 % in zooplankton as compared to about 90 % in pike (Meili, unpubl. data). This supports the hypothesis that methylmercury is accumulated in food chains (6,7,8,17).

Mercury can enter aquatic food chains either from the sediment through benthic animals, or from the water body through plankton organisms or through direct uptake from the water by fishes. The relative importance of these pathways is, however, not known. Moreover, it is not clearly understood in which part of the ecosystem methylation of mercury takes place.

All these findings illustrate the need for further studies on the transition of mercury from the abiotic to the biotic compartments of aquatic ecosystems.

REFERENCES

1 Andersson B., Hedenmark M., Persson K. (1985): Mercury concentration in lake sediments and muscle tissue of pike from some areas in central Sweden. (In Swedish) - Nat. Swed. Env. Prot. Board, Report PM 1984.

2 Andersson, T, Nilsson Å, Håkanson L. & Brydsten L. (1987): Mercury in Swedish lakes (in Swedish, English summary). - Nat. Swed. Env. Prot. Board, Report 3291, 92 p.
3 Andersson, P. & Kärrhage, P. (1984): Effects of liming on the mercury content of fish in the Åva water system. (In Swedish) - Nat. Swed. Env. Prot. Board, Report PM 1771,46p.
4 Björklund, I., Borg, H. & Johansson K. (1984): Mercury in Swedish lakes - regional distribution and causes. - Ambio 13:118-121.
5 El-Daushy F. & Johansson K. (1983): Radioactive lead-210 and heavy metal analyses in four Swedish lakes. Ecol. Bull. 35:555-570.
6 Fagerström, T. (1977): Body weight, metabolic rate, and trace substance turnover in animals. - Oecologia (Berl.) 29:99-104.
7 Fagerström, T., Åsell, B. & Jernelöv, A. (1974): Model for accumulation of methyl mercury in northern pike Esox lucius. - Oikos 25:14-20.
8 Grahn, O., Hultberg, H. & Jernelöv, A. (1976): Distribution of mercury in three different trophic levels - a review study of three lakes in Western Sweden. (In Swedish) - Swed. Water Air Poll. Res. Inst., Report B 292.
9 Hannerz, L. (1968): Experimental investigations on the accumulation of mercury in water organisms. - Inst. Freshwater Res. Drottningholm, Report 48.
10 Hultberg H. & Hasselrot B. (1981): Mercury in the ecosystem (In Swedish). - National Swedish Power Board; Project Coal, Health and Environment; Inf. 04:33-35.
11 Håkanson, L. (1980): The quantitative impact of pH, bioproduction and Hg-contamination on the Hg-content of fish (pike). - Envir. Pollut. B 1:235-304.
12 Iverfeldt Å. & Johansson K. (1987): Mercury in run-off water from small watersheds. - Verh. Internat. Verein. Limnol. 23 (in press).
13 Iverfeldt Å. & Johansson K. (1987): Transport of mercury in small watersheds in Sweden. - Paper presented at the intern. workshop on Geochemistry and Monitoring, Prague.
14 Jensen S. and Jernelöv A. (1967): Biosynthesis of methylmercury I (in Swedish). - Nordforsk Biocid Inf. 10:4.
15 Jensen S. & Jernelöv A. (1969): Biological methylation of mercury in aquatic organisms. - Nature 223:753-754.
16 Jernelöv, A., Landner, L. & Larsson, T. (1975): Swedish perspectives on mercury pollution. - J. Water Poll. Control Fed. 47 (4):810-822.
17 Jernelöv, A. & Lann, H. (1971):Mercury accumulation in food chains.-Oikos 22:403-406.
18 Johansson K. (1980): Heavy metals in acid woodland lakes (In Swedish, English summary). - Nat. Swed. Env. Prot. Board, Report PM 1359, 70 p.
19 Johansson K. (1983): Mercury in sediments in Swedish forest lakes. - Verh. Internat. Verein. Limnol. 22.
20 Johnels, A.G, Westermark, T., Berg, W., Persson, P.I., & Sjöstrand, B. (1967): Pike (Esox lucius L.) and some other aquatic organisms in Sweden as indicators of mercury contamination in the environment. - Oikos 18:323-333.
21 Kjellström T., Kennedy P., Wallis S., Mantell C. (1986): Physical and mental development of children with prenatal exposure to mercury from fish. - Nat. Swed. Env. Prot. Board, Report 3080, 96p.
22 Lee Y.H. (1986): Determination of methyl- and ethylmercury in natural waters at sub-nanogram per liter using SCF-adsorbent preconcentration procedure. Intern. J. Envir. Anal. Chem., 1987 (in print).
23 Lindqvist, O., Jernelöv, A., Johansson, K. & Rodhe, H. (1984): Mercury in the Swedish environment. Global and local sources. - Nat. Swed. Env. Prot. Board, Report PM 1816,105p.
24 Lindqvist O., Johansson K., Timm B. & Hovsenius G. (1984b): Mercury, occurrence and turnover in the environment. Research program 1984-1989. (In Swedish, English summary). - Nat. Swed. Env. Prot. Board, Report PM 1912, 28 p.
25 Lindqvist O., Johansson K. & Timm B. (1986): Mercury, occurrence and turnover in the environment. Progress report Nov. 1986. - Nat. Swed. Env. Prot. Board, Report3265, 43p.
26 Meili M. & Parkman H. (1987): Seasonal mercury accumulation patterns in macroplankton. - Verh. Internat. Verein. Limnol. 23 (in press).
27 Monitor 1986 (1986): Acid and acidified waters.-Nat. Swed. Env. Prot. Board, Inf. 1986.
28 Westöö G. (1966): Methylmercury compounds in fish, identification and determination. - Acta Chem. Scand. 20:2131-2137.

EFFECTS OF HYDROLOGY ON THE REACIDIFICATION OF THE LIMED LAKE GÅRDSJÖN

U. Nyström and H. Hultberg
Swedish Environmental Research Institute
P.O. Box 47086, S-402 58 Gothenburg, Sweden

Summary

The role of the hydrology for the reacidification process in a limed lake was shown by studying the alkalinity, pH and calcium concentrations in the surficial water and in depth profiles in the lake. The situations with high discharges during circulation periods differ sharply from high discharges events during ice-covered periods. The surficial layering of acidic water inflow during the ice-covered period may last all over the winter season. This may cause toxic conditions to biota over shallow bottoms.

1. INTRODUCTION

In Sweden liming operations in lakes and whole catchments have been an important management practice since mid 1970, when governmental subsidiaries became available. Today more than three thousand lakes and large river systems have been limed (1). During the winter 1985, however, only 50% of the limed lakes had an alkalinity >0.05 meq.L^{-1} and negative biological effects may have occurred in the reacidified lakes. The acid Lake Gårdsjön and its subcatchments have been studied by more than twenty scientists since 1979 (2). These studies in hydrology, chemistry and biology form the background to the lime treatment in Lake Gårdsjön (3). The main objectives with lime treatment of the lake were to study the effects on the flora, fauna and nutrients as P, N and org-C as well as metals like Al, Mn and Fe.

2. SITE DESCRIPTION AND HYDROLOGY

The Lake Gårdsjön catchment (2.11 km^2, 113-170 m a.s.l) is situated 14 km from the Swedish west coast and is exposed to extensive sea salt input. The lake surface is 31.2 ha and the surrounding land area is 74.3 ha. 50% of the catchment is located upstream the Lake Gårdsjön inlet. The catchment is underlain by crystalline bedrock and the soils are dominated by thin iron podsols. Clearcuts are characteristic of the land areas upstream of Lake Gårdsjön while Norway spruce and Scots pine cover about 78% of the land areas surrounding Lake Gårdsjön (4).

The pH of the lake decreased from 6.25 in 1949 to 4.7 in 1970 (summer values). Due to increased acidity and acid surges with increased concentrations of toxic aluminium species, almost all fishes had disappeared by 1973. Acidification had also affected plankton, benthic invertebrates, benthic algae and submerged macrophytes with prolific growth of Sphagnum on the lake bottoms (5).

The hydrological regime in this part of Sweden is characterized by low discharge during summer and two large maxima, one caused by snowmelt in spring and the other caused by rainfall in late autumn. This is commonly referred to as the Atlantic rain and snow regime (6). The most frequent wind direction is from south and southwest, which makes the region strongly influenced by orographic precipitation. The long-term water balance is: P=1070, Q=520 and E=550 $mm.yr^{-1}$. During the 3.3 year period after liming, the total runoff was 1760 mm (7). This resulted in a "yearly" average runoff during this time of 530 mm which corresponds almost exactly to the long-term average, 520 $mm.yr^{-1}$. This gives Lake Gårdsjön a turnover time of 1.36 yr and a water exchange of 0.74 yr^{-1}.

3. LIMING

The objective of the liming in the acid Lake Gårdsjön (pH = 4.7) was to study effects on lake chemistry and biology. As the upstream reference lakes should not be affected by the liming, the limestone had to be spread in Lake Gårdsjön exclusively. This means that a more strategical upstream liming could not be carried out.

Acid runoff and seepage from the Lake Gårdsjön adjacent land areas as well as by the main inlet stream may affect major parts of the littoral areas by acid water (pH = 3.8-4.7) and high aluminium concentrations (0.3-1.5 mg L^{-1}). The limestone was therefore concentrated to the littoral areas (80%) and a smaller amount over deep areas (20%). The treatment started on April 28 and was finished by May 1, 1982. The limestone was distributed from a float after mixing with lake water. The total amount spread corresponds to 23.3 mg Ca L^{-1} in the lake water.

4. EFFECTS ON WATER CHEMISTRY

To describe the effects on water chemistry in brief, some easily identified and important events are discussed. The letters within brackets in the text refer to corresponding letters in the Figures 1-3.

A sharp increase in pH occurred immediately after liming. The Ca and alkalinity were still increasing when pH was 7.5 and reached the highest concentrations in October 1982 (A). H^+, Ca and HCO_3 were the only major ions affected by the liming.

The dissolution of limestone and release from sediments will counteract a concentration decrease. When most of the calcium is dissolved, i.e. there is no more easily accessible calcium left, the reacidification process will be more similar to a pure dilution process. This process may be described by a simple differential equation, which has the following solution

$$C(W) = e^{-W} \cdot (C_s - C_b) + C_b$$

where C(W) is the concentration, W is the number of water exchange (= lake volumes) since the start of the dilution computation, C_s = concentration in the water at the start of the computation of the dilution and C_b is the background concentration. The concentration is hence treated as a function of cumulative runoff (here in terms of water exchanges) rather than time, to get a stationary state condition. Ca concentrations in the outlet (surface concentration), from profiles (mean concentration in the lake) and eventually pure dilution computations from six different events show the hydrological role of the acidification process.

In the autumn 1982, during a circulation period, very high discharges caused an even dilution in the whole lake. This situation is shown by the three curves close together (B). During this time, mid November 1982 to mid January 1983, 20% of the added amount of Ca left the lake.

During the winter 1983/84 high discharges caused by temperature above 0^0C and heavy rain or sleet, occurred when the lake was ice-covered. This situation caused instead a tremendous surficial dilution which did not stop until after snowmelt (C). This may be typical, as the most effective mixing current, the wind action, is impeded by the ice-cover.

At the snowmelt in 1985, the limestone dissolution had almost ceased and the reacidification was governed by the hydrology of the entire lake. The alkalinity and pH dropped dramaticly, the pH from 6.6 to 5.3. This probably caused toxic conditions to biota over shallow bottom areas all along the

Lake Gårdsjön shores (D). At the end of the period about 80% of the added limestone had dissolved, but only 18% of the dissolved Ca was still in the lake water mass.

Acid episode studies in other limed lakes, e.g. Lake Hällsjön with a very short turnover time (0.24 y) have shown even more pronounced surficial layering of acidic inflow (8). These studies also showed some short events of bottom and intermediate layering of acidic water, mainly due to temperature conditions.

4. CONCLUSIONS

This treatment shows clearly that although a high mean pH in the lake water mass, toxic conditions to biota over shallow bottoms may arise due to the hydrology. To prevent this phenomenon a more biologic strategical liming is important. This includes upstream liming, liming on discharge areas and liming over shallow bottoms in combination with liming of the lake water mass.

REFERENCES

(1) Monitor (1986). Sura och försurade vatten (Acid and acidified waters). SNV, Solna, Sweden.

(2) Ecological Bulletin 37 (1985). Lake Gårdsjön - An Acid Forest Lake and its Catchment. Andersson & Olsson (Eds.)

(3) Hultberg, H. and Nyström, U. (1987). The role of hydrology in treatment duration and reacidification in the limed Lake Gårdsjön (manuscript).

(4) Olsson, B., Hallbäcken, L., Johansson, S., Melkerud, P-A, Nilsson, S.J. and Nilsson, T. (1985). The Lake Gårdsjön area - physiographical and biological features. Ecol. Bull. (Stockholm) 37, 10-28.

(5) Hultberg, H. (1985b). Changes in fish populations and water chemistry in the Lake Gårdsjön and neighbouring lakes during the last century. Ecol. Bull. (Stockholm) 37, 64-72.

(6) Melin, R. (1970). Hydrologi i Norden (Hydrology in Scandinavia). Svenska Utbildningsförlaget Liber, Stockholm. 207 pp. (in Swedish).

(7) Johansson, S. and Nilsson, T. (1987). Lake Gårdsjön - the hydrological balance 1979-1986 (manuscript).

(8) Hasselrot, B., Andersson, B.I., Alenäs, I. and Hultberg, H. (1987). Response of limed lakes to episodic acid events in southwestern Sweden. Water, Air, and Soil Pollut. 32, 341-362.

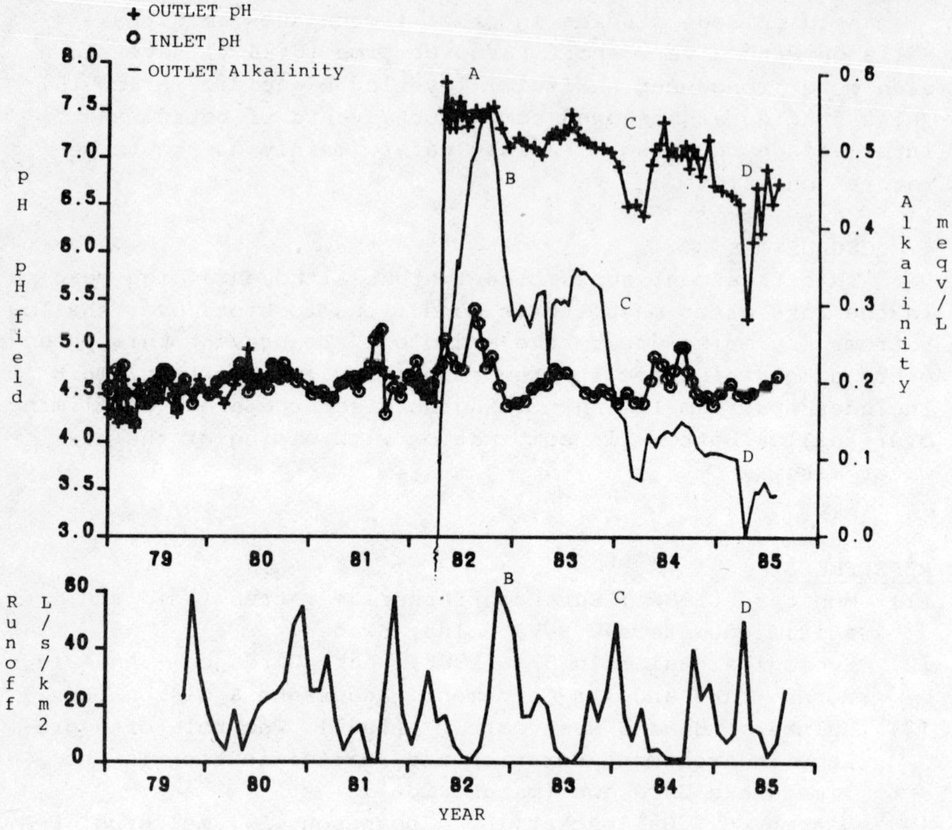

<u>Figure 1</u>. pH of Lake Gårdsjön 1979-85 in the inlet (ө), and outlet (+), and the alkalinity in the outlet (–). Monthly mean runoff, in L s^{-1} km^{-2}, during the monitored hydrological years 1979/80 - 1984/85 at the Lake Gårdsjön outlet.

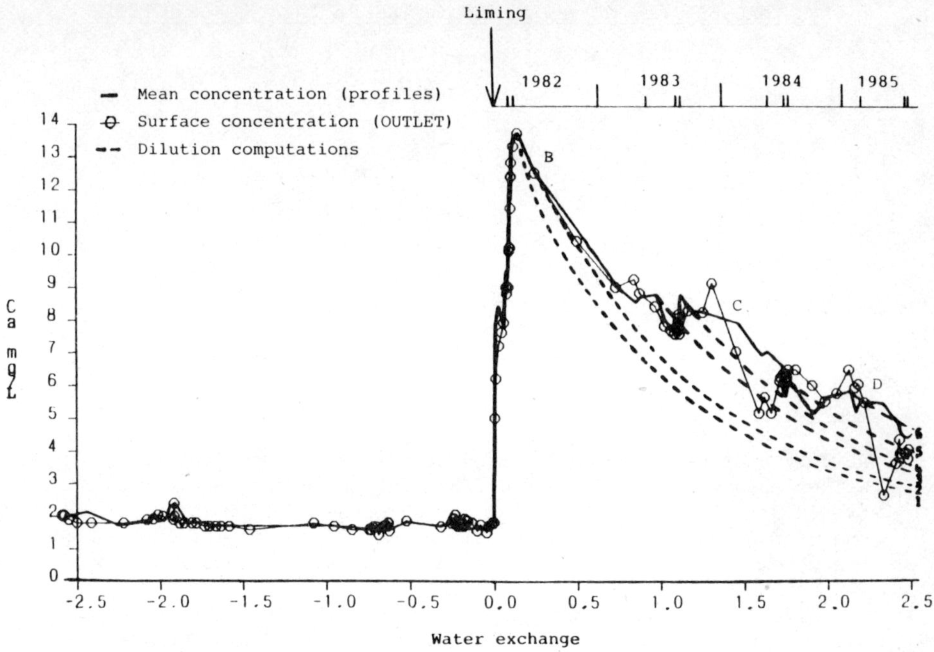

Figure 2. Calcium concentrations, in mg L^{-1}, versus water exchanges, W, in the Lake Gårdsjön. Liming occurred at W = 0. Calcium concentrations in the outlet (⊖) and mean in lake (–). Dotted lines 1-6 are the dilution computations. Calendar years with quarters are also shown.

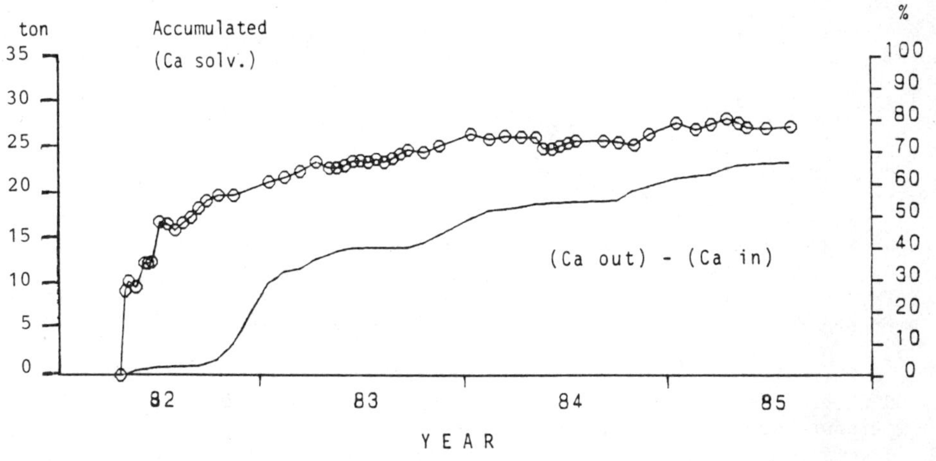

Figure 3. Accumulated dissolved calcium in the Lake Gårdsjön (background excluded) (⊖) and accumulated dissolved calcium lost by the outlet (–). Both in ton and percent of the amount added.

pH CHANGES OVER TWO CENTURIES IN THREE DUTCH MOORLAND POOLS

H. VAN DAM
Research Institute for Nature Management, Leersum

B. VAN GEEL
Hugo de Vries-Laboratory, University of Amsterdam

A. VAN DER WIJK
Centre for Isotope Research, Groningen

M.D. DICKMAN
Dept. of Biological Sciences, Brock University, St. Catharines (Ont.)

Summary

Acidity changes in three isolated perched water moorland pools in The Netherlands were reconstructed from diatoms in sediment cores and interpreted with the help of documentary evidence. Pollen and other micro- and macrofossils were used for dating and to obtain information about the development of vegetation at the study sites. Also Pb-210 was used for dating.

The pools studied showed considerably larger pH changes than acidifying European and American lakes. The pH increased from 4-5 in the first half of the 19th century to c. 6 around 1900, due to eutrophication by external sources, and decreased to recent values between 4 and 5, when eutrophication stopped and acid deposition increased.

Documented evidence of human influence upon the study sites, combined with palaeoecological techniques was found to be necessary to obtain proper conclusions with studies on the history of surface water acidification.

1. INTRODUCTION

Moorland pools in The Netherlands, Belgium and Northern Germany have been acidified over the last sixty years, as appeared from comparison of old and recent data on chemistry, micro- and macrophytes (1,2,3). No information is available about acidity changes in previous centuries.

As data from direct measurements are lacking, diatoms from sediments are used in this paper to reveal the long-term pH history of some of these pools. It is to be expected that the pools, close to major emission sources (e.g. the Ruhr area), were acidified earlier than lakes in remote areas in Northern Europe and Northern America, which started to acidity in the mid 19th century (4).

This paper summarizes the results which are reported elsewhere (5,6) and where details about the study sites, methods, and full reference to the literature are to be found.

2. STUDY SITES

Three isolated moorland pools, differing in acidity, were selected for coring (Table I, Fig. 1 I). They are situated in areas with unconsolidated sands and podzols and have perched water tables. Apart from some bathing activities in Gerritsfles and Kliplo no direct human impact on the pools is known from the present century.

Table I. Selected physical and chemical variables of study sites. Mean values of quarterly measurements (Kliplo May 1981 - Feb. 1985, other pools Aug. 1979 - Feb. 1985) (7).

	Goorven	Kliplo	Gerritsfles
Northern latitude	51°34'	52°50'	52°10'
Eastern longitude	5°13'	6°26'	5°49'
Area (ha)	2.35	0.62	6.78
Maximum depth (m)	1.85	1.14	1.24
Mean depth (m)	0.62	0.82	0.68
Colour (mg Pt l^{-1})	13	30	9
H^+ (ueq l^{-1})	92 (4.0)	4 (5.4)	33 (4.5)
NH_4^+ (ueq l^{-1})	173	41	99
Ca^{2+} (ueq l^{-1})	158	70	82
Al^{3+} (ueq l^{-1})	89	3	24
HCO_3^- (ueq l^{-1})	6	33	12
SO_4^{2-} (ueq l^{-1})	620	130	277

The sand dunes around Goorven were planted with Scots pines (*Pinus sylvestris*) in 1840. *Nymphaea alba* and *Juncus bulbosus* are the most important aquatic macrophytes.

The formerly open landscape near Kliplo (Fig. 1 H) was partly afforested with pines since 1906. The open water is dominated by dense stands of *Potamogeton natans*.

Gerritsfles is in a rather open landscape of sand dunes and Molinia grassland, which are partly invaded by Scots pines. These were planted in the area since 1898 (Fig. 1 A-G). The bottom of the pool is nearly completely covered with *Sphagnum denticulatum (= S. crassicladum)*.

3. METHODS

Plankton tow diatoms were gathered quarterly in Kliplo from 1981 through 1984 and in the other pools from 1979 through 1984. Plankton samples from the periods 1916-1929 and 1948-1964 were taken from museum collections. In each sample 400 diatom valves were counted. The diatom-inferred pH was calculated as a mean of the pH-optima of the species, weighted by the relative abundance of the species (8). Historical pH-measurements were searched in the literature and for each of the above mentioned periods mean values of measured and diatom-inferred pH were calculated.

Bottom cores were taken in 1985 and cut in 1-cm sections. Diatom-inferred pH values were calculated as for the plankton tow samples. Pollen slides were prepared and pollen and other remains of organisms were counted until a pollen sum of c. 300 pollen was reached. Pollen were used for dating and to describe changes in the vegetation at and near the study sites. Pb-210 was also used for dating (9).

Historical data on human impact were obtained from literature, archives, and old topographical maps.

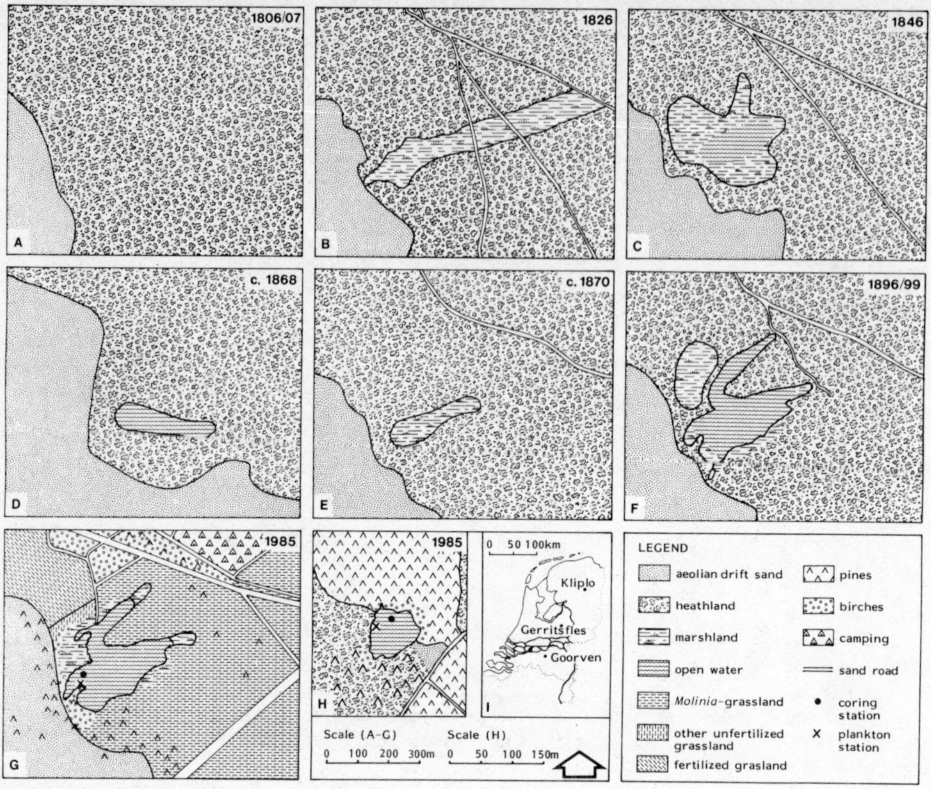

Fig. 1. Historical development of Gerritsfles according to various topographical maps, situation of Kliplo (H) and location of sampling sites in The Netherlands (I).

4. RESULTS AND DISCUSSION

At all stations the pH decreased from c. 1900 to 1985 (Fig. 2). This appears from direct measurements of the pH and from diatom-inferred pH values of the plankton and sediment samples.

In Goorven the pH increased from c. 1810 to c. 1910. This pool developed to a bog after its formation in the Late-Glacial period. Until the early 19th century the peat was excarvated as a fuel by farmers and an acid oligotrophic pool was formed. Subsequently, the pH increased by inflow of agricultural drainage water. This stopped at the turn of the century.

The changes in Gerritsfles have been more complex than in the previous pools. Integration of data on topographical development (Fig. 1), pollen (Fig. 3), diatoms and unpublished data on management at the site leads to an explanation of these changes. In 1806/07 the coring site probably existed as a wet heath with *Erica tetralix* as the most prominent macrophyte. The presence of aerophilous diatoms indicates temporary dry conditions. The high sand percentage indicates a high activity of the aeolian drift sand, west of the coring site (Fig. 3). The excrements of sheep, crossing the wet heath, caused a slight enrichment with nutrients and buffering material and, consequently, a slight increase of the pH (Fig. 2). During the first half of the 19th century the site changed to a more or less permanent pool with a fluctuating water level, as indicated by aerophilous diatoms.

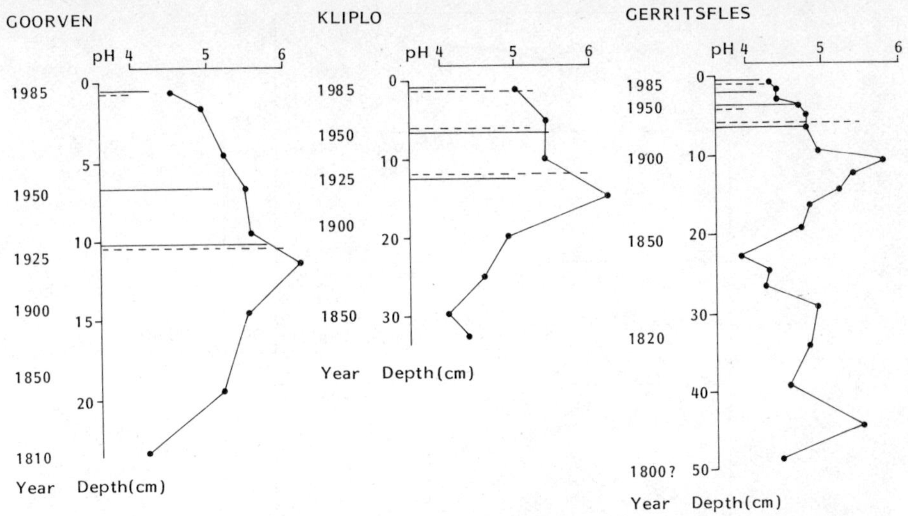

Fig. 2. pH-depth curves. Dating with Pb-210 and pollen. Solid horizontal lines: plankton diatom-inferred pH; broken horizontal lines: direct pH measurements.

In this stage the pool was too deep for the sheep to cross and pollutions with excrements stopped. An acid, oligotrophic pool was formed. After 1850 over 1000 sheep were washed in the pool each year, which caused an increase of the pH and meso-eutraphent plants (e.g. *Luronium natans, Potamogeton natans*) were able to grow. After finishing sheep washing around 1900, the pool became less eutrophic and the pH declined. The relative abundance of *Eunotia exigua*, a diatom indicating the presence of mineral acids, increased particularly after 1950, due to atmospheric acid deposition.

5. CONCLUSIONS

1. The pools studied showed considerably larger pH-changes than acidifying European and American lakes (4).

2. It is not possible to determine when acidification by acid atmospheric deposition began at the study sites, as acidification was caused both by acid atmospheric deposition and discontinuation of eutrophication in the 19th century.

3. At least in the intensively cultivated region of Western Europe documented evidence of human influence upon study sites is a prerequisite to obtain proper conclusions about the history of surface water acidification.

REFERENCES

(1) VAN DAM, H. et al. (1981). Impact of acidification on diatoms and chemistry of Dutch moorland pools. Hydrobiologia 83:425-459.

(2) ROELOFS, J.G.M. (1983). Impact of acidification and eutrophication on macrophyte communities in soft waters in The Netherlands. Aquatic Botany 17:139-155.

(3) VAN DAM, H. and BEIJAARS, K. (1984). Nachweis von Versauerung in West-Europäischen kalkarmen stehenden Gewässern durch Vergleich von alten und rezenten Kieselalgenproben. In: J. Wieting et al. (Eds). Gewässer-

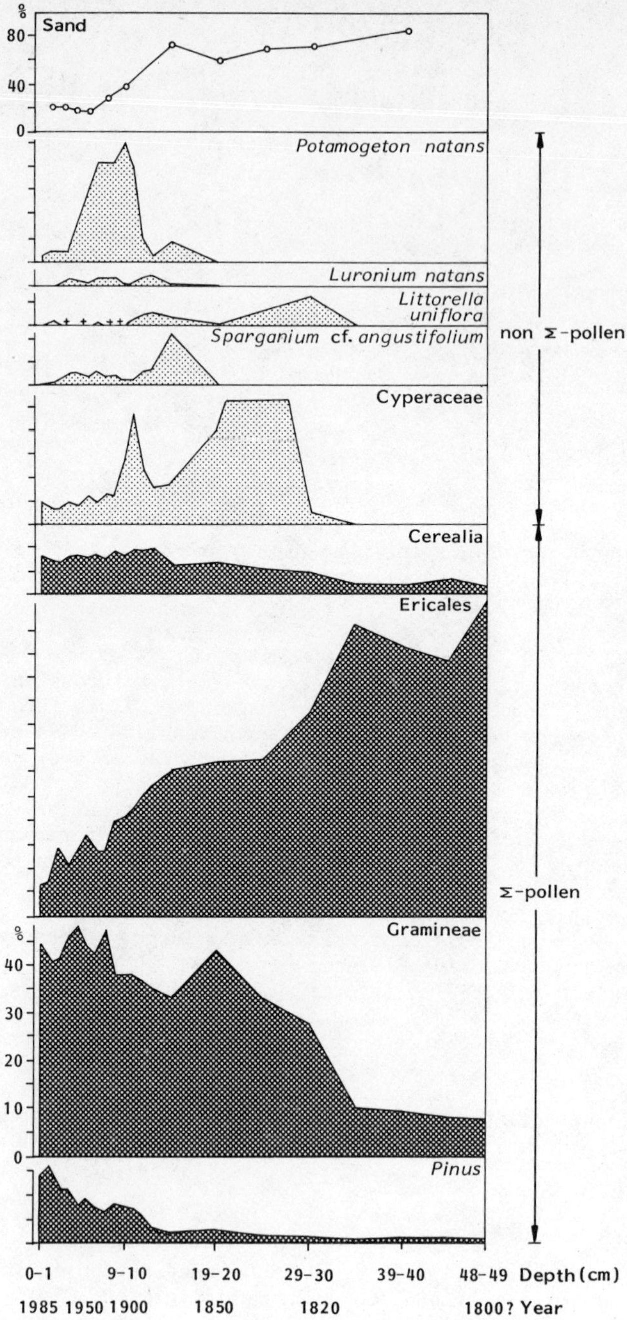

Fig. 3 Selected pollen curves from Gerritsfles.

versauerung in der Bundesrepublik Deutschland, 184-188. Umweltbundesamt and Schmidt, Berlin.
(4) BATTARBEE, R.W. and CHARLES, D.F. (1986). Diatom-based pH reconstruction of acid lakes in Europe and North America. Water, Air and Soil Pollution 30:347-354.
(5) DICKMAN, M.D. et al. (1987). Acidification of a Dutch moorland pool, a palaeolimnological study. Archiv für Hydrobiologie (in press).
(6) VAN DAM, H. et al. (submitted). Palaeolimnological and documented evidence for alkalization and acidification of two moorland pools (The Netherlands). Review of Palaeobotany and Palynology.
(7) VAN DAM, H. (1987). Verzuring van vennen: een tijdsverschijnsel (Acidification of moorland pools: a process in time). Doctoral thesis, Agricultural University. Wageningen. 175p.
(8) TER BRAAK, C.J.F. and VAN DAM, H. (in prep.) Calibration methods for inferring pH from diatoms.
(9) VAN DER WIJK, A. and MOOK, W.G. (1987). 210 Pb-dating in shallow moorland pools. Geologie en Mijnbouw 66 (in press).

DOCUMENTATION OF AREAS POTENTIALLY INCLINED TO WATER ACIDIFICATION IN THE FEDERAL REPUBLIC OF GERMANY

A. Hamm*, J. Wieting, P. Schmitt* and R. Lehmann*

*Bavarian Institute for Water Research Munich - Test Plant Wiedenbach

Federal Environmental Agency Berlin

Summary

Maps about areas in the Federal Republic of Germany, which are potentially dangered by acidification are prepared, based on the soil and rock conditions and land use. These are compared with the actial low pH-(<pH 6.0) values in surface waters of the Federal Republic of Germany.

1. INTRODUCTION

The situation of water acidification is strongly differentiated in the Federal Republic of Germany, as geology, soil properties, land utilization and pollutants charged into the air are extremely varied. With respect to geology, areas sensitive to acidification are primarily found in secondary mountain chains with carbonate-free or low-carbonate rocks (granite-gneiss areas; sandstone, slate). Moorland and sand areas, as those primarily common in the lowlands of Northern Germany are also sensitive to water acidification. In order to gain a more exact survey of the areas of the Federal Republic of Germany sensitive to acidification, various characteristics of rock and soil which are exposed to acid deposition are examined.

2. MAP-CONCEPTION

The Bavarian Institute for Water Research in cooperation with the Federal Environmental Agency developed maps of bases-supplies in the soil (Base Map I) and the carbonate content of rocks (Base Map II) with regard to agricultural land uses and combined these two maps to form a Synthesis Map for areas potentially inclined to water acidification.
In Base Map I, the various buffering capacities of soils according to the base supply is divided into the following four groups:

areas with low base supply in the soil (1)
areas with low to average base supply (2)
areas with average to sufficient base supply (3)
areas which have base supply sufficient enough so that acidification of waters is not expected. (4)

The parameter "Supply of Bases in Soil" was chosen in order to determine sensitivity and buffering capacity of an individual soil type with respect to a specific amount of acid deposition on a given surface area. The supply of bases is also a comparable indicator which is available for all soil types. Of the four soil groups, the appearance of acidic waters is primarily expected in the areas indicated as having soils containing low to average base supplies (groups (1) and (2)).

<u>In Base Map II</u>, exposed rock is categorized into 3 groups according to its carbonate content:

low-carbonate or carbonate-free areas; (1)
areas with average carbonate content; (2)
carbonate-rich areas where acidification of waters is not expected. (3)

The categorization was chosen because the carbonate content has been determined to be the most important factor in the buffering of the various acidic levels in exposed rock. In areas containing small volumes of many differnt types of rock, the carbonate content fluctuates drastically, and thus the mean values were used to categorize such rock into one of the three groups described above.

Base Maps I and II were combined and a Synthesis Map (Fig. 1) which also takes agricultural land use factors into account was developed. In this map, areas with different levels of susceptibility to water acidification were identified. The following three degrees of susceptibility were revealed:

Soil (Supply of Bases)	Rock (Carbonate Content)	Potential for Water Acidification
low (1)	Carbonate free to low carbonate (1)	very high (waters very endangered)
low (1)	Carbonate content (2)	high (waters endangered)
low to average (2)	Carbonate free to low carbonate (1)	
low (1)	Carbonate-rich (3)	low
low to average (2)	Carbonate content (2) average	waters slightly endangered)

In mapping an area's potential for water acidification, the soil parameter carried more weigth, as soils come into direct contact with sub-surface runoff and heavily influence the acidity level of the waters.

The 3 categories for areas which are potentially endangered by water acidification are indicated by the pale tones. Forest agricultural land uses exists are represented by the darker tones.

As can be seen on the Synthesis Map, the waters in the mountain areas and lowlands of Northern Germany are the most susceptible to acidification. In mapping these areas, it was not possible to distinguish between those areas inclined to water acidification due to acid deposition resulting from anthropogenic sources and those which are inclined due to naturally occurring acidic conditions.

3. COMPARISON OF PH-VALUES WITH AREAS POTENTIALLY ENDANGERED BY ACIDIVICATION

In comparing the Synthesis Map with the actual pH-values of surface waters (pH < 6.0) (Fig. 2), an obious regional distribution of pH-values can be seen. Such pH-values have been measured primarily in endangered areas where the buffer capacity has been somewhat exhausted by the acidic surface waters. The following three pH-ranges are represented in Fig. 2:

pH-Range	Symbol	Color
under 4.3	Δ	red
4.3 - 5.0	□	yellow
5.0 - 6.0	o	green

Only the lowest measured pH-values for the time period between 1977 and 1985 could be mapped. The selected increments used in the mapping of the pH-values and the maximum value of pH 6 is hydrochemically and hydrobiologically justified. The upper limit of pH 6 was selected because, at lower pH-values, the first effects on water organisms and changes in the species makeup are first noticeable.

The cut-off value of pH 5.0 was chosen because under pH 5.0 the carbonic acid buffer system no longer functions. Furthermore, pH-values in this range are associated with the release of aluminum from the soil to the water. Waters in this pH range can ce lethally toxic for many water organisms, e.g. river trout in mountain brooks.

The cut-off value of pH 4.3 was used because it represents the titration endpoint of the measure of acidic capacity, K_s. At this and lower pH-values, only more acid-tolerant organisms are to found in the waters.

REFERENCES

ANONYMOUS, Kartierung der zur Gewässerversauerung neigenden Gebiete in der Bundesrepublik Deutschland sowie des aktuelllen Standes der pH-Wert-Situation (< pH 6,0) in Oberflächengewässern. Texte Umweltbundesamt 1987 (in press)

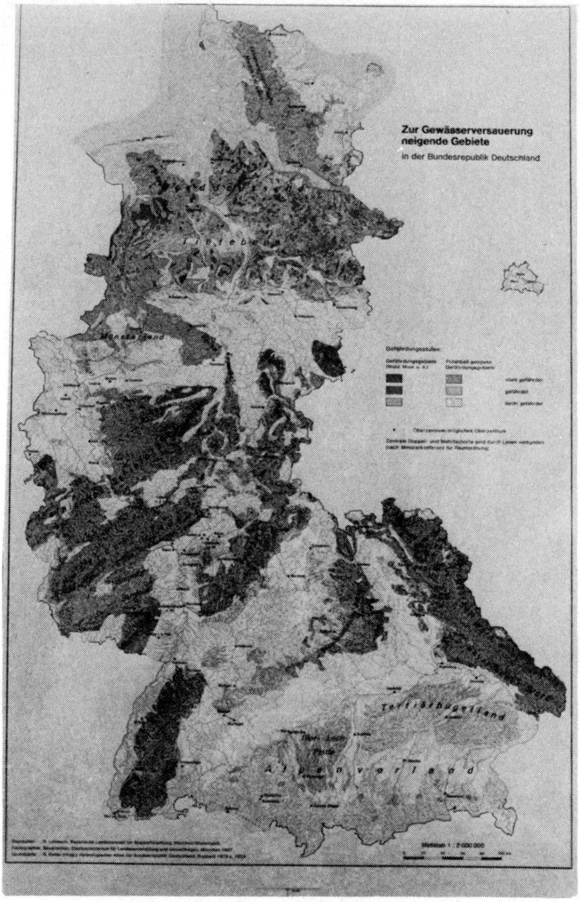

Figure 1 : Map of areas potentially inclined to water acidification in GER-synthesis of base map 1 and base map 2 with regard to agricultural land uses.

red : very endangered
orange : endangered
yellow : slightly endangered

pale tones : principally endangered due to type of rock, but buffering influences by land uses exist ;
darker tones : forest and other areas, actually endangered

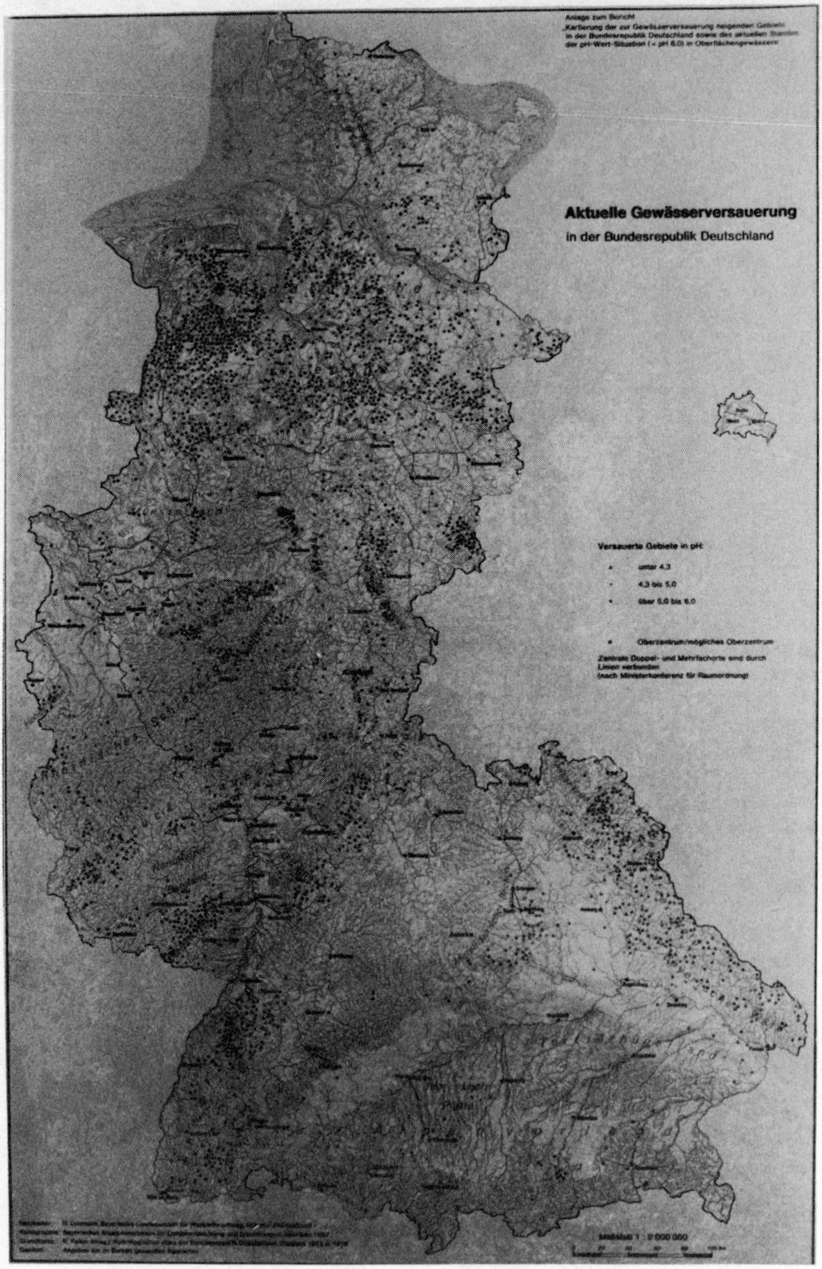

Figure 2 : Map of the actual pH-values (< pH 6.0) of surface waters in GER
Minimal measures pH-values between 1980-85.

red : < pH 4.3
yellow : pH 4.3 - 5.0
green : pH 5.0 - 6.0

INTERNAL OR EXTERNAL TOXICITY OF ALUMINIUM IN FISH EXPOSED TO ACID WATER

H.E. WITTERS (*,γ), J.H.D. VANGENECHTEN (*), S. VAN PUYMBROECK(*) and O.L.J. VANDERBORGHT (*,γ)

(*) Belgian Nuclear Energy Study Centre (SCK/CEN), Biology Department, Mineral Metabolism Laboratory, Boeretang 200, B-2400 Mol, Belgium
(γ) University of Antwerp, Biology Department, Universiteitsplein 1, B-2610 Wilrijk, Belgium

SUMMARY

Aluminium in acidified waters is highly toxic to aquatic organisms, inducing high mortality preceeded by stress on physiological processes such as osmoregulation, acid-base balance, respiration and haematology. However the mechanisms of the toxic action of Al are poorly understood. It is not yet known whether Al excerts its toxic action at the external side of the gill epithelium or whether Al is taken up by the fish. Experiments in which adult rainbow trout were exposed to elevated Al-concentrations (0.2 mg/l) in acid water (pH 5.0), gave evidence that Al is taken up by the fish. The blood Al-content was significantly higher compared to fish at control or acid conditions (0.213 μg/ml $\pm$ 0.040 versus 0.134 μg/ml $\pm$ 0.023; mean $\pm$95% conf. limits). A high rate of death was observed in fish kept in acid water with 0.2 mg Al/l. Furthermore significantly increased net branchial Na and Cl loss, decreased plasma Na and Cl levels and increased haematocrit values were measured. In further experiments acid exposed fish were injected intravenously with AlCl3. The evaluation of the ion and acid-base balance and the haematology in the Al-injected fish showed no physiological disturbances over a 4 day period.
From these experiments we conclude that the cause of the physiological disturbances in fish kept in acid water with elevated Al-concentrations can be situated at or in the outer side of the gill surface rather than via internal toxicity after Al-uptake into the fish.

INTRODUCTION

The decline of fish populations in lakes with pH-levels, higher than those which were indicated as toxic, encouraged the investigations on the combined effects of pH and Al in aquatic organisms. Geochemical and physico-chemical studies indeed pointed to an acid-induced leaching of Al from the soil and a subsequent increase of its concentration in acidified surface waters (1,2). Several mortality studies showed that fish loss at pH-levels around pH 4.7 to pH 5.5 could primarily be attributed to the presence of Al rather than to the proton concentration in the water (3,4). Physiological research has proven that Al exacerbates the effects of acid toxicity in aquatic invertebrates (5,6) and fishes (7,8,9), and that the failure of ion regulation may be the main cause of death. But other physiological processes such as respiration and haematology may be disturbed due to an impaired gas exchange at the gill surface, where a thick mucuslayer is formed.

It is not clear in which way Al is interfering with the physiological processes of Al-exposed fish. Wood & McDonald (9) observed that the decrease of the plasma and whole body ion levels is related to the increase of the Al-accumulation at the gill surfaces. Some speculations were made on the role and the interactions of Al at the outer gill surface. Due to the less acid branchial micro-environment (NH3-excretion), Al is supposed to occur as cationic Al-hydroxides which will bind to the organic ligands at the gill surface, thereby acting as an irritant and stimulating the excessive excretion of mucus. A thick mucus layer enhances the diffusion distance and resistance, thereby interfering with normal ion and gas exchange processes (9). On the other hand evidence exists that Al enters into the fish and accumulates in various tissues such as bone, muscle, liver and kidney (10, 11). It is not clear in which way Al can enter the fish and whether this Al has any physiological or metabolic role.

The experiments presented here, were performed to investigate 1) whether Al-exposed fish have increased Al-concentrations in the blood and 2) whether Al injected in the blood yields similar physiological disturbances comparable to those induced by external Al.

MATERIALS AND METHODS

Adult rainbow trout (160-330 g), obtained from a hatchery in France, were kept in stocking tanks with running tap water (table 1) for about 1 week. Thereafter they were acclimated to a temperature of 8°C and a constant day/night rhythm (14 hr/10 hr) in a closed system of 150 l (waterrecirculation rate = 1000 l/hr), filled with control water of pH 6.8 for 8-10 days.

TABLE 1: pH and average ion composition (mg/l) ± 95% confidence interval, in the experimental solutions.

	pH	Na	Cl	Ca	Al
TAP WATER	7.2	7.9	7.6	32.5	*
CONTROL	6.8	9.6 ±0.4	8.6 ±0.5	1.1 ± 0.1	*
ACID	5.0	9.4 ±0.3	8.7 ±0.2	1.0 ± 0.2	*
ACID+Al	5.0	9.4 ±0.3	10.1 ±0.6	1.0 ± 0.1	0.20 ±.05

(*) values lower than 0.03 mg/l = detection limit of Al by plasma emission spectrometry

Afterwards they were acclimated to acid water of pH 5.0 (addition of a 0.1 N mixture of H_2SO_4/HNO_3 (2 vol./1 vol.)) during a period of 10-15 days. The ionic composition of the different watertypes is shown in table 1. During both treatments the water was changed daily to minimize the accumulation of excretory products. During the experimental period when fish were exposed to acid water with Al, a recirculation (1000 l/hr) system combined with a flow through (25 l/hr) system was used in order to keep the Al-concentration between 0.200 and 0.120 mg Al/l. The labile monomeric form of Al dominated (95-100%) during the whole course of the experiment (8).

During the experimental period fish were catheterized in the dorsal aorta and/or in the urinary papilla (7). The measurement of the net ion fluxes (branchial and renal) and the measurement of parameters in water, plasma, urine and blood were performed as described in Witters (7).

In a first set of experiments fish were exposed to acid water (pH 5) with 0.200 mg Al/l for 1-10 days to allow the determination of the total Al-content in the blood. As 1 to 2 ml of blood was required to determine the Al-concentration only one blood sample was taken from each fish. The blood sample was pipetted and weighed in a teflon high pressure decomposition vessel. An equal volume of nitric acid (65%) was added. The sample was mineralized in a furnace for 5 hours at 120°C. After dilution with bidistilled water, blood Al content was measured by plasma emission spectrometry.

In another series of experiments fish were injected with Al ($AlCl_3$ in physiological saline) and results were compared with sham-injected fish and with Al-exposed fish. All fish were exposed to acid water (pH 5.0) during these experimental treatments. Two groups of fish were injected with Al, either with a low dose (0.300 µg/ml blood) comparable to the dose measured in externally exposed fish (results from first set of experiments) or a higher dose (1.050 µg/ml blood). The amount to be injected via the dorsal catheter was calculated on the basis of an extracellular fluid volume of 20% of the total body weight in trout (12,13). The sham-injected fish were injected with a solution of physiological saline only. The physiological parameters in the fish were measured one day before the Al-injection or the Al-exposure, thus during acid exposure alone and subsequently 6 hours after the start of the experimental treatment and then each of the 3 following days.

RESULTS

Figure 1 illustrates the Al-concentration in the blood of Al-exposed fish, in comparison with fish at control and acid conditions. A correlation analysis (14) on the blood Al-content of Al-exposed fish (fig.1) indicated no significant correlation with exposure time ($r=0.337$, d.f.=20, $P>0.1$). Therefore these data were pooled and the mean value was compared with the average blood Al-concentration in control- and acid-exposed fish. Calculation (F-test) revealed that the variances of the mean of both populations were not different (15). Therefore the student t-test could be used to compare the means. As we may assume blood Al-content in Al-exposed fish to be higher than in controls, a one-tailed test was made (15). The comparison indicated a significantly ($P<0.05$) increased blood Al-concentration in Al-exposed fish: 0.134 µg/ml $\pm$ 0.023 in controls and 0.213 µg/ml $\pm$ 0.040 in Al-exposed fish (mean value $\pm$ 95% confidence limits).

Mortality was highest in the Al-exposed fish compared to the sham or

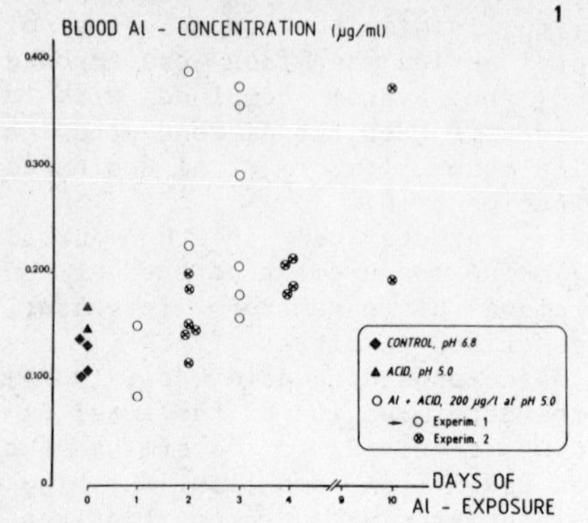

FIGURE 1 : The blood Al-concentration in fish at different experimental conditions. Values for individual fishes are presented.

Al-injected fish. In the group of Al-exposed fish 4 fish died the second day, the remaining two fish died the third day of Al-exposure. In the 3 other experimental treatments (sham injected, low Al-injected and high Al-injected) no mortalities were observed. Figure 2,3 and 4 illustrate the branchial net Na-flux, the plasma Cl-concentration and the haematocrit-value. External Al-exposure seriously disturbed these physiological parameters, as well as the branchial net Cl-flux (not shown) and the plasma Na-concentration (not shown). Blood pH, plasma glucose, plasma PO4 levels and renal ion losses (not shown) were not significantly affected by any of the treatments.

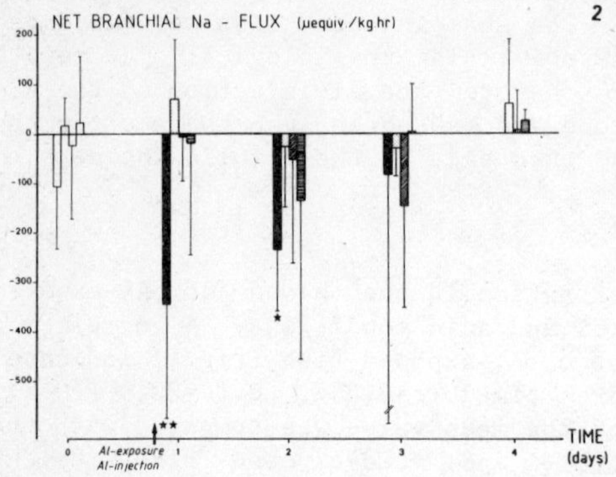

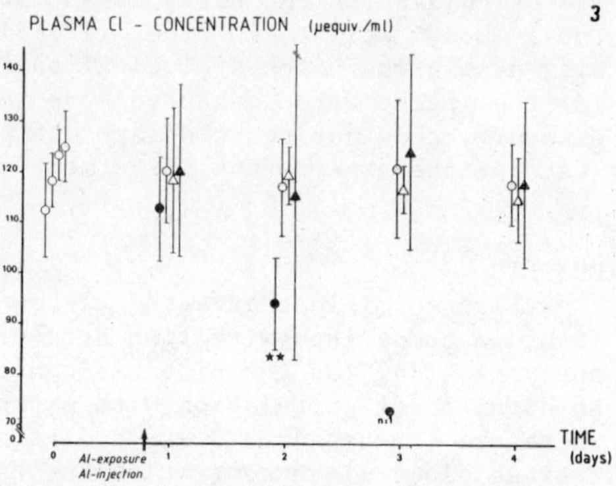

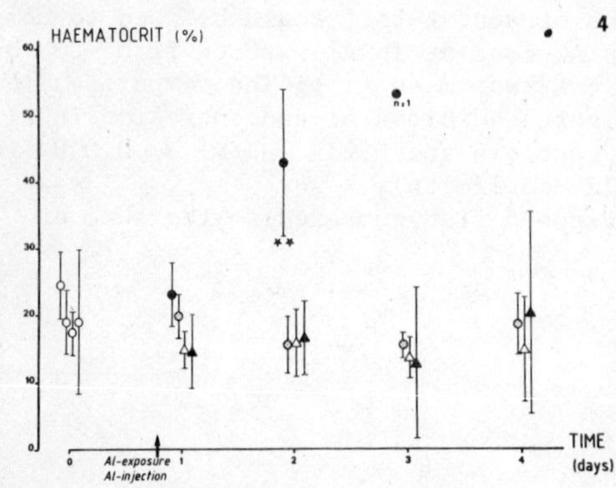

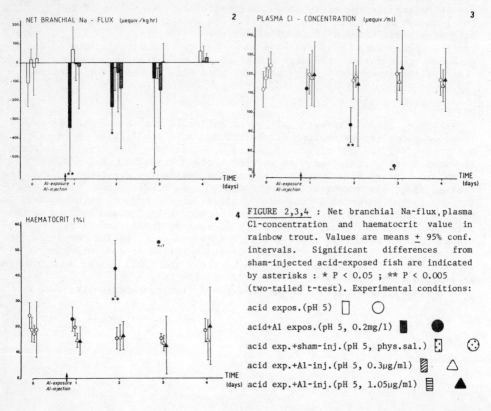

FIGURE 2,3,4 : Net branchial Na-flux, plasma Cl-concentration and haematocrit value in rainbow trout. Values are means ± 95% conf. intervals. Significant differences from sham-injected acid-exposed fish are indicated by asterisks : * $P < 0.05$; ** $P < 0.005$ (two-tailed t-test). Experimental conditions:

acid expos.(pH 5)

acid+Al expos.(pH 5, 0.2mg/l)

acid exp.+sham-inj.(pH 5, phys.sal.)

acid exp.+Al-inj.(pH 5, 0.3µg/ml)

acid exp.+Al-inj.(pH 5, 1.05µg/ml)

DISCUSSION

The Al-content in the blood of Al-exposed fish (fig.1) is of the same order of magnitude as serum Al-levels (range 0.070-0.350 μg/ml) in renal dialysis patients (16,17). In these patients Al is bound to plasma proteins thereby occupying binding sites of transferrin (18). It interferes with biochemical processes requiring other ions such as Fe, Mg and Ca (19). But nothing is known about this type of intoxication-processes in fish.

Measurements of blood Al-concentrations in Al-injected fish showed that these were of comparable magnitude for at least 6 hours as in fish, which were externally exposed to Al (fig.1). However we could not detect any effect of these high blood Al-concentrations on branchial NaCl fluxes, on plasma NaCl concentrations or on the haematocrit value. One day after Al-injection, blood Al-levels were returned to control levels. We momentarily have no measurements on the clearance rate of this metal nor on the possible retention in other organs.

In some experimental studies, Al was measured in bone, kidney, liver, and muscle (10, 11), but it is unclear how Al is taken up by the fish. Since high levels of Al are measured in the gills, probably due to adsorption in the mucuslayer (8,9,10,11), Al-uptake may occur via the gills. We however were not able to find any significant correlation between blood Al-concentration (fig.1) and the Al-content in the gills of these fish.

From these experiments we conclude that the observed ionoregulatory disturbance during exposure of fish to Al in acid water are principally mediated by external interaction of Al with the gill epithelium, as was recently proposed by Wood & McDonald (9). The experiments further indicate that elevated Al-concentrations occur in the blood of Al-exposed. Up to now we have no indications that they influence ionoregulatory mechanisms.

ACKNOWLEDGEMENT

We acknowledge the financial support provided by the Commission of the European Communities, DGXII, contract nr. ENV-836B. The work was carried out in the Mineral Metabolism Laboratories of the Belgian Nuclear Energy Centre.

REFERENCES

(1) CRONAN, C.S. and SCHOFIELD, C.L. (1979). Aluminium leaching respons to acid precipitation: effects of high elevation watersheds in the Northeast. Science 204, 304-306

(2) DICKSON, W. (1980) Properties of acidified waters. In: Proc. Int. Conf. Ecol. Impact Acid Precip., Norway 1980, SNSF project, ed. by Drablos, D. & Tollan, A., 75-83.

(3) SCHOFIELD, C.L. and TROJNAR, J.R. (1980). Aluminium toxicity to brook trout (Salvelinus fontinalis) in acidified waters. In: Polluted rain, ed. by Miller, M. & Morrow, P., Plenum Press, New York, 341-362.

(4) BAKER, J.P. and SCHOFIELD, C.L. (1982). Aluminium toxicity to fish in acidic waters. Water Air Soil Pollut. 18, 289-309.

(5) WITTERS, H.; VANGENECHTEN, J.H.D.; VAN PUYMBROECK, S. and VANDERBORGHT, O.L.J. (1984). Interference of aluminium and pH on the Na-influx in an aquatic insect Corixa punctata (Illig.). Bull. Environ. Contam. Toxicol. 32, 575-579.

(6) HAVAS, M. and LIKENS, G.E. (1985). Changes in Na-22 influx and outflux in Daphnia magna (Straus) as a function of elevated Al-concentrations in soft water at low pH. Proc. Natl. Acad. Sci. USA 82, 7345-7349.
(7) WITTERS, H.E. (1986). Acute acid exposure of rainbow trout, Salmo gairdneri Richardson: effects of aluminium and calcium on ion balance and haematology. Aquatic Toxicology 8, 197-210.
(8) WITTERS, H.E.; VANGENECHTEN, J.H.D.; VAN PUYMBROECK, S. and VANDERBORGHT, O.L.J. (1987). Ionoregulatory and haematological responses of rainbow trout Salmo gairdneri Richardson to chronic acid and aluminium stress. Annls. Soc. r. zool. Belg., 117, suppl. 1, proc."Ecophysiology of Acid Stress in Aquatic Organisms", Antwerp 1987,ed. by Witters, H. & Vanderborght, O., 411-420.
(9) WOOD, C.M. and McDONALD, G. (1987). The physiology of acid/aluminium stress in trout. Annls. Soc. r. zool. Belg., 117, suppl. 1, proc."Ecophysiology of Acid Stress in Aquatic Organisms", Antwerp 1987, ed. by Witters, H. & Vanderborght, O., 399-410.
(10) KARLSSON-NORRGREN, L.; DICKSON, W.; LJUNGBERG, O. and RUNN, P. (1986). Acid water and aluminium exposure: gill lesions and aluminium accumulation in farmed brown trout, Salmo trutta L.. Journal of Fish Diseases 9, 1-9.
(11) WITTERS, H.E. The Al-concentration in some tissues of Al-exposed fish. Unpublished results.
(12) HOAR, W. and RANDALL, D. (1969). Fish physiology. Vol I. Excretion, ion regulation and metabolism. 455 p.
(13) MILLIGAN, C.L. and WOOD, C.M. (1982). Disturbances in haematology, fluid volume distribution and circulatory function associated with low environmental pH in the rainbow trout, Salmo gairdneri. J. Exp. Biol. 99, 397-415.
(14) SOKAL, R.R. and ROHLF, F.J. (1969). Biometry. The principles and practice of statistics in biological research. 776 p.
(15) DIEM, E. (1962). Documenta Geigy. Scientific tables. 6th edition. 778 p.
(16) DE WOLFF, F.A. (1985). Toxicological aspects of aluminium poisoning in clinical nephrology. Clinical Nephrology 24 (1), S9-S14.
(17) WINNEY, R.J., COWIE, J.F. and ROBSON, J.S. (1985). What is the value of plasma/serum aluminium in patients with chronic renal failure? Clinical Nephrology 24 (1), S2-S8.
(18) TRAPP, G.A. (1983). Plasma aluminium is bound to transferrin. Life Sci. 33, 311-316.
(19) VAN DE VYVER, F.L. and M. E. DE BROE (1985). Aluminium in tissues. Clinical Nephrology, 24 (1), S37-S57.

CLOSING SESSION

GENERAL CONCLUSIONS AND RECOMMENDATIONS

SEANCE DE CLOTURE

CONCLUSIONS GENERALES ET RECOMMANDATIONS

SCHLUßSITZUNG

ALLGEMEINE SCHLUSSFOLGERUNGEN UND EMPFEHLUNGEN

Chairmen : Ph. BOURDEAU, Commission of the European Communities, Brussels, Belgium
L. CHABASON, Ministère de l'Environnement, Paris, France

CONCLUSIONS GENERALES

Maurice MULLER
Ministère de l'Environnement, S.R.E.T.I.E
Boulevard du Général Leclerc 14
92524 Neuilly-sur-Seine, FRANCE

Les conférences de synthèse, les communications verbales et les posters présentés au cours des cinq sessions du Symposium permettent de tirer un certain nombre de conclusions concernant les connaissances scientifiques et les recherches à poursuivre ou à entreprendre sur les écosystèmes perturbés. Ces conclusions reprennent pour l'essentiel les notes de synthèse rédigées "à chaud" par les rapporteurs à l'issue de chaque session.

1. LES CLIMATS DE POLLUTION EN EUROPE. LES DEPOTS DANS LES ECOSYSTEMES

1) Les climats de pollution

a) Les différentes ambiances de pollution auxquelles sont soumis les écosystèmes en Europe sont encore mal connues. Il faut donc multiplier dans les sites ruraux les mesures des paramètres de pollution et des facteurs environnementaux -en particulier climatiques- qui leur sont associés.

b) Les biologistes doivent indiquer aux chimistes et aux physiciens de l'atmosphère de quels types d'informations ils ont besoin pour l'étude des effets de la pollution : fréquence, intensité, périodes d'apparition des pointes de pollution, etc., à mettre en parallèle avec l'activité biologique des écosystèmes étudiés.

2) Les dépôts

a) Il faut avoir beaucoup plus d'informations sur les dépôts secs et sur les dépôts occultes (induits par les brouillards, les brumes, les nuages bas, la rosée, ...), ce qui nécessite une meilleure connaissance des flux de polluants et des processus d'échange près des surfaces considérées.

Les méthodes de mesure correspondantes ne pourront se développer qu'à partir d'une réflexion pluridisciplinaire, associant micrométéorologistes, physico-chimistes et biologistes.

b) L'étude des cycles ioniques en forêt (c'est-à-dire des échanges d'ions entre compartiments air, végétation, sol) devrait permettre de mieux estimer les apports respectifs par dépôts secs et par dépôts humides.

c) Les charges critiques en azote, en soufre et en protons H^+ ont été estimées d'après des critères de qualité des sols, des eaux souterraines et des lacs. Cette notion intéressante doit être affinée en tenant compte du cycle naturel des éléments dans les écosystèmes et de la possibilité d'accumulation et d'effets à long terme des polluants.

2. EFFETS DE LA POLLUTION DE L'AIR SUR LES CULTURES AGRICOLES

a) Grâce au développement récent des connaissances sur l'importance des interactions entre pollution de l'air et autres stress environnementaux (biotiques ou abiotiques), il est clair maintenant que les concentrations de polluants susceptibles de provoquer des dommages sur les cultures agricoles sont bien plus basses que celles auxquelles on pensait auparavant (par exemple, le seuil de concentration de SO2 jugé critique a diminué d'un facteur 10 au cours des 20 dernières années).

En outre, ces interactions ont le plus souvent été mises en évidence par des expérimentations à court terme. Pour des plantes à croissance plus lente, des concentrations encore plus faibles pourraient s'avérer nocives à long terme.

b) Il est nécessaire de mener des expérimentations complémentaires, utilisant des techniques de fumigation différentes : expérimentations en chambres closes, en chambres à ciel ouvert et expérimentations en plein-champ. Dans tous les cas, il faut déterminer "l'effet chambre" lui-même.

3. EFFETS DE LA POLLUTION DE L'AIR SUR LES ARBRES ET LES ECOSYSTEMES FORESTIERS

1) Les effets sur les parties aérienne et souterraine des arbres et sur les sols forestiers :

a) Les recherches sur les causes et les mécanismes des altérations des forêts sont moins avancées que les recherches sur les cultures agricoles. Il faut donc poursuivre l'effort dans ce domaine en :

- développant les connaissances sur la physiologie fondamentale des arbres forestiers : quel est le fonctionnement physiologique normal d'un arbre ?
- développant des recherches pluridisciplinaires sur les arbres et la pollution.

Pour des raisons d'efficacité, l'accent devrait être mis d'abord sur un nombre limité d'aspects jugés les plus importants en l'état actuel des connaissances, à savoir :

. les effets interactifs des polluants avec la sécheresse (stress hydrique), la résistance au gel et les carences minérales (stress ionique)
. la translocation des éléments nutritifs (interactions entre effets sur le feuillage et effets sur le système racinaire des arbres)
. les effets des protons H^+ (dépôts acides)
. les changements à long terme au niveau des cuticules
. les aspects hormonaux.

Il faut à chaque fois tenir compte de la dimension "sol" dans ces études !

b) Les chambres à ciel ouvert sont à utiliser non seulement pour vérifier que les concentrations ambiantes de polluants sur des sites particuliers sont susceptibles de provoquer des effets visibles sur les arbres ou des réponses de croissance, mais aussi pour mettre au point et conduire des expériences destinées à tester des hypothèses mécanistiques et à identifier des effets de mélanges de polluants dans des conditions précises (par exemple : brouillard acide + ozone).

c) Il faut étudier plus à fond les processus et les conséquences de l'augmentation des apports d'azote dans les plantes, les sols et les écosystèmes forestiers, par référence au cycle "naturel" de cet élément et à sa charge supposée critique.

d) Il s'avère indispensable de développer au fur et à mesure de l'avancement des recherches, des modèles conceptuels du fonctionnement des écosystèmes forestiers. Ces modèles doivent inclure la possibilité de l'apparition d'une nouvelle période de déclin de la forêt, par exemple après une succession d'années de sécheresse. En effet, il ne faut jamais perdre de vue que l'environnement n'est pas statique, mais qu'il peut évoluer sur une période de plusieurs années.

2) Symptomatologie

a) Les symptômes de dépérissement déjà observés, ainsi que ceux, plus spécifiques, qu'il faudrait arriver à détecter le plus tôt possible et à suivre dans le temps, qu'ils soient macroscopiques, microscopiques ou physiologiques, ne pourront être interprétés en terme de diagnostic que si l'on connaît bien les facteurs écologiques qui interviennent dans la vitalité et la physiologie des arbres étudiés.

b) Il est essentiel de connaître la chronologie du développement des symptômes pour comprendre les relations entre symptomatologie et pollution et les inter-relations entre différents types de symptômes. D'où la nécessité d'observer le plus finement possible l'apparition puis l'évolution sur plusieurs années des symptômes dans quelques placettes-types judicieusement choisies.

3) Approches méthodologiques

a) Tous les systèmes de fumigation expérimentale en chambre actuellement utilisés (chambres fermées ou chambres ouvertes) introduisent des modifications de l'environnement de la plante étudiée. Il est nécessaire de bien caractériser ces modifications par rapport à l'environnement réel de la plante en plein champ.

En particulier, les microclimats réels doivent être connus dans tous les systèmes.

b) De même, il faudrait définir un protocole commun de caractérisation du climat de pollution réel dans les chambres d'expérimentation.

c) Enfin, le problème de l'extrapolation des résultats d'un système de fumigation à un autre et à la situation réelle en plein champ doit être réglé.

4. EFFETS DE LA POLLUTION DE L'AIR SUR LES ECOSYSTEMES AQUATIQUES

a) Il est clair que l'acidification progressive (depuis plus de 50 ans) des eaux de surface, en particulier des lacs dans les pays nordiques et de certains lacs d'altitude en Europe Centrale, trouve son origine dans l'excès des dépôts atmosphériques acides liés au développement industriel.

Cette acidité, associée à la toxicité de l'aluminium libéré en quantités importantes, entraîne des perturbations dans la biologie des eaux de surface, mais on ne connaît pas encore tous les mécanismes impliqués.

b) La réduction des émissions de polluants précurseurs d'acidité devrait donc avoir des effets bénéfiques à plus ou moins long terme. Mais pour le moment, les réductions des rejets de SO2 opérées ces dernières années n'ont pas encore eu les effets escomptés.

c) Certaines pratiques liées à l'aménagement du territoire et à la gestion des terres (en particulier, les plantations de conifères) peuvent intervenir avec la pollution dans le processus d'acidification, mais il n'est pas prouvé qu'elles aient un effet à elles seules.

De toutes façons, il est très difficile de mesurer l'impact des pratiques sylvicoles sur l'acidification des milieux aquatiques.

d) Il faut développer la surveillance biologique des milieux aquatiques.

5. MESURES PREVENTIVES ET CURATIVES

a) Des recherches sont effectuées pour mettre au point des méthodes de restauration des écosystèmes perturbés. Des résultats intéressants ont été obtenus. Exemples : le chaulage des lacs, l'apport calcique dans les sols forestiers, etc...

b) Il faut cependant s'assurer que la généralisation de ces pratiques n'entraîne pas des effets de retour indésirables à long terme.

SUMMING-UP OF THE SYMPOSIUM

P.J.W. SAUNDERS
Department Environment London, United Kingdom

Chairman of the concerted Action
"Effects of Air Pollution on Terrestrial and Aquatic Ecosystems" (COST 612)

1. Chairmen Chabason and Bourdeau; Maurice Muller has set the scene admirably for me to take an holistic - even a "communitaire" view of our proceedings on progress and prospects in European air pollution effects research. It is inevitably provisional because we have yet to digest the full import of the numerous important and excellent papers, posters and discussions. Nevertheless, as Philip Bordeau and Maurice Muller indicated, some clear conclusions have already emerged.

2. It might be useful for me to place these conclusions in a unifying framework along the lines of a conceptual model or system which has been alluded to by several speakers. The model is based upon the environment and its problems. It has three distinct compartments - the atmosphere, the terrestrial ecosystem and the aquatic ecosystem. Each can be broken down into various sub-compartments. For example, the terrestrial system contains plants, soil and soil waters. Between the compartments and sub-compartments of such models, there are pathways in both directions involving various mechanisms controlling fluxes of elements, energy and pollutants. The model is dynamic within the bounds of natural variation but human intervention in any pathway or compartment has a series of repercussions which are reflected in changes of varying magnitude throughout the system.

3. The conceptual model is, happily, a familiar one having almost identical analogues in models used to study geochemical processes, nutrient cycles and radioactive and other pollutants. This emphasises the dynamic interplay and association of our work on atmospheric pollution with the cycling and fluxing of essential nutrients and elements in ecosystems. Nowhere is this more apparent than in work on the problems of nitrogen deposition and its effects upon vegetation, soils and aquatic systems.

4. I suggest we develop conceptual models as tools to identify critical issues where Community research is needed to remedy gaps in knowledge about the pathways, sinks and effects of atmospheric pollutants. Maurice Muller has indicated many such issues. For example, general calls for more research into sub-cellular mechanisms of damage, physiological responses to stresses, and ecosystem changes in response to pollution have been made by several speakers. Such work will be of most use, however, if targetted carefully to answer specific questions relating to the practical problems at hand. This is particularly true of work on sub-lethal stress where our priority must be the discrimination, at sub-cellular and higher levels, of pollution responses from natural stress reactions, at the earliest possible stage of exposure. This is why our work on early diagnosis is so important.

5. For those of you advocating more basic research, I would say that targetted research has its own rewards. Time and again, research into specific and applied problems reveals gaps in fundamental knowledge which must be remedied to answer the questions in hand !

6. Conceptual models also allow us to explore the interfaces between the various disciplines and come programmes of European research. The specialists in plant and soil research appear to be cooperating effectively in most respects as are those in soil and water research, and in land use and water quality modelling. The pressure to improve coordination must, however, be maintained.

7. It is clear on the other hand that the atmosphere chemists and physicists are, with few exceptions, not communicating effectively with the effects specialists on such matters as common sampling and measurement objectives and methods. It is time, I believe, for COST 611 to review its modus operandi, and to form effective links with COST 612. The climate dimension of the EC Climatology programme must also be included in this exercice.

8. The session on fumigation systems reflected excellent cooperation between those using differing approaches and systems. Philip Ineson showed, however, that there is an untapped field for cooperation with soil scientists; cooperation which will maximise effective use of expensive facilities.

9. Great progress has been made in hydrological modelling but there appears to be a need for a European forum or focus that brings together work on atmospheric deposition, land use/management, and water yield and quality. Thought must also be given to bringing the experts in ecoystems together with those developing remedial measures, and those responsible for system management. Extreme remedial measures which cause drastic if temporary change in ecosystems are not the answer ! Several speakers showed that we need to be more subtle in using our extensive knowledge of the environment to achieve cost-effective and lasting solutions to such problems as lake and river acidification.

10. We are still not in a position where we can give reasonable estimates of the quantitative effects of atmospheric pollution upon European crops, forests and aquatic systems. COST 612 may well need to review the position formally in two or three years time, with the aim of producing authoritative impact evaluations.

11. The coordination and focusing of future research requires the development of common protocols, of information and personnel exchange mechanisms, of critical masses of expertise in specific fields of research, and of expensive facilities. We can, indeed are, moving towards optimum utilisation of expertise and facilities. But, the pressure must be maintained and the rate of progress will be highly dependent upon available resources in the forms of money and qualified manpower. The Commission and COST 612 will be analysing the results of this Symposium with these points in mind to determine precise course of action over the next few years.

12. Finally, I would like to give my personal thanks to my many colleagues in the Commission and in COST 612 and its working parties, for all the work they have done to make this symposium such a success.

GENERAL CONCLUSIONS AND THANKS

Ph. BOURDEAU
Director, Commission of the European Communities
Directorate-general for Science, Research and Development
Brussels, Belgium

This gathering, which was preceded by a series of specialized preparatory meetings, has provided the opportunity to assess the considerable progress achieved in the EC, in understanding the effects of air pollution on terrestrial and aquatic ecosystems since the first Symposium on this subject held in Karlsruhe, in September 1983.

We have come a long way, as summed up by Maurice MULLER, Peter SAUNDERS, but much still remains to be done.

Major advances have been made in the physico-chemistry of air pollutants, in particular on oxidants, in measurements of wet and dry deposition, in the characterization and mechanisms of effects on crops, forests, water and buildings.

I should like to mention some of the significant items which emerged from the presentations and discussions : the ecological effects of low doses of air pollutants, the relative importance of acid and oxidizing chemicals, the long-term impact of acidic substances on soils, the need to integrate studies of terrestrial and aquatic systems, the importance of interactions between natural and man-made stresses, the reversibility of lake acidification and the progressive development of sophisticated equipment (closed and open-top chambers) for measuring effects on plants.

There is ample evidence, in my opinion, to justify a reduction of emissions. As far as further research is concerned, three points have come through forcefully : first, the desire, if not the absolute necessity to develop research plans within a unifying conceptual model as proposed by Peter SAUNDERS ; second, the need to encourage work in the basic disciplines of ecosystem ecology, tree physiology and soil science ; third, the reinforcement of multidisciplinary cooperation among scientists concerned with this issue, not only within the European Community but also between the EFTA countries, the US and Canada, and possibly Eastern Europe.

In conclusion, I wish to express our most sincere thanks to those who have helped so much to organise this Symposium in Grenoble...

Nos sincères remerciements s'adressent en particulier à Monsieur le Ministre de l'Environnement A. CARIGNON et à ses collaborateurs, tant au Cabinet qu'à l'Administration, qui ont permis la réalisation de la conférence. Notre gratitude va également aux départements et organismes qui ont participé au succès de la réunion : le Ministère de l'Agriculture, l'Institut National de la Recherche Agronomique et l'Office National des Forêts, la Ville de Grenoble, la Préfecture de l'Isère, le Parc National des Ecrins et Grenoble Alpes Congrès.

INDEX OF AUTHORS